Springer Collected Works in Mathematics

For further volumes:
http://www.springer.com/series/11104

August 1993
near Vancouver B.C.
(photo by W. Casselman)

Armand Borel

Oeuvres - Collected Papers IV

1983–1999

Reprint of the 2001 Edition

Springer

Armand Borel (1923–2003)
Princeton University
Princeton, NJ
USA

ISSN 2194-9875
ISBN 978-3-642-30717-1 (Softcover)
978-3-540-67640-9 (Hardcover)
DOI 10.1007/978-3-642-41240-0
Springer Heidelberg New York Dordrecht London

Library of Congress Control Number: 2012954381

Mathematics Subject Classification (2000): 01A75, 01A60, 11Fxx, 19F27, 20G15, 55N33

Printed on acid-free paper

Springer is part of Springer Science+Business Media (www.springer.com)

Préface

Ce volume contient mes articles publiés, seul ou en collaboration, de 1983 à 1999. Comme dans les trois volumes précédents, certains d'entre eux sont accompagnés de commentaires ou corrections, placés en fin de volume. J'ai aussi profité de cette occasion pour en ajouter quelques-uns aux articles parus avant 1983.

Outre des articles de recherche ou des exposés de développements récents, on y trouvera des souvenirs personnels, quelques considérations sur les mathématiques en général et plusieurs articles de nature historique.

Je suis très reconnaissant à la Maison Springer, pour cette publication, réalisée avec sa compétence habituelle.

Princeton, Décembre 1999 A. Borel

Curriculum vitae

Né à la Chaux-de-Fonds, Neuchâtel, Suisse le 21 mai 1923
Marié le 8 mai 1952 à Gabrielle Aline, née Pittet
Enfants: Dominique et Anne

Études à la section de Mathématiques et Physique de l'École Polytechnique Fédérale (E.P.F.), Zürich, 1942–1947
Diplôme en mathématiques au printemps 1947
Assistant à l'E.P.F., 1947–1949
Boursier du C.N.R.S., Paris, 1949–1950
Remplaçant du professeur d'algèbre à l'Université de Genève, 1950–1952
Doctorat ès Sciences, Université de Paris, 1952
Membre de l'Institute for Advanced Study, Princeton, 1952–1954
Visiting lecturer, University of Chicago, 1954–1955
Professeur à l'E.P.F., Zürich, 1955–1957, 1983–1986
Professeur à l'Institute for Advanced Study, 1957–1993, émérite 1993–

Invitations de un mois ou plus:
University of Mexico, été 1953, été 1979
MIT, Cambridge, Mass., printemps 1958, automne 1969
Tata Institute of Fundamental Research, Bombay, jan.–mars 1961, jan. 1968, déc. 1983–mars 1994, dec. 1989–jan. 1990, déc 1991–jan. 1992, oct. 1993–jan. 1994, oct. 1995–jan. 1996, fév. 1999
Université de Paris, janvier–juin 1964
Université de Genève, printemps 1966
Cours à Columbia University, New York, printemps 1968
University of Utrecht, The Netherlands, printemps 1971
University of Buenos Aires, Argentina, août 1973
Institut des Hautes Études Scientifiques, Bures-sur-Yvette, automne 1973, juin 1978
Cours à Princeton University, octobre 1974–janvier 1975
University of California at Berkeley, printemps 1975
University of Chicago, printemps 1976
University of Washington, Seattle (Walker Ames Professorship), été 1976
Collège de France, Paris, mai 1977
University of Amsterdam, mai 1978
Yale University, automne 1978
Mathematics Institute, Academia Sinica, Beijing, Chine, mai–juin 1981, mai 1993, mai 1998

University of California at San Diego, mars 1998
University of Hong Kong, mars–juin 1999

Membre du Comité de rédaction:
Annals of Mathematics 1962–1979;
Inventiones Math., 1979–1993;
Comm. Math. Helv. 1984–1996;
Mathematical Research Letters, 1994–

Membre, Comité Consultatif, Congrès International des Mathématiciens, Moscou 1966
Président, Comité Consultatif, Congrès International des Mathématiciens, Helsinki 1978

Dr. h.c., Université de Genève, 1972
Membre, American Academy of Arts and Sciences, 1976
Membre étranger, Académie finlandaise des Arts et Sciences, 1978
Associé étranger, Académie des Sciences, Paris, 1981
Membre étranger, American Philosophical Society, 1985
Membre, National Academy of Sciences, USA, 1987
Honorary fellow, Tata Institute of fundamental research, Bombay, India 1990
Membre étranger, Academia Europaea 1995

Médaille Brouwer, Société mathématique de Hollande, 1978
Steele prize, American Mathematical Society, 1991
Prix Balzan, 1992

Membre des sociétés mathématiques de Suisse, de France et des Etats-Unis

Table des matières

Volume IV

121.

(with H. Garland)

Laplacian and the Discrete Spectrum of an Arithmetic Group

Amer. J. Math. **105** (1983) 309–335

To André Weil, on his 77th birthday

Let G be the group of real points of a connected semi-simple algebraic group defined over $\mathbf{Q}$ and Γ an arithmetic subgroup of G (these assumptions are made for convenience in the introduction but will be somewhat relaxed, see 2.1). Let K be a maximal compact subgroup of G and $X = G/K$. Let (σ, E) be a finite dimensional representation of G.

Using a K-invariant admissible (see 5.1) scalar product on E and a G-invariant metric on X, one defines a scalar product on the space $A^\infty(M, \tilde{E})$ of forms with values in the local system $\tilde{E}$ defined by (σ, E). Let $A(M; \tilde{E})_2$ be the space of square integrable forms with measurable coefficients. Our initial objective was to prove that the space $\mathcal{H}^*_{(2)}(M; \tilde{E})$ of square integrable $\tilde{E}$-valued harmonic forms on $\Gamma\backslash X$ is finite dimensional. We shall in fact consider more generally L^2-eigenforms of the Laplace-Beltrami operator Δ_X, or rather, of its natural self-adjoint closure $\bar{\Delta}_X$ (see 5.4) and prove (5.5):

Theorem 1. *The eigenvalues of $\bar{\Delta}_X$ on the space of $\tilde{E}$-valued square integrable q-forms on $\Gamma\backslash X$ have finite multiplicities and tend to infinity.*

If $\Gamma\backslash X$ is compact, this is standard theory of elliptic operators, so our main concern will be the non-compact case.

There is a well-known way (briefly recalled in 5.4) to identify $A^\infty(\Gamma\backslash X; \tilde{E})$ with a space of differential forms on $\Gamma\backslash G$ with coefficients in $C^\infty(\Gamma\backslash G) \otimes E$, in fact with the space of cochains of the relative Lie algebra complex $C^*(\mathfrak{g}, K; C^\infty(\Gamma\backslash G) \otimes E)$. Similarly, $A(\Gamma\backslash X; E)_2$ can be identified to the space of exterior forms with coefficients in $L^2(\Gamma\backslash G) \otimes E$. Furthermore a formula of Kuga's (see 5.1(3)) expresses the action of Δ_X in terms of the Casimir operator C acting on the coefficients. This leads us to study eigenfunctions of C on the space $L^2(\Gamma\backslash G)^\infty$ of smooth vectors in $L^2(\Gamma\backslash G)$ (smooth in the sense of representation theory, $L^2(\Gamma\backslash G)$ being viewed as a

---

Manuscript received October 8, 1982.

unitary module via the right regular representation) or, equivalently, of eigenfunctions of the (self-adjoint) closure $\bar{C}$ of C acting on $L^2(\Gamma\backslash G)^\infty$. We shall prove (4.1):

THEOREM 2. *Fix a K-type r. Then the eigenvalues of $\bar{C}$ on the space $L^2(\Gamma\backslash G)_r$ of L^2-functions on $\Gamma\backslash G$ transforming according to r under K have finite multiplicities and $\to -\infty$. The corresponding eigenfunctions belong to the discrete spectrum and are automorphic forms.*

We shall also consider the eigenvalue problem for a closely related elliptic operator Δ, which can be viewed as a Laplacian on $\Gamma\backslash G$ with respect to a Riemannian metric induced from a left invariant metric on G. For a given K-type, the eigenspaces of C and Δ are the same (with different eigenvalues in general, though). However it turns out that for Δ we need not fix a K-type (whereas it is essential for C). This is also connected with the compactness of convolution operators on the discrete spectrum. We shall also show (4.6)

1 THEOREM 3. (i) *The convolution $*\alpha$ by $\alpha \in L^1(G)$ on the discrete spectrum is a compact operator.*

(ii) *The eigenvalues of Δ on $L^2(\Gamma\backslash G)$ have finite multiplicities and $\to \infty$. Its eigenspaces are contained in the discrete spectrum.*

The proof of Theorem 2 is first carried out for the cuspidal spectrum. The main point is a lemma (1.8) involving a derivative norm and convolution by functions $\varphi_\epsilon (0 < \epsilon \leq 1)$ forming an approximate identity, directly suggested by Prop. 2, Chap. III, Section 5 of [11]. This is valid for any unimodular Lie group and any discrete subgroup and is discussed in Section 1. This implies readily that the eigenvalues of Δ on a closed G-invariant subspace U of $L^2(\Gamma\backslash G)$ have finite multiplicities and $\to \infty$ if the operators $*\varphi_\epsilon$ are compact on U (1.11).

Section 2 describes our assumptions on G and Γ for the remainder of this paper and assembles some definitions, notation and facts about Casimir and Laplace operators and their eigenspaces. Section 3 reviews briefly, in a form convenient for us, some of the fundamental results on the spectral decomposition of $L^2(\Gamma\backslash G)$ due to R. P. Langlands [10]. Section 4 starts with the proof of Theorem 2. There are two steps: for the cuspidal spectrum, the theorem follows from 1.11 and from the fact that $*\alpha$ ($\alpha \in C_c^\infty(G)$) is a compact (even trace class) operator on the cuspidal spectrum (Theorem of Gelfand and Piatetski-Shapiro). From there we proceed by induction, using the results recalled in Section 3. We then draw some consequences, in particular Theorem 3 above and the fact that, given a K-type r, and a non-constant polynomial P, the operator $P(C)$ has a bounded inverse in $L^2(\Gamma\backslash G)_r$ outside the sum of finitely many of the elementary summands provided by the Langlands decomposition (4.2). Some open problems suggested by these results are mentioned in 4.7. Section 5 gives some applications to relative Lie algebra cohomology with respect to $L^2(\Gamma\backslash G)^\infty \otimes E$ (5.3, 5.6) and deduces Theorem 1 from Theorem 2. We refer to a forthcoming joint paper of the first named author and W. Casselman: "L^2-cohomology of locally symmetric manifolds of finite volume," for applications to the L^2-cohomology of $\Gamma\backslash X$.

For G of **Q**-rank one, the finite multiplicity assertion for the eigenvalues in Theorems 1 and 2 was announced in [9], (It was also stated there that they had no finite accumulation point, but the argument alluded to in footnote 2, p. 809, was based on a misunderstanding.) This was the starting point of the present paper. Theorems 2 and 5.5, 5.6 were stated in [2]. The proofs were described in seminars held at the University of Washington (Summer 1976) and the Institute (1979–80) by the first named author. The latter seminar also included a discussion of Theorem 3. A considerable simplification of our original argument was then proposed by N. Wallach and has been adopted here (see 4.4, 4.5). Our proof involved the use of the heat kernel; discussions on the latter with R. Beals had been extremely useful to us. We would also like to thank R. P. Langlands for his help in going from the cuspidal to the discrete spectrum. Another proof of Theorem 2 is given in [13: 7.6]. Also, H. Donnelly has established a quantitative sharpening of Theorem 2 for the cuspidal spectrum (stated in 4.7), first in **Q**-rank one [5] and then in general [6].

1. The derivative norm on a Lie group. In this section, G is a unimodular Lie group with finitely many connected components, dg a Haar measure on G, K a compact subgroup and Γ a discrete subgroup of G.

1.0. *Some notation.*

1. The right translation on G or $\Gamma\backslash G$ by $g \in G$ or its effect on functions is denoted $r_g : r_g f(x) = f(x\cdot g)$.

2. The Lie algebra of a Lie group $G, H, \ldots$ is denoted by the corresponding l.c. German letter. $U(\mathfrak{g})$ is the universal enveloping algebra of $\mathfrak{g}$ over **C** and $Z(\mathfrak{g})$ the center of $U(\mathfrak{g})$.

3. As usual, $\hat{K}$ denotes the set of equivalence classes of irreducible (finite dimensional) representations of K. If (π, V) is a unitary K-module and $r \in \hat{K}$, then V_r is the space of K-finite vectors of type r [16: 4.4.2]. If (π, V) is a unitary G-module, then V^∞ is the space of differentiable vectors.

4. If u and v are positive functions on a set A, we write $u \prec v$ if there exists $C > 0$ such that $u(a) \leq C \cdot v(a)$ for all $a \in A$ and $u \asymp v$ if $u \prec v$ and $v \prec u$.

1.1. The Lie algebra $\mathfrak{g}$ of G is identified with left invariant vector fields on G, or also with the induced vector fields on $\Gamma \backslash G$. Let

$$\{x_1, \ldots, x_n\}, \qquad (n = \dim G) \tag{1}$$

be a basis of $\mathfrak{g}$. On $C_c^\infty(\Gamma \backslash G)$ we consider the usual scalar product

$$(u, v) = \int_{\Gamma \backslash G} u \cdot \bar{v} dg, \tag{2}$$

the scalar product

$$(u, v)_1 = (u, v) + \sum_i (x_i u, x_i u), \tag{3}$$

the associated norms

$$\| u \| = (u, u)^{1/2}, \qquad \| u \|_1 = (u, u)_1^{1/2},$$

and denote $H = L^2(\Gamma \backslash G)$, resp. H_1, the completion of $C_c^\infty(\Gamma \backslash G)$ with respect to $\| \cdot \|$ (resp. $\| \cdot \|_1$).

1.2. The space H (resp. H_1) can also be defined as the completion of the space $C_{0,2}^\infty(\Gamma \backslash G)$ (resp. $C_{1,2}^\infty(\Gamma \backslash G)$) of smooth functions which have finite $\| \cdot \|$ (resp. $\| \cdot \|_1$) norm. In fact, we may (and do) view the x_i as defining at each point an orthonormal frame with respect to a left invariant metric on G or the induced metric ds^2 on $\Gamma \backslash G$ and dg as the Riemannian volume element. These metrics are *complete*. As a consequence, by using the functions σ_r of [1: Section 1] one sees that if u and xu are L^2, then xu is the limit of the functions $x\sigma_r u$ $(x \in \mathfrak{g})$, and our assertion follows. We note also that $C_{1,2}^\infty$ contains H^∞. Let

$$\Delta = -\Sigma x_i^2 \tag{1}$$

be the Laplacian on $\Gamma \backslash G$ with respect to ds^2. Since we have

$$(x^2 u, u) = -(xu, xu), \qquad (x \in \mathfrak{g}, u \in H^\infty), \tag{2}$$

[1: 1.4], we can also write

$$(3)\qquad (u, u)_1 = (u, u) + (\Delta u, u), \qquad (u \in H^\infty).$$

1.3. Lemma. *The identity map of $C_c^\infty(\Gamma\backslash G)$ extends to a continuous injective map $j: H_1 \to H$.*

Since $(u, u) \leq (u, u)_1$, the existence of a continuous map j is clear. Let $u \in H$ and $u_j \in C_c^\infty(\Gamma\backslash G)$ which tend to u in H_1 and assume that $u_j \to 0$ in H. To prove injectivity, we have to show that $u_j \to 0$ in H_1 and for this it suffices to prove that $xu_j \to 0$ in H for all $x \in \mathfrak{g}$. But xu_j is a Cauchy sequence in H since u_j has a limit in H_1 and we have

$$(xu_j, v) = -(u_j, xv) \to 0 \quad (j \geq 1;\ v \in C_c^\infty(\Gamma\backslash G))$$

whence our assertion.

1.4. In the sequel, we identify H_1 to a subspace of H using j. Thus an element $u \in H$ belongs to H_1 if and only if there exists $u_j \in C_{1,2}^\infty(\Gamma\backslash G)$ ($j = 1, \ldots$) which converges to u in H and such that xu_j converges in H to an element which is then equal to xu by definition ($x \in \mathfrak{g}$). By 1.3, this limit is indeed independent of the choice of the u_j; it is the strong derivative of u in the sense of Friedrichs.

1.5. Lemma. *The subspace H_1 of H is invariant under the right regular representation. The induced representation of G on H_1 is continuous with respect to the norm $\|\cdot\|_1$.*

The relations

$$(1)\qquad r_g(x\cdot u) = \mathrm{Ad}\, g(x)\cdot(r_g u),$$

$$(2)\qquad x(r_g u) = r_g(\mathrm{Ad}\, g^{-1}(x)\cdot u), \qquad (x \in \mathfrak{g};\ u \in C_{1,2}^\infty(\Gamma\backslash G),\ g \in G)$$

imply immediately that $C_{1,2}^\infty(\Gamma\backslash G)$ is invariant under G and that if the sequence $u_j \in C_{1,2}^\infty(\Gamma\backslash G)$ is such that $x\cdot u_j$ converges in H for each $x \in \mathfrak{g}$, then the same is true for $\{r_g \cdot u_j\}$ ($g \in G$). The first assertion then follows from the characterization of H_1 given in 1.4. Write

$$(3)\qquad \mathrm{Ad}\, g^{-1}(x_i) = \sum_j c_i^j(g) x_j, \qquad (i = 1, \ldots, n),$$

and let

$$c(g) = \sum_{i,j} c_i^j(g)^2. \tag{4}$$

It is a smooth function on G. For g, $h \in G$, we can write

$$x_i r_g u - x_i r_h u = \sum_j \{c_i^j(g)(r_g x_j u - r_h x_j u) + (c_i^j(g) - c_i^j(h)) r_h x_j u\}.$$

Therefore

$$\|x_i r_g u - x_i r_h u\|^2 \leq \sum_j \{c(g)\|r_g x_j u - r_h x_j u\|^2 + |c_i^j(g) - c_i^j(h)|^2 \|r_h x_j u\|^2\}.$$

$$\|r_g u - r_h u\|_1^2 \leq n \cdot c(g) \sum_j \|r_g x_j u - r_h x_j u\|^2 + \sum_{i,j} |c_i^j(g) - c_i^j(h)|^2 \|r_h x_j u\|^2 + \|r_g u - r_h u\|^2.$$

This implies immediately that $g \mapsto r_g u$ is a continuous map of G into H_1. It is standard that this yields the second assertion.

1.6. *Construction of an approximate indentity.* A family of smooth functions (φ_ϵ) on G $(0 < \epsilon \leq 1)$ is said to be an *approximate indentity* if

(1) $\varphi_\epsilon \geq 0, \quad \varphi_\epsilon \in C_c^\infty(G), \quad \int_G \varphi_\epsilon \cdot dg = 1, \quad (0 < \epsilon \leq 1)$

(2) $\operatorname{Supp} \varphi_\epsilon \to \{1\}$ as $\epsilon \to 0$.

We want to construct such a family whose elements are in addition K-invariant and symmetric, i.e., satisfy

(3) $u(kg) = u(gk) \quad (k \in K), \quad u(g) = u(g^{-1}) \quad (g \in G)$.

Let U_0 be an $\mathrm{Ad}_\mathfrak{g}\, K$-invariant relatively compact symmetric ($U_0 = -U_0$) neighborhood of the origin in $\mathfrak{g}$ such that the exponential mapping sends U_0 homeomorphically onto a neighborhood V_0 of 1 in G; let U_1 be a symmetric $\mathrm{Ad}_{\mathfrak{g}}\, K$-invariant neighborhood of 0 in U_0 whose closure is contained in U_0 and $V_1 = \exp U_1$. Then V_0 and V_1 are K-invariant and symmetric. We choose $\psi \in C_c^\infty(U_0)$, with support in U_1, invariant under K, symmetric, ≥ 0 and such that $\int \psi(X)dX = 1$, where dX is Lebesgue measure on $\mathfrak{g}$. Set

(1) $$\psi_\epsilon = \epsilon^{-n}\psi(x/\epsilon), \qquad (0 < \epsilon \leq 1).$$

We have

(2) $$\operatorname{Supp}\psi_\epsilon \subset \epsilon\cdot\operatorname{Supp}\psi \subset \epsilon\cdot U_1 \to \{0\}, \qquad \text{as} \quad \epsilon \to 0.$$

(3) $$\int \psi_\epsilon dX = 1 \quad \text{and} \quad \psi_\epsilon \text{ is } K\text{-invariant, symmetric.}$$

Let us denote by dY the measure on U_0 transported from dg by $\exp^{-1}$. There exists then $C \geq 1$ such that

(4) $$C^{-1}dX \leq dY \leq CdX \quad \text{on} \quad U_1.$$

Let φ_ϵ' be the function which is equal to $\psi_\epsilon \circ \exp^{-1} | V_0$ on U_0 and is zero outside U_0. Set

(5) $$\delta_\epsilon = \int_G \varphi_\epsilon' dg \quad \text{and} \quad \varphi_\epsilon = \varphi_\epsilon'\delta_\epsilon^{-1}.$$

Then $\{\varphi_\epsilon\}$ $(0 < \epsilon \leq 1)$ is an approximate identity, consisting of K-invariant and symmetric functions. In particular

(6) $$V_\epsilon = \operatorname{Supp}\varphi_\epsilon \subset \exp \epsilon U_1 \to \{1\}, \qquad \text{as} \quad \epsilon \to 0.$$

1.7. LEMMA. *(i) Let $v_\epsilon = \int_{V_\epsilon} dX$. Then*

(i) $v_\epsilon \asymp \epsilon^n$, *and* $\delta_\epsilon \asymp 1$ $(0 < \epsilon \leq 1)$,
(ii) $\int \varphi_\epsilon(g)^2 dg \asymp \epsilon^{-n}$ $(0 < \epsilon \leq 1)$,
(iii) $\|f * \varphi_\epsilon - f\|^2 < \epsilon^{-n}\int_{V_\epsilon} \|r_y f - f\|^2 dy$ $(0 < \epsilon \leq 1; f \in L^2(\Gamma\backslash G))$.

(i) The relation

$$\int_{V_\epsilon} dg \asymp \int_{\ln(V_\epsilon)} dX = \int_{\epsilon \operatorname{Supp}\psi} dX = \epsilon^n \int_{\operatorname{Supp}\psi} dX,$$

proves the first part of (i). We have

$$\delta_\epsilon = \int_G \varphi'_\epsilon dg = \int_{U_0} \psi_\epsilon dY \asymp \int_{\mathfrak{g}} \psi_\epsilon dX = 1.$$

whence the second part of (i). We have

$$\int_G \varphi_\epsilon(g)^2 dg \asymp \int_G \varphi'_\epsilon(g)^2 dg \asymp \int_{\mathfrak{g}} \psi_\epsilon^2 dX = \epsilon^{-2n} \int_{\mathfrak{g}} \psi(X/\epsilon)^2 dX$$

$$= \epsilon^{-n} \int_{\mathfrak{g}} \psi(X)^2 dX \asymp \epsilon^{-n},$$

whence (ii). By definition

$$\|f * \varphi_\epsilon - f\|^2 = \int_{\Gamma\backslash G} dx \left| \int_G f(xy)\varphi_\epsilon(y^{-1})dy - f(x)\right|^2.$$

We have $\varphi_\epsilon(y^{-1}) = \varphi_\epsilon(y)$ by symmetry, and $\int_G \varphi_\epsilon(g)dg = 1$ by construction, hence

$$\|f * \varphi_\epsilon - f\|^2 = \int_{\Gamma\backslash G} dy \left| \int_G \varphi_\epsilon(y^{-1})(f(xy) - f(x))dx\right|^2.$$

The inner integral is in fact an integral over V_ϵ. By the Schwarz inequality:

$$\left|\int_{V_\epsilon} \varphi_\epsilon(y)(f(xy) - f(x))dx\right|^2 \leq \left(\int_{V_\epsilon} \varphi_\epsilon(y)^2 dy\right)\cdot\left(\int_{V_\epsilon} |f(xy) - f(x)|^2 dy\right).$$

We get therefore

$$\|f * \varphi_\epsilon - f\|^2 \prec \epsilon^{-n} \int_{\Gamma\backslash G} dx \int_{V_\epsilon} |f(xy) - f(x)|^2 dy$$

$$= \epsilon^{-n} \int_{V_\epsilon} dy \int_{\Gamma\backslash G} |(r_y f - f)(x)|^2 dx = \epsilon^{-n} \int_{V_\epsilon} dy \, \|r_y f - f\|^2.$$

As remarked in the introduction, the following lemma is the main point of this section and is a transcription to our setting of Prop. 2, p. 175 in [11].

1.8. LEMMA. *There exists a constant $D > 0$ such that*

(1) $\quad \|f * \varphi_\epsilon - f\| \leq D \cdot \epsilon \cdot \|f\|_1, \qquad$ *for all* $\epsilon \in (0, 1]$ *and all* $f \in H_1$.

Assume first that (1) is already proved for all $f \in C_c^\infty(\Gamma\backslash G)$. Given $f \in H_1$, we can find $f_n \in C_c^\infty(\Gamma\backslash G)$ such that $f_n \to f$ in H_1. Fix $\epsilon > 0$. For all n's we have

$$\|f * \varphi_\epsilon - f\| - D\epsilon\|f\|_1 \leq \|(f - f_n) * \varphi_\epsilon - (f - f_n)\|$$
$$+ \|f_n * \varphi_\epsilon - f_n\| - D\epsilon\|f\|_1.$$

But $* \varphi_\epsilon$-Id is a continuous operator on H, hence there exists $d_\epsilon > 0$ such that

$$\|(f - f_n) * \varphi_\epsilon - (f - f_n)\| \leq d_\epsilon \|f - f_n\|.$$

We get therefore, using (1) on $C_c^\infty(\Gamma\backslash G)$:

$$\|f * \varphi_\epsilon - f\| - D\epsilon\|f\|_1 \leq d_\epsilon\|f - f_n\| + D\epsilon(\|f_n\|_1 - \|f\|_1).$$

Since the right side $\to 0$ as $n \to \infty$, this proves the lemma for f. There remains to prove (1) for $f \in C_c^\infty(G)$. By 1.7(iii) it suffices to show the existence of $D_1 > 0$ such that

(2) $\quad \int_{V_\epsilon} \|r_y f - f\|^2 dy \leq D_1 \epsilon^{n+2} \|f\|_1^2, \qquad (0 < \epsilon \leq 1; f \in C_c^\infty(\Gamma\backslash G)).$

Let $Y = \ln y$. Then

$$f(xy) - f(x) = \int_0^1 \frac{d}{dt} f(xe^{tY})dt = \int_0^1 Yf(xe^{tY})dt.$$

We can write

$$Y = \sum_j a_j(y)X_j,$$

where the a_j are smooth functions of y and $j = 1, \ldots, n$. Hence

$$f(xy) - f(x) = \sum_j a_j(y) \int_0^1 X_j f(xe^{tY})dt;$$

$$|f(xy) - f(x)|^2 \leq n \sum_j a_j(y)^2 \left| \int_0^1 X_j f(xe^{tY})dt \right|^2$$

$$\int_{V_\epsilon} dy \, \|r_y f - f\|^2 \leq n \cdot \sum_j \int_{V_\epsilon} dy a_j(y)^2 \int_{\Gamma\backslash G} dx \int_0^1 dt \, |X_j f(xe^{tY})|^2$$

$$= n \sum_j \left(\int_{V_\epsilon} dy a_j(y)^2 \right) \left(\int_0^1 dt \int_{\Gamma\backslash G} dx \, |X_j f(xe^{tY})|^2 \right)$$

(3)
$$\int_{V_\epsilon} dy \, \|r_y f - f\|^2 \leq n \sum_j \left(\int_{V_\epsilon} dy a_j(y)^2 \right) \|X_j f\|^2.$$

To prove (2), it suffices then to show that the integral on the right-hand side is $< \epsilon^{n+2}$. But

$$\int_{V_\epsilon} dy a_j(y)^2 \asymp \int_{\epsilon \, \mathrm{Supp}\, \psi} dY a_j(Y)^2, \qquad (0 < \epsilon \leq 1).$$

Set $Y = \epsilon \cdot Z$. Then

$$\int_{\epsilon \, \mathrm{Supp}\, \psi} dY a_j(Y)^2 \asymp \epsilon^n \int_{\mathrm{Supp}\, \psi} dX \cdot a_j(\epsilon X)^2 = \epsilon^{n+2} \int_{\mathrm{Supp}\, \psi} dX \cdot a_j(X)^2 < \epsilon^{n+2}.$$

1.9. For $\lambda \in \mathbf{C}$, let

(1)
$$H_\lambda^\infty = \{u \in H^\infty \mid \Delta u = \lambda u\}.$$

Since Δ is symmetric, and ≥ 0 on H^∞, we have

(2)
$$H_\lambda^\infty = 0 \quad \text{if} \quad \lambda \notin \mathbf{R}, \qquad \text{or} \quad \lambda \in \mathbf{R}, \qquad \lambda < 0.$$

(3)
$$(H_\lambda^\infty, H_\mu^\infty) = 0 \quad \text{if} \quad \lambda \neq \mu,$$

which, together with 1.2(3), implies:

(4) $$(H_\lambda^\infty, H_\mu^\infty)_1 = 0 \quad \text{if} \quad \lambda \neq \mu.$$

We set

(5) $$H_{[\lambda]}^\infty = \sum_{\mu \leq \lambda} H_\mu^\infty.$$

1.10. Proposition. *Fix* $\lambda \in \mathbf{R}$, $\lambda \geq 0$.

(i) *We have*

(1) $$(u, u)_1 \leq (1 + \lambda)(u, u), \qquad \textit{for all} \quad u \in H_{[\lambda]}^\infty.$$

(ii) *There exists a constant* $D'_\lambda > 0$ *such that*

(2) $$\| u * \varphi_\epsilon - u \| \leq D'_\lambda \cdot \epsilon \cdot \| u \|, \qquad \textit{for all} \quad u \in H_{[\lambda]}^\infty.$$

For $u \in H_\mu^\infty$, $(\mu \leq \lambda)$, the relation (1) is a consequence of 1.2(3). In the general case, it then follows from 1.9(3) and 1.9(4).

We get then (ii) from (i) and 1.8, with $D'_\lambda = (1 + \lambda)D$.

1.11. Corollary. *Let U be a closed G-invariant subspace of H on which the operators* $*\varphi_\epsilon$ $(0 < \epsilon \leq 1)$ *are compact. Then* $U_{[\lambda]}^\infty = U \cap H_{[\lambda]}^\infty$ *is finite dimensional.*

The relation 1.10(2) is also valid for u in the closure $Cl(H_{[\lambda]}^\infty)$ of $H_{[\lambda]}^\infty$ and shows that on that space, the identity mapping into H is a uniform limit of the $*\varphi_\epsilon$. Therefore, on $Cl(U_{[\lambda]}^\infty)$, the identity mapping into H is a uniform limit of compact operators, hence is compact; consequently, $Cl(U_{[\lambda]}^\infty)$ is finite dimensional.

Remark. The operator Δ acting on H^∞ is essentially self-adjoint [16: 4.4.4.3, p. 268]. Let then $\bar{\Delta}$ be its closure and N_λ the λ-eigenspace of $\bar{\Delta}$ in its domain. It is closed in H. Then 1.9, 1.10, 1.11 and their proofs remain valid if we replace H_λ^∞ by H_λ. However, we shall not need this.

2. Assumptions and conventions. Casimir elements and admissible Laplacians.

2.1. *Assumptions on G.* Our main case of interest is when G is the group of real points of a connected semi-simple $\mathbf{Q}$-group $\mathcal{G}$ and Γ is an arithmetic subgroup. However, as is often the case in the theory of automorphic forms, the results are valid in somewhat greater generality and it is also convenient to enlarge the framework for induction purposes.

We impose on G the assumptions made in [3]: G is a reductive real Lie group with finitely many connected components having a quotient by a finite group which is of finite index in the group of real points $\mathcal{G}'(\mathbf{R})$ of a connected reductive $\mathbf{R}$-group $\mathcal{G}'$ and Γ is a discrete subgroup of finite covolume whose intersection with the center of G is cocompact. Then Γ is commensurable with the product of its intersections with Z and with the derived group $\mathfrak{D}G^0$ of the (topological) identity component G^0 of G. We shall further assume that G has no compact semi-simple normal subgroup of strictly positive dimension. For the theorems we are concerned with in this paper, there is no harm in replacing Γ by a subgroup of finite index. We can therefore always reduce us to the case where G is connected. In that case we can write $\mathfrak{D}G$ as an almost direct product of subgroups G_i such that $\Gamma \cap \mathfrak{D}G$ is commensurable with the product of the $\Gamma_i = \Gamma \cap G_i$ and that Γ_i is irreducible in G_i [12: Section 5]. By the results of Margoulis (see [15], e.g.), there are three possibilities for G_i: either G_i has real rank one, or Γ_i is cocompact and arithmetically defined over some number field or Γ_i is not cocompact and G_i has a unique $\mathbf{Q}$-structure with respect to which Γ_i is arithmetic.

2.2. We let K be a maximal compact subgroup of G and $X = G/K$. Furthermore θ is the Cartan involution of G with respect to K and $\mathfrak{g} = \mathfrak{k} \oplus \mathfrak{p}$ the associated Cartan decomposition of $\mathfrak{g}$. If $P = MAN$ is a Langlands decomposition of a cuspidal subgroup, M and A are assumed to be θ-stable. As usual, $\Gamma_P = \Gamma \cap P$. We have $\Gamma_N = \Gamma \cap N = \Gamma \cap AN$ and $\Gamma_M = \Gamma_P/\Gamma_N$ is identified to a subgroup of M via the projection $P \to P/AN \simeq M$.

We note that it cannot be ruled out that M has a non-trivial compact normal semi-simple factor, so that our assumption on G is not strictly hereditary. However this is harmless: again Γ is commensurable with the product of its intersections with the center of M and the derived group $\mathfrak{D}M^0$ of M^0 and there is a decomposition of $\mathfrak{D}M^0$ similar to the one given above for $\mathfrak{D}G^0$, except that in the second case there may be some semi-simple compact factor.

2.3. *Casimir elements and admissible Laplacians.* For brevity, we say that a real valued bilinear symmetric form $F(x, y)$ on $\mathfrak{g}$ is *admissible* if it can be written

(1) $$F(x, y) = \mathrm{Tr}(\tau(x) \circ \tau(y)), \qquad (x, y \in \mathfrak{g}),$$

where $\tau: G \to \mathbf{GL}_N(\mathbf{R})$ is a finite dimensional real representation of G with finite kernel which maps K into $\mathbf{O}(n)$ and $\mathfrak{p}$ into symmetric matrices. Such a form is non-degenerate, invariant under G and θ, therefore under all Cartan involutions of G and satisfies

(2) $$F(\mathfrak{k}, \mathfrak{p}) = 0,$$

$F|\mathfrak{k}$ (resp. $F|\mathfrak{p}$) is negative (resp. positive) non-degenerate.

We let $C_{F,\mathfrak{g}}$ or $C_\mathfrak{g}$ or simply C, depending on the risks of confusion, be the Casimir element in $U(\mathfrak{g})$ associated to F, namely

(3) $$C_{F,\mathfrak{g}} = \Sigma\, x_i \cdot x_i',$$

where (x_i) is a basis of $\mathfrak{g}$ and (x_i') the dual basis with respect to F.

If H is a reductive subgroup of G, then the restriction of F to $\mathfrak{h}$ is admissible.

The *admissible Laplacian* (associated to F), denoted $\Delta_{F,\mathfrak{g}}$, $\Delta_\mathfrak{g}$ or Δ, is then by definition

(4) $$\Delta_{F,\mathfrak{g}} = 2C_{F,\mathfrak{k}} - C_{F,\mathfrak{g}}.$$

Viewed as an element of $U(\mathfrak{g})$, it commutes with $U(\mathfrak{k})$ and is fixed under the extension of $\mathrm{Ad}\, k$ to $U(\mathfrak{g})$ for all $k \in K$. In the sequel, we let (x_i), $(1 \leq i \leq n = \dim G)$ be a basis of $\mathfrak{g}$ whose first m (resp. last $n - m$) elements span $\mathfrak{p}$ (resp. $\mathfrak{k}$) and form there an orthonormal basis with respect to F (resp. $-F$). We have then

(5) $$C_{F,\mathfrak{g}} = \sum_1^m x_i^2 - \sum_{m+1}^n x_j^2,$$

(6) $$C_{F,\mathfrak{k}} = -\sum_{m+1}^n x_j^2 = \Delta_{F,\mathfrak{k}},$$

(7) $$\Delta_{F,\mathfrak{g}} = -\sum_1^n x_i^2.$$

In the sequel, F is fixed and will usually be omitted from the notation. We assume then that if M is a reductive subgroup of G, the elements $C_{\mathfrak{m}}$ and $\Delta_{\mathfrak{m}}$ are also associated to F.

If (π, V) is a continuous G-module and $C_{\mathfrak{g}}$ acts as a scalar multiple of Id. on V^∞, then we denote the eigenvalue of $C_{\mathfrak{g}}$ on V^∞ by one of the symbols $c_{\mathfrak{g}}(\pi)$, $c(\pi)$, $c_{\mathfrak{g}}(V)$ or $c(V)$.

2.4. The Casimir element operating on $L^2(\Gamma\backslash G)^\infty$ is symmetric. Its eigenvalues are real and the eigenspaces corresponding to different eigenvalues are orthogonal. Given a K-type r, we let

$$(1) \qquad A^\infty_{\lambda,r} = \{u \in L^2(\Gamma\backslash G)^\infty_r \mid Cu = \lambda u\}$$

$$(2) \qquad A^\infty_{[\lambda],r} = \sum_{\mu \geq \lambda} A^\infty_{\mu,r}$$

$$(3) \qquad B^\infty_{\lambda,r} = \{u \in L^2(\Gamma\backslash G)^\infty_r \mid \Delta u = \lambda u\},$$

$$(4) \qquad B^\infty_{[\lambda],r} = \sum_{\mu \leq \lambda} B^\infty_{\mu,r}.$$

Let $c_r \in \mathbf{R}$ be such that $C_{\mathfrak{k}} = c_r \cdot \text{Id.}$ in a representation of K of type r. Then

$$(5) \qquad A^\infty_{\lambda,r} = B^\infty_{2c_r-\lambda} \quad \text{and} \quad A^\infty_{[\lambda],r} = B^\infty_{[2c_r-\lambda],r}.$$

Of course $B^\infty_{\lambda,r} = 0$ for $\lambda < 0$, hence $A^\infty_{\lambda,r} = 0$ if $\lambda > 2c_r$. Let finally

$$(6) \qquad B^\infty_\lambda = \{u \in L^2(\Gamma\backslash G)^\infty \mid \Delta u = \lambda u\},$$

$$(7) \qquad B^\infty_{[\lambda]} = \sum_{\mu \leq \lambda} B^\infty_\mu.$$

2.5. The operators C and Δ acting on $L^2(\Gamma\backslash G)^\infty$ are essentially self-adjoint (see [16: p. 268–9]). Let $\bar{C}$ and $\bar{\Delta}$ denote their closures in $L^2(\Gamma\backslash G)$. On $L^2(\Gamma\backslash G)_r$, the operators $\bar{C}$ and $\bar{\Delta}$ differ by a scalar multiple of the identity, hence they have the same domain. We let

$$(1) \qquad A_{\lambda,r} = \{u \in \operatorname{dom} \bar{C} \cap L^2(\Gamma\backslash G)_r \mid \bar{C}u = \lambda u\}$$

$$(2) \qquad B_{\lambda,r} = \{u \in \operatorname{dom} \bar{\Delta} \cap L^2(\Gamma\backslash G)_r \mid \bar{\Delta}u = \lambda u\} \quad (\lambda \in \mathbf{R}, r \in \hat{K})$$

and, as before,

$$(3)\qquad A_{[\lambda],r} = \sum_{\mu \geq \lambda} A_{\lambda,r}, \qquad B_{[\lambda],r} = \sum_{\mu \leq \lambda} B_{\lambda,r} \quad (\lambda \in \mathbf{R};\ r \in \hat{K}).$$

$$(4)\qquad B_\lambda = \{u \in \operatorname{dom} \bar{\Delta} \mid \bar{\Delta} u = \lambda u\}, \qquad B_{[\lambda]} = \sum_{\mu \leq \lambda} B_\mu .$$

Then 2.4(5) is also valid for the spaces in (3), (4). In fact we have:

2.6. Lemma. *The space $A^\infty_{\lambda,r}$ (resp. $B^\infty_{\lambda,r}$) is dense in $A_{\lambda,r}$ (resp. $B_{\lambda,r}$) ($\lambda \in \mathbf{R}$, $r \in \hat{K}$).*

We have only to prove that for $A_{\lambda,r}$. Since $C \in Z(\mathfrak{g})$, the convolution $*\alpha$ ($\alpha \in C^\infty_c(G)$) commutes with C on $L^2(\Gamma\backslash G)^\infty$. From this it follows that it commutes with $\bar{C}$ on dom $\bar{C}$. In fact, if $u \in \operatorname{dom} \bar{C}$, and $u_j \in L^2(\Gamma\backslash G)^\infty$ ($j \in \mathbf{N}$) are such that $u_j \to u$ and Cu_j has a limit v, which is then $\bar{C}u$ by definition, we have

$$u_j * \alpha \to u * \alpha \quad \text{and} \quad C(u_j * \alpha) = Cu_j * \alpha \to v * \alpha,$$

which means that $(\bar{C}u) * \alpha = \bar{C}(u * \alpha)$. If α is moreover K-invariant, then it leaves $A_{\lambda,r}$ stable. Let now $\{\alpha_j\}$ be a Dirac sequence consisting of K-invariant functions in $C^\infty_c(G)$ (e.g., the functions φ_ϵ of Section 1, where ϵ runs through a sequence tending to zero). Then

$$u_j = u * \alpha_j \in A^\infty_{\lambda,r} \quad \text{and} \quad u_j \to u.$$

3. Decomposition of $L^2(\Gamma\backslash G)$.

3.1. We recall here some basic facts about the spectral decomposition of $L^2(\Gamma\backslash G)$. We denote by $L^2(\Gamma\backslash G)_d$ the *discrete spectrum*, i.e., the span of the irreducible closed G-submodules of $L^2(\Gamma\backslash G)$, by $L^2(\Gamma\backslash G)_{ct}$ the *continuous spectrum*, i.e., the orthogonal complement of the discrete spectrum and finally by $L^2(\Gamma\backslash G)_{\text{res}}$ the *residual spectrum*, i.e., the orthogonal complement in the discrete spectrum of the cuspidal spectrum ${}^0L^2(\Gamma\backslash G)$. The discrete spectrum is a countable Hilbert direct sum of irreducible G-modules with finite multiplicities.

3.2. Let *P be a proper cuspidal subgroup and $P = MAN$ the* Langlands decomposition of P. Let V be an irreducible unitary representation of M. For $\mu \in \mathfrak{a}^*_c$, let $\mathbf{C}_\mu$ denote the one-dimensional representation of A

with character μ. As usual, the tensor product $V \otimes \mathbf{C}_\mu$ is viewed as a representation of P on which N acts trivially. We let

$$I_{P,V,\mu} = \mathrm{Ind}_P^G(V \otimes C_{\rho_P + i\mu}) \tag{1}$$

be the induced representation from P to G from $V \otimes C_{\rho_P + i\mu}$. It is unitary if $\mu \in \mathfrak{a}^*$, our only case of interest in this paper.

We recall that the space of $I_{P,V,\mu}$ can be identified as a K-module with the space of $\mathrm{Ind}_{K_M}^K(V)$. In particular, the K-types of $I_{P,V,\mu}$ are independent of μ and if $r \in \hat{K}$ appears in $I_{P,V,\mu}$, then, by Frobenius reciprocity, one element of the set r_M of K_M-types occurring in r must be present in V. Furthermore there exists a constant e_P depending only on G and P such that

$$c_{\mathfrak{g}}(I_{P,V,\mu}) = e_P + c_{\mathfrak{m}}(V) - (\mu, \mu), \qquad (\mu \in \mathfrak{a}^*). \tag{2}$$

3.3. We let

$$L_{P,V} = \int_{\mathfrak{a}^{*+}}^{\oplus} I_{P,V,\mu} d\mu \tag{1}$$

be the continuous integral of the $I_{P,V,\mu}$ over the positive Weyl chamber $\mathfrak{a}^{*+}$ defined by the set $\Phi(P, A)$ of roots of P with respect to A. Thus $L_{P,V}$ is the space of L^2-cross sections, with respect to the Lebesgue measure $d\mu$, of a "Hilbert bundle" over $\mathfrak{a}^{*+}$ whose fibre over μ is $I_{P,V,\mu}$.

We note that if J is a finite set of K-types, then $I_{P,V,\mu,J}$ may be canonically identified as a K-module to some fixed finite dimensional space U (see 3.2) so that

$$L_{P,V,J} \cong L^2(\mathfrak{a}^{*+}, U). \tag{2}$$

If $f \in L^\infty_{P,V,J}$, $z \in Z(\mathfrak{g})$, then zf is given by

$$(z \cdot f)(\mu) = z(I_{P,V,\mu}) f(\mu) \quad (\mu \in \mathfrak{a}^{*+}). \tag{3}$$

3.4. Let $\mathfrak{C}$ be the set of pairs (P, V), where P runs through a representative set of Γ-conjugacy classes of proper cuspidal subgroups and V through a minimal spanning set of irreducible M-submodules in $L^2(\Gamma_M \backslash M)_d$.

The theory of Eisenstein series supplies a G-isomorphism of $L_{P,V}$ ($(P, V) \in \mathfrak{C}$) onto a closed subspace of $L^2(\Gamma\backslash G)_{ct}$. We shall call such subspaces and the irreducible constituents of $L^2(\Gamma\backslash G)_d$ *elementary subspaces*. If $J \subset \hat{K}$ is finite, then the sum of isotypic components of K-types belonging to J of an elementary subspace will also be called an elementary subspace of $L^2(\Gamma\backslash G)_J$. The following facts will be used.

(i) The continuous spectrum is a countable Hilbert direct sum of elementary subspaces $L_{P,V}$ ($(P, V) \in \mathfrak{C}$), where the (P, V) runs through a representative system under a suitable equivalence relation.

(ii) Let $\hat{\bigoplus}_{s\in S} H_s$ be a Hilbert direct sum decomposition of $L^2(\Gamma\backslash G)_{\text{res}}$ into closed irreducible G-invariant subspaces. There exists $(P_s, V_s) \in \mathfrak{C}$ and $x_s \in \mathfrak{a}$ such that

$$c_{\mathfrak{g}}(H_s) = c_{\mathfrak{m}}(V_s) + d_P - (x_s, x_s), \qquad (s \in S), \tag{1}$$

(see 2.3 for the notation). Moreover the x_s are bounded independently of V_s, the constant d_P depends only on P and if H_s contains a K-type r, then V_s contains a K_M-type in r_M. (These properties come from the fact that H_s is the quotient of a generalized principal series representation.)

These facts are proved in [10], although (i) is not explicitly stated there (except in adelic form in Appendix II). A more detailed discussion, including (i) explicitly, is given in [13].

4. Finiteness theorems.

4.1. Theorem. *Let $\lambda \in \mathbf{R}$ and $r \in \hat{K}$. Then the spaces $A_{[\lambda],r}$ and $B_{[\lambda],r}$ (see 2.6) are finite dimensional and contained in the sum of finitely many constituents of the discrete spectrum. They are annihilated by an ideal of $Z(\mathfrak{g})$ of finite codimension and consist of automorphic forms. The eigenvalues of $\bar{C}$ (resp. $\bar{\Delta}$) in $L^2(\Gamma\backslash G)_r$ have finite multiplicities and tend to $-\infty$ (resp. $+\infty$).*

The last assertion obviously follows from the first one for arbitrary λ's. To prove the first one it suffices, by 2.4, 2.6, to show that either $A^{\infty}_{[\lambda],r}$ or $B^{\infty}_{[\lambda],r}$ is finite dimensional.

(i) Let U be a G-invariant closed subspace of $L^2(\Gamma\backslash G)$ on which the convolution operators $*\alpha$ ($\alpha \in C_c^\infty(G)$ are compact. Then 1.11 holds and shows that

$$\dim(B^{\infty}_{[\lambda]} \cap U) < \infty. \tag{1}$$

We have then also, in view of 2.4

$$\dim(B^{\infty}_{[\lambda],r} \cap U),\ \dim(A^{\infty}_{[\lambda],r} \cap U) < \infty. \tag{2}$$

These spaces are then equal to their closures. By a well-known theorem of Gelfand and Piatetski-Shapiro [10: p. 41], our assumption is fulfilled by the cuspidal spectrum ${}^0L^2(\Gamma\backslash G)$. In particular this proves the first assertion when Γ is cocompact, i.e., when G has no proper cuspidal subgroups. We then proceed by induction on dim G and assume G has a proper cuspidal subgroup.

(ii) We now show that $L^2(\Gamma\backslash G)^{\infty}_{ct,r}$ does not contain any non-zero eigenfunction of the Casimir operator C; in fact, slightly more generally, let us prove that, given a non-constant polynomial $P[X] \in \mathbf{C}[X]$, every element of $L^2(\Gamma\backslash G)^{\infty}_{ct}$ annihilated by $P(C)$ is zero. If u belongs to an elementary subspace $L_{P,V}$ (see 3.3), then

$$(P(C)u)(\mu) = P(C(I_{P,V,\mu}))\cdot u(\mu), \qquad (\mu \in \mathfrak{a}^{*+}.$$

But it follows from 3.2(2) that the set of μ's for which $P(C(I_{P,V,\mu}) = 0$ has measure zero. Hence $u = 0$. But any $u \in L^2(\Gamma\backslash G)^{\infty}_{ct}$ annihilated by $P(C)$ is a convergent sum (in L^2) of such elements belonging to elementary subspaces. This proves are assertion.

(iii) We know now that $A^{\infty}_{[\lambda],r}$, hence also $B^{\infty}_{[\lambda]\cdot r}$, belongs to the discrete spectrum. Furthermore, by (i), its intersection with the cuspidal spectrum is finite dimensional. There remains to consider the intersection of $A^{\infty}_{[\lambda],r}$ with the residual spectrum. Assume that $u \in H^{\infty}_{s,r}$ is a non-zero eigenfunction of C with eigenvalue $\mu \geq \lambda$. Then 3.4(1) shows that $c_m(V_s)$ is also bounded from below. Moreover, as recalled in 3.4, V_s contains a K_M-type belonging to r_M. Since this set is finite, the induction assumption, applied to M, shows that there are only a finite number of possibilities for V_s, hence ultimately for s. As a consequence

$$A^{\infty}_{[\lambda],r} \cap L^2(\Gamma\backslash G)_{\text{res}}$$

is contained in finitely many irreducible closed invariant subspaces of the residual spectrum, hence it is finite dimensional. This then shows that $A^{\infty}_{[\lambda],r}$ and $B^{\infty}_{[\lambda],r}$ are finite dimensional and contained in finitely many constituents of the discrete spectrum. The theorem follows.

4.2. Theorem. *Let J be a finite set of K types, $P(X)$ a non-constant polynomial in one indeterminate and c a strictly positive real number. Fix a Hilbert sum decomposition $L^2(\Gamma\backslash G) = \bar{\oplus}_i W_i$ of $L^2(\Gamma\backslash G)$ into elementary subspaces* (3.4). *Then there exists a closed subspace U of $L^2(\Gamma\backslash G)_J$, whose orthogonal complement is a finite sum of spaces $W_{i,J}$, such that $P(C_\mathfrak{g})$ has an inverse on U^∞ which extends to a bounded operator on U, whose norm is $\leq c$.*

We may write $P(X)$ as a product of linear factors $(X + b)$ by a non-zero constant. Therefore it suffices to consider the case where $P(X) = X + b$ for some $b \in \mathbf{C}$.

Write $b = b_0 + ib_1$ $(b_0, b_1 \in \mathbf{R})$. In $L^2(\Gamma\backslash G)_{d,J}$ there are by 4.1 only finitely many elementary subspaces $W_{i,J}$ on which

$$c_\mathfrak{g}(W_i) > -b_0 - |c^{-2} - b_1^2|^{1/2}. \tag{1}$$

Let U_1 be their orthogonal complement in $L^2(\Gamma\backslash G)_{d,J}$. On any elementary subspace W_j such that $W_{j,J}$ is $\neq 0$, contained in U_1, we have

$$c_\mathfrak{g}(W_j) + b_0 \leq -|c^{-2} - b_1^2|^{1/2},$$

$$(c_\mathfrak{g}(W_j) + b_0)^2 \geq |c^{-2} - b_1^2| \geq c^{-2} - b_1^2,$$

$$|c_\mathfrak{g}(W_j) + b| \geq c^{-1}$$

$$|c_\mathfrak{g}(W_j) + b|^{-1} \leq c;$$

hence $(C_\mathfrak{g} + bId)^{-1}$ is bounded, with norm $\leq c$, on each such $W_{j,J}$ and therefore also on their orthogonal sum U_1.

Consider now the elementary subspaces $W_i = L_{P,V}$ contained in the continuous spectrum and satisfying:

$$L_{P,V,J} \neq 0 \quad \text{and} \quad c_\mathfrak{m}(V) + e_P + b_0 > -|c^{-2} - b_1^2|^{1/2}, \tag{2}$$

with e_P as in 3.2(2). The first condition implies that V contains one of the K_M-types in J_M (3.2). Therefore 4.1 shows that there are only finitely many such spaces for a given P. Since the P's under consideration are finite in number, we see that altogether there are only finitely many elementary subspaces W_i in $L^2(\Gamma\backslash G)_{ct}$ satisfying (2). Let then U_2 be the orthogonal complement in $L^2(\Gamma\backslash G)_{ct,J}$ of the sum of the J-isotypic subspaces of the W_i's. For any $L_{P,V}$ with $L_{P,V,J} \neq 0$, contained in U_2, we have

$$c_{\mathfrak{m}}(V) + e_P + b_0 \leq -|c^{-2} - b_1^2|^{1/2}, \tag{3}$$

and therefore, by 3.2(2)

$$\begin{aligned} c_{\mathfrak{g}}(I_{P,V,\mu}) + b_0 &= -(\mu, \mu) + c_{\mathfrak{m}}(V) + e_P + b_0 \\ &\leq -|c^{-2} - b_1^2|^{1/2}, \qquad (\mu \in \mathfrak{a}^*), \end{aligned}$$

$$(c_{\mathfrak{g}}(I_{P,V,\mu}) + b_0)^2 \geq |c^{-2} - b_1^2| \geq c^{-2} - b_1^2;$$

as above, we conclude that

$$|c_{\mathfrak{g}}(I_{P,V,\mu}) + b|^{-1} \leq c, \qquad (\mu \in \mathfrak{a}^*). \tag{4}$$

Then the assignment $f \mapsto Uf$, where Uf is given by

$$Uf(\mu) = (c_{\mathfrak{g}}(I_{P,V,\mu}) + b)^{-1} \cdot f(\mu), \qquad (\mu \in \mathfrak{a}^*), \tag{5}$$

defines a bounded operator on $L_{P,V,J}$, with norm $\leq c$, which is an inverse to $C_{\mathfrak{g}} + bId$ on $L_{P,V,J}$ in view of 3.3(3). The theorem follows, with $U = U_1 \oplus U_2$.

4.3. Corollary. *Let $f \in L^2(\Gamma\backslash G)$ be K-finite and $C_{\mathfrak{g}}$-finite. Then f is Z-finite, contained in the sum of finitely many elementary subspaces of the discrete spectrum. In particular it is an automorphic form and belongs to $L^2(\Gamma\backslash G)^\infty$.*

It is understood here that if $Q(X) \in \mathbf{C}[X]$ is a polynomial in one indeterminate then $Q(C_{\mathfrak{g}}) \cdot f$ is defined *a priori* as a distribution. It is said to be $C_{\mathfrak{g}}$-finite if the space of these distributions, as $Q(X)$ runs through $\mathbf{C}[X]$, is finite dimensional. This is equivalent to requiring the existence of a non-constant $P(X) \in \mathbf{C}[X]$ such that $P(C_{\mathfrak{g}}) \cdot f = 0$. If that is the case, then, as is well known and easy to check, f is also annihilated by a non-constant polynomial in the Laplacian $\Delta_{\mathfrak{g}}$ (2.3(4)). Since the latter is elliptic, $f \in C^\infty(\Gamma\backslash G)$ and so $Q(C_{\mathfrak{g}}) \cdot f$ is just obtained by differentiation. It belongs to $C^\infty(\Gamma\backslash G)$, but not necessarily to L^2. It is however a consequence of 4.3 that it does so for any $Q(X)$.

The second assertion of 4.3 is a consequence of the first one. The first one follows from the slightly more precise:

(*) *Let $P(X)$ be a non-constant polynomial and J a finite set of K-types.*
2 *Then the space $W = \{f \in L^2(\Gamma\backslash G)_J \mid P(X)f = 0\}$ is finite dimensional, contained in finitely many elementary subspaces of the discrete spectrum and belongs to $L^2(\Gamma\backslash G)_d^\infty$.*

It suffices to prove that W is contained in finitely many elementary subspaces in $L^2(\Gamma\backslash G)_d$ since H_J is finite dimensional, Z-finite and equal to H_J^∞ for every irreducible unitary G-module H.

By 4.1(ii), we already know that $L^2(\Gamma\backslash G)_{ct}^\infty$ does not contain any non-zero element annihilated by $P(C)$. If $f \in L^2(\Gamma\backslash G)_{ct}$ is a zero of $P(C)$, then so is $f * \alpha$ for any $\alpha \in C_c^\infty(G)$. But $f * \alpha \in L^2(\Gamma\backslash G)_{ct}^\infty$ hence $f * \alpha = 0$. Since f is a L^2-limit of such elements, we see that $W \subset L^2(\Gamma\backslash G)_d$.

By 4.2 there exists a closed invariant subspace U of $L^2(\Gamma\backslash G)_{d,J}$, whose complement U_1 is the sum of finitely many elementary subspaces, such that $P(C)$ is bijective on U^∞. It suffices to show that $W \subset U_1$. Obviously, $W = W \cap U \oplus W \cap U_1$, therefore there remains to prove that $W \cap U = (0)$. But this is clear since $(W \cap U)^\infty = W^\infty \cap U^\infty$ is dense in $W \cap U$ and $P(C)$ is bijective on U^∞.

Remark. 4.3 is an analogue on $\Gamma\backslash G$ of a result of K. Okamoto [14] on G and was suggested by it: a square integrable function on a connected semi-simple Lie group G with finite center which is K-finite on both sides and is an eigenfunction of C belongs to finitely many elements of the discrete series. (In fact, it is proved there under an assumption (A), but that is now known to be true in general.)

4.4. Our next goal is to establish: (a) similar results for the space $B_{[\lambda]}$; (b) the compactness of certain convolution operators $*\alpha$. Originally, we had proved first (a) and then deduced (b) for fast decreasing functions. N. Wallach showed us how to derive (b) directly from 4.1, for any $\alpha \in L^1(G)$ in fact, and we give his proof. Then (a) will follow from 1.11 and 4.1.

We recall that if (π, V) is a unitary representation, then the operator $\pi(\alpha)$, $(\alpha \in L^1(G))$, defined by

$$\pi(\alpha)\cdot v = \int_G \alpha(g)\pi(g)\cdot v dg,$$

is bounded, with norm bounded by the L^1-norm of α.

In particular, if $\alpha_j \to \alpha$ in $L^1(G)$, then $\pi(\alpha_j) \to \pi(\alpha)$ uniformly.

4.5. Lemma (N. Wallach). *Let (π, V) be a unitary representation. Assume that V has a basis consisting of K-finite eigenfunctions of C and that, for any given K-type r, the eigenvalues of C in V_r^∞ have finite multiplicities and tend to $-\infty$. Then $\pi(\alpha)$ is a compact operator for any $\alpha \in L^1(G)$.*

Since the elements of $L^1(G)^\infty$ are dense in $L^1(G)$ and a uniform limit of compact operators is compact, we may assume $\alpha \in L^1(G)^\infty$. It is then an L^1-limit of finite linear combinations of its isotypic components α_r $(r \in \hat{K})$. Hence we may assume that $\alpha = \alpha_r \in L^1(G)_r^\infty$ for some $r \in \hat{K}$. Then $\pi(\alpha)$ maps V into V_r and annihilates V_s for $s \in \hat{K}, s \neq r$. It suffices to show that its restriction of V_r is compact. By assumption, V_r has a basis (e_i) $(i = 1, 2, \ldots)$ consisting of eigenvectors of C and we can arrange that if λ_i is the eigenvalue of e_i, then $\lambda_i \to -\infty$. Let k be a strictly positive integer. Then

$$\pi(C^k\alpha)\cdot e_i = \pi(\alpha)\cdot\pi(C^k)\cdot e_i = \lambda_i^k\pi(\alpha)e_i,$$

therefore, since $\pi(C^k\alpha)$ is bounded

$$\|\pi(\alpha)e_i\| = |\lambda_i|^{-k}\|\pi(C^k\alpha)\cdot e_i\| \leq |\lambda_i|^{-k}\|e_i\|\,\|\pi(C^k\alpha)\|.$$

Since $|\lambda_i| \to \infty$, this proves the lemma.

4.6. Theorem. (i) *For any $\alpha \in L^1(G)$, the operator $*\alpha$ is compact on $L^2(\Gamma\backslash G)_d$,*

(ii) *For every $\lambda \in \mathbf{R}$, the space $B_{[\lambda]}$ (see 2.5) is finite dimensional, contained in finitely many irreducible G-invariant subspaces of the discrete spectrum and consists of automorphic forms.*

(i) follows from 4.5 and 4.1.

Since Δ commutes with K, the space $B_{[\lambda]}$ is stable under the projection of $L^2(\Gamma\backslash G)$ onto its K-isotypic subspaces $L^2(\Gamma\backslash G)_r$ $(r \in \hat{K})$. Moreover, the (algebraic) direct sum of the K-finite vectors being dense in $L^2(\Gamma\backslash G)$ [16: 4.4.2], the same is true in $B_{[\lambda]}$. But each $B_{[\lambda],r}$ $(r \in \hat{K})$ is finite dimensional, equal to $B_{[\lambda],r}^\infty$ and contained in the discrete spectrum. As a consequence, $B_{[\lambda]}$ also belongs to the discrete spectrum and $B_{[\lambda]}^\infty$ is dense in $B_{[\lambda]}$. By (i) we can apply 1.11 to $B_{[\lambda]}^\infty$, hence the latter is finite dimensional. It is then a finite sum of subspaces $B_{[\lambda],r}$ $(r \in \hat{K})$ and the last assertion of (ii) follows from 4.1.

4.7. Let ${}^0A_{[\lambda],r}$ be the intersection of $A_{[\lambda],r}$ with the cuspidal spectrum and $N(\lambda)$ its dimension. In [6] H. Donnelly proves the inequality

$$\overline{\lim_{\lambda \to -\infty}} \frac{N(\lambda)}{d/2} \leq (4\pi)^{-d/2} \frac{\mathrm{Vol}(\Gamma\backslash X}{\Gamma(d/2 + 1)} \dim r,$$

where $d = \dim X$. Recall that in the compact case this would be an equality. For $\mathbf{SL}_2(\mathbf{Z})$ and some congruence subgroups, there is also equality. It is not known whether this inequality also holds for the discrete spectrum, or to what extent it may even be an equality as in the compact case.

If $\alpha \in C_c^\infty(G)$, then $*\alpha$ is not only a compact but also a trace class operator on the cuspidal spectrum. It is known to be a trace class operator in the discrete spectrum if G has real rank one. Whether this holds true in general is an open question.

5. Applications to relative Lie algebra cohomology and harmonic forms. In this section we assume some familiarity with relative Lie algebra cohomology [3: I] and its relations with the de Rham cohomology on $\Gamma\backslash X$ [3: VII].

5.1. Let (π, V) be a unitary representation of G. Then there is on the complex of relative Lie algebra cohomology

(1) $$C^*(\mathfrak{g}, K; V^\infty \otimes E) = \mathrm{Hom}_K(\Lambda(\mathfrak{g}/\mathfrak{k}), V^\infty \otimes E),$$

a natural scalar product with respect to which one defines an adjoint ∂ to d and a Laplace operator $\Delta = d\partial + \partial d$ [3: II]. If

(2) $$\eta = \sum_I \eta_I \omega^I \in C^i(\mathfrak{g}, K; V^\infty \otimes E),$$

where the ω^I are exterior products of elements of a basis of $\mathfrak{g}/\mathfrak{k} = \mathfrak{p}$ and $\eta_I \in V^\infty \otimes E$, we have, by Kuga's formula [3: II, 2.5]

(3) $$(\Delta\eta)_I = (C(E) - C(V))\eta_I.$$

It is understood that the scalar product on $\mathfrak{g}/\mathfrak{k}$ is defined by the form F which underlies the definition of $C_\mathfrak{g}$ (see 2.3). It is also assumed that the K-invariant scalar product on E is "admissible," i.e., the space $\mathfrak{p}$ is represented by self-adjoint operators. This can always be achieved.

In the sequel, we let J denote the set of K-types occurring in $\Lambda(\mathfrak{g}/\mathfrak{k}) \otimes E^*$. Since (1) can also be written

(4) $$C^*(\mathfrak{g}, K; V^\infty \otimes E) = \mathrm{Hom}_K(\Lambda(\mathfrak{g}/\mathfrak{k}) \otimes E^*, V^\infty),$$

we see that

(5) $$C^*(\mathfrak{g}, K; V^\infty \otimes E) \subset \mathrm{Hom}_K(\Lambda(\mathfrak{g}/\mathfrak{k}), V_J^\infty \otimes E).$$

The following lemma is well-known.

5.2. LEMMA. *We have $H^*(\mathfrak{g}, K; V^\infty \otimes E) = 0$ if either V has no K-type in common with $\Lambda^*(\mathfrak{g}/\mathfrak{k}) \otimes E^*$ or if $C(E) - C(V)$ has a bounded inverse on $V_J^\infty \otimes E$.*

5.1(4) shows that, if our first condition is fulfilled, then the complex itself is already zero. Assume now that $(C(E) - C(V))^{-1}: V_J^\infty \otimes E \to V_J \otimes E$ extends to a bounded operator. Then it maps bijectively $V_J^\infty \otimes E$ onto itself and 5.1(3)(5) show that Δ has an inverse on $C^*(\mathfrak{g}, K; V^\infty \otimes E)$, which necessarily commutes with d. If now $\eta \in C^*(\mathfrak{g}, K; V^\infty \otimes E)$ is closed, then so is $\mu = \Delta^{-1} \cdot \eta$ and we have

$$\eta = \Delta\mu = (d\partial + \partial d)\mu = d\partial\mu,$$

hence η is a coboundary.

5.3. THEOREM. *There exists a finite set of mutually orthogonal elementary subspaces U_i $(i \in I)$ of $L^2(\Gamma\backslash G)$ such that*

(1) $$H^*(\mathfrak{g}, K; L^2(\Gamma\backslash G)^\infty \otimes E) = \bigoplus_{i \in I} H^*(\mathfrak{g}, K; U_i^\infty \otimes E).$$

We can write $L^2(\Gamma\backslash G) = Q \oplus R$, where Q is a Hilbert orthogonal sum of elementary subspaces containing (resp. not containing) a K-type in J. We have

(2)
$$H^*(\mathfrak{g}, K; L^2(\Gamma\backslash G)^\infty \otimes E) = H^*(\mathfrak{g}, K; Q^\infty \otimes E) \oplus H^*(\mathfrak{g}, K; R^\infty \otimes E),$$

and, by 5.2, the second term in (2) is zero. By 4.2, Q_J contains a closed G-invariant subspace U, spanned by elementary subspaces, on which $(c_\mathfrak{g}(E)I - C_\mathfrak{g})^{-1}$ extends as a bounded operator, and whose complement is a finite orthogonal sum of elementary subspaces U_{iJ} $(i \in I)$. Then (1) follows from 5.2.

5.4. *Eigenforms of the Laplacian on* $\Gamma\backslash X$. Let $\tilde{E}$ be the local system on $\Gamma\backslash X$ defined by E, viewed as a Γ-module. Let $\nu: G \to X$ be the canonical projection and $A^\infty(\Gamma\backslash X; \tilde{E})$ the complex of smooth $\tilde{E}$-valued differential forms on $\Gamma\backslash X$. The map which assigns to $\eta \in A^\infty(\Gamma\backslash X; E)$ the form η^0 on $\Gamma\backslash G$ defined by

$$\eta^0(g) = \sigma(g)^{-1}(\eta \circ \nu)(g),$$

yields an isomorphism of graded differential modules

$$(2) \qquad \nu^*: A^\infty(\Gamma\backslash X; \tilde{E}) \tilde{\to} C^*(\mathfrak{g}, K; C^\infty(\Gamma\backslash G) \otimes E),$$

as was pointed out first by Y. Matsushima and S. Murakami (see [3: VII] for details and references). This map also extends to give an isomorphism of graded vector spaces

$$(3) \qquad \begin{aligned} A_{ms}(\Gamma\backslash X; \tilde{E}) &\cong C^*(\mathfrak{g}, K; C_{ms}(\Gamma\backslash G) \otimes E) \\ &= \mathrm{Hom}_K(\Lambda(\mathfrak{g}/\mathfrak{k}), C_{ms}(\Gamma\backslash G) \otimes E), \end{aligned}$$

where A_{ms} and C_{ms} refer to forms with measurable coefficients and to measurable functions respectively. Once an admissible product on E and an invariant metric on $\Gamma\backslash G$ are chosen, this map sends the space $A(\Gamma; E)_2$ of L^2-forms onto $C^*(\mathfrak{g}, K; L^2(\Gamma\backslash G) \otimes E)$. Moreover, the differential d_X, the codifferential ∂_X and the Laplacian Δ_X on $\Gamma\backslash X$, defined first say on the space $A_c^\infty(\Gamma\backslash X; \tilde{E})$ of compactly supported smooth forms, go over to the d, ∂ and Δ considered above. In particular, via these identifications, Δ_X is expressed in terms of the Casimir operator by 5.1(3).

Let $\bar{d}_X$ and $\bar{\partial}_X$ be the closures of d_X and ∂_X operating on $A_c^\infty(\Gamma\backslash X; \tilde{E})$. Since $\Gamma\backslash X$ is complete, these are the same as the strong or weak closures; moreover, Δ_X operating on $A_c^\infty(\Gamma\backslash X; \tilde{E})$ is essentially self-adjoint and its closure $\bar{\Delta}_X$ is equal to $\bar{d}_X\bar{\partial}_X + \bar{\partial}_X\bar{d}_X$, in the operator sense. (For all this, see [4; 7; 8].) View now $\bar{\Delta}_X$ as an operator on $C^*(\mathfrak{g}, K; L^2(\Gamma\backslash G) \otimes E)$ via ν^*. Then $\bar{\Delta}_X$ is also the closure of Δ_X operating on any larger space than compactly supported smooth forms on which it is still symmetric. As such, we can take $C^*(\mathfrak{g}, K; L^2(\Gamma\backslash G)^\infty \otimes E)$. Then we see from 5.1(3) that the domain of $\bar{\Delta}_X$ is $C^*(\mathfrak{g}, K; \mathrm{dom}\ \bar{C}_\mathfrak{g} \otimes E)$, where, as earlier, $\bar{C}_\mathfrak{g}$ is the (self-adjoint) closure of $C_\mathfrak{g}$ operating on $L^2(\Gamma\backslash G)^\infty$.

By 5.1(5), we also have

$$(4)\qquad C^*(\mathfrak{g}, K; \operatorname{dom} \bar{C}_{\mathfrak{g}} \otimes E) \subset \operatorname{Hom}_K(\Lambda(\mathfrak{g}/\mathfrak{k}), (\operatorname{dom} \bar{C}_{\mathfrak{g}})_J \otimes E),$$

and Kuga's formula, which extends to $\bar{\Delta}_X$ and $\bar{C}_{\mathfrak{g}}$ for a given K-type, shows that the space of eigenforms with eigenvalue λ for $\bar{\Delta}_X$ becomes identified to a subspace of a finite sum of spaces

$$\operatorname{Hom}_K(\Lambda(\mathfrak{g}/\mathfrak{k}), A_{2c(r)-\lambda} \otimes E), \qquad (r \in J).$$

Therefore we derive from 4.1:

5.5. THEOREM. *Let $\mathcal{H}^q_\lambda$ be the space of square integrable $\tilde{E}$-valued q-eigenforms of $\bar{\Delta}_X$ on $\Gamma\backslash X$ with eigenvalue λ and $\mathcal{H}^q_{[\lambda]} = \Sigma_{\mu \leq \lambda} \mathcal{H}^q_\mu$. Then $\mathcal{H}^q_{[\lambda]}$ is finite dimensional. If $\eta \in \mathcal{H}^q_\lambda$, then its lift $\nu^*(\eta)$ has coefficients which are automorphic forms contained in $L^2(\Gamma\backslash G)^\infty_d \otimes E$.*

For the space $\mathcal{H}^*_{(2)}(\Gamma\backslash X; \tilde{E})$ of $\tilde{E}$-valued square integrable harmonic forms this can be made more precise.

5.6. PROPOSITION. *Let $L^2(\Gamma\backslash G)_d = \hat{\oplus}_{i\in N} H_i$ be a Hilbert orthogonal direct sum decomposition of the discrete spectrum into closed irreducible G-invariant subspaces. Let I be the set of $i \in \mathbf{N}$ for which $H_{i,J} \neq 0$ and $c(H_i) = c(E)$. Then I is finite and we have*

$$(1)\qquad \mathcal{H}^*_{(2)}(\Gamma\backslash X; \tilde{E}) = \oplus_{i\in I} \operatorname{Hom}_K(\Lambda\mathfrak{g}/\mathfrak{k} \otimes E^*, H_i^\infty)$$

$$= \oplus_{i\in I} H^*(\mathfrak{g}, K; H_i^\infty \otimes E)$$

$$= H^*(\mathfrak{g}, K; L^2(\Gamma\backslash G)^\infty_d).$$

For $i \in I$, the operators d, ∂ and Δ are identically zero on $C^*(\mathfrak{g}, K; H_i^\infty \otimes E)$ (see [3: II, 3.1]). This proves the second equality and shows that the second term of (1) is contained in the first one, or more precisely in the image of the latter by ν^*. By our first assumption each summand in the second term is $\neq 0$, hence the sum is finite since $\mathcal{H}^*_{(2)}(\Gamma\backslash X; \tilde{E})$ is finite dimensional. (In fact, there is also a representation theoretic reason for this: the cohomology being non-zero, H_i must have the same infinitesimal character as E, hence there are only finitely many possibilities for the isomorphism class of H_i and therefore for H_i itself since the discrete spectrum has finite multiplicities.)

If $\eta \in \mathcal{H}^*_{(2)}(\Gamma\backslash X; E)$, then $\nu^*(\eta) \in C^*(\mathfrak{g}, K; A^\infty_{c(E),J} \otimes E)$ by the remarks made in 5.4, hence $\nu^*(\eta)$ is contained in the second term of (1). This proves the first equality in (1). If now $i \notin I$, then $H^*(\mathfrak{g}, K; H_i^\infty \otimes E) = 0$ by 5.2, so that the last equality is a consequence of 4.2 and 5.2.

THE INSTITUTE FOR ADVANCED STUDY
PRINCETON, NEW JERSEY

YALE UNIVERSITY
NEW HAVEN, CONNECTICUT

REFERENCES

[1] A. Borel, Stable real cohomology of arithmetic groups, *Annales E.N.S.* Paris (4) 7 (1974), 235–272.

[2] ———, Cohomology of arithmetic groups, *Proc. Internat. Congr. Math.* 1 (1974), 435–442.

[3] ———, and N. Wallach, Continuous cohomology, discrete subgroups and representations of reductive groups, *Ann. of Math. Studies*, No. 94, Princeton Univ. Press, Princeton, N.J., 1980.

[4] J. Cheeger, On the Hodge theory of Riemannian pseudomanifolds, *Proc. Sympos. Pure Math.*, 36 (1980), 91–146. A.M.S. Providence R.I.

[5] H. Donnelly, On the point spectrum for finite volume symmetric spaces of negative curvature, *Comm. Partial Diff. Equations* 6 (1981), 963–992.

[6] ———, On the cuspidal spectrum for finite volume symmetric spaces, *J. Differential Geometry* 17 (1982), 239–253.

[7] M. Gaffney, A special Stokes theorem for Riemannian manifolds, *Annals of Math.* (2) 60 (1954), 140–145.

[8] ———, Hilbert space methods in the theory of harmonic integrals, *Trans. A.M.S.* 78 (1955), 426–444.

[9] H. Garland, The spectrum of noncompact G/Γ and the cohomology of arithmetic groups, *Bull. A.M.S.* 75 (1969), 807–811.

[10] R. P. Langlands, On the functional equations satisfied by Eisenstein series, *Lecture Notes in Math.*, vol. 544, Springer Verlag, Berlin and New York, 1976.

[11] R. Narasimhan, *Lectures on Topics in Analysis*, Mimeographed Notes, Tata Institute of Fundamental Research, Bombay, 1965.

[12] M. S. Raghunathan, *Discrete Subgroups of Lie Groups*, Erg. Math. u.i. Grenzgeb. 68, Springer Verlag, Berlin-New York-Heidelberg 1972.

[13] M. S. Osborne and G. Warner, *The Theory of Eisenstein Systems*, Academic Press, New York, 1981.

[14] K. Okamoto, On induced representations, *Osaka M. J.* 4 (1967), 85–94.

[15] J. Tits, Travaux de Margulis sur les sous-groupes discrets de groupes de Lie, *Sém. Bourbaki* 1975–76, Exp. 482, *Lecture Notes in Mathematics* 567, Springer Verlag, 1977.

[16] G. Warner, *Harmonic Analysis on Semi-Simple Lie Groups* I, Grund. Math. Wiss. 188, Springer Verlag, Berlin-New York-Heidelberg, 1972.

122.

L^2-Cohomology and Intersection Cohomology of Certain Arithmetic Varieties

in *E. Noether in Bryn Mawr*, Springer-Verlag (1983) 119–131

Introduction

L^2-cohomology of non-compact manifolds and intersection cohomology are two fairly recent topics. In principle, they are independent from one another, one belonging to differential geometry or analysis, the other to algebraic topology and algebraic geometry. However, some interesting relations between the two appeared almost from the beginning. To what extent they are special cases of rather general phenomena is not known at present, but, for a number of people, the evidence is tantalizing enough to make it appear worthwhile to try to investigate this further. From a broader point of view, one can view this as an attempt to add a new chapter to a topic with an already long history: the connections between differential geometry and topology or, more specifically, the representation of topological invariants by means of analytical objects.

A first well-known example is the isomorphism of the first complex cohomology group $H^1(M; \mathbf{C})$ of a compact Riemannian surface M with the direct sum $S_2(M) + \bar{S}_2(M)$ of the spaces of holomorphic and of antiholomorphic differentials. A far-reaching generalization is given by the Hodge theorem which, for a compact smooth Riemannian manifold M, provides an isomorphism between the ith cohomology group $H^i(M; \mathbf{C})$ and the space $\mathscr{H}^i(M)$ of harmonic i-forms ($i = 0, 1, \ldots$). If M is kählerian, in particular, if it is a projective variety, this can be considerably refined by the consideration of the (p, q)-types and the introduction of a Hodge structure on $H^{\cdot}(M; \mathbf{Q})$. The research alluded to above may then be viewed as an attempt to provide similar connections when M is a singular space, in particular, a projective variety. Roughly speaking, if Σ is the singular locus, one tries to establish an isomorphism between the space of L^2-harmonic forms of $M - \Sigma$ with respect to a suitable metric, and the middle intersection cohomology

* Institute for Advanced Study, School of Mathematics, Princeton, NJ 08540, U.S.A.

of M, in the sense of M. Goresky and R. MacPherson [10], [11]. So far, this has been done notably for conical singularities and conical metrics (J. Cheeger [7]), variations of Hodge structures on punctured curves (S. Zucker [14] and for the minimal compactifications of an arithmetic quotient of a bounded symmetric domain in the so-called **Q**-rank one case (i.e. when the singular locus consists of one closed smooth subvariety of pure dimension). Those are the "certain arithmetic varieties" of our title. The main goal of this paper is to discuss this last case. To make the paper a bit more self-contained, we shall also recall some basic notions and facts on L^2-cohomology and on middle intersection cohomology, the latter, however, only for spaces with one singular stratum. For a more general discussion of those topics and of their proven or conjectured relationships, we refer to [8].

1. L^2-Cohomology

1.1. Let M be a Riemannian manifold. We let $A^{\cdot}(M)$ denote the space of smooth complex-valued differential forms on M. The metric defines at each point $x \in M$ a scalar product $(\ ,\ )_x$ on the exterior algebra $\Lambda T^*(M)_x$ of the cotangent space $T^*(M)_x$ at x, whence a, possibly infinite, scalar product

(1) $$(\mu, \eta) = \int_M (\mu_x, \eta_x)_x\, dv \qquad (\mu, \eta \in A^{\cdot}(M)),$$

where dv is the Riemannian volume element. We let $A^{\cdot}_2(M)$ be the space of η for which (η, η) is finite and $A^{\cdot}_{(2)}(M)$ the space of η such that $\eta, d\eta \in A^{\cdot}_2(M)$. The space $A^{\cdot}_{(2)}(M)$ is stable under the exterior differentiation d and the space $H^{\cdot}_{(2)}(M; \mathbf{C})$ of L^2-cohomology of M with complex coefficients is by definition the cohomology of $A^{\cdot}_{(2)}(M)$. Let as usual $\partial\colon A^p(M) \to A^{p-1}(M)$ be the boundary operator. It is formally adjoint to d in the sense that we have

(2) $$(\mu, \partial\eta) = (d\mu, \eta) \quad (\mu, d\mu, \eta, \partial\eta \in A^{\cdot}_2(M),\ \mu \text{ or } \eta \text{ of compact support}).$$

Let $\mathscr{H}^{\cdot}_{(2)}(M; \mathbf{C})$ be the space of L^2-harmonic forms, i.e. of the $\eta \in A^{\cdot}_2(M)$ which satisfy $d\eta = \partial\eta = 0$. There are natural homomorphisms

(3) $$\mathscr{H}^{\cdot}_{(2)}(M, \mathbf{C}) \xrightarrow{\alpha} H^{\cdot}_{(2)}(M; C) \xrightarrow{\beta} H^{\cdot}(M; \mathbf{C}),$$

which are isomorphisms if M is compact, but are neither surjective nor injective in general. If M is complete, our main case of interest in this paper, then α is injective (see, for example, [2]) and (2) is valid for $\mu, d\mu, \eta, \partial\eta \in A^{\cdot}_2(M)$ (a result due originally to M. Gaffney, see, for example, [2] for a proof).

1.2. The cokernel of α is either zero or infinite dimensional. Thus if we wish to compare L^2-cohomology with some topological invariant which is known to be finite dimensional, a prerequisite is that $\mathscr{H}^{\cdot}_{(2)}(M; \mathbf{C})$ be equal to $H^{\cdot}_{(2)}(M; \mathbf{C})$ and be finite dimensional. One may then wonder why one should look at $H^{\cdot}_{(2)}$ at all, rather than limit oneself to $\mathscr{H}^{\cdot}_{(2)}$. The point is that, in order to try to set up such isomorphisms, one often goes over to sheaves of germs of forms. This can be done easily enough with $A^{\cdot}_{(2)}(M)$ (see below for some illustrations), but hardly with harmonic forms.

1.3. A theorem of K. Kodaira asserts that β and $\beta \circ \alpha$ have the same image. When $M = \Gamma\backslash X$ is a non-compact quotient of a negatively curved symmetric space X by an arithmetic group, for instance when $N = \mathbf{SL}_n(\mathbf{Z})\backslash\mathbf{SL}_n(\mathbf{R})/\mathbf{SO}(n)$, it has proved very useful to show that $\beta \circ \alpha$ is an isomorphism in some range, which can be described in terms of roots [2], [3]. More recently, S. Zucker has shown the existence of such a range in which β itself is an isomorphism [15].

1.4. Let E be a local system on M defined by a finite dimensional unitary representation (τ, E_0) of the fundamental group of M. There is then a scalar product on $\Lambda T^*(M)_x \otimes E_x$ and the formula 1.2(1) again defines a scalar product on the space $(M; E)$ of E-valued smooth differential forms. We let $A^{\cdot}_{(2)}(M; E)$ be the complex $A^{\cdot}$ of smooth E-valued forms η such that η and $d\eta$ are square integrable. Its cohomology is the space $H^{\cdot}_{(2)}(M; E)$ of L^2-cohomology of M with coefficients in E. The previous remarks extend obviously to this case.

1.5. We refer to [7] for Hilbert space definitions of L^2-cohomology.

2. Middle Intersection Cohomology for Spaces with a Smooth Singular Set

2.1. Let V be a locally compact space and S a closed subspace. We assume that $V - S$ is a smooth orientable n-manifold and that S is a manifold of pure *even* codimension $2m$, the "singular stratum" of V. Moreover, for $x \in S$, it is assumed that there exists a closed submanifold L_x of $V - S$, called the link of x, such that x has in V a fundamental set of neighborhoods U homeomorphic to a product of a neighborhood B of x in S by a cone $c(L_x)$ over L_x with vertex x. There are some further regularity assumptions which we do not spell out, such as the existence of a PL-structure on V, L_x for which S is a subcomplex and $U \xrightarrow{\sim} B \times c(L_x)$ is PL, etc. cf. [10], [11].

2.2. Let k be a field $C. = C.(V; k)$ be the inductive limit of the simplicial chains with coefficients in k for triangulations compatible with the given PL structure (the chains are allowed to be infinite if V is not compact). Then $IC.(V; k)$ is the space of i-chains ξ satisfying

(1) $$\dim(\xi \cap S) \leq i - m - 1, \qquad \dim(\partial\xi \cap S) \leq i - 1 - m - 1,$$

and the ith *middle intersection homology space* is

(2) $$IH_i(V; k) = H_i(IC.(V; k)).$$

To understand this, one should write the $-m - 1$ on the right-hand sides in (1) as $-2m + (m - 1)$. The difference $i - 2m$ would be the dimension of $\xi \cap S$ if ξ would intersect S in general position. If we replace $m - 1$ by 0 then, by using dual cell subdivisions, one sees that one gets a complex defining cohomology (in complementary dimensions), at any rate if the links are connected [10]. If we replace $m - 1$ by $2m$, then there is no condition at all and we get homology. In fact, we already do so if we replace $m - 1$ by $2m - 2$ and assume again the links to be connected [10]. The $m-1$ is the "perversity." It could be replaced by any integer between 0 and $2m-2$ and this would lead to the definition of other intersection homology groups. By choosing $m-1$, we are in the middle of the possible perversities (or also in some sense half-way between homology and cohomology), whence the expression "middle intersection homology." The homology groups $IH_i(V;k)$ are topological invariants and, moreover, finite dimensional if V is compact.

2.3. If $i \leq m$, the right-hand side of the first relation in 2.2(2) is ≤ -1, hence ξ must have its support in $V-S$. If $i > n-m$, then the right-hand side is $\geq n-2m$, which is the dimension of S, hence the condition imposed on ξ itself is vacuous. From this and similar remarks on the second condition in 2.2(1) we see that if V is compact, then

(1) $$IH_i(V;k) = H_i(V;k) \quad \text{if } i > n-m,$$

(2) $$IH_i(V;k) = H_i(V-S) \quad \text{if } i \leq m-1,$$

(3) $$IH_m(V;k) \text{ is a quotient of } H_m(V-S;k),$$

(4) $$IH_{n-m}(V;k) \text{ is a subspace of } H_{n-m}(V;k).$$

If now $n = 2m$, and $i = m$, then the boundaries in IC_i are those of arbitrary chains while a cycle in IC_m is any cycle with support in $V-S$. As a consequence

(5) $$IH_m(V;k) = \mathrm{Im}((H_m(V-S;k) \to H_m(V;k)) \quad \text{if } n = 2m.$$

This is also true if V is not compact, provided ordinary homology is understood with infinite chains.

2.4. One can also define similarly intersection homology with respect to a local system E of finite dimensional vector spaces over k on $V-S$. Note that any simplex in $IC_.$ is not contained in S, and this allows one easily to define the complex $IC_.(V;E)$ of intersection chains with coefficients in E, and the corresponding middle intersection homology groups $IH_.(V;E)$. By taking finite chains one defines intersection homology with compact supports.

2.5. It will be more convenient for us to use a cohomological grading. We set, by definition

(1) $$IC^i(V;E) = IC_{n-i}(V;E), \qquad IH^i(V;E) = IH_{n-i}(V;E) \qquad (i \in \mathbf{N}),$$

and similarly for compact supports. If S is empty, this means that we take Poincaré duality, when stated as an isomorphism between homology and cohomology, as a definition! However, Poincaré duality becomes again a theorem when formulated as a pairing between cohomology groups of complementary dimensions.

2.6. *Poincaré duality.* One main property of the middle intersection cohomology is that it satisfies Poincaré duality: There exists a perfect pairing

$$IH_c^i(V;E) \times IH^{n-i}(V;E^*) \to k \qquad (i \in \mathbf{N}),$$

where E^* is the local system contragredient to E.

2.7. *Local groups.* By using locally finite simplicial chains, we can define a sheaf $\mathscr{I}\mathscr{C}^*(V; E)$ of germs of middle intersection cochains. It is fine. The stalks of the derived sheaf $\mathscr{H}(\mathscr{I}\mathscr{C}^{\cdot})$ are given by

(1) $$\mathscr{H}^q(\mathscr{I}\mathscr{C}^{\cdot})_x = \begin{cases} E & \text{if } q = 0, \\ 0 & \text{if } q \geq 1. \end{cases} \qquad (x \in V - S);$$

(2) $$\mathscr{H}^q(\mathscr{I}\mathscr{C}^{\cdot})_x = \begin{cases} H^q(L_x; E) & \text{if } q < m, \\ 0 & \text{if } q \geq m. \end{cases} \qquad (x \in S)$$

For $x \in S$, let

$$H^q_x(\mathscr{I}\mathscr{C}^{\cdot}) = \varinjlim_U IH^q_c(U; E) \qquad (q \in \mathbf{N}),$$

where U runs through a fundamental set of neighborhoods of x. Note the place of the subscript x. A more orthodox notation for the left-hand side would be $H^q(f^!_x \mathscr{I}\mathscr{C}^{\cdot})$, where f_x is the inclusion of x in X. We also have

(3) $$\mathscr{H}^q_x(\mathscr{I}\mathscr{C}^{\cdot}) = \begin{cases} 0 & (q \leq n - m), \\ H^{2m-1-n+q}(L_x; E) & (q > n - m). \end{cases}$$

2.8. These properties essentially characterize the intersection cohomology sheaf, up to quasi-isomorphism. To be more precise let $\mathscr{S}$ be a differential graded sheaf on V. Assume

(i) $\mathscr{S}$ is fine, graded by positive degree $i\mathscr{S}|_{V-S}$ is quasi-isomorphic to the locally constant sheaf $\mathscr{E}$ defined by E. The stalks of $\mathscr{H}\mathscr{S}$ on S form a locally constant sheaf on S.

(ii) The stalks of $\mathscr{H}\mathscr{S}$ satisfy 2.7(1) and either

(1) 2.7(2), *where the isomorphisms for $q < m$ satisfy a certain naturality condition, the "attachment condition,"*

or

(2) $$\mathscr{H}^q(\mathscr{S})_x = 0 \quad \text{for } q \geq m \qquad \text{and} \qquad \mathscr{H}^q_x(\mathscr{S}) = 0 \quad \text{for } q \leq n - m,$$

then $\mathscr{S}$ is quasi-isomorphic to $\mathscr{I}\mathscr{C}^{\cdot}(V; E)$. In particular

(3) $$H^{\cdot}(\Gamma(\mathscr{S})) = IH^{\cdot}(V; E), \qquad IH^{\cdot}(\Gamma_c(\mathscr{S})) = IH^{\cdot}_c(V; E),$$

where Γ and Γ_c refer to continuous and compactly supported continuous sections, respectively. Here $\mathscr{H}_x(\mathscr{S})$ is defined by

(4) $$\mathscr{H}_x(\mathscr{S}) = \varinjlim_U H^i(\Gamma_c(\mathscr{S})).$$

In fact, this characterization holds also for $\mathscr{S}$ not fine, but one has then to use hypercohomology.

2.9. We have assumed $V - S$ to be orientable for convenience. If not, one has to twist by the orientation sheaf at appropriate places. For the proofs of all of the above statements, in much greater generality, see [11].

3. A Method to Relate L^2-Cohomology and Intersection Cohomology

3.1. We keep the setup of Section 2. Assume we are given a Riemannian metric on $M = V - S$ and are interested in knowing whether $H^{\cdot}_{(2)}(M; E) = IH^{\cdot}(V; E)$. We describe one way to try to go about it.

Consider on V the presheaf which associates to an open subset U the complex $A^{\cdot}_{(2)}(M \cap U; E)$. This presheaf is in fact a sheaf $\mathscr{A}^{\cdot}$, and the space of its continuous sections over V is just $A^{\cdot}_{(2)}(M; E)$. *Assume it is fine.* Then, in order to prove that $H^{\cdot}_{(2)}(M; E) = IH^{\cdot}(V; E)$, it suffices to show that $\mathscr{A}^{\cdot}$ is quasi-isomorphic to $\mathscr{IC}^*$ and for this it is enough to see that the stalks of the derived sheaf $\mathscr{H}\mathscr{A}^{\cdot}$ satisfy one of the sets of requirements in 2.8. The conditions 2.8(i) and 2.7(1) are obvious, so what remains is to study the derived sheaf $\mathscr{H}\mathscr{A}^{\cdot}$ at points of S. By construction, $H^{\cdot}(\Gamma(\mathscr{A}\,|\,U)) = H^{\cdot}_{(2)}(U \cap M; E)$, therefore

$$\mathscr{H}^{\cdot}(\mathscr{A}^{\cdot})_x = \varinjlim_U H^{\cdot}_{(2)}(U \cap M; E), \tag{1}$$

where U runs through a fundamental set of neighborhoods of x in V. The space $\Gamma_c(\mathscr{A}\,|\,U)$ may be identified with the space $A^{\cdot}_{(2),rc}(U \cap M; E)$ of elements in $A^{\cdot}_{(2)}(U \cap M; E)$ whose support is relatively compact in V. We have therefore

$$\mathscr{H}^{\cdot}_x(\mathscr{A}^{\cdot}) = \varprojlim_U H^{\cdot}(A^{\cdot}_{(2),rc}(U \cap M; E)). \tag{2}$$

In practice, these inductive or projective systems are essentially constant, at any rate for a suitable fundamental set of neighborhoods. Therefore the main burden is to prove either that $H^{\cdot}_{(2)}(U \cap M; E)$ satisfies 2.7(2) or that

$$H^q_{(2)}(U \cap M; E) = 0 \quad \text{for } q \geq m \qquad \text{and} \tag{3}$$
$$H^r(A^{\cdot}_{(2),rc}(U \cap M; E)) = 0 \quad \text{for } r \leq n - m.$$

The impression I have gained so far is that the crucial test is whether the first part of (3) can be proved. If so, it is not unlikely that the argument will also yield 2.7(2) for $i < m$ and that some dual version of it will give the second part of (3). Each of these two suffices to complete the proof of the isomorphism.

3.2. Cheeger [7] considers (not exclusively) the case where $n = 2m$, i.e. where S consists of isolated points and the metric around $x \in S$ is "conical": In this case $U = c(L)$ is the cone over the link. If r is a coordinate along the generating lines of the cone, with x at the origin, then the metric is of the form $ds^2 = dr^2 + r^2\, ds_L^2$. It is not complete. The gist of the matter is indeed the computation of the L^2-cohomology of $c(L) - \{x\}$.

3.3. In [4], V is a complete smooth complex curve and the metric around $x \in S$ is the Poincaré metric of a punctured disc. If we view a neighborhood of x as a cone $c(L)$ over a circle L with coordinate t, the metric around x can be written

$$ds^2 = r^{-2} \cdot dr^2 + e^{-2/r} \cdot dt^2.$$

It is complete. If now E represents a variation of Hodge structure on M and $j: M \hookrightarrow V$ is the inclusion, then [14] proves an isomorphism

$$H^{\cdot}(V; j_* E) = H^{\cdot}_{(2)}(M; E). \tag{1}$$

It was seen later that the left-hand side is indeed middle intersection cohomology, so that this result could also be viewed as another example of an isomorphism between the latter and L^2-cohomology.

4. Locally Symmetric Varieties

4.1. In the sequel, $\mathscr{G}$ is a semi-simple connected **Q**-group, almost simple over **Q**. Let $G = \mathscr{G}(\mathbf{R})$ be the group of real points of $\mathscr{G}$ and K a maximal compact subgroup of G. We assume that $X = G/K$ is a bounded symmetric domain. Let Γ be an arithmetic subgroup of G. Then $M = \Gamma\backslash X$ has a canonical structure of quasi-projective variety and a minimal projective compactification [1], to be denoted V. To simplify we assume that Γ is neat (the subgroup of $\mathbf{C}^*$ generated by the eigenvalue of any element of Γ is torsion-free). In particular, Γ is torsion-free, hence $M = \Gamma\backslash X$ is a complex manifold. We assume, moreover, that G has **Q**-rank one (the maximal **Q**-split tori of $\mathscr{G}$ have dimension one). This implies first that M is not compact, hence $M \neq V$ and second that $S = V - M$ is itself a smooth projective variety. In fact, its connected components are smooth compact arithmetic quotients of lower-dimensional bounded symmetric domains, the "rational boundary components" [1].

4.2. We give here some examples.

(1) Let F be a totally real field, d its degree over **Q** and $\mathfrak{o}_F$ the ring of integers of F. Take for G the group $R_{k/\mathbf{Q}}\mathbf{SL}_2$ obtained by restriction of scalars from $\mathbf{SL}_2$ viewed as a k-group. Then $G = \mathbf{SL}_2(\mathbf{R})^d$, and X is the product H^d of d copies of the upper halfplane H. Take for Γ a subgroup of finite index of $\mathbf{SL}_2\, \mathfrak{o}_F$. Then $\Gamma\backslash X$ is a Hilbert–Blumenthal variety and V is obtained by compactifying M with finitely many points.

(2) Let k be a quadratic imaginary field, h a hermitian form over k^n of Witt index one and of signature (p, q) over **C** $(p \geq q \geq 1)$. This defines a **Q**-form of $\mathbf{SL}_n$ such that $G = \mathbf{SU}(p, q)$. In that case, X may be identified with an open subset of the Grassmannian $\mathbf{G}_{p,n}$ of p planes in $\mathbf{C}^n$ and, similarly, S with an open subset of $\mathbf{G}_{p-1,n-2}$. For $q = 1$, X is the unit ball in $\mathbf{C}^p$. In this case, Γ is a subgroup of finite index of the group of units of h.

(3) The group G is a **Q**-form of **Q**-rank one the exceptional group $\mathbf{E}_6$ of type ${}^2\mathbf{E}^{16}_{6,2}$ over **R** in the notation of [12]. Then X is 16-dimensional and each connected component of S is a compact quotient of the unit ball in $\mathbf{C}^5$.

4.3. The space M has a natural Riemannian metric, stemming from a G-invariant metric on X or from the Bergmann metric in any realization of X as a bounded domain. It is complete.

On M we consider a local system E associated to a Γ-module which is an irreducible rational representation E_0 of $\mathscr{G}$. Then E is not unitary with respect to Γ, unless it is trivial, so that 1.4 is not directly applicable. However, it is well-known and easily checked that $A^{\cdot}(M; E)$ can also be viewed as the space of smooth sections of a K-bundle over $\Gamma\backslash X$, with typical fiber $\Lambda T^*(M)_0 \otimes E_0$, whence the existence of a scalar product. Hence $H^{\cdot}_{(2)}(M; E)$ is again well-defined.

4.4. Theorem. *We keep the notation and assumptions of* 4.1. *Then $H^{\cdot}_{(2)}(M; E)$ is isomorphic to $IH^{\cdot}(V; E)$.*

In [15], S. Zucker conjectures that this theorem is true without any restriction on the **Q**-rank and also checks it in some cases where the **Q**-rank is one, in particular for Hilbert–Blumenthal varieties.

The validity of this conjecture in that last case has been used by Brylinski and Labesse [6] to express the Hasse–Weil zeta function of a canonical model of V, with respect to middle intersection cohomology, or rather to its analogue in the étale cohomology setting, in terms of automorphic L-functions. In fact, a prime reason of interest in this conjecture is the hope that, if true, it will help to extend such results to other arithmetic quotients of bounded symmetric domains.

5. Sketch of the Proof of Theorem 4.4.

5.1. For $\nu \in \mathbf{R}$ and f a function on the half line $\mathbf{R}^+ = [0, \infty)$ we define a "ν-norm" $\|\cdot\|_\nu$ by:

(1) $$\|f\|_\nu^2 = \int_0^\infty |f(r)|^2 e^{2\nu r}\, dr.$$

By $H^{\cdot}(\mathbf{R}^+, \nu)$ we denote the cohomology of the complex $C^{\cdot}_\nu$, where C^0_ν consists of the smooth functions such that f, df/dr have finite ν-norms and C^1_ν of the one-forms $f \cdot dr$, where f is smooth with finite ν-norm. Then, as a special case of 3.69 in [15], we have:

(2) $$H^0(\mathbf{R}^+, \nu) = \begin{cases} 0 & \text{if } \nu \geq 0, \\ \mathbf{C} & \text{if } \nu < 0, \end{cases} \qquad \dim H^1(\mathbf{R}^+, \nu) = \begin{cases} 0 & \text{if } \nu \neq 0, \\ \infty & \text{if } \nu = 0. \end{cases}$$

Let now $C^{\cdot}_{\nu,c}$ be the complex defined as above, except that we consider only functions which vanish in a neighborhood of the origin. Let $H^{\cdot}_c(\mathbf{R}^+, \nu)$ be the cohomology of $C^{\cdot}_{\nu,c}$. Then we have obviously

(3) $$H^0_c(\mathbf{R}^+, \nu) = 0;$$

a computation similar to the one yielding (2) shows:

(4) $$\dim H^1_c(\mathbf{R}^+, \nu) = \begin{cases} 0 & \text{if } \nu < 0, \\ \infty & \text{if } \nu = 0, \\ 1 & \text{if } \nu > 0. \end{cases}$$

These results remain true if $\mathbf{R}^+$ is replaced by $R^{\geq t} = [t, \infty)$ and in fact the maps

$$H^{\cdot}(\mathbf{R}^+, \nu) \to H^{\cdot}(\mathbf{R}^{\geq t}, \nu) \quad \text{and} \quad H_c(\mathbf{R}^{\geq t}, \nu) \to H_c(\mathbf{R}^+, \nu),$$

defined by restriction and inclusion are isomorphisms ($t > 0$).

5.2. We now go on in the setting of Section 4 and want to use the approach outlined in Section 3. We can start because the L^2-sheaf $\mathscr{A}$ on V is indeed fine: in [15], this is proved for certain Satake compactifications but the argument applies to all, and V is one. We have then to consider the L^2-cohomology around a point $x \in S$. A typical neighborhood U of x in V can be written $U = B \times c(L)$, where B is a ball in S with center x and L the link of x. Then

(1) $$U \cap M = B \times c(L)^0, \quad \text{where} \quad c(L)^0 = c(L) - \{x\}.$$

The metric on $U \times M$ is the product metric and a Künneth rule [15: 2.36] allows one to discard U. We are therefore reduced to considering $H^{\cdot}_{(2)}(c(L)^0; E)$. We shall write $c(L)^0$ as

(2) $$c(L)^0 = \mathbf{R}^+ \times L,$$

with the understanding that x is at infinity. With respect to the metric, this is an orthogonal decomposition but the space is by far not a metric product: The metric on $\{r\} \times L$ depends on r in a rather complicated way. It is warped [15].

The space L is fibered over a compact quotient Y of a symmetric space (not complex in general), with fibre a compact nilmanifold. To be more precise, let P be the group of real points of a minimal parabolic $\mathbf{Q}$-subgroup $\mathscr{P}$ of $\mathscr{G}$ and let $P = N.A.{}^0M$ be its usual Langlands decomposition: N is the unipotent radical of P and ${}^0M.A$ a Levi subgroup. Under the projection $\sigma: P \to P/N \simeq {}^0M.A$, the group A maps onto the identity component of the group of real points of a maximal (one-dimensional) $\mathbf{Q}$-split torus of P/N and ${}^0M$ is defined as the intersection of the kernels of certain characters. We have $Z_G(A) = {}^0M.A = {}^0M \times A$. Let $\Gamma_P = \Gamma \cap P$ and $\Gamma_N = \Gamma \cap N$. Then $\Gamma_P \subset {}^0M.N$. We assume ${}^0M$ and A to be stable under the Cartan involution θ associated to K. The rational boundary components correspond to the Γ-conjugacy classes of minimal parabolic $\mathbf{Q}$-subgroups of G. We assume P to be associated to the boundary component S_0 containing x. There exists a decomposition ${}^0M^0 = H.Z$ of the identity component of ${}^0M$ into an almost direct product of two normal $\mathbf{Q}$-subgroups H, Z, also stable under θ, with H semi-simple, such that the symmetric space $X_H = H/K_H$ ($K_H = K \cap H$)) is a bounded domain and the universal covering of S_0 and that the link L admits $N.Z/K_Z$ ($K_Z = K \cap Z$) as a universal covering. Let

$$\Gamma_1 = \Gamma \cap Z.A.N., \qquad \Gamma_Z = \Gamma_1/\Gamma_N, \qquad \Gamma_H = \Gamma_P/\Gamma_1.$$

Then $\Gamma_1 \subset Z.N$ and we have:

$$S_0 = \Gamma_H \backslash X_H \quad \text{and} \quad L = \Gamma_1 \backslash N.Z/K_Z.$$

This shows in particular that L is fibered over $Y = \Gamma_Z \backslash X_Z$, with typical fiber $\Gamma_N \backslash N$. Note that Γ_Z and Γ_H are torsion free since Γ is assumed to be neat.

We identify $\mathbf{R}$ to the Lie algebra of A and therefore A to the multiplicative group of strictly positive numbers. Let $A^+ = \exp \mathbf{R}^+ = \{r \in \mathbf{R}, r \geq 1\}$. We view $U \cap M$ as a fiber bundle over $Y \times A^+ = Y \times \mathbf{R}^+$, with typical fiber $\Gamma_N \backslash N$.

5.3. By a result of van Est [13] (proved first by K. Nomizu for trivial coefficients), we have

$$H^{\cdot}(\Gamma_N \backslash N; E) = H^{\cdot}(\mathfrak{n}; E), \tag{1}$$

where the right-hand side refers to Lie algebra cohomology, $\mathfrak{n}$ being the Lie algebra of N.

In ordinary cohomology, it is known that the spectral sequence of the fibration of L over Y degenerates at E_2, therefore

$$H^{\cdot}(L; E) = H^{\cdot}(Y; H^{\cdot}(\mathfrak{n}; E)) = H^{\cdot}(Y \times \mathbf{R}^+; H^{\cdot}(\mathfrak{n}; E) = H^{\cdot}(c(L)^0; E). \tag{2}$$

Note that, since X_Z is contractible to a point, we have

$$H^{\cdot}(Y; H^*(\mathfrak{n}; E)) = H^{\cdot}(\Gamma_Z; H^{\cdot}(\mathfrak{n}; E)), \tag{3}$$

where the right-hand side denotes Eilenberg–MacLane cohomology. In the L^2-case, there is a partial analogue, namely:

$$H^{\cdot}_{(2)}(c(L)^0; E) = H_{(2)}(Y \times \mathbf{R}^+; H^{\cdot}(\mathfrak{n}; E)). \tag{4}$$

However, the factor $\mathbf{R}^+$, which can be ignored in the topological case, plays an important role in the L^2-setting.

According to a theorem of B. Kostant (recalled in [5: III] or also in [15]) there is a multiplicity-free decomposition of $H^*(\mathfrak{n}; E)$ into irreducible rational ${}^0M.A$ modules (also irreducible under the identity component ${}^0M^0A$ of ${}^0M.A$). Write then

$$H^i(\mathfrak{n}; E) = \bigoplus_{1 \leq j \leq n_i} F_{i,j} \qquad (i \in \mathbf{N}), \tag{5}$$

where $F_{i,j}$ is such a module. Then $F_{i,j}$ is the tensor product of an irreducible ${}^0M$-module by a one-dimensional A-module. Let $\nu(i, j)$ be the real number such that e^r acts on $F_{i,j}$ by $e^{\nu(i,j)r}$. The $\gamma(i, j)$'s and, more precisely, the highest weights of the $F_{i,j}$ are explicitly described by Kostant's theorem. This information plays a basic role in the proof. However, since we leave out the computations, we do not need it explicitly here and omit it. Let $\rho \in \mathbf{R}$ be defined by:

$$e^{2\rho . r} = \det \operatorname{Ad} e^r | \mathfrak{n} \qquad (r \in \mathbf{R}). \tag{6}$$

Then we have

$$H^{\cdot}(c(L)^0; E) = \bigoplus_{i,j} H^{\cdot}(\mathbf{R}^+, \nu(i, j) - \rho) \otimes (H^{\cdot}(\Gamma_Z; F_{i,j})[-i]). \tag{7}$$

(As usual, if $C^{\cdot}$ is a graded complex and $i \in \mathbf{Z}$, then $C^{\cdot}[i]$ is the graded complex defined by $(C^{\cdot}[i])^j = C^{j+i}$ $(j \in \mathbf{Z})$.) This equality, in a different notation, is contained in 3,vi of [15].

On the other hand, we can write, by (2), (3), (5),

$$H^{\cdot}(L; E) = \bigoplus_{i,j} H^{\cdot}(\Gamma_Z; F_{i,j})[-i], \tag{8}$$

or, more explicitly,

$$H^q(L; E) = \bigoplus_{i,j} H^{q-i}(\Gamma_Z; F_{i,j}) \qquad (q \in \mathbf{N}). \tag{9}$$

5.4. We get a fundamental system of neighborhoods $\{U_t\}$ of x just by replacing in the above $\mathbf{R}^+$ by $\mathbf{R}^{\geq t}$ (t tending to infinity). This does not change the ordinary cohomology of $U \cap M$ (obviously) or the L^2-cohomology and the groups $H^{\cdot}(A^{\cdot}_{(2),rc}(U \cap M; E))$ of 3.1 by the end remark of 5.1. We have therefore to check the conditions of 2.8 directly for $U \cap M$.

The first step is to make sure that no infinite dimensional cohomology occurs. In view of 5.1(2) and 5.3(7), this amounts to proving that $H^{\cdot}(\Gamma_Z; F_{i,j}) = 0$ whenever $v(i,j) = \rho$. This can be shown *a priori* from the results of Section 1 in [4]. We assume this has been done. Then we can write

$$H^q(c(L)^0; E) = \bigoplus_{i,j} H^0(\mathbf{R}^+, v(i,j) - \rho) \otimes H^{q-i}(\Gamma_Z; F_{i,j}) \qquad (q \in \mathbf{N}). \tag{10}$$

In view of 5.1, the vanishing part of 2.7(2) is equivalent to:

(*) *Let $q \geq m$, $i,j \in \mathbf{N}$ and $j \leq n_i$. If $v(i,j) < \rho$, then $H^{q-i}(\Gamma_Z; F_{i,j}) = 0$.*

Note that, apart from the attachment condition, which causes no difficulty here, 2.7(2) for $q < m$ is equivalent to

(**) *Let $q < m$, and $i \in \mathbf{N}$. If $H^{q-i}(\Gamma_Z; F_{i,j}) \neq 0$, then $v(i,j) < \rho$.*

We concentrate here on (*). It is first rather easily seen that the proof can be reduced to the case where G is absolutely almost simple; then X is an irreducible bounded symmetric domain. From that point on, however, I have to proceed case by case using J. Tits' classification [12].

A fairly simple case is when Y is a point. Then the formulas (9), (10) simplify to

$$H^q(L; E) = \bigoplus_{1 \leq j \leq n_q} F_{q,j}; \tag{11}$$

$$H^q_{(2)}(c(L)^0; E) = \bigoplus_{1 \leq j \leq n_q} H^0(\mathbf{R}^+; v(q,j) - \rho) \otimes F_{q,j}. \tag{12}$$

In this case (*) follows rather simply from some root and Weyl group considerations and the explicit description of the highest weight of the $F_{i,j}$'s.

There are however some cases where Y is not a point. Some information about the cohomology of Γ_Z is then needed. Since Γ_Z is cocompact, the space $L^2(\Gamma_Z \backslash Z)$ is a countable Hilbert direct sum of irreducible unitary Z-modules, say

$$L^2(\Gamma_Z \backslash Z) = \widetilde{\bigoplus_{i \in I}} H_i, \tag{13}$$

and we have, for a finite dimensional Z-module F:

$$H^{\cdot}(\Gamma_Z; F) = \bigoplus_{i \in I} H^{\cdot}(\mathfrak{z}, K_Z; H_i^\infty \otimes F). \tag{14}$$

(See [5: VII, 5.2]; the right-hand side refers to relative Lie algebra cohomology. Only finitely many summands are non-zero.)

Given a finite dimensional Z-module F, let $c(Z, F)$ be the greatest integer q such that there exists some irreducible unitary Z-module H for which

$$H^q(\mathfrak{z}, K_Z; H^\infty \otimes F) \neq 0.$$

For instance, if $Z = \mathbf{SL}_2(\mathbf{R})$, then $c(Z, \mathbf{C}) = 2$ and $c(Z, F) = 1$ for every irreducible non-trivial F. Now if $q - i > c(Z, F)$ then $H^{q-i}(\Gamma_Z; F) = 0$. Therefore (*) is implied by

(***) *Let $q \geq m$ and $i, j \in \mathbf{N}$, with $j \leq n_i$; assume that $v(i, j) < \rho$. Then $q > c(Z, F_{i,j}) + i$.*

This condition can be verified in the remaining cases. The most interesting is the one labeled ${}^2\mathbf{A}^1_{n,d}$ in [12]. In this case $d/n + 1$ and, over $\mathbf{Q}$,

$$G = \mathbf{SU}_{(n+1)/d}(D, h),$$

where D is a central division algebra of the second kind over a quadratic imaginary field and h a non-degenerate hermitian form of index one on $D^{(n+1)/d}$. Then, $Z = \mathbf{SL}_d(\mathbf{C})$ and the necessary information on $c(Z, F_{i,j})$ is contained in Enright's paper [9].

References

[1] W. Baily and A. Borel. Compactification of arithmetic quotients of bounded symmetric domains. *Ann. Math.*, **84** (1966), 442–528.

[2] A. Borel. Stable real cohomology of arithmetic groups. *Ann. E.N.S., Paris* (4), 7 (1974), 235–272.

[3] ———. Stable real cohomology of arithmetic groups, II. *Manifolds and Lie Groups*. Papers in honor of Y. Matsushima. Progress in Math., 14. Birkhäuser: Boston, 1981, pp. 21–55.

[4] ——— and W. Casselman. L^2-cohomology of locally symmetric manifolds of finite volume. (To appear).

[5] ——— and N. Wallach. *Continuous Cohomology, Discrete Subgroups and Representations of Reductive Groups*. Annals of Math. Studies, 94. Princeton University Press: Princeton, NJ, 1980.

[6] J. L. Brylinski and J. P. Labesse. Cohomologie d'intersection et fonctions L de certaines variétés de Shimura. Preprint.

[7] J. Cheeger. On the Hodge theory of Riemannian pseudomanifolds. *Proc. Symp. Pure Math.*, 36. R.M.S.: Providence, R.I., 1980, pp. 91–146.

[8] ———, M. Goresky, and R. MacPherson. L^2-cohomology and intersection homology. In *Seminar on Differential Geometry*, edited by S. T. Yau. Annals of Math. Studies, 102. Princeton University Press: Princeton, N.J., 1982, pp. 303–340.

[9] T. Enright. Relative Lie algebra cohomology and unitary representations of complex Lie groups. *Duke Math. J.*, **46** (1979), 513–525.

[10] M. Goresky and R. MacPherson. Intersection homology theory. *Topology*, **19** (1980), 135–162.

[11] ———. Intersection homology theory, II. (To appear in *Invent. Math.*)

[12] J. Tits. Classification of algebraic semisimple groups. *Proc. Symp. Pure Math.*, 9. A.M.S.: Providence, R.I., 1966, pp. 33–62.

[13] W. T. van Est. A generalization of the Cartan Leray spectral sequence, I, II. *Proc. Koninkl. Ned. Ad. v. Wet.-Amsterdam, Series A*, **61** (1958), 399–413.

[14] S. Zucker. Hodge theory with degenerating coefficients: L^2-cohomology in the Poincaré metric. *Ann. Math.*, **109** (1979), 415–476.

[15] ———. L^2-cohomology of warped products and arithmetic groups. *Invent. Math.* **70** (1982), 169–218.

123.

On Free Subgroups of Semi-simple Groups

Enseign. Math. (2) **29** (1983) 151–164

In [8], P. Deligne and D. Sullivan show that an odd-dimensional sphere $S^{2n+1}(n \geq 1)$ admits a non-commutative free group of isometries which acts freely. This answers a question raised in [7] (for n even, it was settled there for n odd), recalled in [11], and motivated by the fact that this property implies a strong form of the Hausdorff-Banach-Tarski paradox [6] (see §4). The present paper grew out of the attempt to extend this result to homogeneous spaces of compact semi-simple Lie groups having zero Euler characteristic. More generally we shall prove:

THEOREM A. *Let* U *be a non-trivial connected semi-simple compact Lie group. Then* U *contains a non-commutative free subgroup* Γ *with the following property: for any proper closed subgroup* V, *any element* $\gamma \in \Gamma - \{1\}$, *acting by left translations on* U/V, *has exactly* $\chi(U/V)$ *fixed points, where* $\chi(U/V)$ *is the Euler characteristic of* U/V.

In particular, Γ acts freely if $\chi(U/V) = 0$. Note that since every translation by an element of U is homotopic to the identity, the number of fixed points is the smallest possible in view of the Lefschetz fixed point theorem. The proof shows in fact that there are "many" such subgroups: given $m \in \mathbf{N}$, the set of m-tuples of elements in U which do not generate freely a free subgroup with the property mentioned in the theorem is contained in a set of Haar measure zero in U^m.

The result of [6] alluded to above also extends to actions of free groups *with commutative isotropy groups* (called "locally commutative" in [6]). This suggests looking for such actions in case $\chi(U/V) \neq 0$. We shall see indeed in §4, by a completely different argument, that they always exist (see Theorem 3). This in particular answers a question of T. J. Dekker for S^4 [6].

Let w be a reduced non-trivial word in m letters and their inverses. It defines an obvious map $f_w : U^m \to U$. The main step to prove Theorem A is to show that f_w is a dominant map. In particular Im f_w contains a non-empty open set. Furthermore, $\chi(U/V)$ can be described purely in Lie group terms, by a theorem of Hopf-Samelson [10], (recalled below). This suggests proving more general results for semi-simple algebraic groups, and deriving the above ones as special cases. We shall do so and show first

THEOREM B. *Let* G *be an algebraic connected semi-simple group. Let* $m \in \mathbf{N}$, *and* $f_w : G^m \to G$ *be the map associated to a non-trivial element* w *in the free group on* m *letters. Then* $\operatorname{Im} f_w$ *is not contained in any proper subvariety* $(m \geqslant 2)$.

The proof is by induction on dim G and uses a variant of the key idea of [8]. In this statement, we have implicitly identified an algebraic group H with the group $H(\Omega)$ of its points in a "universal field" Ω, i.e., an algebraically closed extension of infinite transcendence degree of a prime field. Assume now that G is defined over a field K of infinite transcendence degree. It follows from [17], and was well-known over $\mathbf{R}$ or $\mathbf{C}$, that $G(K)$ contains many non-commutative free subgroups, in fact that m "sufficiently general" elements are free generators of a subgroup ($m \in \mathbf{N}$). Theorem B implies a sharpening of that assertion, namely the existence of non-commutative free subgroups in $G(K)$ all of whose elements, except the identity, are outside a given proper subvariety (or even outside a countable union of proper subvarieties defined over a common field of finite transcendence degree over the prime field, see Theorem 2 for the precise statement). As an application, we deduce

THEOREM C. *Let* K *be a field of infinite transcendence degree over its prime field and assume* G *to be defined over* K. *Then there exists* $g = (g_i) \in G(K)^m$ *whose components* g_i *freely generate a subgroup* Γ *of* $G(K)$ *such that every* $\gamma \in \Gamma - \{1\}$ *is regular and generates a Zariski-dense subgroup of the unique maximal torus* T_γ *containing it.*

In fact, there are many such g's. In some sense, a "generic" $g \in G(K)^m$ always gives rise to such a subgroup. If $K = \mathbf{R}, \mathbf{C}$ or is a p-adic field, the set of such g's is dense in the ordinary topology.

Given a closed subgroup H of G, set $\chi(G, H) = 0$ if H does not contain any maximal torus of G. If it does, and T is one, then set $\chi(G, H) = [N_G(T) : N_H(T)]$. Then Theorem C implies the:

COROLLARY. *Every* $\gamma \in \Gamma - \{1\}$, *acting by left translations on* $G(K)/H(K)$, *has at most* $\chi(G, H)$ *fixed points.*

In particular, Γ acts freely if H does not contain any maximal torus of G. Assume now that it contains one, say T_0, which we may assume to be defined over K. Assume further that all maximal K-tori of H are conjugate under $H(K)$ and set

$$\chi(G(K), H(K)) = [N_{G(K)}(T_0) : N_{H(K)}(T_0))] .$$

Then we shall see that γ acts freely on $G(K)/H(K)$ if T_γ is not conjugate to T_0 under $G(K)$ and has $\chi(G(K), H(K))$ fixed points otherwise.

The condition on H and the second alternative hold either if K is algebraically closed or if $K = \mathbf{R}$ and $G(K)$ is compact. In that last case, $\chi(G(K), H(K)) = \chi(G(K)/H(K))$ by [10], and Theorem A follows.

I wish to thank D. Sullivan for having sent me a preprint of [8], which was the starting point of the present paper, and D. Kazdhan and G. Prasad for having pointed out two errors in a previous proof of Theorem B for $\mathbf{SL}_n$.

Notation and conventions. In the sequel, G is a connected semi-simple algebraic group over some groundfield, and p the characteristic of the groundfield. For unexplained notation and notions on linear algebraic groups, we refer to [1]. In particular, in such a group, the word "torus" is meant as in [1], i.e., refers to a connected linear algebraic group which is isomorphic to a product of $\mathbf{GL}_1$'s. In a compact group however it means a topological torus (product of circle groups).

If H is a group, and A, B are subsets of H, then

$$^{B}A = \{bab^{-1} \mid a \in A, b \in B\},\ N_H(A) = \{h \in H \mid hAh^{-1} = A\},$$

$$\mathrm{Tr}_H(A, B) = \{h \in \mathrm{H} \mid h.A.h^{-1} = B\}.$$

If Γ acts on a space X, the *isotropy group* of Γ at x is

$$\Gamma_x = \{\gamma \in \Gamma \mid \gamma \cdot x = x\}.$$

We recall that a morphism $f : X \to Y$ of irreducible algebraic varieties is *dominant* if its image is not contained in any proper algebraic subvariety. If so, then Im f contains a Zariski-dense open subset of Y [1: AG 10.2]. If the groundfield has characteristic zero, then, since f is separable, the differential of f has maximal rank on some non-empty Zariski open subset of X [1: AG, 17.3].

§1. Proof of Theorem B

Let m be an integer ≥ 2. Let $w = w(X_1, ..., X_m)$ be a non-trivial element in the free group $F(X_1, ..., X_m)$ on m letters X_i, i.e., a non-trivial reduced word in the X_i's, with non-zero integral exponents [3: I.81, Prop. 7]. Then given a group H, the word w defines a map $f_w : H^m \to H$ by the rule

$$(1) \qquad f_w(\{h_1, ..., h_m\}) = w(h_1, ..., h_m), \qquad (h_i \in H;\ 1 \leq i \leq m).$$

If H is an algebraic group, then f_w is a morphism of algebraic varieties which is defined over any field of definition for H. In the case where $H = G$ we want to prove

THEOREM 1. *The map* $f_w: G^m \to G$ *is dominant.*

This is a geometric statement. To prove it, we shall identify G with $G(\Omega)$, where Ω is some universal field. We have then to prove that $f_w(G(\Omega)^m)$ is Zariski-dense in $G(\Omega)$.

The Zariski closure Z of $\operatorname{Im} f_w$ is irreducible (since G^m is) and is invariant under conjugation, since $\operatorname{Im} f_w$ is obviously so. Since the semi-simple elements of G are Zariski-dense, and all conjugate to elements in some fixed maximal torus T, it suffices to show that $Z \supset T$.

a) We first consider the case where $G = \mathbf{SL}_n (n \geqq 2)$. Let us prove that $G(\Omega)$ contains a Zariski-dense subgroup H, no element of which, except for the identity, has an eigenvalue equal to one. This statement and its proof were directly suggested by [8].

One can find an infinite field L of the same characteristic as Ω over which there exists a central division algebra D of degree n^2. We may for example take for L a local field (see e.g. XIII, §3, Remarque p. 202 in [14]). We may assume $L \subset \Omega$. Let $\mathscr{D}^1$ be the algebraic group over L whose points in a commutative L-algebra M are the elements of reduced norm one in $D \otimes_L M$. Then $\mathscr{D}^1$ is an anisotropic L-form of $\mathbf{SL}_n$. Of course, D splits over Ω and the isomorphism $D \otimes_L \Omega = \mathbf{M}_n(\Omega)$ yields an isomorphism of $\mathscr{D}^1(\Omega)$ onto $G(\Omega)$. We let H be the image of $D^1 = \mathscr{D}^1(L)$ under such an isomorphism. The group H is Zariski-dense since L is infinite. The fact that any $h \in H - \{1\}$ has no eigenvalue equal to one is then proved as in [8]: the element $h - 1$ is a non-zero element of D, hence is invertible, hence has no eigenvalue zero and therefore h has no eigenvalue one. This proves our assertion. Let p_0 be the characteristic exponent of Ω ($p_0 = 1$ if char $\Omega = 0$ and $p_0 = \text{char}\,\Omega$ otherwise). If $p_0 = 1$, then H consists of semi-simple elements; if not, then $h^q (q = p_0^{n-1})$ is semi-simple for any $h \in G$. Let $f_w^q: G^m \to G$ be defined by $f_w^q(g) = f_w(g)^q$. Then $f_w^q(H)$ consists of semi-simple elements. Let Z_q be the Zariski closure of $\operatorname{Im} f_w^q$. Since $x \mapsto x^q$ is dominant, we have shown:

(*) *Let* V *be the set of semi-simple elements in* $G(\Omega)$ *which have no eigenvalue equal to one. Then* $\{1\} \cup (V \cap \operatorname{Im} f_w^q)$ *is Zariski-dense in* Z_q.

We now prove the theorem for $\mathbf{SL}_n (n \geqq 2)$ by induction on n. It suffices to show that f_w^q is dominant and, for this, that $Z_q \supset T$. Let $n = 2$. The group $\mathbf{SL}_2$ has dimension three and the conjugacy classes of non-central elements have dimension two. If $Z_q \neq G$, then $\dim Z_q \leqq 2$ and Z_q is contained in the union of the set U of unipotent elements of G and of finitely many conjugacy classes of semi-simple elements $\neq 1$. Those are closed, disjoint from U. Since Z_q is irreducible and contains 1, it should then be contained in U. On the other hand,

$Z_q \neq \{1\}$ since G contains non-commutative free subgroups, as follows from [17] (see also Remark 1 below). We then get

$$\{1\} \underset{\neq}{\subset} Z_q \subset U,$$

but this contradicts (*), whence the Theorem for $\mathbf{SL}_2$.

Assume now $n > 2$ and our assertion proved up to $n - 1$. This implies in particular that Z_q contains all subgroups of G isomorphic to $\mathbf{SL}_{n-1}$, hence that $Z_q \cap T$ contains the subtori of T of codimension one consisting of the elements of T which have at least one eigenvalue equal to one. Call Y their union. Assume that $Z_q \cap T \neq T$. Then we may write $Z_q \cap T = Y \cup Y'$, where Y' is a proper algebraic subset of T not containing any irreducible component of Y. Let Q be the Zariski-closure of the set ${}^G Y'$ of conjugates of elements of Y'. We claim that $Y \not\subset Q$. In fact, the subsets Y and Y' are stable under the Weyl group $W = N(T)/T$ (which may be identified with the group of permutations of the basic vectors of Ω^n). Let $J \subset \Omega[T/W]$ be the ideal of Y'. The algebra $\Omega[T/W]$ is isomorphic, under the restriction mapping, to the algebra S of regular class functions on G [16]. Let J' be the ideal of S corresponding to J under this isomorphism and R the variety of zeroes of J'. We have then $Q \subset R$, but $Y \not\subset R$, whence $Y \not\subset Q$.

The difference $Y' - (Y \cap Y')$ contains a conjugate of every semi-simple element of Z_q not having any eigenvalue equal to one. Therefore (*) implies that $Z_q = \{1\} \cup Q$. But this contradicts the fact that $Y \not\subset Q$. Therefore $T \subset Z_q$ and the theorem is proved for $\mathbf{SL}_n$.

b) In the general case we use induction on dim G. If $\mu : G' \to G$ is an isogeny, then the theorem for G' implies it for G, hence we may assume G to be simply connected. It is then a direct product of almost simple groups, whence also a reduction to the case where G is almost simple. By a), it suffices to consider the case where G is not isomorphic to $\mathbf{SL}_n$ for any n. But then it contains a proper connected semi-simple subgroup H of maximal rank (see lemma below). By induction Z contains a maximal torus of H, hence one of G, and therefore T.

We have just used the following lemma:

LEMMA 1. *Assume G to be almost simple, and not isogeneous to $\mathbf{SL}_n$ for any n. Then G contains a proper connected semi-simple subgroup of maximal rank.*

For convenience, we may assume G to be isomorphic to its adjoint group. Let $\Phi = \Phi(G, T)$ be the root system of G with respect to T and $\Delta = \{\alpha_1, ..., \alpha_l\}$ a basis of Φ. Since G is adjoint, Δ is also a basis of the group $X^*(T)$ of rational characters of T. Let d be the dominant root and write

$$d = \sum_{i=1}^{i=l} d_i \alpha_i .$$

The d_i's are strictly positive integers. By assumption, Φ is not of type $\mathbf{A}_m$ for any m. Therefore, by the classification of root systems, one of the d_i's is prime (see e.g. [4]). Say $d_1 = q$, with q prime. Let Ψ be the set of elements in Φ which, when expressed as linear combination of simple roots, have either 0 or $\pm q$ as coefficient of α_1. This is a closed set of roots. In fact, it is a root system with basis $\alpha_2, \ldots, \alpha_l$ and $-d$ [2]. We claim that there exists a closed connected subgroup H of G containing T with root system Ψ.

Let first $q \neq$ char. K. Then there is an element $t \in T$, $t \neq 1$, such that

$$d(t) = \alpha_i(t) = 1 , \qquad (i = 2, \ldots, l) .$$

It has order q, and Ψ is the set of roots which are equal to one on t. Then the identity component of the centralizer of t satisfies our condition.

Let now $q =$ char. Ω. Let $\mathfrak{t}$ be the Lie algebra of T and $\mathfrak{u}$ be the subspace of $\mathfrak{t}$ which annihilates the differentials $d\alpha_i$ of the roots α_i $(i = 2, \ldots, l)$. It is one dimensional and does not annihilate $d\alpha_1$ (since, as recalled above, Δ is a basis of $X^*(T)$, hence the $d\alpha_i (1 \leqq i \leqq l)$ form a basis of the dual space to $\mathfrak{t}$). Of course, the differential of any $\lambda \in X^*(T)$ which is divisible by q in $X^*(T)$ is identically zero on $\mathfrak{t}$. It follows then that

$$\Psi = \{\alpha \in \Phi \mid d\alpha(\mathfrak{u}) = 0\} .$$

Let $\mathfrak{g}$ be the Lie algebra of G and

$$\mathfrak{g}_\alpha = \{x \in \mathfrak{g} \mid \mathrm{Ad}\, t(x) = \alpha(t) \cdot x (t \in T)\} , \qquad (\alpha \in \Phi) ,$$

be the (1-dimensional) eigenspace of T corresponding to α[1, §14]. The previous relation implies that

$$\mathfrak{z}_{\mathfrak{g}}(\mathfrak{u}) = \mathfrak{t} \oplus \bigoplus_{\alpha \in \Psi} \mathfrak{g}_\alpha .$$

By [1: §14] the Lie algebra of the centralizer

$$Z_G(\mathfrak{u}) = \{g \in G \mid \mathrm{Ad}\, g(x) = x, (x \in \mathfrak{u})\} ,$$

of $\mathfrak{u}$ in G is equal to $\mathfrak{z}_{\mathfrak{g}}(\mathfrak{u})$; therefore $Z_G(\mathfrak{u})$ is a semi-simple subgroup satisfying our conditions.

Remarks. 1) We have used [17] only for $\mathbf{SL}_2(\Omega)$, but it is possible to bypass [17] in this case and make our proof, and the whole paper, independent of [17].

We need only to prove that $SL_2(\Omega)$ contains a non-commutative free subgroup F. If Ω has characteristic zero, we may take any torsion-free subgroup of $SL_2(\mathbf{Z})$. Let now $p = \operatorname{char} \Omega$ be >0. Then, by the arithmetic method, using division quaternion algebras over global fields, we can construct a discrete cocompact subgroup of $SL_2(L)$, where L is a local field of characteristic p (cf. A. Borel-G. Harder, *Crelle J. 298* (1978), 53-74). The latter has a torsion-free subgroup F of finite index (H. Garland, *Annals of Math. 97* (1973), 375-423) which is then free, since it acts freely on a tree, namely the Bruhat-Tits building of $SL_2(L)$.

2) For any non-zero $n \in \mathbf{Z}$, the power map $g \mapsto g^n$ is dominant (because it is surjective on any maximal torus [1 : 8.9]), hence Theorem 1 is obvious if the sum of the exponents of one letter in the word w is not zero. (See [11] for a similar remark in the context of compact groups.)

3) If U and V are non-empty open subsets in a connected algebraic group H, then $H = U \cdot V$ [1 : 1.3]. It follows then from Theorem 1 that if w, w' are two words in two letters, say, then the map $G^4 \to G$ defined by

$$f(g_1, g_2, g_3, g_4) = w(g_1, g_2) \cdot w'(g_3, g_4),$$

is surjective. For instance, every element of $G(\Omega)$ is the product of two commutators. However, the map f_w itself is not always surjective; for instance $x \mapsto x^2$ is not surjective in $SL_2(\mathbf{C})$, as pointed out in [11].

4) If $K = \mathbf{C}$, then Theorem 1 implies that $\operatorname{Im} f_w$ contains a dense open set in the ordinary topology. If G is defined over $\mathbf{R}$, then Theorem 1 also shows that $f_w(G(\mathbf{R}))$ contains a non-empty subset of $G(\mathbf{R})$ which is open in the ordinary topology. However it may not be dense. For instance, it is pointed out in [11] that for SU_2, the image of the map defined by $[x^2, yxy^{-1}]$ omits a neighborhood of -1; however this map is surjective in SO_3.

It seems that little is known about the image of f_w, even over $\mathbf{R}$ or $\mathbf{C}$. A general fact however is that the commutator map is surjective in any compact connected semi-simple Lie group [9].

§2. Free subgroups with strongly regular elements

1. In the sequel, K is a field of infinite transcendence degree over its prime field. We shall need the following lemma:

Lemma 2. *Let X be an irreducible unirational K-variety. Let L be a finitely generated subfield of K containing a field of definition of X, and $V_i (i \in \mathbf{N})$ a sequence of proper irreducible algebraic subsets of X defined over an* *algebraic closure* $\bar{L}$ *of* L. *Then* $X(K)$ *is not contained in the union of the* $V_i \cap X(K)$, $(i \in \mathbf{N})$.

By definition of unirationality, there exists for some $n \in \mathbf{N}$ a dominant K-morphism $\varphi : \mathbf{A}^n \to X$, where $\mathbf{A}^n$ denotes the affine n-dimensional space.

This map is already defined over some finitely generated extension of L. Replacing L by the former, we may assume φ to be defined over L, hence $\varphi^{-1}(V_i)$ to be defined over $\bar{L}$. It is a proper algebraic subset since φ is dominant. This reduces us to the case where $X = \mathbf{A}^n$. But then any point whose coordinates generate over $\bar{L}$ a field of transcendence degree n will do.

THEOREM 2. *Assume* G *to be defined over* K. *Let* $\mathscr{V} = \{V_i\}$ $(i \in \mathbf{N})$ *be a family of proper subvarieties of* G, *all defined over an algebraic closure* $\bar{L}$ *of a finitely generated subfield* L *of* K *over which* G *is also defined. Then* $G(K)$ *contains a non-commutative free subgroup* Γ *such that no element of* $\Gamma - \{1\}$ *is contained in any of the* V_i*'s. Given* $m \geqq 2$, *the set of* m*-tuples which freely generate a subgroup having this property is Zariski dense in* G^m.

We may (and do) assume that the identity element is contained in one of the V_i's.

Let w and f_w be as in §1. Then f_w is defined over L hence $f_w^{-1}(Z)$ is defined over $\bar{L}$ for every $Z \in \mathscr{V}$ and is a *proper* algebraic subset by Theorem 1. The sets $f_w^{-1}(Z)$, as w runs through all the non-trivial reduced words (in m letters and their inverses) and Z through $\mathscr{V}$, form then a countable collection of proper algebraic subsets, all defined over $\bar{L}$. But G, hence G^m, is a unirational variety over any field of definition of G [1: 18.2]. Lemma 2 implies therefore the existence of $g = (g_i) \in G(K)^m$ not belonging to any of these subsets. Then the g_i's are free generators of a subgroup which satisfies our conditions. In fact, we see that we can take for g any point of $G(K)^m$ which is generic over $\bar{L}$ and, since $\bar{L}$ has finite transcendence degree over the prime field, such points are Zariski-dense. This establishes the second assertion.

Remark. If $G = \mathbf{SO}_{2n}$ (resp. $\mathbf{SO}_{2n+1}$), this shows for instance the existence of a free subgroup Γ, no element of which except 1 has the eigenvalue 1 (resp. the eigenvalue 1 with multiplicity >1).

2. Any semi-simple element x of G is contained in a maximal torus [1: 11.10]; x is called regular if it is contained in exactly one maximal torus. We shall say that x is *strongly regular* if it is not contained in any non-maximal torus, i.e., if the cyclic group generated by x is Zariski-dense in a maximal torus.

The following result contains Theorem C of the introduction.

COROLLARY 1. *Assume G to be defined over K. Then $G(K)$ contains a non-commutative free subgroup Γ all of whose elements $\neq 1$ are strongly regular. Given $m \geqq 2$, the set of m-tuples $(g_i) \in G(K)^m$ which generate freely a subgroup with that property is Zariski dense in G^m.*

The field K contains a field of definition L of G which is finitely generated over its prime field. Let $\bar{L}$ be an algebraic closure of L in our universal field Ω. Then the subfield generated by $\bar{L}$ and K has infinite transcendence degree over $\bar{L}$. Let S be the set of singular elements of G (i.e., of elements $g \in G$ such that Ad g has the eigenvalue one with multiplicity $>$ rk G). It is algebraic, defined over $\bar{L}$. Fix a maximal L-torus T of G [1: 18.2]. Every proper closed subgroup of T is contained in the kernel of a rational character [1: 8.2]. The characters are all defined over a finite separable extension L' of L [1: 8.11] and form a countable set. For $\lambda \in X^*(T)$, $\lambda \neq 1$, let $T_\lambda = \ker \lambda$, and V_λ the Zariski-closure of ${}^G T_\lambda$. The V_λ and S form a countable set $\mathscr{V}$ of proper algebraic subsets of G which are all defined over $\bar{L}$.

Our assertion is now a special case of the Theorem.

3. We can now prove the Corollary in the introduction. Let Ω be an algebraically closed extension of K. Since $G(K)/H(K)$ may be identified to an orbit of $G(K)$ in $G(\Omega)/H(\Omega)$ it suffices to show:

COROLLARY 2. *Assume K to be algebraically closed. Then every $\gamma \in \Gamma - \{1\}$, operating by left translations on $G(K)/H(K)$, has exactly $\chi(G, H)$ fixed points.*

For $\gamma \in \Gamma - \{1\}$, let F_γ be the fixed point set of γ in $G(K)/H(K)$, and let T_γ be the maximal torus in which the cyclic group generated by γ is dense. Clearly, F_γ is also the set of fixed points of $T_\gamma(K)$. Thus, if F_γ is non-empty, then T_γ is conjugate to a subgroup of H, and H has maximal rank. Assume this is the case and let T_0 be a maximal K-torus of H. Since the maximal tori of H (or G) are conjugate, it is elementary that F_γ may be identified with $\mathrm{Tr}(T_0, T_\gamma)/N_H(T_0)$. But, if $x \in \mathrm{Tr}(T_0, T_\gamma)$, then $\mathrm{Tr}(T_0, T_\gamma) = x \cdot N_G(T_0)$, whence the Corollary.

4. We now generalize slightly the Corollary in case H contains a maximal torus of G, dropping again the assumption that K is algebraically closed. Assume instead

(*) *The maximal K-tori of H are conjugate under $H(K)$.*

If T_0 is a maximal K-torus of H, we then set

$$\chi(G(K), H(K)) = [N_{G(K)}(T_0) : N_{H(K)}(T_0)] .$$

If K is algebraically closed, then (*) is fulfilled and $\chi(G(K), H(K))$ is our previous $\chi(G, H)$. We again set $\chi(G(K), H(K)) = 0$ if H does not contain any maximal torus of G.

COROLLARY 3. *Let* Γ *be as in Theorem 2. Let* H *be a closed* K-*subgroup of maximal rank and assume* (*) *to be satisfied. Then* $\gamma \in \Gamma - \{1\}$ *acts freely if* T_γ *is not conjugate under* $G(K)$ *to* T_0 *and has* $\chi(G(K), H(K))$ *fixed points otherwise.*

The argument is the same as before: F_γ is also the set of fixed points of T_γ. The latter is defined over K. If $F_\gamma \neq \emptyset$, then there exists $x \in G(K)$ such that ${}^xT_\gamma \in H$, hence by (*),

$$\mathrm{Tr}_{G(K)}(T_0, T_\gamma) \neq \emptyset ,$$

and we have, as above, bijections

$$F_\gamma = \mathrm{Tr}_{G(K)}(T_0, T_\gamma)/N_{H(K)}(T_0) = N_{G(K)}(T_0)/N_{H(K)}(T_0) .$$

5. (i) If $K = \mathbf{R}, \mathbf{C}$ or also is a non-archimedean local field with finite residue field, then $G(K)$, endowed with the topology stemming from K, is a Lie group over K, and in particular is a locally compact topological group. In that case, we can use in Theorem 2 a category argument instead of Lemma 2: the $f_w^{-1}(Z)$, being proper algebraic subsets, have no interior point, the intersection of their complement is then dense by Baire's theorem, whence the last assertion of Theorem 2 with "Zariski-dense" replaced by "dense in the K-topology".

(ii) In [4] it is asked whether the hyperbolic n-space admits a non-commutative free group of isometries which acts freely. More generally, one has the

PROPOSITION. *Let* S *be a connected semi-simple non-compact Lie group with finite center,* U *a maximal compact subgroup of* L *and* $X = L/U$ *the symmetric space of non-compact type of* S. *Then* S *contains a non-commutative free subgroup which acts freely on* X.

If $\mathrm{rk}\, S \neq \mathrm{rk}\, U$, this could be deduced from Corollary 2. However, the existence of one such subgroup can be proved much more directly in all cases: if $S = \mathbf{SL}_2(\mathbf{R})$ or $\mathbf{PSL}_2(\mathbf{R})$, then we may take for Γ a free subgroup of finite index in $\mathbf{SL}_2(\mathbf{Z})$ or $\mathbf{SL}_2(\mathbf{Z})/\{\pm 1\}$. If S is of dimension >3, then it contains a copy of $\mathbf{SL}_2(\mathbf{R})$ or of $\mathbf{PSL}_2(\mathbf{R})$, and therefore a *discrete* non-commutative free subgroup Γ. No element $\gamma \in \Gamma - \{1\}$ is contained in a compact subgroup of S, hence Γ acts freely on X.

A similar argument would be valid over a non-archimedean local field K for the Bruhat-Tits buildings attached to semi-simple K-groups.

§3. Compact groups. Proof of Theorem A.

1. Let U be a compact Lie group. Then we may view U as the group $G(\mathbf{R})$ of real points of an algebraic group G defined over $\mathbf{R}$ [5]. Furthermore, the maximal (topological) tori of U are the groups $T(\mathbf{R})$, where T runs through the maximal $\mathbf{R}$-tori of G. They are conjugate under inner automorphisms of U. Corollary 1 to Theorem 2 insures the existence of a non-commutative free subgroup Γ of U such that every $\gamma \in \Gamma - \{1\}$ is strongly regular, i.e., generates a dense subgroup of a maximal torus of U. If now V is a closed subgroup of U, then, by [10], $\chi(U/V) = 0$ if V does not contain a maximal torus of U, and is equal to $[N_U(T) : N_V(T)]$ if V contains a maximal torus T of U. By the results just recalled, we may write $V = H(\mathbf{R})$, where H is an algebraic $\mathbf{R}$-subgroup of G, the condition (*) of §2 is satisfied, and any maximal torus of U is conjugate to T. Theorem A now follows from Corollaries 1 and 3 to Theorem 2.

2. The results of this paper, specialized to compact Lie groups, can of course be proved more directly, in the framework of the theory of compact Lie groups, without recourse to the theory of algebraic groups. For the benefit of the reader mainly interested in that case, we sketch how to modify the above arguments.

The main point is again to prove Theorem 1, where now G stands for a non-trivial compact connected semi-simple Lie group. In part a) of the proof, the role of $\mathbf{SL}_n$ is taken by $\mathbf{SU}_n$. If $n = 2$, G contains non-commutative free subgroups. If $n > 2$, the argument is the same except that now we take for D, exactly as in [8], a division algebra with an involution of the second kind and identify $\mathbf{SU}_n$ to $(D \otimes_L \mathbf{R})^1$, where L is the fixed field, in the center of D, of the given involution of D. In part b), we use the fact that if G is simple, not locally isomorphic to $\mathbf{SU}_n$, then it contains a proper closed connected semi-simple subgroup of maximal rank, for which we can refer directly to [2] (the proof of Lemma 1 was in fact just an adaptation to algebraic groups of the one in [2]).

Then, as pointed out in section 5 of §2, a simple category argument yields Theorem 2, whence also Corollary 1 to Theorem 2 and Theorem A.

§4. Free group actions with commutative isotropy groups

1. Let Γ be a non-commutative free group acting on a set X. Assume that Γ acts freely, or more generally, that the isotropy groups $\Gamma_x (x \in X)$ are commutative (hence cyclic), and that at least one is reduced to $\{1\}$. Then the decomposition theorem 2.2.1, 2.2.2 of [6] implies in particular the following: given $n \geqq 2$, there exists a partition of X into $2n$ subsets X_i and elements $\gamma_i \in \Gamma (1 \leqq i \leqq 2n)$ such that X is the disjoint union of $\gamma_i X_i$ and $\gamma_{n+i} X_{n+i} (i \leqq i \leqq n)$. If we view the operations of

Γ as congruences, this shows that X is equivalent to the union of n copies of itself via finite congruences. The existence of such partitions of S^2 was proved first by R. M. Robinson [13].

This then leads to the problem of finding actions of free groups with commutative isotropy groups in cases where free actions are ruled out. We now prove some results pertaining to that question.

2. Consider first the case of $S^n = SO_{n+1}/SO_n$. The problem is then to find a free non-commutative subgroup Γ of SO_{n+1} such that no two non-commutative elements of Γ are contained in a conjugate of SO_n, i.e., have a common non-zero fixed vector. In [6], this is shown for $n \geqq 2$, but $n \neq 4$. We want to give an alternate proof which also covers that last case. For n odd, there is even a Γ such that no element $\neq 1$ has an eigenvector, as follows from the remark to Theorem 2. So assume n even. If $n = 2$, then the isotropy groups of SO_3 itself on S^2 are commutative, hence any non-commutative free subgroup will do. Assume $n > 2$. The group SO_3 has an (absolutely) irreducible real representation of degree $n + 1$; it can e.g. be realized in the space of spherical harmonics in $\mathbf{R}^3$ of degree $n/2$. Let H be the image of SO_3 in SO_{n+1} under such a representation and let Γ be a free non-commutative subgroup of H. Then any two non-commuting elements of Γ generate a dense subgroup of H, hence do not have a common non-zero proper invariant subspace of $\mathbf{R}^{n+1}$; in particular they have no common fixed vector, whence our assertion.

Example. For the sake of definiteness, we indicate one explicit example in the case $n = 4$.

Let $\alpha, \beta \in (0, 2\pi)$ be two angles such that the rotations of angle α and β of $\mathbf{R}^3$ around two perpendicular axes freely generate a free subgroup $F_{\alpha,\beta}$ of SO_3. We may take e.g. $\alpha = \beta$, where α is such that $\cos \alpha$ is transcendental [7]. Let $\{e_1, ..., e_5\}$ be the canonical basis of $\mathbf{R}^5$. Let $A_\alpha \in SO_5$ be the transformation which is a rotation of angle 2α in the plane $[e_4, e_5]$ spanned by e_4 and e_5 and which is the rotation of angle 4α around the axis spanned by $(3^{1/2}, 0, 1)$ in $[e_1, e_2, e_3]$. Let B_β the element of SO_5 which fixes e_3 and is a rotation of angle 2β (resp. 4β) in the plane $[e_2, e_4]$ (resp. $[e_1, e_5]$). Then A_α and B_β freely generate an irreducible subgroup of SO_5, whose closure is isomorphic to SO_3 and which is therefore locally commutative on S^4.

In fact, in suitable coordinates, this group is just the image of the group $F_{\alpha,\beta}$ under the five-dimensional irreducible representation of SO_3. The easy computations showing this are left to the reader.

3. The above argument extends in the general case to the following sharpening of Theorem A in the case of non-zero Euler characteristic.

THEOREM 3. *Let U be a compact connected non-trivial semi-simple Lie group. Then U contains a non-commutative free subgroup Γ whose elements $\gamma \neq 1$ are regular and such that, for any proper closed subgroup V of maximal rank of U, the isotropy groups $\Gamma_x(x \in U/V)$ of Γ on U/V are commutative and any $\gamma \in \Gamma - \{1\}$ has exactly $\chi(U/V)$ fixed points.*

Proof: First we carry an easy reduction to the case where U is simple and V connected. Let U' be the quotient of U by its center, $\pi : U \to U'$ the natural projection and $V' = \pi(V)$. The isotropy groups of U on U'/V' contain the isotropy groups on U/V, hence we may assume that U has center reduced to the identity. Let V^0 be the identity component of V. Any isotropy group of Γ on U/V contains an isotropy group on U/V^0 as a subgroup of finite index. Both are therefore simultaneously commutative or not commutative. So we may assume V to be connected. Now U is a direct product of simple groups and V, being of maximal rank, is the direct product of its intersections with the simple factors of U [2], whence our reduction.

We now prove the theorem in this case except for the last assertion on the number of fixed points.

If $U = \mathbf{SO}_3$, then any proper closed subgroup has a commutative subgroup of finite index, and any element $\neq 1$ is regular. Therefore we may take for Γ any non-commutative free subgroup. Assume now that $U \neq \mathbf{SO}_3$, hence $\dim U > 3$. Then U has a closed subgroup H, isomorphic to $\mathbf{SO}_3$, which contains regular elements of U and is not contained in any proper subgroup of maximal rank [15: §12]. (This subgroup is called "principal" in [15].) Then any element of infinite order in H is regular in U. In particular any element $\gamma \neq 1$ in a free non-commutative subgroup Γ of H is regular. Moreover any two non-commuting elements of Γ generate a dense subgroup of H. If they were contained in a conjugate of V, then so would H, whence a contradiction.

There remains to see that every $\gamma \in \Gamma - \{1\}$ has exactly $\chi(U/V)$ fixed points on U/V. Let S_γ be the closure of the subgroup of H generated by γ. It is a one-dimensional torus, almost all of whose elements are regular in U. Fix a maximal torus T_0 of V, hence of U. If $x, y \in U$ are such that ${}^xS, {}^yS \subset T_0$, then the inner automorphism by $x \cdot y^{-1}$, which brings ${}^yS_\gamma$ onto ${}^xS_\gamma$, must leave T_0 stable since ${}^xS_\gamma$ contains regular elements, i.e., $x \cdot y^{-1} \in N_U(T_0)$. From this we see again that there is a natural bijection between the fixed point set of γ and $N_U(T_0)/N_V(T_0)$, and our assertion follows as in section 4 of §2.

4. The same argument is valid for a complex semi-simple Lie group, using a principal three-dimensional subgroup, or also over any algebraically closed groundfield. Over a field K of infinite transcendence degree over its prime field,

one would have to assume the existence of a principal three-dimensional subgroup which is defined over K.

5. We note finally that if $\Gamma \subset G(K)$ satisfies the conditions of Corollary 1 to Theorem 2 and if H is a subgroup of maximal rank of G whose identity component is solvable, then for any $x \in G(K)/H(K)$, the isotropy group Γ_x is commutative, since its intersection with the isotropy group of x in $G(K)$ is on one hand free, as a subgroup of Γ, and on the other hand contains a solvable normal subgroup of finite index, since $H(K)$ does.

REFERENCES

[1] Borel, A. *Linear Algebraic Groups.* (Notes by H. Bass), Benjamin, New York 1969.

[2] Borel, A. et J. de Siebenthal. Les sous-groupes fermés de rang maximum des groupes de Lie clos. *Comm. Math. Helv. 23* (1949), 200-220.

[3] Bourbaki, N. *Algèbre 1, 2, 3.* Hermann, Paris 1970.

[4] —— *Groupes et algèbres de Lie, Chap. 4, 5, 6.* Masson, Paris 1981.

[5] Chevalley, C. *Théorie des Groupes de Lie III : Groupes algébriques.* Hermann, Paris 1955.

[6] Dekker, T. J. Decompositions of sets and spaces I, II, III. *Indag. Math. 18* (1956), 581-595, *19* (1957), 104-107.

[7] —— On free groups of motions without fixed points. *ibid. 20* (1958), 348-353.

[8] Deligne, P. and D. Sullivan. Division algebras and the Hausdorff-Banach-Tarski paradox. *L'Enseignement Mathématique 29* (1983), 145-150.

[9] Goto, M. A theorem on compact semi-simple groups. *Jour. Math. Soc. Japan 1* (1949), 270-272.

[10] Hopf, H. und H. Samelson. Ein Satz über die Wirkungsräume geschlossener Liescher Gruppen. *Comm. Math. Helv. 13* (1940), 240-251.

[11] Mycielski, J. Can one solve equations in groups? *Amer. Math. Monthly 84* (1977), 723-726.

[12] —— Equations unsolvable in $GL_2(C)$ and related problems. *ibid. 85* (1978), 263-265.

[13] Robinson, R. M. On the decomposition of spheres. *Fund. Math. 34* (1947), 226-260.

[14] Serre, J.-P. *Corps locaux.* Act. Sci. Ind. 1296, Hermann éd., Paris 1962.

[15] de Siebenthal, J. Sur les sous-groupes fermés connexes d'un groupe de Lie clos. *Comm. Math. Helv. 25* (1951), 210-256.

[16] Steinberg, R. Regular elements of semi-simple algebraic groups. *Publ. Math. I.H.E.S. 25* (1965), 49-80.

[17] Tits, J. Free subgroups in linear groups. *Journal of Algebra 20* (1972), 250-270.

(Reçu le 8 mars 1982)

Armand Borel
The Institute for Advanced Study
Princeton, N.J. 08540
USA

124.

Cohomology and Spectrum of an Arithmetic Group

Proc. of a Conference on Operator Algebras and Group Representations,
Neptun, Rumania (1980), Pitman (1983) 28–45

The study of the cohomology of a discrete subgroup Γ of finite covolume of a semi-simple Lie group G has been carried out during these last twenty years or so from several points of view, with various techniques. In order to relate to the extent possible to one of the main themes of the conference, I shall in this paper emphasize mainly one of those points of view, namely the relationship with the 'spectrum of Γ'. By the latter is meant first of all the space $L^2(\Gamma\backslash G)$ of square integrable functions on the $\Gamma\backslash G$ of right cosets of Γ in G, viewed as usual as a unitary G-module via right translations, but also more generally some non-unitary representations of G occurring on $\Gamma\backslash G$ (possibly only as subquotients). In the lecture, I tried to give a reasonably comprehensive survey. However, in this paper the presentation will be more unbalanced since I shall avail myself of the literature and treat more cursorily topics for which I can give references. Although some overlap is unavoidable, this paper is to be viewed as a sequel to [4; 5; 6; 9].

1. Generalities. Statement of a problem

1.1. As usual we assume G to have finitely many connected components, and the identity component of G to have finite centre. In fact, for our purposes, there is no loss in generality in assuming G to be linear. It is then of finite index in the group of real points of a linear algebraic group defined over the reals. Let K be a maximal compact subgroup of G and $X = G/K$. Then X is a symmetric space with negative curvature, simply connected, complete, hence diffeomorphic to Euclidean space.

1.2. Let (σ, E) be a finite dimensional complex representation of G. Our main object of concern is the Eilenberg–MacLane cohomology space $H^*(\Gamma; E)$, where Γ is a discrete subgroup of G of finite covolume, i.e., such that $\Gamma\backslash G$, endowed with the measure coming from a Haar measure of G, has finite volume. The group Γ operates properly on X (i.e., given a compact subset $C \subset X$, the set of $\gamma \in \Gamma$ such that $\gamma \cdot C \cap C \neq \emptyset$ is finite), and freely if it is torsion-free. In the latter case $\Gamma\backslash X$ is an Eilenberg-MacLane space $K(\Gamma; 1)$, hence

$$H^*(\Gamma; E) = H^*(\Gamma\backslash X; \tilde{E}), \tag{1}$$

where $\tilde{E}$ denotes the local system defined by E on X. Even if Γ has torsion, this remains true, with $\tilde{E}$ standing for a suitable sheaf [9: VII].

Denote by $T_x(X)$ the tangent space to X at $x \in X$. Let $A^q(X; E)$ denote the space of smooth E-valued differential q-forms on X ($q \in \mathbb{N}$) and $A^*(X, E)$ the direct sum of the $A^q(X, E)$, endowed with exterior differentiation. As usual, $g \in G$ operates on $A^q(X; E)$ by the rule

$$(g \circ \omega)(x; X_1, \ldots, X_q) = \sigma(g)(\omega(g^{-1} \cdot x; g^{-1} \cdot X_1, \ldots, g^{-1} X_q)) \tag{2}$$

($g \in G$; $x \in X$, $X_i \in T_x(X)$, $i = 1, \ldots, q$). For any subgroup H of G, let $A^*(X, E)^H$ be the space of H-invariant elements in $A^*(X; E)$. Then, as is well known (see e.g. [9: VII]) by a mild generalization of de Rham's theorem

$$H^*(\Gamma; E) \cong H^*(A^*(X; E)^\Gamma). \tag{3}$$

From the beginning, this relation was used to study the left-hand side in the cocompact case by methods of differential geometry, in particular by means of Hodge theory and techniques inspired by the work of Bochner-Yano (see e.g. [15; 16]). Gradually, it was realized that this could be translated into representation theoretic terms, and this is the point of view adopted in [9]. There it is moreover assumed that Γ is *cocompact*; i.e., $\Gamma\backslash G$ or equivalently $\Gamma\backslash X$ is compact. Here we shall therefore be mainly concerned with the non-cocompact case, in particular with non-cocompact arithmetic subgroups.

1.3. The canonical projection $\pi: G \to X$ induces a projection $\Gamma\backslash G \to \Gamma\backslash X$ also denoted π. The lifting of differential forms by π induces an isomorphism

$$A^*(X, E)^\Gamma \xrightarrow{\sim} C^*(\mathfrak{g}, K; C^\infty(\Gamma\backslash G) \otimes E), \tag{1}$$

which maps exterior differentiation to the differential in Lie algebra cohomology (see 1.4), whence the isomorphism

$$H^*(\Gamma; E) = H^*(\mathfrak{g}, K; C^\infty(\Gamma\backslash G) \otimes E) \tag{2}$$

(see [9: VII]). This formula was proved for the first time, under some slightly more restrictive assumptions, in [16].

1.4. We recall that if V is a smooth G-module, then

$$C^*(\mathfrak{g}, K; V) = \mathrm{Hom}_K(\Lambda(\mathfrak{g}/\mathfrak{k}), V), \tag{1}$$

and the differential $d: C^q \to C^{q+1}$ is given by

$$df(x_0, \ldots, x_q) = \sum_i (-1)^i x_i \cdot f(x_0, \ldots, \hat{x}_i, \ldots, x_q), \qquad (x_i \in \mathfrak{g}/\mathfrak{k}, i = 0, \ldots, q), \tag{2}$$

where ^ indicates omission of the argument underneath. [There is a simplification in (2) due to the fact that $(\mathfrak{g}, \mathfrak{k})$ is a symmetric pair; see [9: §1] for the general formula.]

The space $V_{(K)}$ of K-finite vectors in V is dense, stable under $\mathfrak{g}$, and is a so-called $(\mathfrak{g}, K)$-module [9: 0, 2.5]. Obviously, the right-hand side of (1) does not change if V is replaced by $V_{(K)}$, and the above definition makes sense more generally for $(\mathfrak{g}, K)$-modules [9: I, 2.2].

1.5. Let again (π, V) be a smooth G-module and assume there is an intertwining operator $A: V \to C^\infty(\Gamma \backslash G)$. Then A yields in a natural way a map

$$C^*(\mathfrak{g}, K; V \otimes E) \to C^*(\mathfrak{g}, K; C^\infty(\Gamma \backslash G) \otimes E), \tag{1}$$

and an induced derived map

$$H(A): H^*(\mathfrak{g}, K; V \otimes E) \to H^*(\mathfrak{g}, K; C^\infty(\Gamma \backslash G) \otimes E) = H^*(\Gamma; E).$$

Our main problem here is to discuss this map for some choices of A and V. So far, it has been studied mainly in two cases:

(a) $V \subset L^2(\Gamma \backslash G)$. This leads to the study of stable cohomology, of classes represented by square integrable harmonic forms, and also of the L^2-cohomology of the space $\Gamma \backslash X$ [3; 4; 6]. I have few results to add to those announced earlier and since I cannot give their proofs here, I shall be rather brief.

(b) V is a principal series representation stemming from a representation of a Levi subgroup of some proper parabolic cuspidal subgroup, contained in the space of cusp forms, and A is defined by analytic continuation of Eisenstein series. It leads to classes having a non-trivial restriction to the boundary (in the 'manifold with corners' picture recalled below), which are often not square integrable.

Section 2 is devoted to (b), Section 3 to (a), and Section 4 to some related results and problems. For convenience, we shall assume Γ to be arithmetic with respect to some given $\mathbb{Q}$-structure on G [2], although we could adopt the set-up of [14]. We end this section with a few remarks to give an idea of how (a) and (b) may contribute to the description of $H^*(\Gamma; E)$.

1.6. Assume Γ to be arithmetic and torsion-free. Then $Y = \Gamma\backslash X$ may be identified to the interior of a compact manifold with corners $\bar{Y} = \Gamma\backslash\bar{X}$, where $\bar{X}$ is a suitable completion of X on which Γ operates freely [8]. The space Y is a deformation retract of $\bar{Y}$. Let $H_c(Y; E)$ be the cohomology with compact supports of Y and $H_!(Y; E)$ the image of H_c into H^* under the natural map. We have an exact sequence

$$\cdots \to H_c^q(Y; \tilde{E}) \to H^q(\Gamma: \tilde{E}) \xrightarrow{r} H^q(\partial\bar{Y}; \tilde{E}) \to H_c^{q+1}(Y; \tilde{E}) \to \cdots$$

where r refers to restriction.

Let $H_{(2)}(\Gamma; E)$ be the subspace of $H^*(\Gamma; E)$ represented by closed square integrable forms. It contains $H_!$. By a theorem of Kodaira, it is also the space of classes represented by a square integrable harmonic form (see e.g. [17]). Thus (a) in Section 1.5 gives information on $H_{(2)}$. In (b) we may also sometimes obtain L^2-classes, when one has to take residues of Eisenstein series. But the main aim of (b) is to get hold of, classes not in the kernel of the restriction mapping r, and many of these are not square integrable.

2. Eisenstein series and cohomology

In this section, we want to discuss the use of Eisenstein series to construct cohomology classes. This approach was initiated by G. Harder [11; 12] and pursued by J. Schwermer [18; 19; 20]. Our formulation is representation-theoretic, hence formally different from that of Harder, which is differential-geometric, but it is in substance equivalent to it. Unfortunately, we need the whole paraphernalia of parabolic subgroups. For unexplained notation and notions, we refer the reader to the surveys [1; 2; 21], where he will also find many earlier references.

2.1. Let P be a parabolic $\mathbb{Q}$-subgroup of G, N its unipotent radical and $\tau: P \to P/N$ the canonical projection. By split component of P we mean here the unique subgroup A of the radical of P which is stable under the Cartan involution θ associated to K and which is mapped by τ onto the $\mathbb{Q}$-split component of P/N, i.e., the (topological) identity component of the maximal central $\mathbb{Q}$-split torus of P/N (see e.g. [8]). We let then $M = Z(A)$ and ${}^\circ M$ the usual complement of A ($\tau({}^\circ M)$ is the intersection of the kernels of the squares of the rational characters defined over $\mathbb{Q}$ of P/N). We have then $P = M \cdot N$ (semi-direct) and $M = {}^\circ M \times A$. We fix a Cartan subalgebra $\mathfrak{b}$ of the Lie algebra ${}^\circ\mathfrak{m}$ of ${}^\circ M$.

Then $\mathfrak{h}=\mathfrak{b}\oplus\mathfrak{a}$ is a Cartan subalgebra of $\mathfrak{g}$. Let $W(\mathfrak{g}_c,\mathfrak{h}_c)$ or W be the Weyl group of $\mathfrak{g}_c$ with respect to $\mathfrak{h}_c$ and $\Phi=\Phi(\mathfrak{g}_c,\mathfrak{h}_c)$ the set of roots of $\mathfrak{g}_c$ with respect to $\mathfrak{h}_c$. We fix an ordering on the roots such that the weights of $\mathfrak{h}_c$ in $\mathfrak{p}$ are positive and set, as in [9]:

$$\Phi_M=\Phi(\mathfrak{m}_c,\mathfrak{h}_c),\; W_M=W(\mathfrak{m}_c,\mathfrak{h}_c),$$
$$W^P=\{w\in W \mid w^{-1}(\Phi_M^+)\subset\Phi^+\}.$$

As usual, ρ denotes half the sum of the positive roots.

2.2. We shall denote by E_λ an irreducible finite dimensional representation of G whose infinitesimal character is χ_λ, in Harish-Chandra's parametrization (if $\lambda\in\mathfrak{h}_c^*$ is dominant, then $\lambda-\rho$ is the highest weight). Similarly F_μ will denote an irreducible finite dimensional representation of M with infinitesimal character, χ_μ ($\mu\in\mathfrak{h}_c^*$). We warn the reader that this is a change with respect to [9] which will cause some minor formal differences between some statements here and those in the references given for their proofs. We shall assume that λ is *dominant*.

2.3. ***Cohomology of a face of the boundary.*** The boundary $\partial\bar{Y}$ (see (1.6)) is a union of finitely many faces $e'(P)$ associated to the parabolic $\mathbb{Q}$-subgroups modulo conjugacy by Γ. Let (P, A) be as above. Set $\Gamma_P=\Gamma\cap P$. It is contained in ${}^\circ P={}^\circ M\cdot N$. Let σ be the canonical isomorphism $M\tilde{\to}P/N$ defined by the projection $\tau: P\to P/N$ and $\Gamma_M=\tau(\Gamma_P)$. If M is defined over $\mathbb{Q}$, then $\sigma^{-1}(\Gamma_M)$ is an arithmetic subgroup of ${}^\circ M$ which contains $\Gamma\cap M$ as a subgroup of finite index.

Since M is θ-stable by assumption, the group $K_P=K\cap P$ is contained in ${}^\circ M$ and is a maximal compact subgroup of P, M and ${}^\circ M$. We denote it also K_M. Let $X_M={}^\circ M/K_M$. Then $e'(P)$ is fibred over $Y_M=\sigma^{-1}(\Gamma_M)\backslash X_M$, with fibre the compact nilmanifold $\Gamma_N\backslash N$, where $\Gamma_N=\Gamma\cap N$. It is known that the spectral sequence of this fibration, for cohomology in characteristic zero, degenerates at E_2. The proof given by Harder in a special case (12) extends readily to the general case. The space $H^*(\mathfrak{n}; E_\lambda)$ is in a natural way an M-module; by a theorem of B. Kostant recalled in [9; III, 3.1], we have:

$$H^q(\mathfrak{n}; E_\lambda)=\bigoplus_{s\in W^P,\, l(s)=q} F_{s\lambda}. \tag{1}$$

2.4. ***Cuspidal cohomology of a face.*** We shall in fact be mainly interested in the so-called 'cuspidal' part of the cohomology of the face $e'(P)$, to be defined now. We let ${}^\circ L(\Gamma_M\backslash{}^\circ M)$ be the space of square integrable cuspidal functions on $\Gamma_M\backslash{}^\circ M$. (See [13; 14]: a function is cuspidal if its constant terms with respect to all proper parabolic $\mathbb{Q}$-groups vanish. The definition of a constant term is recalled in 2.5 below.) By a basic theorem due to I. M. Gelfand and I. Piateckii-Shapiro (see [13; 14]), the space ${}^{\circ}L(\Gamma_M\backslash{}^{\circ}M)$ decomposes into a Hilbert direct sum of irreducible ${}^{\circ}M$-modules with finite multiplicities. We may therefore write

$$ {}^{\circ}L(\Gamma_M\backslash{}^{\circ}M) = \widetilde{\bigoplus_{\pi\in{}^{\circ}\hat{M}}} H_\pi, \tag{1}$$

where H_π is the *isotypic* subspace of type π, hence a finite direct sum of copies of π. By [4; 6], the natural map

$$H^*({}^{\circ}\mathfrak{m}, K_M; {}^{\circ}L(\Gamma_M\backslash{}^{\circ}M)^\infty\otimes F)\to H^*(\Gamma_M; F) \tag{2}$$

is injective, for any finite dimensional ${}^{\circ}M$-module F. For $\pi\in{}^{\circ}\hat{M}$ and $s\in W^P$, let then

$$F^*_{\pi,s\lambda} = H^*({}^{\circ}\mathfrak{m}, K_M; H_\pi\otimes F_{s\lambda\,|\,\mathfrak{b}_c})[l(s)], \tag{3}$$

where the $[l(s)]$ indicates a shift in grading, namely

$$F^q_{\pi,s\lambda} = H^{q+l(s)}({}^{\circ}\mathfrak{m}, K_M; H_\pi\otimes F_{s\lambda\,|\,\mathfrak{b}_c}). \tag{4}$$

This space can be non-zero only if $H_\pi\neq 0$, if the representation H_π has a K_M-type in common with

$$\Lambda({}^{\circ}\mathfrak{m}/\mathfrak{k}_M)\otimes F_{-s\lambda\,|\,\mathfrak{b}_c},$$

and if

$$\chi_\pi = \chi_{-s\lambda\,|\,\mathfrak{b}_c}. \tag{5}$$

We shall say that π, s and λ are *compatible* if the latter condition is fulfilled. We note that in order to check it, we need not take $s\in W^P$ since if it is true for some $s\in W$, then it also holds for every $t\in W_M\cdot s$.

The direct sum of these spaces is contained in $H^*(e'(P); E_\lambda)$. By definition this is the cuspidal cohomology

$$H^*_{\text{cusp}}(e'(P): E_\lambda) = H^*_{\text{cusp}}(\Gamma_P; E_\lambda), \tag{6}$$

of Γ_P or $e'(P)$.

2.5. *The restriction map* $\Gamma_P: H^*(\Gamma; E_\lambda)\to H^*(\Gamma_P; E_\lambda)$. In the manifold with corners setting (1.6), it can be viewed as the restriction map $H^*(\bar{Y}; E_\lambda)\to H^*(e'(P): E_\lambda)$. It can also be interpreted directly in $\Gamma\backslash X$. In fact, let

$$\sigma_P: \Gamma_P\backslash X\to\Gamma\backslash X$$

be the natural projection. The manifold $\Gamma_P\backslash X$ is canonically isomorphic to the product $A\times e'(P)$. The submanifolds $a\times e'(P)$ $(a\in A)$ are homologous to one another and r_P may be identified the composition of

σ_P^* with the restriction to any one of them. We consider families of such manifolds when $a \to \infty$ in the sense that $a^\alpha \to \infty$ for every root α of (P, A), in sign $a \to_P \infty$.

To study Γ_P, we can make use of the theory of the constant term. Recall that if f is a continuous (locally integrable would do) function on $\Gamma\backslash X$, then its constant term f_P with respect to P is given by

$$f_P(x) = \int_{\Gamma_N\backslash N} f(xn)\, dn, \tag{1}$$

where the invariant measure dn gives total mass one to $\Gamma_N\backslash N$. If f is an automorphic form, or a bit more generally, if it is of uniform moderate growth (see 4.2) and K-finite, then $f - f_P$ is fast decreasing on $a \times \omega$ for any compact set in $e'(P)$ if $a \to_P \infty$. We can similarly define the constant term of a differential form. It is then a differential form on $\Gamma_P\backslash X$. The formation of the constant term obviously commutes with exterior differentiation. Furthermore, it the coefficients of ω are of the type just mentioned, then $\omega - \omega_P$ decreases fast. It follows that if Z is a compact cycle on $e'(P)$, then the integral of $\omega - \omega_P$ on $a \cdot Z$ is fast decreasing as $a \to_P \infty$. On the other hand, it is also independent of $a \in A$ since $a \cdot Z$ and $a' \cdot Z$ are homologous for $a, a' \in A$. Therefore this integral vanishes, and we may replace ω by its constant term ω_P in order to study the effect of r_P.

2.6. The purpose of the construction outlined below is to get classes represented by automorphic forms with a non-zero restriction to the boundary and, ultimately, to get hold of cross-sections to such restriction maps. In fact we shall be concerned only with the cuspidal part of $H^*(e'(P); E)$ and so in fact consider a restriction map:

$$r_P : H^*(\Gamma; E_\lambda) \to H^*_{\mathrm{cusp}}(e'(P); E_\lambda). \tag{1}$$

At first, one may ask whether it makes good sense to speak of such a map, since the target space is only a subspace of $H^*(e'(P); E_\lambda)$. Since the classes constructed below automatically restrict to cuspidal classes, this need not worry us. In fact, an unpublished theorem of R. P. Langlands, coupled with 4.3 below, yields a natural complement to the cuspidal cohomology, and indeed allows one to define a restriction to cuspidal cohomology by projection modulo that complement. Our classes restrict trivially to the latter, so that our considerations are consistent with such a definition of r_P in (1). However, the theory of Eisenstein series makes it clear that one should not consider one face at a time, but more symmetrically a set of representatives $\mathscr{C}$ of Γ-conjugacy classes in a full set of associated parabolic $\mathbb{Q}$-subgroups, and study the direct sum $r_{\mathscr{C}}$ of the restriction maps r_P

$$r_{\mathscr{C}} : H^*(\Gamma; E) \to \bigoplus_{P \in \mathscr{C}} H^*_{\text{cusp}}(e'(P); E).$$

It is only for such a map that one may hope to construct a cross-section via Eisenstein series. In the case of $\mathbb{Q}$-rank one, this is done in [12]. Note that in that case $e'(P)$ is compact, hence all its cohomology is cuspidal.

2.7. ***Induced representations.*** Let (τ, V) be a unitary representation of $°M$. For $\Lambda \in \mathfrak{a}^*_c$, let $\mathbb{C}_\Lambda$ denote $\mathbb{C}$ acted upon by A

$$a(x) = e^{\Lambda(\log a)} \cdot x = a^\Lambda \cdot x \qquad (x \in \mathbb{C}). \tag{1}$$

We let $I_{P,V,\Lambda}$ or $I_{V,\Lambda}$ be the representation induced from $V \otimes \mathbb{C}_{\Lambda+\rho}$ in the smooth sense (see e.g. [9: III]); it is preunitary if $\Lambda \in i \cdot \mathfrak{a}^*$. In the discussion of Eisenstein series, it is often convenient to view the $I_{V,\Lambda}$ as a family of representations acting on a fixed space. To this effect, one starts from the space

$$U_{V,0} = C^\infty(\Gamma_P(NA \backslash G) \tag{2}$$

of smooth functions on the coset space $\Gamma_P NA \backslash G$ [this makes good sense since Γ_P normalizes $A \cdot N$, and A normalizes N], and sets:

$$(I_{V,\Lambda}(g)f)(x) = f(x \cdot g) \cdot a(xg)^{\Lambda+\rho} \cdot a(x)^{-(\Lambda+\rho)}, \qquad (x, g \in G; f \in U_{V,0}), \tag{3}$$

where $a(x)$ denotes the A-component of $x \in G$, when the latter is written using the decomposition $G = °M \cdot A \cdot N \cdot K$ (which is not direct, but nevertheless determines $a(x)$ uniquely).

We assume now that V is irreducible or a finite sum of equivalent irreducible representations. It then has an infinitesimal character χ_V. Let $\mu \in \mathfrak{b}^*_c$ be dominant and such that $\chi_V = \chi_\mu$. Then, by Theorem 3.3 in [9: III]

$$H^*(\mathfrak{g}, K; I_{V,\Lambda} \otimes E_\lambda) = 0 \quad \text{if} \quad -(\mu + \Lambda) \notin W^P(\lambda), \tag{4}$$

and

$$H^*(\mathfrak{g}, K; I_{V,\Lambda} \otimes E_\lambda) = H^*(°\mathfrak{m}, K_M; V F_{s\lambda|\mathfrak{b}})[l(s)] \otimes \Lambda \mathfrak{a}^*_c, \tag{5}$$

if there exists $s \in W^P$, necessarily unique, such that

$$-(\mu + \Lambda) = s\lambda; \tag{6}$$

this last condition can also be written

$$-\mu = s\lambda\,|_{\mathfrak{b}'_c}, \qquad -\Lambda = s\lambda\,|_{\mathfrak{a}}. \tag{7}$$

Return now to the notation of 2.3. If $V = H_\pi$ $(\pi \in °\hat{M})$, then we write accordingly $I_{\pi,\Lambda}$ for $I_{V,\Lambda}$. Then, if $s \in W^P$ satisfies (6), the equality (5) can be written

$$H^*(\mathfrak{g}, K; I_{\pi,-s\lambda|\mathfrak{a}} \otimes E_\lambda) = F^*_{\pi,s} \otimes \Lambda\mathfrak{a}^*_c. \tag{8}$$

2.8. We now consider a second p-pair (P', A') similar to (P, A) and denote by $M', {}^\circ M', \mathfrak{b}', \mathfrak{h}'$, etc. the analogues of $M, {}^\circ M, \mathfrak{b}, \mathfrak{h}$, etc. (2.1). We assume that (P', A') is *associated to* (P, A), i.e., that the set $W(A, A')$ of isomorphisms of A onto A' induced by inner automorphisms of G defining a $\mathbb{Q}$-isomorphism of $\tau(M)$ onto $\tau'(M')$ is not empty. If $n \in G$ represents an element $w \in W(A, A')$ then ${}^n({}^\circ M) = {}^\circ M'$, hence n induces a bijection of ${}^\circ\hat{M}$ on to ${}^\circ\hat{M}'$. Any other representative n' of w is of the form $n' = n \cdot q$ with $q \in Z(A)$. Since ${}^\circ M$ acts trivially on ${}^\circ\hat{M}$ by inner automorphisms, we see that this bijection depends only on w. We denote by ${}^w\pi$ the image of $\pi \in {}^\circ\hat{M}$. If $A = A'$, we also write $W(A)$ for $W(A, A)$. The group $W(A)$ is the quotient of the normalizer $N(A)$ of A by the centralizer $Z(A)$ of A. We note the obvious relation

$$w \cdot W(A) = W(A, A') = W(A') \cdot w \qquad (w \in W(A, A')). \tag{1}$$

Let $\mathfrak{b}'$ be a Cartan subalgebra of ${}^\circ\mathfrak{m}'$ and $\mathfrak{h}' = \mathfrak{b}' \otimes \mathfrak{a}'$. Let $W(\mathfrak{h}_c, \mathfrak{h}'_c)$ be the set of isomorphisms $\mathfrak{h}_c \to \mathfrak{h}'_c$ defined by inner automorphisms of $\mathfrak{g}_c$ which map $\mathfrak{h}_c$ onto $\mathfrak{h}'_c$. As is well known, any element of $W(A, A')$ is induced by one of $W(\mathfrak{h}_c, \mathfrak{h}'_c)$; if w_0 is one such element, any other one is of the form $t \cdot w_0$ or $w_0 \cdot s$ ($s \in W_M, t \in W_{M'}$) and the double coset $W_{M'} \cdot w_0 \cdot W_M$ is independent of w_0. The cosets $W_{M'} \cdot w_0$ ($w \in W(A, A')$) in $W(\mathfrak{h}_c, \mathfrak{h}'_c)$ are then well defined, and clearly disjoint. By abuse of notation, we shall denote their union by $W_{M'} \cdot W(A, A')$. The set $W(A, A') \cdot W_M$ is defined similarly. If $\pi \in {}^\circ\hat{M}$ has infinitesimal character χ_μ, then ${}^w\pi$ has infinitesimal character $\chi_{\mu'}$, with $\mu' = t^*(\mu)$ for any $t \in W(\mathfrak{h}_c, \mathfrak{h}'_c)$ representing w.

2.9. Let V now be in ${}^\circ L(\Gamma_M \backslash M)^\infty$. The construction of Eisenstein series can be viewed as yielding an intertwining operator (of $(\mathfrak{g}, K)$-modules)

$$E(\Lambda): I_{V,\Lambda,(K)} \to C^\infty(\Gamma \backslash G), \tag{1}$$

whose image is in fact in the space $\mathcal{A}(\Gamma \backslash G)$ of automorphic forms on $(\Gamma \backslash G)$. This operator is first defined for $\operatorname{Re} \Lambda > \rho$, and holomorphic in that tube. It admits a meromorphic extension to $\mathfrak{a}^*_c$ [13; 14]. In the sequel, we shall drop the explicit mention of K-finite vectors, it being understood that our intertwining operators are defined on spaces of K-finite vectors and are equivariant for the $(\mathfrak{g}, K)$-module structure only. For $w \in W(A, A')$, there is furthermore an intertwining operator

$$C(w, \Lambda): I_{P,V,\Lambda} \to I_{P',{}^wV,w\Lambda}, \tag{2}$$

which is also holomorphic for $\operatorname{Re}\Lambda > \rho$, and admits a meromorphic continuation to $\mathfrak{a}_c^*$ [13; 14].

Let $E(\Lambda)_{P'}$ be $E(\Lambda)$ followed by taking the constant term with respect to P'. Then, if $C(w, \Lambda)$ is holomorphic at Λ for all w, we have

$$E(\Lambda)_{P'}(f) = \sum_{w \in W(A, A')} C(w, \Lambda) \cdot f. \tag{3}$$

We get then a natural map

$$H(C(w, \Lambda)) : H^*(\mathfrak{g}, K; I_{P,\pi,\Lambda}) \to H(\mathfrak{g}, K; I_{P', w_\pi, w\Lambda}). \tag{4}$$

The left-hand side of (4) is zero unless there exists $s \in W^P$ which satisfies 2.7(6) and $\Lambda = -s\lambda|_{\mathfrak{a}}$. If $C(w, \Lambda)$ is holomorphic at $-s\lambda|_{\mathfrak{a}}$, we have then by 2.7(8) a linear map

$$H(C(w, -s\lambda|_{\mathfrak{a}})) : F^*_{\pi,s} \otimes \Lambda\mathfrak{a}_c^* \to F^*_{w_\pi, ws} \otimes \Lambda\mathfrak{a}'_{c}. \tag{5}$$

By making explicit the various maps involved, we can see that $H(C(w, -s\lambda|_{\mathfrak{a}})$ is the tensor product of a linear map

$$c^*_{w,s,\lambda} : F^*_{\pi,s} \to F^*_{w_\pi, ws}, \tag{6}$$

by the natural map $\Lambda\mathfrak{a}_c^* \to \Lambda\mathfrak{a}_c'^*$ transpose of w^{-1}. We have $\chi_{w_\pi} = \chi_{-w(s\lambda|_{\mathfrak{b}'})}$, hence there is a natural injection

$$j_{P'} : F^*_{w_\pi, ws} \to H^*_{\text{cusp}}(e'(P'); E_\lambda). \tag{7}$$

Note (November 1982): The reason given for the existence of $c^*_{w,s,\lambda}$ in (6) is not correct. However, W. Casselman and B. Speh have independently shown that the space $F^*_{\pi,s} \oplus \Lambda^i\mathfrak{a}_c^*$ in 2.7(8) is in the kernel of the map into $H^*(\Gamma; E)$ for $i \geq 1$; this yields the same consequences.

2.10. Theorem. *Let (P, A) and (P', A') be associate p-pairs as above. Let $\pi \in {}^\circ\hat{M}$, $s \in W^P$ and the irreducible representation E_λ of G be compatible (2.4). Assume that the operators $C(w, \Lambda)$ are holomorphic at $\Lambda = -s\lambda|_{\mathfrak{a}}$ for all $w \in W(A, A')$. Then the image of the composition $\nu \circ \mu$:*

$$F^*_{\pi,s} \xrightarrow{\mu} \bigoplus_{w \in W(A,A')} F^*_{w_\pi, ws} \xrightarrow{\nu} H^*_{\text{cusp}}(e'(P'); E_\lambda), \tag{1}$$

where $\mu = \oplus c^*_{w,s,\lambda}$ and ν is the sum of the $j_{P'}$, is contained in the image under $r_{P'}$ of a subspace of $H^*(\Gamma; E_\lambda)$ represented by harmonic forms.

Idea of proof. Let $T = E(-s\lambda|_A)$. It intertwines $I = I_{P,\pi-s\lambda|_{\mathfrak{a}}}$ and $C^\infty(\Gamma\backslash G)$. Let J be its image. The space J is in fact contained in the space $\mathcal{A}(\Gamma\backslash G)$ of automorphic forms on $\Gamma\backslash G$. Then T induces maps

$$C^*(\mathfrak{g}, K; I \otimes E) \to C^*(\mathfrak{g}, K; J \otimes E) \to C^*(\mathfrak{g}, K; C^\infty(\Gamma\backslash G) \otimes E), \tag{2}$$

$$H^*(\mathfrak{g}, K; I \otimes E) \to H^*(\mathfrak{g}, K; J \otimes E) \to H^*(\Gamma; E). \tag{3}$$

Let ω be a closed form in $C^*(\mathfrak{g}, K; I \otimes E_\lambda)$. In order to study the effect of r_P on the cohomology class of ω, we may by 2.5 replace ω by its constant term $\omega_{P'}$. By 2.9(3), the form ω_P may be viewed as an element of the sum of the spaces

$$C^*(\mathfrak{g}, K; I_{P', {}^w\pi, -ws\lambda|_{\mathfrak{a}}}), \qquad (w \in W(A, A')). \tag{4}$$

The proofs of 2.5 and 3.3 in [9: III] yield respectively the equalities

$$C^*(\mathfrak{p}', K_{P'}; H_{w_\pi} \otimes \mathbb{C}_{-ws(\lambda+P)_{\mathfrak{a}'}} \otimes E_\lambda) = C(\mathfrak{g}, K; I_{P', {}^w\pi, -s\lambda|_{\mathfrak{a}'}} \otimes E_\lambda), \tag{5}$$

$$H^*(\mathfrak{p}', K_{P'}; H_{w_\pi} \otimes \mathbb{C}_{-ws\lambda|_{\mathfrak{a}'}} \otimes E_\lambda) = F^*_{w_\pi, ws} \otimes \Lambda \mathfrak{a}'_c, \tag{6}$$

where $K_{P'} = K \cap P'$. Going from the right to the left in (5) corresponds to viewing X as a quotient of P' rather than G, i.e., to identifying X with $P'/K_{P'}$. From this, one sees that the map

$$H^*(\mathfrak{p}', K_{P'}; H_{w_\pi} \otimes \mathbb{C}_{-w(s\lambda+\rho)|_{\mathfrak{a}}} \otimes E_\lambda \to H^*(e'(P'); E_\lambda) \tag{7}$$

defined by the restriction to a submanifold $a \cdot e'(P') \subset U_t$ (notation of 2.5) amounts to the natural inclusion

$$F^*_{w_\pi, -ws\lambda} \to H^*_{\mathrm{cusp}}(e'(P'), E_\lambda), \tag{8}$$

hence the image of the class $[\omega]$ of ω under $\nu \circ \mu$ in (1) is that obtained from the image of the class of $T\omega$ in $H^*(\Gamma; E)$ by restriction to $H^*(\Gamma_{P'}; E)$.

To complete the proof, it suffices to show that the image of $H(T)$ consists of harmonic forms. Let $\mu \in \mathfrak{b}_c^*$ be dominant such that $\chi_\pi = \chi_\mu$. Then the infinitesimal character of I is $\chi_{\mu+\Lambda}$. This is also the infinitesimal character of J. Let C be the Casimir operator. By Kuga's formula, the action of Δ on $\omega \in C^*(\mathfrak{g}, K; J \otimes E)$ is obtained by applying $-C(J) \otimes 1 + 1 \otimes C(E)$ to $J \otimes E$. But if a representation has infinitesimal character χ_ν, then the Casimir operator is $((\nu, \nu) - (\rho, \rho))$, where $(\,,\,)$ refers to the scalar product associated to the invariant form on $\mathfrak{g}$ used to define C. In our case $\mu + \Lambda = -s\lambda$, hence

$$C(J) = (s\lambda, s\lambda) - (\rho, \rho) = (\lambda, \lambda) - (\rho, \rho) \quad \text{and} \quad C(E) = (\lambda, \lambda) - (\rho, \rho)$$

whence our assertion.

2.11. ***Remarks.*** (1) We see that $C(\mathfrak{g}, K; J \otimes E_\lambda)$ consists of harmonic forms. But we do not know whether all cohomology classes are represented by forms which are both closed and coclosed.

(2) Let now $P' = P$. Then $1 \in W(A, A')$. Since $C(1, \Lambda)$ is the identity, the first map μ in (1) is injective. However, the various spaces $F^*_{w_\pi, ws}$ may not be disjoint, hence we cannot assert in general that r_P is injective.

There is, however, one case where this is so and we shall devote Proposition 2.13 to it.

(3) The $C(w, \Lambda)$ satisfy a functional equation [13; 14], in particular

$$C(wt, \Lambda) = C(w, t\Lambda) \cdot C(t, \Lambda), \qquad (t \in W(A, A),\ w \in W(A, A')). \qquad (1)$$

Together with 2.8(1), this shows that if we replace $F^*_{\pi,s}$ by $F^*_{t\pi,ts}$ in 2.10(1) for any $t \in W(A)$, then the target space and image of μ are unchanged.

2.12. Lemma. *Assume that π, s, λ are compatible and that $s\lambda|_A = 0$ and $F^*_{\pi,s} \neq 0$. Then* $\dim N$ *is even,* $l(s) = \dim N/2$, *and the representations ${}^w\pi$ $(w \in W(A, A'))$ have distinct infinitesimal characters. In particular, the subspaces $F^*_{w_\pi, ws}$ of $H(e'(P'); E_\lambda)$ are linearly independent.*

The representation $I_{\pi,0}$ is unitary hence $F^*_{\pi,s} \otimes \Lambda \mathfrak{a}^*_c$ satisfies Poincaré duality. The same is true for $H^*({}^\circ\mathfrak{m}, K_M; H_\pi \otimes F_{s\lambda|\mathfrak{b}_c})$. Let c and c' be the smallest and biggest degrees in which the latter is $\neq 0$. Then $c + c' = \dim X_M$. Similarly, using 2.7(5), we get

$$\dim X = c + c' + 2l(s) + \dim \mathfrak{a}.$$

Since $\dim X = \dim X_M + \dim \mathfrak{a} + \dim N$, the first assertion follows. Moreover, $s\lambda|\,\mathfrak{b}_c = s\lambda$ is a regular element. Therefore the transforms of $s\lambda|_{\mathfrak{b}_c}$ under $W_{M'} \cdot W(A, A')$, where $W_{M'} \cdot W(A, A')$ is embedded in $W(\mathfrak{h}_c, \mathfrak{h}'_c)$ as in 2.8 are all distinct. In particular $ws\lambda|_{\mathfrak{b}'_c}$ and $w's\lambda|_{\mathfrak{b}'_c}$ belong to different $W_{M'}$-orbits if $w \neq w'$ $(w, w' \in W(A, A'))$, hence the ${}^w\pi$ have distinct infinitesimal characters. This implies the last assertion of the lemma.

2.13. Proposition. *Assume that π, λ, s are compatible and that $s\lambda|_A = 0$. Then the map $\nu \circ \mu$ of 2.10 is defined and is injective.*

The operators $E(\Lambda)$ and $C(w, \Lambda)$ are all holomorphic on the 'imaginary axis' $\operatorname{Re} \Lambda = 0$ and the $C(w, \Lambda)$ are unitary there [13; 14]. Thus μ is well defined; since the $F^*_{w_\pi, ws}$ are linearly independent by 2.12, it also follows that ν is injective, whence the proposition.

2.14. Let $F^*_{P,(\pi,s)}$ be the direct sum of the spaces $F^*_{w_\pi, ts}$ $(t \in W(A, A))$. These spaces are linearly independent and if $\{(P_i, A_i)\}_{i \in I}$ is a set of representative modulo Γ of associated p-pairs, then 2.13 provides an injective map

$$F^*_{P,(\pi,s)} \to \bigoplus_i H^*_{\text{cusp}}(e'(P_i), E_\lambda),$$

whose image is contained in that of $H^*(\Gamma; E_\lambda)$ under the sum of the maps

r_{P_i}. As in 2.11(3), its image would not change if $F^*_{P,(\pi,s)}$ were replaced by $F^*_{P',(w_\pi,ws)}$, where (P', A') is one of the (P_i, A_i) and $w \in W(A, A')$. It seems reasonable to conjecture that this image is equal to that of $H^*(\Gamma; E)$ under restriction. If G has $\mathbb{Q}$-rank one, this follows from Poincaré duality (see [12], case $\lambda = -\rho$ on pp. 156–158).

If $G = \mathbb{SL}_3$ and Γ is a (neat) congruence subgroup of $\mathbb{SL}_3(\mathbb{Z})$, this has been proved for the proper maximal parabolic $\mathbb{Q}$-subgroups by J. Schwermer (unpublished).

2.15. We have limited ourselves to the case where the $C(w, \Lambda)$ are holomorphic at $-s\lambda|_{\mathfrak{a}}$. However, in particular when E_λ is the trivial representation, the $C(w, \Lambda)$ may very well have poles at such points. In this case, we may have to take residues of these operators and of Eisenstein series. The latter will in general intertwine only a quotient of the principal series $I_{P,\pi,\Lambda}$ and $C^\infty(\Gamma\backslash G)$, and the situation is much harder to describe. In fact, in most cases, it cannot be investigated thoroughly because of lack of knowledge of the operators $C(w, \Lambda)$. We refer to [11; 12] for some cohomological results involving residues.

3. Square integrable classes

3.1. In this section, we survey rather briefly some results and problems concerning L^2-classes. As before, we shall consider cohomology with coefficients in a finite dimensional irreducible G-module (σ, E). Sometimes, only the case $E = \mathbb{C}$ is considered in the references given below, but the extension to the more general one is easy.

3.2. The space $L^2(\Gamma\backslash G)$ is the direct sum of the discrete spectrum $L^2(\Gamma\backslash G)_d$ and the continuous spectrum $L^2(\Gamma\backslash G)_{ct}$.

As stated in [4], and is proved in [7], the space $\mathcal{H}_2$ of harmonic E-valued L^2-forms is finite dimensional, and is equal to $H^*(\mathfrak{g}, K; L^2(\Gamma\backslash G)_d \otimes E)$. The space $L^2(\Gamma\backslash G)_d$ is a Hilbert direct sum of irreducible subspaces with finite multiplicities. Let H_π denote the isotypic subspace corresponding to $\pi \in \hat{G}$. Then it can be proved that $H^*(\mathfrak{g}, K; L^2(\Gamma\backslash G)_d \otimes E)$ is a finite algebraic direct sum

$$H^*(\mathfrak{g}, K; L^2(\Gamma\backslash G)_d \otimes E) = \bigoplus_{\pi \in \hat{G}} H^*(\mathfrak{g}, K; H_\pi \oplus E). \tag{1}$$

Moreover, for a summand to be non-zero, it is necessary that $\chi_\pi = \chi_{E^*}$, [9: I, 5.3], in which case the exterior differential on $C^*(\mathfrak{g}, K; H_\pi \otimes E)$ is identically zero and $C^*(\mathfrak{g}, K; H_\pi \otimes E)$ consists of harmonic forms [9: II, 3.1]. Since H is unitary, it follows moreover that harmonic forms are closed and coclosed [9: II, 2.4], hence

$$H^*(\mathfrak{g}, K; H_\pi \otimes E) = \mathrm{Hom}_K(\Lambda^* \mathfrak{g}/\mathfrak{k}; H_\pi \otimes E). \tag{2}$$

The relation (1) implies in particular that the known results in relative Lie algebra cohomology with coefficients in irreducible unitary representations apply to the square integrable harmonic forms. Furthermore, one can prove the existence of ranges of dimensions in which all cohomology classes have L^2-representatives or even in which $\mathcal{H}_2$ is isomorphic to the cohomology, and get then information on $H^*(\Gamma; E)$ itself (see [3; 6] for examples in the non-cocompact case, and [9] or the earlier work of Matsushima, Murakami, Parthasarathy and others referred to there in the cocompact case).

3.3. The first finiteness result above is in fact part of a more general statement concerning eigenvalues of Casimir operators or Laplace operators. The one given in [4] can also be derived from a more general one proved in [7]. It states that if Δ is the Laplace operator on $\Gamma\backslash G$ associated to a metric coming from a left invariant Riemannian metric on G, then the eigenvalues of Δ on $L^2(\Gamma\backslash G)_d$ have finite multiplicities, and tend to $+\infty$. From this one can further deduce that if $\alpha \in C_c^\infty(G)$, then convolution by α is a compact operator on $L^2(\Gamma\backslash G)_d$. It is not known, however, whether it is of Hilbert–Schmidt class. Another open problem is whether there are better results on the asymptotic behaviour of eigenvalues. For instance is the H. Weyl estimate for the eigenvalues of the Laplacian on a compact manifold still valid in our case? As far as I know, this has been answered (positively) for $\mathbb{SL}_2(\mathbb{Z})$ and some of its congruence subgroups (cf. [22] for a survey).

3.4. Let $A_{(2)}(\Gamma\backslash X; E)$ denote the set of smooth forms η such that η, $d\eta$ and are square integrable. This is a subcomplex of $A^*(\Gamma\backslash X)$. By definition, its cohomology is the L^2-cohomology $H_{(2)}(\Gamma\backslash X; E)$ of $\Gamma\backslash X$. By going to a suitable closure $\bar{d}$ of the exterior differential d, one can also give a Hilbert space definition of L^2-cohomology [10]. Since $\Gamma\backslash X$ is complete, a harmonic L^2-form is not the coboundary of an L^2-form, hence the natural map $\mathcal{H}_2 \to H_{(2)}$ is injective (see, e.g., [3]). It is surjective if $\bar{d}$ has closed range, in particular if $H_{(2)}$ is finite dimensional. It can be shown that

$$H_{(2)}(\Gamma\backslash X; E) = H^*(\mathfrak{g}, K; L^2(\Gamma\backslash G)^\infty \otimes E). \tag{1}$$

In view of 3.1, it follows that $\mathcal{H}_2 = H_2$ if and only if

$$H(\mathfrak{g}, K; L^2(\Gamma\backslash G)_{ct}^\infty \otimes E) = 0. \tag{2}$$

Now according to [14], $L^2(\Gamma\backslash G)_{ct}$ is a Hilbert direct sum of continuous integrals. Each of those is a continuous sum of unitarily induced representations $I_{P,\pi,i\mu}$ $(\mu \in \mathfrak{a}_*)$, where π runs through the discrete spectrum $L^2(\Gamma_M \backslash {}^\circ M)_d$ (see Section 2 for the notation). The integral is to be taken over a Weyl chamber C in $\mathfrak{a}^*$.

Therefore, we are reduced to evaluate

$$H^*\left(\mathfrak{g}, K; \left(\int_C I_{P,\pi,i\mu}\, d\mu\right) \otimes E\right). \tag{3}$$

This can be done by a computation similar to that of 3.3 in [9: III], with $E = E_\lambda$ as in Section 2. One finds this space is zero-dimensional unless $s\lambda\,|_{\mathfrak{a}} = 0$. If that is the case, then, as remarked in the proof of 2.12, $\mathfrak{b}$ contains an element regular in $\mathfrak{g}$. If there is no P satisfying this latter condition, then (2) holds. Otherwise, this cohomology space is given by the right-hand side of 2.7(8), with $\Lambda\mathfrak{a}_c^*$ replaced by

$$H^*\left(\mathfrak{a}; \int_C \mathbb{C}_{i\mu}\, d\mu\right). \tag{4}$$

It can easily be shown that this space is infinite dimensional in dimensions $q \in [1, \dim \mathfrak{a}]$ and is zero otherwise. As a consequence, if $F^*_{\pi,s} \neq 0$, then the L^2-cohomology is infinite dimensional. This happens in many cases [6].

3.5 (*added November 1982*). The argument underlying Section 4 in [6] proves a weaker statement. However, in a paper written jointly with W. Casselman, 'L^2-cohomology of locally symmetric manifolds of finite volume' (to appear), it is shown that the conclusion of Theorem 4 holds if no proper parabolic $\mathbb{Q}$-subgroup of G contains a fundamental parabolic subgroup of G. This condition is in particular fulfilled if $\operatorname{rk} G = \operatorname{rk} K$.

4. Further remarks and problems

1 **4.1.** We have emphasized an analytic approach to the study of cohomology, via automorphic forms or automorphic representations. But even in the cocompact case, many results have been obtained by more geometric methods. Moreover, the method outlined in Section 2, even if one knew enough on the $C(w, \Lambda)$ to carry it out in full, would not give a full account of $H^*(\Gamma; E)/H_!(\Gamma; E)$. In fact, by construction, it yields mostly classes whose restriction to some faces are not cohomologous to zero. However, except in $\mathbb{Q}$-rank one, where $\partial \bar{Y}$ is a disjoint union of faces, it seems clear that $H^*(\Gamma; E)$ will contain elements not having this property.

In fact, the closures of the faces $e'(P)$ form a closed cover of the boundary, whose nerve is the quotient of the Tits building of proper parabolic $\mathbb{Q}$-subgroups of G by Γ, acting by conjugation, and the cohomology of the boundary may be investigated by means of the Leray spectral sequence of this cover. The classes which restrict non-trivially to some faces will contribute to the terms of base degree zero in the E_∞-term of that spectral sequence. But there may be elements with non-zero base degree and one has to expect that some of those may also be in the image of $H^*(\Gamma; E)$. I should say, however, that no concrete examples of such classes, which I have come to call 'ghost classes', have been given so far. The structure of the spectral sequence seems to indicate that if one wishes to construct such classes by analytic means, one should have some 'lifting' procedure which would associate to the cohomology of a face, or some part of it, elements of higher dimension in $H^*(\Gamma; E)$. I do not know whether this is possible in any systematic way.

2 **4.2.** In [12] it is shown that if G has $\mathbb{Q}$-rank one, every cohomology class can be represented by a closed and coclosed (hence harmonic) automorphic form. It is natural to ask whether this holds true in general. The first question would be whether the natural inclusion $\mathscr{A}(\Gamma\backslash G) \to C^\infty(\Gamma\backslash G)$ induces an isomorphism in relative Lie algebra cohomology. Now, in order to be an automorphic form, a function $f \in C^\infty(\Gamma\backslash G)$ has to satisfy three conditions (see, e.g., [13, 14]):

(a) f is K-finite (on the right).
(b) f is $\mathfrak{Z}$-finite: the $z \cdot f$, where z runs through the centre of the universal algebra $U(\mathfrak{g})$ of $\mathfrak{g}$, span a finite dimensional space.
(c) f is of moderate growth.

As is well-known, these conditions imply that (c) can be replaced by

(c′) f is of 'uniform moderate growth',

By this it is meant that f and all its derivatives $x \cdot f$ $(x \in U(\mathfrak{g}))$ are of moderate growth *with the same exponent.* Let $C^\infty_{\mathrm{umg}}(\Gamma\backslash G)$ be the space of $f \in C^\infty(\Gamma\backslash G)$ which satisfy (c′). Then one can prove

4.3. Theorem. *The inclusion* $C^\infty_{\mathrm{umg}}(\Gamma\backslash G) \to C^\infty(\Gamma\backslash G)$ *induces an isomorphism of* $H^*(\mathfrak{g}, K; C^\infty_{\mathrm{umg}}(\Gamma\backslash G) \otimes E)$ *onto* $H^*(\Gamma; E)$.

The technique is similar to the one which yields 3.3(1). One uses a spectral sequence in relative Lie algebra cohomology, and a homotopy operator on $A^*(\Gamma\backslash G; E)$ given by convolution with a compactly supported smooth K-invariant function. As usual (see 1.4) we may replace $C^\infty_{\mathrm{umg}}(\Gamma\backslash G)$ by its subspace of K-finite functions; therefore we may compute cohomology using forms whose coefficients satisfy (a) and (c′), and the main open question is whether one can also get (b) fulfilled.

Note (November 1982). The proofs of 3.3(1) and Theorem 4.3 are contained in a forthcoming paper, 'Regularization theorems in Lie algebra cohomology. Applications'.

References

1. Borel, A. Linear algebraic groups. In *Algebraic Groups and Discontinuous Subgroups. Proc. Symp. Pure Math.*, **9** (1966), 3–19, A.M.S., Providence, R.I.
2. Borel, A. Reduction theory for arithmetic groups. In *Algebraic Groups and Discontinuous Subgroups. Proc. Symp. Pure Math.*, **9** (1966), 20–25, A.M.S., Providence, R.I.
3. Borel, A. Stable real cohomology of arithmetic groups. *Ann. Sci. E.N.S.*, Paris, **7** (4) (1974), 235–272.
4. Borel, A. Cohomology of arithmetic groups. *Proc. I.C.M.*, Vancouver, 1974, Vol. 1, 435–442.
5. Borel, A. Cohomologie de sous-groupes discrets et représentations de groupes semi-simples. *Astérisque*, **32–33** (1976), 73–112.
6. Borel, A. Stable and L^2-cohomology of arithmetic groups. *Bull. A.M.S. N.S.*, **3** (1980), 1025–1027.
7. Borel, A. and Garland, H. Laplacian and discrete spectrum of an arithmetic group. *Amer. J. Math.* (to appear).
8. Borel, A. and Serre, J.-P. Corners and arithmetic groups. *Comment. Math. Helv.*, **48** (1973), 436–491.
9. Borel, A. and Wallach, N. Continuous cohomology, discrete subgroups and representations of reductive groups. *Ann of Math. Studies*, **94**, Princeton University Press, Princeton, 1980.
10. Cheeger, J. On the Hodge theory of Riemannian manifolds. In *Geometry of the Laplace operator. Proc. Symp. Pure Math.*, **36** (1980), 91–146, A.M.S., Providence, R.I.
11. Harder, G. On the cohomology of $\mathbb{SL}_2(\mathbb{Q})$. In *Lie Groups and Their Representations*, I. M. Gelfand, ed., A. Hilger, London, 1975, 139–150.
12. Harder, G. On the cohomology of discrete arithmetically defined groups. *Proc. Int. Colloquium, Bombay, 1973*, Oxford University Press, 1975, 129–160.
13. Harish-Chandra, Automorphic forms on semi-simple Lie groups. Notes by J. G. M. Mars. *Lecture Notes in Math.*, **62**, Springer-Verlag, 1968.
14. Langlands, R. P. On the functional equation satisfied by Eisenstein series. *Lecture Notes in Math.*, **544**, Springer-Verlag, 1976.

15. Matsushima, Y. and Murakami, S. On vector bundle valued harmonic forms and automorphic forms on symmetric manifolds. *Ann. of Math.*, **78** (2) (1963), 365–416.
16. Matsushima, Y. and Murakami, S. On certain cohomology groups attached to hermitian symmetric spaces. *Osaka J. Math.*, **2** (1965), 1–35.
17. deRham, G. Variétés différentiables. *Act. Sci. Ind.*, **1222,** Hermann, Paris, 1960.
18. Schwermer, J. Sur la cohomologie des sous-groupes de congruence de $\mathbb{SL}_3(\mathbb{Z})$. *C.R. Acad. Sci. Paris Sér. A*, **283** (1976), 817–820.
19. Schwermer, J. Eisensteinreihen und die Kohomologie von Kongruenzuntergruppen von $\mathbb{SL}_n(\mathbb{Z})$. *Bonner Math. Schriften*, **99** (1977), Universität Bonn.
20. Schwermer, J. Sur la cohomologie de $\mathbb{SL}_n(\mathbb{Z})$ à l'infini et les séries d'Eisenstein. *C.R. Acad. Sci. Paris Sér. A*, **289** (1979), 413–415.
21. Springer, T. A. Reductive groups. In *Automorphic Forms, Representations and L-functions. Proc. Symp. Pure Math.*, **33** (1979), Part 1, 3–27.
22. Venkov, A. B. Spectral theory of automorphic functions, the Selberg zeta-function, and some problems of analytic number theory and mathematical physics. *Uspekhi Mat. Nauk*, **34,** 3 (1979), 69–135, *Russian Math. Surveys*, **34,** 3 (1979), 79–153.

A. Borel
The Institute for Advanced Study
School of Mathematics
Princeton
New Jersey 08540
USA

Received December, 1980

125.

Regularization Theorems in Lie Algebra Cohomology. Applications

Duke Math. J. 50 (1983) 605–623

In computing the cohomology of a complex, it is sometimes useful to be able to replace the given complex by a smaller one without altering the cohomology. This paper proves some theorems of this type in the framework of Lie algebra or relative Lie algebra cohomology with coefficients in infinite dimensional modules. The subcomplex will usually consist of smooth vectors in the representation theoretic sense, and the passage from one to the other will involve smoothing or regularization operators, whence our title. For motivation, we first describe three applications, the first two of which are at the origin of this paper. We let G be a connected semi-simple Lie group, K a maximal compact subgroup of G, $\mathfrak{g}, \mathfrak{k}$ their Lie algebras, $X = G/K$ and Γ a torsion-free discrete subgroup of finite covolume, not cocompact since otherwise the problems considered here do not arise.

A. The cohomology $H^{\cdot}(\Gamma; \mathbf{C})$ of Γ with complex coefficients can be expressed as the cohomology of the complex $A^{\infty}(\Gamma\backslash X; \mathbf{C})$ of smooth complex valued differential forms on $\Gamma\backslash X$. Lifting those forms on $\Gamma\backslash G$ yields an isomorphism of $A^{\infty}(\Gamma\backslash X; \mathbf{C})$ with the relative Lie algebra complex $C^{\cdot}(\mathfrak{g}; \mathfrak{k}, C^{\infty}(\Gamma\backslash G))$ with coefficients in the space of complex valued smooth functions on $\Gamma\backslash G$. One wishes to replace $C^{\infty}(\Gamma\backslash G)$ by a smaller space. In [2] it is shown that we can use the space $C^{\infty}_{mg}(\Gamma\backslash G)$ of smooth functions which, together with their derivatives with respect to left invariant differential operators, have moderate growth. Here we shall see that we can reduce it further to the space $C^{\infty}_{umg}(\Gamma\backslash G)$ of functions of uniform moderate growth (i.e., the exponent limiting the growth on a Siegel set can be chosen independently of the derivatives). An application of this to the decomposition of $H^{\cdot}(\Gamma; \mathbf{C})$ was pointed out by R. P. Langlands several years ago (3.4). Actually, one would like to be able to replace $C^{\infty}_{umg}(\Gamma\backslash G)$ by automorphic forms [3; 7], but this seems to be a much harder step.

B. The L^2-cohomology of $\Gamma\backslash X$ is the cohomology of the complex $A^{\infty}_{(2)}(\Gamma\backslash X)$ of smooth forms η forms on $\Gamma\backslash X$ such that η and $d\eta$ are square integrable. It can be identified to a subcomplex $C^{\cdot}_{(2)}$ of $C^{\cdot}(\mathfrak{g}, \mathfrak{k}; C^{\infty}(\Gamma\backslash G))$ which contains $C^{\cdot}(\mathfrak{g}, \mathfrak{k}; L^2(\Gamma\backslash G)^{\infty})$. We want to prove that $H^{\cdot}(\mathfrak{g}, \mathfrak{k}; L^2(\Gamma\backslash G)^{\infty}) \to H^{\cdot}(C^{\cdot}_{(2)})$ is an isomorphism. This then gives some means to compute the latter. For applications, see [4].

C. The complexes $C^{\cdot}(\mathfrak{g}, \mathfrak{k}; L^2(\Gamma\backslash G)^{\infty})$ and $C^{\cdot}_{(2)}$ are embedded in the graded Hilbert space $C^{\cdot}(\mathfrak{g}, \mathfrak{k}; L^2(\Gamma\backslash G)) = \mathrm{Hom}_{\mathfrak{k}}(\Lambda(\mathfrak{g}/\mathfrak{k}), L^2(\Gamma\backslash G))$. The closures of d operating on $C^{\cdot}_{(2)}$ and on $C^{\cdot}(\mathfrak{g}, \mathfrak{k}; L^2(\Gamma\backslash G)^{\infty})$ are the same. We shall prove that

Received November 22, 1982.

the natural map $H^{\cdot}(\mathfrak{g},\mathfrak{k}; L^2(\Gamma\backslash G)^\infty) \to H^*(\operatorname{dom} \bar{d})$ is an isomorphism. This is another way to look at B, but it leads to a different type of generalization.

The simplest way to get a function of uniform moderate growth from one of moderate growth is to convolve it with a smooth function $\alpha \in C_c^\infty(G)$ of compact support. The convolution is a sum of translations by elements of G, which are all homotopic to the identity. This suggests then in case A to try to use $*\alpha$ as a homotopy operator. It is indeed one (under some mild conditions), but on $A^\infty(\Gamma\backslash G; \mathbf{C})$, i.e., on $C^{\cdot}(\mathfrak{g}; C^\infty(\Gamma\backslash G))$, and not on $C^{\cdot}(\mathfrak{g},\mathfrak{k}; C^\infty(\Gamma\backslash G))$, which it does not leave stable: the latter is defined by two conditions; one is $\mathfrak{k}$-invariance, and this can be taken care of by assuming α to be K-invariant; the other one however, the vanishing of inner products i_x $(x \in \mathfrak{k})$ is definitely not compatible with convolution and there lies our problem. In order to get around that difficulty, we proceed in three steps.

(1) Prove that $H^{\cdot}(\mathfrak{g}; C_{umg}^\infty(\Gamma\backslash G)) \to H^{\cdot}(\mathfrak{g}; C_{mg}^\infty(\Gamma\backslash G))$ is an isomorphism using a homotopy operator given by $*\alpha$ for a suitable $\alpha \in C_c^\infty(G)$.

(2) Construct relative Lie algebra spectral sequences (E_r') and (E_r) abutting to $H^{\cdot}(\mathfrak{g}; C_{umg}^\infty(\Gamma\backslash G))$ and $H^{\cdot}(\mathfrak{g}; C_{mg}^\infty(\Gamma\backslash G))$ respectively and in which

$$E_2'^{p,q} = H^p(\mathfrak{g},\mathfrak{k}; C_{umg}^\infty(\Gamma\backslash G) \otimes H^q(\mathfrak{k})), \qquad E_2^{p,q} = H^p(\mathfrak{g},\mathfrak{k}; C_{mg}^\infty(\Gamma\backslash G) \otimes H^q(\mathfrak{k})).$$

(3) Use (1) and the comparison theorem for spectral sequences to deduce that $H^{\cdot}(\mathfrak{g},\mathfrak{k}; C_{umg}^\infty(\Gamma\backslash G)) \to H^{\cdot}(\mathfrak{g},\mathfrak{k}; C_{mg}^\infty(\Gamma\backslash G))$ is an isomorphism.

For B and C, steps (1) and (3) are the same, but there are some further technical difficulties in (2): in B, the complex $C_{(2)}^{\cdot}$ is not a Lie algebra complex, but only a subcomplex of one; in C, the complex $\operatorname{dom} \bar{d}$ is not embedded in a Lie algebra complex. However, it contains a dense subcomplex which is one. Also our spaces are infinite dimensional. These technicalities prevent us from just quoting the known existence theorem, say in [10; 11], for the spectral sequence of a Lie algebra modulo a reductive Lie subalgebra. We shall prove it in §1 in the degree of generality required here (see 1.6, 1.10).

§2 establishes the general results in the context of Lie algebra cohomology. There are two cases. In one we start from a differentiable G-module V and consider the inclusion of two subcomplexes of $C^{\cdot}(\mathfrak{g}; V)$ (2.5); in the other we let W be a continuous G-module which is a Hilbert space, E a finite dimensional G-module, $\bar{d}$ the closure in $C^{\cdot}(\mathfrak{g}; W \otimes E)$ of d operating on $C^{\cdot}(g; W^\infty \otimes E)$ and study the inclusion of the latter space in $\operatorname{dom} \bar{d}$ (2.7).

§3 is devoted to the applications, which are stated and proved under more general assumptions than above, with respect to G, K, Γ and the coefficients. A and B are special cases of (2.5) and C of (2.7).

In this paper, we consider only the global case, where a group operates. However, similar regularization methods may be applied to the study of L^2-cohomology "at infinity," i.e., of certain subsets of $\Gamma\backslash X$ which are relatively compact in a given Satake compactification, as is already indicated in [6, 7] and will also be shown elsewhere by the author.

Theorem 1.6 and A, B were discussed in a seminar at the Institute in 1979–80. A was announced in [3] and B in [1]. Another proof of C, for reductive groups, is contained in [4]. A sketch of the proof of B was given in a Conference on Shimura varieties in Vancouver (1981).

§1. Spectral sequences in Lie algebra cohomology.

1.1. *Notation and conventions.* With only minor deviations, they are those of [5: I, §1], with which we assume familiarity. In particular, F is a field of characteristic zero, $\mathfrak{g}$ a finite dimensional Lie algebra over F and $\mathfrak{k}$ a reductive subalgebra of $\mathfrak{g}$. We fix a basis (e_r) of $\mathfrak{g}$ $(r = 1, \ldots, n)$ and let (e^r) be the dual basis of $\mathfrak{g}^*$. We write θ_r for θ_{e_r} $(1 \leqslant r \leqslant n)$.

(π, V) is a $(\mathfrak{g}, \mathfrak{k})$-module over a field $F' \supset F$, and $C^\cdot(\mathfrak{g}; V)$ the cochain complex for Lie algebra cohomology. We shall write it as

$$C^\cdot(\mathfrak{g}; V) = V \otimes \Lambda \mathfrak{g}^*.$$

We recall the formulae

$$\theta_x = di_x + i_x d, \qquad (x \in \mathfrak{g}) \tag{1}$$

$$d = 1 \otimes d_{\mathfrak{g}} + \sum_r \pi(e_r) \otimes \epsilon(e^r), \tag{2}$$

where $\epsilon(\ )$ denotes the exterior product, and

$$d_{\mathfrak{g}} = (1/2) \sum_r \epsilon(e^r) \cdot \theta_r. \tag{3}$$

We can write

$$\theta_t(e_s) = [e_t, e_s] = \sum_r c^r_{ts} e_r, \tag{4}$$

hence

$$\theta_t(e^s) = \sum_r c^s_{rt} e^r. \tag{5}$$

1.2. Given a $\mathfrak{k}$-module U, we let

$$U^{\mathfrak{k}} = \{u \in u \mid \theta_x u = 0 \text{ for all } x \in \mathfrak{k}\}. \tag{1}$$

We note that if (U, d) is a differential complex on which $\mathfrak{k}$ operates, then 1.1(1) shows that the natural operation of $\mathfrak{k}$ on $H^\cdot(U)$ is trivial. If now U is a $(\mathfrak{k}, \mathfrak{k})$ module, then, by full reducibility, to go over to $\mathfrak{k}$-invariants is an exact functor, hence

$$H^\cdot(U^{\mathfrak{k}}) = H^\cdot(U)^{\mathfrak{k}} = H^\cdot(U). \tag{2}$$

1.3. For brevity, we shall say that a subspace U of $C^{\cdot}(\mathfrak{g}; V)$ is *fully* $\mathfrak{k}$*-stable* if it is invariant under θ_x, i_x for $x \in \mathfrak{k}$ and $\epsilon(y)$ for $y \in \mathfrak{k}^*$. We then define the *basic subspace* U_B of U by

$$U_B = \{u \in U;\ \theta_x = i_x u = 0,\ (x \in \mathfrak{k})\}. \tag{1}$$

If U is a subcomplex, then U_B is also one (use 1.1(1)), the *basic subcomplex* of U. By 1.1(1) a subcomplex is fully $\mathfrak{k}$-stable if it is stable under i_x $(x \in \mathfrak{k})$ and $\epsilon(y)$ $(y \in \mathfrak{k}^*)$.

1.4. We choose once and for all a $\mathfrak{k}$-invariant complement $\mathfrak{m}$ to $\mathfrak{k}$ in $\mathfrak{g}$. We assume that

$$(e_i)\ (i = 1, \ldots, m) \qquad (\text{resp. } (e_a),\ (a = m+1, \ldots, n))$$

is a basis of $\mathfrak{m}$ (resp. $\mathfrak{k}$).

In the rest of §1 *indices* i, j, k *run from* 1 *to* m, *indices* a, b, c, d *from* $m+1$ *to* n *and indices* r, s, t *from* 1 *to* n.

We have a natural bigrading of $C^{\cdot} = C^{\cdot}(\mathfrak{g}; V)$ given by

$$C^{p,q} = V \otimes \Lambda^p \mathfrak{m}^* \otimes \Lambda^q \mathfrak{k}^*. \tag{1}$$

We let

$$F^p = F^p C^{\cdot} = \sum_{p' \geq p, q} C^{p',q}, \qquad (p \in \mathbf{N}) \tag{2}$$

$$F^{p,q} = F^p \cap C^{p+q}, \qquad (p, q \in \mathbf{N}). \tag{3}$$

We claim that $dF^p \subset F^p$ and that F^p is fully $\mathfrak{k}$-stable ($p \in \mathbf{N}$). Stability under i_x or $\epsilon(y)$ $(x \in \mathfrak{k}, y \in \mathfrak{k}^*)$ is clear. As regards d, it is enough to see that $1 \otimes d_{\mathfrak{g}}$ leaves F^p stable. For this, it suffices to check that $\theta_a(e^i) \in F^1$. This follows from 1.1(5) and the equality $c^i_{r,a} = 0$ for $r > m$. Stability under θ_x then follows from 1.1(1). We let further

$$N^p = V \otimes \Lambda^p \mathfrak{m}^*, \qquad (p \in \mathbf{N}). \tag{4}$$

We have in particular

$$C^p(\mathfrak{g}, \mathfrak{k}; V) = N^{p,\mathfrak{k}}, \qquad (p \in \mathbf{N}). \tag{5}$$

1.5. The θ_x $(x \in \mathfrak{g})$ define a representation of $\mathfrak{g}$ into $C^{\cdot}(\mathfrak{g}; V)$ which commutes with d. For $y \in U(\mathfrak{g})$, the universal enveloping algebra of $\mathfrak{g}$, we also denote by θ_y the endomorphism of $C^{\cdot}(\mathfrak{g}; V)$ associated to y under that representation.

Since $\mathfrak{k}$ and $\mathfrak{m}$ are $\mathfrak{k}$-invariant, each space $C^{p,q}$ is $\mathfrak{k}$-invariant and the restriction of θ_x $(x \in \mathfrak{k})$ to $C^{p,q}$ is the same action as that stemming from the $\mathfrak{k}$-module structure. More formally, $[\mathfrak{k}, \mathfrak{m}] \subset m$ and $[\mathfrak{k}, \mathfrak{k}] \subset \mathfrak{k}$ imply

$$c^i_{b,a} = c^b_{i,a} = 0, \tag{1}$$

hence

$$\theta_a e^i = \sum_j c^i_{j,a} e^j; \qquad \theta_a e^b = \sum_c c^b_{c,a} e^c. \tag{2}$$

1.6. THEOREM. *Let U be a fully $\mathfrak{k}$-invariant graded subcomplex of $C^\cdot(\mathfrak{g}; V)$. Let $F^pU = F^p \cap U$. Then there is a spectral sequence (E_r) associated to the filtration (F^pU), abutting to $H^\cdot(U)$, with the following properties:*

(i) *There is a natural isomorphism*

$$E_0^{p,q} = (U_B \cap N^p) \otimes C^q(\mathfrak{k}),$$

such that

$$d_0 = (-1)^{p+1} I \otimes d_{\mathfrak{k}} \qquad \text{on } E_0^{p,q}, \qquad (p, q \in \mathbf{N}).$$

(ii) $$E_1^{p,q} \cong E_1^{p,q,\mathfrak{k}} = U_B^p \otimes H^q(\mathfrak{k}), \qquad (p, q \in \mathbf{N})$$

(iii) $$E_2^{p,q} = H^p(U_B) \otimes H^q(\mathfrak{k}), \qquad (p, q \in \mathbf{N}).$$

The assumption of full $\mathfrak{k}$-stability implies immediately

$$U = (N^\cdot \cap U) \otimes C^\cdot(\mathfrak{k}), \tag{1}$$

hence

$$U^p / U^{p+1} = (N^p \cap U) \otimes C^\cdot(\mathfrak{k}). \tag{2}$$

$$U_B = (N^\cdot \cap U)^{\mathfrak{k}}. \tag{3}$$

We have now to compute $du \bmod F^{p+1}$ for $u \in U^p$. By 1.2, we may replace U by $U^{\mathfrak{k}}$ and therefore assume

$$\theta_x u = 0, \qquad (x \in \mathfrak{k}). \tag{4}$$

Using the bigrading 1.4(1), we can write

$$u = \sum_{\mu \in I} u_\mu a_\mu \otimes b_\mu, \qquad \Big(u_\mu \in V;\ a_\mu \in \sum_{p' > p} \Lambda^{p'} \mathfrak{m}^*;\ b_\mu \in \Lambda \mathfrak{k}^*, (\mu \in I)\Big). \tag{5}$$

In d, as given by 1.1(2), we may neglect the terms $\pi(e_i) \otimes \epsilon(e^i)$, since they strictly increase the filtration. We have then, by 1.1(2), (3),

$$2du = \Big(1 \otimes \sum \epsilon(e^r)\theta_r + 2\sum \pi(e_a) \otimes \epsilon(e^a)\Big)u, \qquad \bmod F^{p+1}.$$

But, by (4):

$$0 = \theta_a u = (\pi(e_a) \otimes 1 + 1 \otimes \theta_a)u,$$

hence

$$2du = \left(1 \otimes \sum \epsilon(e^i)\theta_i + \sum \pi(e_a) \otimes \epsilon(e^a)\right)u,$$

which can be written

$$2du = \sum_{\mu,i} u_\mu e^i \wedge \theta_i(a_\mu \wedge b_\mu) + \sum_{\mu,a} \pi(e_a) \cdot u_\mu e^a \wedge a_\mu \wedge b_\mu .$$

Modulo F^{p+1}, the first sum is equal to

$$\sum_{\mu,i} u_\mu e^i \wedge \theta_i(a_\mu) \wedge b_\mu .$$

We claim that we have

$$\sum_i e^i \wedge \theta_i v = \sum_a e^a \wedge \theta_a v \qquad \text{mod } F^{p+1} \quad \text{for} \quad v \in \Lambda^p \mathfrak{m}^*, \quad (p \in \mathbf{N}). \tag{6}$$

Since the θ_x $(x \in \mathfrak{g})$ are derivations with respect to the exterior product, it suffices to show this for $p = 1$ and $v = e^j$ $(j = 1, \ldots, m)$. The left-hand side is equal to

$$\sum_{i,s} e^i \wedge c^j_{s,i} e^s = \sum_{i,a} e^i \wedge c^j_{a,i} e^a, \qquad \text{mod } F^2,$$

and the right-hand side to

$$\sum_a e^a \wedge \theta_a e^j = \sum_{a,s} e^a \wedge c^j_{s,a} e^s.$$

Since $c^j_{b,a} = 0$, the right-hand side is then also equal to

$$\sum_{a,i} e^a \wedge c^j_{i,a} e^i,$$

and (6) follows. We can now write

$$2du = \sum_{\mu,a} (u_\mu e^a \wedge \theta_a(a_\mu) \wedge b_\mu + \pi(e_a) u_\mu \cdot e^a \wedge a_\mu \wedge b_\mu), \qquad \text{mod } F^{p+1}.$$

Condition (4) for $x = e_a$ then implies

$$2du = -\sum u_\mu e^a \wedge a_\mu \wedge \theta_a(b_\mu).$$

The image of du in E_0 is $d_0\tilde{u}$, where $\tilde{u}$ is the image of u, hence

$$2d_0\tilde{u} = (-1)^{p+1}\sum u_\mu a_\mu \otimes \Big(\sum_a e^a \wedge \theta_a\Big) b_\mu$$

$$d_0\tilde{u} = (-1)^{p+1}\sum u_\mu a_\mu \otimes d_{\mathfrak{k}} b_\mu, \tag{7}$$

which concludes the proof of (i).

We get then, using (1):

$$E_1^{p,q} = (N^p \cap U) \otimes H^q(\mathfrak{k}), \qquad (p,q \in \mathbf{N}).$$

But 1.2, applied to $F^pU/F^{p+1}U$, yields

$$E_1^{p,q} = (E_1^{p,q})^{\mathfrak{k}}, \qquad (p,q \in \mathbf{N}).$$

Since $\mathfrak{k}$ acts trivially on its own cohomology, we obtain

$$E_1^{p,q} = (N^p \cap U)^{\mathfrak{k}} \otimes H^q(\mathfrak{k}).$$

In view of (3), this proves (ii).

In order to establish (iii), we need to show that d_1 is just $(d \mid U_B) \otimes 1$. For this, we take representatives and compute $du \bmod F^{p+2}$. Write

$$u = \sum_\mu u_\mu a_\mu \wedge b_\mu \qquad (u_\mu a_\mu \in U_B^p,\ b_\mu \in \Lambda\mathfrak{k}^*,\ d_{\mathfrak{k}} b_\mu = 0). \tag{8}$$

Then

$$du = \sum_\mu d(u_\mu a_\mu) \otimes b_\mu + (-1)^p u_\mu a_\mu \otimes d_{\mathfrak{g}} b_\mu .$$

It is now enough to show that the second term on the right-hand side is contained in F^{p+2}. We have

$$2db_\mu = \sum_a e^a \wedge \theta_a b_\mu + \sum_i e^i \wedge \theta_i b_\mu .$$

Since $d_{\mathfrak{k}} b_\mu = 0$, it suffices to show

$$(1/2)\sum_a e^a \wedge \theta_a b_\mu = d_{\mathfrak{k}} b_\mu, \tag{9}$$

$$\sum_i e^i \wedge \theta_i b_\mu \in F^2. \tag{10}$$

For (9), this follows from 1.1(3) and 1.5. Consider now $\theta_i(e^{a_1} \wedge \cdots \wedge e^{a_q})$. We have

$$\theta_i e^b = \sum_r c_{r,i}^b e^r,$$

and $c_{r,i}^a = 0$ for $r > m$ by 1.5(1). Therefore $\theta_i(e^b) \subset F^1$, and (10) follows.

From now on $F = \mathbf{R}$, $F' = \mathbf{C}$. We let G be a Lie group with finite component group and Lie algebra $\mathfrak{g}$, and assume that $\mathfrak{k}$ is the Lie algebra of a compact subgroup K of G.

1.7. We want to discuss an extension of 1.6 needed for C in the introduction. We choose a K-invariant scalar product on $\mathfrak{g}$ and endow $\mathfrak{g}^*$, $\Lambda\mathfrak{g}$ and $\Lambda\mathfrak{g}^*$ with the canonically associated scalar products. (π, V) is a continuous representation of G in a Hilbert space V. The space $C^{\cdot}(\mathfrak{g}; V)$ is defined as in 1.1. It is viewed as the graded Hilbert space product of $\Lambda\mathfrak{g}^*$ and V. *A priori*, there is no differential on it, but there is one, given by 1.1(2), (3) on $C^{\cdot}(\mathfrak{g}; V^{\infty})$ and $C^{\cdot}(\mathfrak{g}; V^{\infty}_{(K)})$. We shall see in 2.4 that these two complexes have the same cohomology, but this is not needed here.

1.8. LEMMA. *The differential d operating on $C^{\cdot}(\mathfrak{g}; V^{\infty})$ has a closure $\bar{d}$ in $C^{\cdot}(\mathfrak{g}; V)$. It is equal to the closure of d operating on $C^{\cdot}(\mathfrak{g}; V^{\infty}_{(K)})$. Given $u \in U = \operatorname{dom} \bar{d}$ there exists a sequence $u_m \in C^{\cdot}(\mathfrak{g}; V^{\infty}_{(K)})$ such that $u_m \to u$, du_m has a limit and $\theta_y u_m \to \theta_y u$ for all $u \in U(\mathfrak{k})$. The space $U_{(K)}$ is fully $\mathfrak{k}$-stable and K-invariant.*

There is a continuous representation $(\tilde{\pi}, \tilde{V})$ of G, with $\tilde{V} = V$ canonically, such that

$$((\pi(g)u, \tilde{\pi}(g)\tilde{v}) = (v, \tilde{v}), (v, \tilde{v} \in V), g \in G). \tag{1}$$

[This is the complex conjugate contragredient representation, viewed as a representation on V via the canonical identification of the conjugate complex dual of V to V by means of the scalar product on V.] We have then

$$(\pi(x)v, \tilde{v}) + (v, \pi(x)\tilde{v}) = 0, \qquad (v \in V^{\infty}, \tilde{v} \in \tilde{V}^{\infty}, g \in G). \tag{2}$$

For $x \in \mathfrak{g}^*$, let us also denote by $\tilde{i}_x$ the adjoint to $\epsilon(x)$ on $\Lambda\mathfrak{g}^*$. [If we identify $\Lambda\mathfrak{g}^*$ to its dual via the chosen scalar product, then $\tilde{i}_x$ comes from the interior product i_x.] Then it is clear from 1.1(2), (3) that $-\tilde{\partial}$, where

$$\tilde{\partial} = (1/2)\sum_r \theta_r \circ \tilde{i}_{e'} + \sum_r \tilde{\pi}(e_r) \otimes \tilde{i}_{e'} \tag{3}$$

is a formal adjoint to d, whose domain contains $V^{\infty} \otimes \Lambda\mathfrak{g}^*$. Since this domain is dense, this proves the existence of $\bar{d}$. The differential d commutes with K (in fact with G), hence with the projections onto isotypic subspaces. Since any smooth vector is limit of its Fourier series with respect to K-types, the second assertion follows.

Assume now that $u \in U_{(K)}$. Let $u_m \in C^{\cdot}(\mathfrak{g}; V^{\infty}_{(K)})$ be such that $u_m \to u$ and du_m has a limit v. The element u is K-finite. There exists therefore $\alpha \in C_c^{\infty}(K)$, which is K-invariant, such that

$$u = \pi_K(\alpha) \,.\, u = \int_K \alpha(k)\pi(k)u \, dk. \tag{4}$$

Then

$$u'_m = \pi_K(\alpha)u_m \to \pi_K(\alpha)\cdot u = u \quad \text{and} \quad du'_m \to \pi_K(\alpha)\cdot v. \tag{5}$$

Moreover, if $y \in U(\mathfrak{k})$ and if we set $\beta = \theta_y \alpha$, then

$$\theta_y \cdot u'_m = \pi_{K,\beta}\cdot u'_m \to \pi_{K,\beta}\cdot u = (\theta_y \pi_{K,\alpha})\cdot u = \theta_y(\pi_{K,\alpha}\cdot u) = \theta_y \cdot u. \tag{6}$$

That dom $\bar{d}$ is invariant under K and $\epsilon(x)$ for any $x \in \mathfrak{g}^*$ is clear. By what has just been proved, it is also invariant under the θ_x for $x \in \mathfrak{k}$. Then 1.1(2) implies also the invariance under i_x ($x \in \mathfrak{k}$). This concludes the proof of the lemma.

Remark. The operator $\tilde{\partial}$ is essentially the differential in homology, but identified to an operator on $C^{\cdot}(\mathfrak{g}; V)$ by means of the scalar product, whereas the natural domain of definition of the boundary operator is $C_{.}(\mathfrak{g}; W) = W \otimes \Lambda\mathfrak{g}$, where (τ, W) is a $\mathfrak{g}$-module. In that case, it is defined by

$$\partial = (1/2)1 \otimes \sum_r \theta_r \circ i_{e^r} + \sum_r \tau(e_r) \otimes i_{e^r}. \tag{7}$$

It is clear that d on $C^{\cdot}(\mathfrak{g}; V)$ and $-\partial$ on $C_{.}(\mathfrak{g}; V^*)$, where V^* is the dual $\mathfrak{g}$-module to V, are adjoint of one another with respect to the canonical pairing between $C^{\cdot}(\mathfrak{g}; V)$ and $C_{.}(\mathfrak{g}; V^*)$. See [11] for homology with trivial coefficients.

1.9. Let F^p, N^p, and $C^{p,q}$ be as in 1.4. Since U is fully $\mathfrak{k}$-stable, we again have

$$u = (N^{\cdot} \cap U) \otimes \Lambda\mathfrak{k}, \qquad U^{p,q} = U \cap C^{p,q} = (N^p \cap U) \otimes \Lambda\mathfrak{k}^*. \tag{1}$$

LEMMA. *We have $U = \bigoplus U^{p,q}$. Given $u \in U^{p,q}$, we may find elements in $C^{p,q} \cap C^*(\mathfrak{g}; V)$ which satisfy* 1.8.

Assume $u \in U$ to be homogeneous and let d be its degree. We can write

$$u = \sum_{p+q=d} u^{p,q}, \qquad (u^{p,q} \in C^{p,q}; p, q \in \mathbf{N}). \tag{2}$$

Let u_m be as in 1.8, of degree d, and write similarly

$$u_m = \sum_{p+q=d} u_m^{p,q}, \qquad (u_m^{p,q} \in C^{p,q} \cap C^{\cdot}(\mathfrak{g}; V); p, q \in \mathbf{N}). \tag{3}$$

Then

$$u_m^{p,q} \to u^{p,q}, \qquad \theta_x u_m^{p,q} \to \theta_x u^{p,q} \qquad (x \in U(\mathfrak{k}); p, q \in \mathbf{N}). \tag{4}$$

The first assertion is obvious and the second one follows from the fact that the spaces $C^{p,q}$ are $\mathfrak{k}$-invariant (same argument as in 1.5). There remains to see that $du_m^{p,q}$ has a limit. This is not a priori clear since d is not homogeneous with respect to the bidegree (p, q). Let p_0 be the minimum of the p's for which $u_m^{p,q} \neq 0$ for arbitrarily large m's. By taking iterated interior products by q_0 elements i_x ($x \in \mathfrak{k}$) and then again exterior products by q_0 elements $\epsilon(x)$ ($x \in \mathfrak{k}$), and using

1.1(1), one sees that $du_m^{p,q}$ has a limit for $(p,q) = (p_0, q_0)$. One then proceeds by induction on p_0.

Remark. We can also express this by saying that $U^{p,q}$ is the closure of d operating on $C^{p,q}(\mathfrak{g}; V)$. Then the same is true for F^pU. Let (b_μ) be the standard basis of $\Lambda^q\mathfrak{k}^*$, made of exterior products of the elements e^a. We can always write,

$$u^{p,q} = \sum_\mu a_\mu \otimes b_\mu, \qquad (a_\mu \in N^p \cap U),$$

$$u_m^{p,q} = \sum_\mu a_{m,\mu} \otimes b_\mu, \qquad (a_{m,\mu} \in N^p \cap C^\cdot(\mathfrak{g}; V))$$

and the previous argument shows that for each μ, the $a_{m,\mu}$ and a_μ satisfy 1.8. Thus the full $\mathfrak{k}$-stability allows us to argue componentwise with respect to the bidegree (p,q), although d is not homogeneous.

1.10. Proposition. *Theorem* 1.6 *holds for U and d.*

We review the steps of the proof of 1.6 and indicate how to modify them. As already remarked, we again have 1.6(1), whence also 1.6(2), (3). In order to compute $\bar{d}u \bmod F^{p+1}$ for $u \in U^p$, we choose u_m as in 1.8 and consider $\lim du_m$. By the above remark, we may assume $u_m \in U^p$. In decomposing u as in 1.6, we may assume a_μ and b_μ to be independent of m. The computations of 1.6 remain valid for u_m. The element d_0u_m is given by (7) and we get $\bar{d}_0u$ just by letting $u_{m,\mu} \to u_\mu$, and (7) for $\bar{d}_0u$ follows. This yields (i); then (ii) follows as in 1.6.

To prove (iii) we start again with (8) and then can also write

$$u_m = \sum_\mu u_{m\mu}a_\mu \otimes b_\mu, \qquad (u_{m,\mu}a_\mu \in U_B^p,\ b_\mu \in \Lambda\mathfrak{k}^*;\ d_\mathfrak{k}b_\mu = 0)$$

and get

$$du_m = \sum_\mu d(u_{m,\mu}a_\mu) \otimes b_\mu, \qquad \bmod F^{p+2}.$$

Passing to the limit, we obtain

$$\bar{d}u = \sum_\mu \bar{d}(u_\mu a_\mu) \otimes b_\mu,$$

which proves (iii).

§2. Regularization theorems.

2.1. The identity components of G and K are denoted G° and K° respectively. We fix a left Haar measure dg on G and set

$$I(\alpha) = \int_G \alpha(g)\,dg, \qquad (\alpha \in C_c^\infty(G)). \tag{1}$$

V is now a locally complete topological vector space, (π, V) a *differentiable* representation of G on V. The space $V_{(K)}$ of K-finite vectors in V is then a $(\mathfrak{g}, K)$-module.

We view $C^{\cdot}(\mathfrak{g}; V)$ as a G-module under the tensor product of π and of the coadjoint representation. It is also differentiable. Its differential is given by the θ_x $(x \in \mathfrak{g})$ and we shall also denote by θ the global representation, by θ_g or $\theta(g)$ the action of $g \in G$ and by θ_α that of $\alpha \in C_c^\infty(G)$. If α is K-invariant, then it leaves $C^{\cdot}(\mathfrak{g}; V)_{(K)}$ stable. It follows from 1.1(2), (3) that d commutes with G, hence also with the θ_α's.

All this applies of course to $C^{\cdot}(\mathfrak{g}; V)$ viewed as a K-module. In particular d commutes with the operators $\theta_{K,\beta}$ $(\beta \in C_c^\infty(K))$, where $\theta_{K,\beta}$ is defined by

$$\theta_{K,\beta}(\eta) = \int_K \beta(k)\theta_k \cdot \eta \, dk, \qquad (\eta \in C^*(\mathfrak{g}; V)), \tag{2}$$

(dk is the Haar measure of total measure one). Among those operators are the projectors on the K-isotypic subspaces, hence d leaves stable those spaces and therefore also $C^{\cdot}(\mathfrak{g}; V)_{(K)}$.

2.2. A subspace U of $C^{\cdot}(\mathfrak{g}; V)$ is *fully K-stable* if it is fully $\mathfrak{k}$-stable (1.3) and K-stable. Full $\mathfrak{k}$-stability is then equivalent to full K°-stability. If U is fully K-stable, then the basic subspace (subcomplex if U is one) U_B of U is

$$U_B = \{u \in U, \theta_k u = u \ (k \in K), i_x u = 0, (x \in \mathfrak{k})\}. \tag{1}$$

The basic subspace U_B° in the sense of 1.3 is then the basic subspace in the present sense with respect to K°. It contains U_B; in fact there is a natural operation of K/K° on U_B° and

$$U_B = (U_B^\circ)^{K/K^\circ}. \tag{2}$$

In case these are subcomplexes, we then also have

$$H^{\cdot}(U_B) = (H^{\cdot}(U_B^\circ))^{K/K^\circ}. \tag{3}$$

The elements of U_B are K-fixed, hence K-finite and those of U_B° are also K-finite. Therefore U and $U_{(K)}$ have the same basic subspace. If W is a G-stable subspace of V, then $C^{\cdot}(\mathfrak{g}; W)$ is fully K-stable and we have:

$$C^{\cdot}(\mathfrak{g}, W)_B = C^{\cdot}(\mathfrak{g}, K; W). \tag{4}$$

2.3. *Homotopy operators*. We fix a symmetric K-invariant relatively compact open exponential neighborhood C_0 of 1 in G, let ln denote the inverse of the exponential map and set $D_0 = \ln(C_0)$. We also fix a left invariant Haar measure dg on G. Given $\alpha \in C_c^\infty(G)$ with support in C_0 we want to construct a linear map E_α of $C^\cdot(\mathfrak{g}; V)$ into itself, decreasing the degree by one, and such that

$$\theta_\alpha\eta - I(\alpha)\eta = dE_\alpha\eta + E_\alpha d\eta, \qquad (\eta \in C^\cdot(\mathfrak{g}; V)). \tag{1}$$

In particular, θ_α is a homotopy operator if $I(\alpha) = 1$.

Let $x \in \mathfrak{g}$, $g = \exp x$, and η a p-form. We have

$$(\theta_g\eta - \eta) = \int_0^1 dt \frac{d}{ds}\left(\theta(e^{(t+s)x}\eta)\right)_{s=0}$$

$$\theta_g\eta - \eta = \int_0^1 dt\, \theta(e^{tx})\theta_x\eta$$

and therefore, since $\theta_x = di_x + i_x d$,

$$\theta_g\eta - \eta = dE_g\eta + E_g d\eta, \tag{2}$$

where

$$E_g\eta = \int_0^1 dt\, \theta(e^{tx})i_x\eta, \qquad (x \in \mathfrak{g}; g = \exp x). \tag{3}$$

Let now $\alpha \in C_c^\infty(C_0)$. Then

$$\theta_\alpha\eta - I(\alpha)\eta = \int_G \alpha(g)(\theta_g\eta - \eta)\, dg = dE_\alpha\eta + E_\alpha d\eta, \tag{4}$$

where

$$E_\alpha\eta = \int_G \alpha(g)E_g\eta\, dg = \int_G \alpha(g)\, dg \int_0^1 \theta(e^{t\ln g})i_{\ln g}\eta \cdot dt. \tag{5}$$

2.4. Lemma. *Let U be a K-stable subcomplex of $C^\cdot(\mathfrak{g}; V)$. Then the inclusions $U^{K^\circ} \to U_{(K)} \to U$ induce isomorphisms in cohomology.*

We shall apply 2.3, but viewing U as a K°-module. Fix a finite set of elements $(x_i)_{i \in I}$ of K such that $(C_0 \cdot x_i)$ is a covering of K° and let (μ_i) be an associate partition of unity. Let $\nu_i = r_{x_i}\mu_i$, i.e.,

$$\nu_i(x) = \mu_i(x \cdot x_i), \qquad (x \in K^\circ; i \in I). \tag{1}$$

We have then $I(\nu_i) = I(\mu_i)$, $(i \in I)$ hence

$$\sum_i I(\nu_i) = \sum I(\mu_i) = 1 \tag{2}$$

$$\sum \theta_{\mu_i}\eta = \bar{\eta} = \int_{K^\circ} \theta_k\eta\, dk, \qquad (\eta \in U). \tag{3}$$

Let $\eta \in U$ be closed. Since ν_i has support in C_0, (2) and 2.3(1) shows that

$$\eta \sim \sum \theta_{\nu_i}\eta$$

where $\sim$ means cohomologous to. But the exponential map on K° is surjective, hence 2.3(2) is valid for any $k \in K^\circ$ and we get

$$\eta \sim \sum \theta_{\nu_i}\eta \sim \sum \theta_{x_i}\theta_{\mu_i}\eta = \sum \theta_{\mu_i}\eta = \bar{\eta}.$$

This shows that the derived maps of the inclusions $U^{K^\circ} \to U_{(K)}$ and $U^{K^\circ} \to U$ are surjective. If now $\eta \in U^{K^\circ}$ and $\mu \in U$ are such that $\eta = d\mu$, then we have

$$\eta = \bar{\eta} = (\overline{d\mu}) = d\bar{\mu}$$

since $\mu \mapsto \bar{\mu}$ commutes with d, and the injectivity follows.

2.5. THEOREM. *Let (π, V) be as in 2.1. Let U be a G-stable subcomplex of $C^\cdot(\mathfrak{g}; V)$ and U' a G-stable subcomplex of U containing $\theta(C_c^\infty(G))U$. Assume U and U' to be stable under the operators E_α ($\alpha \in C_c^\infty(G)$).*

(i) *The maps $j^*: H^\cdot(U') \to H^\cdot(U)$ and $H^\cdot(U'_{(K)}) \to H^\cdot(U_{(K)})$ associated to inclusions are isomorphisms.*

(ii) *If $U_{(K)}$ and $U'_{(K)}$ are fully K-stable (2.2), then the inclusion $U'_B \to U_B$ of the basic complexes (2.2) is an isomorphism in cohomology.*

(i) Let $\alpha \in C_c^\infty(C_0)$ and assume $I(\alpha) = 1$. If $\eta \in U$ is closed, then we have $\theta_\alpha(u) = u + dE_\alpha u$. Since $E_\alpha u \in U$, this shows that j^* is surjective. Let $\eta \in U'$ be closed. If there exists $\eta \in U$ such that $\eta = d\mu$, then $\theta_\alpha(\eta) = d\theta_\alpha\mu$ is a coboundary in U' and the previous argument shows that $\theta_\alpha\mu$ is cohomologous to η within U'. Hence j^* is injective.

If we take α to be K-invariant, then the same proof is valid for the subcomplexes of K-finite elements.

By 1.6, there exist spectral sequences E'_r and E_r abutting to $H^\cdot(U'_{(K)})$ and $H^\cdot(U_{(K)})$ respectively, in which

$$E_2'^{p,q} = H^p(U_B'^\circ) \otimes H^q(\mathfrak{k}) \qquad E_2^{p,q} = H^p(U_B^\circ) \otimes H^q(\mathfrak{k}), \qquad (p, q \in \mathbf{N})$$

(where the superscript $^\circ$ refers to the basic subcomplexes in the sense of 1.3, see 2.2). The inclusion $U'_{(K)} \to U_{(K)}$ is compatible with the filtration and gives rise to a homomorphism (j_r^*) of spectral sequences. It is clear that $j_2^* = j_B^* \otimes I$, where j_B^* is induced from the inclusion of the basic complexes. By (i), j_∞^* is an isomorphism. Then j_B^* is an isomorphism by the comparison theorem for spectral sequences (see [12: 5.3] or [14: XI, §11]).

This proves (ii) for $K = K^\circ$. In the general case, j_B^* commutes with the natural action of K/K° and we just have to use 2.2(3).

2.6. We now come to the situation described in 1.7–1.9 and assume that $V = W^\infty \otimes E$, where W is a Hilbert space as well as a continuous G-module and

E a finite dimensional G-module. We also assume E to be endowed with a K-invariant scalar product and view again

$$C^{\cdot}(\mathfrak{g}; W \otimes E) = W \otimes E \otimes \Lambda\mathfrak{g}^*$$

as a graded Hilbert space. By 1.8, we know that d operating on $C^{\cdot}(\mathfrak{g}; V_{(K)})$ has a closure $\bar{d}$ and that if $U = \operatorname{dom} \bar{d}$, then $U_{(K)}$ is fully K-stable. We let again U_B be the basic subcomplex of U or $U_{(K)}$ (2.2).

2.7. THEOREM. *We keep the notation and assumptions of 2.6. Then the inclusions $C^{\cdot}(\mathfrak{g}; V_{(K)}) \to U_{(K)}$ and $C^{\cdot}(\mathfrak{g}, K; V) \to U_B$ induce isomorphisms in cohomology.*

Let $\alpha \in C_c^\infty(C_0)$ be as in 2.3. We note first that the operator E_α in (1) is bounded. To see that it suffices, in view of 2.3(5), to show that the right-hand side of 2.3(3) is bounded, with a norm bounded by some constant when x varies in a compact subset of $\mathfrak{g}$. But this is obvious. Therefore E_α extends to an operator on U. Using 2.3(1) we see that $E_\alpha(U) \subset U$ and that

$$\theta_\alpha \eta - I(\alpha)\eta = \bar{d}E_\alpha \eta + E_\alpha \bar{d}\eta, \qquad (\eta \in U).$$

The first assertion is then proved exactly as in 2.5. The second one also, except that we have to invoke 1.10 for the existence of the spectral sequence (E_r) for U.

§3. Applications. In this section, we discuss (A), (B), (C) of the introduction, under more general assumptions.

A. *Cohomology with growth conditions.*

3.1. We adopt here the framework of [2]. G is the group of real points of a connected reductive Q-group $\mathcal{G}$ without nontrivial rational characters defined over Q and Γ is an arithmetic subgroup. We let $C_{mg}^\infty(\Gamma \backslash G)$ be the space of smooth functions on $\Gamma \backslash G$ which, together with their $U(\mathfrak{g})$-derivatives, have moderate growth [2: 3.2]. Moreover, $C_{umg}^\infty(\Gamma \backslash G)$ is the space of smooth functions of *uniform moderate growth*: the exponent which bounds the growth on a Siegel set can be chosen to be the same for all $U(\mathfrak{g})$-derivatives. If f is a function of moderate growth, then $f * \alpha$ ($\alpha \in C_c^\infty(G)$) is of uniform moderate growth. K is a maximal compact subgroup of G and $X = G/K$. Finally, (σ, E) is a finite dimensional G-module. We use the standard identification.

$$A^\infty(\Gamma \backslash X; \tilde{E}) = C^{\cdot}(\mathfrak{g}, K; C^\infty(\Gamma \backslash G) \otimes E),$$
$$A^\infty(\Gamma \backslash G; E) = C^{\cdot}(\mathfrak{g}; C^\infty(\Gamma \backslash G) \otimes E \qquad (1)$$

[5: VII], and let

$$\begin{aligned} A^{\infty}_{mg}(\Gamma\backslash X;\tilde{E}) &= C^{\cdot}(\mathfrak{g},K;C^{\infty}_{mg}(\Gamma\backslash G)\otimes E);\\ A^{\infty}_{mg}(\Gamma\backslash G;E) &= C^{\cdot}(\mathfrak{g};C^{\infty}_{mg}(\Gamma\backslash G)\otimes E)\end{aligned}\qquad(2)$$

$$\begin{aligned} A^{\infty}_{umg}(\Gamma\backslash X;\tilde{E}) &= C^{\cdot}(\mathfrak{g},K;C^{\infty}_{umg}(\Gamma\backslash G)\otimes E);\\ A^{\infty}_{umg}(\Gamma\backslash G;E) &= C^{\cdot}(\mathfrak{g};C^{\infty}_{umg}(\Gamma\backslash G)\otimes E).\end{aligned}\qquad(3)$$

Then the first term in (2) (resp. (3)) is the basic subcomplex of the second one. We have

$$H^{\cdot}(\Gamma\backslash G;\tilde{E}) = H^{\cdot}(\mathfrak{g};C^{\infty}(\Gamma\backslash G)\otimes E)\qquad(4)$$

$$H^{\cdot}(\Gamma\backslash X;\tilde{E}) = H^{\cdot}(\mathfrak{g},K;C^{\infty}(\Gamma\backslash G)\otimes E).\qquad(5)$$

3.2. THEOREM. *The inclusions*

$$A^{\infty}_{umg}(\Gamma\backslash G;\tilde{E})\to A^{\infty}_{mg}(\Gamma\backslash G;\tilde{E})\to A^{\infty}(\Gamma\backslash G;\tilde{E}),\qquad(1)$$

$$A^{\infty}_{umg}(\Gamma\backslash X;\tilde{E})\to A^{\infty}_{mg}(\Gamma\backslash X;\tilde{E})\to A^{\infty}(\Gamma\backslash X;\tilde{E})\qquad(2)$$

induce isomorphisms in cohomology.

In view of 3.1(1), (2), (3), and 2.4 this amounts to showing that the inclusions

$$C^{\cdot}(\mathfrak{g};C^{\infty}_{umg}(\Gamma\backslash G)_{(K)}\otimes E)\xrightarrow{\mu}C^{\cdot}(\mathfrak{g};C^{\infty}_{mg}(\Gamma\backslash G)_{(K)}\otimes E)\xrightarrow{\nu}C^{\cdot}(\mathfrak{g};C^{\infty}(\Gamma\backslash G)_{(K)}\otimes E),\qquad(3)$$

$$C^{\cdot}(\mathfrak{g},K;C^{\infty}_{umg}(\Gamma\backslash G)\otimes E)\xrightarrow{\mu_B}C^{\cdot}(\mathfrak{g},K;C^{\infty}_{mg}(\Gamma\backslash G)\otimes E)\xrightarrow{\nu_B}C^{\cdot}(\mathfrak{g},K;C^{\infty}(\Gamma\backslash G)\otimes E),\qquad(4)$$

induce isomorphisms in cohomology.

We start from the fact that ν_B^* is an isomorphism [2: 3.4]. The homomorphism of spectral sequences (1.6) induced by ν is then an isomorphism at the E_2 level, hence also at the E_∞ level. This shows that ν^* is an isomorphism.

To prove that μ^* and μ_B^* are isomorphisms, we have to check that 2.5 applies, with

$$U'_{(K)} = C^{\cdot}(\mathfrak{g};C^{\infty}_{umg}(\Gamma\backslash G)_{(K)}\otimes E),\qquad U_{(K)} = C^{\cdot}(\mathfrak{g};C^{\infty}_{mg}(\Gamma\backslash G)_{(K)}\otimes E),$$

$$V = C^{\infty}(\Gamma\backslash G)_{(K)}\otimes E.$$

The representation π is then the tensor product of the right regular representation by σ. For $\alpha \in C_c^\infty(G)$, the endomorphism $\pi(\alpha)$ is essentially convolution. More precisely, if $u \in C^\infty(\Gamma\backslash G)$, $e \in E$, (e_i) is a basis of E and we write $\sigma(g)\cdot e = \sum_i c_i(g)e_i$, then

$$\pi(\alpha)(u \otimes e) = \sum_i u * \alpha_i \otimes e_i,$$

where α_i is given by $\alpha_i(g) = \alpha(g^{-1})c_i(g^{-1})$. Therefore $\pi(\alpha)$ maps $C_{mg}^\infty(\Gamma\backslash G) \otimes E$ into $C_{umg}^\infty(\Gamma\backslash G) \otimes E$. Then the first assumption of 2.5 is satisfied. The conditions of moderate growth or uniform moderate growth are clearly invariant under interior products; therefore, U and U' are stable under the E_g's $(g \in G)$. Moreover, these conditions are, by definition, uniform on compact sets, whence also the invariance under the E_α's $(\alpha \in C_c^\infty(G))$. Therefore the second assumption of 2.5 also holds. Finally full K-stability is automatic, since we deal with Lie algebra subcomplexes (2.2).

3.3. *Remark.* This theorem is also valid if Γ is only assumed to have finite covolume, G is in the usual reductive class [5] and the intersection of Γ with the center Z of G is cocompact in Z. Arithmeticity intervened here, because we started from 3.4 in [2], whose proof makes use of the compactification of $\Gamma\backslash X$ by a manifold with corners, which has been explicitly carried out in the literature only in the arithmetic case. However, its existence (as a manifold with boundary) can be deduced in the real rank one case from the reduction theory of H. Garland and M. S. Raghunathan (see [14: XIII] for a generalization) and the results of G. A. Margulis on the arithmeticity of discrete subgroups of finite covolume allow one to reduce oneself to a product of arithmetic and real rank one situations (after passing to a subgroup of finite index, but this is harmless for the kind of theorems proved here).

3.4. We use here the terminology of [13]. Let $\mathscr{P}$ be a set of representatives of the associated classes of cuspidal subgroups of G (or of parabolic Q-subgroups under our original assumptions). For $P \in \mathscr{P}$, let $C_{umg}^\infty(\Gamma\backslash G)_{(P)}$ be the space of functions of uniform moderate growth which are negligible outside P (for $Q \in \mathscr{P}$, $Q \neq P$, the constant term f_Q of f is orthogonal to cusp forms on $\Gamma_M\backslash M$, where $P = NAM$ is a Langlands decomposition of P). Then Langlands showed some years ago (unpublished) that $C_{umg}^\infty(\Gamma\backslash G)$ is the direct sum of the $C_{umg}^\infty(\Gamma\backslash G)_{(P)}$. In view of 3.2 this yields a direct sum decomposition

$$H^\cdot(\Gamma; E) = \bigoplus_{P \in \mathscr{P}} H^\cdot(\Gamma; E)_{(P)}, \tag{1}$$

where, by definition,

$$H^\cdot(\Gamma; E)_{(P)} = H^\cdot(\mathfrak{g}, K; C_{umg}^\infty(\Gamma\backslash G)_{(P)} \otimes E), \qquad (P \in \mathscr{P}). \tag{2}$$

For $P = G$, the corresponding summand is the cuspidal cohomology. It is already known to be a direct summand [2: 5.3, 5.5]. The relation (1) provides moreover a direct complement to it.

B. *Square integrable cohomology.*

3.4. For the remainder of the paper, unless otherwise stated, G and K are as in §2, $X = G/K$ and Γ is any discrete subgroup of G. Again, (σ, E) is a finite dimensional representation of G. We let $\tilde{E}$ be the local system on $\Gamma\backslash G$ or $\Gamma\backslash X$ defined by E. We assume $M = \Gamma\backslash X$ or $\Gamma\backslash G$ endowed with a Riemannian metric which stems from a metric on G which is left invariant under G and right invariant under K. It is complete. We let $A^\infty_{(2)}(M;\tilde{E})$ be the space of $\eta \in A^\infty(M;\tilde{E})$ such that η and $d\eta$ are square integrable. Then $H^\cdot(A^\infty_{(2)}(M;\tilde{E})) = H^\cdot_{(2)}(M;E)$ is the L^2-cohomology space of M with coefficients in $\tilde{E}$.

The Riemannian volume element on $\Gamma\backslash G$ comes from a left invariant Haar measure, with respect to which $L^2(\Gamma\backslash G)$ is defined. The group G operates on $L^2(\Gamma\backslash G)$ by right translations. If it is not unimodular, this representation is not unitary, but of course continuous. Any element in $L^2(\Gamma\backslash G)^\infty$ is a smooth function and all its $U(\mathfrak{g})$-derivatives are L^2. We have obviously

$$\left(L^2(\Gamma\backslash G)\otimes E\right)^\infty = L^2(\Gamma\backslash G)^\infty \otimes E.$$

We identify $A^\infty_{(2)}(\Gamma\backslash X;\tilde{E})$ and $A^\infty_{(2)}(\Gamma\backslash G;\tilde{E})$ with subcomplexes of $C^\cdot(\mathfrak{g}; C^\infty(\Gamma\backslash G)\otimes E)$ and $C^\cdot(\mathfrak{g},K; C^\infty(\Gamma\backslash G)\otimes E)$. They respectively contain $C^\cdot(\mathfrak{g}; L^2(\Gamma\backslash G)^\infty\otimes E)$ and $C^\cdot(\mathfrak{g},K; L^2(\Gamma\backslash G)\otimes E)$.

3.5. THEOREM. *The inclusions* $C^\cdot(\mathfrak{g}; L^2(\Gamma\backslash G)^\infty\otimes E)\to A^\infty_{(2)}(\Gamma\backslash G;\tilde{E})$ *and* $C^\cdot(\mathfrak{g},K; L^2(\Gamma\backslash G)^\infty\otimes E)\to A^\infty_{(2)}(\Gamma\backslash X;\tilde{E})$ *induce isomorphisms in cohomology.*

By 2.4, we may replace the first two complexes by the subspaces of K-finite elements. We want again to use 2.5 with

$$V = C^\infty(\Gamma\backslash G)\otimes E,\quad U = A^\infty_{(2)}(\Gamma\backslash G;\tilde{E})\quad\text{and}\quad U' = C^\cdot\left(\mathfrak{g}; L^2(\Gamma\backslash G)^\infty\otimes E\right).$$

For this, the main point is to check that $U_{(K)}$ is fully K-stable. The other conditions are clearly satisfied.

$U_{(K)}$ is in fact contained in $C^\cdot(\mathfrak{g}; L^2(\Gamma\backslash G)\cap C^\infty(\Gamma\backslash G)_{(K)}\otimes E)$. The representation of K on $L^2(\Gamma\backslash G)\otimes E$ is unitary hence all elements of $L^2(\Gamma\backslash G)_{(K)}\otimes E$ are smooth vectors for the K-module structure. This means in particular that all their $U(\mathfrak{k})$-derivatives are also L^2 and implies that if $\eta\in C^\cdot(\mathfrak{g}; L^2(\Gamma\backslash G)\cap C^\infty(\Gamma\backslash G)_{(K)}\otimes E)$, then so is $\theta_y\eta$ for any $y\in U(\mathfrak{k})$. Since these endomorphisms commute with d, it follows that $U_{(K)}$ is also stable under θ_y. Let $x\in\mathfrak{g}^*$ and $\eta\in U_{(K)}$. Then the coefficients of $x\wedge\eta$ are linear combinations of those of η and the coefficients of

$$d(x\wedge\eta) = dx\wedge\eta - x\wedge d\eta$$

are linear combinations of those of η and of $d\eta$. Hence $\eta \in U_{(K)}$ implies $x \wedge \eta \in U_{(K)}$. Let now $x \in \mathfrak{k}$. Then the coefficients of $i_x\eta$ are linear combinations of those of η, hence $i_x\eta$ is square integrable. We have

$$di_x\eta = \theta_x\eta - i_x d\eta$$

hence $di_x\eta$ is also L^2, i.e., $i_x\eta \in U_{(K)}$.

The theorem now is a special case of 2.5.

3.6. Let again M be either $\Gamma\backslash X$ or $\Gamma\backslash G$ and let $A_{(2)}(M;\tilde{E})$ be the space of E-valued square integrable forms on M with measurable coefficients. We denote by d_M the exterior differentiation on M. The differential d_M, operating on $A^{\infty}_{(2)}(M;\tilde{E})$, has a closure $\bar{d}_M$ in $A_{(2)}(M;\tilde{E})$ and the inclusion

$$A^{\infty}_{(2)}(M;\tilde{E}) \to \operatorname{dom} \bar{d}_M \tag{1}$$

is an isomorphism in cohomology [8]. A natural extension of the isomorphism 3.1(1) yields graded vector space isomorphisms:

$$\begin{aligned} A_{(2)}(\Gamma\backslash X;\tilde{E}) &\xrightarrow{\sim} C^{\cdot}(\mathfrak{g}, K; L^2(\Gamma\backslash G) \otimes E), \\ A_{(2)}(\Gamma\backslash G;\tilde{E}) &\xrightarrow{\sim} C^{\cdot}(\mathfrak{g}; L^2(\Gamma\backslash G) \otimes E). \end{aligned} \tag{1}$$

We claim that they map $\operatorname{dom} \bar{d}_{\Gamma\backslash X}$ and $\operatorname{dom} \bar{d}_{\Gamma\backslash G}$ onto $\operatorname{dom} \bar{d}$, hence bring $\bar{d}_{\Gamma\backslash X}$ and $\bar{d}_{\Gamma\backslash G}$ onto $\bar{d}$. This follows from the fact that $\bar{d}_M$ is also the closure of d operating on smooth compactly supported forms because M is complete (Theor. of Gaffney, see [4; 8] for references), as was already remarked in [4] in the first case. [Actually, in the second case, this can be seen more simply by using a Dirac sequence of K-invariant functions in $C_c^{\infty}(G)$.] It follows from Prop. 2.7 that the inclusions

$$C^{\cdot}(\mathfrak{g}, K; L^2(\Gamma\backslash G)^{\infty} \otimes E) \to \operatorname{dom} \bar{d}_{\Gamma\backslash X}, \qquad C^{\cdot}(\mathfrak{g}; L^2(\Gamma\backslash G)^{\infty} \otimes E) \to \operatorname{dom} \bar{d}_{\Gamma\backslash G},$$

induce isomorphisms in cohomology. This includes in particular C in the introduction.

References

1. A. Borel, *Stable and L^2-cohomology of arithmetic groups*, Bull. Amer. Math. Soc. (N.S.) **3** (1980), 1025–1027.
2. ———, "Stable real cohomology of arithmetic groups II" in *Manifolds and Lie Groups*, Papers in honor of Y. Matsushima, Progress in Math. 14, Birkhäuser Boston, 1981.
3. ———, *Cohomology and spectrum of an arithmetic group*, Proc. Conf. on Group Representations and Operator Algebras, Neptun, Rumania, 1980. To appear.
4. ——— and W. Casselman, *L^2-cohomology of locally symmetric manifolds of finite volume*, Duke Math. J. **50** (1983), 625–647
5. ——— and N. Wallach, *Continuous Cohomology, Discrete Subgroups and Representations of Reductive Groups*, Ann. of Math. Studies **94**, Princeton Univ. Press, Princeton, NJ, 1980.

6. J. L. Brylinski and J. P. Labesse, *Cohomologie d'intersection et fonctions L de certaines variétés de Shimura*, preprint, 1982.
7. W. Casselman, *L^2-cohomology of real rank one groups*, Proc. Conf. in Utah, 1982, to appear.
8. J. Cheeger, On the Hodge theory of Riemannian pseudomanifolds, Proc. Sympos. Pure Math. **36**, Amer. Math. Soc., Providence, RI, 1980, pp. 91–146.
9. W. Greub, S. Halperin, R. Vanstone, *Connections, Curvatuve and Cohomology*, Vol. III, Academic Press, New York, 1976.
10. G. Hochschild and J.-P. Serre, *Cohomology of Lie algebras*, Annals of Math. (2) **57** (1953), 591–603.
11. J.-L. Koszul, *Homologie et cohomologie des algèbres de Lie*, Bull. Soc. Math. France **78** (1950), 65–127.
12. T. Kudo and S. Araki, *Topology of H_n-spaces and H-squaring operations*, Memoirs Fac. Sci. Kyusyu Univ., Ser. A, Vol. **10** (1956), 85–120.
13. R. P. Langlands, On the Functional Equations Satisfied by Eisenstein Series, Lecture Notes in Math. **544**, Springer-Verlag, Berlin and New York, 1976.
14. S. MacLane, Homology, Grund. Math. Wiss. **114**, Springer-Verlag, Berlin, Heidelberg, New York, 1963.
15. M. S. Raghunathan, Discrete Subgroups of Lie Groups, Ergeb. d. Mathematik **68**, Springer-Verlag, 1972.

School of Mathematics, The Institute for Advanced Study, Princeton, New Jersey 08540

126.

(with W. Casselman)

L^2-Cohomology of Locally Symmetric Manifolds of Finite Volume

Duke Math. J. **50** (1983) 625–647

The L^2-cohomology space $H^{\cdot}_{(2)}(M; \mathbf{C})$ of a Riemannian manifold M may be defined as the cohomology of the complex $A^{\infty}_{(2)}(M)$ of complex valued smooth differential forms ω such that ω and $d\omega$ are square integrable (with respect to the scalar product associated to the given metric). In this paper, we consider the case where $M = \Gamma\backslash X$ is the quotient of the symmetric space X of maximal compact subgroups of a real linear semi-simple Lie group G by a discrete subgroup of finite covolume. If M is compact, then $H^*_{(2)}(M; \mathbf{C}) = H^*(M; \mathbf{C})$, so we shall always assume M to be noncompact. In that case, the relationship with the ordinary cohomology is more tenuous, but L^2-cohomology has become of interest in other contexts as well, in particular via its conjectured or proven relationships with intersection cohomology [6; 7; 8; 9; 15]. In fact, in view of such applications, we shall more generally consider the L^2-cohomology space $H^{\cdot}_{(2)}(M; \tilde{E})$ with respect to a local system $\tilde{E}$ on $\Gamma\backslash X$ associated to a finite dimensional representation (r, E) of G (endowed with an admissible scalar product).

Let $\mathscr{H}^{\cdot}_{(2)}(M; \tilde{E})$ be the space of harmonic square integrable $\tilde{E}$-valued harmonic forms. There is a natural map $j : \mathscr{H}^{\cdot}_{(2)}(M; \tilde{E}) \to H^{\cdot}_{(2)}(M; \tilde{E})$. In our case, it is injective since M is complete; moreover $\mathscr{H}^{\cdot}_{(2)}(M; \tilde{E})$ is finite dimensional [1, 4]. Then it is easily seen that $H^{\cdot}_{(2)}(M; \tilde{E})$ is finite dimensional if and only if j is bijective. Our main result is a sufficient condition for finite dimensionality, namely

THEOREM A. *The space $H^{\cdot}_{(2)}(\Gamma\backslash X; \tilde{E})$ is finite dimensional if no proper cuspidal parabolic subgroup of G contains a Cartan subgroup of a maximal compact subgroup K of G; in particular if* $\operatorname{rank} G = \operatorname{rank} K$.

(See 4.5.) Recall that a parabolic subgroup P of G is cuspidal if $\Gamma \cap R_uP$ is cocompact in R_uP, where R_uP is the unipotent radical of P. (If G is algebraic, defined over $\mathbf{Q}$ and Γ is arithmetic, then P is cuspidal if and only if it is also defined over $\mathbf{Q}$.) It may well be that there is a converse if we allow passing to a subgroup of finite index, and if moreover E, assumed to be irreducible, is equivalent to its complex conjugate contragredient representation. In fact, if the latter condition is fulfilled and G has a minimal cuspidal subgroup which is fundamental (see 1.7), then Γ always has a subgroup of finite index such that $H^{\cdot}_{(2)}(\Gamma\backslash X; \tilde{E})$ is infinite dimensional and a (n expected) generalization of a result

Received November 22, 1982.

of de George and Wallach [10] would yield this whenever G has a proper cuspidal subgroup which is fundamental (see 4.6, 4.7 for a discussion and some examples, among them odd-dimensional hyperbolic manifolds).

To prove these results we start from the fact that

$$H^{\cdot}_{(2)}(\Gamma\backslash X; \tilde{E}) \cong \mathrm{Ext}^{\cdot}_{(\mathfrak{g},K)}(E^*, L^2(\Gamma\backslash G)^{\infty}). \tag{1}$$

This was announced in [2] and is proved in [3], under more general assumptions on G and Γ, by use of smoothing homotopy operators and spectral sequences. In §5, we give a direct proof, based on the results of [4] and [11] on $L^2(\Gamma\backslash G)$, of an equivalent statement. The $\mathrm{Ext}^{\cdot}$ group with respect to the discrete spectrum gives the space of L^2-harmonic forms and is finite dimensional [1, 4], so that we have to show that the cohomology with respect to the continuous spectrum is zero. It is proved in [4] to be a finite sum of groups $\mathrm{Ext}^{\cdot}_{(\mathfrak{g},K)}(E^*, L^{\infty}_{P,V})$, where $L_{P,V}$ is a direct integral of unitarily induced principal series representations (4.3). Those $\mathrm{Ext}^{\cdot}$ groups are discussed in §3. The nonvanishing implies some strong conditions on the infinitesimal character of E. Via some relations between complex conjugation, opposition involutions and Cartan involution which are established in §1, these imply that P contains a fundamental parabolic subgroup (and conversely). See 3.6, which also proves similar relations in the case of a single principal series representation. From this Theorem A follows easily.

In this paper, we confine ourselves to the global L^2-cohomology. Since it is either infinite dimensional or equal to the space of square integrable harmonic forms, one may wonder whether there is much point in not limiting oneself *a priori* to the latter. However, a main motivation for us is that similar techniques allow one to study L^2-cohomology at infinity, with respect to a Satake compactification and to investigate Zucker's conjecture [15:6.20], as is already apparent from [6, 7, 15] and will be illustrated further elsewhere. Obviously, this case cannot be reduced to the consideration of harmonic forms.

A statement stronger than Theorem A was announced in [2], but the author subsequently noticed a gap in the proof. The second named author then provided a lemma which, combined with the other results of [2], showed finite dimensionality at least when G and K have the same rank. This paper grew out of some discussions about various variants and generalizations of that lemma, which are contained in §1.

In this paper, G denotes a group of finite index in the group of real points of a connected semi-simple algebraic group $\mathcal{G}$ defined over $\mathbb{R}$, *K is a maximal compact subgroup of G and $X = G/K$. Moreover, (r, E) is an irreducible finite dimensional rational representation of $\mathcal{G}$. Beginning with §2, $\mathfrak{g}$ is the Lie algebra of G.*

§1. Opposition involution, complex conjugation and fundamental parabolic subgroups.

1.1. In this section, $\mathfrak{g}$ is a real, semi-simple Lie algebra, $\mathfrak{h}$ a Cartan subalgebra, $\mathfrak{k}$ a maximal compact subalgebra and $\mathfrak{g}_c = \mathfrak{g} \otimes_{\mathbb{R}} \mathbb{C}$. Furthermore $\mathfrak{g}_u$

(resp. $\mathfrak{g}_{sp}$) is a compact (resp. split) form of $\mathfrak{g}_c$; we let $\operatorname{Dyn}\mathfrak{g}_c$ be the Dynkin diagram of the root system $\Phi = \Phi(\mathfrak{g}_c, \mathfrak{h}_c)$ and $W = W(\mathfrak{g}_c, \mathfrak{h}_c)$ the Weyl group of $\mathfrak{g}_c$ with respect to $\mathfrak{h}_c$. We recall that the automorphism group $\operatorname{Aut}(\operatorname{Dyn}\mathfrak{g}_c)$ of $\operatorname{Dyn}\mathfrak{g}_c$ is canonically isomorphic to the quotient of the automorphism group $\operatorname{Aut}(\Phi)$ of Φ by W, and also to the group $\operatorname{Aut}(\Phi, \Delta)$ of automorphisms of Φ leaving a given basis Δ stable. Given $\psi \in \operatorname{Aut}\operatorname{Dyn}\mathfrak{g}_c$, and such an isomorphism, we shall denote by ψ_B the corresponding element in $\operatorname{Aut}(\Phi, \Delta)$. This automorphism also defines a bijection of the set $P(\Phi)^+_\Delta$ of the dominant weights for the ordering associated to Δ as well as automorphisms of $\mathfrak{h}_c$ or $\mathfrak{h}_c^*$, to be denoted in the same way.

Let $\iota_\mathfrak{g} \in \operatorname{Aut}(\operatorname{Dyn}\mathfrak{g}_c)$ be the "opposition involution." We have then

$$\iota_{\mathfrak{g},\Delta} = -w_{\mathfrak{g},\Delta}, \tag{1}$$

where $w_{\mathfrak{g},\Delta}$ is the longest element of W, for the notion of length defined by the simple reflections to the elements of Δ. We recall that, as a permutation of $P(\Phi)^+_\Delta$, the transformation $\iota_{\mathfrak{g},B}$ maps the highest weight of an irreducible $\mathfrak{g}_c$-module into that of the contragredient one. As is known,

$$\iota_\mathfrak{g} = 1 \Leftrightarrow \mathfrak{g}_{sp} \text{ and } \mathfrak{g}_u \text{ are inner forms of one another.} \tag{2}$$

The complex conjugation of $\mathfrak{g}_c$ with respect to $\mathfrak{g}$ defines an involution of $P(\Phi)$, whence an automorphism $\gamma_\mathfrak{g}$ of order $\leqslant 2$ of $\operatorname{Dyn}\mathfrak{g}_c$. If $\mathfrak{g}'$ is another real form of $\mathfrak{g}$, then

$$\gamma_\mathfrak{g} = \gamma_{\mathfrak{g}'} \Leftrightarrow \mathfrak{g} \text{ and } \mathfrak{g}' \text{ are inner forms of one another.} \tag{3}$$

The permutation $\gamma_{\mathfrak{g},\Delta}$ of $P(\Phi)^+_\Delta$ onto itself transforms the highest weight of an irreducible $\mathfrak{g}_c$-module $\pi : \mathfrak{g}_c \to \mathfrak{gl}_n(\mathbf{C})$ into that of the complex conjugate one (with respect to $\mathfrak{g}$), defined by $\bar{\pi}(x) = \overline{\pi(\bar{x})}$, where $x \mapsto \bar{x}$ is the complex conjugation of $\mathfrak{g}_c$ with respect to $\mathfrak{g}$. In particular, $\bar{\pi}(x) = \overline{\pi(x)}$ if $x \in \mathfrak{g}$.

1.2. Our objective in §1 is to study some properties of

$$\tau_\mathfrak{g} = \iota_\mathfrak{g} \circ \gamma_\mathfrak{g}. \tag{1}$$

Clearly $\tau_{\mathfrak{g},\Delta}$, acting on $P(\Phi)^+_\Delta$, transforms the highest weight of an irreducible finite dimensional representation (π, E) into that of the complex conjugate contragredient one $(\bar{\pi}^*, \bar{E}^*)$.

It will be convenient to extend the previous definitions to the case of real algebraic reductive Lie algebras. Let then $\mathfrak{l} = \mathfrak{g} \oplus \mathfrak{z}$, where $\mathfrak{z}$ is the Lie algebra of the group of real points of a torus $\mathcal{Z}$ defined over $\mathbf{R}$. We have the decomposition $\mathfrak{z} = \mathfrak{z}_d \oplus \mathfrak{z}_u$, where $\mathfrak{z}_d$ is $\mathbf{R}$-split and $\mathfrak{z}_u$ is the Lie algebra of the maximal compact subgroup of Z. Let $X \subset \mathfrak{z}_c^*$ be the lattice of the differentials of the rational characters of $\mathcal{Z}$. If $\lambda \in X$, then $-\lambda$ is the differential of the contragredient character. Accordingly we set $\iota_\mathfrak{z} = -1$, viewed as automorphism of X, $\mathfrak{z}_c$ or $\mathfrak{z}_c$. Any such λ is purely imaginary on $\mathfrak{z}_u$ and real on $\mathfrak{z}_d$. We then let $\gamma_\mathfrak{z}$ be -1 on $\mathfrak{z}_u$ and the identity on $\mathfrak{z}_d$. Then $\tau_\mathfrak{z} = -\gamma_\mathfrak{z}$, viewed as an automorphism of X, transforms the differential of an irreducible rational representation of $\mathscr{G}$ into that of the contragredient complex conjugate one. We let then

$$z_{\mathfrak{l},\Delta} = \iota_{\mathfrak{g},\Delta} + \iota_{\mathfrak{z}}, \qquad \gamma_{\mathfrak{l},\Delta} = \gamma_{\mathfrak{g},\Delta} + \gamma_{\mathfrak{z}}, \qquad \tau_{\mathfrak{l},\Delta} = \tau_{\mathfrak{z},\Delta} + \tau_{\mathfrak{g}}.$$

It is then again true that $\tau_{\mathfrak{l},B}$ transforms the highest weight of a rational irreducible representation of $\mathfrak{l}$ into that of the contragredient complex conjugate one. Now if V is a unitary representation of a connected group L with Lie algebra $\mathfrak{l}$, then V is equivalent to its contragredient complex conjugate. Therefore:

1.3. LEMMA. *If the infinitesimal character χ_λ (λ dominant) of an irreducible finite dimensional representation of $\mathfrak{l}$ is equal to that of a unitary representation of connected group with Lie algebra $\mathfrak{l}$, then $\tau_{\mathfrak{l},B}(\lambda) = \lambda$.*

1.4. Let θ be the Cartan involution of $\mathfrak{l}$ with respect to $\mathfrak{k} \oplus \mathfrak{z}_u$. Its restriction to $\mathfrak{g}$ is the Cartan involution of $\mathfrak{g}$ with respect to $\mathfrak{k}$. If $\mathfrak{h}$ is θ-stable, then θ defines an automorphism of Dyn $\mathfrak{g}_c$; however, that automorphism may depend on the conjugacy class of $\mathfrak{h}$ in $\mathfrak{g}$. We shall consider it only when $\mathfrak{h}$ is fundamental and θ-stable, i.e., when $\mathfrak{h}$ contains a Cartan subalgebra $\mathfrak{t}$ of $\mathfrak{k}$. We recall that $\mathfrak{t}$ always contains elements which are regular in $\mathfrak{g}$, hence $\mathfrak{h} = \mathfrak{z}_{\mathfrak{g}}(\mathfrak{t})$ is a θ-stable Cartan subalgebra.

1.5. PROPOSITION. *Assume that $\mathfrak{h}$ is θ-stable and fundamental and let Δ be a basis of Φ such that the corresponding positive Weyl chamber contains an element of $\mathfrak{k} \cap \mathfrak{h}$ regular in $\mathfrak{g}$. Then $\tau_{\mathfrak{l},\Delta}$ leaves $\mathfrak{h} \oplus \mathfrak{z}$ stable and is equal to the restriction of θ on $\mathfrak{h} \oplus \mathfrak{z}$.*

The Cartan involution θ_0 of $gl_n(\mathbf{C})$ with respect to su_n is $x \mapsto -{}^t\bar{x}$. If $\pi : \mathfrak{l} \to gl_n(\mathbf{C})$ is a rational representation, then it is always possible, by conjugation in $gl_n(\mathbf{C})$, to arrange that $\pi(\mathfrak{g})$, is stable under θ_0 and $\pi(\mathfrak{k}) = su_n \cap \pi(\mathfrak{g})$. For $x \in \mathfrak{h} \oplus \mathfrak{z}$, we have then

$$\pi(\theta(x)) = \theta_0(\pi(x)) = -{}^t\bar{\pi}(x).$$

If now π is irreducible, this means that θ transforms the highest weight of π into that of $\bar{\pi}^*$. Since this is true for any such π, the proposition follows.

1.6. COROLLARY. (a) *Let π be an irreducible rational representation of $\mathfrak{l}_c$ and λ its highest weight. Then $\pi \sim \tilde{\pi}^*$ if and only if $\theta(\lambda) = \lambda$.*

(b) *The following three conditions are equivalent*: (i) $\tau_{\mathfrak{g}} = 1$; (ii) $\mathfrak{g}$ *and* $\mathfrak{g}_u$ *are inner forms of one another*; (iii) rk $\mathfrak{g}$ = rk $\mathfrak{k}$. *They imply that every simple factor of* $\mathfrak{g}$ *is absolutely simple.*

(c) *Let $\mathfrak{h}_d$ be the* $\mathbf{R}$*-split part of $\mathfrak{h}$ and $\mathfrak{u} = \mathfrak{z}(\mathfrak{h}_d)$. Then*

$$\Phi(\mathfrak{u}_c, \mathfrak{h}_c) = \{\alpha \in \Phi \mid \tau_{\mathfrak{g},\Delta}(\alpha) = \alpha\}.$$

(a) follows from the proposition and the end remark of 1.2.

(b) In view of the proposition, $\tau_{\mathfrak{g},\Delta} = 1$ if and only if $\mathfrak{t} = \mathfrak{k}$, whence the equivalence of (i) and (iii). Now if a real simple Lie algebra is not absolutely simple, the rank of a maximal compact subalgebra is half the rank of the algebra. Therefore (iii) implies that every simple factor of $\mathfrak{g}$ is absolutely simple. Finally, the equivalence of (ii) and (iii) is standard.

(c) By 1.5, $\tau_{\mathfrak{g},\Delta}$ is the identity on $\mathfrak{t}$ and $-Id.$ on $\mathfrak{h}_d$. Therefore a root α is fixed under $\tau_{\mathfrak{g},\Delta}$ if and only if it is zero on $\mathfrak{h}_d$, i.e., if and only if it belongs to $\Phi(\mathfrak{u}_c, \mathfrak{h}_c)$.

1.7. A Cartan subalgebra is said to be *maximally compact* or *fundamental* if it contains a Cartan subalgebra $\mathfrak{t}$ of a maximal compact subalgebra of $\mathfrak{g}$. The maximally compact Cartan subalgebras form one conjugacy class with respect to $\mathrm{Ad}\,\mathfrak{g}$.

A parabolic subalgebra $\mathfrak{p}$ is *fundamental if* any split component of $\mathfrak{p}$ is the split part of a maximally compact Cartan subalgebra of $\mathfrak{g}$. It is proper if and only if $\mathrm{rk}\,\mathfrak{g} \neq \mathrm{rk}\,\mathfrak{k}$. The fundamental parabolic subalgebras form one associate class; their Levi subalgebras are the centralizers of the split parts of the maximally compact Cartan subalgebras of $\mathfrak{g}$ [5:III].

1.8. PROPOSITION. *Let $\mathfrak{q}$ be a parabolic subalgebra of $\mathfrak{g}$ and $\mathfrak{l}$ a Levi subalgebra of $\mathfrak{q}$. Then the following conditions are equivalent*:

(i) $\mathfrak{q}$ *contains a fundamental parabolic subalgebra*;

(ii) $\mathfrak{q}$ *contains a Cartan subalgebra of* $\mathfrak{k}$;

(iii) $\mathfrak{q}$ *contains an element of* $\mathfrak{k}$ *regular in* $\mathfrak{g}$;

(iv) $\tau_{\mathfrak{l},\Delta}$ *fixes an element regular in* $\mathfrak{g}_c$.

The implications (i)$\Rightarrow$(ii)$\Rightarrow$(iii) are obvious from 1.7. Moreover, (iii)$\Rightarrow$(iv) follows from 1.5. There remains to prove that (iv)$\Rightarrow$(i).

We assume, as we may that $\mathfrak{l}$ is θ-stable. Let $\mathfrak{s}$ be a Cartan subalgebra of $\mathfrak{l} \cap \mathfrak{k}$. Then $\mathfrak{c} = \mathfrak{z}_{\mathfrak{l}}(\mathfrak{s})$ is a Cartan subalgebra of $\mathfrak{l}$, hence of $\mathfrak{g}$, in which $\mathfrak{s}$ is maximal compact, i.e., we can write $\mathfrak{c} = \mathfrak{s} \oplus \mathfrak{r}$ with $\mathfrak{r}$ split and θ-stable. But the restriction of θ to $\mathfrak{l}$ is a Cartan involution of $\mathfrak{l}$ and, by 1.5, its restriction to $\mathfrak{c}$ coincides with $\tau_{\mathfrak{l}}$. In view of the assumption (iv), it follows that $\mathfrak{s}$ contains an element regular in $\mathfrak{g}$. But then $\mathfrak{c} = \mathfrak{z}_{\mathfrak{g}}(\mathfrak{s})$. In particular, $\mathfrak{c}$ contains any commutative subalgebra containing $\mathfrak{s}$; therefore $\mathfrak{s}$ is a Cartan subalgebra of $\mathfrak{k}$, the Cartan subalgebra $\mathfrak{c}$ is maximally compact and $\mathfrak{z}(\mathfrak{r})$ is a Levi subalgebra of a fundamental parabolic subalgebra (1.7). Since $\mathfrak{r}$ also contains a split component $\mathfrak{a}$ of $\mathfrak{q}$, we have

$$\mathfrak{q} \supset \mathfrak{z}(\mathfrak{a}) \supset \mathfrak{z}(\mathfrak{r}).$$

Let $\mathfrak{a}'$ be a maximal split commutative subalgebra of $\mathfrak{g}$ contained in $\mathfrak{z}(\mathfrak{r})$ and $\mathfrak{p}_0$ a minimal parabolic subalgebra of $\mathfrak{g}$ contained in $\mathfrak{q}$, with Levi subalgebra $\mathfrak{z}(\mathfrak{a}')$. Then $\mathfrak{z}(\mathfrak{r})$ and the nil-radical of $\mathfrak{p}_0$ generate a parabolic subalgebra of $\mathfrak{g}$ contained in $\mathfrak{q}$ with Levi subalgebra $\mathfrak{z}(\mathfrak{r})$, which is therefore fundamental.

§2. Relative Lie algebra cohomology and Casimir operators. We assemble here some known or easy facts pertaining to cohomology and the Casimir operator.

2.1. Given a nondegenerate invariant bilinear form B on $\mathfrak{g}$, there is defined a Casimir element in the center of the universal enveloping algebra $U(\mathfrak{g})$ of $\mathfrak{g}$ (see e.g., [5:II, 1.3] or [4]), to be denoted $C_{B,\mathfrak{g}}$ or simply $C_{\mathfrak{g}}$ or C, if B is understood. We always assume that, once B is chosen for $\mathfrak{g}$, the Casimir element $C_{\mathfrak{m}}$ for a reductive subalgebra $\mathfrak{m}$ is also defined using B. If (π, U) is a continuous representation of G such that $C(\pi)$ is a scalar multiple of the identity on U^∞, then the eigenvalue of C on U^∞ is denoted by either of the symbols $c(\pi), c(U), c_{\mathfrak{g}}(\pi), c_{\mathfrak{g}}(U)$.

If (π, V) is a unitary representation of G, then the Laplace operator Δ on $C^{\cdot}(\mathfrak{g}, K; V^\infty \otimes E)$ is given componentwise by Kuga's formula

$$(\Delta\eta)_I = (c(\sigma) - c(\pi))\eta_I, \tag{1}$$

[5:II, 2.5]. Here we write as usual

$$\eta = \Sigma_I \eta_I \omega^I,$$

with I a set of indices, $\eta_I \in V^\infty \otimes E$ and ω^I the exterior product of the elements ω^i $(i \in I)$, where (ω^i) is a basis of $\mathfrak{g}^*$.

2.2. Let now (π, V) be irreducible. Then $C(\pi)$ is multiplication by a scalar $c(\pi)$. We recall that $\operatorname{Ext}^{\cdot}_{(\mathfrak{g},K)}(E^*, V^\infty) = 0$ if either $c(\pi) \neq c(\sigma)$ or π and σ^* have distinct infinitesimal characters $\chi_\pi, \chi_{\sigma^*}$ and that if $c(\sigma) = c(\pi)$, then d, ∂ and Δ are identically zero, and

$$\operatorname{Ext}^{\cdot}_{(\mathfrak{g},K)}(E^*, V) = C^{\cdot}(\mathfrak{g}, K; V^\infty \otimes E), \tag{1}$$

consists of harmonic forms (for all this, see [5:I, II]). In particular, we see that

$$C^{\cdot}(\mathfrak{g}, K; V^\infty \otimes E) \neq 0 \quad \text{and} \quad c(\pi) = c(\sigma) \Rightarrow \chi_\pi = \chi_{\sigma^*}. \tag{2}$$

We note also that

$$E \not\sim \bar{E}^* \Rightarrow \operatorname{Ext}_{(\mathfrak{g},K)}(E^*, V) = 0, \tag{3}$$

since $E \not\sim \bar{E}^*$ implies that E^* and V do not have the same infinitesimal character. If λ is the highest weight of E in the set up of 1.5, then the condition $E \not\sim \bar{E}^*$ is equivalent to $\theta\lambda \neq \lambda$ by 1.3, 1.5. In this form, (3) is proved in [5:II, 6.12].

2.3. Finally let us recall that if (π, V) is any continuous representation, then we may also write

$$C^{\cdot}(\mathfrak{g}, K; V^\infty \otimes E) = \operatorname{Hom}_K(\Lambda^{\cdot}(\mathfrak{g}/\mathfrak{k}) \otimes E^*, V^\infty), \tag{1}$$

therefore

$$C^{\cdot}(\mathfrak{g}, K; V^\infty \otimes E) \subset \operatorname{Hom}_K(\Lambda(\mathfrak{g}/\mathfrak{k}), V_I^\infty \otimes E), \tag{2}$$

where I is the set of K-types occurring in $\Lambda(\mathfrak{g}/\mathfrak{k}) \otimes E^*$. Note that the right hand side may not be a complex.

§3. The relative Lie algebra cohomology with respect to the direct integral of induced representations.

3.1. We first consider direct integrals of one-dimensional representations of a vector group. Let then A be a vector group, $\mathfrak{a}$ its Lie algebra. For $D \subset \mathfrak{a}^*$ of strictly positive measure and $\lambda \in \mathfrak{a}^*$, we consider the direct integral

$$I_{\lambda,D} = \int_D^{\oplus} \mathbf{C}_{\lambda + i\mu}\, d\mu. \tag{1}$$

$I_{\lambda,D}$ is then the space of square integrable functions on D, where $a \in A$ acts by

$$(a.f)(\mu) = a^{\lambda + i\mu} f(\mu). \qquad (\mu \in D).$$

We note that we have an obvious isomorphism

$$I_{\lambda,D} = \mathbf{C}_\lambda \otimes I_{0,D}. \tag{2}$$

If $\alpha \in C_c^\infty(A)$, then let $\check{\alpha}$ be its Fourier transform, given by

$$\check{\alpha}(\lambda + i\mu) = \int_A a^{\lambda + i\mu} \alpha(a)\, da. \tag{3}$$

Then α operates on $I_{\lambda,D}$ by

$$(\alpha(f))(i\mu) = \check{\alpha}(\lambda + i\mu) f(i\mu). \tag{4}$$

The Gårding vectors are therefore finite linear combinations of functions of the type $\check{\alpha}.f$ ($f \in I_{\lambda,D}$, $\alpha \in C_c^\infty(A)$).

3.2. Proposition. *We keep the notation of* 3.1.

(i) *If* $\lambda \neq 0$, *or if* D *is in the complement of some neighborhood of the origin in* $i\mathfrak{a}^*$, *then* $H^\cdot(\mathfrak{a}; I_{\lambda,D}) = 0$.

(ii) *If* $\lambda = 0$ *and there is a neighborhood* U *of the origin such that* $U \cap D$ *contains the intersection of* U *with an open cone with vertex as the origin, then*

$$H^0(\mathfrak{a}; I^\infty_{\lambda,D}) = 0 \textit{ and } H^i(\mathfrak{a}; I^\infty_{\lambda,D}) \textit{ is infinite dimensional for } 1 \leqslant i \leqslant \dim \mathfrak{a}. \tag{1}$$

(i) We remark first that if $D = D_1 \cup D_2$ is the union of two measurable subsets whose intersection has measure zero, then

$$H^\cdot(\mathfrak{a}; I^\infty_{\lambda,D}) = H^\cdot(\mathfrak{a}; I^\infty_{\lambda,D}) + H^\cdot(\mathfrak{a}; I^\infty_{\lambda,D}). \tag{2}$$

It follows for instance from Kuga's formula that the Laplacian Δ_μ on $C^\cdot(\mathfrak{a}; \mathbf{C}_{\lambda+i\mu})$ is multiplication by $\langle\lambda,\lambda\rangle + \langle\mu,\mu\rangle$. Thus if $(\lambda,\mu) \neq (0,0)$ Δ_μ has an inverse bounded in norm by $(\langle\lambda,\lambda\rangle + \langle\mu,\mu\rangle)^{-1}$. Therefore, if either $\lambda \neq 0$ or if there exists a constant $c > 0$ such that $|\mu| \geqslant c$ for $\mu \in D$, then Δ^{-1} is a bounded operator. Hence (i) follows from [4:5.2]

(ii) From now on $\lambda = 0$. In view of (2), we may assume that D is contained in a ball of any prescribed non zero radius around the origin. Then there exists $\alpha \in C_c^\infty(A)$ such that $|\check{\alpha}| \geqslant 1$ on D. It follows then from the end remarks of 3.1 that all elements of $I_{0,D}$ are smooth vectors (see also 5.6).

Let (e_j) be a basis of $\mathfrak{a}$, (ω^j) the dual basis of $\mathfrak{a}^*$ and x^j the corresponding coordinates. If

$$\eta = \sum_J \eta_J \omega^I \in C^q(\mathfrak{a}; I_{0,D}), \tag{3}$$

then

$$(d\eta)_I = \sum_j \pm x^{i_j} \cdot \eta_{I-\{i_j\}}, \qquad \text{where } I = \{i_1, \ldots, i_q, i_{q+1}\}. \tag{4}$$

Let first $q = 0$. Then $d\eta = 0$ implies that $x^j \cdot \eta = 0$ for all j, hence $\eta = 0$. Let $r^2 = \langle \mu, \mu \rangle$. Using polar coordinates, one sees immediately that if $\eta = d\mu$ ($\mu \in C^{q-1}(\mathfrak{a}; I)$), then η / r must have coefficients which are square integrable. Let $\sigma = \omega^1 \wedge \cdots \wedge \omega^{q-1}$ ($\sigma = 1$ if $q = 0$), and $\eta = d\sigma$. Consider

$$\eta_\alpha = \eta \cdot r^{-\alpha} \qquad (n/2 - 1 \leqslant \alpha < n/2;\ n = \dim \mathfrak{a}). \tag{5}$$

Each form η_α is closed. A simple computation shows that $\eta_\alpha / r \notin L^2$, hence η_α is not a coboundary. In order to prove (ii), for $q \geqslant 1$, it is then enough to show that the η_α for different α are linearly independent modulo coboundaries. Now the linear dependence of the η_α is the same as that of the functions $r^{-\alpha}$ for r small, i.e., of the functions $e^{x\alpha}$ for x large. Hence they are linearly independent. If we have a relation

$$\sum_1^m c'_{\alpha_i} \eta_{\alpha_i} = d\mu, \qquad \big(\mu \in C^{q-1}(\mathfrak{a}; I_{0,D});\ c_{\alpha_i} \neq 0\ (1 \leqslant i \leqslant m)\big),$$

then

$$\int_0^a \Big| \sum c'_{\alpha_i} r^{-\alpha_i - 1} \Big|^2 r^{n-1}\, dr < \infty.$$

We may assume $\alpha_1 > \cdots > \alpha_m$. Then this relation implies

$$\int_0^a r^{n-1} / r^{2\alpha_1 + 2}\, dr < \infty,$$

which contradicts the assumption $\alpha_1 \geqslant (n/2) - 1$.

3.3. We collect here some notation about parabolic subgroups and some facts on induced representations.

(a) We assume $\mathcal{G}$ to be defined over a subfield F of $\mathbb{R}$. Let P be its proper parabolic F-subgroup of G, A a maximal F-split component of its radical and $P = M.A.N$ its standard decomposition: N is the unipotent radical, M the intersection of the kernels of the squares of the rational characters defined over F

of $M.A$, and $M.A = M \times A = Z_G(A)$. We use the notation of [5], (except that we write M and $M.A$ for ${}^0M$ and M), adapted to this more general case (there $F = \mathbf{R}$). In particular $\mathfrak{b}$ is a Cartan subalgebra of $\mathfrak{m}$, hence $\mathfrak{h} = \mathfrak{b} \oplus \mathfrak{a}$ one of $\mathfrak{g}$. We fix an ordering on $\Phi = \Phi(\mathfrak{g}_c, \mathfrak{h}_c)$ compatible with $\Phi(P, A)$ and let $W = W(\mathfrak{g}_c, \mathfrak{h}_c)$. Then $W_M = W(\mathfrak{m}_c, \mathfrak{b}_c) \subset W$ and we have $W = W_M . W^P$ (uniquely), where

$$W^P = \left\{ w \in W \mid w^{-1}(\Phi_M^+) \subset \Phi^+ \right\}. \tag{1}$$

Finally, $2\rho_P$ is the character of A on $\mathfrak{n}$. It is the restriction to A of

$$2\rho_{\mathfrak{g}} = 2\rho = \sum_{\alpha \in \Phi^+} \alpha. \tag{2}$$

We denote by $F_\nu (\nu \in \mathfrak{b}_c^*)$ a finite dimensional irreducible representation of M which is also irreducible for $\mathfrak{m}_c$, and has infinitesimal character χ_ν (if ν is dominant, its highest weight is then $\nu - \rho_{\mathfrak{m}}$).

(b) Let V be an irreducible unitary M-module. We consider the induced representation

$$I_{P,V,\mu} = \mathrm{Ind}_P^G(V \otimes \mathbf{C}_{\rho + i\mu}) \qquad (\mu \in \mathfrak{a}^*) \tag{3}$$

Let $K_P = K \cap P$. It is a maximal compact subgroup of P. If we assume (as we may) K such that A is stable under the Cartan involution θ associated to K, then $K_P = K \cap M$ is maximal compact in M. We recall that, as a K-module, $I_{P,V,\mu}$ is independent of μ and can be viewed as the induced representation $\mathrm{Ind}_{K_M}^K(V)$. In particular, by Frobenius reciprocity, if a K-type τ occurs in $I_{P,V,\lambda}$, then one of the K_M-types of $\tau|_{K_M}$ occurs in V.

(c) We recall that there is a constant d_P depending only on P such that

$$c_{\mathfrak{g}}(I_{P,V,\lambda}) = d_P + c_{\mathfrak{m}}(V) - (\lambda, \lambda), \qquad (\lambda \in \mathfrak{a}^*). \tag{4}$$

(d) Fix a measurable subset D of $\mathfrak{a}^*$. (In the applications here, it will be the positive Weyl chamber $\mathfrak{a}^{*+}$ defined by $\Phi(P, A)$.) We let

$$L_{P,V,D} = L_{P,V} = \int_D^{\oplus} I_{P,V,\mu}\, d\mu, \tag{5}$$

be the direct integral over D of the $I_{P,\mu}$ $(\mu \in D)$. Its K-types are the same as those of $I_{P,V,\mu}$ for any μ.

3.4. Theorem. *We keep the previous notation and let $\lambda \in \mathfrak{h}_c^*$ be dominant and such that χ_λ is the infinitesimal character of E. Assume that* $\mathrm{Ext}^{\cdot}_{(\mathfrak{g},K)}(E^*, L^\infty_{P,V}) \neq 0$. *Then*

(i) *There exists $s \in W^P$ such that $s(\lambda) | A = 0$ and that $\chi_{V^*} = \chi_{s(\lambda)|\mathfrak{b}}$. Such an s is unique. Moreover* $2l(s) = \dim N$.

(ii) $\mathrm{Ext}^{\cdot}_{(\mathfrak{g},K)}(E^*, L^\infty_{P,V}) = (\mathrm{Ext}^{\cdot}_{(\mathfrak{m},K_M)}(F^*_{s\lambda|\mathfrak{b}_c}, V) \otimes H^{\cdot}(\mathfrak{a}; \int_D^{\oplus} C_{i\mu}\, d\mu))[-l(s)]$.

(iii) *P contains a fundamental parabolic subgroup.*

Note that (i) and (ii) are quite similar to Theorem 3.3 in [5], which computes the cohomology with respect to an induced representation. As we shall see, the proof is analogous.

From basic facts about direct integrals, we have

$$\int_D^{\oplus} I_{V,\mu}\, d\mu = I_P^G\left(\int_D^{\oplus} (V \otimes \mathbf{C}_{i\mu+\rho})\, d\mu\right) = I_P^G\left(\mathbf{C}_\rho \otimes \int_D^{\oplus} (V \otimes C_{i\mu})\, d\mu\right). \tag{1}$$

By an analogue of Shapiro's lemma [5:III, 25] we have then

$$\operatorname{Ext}^{\cdot}_{(\mathfrak{g},K)}(E^*, L^{\infty}_{P,V}) = H^{\cdot}\left(\mathfrak{p}, K_P; \mathbf{C}_\rho \otimes \left(\int_D^{\oplus} (V \otimes \mathbf{C}_{i\mu})\, d\mu\right)^{\infty}\right). \tag{2}$$

We consider the Hochschild–Serre type spectral sequence (E_r) in relative Lie algebra cohomology for $\mathfrak{p}$ mod $\mathfrak{n}$ [5:I, §6]. It abuts to the right hand side of (2) and its E_2 term is given by

$$E_2^{p,q} = H^{\cdot}\left(\mathfrak{m} \oplus \mathfrak{a}, K_M; H^q(\mathfrak{n}; \mathrm{E}) \otimes \mathbf{C}_{\rho_P} \otimes \left(\int_D^{\oplus} (V \otimes \mathbf{C}_{i\mu})\, d\mu\right)^{\infty}\right). \tag{3}$$

By the theorem of Kostant giving the decomposition of $H^{\cdot}(\mathfrak{n}; E)$ as a $M.A$ module [5:III, 3.1] we have

$$H^q(\mathfrak{n}; E) = \bigoplus_{s \in W^P;\, l(s) = q} F_{s\lambda}, \qquad (q \in \mathbf{N}). \tag{4}$$

This allows one to decompose $E_2^{\cdot,q}$ into a direct sum of pieces $Q_s^{\cdot}$ indexed by W^P, where

$$Q_s^{\cdot} = H^{\cdot}\left(\mathfrak{m} \oplus \mathfrak{a}, K_M; F_{s\lambda} \otimes \mathbf{C}_{\rho_P} \otimes \left(\int (V \otimes \mathbf{C}_{i\mu})\, d\mu\right)^{\infty}\right) \in E_2^{\cdot, l(s)}. \tag{5}$$

On the other hand, we have a canonical isomorphism

$$\int_D^{\oplus} (V \otimes \mathbf{C}_{i\mu})\, d\mu = V \hat{\otimes} \int_D^{\oplus} \mathbf{C}_{i\mu}\, d\mu, \tag{6}$$

where $\hat{\otimes}$ indicates a Hilbert tensor product. We have only to consider the K_M-types which occur in $\Lambda^{\cdot}(\mathfrak{m} + \mathfrak{a})$. They form a finite set F; hence

$$\left(E \otimes V \otimes \mathbf{C}_{\rho_P} \hat{\otimes} \int_D^{\oplus} \mathbf{C}_{i\mu}\, d\mu\right)_F = (E \otimes V \otimes \mathbf{C}_{\rho_P})_F \otimes \int_D^{\oplus} \mathbf{C}_{i\mu}\, d\mu, \tag{7}$$

where we could replace $\hat{\otimes}$ by an ordinary tensor product since the first factor on the right hand side is finite dimensional. The highest weight of $E_{s\lambda}$ is $s\lambda - \rho$. So we can write

$$F_{s\lambda} = F_{s\lambda \mid \mathfrak{b}_c} \otimes \mathbf{C}_{(s\lambda - \rho) \mid A} \tag{8}$$

as a tensor product of the irreducible representation of M with infinitesimal character $\chi_{s\lambda \mid \mathfrak{b}_c}$ and the one-dimensional representation of A with character $(s\lambda - \rho) \mid A = s\lambda \mid A - \rho_P$. By the Künneth rule [5:I, 1.3], we get

$$Q_s^{\cdot} = H^{\cdot}(\mathfrak{m}, K_M; V \otimes F_{s\lambda \mid \mathfrak{b}_c}) \otimes H^{\cdot}\left(\mathfrak{a}; \left(\int_D^{\oplus} \mathbf{C}_{s\lambda \mid \mathfrak{a} + i\mu} \, d\mu\right)^{\infty}\right). \tag{9}$$

(We have used 3.1(2).) We recall that

$$Q_s^p \subset E_2^{p, l(s)} \qquad (p \in \mathbf{N}).) \tag{10}$$

If $s\lambda \mid \mathfrak{a} \neq 0$, then the second factor is zero by 3.2 and if the second equality of (i) is not true, then the first factor is zero (by Wigner's lemma [5:I, 5.3]). The two conditions of (i) are then fulfilled. Since they determine $s \in W^P$ uniquely (see [5], p. 95), this proves (i). As a consequence, $Q_s^{\cdot} \neq 0$ for at most one value of s and $E_2^{p,q}$ is not zero at most for $q = l(s)$. Therefore $E_2 = E_\infty$ and the equality in (ii) follows.

Consider now $\mathrm{Ext}^{\cdot}_{(\mathfrak{g},K)}(E^*, I^{\infty}_{P,V,0})$. From [5:III, 3.3] we see that since (i) holds, we have

$$\mathrm{Ext}^{\cdot}_{(\mathfrak{g},K)}(E^*, I^{\infty}_{P,V,0}) = (\mathrm{Ext}^{\cdot}_{(\mathfrak{m},K_M)}(F^*_{s\lambda \mid \mathfrak{b}_c}, V^{\infty}) \otimes \Lambda^{\cdot} \mathfrak{a}_c^*)[-l(s)]. \tag{11}$$

The representations $I_{P,V,0}$ of G and V of M are unitary, admissible. Therefore, the two $\mathrm{Ext}^{\cdot}$ groups satisfy Poincaré duality [5:II, 3.4]. Let c and C (resp. b and B) the smallest and biggest integers for which the first (resp. second) Ext^q is not zero. Then

$$c + C = \dim X, \qquad b + B = \dim M / K_M. \tag{12}$$

By (11), we have

$$c = b + l(s), \qquad C = B + \dim \mathfrak{a} + l(s). \tag{13}$$

But

$$\dim X = \dim M / K_M + \dim \mathfrak{a} + \dim N. \tag{14}$$

The second assertion of (ii) now follows from (12), (13), (14).

The second equality of (i) shows that $F_{s\lambda \mid \mathfrak{b}_c}$ has the same infinitesimal character as a unitary representation, hence, by 1.3:

$$\tau_{\mathfrak{m}}(s\lambda \mid \mathfrak{b}) = s\lambda \mid \mathfrak{b}. \tag{15}$$

(We use the fact that $s(\lambda) \mid \mathfrak{b}$ is dominant.) However, since $s\lambda \mid \mathfrak{a} = 0$, the element $s\lambda \mid \mathfrak{b}$ may be identified to a regular element of $\mathfrak{g}$, hence (iii) follows from 1.8.

3.5. COROLLARY. (i) *If D is contained in the complement of a neighborhood of the origin, then* $\mathrm{Ext}^{\cdot}_{(\mathfrak{g},K)}(E^*, L^{\infty}_{P,V}) = 0$.

(ii) *Assume that D contains the interior of a cone (with interior points and vertex at the origin). Then* $\mathrm{Ext}^{\cdot}_{(\mathfrak{g},K)}(E^*, L^{\infty}_{P,V}) \neq 0$ *if and only if* (i) *is fulfilled and* $\mathrm{Ext}^{\cdot}_{(\mathfrak{m},K_M)}(F^*_{s\lambda|\mathfrak{b}}, V^{\infty}) \neq 0$. *In that case, it is infinite dimensional in exactly the degrees of the form* $q + i + (\dim N)/2$ *where* $1 \leqslant i \leqslant \dim \mathfrak{a}$ *and q runs through the integers such that* $\mathrm{Ext}^q_{(\mathfrak{m},K)}(F^*_{s\lambda|\mathfrak{b}_c}, V^{\infty}) \neq 0$.

(i) follows from 3.4 and 3.2.

(ii) By 3.2, the last direct integral in 3.4(ii) is infinite dimensional precisely in the degrees $i \in [1, \dim \mathfrak{a}]$. Then our assertion follows from 3.4.

3.6. THEOREM. *Let $\lambda \in \mathfrak{h}^*_c$ be dominant regular and P be a proper parabolic subgroup of G. Then the following conditions are equivalent*:

(i) *There exists $V \in \hat{M}$ such that* $\mathrm{Ext}^{\cdot}_{(\mathfrak{g},K)}(E^*, I_{P,V,0}) \neq 0$.

(ii) *There exists $V \in \hat{M}$ such that* $\mathrm{Ext}^{\cdot}_{(\mathfrak{g},K)}(E^*, L_{P,V}) \neq 0$.

(iii) *There exists $s \in W^P$ such that $s\lambda|\mathfrak{a} = 0$ and $\tau_{\mathfrak{m}}(s\lambda|\mathfrak{b}) = s\lambda|\mathfrak{b}$.*

(iv) *P contains a fundamental parabolic subgroup and $\tau_{\mathfrak{g}}(\lambda) = \lambda$.*

It is understood here that the D underlying the definition of $L_{P,V}$ contains an open cone with vertex at the origin, as in 3.5(ii) and that $\tau_{\mathfrak{g}}$ and $\tau_{\mathfrak{m}}$ stand for $\tau_{\mathfrak{g},\Delta}, \tau_{\mathfrak{m},\Delta'}$ where Δ is the basis of Φ defining the order chosen in 3.3(a) and $\Delta' = \Delta \cap \Phi_M$.

By 3.4 (resp. [5:III, 3.3], (i) (resp. (ii)) implies the existence of $s \in W^P$ such that $s\lambda|\mathfrak{a} = 0$ and then, in view of 3.4(ii) (resp. 3.4(11)) we have $\mathrm{Ext}^{\cdot}_{\mathfrak{m},K_M}(F^*_{s\lambda|\mathfrak{b}}, V) \neq 0$. By 1.2, 1.3, this last condition is equivalent to $\tau_{\mathfrak{m}}(s\lambda|\mathfrak{b}) = s\lambda|\mathfrak{b}$. This shows that (i)$\Leftrightarrow$(ii) and that these conditions imply (iii).

(iii)$\Rightarrow$(ii), (i). Given a finite dimensional irreducible representation F of M, which is equivalent to its contragredient complex conjugate, there always exists $V \in \hat{M}$ such that $\mathrm{Ext}^{\cdot}_{(\mathfrak{m},K_M)}(F, V) \neq 0$. One can take in particular a fundamental series representation. This can be deduced from [14]. This being granted, (iii)$\Rightarrow$(ii) follows from 3.5(ii) and (iii)$\Rightarrow$(i) from 3.4(11).

(iii)$\Rightarrow$(iv). The element $s\lambda$ is regular. If $s\lambda|\mathfrak{a} = 0$, then $s\lambda|\mathfrak{b}$ is regular in $\mathfrak{g}$. Since it is fixed under $\tau_{\mathfrak{m}}$, the first part of (iii) follows from 1.8. We may now assume that $\mathfrak{b}$ contains a Cartan subalgebra $\mathfrak{t}$ of $\mathfrak{k}$, hence that $\mathfrak{h} = \mathfrak{z}_{\mathfrak{g}}(\mathfrak{t})$ and $\mathfrak{b} = \mathfrak{z}_{\mathfrak{m}}(\mathfrak{t})$. By the second assumption of (iii) and 1.5, there exists then $w \in W_M$ such that $\theta(ws\lambda|\mathfrak{b}) = ws\lambda|\mathfrak{b}$. Since w acts trivially on $\mathfrak{a}$ and $s\lambda|\mathfrak{a} = 0$, we have also $\theta(ws\lambda) = ws\lambda$. Proposition 1.5 now implies that λ is fixed under $\tau_{\mathfrak{g}}$.

(iv)$\Rightarrow$(iii). We may assume that the $\mathfrak{t}$ chosen above belongs to the θ-stable Levi subalgebra of a fundamental parabolic subgroup $Q \subset P$. We have then $\mathfrak{h} = \mathfrak{t} + \mathfrak{a}_Q$, where $\mathfrak{a}_Q$ is a split component of the Lie algebra $\mathfrak{q}$ of Q. In view of our assumption and 1.5, there exists $u \in W$ such that $\theta(u\lambda) = u\lambda$. Since θ leaves $\mathfrak{b}$ stable and is -1 on $\mathfrak{a}_Q$, in particular on $\mathfrak{a}$, we get

$$u\lambda|\mathfrak{a} = 0, \qquad \theta(u\lambda|\mathfrak{b}) = u\lambda|\mathfrak{b}.$$

We can write uniquely $u = w.s$ with $w \in W_M$ and $s \in W^P$. Since w acts trivially on $\mathfrak{a}$ and $u\lambda|\mathfrak{a} = 0$, we also have $s\lambda|\mathfrak{a} = 0$. Since $s \in W^P$, the element $s\lambda|\mathfrak{b}$ is dominant. The relation $\theta(ws\lambda|\mathfrak{b}) = ws\lambda|\mathfrak{b}$ and 1.5 then imply that $\tau_{\mathfrak{m}}$ fixes $s\lambda|\mathfrak{b}$.

3.7. COROLLARY. *Let P be fundamental and $E_\lambda \sim \bar{E}_\lambda^*$. Then there exists a discrete series representation V of M such that* $\mathrm{Ext}^q_{(\mathfrak{g},K)}(E^*, L^\infty_{P,V})$ *is infinite dimensional in exactly the degrees*

$$q = (\dim X - \dim A)/2 + i, \qquad (1 \leqslant i \leqslant \dim \mathfrak{a}). \tag{1}$$

By 3.6, there exists $s \in W^P$ such that $s\lambda \mid \mathfrak{a} = 0$ and $\tau_{\mathfrak{m}}(s\lambda \mid \mathfrak{b}) = s\lambda \mid \mathfrak{b}$. In fact, P being fundamental, we have anyhow $\tau_{\mathfrak{m}} = 1$ by 1.5, since $\mathrm{rk}\, M = \mathrm{rk}\, K_M$. This last fact also implies the existence of a discrete series representation V of M such that $\mathrm{Ext}^q_{(\mathfrak{m},K_M)}(F^*, V^\infty) \neq 0$. We have then

$$\mathrm{Ext}^i_{(\mathfrak{m},K_M)}(F^*, V) = \begin{cases} 0, & \text{if } i \neq (\dim M/K_M)/2, \\ \mathbf{C}, & \text{if } i = (\dim M/K_M)/2. \end{cases}$$

(See [5:II, 5.3].) Then, by 3.4, 3.5, $\mathrm{Ext}^q_{(\mathfrak{g},K)}(E^*, L^\infty_{P,V})$ is infinite dimensional exactly for

$$q = (\dim N + \dim(M/K_M)/2 + i, \qquad (1 \leqslant i \leqslant \dim \mathfrak{a}). \tag{2}$$

In view of 3.4(14), this proves (1).

§4. L^2-cohomology and relative Lie algebra cohomology.

4.1. From now on Γ is a discrete non cocompact subgroup of G of finite covolume. We may (and do) assume that G has no nontrivial normal compact connected subgroup. If Γ' is a normal subgroup of finite index of Γ, then Γ/Γ' operates on $H^{\cdot}_{(2)}(\Gamma'\backslash X; E)$ and

$$H^{\cdot}_{(2)}(\Gamma\backslash X; \tilde{E}) = H^{\cdot}_{(2)}(\Gamma'\backslash X; \tilde{E})^{\Gamma/\Gamma'}. \tag{1}$$

Now the theorems we want to prove are of two types: finite dimensionality or existence of a subgroup of finite index for which the L^2-cohomology is infinite dimensional. We may therefore always replace Γ by a subgroup of finite index. We could therefore also assume G to be connected and then Γ irreducible [13: 5.20]. According to the results of Margoulis, then either G has real rank one or G admits a (unique) $\mathbf{Q}$-structure with respect to which Γ is arithmetic. This shows that the conditions assumed in [11] are always satisfied. Furthermore there is always a subfield F of $\mathbf{R}$ ($\mathbf{R}$ itself in the first case, $\mathbf{Q}$ in the second one) such that the Langlands decomposition $P = M.A.N$ of a cuspidal parabolic subgroup coincides with its decomposition relative to F where A is a maximal F-split component, which was our starting point in §3. The results of §3 are therefore available to us.

4.2. Let $A^\infty(X; E)$ be the space of smooth E-valued exterior differential forms on X. Then, for any discrete subgroup Γ, there is a canonical isomorphism

$$A^\infty(X; E)^\Gamma = C^{\cdot}(\mathfrak{g}, K; C^\infty(\Gamma\backslash G) \otimes E) \tag{1}$$

(see [5:VII]). If $\nu: G \to G/K$ is the canonical projection, this isomorphism is obtained by associating to $\varphi \in A$ the form $\varphi^0 = \tau(g^{-1})(\varphi \circ \nu)(g)$. On the right hand side there is a natural scalar product [5:II]. We then have

$$A^{\infty}_{(2)}(\Gamma\backslash X; \tilde{E})$$
$$= \{\eta \in C^{\cdot}(\mathfrak{g}, K; C^{\infty}(\Gamma\backslash G) \otimes E \mid \eta, d\eta \text{ are square integrable}\}. \quad (2)$$

Let $L^2(\Gamma\backslash G)^{\infty}$ be the space of smooth vectors in $L^2(\Gamma\backslash G)$, viewed as a unitary G-module under right translations. It is a $(\mathfrak{g}, K)$-module, hence we may consider the relative Lie algebra complex $C^*(\mathfrak{g}, K; L^2(\Gamma\backslash G)^{\infty} \otimes E)$. Since the elements of $L^2(\Gamma\backslash G)^{\infty}$ are smooth functions, we have a natural inclusion

$$C^{\cdot}(\mathfrak{g}, K; L_2(\Gamma\backslash G)^{\infty} \otimes E) \to C^{\cdot}(\mathfrak{g}, K; C^{\infty}(\Gamma\backslash G) \otimes E), \quad (3)$$

whose image is contained in $A^{\infty}_{(2)}(\Gamma\backslash X; \tilde{E})$. By a result announced in [2], proved in [3], the inclusion:

$$C^{\cdot}(\mathfrak{g}, K; L^2(\Gamma\backslash G)^{\infty} \otimes E) \to A^{\cdot}_{(2)}(\Gamma\backslash X; \tilde{E}) \quad (4)$$

induces an isomorphism in cohomology. Otherwise stated, we have a natural isomorphism

$$H^{\cdot}_{(2)}(\Gamma\backslash X; \tilde{E}) = \mathrm{Ext}^{\cdot}_{(\mathfrak{g},K)}(E^*, L^2(\Gamma\backslash G)^{\infty}). \quad (5)$$

An equivalent result will be proved in §5. In the remainder of this section, we concern ourselves with the right hand side of (5).

We write $L^2(\Gamma\backslash G)$ as the direct sum of the discrete spectrum $L^2(\Gamma\backslash G)_d$ and of its orthogonal complement, the so-called continuous spectrum $L^2(\Gamma\backslash G)_{ct}$. We have then a first obvious reduction:

$$\mathrm{Ext}^{\cdot}_{(\mathfrak{g},K)}(E^*, L^2(\Gamma\backslash G)^{\infty})$$
$$= \mathrm{Ext}^{\cdot}_{(\mathfrak{g},K)}(E^*, L^2(\Gamma\backslash G)^{\infty}_d) \oplus \mathrm{Ext}^{\cdot}_{(\mathfrak{g},K)}(E^*; L^2(\Gamma\backslash G)^{\infty}_{ct}). \quad (6)$$

Before pursuing this further, we recall some results of [11] on the continuous spectrum.

4.3. Let P be a proper cuspidal parabolic subgroup and $P = M.A.N$ a Langlands decomposition. Let $\Gamma_P = \Gamma \cap P$ and Γ_M the image of Γ_P under the natural projection $\tau: P \to P/A.N = M$. We identify it to a subgroup of M. The group Γ_M is discrete in M, of finite covolume. The group M is reductive, not necessarily semi-simple. However, if Z is the center of its identity component, then Z is central in M and $Z \cap \Gamma_M$ is cocompact in Z.

Let V be an irreducible constituent of $L^2(\Gamma\backslash M)_d$, and let

$$L_{P,V} = \int^{\oplus}_{\mathfrak{a}^{*+}} I_{P,\mu}\, d\mu. \quad (1)$$

(notation of 3.3(d)), where $\mathfrak{a}^{*+}$ is the positive Weyl chamber defined by $\Phi(P,A)$. The theory of Eisenstein series [11] provides a G-isomorphism of $L_{P,V}$ onto a closed subspace of $L^2(\Gamma\backslash G)_{ct}$, and the latter is a countable Hilbert direct sum of such subspaces, to be called *elementary subspaces*. (Each $L_{P,V}$ occurs, but the sum, as P and V vary, is not direct; one has to restrict the P's to a representative system of associate classes modulo Γ-conjugacy.) This theorem is implicitly proved in [11], although it is stated there only in adelic form [11: Appendix II]. It is dealt with more explicitly in [12].

We denote by $\mathscr{C}$ the set of pairs (P, V), where P runs through a set of Γ-conjugacy classes of proper cuspidal subgroups and V through a minimal spanning set of irreducible constituents of the discrete spectrum of Γ_M.

4.4. Proposition. (i) *The space $\mathscr{H}^{\cdot}_{(2)}(\Gamma\backslash X;\tilde{E})$ is finite dimensional. There exists a finite set (H_i) $(i \in S)$ of mutually orthogonal closed irreducible G-invariant subspaces of $L^2(\Gamma\backslash G)_d$ such that*

$$\mathscr{H}^{\cdot}_{(2)}(\Gamma\backslash X;E) = \mathrm{Ext}^{\cdot}_{(\mathfrak{g},K)}\big(E^*, L^2(\Gamma\backslash G)^{\infty}_{d}\big) = \bigoplus_{i\in S}\mathrm{Ext}^{\cdot}_{(\mathfrak{g},K)}(E^*, H_i^{\infty}). \quad (1)$$

(ii) *There exists a finite set $T \subset \mathscr{C}$ such that the $L_{P,V}$ $((P,V)\in T)$ are mutually orthogonal and*

$$\mathrm{Ext}^{\cdot}_{(\mathfrak{g},K)}\big(E^*, L^2(\Gamma\backslash G)^{\infty}_{ct}\big) = \bigoplus_{(P,V)\in T}\mathrm{Ext}^{\cdot}_{(\mathfrak{g},K)}(E^*, L^{\infty}_{P,V}). \quad (2)$$

This is proved in [4]. See 5.3 and 5.7 there. Combined with 4.3 and 3.4, it yields:

4.5. Theorem. *The space $H^{\cdot}_{(2)}(\Gamma\backslash X;\tilde{E})$ is finite dimensional if and only if $\mathrm{Ext}^{\cdot}_{(\mathfrak{g},K)}(E^*, L_{P,V}) = 0$ for every $(P,V)\in\mathscr{C}$ (see 4.3 for $\mathscr{C}$).*

(ii) *If no proper cuspidal subgroup of G contains a fundamental parabolic subgroup, then $H^{\cdot}_{(2)}(\Gamma\backslash X;\tilde{E})$ is finite dimensional (and equal to the space $\mathscr{H}^{\cdot}_{(2)}(\Gamma\backslash X;\tilde{E})$ of L^2-harmonic forms).*

4.6. We shall discuss the converse and examples of infinite dimensional cohomology in the case where G has a proper cuspidal subgroup which is a fundamental parabolic subgroup, and where E is equivalent to its complex conjugate contragredient representation.

Let $P = M.A.N$ be such a group. Then $\mathrm{rk}\, M = \mathrm{rk}\, K_M$ hence M has a discrete series, and we can always find one such that $\mathrm{Ext}^{\cdot}_{(\mathfrak{g},K)}(E^*, L_{P,V}) \neq 0$ (3.7). The question is therefore whether such a representation occurs in $L^2(\Gamma_M\backslash M)$. Consider all the subgroups of finite index of G. Let $\mathscr{S}$ be the set of the corresponding subgroups of M. The intersection of these subgroups is reduced to $\{1\}$. If the following assertion:

Any discrete series representation of M with infinitesimal characters $\chi_{s\lambda|\mathfrak{b}}$ occurs in $L^2(\Sigma\backslash M)_d$ for some $\Sigma\in\mathscr{S}$. (*)

is true, then $H_{(2)}^{\cdot}(\Gamma\backslash X;\tilde{E})$ is infinite dimensional and this particular $\mathrm{Ext}^{\cdot}$ contributes to infinite dimensionality in the degrees given by 3.7(1).

The assertion (*) is rather widely believed to be true in general (i.e., for M reductive with compact center and Σ of finite covolume). So far it is known to hold at least in the following cases:

(a) M is compact (i.e., P is a minimal parabolic subgroup of G). It suffices then to take $\Sigma = \{1\}$, because $L^2(\Sigma\backslash M) = L^2(M)$ contains every irreducible (necessarily finite dimensional) continuous representation of M.

As an example, let $G = \mathbf{SO}(n, 1)$, with n odd. Then X is the hyperbolic n-space. If $\Gamma_M = \{1\}$, and $E = \bar{E}^*$, then $H_{(2)}^i(\Gamma\backslash X;\tilde{E})$ is infinite dimensional if (and only if) $i = (n+1)/2$.

(b) $\Gamma_M\backslash M$ is compact, (i.e., P is also minimal among cuspidal subgroups, or percuspidal in the terminology of [11]). In this case, (*) is proved in [10].

(c) The discrete series representations with infinitesimal character $\chi_{s(\lambda)\,|\,\mathfrak{b}_c}$ are integrable.

This can be seen using Poincaré series: Start with f on M belonging to such a discrete series D and of minimal K-type, and consider the Poincaré series

$$P_f(x) = \Sigma_{\gamma\in\Sigma} f(\gamma x), \qquad (x \in M; \Sigma \in \mathscr{S}). \tag{1}$$

It is well known to converge absolutely and locally uniformly and to represent a cuspidal automorphic form. If $P_f \neq 0$ its M-translates on $\Sigma\backslash M$ are contained in the sum of finitely many copies of D belonging to $L^2(\Gamma\backslash M)_d$. Now one sees easily from the convergence proof of P_f that by taking a suitably small $\Sigma \in \mathscr{S}$ one can arrange to have $P_f \neq 0$ if $f \neq 0$.

Now if $\mu \in \mathfrak{b}_c^*$ is sufficiently regular, then any discrete series with infinitesimal character χ_μ is integrable. It follows therefore that if the highest weight of E is sufficiently regular (e.g., if it is a high multiple of ρ), then the L^2-cohomology is infinite dimensional.

(d) The condition (*) is also satisfied if $M^0 = \mathbf{SL}_2(\mathsf{R})$ and $\Gamma = \mathbf{SL}_2(\mathsf{Z})$ for $\mathscr{S}$ the set of congruence subgroups, because $L^2(\Sigma\backslash\mathbf{SL}_2(\mathsf{R}))_d$ contains the two lowest discrete series representations if the genus of Σ is ≥ 1, and any other discrete series is integrable. Therefore, if $(\Gamma_M \cap M^0)\backslash M^0$ can be written as a direct product of quotients $\Gamma_i\backslash M_i$ where either Γ_i is cocompact or $M_i = \mathbf{SL}_2(\mathsf{R})$ and Γ_i is a congruence subgroup of $\mathbf{SL}_2(\mathsf{Z})$, then (*) also holds.

As an example, take $G = \mathbf{SL}_n(\mathsf{R})$ and $\Gamma = \mathbf{SL}_n(\mathsf{Z})$. Then $\operatorname{rk} G = n-1$ and $\operatorname{rk} K = [n/2]$. Let $P = M.A.N$ be a fundamental parabolic subgroup. Then M^0 is a product of $[n/2]$ copies of $\mathbf{SL}_2(\mathsf{R})$ and Γ_{M^0} the product of $[n/2]$ copies of $\mathbf{SL}_2(\mathsf{Z})$. Hence we get infinite dimensionality in (at least) the degrees

$$(n-1)/2 + i, \qquad (1 \leq i \leq [n/2]).$$

4.7. In 4.6, we have stressed discrete series representations because these are practically the only ones for which there are some theorems proving they occur in the discrete spectrum. But (*) is only a sufficient condition for infinite dimensionality. Given $s \in W^P$ such that $s(\lambda)|\mathfrak{a} = 0$, a weaker but still sufficient one is:

There exists a constituent V of $L^2(\Sigma \backslash M)_d$ such that (**)
$\mathrm{Ext}^{\cdot}_{(\mathfrak{m},K_M)}(F^*_{s(\lambda)\,|\,\mathfrak{b}_c}, V^\infty) \neq 0$ *for some $\Sigma \in \mathscr{S}$.*

In case (b) of 4.6 this can be proved directly without recourse to [10]: In fact, assume $\Sigma \subset M$ to be discrete, cocompact, and torsion free. Let

$$L^2(\Sigma \backslash M) = \widetilde{\bigoplus_{i \in I}} V_i,$$

be a decomposition of $L^2(\Sigma \backslash M)$ into irreducible M-modules. Then [5:VII, 5.2]:

$$H^{\cdot}(\Gamma; E) = \bigoplus_{i \in I} \mathrm{Ext}^{\cdot}_{(\mathfrak{m},K_M)}(F^*, V_i). \tag{1}$$

Since Σ is torsion free we have, with χ denoting now the Euler characteristic

$$\chi(H^{\cdot}(\Gamma; E)) = (\dim E).\chi(H^{\cdot}(\Gamma; \mathbf{C})).$$

But $\operatorname{rk} M = \operatorname{rk} K_M$, hence $\chi(H^{\cdot}(\Gamma; \mathbf{C})) \neq 0$, and at least one of the terms in (1) has to be nonzero.

Remark. In a recent preprint: "L^2-index and the Selberg trace formula," D. Barbasch and H. Moscovici show that (**) holds for real rank one groups with a discrete series. Hence (**) is also true if $(\Sigma \cap M^0)\backslash M^0$ can be written as a product of factors $\Gamma_i \backslash M_i$ where either M_i has real rank one or Γ_i is cocompact, (and M has a discrete series, as was assumed in 4.6).

§5. Appendix: Reduction to $\mathrm{Ext}^{\cdot}_{(\mathfrak{g},K)}(E^*, L^2(\Gamma \backslash G)^\infty)$.
In this section J denotes the set of K-types in $\Lambda(\mathfrak{g}/\mathfrak{k}) \otimes E^*$.

5.1. Let (π, V) be a unitary representation of G. The relative Lie algebra complex $C^{\cdot}(\mathfrak{g}, K; V^\infty \otimes E)$ is contained and dense in the graded Hilbert space

$$C^{\cdot}(\mathfrak{g}, K; V \otimes E) = \mathrm{Hom}_K(\Lambda^{\cdot}(\mathfrak{g}/\mathfrak{k}), V \otimes E). \tag{1}$$

We denote by $\bar{d}$ the closure in that space of d acting on $C^{\cdot}(\mathfrak{g}, K; V^\infty \otimes E)$ (well defined since ∂ is a densely defined formal adjoint to d) and let

$$C^{\cdot}_{(2)}(\mathfrak{g}, K; V \otimes E) = \operatorname{dom} \bar{d}. \tag{2}$$

Thus $\eta \in \operatorname{dom} \bar{d}$ if and only if there exists a sequence (η_j) of elements in $C^{\cdot}(\mathfrak{g}, K; V^\infty \otimes E)$ converging to η and such that the $d\eta_j$ converge to some element, which is then $\bar{d}\eta$ by definition. The space $\operatorname{dom} \bar{d}$ is a differential graded complex with differential $\bar{d}$ and we denote its cohomology space by

$$H^{\cdot}_{(2)}(\mathfrak{g}, K; V \otimes E) \quad \text{or} \quad \mathrm{Ext}^{\cdot}_{(\mathfrak{g},K)}(E^*, V). \tag{3}$$

It is proved in [3] that the natural homomorphism

$$j^* : \mathrm{Ext}^{\cdot}_{(\mathfrak{g},K)}(E^*, V^{\infty}) \to \mathrm{Ext}^{\cdot}_{(\mathfrak{g},K)}(E^*, V), \tag{4}$$

induced by inclusion is an isomorphism in general. We shall prove it here when $V = L^2(\Gamma\backslash G)$ and in a few related special cases. Since moreover there is a natural isomorphism

$$H^{\cdot}_{(2)}(\Gamma\backslash X; \tilde{E}) = \mathrm{Ext}^{\cdot}_{(\mathfrak{g},K)}(E^*, V), \tag{5}$$

(see 5.6), this will establish 4.2(5), (or (1) in the introduction).

We note that 2.3(2) also implies

$$C^{\cdot}_{(2)}(\mathfrak{g}, K; V \otimes E) \subset \mathrm{Hom}_K(\Lambda(\mathfrak{g}/\mathfrak{k}), V_J \otimes E). \tag{6}$$

5.2. Lemma. *Assume that $(C - c(E)I)$ has a bounded inverse on $V_J \otimes E$. Then* $\mathrm{Ext}^{\cdot}_{(\mathfrak{g},K)}(E^*, V) = 0$.

Since $(C - c(E)I)$ maps $V_J^{\infty} \otimes E$ into itself, it is clear that $(C - c(E)I)$ and $(C - c(E)I)^{-1}$ are inverse bijections of $V_j^{\infty} \otimes E$. From 5.1(6) and Kuga's formula, we see then Δ^{-1} also has a bounded inverse on $C^{\cdot}(\mathfrak{g}, K; V \otimes E)$ and that Δ and Δ^{-1} are inverse bijections of $C^{\cdot}(\mathfrak{g}, K; V^{\infty} \otimes E)$ onto itself. Moreover Δ^{-1} commutes with d and ∂ on the latter space, since Δ does so. We have then

$$I = d(\Delta^{-1}\partial) + (\Delta^{-1}\partial)d, \qquad \text{on } C^{\cdot}\left(\mathfrak{g}, K; L^2(\Gamma\backslash G)^{\infty} \otimes E\right). \tag{1}$$

We want to prove

The operator $\Delta^{-1}\partial$ on $C^{\cdot}(\mathfrak{g}, K; V^{\infty} \otimes E)$ extends to a bounded operator on $C^{\cdot}(\mathfrak{g}, K; V \otimes E)$, whose image is contained in $\mathrm{dom}\,\bar{d}$. (*)

If this is granted, then (1), with d replaced by $\bar{d}$, remains valid on $\mathrm{dom}\,\bar{d}$ and shows that if $\eta \in \mathrm{dom}\,\bar{d}$ is closed, then $\eta = \bar{d}(\Delta^{-1}\partial\eta)$ is a coboundary, which proves the lemma. There remains to check (*). We have

$$\begin{aligned}(u, \Delta^{-1}u) &= (\Delta^{-1}d\partial u, \Delta^{-1}u) + (\Delta^{-1}\partial du, \Delta^{-1}u), \qquad (u \in C^*(\mathfrak{g}, K; V^{\infty} \otimes E))\\ &= (d\Delta^{-1}\partial u, \Delta^{-1}u) + (\partial\Delta^{-1}du, \Delta^{-1}u)\\ &= \|\Delta^{-1}\partial u\|^2 + \|\Delta^{-1}du\|^2,\end{aligned}$$

hence

$$\|\Delta^{-1}\partial u\|^2 + \|\Delta^{-1}du\|^2 \leqslant (u, \Delta^{-1}u) \leqslant \|\Delta^{-1}\|\,\|u\|^2,$$

which shows that $\Delta^{-1}\partial$ and $\Delta^{-1}d$ extend to bounded operators.

Similarly, since $d\partial u$ and ∂du ($u \in C^*(\mathfrak{g}, K; V^\infty \otimes E)$) are orthogonal, we have

$$\|u\|^2 = \|\Delta^{-1}d\partial u\|^2 + \|\Delta^{-1}\partial du\|^2 \qquad (u \in C^*(\mathfrak{g}, K; V^\infty \otimes E),$$

which shows that $d(\Delta^{-1}\partial)$ and $\partial(\Delta^{-1}d)$ extend to bounded operators. Therefore, if

$$u_j \to u \qquad (u_j \in C^\cdot(\mathfrak{g}, K; V^\infty \otimes E), j = 1, 2, \dots; u \in C^\cdot(\mathfrak{g}, K; V \otimes E)),$$

then $\Delta^{-1}\partial u_j \in C^*(\mathfrak{g}, K; V^\infty \otimes E)$, as remarked above, and

$$\Delta^{-1}\partial u_j \to \Delta^{-1}\partial u, \qquad d(\Delta^{-1}\partial u_j) \to (d\Delta^{-1}\partial)u,$$

which, by definition, means that we have

$$\Delta^{-1}\partial u \in \operatorname{dom} \bar{d}, \qquad \bar{d}((\Delta^{-1}\partial)u) = (d\Delta^{-1}\partial)(u).$$

5.3. An analogous, but simpler, argument shows that the lemma is also true if V is replaced by V^∞ (see 5.2 in [4]). Thus j^* of 5.1(4) is certainly an isomorphism under the assumption of the lemma, since both spaces are null.

5.4. PROPOSITION. *The natural homomorphism*

$$j^* : \operatorname{Ext}^\cdot_{(\mathfrak{g},K)}\big(E^*, L^2(\Gamma\backslash G)^\infty\big) \to \operatorname{Ext}^\cdot_{(\mathfrak{g},K)}\big(E^*, L^2(\Gamma\backslash G)\big),$$

is an isomorphism.

There exists a finite orthogonal decomposition of $L^2(\Gamma\backslash G)$ into closed G-invariant subspaces

$$L^2(\Gamma\backslash G) = \bigoplus_{i=0}^{i=m} V_i,$$

such that $(C - c(E)I)$ has a bounded inverse on $V_{0,J}$ and the V_i ($i \geqslant 1$) are elementary subspaces [4 : 4.2]. It then suffices to prove our claim for each V_i. For V_0, this is taken case of by 5.3. If V_i is in the discrete spectrum, then $V_{i,J}$ is finite dimensional and consists of smooth vectors so that in that case

$$C^\cdot(\mathfrak{g}, K; V_i^\infty \otimes E) = C^\cdot_{(2)}(\mathfrak{g}, K; V_i \otimes E)$$

and our assertion is obvious. There remains to consider the case where V_i is in the continuous spectrum, i.e., is a continuous integral $L_{P,V,D}$ as in 3.3(d), (with

$D = \mathfrak{a}^{*+}$ but this is irrelevant here). To settle this case, it suffices to prove:

5.5. LEMMA. (i) *There exists an open set U of $\mathfrak{a}^*$, whose complement is compact, such that $(C - c(E)I)$ has a bounded inverse on $L_{P,V,U}$.*

(ii) *If B is a relatively compact subset of $\mathfrak{a}^*$ and F a finite set of K-types, then $L_{P,V,B,F} = L^\infty_{P,V,B,F}$.*

Indeed, assume this for the moment. Let $U' = D \cap U$ and $U'' = D \cap (\mathfrak{a}^* - U)$. We have then

$$\mathrm{Ext}^{\cdot}_{(\mathfrak{g},K)}(E^*, L_{P,V,D}) = \mathrm{Ext}^{\cdot}_{(\mathfrak{g},K)}(E^*, L_{P,V,U'}) + \mathrm{Ext}^{\cdot}_{(\mathfrak{g},K)}(E^*, L_{P,V,U''}). \quad (1)$$

$$\mathrm{Ext}^{\cdot}_{(\mathfrak{g},K)}(E^*, L^\infty_{P,V,D}) = \mathrm{Ext}^{\cdot}_{(\mathfrak{g},K)}(E^*, L^\infty_{P,V,U'}) + \mathrm{Ext}^{\cdot}_{(\mathfrak{g},K)}(E^*, L^\infty_{P,V,U''}). \quad (2)$$

By 5.2 and (i), the two first terms on the right hand sides of (1), (2) are zero and, by 5.1(6) and (ii), the complexes defining the two other terms are the same. There remains then to prove the lemma.

By 3.3(3), there exists a constant d such that

$$c(I_{P,V,\lambda}) - c(E) = -(\lambda, \lambda) + d, \qquad (\lambda \in \mathfrak{a}^*).$$

Fix a constant $a > 0$ and let

$$U = \{\lambda \in \mathfrak{a}^* \mid |c(I_{P,V,\lambda}) - c(E)| \geq a^{-1}\}.$$

Then the complement of U is clearly compact and the equality

$$((C(L_{P,V}) - c(E)I)^{-1} \cdot u)(\mu) = (c(I_{P,V,\mu}) - c(E))^{-1} \cdot u(\mu),$$

where

$$u \in L_{P,V,U},$$

shows that $(C(L_{P,V,U}) - c(E))^{-1}$ has norm $\leq a$ on $L_{P,V,U}$. This proves (i).

(ii) This is no doubt well known, but we do not know of a reference. By compactness, it suffices to show that every point $\lambda \in \mathfrak{a}^*$ has a neighborhood B for which our equality is true. For $\alpha \in C_c^\infty(G)$, let us denote by $\pi_\mu(\alpha)$ the effect of α on $I_{P,V,\mu}$ ($\mu \in \mathfrak{a}^*$). If α is K-invariant, then $\pi_\mu(\alpha)$ leaves stable the K-isotypic subspaces $I_{P,V,\mu,r}$ ($r \in \hat{K}$). We already recalled (3.3(b)) that we have an identification

$$L_{P,V,D,F} = L^2(D; Q),$$

where $Q \cong I_{P,V,\mu,F}$ for any μ and is finite dimensional. The $\pi_\mu(\alpha)$, ($\mu \in D$) can be viewed as endomorphisms of Q which depend continuously, in fact smoothly, on μ. We can also always find α so that $\pi_\lambda(\alpha)$ is arbitrarily close to 1 on Q, in particular is invertible there. We then choose B so that $\pi_\mu(\alpha)$ has an invers-

bounded in norm by some constant on B. If now $u \in L^2(B, Q)$, then the function v given by

$$v(\mu) = \pi_\mu(\alpha)^{-1} \cdot u(\mu) \qquad (\mu \in B)$$

belongs to $L^2(B, Q)$ and therefore

$$u = \pi_{L_{P,V,B}}(\alpha) v$$

is a Gårding vector, hence is differentiable.

5.6. PROPOSITION. *There is a canonical isomorphism*

$$H^{\cdot}_{(2)}(\Gamma\backslash X; \tilde{E}) \cong \mathrm{Ext}^{\cdot}_{(\mathfrak{g},K)}\big(E^*, L^2(\Gamma\backslash G)\big). \tag{1}$$

To avoid any confusion, we denote here by d_X the exterior differentiation on $\Gamma\backslash X$.

As in the introduction, let $A^\infty_{(2)}(\Gamma\backslash X; \tilde{E})$ be the complex of smooth $\tilde{E}$-valued forms ω on $\Gamma\backslash X$ such that ω and $d\omega$ are square integrable. Then, by definition

$$H^{\cdot}_{(2)}(\Gamma\backslash X; \tilde{E}) = H^{\cdot}(A^\infty_{(2)}(\Gamma\backslash X; \tilde{E})). \tag{1}$$

Let now $A(\Gamma\backslash X; \tilde{E})_2$ be the Hilbert space of square integrable $\tilde{E}$-valued forms with measurable coefficients and $\bar{d}_X$ the closure on $A(\Gamma\backslash X; \tilde{E})_2$ of d_X operating on $A^\infty_{(2)}(\Gamma\backslash X; \tilde{E})$. Then $\mathrm{dom}\,\bar{d}_X$ is a differential graded complex with differential $\bar{d}_X$. By [8:§8] the inclusion

$$A^\infty_{(2)}(\Gamma\backslash X; \tilde{E}) \to \mathrm{dom}\,\bar{d}_X \tag{2}$$

induces an isomorphism in cohomology. To establish (1), it suffices therefore to check that:

$$H^{\cdot}(\mathrm{dom}\,\bar{d}_X) = \mathrm{Ext}^{\cdot}_{(\mathfrak{g},K)}\big(E^*, L^2(\Gamma\backslash G)\big). \tag{3}$$

By lifting forms from $\Gamma\backslash X$ to $\Gamma\backslash G$, one defines an isomorphism of differential graded complexes

$$\nu^* : A^\infty(\Gamma\backslash X; \tilde{E}) = C^{\cdot}(\mathfrak{g}, K; C^\infty(\Gamma\backslash G) \otimes E), \tag{4}$$

[5:VII]. In particular, ν^* brings d_X onto d. The same construction provides an isomorphism of graded vector spaces

$$\nu^* : A^{\cdot}(\Gamma\backslash X; \tilde{E})_2 = C^{\cdot}\big(\mathfrak{g}, K; L^2(\Gamma\backslash G) \otimes E\big). \tag{5}$$

We have to show that

$$\nu_* \circ \bar{d}_X \circ \nu^{*-1} = \bar{d}. \tag{6}$$

By definition, $\nu^* \circ \bar{d}_X \circ \nu^{*-1}$ is the closure of $\bar{d}$ operating on $\nu^*(A^\infty_{(2)}(\Gamma\backslash X; \tilde{E})$, whereas $\bar{d}$ is the closure of d operating on the smaller space $C^\cdot(\mathfrak{g}, K; L^2(\Gamma\backslash G)^\infty \otimes E)$. However, $\Gamma\backslash X$ is complete. Therefore, by a result proved originally by M. Gaffney (see [4; 8; 9] for references) $\bar{d}_X$ is also the closure of d_X operating on the space $A^\infty_c(\Gamma\backslash X; \tilde{E})$ of smooth compactly supported $\tilde{E}$-valued forms. Hence the left hand side of (6) is also the closure of d operating on

$$\nu^*(A^\infty_c(\Gamma\backslash X; \tilde{E})) = C^\cdot(\mathfrak{g}, K; C^\infty_c(\Gamma\backslash G) \otimes E).$$

This closure is then also that of d operating on any intermediate stable subspace: as such we may take $C^\cdot(\mathfrak{g}, K; L^2(\Gamma\backslash G)^\infty \otimes E)$, which proves (6). This implies (3).

5.7. The relationships between the various isomorphisms considered here can be summarized as follows. We have natural inclusions

$$\begin{array}{ccc} C^\cdot\left(\mathfrak{g}, K; L^2(\Gamma\backslash G)^\infty \otimes E\right) & \xrightarrow{\alpha} & C^\cdot_{(2)}\left(\mathfrak{g}, K; L^2(\Gamma\backslash G) \otimes E\right) \\ \downarrow \beta & & \downarrow \gamma \\ A^\infty_{(2)}(\Gamma\backslash X; \tilde{E}) & \xrightarrow{\delta} & \operatorname{dom} \bar{d}_X \end{array} \tag{1}$$

(where the two bottom spaces have been identified to their images under ν^*). We have just seen that γ is an isomorphism. As recalled above, the cohomology map $H^\cdot(\delta)$ induced by δ is an isomorphism. Therefore $H^\cdot(\alpha)$ is an isomorphism if and only if $H^\cdot(\beta)$ is. Here we have proved that $H^\cdot(\alpha)$ is one. As stated in 4.2(5), the result of [2, 3] quoted there is in fact a direct proof that $H^\cdot(\beta)$ is an isomorphism.

References

1. A. Borel, *Cohomology of arithmetic groups*, Proc. Internat. Congr. Math. Vancouver **1** (1974), 435–442.
2. ———, *Stable and L^2-cohomology of arithmetic groups*, Bull. Amer. Math. Soc. **3** (1980), 1025–1027.
3. ———, *Regularization theorems in Lie algebra cohomology. Applications*, Duke Math. J. **50** (1983), 605–624.
4. A. Borel and H. Garland, *Laplacian and discrete spectrum of an arithmetic group*, Amer. J. Math. **105** (1983), 309–335
5. A. Borel and N. Wallach, *Continuous Cohomology, Discrete Subgroups and Representations of Reductive Groups*, Ann. of Math. Studies **94**, Princeton Univ. Press, Princeton, NJ, 1980.
6. J. L. Brylinski and J. P. Labesse, *Cohomologie d'intersection et fonctions L de certaines variétés de Shimura*, preprint, 1982.
7. W. Casselman, *L^2-cohomology of real rank one groups*, Proc. Conf. in Utah (1982), to appear.
8. J. Cheeger, *On the Hodge theory of Riemannian pseudomanifolds*, Proc. Sympos. Pure Math. **36**, Amer. Math. Soc., Providence, RI, 1980, pp. 91–146.
9. J. Cheeger, M. Goresky and R. MacPherson, "L^2-cohomology and intersection homology of singular varieties," in *Seminar in Differential Geometry*, S. T. Yau, ed., Annals of Math. Studies **102**, Princeton Univ. Press, 1982, pp. 303–340.
10. S. de George and N. Wallach, *Limit formulas for multiplicities in $L^2(\Gamma\backslash G)$*, Annals of Math. **107** (1978), 133–150.

11. R. P. LANGLANDS, *On the Functional Equations Satisfied by Eisenstein Series*, Lecture Notes in Math. **544**, Springer-Verlag, Berlin and New York, 1976.
12. S. OSBORNE AND G. WARNER, *The Theory of Eisenstein Systems*, Academic Press, 1981.
13. M. S. RAGHUNATHAN, *Discrete Subgroups of Lie Groups*, Ergeb. d. Mathematik **68**, Springer Verlag, 1972.
14. D. VOGAN AND G. ZUCKERMAN, *Unitary representations with non-zero cohomology*, preprint.
15. S. ZUCKER, *L^2-cohomology of warped products and arithmetic groups*, Inv. Math. **70** (1982), 169–218.

BOREL: THE INSTITUTE FOR ADVANCED STUDY, PRINCETON, NEW JERSEY 08540.
CASSELMAN: MATHEMATICS DEPARTMENT, THE UNIVERSITY OF BRITISH COLUMBIA, VANCOUVER V6T 1W5, CANADA.

The L^2-Cohomology of Negatively Curved Riemannian Symmetric Spaces

Ann. Acad. Sci. Fenn. Ser. A, I. Math 10 (1985) 95–105

Let G be a connected linear semi-simple Lie group, K a maximal compact sub group of G. As is well-known, the quotient space $X=G/K$ is homeomorphic to euclidean space and, endowed with a G-invariant metric, is a Riemannian symmetric space with negative curvature without flat component and any such space can be obtained in this way. We fix an irreducible finite dimensional representation (r, E) of G. Our object of interest in this paper is the L^2-cohomology space $H^{\cdot}_{(2)}(X; E)$ of X with respect to E. It can be defined first as the cohomology of the complex $A^{\cdot}_{(2)}(X; E)$ of E valued smooth differential forms η on X such that η and $d\eta$ are square integrable, where d is exterior differentiation. To get a Hilbert space definition, we may consider the completion $\bar{A}^{\cdot}_{(2)}(X; E)$ of $A^{\cdot}_{(2)}(X; E)$ with respect to the square norm $(\eta, \eta)+(d\eta, d\eta)$, and the graph closure or strong closure $\bar{d}$ of d. It is known that the inclusion $A^{\cdot}_{(2)}(X; E)\to\bar{A}^{\cdot}_{(2)}(X; E)$ induces an isomorphism in cohomology [6]. The group G operates on these complexes and hence on the cohomology. In the Hilbert space definition, $H^{\cdot}_{(2)}(X; E)$ appears as the quotient of the closed subspace of the cocycles in $\bar{A}_{(2)}(X; E)$ by the image of $\bar{d}$. Therefore, if $\bar{d}$ has a closed range, then $H^{\cdot}_{(2)}(X; E)$ has a natural Hilbert space structure and yields a unitary representation of G. Our first objective is to prove that this is the case when G and K have the same rank and to identify the representations thus obtained. We shall prove

Theorem A. *Let* $m=(\dim X)/2$ *and assume that* $\operatorname{rk} G=\operatorname{rk} K$. *Then*
(i) *The range of* $\bar{d}$ *is closed. We have*

$$H^i_{(2)}(X; E) = 0, \quad \text{if} \quad i \neq m. \tag{1}$$

(ii) *The* G-*space* $H^m_{(2)}(X; E)$ *is the direct sum of the discrete series representations of* G *having the same infinitesimal character as* (r, E).

The proof of (ii) shows in fact that $H^m_{(2)}(X; E)$ may be identified with the space of square integrable harmonic m-forms. Interpreted in this way, (ii) is quite reminiscent of some characterizations of the discrete series as spaces of harmonic square integrable sections of certain K-bundles over X (see e.g. [10]).

We shall also obtain some information in the case of unequal ranks:

Theorem B. *Assume that* $l_0 = \mathrm{rk}\, G - \mathrm{rk}\, K$ *is not zero.*

(i) *If* E *is not equivalent to its contragredient complex conjugate* $\bar{E}^*$, *then* $H^{\cdot}_{(2)}(X; E) = 0$.

(ii) *If* $E \sim \bar{E}^*$, *then* $\bar{d}$ *does not have closed range, and* $H^i_{(2)}(X, E)$ *is infinite dimensional at least for* $i \in (m-(l_0/2), m+(l_0/2)]$.

Our starting point is a regularization theorem of [1] which yields a canonical isomorphism

$$\mathrm{Ext}^{\cdot}_{(\mathfrak{g}, K)}(E^*, L^2(G)^\infty) \xrightarrow{\sim} H^{\cdot}_{(2)}(X; E), \tag{1}$$

where the left-hand side refers to $\mathrm{Ext}^{\cdot}$ in the category of $(\mathfrak{g}, K)$-modules (cf. [5:I]) and $L^2(G)$ is viewed as a G-module via the right regular representation. We may then investigate the left-hand side using the results of Harish-Chandra [8] on $L^2(G)$. This reduces us to the computation of $\mathrm{Ext}^{\cdot}_{(\mathfrak{g}, K)}(E^*, L^\infty_{P,\omega})$, where the $L_{P,\omega}$ are the direct summands of $L^2(G)$ given by [8]. Those are defined in Section 1, and the computations performed in Section 2. Theorems A and B are proved in Section 3.

This procedure is quite similar to the study of $H^{\cdot}_{(2)}(\Gamma\backslash X; E)$ in [2], where Γ is a discrete subgroup of finite covolume of G. In fact, (1) above is also valid if X and G are replaced by $\Gamma\backslash X$ and $\Gamma\backslash G$ (for any discrete $\Gamma \subset G$). Modulo a result of [3] (whose role is played here by 1.4), we are again reduced to the discussion of $\mathrm{Ext}^{\cdot}$-groups with respect to some elementary subspaces of $L^2(\Gamma\backslash G)$ which are given by Langlands' results [11]. In short, [2] and the present paper correspond to the two cases where extensive information on $L^2(\Gamma\backslash G)$ is available.

Some notation. The Lie algebra of a Lie group $A, G, \ldots$ is denoted by the corresponding lower case German letter $\mathfrak{a}, \mathfrak{g}, \ldots$.

A reductive group is always meant to satisfy the conditions of [5: 0, 3.1]. In particular, it belongs to Harish-Chandra's class [7].

The space of smooth vectors of a continuous representation (π, V) of a Lie group L is denoted V^∞. If the center $\mathscr{Z}$ of the universal enveloping algebra of L acts by scalars on V^∞, we denote by χ_π or χ_V the character of $\mathscr{Z}$ thus obtained, the so-called infinitesimal character of π.

The contragredient of a representation (π, V) is denoted (π^*, V^*).

The set of equivalence classes of irreducible unitary (resp. square integrable) representations of the reductive group L with compact center is denoted $\hat{L}$ (resp. $\hat{L}_d$). If L is compact and F a finite subset of L, then, for any continuous L-module V, we let V_F denote the sum of the isotypic subspaces V_τ $(\tau \in F)$.

1. The decomposition of $L^2(G)$

In this section, we recall some of the fundamental results of Harish-Chandra [8] on the spectral decomposition of $L^2(G)$, in a form adapted to our needs.

1.1. Let (P, A) be a p-pair (cf. [7] or [5: 0, 3.4]) and $P=N_P A_P M_P$ or simply $P=NAM$ the associated Langlands decomposition of P. In particular, N is the unipotent radical of P, A is a split component of the radical of P and the centralizer $Z(A)$ of A in G is the direct product of A and M. For $\lambda \in \mathfrak{a}_c^*$, we denote by C_λ the one-dimensional representation of A, where $a\in A$ acts by multiplication by $a^\lambda = \exp \lambda(\log a)$. Given $(\omega, V_\omega)\in \hat{M}_d$ and $\lambda\in \mathfrak{a}_c^*$, we view as usual $V_\omega \otimes C_\lambda$ as a representation of P on which N acts trivially. Let

$$(1) \qquad I_{P,\omega,i\mu} = \mathrm{Ind}_P^G(V_\omega \otimes C_{\varrho_P + i\mu}) \quad (\omega\in \hat{M}_d;\ \mu \in \mathfrak{a}_c^*)$$

where ϱ_P is defined by

$$a^{2\varrho_P} = \det \mathrm{Ad}\, a|_{\mathfrak{n}} \quad (a\in A).$$

It is unitary if $\mu\in\mathfrak{a}^*$, our only case of interest in this paper.

We shall assume that A and M are stable under the Cartan involution of G associated to K. In particular, $K\cap M$ is a maximal compact subgroup of M and P. We recall that the K-module structure of $I_{P,\omega,i\mu}$ is "independent of μ", i.e., there exists a canonical K-equivariant isomorphism of Hilbert space of $I_{P,\omega,i\mu}$ on a fixed K-module $U_{(\omega)}$, namely $U_{(\omega)} = \mathrm{Ind}_{K\cap M}^K(V_\omega)$, where V_ω is viewed as a $(K\cap M)$-module. In particular, the K-types of $I_{P,\omega,i\mu}$ are independent of μ.

1.2. Recall that a parabolic subgroup P is *cuspidal* if M has the same rank as its maximal compact subgroups, *fundamental* if M contains a Cartan subgroup of K. The group G is its own fundamental parabolic subgroup if and only if G and K have the same rank. According to [8], there exists a finite set S of non-conjugate cuspidal parabolic subgroups of G, containing exactly one fundamental parabolic subgroup, with the following properties:

$$(1) \qquad L^2(G) = \tilde{\bigoplus}_{P\in S} L_P,$$

with

$$(2) \qquad L_P = \tilde{\bigoplus}_{\omega\in \hat{M}_{P,d}} L_{P,\omega},$$

$$(3) \qquad L_{P,\omega} = \int_{\mathfrak{a}^*}^{\oplus} I_{P,\omega,i\mu}^* \hat{\otimes} I_{P,\omega,i\mu}\, d\mu_\omega.$$

Here $d\mu_\omega$ is a certain measure (the Plancherel measure), which is the product of an analytic function by the Lebesgue measure, $\tilde{\oplus}$ stands for a Hilbert direct sum and $\hat{\otimes}$ for the usual Hilbert space completion of the algebraic tensor product of two Hilbert spaces.

The action of G by left (resp. right) translations is given by the natural action in (3) on the first (resp. second) factor of the integrand. If $P=G$, then (3) can be written more simply as

(4) $$L_{P,\omega} = V_\omega^* \hat{\otimes} V_\omega d_\omega,$$

where ω runs through $\hat{G}_d$ and d_ω is the formal degree of ω.

The spaces $L_{P,\omega}$ will be called *elementary subspaces of* $L^2(G)$.

1.3. *Casimir operators.* We fix an admissible trace form on $\mathfrak{g}$ [3: 2.3], say the Killing form if $\mathfrak{g}$ is semi-simple, and for a reductive subalgebra $\mathfrak{m}$ of $\mathfrak{g}$, denote by $C_\mathfrak{m}$ the Casimir operator associated to the same trace form. If $C_\mathfrak{m}$ acts by a scalar multiple of the identity on the space H^∞ of smooth vectors of a continuous representation (π, H_π) of a reductive subgroup M of G with Lie algebra $\mathfrak{m}$, we denote by $c(\pi)$ or $c(H_\pi)$ the eigenvalue of $C_\mathfrak{m}$. We recall that $C_\mathfrak{g}$ acts by a scalar multiple of the identity on $I^\infty_{P,\omega,i\mu}$ and that there exists a constant e_P, depending only on G and P, such that

(1) $$c(I_{P,\omega,i\mu}) = e_P + c_\mathfrak{m}(V_\omega) - (\mu, \mu) \quad (\omega \in \hat{M}_d;\ \mu \in \mathfrak{a}^*).$$

1.4. Lemma. *Let J be a finite subset of $\hat{K}$. Then we can write $L^2(G)$ as a direct sum of two G-stable subspaces Q, R such that Q is the sum of finitely many elementary subspaces and $R_J = 0$.*

By 1.2(1), it suffices to prove the existence of such a decomposition for a space L_P $(P \in S)$. Let J_P be the set of $(K \cap M)$-types occurring in the restriction to $K \cap M$ of the elements $\tau \in J$. It is finite. Let $U_{(\omega)}$ be as in 1.1. By Frobenius reciprocity, $U_{(\omega),J} \neq 0$ implies $V_{\omega,J_P} \neq 0$. It is known that there are only a finite number of $\omega \in \hat{M}_d$ containing a given $(K \cap M)$-type: this follows from the description of the $(K \cap M)$-types in a discrete series representation given by Blattner's formula [9], which in particular shows the existence of a single minimal $(K \cap M)$-type with multiplicity one. As a consequence the set of $L_{P,\omega}$ with a non-trivial J-component is finite. We let then Q be their direct sum and R the orthogonal complement of Q in L_P.

2. Relative Lie algebra cohomology with respect to an elementary subspace

2.1. We consider in this section the cohomology space $\mathrm{Ext}^{\cdot}_{(\mathfrak{g},K)}(E^*, L^\infty_{P,\omega})$, where $L_{P,\omega}$ is viewed as a G-module via right translations (1.2). It is therefore the cohomology of the complex

(1) $$C^{\cdot}(\mathfrak{g}, K; L^\infty_{P,\omega} \otimes E) = \mathrm{Hom}_K(\Lambda^{\cdot}\mathfrak{g}/\mathfrak{k}, L^\infty_{P,\omega} \otimes E) = \mathrm{Hom}_K(\Lambda^{\cdot}\mathfrak{g}/\mathfrak{k} \otimes E^*, L^\infty_{P,\omega}).$$

If J is the (finite) set of K-types occurring in $\Lambda^{\cdot}(\mathfrak{g}/\mathfrak{k}) \otimes E^*$, we have therefore

(2) $$C^{\cdot}(\mathfrak{g}, K; L^\infty_{P,\omega} \otimes E) \subset \mathrm{Hom}_K(\Lambda^{\cdot}\mathfrak{g}/\mathfrak{k} \otimes E^*, L^\infty_{P,\omega,J}) = \mathrm{Hom}_K(\Lambda^{\cdot}\mathfrak{g}/\mathfrak{k}, L^\infty_{P,\omega,J} \otimes E).$$

The action of G by left translations is an automorphism of this complex and goes over to the cohomology. This complex is contained in the graded Hilbert space

$\mathrm{Hom}_K(\Lambda^*\mathfrak{g}/\mathfrak{k}, L_{P,\omega}\otimes E)$ (where, as usual, E is endowed with an "admissible" scalar product, i.e., one which is invariant under K and with respect to which the orthogonal complement of $\mathfrak{k}$ in $\mathfrak{g}$ is represented by self-adjoint operators). If $\bar{d}$ is the closure of d, then the inclusion $C^{\cdot}(\mathfrak{g}, K; L^{\infty}_{P,\omega}\otimes E)\to\mathrm{dom}\,\bar{d}$ is an isomorphism in cohomology [1: 2.7].

2.2. We first consider the case of a discrete series representation, i.e., where $P=G$ and $L_{P,\omega}=V^*_\omega\hat{\otimes}V_\omega$, with $\omega\in\hat{G}_d$. Since $V_{\omega,J}$ is finite dimensional and consists of smooth vectors we have then

$$(V^*_\omega\hat{\otimes}V_\omega)^\infty_J=V^*_\omega\otimes V_{\omega,J},$$

(note that $\hat{\otimes}$ has been replaced by $\otimes$), whence

$$C^{\cdot}(\mathfrak{g}, K; L^\infty_{P,\omega}\otimes E)=V^*_\omega\otimes C^{\cdot}(\mathfrak{g}, K; V^\infty_\omega\otimes E).$$

Since the second factor on the right-hand side is finite dimensional, we see that $d=\bar{d}$ and that

$$\mathrm{Ext}^{\cdot}_{(\mathfrak{g},K)}(E^*, L^\infty_{P,\omega})=V^*_\omega\otimes\mathrm{Ext}^{\cdot}_{(\mathfrak{g},K)}(E^*, V^\infty_\omega),$$

this isomorphism being G-equivariant, G operating through the given representation on the first factor of the right-hand side, and trivially on the second factor. But the value of the second factor is well-known [5: II, 5.3] (see 2.9), therefore we get

2.3. Proposition. *Let $P=G$ and $\omega\in\hat{G}_d$. Then $\bar{d}$ has closed range. We have $\mathrm{Ext}^i_{(\mathfrak{g},K)}(E^*, L^\infty_{P,\omega})=0$ if $i\neq m=(\dim X)/2$ or $\chi_\omega\neq\chi_{r^*}$. If $\chi_\omega=\chi_{r^*}$, then*

(1) $$\mathrm{Ext}^m_{(\mathfrak{g},K)}(E^*, L^\infty_{P,\omega})=V^*_\omega.$$

2.4. Assume now that $P\neq G$, hence that $L_{P,\omega}$ is a direct integral of induced representations. We have

(1) $$L^\infty_{P,\omega,J}=\Big(\int^\oplus_{\mathfrak{a}^*}(I^*_{P,\omega,i\mu}\hat{\otimes}I_{P,\omega,i\mu,J})\,d\mu_\omega\Big)^\infty.$$

But $I_{P,\omega,i\mu,J}$ is finite dimensional, so that we may again replace $\hat{\otimes}$ by $\otimes$ and write

(2) $$L^\infty_{P,\omega,J}=\Big(\int^\oplus_{\mathfrak{a}^*}I^*_{P,\omega,i\mu}\otimes I_{P,\omega,i\mu,J}\,d\mu_\omega\Big)^\infty.$$

2.5. We shall have to use some results of [5: III] on cohomology with respect to $I_{P,\omega,i\mu}$. We recall them here, together with some of the relevant notation. We fix a Cartan subalgebra $\mathfrak{b}$ of $\mathfrak{m}$, let $\mathfrak{h}=\mathfrak{b}\oplus\mathfrak{a}$, fix an ordering on the set $\Phi(\mathfrak{g}_c, \mathfrak{h}_c)$ of roots of $\mathfrak{g}_c$ with respect to $\mathfrak{h}_c$ compatible with $\Phi(P, A)$ and let W^P be the usual canonical set of representatives of right classes of the Weyl group $W(\mathfrak{g}_c, \mathfrak{h}_c)$ of $\mathfrak{g}_c$ with respect to $\mathfrak{h}_c$ modulo the Weyl group $W(\mathfrak{m}_c\oplus\mathfrak{a}_c, \mathfrak{h}_c)$ of $\mathfrak{m}_c\oplus\mathfrak{a}_c$ with respect to $\mathfrak{h}_c$. Furthermore let $\lambda-\varrho$ be the highest weight of r, where $\lambda\in\mathfrak{h}^*_c$ is dominant and 2ϱ is the sum of the positive roots. Since μ is real, [5: III, 3.3] shows that for

$\mathrm{Ext}^{\cdot}_{(\mathfrak{g},K)}(E^*, I^{\infty}_{P,\omega,i\mu})$ not to be zero, first there must exist $s \in W^P$ such that

$$s(\lambda)|_{\mathfrak{a}^*} = 0, \ \chi_{-s(\lambda)}|_{\mathfrak{b}_c} = \chi_\omega; \tag{1}$$

this condition is independent of μ and satisfied by at most one $s \in W^P$. Furthermore, we must also have

$$\mu = 0. \tag{2}$$

By assumption, P is cuspidal, hence for the first equality of (1) to hold, it is necessary that P be fundamental [5: III, 5.1].

2.6. Lemma. *Assume that* $\mathrm{Ext}^{\cdot}_{(\mathfrak{g},K)}(E^*, I^{\infty}_{P,\omega,0})=0$. *Then*

$$\mathrm{Ext}^{\cdot}_{(\mathfrak{g},K)}(E^*, L^{\infty}_{P,\omega}) = 0. \tag{1}$$

Recall that the K-types of $I^{\infty}_{P,\omega,i\mu}$ are independent of μ. If $I_{P,\omega,i\mu,J}=0$ for some μ, then it is so for all μ's and $C^{\cdot}(\mathfrak{g}, K; L_{P,\omega} \otimes E)=0$, which obviously yields (1). Assume now that $I_{P,\omega,i\mu,J} \neq 0$, hence that

$$C^{\cdot}(\mathfrak{g}, K; I^{\infty}_{P,\omega,i\mu} \otimes E) \neq 0 \quad (\mu \in \mathfrak{a}^*). \tag{2}$$

In view of the assumption of 2.6 and of the results recalled in 2.5, we have

$$\mathrm{Ext}^{\cdot}_{(\mathfrak{g},K)}(E^*, I^{\infty}_{P,\omega,i\mu}) = 0 \quad (\mu \in \mathfrak{a}^*). \tag{3}$$

By [5: II, 3.1], we deduce from (2) and (3):

$$c(I_{P,\omega,i\mu}) - c(E) \neq 0 \quad (\mu \in \mathfrak{a}^*). \tag{4}$$

It follows from 1.3(1) that, given a constant $d>0$, there exists a compact set $D \subset \mathfrak{a}^*$ such that $|c(I_{P,\omega,i\mu}) - c(E)| \geqq d$ outside D. Since $c(I_{P,\omega,\lambda})$ is a continuous function of λ, we see that there exists $c>0$ such that

$$|c(I_{P,\omega,i\mu}) - c(E)| \geqq c \quad \text{for all} \quad \mu \in \mathfrak{a}^*. \tag{5}$$

The constant $c(I_{P,\omega,i\mu})$ is also the eigenvalue of the Casimir operator on $(I^*_{P,\omega,i\mu} \hat{\otimes} I_{P,\omega,i\mu})^\infty$, it being understood that G acts only on the second factor. The Casimir operator $C_\mathfrak{g}$ operates therefore on $L^{\infty}_{P,\omega}$ by the rule

$$C_\mathfrak{g} f(\mu) = c(I_{P,\omega,i\mu}) f(\mu), \quad (f \in L^{\infty}_{P,\omega}; \mu \in \mathfrak{a}^*).$$

From (5), we see then that $(C_\mathfrak{g} - c(E) \cdot I)$ has a bounded inverse on $L^{\infty}_{P,\omega}$. Therefore 2.5 follows from 5.2 in [3].

2.7. Proposition. *Assume that P is not fundamental. Then*

$$\mathrm{Ext}^{\cdot}_{(\mathfrak{g},K)}(E^*, L^{\infty}_{P,\omega})=0.$$

In fact, as recalled in 2.5, the assumption of 2.6 is satisfied if P is not fundamental.

2.8. Proposition. *Assume P to be fundamental. Then either*

$$\mathrm{Ext}^{\cdot}_{(\mathfrak{g},K)}(E^*, L^{\infty}_{P,\omega})=0$$

or $\bar{d}$ *does not have closed range.*

If 2.5(1) is not fulfilled, then $\mathrm{Ext}^{\cdot}_{(\mathfrak{g},K)}(E^*, L^{\infty}_{P,\omega})=0$ by 2.6. Assume then 2.5(1) to hold. Let

(1) $$L'_{P,\omega} = \int^{\oplus}_{\mathfrak{a}^*} I_{P,\omega,i\mu}\, d\mu_{\omega}.$$

As recalled in 1.1, there is a canonical K-isomorphism

(2) $$\alpha: I_{P,\omega,i\mu} \xrightarrow{\sim} U = \mathrm{Ind}^{K}_{K\cap M}(V_{\omega}) \quad (\mu\in\mathfrak{a}^*).$$

From this we get a canonical injective $(K\times G)$-homomorphism

(3) $$U_{(K)}\otimes L'_{P,\omega} \to L_{P,\omega}$$

where $U_{(K)}$ is the space of K-finite vectors in U, and a K-equivariant homomorphism

(4) $$\beta\colon \mathrm{Ext}^{\cdot}_{(\mathfrak{g},K)}(E^*, U_{(K)}\otimes L'^{\infty}_{P,\omega}) = U_{(K)}\otimes \mathrm{Ext}^{\cdot}_{(\mathfrak{g},K)}(E^*, L'^{\infty}_{P,\omega}) \to \mathrm{Ext}^{\cdot}_{(\mathfrak{g},K)}(E^*, L^{\infty}_{P,\omega}).$$

We claim that β is injective. For any $\tau\in\hat{K}$, the space U_{τ} is finite dimensional. Let us denote by ${}_{\tau}L_{P,\sigma}$ the isotypic component of type τ for the left action of K, i.e., on the factors $I^*_{P,\omega,i\mu}$. It is clear from the definitions that α induces a $(K\times G)$-isomorphism of $U_{\tau}\otimes L'_{P,\omega}$ onto ${}_{\tau}L_{P,\omega}$. We have an isomorphism

$$\mathrm{Ext}^{\cdot}_{(\mathfrak{g},K)}(E^*, U_{(K)}\otimes L'^{\infty}_{P,\omega}) = \oplus_{\tau\in\hat{K}}\, \mathrm{Ext}^{\cdot}_{(\mathfrak{g},K)}(E^*, U_{\tau}\otimes L'_{P,\omega}).$$

Let $\eta\in C^{\cdot}(\mathfrak{g}, K; U_{(K)}\otimes L'^{\infty}_{P,\omega}\otimes E)$ be a cocycle. We may write

$$\eta = \textstyle\sum_{\tau\in\hat{K}}\eta_{\tau},$$

where η_{τ} is a cocycle in $C^{\cdot}(\mathfrak{g}, K; U_{\tau}\otimes L'^{\infty}_{P,\omega}\otimes E)$. Let F be the set of τ's for which $\eta_{\tau}\neq 0$. It is finite. Let $\varphi_{\tau}\in C^{\infty}_c(K)$ be the function which defines the projector of any continuous K-module onto its τ-isotypic component and let $\varphi_F=\sum_{\tau\in F}\varphi_{\tau}$. Assume now that $\alpha(\eta)=d\mu$ for some $\mu\in C^{\cdot}(\mathfrak{g}, K; L^{\infty}_{P,\omega}\otimes E)$. Then we have

$$\alpha_F * \eta = \eta = d(\alpha_F * \mu),$$

since the operation of K on the left commutes with differentiation. The element $\alpha_F*\mu$ is contained in $C^{\cdot}(\mathfrak{g}, K; {}_{F}L^{\infty}_{P,\omega}\otimes E)$, which can be identified to the image under α of $C^{\cdot}(\mathfrak{g}, K; U_F\otimes L'^{\infty}_{P,\omega}\otimes E)$. Therefore η is already cohomologous to zero in the latter space, which proves our contention.

For any finite dimensional subspace W of $U_{(K)}$ we have

(5) $$\mathrm{Ext}^{\cdot}_{(\mathfrak{g},K)}(E^*, W\otimes L'^{\infty}_{P,\omega}) = W\otimes \mathrm{Ext}^{\cdot}_{(\mathfrak{g},K)}(E^*, L'^{\infty}_{P,\omega}).$$

By the same proof as that of 3.4 in [2], one shows:

(6) $$\mathrm{Ext}^{\cdot}_{(\mathfrak{g},K)}(E^*, L'^{\infty}_{P,\omega}) = \left(\mathrm{Ext}^{\cdot}_{(\mathfrak{m},K\cap M)}(F^*_{s\lambda|\mathfrak{b}_c}, V^{\infty}_{\omega})\otimes H^{\cdot}\left(\mathfrak{a}, \int^{\oplus}_{\mathfrak{a}^*} C_{i\mu}\, d\mu_{\omega}\right)\right)[-(\dim N)/2]$$

where $\lambda \in \mathfrak{a}_c^*$ is dominant such that $\chi_\lambda = \chi_r$. [The only difference is that $d\mu_\omega$ replaces the Lebesgue measure, but this does not affect the argument.] Since we assume the left-hand side to be non-zero, the first factor of the right-hand side is not zero. We claim that, as in [2: 3.2], we have

$$H^0\left(\mathfrak{a}; \int_{\mathfrak{a}^*}^{\oplus} C_{i\mu}\, d\mu_\omega\right) = 0, \tag{7}$$

$$\dim H^i\left(\mathfrak{a}; \int_{\mathfrak{a}^*}^{\oplus} C_{i\mu}\, d\mu_\omega\right) = \infty \quad (i = 1, \ldots, \dim \mathfrak{a}),\ \bar{d} \textit{ is not closed.} \tag{8}$$

The group P is fundamental, therefore $d\mu_\omega$ is the product of the Lebesgue measure $d\mu$ by a polynomial, say R, which is strictly positive on the regular elements [8: § 24, Theorem 1]. From this (7) follows as in *loc. cit.* As regards (8), it is enough to prove it if the direct integral is taken over some measurable set D of strictly positive measure. Take for instance for D the positive Weyl chamber. Let $R^{1/2}$ be the positive square root of R on D. Then $\varphi \mapsto R^{1/2}\varphi$ defines an equivariant isomorphism

$$\int_D^{\oplus} C_{i\mu}\, d\mu_\omega \xrightarrow{\sim} \int_D^{\oplus} C_{i\mu}\, d\mu \tag{9}$$

which reduces us to [2: 3.2]. Since the map β of (4) is injective, 2.8 follows.

Remarks. (1) The first factor on the right-hand side of (6) is non-zero only in the middle dimension $m_0 = (\dim M/(K_M \cap M))/2$ [5: II, 5.3]. Let moreover $l_0 = \dim A$ (i.e., $l_0 = \mathrm{rk}\, G - \mathrm{rk}\, K$). Since $2m = 2m_0 + l_0 + \dim N$, we get:

$$\dim \mathrm{Ext}^i_{(\mathfrak{g}, K)}(E^*, L'^{\infty}_{P,\omega}) = \begin{cases} \infty & i \in (m-(l_0/2), m+(l_0/2)] \\ 0 & i \notin (m-(l_0/2), m+(l_0/2)], \end{cases}$$

assuming that $\mathrm{Ext}^{\cdot}_{(\mathfrak{g}, K)}(E^*, L'^{\infty}_{P,\omega}) \neq 0$.

(2) I do not know whether β is also surjective. In particular, is $\mathrm{Ext}^i_{(\mathfrak{g}, K)}(E^*, L'^{\infty}_{P,\omega})$ zero outside the interval $(m-(l_0/2), m+(l_0/2)]$?

(3) We already pointed out that the proof of 3.4 in [2] is also valid if the $L_{P,V}$ there is replaced by our $L'_{P,\omega}$. In the same way, 3.5 and 3.6 in [2] and their proofs also hold under that change. In particular, the implication (iv)$\Rightarrow$(ii) of 3.6 shows that $\mathrm{Ext}_{(\mathfrak{g}, K)}(E^*, L'^{\infty}_{P,\omega}) = 0$ if E is not equivalent to $\bar{E}^*$, for any $\omega \in \hat{M}_d$.

In case $E \sim \bar{E}^*$, we want now to show the existence of some $\omega \in \hat{M}_d$ for which $\mathrm{Ext}_{(\mathfrak{g}, K)}(E^*, L'^{\infty}_{P,\omega}) \neq 0$. For this we need to bring a complement to 5.5, 5.7 of [5: II, § 5].

2.9. Remark on [5: II, § 5]. In 5.3, *loc. cit.* it is proved that if M is a connected linear semi-simple group, L a maximal compact subgroup of M, F an irreducible dimensional representation of M, and V a discrete series representation of M, then

$$\mathrm{Ext}^{\cdot}_{(\mathfrak{m}, L)}(F, V) = 0 \quad \text{if} \quad \chi_V \neq \chi_F; \tag{1}$$

$$\dim \mathrm{Ext}^i_{(\mathfrak{m}, L)}(F, V) = \delta_{i,q} \quad (q = (\dim M/L)/2,\ i \in \mathbf{Z}) \quad \text{if} \quad \chi_V = \chi_F. \tag{2}$$

It is then shown (5.5, 5.7) that if M is reductive, with compact center, and F, V are as before, then

$$\dim \operatorname{Ext}^i_{(\mathfrak{m}, L)}(F, V) \leqq \delta_{i,q} \quad (i \in \mathbf{Z}). \tag{3}$$

We want now to point out that if $M = M_P$ as above, with P cuspidal in G, then, given F, there exists V in the discrete series of M such that

$$\operatorname{Ext}^q_{(\mathfrak{m}, L)}(F, V) = \mathbf{C}. \tag{4}$$

Let first M be any connected reductive group with compact center. It is then the almost direct product of a semi-simple group M' by a torus T. The representation F is the tensor product of an irreducible representation F' of M' by a one-dimensional representation $\mathbf{C}_\lambda$ of T. Fix a discrete series representation V' of M' with infinitesimal character equal to $\chi_{F'}$. Then, by known results on the L-weights of V', the characters of $T \cap M'$ given by V' and F' are the same. Therefore $V' \otimes \mathbf{C}_\lambda$ is also a representation of M, hence an element V of the discrete series of M. Using the Künneth rule, one sees immediately that

$$\operatorname{Ext}^{\cdot}_{(\mathfrak{m}, L)}(F, V) = \operatorname{Ext}^{\cdot}_{(\mathfrak{m}', L \cap M')}(F', V'), \tag{5}$$

and we are reduced to (2) above, taking into account the fact that

$$M/L = M'/(M' \cap L).$$

Let now $M = M_P$, with P cuspidal in G. We claim that M is the direct product of M^0 by a finite elementary abelian 2-group, say Z. By our standing assumption G is linear. Let G_c be its complexification. It is an algebraic $\mathbf{R}$-group. The group P is of finite index in the group of real points of the parabolic $\mathbf{R}$-subgroup $\mathscr{P}$ of G_c with Lie algebra $\mathfrak{p}_c$, and A is the identity component, in ordinary topology, of the group of real points $\mathscr{A}(\mathbf{R})$ of the maximal $\mathbf{R}$-split torus $\mathscr{A}$ of the radical of $\mathscr{P}$ with Lie algebra $\mathfrak{a}_c$. The group $\mathscr{A}(\mathbf{R})$ is the direct product of A by an elementary abelian 2-group Z_0, the group of elements of order $\leqq 2$ of $\mathscr{A}(\mathbf{R})$. By a result of Matsumoto (see [4: § 14]), R meets every connected component of $\mathscr{P}(\mathbf{R})$. Since $N \cdot A$ is connected, and Z_0 centralizes A, this implies immediately our assertion, with $Z = Z_0 \cap M$.

The representation F is the tensor product of an irreducible representation F^0 of M^0 by a one-dimensional representation $\mathbf{C}_\sigma$ of Z. By the previous argument we may find a discrete series representation V^0 of M^0 such that

$$\operatorname{Ext}^{\cdot}_{(\mathfrak{m}, L^0)}(F^0, V^0) \neq 0 \tag{6}$$

where $L^0 = M^0 \cap L$. We then take $V = V^0 \otimes \mathbf{C}_\sigma$. Since Z is central, it acts trivially on $\Lambda^{\cdot}\mathfrak{m}/\mathfrak{l}$, from which it follows that

$$C^{\cdot}(\mathfrak{m}, L; F \otimes V) = C^{\cdot}(\mathfrak{m}, L^0; F^0 \otimes V^0), \tag{7}$$

whence

(8) $$\operatorname{Ext}^{\cdot}_{(\mathfrak{m}, L)}(F, V) = \operatorname{Ext}^{\cdot}_{(\mathfrak{m}, L^0)}(F^0, V^0),$$

and our assertion.

2.10. Remark. We take this opportunity to correct an oversight in [2]: In the proof of 3.7, we apply [5: II, 5.3] to $\operatorname{Ext}^{\cdot}_{(\mathfrak{m}, K\cap M)}(F^*, V)$ although M is not necessarily connected semi-simple. But $M = M_P$ with P fundamental and G is semi-simple, linear, so that the above holds. Also F^* there stands for $F^*_{s\lambda|\mathfrak{b}_c}$.

3. Proof of Theorems A and B

3.1. By [1: 3.5], there is a canonical inclusion

(1) $$C^{\cdot}(\mathfrak{g}, K; L^2(G)^{\infty} \otimes E) \to A^{\cdot}_{(2)}(K; E),$$

which induces an isomorphism in cohomology. Let us denote by d (resp. d_X) the differential on the left (resp. right)-hand side. The left-hand side is contained in the graded Hilbert space

(2) $$C^{\cdot}(\mathfrak{g}, K; L^2(G) \otimes E) = \operatorname{Hom}_K(\Lambda^{\cdot}\mathfrak{g}/\mathfrak{k}, L^2(G) \otimes E).$$

Let $\bar{d}$ be the closure of d. Then (1) extends to an isomorphism of the graded Hilbert space $C^{\cdot}(\mathfrak{g}, K; L^2(G) \otimes E)$ onto the space of L^2-forms on X with measurable coefficients, which maps dom $\bar{d}$ onto dom $\bar{d}_X$ and $\bar{d}$ onto $\bar{d}_X$ [1: 3.6]. We are therefore reduced to the discussion of $\operatorname{Ext}^{\cdot}_{(\mathfrak{g}, K)}(E^*, L^2(G)^{\infty})$ and of the range of $\bar{d}$.

3.2. As in 2.1, let J be the set of K-types occurring in $\Lambda^{\cdot}(\mathfrak{g}/\mathfrak{k}) \otimes E$. By 1.4, we can write $L^2(G) = Q \oplus R$, where Q is a sum of finitely many elementary subspaces, R the orthogonal complement to Q and $R_J = 0$. In view of 2.1(2), which is valid for any continuous G-module, we have then

(1) $$\operatorname{Ext}^{\cdot}_{(\mathfrak{g}, K)}(E^*, L^{\infty}_{P,\omega}) = \operatorname{Ext}^{\cdot}_{(\mathfrak{g}, K)}(E^*, R^{\infty}) = 0 \quad (L_{P,\omega} \subset R),$$

since the complexes which give rise to these cohomology spaces are already zero. We have therefore

(2) $$\operatorname{Ext}^{\cdot}_{(\mathfrak{g}, K)}(E^*, L^2(G))^{\infty} = \oplus_{P \in S, \omega \in \hat{M}_{P,d}} \operatorname{Ext}^{\cdot}_{(\mathfrak{g}, K)}(E^*, L^{\infty}_{P,\omega}),$$

where the sum on the right-hand has at most finitely many non-zero terms. We can now use the results of Section 2.

3.3. Let first G and K be of equal rank. Then G is its own fundamental parabolic subgroup and has a discrete series. Theorem A now follows from 3.2(2) and 2.3, 2.7.

3.4. Let now $l_0 = \operatorname{rk} G - \operatorname{rk} K$ be $\neq 0$. By 2.7 we may, on the right-hand side, restrict the summation to the unique fundamental parabolic subgroup of G contained in S. Since $L_{P,\omega}$ is unitary, it is standard that $\mathrm{Ext}^{\cdot}_{(\mathfrak{g},K)}(E^*, L^{\infty}_{P,\omega})=0$ if $E \nsim \bar{E}^*$, which proves Theorem B in that case. So assume $E \sim \bar{E}^*$. In view of the injectivity of β in 2.8 and of the remark to 2.8, it suffices, to conclude the proof of Theorem B, to show the existence of $\omega \in \hat{M}_{P,d}$ such that $\mathrm{Ext}^{\cdot}_{(\mathfrak{g},K)}(E^*, L'_{P,\omega}) \neq 0$. We use the notation of 2.5. Since P is fundamental, there exists $s \in W^P$ such that 2.5(1) is satisfied [2: 3.6]. Then $\mathrm{Ext}^{\cdot}_{(\mathfrak{g},K)}(E^*, L'^{\infty}_{P,\omega})$ is given by 2.8(6), and it is therefore enough to show the existence of $\omega \in \hat{M}_{P,d}$ such that $\mathrm{Ext}^{\cdot}_{(\mathfrak{m},K\cap M)}(F^*_{s\lambda|\mathfrak{b}_c}, V^{\infty}_{\omega}) \neq 0$. But this follows from 2.9.

References

[1] BOREL, A.: Regularization theorems in Lie algebra cohomology. Applications. - Duke Math. J. 50, 1983, 605—623.

[2] BOREL, A., and W. CASSELMAN: L^2-cohomology of locally symmetric manifolds of finite volume. - Duke Math. J. 50, 1983, 625—647.

[3] BOREL, A., and H. GARLAND: Laplacian and the discrete spectrum of an arithmetic group. - Amer. J. Math. 105, 1983, 309—335.

[4] BOREL, A., et J. TITS: Groupes réductifs. - Inst. Hautes Études Sci. Publ. Math. 27, 1965, 55—151.

[5] BOREL, A., and N. WALLACH: Continuous cohomology, discrete subgroups, and representations of reductive groups. - Ann. of Math. Stud. 94, Princeton Univ. Press, Princeton, N. J., 1980, 1—387.

[6] CHEEGER, J.: On the Hodge theory of Riemannian pseudomanifolds. - Proceedings of Symposia in Pure Mathematics 36, American Mathematical Society, Providence, R. I., 36, 1980, 91—146.

[7] HARISH-CHANDRA: Harmonic analysis on real reductive groups, I. - J. Funct. Anal. 19, 1975, 104—204.

[8] HARISH-CHANDRA: Harmonic analysis on real reductive groups, III. The Maass—Selberg relations and the Plancherel formula. - Ann. of Math. 104, 1976, 117—201.

[9] HECHT, H., and W. SCHMID: A proof of Blattner's conjecture. - Invent. Math. 31, 1976, 129—154.

[10] HOTTA, R.: On realization of the discrete series for semisimple Lie groups. - J. Math. Soc. Japan 23, 1971, 384—407.

[11] LANGLANDS, R. P.: On the functional equations satisfied by Eisenstein series. - Lecture Notes in Mathematics 544. Springer-Verlag, Berlin—Heidelberg—New York, 1976, 1—337.

Institute for Advanced Study
School of Mathematics
Princeton, New Jersey 08540
USA

Eidgenössische Technische Hochschule
CH—8092 Zürich
Switzerland

Received 7 November 1983

130.

On Affine Algebraic Homogeneous Spaces

Archiv d. Math. 45 (1985) 74–78

1. Affine homogeneous spaces. In this Note, algebraic varieties are reduced, defined over an algebraically closed groundfield k. We let G denote a connected algebraic affine algebraic group, X an algebraic homogeneous space of G and H an isotropy group of G on X. We are interested in the case where X is an affine variety or is isomorphic to an affine space. We shall prove:

1.1 Theorem. (i) *If X is an affine variety, then* $\dim R_u G \geqq \dim R_u H$.
1 (ii) *If X is isomorphic to an affine space, then $R_u G$ is transitive on X.*

Here and below we use in general the notation of [2]. In particular, if M is an affine algebraic group and N a closed subgroup, then M^0 is the identity component of M and M/N the quotient variety of left cosets of M modulo N. By definition, we let the unipotent radical $R_u M$ of M be equal to that of M^0.

The second assertion answers a question raised by A. H. Andersen. The first one implies that if G is reductive and G/H is affine, then H^0 is reductive. This is known (as well as a converse: G/H is affine if G and H are reductive) and several proofs have been given (see [4; 8]). The first valid in arbitrary characteristic was a transcription in étale cohomology of the topological argument given over $\mathbb{C}$ in [3], which was pointed out to me at the time by A. Grothendieck and is alluded to in [1: 7.10]. We shall use here a dual argument making use of étale cohomology with proper supports. Either version is very simple, the only problem being to find references for some foundational material on the cohomology theory used. This is a bit easier for étale cohomology with proper supports than for the usual one, whence the shift to the former.

We fix a prime l prime to the characteristic p of k. Given a smooth variety Y, we let $H_c^{\cdot}(Y)$ denote the group of étale cohomology of Y with proper supports and coefficients in $\mathbb{Z}/\ell\mathbb{Z}$ (cf. Section 2 for references) and set

$$(1) \qquad m(Y) = \min\{j \mid H_c^j(Y) \neq 0\}.$$

In Section 2 we shall recall some basic facts on étale cohomology and prove (see 2.8)

$$(2) \qquad m(X) = \dim X + \dim R_u G - \dim R_u H.$$

From this we deduce

1.2 Proposition. (i) *If* $m(X) \geqq \dim X$, *then* $\dim R_u G \geqq \dim R_u H$.
(ii) *If* $m(X) = 2 \cdot \dim X$, *then* $R_u G$ *is transitive on* X.

Proof. (i) follows trivially from 1.1 (2). Under the assumption of (ii), 1.1 (2) implies

$$\dim X = \dim R_u G - \dim R_u H.$$

Now the orbit of the origin in X under $R_u G$ is the image under a bijective morphism of $R_u G/(R_u G \cap H)$. Since $(R_u G \cap H)^0 \subset R_u H$, the orbit has then dimension $\geqq \dim R_u G - \dim R_u H$, hence $R_u G$ is transitive.

1.3 If now X is an affine variety, then $m(X) \geqq \dim X$ and if X is an affine space, then $m(X) = 2 \cdot \dim X$, (cf. 2.2, 2.3). Therefore 1.1 is a consequence of 1.2. There remains then to prove 1.1 (2). Before going over to étale cohomology, we draw some consequences of 1.2 and of some known facts.

1.4 Theorem. *The following conditions on the homogeneous space* X *are equivalent:*

(i) X *is isomorphic to an affine space.*
(ii) $m(X) = 2 \cdot \dim X$.
(iii) X *is homogeneous with respect to a connected unipotent group.*

Proof. That (i) ⇒ (ii) is recalled in 2.3. The implication (ii) ⇒ (iii) follows from 1.2. Finally, (iii) ⇒ (i) follows from Theorem 5, p. 119 in [9].

1.5 Theorem. *Let* G *be reductive. Then the following conditions are equivalent:*

(i) H^0 *is reductive.*
(ii) X *is an affine variety.*
(iii) $m(X) \geqq \dim X$.

We note first that we can replace X by G/H. In fact if $\alpha: Y \to Z$ is a bijective morphism of smooth varieties, then Y is affine if and only if Z is so (see e.g. [2; AG, 18.31]) and $m(Y) = m(Z)$ (2.4).

That (i) ⇒ (ii) is well-known, and proved e.g. in [4; 8]. From 2.2, we have that (ii) ⇒ (iii). Finally (iii) ⇒ (i) by 1.2.

1.6 Remark. Let H be unipotent, connected. By 1.2, the condition $\dim H \leqq \dim R_u G$ is necessary for G/H to be affine. However it is not sufficient, as is shown by Example 5.5 in [4]. On the other hand, the condition $H \subset R_u G$ is sufficient for G/H to be affine, as follows from 4.6 in [4], but the same example 5.5 shows that this condition is not necessary. A group theoretical characterisation of the unipotent H's for which G/H is affine does not seem to be known (except in characteristic zero for H of dimension one [4: 5.3]).

2. Étale cohomology with proper supports.

2.1 As before, l is a prime number, prime to the characteristic of k. Let M be a smooth variety, d its dimension and F a torsion sheaf of $\mathbb{Z}/(l)$-modules on M. We let $H^i(M; F)$

(resp. $H^i_c(M;F)$) denote the i-th étale cohomology group (resp. with proper supports) of M with coefficients in F, and $H^{\cdot}(M;F)$ (resp. $H^{\cdot}_c(M;F)$) the direct sum of these groups ($i \in \mathbb{Z}$). *If* F is the constant sheaf with stalk $\mathbb{Z}/(l)$, we simply write $H^i(M)$ and $H^i_c(M)$. We refer to [5; 6] for a discussion of these groups and recall here the properties needed in this Note.

2.2 The group $H^i(M;F)$ vanish for $i \notin [0, 2d]$ (see [6: VI, 1.1]). If M is an affine variety, they vanish for $i > d$. [6: VI, 7.2]. Let $\mathcal{O}$ be the orientation sheaf of M [5: VI, n° 1], i.e. the tensor product of d copies of the sheaf μ_l of l-th roots of unity on M. There exists a non-degenerate pairing

$$H^i_c(M) \times H^{2d-i}(M;\mathcal{O}) \to \mathbb{Z}/(l) \quad (i \in \mathbb{Z}) \tag{1}$$

[6: VI, 11.2]. We have therefore in particular

$$H^i_c(M) = 0 \quad \text{for } i < d \quad \text{if } M \text{ is affine,} \tag{2}$$

$$H^{2d}_c(M) = \mathbb{Z}/(l) \quad \text{if } M \text{ is orientable.} \tag{3}$$

2.3 If $M = \mathbb{A}^d$ is isomorphic to the affine space of dimension d, then

$$H^i_c(M) = 0 \quad i \neq 2d \quad \text{and } H^{2d}_c(M) = \mathbb{Z}/(l). \tag{1}$$

For $d = 1$ this follows from the long exact sequence in cohomology with proper supports [6: III, 1.13] of the projective line $\mathbb{P}^1$ modulo a point and the known results on $H^{\cdot}(\mathbb{P}^1)$ [6: III. 3.5]. For a general d, one may then use the Künneth rule [6: VI, 8.5]. We could also derive this from 2.2 and the equality $H^i(\mathbb{A}^d) = 0$ for $i > 0$, which follows from [6: 4.20].

2.4 Let N be a smooth variety and $f: M \to N$ a bijective morphism. Then f induces isomorphisms $f^{\cdot}: H^{\cdot}(N) \to H^{\cdot}(M)$ and $f^{\cdot}_c: H^{\cdot}_c(N) \to H^{\cdot}_c(M)$.

In fact, in this case, by the proper base change theorem [6: VI, 2.3] the stalk of the Leray sheaf $Rf_* \mathbb{Z}/(l)$ (resp. $R_c f_* \mathbb{Z}/(l)$) is the cohomology of the fibre, hence reduces to $\mathbb{Z}/(l)$, in dimension zero.

2.5 Proposition. *Let (E_r) be the spectral sequence of the canonical projection $\pi: G \to G/H$, for étale cohomology with proper supports.*

(i) *The term $E_2^{p,q}$ contains a subspace equal to $H^p_c(G/H) \otimes H^q_c(H^0)$ $(p, q \in \mathbb{N})$.*
(ii) *We have the equality $m(G) = m(G/H) + m(H)$.*

Proof. The spectral sequence of π abuts to $H^{\cdot}_c(G)$ and its term E_2 is given by

$$E_2^{p,q} = H^p_c(G/H; R^q_c \pi_* \mathbb{Z}/(l)) \quad (p, q \in \mathbb{N}) \tag{1}$$

where $R^q_c \pi_* \mathbb{Z}/(l)$ is the q-th Leray sheaf of π, for the cohomology under consideration. (The notation $R^q_c \pi_*$ is borrowed from [6]. In [5] $R^q \pi_!$ is used.) By the proper base change theorem [6: VI, 2.3]:

$$(R^q_c \pi_* \mathbb{Z}/(l))_y = H^q_c(\pi^{-1} y), \quad (y \in G/H; q \in \mathbb{N}).$$

The fibration of G by H is locally trivial in the étale topology and has H as structural group, acting on itself by right translations. Therefore the stalks of the Leray sheaf form a locally constant system, with fibre $H_c^{\cdot}(H)$ and structural group the action of H induced by right translations. This action is trivial for H^0. The space $H_c^{\cdot}(H)$ is a direct sum of the cohomology of the connected components xH^0, which are permuted by H/H^0, so that we may identify $H_c^{\cdot}(H)$, as an H/H^0-module, with $\mathrm{Ind}_{H^0}^{H}(H_c^{\cdot}(H^0))$, where $H_c^{\cdot}(H^0)$ is viewed as a trivial H^0-module. It contains therefore a subspace isomorphic to $H_c^{\cdot}(H^0)$ invariant under H/H^0. The latter defines a constant subsheaf of $R_c^q \pi_* \mathbb{Z}/(l)$ with stalk $H_c^{\cdot}(H^0)$. Then (i) follows from (1).

(ii) That $m(G) \geqq m(G/H) + m(H)$ is obvious from the spectral sequence. On the other hand, by the standard minimal cocycle argument, if $a = m(G/H)$ and $b = m(H)$, then $E_2^{a,b} = E_\infty^{a,b}$. Since $E_2^{a,b} \neq 0$ by (i), this proves the reverse inequality.

2.6 Proposition. *We have the equality*

(1) $$m(G) = \dim G + \dim R_u G.$$

Proof. Let $L = G/R_u G$. By 2.5 and 2.3

(2) $$m(G) = m(L) + 2 \cdot \dim R_u G.$$

Since $\dim G = \dim L + \dim R_u G$, there remains therefore to see that $m(L) = \dim L$. Let B be a minimal parabolic subgroup of L and T a maximal torus of B. We have then

(3) $$m(B) = m(T) + 2 \cdot \dim R_u B = \dim L.$$

On the other hand, since L/B is a projective variety, $m(L/B) = 0$, and 2.5 gives $m(L) = m(B)$. Together with (3), this proves our assertion.

2.7 Remark. In the reductive case, a more precise result follows from [7]: If G_0 is a smooth group scheme over $\mathbb{Z}$ such that $G = G_0 \times k$, and K is a maximal compact subgroup of $G_0(\mathbb{C})$, then $H^{\cdot}(G)$ is canonically isomorphic to $H^{\cdot}(K; \mathbb{Z}/(l))$, where the latter refers to singular cohomology, say [7: 5.2]. But K is a connected compact orientable manifold whose topological dimension is equal to the dimension of G. By applying Poincaré duality to both groups, we get

$$H^i(K; \mathbb{Z}/(l)) = H_c^{n+i}(G) \quad (i \in \mathbb{N}; n = \dim G),$$

which implies in particular that $m(G) = n$.

2.8 Proof of 1.1 (2). By 2.4, we may replace X by G/H. We have then

$$m(G/H) = m(G) - m(H)$$

by 2.5 (ii). Using 2.6, we get

$$m(G/H) = \dim G - \dim H + \dim R_u(G) - \dim R_u H$$

and 1.1 (2) follows.

References

[1] A. BOREL, Introduction aux groupes arithmétiques. Act. Sci. Ind. **1341**, Paris 1969.
[2] A. BOREL, Linear algebraic groups (Notes by H. Bass). Math. Lecture Notes Series, New York 1969.
[3] A. BOREL and HARISH-CHANDRA, Arithmetic subgroups of algebraic groups. Ann. of Math. **75**, 485–535 (1962).
[4] E. CLINE, B. PARSHALL and L. SCOTT, Induced modules and affine quotients. Math. Ann. **230**, 1–14 (1947).
[5] P. DELIGNE, Cohomologie étale, les points de départ (rédigé par J-F. Boutot). In: Séminaire de Géométrie algébrique SGA 41/2. LNM **569**, Berlin 1977.
[6] J. S. MILNE, Étale cohomology. Princeton 1980.
[7] M. RAYNAUD, Modules projectifs universels. Invent. Math. **6**, 1–26 (1968).
[8] R. W. RICHARDSON, Affine coset spaces of affine algebraic groups. Bull. London Math. Soc. **91**, 38–41 (1977).
[9] M. ROSENLICHT, Questions of rationality for solvable algebraic groups over nonperfect fields. Ann. Mat. Pura Appl. (IV) **61**, 97–120 (1963).

Eingegangen am 2. 8. 1984

Anschrift des Autors:

A. Borel
ETH Zürich, Mathematik
Rämistr. 101
CH-8092 Zürich

131.

(with W. Casselman)

Cohomologie d'intersection et L^2-cohomologie de variétés arithmétiques de rang rationnel 2

C. R. Acad. Sci. Paris **301** (1985) 369–373

1 Dans cette Note, G désigne le groupe des points réels d'un Q-groupe connexe presque simple sur Q, de rang rationnel $\mathrm{rg}_{\mathbf{Q}}(\mathscr{G}) \geqq 1$ et Γ un sous-groupe arithmétique net ([3], § 17) de $\mathscr{G}$. On suppose que le quotient $X = G/K$ de G par un sous-groupe compact maximal K est un domaine borné symétrique. Soient $V = \Gamma \backslash X$ et V^* la compactification (projective normale) de V construite dans [2]. La résultat principal de cette Note (théorème 1) affirme en particulier que la L^2-cohomologie de V, définie à partir d'une métrique riemannienne G-invariante sur X, est naturellement isomorphe à la cohomologie d'intersection intermédiaire [10] de V^*, lorsque $\mathrm{rg}_{\mathbf{Q}}(\mathscr{G}) \leqq 2$.

TOPOLOGY. — Intersection cohomology and L^2-cohomology of arithmetic varieties of rational rank 2.

In this Note, G *denotes the group of real points of a connected almost* **Q**-*simple* **Q**-*group* $\mathscr{G}$, *of* **Q**-*rank* $\mathrm{rk}_{\mathbf{Q}}(\mathscr{G}) \geqq 1$, *and* Γ *a neat* ([3], § 17) *arithmetic subgroup of* $\mathscr{G}$ *contained in* G. *We assume the quotient* $X = G/K$ *of* G *by a maximal compact subgroup to be a bounded symmetric domain. Let* $V = \Gamma \backslash X$ *and* V^* *the (projective normal) compactification of* V *constructed in* [2]. *The main result of this Note (Theorem* 1) *asserts, among other things, that the* L^2-*cohomology of* V, *defined by means of a* G-*invariant Riemannian metric on* X, *is naturally isomorphic to the middle intersection cohomology* [10] *of* V^* *if* $\mathrm{rk}_{\mathbf{Q}}(\mathscr{G}) \leqq 2$.

1. Soit E l'espace d'une représentation complexe de dimension finie de G, muni d'une structure hermitienne admissible ([8], VII, 2.2). On note aussi E le système local sur V qui lui est associé. On rappelle que V^* est le quotient d'un espace X^* contenant X comme ouvert dense, sur lequel l'action de Γ dans X se prolonge continûment. L'espace X^* est réunion disjointe de X et de « composantes à la frontière rationnelles ». Ces dernières correspondent canoniquement aux **Q**-sous-groupes propres paraboliques maximaux de $\mathscr{G}$. Soit P l'intersection de G avec un tel sous-groupe. La composante à la frontière X_P associée à P est un espace homogène de P et un domaine borné symétrique. L'image $\Gamma(X_P)$ de $\Gamma(P) = \Gamma \cap P$ dans $\mathrm{Aut}\, X_P$ par $\psi : P \to \mathrm{Aut}\, X_P$ est un sous-groupe arithmétique net. L'espace V^* est réunion disjointe de V et de sous-variétés $V_P = \Gamma(X_P) \backslash X_P$ localement fermées correspondant biunivoquement aux classes de conjugaison de Γ dans l'ensemble des **Q**-sous-groupes propres paraboliques maximaux. Les V_P sont les composantes connexes des strates d'une stratification définissant une structure de pseudo-variété sur V^* et possédant $\mathrm{rg}_{\mathbf{Q}}(\mathscr{G})$ strates (propres). On dira que la strate de plus grande dimension (en dehors de V) est de niveau un et, pour $i > 1$, que la plus grande state du complément dans $V^* - V$ des strates de niveaux $< i$ est de niveau i. Le niveau maximum est donc le **Q**-rang de $\mathscr{G}$.

On note $\mathscr{I}\mathscr{C}^{\cdot}(E)$ un complexe de faisceaux sur V^* définissant la cohomologie d'intersection intermédiaire de V^*, au sens de Goresky-MacPherson, à coefficients dans E [10].

2. Étant donnée une variété riemannienne indéfiniment différentiable M et un système local F d'espaces vectoriels hermitiens sur M, on désigne par $\Omega^{\cdot}_{(2)}(M; F)$ le complexe des formes différentielles η indéfiniment différentiables sur M, à coefficients dans F, telles que η et $d\eta$ soient de carré intégrable. Par définition, $H^{\cdot}(\Omega^{\cdot}_{(2)}(M; F))$ est la L^2-cohomologie de M, à valeurs dans F.

Fixons sur X une métrique riemannienne G-invariante. Cela définit une métrique riemannienne complète sur V. On note $\mathscr{L}^{\cdot}(E)$ le préfaisceau différentiel sur V^* qui associe à un ouvert U le complexe $\Omega^{\cdot}_{(2)}(U \cap V; E)$ et $\mathscr{L}^{\cdot}_{\mathrm{loc}}(E)$ le complexe de faisceaux associé. Il est fin [13].

0249-6321/85/030100369 $ 2.00 © Académie des Sciences

THÉORÈME 1. — *Le complexe de faisceaux* $\mathscr{L}^{\bullet}_{\mathrm{loc}}(E)$ *est quasi-isomorphe à* $\mathscr{I}\mathscr{C}^{\bullet}(E)$ *le long des strates de niveaux* $i=1, 2$. *En particulier* $\mathscr{L}^{\bullet}_{\mathrm{loc}}(E)$ *est quasi-isomorphe à* $\mathscr{I}\mathscr{C}^{\bullet}(E)$ *si de plus* $\mathrm{rg}_{\mathbf{Q}}(\mathscr{G})\leqq 2$.

Dans [13], S. Zucker a conjecturé que $\mathscr{L}^{\bullet}_{\mathrm{loc}}(E)$ est quasi-isomorphe à $\mathscr{I}\mathscr{C}^{\bullet}(E)$ sans restriction sur $\mathrm{rg}_{\mathbf{Q}}(\mathscr{G})$ et l'a vérifié dans quelques cas où $\mathrm{rg}_{\mathbf{Q}}(\mathscr{G})=1$. La validité de cette conjecture sous cette dernière condition a été annoncée dans [4]. Dans [14], Zucker démontre l'énoncé précédent lorsque $G=\mathbf{Sp}(2n, \mathbf{R})$, E quelconque, et $G=\mathbf{SU}(p, 2)$, $E=\mathbf{C}$.

Dans la suite de cette Note, nous indiquons quelques-uns des résultats intermédiaires intervenant dans la démonstration de ce théorème. Jusqu'à nouvel avis, nous revenons aux hypothèses générales du début de cette Note.

3. RÉGULARISATION. — Soit $\mathscr{L}^{\bullet}_{\infty}(E)$ le préfaisceau différentiel gradué sur V^* qui associe à tout ouvert U de V^* le complexe :

$$\mathrm{Hom}^{\bullet}_{K}(\Lambda^{\bullet}(\mathfrak{g}/\mathfrak{k}), L^{2,\infty}(\pi^{-1}(U\cap V))\otimes E),$$

où l'on pose, pour tout ouvert Y de $\Gamma\backslash G$:

$$L^{2,\infty}(Y)=\{f\in C^{\infty}(Y)\,|\,R_X f\in L^2(Y), (X\in U(\mathfrak{g}))\}.$$

En utilisant un procédé bien connu ([8], VII), on peut identifier canoniquement $\Omega^{\bullet}_{(2)}(U\cap V; E)$ à un sous-espace de $\mathrm{Hom}_K(\Lambda(\mathfrak{g}/\mathfrak{k}), L^2(\pi^{-1}(U\cap V))\otimes E)$, donc $\mathscr{L}^{\bullet}_{\infty}(E)$ à un sous-faisceau différentiel gradué de $\mathscr{L}^{\bullet}(E)$.

Soit $\mathscr{L}^{\bullet}_{\infty,\mathrm{loc}}(E)$ le faisceau différentiel gradué associé à $\mathscr{L}^{\bullet}_{\infty}(E)$. Il est contenu dans $\mathscr{L}^{\bullet}_{\mathrm{loc}}(E)$. Le procédé de régularisation de [5], convenablement localisé, permet de démontrer la :

PROPOSITION 1. — *L'inclusion* $\mathscr{L}^{\bullet}_{\infty,\mathrm{loc}}(E)\to\mathscr{L}^{\bullet}_{\mathrm{loc}}(E)$ *est un quasi-isomorphisme.*

4. AUTODUALITÉ. — Le complexe $\mathscr{D}^{\bullet}(E)=\mathscr{L}^{\bullet}_{\infty}(E)(U\cap V)$ est un complexe borné d'espaces de Fréchet réflexifs. Soit $\mathscr{C}^{\bullet}$ le complexe dual, tel que $\mathscr{C}^{i}$ soit le dual fort de $\mathscr{D}^{n-i}(E)$ ($n=\dim_{\mathbf{R}}X$, $i\in\mathbf{N}$), les différentielles de $\mathscr{C}^{\bullet}$ étant les transposées de celles de $\mathscr{D}^{\bullet}(E)$. Soient d'autre part $\mathscr{D}^{\bullet}_{c}(E)$ le sous-complexe de $\mathscr{D}^{\bullet}(E)$ formé des éléments dont le support est l'intersection de V avec un compact de U et E^* la représentation contragrédiente de E. Le produit extérieur de formes différentielles et l'accouplement de E et E^* permettent de définir une inclusion $\mathscr{D}^{\bullet}_{c}(E^*)\to\mathscr{C}^{\bullet}$. On a la :

PROPOSITION 2. — *L'inclusion* $\mathscr{D}^{\bullet}_{c}(E^*)\to\mathscr{C}^{\bullet}$ *est un quasi-isomorphisme.*

On en déduit en particulier que si $H^i(\mathscr{D}^{\bullet}(E))=0$ pour $i\geqq c$ ($c\in\mathbf{N}$), alors $H^j(\mathscr{D}^{\bullet}_{c}(E^*))=0$ pour $j\leqq n-c$, ce qui nous suffira. Plus généralement, si $H^i(\mathscr{D}^{\bullet}(E))$ est de dimension finie pour tout i, alors $H^{n-i}(\mathscr{D}^{\bullet}_{c}(E^*))$ est le dual de $H^i(\mathscr{D}^{\bullet}(E))$ pour tout i.

5. Par définition, $\mathscr{L}^{\bullet}_{\mathrm{loc}}(E)$ est borné, nul en degrés $i<0$ et quasi-isomorphe à E sur V. On voit facilement que son faisceau dérivé $\mathscr{H}^{\bullet}\mathscr{L}^{\bullet}_{\mathrm{loc}}(E)$ est localement constant le long des V_P. De plus il suit de la dernière remarque du n° 4 que si $x\in V_P$, la nullité de $\mathscr{H}^{i}(\mathscr{L}^{\bullet}_{\mathrm{loc}}(E)\,x)$ pour $i\geqq\mathrm{cd}(V_P)$ entraîne celle de $\mathscr{H}^{j}(D_X\mathscr{L}^{\bullet}_{\mathrm{loc}}(E)_x)$ pour $j\leqq n-\mathrm{cd}(V_P)$, où D_X désigne le passage au dual de Verdier. Comme ces conditions caractérisent $\mathscr{I}\mathscr{C}^{\bullet}(E)$ à quasi-isomorphisme près ([6], [11]) il suffit, pour démontrer la conjecture de Zucker, d'établir la première condition d'annulation pour E et E^*, donc de prouver :

$$H^i(U\cap V; \mathscr{L}^{\bullet}_{\infty}(E))=0 \qquad (i\geqq\mathrm{cd}(V_P)), \tag{1}$$

pour un système fondamental de voisinages U de $x\in V_P$ dans V^*, quels que soient x, V_P et E, où $\mathbf{H}^i$ indique l'hypercohomologie.

6. Fixons un Q-sous-groupe parabolique propre maximal $\mathscr{P}$, et soit $\mathscr{P}=\mathscr{M}.\mathscr{N}$ une décomposition de Levi de $\mathscr{P}$. Notons A la composante neutre du groupe des points réels du tore déployé sur Q maximal du centre de $\mathscr{M}$. Il existe dans le centre de l'algèbre de Lie $\mathfrak{n}$ de N un cône positif homogène autoadjoint C_P stable et homogène sous l'action (adjointe) de P [1]. Soient $\varphi: P \rightarrow \operatorname{Aut} C_P$ l'homomorphisme ainsi défini et G_P son noyau. Il est défini sur Q. Le groupe P possède un Q-sous-groupe distingué Z_P, d'intersection finie avec G_P, qui est cocompact dans le noyau de $\psi: P \rightarrow \operatorname{Aut} X_P$ ([1], [2]). Soient encore $L_P=Z_P/N$ et $K(L_P)$ un sous-groupe compact maximal de L_P. On a donc $C_P=L(P)/K(L_P)$. Pour toute partie $\mathscr{Q}$ de C_P notons $\mathscr{Q}(L_P)$ et $\mathscr{Q}(Z_P)$ ses images réciproques dans L_P et Z_P respectivement. On peut écrire $U \cap V$ (*cf.* supra) comme le produit d'un voisinage W de x dans V_P par un quotient $\Gamma(Z_P) \diagdown \mathscr{C}(Z_P)/K(Z_P)$, où $\mathscr{C}$ est une core ([1], p. 120) convenable dans C_P. On a aussi :

(1) $$H^{\cdot}(U \cap V;\mathscr{L}^{\cdot}_{\infty}(E))=H^{\cdot}(\mathfrak{z}_P, K(Z_P); L^{2,\infty}(\Gamma(Z_P)\diagdown \mathbf{C}(Z_P)\otimes \mathbf{C}(\rho_P)\otimes E)).$$

L'ouvert $U \cap V$ est fibré de fibre type $(\Gamma \cap N) \diagdown N$. On prouve d'abord, par une généralisation d'un argument de van Est, que l'on peut remplacer les coefficients dans le deuxième membre de (1) par $L^{2,\infty}(\Gamma(L_P)\diagdown \mathscr{C}(L_P))\otimes E$. Ensuite, en utilisant l'existence de représentants harmoniques pour $H^{\cdot}(\mathfrak{n}; E)$, on montre que la suite spectrale en cohomologie d'algèbre de Lie relative mod $\mathfrak{n}$ dégénère, d'où :

Proposition 2. — *On a l'égalité :*

(2) $$H^{\cdot}(U \cap V;\mathscr{L}^{\cdot}_{\infty}(E))=H^{\cdot}(\mathfrak{l}_P, K(L_P); L^{2,\infty}(\Gamma(L_P)\diagdown \mathscr{C}(L_P))\otimes \mathbf{C}(\rho_P)\otimes H^{\cdot}(\mathfrak{n}; E)),$$

où $\mathfrak{l}_P$ *est l'algèbre de Lie* L_P, $\mathbf{C}(\rho_P)$ *la représentation de degré* 1 *de* L_P *définie par* ρ_P *et* $\Gamma(L_P)=(\Gamma \cap Z_P)/(\Gamma \cap N)$.

(Selon l'usage, $\rho_P^2(l)=\operatorname{Tr}(\operatorname{Ad} l \mid \mathfrak{n})$ pour $l \in L_P$.)

Des analogues de la proposition 2 et de l'autodualité, dans un contexte différent, figurent dans [13] et [14].

7. Désignons par $A_2(\Gamma(L_P)\diagdown \mathscr{C}(L_P))$ l'espace des formes automorphes sur $\Gamma(L_P)\diagdown L_P$ qui sont de carré intégrable sur $\Gamma(L_P)\diagdown \mathscr{C}(L_P)$. On a :

Théorème 2. — *Supposons* X_P *de niveau* $\leqq 2$. *Alors l'inclusion de* $A_2(\Gamma(L_P)\diagdown \mathscr{C}(L_P))$ *dans* $L^{2,\infty}(\Gamma(L_P)\diagdown \mathscr{C}(L_P))$ *induit un isomorphisme :*

$$H^{\cdot}(\mathfrak{z}_P, K(Z_P), A_2(\Gamma(L_P)\diagdown \mathscr{C}(L_P)) \xrightarrow{\sim} H^{\cdot}(U \cap V; \mathscr{L}^{\cdot}_{\infty}(E)).$$

Ce théorème est l'analogue d'un résultat de [10] sur la cohomologie ordinaire des groupes arithmétiques de rang rationnel 1. Les deux démonstrations sont du reste apparentées. En particulier, celle du théorème 2 repose aussi sur un théorème du type Paley-Wiener en théorie des formes automorphes.

Désignons par $\operatorname{Log} A_P$ l'algèbre des fonctions sur A_P, qui, via l'exponentielle, proviennent de polynômes sur $\mathfrak{a}_P$. On voit aisément que $H^0(\mathfrak{a}_P; \operatorname{Log} A_P)=\mathbf{C}$ et que $H^i(\mathfrak{a}_P; \operatorname{Log} A_P)=0$ pour $i \geqq 1$. On peut analyser la structure de $A_2(\Gamma(L_P)\diagdown \mathscr{C}(L_P))$ à l'aide de la théorie des séries d'Eisenstein, convenablement complétée. Cela permet de déduire le théorème 1 du théorème 2 par application du théorème 3 à plusieurs situations différentes.

Théorème 3. — *Soient* H *une représentation unitaire irréductible de* L_P *et* χ *un caractère de* A_P *qui est* <1 *sur la chambre de Weyl positive* $A_P^+=A_P \cap \rho_P^{-1}((1, \infty))$. *Alors :*

(3) $$H^j(\mathfrak{z}_P, K(Z_P); H \otimes \mathbf{C}(\chi . \rho_P)\otimes \operatorname{Log} A_P \otimes E)=0 \quad si \quad j \geqq \operatorname{cd}(V_P).$$

Posons $L=L_P$, $Z=Z_P$, supposons $\mathscr{C}$ assez petit pour que $\Gamma(L)\backslash\mathscr{C}(L)$ soit contenu dans $\Gamma(L)\backslash\rho_P^{-1}((0,\infty))$ et soit $A_{2,\text{dis}}(\Gamma(L)\backslash\mathscr{C}(L))$ le sous-espace des formes automorphes sur $\Gamma(L)\backslash L$ qui sont de carré intégrable sur $\Gamma(L)\backslash\mathscr{C}(L)$. Le $(\mathfrak{l}/\mathfrak{a}, K(L))$-module des formes automorphes de carré intégrable sur $\Gamma(L)A_P\backslash L$ est somme directe de sous-modules irréductibles unitaires H_i $(i\in I)$. Comme $L\cong L/A_P\times A_P$, on peut écrire :

$$A_{2,\text{dis}}(\Gamma(L)\backslash\mathscr{C}(L))=\oplus_{i\in I},\ H_i\otimes C(\chi)\otimes \operatorname{Log} A_P,$$

où χ parcourt les caractères qui sont <1 sur A_P^+. Sans restriction sur le niveau, le théorème 3 implique :

COROLLAIRE. — *On a :*

$$H^q(\mathfrak{z}, K(Z); A_{2,\text{dis}}(\Gamma(L)\backslash\mathscr{C}(L))\otimes C(\rho_P)\otimes \operatorname{Log} A_P\otimes E)=0 \quad \textit{si} \quad q\geqq \operatorname{cd}(V_P).$$

Si V_P est de niveau 1, le groupe $\Gamma(L)$ est cocompact dans L/A et l'espace $A_{2,\text{dis}}(\Gamma(L)\backslash\mathscr{C}(L))$ est égal à $A_2(\Gamma(L)\backslash\mathscr{C}(L))$, ce qui établit le théorème 1 en niveau 1. Si V_P est de niveau 2, alors on peut montrer, en utilisant notamment l'équation fonctionnelle des séries d'Eisenstein, que la cohomologie à coefficients dans le quotient du deuxième espace par le premier s'identifie à une partie de la L^2-cohomologie des strates de niveau 1 contenant V_P dans leur adhérence, à coefficients dans un système local convenable, aussi justiciable du théorème 3. L'annulation cherchée résulte alors d'une double application de ce dernier.

8. A l'aide de différentes réductions, on déduit le théorème 3 lui-même d'un théorème d'annulation qui ne fait plus intervenir de **Q**-structure. Pour l'énoncer, nous abandonnons certaines des conventions adoptées jusqu'ici, supposons G presque simple sur **R** et P parabolique propre maximal sur **R**. C'est donc le normalisateur d'une composante à la frontière X_P dans la compactification de Satake de X sous-jacente à la construction de V ([2], § 1). On a alors :

THÉORÈME 4. — *Soit* H *une représentation unitaire irréductible de* L. *Alors :*

$$H^j(\mathfrak{l}, K(L); H\otimes H^i(\mathfrak{n}; E))=0 \quad \textit{si} \quad i<(\dim N)/2 \quad \textit{et} \quad 2(j+i)\geqq \dim X-\dim X_P.$$

La démonstration fait intervenir la décomposition de Kostant de $H^*(\mathfrak{n}; E)$, la proposition 2.6 de [9] et, de façon essentielle, les résultats de [12].

9. REMARQUES. — En fait les démonstrations des résultats énoncés dans les n°s 3 à 8 n'utilisent pas toujours pleinement l'hypothèse que X et X_P ont des structures complexes invariantes. Il en résulte notamment que le théorème 1 vaut aussi dans des cas où l'on suppose que G et K d'une part, G_P et $K(G_P)$ d'autre part, ont le même rang réel, conformément à une généralisation de la conjecture de Zucker proposée par l'un de nous, et mentionnée dans [14].

Remise le 20 mai 1985.

RÉFÉRENCES BIBLIOGRAPHIQUES

[1] A. ASH, D. MUMFORD, M. RAPOPORT et Y. TAI, *Smooth compactifications of locally symmetric varieties*, Math. Sc. Press, 1975.

[2] W. BAILY et A. BOREL, Compactifications of arithmetic quotients of bounded symmetric domains, *Ann. Math.*, 84, 1966, p. 442-528.

[3] A. BOREL, *Introduction aux groupes arithmétiques*, Hermann, Paris, 1969.

[4] A. BOREL, L^2-cohomology and intersection cohomology of certain arithmetic varieties, in *E. Noether in Bryn Mawr*, Springer, 1983, p. 119-131.

[5] A. Borel, Regularization theorems in Lie algebra cohomology. Applications. *Duke Math. J.*, 50, 1983, p. 605-623.

[6] A. Borel, Sheaf theoretic intersection cohomology, in *Progress in Math.*, 50, Birkhäuser-Boston, 1984, p. 47-182.

[7] A. Borel et W. Casselman, L^2-cohomology of locally symmetric varieties of finite volume, *Duke Math. J.*, 50, 1984, p. 625-647.

[8] A. Borel et N. Wallach, *Continuous cohomology, discrete subgroups and representations of reductive groups, Ann. Math. Studies*, Princeton University Press, 1980.

[9] W. Casselman, L^2-cohomology for groups of real rank one, in *Representation theory of reductive groups, Progress in Math.*, 40, Birkhaüser, Boston, 1983, p. 69-82.

[10] W. Casselman, Automorphic forms and a Hodge theory for congruence subgroups of $SL_2(Z)$, in *Lie groups representations II, Springer Lecture Notes in Math.*, n° 1041, 1983, p. 103-140.

[11] M. Goresky et R. MacPherson, Intersection homology II, *Inv. Math.*, 71, 1983, p. 77-129.

[12] D. Vogan et G. Zuckerman, Unitary representations with non-zero cohomology, *Comp. Math.*, 53, 1984, p. 51-90.

[13] S. Zucker, L^2-cohomology of warped products and arithmetic groups, *Inv. Math.*, 70, 1982, p. 169-218.

[14] S. Zucker, L^2-cohomology and intersection homology of locally symmetric varieties II, preprint, 1984.

A. B. : *The Institute for Advanced Study, Princeton*, N.J. 08540, *U.S.A.*
F.I.M., E.T.H.-Zentrum, 8092 *Zurich, Suisse;*
W. C. : *Department of Mathematics, University of British Columbia, Vancouver* (*V6T 1W5*), *B.C., Canada.*

132.

Hermann Weyl and Lie Groups

in *Hermann Weyl 1885–1985*. K. Chandrasekharan ed., Springer-Verlag 1986, 53–82

During the first thirteen years or so of his scientific career, H. Weyl was concerned mostly with analysis, function theory, differential geometry and relativity theory. His interest in representations of Lie groups or Lie algebras[1] and Invariant Theory grew out of problems raised by the mathematical underpinning of relativity theory:

> "...but for myself I can say that the wish to understand what really is the mathematical substance behind the formal apparatus of relativity theory led me to the study of representations and invariants of groups;"

[W 147: p. 400]. This involvement, at first somewhat incidental, turned rapidly into a major one. Weyl soon mastered, furthered and combined existing techniques, and within two years announced in 1924 a number of basic contributions.

The first serious encounter with Lie group theory arose out of a problem considered in the 4th edition of „Raum Zeit, Materie" [W I] on the nature of the metric in space-time, an analogue in this context of the Helmholtz-Lie space problem, which aims at characterizing the orthogonal group by means of some general mobility axioms. As you know, the mathematical framework of general relativity is a 4-manifold, say M, endowed with a Riemannian metric (of Lorentz type). The latter, in particular, assigns to $x \in M$ a quadratic form on the tangent space $T_x(M)$ at M. Riemann himself had already alluded to the possibility of considering metrics associated to biquadratic forms or more general functions. H. Weyl investigated, for manifolds of arbitrary dimension n, whether it was possible to start from a general notion of congruence, defined at each point by a closed subgroup G of automorphisms of the tangent space at x, belonging to a given conjugacy class of closed subgroups of $\mathbf{GL}_n(\mathbf{R})$,[2] and deduce from some general axioms that it would be conjugate to the orthogonal group of a (possibly indefinite) metric. This would then prove that the notion of congruence was associated to a Riemannian metric of some index. The problem was reduced to showing that the complexification of the Lie algebra of G was

* A Hermann Weyl Centenary Lecture, ETH Zürich, November 7, 1985.

conjugate in $\mathfrak{gl}_n(\mathbf{C})$ to that of the complex orthogonal group. [Originally, it dealt with real Lie algebras. However, since only the quadratic character of the metric, and not the index, was at issue, it was sufficient to consider their complexifications.][3] In [W I], H. Weyl stated he could prove it for $n=2, 3$ and, shortly afterwards, published a general proof in [W 49]. It was a rather delicate and long case-by-case argument, which H. Weyl himself likened to mathematical tightrope dancing („mathematische Seiltänzerei").

E. Cartan read about the problem in the French translation of [W I], published in 1922, and lost no time in providing a general proof in the framework of his determination of the irreducible representations of simple Lie algebras [C 65]. It was more general and more natural than Weyl's, even than Weyl's somewhat streamlined argument in [W II]. The comparison between them appears to have been a strong incentive for Weyl to delve into E. Cartan's work. Whether he had other reasons I do not know; at any rate he did so not long after, with considerable enthusiasm, as he was moved to write later on to E. Cartan (3.22.25):

> „Seit der Bekanntschaft mit der allgemeinen Relativitätstheorie hat mich nichts so ergriffen und mit Begeisterung erfüllt wie das Studium Ihrer Arbeiten über die kontinuierlichen Gruppen."

[W I] makes ample use of tensor calculus. A student of Weyl considered the problem of showing that the usual tensors form a family characterised by some natural conditions. This eventually amounted to proving that all the Lie group homomorphisms of $\mathbf{GL}_n(\mathbf{R})^+$ into $\mathbf{GL}_n(\mathbf{R})$, where $\mathbf{GL}_n(\mathbf{R})^+$ is the group of elements in $\mathbf{GL}_n(\mathbf{R})$ with positive determinant, are compositions of inner automorphisms, passage to the contragredient, and sums of maps $M \mapsto |\det M|^\alpha$ $(\alpha \in \mathbf{R})$ [Wn].[4] About thirty years later, that former student told me that, at the time, Weyl was clearly broadening his interest in representations of semisimple Lie groups, and had suggested that he work further along those lines. He even felt he might have shared some of the excitement to come, had he done so, but he had preferred to go back to his main interest, analysis,[5] and became indeed a well-known analyst: I was talking about Alexander Weinstein.

Another push into Lie groups came, in a way, again from the tensor calculus in [W I], but for a completely different reason. In 1923 E. Study, a wellknown expert in Invariant theory for over thirty years, published a book on Invariant Theory [St 2]. In a long foreword, he complained that Invariant Theory, in particular the so-called symbolic method to generate invariants, had been all but forgotten and that several mathematicians did less by other methods than would be possible using it. Among those was H. Weyl, identified by a quote, criticized for his treatment of tensors. Apparently H. Weyl was stung by this. This can already be seen by the rather sharply worded footnote in his answer to Study, which makes up the first part of [W 60], but it was also well remembered about 25 years later by one of his colleagues here at the time, M. Plancherel, who mentioned it to me then as an example of the extraordinary ability H. Weyl had, shared only by J. von Neumann among the mathematicians he had known, to get into a new subject and bring an important contribution to it within a few months. In fact H. Weyl published two papers on Invariant Theory in 1924, [W 60: I], [W 63], which brings me to the achievements already alluded to before.

For the sake of the discussion I shall divide them and more generally Weyl's output in this area into two parts, one concerned with linear representations of semisimple Lie groups, complex or compact, and semisimple Lie algebras, which operates with Lie algebra techniques and transcendental means, and is tied up to the real or complex numbers, and one concerned with Invariant Theory and representations of classical groups, initiated by [W 60, W 63], in which Weyl wears an algebraist's hat. This will be convenient to me as an organisational principle, but is, of course, to some extent artificial and should not be construed as a sharp division. In the following years, H. Weyl wrote a number of papers on both, but remained actively interested for a longer time in the second one, and I shall discuss it later.

At that time, *the* outstanding contribution to the former one was the work of E. Cartan which H. Weyl was discovering [C 5, C 37]. A second stimulus was provided by two papers of I. Schur [Sc 2] about representations of the special orthogonal group $\mathbf{SO}_n$, (as well as of the full orthogonal group $\mathbf{O}_n$, but I shall limit myself to the former) and invariants for $\mathbf{SO}_n(\mathbf{C})$, in which he, in particular, extended the theory of characters and orthogonality relations known for finite groups by the work of Frobenius and Schur done at the turn of the century. For later reference also, let me recall briefly some features of the latter.

Let G be a finite group, $\hat{G}$ the set of equivalence classes of irreducible complex representations of G, and $F(G)$ the space of complex valued functions on G. On $F(G)$ there is a natural finite dimensional Hilbert space structure, with scalar product given by

(1) $$(f,g)=|G|^{-1}\sum_{g\in G} f(x)\bar{g}(x) \qquad (f,g\in F(G))$$

where $|G|$ is the order of G. Given an irreducible representation $\pi: G\to \mathbf{GL}_n(\mathbf{C})$ by matrices $(a^i_j(g))$, let V_π be the vector subspace of $F(G)$ generated by the coefficients (a^i_j). It depends only on the equivalence class $[\pi]\in\hat{G}$ of π (and can be defined more intrinsically) and will also be labelled by $[\pi]$. Moreover, we have the orthogonal decomposition

(2) $$F(G)=\bigoplus_{\pi\in\hat{G}} V_\pi,$$

and the a^i_j form a basis of V_π (orthogonal if π is unitary). The space $F(G)$ is a $G\times G$ module via left and right translations, and the V_π are the irreducible $G\times G$ submodules. If E_π is a representation space for π, then $V_\pi\cong \operatorname{End} E_\pi$, hence is isomorphic to $E_{\pi^*}\otimes E_\pi$ as a $G\times G$ module, where π^* is the contragredient representation to π. As a G-module under right (or left) translations $F(G)$ is the regular representation of G, and V_π is isomorphic to the direct sum of d_π copies of π, (or π^*), where d_π is the degree of π.

Let χ_π be the character of π.[6] It belongs to the space

$$(3) \qquad C(G)=\{f\in F(G) | f(xyx^{-1})=f(y) \quad (x,y\in G)\}$$

of class functions on G; we have the orthogonality relations

$$(4) \qquad |G|^{-1}(\chi_\pi,\chi_{\pi'})=\delta_{\pi,\pi'} \quad (\pi,\pi'\in\hat{G})$$

and the χ_π $(\pi\in\hat{G})$ form an orthogonal basis of $C(G)$.

A very simple consequence of the orthogonality relations is that the average over G of the character of a given representation π gives the dimension of the space of fixed vectors.

The key for the extension of these results to orthogonal groups was provided by a paper of Hurwitz [Hu]. The main concern there was the invariant problem for $\mathbf{SL}_n(\mathbf{C})$ and $\mathbf{SO}_n(\mathbf{C})$: given a finite dimensional holomorphic representation of one of these groups on a space E, show that the invariant polynomials on E form a finitely generated algebra. The known averaging procedure for finite groups to prove this could not be directly applied since $\mathbf{SO}_n(\mathbf{C})$ or $\mathbf{SL}_n(\mathbf{C})$ were not "bounded", but it could be to their "bounded" subgroups $\mathbf{SO}_n$ and $\mathbf{SU}_n$ and that was sufficient, since a holomorphic function on $\mathbf{SO}_n(\mathbf{C})$ [resp. $\mathbf{SL}_n(\mathbf{C})$] is completely determined by its restriction to $\mathbf{SO}_n$ (resp. $\mathbf{SU}_n$), in the same way as a holomorphic function on an connected open set of $\mathbf{C}$ is determined by its restriction to a line. This was the first instance of what H. Weyl first called the "unitary restriction" („unitäre Beschränkung"), and later [W VI] "unitarian trick".

I. Schur, whose initial motivation was also Invariant Theory, extended the orthogonality relations to $\mathbf{SO}_n$, drew the consequence about the average of a character to compute the dimension of the spaces of invariants in a representation of $\mathbf{SO}_n(\mathbf{C})$, using Hurwitz's integration device, and then went over the determination of all the continuous irreducible representations of $\mathbf{SO}_n$, their characters and dimensions. He also checked that these representations were in fact analytic and, even more, rational in the sense that their coefficients could be expressed as polynomials in the entries of the elements of $\mathbf{SO}_n$. Also full reducibility of finite dimensional representations could be established as in the finite group case, by the construction of an invariant positive non-degenerate hermitian form, where the averaging over the group was replaced by an integration.

H. Weyl was now ready to strike. He first extended Schur's method to $\mathbf{SL}_n(\mathbf{C})$ and the symplectic group $\mathbf{Sp}_{2n}(\mathbf{C})$ [W 62] and then almost immediately afterwards combined the Hurwitz-Schur and the Cartan approaches in an extraordinary synthesis, announced first in the form of a letter to Schur [W 61]. Until Weyl came on the scene, neither did Cartan know about the work of

Hurwitz and Schur nor did Schur about E. Cartan's, as can be seen from the introduction to [Sc 2] in the latter case, from a letter of E. Cartan to H. Weyl (3.1.25) in the former case. I. Schur expressed his admiration for the results of H. Weyl and shortly afterwards suggested that Weyl write them up and publish them in Math. Zeitschrift, which was soon done [W 68].

In these papers H. Weyl first discusses separately the series of classical groups: $\mathbf{SL}_n(\mathbf{C})$, $\mathbf{SO}_n(\mathbf{C})$, and $\mathbf{Sp}_{2n}(\mathbf{C})$, and then sets up the general theory. A first main goal is to prove the full reducibility of the finite dimensional representations of a complex semisimple Lie algebra, a problem which E. Cartan had hardly alluded to in print before 1925. To be more precise, E. Cartan had in [C 37] given in principle a construction of all irreducible representations of a given simple algebra[7] and had just not considered more general ones. However, H. Weyl pointed out that, as far as he could see, an argument of E. Cartan at one important point could be justified only if full reducibility were available.[8]

The key point was to show that the "unitary restriction" could be applied in the general situation. This was done in two steps: first H. Weyl showed that a given complex semi-simple Lie algebra $\mathfrak{g}$ has a "compact real" form $\mathfrak{g}_u$, i.e., a real Lie subalgebra such that $\mathfrak{g} = \mathfrak{g}_u \otimes_{\mathbf{R}} \mathbf{C}$, on which the restriction of the Killing form is negative nondegenerate.[9] For instance, if $\mathfrak{g} = \mathfrak{sl}_n(\mathbf{C})$ one can take for $\mathfrak{g}_u$ the Lie algebra of $\mathbf{SU}_n$. To this effect, H. Weyl had first to outline the general theory of semisimple Lie algebras, for which the only sources until then were the papers of W. Killing and Cartan's Thesis, all extremely hard to read, and then prove the existence of $\mathfrak{g}_u$ by a subtle argument using the constants of structure. Already this exposition, which among other things, stressed the importance of a finite reflection group (S), later called the Weyl group, was a landmark and for many years the standard reference. But there it was really only preliminary material. Identify $\mathfrak{g}$ to a subalgebra of $\mathfrak{gl}(\mathfrak{g})$ by the adjoint representation, which is possible since $\mathfrak{g}$, being semisimple, is in particular centerless, and let G^0 be the complex subgroup of $GL(\mathfrak{g})$ with Lie algebra $\mathfrak{g}$. It leaves invariant the Killing form $K_{\mathfrak{g}}$ defined by $\mathscr{K}_{\mathfrak{g}}(x, y) = \mathrm{tr}(ad x \circ ad y)$, which is nondegenerate by a result of E. Cartan. Let then G_u^0 be the real Lie subgroup of G^0 (viewed now as a real Lie group) with Lie algebra $\mathfrak{g}_u$. Since the restriction K_u of $K_{\mathfrak{g}}$ to $\mathfrak{g}_u$ is negative nondegenerate, it can be viewed as a subgroup of the orthogonal group of K_u, hence is compact.[10]

This is a situation to which the Schur-Hurwitz device can be applied, therefore any finite dimensional representation π of $\mathfrak{g}$ which integrates to one of G^0 is fully reducible. However, in general, a representation π of $\mathfrak{g}$ will integrate to one not of G^0, but of some covering group G_π of G^0. In the latter, there is a closed real analytic subgroup $G_{\pi,u}$ with Lie algebra $\mathfrak{g}_u$, and the Schur-Hurwitz method will be applicable only if $G_{\pi,u}$ is itself compact, i.e., is a finite covering of G_u^0, which is equivalent to $G_{\pi,u}$ having a finite center. This H. Weyl shows at one stroke for all possible $G_{\pi,u}$ by proving more generally that G_u^0 has a finite fundamental group, i.e. that its universal covering has finite center.[11] This first of all yields the full reducibility, but also sets the stage to extend the character theory of Schur for $\mathbf{SO}_n$ to all compact semi-simple groups. All this is by now so standard that a detailed summary is surely superfluous. I shall content myself with some remarks mainly to comment on some of the work which arose from it. Let then K be a compact connected semisimple Lie group and T a maximal torus. It is unique up to conjugacy, and Weyl shows that it meets every conjugacy class. Its Lie algebra $\mathfrak{t}$ is a Cartan subalgebra of the Lie algebra $\mathfrak{k}$ of K. Its character group $X(T)$ is a free abelian group of rank equal to $l=\dim\mathfrak{t}$. A *root* is defined globally as a non-trivial character of T in $\mathfrak{k}_{\mathbf{C}}=\mathfrak{k}\bigotimes_{\mathbf{R}}\mathbf{C}$ with respect to the adjoint representation of K. The Weyl group W is the group of automorphisms of $X(T)$ induced by inner automorphisms of K. It is generated by the reflections s_α, where α is a root and s_α the unique involution of $X(T)$ having a fixed point set of corank one, mapping α to $-\alpha$ and leaving stable the set of roots. Since T meets every conjugacy class in K, it suffices to describe the restrictions to T of the irreducible characters of K. At first these are finite sums of characters of T, i.e. trigonometric polynomials, invariant under W. To describe them, we consider the trigonometric sums

$$A(\lambda)=\sum_{w\in W}(\det w)w\cdot\lambda,\qquad (\lambda\in X(T)).$$

The sum $A(\lambda)$ is skew invariant with respect to W, equal to zero if λ is fixed under a reflection s_α in W and depends only on the orbit $W\cdot\lambda$ of λ. Fix a closed convex cone C in $X(T)_{\mathbf{R}}=X(T)\bigotimes_{\mathbf{Z}}\mathbf{R}$ which is a fundamental domain for W, call λ dominant if it belongs to C and say that a root α is positive if it lies in the half-space bounded by the fixed point set of s_α and containing C. Let 2ϱ be the sum of the positive roots. We have Weyl's denominator formula

$$A(\varrho)=\prod_{\alpha>0}(\alpha^{1/2}-\alpha^{-1/2}). \tag{5}$$

Weyl shows that for every continuous class function f on K we have

$$\int_K f(k)dk=\int_T f(t)\mu(t)dt, \tag{6}$$

where dk (resp. dt) is the invariant measure on K (resp. T) with mass 1 and

$$\mu=|W|^{-1}|A(\varrho)|^2,\qquad (|W| \text{ of order of } W), \tag{7}$$

a point which Schur had singled out in his praise of Weyl's results. From this and the orthogonality relations for characters of K or of T Weyl deduces that every irreducible character is of the form

$$\chi_\pi=A(\lambda_\pi+\varrho)\cdot A(\varrho)^{-1} \tag{8}$$

where λ_π is dominant.[12] He does not show however, but derives from E. Cartan's work, that every dominant λ occurs in this way.[13] I shall soon come back to this problem. He also deduces from (8) a formula for the degree of π. Both (8) and this formula were not at all to be seen from Cartan's construction of irreducible representations.

Among many things, these papers mark the birthdate of the systematic global theory of Lie groups. The original Lie theory, created in 1873, was in principle local, but during these first fifty years, global considerations were not ruled out, although the main theorems were local in character. However, a striking feature here was that algebraic statements were proved by global arguments which, moreover, seemed unavoidable at the time.[14] H. Weyl had not bothered to define the concepts of Lie group or of universal covering (the latter being already familiar to him in the context of Riemann surfaces).[15] He had just taken them for granted, but could of course lean on the examples of the classical groups, which had been known global objects already in the early stages of the Lie theory.

These papers had a profound impact on E. Cartan. He had first known the results of H. Weyl through the announcements [W 61, W 62], in which the general case was only cursorily discussed. His first reaction [C 81] was to show that, given the Hurwitz device, the use of "analysis situs" could be avoided by means of older results of his.[16] However, once the full papers were published, his outlook changed and from then on, the global point of view and analysis situs were foremost in his mind. He began to supplement his earlier work on Lie algebras with a systematic study of global properties of Lie groups. In [C 103, C 113] he developed the geometry of singular elements in a compact semisimple Lie group, used the Weyl group systematically, to the extent even of deriving some basic properties of compact Lie groups or Lie algebras from results on Euclidean reflection groups. The scope of these investigations was further increased when he began to look from this point of view at a problem originating in differential geometry he was involved with at the time: the study of a class of Riemannian manifolds he later (from [C 117] onwards) called symmetric spaces. These spaces were originally, by definition, those in which the Riemannian curvature tensor is invariant under parallelism. They are locally Riemannian products of irreducible ones. Cartan had classified them [C 93, C 94] and seen, with considerable astonishment, that this classification was essentially the same as that of the real forms of complex simple Lie algebras he had carried out ten years before [C 38]. Up to that point, Cartan's investigations had been really local, although this was a tacit rather than an explicit assumption. In present day parlance we would say he had classified isomorphy classes of local irreducible symmetric spaces. However, he had soon recognized that these spaces were also characterized by the condition that the local symmetry at a point x, i.e. the local homeomorphism which flips the geodesics through x, is isometric [C 93: Nr. 14], which of course led to his choice of terminology. There also Cartan adopted systematically a global point of view, put in the foreground a global version of this second condition (each point is isolated fixed point of a global involutive isometry), and developed a theory of semisimple groups and symmetric spaces in which Lie group theory and differential geometry were beautifully combined (see, e.g., [C 107, 116]). A particularly striking example is his proof of the existence and conjugacy of maximal compact subgroups by means of a fixed-point theorem in Riemannian geometry [C 116]. Nowadays, all this has been streamlined, in the sense that group theoretical (resp. differential geometric) results have been given group theoretical (resp. differential geometric) proofs. Such an evolution is unavoidable and has many advantages, but sometimes loses some of the freshness and suggestive power of the original approach. It seems to me still fascinating to watch Cartan explore this new territory and display, as he once put it [C 105]:

"toute la variété des problèmes que la Théorie des Groupes et la Géométrie, en s'appuyant mutuellement l'une sur l'autre, permettent d'aborder et de résoudre."

In [C 111] he also points out that, since the set of singular elements in a compact semisimple group has codimension 3, H. Weyl's homotopy argument for the finiteness of the fundamental group (cf. [11]) can also be pushed to show that the second Betti number is always zero (in fact, what he sketches leads to a proof that the second homotopy group is zero). He then sets up a program to compute the Betti numbers of compact Lie groups or their homogeneous spaces by means of closed differential forms, conjecturing on that occasion the theorems which de Rham was soon going to prove. Using invariant differential forms, he reduced these computations to purely algebraic problems [C 118]. This led later to the cohomology of Lie algebras.

We now come back to the problem of showing that every dominant character λ occurs in (8) as the highest weight of an irreducible representation. Already in a footnote to [W 61], H. Weyl had stated that this "completeness" can be proved by transcendental means. In [W 68: III, § 4] he is more precise but also more circumspect: the problem is to decompose the regular representation of a compact group, but there are serious technical difficulties, and this method is maybe not really worth pursuing until simplifications are available since the result is known anyway from Cartan.[17] However, whatever difficulties there were, they were soon surmounted in the paper written jointly with his student F. Peter [W 73], which not only proves the sought for completeness, but is very broadly conceived and is to be viewed as the foundational paper for harmonic analysis on compact topological groups. In the simplest case, that of the circle group, this boiled down to the study of trigonometric series but even there, the group theoretical point of view was new.[18] We now refer for comparison to the earlier discussion of finite groups. Since G is now a compact Lie group, the authors replace $F(G)$ by the space $L^2(G)$ of square integrable functions with respect to a fixed invariant measure. It defines, via right translations, a unitary representation of G. The completeness problem is to show that the algebraic direct sum F of the spaces V_π, defined as above, with π running through the equivalence classes of irreducible finite dimensional continuous representations, is dense in $L^2(G)$, (Peter-Weyl theorem). Indeed, if that is the case, then the characters will form an orthonormal basis of the space of measurable class functions, and this will imply easily that any expression $A(\lambda+\varrho)A(\varrho)^{-1}$ $(\lambda \geqq 0)$ does occur as a character in the right-hand side of (8). To prove this density, the authors introduce the convolution algebra $C^*(G)$ of the continuous functions on G. They show that any finite dimensional representation π yields also one of $C^*(G)$ by the formula

$$\pi(\alpha) = \int_G \alpha(x)\pi(x)dx$$

and view $\pi(\alpha)$ as a Fourier coefficient of α. The convolution $\alpha*$ by α is an integral operator of Hilbert-Schmidt type, with kernel $k(x, y) = \alpha(x \cdot y^{-1})$, which is self adjoint if $\alpha = \tilde{\alpha}$, where $\tilde{\alpha}(x) = \bar{\alpha}(x^{-1})$. This operator commutes with right translations hence its eigenspaces are invariant under G operating on the right. The authors consider in particular the operator associated to $\alpha * \tilde{\alpha}$. An extension of E. Schmidt's theory of eigenvalues and eigenspaces for such integral operators on an interval shows that $(\alpha * \tilde{\alpha}) *$ has a non-zero eigenvalue if $\alpha \neq 0$ and that the corresponding eigenspace is necessarily finite dimensional. All such eigenspaces are then contained in F. Using the operators so associated to an "approximate identity", (a sequence of positive functions, whose supports tend to $\{1\}$ and whose integral is one), they show that every finite dimensional irreducible representation of G occurs in this way (up to equivalence) and that every continuous function is a uniform limit of elements of F. This implies in particular that F is dense. As a further consequence, the finite dimensional representations separate the elements of G and the continuous class functions separate the conjugacy classes in G. Apart from some technical simplifications, such as the use of the spectral theorem for completely continuous operators (as done first in [Wi 1: § 21]), this is pretty much the way it is presented today and it made a deep impression at the time it was published. Weyl himself viewed it as one of the most interesting and surprising applications of integral equations [W 80: p. 196]. It was immediately extended to homogeneous spaces of compact Lie groups by E. Cartan, with emphasis on symmetric spaces [C 117], thus supplying in particular a group theoretical framework to the theory of certain special functions, such as spherical harmonics, and then again by H. Weyl [W 98].

H. Weyl had seen that the same approach would also yield the main approximation theorem in H. Bohr's theory of almost periodic functions on the line [W 71, 72]. After Haar showed the existence of an invariant measure on any locally compact group, and noted that this allowed one to generalize [W 73] to compact groups without further ado, J. von Neumann extended to groups S. Bochner's definition of Bohr's almost periodic functions (cf. [Wi 1: §§ 33, 41]) and this led to what H. Weyl called the "culminating point of this trend of ideas" [W VI: p. 193], providing the natural domain of validity for the arguments of the Peter-Weyl theory, but he pointed out immediately its limitations by quoting a result of Freudenthal, to the effect that a group whose points are separated by almost periodic functions is the product of a compact group by the additive group of a vector space, i.e. one does not get more than the two cases initially considered. Therefore, an extension of this theory to other non-compact groups, in particular non-compact semisimple groups, would have to be based on quite different ideas and Weyl never tried his hand at it. Still his work has exerted a significant influence on its development. First the obvious one: the character formula, the Peter-Weyl theorem, the use of a suitable convolution algebra of functions have been for all a pattern, a model. The results of Harish-Chandra on the discrete series for instance, albeit much harder to prove, bear a considerable formal analogy with them. But, less obviously maybe, Weyl was also of help via his work on differential equations [W 8], which gave Harish-Chandra a crucial hint in his quest for an explicit form of the Plancherel measure. In the simplest case, that of spherical functions for $\mathbf{SL}_2(\mathbf{R})$, (or real rank-one groups) the problem reduces essentially to the spectral theory of an ordinary differential equation on the line, with eigenfunctions depending on a real parameter λ. It is the reading of [W 8] which suggested to Harish-Chandra that the measure should be the inverse of the square modulus of a function in λ describing the asymptotic behaviour of the eigenfunctions [HC: II, p. 212], and I remember well from seminar lectures and conversations that he never lost sight of that principle, which is confirmed by his results in the general case as well.[19]

Around 1927, H. Weyl got involved with the applications of group representations to quantum mechanics. His first paper [W 75] contains notably some suggestions or heuristic arguments which also led to new developments in unitary representations of non-compact Lie groups, but of a quite different nature from those mentioned above. Weyl proposes that the spectral theorem should allow one to associate to an unbounded hermitian operator A on a Hilbert space H a one-parameter group $\{\exp itA\}$ $(t \in \mathbf{R})$ of unitary transformations of H having iA as an infinitesimal generator, as is well known to be the case in the finite dimensional case. Then, given operators satisfying the Heisenberg commutation relations

$$(9) \qquad [P_j, Q_k] = \delta_{jk}, \qquad [P_j, P_k] = [Q_j, Q_k] = 0, \qquad (1 \leqq j, k \leqq n),$$

Weyl views these relations as defining a Lie algebra and considers the associated group N of unitary transformations generated by the $\exp itU$, where U runs through the real linear combination of the P_j and Q_k. To (9) correspond commutator relations in N, which have since been known as the "Weyl form" or "integrated form" of (9), (and also prove that N is a "Heisenberg group"). He then gives some heuristic arguments to prove the uniqueness (up to equivalence) of an irreducible unitary representation of N with a given central character, out of which follows the uniqueness of the Schrödinger model for the canonical variables. Rigorous treatments of these suggestions led to the Stone theorem on one-parameter groups of unitary representations [So], which soon became a foundational result in unitary representations of non-compact Lie groups, and to the Stone-von Neumann uniqueness theorem [Ne] [So], itself a fundamental result and the source of many further developments (for all this see [Ho 4] and [M]).

Weyl soon provided a systematic and impressive exposition in his book "Gruppentheorie und Quantenmechanik" [W V]. I shall not attempt to discuss its importance in physics and shall go on confining myself mostly to Lie groups. As far as Weyl was concerned, the main mathematical contribution stemming from it is the paper on spinors [W 105], written jointly with R. Brauer. Infinitesimally, the spinor representations had already been described by Cartan in 1913 [C 37], by their weights. But [W 105] gave a global definition, based on the use of the Clifford algebra, itself suggested by Dirac's formulation of the equations for the electron. However, the most unexpected fall-out originated with a physicist, H. Casimir, and led to the first algebraic proof of the complete reducibility theorem. In the representations of $\mathfrak{g} = \mathfrak{sl}_2(\mathbf{C})$, or equivalently $\mathfrak{so}_3(\mathbf{C})$, an important role in the quantum theoretic applications is played by a polynomial of second degree in the elements of $\mathfrak{g}$, which represents the "square of the magnitude of the moment of momentum" [W V], p. 156 (or p. 179 in the English version), the sum of the squares of the infinitesimal rotations around the coordinate axes. It commutes with all of $\mathfrak{g}$, hence is given by a scalar in any irreducible representation: this yields an important quantum number $j(j+1)$, in the representation of degree $2j+1$ ($2j \in \mathbf{N}$). Casimir was struck by this commutation property and defined in 1931 an analogous operator for a general semisimple Lie algebra, called later on, and maybe in [W III] for the first time, the Casimir operator, and indicated how it would allow one to derive the Peter-Weyl theorem from results about self-adjoint elliptic operators [Cs]. A year later, he noticed that in the case of $\mathfrak{sl}_2$, it could be used to give a purely algebraic proof of full reducibility, which was later extended to the general case by B. L. van der Waerden, using the general Casimir operator [CW]. As we shall soon see, H. Weyl was quite concerned at the time with finding algebraic proofs of results obtained first in a transcendental way, but in a different context, and it seems that this problem was not anymore of much interest to him,[20] although he had concluded his first announcement [W 62] by saying that an algebraic proof would be desirable and had suggested, at the time of the lectures [W 3], that it would be worthwhile to develop a purely algebraic theory of Lie algebras, valid at least over arbitrary fields of characteristic zero, a suggestion which was picked up by N. Jacobson (see [J 1]) and had a considerable impact on his research interests.

Although I have spoken at some length of the Math. Zeitschrift papers, I have not yet exhausted their content and my survey has been incomplete on at least two counts. Making up for it will provide a bridge towards the more algebraic aspects of Weyl's work.

In Chapter I of [W 68], devoted to the representations of $\mathbf{SL}_n(\mathbf{C})$ and $\mathbf{GL}_n(\mathbf{C})$, Weyl not only combines Hurwitz-Schur and Cartan, but also relates the results to older ones going back to Schur's Thesis [Sc 1]: After having determined all the holomorphic irreducible representations of $\mathbf{SL}_n(\mathbf{C})$, he points out that the matrix coefficients are in fact polynomials in the matrix entries, and that these representations are the irreducible constituents of the representations of $\mathbf{SL}_n(\mathbf{C})$ in the tensor algebra over $\mathbf{C}^n$. They are therefore just the tensor spaces, described by means of symmetry conditions on the coefficients. In this he sees the "group theoretical foundation of tensor calculus," a point important enough for him to make it the title of this Chapter and of the announcement [W 62]. Moreover, Schur had given a direct algebraic construction of those, setting up the well-known correspondence with representations of the symmetric groups, stated in terms of Young diagrams, which also yielded an algebraic proof of full reducibility. This example of an algebraic treatment is one to which he will come back repeatedly and which he will try to extend to other classical groups. It later became of even greater interest to him in view of its applications to quantum theory [W V: Ch. V].

The second point is "Invariant Theory". In broad terms its general problem is, given a group G and a finite dimensional representation of G in a vector space V, to study the polynomials on V which are invariant under G (or sometimes only semi-invariant, i.e. multiplied by a constant under the action of a group element). The questions which are usually asked are whether the ring of invariants is finitely generated (first main theorem) and if so, whether the ideal of relations between elements in a generating set is finitely generated (second main theorem). In concrete situations, one will want of course an explicit presentation of the ring of invariants in terms of generators and relations. One may also look for the dimension of the space of homogeneous invariants of a given dimension (the "counting of the number of invariants"). Such a formulation, however, where G and V are free, emerged at a later stage of the theory, as an abstraction from the classical invariant theory, to which I shall come in a moment, which focuses on very specific instances of these questions.[21]

As remarked earlier, such problems were at the origin of the papers of Hurwitz [Hu] and Schur [Sc 2], and it was to be expected they would also be very much in Weyl's mind, even independently of the Study incident. Indeed, he points out at the end of [W 68] that the unitary restriction method now allows one to prove the first main theorem for all semisimple groups, which provides, for the first time, a natural group theoretical domain of validity for it. In addition, following Schur, he notes that the dimension of the space of fixed vectors in a given representation π is given by integrating over G the character χ_π of the representation. He pursues this in [W 69, 70] where he also states that, more generally, the multiplicity of an irreducible representation σ in π is given by the integral $\int \chi_\pi(g)\overline{\chi_\sigma(g)}dg$, where dg is the invariant measure with total mass 1, and applies this to a number of classical cases. Although he also stresses in [W 68] the superiority of the integration method over the traditional procedures, based on differentiation operators of the kind of Cayley's Ω-process, he soon became preoccupied with finding an algebraic framework for the main results of [W 68] pertaining to classical groups, which would encompass the classical invariant theory.

We have now come close to the two main themes of this second, more algebraic, part of Weyl's work, which culminates in his book on Classical Groups [W VI]:

> "...The task may be characterized precisely as follows: with respect to the assigned group of linear transformations in the underlying vector space, to decompose the space of tensors of given rank into its irreducible invariant subspaces. ...Such is the problem which forms one of the mainstays of this book, and in accordance with the algebraic approach its solution is sought for not only in the field of real numbers on which analysis and physics fight their battles, but in an arbitrary field of characteristic zero. However, I have made no attempt to include fields of prime characteristic."

After having briefly explained that the determination of representations logically precedes the search for algebraic invariants, he goes on to say:

> "...My second aim, then, is to give a modern introduction to the theory of invariants. It is high time for a rejuvenation of the classic invariant theory, which has fallen into an almost petrified state."

All this seems rather clear, but I dare say I am not the only one to have found the book of rather difficult access. According to the introduction, the program is first to decompose tensor representations and then to derive Invariant Theory. But Weyl apparently could not do this for all classical groups, algebraically and in that order, so that the itinerary between the two is more sinuous, starting in fact with invariant theory. It would seem also that, in spite of some occasional, rather pungent, comments on the symbolic method, Weyl had found Invariant Theory and some of its specific techniques of independent interest, since he gives them prominent billing in [W VI] and had also devoted to them a course here (of which a 13 page outline can be found in the Weyl Archives) and at the Institute for Advanced Study [W IV]. In addition, Weyl could not realize his program fully algebraically and did not refrain from introducing and using the integration method, whether to realize his immediate goals or in its own right. This rather subtle interplay between various points of view does, of course, broaden the horizon of "the humble who want to learn" for whom the book is "primarily meant" [W VI: viii], but does not make it easier for them to get a clear picture of its organization. This somewhat tentative character may also have been felt by Weyl when he wrote [W 117]:

> "At present I have come to a certain end, or at least to a certain halting point, from which it seems profitable to look back upon the track so far pursued, and this is what I have tried to do in my recent book, *The Classical Groups, their Invariants and Representations*."

Reflecting upon the greater finality of the transcendental results as compared to the algebraic ones, with the added wisdom of fifty or so more years, one may observe that in the former case Weyl already had the natural framework and all the necessary tools at his disposal, but not so in the latter; in particular, the point of view of linear algebraic groups, or the use of the universal enveloping algebras of Lie algebras, which became essential, were not part of his vision of future developments, as described at the end of [W 117] where he forecast

"...a similar book dealing comprehensively with the representations and invariants of all semi-simple Lie algebras in an arbitrary characteristic."

I shall try shortly to give some idea of the content of [W VI] and some of the developments to which it has led. Before that, however, I should backtrack and say something about the already often mentioned Classical Invariant Theory and the earlier work of Weyl pertaining to it.

A typical example is the search of invariants of p vectors for $G = \mathbf{SL}_n(\mathbf{C})$, $\mathbf{SO}_n(\mathbf{C})$, i.e. of polynomials in the coordinates (x_{ij}) $(i=1, \ldots, n; j=1, \ldots, p)$ of p vectors $\mathbf{x}_j$ $(j=1, \ldots, p)$ homogeneous in the coordinates of each $\mathbf{x}_j$, which are invariant under G. In other words, one is looking for the fixed vectors in the tensor product of p copies of the polynomial algebra over $\mathbf{C}^n$. In the case of $\mathbf{SL}_n(\mathbf{C})$ or $\mathbf{GL}_n(\mathbf{C})$ one wants more generally the invariant of p vectors and q covectors $(\mathbf{y}_k)$ $(k=1, \ldots, q)$. The first main theorem in this last case says that the ring of invariants is generated by the products $\langle \mathbf{x}_j, \mathbf{y}_k \rangle$, to which one should add the determinants in n of the p vectors or q covectors in the case of $\mathbf{SL}_n$. For $\mathbf{O}_n$ there is no need to add covectors, the invariants of p vectors are generated by the scalar products $(\mathbf{x}_j, \mathbf{x}_k)$. Also a generating set of relations between these elements was given. Let me now limit myself to the case of p vectors. To prove such theorems one uses differentiation operators which commute with G, hence transform invariants to invariants, and which allow one to decrease the degree in certain variables and to carry out induction proofs. The first ones are the polarisation operators $D_{ik} = \Sigma_j x_{ji} \partial / \partial x_{jk}$. Another fundamental one is Cayley's operator

$$\Omega = \Sigma (\operatorname{sgn} \sigma) \partial^n / \partial x_{\sigma(1),1}, \ldots, \partial x_{\sigma(n),n},$$

where σ runs through the symmetric group $\mathfrak{S}_n$ in n letters. This operator commutes with the D_{ij} for $i \neq j$, but not with the D_{ii}. Capelli [Ca 1] showed that $H = \det(x_{ij}) \det(\partial / \partial x_{ij})$ (for $p = n$) does commute with all D_{ij} and gave an expression for H as a determinant in those, the so-called Capelli identity, which is the main formal tool in much of the theory.[22] There are of course many variants of this problem. One may, e.g., replace the identity representation of G by a symmetric power, hence consider the invariants of p homogeneous forms of a given degree. The "symbolic method" reduces in principle the form problem to the search of multilinear invariants, i.e. to the fixed points in tensor powers of the identity representation. In the nineteenth century, such procedures to generate invariants out of a given one were often used to try to check the validity of the first and second main theorems. In that function, they were discredited by Hilbert's work on Invariant Theory, which "almost kills the whole subject. But its life lingers on, however flickering, during the next decades" [W VI: p. 27, 28].

In his answer to Study, in [W 60], Weyl goes back to Capelli, provides a new proof of Capelli's identity and then establishes the first main theorem for $\mathbf{SL}_n$, $\mathbf{SO}_n$, which were known, as well as for $\mathbf{Sp}_{2n}$, which had not been considered before. He also discusses the determination of invariants for some non-simple groups such as the group of euclidean motions or the subgroups of $\mathbf{SL}_n$, or $\mathbf{GL}_n$, leaving invariant a strictly increasing sequence of proper subspaces (now called parabolic subgroups, *see* [24]). [W 63] centers on the use of the symbolic method.

After some contributions to the program outlined above, Weyl publishes his book [W VI]. It starts with an exposition of the first and second main theorems for invariants of vectors and covectors for the classical groups. A central point of the book, or at any rate of its algebraic part, is the double commutant theorem, proved first in [W 107], which is the main principle on which Weyl organizes the discussion of the decomposition of tensor representations. Let A be a subalgebra of the algebra End(E) of endomorphisms of a finite dimensional vector spaçe E over a field K of characteristic zero and let A' be its commuting algebra in End V. Assume V is a fully reducible A-module. Then V is also fully reducible under A', and A is the commuting algebra of A'. In particular, V decomposes into a direct sum of $A' \times A$ irreducible submodules. Any such is the tensor product $U_\sigma \otimes U'_{\sigma'}$, of an irreducible A-module by an irreducible A'-module, whence a correspondence between some irreducible representations σ of A and σ' of A', which is particularly nice of no σ or σ' occurs twice in these pairs. The assumption of full reducibility under A is in particular fulfilled if A is the enveloping algebra of a finite group (even in positive characteristic prime to the order of that group).

The origin of this theorem, and its most perfect illustration, is the reciprocity between irreducible representations of the symmetric groups $\mathfrak{S}_p$ and the irreducible subspaces of the tensor representations of $\mathbf{GL}_n$, discussed first in Schur's Thesis [Sc 1] (for $K = \mathbf{C}$). Let V be the tensor product of p copies of K^n. It is operated upon by $\mathbf{GL}_n$, via the p-th tensor power of the identity representation and by $\mathfrak{S}_p$, via the permutations of the factors and these operations commute. Let A' and A be the corresponding enveloping algebras. Schur proves that A' is the centralizer of A and deduces from this full reducibility. He also shows that the correspondence (σ, σ') is bijective and determines the characters of the σ'. After having discussed this case, Weyl goes over to the orthogonal and symplectic groups. However, this has to be more roundabout since there is no finite group on the other side. Weyl had made the first steps in that direction in [W 96]; here he avails himself also of some results of R. Brauer [Br 2]. He considers simultaneously the enveloping algebra of the group under consideration in a tensor representation and its commuting algebra A'. None of them is a priori known to be fully reducible, and the interplay between information gained successively on each of them is rather subtle. An important fact is that the description of A' is equivalent to the first main theorem. However, in the case of $\mathbf{gl}_n$, this relationship can be used conversely to prove the first main theorem [Br 2].

Weyl's next goal is the determination of the characters of the irreducible constituents of the tensor representations. Lacking an algebraic method valid for all classical groups, he turns to the transcendental one of [W 68], based on integration. For $\mathbf{GL}_n$, however, algebraic treatments were available. I already mentioned one by Schur [Sc 1]. Here Weyl follows a slightly earlier one due to Frobenius. He then goes over to more general aspects of Invariant Theory, which he discusses both from the algebraic and the transcendental points of view.

A supplement to the main text in the second edition includes some complements published first in [W 122]. In particular, it points out that a subalgebra of a matrix algebra over a subfield of $\mathbf{R}$ which is stable under transposition is fully reducible, which gives an algebraic proof of full reducibility in such cases.

Obviously not an easy book.[23] Its results were not as spectacular and clear cut as those of the Math. Zeitschr. and M. Annalen papers and, not surprisingly, it did not have such an immediate impact, but this was really only a question of time, and its influence has been felt more and more during these last fifteen years or so.

First of all, its treatment of Classical Invariant Theory became the standard reference and made it available to potential users, whether specialists or not. This led to further applications but also to improvements of the theory, already over $\mathbf{C}$. As a first example, [ABP] provides a new proof of the determination of the tensor invariants for $\mathbf{O}_n$ directly from those for $\mathbf{GL}_n$, bypassing the use of the Capelli identity. The application the authors had in mind was to a new proof of the Atiyah-Singer theorem by the heat equation method. As a second example, M. Artin [A] was led to consider invariants of r matrices and conjectured they would be generated by traces. This was soon deduced from the classical theory by Procesi [P 1], who more generally described the invariants of p vectors, q covectors, and r matrices, results which were then used to determine the rational Waldhausen K-theory of simply connected spaces [DHS].

As a second type of developments, let me discuss some in which the restriction on the characteristic of the groundfield was lifted, even beyond what Weyl could envisage. Progress was made almost simultaneously in two directions. On the one hand, the first main theorem was extended to all semisimple (even reductive) linear algebraic groups over an algebraically closed groundfield K of arbitrary characteristic.[24] In characteristic zero, this was essentially Weyl's theorem mentioned earlier (modulo a harmless reduction to $\mathbf{C}$). But this proof was based on the full reducibility, which is false in positive characteristic, so it could not be extended directly. Motivated by his Geometric Invariant Theory, D. Mumford proposed a weaker notion, later called "geometric reductivity": Given a rational representation $G \to GL(V)$ and a pointwise fixed line $D \subset V$, there should exist a homogeneous hypersurface W meeting D only at the origin and invariant under G. (In case of full reducibility, W could be a hyperplane.) and conjectured that every semisimple algebraic group would satisfy this condition. Soon after, M. Nagata showed that it would imply the first main theorem [Na]. The validity of the latter was then assured when W. Haboush proved geometric reductivity in general [Ha], a few years after Seshadri had established it for $\mathbf{GL}_2$.

On the other hand, by introducing new methods in combinatorics, [DRS] gave a characteristic free proof of the first main theorem for $\mathbf{GL}_n$. This was quickly seized upon by de Concini and Procesi to extend the first and second main theorems to classical groups over a commutative ring A, subject only to the condition that a polynomial in n indeterminates with coefficients in A which is zero on A^n is identically zero [CP]. In particular, this applies to any algebraically closed groundfield or to $\mathbf{Z}$.

At this point, although it is one more generation removed from Weyl, I feel it is natural to mention a further and very extensive characteristic-free theory, in which [CP] is an essential tool, the "standard monomial theory" of Seshadri et al. In its geometric form, it aims at giving canonical bases for spaces of sections of homogeneous line bundles on flag varieties. As far as I know, it does not yet work in full generality but in a vast class of cases. In characteristic zero, when it does, it allows one to give a canonical basis of an irreducible representation; more explicitly, it yields a combinatorial procedure to single out an irreducible representation with a given highest weight in the tensor product of fundamental representations in which Cartan looked for it (see [7]), assuming this is done for the fundamental ones which occur. This generalizes work by W. V. D. Hodge for $\mathbf{SL}_n$, who used it to give explicit equations for Schubert varieties and applied it to enumerative geometry, and it has similar applications and goals in the general case. See the survey [Se] by Seshadri.

As my final item, I now turn to a development of which [W VI], or more specifically the bicommutant theorem, has been one of the foster parents, namely R. Howe's theory of reductive pairs.

In the early seventies, some remarkable correspondences were set up between certain families of irreducible unitary representations of some pairs of real reductive groups. Knowing [W VI], Howe was led to think that the proper framework for such correspondence was a generalization of the situation of the bicommutant theorem: start from a real semisimple group G and a so-called "reductive pair" (H, H') in G, i.e. two reductive subgroups H, H' each one of which is the centralizer of the other. Let now π be an irreducible representation of G, say first finite dimensional, to avoid analytical difficulties. Then it decomposes into irreducible $H \times H'$ submodules, each of which is a tensor product of an irreducible representation σ of H by one σ' of H', whence again some relationship between the irreducible representations of H and H' which occur. If now π is unitary, infinite dimensional, then the decomposition into $H \times H'$ modules presents all kinds of difficulties. They become more manageable if one of the groups is compact. Also one may anyhow restrict one's attention to the closed irreducible subspaces. It turned out that the correspondences mentioned earlier could be given a natural explanation by viewing the two groups in question as a reductive pair in a symplectic group $\mathbf{Sp}_{2n}(\mathbf{R})$ or rather by going over to a two-fold covering of $\mathbf{Sp}_{2n}(\mathbf{R})$ and taking for π the so-called oscillator or metaplectic representation. The choice of that particular representation had been suggested by Weil's use of it, over local and global fields, in his group-theoretical treatment of θ-functions [Wi 2]. Accordingly, this principle has had many applications and variants to groups over local or global fields. In the latter case it leads to correspondences built from θ-series between spaces of automorphic forms. This is one of the most fruitful principles in representation theory of, and automorphic forms for, classical groups, which has suggested many problems and has been confirmed by many special results. See [Ho 3] for a survey of theorems and conjectures and [Ho 5] for further results. At this point, we seem again to be far removed from Weyl and obviously such developments are not direct offsprings of his work. But the importance of [W VI] as one of the main influences on the genesis of this general principle has been stressed by Howe himself, who has also applied it to groups over finite fields [Ho 1] and then to Invariant Theory itself [Ho 2]. There Howe generalizes the classical theory to the determination of "superinvariants," i.e. of invariants in the tensor products of symmetric and exterior powers, and proposes a general recasting of the whole theory from that point of view.[25]

With this I conclude my attempt to give an idea of Weyl's work on Lie groups and of its repercussions. As you can see, those were felt in a broad range of topics in analytical, differential geometric, topological or algebraic contexts and took many forms: general theorems or specific results on special cases, clear cut statements as well as less sharply delineated suggestions or guiding principles, mirroring the many-sidedness of Weyl's output and outlook.

Early in this lecture, I quoted from a letter to Cartan in which Weyl expresses his admiration for Cartan's work on continuous groups. He goes on to say, commenting on the results announced in [W 61]:

„Meinen gegenwärtigen Anteil an dieser Theorie schätze ich gar nicht besonders hoch; ich komme mir eigentlich nur vor wie der zufällige Treffpunkt von Ihnen und Schur."

This is of course putting it very mildly. Not only much more than chance was needed to produce such a synthesis, but Weyl had to be a meeting ground for, and to combine, not only Schur and Cartan, but invariant theory, topology and functional analysis as well. At that time, no one else was conversant with all of these; in fact, except for Schur, with hardly more than one. Although I limited myself to a rather sharply circumscribed and quantitatively minor part of Weyl's work, this already provides a demonstration of, a practical lesson in, the unity of mathematics, given to us by a man whose mind was indeed a meeting ground for most of mathematics and mathematical physics.

Notes

[1] (53) As a rule, I shall use current terminology rather than the one of the original papers and content myself with some occasional historical remarks about the latter.

The term "Lie algebra" appears first in [W III], and was suggested by N. Jacobson, but the concept was present very early in the theory. Until then, the usual terminology was "infinitesimal group," or sometimes just "group," (or "abstract group" in [W II]). Obviously, H. Weyl felt at the time more comfortable with the latter, since he reverts almost exclusively to it, after having introduced "Lie algebra" as an alternate to infinitesimal group in a formal definition, and later uses only "subgroup" and "invariant subgroup" for the present-day "subalgebra" and "ideal."

"Lie group" was introduced by E. Cartan around 1930 (see in particular [C 128]). For Lie (disregarding the distinction between local and global), and until then, they were the "finite and continuous groups." However, E. Cartan uses there the latter expression for what we would call locally euclidean groups, and requires twice differentiability for Lie groups.

[2] (53) Some notation:

If K is a field, $\mathbf{M}_n(K)$ is the algebra of $n \times n$ matrices with coefficients in K, and $\mathbf{GL}_n(K)$ [resp. $\mathbf{SL}_n(K)$] the group of invertible (resp. determinant one) elements of $\mathbf{M}_n(K)$. The transpose of $X \in \mathbf{M}_n(K)$ is denoted ${}^tX$.

$\mathbf{Sp}_{2n}(K)$ is the symplectic group, i.e. the subgroup of elements of $\mathbf{GL}_{2n}(K)$ leaving invariant the standard antisymmetric bilinear form $\Sigma_1^n x_i y_{n+i} - x_{n+i} y_i$.

The terminology "symplectic group" is introduced in [W VI]. Earlier, for $K = \mathbf{C}$, Weyl had used "complex group," as a shorthand for group leaving a complex of lines invariant, which goes back to S. Lie (except for the fact that Lie's classical groups were groups of projective transformations).

$\mathbf{U}_n = \{A \in \mathbf{GL}_n(\mathbf{C}) \mid {}^t\bar{A} \cdot A = I\}$ is the unitary group on $\mathbf{C}^n$ and $\mathbf{SU}_n = \mathbf{U}_n \cap \mathbf{SL}_n(\mathbf{C})$.

$\mathbf{O}_n = \{A \in \mathbf{GL}_n(\mathbf{R}) \mid {}^tA \cdot A = I\}$ is the orthogonal group on $\mathbf{R}^n$ and $\mathbf{SO}_n = \mathbf{O}_n \cap \mathbf{SL}_n(\mathbf{R})$ the special orthogonal group, $\mathbf{O}_{p,q}$ the subgroup of $\mathbf{GL}_n(\mathbf{R})$ leaving $\Sigma_1^p x_i^2 - \Sigma_{p+1}^n x_j^2$ invariant $(n = p + q)$.

$\mathbf{O}_n(\mathbf{C}) = \{A \in \mathbf{GL}_n(\mathbf{C}) \mid {}^tA \cdot A = I\}$ and $\mathbf{SO}_n(\mathbf{C}) = \mathbf{SL}_n(\mathbf{C}) \cap \mathbf{O}_n(\mathbf{C})$.

The Lie algebra of a complex or real Lie group is denoted by the corresponding German letters: $\mathfrak{gl}_n$; $\mathfrak{gl}_n(\mathbf{R})$, $\mathfrak{gl}_n(\mathbf{C})$, $\mathfrak{so}_n$, ...

(3) (54) E. Cartan [C 65] translated the problem in the formalism of affine connections he was developing at the time, which made Weyl's transition from half-philosophical considerations to the actual mathematical problem easier to grasp for some. In present day terminology, the problem can be stated geometrically as follows: Let G be a closed subgroup of $\mathbf{SL}_n(\mathbf{R})$. Let M be a smooth manifold of dimension n. Choose in a neighborhood U of a point x a trivialisation of the bundle P of frames, i.e. n smooth everywhere linearly independent vector fields (e_i) (or, equivalently, as Weyl and Cartan express it, a set of n everywhere linearly independent one-forms ω_i). Let Q be the fibre

bundle over U whose fiber at y is the set of G-transforms of $\{e_i(y)\}$, *where* $\{e_i(y)\}$ is used to identify the tangent space $T_y(M)$ with $\mathbf{R}^n$. It is assumed that for every choice of the e_i's, the principal G-bundle Q has one and only one torsion-free affine connection (the existence is Axiom I and the uniqueness Axiom II). Then the complexification of the Lie algebra $\mathfrak{g}$ of G is conjugate in $\mathfrak{gl}_n(\mathbf{C})$ to $\mathfrak{so}_n(\mathbf{C})$.

Algebraically, this amounts to the following problem. Let G and $\mathfrak{g}$ be as before and

$$(C^i_{j,s}) \qquad (1 \leqq i, j \leqq n; 1 \leqq s \leqq m = \dim G)$$

be a basis of $\mathfrak{g}$. Assume that for any set of $m \cdot n^2$ real numbers $a^i_{j,s}$ one can find a unique set of $n \cdot m$ constants b^s_k $(s=1, \ldots, m; k=1, \ldots, n)$ such that the differences $a^i_{j,k} - \Sigma_s C^i_{j,s} a^s_k$ are symmetric in i, j. Then the complexification of $\mathfrak{g}$ is conjugate to $\mathfrak{so}_n(\mathbf{C})$.

In fact, Cartan solves a somewhat more general problem and shows that already the existence of the b^s_k forces $\mathfrak{g}$ to belong to a very small list.

For the sake of completeness, I should point out that Weyl had broached this question in 1919 already, in his comments to Riemann's Habilitationsschrift [Ri], but rather briefly, in a differential geometric way. It was noticed later that the problem stated there was in fact a variant of the one of [W I], but that Cartan's solution to the latter also gave one to the former. See [F] for a discussion, references, and another variation on the problem, and [K] for a hermitian analogue.

[4] (54) The main point is to find the automorphisms of $\mathbf{SL}_n(\mathbf{R})$. This was essentially reduced to a Lie algebra problem: to show that all derivations of the Lie algebra $\mathfrak{sl}_n(\mathbf{R})$ are inner. Cartan had already proved this fact in his Thesis [C 5: p. 137], for all semisimple Lie algebras (over $\mathbf{C}$ really, his standing assumption there, but this anyhow implies the result over $\mathbf{R}$). Prompted by Weinstein's result, he determined the structure of the group of automorphisms of all complex simple Lie algebras in [C 82].

[5] (54) He did not have to go far to find a topic and soon studied a problem in two-dimensional hydrodynamics suggested to him, together with a method of attack, by H. Weyl (Math. Zeitschr. **19** (1924), 265–275), a problem on which Weyl subsequently also wrote a paper [W 76].

[6] (56) This was called "characteristic" at the time. H. Weyl introduces "character" from about 1927 on, first as an alternate possibility, and then later shifted to it. In [W V] he suggests it in order to avoid using characteristic, which has another meaning pertaining to eigenvalues or eigenspaces of linear transformations, but reverses in Chap. V characteristic for finite groups and characters for continuous groups. In a lecture in French [W 74] he had used "caractère".

[7] (57) These results are to be found in many textbooks, e.g., [Bo 2], [J]. I shall recall them later in an equivalent global formulation, favored by Weyl. For the sake of completeness I give here a capsule review of some salient features in the framework of E. Cartan.

Let $\mathfrak{g}$ be a complex semisimple Lie algebra, $\mathfrak{h}$ a Cartan subalgebra. A linear form λ on $\mathfrak{h}$ is a *weight* of a representation $\pi:\mathfrak{g}\to\mathfrak{gl}(V)$ if

$$V_\lambda=\{v\in V\mid \pi(h)v=\lambda(h)\,(h\in\mathfrak{h})\}\neq 0.$$

The space V is always the direct sum of the V_λ. The weights of all finite dimensional representations generate a lattice P in the smallest $\mathbf{Q}$-subspace $\mathfrak{h}^*_{\mathbf{Q}}$ of $\mathfrak{h}^*$ spanned by them, which is a $\mathbf{Q}$-form of $\mathfrak{h}^*$. A nonzero weight of the adjoint representation is a *root*. The roots generate a sublattice Q of P. Let $\mathfrak{h}^*_{\mathbf{R}}$ be the real span of P. For each root α there is a unique automorphism s_α of order 2 of $\mathfrak{h}^*_{\mathbf{R}}$ leaving the set of roots stable, transforming α to $-\alpha$ and having a fixed point set of codimension one. Fix a set Δ of "simple roots", i.e. $l=\dim\mathfrak{h}$ linearly independent roots such that any other root is a integral combination of the $\alpha\in\Delta$ with coefficients of the same sign, and call positive those with positive coefficients. Introduce a partial ordering among the weights by saying that $\lambda\geqq\mu$ if $\lambda-\mu$ is a positive linear combination of simple roots. Then $\lambda\in P$ is said to be dominant if $\lambda\geqq s_\alpha\lambda$ for all simple α's. Let P^+ be the set of dominant weights. [The group W of automorphisms of $\mathfrak{h}^*_{\mathbf{R}}$ generated by the s_α is one realization of the group (S) introduced by Weyl in [W 68] and called later the Weyl group. A weight can also be defined as dominant if it belongs to a suitable fundamental domain C of W, namely the intersection of the half spaces E_α $(\alpha\in\Delta)$, where E_α is the half-space bounded by the fixed point set of s_α and containing α.]

Let π be irreducible. Cartan shows first it has a unique highest weight λ_π, i.e. a weight which is > than any other weight. This weight has multiplicity one. The main result of [C 37] is that any $\lambda\in P^+$ is the highest weight of one and only one (up to equivalence) irreducible representation. The dominant weights are linear combinations with positive integral coefficients of the so-called fundamental ones ω_i $(1\leqq i\leqq l)$, where $l=\dim\mathfrak{h}$ is the rank of $\mathfrak{g}$. The strategy of E. Cartan to prove this is first to exhibit a representation V_i for each fundamental dominant weight ω_i and then to locate an irreducible representation having a given dominant weight $\lambda=\Sigma c_i\cdot\omega_i$ as its highest weight as the smallest $\mathfrak{g}$-submodule in the tensor product $V_1^{c_1}\otimes\ldots\otimes V_l^{c_l}$ of c_1 copies of $V_1,\ldots,c_l$ copies of V_l containing $V_{\omega_1}^{c_1}\otimes\ldots\otimes V_{\omega_l}^{c_l}$. It is this last point that H. Weyl criticized (see below).

[8] (57) Cartan agreed there was a gap, but in fact he could have easily bypassed full reducibility to realize his goal at that point, namely to show the existence of an irreducible representation with a given dominant weight as its highest weight. For this, a slight refinement of his argument would have sufficed, as is customarily done nowadays in the algebraic proof found later, or to show the existence of an irreducible highest weight module in situations where full reducibility is not true or not known such as linear reductive groups in positive characteristic or Kac-Moody Lie algebras: One takes the smallest $\mathfrak{g}$-submodule containing the given highest weight vector, shows that it has a greatest proper invariant subspace and then divides out by the latter.

Since E. Study was somewhat of a villain in the 1923 incident related earlier, let me add as a counterpart that he was well aware of this problem around 1890 and had brought the first contribution to it. In fact, S. Lie reports in [L: 785-8] that Study has proven full reducibility (phrased however differently, in terms of projective representations) for $\mathfrak{sl}_2$ in an unpublished manuscript and that it was quite sure it would be true more generally for $\mathfrak{sl}_n$. In a letter to S. Lie (December 31, 1890), referred to in [Hw], Study even goes as far as conjecturing it should hold for all simple or semisimple Lie algebras. To both of them, this was an important problem. The manuscript was not published, apparently because the proof appeared too complicated and simplifications were hoped for.

In [C 5: p. 134] E. Cartan refers to Lie's summary of Study's results and comments that it would be easy to prove all those pertaining to $\mathfrak{sl}_2$ using some results proved before. In [Hw], the author surmises that Cartan must have had the full reducibility in mind, and not just the determination of the irreducible representations of $\mathfrak{sl}_2$, which was due to Lie anyway. Prompted by this remark, which T. Hawkins reminded me of in a useful correspondence on a first draft of this paper, for which I am glad to thank him, I checked that this is indeed the case: Cartan's arguments from line 6, p. 100 to line 9, p. 102 do provide all the ingredients for an algebraic proof, but this consequence is not explicitly stated since Cartan's goal at that point is to supply a complete proof to a theorem announced earlier by F. Engel, namely that every non-solvable Lie algebra contains a subalgebra isomorphic to $\mathfrak{sl}_2$. (As pointed out on p. 103, the original argument by Engel was not complete, but he also published another proof in 1893.) However, Cartan casually uses full reducibility later (p. 116, lines 13–23), referring to the previous Chapter, showing that he felt he had indeed proved it there.

The next reference in print to that problem by E. Cartan I am aware of, explicit this time, occurs in a footnote to a 1925 paper [C 80: p. 30], added shortly before publication, after Cartan had seen Weyl's announcements [W 61, 62]. Part of this paper is devoted to the decomposition into irreducible tensors of the curvature and torsion tensors of an affine connection. Cartan states in that footnote that at the time he was writing the paper (December 1922) he viewed the full reducibility of tensor representations of semisimple Lie algebras as likely. At about the same time he wrote in [C 81] that this full reducibility had seemed to be extremely probable.

I do not know of any other statement by E. Cartan (expect for a similar one in a letter to Weyl) referring to that problem, which indicates that he may have thought about it between his Thesis and Weyl's work. See [Hw] for a detailed survey of the early stages of the representation theory of Lie algebras.

[9] (57) E. Cartan had checked this case by case in [C 38], but, as he wrote later to Weyl (3.28.1925), without looking for it, without realizing its importance, while determining all real forms of complex simple Lie algebras.

[10] (57) H. Weyl takes this for granted. However, as pointed out later by E. Cartan [C 116: p. 7], one has to know also that G_u is a closed subgroup of the

orthogonal group of B_u, but this follows immediately from the fact that G_u is the identity component of the group of automorphisms of $\mathfrak{g}_u$, (see [4] above).

[11] (57) In view of its importance and novelty, I'll try to sketch Weyl's argument, using some concepts introduced in the next 10 lines of the main text, with $K = G_u^0$. Let us say that $t \in T$ is regular if T is the identity component of its centralizer, singular otherwise. Using the fact that the roots come in pairs $\pm\alpha$, Weyl shows that the set of singular elements has codimension three, hence that any loop based on the identity is homotopic to one containing only regular elements, except for the base point. Any regular element is conjugate to one and only one element contained in a fundamental domain C_0 of W in T. From this it can be shown that the loop is homotopic to one contained in C_0. Under the exponential map C_0 is the image of a fundamental domain C_1 of the group of affine transformations of $\mathfrak{t}$ generated by W and the translations by the kernel of the exponential map. This is a simplex if G_u^0 is simple. Then a loop in C_0 is the image under the exponential map of a path in C_1 going from the origin to one vertex mapping to the identity and its homotopy class is determined by that vertex, whence the sought for finiteness.

[12] (58) There is a quibbling point which might bother the fastidious reader: 2ϱ is a well defined character of T, but not necessarily ϱ itself, hence $\lambda + \varrho$ in (8) might not be a character. However, it is if K is simply connected. We could make that assumption to be safe, but it is not necessary because the ambiguities of numerator and denominator cancel out in (8) and $|A(\varrho)|^2$ is anyhow well defined.

[13] (58) The differential of λ_π, multiplied by $\sqrt{-1}$, is then the highest weight of the representation of the complexification $\mathfrak{g} = \mathfrak{k}_{\mathbf{C}}$ of $\mathfrak{k}$ defined by the differential $d\pi$ of π, in the sense of E. Cartan. To see this and relate the concepts named similarly here and in [7], note that $\mathfrak{t}_C = \mathfrak{h}$ is a Cartan subalgebra of $\mathfrak{g}$ and that the map which assigns to $\lambda \in X(T)$ the linear form $i \cdot d\lambda$ on $\mathfrak{h}$ extends to an isomorphism of $X(T)_{\mathbf{R}}$ onto $\mathfrak{h}^*_{\mathbf{R}} = i \cdot \mathfrak{t}^*$ which commutes with the Weyl group and maps roots to roots and $X(T)$ to a lattice sitting between Q and P, (equal to Q if K is the adjoint group, to P if K is simply connected).

[14] (59) In his answer to the letter where Cartan outlines the argument sketched in [16] below, H. Weyl remarks that the starting point and general framework of both approaches are quite similar and adds:

> "Insbesondere entnehme ich aus Ihren Mitteilungen auch, daß es offenbar einigermaßen aussichtslos ist, den Satz von der vollen Reduzibilität zu beweisen, ohne den infinitesimalen Ansatz zu verlassen,"

echoed by Cartan in his next letter (3.28.25):

> "La difficulté, je n'ose dire l'impossibilité, de trouver une démonstration directe ne sortant pas du domaine strictement infinitésimal montre bien la nécessité de ne sacrifier aucun des deux points de vue...".

[15] (59) Formal definitions were provided soon after by O. Schreier, Abh. math. Sem. Hamburg **4** (1926), 15–32, **5** (1927), 233–244, and then by E. Cartan in [C 128], the first exposition of Lie groups and homogeneous spaces from the global point of view.

[16] (59) "sans être obligé de se livrer à des considérations d'analysis situs, toujours délicates," as he wrote to H. Weyl (3.1.25). First the existence of a compact form $\mathfrak{g}_u$ had been checked case by case in his determination of all the real forms [C 38], as recalled in [9]. Second, in order to apply the integration method, it was not really needed to know that the universal covering of G_u^0 was compact, but only that any covering $G_{\pi,u}$ provided by a linear representation π had a finite center. For this he gives an argument based on his discussion of the weights of irreducible representations [C 37]. Let $\mathfrak{t}$ be a Cartan subalgebra of $\mathfrak{g}_u$ and $T_{\pi,u}$ the integral group associated to it in the given representation π. E. Cartan's argument is that the weights of π in his sense are rational linear combinations of roots (because Q has finite index in P, in the notation of [7]), and therefore, if say, k is a common denominator, then at most k elements of $T_{\pi,u}$ may lie over a given element of the image of $T_{\pi,u}$ in the adjoint group. As I understand it, for this argument to be valid one would have to know that every element of $G_{\pi,u}$ is conjugate to one in $T_{\pi,u}$. This is indeed true and contained in [W 68], but its proof uses compactness and I do not see that it was already available to Cartan at that point.

[17] (60) He also speaks of promising "Ansätze" of Cartan, which would bring noteworthy simplifications, known to him through letters, but I did not find any in the correspondence I have seen.

[18] (60) Weyl did not hesitate to claim it was the better point of view [W 74]:

> "Et même dans ce cas particulier, notre méthode est supérieure aux méthodes anciennes et classiques de la théorie des séries de Fourier, car elle permet, comme je le crois, de se rendre compte pour la première fois des véritables raisons de la validité de la formule de Parseval. J'en vois aussi une confirmation dans le fait qu'elle put être appliquée aussi sans modification au cas traité dernièrement par H. Bohr des fonctions *presque périodiques*..."

[19] (62) In [HC: IV], it appears in the relation between the function $\mu_\omega(\nu)$, which gives the Plancherel measure for a continuous family of unitarily induced principal series representations and the c-functions, which govern the asymptotic behavior of Eisenstein integrals (see Corollary, p. 144).

[20] (63) In fact, [Cs] is just mentioned in [W III: II, p. 33] with no attempt to give an idea of the proof, even in the case of $\mathfrak{sl}_2$, which should have pleased Weyl in view of its origin, although [W III: I] describes the Casimir operator (p. 52) and its use to prove the Peter-Weyl theorem (pp. 100–102).

Soon R. Brauer provided another algebraic proof [Br 2], also using the Casimir operator, but in a quite different and simpler way, to which the one of [Bo 1] is very closely related, followed a few years later by J. H. C. Whitehead, whose argument, cohomological in spirit, was a boost to the development of the cohomology theory of Lie algebras (see [J]).

The Casimir operator was the first example of an operator in the center $\mathscr{Z}(\mathfrak{g})$ of the universal enveloping algebra $\mathscr{U}(\mathfrak{g})$ of a complex semisimple Lie algebra $\mathfrak{g}$. The search for others was taken up again by a physicist, G. Racah, whom I mention here because he wrote two letters to Weyl about it in 1947 and 1949. He determined in fact the degrees of the generators of $\mathscr{Z}(\mathfrak{g})$, see [Ra]. He was hoping that the eigenvalues of these generators would separate the irreducible representations (which is true). This would then allow one to adapt directly to the general case the full reducibility proof for $\mathfrak{sl}_2$, avoiding the technical complications of v. d. Waerden's argument.

In view of the other proofs mentioned above, this was not a goal of interest to the mathematicians but, independently, and for other reasons, Chevalley and Harish-Chandra also determined the structure of $\mathscr{Z}(\mathfrak{g})$ [HC: I, 292–360]. Soon $\mathscr{U}(\mathfrak{g})$ and $\mathscr{Z}(\mathfrak{g})$ became a basic tool for the study of finite or infinite dimensional representations of semisimple Lie groups, the foundational paper in that respect being the one of Harish-Chandra's just quoted. One of the first uses of $\mathscr{U}(\mathfrak{g})$ made there is an algebraic, classification-free, proof of the existence of an irreducible finite dimensional $\mathfrak{g}$ module with a given highest weight. The final step towards the algebraic treatment of all the main results of [W 68] was taken by H. Freudenthal, who supplied an algebraic proof of the character formula, and also one giving the multiplicity of a given weight in an irreducible representation (see [Bo 2], [J]).

[21] (64) Notwithstanding the foreword to [St 2], Study had very early advocated a broadening in this direction of Invariant Theory, following Lie, as can be seen from [St 1], a book which incidentally also describes all irreducible representations of $\mathbf{SL}_3$.

[22] (66) Many mathematicians not steeped in Classical Invariant Theory would concur with [ABP] in viewing the role of Capelli's identity as somewhat "mysterious". It becomes maybe less so, once it is pointed out that we are dealing with an element in the center of a universal enveloping algebra, as this was made very explicit in [Ho 2], on which I shall comment later (see [25]). It is remarkable that, without the terminology, this was quite clearly seen by Capelli himself: In [Ca 3] he writes the commutation relations for the D_{ij}'s [which show that they form a Lie algebra isomorphic to $\mathfrak{gl}_p(\mathbf{C})$]: if we view the p vectors as the p columns of a $n \times p$ matrix then the action of $\mathbf{GL}_n(\mathbf{C})$ is given by left multiplication on $\mathbf{M}_{n,p}(\mathbf{C})$ and the D_{ij}'s span the Lie algebra of $\mathbf{GL}_p(\mathbf{C})$, acting by right multiplication). Capelli then considers its enveloping algebra U, i.e. the associative subalgebra of differential operators on $\mathbf{M}_{n,p}(\mathbf{C})$ it generates, and proves a theorem about its structure, which amounts to say in present day terminology, that it is isomorphic to $\mathscr{U}(\mathfrak{gl}_p)$ and satisfies the Poincaré-Birkhoff-Witt theorem. More precisely, he defines a notion of "irreducible degree," (the usual filtration of $\mathscr{U}(\mathfrak{gl}_p)$] and shows that the associated graded algebra is a polynomial algebra. He adds that H belongs to the center of this algebra, as proved in [Ca 1] and that he had described $n-1$ other operators commuting with the D_{ij}'s in another paper [Ca 2]. Their expressions in terms of the polarization operators provide other "Capelli identities". Later on he showed that these operators are algebraically independent and generate the center of U, (Rend. d. R. Acc. delle Scienze di Napoli (2), VII (1893), 29–38, 155–162). Altogether he had determined the structure of the center of $\mathscr{U}(\mathfrak{gl}_p)$. I thank C. Procesi who told me Capelli had done that, which led me to look for it. See

also [P 2]. Of these papers, H. Weyl quotes only the first one, as far as I could see.

[23] (68) and an illustration of some of the comments of Chevalley and Weil [CWi] on Weyl's mathematical personality:

"Plutôt que de saisir l'idée brutalement au risque de la meurtrir, il aimait bien mieux la guetter dans la pénombre, l'accompagner dans ses évolutions, la décrire sous ses multiples aspects, dans sa vivante complexité. Etait-ce de sa faute si ses lecteurs, moins agiles que lui, éprouvaient parfois quelque peine à le suivre?"

[24] (68) For the convenience of the reader, I recall some definitions pertaining to linear algebraic groups, and refer to either my book (Benjamin, 1969) or the one of J. E. Humphreys' (Graduate Texts in Math., Springer 1975) for more details.

Let K be an algebraically closed groundfield. A linear algebraic group over K is a subgroup G of $\mathbf{GL}_n(K)$ whose elements are all the invertible matrices whose coefficients annihilate a given family of polynomials in n^2 indeterminates, with coefficients in K. It may be viewed as an affine variety with coordinate ring $K[G]$ generated by the matrix entries and the inverse of the determinant. A linear representation $\sigma: G \to \mathbf{GL}_m(K)$ is rational if the coefficients of $\sigma(g)$ belong to $K[G]$. Let G be irreducible as an algebraic variety. It is semi-simple if it has no infinite normal commutative subgroup, reductive if its center consists of semisimple elements. The flag varieties of G are the homogeneous spaces of G which are projective varieties. The isotropy groups of points in those are the parabolic subgroups of G. If $G = \mathbf{SL}_n(K)$ these are the stability groups of the flags, i.e. strictly increasing sequences of subspaces of K^n whose dimension form a given sequence of integers, of which Weyl has studied the invariant theory (in characteristic zero) in [W 60], as already pointed out.

Let now $K = \mathbf{C}$ and G be semisimple. It is also a complex Lie group but the representation theories of G viewed as a complex Lie group or as an algebraic group are essentially the same since any holomorphic finite dimensional representation is automatically rational. Moreover, any irreducible one occurs in a tensor product of copies of the defining representation of G and of its contragredient. This had been shown by Schur [Sc 2] for $\mathbf{SO}_n(\mathbf{C})$ and by Weyl [68] for $\mathbf{SL}_n(\mathbf{C})$ and $\mathbf{Sp}_{2n}(\mathbf{C})$, and now supplies the natural framework for an algebraic theory over more general groundfields. However, this does not hold for $\mathbf{GL}_n(\mathbf{C})$, i.e. for reductive non-semisimple groups, as was already stressed by Weyl in [W 68: I, § 8]: the rational representations of $\mathbf{GL}_n(\mathbf{C})$ are fully reducible [Sc 1], but not all of the holomorphic ones are so.

[25] (70) Let (H, H') be a reductive pair in $\mathbf{GL}_N(\mathbf{C})$ and assume H, H' to be connected. Instead of the enveloping algebras of H and H' consider those B and B' of their Lie algebras $\mathfrak{h}$ and $\mathfrak{h}'$. They are the centralizers of one another in $\mathbf{M}_N(\mathbf{C})$ and are fully reducible. Their intersection is the center of each. Now B and B' are quotients of $\mathscr{U}(\mathfrak{h})$ and $\mathscr{U}(\mathfrak{h}')$ and, by full reducibility, the center of B (resp. B') is the image of $\mathscr{Z}(\mathfrak{h})$, resp. $\mathscr{Z}(\mathfrak{h}')$. This remark is made in [Ho 3] and in particular applied to the Capelli identities, in the case of p vectors in n-space.

There we get back to the situation described in [19]: $N=n\cdot p$, H is $\mathbf{GL}_n(\mathbf{C})$ operating on $\mathbf{M}_{n,p}(\mathbf{C})$ by left translations and H' is $\mathbf{GL}_p(\mathbf{C})$ operating by right translations, B' is isomorphic to $\mathscr{U}(\mathfrak{gl}_p)$ and $B\cap B'$ is a polynomial algebra in the Capelli elements.

References

[A] M. Artin, On Azumaya algebras and finite dimensional representations of rings, J. Algebra **11** (1969), 532–563

[ABP] M. Atiyah, R. Bott, V. Patodi, On the heat equation and the Index Theorem, Inv. Math. **19** (1973), 279–330

[Bo 1] N. Bourbaki, Groupes et Algèbres de Lie I, Hermann, Paris 1971

[Bo 2] N. Bourbaki, Groupes et Algèbres de Lie 7, 8, Hermann, Paris 1975

[Br 1] R. Brauer, Eine Bedingung für vollständige Reduktion von Darstellungen gewöhnlicher und infinitesimaler Gruppen, Math. Zeitschr. **41** (1936), 330–339; C.P.III, 462–471

[Br 2] R. Brauer, On algebras which are connected with the semisimple continuous groups, Annals of Math. **38** (1937), 857–872; C.P.III, 446–461

[Ca 1] A. Capelli, Über die Zurückführung der Cayley'schen Operation Ω auf gewöhnliche Polar-Operationen, Math. Annalen **29** (1887), 331–338

[Ca 2] A. Capelli, Ricerca delle operazioni invariantive fra piu serie di variabili permutabili con ogni altra operazione invariantiva fra le stesse serie, Atti delle Scienze Fis. e Mat. di Napoli (2) I, (1888), 1–17

[Ca 3] A. Capelli, Sur les opérations dans la théorie des formes algébriques, Math. Annalen **37** (1890), 1–37

E. Cartan, papers:
The numbering is the one of the "Oeuvres Complètes" (O.C.), 6 vol. Gauthier-Villars, Paris 1952

[C 5] Sur la structure des groupes de transformations finis et continus, Thèse, Paris, Nony; 2ᵉ édition, Vuibert 1933; O.C. I_1, 137–253

[C 37] Les groupes projectifs qui ne laissent invariante aucune multiplicité plane, Bull. Soc. math., t. **41**, (1913), 53–96; O.C. I_1, 355–398

[C 38] Les groupes réels simples finis et continus, Ann. Éc. Norm., t. **31**, (1914), 263–355; O.C. I_1, 399–491

[C 65] Sur un théorème fondamental de M. H. Weyl, J. Math. pures et appliquées, t. **2**, (1923), 167–192; O.C. III_1, 633–648

[C 80] Sur les variétés à connexion affine et la théorie de la relativité généralisée, Ann. Sci. E.N.S. Paris **42**, (1925), 17–88; O.C. III_2, 921–995

[C 81] Les tenseurs irréducibles et les groupes linéaires simples et semisimples, Bull. Soc. math., t. **49**, (1925), 130–152; O.C. I_1, 531–553

[C 82] Le principe de dualité et la théorie des groupes simples et semi-simples, Bull. Sc. Math. **49** (1925), 361–374; O.C. I_1, 555–568

[C 93] Sur une classe remarquable d'espaces de Riemann, Bull. Soc. math., t. **54**, (1926), 214–264; O.C. I_2, 587–637

[C 94] Sur une classe remarquable d'espaces de Riemann, Bull. Soc. math., t. **55**, (1927), 114–134; O.C. I_2, 639–659

[C 103] La géométrie des groupes simples, Annali di Mat., t. **4**, (1927), 209–256; O.C. I_2, 793–840

[C 105] La théorie des groupes et la géométrie, L'Enseignement math., t. **26**, (1927), 200–225; O.C. I_2, 841–866

[C 107] Sur certaines formes riemanniennes remarquables des géométries à groupe fondamental simple, Ann. Éc. Norm., t. **44**, (1927), 345–467; O.C. I_2, 867–989

[C 111] Sur les nombres de Betti des espaces de groupes clos, C. R. Acad. Sc., t. **187**, (1928), 196–198, O.C. I_2, 999–1001

[C 113] Complément au mémoire "sur la géométrie des groupes simples", Annali di Mat., t. **5**, (1928), 253–260; O.C. I_2, 1003–1010

[C 116] Groupes simples clos et ouverts et géométrie riemannienne, J. Math. pures et appliquées, t. **8**, (1929), 1–33; O.C. I_2, 1011–1043

[C 117] Sur la détermination d'un système orthogonal complet dans un espace de Riemann symétrique clos, Rend. Circ. mat. Palermo, t. **53**, (1929), 217–252; O.C. I_2, 1045–1080

[C 118] Sur les invariants intégraux de certains espaces homogènes clos et les propriétés topologiques de ces espaces, Ann. Soc. Pol. math. **8** (1929), 181–225; O.C. I_2, 1081–1125

[C 128] La théorie des groupes finis et continus et l'analysis situs, Mem. Sci. Math. XLII (1930), Gauthier-Villars; O.C. I_2, 1165–1224

[Cs] H. Casimir, Über die Konstruktion einer zu den irreduziblen Darstellungen halbeinfacher kontinuierlicher Gruppen gehörigen Differentialgleichung, Proc. Kon. Ak. Wet. Amsterdam **34** (1931), 844–846

[Cw] H.L. Casimir, B.L.v.d. Waerden, Algebraischer Beweis der vollständigen Reduzibilität der Darstellungen halbeinfacher Liescher Gruppen, Math. Annalen **111** (1935), 1–12

[CWi] C. Chevalley, A. Weil, Hermann Weyl (1885–1955), Ens. Math. III (1957), 157–187; H. Weyl, Gesammelte Abhandlungen IV, 655–685

[CP] C. De Concini, C. Procesi, A characteristic free approach to invariant theory, Adv. in Math. **21** (1976), 300–354

[DRS] P. Doubilet, G.C. Rota, J. Stein, On the foundations of combinatorial theory, Vol. IX, 185–216, Studies in Applied Math. **53** (1974)

[DHS] W. Dwyer, W.C. Hsiang, R. Staffeldt, Pseudo isotopy and invariant topology I, Topology **19** (1980), 367–385, II, Topology Symposium Siegen 1979, LNM **788** (1980), 418–441, Springer

[F] H. Freudenthal, Zu den Weyl-Cartanschen Raumproblemen, Archiv der Math. **11** (1960), 107–115

[Ha] W. Haboush, Reductive groups are geometrically reductive, Annals of Math. **102** (1975), 67–84

[HC] Harish-Chandra, Collected Papers, 4 vol., Springer 1984

[Ho 1] R. Howe, Invariant theory and duality for classical groups over finite fields with applications to their singular representation theory, preprint

[Ho 2] R. Howe, Remarks on classical invariant theory, preprint

[Ho 3] R. Howe, θ-series and invariant theory, in Automorphic forms, Representations and *L*-functions, Proc. symp. pure math. **33** (1979) Part I, 275–285

[Ho 4] R. Howe, On the role of the Heisenberg group in harmonic analysis, Bull. A.M.S. (N.S.) **3** (1980), 821–843

[Ho 5] R. Howe, Transcending classical invariant theory, preprint

[Hu] A. Hurwitz, Über die Erzeugung der Invarianten durch Integration, Nachr. kön. Ges. Wiss. Göttingen, 1897, 71–90; Mathematische Werke II, 546–564

[Hw] T. Hawkins, Elie Cartan and the prehistory of the theory of representations of Lie algebras, preprint

[J 1] N. Jacobson, Rational methods in the theory of Lie algebras, Annals of Math. **36** (1935), 875–881

[J] N. Jacobson, Lie Algebras, Intersc. tracts pure and applied math. **10**, Interscience Publ. 1962

[K] W. Klingenberg, Eine Kennzeichnung der Riemannschen sowie der Hermiteschen Mannigfaltigkeiten, Math. Zeitschr. **70** (1959), 300–309

[L] S. Lie, Fr. Engel, Theorie der Transformationsgruppen III, Teubner, Leipzig 1893

[M] G. Mackey, Hermann Weyl and the applications of group theory to quantum mechanics, in Exakte Wissenschaften und ihre philosophische Grundlegung – Vorträge des internationalen Hermann Weyl Kongresses, Kiel 1985. P. Lang, Berlin 1986

[Na] M. Nagata, Invariants of a group in an affine ring, J. Math. Kyoto Univ. **3** (1963/64), 369–377

[Ne] J. v. Neumann, Die Eindeutigkeit der Schrödingerschen Operatoren, Math. Annalen **104** (1931), 570–578

[P 1] C. Procesi, The invariant theory of $n \times n$ matrices, Adv. in Math. **19** (1976), 306–381

[P 2] C. Procesi, Sulla formula di Gordan Capelli in La Matematica nella cina antica e di oggi, Università di Ferrara 1979

[Ra] G. Racah, Sulla caratterissione delle representazioni irreducibili dei gruppi semisemplici di Lie, Atti Acc. Naz. Lincei VIII (1950), 108–112

[Ri] B. Riemann, Über die Hypothesen, welche der Geometrie zugrunde liegen, edited and commented by H. Weyl, Springer, Berlin 1919

[Sc 1] I. Schur, Über eine Klasse von Matrizen, die sich einer gegebenen Matrix zuordnen lassen, Dissertation, Berlin 1901; Ges. Abh. I, 1–72

[Sc 2] I. Schur, Neue Anwendungen der Integralrechnung auf Probleme der Invariantentheorie I, Sitzungsber. Pr. Ak. d. Wiss. Berlin 1924, 189–208 II, *ibid.* 297–321 Ges. Abh. II, 440–484

[Se] C.S. Seshadri, Standard monomial theory and the work of Demazure, Advanced Studies in pure math. **1** (1983), 355–384, Kinokunya and North-Holland

[So] M.H. Stone, Linear transformation in Hilbert space III. Operational methods and group theory, Proc. Nat. Ac. Sci. USA **16** (1930), 172–175

[St 1] E. Study, Methoden zur Theorie der ternären Formen, Teubner, Leipzig 1889

[St 2] E. Study, Einleitung in die Theorie der Invarianten linearer Transformationen auf Grund der Vektorrechnung, Vieweg, Braunschweig 1923

[Wi 1] A. Weil, L'intégration dans les groupes topologiques et ses applications, Hermann, Paris 1940

[Wi 2] A. Weil, Sur certains groupes d'opérateurs unitaires, Acta Math. **111** (1964), 143–221. Sur la formule de Siegel dans la théorie des groupes classiques, *ibid.* **113** (1965), 1–87; O.S. III, 1–69, 70–157

[W I] H. Weyl, Raum, Zeit, Materie; Vorlesungen über allgemeine Relativitätstheorie, 4. Auflage, Springer, Berlin 1921

[W II] H. Weyl, Mathematische Analyse des Raumproblems, Springer, Berlin 1923; Wissenschaftliche Buchgesellschaft, Darmstadt 1977

[W III] H. Weyl, The structure and representations of continuous groups, I, Notes by N. Jacobson, II, Notes by R. Brauer, The Institute for Advanced Study, 1934–1935

[W IV] H. Weyl, I, Elementary Theory of Invariants. II, Invariant Theory, Outline by A.H. Clifford, The Institute for Advanced Study, Princeton, 1935–1936

[W V] H. Weyl, Gruppentheorie und Quantenmechanik, Hirzel, Leipzig 1928. English translation (of the second edition) by H.P. Robertson, Dover Publ.

[W VI] H. Weyl, The Classical Groups, their Invariants and Representations, Princeton University Press, 1st edition, 1939, 2nd edition, 1946

H. Weyl, papers:
The numbering is the one of the Gesammelte Abhandlungen (G.A.) 4 Bde., Springer, Berlin 1968

[W 8] Über gewöhnliche Differentialgleichungen mit Singularitäten und die zugehörigen Entwicklungen willkürlicher Funktionen, Math. Annalen **68** (1910), 220–269; G.A. I, 248–297

[W 49] Die Einzigartigkeit der Pythagoreischen Maßbestimmung, Math. Zeitschr. **12** (1922), 114–146; G.A. II, 263–295

[W 60] Randbemerkungen zu Hauptproblemen der Mathematik, Math. Zeitschr. **20** (1924), 131–150; G.A. II, 433–452

[W 61] Zur Theorie der Darstellung der einfachen kontinuierlichen Gruppen. (Aus einem Schreiben an Herrn I. Schur), Sitzungsberichte der Preußischen Akademie der Wissenschaften zu Berlin (1924), 338–345; G.A. II, 453–460

[W 62] Das gruppentheoretische Fundament der Tensorrechnung, Nachrichten der Gesellschaft der Wissenschaften zu Göttingen. Mathematisch-physikalische Klasse (1924), 218–224; G.A. II, 461–467

[W 63] Über die Symmetrie der Tensoren und die Tragweite der symbolischen Methode in der Invariantentheorie, Rendiconti del Circolo Matematico di Palermo **48** (1924), 29–36; G.A. II, 468–475

[W 68] Theorie der Darstellung kontinuierlicher halbeinfacher Gruppen durch lineare Transformationen. I, II, III und Nachtrag, I: Math. Zeitschr. **23** (1925), 271–309; II: Math. Zeitschr. **24** (1926), 328–376; III: Math. Zeitschr. **24** (1926), 377–395; Nachtrag: Math. Zeitschr. **24** (1926), 789–791; G.A. II, 543–647

[W 69] Zur Darstellungstheorie und Invariantenabzählung der projektiven, der Komplex- und der Drehungsgruppe, Acta Mathematica **48** (1926), 255–278; G.A. III, 1–24

[W 70] Elementare Sätze über die Komplex- und die Drehungsgruppe, Nachrichten der Gesellschaft der Wissenschaften zu Göttingen, Mathematisch-physikalische Klasse, (1926), 235–243; G.A. III, 25–33

[W 71] Beweis des Fundamentalsatzes in der Theorie der fastperiodischen Funktionen, Sitzungsberichte der Preußischen Akademie der Wissenschaften zu Berlin, (1926), 211–214; G.A. III, 34–37

[W 72] Integralgleichungen und fastperiodische Funktionen, Math. Annalen **97** (1927), 338–356; G.A. III, 38–57

[W 73] Die Vollständigkeit der primitiven Darstellungen einer geschlossenen kontinuierlichen Gruppe (gem. mit F. Peter), Math. Annalen **97** (1927), 737–755; G.A. III, 58–75

[W 74] Sur la représentation des groupes continus, Ens. Math. **26** (1927), 226–239; G.A. III, 76–89

[W 75] Quantenmechanik und Gruppentheorie, Zeitschr. f. Physik **46** (1927), 1–46; G.A. III, 90–135

[W 76] Strahlbildung nach der Kontinuitätsmethode behandelt, Nach. Ges. Wiss. Göttingen (1927), 227–237; G.A. III, 136–146

[W 79] Der Zusammenhang zwischen der symmetrischen und der linearen Gruppe, Annals of Math. **30** (1929), 499–516; G.A. III, 171–188

[W 80] Kontinuierliche Gruppen und ihre Darstellungen durch lineare Transformationen, Atti Congr. Intern. d. Mat. Bologna 1928, Vol. I, (1929), 233–246; G.A. III, 189–202

[W 96] Über Algebren, die mit der Komplexgruppe in Zusammenhang stehen, und ihre Darstellungen, Math. Zeitschr. **35** (1932), 300–320; G.A. III, 359–379

[W 98] Harmonics on homogeneous manifolds, Annals of Math. **35** (1934), 486–499; G.A. III, 386–399

[W 105] Spinors in *n* dimensions (R. Brauer und H. Weyl), American Journal of Math. **57** (1935), 425–449; G.A. III, 493–516

[W 107] Generalized Riemann Matrices and factor sets, Annals of Math. **37** (1936), 709–745; G.A. III, 534–570

[W 117] Invariants, Duke Math. Jour. **5** (1939), 489–502; G.A. III, 670–683

[W 122] On the use of indeterminates in the theory of the orthogonal and symplectic groups, Amer. Jour. of Math. **63** (1941), 777–784; G.A. III, 670

[W 147] Relativity theory as a stimulus in mathematical research, Proceedings of the American Philosophical Society **93** (1949), 535–541; G.A. IV, 394–400

[Wn] A. Weinstein, Fundamentalsatz der Tensorrechnung, Math. Zeitschr. **16** (1923), 78–91

ETH, 8092 Zürich, Switzerland
and
The Institute for Advanced Study, Princeton, N.J., 08540, U.S.A.

(Received 18 December 1985)

134.

A Vanishing Theorem in Relative Lie Algebra Cohomology

in *Algebraic Groups*, Utrecht 1986, Lect. Notes Math. **1271** (1987), 1–16,
Springer-Verlag

This paper is devoted to the proof of a vanishing theorem for the relative Lie algebra cohomology of a semisimple Lie group with respect to tensor products of irreducible finite dimensional and unitary representations. It plays an essential role in the proofs of Zucker's conjecture in rational ranks 1 and 2 which are sketched in [B] and [BC2] respectively. Since its context is somewhat broader than that of the conjecture, and the techniques used to establish it do not occur in other steps of the proof of Zucker's conjecture in those cases, it seemed best to devote a separate paper to it, although, so far, it is of interest only through this application.

Let L be a connected reductive Lie group with compact center, $\mathfrak{l}$ its Lie algebra, $K(L)$ a maximal compact subgroup of L, E a finite dimensional and H a unitary representation of L, both irreducible. We consider the relative Lie algebra cohomology space $H^{\cdot}(\mathfrak{l},K(L);E\otimes H)$ and are interested in the highest degree in which it is non-zero. Our starting point to study this are the results of Vogan-Zuckerman [VZ]. For L simple, connected and E as above, they give a list of the irreducible unitary representations H of L for which the above cohomology space $H^{\cdot}$ is not identically zero and a description of $H^{\cdot}$. [If L is a complex simple Lie group, viewed as real group, this was done earlier by Enright [E].] However, in our situation, L (which is not necessarily simple, but this is a minor point) is a factor of a Levi subgroup $M.A$ of a maximal proper parabolic subgroup P of a simple Lie group G and the desired vanishing condition involves M,G and the dimensions of various symmetric spaces, so that some work is needed to prove it starting from [VZ]. The first proof, alluded to in [B], was a painful case by case checking, which in fact could not be carried out in all cases involving the two exceptional bounded domains. However, by a happy feature of the classification of $\mathbf{Q}$-simple groups of $\mathbf{Q}$-rank 1 this was not needed to prove the conjecture in the $\mathbf{Q}$-rank 1 case. My goal here is to provide an essentially *a priori* proof of that vanishing theorem. It is still not quite classification free: condition (B) of §1 will have to be checked case by case, but this is of a much more elementary nature (and, I hope, will also be replaced later by a more conceptual argument).

§1 introduces some notation and definitions and the two basic assumptions (A) and (B) of the vanishing theorem. The latter is stated in §2, followed by some brief indications on its applications. §3 recalls some of the results of [VZ] and uses them to reformulate the vanishing theorem in terms of Lie algebra data (3.7(7)). In §4, 3.7(7) is deduced from a slightly stronger inequality 4.4(1). The latter is proved in §5. Finally §6 checks condition (B) in the cases of interest for the applications alluded to above.

§1. *Preliminaries. Notation. Assumptions.*

1.1. Let L be a reductive group, $K(L)$ a maximal compact subgroup of L and $\mathfrak{h}$ a Cartan subalgebra of $\mathfrak{l}$. For $\mu \in \mathfrak{h}_c^*$, let F_μ denote an irreducible finite dimensional representation of L with infinitesimal character χ_μ (i.e. highest weight $\mu-\rho_L$ if μ is dominant). Let $C(L,\mu)$ be the greatest q for which there exists $H \in \hat{L}$ such that $H^q(\mathfrak{l},K(L);H\otimes F_\mu) \neq 0$. If there is no such q, then set $C(L,\mu) = -\infty$. This is the case if and only if $F_\mu \not\cong \bar{F}_\mu^*$ [V; VZ].

1.2. In the sequel G is a connected real simple Lie group, with finite center, K a maximal compact subgroup, θ the associated Cartan involution, E an irreducible representation of G with infinitesimal character χ_λ, P a proper maximal parabolic subgroup and $P = N.A.M.$ a Langlands decomposition of P, with A,M stable under θ. We fix a Cartan subalgebra $\mathfrak{h} = \mathfrak{a} \oplus \mathfrak{h}_M$ invariant under θ, where $\mathfrak{h}_M$ is a *fundamental* Cartan subalgebra of $\mathfrak{m}$. We choose an ordering on the set $\Phi(\mathfrak{g}_c)$ of roots of $\mathfrak{g}_c$ with respect to $\mathfrak{h}_c$ such that

$$\Phi^+|_{\mathfrak{a}} \subset \Phi(P,A) \cup \{0\} \tag{1}$$

where $\Phi(P,A)$ denotes the weights of $\mathfrak{a}$ in $\mathfrak{n}$. We let W^P be the usual set of representatives of $W(\mathfrak{m}_c \oplus \mathfrak{a}_c)\backslash W(\mathfrak{g}_c)$. An element ν of $\mathfrak{a}^*$ is >0 if it is a positive real multiple of some element in $\Phi(P,A)$.

1.3. We assume given a decomposition of M into an almost direct product $M = L.G_P$ of θ-invariant closed subgroups and write accordingly $\mathfrak{h}_M = \mathfrak{h}_L \oplus \mathfrak{h}_{G_P}$. The groups $K(L) = K \cap L$ and $K(G_P) = K \cap G_P$ are maximal compact in L and G_P respectively. The following assumption is basic

$$\text{rk } G = \text{rk } K \quad\text{and}\quad \text{rk } G_P = \text{rk } K(G_P). \tag{A}$$

It follows first that $\mathfrak{h}_{G_P}$ is a Cartan subalgebra of $K(G_P)$. We may write $\mathfrak{h}_L = \mathfrak{a}_0 \oplus \mathfrak{t}_0$, with $\mathfrak{a}_0$ split and $\mathfrak{t}_0$ a Cartan subalgebra of $K(L)$. Then $\mathfrak{t}_0 \oplus \mathfrak{h}_{G_P} = \mathfrak{t}$ is a Cartan subalgebra of a maximal compact subgroup $K(M)$ of M. The centralizer $\mathfrak{z}_\mathfrak{g}(\mathfrak{t})$ of $\mathfrak{t}$ in $\mathfrak{g}$ can be written as $\mathfrak{t} \oplus \mathfrak{z}_1$. We have $\mathfrak{z}_1 \cap \mathfrak{m} = 0$ since $\mathfrak{t}$ contains regular elements of $\mathfrak{m}$. From (A) it follows that $\mathfrak{z}_1$ has a split Cartan subalgebra $(\mathfrak{a}_0 \oplus \mathfrak{a})$ and a compact one. Therefore $\mathfrak{z}_1$ is semi-simple and split over $\mathbb{R}$.

At the very end of the proof, I shall need to use the following assumption, which is easily checked in the cases of interest (see §6) and which may well be true whenever (A) is and G_P is the greatest factor of M satisfying the second equality of (A).

(B) *The intersection of* $\mathfrak{l}$ *with* $\mathfrak{z}_{\mathfrak{g}}(\mathfrak{z}_1)$ *is the Lie algebra of a compact subgroup of* L.

Note that $\mathfrak{z}_1$ contains $\mathfrak{a}_0$. Therefore, if $\mathfrak{a}_0$ is maximal $\mathbb{R}$-split in $\mathfrak{l}$, (e.g. if $\mathfrak{l}$ is a complex semi-simple Lie algebra, viewed as a real Lie algebra), then (B) is fulfilled. It so happens that this stronger condition is often fulfilled.

§2. *The vanishing theorem.*

2.0. *Notation.* For $s \in W^P$ we write $\nu(s\lambda)$ for $s\lambda|\mathfrak{h}_L$. For H reductive, X_H is the quotient of H by a maximal compact subgroup.

2.1. THEOREM. *We keep the previous notation and assume* (A),(B) *of* 1.3 *to hold. Let* $s \in W^P$. *Then*

(1) $$C(L,\nu(s\lambda)) + \ell(s) < (\dim X_G - \dim X_{G_P})/2, \quad \text{if } s\lambda|\mathfrak{a} > 0.$$

2.2. By [C: 2.6] the condition $s\lambda|\mathfrak{a} > 0$ is equivalent to $2\ell(s) < \dim N$. By [K], $F_{s\lambda}$, viewed as a M.A. module, occurs in $H^{\ell(s)}(\mathfrak{n};E_\lambda)$, with multiplicity one, and nowhere else in $H^{\cdot}(\mathfrak{n};E_\lambda)$. We can therefore also write (1) as:

(2) $$H^q(\mathfrak{l},K(L);H^i(\mathfrak{n};E_\lambda)\otimes H) = 0 \quad \text{for } i < (\dim N)/2 \text{ and } q + i \geq m,$$

where

(3) $$m = (\dim X_G - \dim X_{G_P})/2.$$

2.3. We now sketch two (related) applications of 2.1.

(a) Combined with the results of [BC1], 2.1 implies that if Γ is an arithmetic subgroup of L, for some given $\mathbb{Q}$-structure, then

(1) $$H^q(\mathfrak{l},K(L);F_{\nu(s\lambda)} \otimes L^{2,\infty}(\Gamma\backslash L)) = 0 \quad \text{for } q \geq m-\ell(s) \text{ and } \ell(s) < (\dim N)/2.$$

(b) We consider the setup of Zucker's conjecture [B; BC2; Z]. Let $\mathbf{G}$ be a $\mathbb{Q}$-simple connected linear algebraic group, and Γ an arithmetic subgroup. Assume that the symmetric space X of maximal compact subgroups of $G = \mathbf{G}(\mathbb{R})$ is a bounded symmetric domain and let V^* be the minimal compactification of $V = \Gamma\backslash X$ constructed in [BB]. Assume that P is the normalizer of a rational boundary component X_P of maximal dimension. We have an almost direct product decomposition $M = L.G_P$ where L is the greatest subgroup of M acting trivially on X_P, (possibly up to a compact factor), and X_P is the symmetric space of maximal compact subgroups of G_P. Modulo various reductions, 2.1 allows one to prove that the local L^2-cohomology of the intersection with V of a suitable neighborhood U of a point of the image V_P of X_P in V^* vanishes from the complex codimension of X_P on. This is the main point needed to show that the L^2-cohomology sheaf on V^* is homology isomorphic to the middle intersection cohomology sheaf along V_P. In this case, it is a rather direct application of 2.1. For the next stratum (level 2), the situation is much more complicated and 2.1 has to be used in conjunction with results of Casselman on automorphic forms (cf. [BC 2]).

§3. *Results of Vogan-Zuckerman. Reformulation of* 2.1.

3.1. In the sequel, I write $\mathfrak{h}_o$ for $\mathfrak{h}_L$. I recall that under the first assumption of 1.3(A), $-w_G$ gives the effect of complex conjugation on the Dynkin diagram (see [BC1]). It leaves the set $\Delta(\mathfrak{l}_c)$ of simple roots of $\mathfrak{l}_c$ with respect to $\mathfrak{h}_{o,c}$ stable. Also w_G leaves $\mathfrak{a}$ stable. Moreover the transformation $\tau_{\mathfrak{l}}$ (cf. [BC1]), which assigns to a representation its complex conjugate contragredient one, is given by $w_L w_G$. This is the same as $w_M w_G$, because $w_{G_P} \cdot w_G$ is the identity of $\Delta(\mathfrak{g}_P, \mathfrak{h}_{G_P})$ in view of (A). We write also α^* for $w_m w_G \alpha$ $(\alpha \in \Delta(\mathfrak{l}_c))$. We have then

(1) $C(L,\nu(s\lambda)) \neq -\infty \Leftrightarrow (s\lambda,\alpha) = (s\lambda,\alpha^*)$ for all $\alpha \in \Delta(\mathfrak{l}_c) \Leftrightarrow s\lambda|\mathfrak{a}_o = 0$.

3.2. We have $\mathfrak{h}_o = \mathfrak{t}_o \oplus \mathfrak{a}_o$ and $\mathfrak{t}_o$ contains regular elements of $\mathfrak{l}$. We may assume that $\Phi^+(\mathfrak{l}_c)$ is given by $\alpha(it) > 0$ for some regular $t \in \mathfrak{t}_o$. We let also θ denote the Cartan involution of L with respect to $K(L)$. It is indeed the restriction of θ. Its restriction to $\mathfrak{h}_o$ is also the linear transformation defined by $\tau_{\mathfrak{l}}$. The algebra $\mathfrak{b}_c = \mathfrak{h}_{o,c} \oplus_{\alpha>0} \mathfrak{l}_{c,\alpha}$ is θ-stable. A *standard* θ*-stable parabolic subalgebra* $\mathfrak{q}$ of $\mathfrak{l}_c$ is one which contains $\mathfrak{b}_c$ and has a Levi subalgebra $\mathfrak{m}_\mathfrak{q}$ which is the centralizer of some element in $\mathfrak{t}$. It is in particular θ-invariant, and $\mathfrak{m}_{\mathfrak{q},o} = \mathfrak{m}_\mathfrak{q} \cap \mathfrak{l}$ is a real form of $\mathfrak{m}_\mathfrak{q}$. We let $M_{\mathfrak{q},o}$ be the corresponding analytic subgroup of L and $X_\mathfrak{q}$ the symmetric space $M_{\mathfrak{q},o}/K(M_{\mathfrak{q},o})$.

Let $\nu \in \mathfrak{h}_o^*$ be dominant regular. We say that $\mathfrak{q}$ and ν are *compatible* if $\nu - \rho_L$ is the differential of a unitary character of $M_{\mathfrak{q},o}$. This is the case if and only if ν is zero on $\mathfrak{a}_o$ and on the derived algebra of $\mathfrak{m}_\mathfrak{q}$. Given a compatible pair $(\mathfrak{q},\nu)$, [VZ] defines an irreducible representation $A_\mathfrak{q}(\nu-\rho_L)$ of L, which is unitary by [V], such that

(1) $H^\cdot(\mathfrak{l},K(L);A_\mathfrak{q}(\nu-\rho_L)\otimes F_\nu) = H^\cdot(\mathfrak{m}_{\mathfrak{q},o},K(M_{\mathfrak{q},o});\mathbf{C})$, suitably translated.

"Suitably translated" means so that the left-hand side satisfies Poincaré duality. It follows that the top cohomology occurs in dimension $(\dim X_L+\dim X_{\mathfrak{q}})/2$. Moreover, by [VZ], any irreducible unitary representation H such that $H^{\cdot}(\mathfrak{l},K(L);H\otimes F_\nu) \neq 0$ is so obtained. Therefore

(2) $$2.C(L,\nu) = \mathrm{Max}_{\mathfrak{q}}(\dim X_L+\dim X_{\mathfrak{q}}) \quad \text{if} \quad \nu|\mathfrak{a}_o = 0,$$

where $\mathfrak{q}$ runs through the standard θ-stable parabolic subalgebras which are compatible with ν. The biggest possible $M_{\mathfrak{q},o}$ is the one which is generated by $\mathfrak{h}_o$ and a semi-simple group $M'_{\mathfrak{q},o}$ whose roots are all the $\beta \in \Phi(\mathfrak{l}_c)$ such that

(3) $$(\nu,\beta) = (\rho_L,\beta).$$

We can write $\beta = \Sigma\, c_\alpha \alpha$ (with $\alpha \in \Delta(\mathfrak{l}_c)$) and the c_α integers all of the same sign. We have then

(4) $$\textstyle\sum_\alpha c_\alpha(\nu,\alpha) = \sum c_\alpha(\rho_L,\alpha).$$

But $(\rho_L,\alpha) = (\alpha,\alpha)/2$ and, since ν is dominant regular, $(\nu,\alpha) \geq (\alpha,\alpha)/2$. Equality holds, therefore, if and only if

(5) $$c_\alpha \neq 0 \Rightarrow (\nu,\alpha) = (\rho_L,\alpha).$$

The roots of $M'_{\mathfrak{q},o}$ are therefore all the linear combinations of the elements of

(6) $$\Delta_\nu = \{\alpha \in \Delta(\mathfrak{l}_c) \,|\, (\nu,\alpha) = (\rho_L,\alpha)\}.$$

Let us write M_ν, M'_ν, X_ν for $M_{\mathfrak{q},o}, M'_{\mathfrak{q},o}, X_{\mathfrak{q}}$ for this choice of $\mathfrak{q}$. We have

(7) $$\text{If} \quad \nu|\mathfrak{a}_o = 0, \quad \text{then } 2.C(L,\nu) = \dim X_L + \dim X_\nu.$$

Moreover, it is clear that

(8) $$\dim X_\nu \leq \dim X'_\nu + \dim \mathfrak{a}_o,$$

where X'_ν is the symmetric space of maximal compact subgroups of M'_ν.

In all this, it was assumed that $\tau_1\nu = \nu$. Then $\Phi(M'_\nu)$ and Δ_ν are automatically τ_1-stable. However the assumption $\tau_1\nu = \nu$ does not play a role in the implication (5). Generalizing the above slightly, for any dominant regular ν we let Δ_ν be the greatest τ_1-stable subset of $\Delta(\mathfrak{l}_c)$ whose elements are orthogonal to $\nu - \rho_L$ and M'_ν, M_ν the corresponding groups. Then $\Phi(M'_\nu)$ is the greatest τ_1-stable subset of $\Phi(1_c)$ all of whose elements are orthogonal to $\nu - \rho_L$ and Δ_ν is a basis of $\Phi(M'_\nu)$.

3.3. We now come back to 2.1. We have

$$\dim X_G = \dim X_L + \dim X_{G_P} + \dim N + 1.$$

Therefore by 3.2(7), 2.1(1) can be written

(1) $$\ell(s) + (\dim X_L + \dim X_{\nu(s\lambda)})/2 < (\dim X_L + \dim N+1)/2$$

or

(2) $$\ell(s) + \dim X_{\nu(s\lambda)}/2 < (\dim N+1)/2.$$

The map $s \mapsto s' = w_M w_G s$ is an involution of W^P. We have

(3) $$\ell(s') + \ell(s) = \dim N$$

(4) $$s'\lambda|\mathfrak{a} + s\lambda|\mathfrak{a} = 0$$

(5) $F_{\nu(s'\lambda)}$ is complex conjugate contragredient to $F_{\nu(s\lambda)}$.

[This is the s' occurring in the proof of 2.6 in [C].] The relation (2) is equivalent to

$$\ell(s) + \dim X_{\nu(s\lambda)}/2 \leq (\dim N)/2 = (\ell(s)+\ell(s'))/2,$$

hence to

(6) $$\ell(s') - \ell(s) \geq \dim X_{\nu(s\lambda)} \quad \text{if} \quad \ell(s) < (\dim N)/2 \quad \text{and} \quad s\lambda|\mathfrak{a}_o = 0.$$

The left-hand side depends only on s, and we want to prove 2.1 for any dominant regular λ. Therefore we are reduced to showing

(7) *Fix* $s \in W^P$ *such that* $\ell(s) < (\dim N)/2$. *Then* $\ell(s') - \ell(s) \geq \max \dim X_{\nu(s\lambda)}$, *where* λ *runs through the regular dominant weights of* $\mathfrak{g}$ *such that* $s\lambda|\mathfrak{a}_o = 0$.

§4. *Further reductions.*

4.1. Let μ,ν be regular dominant such that $\sigma = \mu - \nu$ is dominant. Then

$$(s\mu,\alpha) \geq (s\nu,\alpha) \quad \text{for} \quad \alpha \in \Delta(1_c).$$

In fact $(s\sigma,\alpha) = (\sigma,s^{-1}\alpha) \geq 0$ since σ is dominant and $s^{-1}\alpha > 0$ by definition of W^P. We also know that $\nu(s\mu)$ is regular dominant (for 1_c), hence

(1) $$(s\mu,\alpha) \geq (s\nu,\alpha) \geq (\rho_L,\alpha) \qquad (\alpha\in\Delta(1_c)).$$

Since $\mu - \rho$ is dominant, this yields in particular

(2) $$(s\mu,\alpha) = (\rho_L,\alpha) \Rightarrow (s\rho,\alpha) = (\rho_L,\alpha) \qquad (\alpha\in\Delta(1_c)),$$

therefore

(3) $$M_{\nu(s\mu)} \subset M_{\nu(s\rho)}, \quad \dim X_{\nu(s\mu)} \leq \dim X_{\nu(s\rho)}.$$

Note that we have not assumed τ_1-stability, and have used the convention made at the end of 3.2. In the sequel we replace the index $\nu(s\rho)$ in $M_{\nu(s\rho)}, M'_{\nu(s\rho)}, X_{\nu(s\rho)}, X'_{\nu(s\rho)}, M'_{\nu(s,\rho)}$ by s. In view of (3) we see that, in order to prove 3.3(7), and hence 2.1, it suffices to establish

4.2. PROPOSITION. *If* $\ell(s) < (\dim N)/2$, *then*

(1) $$\ell(s') - \ell(s) \geq \dim X_s.$$

Remark. It is not clear to me that $\nu(s\rho)$ is τ_1-stable. If it is, then (1) is equivalent to 3.3(7), because $\dim X_s$ is then the maximum of the right-hand side of 3.3(7). Otherwise, it is conceivably stronger. In fact, we shall prove a still slightly stronger inequality, namely

(2) $$\ell(s') - \ell(s) \geq \dim X'_s + \dim \mathfrak{a}_o + 1.$$

4.3. We let $\Phi_{\mathfrak{n}}$ be the set of weights of $\mathfrak{h}_c$ in $\mathfrak{n}_c$. Therefore $\Phi^+ = \Phi^+_{\mathfrak{m}} \amalg \Phi_{\mathfrak{n}}$. Let

(1) $$A_s = \left\{\alpha \in \Phi^+ \mid s^{-1}\alpha > 0\right\}, \quad B_s = \left\{\alpha \in \Phi^+ \mid s^{-1}\alpha < 0\right\}.$$

Then

(2) $$\Phi^+ = A_s \amalg B_s, \quad A_s \supset \Phi^+_{\mathfrak{m}}, \quad \ell(s) = \text{Card } B_s.$$

(3) $s\rho = \rho - \langle B_s\rangle$, where $\langle B_s\rangle$ is the sum of the elements in B_s.

The discussion in [C: 2.6] and standard facts about reduced decompositions show that $B_s \subset B_{s'}$. Let $C_s = B_{s'} - B_s$. Then

$$\ell(s') - \ell(s) = \text{Card } C_s.$$

To prove 2.1, it suffices, in view of 4.2(1), (2), to show:

4.4. PROPOSITION. *We have the inequality*

(1) $$\text{Card } C_s \geq \dim X'_s + \dim \mathfrak{a}_o + 1.$$

§5. *Proof of Proposition* 4.4.

5.0. *Notation*. Let $\alpha,\beta \in \Phi$. We write $\alpha \perp\!\!\!\perp \beta$ if α is strongly orthogonal to β, i.e. if neither $\alpha + \beta$ nor $\alpha - \beta$ is a root. This implies in particular that $(\alpha,\beta) = 0$. We have $\alpha \perp\!\!\!\perp \beta$ if and only if $[\mathfrak{g}_{\pm\alpha},\mathfrak{g}_{\pm\beta}] = 0$. Recall that if $\alpha \perp \beta$ but $\alpha \not\perp\!\!\!\perp \beta$, then $\alpha + \beta$ and $\alpha - \beta$ are roots.

5.1. LEMMA (i) *Let* $\alpha \in \Delta(\ell_c)$. *Then* $(s\rho,\alpha) = (\rho_L,\alpha)$ *if and only if* $s^{-1}\alpha$ *is simple.*

(ii) $\Delta_s = \{\alpha \in \Delta(1_c) | s^{-1}\alpha$ *and* $s^{-1}w_m w_G\alpha$ *are simple*$\}$.

The assertion (ii) follows from (i) and the definition of Δ_s (see 3.1(1) and 3.2(6)).

Proof of (i): Recall that

(1) $$2(\rho,\beta) = (\beta,\beta) \quad \text{if } \beta \text{ is simple}$$

and similarly

(2) $$2(\rho_L,\beta) = (\beta,\beta) \quad \text{if } \beta \in \Delta(1_c).$$

If $s^{-1}\alpha = \beta$ is simple, then

$$2(s\rho,\alpha) = 2(\rho,s^{-1}\alpha) = (\beta,\beta) = (s^{-1}\alpha,s^{-1}\alpha) = (\alpha,\alpha) = 2(\rho_L,\alpha).$$

Assume now $(s\rho,\alpha) = (\rho_L,\alpha)$. Since $s^{-1}\alpha > 0$ we may write $s^{-1}\alpha = \Sigma\, c_\beta\beta$ with β simple, $c_\beta \in \mathbb{N}$ and we have then

(3) $$(s\rho,\alpha) = \Sigma\, c_\beta(\rho,\beta).$$

or equivalently

(4) $$(\alpha,\alpha) = \Sigma\, c_\beta(\beta,\beta).$$

The possible values for the square norms of the roots, suitably normalized, are either 1 or 1 and 2 or 1 and 3. In the first case, (4) shows that $s^{-1}\alpha$ is simple. In the last case, (which occurs only for the Lie type $\mathbf{G}_2$), α should be long, the β with $c_\beta \neq 0$ should be short and $s^{-1}\alpha$ should be a sum of three distinct simple roots, which is absurd since $\mathfrak{g}$ has rank 2. In the second case, if $s^{-1}\alpha$ is not simple, the only possibility is

(5) $$s^{-1}\alpha = \beta_1 + \beta_2, \quad (\beta_1,\beta_2) + (\beta_2,\beta_2) = (\alpha,\alpha).$$

Then β_1,β_2 are short and α long. However if the sum of two simple short roots is a root, it is also short (being the transform of one by the reflection to the other), a contradiction since $s^{-1}\alpha$ is also long.

5.2. We shall write τ for $w_M w_G$ and set $\tau' = -\tau$. Then $s' = \tau.s$. Also, τ' leaves $\Phi_{\mathfrak{n}}$ stable. Both τ and τ' are of order 2. We claim

(1) $$\tau' B_s \cap B_{s'} = \phi.$$

In fact, if $\alpha \in B_s$ then

$$s'^{-1}.\tau'.\alpha = s^{-1}\tau\tau'\alpha = -s^{-1}\alpha > 0.$$

Since $\operatorname{Card}\Phi_{\mathfrak{n}} = \ell(s) + \ell(s') = \operatorname{Card} B_s + \operatorname{Card} B_{s'}$, it follows that $\Phi_{\mathfrak{n}}$ is the disjoint union of $B_{s'}$ and $\tau' B_s$, or also of $B_s, C_s, \tau' B_s$ and that $\tau' C_s = C_s$. We have then the following characterizations of these subsets in terms of the sign of $s^{-1}\alpha$, $s'^{-1}\alpha$ or $s^{-1}\tau'\alpha$:

(2)

$\alpha \in$	$\tau' B_s$	C_s	B_s
$s^{-1}\alpha$	>0	>0	<0
$s'^{-1}\alpha$	>0	<0	<0
$s^{-1}\tau'\alpha$	<0	>0	>0

5.3. LEMMA. (i) *For* $\alpha \in \Phi^+$ *we have the equivalences* $\tau'\alpha = \alpha \Leftrightarrow \alpha|\mathfrak{t} = 0 \Leftrightarrow \alpha \in C_s^{\tau'}$.

(ii) $\Phi(\mathfrak{z}_1) = C_s^{\tau'} \cup (-C_s^{\tau'})$.

(iii) $C_s^{\tau'}$ *contains a basis of* $(\mathfrak{a}_o \oplus \mathfrak{a})^*$.

Proof. The map τ', viewed as transformation of $\mathfrak{h} = \mathfrak{t} \oplus \mathfrak{a}_o \oplus \mathfrak{a}$, is the identity on $\mathfrak{a}_o \oplus \mathfrak{a}$ and -1 on $\mathfrak{t}$. This proves the first equivalence in (i). Since $\mathfrak{t}$ contains regular elements of $\mathfrak{m}$, any positive root identically zero on $\mathfrak{t}$ belongs to Φ_n. Being fixed under τ', such a root must then belong to C_s, since $B_s \cap \tau' B_s = \phi$. This proves (i). The roots of $\mathfrak{z}_1$ are clearly those of $\mathfrak{g}$ which are zero on $\mathfrak{t}$, therefore (i) $\Rightarrow$ (ii). We already pointed out (1.3) that $\mathfrak{z}_1$ is semisimple, and that $\mathfrak{a}_o \oplus \mathfrak{a}$ is a Cartan subalgebra of $\mathfrak{z}_1$. Hence (ii) $\Rightarrow$ (iii).

5.4. LEMMA. *The space* $V_s = \oplus_{\alpha \in C_s} \mathfrak{g}_\alpha$ *is invariant under* $\mathfrak{m}'_s$, *acting by the adjoint representation.*

It suffices to show that $[\mathfrak{g}_\beta, \mathfrak{g}_\alpha] \subset V_s$ if $\beta \in \pm\Delta_s$ and $\alpha \in C_s$. This is clear if $\alpha \perp \beta$. If not, we have to show that if $\alpha + \varepsilon\beta$ is a root, where ε is equal to 1 or to -1, then it belongs to C_s. It is the sum of a positive root α and of $\varepsilon\beta$, with β simple and $\varepsilon = \pm 1$. Therefore it is positive. Its transform

$$s^{-1}(\alpha+\varepsilon\beta) = s^{-1}\alpha + \varepsilon . s^{-1}\beta \tag{1}$$

is also positive for the same reason: $s^{-1}\alpha$ is positive since $\alpha \in A_s$ and $s^{-1}\beta$ is simple by 5.1. Similarly, since $\tau'\alpha \in C_s$, we have $s^{-1}\tau'\alpha > 0$ and

$$s^{-1}\tau'(\alpha+\varepsilon\beta) = s^{-1}\tau'\alpha - \varepsilon s^{-1}\beta^* > 0. \tag{2}$$

But (1), (2) and 5.2(2) imply that $\alpha + \varepsilon\beta \in C_s$.

5.5. LEMMA. *Let* $\Psi = \{\beta \in \Phi^+(\mathfrak{m}'_s) | \beta \not\perp C_s^\tau \}$. *Then*

$$2 \text{ Card } (\psi/\langle\tau'\rangle) + \text{Card } C_s^{\tau'} \leq \text{Card } C_s.$$

[By definition, $\Phi^+(\mathfrak{m}'_s)$ is invariant under τ', hence so is Ψ. It therefore makes sense to speak of the set $\Psi/\langle\tau'\rangle$ of orbits on Ψ of the group $\langle\tau'\rangle = \{1,\tau'\}$ generated by τ'.]

Let $\alpha_1,\ldots,\alpha_q$ be the elements of $C_s^{\tau'}$. Set

$$E_1 = \{\beta \in \Psi | \beta \not\perp \alpha_1\}$$
$$E_i = \{\beta \in \Psi | \beta \perp \alpha_1,\ldots,\beta \perp \alpha_{i-1}, \beta \not\perp \alpha_i\} \quad (2 \leq i \leq q).$$

Then $\psi = \coprod_i E_i$.

If $\beta \in E_i$, then for some $\varepsilon_\beta = \pm 1$, the element $\alpha_i + \varepsilon_\beta\beta$ is a root, and belongs to C_s by 5.4. Moreover, if $\beta = \beta^*$, then both $\alpha_i + \beta$ and $\alpha_i - \beta$ belong to C_s (because if $\alpha_i + \varepsilon_\beta\beta \in C_s$, then $\tau'(\alpha_i+\varepsilon_\beta\beta) = \alpha_i - \varepsilon_\beta\beta \in C_s$.) None of these belongs to $C_s^{\tau'}$, since no root of $\mathfrak{m}_s'$ restricts to zero on $\mathfrak{t}$. It suffices therefore to show that these elements are all distinct, i.e. we are reduced to proving

(*) *Let* $\beta \in E_i$, $\beta' \in E_j$ *and assume that* $\alpha_i + \varepsilon\beta$, $\alpha_j + \varepsilon'\beta' \in C_s$ *and* $\alpha_i + \varepsilon\beta = \alpha_j + \varepsilon'\beta'$. *Then* $i = j$, $\varepsilon = \varepsilon'$ *and* $\beta = \beta'$.

We may assume that $i \leq j$. Assume first that $i < j$. Then we have

(1) $$\alpha_i - \alpha_j = \varepsilon'\beta' - \varepsilon\beta \neq 0.$$

Being fixed under τ', the roots α_i and α_j are zero on $\mathfrak{t}$, hence so is $\varepsilon'\beta' - \varepsilon\beta$. But no root of $\mathfrak{m}$ is zero on $\mathfrak{t}$ (recall that $\mathfrak{t}$ contains regular elements of $\mathfrak{m}$). Therefore $\varepsilon'\beta - \varepsilon\beta$ and $\alpha_i - \alpha_j$ are not roots, and we have

(2) $$(\alpha_i,\alpha_j) \leq 0, \quad (\varepsilon'\beta',\varepsilon\beta) \leq 0.$$

But

$$(\alpha_i+\varepsilon\beta,\alpha_j+\varepsilon'\beta') = (\alpha_i+\varepsilon\beta,\alpha_i+\varepsilon\beta) > 0,$$

therefore

(3) $$(\alpha_i,\alpha_j) + (\alpha_i,\varepsilon'\beta') + (\varepsilon\beta,\alpha_j) + (\varepsilon\beta,\varepsilon'\beta') > 0.$$

By definition $\beta' \perp\!\!\!\perp \alpha_i$, hence a fortiori $(\beta',\alpha_i) = 0$. In view of (2), the relation (3) implies therefore

(4) $$(\varepsilon\beta,\alpha_j) > 0.$$

It follows that $\alpha_j - \varepsilon\beta$ is a root, but then so is $\alpha_i - \varepsilon'\beta'$, which contradicts the definition of β'. This proves (*) when $i < j$.

If now $i = j$. Then $\varepsilon\beta = \varepsilon'\beta'$. Since β and β' are >0, this implies $\varepsilon = \varepsilon'$ and $\beta = \beta'$.

5.6. By 5.3(iii), Card $C_s^{\tau'} \geq \dim \mathfrak{a}_o + 1$. In order to prove 4.4 it suffices therefore to show

(1) $$2\ \mathrm{Card}\ (\psi/\langle\tau'\rangle) \geq \dim X_s'.$$

The dimension of X'_s is the complex dimension of the -1 eigenspace of θ (extended by linearity to $\mathfrak{g}_c$) on $\mathfrak{m}'_{s,c}$. We let σ be the complex conjugation of $\mathfrak{g}_c$ with respect to $\mathfrak{g}$. We have

$$\sigma\beta = -\beta^*, \quad \sigma\beta^* = -\beta \qquad (\beta\in\Phi).$$

Let $\beta \in \Phi^+(\mathfrak{m}'_s)$. We distinguish three cases:

(i) $\beta \neq \beta^*$. Then

$$\mathfrak{g}_\beta + \mathfrak{g}_{\beta*} + \mathfrak{g}_{-\beta} + \mathfrak{g}_{-\beta*}$$

is 4-dimensional and σ stable. θ permutes $\mathfrak{g}_\beta$ and $\mathfrak{g}_{\beta*}$ (resp. $\mathfrak{g}_{-\beta}$ and $\mathfrak{g}_{-\beta*}$) (recall that $\theta|\mathfrak{h}_c$ is defined by *). Therefore its -1 eigenspace there has dimension 2. But $\beta \neq \beta^*$ implies that $\beta|\mathfrak{a}_o \neq 0$, hence that $\beta \not\perp C_s^{\tau'}$ by 5.3(iii). Therefore β belongs to Ψ and gives a contribution 2 to the left-hand side of (1).

(ii) $\beta = \beta^*$, $\beta \in \Psi$. In this case, $\mathfrak{g}_\beta \oplus \mathfrak{g}_{-\beta}$ is σ-stable and θ-stable. The contribution to the dimension of X'_s is at most two, but $\beta \in \Psi$ adds two to the left-hand side of (1).

(iii) $\beta = \beta^*$, $\beta \notin \Psi$. Then again $\mathfrak{g}_\beta + \mathfrak{g}_{-\beta}$ is invariant under σ and θ. The root β is not in Ψ, hence does not contribute to the left-hand side of (1). Therefore we must show that θ is the identity on $\mathfrak{g}_\beta \oplus \mathfrak{g}_{-\beta}$.

By assumption $\beta \perp C_s^{\tau'}$. Then 5.3(ii) shows that $\mathfrak{g}_\beta$, $\mathfrak{g}_{-\beta}$ centralize $\mathfrak{z}_1$. Assumption 1.3(B) then yields the result.

§6. *The condition* (B).

6.1. If L is compact, condition (B) of 1.3 is automatically fulfilled. In this case $C(L,\mu) = 0$ for all μ's and 2.1(1) amounts to the relation

$$2.\ell(s) < \dim N + 1 \quad \text{if} \quad s\lambda|\mathfrak{a} > 0,$$

which already follows from 2.6 in [C]. As pointed out in [B], it can very easily be checked case by case. A proof is also included in [Z].

6.2. As already remarked at the end of §1, condition (B) is satisfied if $\mathfrak{a}_o$ is maximal $\mathbb{R}$-split in $\mathfrak{l}$, i.e. if

$$\text{(1)} \qquad \operatorname{rk} K(L) + \operatorname{rk}_{\mathbb{R}}\mathfrak{l} = \operatorname{rk}\mathfrak{l}.$$

This equality holds in particular if $\mathfrak{l}$ is a complex simple Lie algebra, viewed as a real Lie algebra, (in which case rank $K(L)$ and $\operatorname{rk}_{\mathbb{R}}(L)$ are equal to half of rk (L)), or if L is locally isomorphic to $\mathbf{SO}(2m+1,1)$ $(m\in\mathbb{N})$, in which case $K(L) = \mathbf{O}(2m+1)$ and $rk(L) = m+1$, $rk_{\mathbb{R}}(L) = 1$, $rk\ K(L) = m$.

6.3. For the following, we refer to [BB: §§1,2]. We now consider the irreducible bounded symmetric domains. Let G be a simple non-compact Lie group such that the symmetric space $X = G/K(G)$ of maximal compact subgroups of G is a bounded symmetric domain. Underlying the construction of V^* (see 2.3), there is first a Satake compactification $\bar{X}$ of X. Its boundary is the union of finitely many orbits of G, each one of which is fibered in the so-called real boundary components. If G is given moreover a $\mathbf{Q}$-structure, then the rational boundary components are among the real ones. A real boundary component is characterized by its normalizer, which is a proper maximal parabolic subgroup of G. Let P be one, $P = NAM$ its Langlands decomposition and X_P the associated boundary component. Then the identity component M^0 of M has an almost direct product decomposition $M^0 = L.G_P$, where L is the greatest connected normal subgroup of M^0 acting trivially on X_P and $X_P = G_P/K(G_P)$. [The notation is the one used so far in this paper, itself borrowed from [BC]. In [BB], F stands for the present X_P, $N(F)$ is the normalizer of F, and $Z(F) \cap M^0$ (resp. $G(F)$) stands for L (resp. G_P).] The space X_P is also a bounded symmetric domain, therefore 1.3(A) is satisfied.

6.4. In the sequel we are only concerned with types of Lie algebras. Equalities between Lie groups are therefore meant to be only local isomorphisms of identity components.

Let t be the $\mathbb{R}$-rank of G and ${}_{\mathbb{R}}\Delta = \{\alpha_1,\ldots,\alpha_t\}$ the set of simple $\mathbb{R}$-roots of G. The Dynkin diagram for the $\mathbb{R}$-roots is of type $\mathbf{C}_t$ or $\mathbf{BC}_t$. We use the canonical numbering of the vertices of [BB: 1.2]. The Satake compactification $\bar{X}$ of X is the one associated to the last simple $\mathbb{R}$-root. The proper maximal parabolic subgroups, up to conjugacy, are indexed by $b \in [1,t]$. Let P_b be the one assigned to b. The simple $\mathbb{R}$-roots of L (resp. G_P) are the α_i $(i<b)$ (resp. $i > b$). The $\mathbb{R}$-diagram of L is then of type $\mathbf{A}_{b-1}$. To find out the type of L, one has first to determine the one of $\mathfrak{l}_c$. This amounts essentially to finding the roots in the Dynkin diagram Dyn $\mathfrak{g}_c$ of $\mathfrak{g}_c$ which restrict to the $\mathbb{R}$-roots of L and is easily read off the tables of [T]. Then one has to decide which real form of $\mathfrak{l}_c$ is $\mathfrak{l} = \mathfrak{g} \cap \mathfrak{l}_c$. These are simple computations, which we do not give in detail.

6.5. Up to local isomorphism, the possible G's are, in E. Cartan's notation (see [H: p. 354])

A III, BD I (p=2), C I, D III, E III, E VII .

We now discuss each type separately.

Type A.III. In Tits notation [T: p. 55] this is $^2\mathbf{A}_n$. In this case $L = \mathbf{SL}_{b-1}(\mathbf{C})$ and 6.2 applies.

Type BD I. Here $X = \mathbf{SO}(p,2)/S(\mathbf{O}(2)\times\mathbf{O}(p))$, and $t = 2$. We have to consider the case $b = 2$. Then $L = \mathbf{SO}(p-1,1)$, $K(L) = \mathbf{O}(p-1)$, and the $\mathbb{R}$-rank of L is 1.

If p is even, then 6.2(1) is satisfied.

Let now $p = 2m + 1$ be odd. Then $\mathrm{rk}\, L = \mathrm{rk}\, K(L) = m$ and 6.2(1) does not hold. Some computation is needed.

We assume the coordinates x_i in $\mathbb{R}^n$ $(n=2m+3)$ so chosen that G is the orthogonal group of the form $-x_1^2 + x_2^2 - x_3^2 + \Sigma_4^n x_i^2$. As a maximal $\mathbb{R}$-split torus, we may take the product of the orthogonal groups of the two hyperbolic planes spanned by the basis vectors (e_1,e_2) and (e_3,e_4) respectively. The group L is the subgroup of G leaving e_1 and e_2 fixed. It has the orthogonal group of the last $n - 3$ variables as a maximal compact subgroup. The Lie algebra $\mathfrak{t}$ is the one of a maximal torus of the latter group. It is a Cartan subalgebra of $\mathfrak{l}$, hence $\mathfrak{a}_0 = 0$. The derived algebra $\mathfrak{z}_1$ of the centralizer of $\mathfrak{t}$ is the Lie algebra of the orthogonal group $\mathbf{SO}(1,2)$ of the first three variables. Its centralizer is the orthogonal group of the last $n - 3$ variables and is indeed compact.

Type C I. Here G is the symplectic group $\mathbf{Sp}_{2b}(\mathbb{R})$, $K(G) = \mathbf{U}(n)$, X is the Siegel upper-half space of genus n and $t = n$. For $P = P_b$, we have $G_P \simeq \mathbf{Sp}_{2n-2b}(\mathbb{R})$ and $L = \mathbf{SL}_b(\mathbb{R})$. After permutation of the coordinates we may assume that L is a block matrix with entries $A, {}^tA^{-1}, I_{2n-2b}$, $(A \in \mathbf{SL}_b(\mathbb{R}))$. Let $c = [b/2]$. Let $Z = \begin{pmatrix} 0 & 1 \\ -1 & 0 \end{pmatrix}$. As $\mathfrak{t}_o$ we may take for b even the block matrices with blocks

$$y_1 Z, \ldots, y_c Z, -y_1 Z, \ldots, -y_c Z, 0_{2n-2b} \qquad (y_i \in \mathbb{R}, i=1,\ldots,c)$$

and $\mathfrak{t}$ is the direct sum of $\mathfrak{t}_o$ and of a compact Cartan subalgebra of $\mathbf{Sp}_{2n-2b}\mathbb{R}$, operating on the last $2n - 2b$ coordinates. To this we should add the 2×2 zero matrix in (e_b, e_{2b}) if b is odd. In Z we have a direct product of c (resp. $c+1$) copies D_i of $\mathbf{SL}_2(\mathbb{R})$ if b is even (resp. odd). For $i \in [1,c]$, D_i is the group of matrices

$$(1) \qquad \begin{pmatrix} aI_2 & bI_2 \\ cI_2 & dI_2 \end{pmatrix} \qquad (ad-bc=1)$$

acting on the 4-plane V_i with basis $(e_{2i-1}, e_{2i}, e_{b+2i-1}, e_{b+2i})$. For b odd, D_{c+1} is the unimodular group of the plane V_{c+1} spanned by (e_b, e_{2b}). The centralizer of Z_1 is contained in the centralizer of the product of the D_i's. It suffices to show that the intersection of the latter with L is compact. This intersection obviously leaves the V_i's invariant. On V_{c+1}, it centralizes $\mathbf{SL}_2(\mathbb{R})$, hence consists of scalar matrices cI_2. But the symplectic condition imposes that $c = c^{-1}$, hence $c = \pm 1$. On V_i $(1 \leq i \leq c)$ we have to find the centralizer of the group of matrices (1). Writing an element of it in 2×2 blocks, we see easily that it consists of matrices $\begin{pmatrix} A & 0 \\ 0 & A \end{pmatrix}$ $(A \in \mathbf{GL}_2(\mathbb{R}))$. But we must have in addition $A = {}^tA^{-1}$, hence A is orthogonal.

Type D III. Here $G = \mathbf{SO}^*_{2n}$ (notation of [H]), $K(G) = \mathbf{U}_n$ and $t = [n/2]$. In this case, $L = \mathbf{SL}_b(\mathbb{H})$, where $\mathbb{H}$ is the field of (Hamilton) quaternions. The group L has rank $2b-1$ and $\mathbb{R}$-rank $b-1$. Moreover $K(L)$ is the unitary group on quaternionic b-dimensional space and has therefore rank b. This shows that 6.2(1) holds.

Type E III. Here G is real form $\mathbf{E}_{6,-14}$ of $\mathbf{E}_6$, $K(G)$ is locally isomorphism to $\mathbf{SO}(10) \overset{\circ}{\times} \mathbf{SO}(2)$, and $t = 2$. For $b = 2$, $Z = \mathbf{SO}(7.1)$, and 6.2(1) is satisfied.

Type E VII. Here G is the real form $\mathbf{E}_{7,-25}$ of $\mathbf{E}_7$ and $K(G)$ is locally isomorphic to $\mathbf{E}_6 \times \mathbf{SO}(2)$, (where the compact $\mathbf{E}_6$ is meant). Moreover $t = 3$.

For $b = 2$, we have $L = \mathbf{SO}(9,1)$ and we may again apply 6.2(1).

Let $b = 3$. Then L is the real form $\mathbf{E}_{6,-26}$ of $\mathbf{E}_6$, with maximal compact subgroup of type $\mathbb{F}_4$. It has real rank 2, hence 6.2(1) is again true.

6.6. I have proposed a slight generalization of Zucker's conjecture, in which the underlying symmetric space is not necessarily hermitian, but where all the real boundary components of the Satake compactification to be used satisfy (A). See [Z] for the statement, where the new cases are also enumerated. Apart from some rank 2 groups there are two new series: $\mathbf{SO}(p,q)$ $(1 \leq p \leq q)$, with $p+q$ odd, and the unitary groups $\mathbf{Sp}(p,q)$ $(1 \leq p \leq q)$ of indefinite quaternionic forms.

If $G = \mathbf{SO}(p,q)$, the system of $\mathbb{R}$-roots is of type $\mathbb{B}_p$ and the relevant Satake compactification is associated to the short simple $\mathbb{R}$-root. (Thus, for $p = 2$, it is different from the one considered above.) The $\mathbb{R}$-rank is p, and the types of boundary components are again indexed by $b \in [1,p]$. Given b, the corresponding group G_P is $\mathbf{SO}(p-b,q-b)$ and $L = \mathbf{SL}_b(\mathbb{R})$. The computations to check (B) are the same as in the case C I.

If $G = \mathbf{Sp}(p,q)$, the Satake compactification is associated to the long simple root. Again the $\mathbb{R}$-rank is p and the types of boundary components are indexed by $b \in [1,p]$. For a given b, we have $G_p = \mathbf{Sp}(p-b,q-b)$ and $L = \mathbf{SL}_b(\mathbb{H})$. As we saw above in the case D III, 6.2(1) is satisfied.

6.7. In all this, we have assumed G to be simple. In fact, in the applications, G is the group of real points of a $\mathbf{Q}$-simple algebraic group, hence is not always simple over $\mathbb{R}$ and symmetric space of G is then a product of symmetric spaces of real simple groups. The needed vanishing conditions follow from 2.1 and a suitable Künneth rule.

The Institute for Advanced Study, Princeton, NJ 08540 U.S.A.

References

[BB] W. Baily and A. Borel, *Compactifications of arithmetic quotients of bounded symmetric domains*, Ann. Math., 84, 1966, p. 442-528.

[B] A. Borel, L^2*-cohomology and intersection cohomology of certain arithmetic varieties*, in E. Noether in Bryn Mawr, Springer, 1983, p. 119-131.

[BC1] A. Borel and W. Casselman, L^2*-cohomology of locally symmetric manifolds of finite volume*, Duke Math. J. 50 (1983), p. 625-647.

[BC2] A. Borel et W. Casselman, *Cohomologie d'intersection et* L^2*-cohomologie de variétés arithmétiques de rang rationnel* 2. C. R. Acad. Sci. Paris 301, (1985), p. 369-373.

[C] W. Casselman, L^2*-cohomology for groups of real rank one*, in Representation theory of reductive groups, Progress in Math., 40, Birkhäuser, Boston, 1983, p. 69-82.

[E] T. Enright, *Relative Lie algebra cohomology and unitary representations of complex Lie groups*, Duke Math. J. 47, (1980), p. 1-15.

[H] S. Helgason, Differential Geometry and Symmetric Spaces, Adademic Press 1962.

[K] B. Kostant, *Lie algebra cohomology and the generalized Borel-Weil theorem*, Annals of Math. 74, (1961), p. 329-387.

[T] J. Tits, *Classification of algebraic semisimple groups*, in Algebraic groups and discontinuous subgroups, Proc. Symp. Pure Math. A.M.S. IX, (1966), p. 33-62.

[V] D. Vogan, *Unitarizability of certain series of representations*, Annals of Math., 120, (1984), p. 141-187.

[VZ] D. Vogan and G. Zuckerman, *Unitary representations with non-zero cohomology*, Comp. Math., 53, 1984, p. 51-90.

[Z] S. Zucker, L^2*-cohomology and intersection homology of locally symmetric varieties II*, preprint, 1984.

135.

(with G. Prasad)

Sous-groupes discrets de groupes p-adiques à covolume borné

C. R. Acad. Sci. Paris **305** (1987), 357–362

Discrete subgroups of p-adic groups with bounded covolume

Abstract — *In this Note, k is a p-adic field of characteristic zero, G the group of k-rational points of an almost absolutely simple k-group $\mathscr{G}$, of k-rank $l(\mathscr{G}) \geqq 2$ and Γ a discrete cocompact subgroup of* G. *We state some finiteness properties of the set of such triples for which* G/Γ *has a volume bounded by a given constant, with respect to a suitably normalized Haar measure, when $\mathscr{G}$ or the order q of the residue field $\bar{k}$ of k vary.*

Résumé — Dans cette Note, k est un corps p-adique de caractéristique zéro, G le groupe des points rationnels sur k d'un k-groupe absolument presque simple $\mathscr{G}$, de k-rang $l(\mathscr{G}) \geqq 2$, et Γ un sous-groupe discret cocompact de G. On annonce quelques propriétés de finitude de l'ensemble de ces triples tels que G/Γ ait un volume borné à l'avance, pour une normalisation convenable de la mesure de Haar, lorsque $\mathscr{G}$ ou l'ordre q du corps résiduel $\bar{k}$ de k varient.

1. On note T(G) l'immeuble de Bruhat-Tits de G ([6], [19]). C'est un complexe simplicial sur lequel G opère canoniquement par automorphismes simpliciaux. On désigne par I un sous-groupe d'Iwahori de G ([6], 5.2.6; [19] 3.7). C'est un sous-groupe ouvert compact de G, qui est d'indice fini dans le sous-groupe des éléments de G qui fixent (point par point) une chambre C de T(G); il lui est égal si, par exemple, $\mathscr{G}$ est simplement connexe. Soit μ_T la mesure de Haar sur G, introduite dans [19], 3.7, qui attribue la mesure 1 à I. On note de la même manière la mesure invariante définie par μ_T sur G/Γ.

Par ailleurs ([14], 3.3), G possède une et une seule mesure invariante μ_G, telle que $\mu_G(G/\Gamma)=\chi(\Gamma)$, où $\chi(\Gamma)$ désigne la caractéristique d'Euler-Poincaré de Γ, au sens de C. T. C. Wall. Rappelons que $\chi(\Gamma) \in \mathbf{Q}$, $(-1)^{l(\mathscr{G})}\chi(\Gamma) \geqq 0$, et que $\chi(\Gamma')=[\Gamma:\Gamma']\chi(\Gamma)$ si Γ' est un sous-groupe d'indice fini $[\Gamma:\Gamma']$ de Γ. De plus, si Γ est sans torsion, alors $\chi(\Gamma)$ est la caractéristique d'Euler-Poincaré de la cohomologie rationnelle de Γ. Nous notons μ_{EP} la valeur absolue de la mesure d'Euler-Poincaré. On a donc $\mu_{EP}(G/\Gamma)=|\chi(\Gamma)|$.

Pour comparer μ_{EP} et μ_T, il suffit de connaître la mesure $\mu_{EP}(I)$ de I. Si $\mathscr{G}$ est simplement connexe, cette dernière est la somme d'une série de terme général $[I w I : I]^{-1}$, où w parcourt le groupe de Weyl affine de G ([14], théor. 6). Elle peut être envisagée comme la valeur en un point convenable d'une fonction rationnelle étudiée dans [9] et [16]. On en déduit la :

Proposition 1. — *Il existe une constante $d>1$, telle que l'on ait, quels que soient k et* G :

$$d^{-l(\mathscr{G})}\mu_T \leqq \mu_{EP} \leqq \mu_T. \tag{1}$$

2. Supposons tout d'abord k et G fixés et soit μ une mesure de Haar sur G. Étant donné $c>0$, il n'existe qu'un nombre fini de classes de conjugaison de sous-groupes discrets Γ tels que $\mu(G/\Gamma) \leqq c$ [2]. Notre premier but est d'annoncer un résultat semblable lorsque $\mu=\mu_T$ ou μ_{EP} et que l'on fait varier G ou, dans une certaine mesure, le corps k.

Soit (k', G′, Γ') un deuxième triple du type envisagé ici. S'il existe un isomorphisme $\varphi : k \to k'$, alors on peut considérer le k'-groupe ${}^{\varphi}\mathscr{G}$ obtenu à partir de $\mathscr{G}$ par changement de base et on a un isomorphisme canonique (de groupes topologiques) $\varphi^0 : \mathscr{G}(k) \to {}^{\varphi}\mathscr{G}(k')$

Note présentée par Armand Borel.

0249-6291/87/03050357 \$ 2.00

([4], 1.7). On dira que (k', G', Γ') est isomorphe à (k, G, Γ) s'il existe un isomorphisme $\varphi : k \to k'$ et un k'-isomorphisme $\psi : \mathscr{G}' \to {}^{\varphi}\mathscr{G}$ tels que $\psi(k')$ amène Γ' sur $\varphi^0(\Gamma)$.

THÉORÈME. — *Fixons deux constantes* a, $b>0$. *Alors il n'existe qu'un nombre fini de classes d'isomorphie de triples* (k, G, Γ) *tels que le degré de ramification absolu* $e(k)$ *de* k *soit* $\leqq b$ *et que* $\mu_{EP}(G/\Gamma) \leqq a$ *(resp.* $\mu_T(G/\Gamma) \leqq a$*).*

En fait, la démonstration montre que l'on peut remplacer la constante a par une exponentielle $\exp_q(r.\mathrm{l}(\mathscr{G})^{2-\varepsilon})$, où $\varepsilon>0$ est arbitraire, r est une constante positive dépendant de ε et $\mathrm{l}(\mathscr{G})$ est le rang de $\mathscr{G}$ sur l'extension maximale non ramifiée de k.

Désignons par $\mathscr{C}(T(G))$ ou simplement $\mathscr{C}$ l'ensemble des chambres de $T(G)$. Le groupe Γ opère sur $\mathscr{C}$ et n'a qu'un nombre fini d'orbites. Supposons que Γ opère par automorphismes spéciaux. Le nombre d'éléments de $\mathscr{C}/\Gamma$ est égal au nombre d'orbites dans G/Γ du fixateur G^C de C dans G. Le groupe I est d'indice fini dans G^C, au plus égal à l'ordre du groupe Aut Δ des automorphismes du diagramme de Dynkin local relatif Δ de $\mathscr{G}$ ([17], 3.5.3), donc au plus à $\mathrm{l}(\mathscr{G})+1$. Par conséquent

$$\mathrm{Card}\,\mathscr{C}(T(G))/\Gamma \geqq \mu_T(G/\Gamma).(\mathrm{l}(\mathscr{G})+1)^{-1}.$$

Le théorème ci-dessus, ou plutôt le renforcement qui en suit l'énoncé, entraîne donc le corollaire suivant, où ε, $\mathrm{l}(\mathscr{G})$ et r sont comme plus haut :

COROLLAIRE 1. — *Il n'existe qu'un nombre fini de classes d'isomorphie de triples* (k, G, Γ) *tels que* Γ *soit formé d'automorphismes spéciaux,* $e(k) \leqq b$ *et* $\mathrm{Card}\,\mathscr{C}(T(G))/\Gamma \leqq q^{r\mathrm{l}(\mathscr{G})^{2-\varepsilon}}$.

Soit c un entier $<\mathrm{l}(\mathscr{G})$, et soit F une face de C de codimension c. Le nombre des chambres de $T(G)$ contenant F est borné par l'ordre du groupe des points rationnels sur le corps résiduel $\bar{k}$ d'un $\bar{k}$-groupe réductif $\mathscr{L}$ de $\bar{k}$-rang égal à c. Il n'y a qu'un nombre fini de possibilités pour $\mathscr{L}$, donc il existe un entier $d=d(c)$, ne dépendant que de c, tel que cet ordre soit borné par q^d. Le corollaire 1 entraîne par conséquent le suivant, suggéré par J. Tits :

COROLLAIRE 2. — *Soit* $c \in \mathbb{N}$. *Il n'existe qu'un nombre fini de classes d'isomorphie de triples* (k, G, Γ) *tels que* $\mathrm{l}(\mathscr{G})>c$, $e(k) \leqq b$ *et que* Γ *soit formé d'automorphismes spéciaux et transitif sur l'ensemble des faces de codimension* c *de* $T(G)$ *d'un type donné.*

Si l'on se borne à des sous-groupes Γ sans torsion, la démonstration du théorème est extrêmement aisée : il suffit d'utiliser la proposition 1 et l'inégalité (3) ci-dessous. Dans ce cas $\chi(\Gamma)$ est la somme alternée des nombres de Betti rationnels $b_i(\Gamma)$ de Γ. Or, d'après un théorème de W. Casselman (*cf.* [5], XIII, 2.6 pour une démonstration), établi tout d'abord dans [7] pour des corps résiduels assez grands, on a $b_i(\Gamma)=0$ pour $i \neq 0$, $l(\mathscr{G})$. D'autre part $b_l(\Gamma)$ est la multiplicité de la représentation spéciale de G dans $L^2(G/\Gamma)$ ([7], 10.4 ou [5], *loc.cit.*). On obtient donc le :

COROLLAIRE 3. — *Il n'existe qu'un nombre fini de classes d'isomorphie de triples* (k, G, Γ) *tels que* Γ *soit sans torsion,* $e(k) \leqq b$ *et que la multiplicité de la représentation spéciale de* G *dans* $L^2(G/\Gamma)$ *soit* $\leqq a$.

3. REMARQUES. — (1) Le théorème, pour μ_T, et le corollaire 1 répondent à des questions posées par J.-P. Serre et J. Tits, qui sont à l'origine de ce travail et de [2]. Ces questions leur avaient été suggérées par le résultat de [8], qui fournit une liste explicite, finie, des classes d'isomorphie de triples (k, G, Γ) dans lesquels Γ est formé d'automorphismes spéciaux et agit *transitivement* sur $\mathscr{C}(T(G))$. Cela contient en particulier le corollaire 1 pour $a=1$, sans restriction sur le degré de ramification de k. En fait, nous ne savons pas si l'hypothèse faite sur $e(k)$ dans le théorème ou les corollaires est nécessaire.

(2) Supposons, pour simplifier, $\mathcal{G}$ simplement connexe. Alors $G^C = I$, le groupe agit par automorphismes spéciaux et

$$\text{Card}\, \mathscr{C}\,(T(G))/\Gamma \geqq \mu_T(G/\Gamma). \tag{2}$$

La condition Card $\mathscr{C}/\Gamma \leqq a$ est donc en principe plus forte que $\mu_T(G/\Gamma) \leqq a$. Signalons un cas où nous pouvons en déduire un résultat de finitude sans restriction sur k.

Proposition 2. — *Soit a un entier positif. Il existe un entier $n(a)$ tel que si $n \geqq n(a)$, le groupe $\mathbf{SL}_n(k)$ ne contienne aucun sous-groupe discret ayant au plus a orbites dans $\mathscr{C}(T(\mathbf{SL}_n(k)))$, quel que soit k.*

(3) On peut aussi démontrer un théorème de finitude analogue au théorème pour des sous-groupes cocompacts irréductibles de produits de groupes p-adiques sur des corps de base différents.

4. Le reste de cette Note est consacré à quelques indications sur la démonstration du théorème. Comme $\mu_{EP} \leqq \mu_T$, il suffit de l'établir pour μ_{EP}. Mais en fait, on le prouve tout d'abord pour μ_T, et l'on constate que l'énoncé reste valable si la constante a est remplacée par une exponentielle comme indiqué plus haut; vu (1), cela entraîne le théorème pour μ_{EP}.

Supposons tout d'abord k fixé. Il s'agit de montrer que l'inégalité $\mu_T(G/\Gamma) \leqq a$ ne peut être satisfaite que pour un nombre fini de choix de $\mathcal{G}$ (à k-isomorphisme près) et de Γ (à conjugaison près). Vu le résultat de [2] rappelé ci-dessus, le théorème est vrai pour un ensemble fini de $\mathcal{G}$. On peut donc se borner à étudier les séries de groupes classiques (*cf.* [18] pour la classification) et, dans chaque série on peut omettre les groupes de rang absolu borné à l'avance.

Fixons un sous-groupe d'Iwahori I de G et soit K un sous-groupe compact contenant I. On a $\mu_T(K) = [K : I]$ et la mesure de l'image de K dans G/Γ est le quotient de $[K : I]$ par l'ordre de $\Gamma \cap K$. Pour démontrer le théorème, lorsque G parcourt une série de groupes classiques, il suffit donc de faire voir que $[K_G : I] . A(G)^{-1} \to \infty$ pour $K_G \subset G$ compact ouvert contenant I convenable, A (G) désignant une borne supérieure de l'ordre des sous-groupes finis de G. On voit tout d'abord facilement qu'il existe $K_G \supset I$ tel que

$$\log_q [K_G : I] \geqq \max(l(\mathcal{G})^2/2,\ l(\mathcal{G})) + 1, \tag{3}$$

où, comme précédemment, $l(\mathcal{G})$ est le rang de $\mathcal{G}$ sur l'extension maximale non ramifiée de k. Pour la suite de la démonstration, on distingue trois cas.

5. On suppose que $G = \mathbf{SL}_n(k)$. Un théorème de Jordan (*voir* [15], Satz 200) ramène l'estimation de A (G) à celle d'une borne supérieure de l'ordre des sous-groupes commutatifs finis de G. On en déduit l'existence d'une constante c, indépendante de k et n, telle que

$$\log_2 A(G) \leqq cn^2/\log_2 n + n(2 + \log_2 e(k)) + (n-1) \log_2(q+1). \tag{4}$$

Il s'ensuit que si k est fixé, le quotient $[K_G : I] . A(G)^{-1} \to \infty$. En fait, même le quotient de cette expression par $\exp_q rn^{2-\varepsilon}$ ($\varepsilon > 0$, $r > 0$ dépendant de ε) tend vers l'infini, ce qui entraîne d'une part le renforcement au théorème et d'autre part, vu (1), que $\mu_{EP}(K_G) . A(G)^{-1} \to \infty$, lorsque $G = \mathbf{SL}_n(k)$ et k est fixé.

6. Supposons maintenant que $\mathcal{G}$ soit un groupe linéaire classique, mais non une forme intérieure ${}^1\mathbf{A}_n$ du type $\mathbf{A}_n$. La classification de [18] montre alors qu'il existe un plongement

$G \subset \mathbf{SL}_{8l+12}(k)$. Vu (4), il s'ensuit que l'on a, avec $l=l(\mathscr{G})$ et $a(l)=8l+12$:

(5) $$\log_2 A(G) \leqq c.a(l)^2/\log_2 a(l) + a(l)(2+\log_2(k)) + (a(l)-1)\log_2(q+1),$$

ce qui permet d'arriver aux mêmes conclusions que dans le cas précédent.

On passe de là assez facilement aux groupes localement isomorphes aux groupes classiques considérés dans ce numéro, en utilisant notamment quelques résultats de cohomologie galoisienne.

Jusqu'à présent, nous avons supposé k fixé. Mais les relations (1), (3), (4) et (5) montrent que si k varie, $e(k)$ restant borné, alors $\mu_T(K_G).A(G)^{-1}$ tend vers l'infini avec q, pour autant que $l(G)$ soit assez grand. Pour éliminer cette dernière restriction, il faut encore examiner le cas d'une famille de groupes isomorphes entre eux sur une clôture algébrique de $\mathbf{Q}_p$, le corps k parcourant l'ensemble de ses sous-corps de degré fini sur $\mathbf{Q}_p$ et de degré de ramification absolue fixé. On doit alors utiliser une autre borne supérieure de A (G), qui s'obtient à partir du théorème de Jordan cité plus haut, du fait que tout sous-groupe commutatif fini F de G est contenu dans le normalisateur d'un k-tore maximal [3] et d'un résultat analogue en caractéristique p si F est d'ordre premier à p ([17], II, 5.16).

7. Il reste à étudier le cas où $\mathscr{G}$ est une forme intérieure de $\mathbf{SL}_n$, différente de $\mathbf{SL}_n$. Cela signifie que G est localement isomorphe à $\mathbf{SL}_n(D)$ ($n \geqq 3$), où D est une algèbre à division centrale sur k, de degré $d \geqq 2$. Si $G=\mathbf{SL}_n(D)$, on a $G \subset \mathbf{SL}_{nd^2}(k)$, mais l'estimation de A (G) qui en résulte ne suffit pas, à moins que l'on ne suppose d borné, aussi procédons-nous différemment. On considère tout d'abord le cas où Γ est défini arithmétiquement. Cela signifie qu'il existe un corps de nombres totalement réel F, une place ultramétrique w de F et un F-groupe $\mathscr{H}$ absolument presque simple et anisotrope à l'infini tels que la complétion F_w de F en w soit isomorphe à k et qu'il existe un isomorphisme de G sur $\mathscr{H}(F_w)$ qui applique Γ sur un sous-groupe $\{w\}$-arithmétique de $\mathscr{H}(F)$. Le groupe $\mathscr{H}$ est de type ${}^2A_{nd-1}$. Sa définition fait intervenir une extension quadratique totalement imaginaire E de F dans laquelle w se décompose, et une algèbre à division $\mathfrak{D}$ centrale sur E, de degré $\mathfrak{d}$ sur E, munie d'une involution de deuxième espèce σ dont la restriction à E a F comme ensemble de points fixes. Le groupe $\mathscr{H}$ est alors le groupe spécial unitaire d'une forme σ-hermitienne h sur $\mathfrak{D}^m$, (où m est tel que $m\mathfrak{d}=nd$), qui est anisotrope en toute place réelle de F. Soit $\delta=\det(h) \in F^*$ et soit h' la forme hermitienne sur E^{nd}, relative à l'automorphisme non trivial de E sur F, donnée par la matrice identité d'ordre nd si nd est impair et par la matrice diagonale de coefficients diagonaux $(1, \ldots, 1, \delta)$ si nd est pair, et soit $\mathscr{H}'$ le groupe spécial unitaire de h'. Alors $\mathscr{H}'(F_w)=\mathbf{SL}_{nd}(F_w)$. Par des comparaisons de volumes et en utilisant le fait que les nombres de Tamagawa de $\mathscr{H}$ et $\mathscr{H}'$ sont égaux à 1 [12], on montre l'existence d'un sous-groupe discret Λ de $\mathbf{SL}_{nd}(k)$ tel que

(6) $$\frac{\mu_T(\mathbf{SL}_n(D)/\Gamma)}{\mu_T(\mathbf{SL}_{nd}(k)/\Lambda)} \geqq \frac{(q-1)^{dn}}{(q^d-1)^n} \cdot \prod_{v \in S} (q_v^{nd/2-1}(q_v-1))^{nd(1-d_v^{-1})},$$

où S en désigne l'ensemble (fini) des places ultramétriques $v \neq w$ de F qui se décomposent en E et en lesquelles l'ordre d_v de $\mathfrak{D} \otimes_E F_v$ dans le groupe de Brauer de F_v est >1, et q_v désigne l'ordre du corps résiduel de F_v. Cela nous ramène au cas précédemment traité, et prouve le théorème pour les sous-groupes de $\mathbf{SL}_n(D)$, définis arithmétiquement (n et D variables). D'après G. A. Margulis [10], tout sous-groupe discret cocompact de $\mathbf{SL}_n(D)$ ($n \geqq 3$) est commensurable à un sous-groupe arithmétiquement défini. Le passage à ces sous-groupes se fait alors par un procédé semblable à celui de [1], § 5, utilisant en outre un résultat de J. Rohlfs [13] 2.6 (prouvé pour des groupes déployés, mais dont la démonstration s'étend d'elle-même au cas général, comme cela est remarqué dans [11] 2.6). Il n'y a de nouveau pas grande difficulté à ramener à ce cas celui des groupes isogènes aux groupes $\mathbf{SL}_n(\mathrm{D})$.

8. La comparaison de volumes précédente conduit aussi à la proposition 3 ci-dessous, et, par suite, à une estimation de certains nombres de classes.

Comme $\mathfrak{D}$ admet une involution σ de seconde espèce, $\mathfrak{D} \otimes_{\mathrm{F}} \mathrm{F}_v$ est isomorphe à l'algèbre de matrices $\mathbf{M}_{m\mathfrak{d}}(\mathrm{E}_v)$, (où $\mathrm{E}_v = \mathrm{E} \otimes_{\mathrm{F}} \mathrm{F}_v$), pour toute place v de F qui ne se décompose pas sur E. Soient $\mathrm{S} = \mathrm{S}(\mathfrak{D})$ et q_v, d_v définis comme au paragraphe 7, à cela près que l'on n'exclut pas de place w. Nous supposons $\mathfrak{D}$ non commutative ou, ce qui est équivalent, S non vide. Soit encore **A** l'anneau des adèles de F. Si M est un ensemble fini, on note $|\mathrm{M}|$ son cardinal.

Proposition 3. — *Soit* K *un sous-groupe compact maximal de* $\mathscr{H}(\mathbf{A})$. *Alors il existe un sous-groupe compact maximal* K' *de* $\mathscr{H}'(\mathbf{A})$ *tel que*

$$|\mathscr{H}(\mathrm{F}) \backslash \mathscr{H}(\mathbf{A})/\mathrm{K}| \geqq f^{-1} \prod_{v \in \mathrm{S}} (q_v^{(m\mathfrak{d}/2-1)}(q_v - 1))^{m\mathfrak{d}(1-d_v^{-1})} |\mathscr{H}'(\mathrm{F}) \backslash \mathscr{H}'(\mathbf{A})/\mathrm{K}'|,$$

où $f = f(m\mathfrak{d})$ *est le maximum des ordres des sous-groupes finis de* $\mathbf{SL}_{m\mathfrak{d}}(\mathrm{E})$.

Cela entraîne en particulier que, étant donné un entier positif c, il existe un ensemble fini $\mathscr{D}_c$ d'algèbres à division centrales sur E et un entier positif $r(c)$ tels que l'on ait $|\mathscr{H}(\mathrm{F}) \backslash \mathscr{H}(\mathbf{A})/\mathrm{K}| > c$ pour tout sous-groupe compact maximal K de $\mathscr{H}(\mathbf{A})$ si $m\mathfrak{d} > r(c)$ ou bien $\mathfrak{D} \notin \mathscr{D}_c$.

Note reçue le 15 juin 1987.

Références bibliographiques

[1] A. Borel, Commensurability classes and volumes of hyperbolic 3-manifolds, *Ann. Sc. Norm. super. Pisa*, Cl. Sci., 8 (4) 1981, p. 1-33; 0. III, p. 617-649.

[2] A. Borel, *On the set of discrete subgroups of bounded covolume in a semisimple group*, preprint, 1986.

[3] A. Borel et G. D. Mostow, On semi-simple automorphisms of Lie algebras, *Ann. Math.*, 61, 1955, p. 389-405; 0. I, 460-476.

[4] A. Borel et J. Tits, Homomorphismes « abstraits » de groupes algébriques simples, *Ann. Math.*, 97, 1973, p. 499-571; 0. III, p. 171-243.

[5] A. Borel et N. Wallach, Continuous cohomology, discrete subgroups and representations of reductive groups, *Ann. Math. Stud.*, 94, 1980, Princeton Univ. Press.

[6] F. Bruhat et J. Tits, Groupes réductifs sur un corps local II : Schémas en groupes. Existence d'une donnée radicielle valuée, *Publ. Math. I.H.E.S.*, 60, 1984, p. 199-376.

[7] H. Garland, p-adic curvature and the cohomology of discrete subgroups of p-adic groups, *Ann. Math.*, 97, 1973, p. 375-423.

[8] W. M. Kantor, R. A. Liebler et J. Tits, On discrete chambertransitive automorphism groups of affine buildings, *Bull. A.M.S.*, (N.S.), 16, 1987, p. 129-133.

[9] I. G. Macdonald, The Poincaré series of a Coxeter group, *Math. Ann.*, 199, 1972, p. 161-174.

[10] G. A. Margulis, Discrete groups of motions of manifolds of non positive curvature, *Proc. I. C. M. Vancouver* 1974, 1, 21-34 (in Russian); English translation in *A.M.S. Translations*, 109, 1977, p. 33-45.

[11] G. A. Margulis et J. Rohlfs, On the proportionality of covolumes of discrete subgroups, *Math. Ann.*, 275, 1986, p. 197-205.

[12] J. G. M. Mars, The Tamagawa number of $^2\mathrm{A}_n$, *Ann. Math.*, 89, 1969, p. 557-574.

[13] J. Rohlfs, Die maximalen arithmetisch definierten Untergruppen zerfallender einfacher Gruppen, *Math. Ann.*, 244, 1979, p. 219-231.

[14] J.-P. Serre, Cohomologie des groupes discrets, *Ann. Math. Studies*, 70, 1971, p. 77-168, Princeton Univ. Press; 0. II, p. 593-685.

[15] A. Speiser, *Die Theorie der Gruppen von endlicher Ordnung*, Grund. Math. Wiss., 5, Springer-Verlag.

[16] R. Steinberg, Endomorphisms of linear algebraic groups, *Mem. A.M.S.*, 80, 1968.

[17] T. Springer et R. Steinberg, Conjugacy classes, in *Seminar on algebraic groups and related finite groups; Springer Lect. Notes in Math.*, 131, 1970, p. 167-294.

[18] J. Tits, Classification of algebraic semisimple groups, *Proc. Symp. Pure Math.*, 9, 1966, p. 33-62.

[19] J. Tits, Reductive groups over local fields, *Proc. Symp. Pure Math.*, 33, 1978, Part 1, p. 29-69.

A. B. : *The Institute for Advanced Study, Princeton NJ* 08540, *U.S.A.*;

G. P. : *Tata Institute of Fundamental Research, Homi Bhabha Rd. Colaba, Bombay* 400005, *India*;
Mathematical Sciences Research Institute, 1000 *Centennial Drive, Berkeley, CA* 94720, *U.S.A.*

136.

On the Set of Discrete Subgroups of Bounded Covolume in a Semisimple Group

Proc. Indian Acad. Sci. (Math. Sci.) **97** (1987), 45–52

Abstract. In this note G is a locally compact group which is the product of finitely many groups $\mathscr{G}_s(k_s)(s \in S)$, where k_s is a local field of characteristic zero and $\mathscr{G}_s$ an absolutely almost simple k_s-group, of k_s-rank $\geqslant 1$. We assume that the sum of the r_s is $\geqslant 2$ and fix a Haar measure on G. Then, given a constant $c > 0$, it is shown that, up to conjugacy, G contains only finitely many irreducible discrete subgroups L of covolume $\leqslant c$ (4.2). This generalizes a theorem of H C Wang for real groups. His argument extends to the present case, once it is shown that L is finitely presented (2.4) and locally rigid (3.2).

Keywords. Discrete subgroups; bounded covolume; semisimple group.

1. Introduction

Let first G be a connected semisimple Lie group with finite centre and no compact factors, of $\mathbb{R}$-rank $\geqslant 2$. Fix a Haar measure on G, hence on any quotient of G by a discrete subgroup. The total measure $v(G/L)$ of G/L is called the covolume of L, and will be denoted $c(L)$. By a result of H C Wang [15], given a constant $c > 0$, the number of conjugacy classes of irreducible discrete subgroups for which $c(L) \leqslant c$ is finite. Recently, Serre and Tits asked whether this is true in the p-adic case. In this paper, we show this is indeed the case. In fact, we prove more generally a similar assertion for discrete subgroups of products of semisimple groups over local fields of characteristic zero without compact factors.

This paper answers only the first of several questions of increasing generality pertaining to the finiteness of the number of pairs G, L with L of covolume bounded by a given constant, for a suitable a priori universal normalization of the Haar measure, when G or even the groundfield are allowed to vary. A number of results in that direction have since been obtained jointly with G Prasad and announced in *C R Acad. Sci.* **305** (1987), 357–362. These and others will be proved in two papers now in preparation.

In § 2 we fix our notation and assumptions and review, or extend to the present case, some known properties of discrete subgroups of finite covolume. In § 3 we show that Marguli's superrigidity implies local rigidity. Once this has done, H C Wang's argument may be used without change (§ 4).

2. Preliminaries

2.1. *Notation and assumptions*

S is a finite set. For each $s \in S$, there is given a local field k_s of characteristic zero and almost absolutely simple isotropic k_s-group $\mathscr{G}_s$. We view $G_s = \mathscr{G}_s(k_s)$ as a locally compact topological group, using the topology of k_s, and let G be the product of the G_s. For $T \subset S$, let pr_T be the projection of G onto $G_T = \prod_{s \in T} G_s$.

Let S_∞ (resp. S_f) be the set of s for which k_s is archimedean (resp. non-archimedean), $G_\infty = G_{S_\infty}$, $G_f = G_{S_f}$ and pr_∞ (resp. pr_f) be pr_T for $T = S_\infty$ (resp. S_f).

Let V be the set of places of $\mathbb{Q}$. It is the union of the set of primes and of the infinite place ∞. For $v \in V$, let $\mathbb{Q}_v$ be the completion of $\mathbb{Q}$ at v. We denote by S_v the set of $s \in S$ for which k_s contains $\mathbb{Q}_v$. We also write pr_v for pr_T when $T = S_v$. For $T \subset S_v$, let

$$R\mathscr{G}_T = \prod_{s \in T} R_{k_s/\mathbb{Q}_v} \mathscr{G}_S RG_T = \prod_{s \in T} R_{k_s/\mathbb{Q}_v} \mathscr{G}_s(\mathbb{Q}_v), \quad (T \subset S_v), \tag{1}$$

where $R_{k_s/\mathbb{Q}_v}$ refers to Weil's restriction of scalars [16:I]. The group $R\mathscr{G}_T$ is an algebraic $\mathbb{Q}_v$-group, and $RG_T = R\mathscr{G}_T(\mathbb{Q}_v)$ is in a natural way an analytic group over $\mathbb{Q}_v$. By [16:1·3·2], there is a canonical isomorphism of topological groups

$$G_T = RG_T. \tag{2}$$

We let r_s denote the k_s-rank of $\mathscr{G}_s$ and $r(G)$ the sum of the r_s ($s \in S$). By assumption $r_s \geqslant 1$ for all s.

2.2. A discrete subgroup L of G is "irreducible" if there is no partition $S = A \cup B$ into non-empty subsets such that $(L \cap G_A)(L \cap G_B)$ is of finite index in G.

2.3. *Lemma*

Let L be an irreducible subgroup of finite covolume of G.

(a) *For every $s \in S$, $\mathrm{pr}_s L$ is Zariski dense in $\mathscr{G}_s$.*
(b) *Let $T \subset S_v$. Then $\mathrm{pr}_T L$ is Zariski dense in $R\mathscr{G}_T$. Assume $T \neq S_v$ and v finite. Then the closure H of $\mathrm{pr}_T L$ contains the product of the groups G_s^+ ($s \in T$). In particular, it is open of finite index.*

[G_s^+ denotes the subgroup of G_s generated by the k_s-rational points of the connected unipotent k_s-subgroups of $\mathscr{G}_s$, cf. [3:§6].]

(a) follows from [9:1.12].
(b) The group $\mathrm{pr}_T L$ has the property (S) (cf. [1], [12:5.1]). Therefore its image on the right-hand side of (2) in §2·1 is Zariski-dense by [1] if $v = \infty$, by [14] if v is finite. To prove the second assertion we consider first the case where $T \neq S_v$ is reduced to one place t. By [5:III, §2, no. 2], H is a Lie subgroup of RG_v. If it were not open, then its Lie algebra would be a proper subspace of the Lie algebra of RG_t, which would be invariant under $\mathrm{pr}_T L$, in contradiction with (a). Therefore H is open. But $H \times \prod_{s \neq t} G_s$ is open in G, of finite covolume. Hence H has finite index in G_t. In particular, it is not compact. Then it contains G_t^+ by a theorem of Tits stated in [3:9.10], for which a proof is given in [11].

To go from this case to the general one, the argument is the same as that of Proposition 4.2 in [9]. We repeat it for the sake of completeness.

Let U be a compact open subgroup of $G_{T-\{t\}}$ and $H_U = H \cap (G_t \times U)$. Since the projection of $G_t \times U$ onto G_t is proper, H_U projects onto a closed subgroup H' of G_t, in which $\mathrm{pr}_t L$ is obviously dense. Therefore H' contains G_t^+ and $G_t^+ \times U \subset H_U$. This being true for every U, we see that $G_t^+ = \cap_U (G_t^+ \times U) \subset H$.

2.4. PROPOSITION

Let L be a discrete subgroup of finite covolume of G. Then L is finitely presented.

If a subgroup of finite index of L is finitely presented, then L is itself finitely presented, as is well-known. Therefore we may assume L to be irreducible.

If $S = S_\infty$, the result is known: If one group G_s has $\mathbb{R}$-rank one, this is 13·20 in [12]. If not, then L is arithmetic [8], in which case this is proved e.g. in [2]. Let now $S \neq S_\infty$ and X_f the product of the Bruhat-Tits buildings of the groups G_s for $s \in S_f$. The group L operates on S_f. We claim that X_f is a finite polyhedron mod L. In fact either $S = S_v$, in which case L is discrete and cocompact, or $S \neq S_f$ and then $\mathrm{pr}_f L$ is dense in an open subgroup of finite index (2.3). In both cases our assertion is clear. The isotropy group K_σ in G_f of a face σ of X_f is a compact open subgroup. The isotropy group of σ in L is $L \cap (G_\infty \times K_\sigma)$. Its projection on G_∞ is discrete of finite covolume. By the above, it is finitely presented. Then L is finitely presented by Theorem 4 in [6].

Our next remark is a trivial extension to our situation of a result of Kazhdan–Margulis.

2.5. PROPOSITION

(a) *Let $S = S_f$. Then G has a compact open neighbourhood of 1 which meets any discrete subgroup at the identity only.*

(b) *The group G has a neighbourhood of the identity U such that if L is discrete in G, then a conjugate of L meets U only at 1. In particular $c(L)$ has a strictly positive minimum.*

In case (a) we just have to take for U a torsion-free compact open subgroup, which always exists.

If $G = G_\infty$, then our assertion is a well-known result of Kazhdan–Margulis [7; 12:XI]. Let U_∞ be such a neighbourhood in G_∞ and U_f a torsion-free compact open subgroup of G_f. Then $U = U_\infty \times U_f$ satisfies our condition.

2.6. *Lemma*

Let k be a local field, $\mathscr{H}$ an almost absolutely simple k-group, (σ, E) an absolutely irreducible k-representation of $\mathscr{H}$ and L a finitely generated Zariski dense subgroup of $\mathscr{H}(k)$. Assume that the set of traces $\mathrm{Tr}\,\sigma(x)$ $(x \in L)$ is bounded. Then $\mathscr{H}(k)$ is relatively compact in $GL(E)(k)$.

Let $\bar{k}$ be an algebraic closure of k. By Burnside's theorem, every $A \in \mathrm{End}\, E(\bar{k})$, is a linear combination of elements of $\sigma(H(\bar{k}))$. In turn, those are linear combinations of elements of L, since the latter is assumed to be Zariski-dense. Let then $\{t_i\}$ $(i = 1, \ldots, m = \dim^2 E)$ be a basis of End E consisting of elements of L, and let $\{u_j\}$ be the dual basis with respect to the trace form. We can write

$$u_j = \sum_i d_{ij} t_i, \quad (d_{ij} \in k).$$

We let $\mathscr{H}(k)$ operate on End E via σ by left multiplication. We may identify $\sigma(\mathscr{H}(k))$ with the orbit of the identity. It suffices therefore to see that the representation of L on End (E), defined by left multiplication, has bounded coefficients with respect to the basis

(u_j). This is easy: Let

$$h\cdot u_j = \sum h_{ji} u_i, \quad (h\in L).$$

Multiplying on the right by t_i and taking traces, we get

$$h_{ji} = \mathrm{tr}\,(h\cdot u_j\cdot t_i) = \sum_a d_{aj}\,\mathrm{tr}\,(h\cdot t_a\cdot t_j).$$

Hence $|h_{ji}|$ has a universal bound in view of our assumptions.
Note. This argument just copies a known one of E B Vinberg.

3. Local rigidity

3.1. We fix an irreducible discrete subgroup L of finite covolume of G. Let $\mathbf{a} = (a_1, \ldots, a_m)$ be a generating set for L and $w_k\,(k\in K)$ be a finite set of defining relations for L (2.4). As usual, the set $R(L, G)$ of homomorphisms of L into G is identified with the set of m-tuples $\mathbf{g} = (g_1, \ldots, g_m)\in G^m$ which satisfy the relations $w_k(g_1, \ldots, g_m) = 0\,(k\in K)$. It is a closed subset of G^m, invariant under the group Aut G of automorphisms of G, acting componentwise in particular under the group Int G of inner automorphisms, where $x\in G$ acts by $\mathbf{g}\mapsto (xg_1x^{-1}, \ldots, xg_mx^{-1})$. The group L is said to be *locally rigid* if Int $G(\mathbf{a})$ contains a neighbourhood of $\mathbf{a}$ in $R(L, G)$.

3.2 Theorem. *Let L be as in 3.1 and assume that $r(G)\geq 2$. Then L is locally rigid.*

For $\mathbf{y} = (y_i)\in R(L, G)$, the map $a_i\mapsto y_i\ (i = 1, \ldots, m)$ extends to a homomorphism $\alpha_{\mathbf{y}}$ of L onto the subgroup $L_{\mathbf{y}}$ generated by the components y_i of $\mathbf{y}$.

(a) Let $s\in S$. We claim first that for $\mathbf{y}\in R(L, G)$ sufficiently close to $\mathbf{a}$, the group $\mathrm{pr}_s L_{\mathbf{y}}$ is Zariski-dense in $\mathscr{G}_s$ and not relatively compact in G_s.

Assume the first assertion to be false. There exists then a sequence $\mathbf{y}_i\to\mathbf{a}$ in $R(L, G)$ and, for each i, a proper subalgebra $\mathfrak{q}_i$ of the Lie algebra $\mathfrak{g}_s$ of $\mathscr{G}_s$ such that

$$\mathrm{Ad}\, y_{ij}(\mathfrak{q}_i) = \mathfrak{q}_i \quad (j = 1, \ldots, m).$$

Passing to a subsequence, we may assume the $\mathfrak{q}_i$ to have a constant dimension, say d, and then, again going over to a subsequence, we may assume that the $\mathfrak{q}_i(k_s)$ converge to a subspace $\mathfrak{q}$ in the Grassmannian of d-planes in $\mathfrak{g}_s(k_s)$. By continuity, $\mathrm{Ad}\, a_j$ leaves $\mathfrak{q}$ invariant for all j's, hence also for all $x\in L$, and, by (2.3) for all $x\in G_s$. But $\mathfrak{g}_s(k_s)$ is simple, whence a contradiction.

Write L_i for $L_{\mathbf{y}}$ when $\mathbf{y} = \mathbf{y}_i$. If now the $\mathrm{pr}_s L_i$ were all relatively compact, the traces of the elements of these groups in some linear realization of $\mathscr{G}_s$ would be uniformly bounded and so would be those of the elements of $\mathrm{pr}_s L$. By 2.6, $\mathrm{pr}_s L$ would be relatively compact. This however, would contradict the fact that G_s is not compact and $\mathrm{pr}_s L$ is dense in an open subgroup of G_s if $S\neq\{s\}$ (2·3), or has finite covolume if $S = \{s\}$.

(b) Let $v\in V$ and $T\subset S_v$. We assert now that for $\mathbf{y}\in R(L, G)$ sufficiently close to $\mathbf{a}$, the group $\mathrm{pr}_T L_{\mathbf{y}}$ is Zariski dense in $\mathscr{G}_T$.

If T consists of one place, this is (a). Assume then that T has at least two elements.

Let $\mathcal{M}_y$ be the Zariski closure of $\mathrm{pr}_T L_y$ and m_y the Lie algebra of RM_y. In view of (a), it maps on to the Lie algebra of RG_t for any $t \in T$. Therefore, for any $x \in L_y$ the characteristic polynomial $C(\lambda, \mathrm{Ad}\,\mathrm{pr}_t(x) | L(RG_t))$ divides the characteristic polynomial $C(\lambda, \mathrm{Ad}\,\mathrm{pr}_T(x) | \mathcal{M}_y)$.

Assume (b) to be false. The previous argument, carried out in RG_T, shows the existence of a sequence $\mathbf{y}_i \in R(L, G)$ tending to $\mathbf{a}$, such that the $m_{\mathbf{y}_i}$ have a constant dimension and tend to a proper subspace $\mathfrak{q}$ of the Lie algebra $L(RG_T)$ of RG_T. For every $x \in L$, the elements $\alpha_{\mathbf{y}_i}(x)$ tend to x. Since $\mathrm{pr}_T L$ is Zariski-dense in $\mathcal{G}_T$ (2.3), it follows that $\mathfrak{q}$ is an ideal of $L(RG_T)$, proper by our assumption. It is therefore equal to the Lie algebra of $RG_{T'}$ for some $T' \subset T$, with $T' \neq \phi, T$. Fix $t \in T$. We have seen that $C(\lambda, \mathrm{pr}_t(\alpha_{\mathbf{y}_i}(x)) | L(RG_t))$ divides $C(\lambda, \mathrm{pr}_T(\alpha_{\mathbf{y}_i}(x) | \mathcal{M}_{\mathbf{y}}))$. By continuity, it follows that for all $x \in L$, $C(\lambda, \mathrm{Ad}\,\mathrm{pr}_t(x) | L(RG_t))$ divides $C(\lambda, \mathrm{Ad}\,\mathrm{pr}_T(x) | L(G_{T'}))$. By Zariski-density, this should then be true for any $x \in G_T$. But, for $t \in T$, $t \notin T'$, this is absurd. This contradiction proves (b).

(c) Let $v \in V$ be finite and $T \subset S_v$. We now claim that for $\mathbf{y} \in R(L, G)$ sufficiently close to $\mathbf{a}$, the group $\mathrm{pr}_T L_{\mathbf{y}}$ contains an open subgroup of finite index of RG_T.

We know that $\mathrm{pr}_T L$ is dense in an open subgroup of finite index of RG_T (2.3). We claim that $\mathrm{pr}_T L_{\mathbf{y}}$ is not discrete if $\mathbf{y}$ is close enough to $\mathbf{a}$. If it were discrete, then there would be a sequence $\mathbf{y}_i \to \mathbf{a}$, such that the closure of the union of the $L_{\mathbf{y}_i}$ would contain $\mathrm{pr}_T L$, hence, by (2.3), an open subgroup of finite index of RG_T, but this contradicts 2.5(a). Combined with (b), this shows that $\mathrm{pr}_T L_{\mathbf{y}}$ is dense in an open subgroup H of RG_T for $\mathbf{y}$ close to $\mathbf{a}$. Let $t \in T$ and $T' = T - \{t\}$. The kernel N of the restriction of $\mathrm{pr}_{T'}$ to H is then open in G_t. Since G_t^+ is simple modulo center [12], it follows that $G_t^+ \subset N$. Therefore, H contains the product of the G_t^+ $(t \in T)$. Since G_t^+ has finite index in G_t, our assertion follows.

(d) We assume here that all $\mathcal{G}_s$ are of adjoint type, hence absolutely simple. We fix $\mathbf{y} \in R(L, G)$ close enough to $\mathbf{a}$ so that (b) and (c) hold for $L_{\mathbf{y}}$. We write L' and α for $L_{\mathbf{y}}$ and $\alpha_{\mathbf{y}}$. We want to prove that α extends to an automorphism of G.

Fix s. By [8], there exist $t = t(s) \in S$, a continuous homomorphism $\mu_s : k_t \to k_s$ and a k_s-morphism $v_s : {}^{\mu_s}\mathcal{G}_t \to \mathcal{G}_s$ such that the composition

$$\sigma_s : G \xrightarrow{\mathrm{pr}_t} G_t \xrightarrow{\mu_s^0} {}^{\mu_s}\mathcal{G}_t(\mu_s(k_t)) \xrightarrow{j} {}^{\mu_s}\mathcal{G}_t(k_s) \xrightarrow{v_s(k_s)} G_s \tag{3}$$

extends $\mathrm{pr}_s \circ \alpha : L \to \mathrm{pr}_s L'$.

We claim that in fact μ_s is an isomorphism. The morphism v_s is non-trivial, hence is a k_s-isomorphism, since both groups are absolutely k_s-simple. Therefore $v_s(k_s)$ is an isomorphism of topological groups. The third arrow in (3) is just induced by the inclusion $\mu_s(k_t) \to k_s$. If it were not surjective, then Im j would be a closed subgroup H if infinite index, not discrete, not relatively compact and $\mathrm{pr}_s L'$ would be contained in such a subgroup. The Lie algebra of H in $L(RG_s)$ would be proper, contradicting the Zariski density of $\mathrm{pr}_s(L')$. Therefore, σ_s is a continuous and open surjective homomorphism of G onto G_s, which extends $\mathrm{pr}_s \circ \alpha$, and whose kernel is the product of the factors G_u with $u \neq t$. Let now σ be the product of the σ_s. It is a continuous homomorphism of G into G which extends α. We want to prove that σ is an automorphism.

Now that j is the identity in k_s we see that $v_s(k_s) \circ \mu_s^0$ is an isomorphism of $G_{t(s)}$ onto G_s. It suffices therefore to show that $s \mapsto t(s)$ is a permutation of S, or, equivalently, that no G_s is contained in the kernel of σ.

Of course, $s \mapsto t(s)$ leaves each S_v stable. It is therefore equivalent to show that σ induces an automorphism of G_{S_v} onto itself for every $v \in V$, and for this that no G_s belongs to the kernel of $\sigma | G_v$. Assume to the contrary that $\sigma(G_s) = 1$ for some $s \in S_v$. Then Card $S_v \geqslant 2$. Let $T' = S_v - \{s\}$ and $T = S_v$. The homomorphism σ induces a continuous homomorphism of $G_{T'}$ into G_T hence also of $RG_{T'}$ into RG_T. These groups are analytic over $\mathbb{Q}_v$, hence σ is analytic [5:III, § 8]. The image is then a proper Lie subgroup H, whose projection on each factor RG_s is equal to $RG_s (s \in T')$. Its Lie algebra $\mathfrak{h}$ maps onto the Lie algebra of RG_s under $\mathrm{pr}_s (s \in T')$. On the other hand, $\mathfrak{h}$ should be invariant under $\mathrm{pr}_{T'} L'$, hence under Zariski dense subgroup (see (b)), and should therefore be a proper ideal. But this contradicts the previous surjectivity assertion.

(e) The automorphism σ belongs to the group $A(G)$ of automorphisms of G which are compositions of permutations of factors, field isomorphisms and restrictions to rational points of morphisms of algebraic groups of the various factors. [These are in fact all automorphisms of G, but we need not know that.] In this group the product B of the groups Aut $\mathscr{G}_s(k_s)$ is open of finite index. Therefore Int $G = \Pi_{s \in S}$ Int G_s is also open of finite index in $A(G)$.

(f) We can now prove the theorem. Let $\mathscr{G}'_s = \mathrm{Ad}\, \mathscr{G}_s$ and let $\pi_s : \mathscr{G}_s \to \mathscr{G}'_s$ be the canonical isogeny ($s \in S$). The morphism $\pi_s(k_s) : G_s \to G'_s$ has finite kernel, hence is proper, and its image is open of finite index [3:3.19], therefore the same is true for the product $\pi : G \to G'$ of the $\pi_s(k_s)$. In view of (e), $\pi(G)$ is also open of finite index in $A(G')$. In fact, π is a local homeomorphism.

Since π is proper and with image open of finite index, $L' = \pi(L)$ is discrete, of finite covolume in G'. With the notation and conventions of 3.1, let $\mathbf{a}' \in R(L', G')$ be the point with components $\mathbf{a}'_i = \pi(a_i)$ $(i = 1, \ldots, m)$ in G'^m. By (d), $A(G')(\mathbf{a}')$ contains a neighbourhood of $\mathbf{a}'$ in $R(L, G')$. By homogeneity, $A(G')(\mathbf{a}')$ is open in $R(L', G')$. Since all spaces under consideration are locally compact, countable at infinity, the orbit map $\beta : x \mapsto x \cdot \mathbf{a}'$ induces a homeomorphism of $A(G')/H$, where H is the isotropy group of $\mathbf{a}'$, onto $A(G')(\mathbf{a}')$ [4:VII, App. 1]. In particular, for every neighbourhood U of 1 in $A(G')$, $\beta(U)$ is a neighbourhood of $\mathbf{a}'$ in $R(L', G')$.

The group $\pi(G)$, identified to a subgroup of $A(G')$ via the inclusion of $G' = \mathrm{Int}\, G'$ into $A(G')$, is also open of finite index. If we let it act on $R(L', G')$ via this homomorphism, it follows that $g \mapsto g \cdot \mathbf{a}'$ is an open map.

Fix now an open neighbourhood V of $\mathbf{a}$ in G^m which is mapped homeomorphically onto a neighbourhood V' of $\mathbf{a}'$ by π. Let U be a neighbourhood of 1 in G such that $\mathrm{Int}\, g \cdot \mathbf{a} \subset V$ for $g \in U$. If now $\mathrm{y} \in R(L, G)$ is such that $\pi(\mathrm{y}) \in U \cdot \mathbf{a}'$, then there exists $g \in U$ such that $g \cdot \mathbf{a} = \pi(\mathrm{y})$. It follows that $g \cdot \mathbf{a}$ is an element of V which maps onto $\pi(\mathrm{y})$, hence $g \cdot \mathbf{a} = \mathrm{y}$. Therefore L is locally rigid.

4. Covolumes

4.1. The group G is locally compact, therefore the space of closed subgroups of G, endowed with the topology defined in [4:VIII, § 5] is compact. For any neighbourhood U of the identity, the subspace N_U of discrete subgroups which meet U only at 1 form a compact subset [4:VIII, § 5, no. 3, Prop. 7]. Moreover, the function $L \mapsto c(L)$ is lower semi-continuous [4:VIII, § 5, no. 2, Prop. 4]. In particular if a sequence of elements L_i of N_U tends to L and $c(L_i) \leqslant c$ for all i, then $c(L) \leqslant c$. We recall that $L_i \mapsto L$ if and only if the following condition is fulfilled (*loc. cit.*, no. 6): For any compact set $C \subset G$ and any neighbourhood U of 1 in G, we have

$$L_i \cap C \subset L \cdot U \text{ and } L \cap C \subset L_i \cdot U \text{ for } i \text{ big enough.} \tag{4}$$

4.2 Theorem. *Fix a constant $c > 0$. Assume $r(G) \geqslant 2$. Then the discrete subgroups of G with covolume $c(L) \leqslant c$ form finitely many conjugacy classes.*

In view of 2.3, 2.4, 3.2 and 4.1, the argument of H C Wang in the real case [15] goes over without change. For the sake of completeness, we describe it briefly.

Assume 4.2 to be false. Then we can find an infinite sequence L_i of nonconjugate discrete subgroups with covolume $\leqslant c$. Passing to a subsequence, we may assume that $c(L_i)$ has a limit $b \leqslant c$. Replacing L_i by a conjugate, if necessary, we may assume that $L_i \cap U = 1$, where U is a suitable neighbourhood of 1 in G (2.5). Then a cofinal subsequence of the L_i's has a limit L and $c(L) \leqslant b$ (4.1). We now consider the setup of 3.1. Let $a_1, \ldots, a_m$ be a generating set for L and w_k $(k \in K)$ a finite defining set of relations (2.4). By 4.1, we can find $x_{ij} \in L_i$ such that

$$\lim_{i \to \infty} x_{ij} = a_j \quad (j = 1, \ldots, m).$$

For i big enough, we have then $w_k(x_{il}, \ldots, x_{im}) \in U \cap L_i$, hence $w_k(x_{il}, \ldots, x_{im}) = 1$ $(k \in K)$. Therefore the map $a_j \mapsto x_{ij}$ $(j = 1, \ldots, m)$ extends to a homomorphism of L onto the subgroup L'_i of L_i generated by the x_{ij}'s, and $x_i = (x_{il}, \ldots, x_{im})$ is a point of $R(L, G)$, which comes arbitrarily close to **a** if i is big enough. For such i's, L'_i is by 3.2 conjugate to L under an inner automorphism. In particular $c(L'_i) = c(L)$, and L'_i has finite index in L_i. Then $c(L'_i)/c(L_i)$ is an integer $\geqslant 1$. On the other hand, it is equal to $c(L)/c(L_i)$, hence tends to $c(L)/b$, which is $\leqslant 1$. Altogether we get $c(L'_i) = c(L_i) = c(L)$, hence $L_i = L'_i$ and L_i is conjugate to L The L_i's are therefore pairwise conjugate for i big enough, whence a contradiction.

Remarks

(1) In the real case, the result of [15] is also valid if $r(G) = 1$, provided that G is not locally isomorphic to $SL_2(\mathbb{R})$ or $SL_2(\mathbb{C})$. [In fact, H C Wang ruled only $SL_2(\mathbb{R})$ out, but this was an oversight.] In the p-adic case, this is however false, since a torsion-free discrete subgroup of finite covolume of a simple p-adic group of relative rank one is cocompact and free.

(2) The previous argument also proves the following statement: Let L be a discrete finitely generated subgroup of G which is locally rigid, and $\{L_i\}$ a sequence of discrete subgroups of G which tend to L. If the L_i's are cocompact, then L is also cocompact.

In fact, the above proof shows that L is conjugate to a subgroup of L_i for i big enough.

Of the real groups G under consideration here, without restriction on the $\mathbb{R}$-rank, only groups locally isomorphic to $SL_2(\mathbb{R})$ or $SL_2(\mathbb{C})$ have discrete subgroups of finite covolume which are not locally rigid. On the other hand, they do contain sequences of discrete cocompact subgroups converging to a discrete non-cocompact subgroup of finite covolume: for $SL_2(\mathbb{C})$, such sequences are obtained by Dehn surgery. In $SL_2(\mathbb{R})$, such examples are easy to obtain geometrically: For instance we may consider a sequence of triangle groups (or rather of their subgroups of index two consisting of conformal transformations) with signatures $\pi/a, \pi/b_n, \pi/b_n$ where $a, b_n \in \mathbb{N}, a > 2$, and $b_n \to \infty$. The limit will be the subgroup of conformal transformations in the triangle group $(\pi/a, 0, 0)$. Therefore, for real groups, there is also a converse to the previous statement, so that the failure of local rigidity is necessary and sufficient for the existence of a sequence of discrete cocompact subgroups whose limit is not cocompact.

References

[1] Borel A Density properties of certain subgroups of semisimple groups, *Ann. Math.* **72** (1960) 179–188

[2] Borel A and Serre J-P. Corners and arithmetic groups, *Comm. Math. Helv.* **48** (1973) 436–491

[3] Borel A and Tits J, Homomorphismes "abstraits" de groupes algébriques simples, *Ann. Math.* **97** (1973) 499–571

[4] Bourbaki N, *Intégration* (Paris: Hermann) Chap. VII and VIII (1963)

[5] Bourbaki N, *Groupes et algèbres de Lie*, (Paris: Hermann) Chap. II and III (1972)

[6] Brown K, Presentations for groups acting on simply-connected complexes, *J. Pure Appl. Algebra* **32** (1984) 1–10

[7] Kazhdan D A and Margulis G A, A proof of Selberg's hypothesis, *Math. Sbornik (N.S.)* **75** (1968) 162–168

[8] Margulis A G, Discrete groups of motions of manifolds of non-positive curvature, *AMS Transl.* **109** (1977) 33–45

[9] Prasad G, Strong approximation for semi-simple groups over function fields, *Ann. Math.* **105** (1977) 553–572

[10] Prasad G, Lattices in semi-simple groups over local fields, *Adv. Math. Suppl. Stud.* **6** (1979) 285–356

[11] Prasad G, Elementary proof of a theorem of Bruhat-Tits-Rousseau and of a theorem of Tits, *Bull. Soc. Math. France* **110** (1982) 197–202

[12] Raghunathan M S, *Discrete subgroups of Lie groups*, Erg. d. Math. u. Grenzgeb., **68** (Berlin-Heidelberg-New York: Springer-Verlag) (1970)

[13] Tits J, Algebraic and abstract simple groups, *Ann. Math.* **80** (1964) 313–329

[14] Wang H C, Topics on totally discontinuous groups, in: *Symmetric spaces*, (eds) W M Boothby and G Weiss (New York: Marcel Dekker) 460–487 (1972)

[15] Wang S P, On density properties of *S*-subgroups of locally compact groups, *Ann. Math.* **94** (1971) 325–329

[16] Weil A, *Adeles and algebraic groups*, PM 23, (Boston: Birkhäuser) (1982)

[17] Zimmer E, Ergodic theory and semisimple groups, *Monographs in Mathematics* Vol. 81 (Boston: Birkhäuser) (1984)

137.

(avec G. Prasad)

Valeurs de formes quadratiques aux points entiers

C. R. Acad. Sci. Paris 307 (1988), 217–220

Résumé – D'après G. A. Margulis ([3], [4], [5]), une forme quadratique indéfinie non dégénérée sur $\mathbf{R}^n$ ($n \geqq 3$), non multiple d'une forme rationnelle, prend sur $\mathbf{Z}^n$ des valeurs non nulles arbitrairement petites en valeur absolue. Nous esquissons ici la démonstration d'une généralisation de ce résultat qui met en jeu un corps de nombres k, un ensemble fini S de places de k contenant l'ensemble S_∞ des places archimédiennes de k, et, pour chaque $s \in S$, les valeurs aux points de k^n à coordonnées S-entières d'une forme quadratique non dégénérée isotrope sur k_s^n, où k_s est la complétion de k en s.

Values of quadratic forms at integral points

Abstract – *According to G. A. Margulis ([3], [4], [5]), an indefinite non-degenerate quadratic form on* $\mathbf{R}^n$ ($n \geqq 3$), *which is not a multiple of a rational form, takes on* $\mathbf{Z}^n$ *non-zero values which are arbitrarily small in absolute value. We sketch here the proof of a generalization of this result which involves a number field k, a finite set S of places of k containing the set* S_∞ *of the archimedean ones and, for each* $s \in S$, *the values at the points of* k^n *with S-integral coordinates of a non-degenerate isotropic quadratic form on* k_s^n, *where* k_s *is the completion of k at s.*

1. Dans la suite, k, S, S_∞ et k_s sont comme ci-dessus. De plus $S_f = S - S_\infty$, $\mathfrak{o}_S$ est l'anneau des S-entiers de k, $|\,.\,|_s$ la valeur absolue normalisée de k_s et k_S la somme directe des k_s ($s \in S$). Le plongement diagonal de k dans k_S induit un plongement de k^n dans k_S^n qui envoie $\mathfrak{o}_S^n$ sur un sous-groupe discret cocompact de k_S^n (ce dernier étant muni de sa structure usuelle de groupe commutatif localement compact). Soit F une forme quadratique sur k_S^n, autrement dit une famille $(F_s)_{s \in S}$, où F_s est une forme quadratique sur k_s^n. Si F_0 est une forme quadratique sur k^n, on note $F_{0,s}$ la forme sur k_s^n obtenue à partir de F_0 par extension des scalaires; on dira, par abus de langage, que F est *rationnelle* si elle est de la forme $c.F_0$, avec $c \in k_S^*$, où F_0 est une forme quadratique sur k^n, et que F est *irrationnelle* sinon. La forme F est donc irrationnelle si, et seulement si, il existe $x, y \in k^n$ tels que $F(y) \neq 0$ et $F(x)/F(y) \notin k$.

THÉORÈME A. – *On conserve les notations précédentes. Soit* $F = (F_s)$ *une forme quadratique sur* k_S^n. *On suppose* $n \geqq 3$, *et* F_s *non dégénérée et isotrope pour tout* $s \in S$. *Alors les conditions suivantes sont équivalentes :*

(i) F *est irrationnelle.*

(ii) *Pour tout* $\varepsilon > 0$, *il existe* $x \in \mathfrak{o}_S^n$ *tel que* $0 < \max_s |F_s(x)|_s \leqq \varepsilon$.

(iii) *Pour tout* $\varepsilon > 0$, *il existe* $x \in \mathfrak{o}_S^n$ *tel que* $0 < |F_s(x)|_s \leqq \varepsilon$ *pour tout* $s \in S$.

2. Il est clair que (iii) ⇒ (ii) et élémentaire que (ii) ⇒ (i). D'autre part, si (i) est satisfaite, il est facile de voir qu'il existe un sous-espace $V \subseteqq k^n$ de dimension trois tel que $F|_{V \otimes_k k_S}$ satisfasse aux mêmes hypothèses que F. Il suffit donc de démontrer le théorème A si $n = 3$, ce que nous supposons dans la suite, sauf au paragraphe 6, où cette restriction n'intervient pas.

Supposons $S = S_\infty$ et $k = \mathbf{Q}$. Alors l'implication (i) ⇒ (ii) (conjecture d'Oppenheim) a été établie en particulier par A. Oppenheim si F représente zéro rationnellement et $n \geqq 4$ ([7], [8]), par B. J. Birch, H. Davenport et H. Ridout si $n \geqq 21$, [2] avant de l'être en général par G. A. Margulis ([3], [4], [5]). La généralisation du problème au cas des places archimédiennes d'un corps de nombres a été proposée dans [10] où, suivant [7],

Note présentée par Armand BOREL.

0249-6291/88/03070217 $ 2.00

l'implication (i) $\Rightarrow$ (ii) est prouvée si $n \geqq 5$ et F représente zéro rationnellement. Ici, nous admettons aussi des places finies dans S, suivant une suggestion de G. Faltings.

3. *L'implication* (i) $\Rightarrow$ (ii) *dans le cas archimédien.* — On suppose ici que $S=S_\infty$. Par suite, $\mathfrak{o}_S=\mathfrak{o}$ est l'anneau des entiers de k et $k_S=k\otimes_{\mathbf{Q}}\mathbf{R}$ est un espace vectoriel de dimension $[k:\mathbf{Q}]$ sur $\mathbf{R}$. Dans ce cas, la démonstration suit dans ses grandes lignes celle de Margulis. Nous nous bornerons à indiquer quelques-unes des modifications nécessaires, en supposant le lecteur familier avec l'un des articles ([3], [4], [5]).

Soient $G=SL_3(k_S)=\prod_{s\in S} SL_3(k_s)$ et $\Gamma=SL_3(\mathfrak{o})$, plongé diagonalement dans G. Le groupe Γ est le stabilisateur dans G du réseau $\Lambda_0=\mathfrak{o}^3$ de k_S^3 et l'espace homogène $\Omega=G/\Gamma$ s'identifie à un sous-espace fermé de l'espace des réseaux de k_S^3. Soient encore $H_s=SO(F_s)$ et H le produit des H_s.

Supposons tout d'abord que F ne représente pas zéro rationnellement, i. e., que F ne s'annule pas sur $k^3-\{0\}$. Alors, si (ii) (où la condition « $0<$ » devient superflue) n'est pas satisfaite, le critère de Mahler montre que l'orbite $H(\Lambda_0)$ est relativement compacte dans Ω. Comme dans ([3], [4]) il suffit pour traiter ce cas de prouver le :

Théorème 1. — *Toute orbite relativement compacte de* H *dans* Ω *est compacte.*

Étant donné $s\in S$, on choisit un sous-groupe unipotent $V_{1,s}$ de dimension un de H_s et note $D_{1,s}$ le groupe des points dans k_s d'un tore déployé de H_s normalisant $V_{1,s}$. Soient de plus W_s l'unique sous-groupe unipotent maximal de $G_s=SL_3(k_s)$ contenant $V_{1,s}$ et $V_{2,s}$ son centre. Le normalisateur de $V_{1,s}$ dans G_s est $D_s.V_s$, où $V_s=V_{1,s}.V_{2,s}$ et celui de V_s est $D_s.W_s$. Si s est réelle, on dispose des lemmes géométriques de ([3], [4]), notamment des lemmes 5, 6 et 7. On a besoin d'une version adaptée au cas complexe. Le point de départ est la variante suivante du lemme 5(*b*) de [3], dont la démonstration nous a été suggérée par P. Deligne :

Lemme 1. — *Soient* U *un sous-groupe unipotent connexe de* G_s, $N_s(U)$ *son normalisateur dans* G_s *et* M *un sous-ensemble de* $G_s-N_s(U)$ *dont l'adhérence contient* 1. *Alors* $(N_s(U)\cap\overline{UMU})/U$ *contient l'image de* U *par une application polynomiale non constante.*

Combiné avec les résultats de ([3], [4]), cela permet d'obtenir des analogues des lemmes 5, 6, 7 dont les conclusions fournissent des sous-variétés ou sous-groupes analytiques sur k_s non compacts.

Par ailleurs si U est un sous-groupe unipotent connexe de G, produit de groupes unipotents $U_s\subset G_s$ ($s\in J\subset S$) et C un ensemble compact invariant minimal pour U dans Ω, qui n'est pas une orbite de U, on est amené à considérer le stabilisateur F de C dans $N_G(U)$ et on a besoin du :

Lemme 2. — *Supposons* F *transitif sur* C. *Alors la composante neutre de* F *est un groupe algébrique, unipotent, produit de ses intersections avec les groupes* G_s, *qui sont toutes de même dimension.*

4. Fixons une orbite relativement compacte $H.z$ de H dans Ω. Supposons-la non compacte et soit X un ensemble fermé H-invariant minimal non vide contenu dans l'adhérence Z de Hz. Il existe $d\in D_s$ tel que W_s soit l'ensemble des $x\in G$ pour lesquels $\lim_{j\to\infty} d^j.x.d^{-j}=1$. Cela entraîne d'une part que le groupe d'isotropie dans W_s d'un point $u\in Z$ est réduit à l'identité [11 : 1. 12], donc que l'orbite de u par un sous-groupe de dimension $\geqq 1$ de W_s est non fermée, et d'autre part que toute orbite de D_sW_s dans Ω est dense [1]. En particulier, X ne contient aucun sous-ensemble non vide invariant par W_s. Le premier pas de la démonstration consiste à faire voir que X ne contient aucun

sous-ensemble Y non vide stable par $V_s (s \in S)$. A cet effet, on montre, par récurrence descendante sur le nombre d'éléments d'une partie non vide J de S, qu'il n'y a pas de sous-ensemble invariant par le produit V_J des $V_s (s \in J)$. A l'aide du lemme 2, on prouve tout d'abord que, si $y \in Y$, l'ensemble M des $g \in G - N_G(V_J)$ tels que $g.y \in Y$ contient l'élément neutre dans son adhérence et on utilise alors le lemme 1 et les analogues des lemmes géométriques de ([3], [4]). Cela étant acquis, on considère un sous-ensemble fermé M de X, invariant minimal par rapport au produit V_1 des $V_{1,s}$, et le deuxième pas de la démonstration consiste à faire voir que M est stable par le produit D des groupes D_s. Comme D_s normalise V_1, il suffit pour cela, étant donné $s \in S$, de montrer qu'un sous-ensemble fermé C invariant minimal pour $V_{1,s}$ contenu dans M est invariant par D_s. Les lemmes 1 et 2 permettent de prouver que C est invariant par un sous-groupe analytique sur k_s de $D_s . V_s$ non contenu dans $V_{1,s}$ et le résultat déjà obtenu entraîne facilement que ce sous-groupe n'est autre que D_s. La démonstration du théorème 1 s'achève alors comme dans [3], [4] ou [5].

5. Supposons maintenant que F représente zéro rationnellement. S'il y a une infinité de droites de k^3 sur lesquelles les F_s sont nulles, on voit tout de suite que F est rationnelle. Dans le cas contraire, on utilise le théorème suivant, qui est l'analogue du principal résultat du paragraphe 4 de [5], et se démontre de la même manière.

THÉORÈME 2. — *Soit* Y *la réunion d'un nombre fini de droites de* k^3. *Supposons qu'il existe* $\varepsilon > 0$ *tel que* $\max_s |F(x)|_s \geqq \varepsilon$ *pour tout* $x \in \Lambda_0 - (Y \cap \Lambda_0)$. *Alors* $H.(\Lambda_0)$ *est compacte et* F *est rationnelle.*

6. *L'implication* (i) $\Rightarrow$ (iii). — On considère tout d'abord le cas où il existe s tel que F_s soit multiple d'une forme rationnelle F_0 sur k^n. Soient S' l'ensemble des $s \in S$ tels que $F_s = c_s . F_{0,s} (c_s \in k_s^*)$ et $T = S - S'$. Vu l'hypothèse, T n'est pas vide. Quitte à remplacer F par un multiple, on peut supposer que $F_s = F_{0,s}$ pour $s \in S'$. Notons Q_0 la k-variété et Q_s la k_s-variété définies respectivement par $F_0 = 0$ et $F_s = 0$ $(s \in S)$. On a donc $Q_s(k_s) = Q_0(k_s)$ pour $s \in S'$ mais $Q_s \neq Q_0 \times k_s$ pour $s \in T$. Soient $H_0 = \mathbf{SO}(F_0)$ et Γ un sous-groupe S-arithmétique de H_0, vu comme sous-groupe discret de $H_{0,S} = H_0(k_S)$. L'approximation forte dans le revêtement universel de H_0 et le fait que les groupes $H_0(k_s)$ sont non compacts pour $s \in S'$, entraînent que la projection Γ_T de Γ dans le produit des $H_0(k_t) (t \in T)$ possède un sous-groupe d'indice fini qui est dense dans un sous-groupe ouvert L, lui-même produit de ses sous-groupes $L_t = L \cap H_{0,t} (t \in T)$. Soit M_t l'ensemble des points de $k_t^n - Q_0(k_t)$ qui sont transformés par L_t d'un point de $Q_t(k_t)$. C'est un ouvert non borné de k_t^n, stable par homothéties. Un raisonnement classique de géométrie des nombres permet d'établir le :

LEMME 3. — *Soit* U_s *un voisinage de l'origine dans* $k_s^n (s \in S')$. *Alors il existe* $x \in \mathfrak{o}_S^n$ *tel que* $x \in U_s$ *si* $s \in S'$ *et* $x \in M_t$ *si* $t \in T$.

Soit $\varepsilon > 0$. Choisissons U_s tel que $|F_s(x)|_s < \varepsilon$ pour $x \in U_s (s \in S')$ et soit $x \in \mathfrak{o}_S^n$ satisfaisant aux conditions du lemme 3. On peut trouver $h'_t \in L_t$ tel que $F_t(h'_t . x) = 0$, donc aussi $h_t \in L_t$ tel que $0 < |F_t(h_t . x)|_t \leqq \varepsilon/2$. Il existe alors $\gamma \in \Gamma$ tel que $0 < |F_t(\gamma . x)|_t < \varepsilon$ pour tout $t \in T$. Par construction $F_{0,t}(x) \neq 0$, donc $F_{0,s}(x) \neq 0$ pour $s \in S'$. Par hypothèse, $|F_s(x)|_s < \varepsilon$. Mais $\gamma \in H_0$ et $F_s = F_{0,s}$ si $s \in S'$. Par conséquent $F_s(\gamma . x) = F_s(x)$ $(s \in S')$ et, par suite, $\gamma . x$ satisfait à (iii).

7. Pour terminer la démonstration de (i) $\Rightarrow$ (iii), il reste donc à considérer le cas où aucune forme F_s n'est multiple d'une forme k-rationnelle, et on peut supposer $n = 3$. Il suit de cette hypothèse que le cône $Q_s = 0$ contient au plus un nombre fini de droites

rationnelles sur k ($s \in S$). Nous voulons montrer que la condition (iii) est déjà satisfaite sur $\mathfrak{o}^3$. Les hypothèses présentes impliquent que la restriction de F à $k_{\mathbf{R}} = k \otimes_{\mathbf{Q}} \mathbf{R}$ est irrationnelle et, compte tenu du théorème 2, que la condition (iii) est satisfaite pour tout $s \in S_\infty$ par des $x \in \mathfrak{o}^3$ appartenant à une infinité de droites rationnelles. En particulier, cela établit (iii) si $S = S_\infty$. Fixons $\varepsilon > 0$ et soit $s \in S_f$. Comme $\mathfrak{o}^3$ est un ensemble borné de k_s^3, il existe un entier a_s de k_s tel que $|F_s(a_s . x)|_s < \varepsilon$ pour tout $x \in \mathfrak{o}^3$. Par approximation forte dans k, on peut trouver $a \in \mathfrak{o}$ tel que $|a|_s = |a_s|_s$ pour $s \in S_f$. Soit $b = \max_{s \in S_\infty} |a|_s$. Vu ce qui a déjà été établi, il existe $x \in \mathfrak{o}^3$ tel que $0 \leqq |F_s(x)|_s < \varepsilon / b^2$ pour tout $s \in S$ et $F_s(x) \neq 0$ pour $s \in S_\infty$. Mais on dispose d'éléments x appartenant à une infinité de droites rationnelles et le cône Q_s n'en contient qu'un nombre fini ($s \in S_f$). Par conséquent, on peut aussi trouver un tel x qui n'annule aucune forme F_s, donc tel que $a . x$ satisfasse à (iii).

8. *Remarques sur l'ensemble* $F(\mathfrak{o}_S^n)$. Si s n'est pas complexe, $k_s^* \neq k_s^{*2}$ et k_s^*/k_s^{*2} est un groupe fini. Sous les hypothèses du théorème A, considérons la condition :

(iv) *Soient* $\varepsilon > 0$ *et* c_s *un élément de* k_s^*/k_s^{*2} ($s \in S$). *Alors il existe* $x \in \mathfrak{o}_S^n$ *satisfaisant à* (iii) *et tel que* $F_s(x) \in c_s$ *pour tout* s.

Il est élémentaire que cette condition équivaut à :

(v) *L'ensemble* $F(\mathfrak{o}_S^n)$ *des valeurs de* F *sur* $\mathfrak{o}_S^n$ *est dense dans* k_S.

Nous conjecturons que (iii) ⇒ (iv), donc que les conditions (i) à (v) sont équivalentes sous les hypothèses du théorème A et indiquons pour terminer les cas où, à notre connaissance, cela est démontré.

Supposons tout d'abord que $S = S_\infty$. Cette équivalence est claire si k est purement complexe, puisqu'alors (iii) et (iv) sont identiques. Si $k = \mathbf{Q}$, la condition (iv) signifie qu'il existe des $x \in \mathbf{Z}^n$ sur lesquels F prend des valeurs non nulles, arbitrairement petites en valeur absolue et de signe donné. Dans ce cas, l'implication (iii) ⇒ (iv) pour $n \geqq 3$ a été établie par A. Oppenheim [6] (qui a de plus remarqué qu'elle est fausse en général si $n = 2$). Ce résultat a été généralisé au cas d'un corps de nombres totalement réel par S. Raghavan [9].

Enfin, mentionnons qu'une légère modification du raisonnement du paragraphe 6 montre, sans restriction sur S, que (i) ⇒ (iv) si l'on peut trouver un S′ comme dans le paragraphe 6 qui est formé de places complexes.

Dans tous ces cas, les conditions (i) à (v) sont donc équivalentes.

Note reçue le 31 mai 1988, acceptée le 6 juin 1988.

RÉFÉRENCES BIBLIOGRAPHIQUES

[1] S. G. DANI, Orbits of Horospherical flows, *Duke Math. J.*, 53, 1986, p. 177-188.

[2] H. DAVENPORT et H. RIDOUT, Indefinite quadratic forms, *Proc. London Math. Soc.*, III Ser., 9, 1959, p. 544-555.

[3] G. A. MARGULIS, Formes quadratiques indéfinies et flots unipotents sur les espaces homogènes, *C. R. Acad. Sci. Paris*, 304, série I, 1987, p. 249-253.

[4] G. A. MARGULIS, In definite quadratic forms and unipotent flows on homogeneous spaces, *Semester on Dynamical Systems and Ergodic* Theory, Warsaw, 1986 Banach Center Publications (à paraître).

[5] G. A. MARGULIS, *Discrete subgroups and ergodic theory* (à paraître).

[6] A. OPPENHEIM, Values of quadratic forms (I), *Quarterly Jour. Math.* (2), 4, 1953, p. 54-59.

[7] A. OPPENHEIM, Values of quadratic forms (II), *ibid*, p. 60-66.

[8] A. OPPENHEIM, Values of quadratic forms (III), *Monatshefte f. Math.* (N.F.), 57, 1954, p. 97-101.

[9] S. RAGHAVAN, Values of quadratic forms, *Comm. pure and applied Math.*, 30, 1977, p. 273-281.

[10] S. RAGHAVAN et K. G. RAMANATHAN, On a diophantine inequality concerning quadratic forms, *Göttingen Nachr. Mat. Phys. Klasse*, 1968, p. 251-262.

[11] M. S. RAGHUNATHAN, *Discrete subgroups of Lie groups*, Springer, Berlin-Heidelberg-New York, 1972.

A. B. : *The Institute for Advanced Study, Princeton, NJ* 08540, *U.S.A.;*
G. P. : *Tata Institute of Fundamental Research, Colaba, Bombay* 400005, *India* et *The Institute for Advanced Study, Princeton, NJ* 08540, *U.S.A.*

138.

The School of Mathematics at the Institute for Advanced Study

in *A Century of Mathematics in America*, Part III, (editor P. Duren, with the assistance of R. Askey, H. Edwards, U. Merzbach), Amer. Math. Soc., Providence, R.I., (1989) 119–147

In the late twenties, Abraham Flexner, a prominent figure in higher education, had made an extensive study of universities in the U.S. and Europe and was extremely critical of many features of American universities. In particular, he deplored the lack of favorable conditions for carrying out research. In January 1930, while preparing for publication an expanded version of three lectures he had given in 1928 at Oxford on universities, he saw in the New York Times an article on a meeting of the American Mathematical Society (AMS), in which Oswald Veblen, professor at Princeton University, was quoted as having stated that America still lacks a genuine seat of learning and that American academic work is inferior in quality to the best abroad. He immediately wrote to Veblen, saying there was not the slightest doubt in his mind that both statements were true and hoping that Veblen had been correctly quoted. In his answer, Veblen confirmed these views, described the context of his remarks and wrote in conclusion:

> Here in Princeton the scientific fund which we owe largely to you and your colleagues on the General Education Board, is having an influence in the right direction, and I think our new mathematical building which is going to be devoted entirely to research and advanced instruction will also help considerably. I think my mathematical institute which has not yet found favor may turn out to be one of the next steps. Anyhow it seems to me to fit in with the concept of a seat of learning.

Born in Switzerland, Armand Borel did his undergraduate work at the Federal School of Technology (ETH) in Zürich. He obtained his doctorate degree at the University of Paris in 1952 and then spent two years at the Institute for Advanced Study in Princeton. He has been professor there since 1957.

The first Faculty of the School of Mathematics (minus J. von Neumann) with the second Director. From left to right: J. Alexander, M. Morse, A. Einstein, F. Aydelotte, Director, H. Weyl and O. Veblen.

(Photograph courtesy of the Institute for Advanced Study.)

Here Veblen was alluding first to the efforts, initiated by Fine and pursued with the help of Eisenhart and Veblen, to improve research conditions in his department and to the construction of what became Fine Hall; second to a plan for an "Institute for Mathematical Research" he had outlined and presented (without success) around 1925 to the National Research Council and to the General Education Board of the Rockefeller Foundation. It was to consist of four or five senior mathematicians who would devote themselves entirely to research, their own and that of some younger men, and of some younger mathematicians. Members would be free to give occasional courses for advanced students. It could operate within a university or be entirely independent of any institution.[1]

Shortly before, Flexner had been approached by two gentlemen who were surveying medical education on behalf of two persons who wanted to use part of their fortune to establish and endow a medical college in Newark. Since Flexner was an authority on medical education in the U.S., it was only natural to seek his counsel. He advised against it, explaining why in his opinion there was no real need for a new institution of the type they had in mind. Instead, he showed them the proofs of his book on universities and outlined his plan for an institution of higher learning, where scholars would pursue their researches and interests freely and independently. They were so fascinated by it that they swayed the potential donors, namely Louis Bamberger and his sister, Mrs. Felix Fuld, born Caroline Bamberger, convinced them to look into this possibility and soon introduced them to Flexner. This initiated a series of discussions and a correspondence extending over several months, at the end of which the Bambergers agreed enthusiastically to back up Flexner's plan, on condition that he would be the first director. A certificate of incorporation for a corporation to be known by law as the "Institute for Advanced Study – Louis Bamberger and Mrs. Felix Fuld Foundation" was filed with the state of New Jersey in May 1930 and the New York Times announced in June the creation of an Institute for Advanced Study, to be located in or near Newark, on a gift of $5 million from Louis Bamberger and his sister, Mrs. Felix Fuld. Veblen learned about it for the first time through that press release, although there had been a little further correspondence between the two about the idea of an Institute, but carried out *in abstracto*, at any rate on Veblen's side. He wrote immediately to Flexner that he was greatly pleased and he expressed the wish that this Institute would be located in the Borough or Township of Princeton "*so that you could use some of the facilities of the University and we could have the benefit of your presence.*" This heralded an increasing involvement of Veblen with this project, first as a consultant, then

[1] For this and the development of mathematics in Princeton until WW II, see William Aspray's article in *A Century of Mathematics in America, Part II* (editor, P. Duren, with assistance of R. A. Askey and U. C. Merzbach), Amer. Math. Soc., Providence, R.I., 1989, pp. 195–215.

as a professor having the primary responsibility for the building up of the School of Mathematics.

The Institute was eventually to consist of a few schools, but Flexner decided early on to start first with one in mathematics, because "mathematics is fundamental, requires the least investment in plant or books and he could secure greater agreement upon personnel than in any other field".[2] He began to make extensive inquiries in the U.S. and in Europe as to who would be the best choices for a faculty in mathematics. Among American mathematicians, the two most prominent names were those of George D. Birkhoff and Veblen. Flexner started with the former, on the theory that Veblen was already in Princeton anyhow. An offer was made, at an extremely high salary and accepted in March 1932, but Birkhoff asked to be released eight days later. After further inquiries, Flexner came to the conclusion that: "*If the Princeton authorities agreed willingly and unreservedly, we could not do better than to select Veblen.*" They did so quickly, and Eisenhart telegraphed to Veblen in June:

> Have talked with those concerned and they approve. Congratulate you heartily. Look forward to big things.

1932 was marked by extensive travelling, wide ranging consultations, and discussions, correspondence and negotiations with Veblen, Einstein and Weyl. (Of course, no outside advice was needed in the case of Einstein, and Flexner forged ahead as soon as he understood that he might be interested.) In October two faculty nominations were announced, that of Veblen, already effective October 1st, 1932 and that of Einstein, effective October 1st, 1933 (as well as the nomination of Walther M. Mayer, the then collaborator of Einstein, as an "associate"). It was also announced that the new Institute would be located in or near Princeton (a shift formally proposed in April 1932) and would be housed temporarily at Fine Hall. The school would officially begin its activities in Fall 1933, but in fact, during the academic year 1932–1933, Veblen already conducted a seminar in "Modern Differential Geometry."

It is well-known that Einstein was enthusiastic from the beginning ("Ich bin Feuer und Flamme dafür," he had stated to Flexner) and excessively modest in his financial requirements, but the negotiations were not all that smooth. In 1933 Flexner learned that Einstein had also accepted a professorship in Madrid and one at the Collège de France. Since their residence requirements were minimal (in the former case, nonexistent in the latter), while those of the Institute were for him only from October to April 15, Einstein did not see any incompatibility; on the other hand, if Flexner felt otherwise, he would agree to terminate the arrangement with the Institute.... The Madrid offer also included the right to name a professor and Einstein tried to use it as

[2] A. Flexner, *I remember*, Simon and Schuster, New York, 1940, pp. 359–360.

a leverage to secure a professorship at the Institute for W. Mayer (without success). In summer of 1933, Flexner had asked whether Einstein could arrive soon enough to participate in a general organizational meeting of the members of the school on October 2nd. Einstein felt he could not because this would entail spending one month away from W. Mayer, which would be too detrimental to his work. He arrived on October 17. He was reminded of that when he complained later that he had not been consulted about invitations and stipends. The collaboration with Mayer was over within a few months.

In Europe, the two names of mathematicians mentioned to Flexner above all others were those of G. H. Hardy and H. Weyl. While in Cambridge, Flexner got readily convinced that there was no way to lure Hardy away from Cambridge and he turned his attention to H. Weyl. (Hardy and Einstein, as well as J. Hadamard, had singled out Weyl as the most important appointment to be made from Europe.) Both he and Veblen, who had received an offer in June and was in Europe at the time, began discussing the matter with Weyl. He was interested from the start, in spite of strong misgivings about leaving Germany, and immediately expressed some desiderata about the school. First he thought it was absolutely necessary to add to Einstein, Veblen and himself a younger mathematician, preferably an algebraist. Weyl commented (in a letter to A. Flexner, dated July 30, 1932):

> The reason lies with the plans for filling the three main positions. By his personality, Veblen is certainly the most qualified American one can wish as the guiding spirit in an institution such as the one you have founded. But he is not a mathematician of as much depth and strength as say Hardy. The participation of Einstein is of course invaluable. But he pursues long-range speculative ideas, the success of which no one can vouch for. He comes less under consideration as a guide for young people to problems which have necessarily to be of shorter range. I am of a similar nature, at any rate I am also one who prefers to think by himself rather than with a group and who communicates with others only for general ideas or for a final well-rounded presentation. Therefore I put so much value on having a man of the type of Artin or v. Neumann.[3]

[3]Der Grund liegt mit in der Art der in Aussicht genommenen Besetzung der drei Hauptstellen. Veblen ist zufolge seiner menschlichen Qualitäten sicher der geeignetste Amerikaner, den man sich als führenden Geist in einer solchen Institution wie der von Ihnen gegründeten wünschen kann. Aber er ist doch nicht ein Mathematiker von ähnlicher Tiefe und Stärke wie etwa Hardy. Einsteins Mitwirkung ist natürlich unbezahlbar. Aber er verfolgt spekulative Ideen auf lange Sicht, deren Erfolg niemand verbürgen kann. Als Führer junger Leute zu eigenen, notwendig auf näher gesteckte Ziele gerichteten Problemen kommt er weniger in Betracht. Ich bin von ähnlicher Natur, jedenfalls auch Einer, der lieber einsam als mit einer Gruppe gemeinsam denkt und mitteilsam nur in bezug auf die allgemeinen Ideen oder in der fertigen gerundeten Darstellung. Mit darum lege ich so viel Wert auf einen Mann vom Typus Artin oder v. Neumann.

In fact, this was important enough to Weyl that Flexner included in his official proposal to him: "*the understanding that when the right person has been found, an algebraist of high promise and capacity will be appointed*". Later Weyl also pointed out the necessity for him to be allowed to give now and then regular courses. He was of course assured he would be welcome to do so, and he accepted in principle the offer in December 1932. But then, in three successive telegrams on January 3, 4, and 12, 1933 he withdrew, then accepted "irrevocably" ("unwiderruflich") and withdrew again. Later on he apologized profusely, explaining he had not realized he was suffering from nervous exhaustion. In his last telegram, he had given as his reason that he felt his effectiveness was tied to the possibility of operating in his mother tongue (a worry still faintly echoed in the foreword to his *Classical Groups*). But the deterioration of the conditions in Germany, in particular the passing of laws not only against Jews, but also against Aryans married to Jews (his case) made his leaving Germany all but unavoidable and in the course of the year he accepted a renewed Institute offer and began his activities at the Institute in January 1934.

The year 1933 also saw the addition to the school faculty of James Alexander and John von Neumann. It had been agreed between Eisenhart, Flexner, and Veblen that an offer would be made to either Lefschetz or Alexander, who both wanted the appointment. The choice fell on the latter, for reasons I have not seen stated anywhere. I have heard indirectly that Eisenhart had said he could more easily spare Alexander than Lefschetz. In view of the much greater involvement of the latter in all the activities of the department, this seems rather plausible. It is also well-known that later Lefschetz was not stingy with critical remarks about Veblen or the Institute. (In 1931, Flexner had asked his views first on the desirability, nature and location of an Institute and second on whom he would choose in mathematics, were he asked to do so. His answer to the second question was Veblen, Alexander and himself from Princeton, Morse and Birkhoff from Harvard; from Europe, he would add above all Weyl, but, since he was holding the most prestigious chair in mathematics in the world, there was no chance to attract him.) J. von Neumann had been half-time professor at the University for some time and the University was trying to make other arrangements. Veblen had suggested to offer him a position at the Institute but at first Flexner was reluctant to take a third mathematician from Fine Hall. However, after Weyl redeclined and after a further conference between von Neumann, Eisenhart, Veblen, and Flexner, an offer was made and quickly accepted. It was also agreed that the two institutions would, henceforth, jointly publish (and share the financial responsibility for) the *Annals of Mathematics*, with managing editors Lefschetz (who had been one since 1928) and von Neumann.

The appointment of Marston Morse in 1934, effective January 1st, 1935, brought to six the school faculty, which was to remain unchanged for the next ten years. To have assembled within three years such an outstanding faculty was an extraordinary success by any standard. In a report to the trustees of the Institute in January 1938, Flexner credited for this achievement Veblen and the help received from the University, in particular from L. P. Eisenhart, then dean of the faculty.

It was, of course, a tremendous boost for the development of the school that it could function in the framework of an outstanding department, strongly committed to research, and make full use of its facilities, vastly superior to those of any other mathematics department in the country. President Hibben and Eisenhart felt that the development of the Institute would be mutually beneficial, although the Institute was offering unique conditions for work, superior salaries, and therefore might again be successful in attracting faculty members besides Veblen. But others in the university community apparently had different opinions, so that, after the third appointment from the university faculty, Flexner and some trustees, in particular L. Bamberger, felt they had to assure the university authorities they would not in the future offer positions to Princeton University professors. As far as I can gather from the record available to me, they did so early in 1933 in one conversation with Acting President Duffield, (Hibben was retired by then). Whether this was meant for a limited time or forever, I do not know. I also have no knowledge of an official written statement by the Institute to that effect, nor of one by the University taking cognizance of such a commitment. On the contrary, the only university document of an official character on this matter I know of (prior to 1963, see below) takes a completely different position. To be more precise, L. P. Eisenhart had written to A. Flexner on November 26, 1932:

> I agree with you that the relationship of the Institute and our Department of Mathematics must be thought of as a matter of policy extending over the years. Accordingly I am of the opinion that any of its members should be considered for appointment to the Institute on his merits alone and not with reference to whether for the time being his possible withdrawal from the Department would give the impression that such withdrawal would weaken the Department. For, if this were not the policy, we should be at a disadvantage in recruiting our personnel from time to time. If our Trustees and alumni were disturbed by such a withdrawal, as you suggest, they should meet it by giving us at least as full opportunity to make replacements intended to maintain our distinction. The only disadvantage to us of such withdrawals would arise, if we were hampered in any way in continuing the policy which has brought us to the position which we now occupy. This policy has been to watch the field carefully and try out men of promise at every possible opportunity. If it is to be the policy of the Institute to have

> young men here on temporary appointment, this would enable us to be in much better position to watch the field.
>
> In my opinion the ideas set forth are so important for the future of our Department that it is my intention to present them to the Curriculum Committee of our Board of Trustees at its meeting next month, after I have had an opportunity to discuss them further with you next week.

Accordingly, Eisenhart presented on December 17 to the Curriculum Committee of the Board of Trustees a statement "on certain matters of policy in connection with the relation of Princeton University to the Institute", a copy of which was kindly given to me by A. W. Tucker. One paragraph reproduces in substance, even partly in wording, the first one quoted above. In conclusion, Eisenhart states that he is presenting this statement "with the expectation that you will approve of the position which I have taken...". It was indeed "approved in principle" by the committee. Obviously the latter was empowered to do so and to speak in the name of the Board of Trustees. Had it been solely advisory, Eisenhart could only have asked the committee to recommend to the board that it approve of his position. I am not aware of any other statement by university authorities addressing this question, again prior to 1963.

As already mentioned, Eisenhart was at the time dean of the faculty. Tucker pointed out to me that, in the organization of the University, this position was next in line to the presidency and that there was in fact no president in charge at that time: Hibben had retired in June 1932 and Dodds would be nominated and become president in late spring 1933. During the academic year 1932–1933, there was only an acting president, namely the Chairman of the Board of Trustees, E. D. Duffield, living in Newark, who mainly took care of off-campus, external affairs. Under those circumstances, Eisenhart was in fact addressing the Curriculum Committee as the chief academic officer of the University.

Although Flexner had not mentioned it in his formal report, he was of course acutely aware of another powerful factor for the rapid growth of the Institute, namely the anti-Semitic policies of the Nazi regime, without which the Institute could hardly have attracted Einstein, Weyl, and von Neumann. This was in fact only the beginning of the Institute's involvement with the migration of European scholars to the U.S. It is a well-known fact that Veblen played a prominent role in helping European mathematicians who had to leave Europe to relocate in the United States.[4] He, Einstein, and Weyl,

[4]See in particular the articles by L. Bers, D. Montgomery and N. Reingold in *A Century of Mathematics in America, Part I* (editor, P. Duren, with assistance of R. A. Askey and U. C. Merzbach), Amer. Math. Soc., Providence, R.I., pp. 231–243, pp. 118–129, pp. 175–200, respectively.

through a network of informants, were well aware of many such cases and often aided in a crucial way by offering first a membership, sometimes with a grant from the Rockefeller Foundation.

At the official Institute opening on October 1, 1933, the school already had over twenty visitors. The level of activities was high from the beginning. While emphasizing the importance of the freedom to carry on one's own research, and the opportunity of making informal contacts and arrangements, the early yearly *Bulletins* issued by the IAS list an impressive collection of lectures, courses and seminars. Among those given in the first four years, let me mention: A two-year joint seminar on topology by Alexander and Lefschetz, followed by a two-year joint course on topology, a joint seminar (extended over several years) by Veblen and von Neumann on various topics in quantum theory and geometry, a course and a seminar by H. Weyl on continuous groups (the subject matter of the famous *Lecture Notes* written by N. Jacobson and R. Brauer), followed by a course on invariant theory, courses and seminars by M. Morse in analysis in the large, a two-year course by von Neumann on operator theory, lectures on quantum theory of electrodynamics by Dirac, on class field by E. Noether, on quadratic forms by C. L. Siegel, and on the theory of the positron by Pauli. In 1935 H. Weyl started and for a number of years led a seminar on current literature. There was also of course a weekly joint mathematical club. The membership steadily increased and Veblen could state around 1937 that in Fine Hall there were altogether approximately seventy research mathematicians and an intense activity. This figure included the members and visitors of the University, too. There was no physical separation in Fine Hall between the two groups, which intermingled freely.[5] Many faced the familiar dilemma of having to choose between attending lectures or minding one's own work. There were also some grumblings that all this was too distracting for the graduate students. The trustees, mindful of the financial aspect, were asking for some limitation and even a reduction of the number of members; Veblen apparently was not too receptive. Almost from the start, Princeton had become a world center for mathematics, the place to go to after the demise of Göttingen.

That the Institute had in this way a considerable impact on mathematical research in Europe and in the United States needs hardly any elaboration. Less evident, and maybe less easy to imagine nowadays, is its role in the improvement of the conditions in American universities by the sheer force of the example of an institution providing such exceptional conditions and opportunities to faculty and visitors. In 1938 Flexner was pleased to quote to the trustees from a letter written to him on another matter by the secretary of the AMS, Dean R. G. D. Richardson of Brown University: "... *The Institute*

[5]For many recollections about Fine Hall at this time, see *The Princeton Mathematics Community in the* 1930*s. An Oral History Project*, administered by C. C. Gillespie edited by F. Nebeker, 1985, Princeton University (unpublished, but available for consultation).

has had a very considerable share in the building up of the mathematics to its present level.... Not only has the Institute given ideal conditions for work to a large number of men, but it has influenced profoundly the attitude of other universities."

The School of Mathematics developed along lines certainly consonant with the vision of the founders, as outlined in the first documents, but not identical with it. Underlying the original concept was a somewhat romantic vision of a few truly outstanding scholars, surrounded by a few carefully selected associates and students, pursuing their research free from all outside disturbances, and pouring out one deep thought after another. Einstein, Weyl, and Veblen soon decided they were not quite up to that lofty ideal and that the justification for the Institute would not be just their own work but, even to a much greater extent, to exert an impact on mathematics, in particular mathematics in the United States, chiefly through a vigorous visitors program. The visitors (called "workers" initially, "members" from 1936 on) were to be mathematicians having carried out independent research at least to the level of a Ph.D. and to be considered on the strength of their research and promise, regardless of whether or not they were assured of a position after their stay at the Institute. Furthermore, their interests did not have to be closely connected to those of one of the faculty members. Originally it was intended that the Institute would also have a few graduate students (but no undergraduates) and would grant degrees. It was officially accredited to do so in 1934. But already then, Flexner stated that it had been done because this seemed a wise thing to do, but it would not be a policy of the Institute to grant degrees, earned or honorary. Indeed, it has so far never done so. This view was confirmed in the 1938 issue of the yearly *Bulletin*, which stated that the Institute had discarded undergraduate and graduate departments on the ground that these already existed in abundance.

In short, the School of Mathematics had very early taken in many ways the shape it still has now, albeit on a different scale, at any rate for the visitors program. It was called School of Mathematics, although its most famous member was not a mathematician. In fact, when asked which title he would want to have, Einstein chose Professor of Theoretical Physics. However, it had been understood from the start that the school would also include theoretical physics. Internally, it was sometimes referred to as School of Mathematics and Theoretical Physics and there were always some visitors specifically in theoretical physics. The faculty had contemplated early on the addition of theoretical physicists; in particular Schrödinger was suggested by Weyl in 1934 and then also by Einstein. Dirac was also mentioned. But the director felt that he could not increase the faculty in the school: He was at the time starting two other schools, in economics and politics and in humanistic studies. Moreover, the financial situation caused some worry and he and the trustees felt some caution was called for. Still, Dirac was a visiting professor in 1934–1935 and Pauli the following year. Later, Pauli spent the war years at the Institute and was offered a professorship in 1945. He was interested but felt he could not commit himself before he had gone back at least for a while to Zürich, where his position had been kept open for him. He stayed at the Institute for one more year with the official title of Visiting Professor, but functioning as a professor and chose later to go back definitely to Zürich. The first real expansion in theoretical physics took place under the first half of Oppenheimer's directorship. As theoretical physics grew at the Institute, the two groups operated more and more independently from one another until it was decided, in 1965, to separate them officially by setting up a School of Natural Sciences. In the sequel, "School of Mathematics" will be meant in the narrow sense it has today.

The Institute developed first very informally. As already stated, Flexner relied for mathematics largely on outside advice, mainly that of Veblen. He had to: "*Mathematicians, like cows in the dark, all look alike to me*", he had said to the trustees at the January 1938 meeting. But this was to be an exception. He had already much more input in the setting up of the School in Economics and Politics and he expected fully it would be so in most aspects of the governance of the Institute. The correspondence with Veblen had shown already some differences of opinion on the eventual shape and running of the Institute, but they were not urgent matters at the time and could be overlooked while dealing with the tasks at hand, on which Flexner and Veblen were usually fully and warmly in agreement. However, as the Institute grew, differences of opinion between the director and some trustees on one hand, and the faculty on the other, became more apparent and relevant. The former liked to view the Institute as consisting of three essentially autonomous schools. They were willing to let each one run its own academic affairs; but there was a rather widespread feeling that professors were often conservative, parochial, not really able to see the Institute globally. Besides it was wrong for them to get involved in administrative matters (after all, Flexner had so often heard professors complain about those duties, which take so much precious time away from research and there he was offering them the possibility of having none...). On the other hand, the faculties of the three schools, which had been chosen quite independently and did not know one another, began to meet, to discuss matters of common interest, to compare views and problems and as a consequence to develop some feeling of being parts of one larger body. Understandably, they wanted to have at least a strong consultative voice in important academic matters. This came to a head when Flexner appointed two professors in economics without any faculty consultation. Added to earlier grievances, it led to such an uproar that Flexner had to resign. But, at a more basic level, there was no attempt to reconcile these two rather antagonistic attitudes in order to arrive at a *modus vivendi* offering a better framework to resolve any conflict that might arise again. None did arise under the next director, Frank Aydelotte (1939–1947), who earned the confidence of the faculty by his way of handling Institute matters (but, as a counterpart, less than unanimous approval from the trustees). Some conflict did surface, not to say erupt, under the next two directors, J. Robert Oppenheimer (1947–1966) and Carl Kaysen (1966–1976). Fortunately, except in one case to which I shall have to come back, these disputes had comparatively little visible impact on the workings of the School of Mathematics, as unpleasant and distracting as they were to its faculty, so that with relief I may pronounce these matters as outside the scope of this account and ignore them altogether. To conclude this long digression, let me add that a prolonged, in my opinion largely successful, effort was made over several years and concluded in 1974 to set up some Rules of Governance for handling in an orderly way between trustees, faculty and the director all aspects of the academic business of the Institute. There has been no such crisis under the present director, Marvin L. Goldberger (1986–), nor under the previous one, Harry Woolf (1976–1986).

In the fall 1939, a new chapter in the life of the Institute began with the moving of the Institute into the newly built Fuld Hall, on its own grounds. In preparation for this change, the school had begun to build up a library, aided in this first of all by Alfred Brauer, whom Weyl had taken as his assistant for this purpose. (Brauer did the same later on, on a bigger scale, for the Mathematics Department of the University of North Carolina at Chapel Hill.) In spite of the war, the Institute operated normally, although some professors were engaged in war work, albeit on a somewhat reduced scale. The influx from Europe increased and, again this had a direct bearing on the school: Siegel was given permanent membership, converted to a professorship in 1945. Kurt Gödel, after having been a member for about ten years, became a permanent member in 1946 and a professor in 1953. Why it took so long for Gödel is a matter of some puzzlement. There was of course unanimous admiration for his achievements and some faculty members had long favored giving him a professorship. The reluctance of others reflected doubts not on his scientific eminence, but rather on his effectiveness as a colleague in dealing with school or faculty matters (Siegel has been quoted to me as having said that one crazy man (namely himself) in the school faculty was enough) or on whether they would not be too much of an imposition on him. As a colleague of his in later years, I would say I found that, his remoteness not withstanding, he would acquit himself well of some of the school business, hence that those fears were not all well founded. On the other hand, I have to confess that I found the logic of Aristotle's successor in more difficult affairs sometimes quite baffling.

After the war, the activities of the school and its membership increased gradually. There was a conscious effort to have members from Eastern Europe or East Asia, in particular Poland, China, India. 1946 was also the beginning of the first (and so far only) venture of the Institute outside the realm of purely theoretical work, namely the construction of a computer under von Neumann's leadership. This has been described in considerable detail by H. Goldstine in his book,[6] to which I refer for details. The computer was used for a few years by a group working on meteorology and von Neumann wanted this to become a permanent feature at the Institute. But the faculty did not follow him. Even the faculty members who had a high regard for this endeavour in itself felt that it was out of place at the Institute, especially in view of the fact that there was no related work done at the University. The computer was given to the University in the late fifties.

Of the first faculty, Alexander resigned in 1947, remaining for some time as a member, Einstein became Professor Emeritus in 1946, Veblen in 1950 and Weyl in 1951. Siegel resigned in 1951 to return to Germany. Added to the faculty in 1951 were Deane Montgomery and Atle Selberg, who had been permanent members since 1948 and 1949 respectively, followed in 1952 by Hassler Whitney.

I came to the Institute in the fall of 1952, not knowing really what to expect. The only recommendation I can remember having received was to appear now and then at tea. This may have been prompted by memories of more formal days, but I soon realized that they were not counting heads. Instead, I found a most stimulating atmosphere, many people to talk to, and suggestions came from many sides. Let me indulge in some reminiscences of those good old days, with the tenuous justification that it is not out of order to describe in this paper some of the experiences and impressions of one visiting member.

F. Hirzebruch, whom I had known in 1949 when he spent some time in Zürich, came once to my office to describe the Chern polynomial of the tangent bundle for a complex Grassmannian. It was a product of linear factors and the roots were formally written as differences of certain indeterminates; Hirzebruch proceeded to tell me how to interpret them but he could not finish: they looked to me like roots in the sense of Lie algebra theory and this was just too intriguing for me to listen to any explanation. An extension to generalized flag manifolds suggested itself, but it was not clear at the moment whether this was more than a coincidence and wishful thinking. A few days later however, it became clear it was not and that marked the start of our joint work on characteristic classes of homogeneous spaces, to which we came back off and on over several years. Conversations with D. Montgomery and H. Samelson led to a paper on the ends of homogeneous spaces. A Chinese member, the topologist S. D. Liao, lectured on a theorem on periodic homeomorphisms of homology spheres he had proved using Smith theory. Having the tools of "French topology" at my finger tips, I tried to establish it in that

[6] H. H. Goldstine, *The computer, Part III*, Princeton University Press, Princeton, N.J., 1972.

framework, succeeded and then, by continuation, obtained new proofs of the Smith theorems themselves. This was the beginning of an involvement with the homology of transformation groups. Of much interest to me also was the seminar on groups, let by D. Montgomery, including his lectures on the fifth Hilbert problem, solved shortly before by him, L. Zippin and A. Gleason, and the contacts with H. Yamabe, his assistant that year.

At the University, Kodaira was lecturing on harmonic forms ("a silent movie" as someone had put it. The lectures were perfectly well organized, with everything beautifully written on the blackboard, but given with a very soft, low-pitched voice which was not so easy to understand.) Tate was lecturing on his thesis in Artin's seminar. The topology at the University gravitated around N. Steenrod, and his seminar was the meeting ground of all topologists. Among those was J. C. Moore, whom I had looked for immediately after my arrival with a message from Serre. This was the beginning of extensive discussions, and a friendship which even moved him to put his life and car at stake by volunteering to teach me how to drive.

My discussions with Hirzebruch went beyond our joint project. He was at the time developing the formalism of multiplicative sequences or functors, genera and experimenting with reduced powers, the Todd genus and the signature. In the latter case, this was soon brought to a first completion after Thom's results on cobordism were announced. Sheaf theory, in particular cohomology with respect to coherent sheaves, had been spectacularly applied to Stein manifolds by H. Cartan and J.-P. Serre; Kodaira, Spencer, Hirzebruch were naturally looking for ways to apply such techniques to algebraic geometry. So was Serre, of course. Being in steady correspondence with him, I was in a privileged position to watch the developments on both sides, as well as to serve as an occasional channel of communication. The breakthroughs came at about the same time in spring 1953 (I shall not attempt an exact chronology) and overlapped in part. Serre's first results were outlined in a letter to me, to be found in his *Collected Papers* (I, 243–250, Springer-Verlag, Berlin and New York, 1986); included were the analytic duality and a first general formulation of a Riemann–Roch theorem for *n*-dimensional algebraic manifolds. It was soon followed by the analogue for projective manifolds of the Theorems A and B on Stein manifolds. Spencer and Kodaira gave in particular a new proof of the Lefschetz theorem characterizing the cohomology classes of divisors. Soon came a vanishing theorem, established by Kodaira via differential geometric methods and by Cartan and Serre via functional analysis. Attention focussed more and more on the Riemann–Roch theorem, whose formulation became more precise, still with no proof. During the summer, we parted, I to go to the first AMS Summer Institute, devoted to Lie algebras and Lie groups (6 weeks, about thirty participants, roughly two lectures a day, a leisurely pace unthinkable nowadays) and then to Mexico (where I lectured sometimes in front of an audience of one, but not less than one, as Siegel is rumored to have done once in Göttingen, a rumor which unfortunately I could not have confirmed).[7]

1 Back at the Institute for a second year, I found again Hirzebruch, whose membership had also been renewed. The relationship between roots in the Lie algebra sense and characteristic classes had been made secure, but this whole project had been left in abeyance, there being so much else to do. Now we began to make more systematic computations, using or proving facts of Lie algebra theory and translating them into geometric properties of homogeneous spaces. Quite striking was the equality of the dimension of the linear system on a flag variety associated to a line bundle defined by a dominant weight and of the dimension of the irreducible representation with that given weight as highest weight. Shortly after, I went to Chicago, described this "coincidence" to André Weil, and out of this came shortly what nowadays goes by the name of the Borel–Weil theorem. After I came back, Hirzebruch was not to be seen much for a while, until he emerged with the great news that he thought he had a proof of the Riemann–Roch theorem. This was first scrutinized in private seminars and found convincing. I also provided a spectral sequence to prove a lemma useful to extend the theorem from line bundles, the case treated by Hirzebruch, to vector bundles. A bit later, Kodaira proved that Hodge varieties are projective. All this, and the work of Atiyah and Hodge giving a new treatment of integrals on algebraic curves, completed a sweeping transformation of complex algebraic geometry. Until then, it had been rather foreign to me, with its special techniques and language (generic points and the like). It was quite an experience to see all of a sudden its main concepts, theorems and their proofs all expressed in a more general and much more familiar framework and to witness these dramatic advances. This led me more and more to think about linear algebraic groups globally, in terms of algebraic geometry rather than Lie algebras, an approach on which I would work intensively the following year in Chicago, benefitting also from the presence of A. Weil.

During that second year, I also gave a systematic exposition of Cartan's theory of Riemannian symmetric spaces and got personally acquainted with O. Veblen, on the occasion of a seminar on holonomy groups he was holding in his office. I had of course no idea of his role in the development of the Institute, nor did I know about Flexner and his avowed ambition to create a "paradise for scholars". But I surely had felt it was one, or a very close approximation, so when I was offered a professorship in 1956, I was strongly inclined to accept it. It raised serious questions of course. I realized that, viewed from the inside, with the responsibilities of a faculty member, paradise might not always feel so heavenly. I had also to weigh a very good

[7](Added in proof) B. Devine just drew my attention to the interview of Merrill Flood by A. Tucker in the collection referred to in footnote five above, according to which such an incident did indeed take place once in Fine Hall.

university position (at the ETH in Zürich) with the usual mix of teaching and research against one entailing a "total, almost monastic, commitment to research", (as someone wrote to us much later, while declining a professorship). In fact, the offer had hit me (not too strong a word) while I was visiting Oxford and in a conversation the day before, J. H. C. Whitehead had made some rather desultory remarks about this "mausoleum". To him it was obviously essential to be surrounded by collaborators and students at various levels. I also had to gauge the impact on my family of such a move. But, after some deliberation and discussions with my wife, who left the decision entirely to me, I felt I just could not miss this opportunity.

My professorship started officially on July 1st, 1957, but I was already here in the spring. I found Raoul Bott, with whom I had many common
2 interests. Sometime before, Hirzebruch and I had made some computations on low-dimensional homotopy groups of some Lie groups and, to our surprise, some of our results were contradicting a few of those contained in a table published by H. Toda. There ensued a spirited controversy, in which the homotopists felt at first quite safe. Bott was very interested; he and Arnold Shapiro, also at the Institute at the time, thought first they had another proof of Toda's result on $\pi_{10}(G_2)$, one of the bones of contention, but a bull session disposed of that. Later, Bott and Samelson confirmed our result. Eventually, the homotopists conceded. At the time, I had not understood why Raoul was so interested in those very special results, but I did a few months later when he announced the periodicity to which his name is now attached: Our corrections to Toda's table had removed a few impurities which stood in the way of even conjecturing the periodicity.

There was also a very active group on transformation groups around D. Montgomery who, with the Hilbert fifth problem behind him, had gone back fully to his major interest. My involvement with this topic increased, culminating in a seminar held in 1958–1959.

But I was now a faculty member in mathematics (together with K. Gödel, D. Montgomery, M. Morse, A. Selberg, H. Whitney, as already mentioned, Arne Beurling, who had joined in 1954, and A. Weil from fall 1958 on) and had to have some concerns going beyond my immediate research interests. Foremost were two, the membership and the seminars. As regards the former, it was not just to sit and wait for applicants and select among them, but also of course to seek them out. Weil and I felt that in the fields somewhat familiar to us, a number of interesting people had not come here and I remember that for a few years, in the fall we would make lists at the blackboard of potential nominees and plan various proposals to the group. In this way, in particular, we contributed not insignificantly to the growth of the Japanese contingent of visitors, which soon reached such a size that the housing project was sometimes referred to as "Little Tokyo" and that a teacher at the nursery school found it handy to learn a few (mostly disciplinary) Japanese words.

After a few years however, there was no significant "backlog" anymore and no need to be so systematic. As to the seminars, there were first some standard ones, like the members' seminar and the seminar in groups and topology, led by D. Montgomery. Others arose spontaneously, reflecting the interests of the members or faculty. We felt that the Princeton community owed it to itself also to supply information about recent developments and that beyond the graduate courses offered by the University and the research seminars, there should be now and then some systematic presentations of recent or even not-so-recent developments. In that respect, J.-P. Serre, a frequent fall term visitor during those years, and I organized in fall 1957 two presentations, one on complex multiplication and a much more informal one where we wrestled with Grothendieck's version of the Riemann–Roch theorem. As soon as he arrived, Weil set up a joint University–Institute seminar on current literature, thus reviving the tradition of the H. Weyl seminar, which he had known while visiting the Institute in the late thirties, and had also kept up in Chicago. The rule was that X was supposed to report on the work of Y, Z, with $X \neq Y, Z$. Later on, the responsibility for this seminar was shared with others. It was quite successful for a number of years, but was eventually dropped for apparent lack of interest. As I remember it, it became more and more difficult to find people willing to make a serious effort to report on someone else's work to a relatively broad non-specialized audience. Maybe the increase in the overall number of seminars at the University and the Institute, at times somewhat overwhelming, was responsible for that, I don't know.

During those years, algebraic and differential topology were in high gear in Princeton. In 1957–1958 J. F. Adams was here, at the time he had proved the nonexistence of maps of Hopf invariant one (except in the three known cases). Also Kervaire, while here, proved the non-parallelizability of the n-sphere ($n \neq 1, 3, 7$) and began his joint work with J. Milnor. In fall 1959 Atiyah and Hirzebruch developed here (topological) K-theory as an extraordinary homology theory, after having established the differentiable Riemann–Roch theorem; Serre organized a seminar on the first four chapters of Grothendieck's EGA. During that year, Kervaire, then at NYU came once to me to outline, as a first check, the construction of a ten-dimensional manifold not admitting any differentiable structure! M. Hirsch and S. Smale were spending the years 1958–1960 here, except that Smale went to Brazil in 1960. Soon Hirsch was receiving letters announcing marvelous results, so wonderful that we were mildly wondering to what extent they were due to the exhilarating atmosphere of the Copacabana beach, but they held out. (At the Bonn Tagung in June, as the program was being set up from suggestions from the floor, as usual, the first three topics proposed were the proofs of the Poincaré conjecture in high dimension by Smale and by Stallings and the construction of a nondifferentiable manifold by Kervaire; Bott, freshly arrived and apparently totally unaware of these developments, asked whether this was a joke!)

During these first years at the Institute, my active research interests shifted gradually from algebraic topology and transformation groups to algebraic and arithmetic groups, as well as automorphic forms. That last topic was already strongly represented here by Selberg, and had been before by Siegel. This general area was also one of active interest for Weil, and it soon became a major feature in the school's activities. Without any attempt at a precise history, let me mention a few items, just to give an idea of the rather exciting atmosphere. I first started with two projects on algebraic groups, one with an eye towards reduction theory, on the structure of their rational points over non-algebraically closed fields, the other on the nature of their automorphisms as abstract groups. Some years later, I realized that Tits had proceeded along rather similar lines and we decided to make two joint endeavours out of that. But I was more and more drawn to discrete subgroups, especially arithmetic ones. Rigidity theorems for compact hermitian symmetric spaces, hyperbolic spaces and discrete subgroups were proved by Calabi, Vesentini, while here, Selberg and then Weil. It is also at that time that I proved the Zariski density of discrete subgroups of finite covolume of semisimple groups. Weil was developing the study of classical groups over adeles and of what he christened Tamagawa numbers. I. Satake, while here, constructed compactifications of symmetric or locally symmetric spaces. It became more and more imperative to set up a reduction theory for general arithmetic groups. The Godement conjecture and the construction of some fundamental domain of finite area became prime targets. The first breakthroughs came from Harish-Chandra. I then proved some results of my own; he suggested that we join forces and we soon concluded the work published later in our joint *Annals* paper. This was in summer 1960. The next year and a half I tried alternatively to prove or disprove a conjecture describing a more precise fundamental domain and finally succeeded in establishing it. Combined with the other activities here and at the University, this all made up for a decidedly upbeat atmosphere. But in 1962 rumors began to spread that it was not matched by equally fruitful and harmonious dealings within the faculty. Harish-Chandra, who was spending the year 1961–1962 here, asked me one day, What about those rumors of tremors shaking the Institute to its very foundations? We were indeed embroiled in a bitter controversy, sparked by the school's proposal to offer a professorship to John Milnor, then on the Princeton faculty.

Before we presented this nomination officially, the director had indeed warned us, without being very precise, that there might be some difficulty due to the fact that Milnor was at the University, and we could hardly anticipate the uproar that was to follow. The general principle of offers from one institution to the other and the special case under consideration were heatedly debated in (and outside) two very long meetings (for which I had to produce minutes, being by bad luck the faculty secretary that year). A number of colleagues in physics and historical studies stated that it had always been their understanding that there was some agreement prohibiting the Institute to offer a professorship to a Princeton University colleague. In fact, the historians extended this principle even to temporary memberships. Fear was expressed that such a move would strain our relations with the University, which some already viewed as far from optimal. In between the two meetings, the director produced a letter from the chairman of the Board of Trustees, S. Leidesdorf, referring to a conversation he had participated in between Flexner and the president of the University, in which it had been promised not to make such offers. He viewed it as a pledge, which could be abrogated only by the University.

Those views were diametrically opposite to those of the mathematicians here and at the University, which were in fact quite similar to those of Eisenhart in the letter quoted earlier or in his statement to the curriculum committee, both naturally and unfortunately not known to us at the time. He really had said it all. First of all, the school used to give sometimes temporary memberships to Princeton faculty. This was on a case-by-case basis, not automatic, and it had never occurred to us to rule it out *a priori*. We also felt that our relations with Fine Hall were excellent and would not be impaired by our proposal. In fact D. Spencer had told us right away we should feel free to act. D. Montgomery stated that Veblen had repeatedly told him, in conversations between 1948 and 1960, that there had never been such an agreement. J. Alexander, asked for his opinion, wrote to Montgomery that he had never known of such an agreement (whether gentlemanly or ungentlemanly). He also remembered certain conversations in which an offer to a university professor was contemplated, or feared by some university colleague, conversations which would have been inconceivable, had such an agreement been known. Finally he had "no knowledge of deals that may have been consummated in 'smoke-filled rooms' or of 'secret covenants secretly arrived at.' All this sort of stuff is over my depth." A. W. Tucker, chairman of the University Mathematics Department, consulted his senior colleagues and wrote to A. Selberg, our executive officer, that in their opinion (unanimous, as he confirmed to me recently) the Institute should be free to extend an offer to Milnor. Of course, were he to accept it, this would be a great loss, but any such "restraint of trade" was distasteful to them and could well prove damaging in the long run. It would be much better, they felt, if the University would answer with a counteroffer attractive enough to keep Milnor. The point was repeatedly made that, when two institutions want the services of a given scholar, it is up to the individual to choose, not up to administrators or colleagues to tell him what to do; also, as Eisenhart had already pointed out, that such a blanket prohibition might be damaging to the recruiting efforts of the University.

In the course of the second faculty meeting a colleague in the School of Historical Studies, the art historian Millard Meiss, stated it had indeed been his understanding there was such an agreement; he noted that the mathematicians and his school acted differently with regard to temporary memberships; he felt the rule had been a wise one in the earlier days of the Institute, but was very doubtful it had the same usefulness today. Accordingly, he proposed a motion, to the effect that the faculty should be free to extend professorial appointments to faculty members of Princeton University, with due regards to the interests of science and scholarship, and to the welfare of both institutions. He also insisted that this should occur only rarely. This motion was viewed as so important ("the most important motion I have voted on in the history of the Institute", commented M. Morse) that it was agreed to have the votes recorded by name, with added comments if desired. It was passed by fourteen *yes* against four *no*, with two abstentions.

After this, it would have seemed most logical to take up the matter with the president of the University, R. Goheen, but nothing of the kind was done at the time and the tension just mounted until the trustees meeting in April. There, as we were told shortly afterwards by the director, the Milnor nomination did not even come to the board: The trustees had first reviewed the matter of invitations to Princeton University faculty, with regard to the Meiss motion, and had voted a resolution to the effect that the agreement with Princeton University to refrain from such a practice was still binding.

In this affair we had worked under a further handicap: In those days, it was viewed as improper to talk about a possible appointment with the nominee before he had received the official offer (nowadays, the other way around is the generally accepted custom). Consequently, none of us had ever even hinted at this in conversations with Milnor. But he had heard about it from other sources and it became known that he would have been seriously interested in considering such an offer. The director and the trustees may not have felt so fully comfortable with their ruling after all. At any rate, they soon proposed to offer some long-term arrangement to Milnor, whereby he could spend a term or a year at the Institute during any of the next ten years. This was of course very pleasant for Milnor, and we gave this proposal our blessing, but it fell short of what we had asked for. Finally, eighteen months later, in October 1963, we were informed that, following instructions from the trustees, the director had taken up the matter of general policy with President Goheen in January 1963 and we received a copy of a letter written on January 21, 1963 by President Goheen to the director, outlining one. Although cautious in tone, it allowed one institution to extend an offer to a faculty member of the other, after close consultation "*to the end of matching the interests of the individual with the common interests of the two institutions to the fullest extent possible.*" In conclusion, he urged that "*this agreement supplant any specific or absolute prohibition that we may have inherited from our predecessors.*"

Right after the next trustees meeting the director wrote to Goheen on April 22, in part: "*The Trustees asked me to tell you that they welcome your letter, and that they have asked me to let it be a guide to future policy of the Institute.*" As far as I know, the matter was never reconsidered and this agreement is still in force. At the time we were apprised of this (October 1963), it would have therefore been "legally" possible for us to present again our proposal, although Milnor was still a Princeton faculty member.

But we could not! During 1962–1963, we had asked for two additions to our group; they had been granted and no chair was available to us anymore. How had this come about?

This experience had left strong marks. It was not just the decision of the trustees, but the way the matter had been handled and the breakdown in relations within the faculty (also contributed to by conflicting views on some nominations in the School of Historical Studies), the ruling from on high by the board, without bothering to have a meaningful discussion with us, bluntly disregarding our wishes, as well as those of the faculty as expressed by the Meiss motion, all this chiefly on the basis of a rather flimsy recollection of the chairman of the board, promoted to the status of an irrevocable pledge. Some of us were wondering whether to withdraw entirely into one's own work or to resign, and were sounded out as to their availability. One Chairman, who had for some time wanted to set up a mathematics institute within his own institution, toyed with the idea of making an offer to all of us. We still had the option of making another nomination and there were indeed two or three names foremost on our minds. But just choosing one and presenting it would not suffice to restore our morale. Something more was needed to help us rebound. It was Weil who suggested that we present two nominations instead of just one, as was expected from us. After some discussions, we agreed to do so and nominated Lars Hörmander and Harish-Chandra.

This took the rest of the faculty and the director completely by surprise. The latter did not raise any objection on budgetary grounds. He also made it clear at some point that if granted, this request would have no bearing on faculty size for the other groups. Since our nominations were readily agreed to be scientifically unassailable, it would seem that our proposal would go through reasonably smoothly, but not at all. Our request had been addressed by A. Selberg, still our executive officer, directly to the director and the trustees, bypassing several steps of the standard procedure for faculty nominations, which seemed unpracticable in the climate at the time, and also not fulfilling one requirement in the by-laws. And it is indeed on grounds of procedure that the director and some colleagues raised various objections. There was overwhelming agreement on the necessity of major changes in our procedure for faculty appointments. The question was whether this review should precede or follow the handling of our two nominations. Again, this grew into a full-size debate and we did not know how our proposal would fare at the

April trustees meeting. There, as we were told at the time, the director recommended to postpone the whole matter, but the trustees, after having heard Selberg present our case, voted to grant our request under one condition, namely that a faculty meeting be held to discuss our nominations. This was really only to restore some semblance of formal compliance with the by-laws, and they were anxious that this matter be brought with utmost dispatch to a happy conclusion, so that the Institute would soon regain its strength and some measure of serenity. This meeting was held within a week and the offers were soon extended.

Harish-Chandra accepted quickly, Hörmander after a few months. Finally, this sad episode was behind us. We felt and were stronger than before and could devote ourselves again fully to the business of the school. In fall 1963 there were the usual seminars on members and faculty research interests. Harish-Chandra started a series of lectures, which became an almost yearly feature: every week two hours in a row, most of the time on his own work, i.e., harmonic analysis on reductive groups (real, later also p-adic), documenting in particular his march towards the Plancherel formula. He was not inclined to lecture on other people's work. One year however he did so, he "took off", as he said, viewing it as some sort of sabbatical, and lectured on the first six chapters of Langlands' work on Eisenstein series (then only in preprint form). There were also some seminars on research carried out outside Princeton: I launched one on the Atiyah–Singer index theorem, for non-analysts familiar with all the background in topology. Eventually, R. Palais took the greater load and wrote the bulk of the *Notes* (published in the *Annals of Math. Studies* under his editorship). The following year, there was similarly a "mutual instruction" seminar on Smale's proof of the Poincaré conjecture in dimensions ≥ 5. Still, we felt some imbalance in the composition of the membership and the activities of the school. Of course, there is no statutory obligation for the school membership to represent all the main active fields of mathematics. In any case, in view of the growth of mathematics and of the number of mathematicians, as compared to the practically constant size of the school (the membership size hovering around 50–60 and that of the faculty around 7–8), such a goal was not attainable anymore. Nevertheless, it has always been (and still is) our conviction that the school will fulfill the various needs of its membership best if it offers a wide variety of research interests, and that this is a goal always to keep in mind and worth striving for, even if not fully reachable. For this and other reasons we decided in 1965 to have more direct input in part of the work and composition of the school by setting up a special program now and then. This idea was of course not to have the school fully organized all of a sudden, rather to add a new feature to the mathematical life here, without supplanting any of the others. Such a program was to involve as a rule about a quarter, at most a third, of the membership, with a mix of invited experts and of younger people. It would often be centered on an area not well represented on the faculty, but not obligatorily so. We did not want to refrain from organizing a program in one of our fields of expertise, if it seemed timely to gather a group of people working in it to spend a year here. It was of course expected that such a program would include a number of seminars for experts to foster further progress, but we also hoped it would feature some surveys and introductory lectures aimed at people with peripheral interests, and would also facilitate to newcomers access to the current research and problems. Pushing this "instructional" aspect a bit further, we also decided to have occasionally two related topics, hoping this would increase contacts between them.

The first such program took place in 1966–1967 and was devoted to analysis, with emphasis on harmonic analysis and differential equations. In agreement with the last guideline stated above, the second one (1968–1969) involved two related topics, namely algebraic groups and finite groups. As a focus of interaction, we had in mind first of all the finite Chevalley groups and their variants (Ree and Suzuki groups). They played that role indeed, but so did the Weyl groups and their representations, as can be seen from the *Notes* which arose from this. The third program (1970–1971) centered on analytic number theory.

In 1971, again with an eye to increasing breadth and exposure to recent developments, another activity was initiated here, namely an ongoing series of survey lectures. In the sixties and before, the dearth of expository or survey papers had often been lamented. The *AMS Bulletin* was a natural outlet for such, first of all because the invited speakers for one-hour addresses are all asked to write one. But this did not seem to elicit as many as one could wish and various incentives were tried, with limited success. It had always seemed to me that most of us are cold to the idea of just sitting down to write an expository paper, unless there is an oral presentation first. But the example just mentioned showed that this condition was not always sufficient. Already in my graduate student days, I had been struck by some beautiful surveys in the *Abhandlungen des Math. Sem. Hamburg.* They were usually the outgrowth of a few lectures given there. This suggested to me that one might have a better chance of getting a paper if the prospective author were invited to give some comprehensive exposition in a few lectures, not just one. However I had done nothing to implement such a scheme, just talking about it occasionally, until the 1970 International Congress in Nice. There K. Chandrasekharan, then president-elect of the IMU, told me he wanted to set up a framework for an ongoing series of lectures sponsored by the IMU, to be given at various locations, with the express purpose to engender survey papers. Would I help to organize it? Our ideas were so similar that we quickly agreed on the general format: A broad survey, for non-specialists, given in four to six one-hour lectures, within a week or two. Expenses would be covered, but the real fee would be paid only upon receipt of a manuscript suitable for inclusion in this series. A bit later, I suggested as an outlet for publication the *Enseignement Mathématique*, mainly for two reasons: First, it is in some way affiliated to the IMU, being the official organ of the International Commission for Mathematical Education. Second, it has the rare, if not unique, capacity to publish as a separate monograph, sold independently, any article or collection of articles published in that journal.

The first two such sets of lectures were given at the Institute in the first quarter of 1971, by Wolfgang Schmidt and Lars Hörmander (who was a visitor, too, having resigned from the faculty in 1968), both soon written up and indeed published in the *Enseignement Mathématique*. But a difficulty arose with our third proposal, namely to invite Jürgen Moser, then at NYU, to give a survey on some topics in celestial mechanics. From the point of view of the IMU, these lectures were meant to promote international cooperation. Accordingly, the lecturer was to be from a geographically distant institution, so that the invitation would also foster personal contacts. They felt that we did not need an IMU sponsorship to bring Moser from NYU to the Institute. They certainly had a point. On the other hand, it was also a sensible idea to have such a set of lectures from Moser. In the school, we were really after timely surveys, whether or not they were contributing to international cooperation, while this latter aspect was essential for the IMU. Also, they wanted of course to have such lecture series be given at various places and their budget was limited. Since we planned to have about one or two per year, our requests might well exceed it, so that some difficulties might be foreseen also on that score. We therefore decided to start a series of similar lectures of our own, and to call them the Hermann Weyl Lectures, an ideal label, in view of Weyl's universality: It was a nice touch to be able on many occasions to trace so much of the work described in those lectures to some of his. We planned to publish them as a rule, though not obligatorily, in the *Annals of Mathematics Studies*. Otherwise, the conditions and format of the lectures were to be the same. Our series started indeed with J. Moser's lectures, resulting in an impressive two-hundred page monograph. For a number of years, the H. Weyl lectures were a regular feature here, at the rate of one to two sets per year. As to their original purpose, namely to bring out survey papers, I must regretfully acknowledge that our record is a mixed one, and that the list of speakers who did not contribute any is about as distinguished as that of those who did. Maybe Moser's contribution was a bit daunting, although F. Adams and D. Vogan rose to the challenge, even topping its number of pages (slightly in the former case, largely in the latter). Overall, the high quality of the monographs growing out of the H. Weyl lectures has made the series very worthwhile. Their frequency has declined in recent years. Since we started this, "distinguished" lecture series have sprung up at many places. Also, symposia, conferences and workshops on specific topics have proliferated, often leading to publications containing many surveys or introductory papers. There is indeed nowadays quite a steady flow of papers of this type so maybe the need for our particular series has decreased. One of the nice features of the Institute is that we need not pursue a given activity if we do not feel it fulfills a useful function in the mathematical community. So we may well leave this one in abeyance and revive it whenever we see a good opportunity.

In 1966 C. Kaysen had taken up the directorship and found the school faculty in good shape. He thought that, at least with our group, he would not face requests for new appointments. But we pointed out to him that our age distribution was a bit unfortunate and would later create some problems, with retirements expected in 1975, 1976, 1977, and 1979. Therefore it might be desirable to consider some advance replacements; also that some minimal expansion might be to the good. He agreed. In 1969 Michael Atiyah joined the faculty. Originally, this appointment had been meant to be an expansion, but it was not anymore, after Hörmander had resigned in 1968. Later, we made offers successively to John Milnor and Robert P. Langlands, who came to the faculty in 1970 and 1972 respectively.

In the sixties, considerable progress was made in the general area I had already singled out as a very strong one here: Algebraic groups, arithmetic groups and automorphic forms, number theory, harmonic analysis on reductive groups. Much of it was done here, but also at the University by G. Shimura, and by R. P. Langlands who was there for three years. It continued unabated, or even at an increased pace, after Langlands joined us. This whole general field had become such an active and important part of "core mathematics" that it was all to the good. However, that was not matched by activities of similar scope in other areas and created some imbalance, accentuated by Atiyah's resignation in 1972. For reasons already explained, in our view it was not in the best interest of the school in the long run and to correct it by increasing activities in other areas became a concern. There were two obvious means to try to remedy this: the special programs and new faculty appointments. But they were not available to us during the energy crisis and the immediately following years. The financial situation of the Institute was worrisome and we had not even been authorized to replace Atiyah. Also, we had not been able to take care completely within our ordinary budget of the special programs, which entailed invitations to well-established people. We always had had to get some outside support, besides our standing NSF contract, and that was hard to come by in those years. But we resumed both as soon as it became possible: Enrico Bombieri came to the faculty in 1977 and Shing-Tung Yau in 1980, broadening greatly its coverage. We had also to wait until 1977 for the programs but have had one almost every year from then on.

In 1977–1978, our program was devoted to Fourier integral operators and microlocal analysis with the participation in particular of L. Hörmander and M. Kashiwara. This was again an attempt to increase contacts between two rather different points of view, in this case the classical approach and the more recent developments of the Japanese school around M. Sato. It led to a collection of papers providing a mix of both. The next one was on finite simple groups and brought here a number of the main participants to the collective enterprise to classify the finite simple groups. 1979–1980 was the year of the biggest program to date, on differential geometry and analysis, in particular nonlinear PDE. The number of seminars was somewhat overwhelming. Several were concentrated at the end of the week, so as to make it easier for people in neighboring (in a rather wide sense including New York and Philadelphia) institutions to participate. Roughly speaking, the main activities were subdivided in three parts: differential geometry, minimal submanifolds, and mathematical physics, with seminar coordinators L. Simon for the second one, S. T. Yau for the other two. A remarkable feature of the third one (devoted to relativity, the positive mass conjecture, gauge theories, quantum gravity) was the cooperation between mathematicians and physicists, probably a first here since the early days. Two volumes of *Notes* resulted from this program.

There was none the following year but then, in 1981–1982, we had one on algebraic geometry, at least as big as the previous one. Again, seminars were also attended by visitors from outside, two even coming from Cambridge, Massachusetts: D. Mumford and P. Griffiths would visit every second or third week for two to three days, each to lead one of the main seminars. We had decided to concentrate on the more geometric (as opposed to arithmetic) aspects of algebraic geometry, since we intended to have in 1983–1984 a program on automorphic forms and *L*-functions. But even with that limitation, it was of considerable scope (Hodge theory, moduli spaces, *K*-theory, crystalline cohomology, low-dimensional varieties, etc.). Griffiths' seminar also led to a set of *Notes*. This was again very successful but the evolution of these seminars betrayed a natural tendency, namely to try each time to improve upon the previous one, leading not unnaturally to bigger and bigger programs. As already stated, our original intention had been to add an activity, not to suppress any, and we began to wonder whether these programs, carried out at such a scale, might not hamper somewhat other important aspects of the mathematical life here, such as variety, informality, the opportunity for spontaneous activities and unplanned contacts, quiet work, etc. So we decided to scale them down a bit. Again, this was not meant as a straightjacket; rather, that the initial planning would usually be on a more modest scale. But, if outside interest would lead to a growth beyond our original expectations (as is the case with the present program on dynamical systems), we would of course do our best to accommodate it. We were aided in fact in our general resolve by the emergence of the Mathematical Sciences Research Institute at Berkeley: Big programs are an essential feature there and they have more financial means than we to carry them out. There is no need to compete for size.

S. T. Yau had resigned in 1984 and was soon replaced by Pierre Deligne. The retirements we had warned C. Kaysen about had caught up with us for some time and our group was reduced to six, two fewer than the size we were entitled to at the time, so that we had the possibility of making two appointments. We were anxious to seize this opportunity to catch up with some new major trends in mathematics. There had been some very interesting shifts in the overall balance of research interests, partly influenced by the development of computers, notably towards nonlinear PDE and their applications (with which we had lost first-hand contact after Yau's resignation), dynamical systems, mathematical physics, as well as an enormous increase of the interaction with physicists, the latter visible notably around string theory and conformal field theory (CFT). These last two topics were very strong at the University, but underrepresented here (not only in the faculty, but also in the membership). As a first attempt to improve this situation, I suggested in fall 1985 to E. Witten to give at the Institute a few lectures on string theory aimed at mathematicians. They were very well attended, so that the next logical move was to think about organizing a program in string theory and to ask Witten whether this seemed to him worth pursuing and, if so, whether he would agree to help, first as a consultant and then as a participant. That same year, we made two successful offers to Luis Caffarelli and Thomas C. Spencer, thus increasing considerably our range of expertise in some of the "most wanted" directions.

The first question put to Witten was not entirely rhetorical, given the abundance at the time of conferences and workshops on these topics. But it was agreed after some thought that a year-long program here would have enough features of its own to make it worth trying. A bit later, an expert to whom I had written about it warned that, in view of the usually rather frantic pace of research in physics, this might be all over and passé at the time of the program (1987–1988); but it seemed to us there was enough new mathematics to chew on for slower witted mathematicians to justify such a program on those grounds (later, that expert volunteered to eat his words). Anyway, we went ahead. The program had originated within the School of Mathematics, but the School of Natural Sciences became gradually more involved and eventually contributed to the invitations. In fact, the borderline between the two schools became somewhat blurred, the physicists D. Friedan, P. Goddard and D. Olive being members in mathematics, while the mathematicians G. Segal and D. Kazhdan were invited by the School of Natural Sciences. A primary goal of this program was to increase the contacts between mathematicans and physicists and to help surmount some of the difficulties in communication due to differences in background, techniques, language and goals. Accordingly, we had invited several mathematically minded physicists and some mathematicians with a strong interest in physics, all rather keen to contribute to the dialogue. The program was very intense, too, with an impressive array of seminars, notably many lectures on various versions of CFT, and many discussions in and outside the lecture rooms.

Our last two appointments, succeeded by that of E. Witten in the School of Natural Sciences, have quickly made the Institute a major center of interaction between physics and mathematics and also increased significantly the membership in analysis. Altogether, the school faculty seems to me to be about as broad as can be expected from seven people. I hope it is not just wishful thinking on my part to believe that by its concern for the school and its own work, it is well on its way to maintain a tradition worthy of the vision of the first faculty.

The reader will have noticed that, from the time I came to the Institute, this account is largely based on personal recollections and falls partly under the label of "oral history", with, as a corollary, an emphasis or maybe even an overemphasis on the events or activities I have been involved with or witnessed from close quarters. Even with those, I have not been even-handed at all and this paper makes no claim to offer a balanced and complete record of the school history and of all the work done there.[8] Such an undertaking would have brought this essay to a length neither the editors nor the author would have liked to contemplate. Also absent is any effort to evaluate the impact of the school on mathematics in the U.S. and beyond: How much benefit did visitors gain? How influential has their stay here been on their short-range and long-range activities? What mathematical research was carried out or has originated here? How important has been the presence and work of the faculty? These are some of the questions which come to mind. To try to answer them would again have had an unfortunate effect on the length of this paper. Besides, an evaluation of this sort is more credible if it emanates from the outside, at any rate not solely from an interested party of one. Moreover, as a further inducement for me to refrain from attempting one, two evaluations of relatively recent vintage do exist. First, a report by a 1976 trustee–faculty committee, whose charges were to review the past, evaluate the Institute and provide some guidelines for the future. Its assessment was based in part on the letters of a number of scholars and on the answers (over five hundred from mathematicians) to a questionnaire sent to all past and present members on behalf of that committee. Second, one by a 1986 visiting committee, chaired by G. D. Mostow. Both, though not exempt from

[8] In that connection, let me mention that *A Community of Scholars. The Institute for Advanced Study, Faculty and Members 1930–1980*, published by the Institute for Advanced Study on the occasion of its fiftieth year, contains in particular a list of faculty and members up to 1980 and, for most, of work related to IAS residence.

criticisms, conclude that the School of Mathematics has been successful in many ways. As a brief justification for this claim and without further elaboration, let me finish by quoting from a letter written in 1976 by I. M. Singer to the chairman of the review committee, Martin Segal, who was happy to share it with the committee:

> Their [the members'] stay at the Institute under the guidance of the permanent staff affects their mathematical careers enormously. Their contacts with their peers continue for decades. They leave the Institute, disperse to their universities, and carry with them a deeper understanding of mathematics, higher standards for research, and a sophistication hard to attain elsewhere.
>
> Such was the case when I was here twenty years ago. Last fall when I signed the Visitors' Book I turned the pages to see who was here in 1955–1956. Many are world famous and they are all close professional friends. I notice the same thing happening now with the younger group. Before I came in 1955, the Institute was described to me as I am describing it to you. It remains true now as it has been for the last thirty years.

In preparing this article I benefitted from the use of some archival material. I thank E. Shore and M. Darby at the Institute for their help in dealing with the Institute archives and R. Coleman at the University for having kindly sent me copies of some documents in the University archives. I am also grateful to A. Selberg and A. W. Tucker for having shared with me some of their recollections, and especially to D. Montgomery for having done so in the course of many years of close friendship.

139.

(with G. Prasad)

Finiteness Theorems for Discrete Subgroups of Bounded Covolume in Semi-simple Groups

Publ. Math. I.H.E.S. **69** (1989), 119–171; Addendum, ibid. **71** (1990), 173–177

Introduction

1. It is well-known that a real non-compact simple Lie group not locally isomorphic to $SL_2(\mathbf{R})$ or $SL_2(\mathbf{C})$ has only finitely many conjugacy classes of discrete subgroups of covolumes bounded by a given constant [44]. Motivated by the results of [17], J. Tits asked whether the same would be true for p-adic groups, not only for a single ambient group, but also when the ground field and the group are allowed to vary, with a specific universal normalization of Haar measures. This problem was our starting point. We were naturally led to consider also an analogue for groups over $\mathbf{R}$ or $\mathbf{C}$ and then, as a common generalization, irreducible discrete subgroups of products of simple groups over local fields.

2. In this introduction we shall outline some of the main results obtained so far, referring to §§7, 8 for the most precise assumptions and general statements. We let k be a global field; V, V_∞, V_f respectively be the set of places, of archimedean places, and of nonarchimedean places of k and k_v the completion of k at $v \in V$. Let G be an absolutely almost simple simply connected k-group and G′ be a k-group centrally k-isogenous to G. If $v \in V_f$, we let μ'_v be the Haar measure on $G'(k_v)$ which assigns the volume one to the stabilizer of a chamber in the Bruhat-Tits building of $G(k_v)$. This is (essentially) the normalization proposed by J. Tits, so μ'_v will be called here the Tits measure. If v is archimedean, and k_v is identified with $\mathbf{R}$ or $\mathbf{C}$, then μ'_v is the Haar measure which gives the volume one to a maximal compact subgroup of $R_{k_v/\mathbf{R}}(G')(\mathbf{C})$. (Originally, we had considered the measure associated to the Killing form. This μ'_v was suggested to us by P. Deligne.) For a finite set of places $S \subset V$, we let μ'_S be the Haar measure on $G'_S = \prod_{v \in S} G'(k_v)$ which is the product of the μ'_v's. When $G' = G$, we set $\mu'_S = \mu_S$. Then we have (7.8):

1 *Theorem A. — Let $c > 0$ be given. Assume k runs through the number fields. Then there are only finitely many choices of k, of G'/k of absolute rank $\geqslant 2$ up to k-isomorphism, of a finite set S of places of k containing all the archimedean places, of arithmetic $\Gamma' \subset G'_S$ up to conjugacy, such that $\mu'_S(G'_S/\Gamma') \leqslant c$.*

---

* Supported by the National Science Foundation during 1986-1988 at the Mathematical Sciences Research Institute, Berkeley, and at the Institute for Advanced Study, Princeton.

[The proof will also yield the finiteness of the number of natural equivalence classes of (k, G', S, Γ') in the function field case, under some mild restrictions. We note also that, in view of the arithmeticity results of [23] and [43], irreducible discrete subgroups of finite covolume of simple groups over local fields are of the type considered here under rather general assumptions. This leads to an apparently different formulation of this finiteness theorem. See Remark 7.9.]

3. The starting point of the proof is a formula of [31] for $\mu_S(G_S/\Lambda_0)$, where Λ_0 is a " principal " S-arithmetic subgroup contained in $G(k)$. (The volume formula in [31] involves the Tamagawa number $\tau_k(G)$ of G, which has recently been proved to be equal to one if k is a number field.) To deal with a subgroup of G'_S commensurable with the image of Λ_0, we need an estimate for the index of the latter in its normalizer. This is done by consideration of the first Galois cohomology set with coefficients in the center C of G (or flat cohomology if C is not reduced) via a slight generalization of an exact sequence due to Rohlfs [32] (see §§2, 5). The proof uses number theoretical estimates, in particular some involving discriminants given in §6. These arguments yield first the finiteness of the set of triples (k, G, S) in 7.3. The finiteness of the number of conjugacy classes of Γ' in a given G'_S, which follows from [3] in characteristic zero, is proved in 7.7 with respect to conjugacy under $(\mathrm{Ad}\, G)(k)$. For the proof of the finiteness theorems in §7, we have to know that given a finite subset $\mathscr{R}$ of V, the set of inner forms of G which are k_v-isomorphic to G for all $v \notin \mathscr{R}$ is finite. This is well-known in characteristic zero [5]. A proof in the function field case is supplied in Appendix B.

4. Another possible natural normalization of Haar measures in the nonarchimedean case is the absolute value of the Euler-Poincaré measure introduced by J.-P. Serre in [33]. If $v \in V_\infty$, we may use on $G'(k_v)$ a similar measure, provided $G'(k_v)$ has a compact Cartan subgroup. If this condition is fulfilled for every $v \in V_\infty \cap S$, then the corresponding product measure on G'_S is also a Haar measure. It may be smaller then μ'_S, but by a controllable factor (see 4.4) and the estimates are good enough to ensure that Theorem A also holds for this choice of the Haar measure in these cases except maybe if G is of type A_2. With that *proviso*, it yields therefore the finiteness of the number of (k, G', S, Γ') such that $0 \neq |\chi(\Gamma')| \leqslant c$ where χ is the Euler-Poincaré characteristic in the sense of C. T. C. Wall (see 7.3, 7.8).

5. Earlier results pertaining to p-adic groups in characteristic zero, announced in [4], were proved in a completely different way, by comparing the index of an Iwahori subgroup in a maximal parahoric subgroup with an estimate for the order of finite subgroups of G(K), where K is a nonarchimedean local field of characteristic zero and G is a semi-simple group defined over K. This method does not depend on any information on Tamagawa numbers and allows us to vary G, and also K among local fields having a bounded absolute ramification index. This is the subject matter of §8.

6. The main result of [17] gives an explicit list of triples (F, G, Γ) where F is a nonarchimedean local field, G an absolutely almost simple F-group of F-rank $\geqslant 2$ and Γ a discrete subgroup of G(F) which acts transitively on the chambers of the Bruhat-Tits building of G(F). In particular this set is finite. It is clear from the definition of the Tits measure μ_T of G(F) that, in that case, $\mu_T(G(F)/\Gamma) \leqslant 1$. Therefore this finiteness assertion follows from Theorem A. More generally, we show the finiteness of the number of triples (F, G, Γ) consisting of a nonarchimedean local field F of characteristic zero, an absolutely almost simple F-group G of absolute rank $\geqslant 2$, and a discrete subgroup Γ of G(F) which is transitive on the set of the facets of a given type of the Bruhat-Tits building of G(F). In fact, we shall establish more general results in the S-arithmetic framework (see 7.10, 7.11).

7. Let G be as in 2. Let S be a finite subset of V containing V_∞. A collection $P = (P_v)_{v \in V_f - S}$, where P_v is a parahoric subgroup of $G(k_v)$, is said to be *coherent* if the product of the P_v's by $G_S = \prod_{v \in S} G(k_v)$ is an open subgroup of the adèle group G(A).

It is known that if either k is a number field or G is anisotropic over k, and $(P_v)_{v \in V_f}$ is a coherent collection of parahoric subgroups, then the "class number"

$$\mathfrak{c}(P) = \#\,((G_\infty \times \prod_{v \in V_f} P_v)\backslash G(A)/G(k))$$

is finite (and, by strong approximation, equal to one if G_∞ is non-compact), where $G_\infty = \prod_{v \in V_\infty} G_v$. Arguments similar, in fact in part common, to those of 7.3 and 7.7 yield (see 7.2, 7.6):

Theorem B. — Let $c \in \mathbf{N}$ be given. Then there are, up to natural equivalence, only finitely many number fields k, absolutely almost simple simply connected k-groups G *with* G_∞ *compact, and coherent collections* P *of parahoric subgroups such that* $\mathfrak{c}(P) \leqslant c$.

We thank Moshe Jarden and A. M. Odlyzko for conversations and correspondence on discriminant and class numbers of global fields. We are indebted to J. Tits for having kindly provided more conceptual proofs of two properties of volumes of parahoric subgroups stated in 3.1 and proved in Appendix A, and for his careful reading of the manuscript and his helpful suggestions.

TABLE OF CONTENTS

0. Notation, conventions and preliminaries

In this section, we recall or fix some notation and conventions, often to be used without reference. In addition we prove some facts about global fields (mostly function fields), for which we could not give references.

0.0. As usual $\mathbf{Q}$, $\mathbf{R}$ and $\mathbf{C}$ will denote respectively the fields of rational, real and complex numbers; $\mathbf{Z}$ the ring of rational integers.

The number of elements of a finite set S will be denoted by $\# S$.

0.1. Throughout this paper k is a global field i.e. a number field or the function field of a curve over a finite field, and A is the k-algebra of adèles of k endowed with the usual locally compact topology. Let V be the set of places of k, and V_∞ (resp. V_f) the subset of archimedean (resp. nonarchimedean) places. For a set S of places of k, let $S_f = S \cap V_f$, and $S_\infty = S \cap V_\infty$.

For $v \in V$, k_v denotes the completion of k at v and $|\ |_v$ the normalized absolute value on k_v. For $v \in V_f$, let $\hat{k}_v$ be the maximal unramified extension of k_v; let $\mathfrak{o}_v$ and $\hat{\mathfrak{o}}_v$ be the ring of integers of k_v and $\hat{k}_v$ respectively; let q_v be the cardinality of the residue field of k_v and $v(x)$ the normalized additive valuation of $x \in k_v^\times$. Recall that, for $x \in k_v^\times$,

$$|x|_v = [\mathfrak{o}_v : x\mathfrak{o}_v]^{-1} = q_v^{-v(x)} \quad \text{if } x \in \mathfrak{o}_v,$$
$$|x|_v = [x\mathfrak{o}_v : \mathfrak{o}_v] = q_v^{-v(x)} \quad \text{if } x \notin \mathfrak{o}_v.$$

0.2. Except in §8, G will be an absolutely almost simple, simply connected algebraic group defined over k, $\overline{\mathrm{G}}$ its adjoint group (i.e. the group of its inner automorphisms), $\varphi : \mathrm{G} \to \overline{\mathrm{G}}$ the natural central isogeny and G′ a k-group centrally k-isogeneous to G. We fix a central k-isogeny $\iota : \mathrm{G} \to \mathrm{G}'$ and let $\varphi' : \mathrm{G}' \to \overline{\mathrm{G}}$ be the unique central isogeny such that $\varphi = \varphi' . \iota$; it is defined over k.

Let C be the center of G and C′ that of G′. Let r be the absolute rank of G and for $v \in V_f$, let r_v be its rank over $\hat{k}_v$.

0.3. For a subset $\mathscr{X}$ of V, let $\mathrm{G}_{\mathscr{X}}$ (resp. $\mathrm{G}'_{\mathscr{X}}$) denote the direct product of the $\mathrm{G}(k_v)$ (resp. $\mathrm{G}'(k_v)$), $v \in \mathscr{X}$, if $\mathscr{X}$ is finite, and their restricted direct product if $\mathscr{X}$ is infinite. The group $\mathrm{G}(k)$ (resp. $\mathrm{G}'(k)$) will always be viewed as a subgroup of $\mathrm{G}_{\mathscr{X}}$ (resp. $\mathrm{G}'_{\mathscr{X}}$) in terms of its diagonal embedding.

For $v \in V$ and $\mathscr{X} \subset V$, the homomorphisms $G(k) \to G'(k)$, $G(k_v) \to G'(k_v)$, $G_{\mathscr{X}} \to G'_{\mathscr{X}}$, induced by ι will also be denoted by ι.

0.4. Let S be a finite set of places of k containing all the archimedean ones. We assume that for every nonarchimedean $v \in S$, G is isotropic over k_v, or, equivalently, $G(k_v)$ is noncompact. Let $\mathscr{S} = \mathscr{S}(G)$ be the subset of S consisting of the places v such that G is isotropic over k_v. We assume further that $\mathscr{S}$ is nonempty.

0.5. We shall assume familiarity with the Bruhat-Tits theory of reductive groups over nonarchimedean local fields. All we need is stated in [41], and the proofs of most of the results can be found in [8].

For $v \in V_f$, we shall let X_v denote the Bruhat-Tits building of $G(k_v)$. We recall that $G(k_v)$ acts on X_v by *special* simplicial automorphisms; in particular any simplex stable under an element $g \in G(k_v)$ is pointwise fixed by g.

0.6. Let $\mathscr{K}$ be a compact open subgroup of G_{V-S}. Let $\Lambda = G(k) \cap \mathscr{K}$. Any subgroup of G_S (resp. G'_S) commensurable with Λ (resp. $\iota(\Lambda)$) is called an S-arithmetic subgroup.

Let $G_S \to G_{\mathscr{S}}$ and $G'_S \to G'_{\mathscr{S}}$ be the natural projections. Then any subgroup of $G_{\mathscr{S}}$ (resp. $G'_{\mathscr{S}}$) commensurable with the projection of an S-arithmetic subgroup of G_S (resp. G'_S) will be called an arithmetic subgroup. Arithmetic subgroups are discrete and of finite covolume.

0.7. If K is a number field, $a(K)$ will denote the number of its archimedean places, D_K the absolute value of its discriminant over $\mathbf{Q}$, and h_K its class number.

0.8. If K is a global function field, let $a(K) = 1$, g_K be its genus, q_K be the cardinality of its field of constants, and h_K be its "class number" i.e. the order of the quotient of the group of its divisors of degree zero by the subgroup of principal divisors. Let $D_K = q_K^{2g_K-2}$ in this case. The following bounds for the class number h_K are known.

(1) $$(q_K^{1/2} - 1)^{2g_K} \leqslant h_K \leqslant (q_K^{1/2} + 1)^{2g_K}.$$

For the sake of expository completeness we sketch a proof pointed out to us by Manohar Madan: The zeta-function $\zeta_K(s)$ of K can be written in the form

$$\zeta_K(s) = \frac{P(q_K^{-s})}{(1 - q_K^{-s})(1 - q_K^{1-s})},$$

where P is a polynomial of degree $2g_K$ with integral coefficients, $P(0) = 1$ and $P(1) = h_K$ ([45: Chapter VII, §6, Theorem 4]). According to the "Riemann hypothesis" for curves over finite fields proved by A. Weil (see [1] for an elementary proof), the roots of P have absolute value $q_K^{-1/2}$. This at once implies the above bounds.

It is a well known result of Hermite and Minkowski that (up to isomorphism) there are only finitely many number fields with a given discriminant (see [20: Chapter V,

Theorem 5]). For global function fields the following finiteness assertion holds. Its proof was supplied to us by Moshe Jarden and Dinesh Thakur.

0.9. *Proposition. — For given g and q, there are only finitely many global function fields of genus g and field of constants of cardinality q.*

For its proof we need the following lemma.

0.10. *Lemma. — Let* K *be a global function field of genus g and field of constants* K_0. *Suppose that* K/K_0 *has a prime divisor* P *of degree 1. Then* $K = K_0(x, y)$, *where* (x, y) *satisfy an equation* $f(x, y) = 0$ *with coefficients in* K_0, *of degree at most* $4g$.

Proof. — To each divisor D of K we associate the K_0-vector space

$$\mathscr{L}(D) = \{ x \in K \mid (x) + D \geqslant 0 \}, \quad \text{and set} \quad \dim(D) = \dim(\mathscr{L}(D)).$$

If $g = 0$, then $K = K_0(x)$ with x transcendental over K_0 ([11: §18, Theorem]). So, assume that $g > 0$. By the Riemann-Roch theorem, $\dim(nP) = n + 1 - g$ if $n > 2g - 2$. Hence, $\mathscr{L}((2g-1)P) \subset \mathscr{L}(2gP) \subset \mathscr{L}((2g+1)P)$. Choose

$$x \in \mathscr{L}(2gP) - \mathscr{L}((2g-1)P) \quad \text{and} \quad y \in \mathscr{L}((2g+1)P) - \mathscr{L}(2gP).$$

Then $v_P(x) = -2g$, $v_P(y) = -(2g+1)$ and $(x)_\infty = 2gP$ i.e., $\deg(x)_\infty = 2g$, where v_P is the additive valuation associated with P.

If i and j are integers between 0 and $4g$, then

$$v_p(x^i y^j) = -2gi - (2g+1)j \geqslant -16g^2 - 4g,$$

and therefore, $x^i y^j \in \mathscr{L}((16g^2 + 4g)P)$. As

$$\dim \mathscr{L}((16g^2 + 4g)P) = 16g^2 + 3g + 1 < (4g+1)^2,$$

there exist $a_{ij} \in K_0$, $0 \leqslant i, j \leqslant 4g$, not all zero, such that $\Sigma_{i,j}\, a_{ij}\, x^i y^j = 0$. We prove that $K = K_0(x, y)$.

Note that $[K : K_0(x)] = \deg(x)_\infty$ by the theorem on [11: p. 25]. Therefore, by the above, $[K : K_0(x)] = 2g$. Hence, in order to prove that $K = K_0(x, y)$, it suffices to show that $[K_0(x, y) : K_0(x)] \geqslant 2g$. If we had $[K_0(x, y) : K_0(x)] < 2g$, there would exist $b_{ij} \in K_0$ with $0 \leqslant j \leqslant 2g - 1$, not all 0, such that $\Sigma\, b_{ij}\, x^i y^j = 0$. Hence there would exist distinct pairs (i, j) and (r, s) with $0 \leqslant j, s \leqslant 2g - 1$ such that $v_P(x^i y^j) = v_P(x^r y^s)$. Thus $2gi + (2g+1)j = 2gr + (2g+1)s$. As $2g$ and $2g + 1$ are relatively prime, this would imply that $2g$ divides $s - j$. It would then follow that $s = j$ and $r = i$. This contradiction concludes the proof.

Proof of Proposition 0.9. If $g = 0$, then K is either a rational function field over K_0 or $K = K_0(x, y)$ where (x, y) satisfy a quadratic equation with coefficients in K_0 ([12]). If $g = 1$, then, by a theorem of F. K. Schmidt, K has a prime divisor of degree 1 [9]. So in view of the above lemma we may (and we shall) assume that $g \geqslant 2$.

Denote the unique extension of degree $(2g-2)!$ of K_0 by K'_0. As K has a prime divisor of degree $\leqslant 2g-2$ ([11: p. 52]), $K' = K'_0 K$ has a prime divisor of degree 1. By the preceding lemma, $K' = K'_0(x, y)$, where (x, y) satisfy an equation of degree $\leqslant 4g$ with coefficients in K'_0. There are therefore only finitely many possibilities for K'. For each of these possibilities K is an intermediate field between $K_0(x)$ and K'. Let p be the characteristic of K. Since $[K_0(x) : K_0(x)^p] = p$, the field K' is generated over K_0 by one element [12: Lemma 24.31]. Hence there are only finitely many possibilities for K[21].

0.11. *Lemma. — A global function field* L *contains only finitely many subfields* K *such that* L/K *is a Galois extension.*

Any subfield K of L such that L/K is a Galois extension is the fixed field of a suitable subgroup of the automorphism group of L. Now the lemma follows from the well-known result that the automorphism group of any global function field is finite.

Let K be a global field and n be a positive integer. Let K_n be the subgroup of $K^\times$ consisting of all $x \in K^\times$ such that for every normalized nonarchimedean valuation v of $K^\times$, $v(x) \in n\mathbf{Z}$. Clearly, $K_n \supset K^{\times n}$.

The proof of the following proposition was suggested by Moshe Jarden and Dipendra Prasad.

0.12. *Proposition.* — $\#(K_n/K^{\times n}) \leqslant h_K\, n^{a(K)}$.

Proof. — If K is a number field (resp. global function field), let $\mathscr{P}$ be the group of all fractional principal ideals (resp. principal divisors) of K and $\mathscr{I}$ be the group of all fractional ideals (resp. divisors of degree zero). We shall use multiplicative notation for the group operation in both $\mathscr{I}$ and $\mathscr{P}$. The kernel of the natural map $x \mapsto (x)$ of $K^\times$ onto $\mathscr{P}$ is precisely the group U of units. This gives us our first short exact sequence

(1) $$1 \to U \to K^\times \to \mathscr{P} \to 1.$$

Let $\mathscr{C} = \mathscr{I}/\mathscr{P}$; then the class number h_K equals $\#\mathscr{C}$ and we have a second short exact sequence

(2) $$1 \to \mathscr{P} \to \mathscr{I} \to \mathscr{C} \to 1.$$

Now note first that since $U \cap K^{\times n} = U^n$, (1) gives rise to the following short exact sequence,

(3) $$1 \to U/U^n \to K^\times/K^{\times n} \to \mathscr{P}/\mathscr{P}^n \to 1.$$

As $U \subset K_n$, (3) yields another short exact sequence:

(4) $$1 \to U/U^n \to K_n/K^{\times n} \to (\mathscr{P} \cap \mathscr{I}^n)/\mathscr{P}^n \to 1.$$

On the other hand, let $\mathscr{C}_n$ be the subgroup of all elements of $\mathscr{C}$ whose order is a divisor of n. If for $x \in K^\times$, there exists $I \in \mathscr{I}$ such that $(x) = I^n$, then I is unique. Therefore the map $(x) \mapsto I\mathscr{P}$ induces an isomorphism

(5) $$(\mathscr{P} \cap \mathscr{I}^n)/\mathscr{P}^n \cong \mathscr{C}_n.$$

Combining (4) and (5) we get

$$[K_n : K^{\times n}] = [U : U^n] \# \mathscr{C}_n.$$

Obviously, $\# \mathscr{C}_n \leqslant h_K$. So it suffices to prove that $[U : U^n] \leqslant n^{a(K)}$.

If K is a number field, then by Dirichlet's unit theorem, $U \cong \mu(K) \times \mathbf{Z}^{a(K)-1}$ where $\mu(K)$ is the cyclic group of roots of unity in K ([45:IV, Theorem 9]). Thus U is the direct product of $a(K)$ cyclic groups. From this we conclude that $[U : U^n] \leqslant n^{a(K)}$.

If K is a global function field, then U is the group of non-zero elements of the field of constants (*loc. cit.*). As the latter field is finite, U is cyclic. Therefore, $[U : U^n] \leqslant n$.

1. Remarks on arithmetic subgroups

In this section, for the sake of completeness, we prove in our framework some properties of arithmetic subgroups which are well-known in characteristic zero.

1.1. Let $v \in V_f$. We observe first that the fixed point set F of a compact open subgroup H of $G(k_v)$ on the Bruhat-Tits building X_v of $G(k_v)$ is compact. In fact, H acts continously on the compactification $\overline{X}_v$ of X_v constructed in [6]. If F were not compact, then H would have a fixed point in $\overline{X}_v - X_v$. But there, by construction, the isotropy subgroups are of the form $P(k_v)$, where P is a proper parabolic k_v-subgroup of G, and those subgroups do not contain any open subgroup of $G(k_v)$.

1.2. *Proposition. — Let* Γ' *be an arithmetic subgroup of* $G'_{\mathscr{S}}$. *Then* $\varphi'(\Gamma')$ *is contained in* $\overline{G}(k)$ *and is Zariski-dense. The subgroups* $\Gamma' \cap G'(k)$ *and* $\Gamma' \cap \iota(G(k))$ *are normal subgroups of* Γ'.

The subgroup $\Gamma' \cap \iota(G(k))$ is of finite index in Γ', hence contains a subgroup Γ'_0 which is normal, of finite index, in Γ'. Since $G'(k)$ is contained in the commensurability group of Γ'_0, the latter is Zariski-dense in G', and hence $\varphi'(\Gamma'_0)$ is a Zariski-dense subgroup of $\overline{G}$. For $\gamma' \in \Gamma'$, the element $\varphi'(\gamma')$ normalizes $\varphi'(\Gamma'_0)$, so it is a k-automorphism of G. This implies that $\varphi'(\Gamma') \subset \overline{G}(k)$ and that Γ' normalizes $G'(k)$ and $\iota(G(k))$, hence also $\Gamma' \cap G'(k)$ and $\Gamma' \cap \iota(G(k))$.

1.3. For $v \in V_f$, $\mathrm{Aut}(G(k_v))$, and so in particular $\overline{G}(k_v)$, acts on the building X_v by simplicial automorphisms. In view of 1.2, this allows one to define an action of any arithmetic subgroup of $G'_{\mathscr{S}}$ on X_v $(v \in V_f)$. This will be used in the sequel without further reference.

A compact open subgroup $\mathscr{K}$ of G_{V-S} contains, as a subgroup of finite index, a direct product $\prod_v \mathscr{K}_v$, where, for $v \notin S$, $\mathscr{K}_v$ is a compact open subgroup of $G(k_v)$ which is hyperspecial for all the v's outside some finite subset T of V containing S; see [41: 3.9]. If $\mathscr{K}$ is such a group, then $\Lambda_{\mathscr{K}} = G(k) \cap \mathscr{K}$ is an S-arithmetic subgroup of G_S, and in its natural embedding in G_{V-S}, its closure is $\mathscr{K}$ by strong approximation ([30], [22]).

1.4. *Proposition. — Let* Γ' *be an arithmetic subgroup of* $G'_{\mathscr{S}}$ *and* Λ *be the inverse image in* $G(k)$ *of* $\Gamma' \cap \iota(G(k))$ *under* ι.

(i) *The fixed point set of* Γ' *in* X_v $(v \notin S)$ *is compact, not empty.*

(ii) *For any field extension* K *of* k, *the normalizer of* $\varphi'(\Gamma')$ *in* $\overline{G}(K)^{\mathscr{S}}$ *is contained in* $\overline{G}(k)$ *(*$\overline{G}(k)$ *embedded in* $\overline{G}(K)^{\mathscr{S}}$ *diagonally),* $\varphi'(\Gamma')$ *is of finite index in its normalizer, and the normalizer* $N(\Gamma')$ *of* Γ' *in* $G'_{\mathscr{S}}$ *is arithmetic.*

(iii) Γ' *is contained in only finitely many arithmetic subgroups.*

(iv) *If* Γ' *is maximal, then for* $v \notin S$, *the closure* P_v *of* Λ *in* $G(k_v)$ *is a parahoric subgroup of* $G(k_v)$, $\Lambda = G(k) \cap \Pi_v P_v$, *and* Γ' *is the normalizer of* $\iota(\Lambda)$ *in* $G'_{\mathscr{S}}$.

Proof. — By strong approximation, the projection of Λ in G_{V-S} is dense in a compact open subgroup. Therefore, its fixed point set $\mathscr{F}_v$ in X_v is compact (1.1), nonempty (by the fixed-point theorem of Bruhat-Tits [8: I, 3.2.4]), and reduced to the unique fixed point of a hyperspecial parahoric subgroup P_v for $v \in V - T$, where T is a suitable finite subset of V containing S (1.3). Since $\iota(\Lambda)$ is of finite index in Γ', the group of automorphisms of X_v (for $v \notin S$) determined by Γ' is relatively compact, therefore its fixed point set F_v is not empty; F_v is obviously contained in $\mathscr{F}_v$ and so in particular it is compact and (i) is proved.

By 1.2, $\varphi'(\Gamma')$ is contained and Zariski-dense in $\overline{G}(k)$. Therefore its normalizer in $\overline{G}(K)^{\mathscr{S}}$ is contained in $\overline{G}(k)$ and so it coincides with the normalizer $N(\varphi'(\Gamma'))$ of $\varphi'(\Gamma')$ in $\overline{G}(k)$. Obviously, F_v is stable under the natural action of $N(\varphi'(\Gamma'))$ on X_v. Hence, for all $v \notin S$, $N(\varphi'(\Gamma'))$ is a relatively compact subgroup of $\overline{G}(k_v)$. From this we conclude that $N(\varphi'(\Gamma'))$ is a discrete subgroup of $\overline{G}_{\mathscr{S}} := \Pi_{v \in \mathscr{S}} \overline{G}(k_v)$, and as it contains $\varphi'(\Gamma')$, which is a discrete subgroup of $\overline{G}_{\mathscr{S}}$ of finite covolume, the index of $\varphi'(\Gamma')$ in it is finite*. This implies in particular that the normalizer $N(\Gamma')$ of Γ' in $G'_{\mathscr{S}}$ is arithmetic, which proves (ii).

For $v \in T - S$, let $\mathscr{P}_v$ be the (finite) set of parahoric subgroups of $G(k_v)$ which fix some facet contained in $\mathscr{F}_v$. For $P = \Pi_{v \notin S} P_v$, where $P_v \in \mathscr{P}_v$ if $v \in T - S$, and P_v is the hyperspecial parahoric subgroup as above if $v \in V - T$, let $\Lambda_P = G(k) \cap P$, $\Lambda'_P = \iota(\Lambda_P)$ and $N(\Lambda'_P)$ be the normalizer of Λ'_P in $G'_{\mathscr{S}}$. As (by (i)) any arithmetic subgroup containing Γ' has a fixed point in $\mathscr{F}_v$, $v \notin S$, it is contained in the normalizer of Λ'_P for a suitable P. Since according to (ii), $N(\Lambda'_P)$ itself is an arithmetic subgroup, it follows that $\Gamma' = N(\Lambda'_P)$ for some P if Γ' is maximal. This proves (iv). Also, since there are only finitely many P's and, for each P, $[N(\Lambda'_P) : \Gamma']$ is finite, we conclude that the arithmetic subgroups of $G'_{\mathscr{S}}$ containing Γ' are finite in number, which proves (iii).

1.5. The group Λ defined in 1.4 (iv) will be called the *principal* S*-arithmetic subgroup* determined by the coherent collection $P = (P_v)_{v \in V_f - S}$ of parahoric subgroups. We shall also say that Λ and $\Gamma' = N(\iota(\Lambda))$ are *associated* to P.

---

* For a different proof, see §1.5 in Lattices in semi-simple groups over local fields by G. Prasad, *Advances in Math. Studies in Algebra and Number Theory*, Academic Press (1979).

2. The action of the first Galois cohomology group of the center of G on Δ_v

2.1. For v nonarchimedean, let T_v be a maximal $\hat{k}_v$-split torus of G which is defined over k_v and contains a maximal k_v-split torus of G; according to the Bruhat-Tits theory such a torus exists. Let $\hat{I}_v$ be an Iwahori subgroup of $G(\hat{k}_v)$ defined over k_v (i.e., stable under the Galois group of $\hat{k}_v/k_v$) such that the chamber in the Bruhat-Tits building of $G/\hat{k}_v$ fixed by $\hat{I}_v$ lies in the apartment corresponding to T_v, and let $I_v = \hat{I}_v \cap G(k_v)$. Let $\hat{\Delta}_v$ be the basis of the affine root system of $G/\hat{k}_v$ relative to T_v, determined by the Iwahori subgroup $\hat{I}_v$, and Δ_v be the basis of the affine root system of G/k_v relative to the maximal k_v-split torus contained in T_v, determined by the Iwahori subgroup I_v of $G(k_v)$.

2.2. Any subset $\Theta_v \subset \Delta_v$ determines a parahoric subgroup P_{Θ_v}, of $G(k_v)$, containing I_v (which is assigned to the empty set); moreover any parahoric subgroup of $G(k_v)$ is conjugate to a unique subgroup of the form P_{Θ_v}. A parahoric subgroup of $G(k_v)$ which is conjugate to P_{Θ_v} is said to be of *type* Θ_v.

$\mathrm{Aut}(G(k_v))$, and so in particular $\overline{G}(k_v)$, acts on the set of parahoric subgroups of $G(k_v)$, and there is a homomorphism

$$\xi_v : \overline{G}(k_v) \to \mathrm{Aut}(\Delta_v)$$

such that for $g \in \overline{G}(k_v)$, the conjugate of P_{Θ_v} $(\Theta_v \subset \Delta_v)$ under g is a parahoric subgroup of type $\xi_v(g)\,(\Theta_v)$.

There is a similar homomorphism

$$\hat{\xi}_v : \overline{G}(\hat{k}_v) \to \mathrm{Aut}(\hat{\Delta}_v).$$

Furthermore, ξ_v (resp. $\hat{\xi}_v$) is trivial on $\varphi(G(k_v))$ (resp. $\varphi(G(\hat{k}_v))$). Let Ξ_v (resp. $\hat{\Xi}_v$) be its image.

2.3. *Lemma.* — *Let* $g \in \overline{G}(k_v)$.

(i) *If* $\hat{\xi}_v(g)$ *is trivial, then so is* $\xi_v(g)$. *In particular,* Ξ_v *is a subquotient of* $\hat{\Xi}_v$.
(ii) *Assume* G *to be quasi-split over* k_v. *If* $\xi_v(g) = 1$, *then* $\hat{\xi}_v(g) = 1$.

Proof. — (i) The first assertion follows immediately from the fact that two parahoric subgroups of $G(\hat{k}_v)$ which are defined over k_v are conjugate in $G(\hat{k}_v)$ if, and only if, their intersections with $G(k_v)$ are conjugate in $G(k_v)$ [8: II, Proposition 5.2.10 (ii)].

(ii) Assume G to be quasi-split over k_v. If it does not split over $\hat{k}_v$, then it is *residually split* over k_v; $\hat{\Delta}_v$ then has a natural identification with Δ_v and the second assertion of the lemma is obvious. We assume therefore that G splits over $\hat{k}_v$. Then $G(k_v)$ has a hyperspecial parahoric subgroup ([41:1.10.2]) to which corresponds a hyperspecial vertex of $\hat{\Delta}_v$. If $\xi_v(g) = 1$, then this vertex is fixed under $\hat{\xi}_v(g)$. But, by [16: 1.8], the group $\hat{\Xi}_v$ operates freely on the set of hyperspecial vertices of $\hat{\Delta}_v$. Therefore, $\hat{\xi}_v(g) = 1$, whence (ii).

Remark. — The conclusion of (ii) may fail if G is not quasi-split over k_v; it fails, for example, if G is anisotropic over k_v or if it is an inner form of type D_r whose k_v-rank is $r-2$.

2.4. Let K be a field and H be an affine algebraic group-scheme over K. If K is of characteristic zero, then $H^1(K, H)$ denotes as usual the first Galois cohomology set with coefficients in H. If K is of positive characteristic, then we let it stand for the set denoted $\check{H}^1(\mathrm{Spec}(K)_{f1}, H)$ in [25: III, §§3, 4], or $H^1_f(K, H)$, $H^1(K, H)$, in [34]. If H is commutative, similar groups are defined in all positive degrees. The usual exact sequence in Galois cohomology associated to a short exact sequence of group schemes is also available [25: III, Prop. 4.5] as well as the long exact cohomology sequence associated to a short exact sequence of commutative group schemes [34]. Moreover, if H is smooth, then these two cohomology sets are canonically isomorphic [25: III, Theorem 3.9]. (It is assumed there that the group-scheme is commutative, and the assertion is proved for cohomology groups in any degree $i \geqslant 0$, but this assumption is not used for $i = 1$, as is tacitly understood later in 4.8.) From this it follows that we need not distinguish between the two cases in our discussion below of cohomology with coefficients in C.

2.5. Let C be the center of G. It is k-isomorphic to the center of the unique simply connected, quasi-split *inner* k-form $\mathscr{G}$ of G.

The natural central k-isogeny $\varphi : G \to \overline{G}$ gives rise to the following commutative diagram with exact rows:

$$\begin{array}{ccccccccccc}
1 & \longrightarrow & C(k) & \longrightarrow & G(k) & \xrightarrow{\varphi} & \overline{G}(k) & \xrightarrow{\delta} & H^1(k, C) & \longrightarrow & H^1(k, G) \\
 & & \downarrow & & \downarrow & & \downarrow & & \downarrow & & \downarrow \\
1 & \longrightarrow & C(k_v) & \longrightarrow & G(k_v) & \xrightarrow{\varphi} & \overline{G}(k_v) & \xrightarrow{\delta_v} & H^1(k_v, C) & \longrightarrow & H^1(k_v, G), \\
 & & \downarrow & & \downarrow & & \downarrow & & \downarrow & & \downarrow \\
1 & \longrightarrow & C(\hat{k}_v) & \longrightarrow & G(\hat{k}_v) & \xrightarrow{\varphi} & \overline{G}(\hat{k}_v) & \xrightarrow{\hat{\delta}_v} & H^1(\hat{k}_v, C) & \longrightarrow & H^1(\hat{k}_v, G).
\end{array}$$

Since for any nonarchimedean v, $H^1(k_v, G)$ and $H^1(\hat{k}_v, G)$ vanish, ([19], [8: III]; [38]) δ_v and $\hat{\delta}_v$ are both surjective, therefore we have a commutative diagram

$$(1) \qquad \begin{array}{ccc}
\overline{G}(k_v)/\varphi(G(k_v)) & \xrightarrow[\cong]{\delta_v} & H^1(k_v; C) \\
\downarrow & & \downarrow \\
\overline{G}(\hat{k}_v)/\varphi(G(\hat{k}_v)) & \xrightarrow[\cong]{\hat{\delta}_v} & H^1(\hat{k}_v; C)
\end{array}$$

As ξ_v and $\hat{\xi}_v$ are trivial on $\varphi(G(k_v))$ and $\varphi(G(\hat{k}_v))$ respectively, they induce homomorphisms

$$(2) \qquad H^1(k_v, C) \to \mathrm{Aut}(\Delta_v), \qquad H^1(\hat{k}_v, C) \to \mathrm{Aut}(\hat{\Delta}_v),$$

also to be denoted ξ_v and $\hat{\xi}_v$ respectively.

2.6. Let $\varepsilon = \varepsilon(G) = 2$ if G is of type D_r with r even, and let it be 1 otherwise. Let $n = n(G) = r + 1$ if G is of type A_r; $n = 2$ if G is of type B_r, C_r (r arbitrary), or D_r (with r even), or E_7; $n = 3$ if G is of type E_6; $n = 4$ if G is of type D_r with r odd; $n = 1$ if G is of type E_8, F_4 or G_2.

Let μ_n^ε be the kernel of the endomorphism $m_n : x \mapsto x^n$ of $(GL_1)^\varepsilon$. If G splits over some field K, then C is isomorphic to μ_n^ε over K. For any field K, $H^1(K, \mu_n^\varepsilon)$ is canonically isomorphic to $(K^\times/K^{\times n})^\varepsilon$.

Let now $v \in V_f$ be such that G splits over $\hat{k}_v$. Then C is isomorphic to μ_n^ε over $\hat{k}_v$. Moreover, it is known [16: 1.8] that $\hat{\Xi}_v$ is isomorphic to $(\mathbf{Z}/n\mathbf{Z})^\varepsilon$. The second assertion of 2.3 (i) then shows that the order of Ξ_v is a divisor of n^ε. We identify C with μ_n^ε in terms of a fixed $\hat{k}_v$-isomorphism $\theta : C \to \mu_n^\varepsilon$. This then provides an identification of $H^1(\hat{k}_v, C)$ with $(\hat{k}_v^\times/\hat{k}_v^{\times n})^\varepsilon$, with respect to which we have:

2.7. *Proposition. — Let v be a nonarchimedean place of k such that* G *splits over* $\hat{k}_v$. *Then the kernel of* ξ_v *is the subgroup* $(\hat{\mathfrak{o}}_v^\times \hat{k}_v^{\times n}/\hat{k}_v^{\times n})^\varepsilon$ *of* $(\hat{k}_v^\times/\hat{k}_v^{\times n})^\varepsilon$.

Proof. — Let us write $\mathscr{C}$ for $(GL_1)^\varepsilon$. Let $H = (\mathscr{C} \times G)/C_\theta$, where

$$C_\theta = \{(\theta(x), x^{-1}) \mid x \in C\},$$

and Z be a maximal $\hat{k}_v$-split torus of H; it is a maximal torus of H and $T := G \cap Z$ is a maximal torus of G. Since in a split torus, every subtorus is split and a direct factor, there exists a $\hat{k}_v$-subtorus D of Z, of dimension ε, such that $Z = D \times T$, hence such that $H = D \ltimes G$ is a semi-direct product of D and the normal subgroup G. Let p be the projection of H onto D. Then we have a sequence of isomorphisms:

$$(1) \qquad H^1(\hat{k}_v, C) \xrightarrow{\delta_v^{-1}} \overline{G}(\hat{k}_v)/\varphi(G(\hat{k}_v)) \xrightarrow{\cong} H(\hat{k}_v)/\mathscr{C}(\hat{k}_v)\, G(\hat{k}_v) \xrightarrow{\cong} D(\hat{k}_v)/D(\hat{k}_v)^n.$$

We extend φ to a homomorphism of H onto $\overline{G}$, also denoted φ. Its kernel is precisely $\mathscr{C}$. Since the latter is $\hat{k}_v$-split, the homomorphism $H(\hat{k}_v) \to \overline{G}(\hat{k}_v)$ is surjective (Hilbert's Theorem 90). The inverse image in $H(\hat{k}_v)$ of $\varphi(G(\hat{k}_v))$ is $\mathscr{C}(\hat{k}_v)\, G(\hat{k}_v)$, whence the second isomorphism in (1). The kernel of $p : H(\hat{k}_v) \to D(\hat{k}_v)$ is $G(\hat{k}_v)$. By restriction to $\mathscr{C}$, p defines a $\hat{k}_v$-morphism

$$\mathscr{C} = (GL_1)^\varepsilon \to D$$

whose kernel is C, and with a suitable identification of D with $(GL_1)^\varepsilon$, this homomorphism is the homomorphism m_n defined above. Therefore, the image of $\mathscr{C}(\hat{k}_v)$ under p is $D(\hat{k}_v)^n$. This yields the third isomorphism in (1). The aforementioned identification of D with $(GL_1)^\varepsilon$ gives an identification of $D(\hat{k}_v)/D(\hat{k}_v)^n$ with $(\hat{k}_v^\times/\hat{k}_v^{\times n})^\varepsilon$, and with this identification, the isomorphism $H^1(\hat{k}_v, C) \to (\hat{k}_v^\times/\hat{k}_v^{\times n})^\varepsilon$, induced by θ, is the composite of the three isomorphisms in (1).

The composite $\xi_v . \varphi$ defines a homomorphism $H(\hat{k}_v) \to \mathrm{Aut}(\hat{\Delta}_v)$, which is trivial on $\mathscr{C}(\hat{k}_v)\, G(\hat{k}_v)$ and also on the maximal bounded subgroup Z_c of $Z(\hat{k}_v)$ (see [41: 2.5]).

But it is obvious that in the identification of $D(\hat{k}_v)/D(\hat{k}_v)^n$ with $(\hat{k}_v^\times/\hat{k}_v^{\times n})^e$, the image of the maximal bounded subgroup of $D(\hat{k}_v)$ in $D(\hat{k}_v)/D(\hat{k}_v)^n$ is $(\hat{\mathfrak{o}}_v^\times \hat{k}_v^{\times n}/\hat{k}_v^{\times n})^e$. This shows that in the identification of $H^1(\hat{k}_v, C)$ with $(\hat{k}_v^\times/\hat{k}_v^{\times n})^e$, the kernel of $\hat{\xi}_v$ contains $(\hat{\mathfrak{o}}_v^\times \hat{k}_v^{\times n}/\hat{k}_v^{\times n})^e$. As the groups $\hat{\Xi}_v$ and $(\hat{k}_v^\times/\hat{\mathfrak{o}}_v^\times \hat{k}_v^{\times n})^e$ have equal order ($= n^e$, see 2.6), the kernel of $\hat{\xi}_v$ cannot be bigger. This proves the proposition.

2.8. For $v \notin S$, let P_v be a parahoric subgroup of $G(k_v)$ such that $G_S . \prod_{v \notin S} P_v$ is an open subgroup of $G(A)$. Let $\Theta_v (\subset \Delta_v)$ be the type of P_v. Let $\Lambda = G(k) \cap \prod_{v \notin S} P_v$ and $\Lambda' = \iota(\Lambda)$. In the sequel, we shall view Λ and Λ' as arithmetic subgroups of $G_{\mathscr{S}}$ and $G'_{\mathscr{S}}$ respectively. Let Γ' be the normalizer of Λ' in $G'_{\mathscr{S}}$; it is an arithmetic subgroup of $G'_{\mathscr{S}}$ (see 1.4 (ii)). We recall that $\varphi'(\Gamma')$ is contained in $\overline{G}(k)$; see 1.2. Hence the natural homomorphism $\delta : \overline{G}(k) \to H^1(k, C)$, whose kernel is $\varphi(G(k))$, induces a homomorphism

$$\partial : \Gamma'/\Lambda' \to H^1(k, C).$$

Let Ξ_v be as in 2.2, and let Ξ be the direct sum of the Ξ_v, $v \notin S$. Then Ξ acts on $\Delta := \prod_{v \notin S} \Delta_v$. Let $\Theta = \prod_{v \notin S} \Theta_v (\subset \Delta)$; let Ξ_Θ be the stabilizer of Θ in Ξ, and Ξ_{Θ_v} that of Θ_v in Ξ_v.

For $c \in H^1(k, C)$, let c_v denote the cohomology class in $H^1(k_v, C)$ determined by c. The maps ξ_v's induce a map $\xi : H^1(k, C) \to \Xi$ given by $\xi(c) = (\xi_v(c_v))_{v \in V - S}$ ($c \in H^1(k, C)$). Let

$$H^1(k, C)_\Theta = \{ c \in H^1(k, C) \mid \xi(c) \in \Xi_\Theta \},$$

$$H^1(k, C)'_\Theta = \{ c \in H^1(k, C)_\Theta \mid c_v \in \delta_v \varphi'(G'(k_v)) \quad \text{for all } v \in \mathscr{S} \}$$

and

$$\delta(\overline{G}(k))'_\Theta = \delta(\overline{G}(k)) \cap H^1(k, C)'_\Theta.$$

Let $\gamma' \in \Gamma'$. Then $\varphi'(\gamma')$ belongs to $\overline{G}(k)$ (1.2), and it stabilizes Λ, hence also P_v for all $v \notin S$. Therefore, $\delta_v \varphi'(\gamma') \in \Xi_v$. This shows that ∂ maps Γ'/Λ' into $\delta(\overline{G}(k))'_\Theta$.

In the notation introduced above we have:

2.9. *Proposition. — The following sequence is exact*

$$1 \to \Big(\prod_{v \in \mathscr{S}} C'(k_v)\Big)/(C'(k) \cap \Lambda') \to \Gamma'/\Lambda' \xrightarrow{\partial} \delta(\overline{G}(k))'_\Theta \to 1.$$

Apart from minor modifications, the above proposition is due to J. Rohlfs when G is k-split [32]. It was already remarked in [24] that the proof of [32] goes over without change to the more general case if k is a number field. Since our context is slightly more general (for example, we allow k to be of positive characteristic), we repeat the proof.

We begin by showing that ∂ is surjective. Let $c \in \delta(\overline{G}(k))'_\Theta$, and $g \in \overline{G}(k)$ be such that $\delta(g) = c$. Then the parahoric subgroup $g(P_v)$ is of the same type as P_v ($v \notin S$). There exist therefore $h_v \in G(k_v)$ such that $g(P_v) = h_v P_v h_v^{-1}$. Moreover, for v outside a finite set T of places containing S, we have $g(P_v) = P_v$ and so $h_v \in P_v$. By strong approximation ([30], [22]), we can find an $h \in G(k)$ such that for all $v \notin T$, $h \in P_v$, and which is so close to h_v, for $v \in T - S$, that $h_v P_v h_v^{-1} = hP_v h^{-1}$. This last equality is then true for all $v \notin S$. Therefore, $\varphi(h)^{-1} g$ stabilizes P_v for all $v \notin S$. Also,

$$\delta(\varphi(h)^{-1} g) = \delta(g) = c.$$

As $c \in \delta(\overline{G}(k))'_\Theta$, there is, for $v \in \mathscr{S}$, a $\gamma'_v \in G'(k_v)$ such that $\varphi'(\gamma'_v) = \varphi(h)^{-1} g$. Then since $\varphi(h)^{-1} g$ stabilizes P_v for all $v \notin S$, the element $\gamma' = (\gamma'_v)_{v \in \mathscr{S}}$ belongs to Γ'. Therefore ∂ is surjective. If now $\delta\varphi'(\gamma') = 1$, then $\varphi'(\gamma') = \varphi(g)$ for some $g \in G(k)$ and, consequently, $\gamma' \in \iota(g) . \Pi_{\mathscr{S}} C'(k_v)$. From this the exactness on the left follows.

2.10. As $\delta(\overline{G}(k))'_\Theta \subset H^1(k, C)'_\Theta$, Proposition 2.9 gives the following exact sequence:

$$1 \to (\prod_{v \in \mathscr{S}} C'(k_v))/(C'(k) \cap \Lambda') \to \Gamma'/\Lambda' \xrightarrow{\partial} H^1(k, C)'_\Theta.$$

Now let $\quad H^1(k, C)_\xi = \{c \in H^1(k, C) \mid \xi(c) = 1\}$,

and $\quad H^1(k, C)'_\xi = \{c \in H^1(k, C)_\xi \mid c_v \in \delta_v \varphi'(G'(k_v))\}$.

It is obvious that

$$(*) \qquad \# H^1(k, C)'_\Theta \leqslant \# H^1(k, C)'_\xi . \prod_{v \in V - S} \# \Xi_{\Theta_v},$$

and since $\# C'(k_v) \leqslant n^\epsilon$ for all v, we conclude from the above exact sequence that

$$[\Gamma' : \Lambda'] \leqslant \# \prod_{v \in \mathscr{S}} C'(k_v) . \# H^1(k, C)'_\xi . \prod_{v \in V - S} \# \Xi_{\Theta_v}$$

$$\leqslant n^{\epsilon \# \mathscr{S}} . \# H^1(k, C)_\xi . \prod_{v \in V - S} \# \Xi_{\Theta_v}.$$

3. Lower bound for the covolumes of arithmetic subgroups

We shall continue to use the notation introduced in §§0 and 2.

3.1. In the sequel, we shall use the fact that for $v \in V_f$ a parahoric subgroup P_v^m of $G(k_v)$ of maximal volume is necessarily *special*. We also need to know that if P_v is a parahoric subgroup of $G(k_v)$, of type Θ_v, such that $P_v^m \cap P_v$ contains an Iwahori subgroup I_v, then

$$(*) \qquad [P_v^m : I_v] \geqslant [P_v : I_v] (\# \Xi_{\Theta_v}).$$

This could be rather laboriously checked case by case, using the " reduction mod p " of the parahoric subgroup P_v (see 3.5, 3.7 in [41]) to compute the index of an Iwahori subgroup it contains. More conceptual proofs are given in Appendix A.

3.2. Let Γ' be a maximal arithmetic subgroup of $G'_{\mathscr{S}}$, $\Lambda' = \Gamma' \cap \iota(G(k))$ and Λ be its inverse image in $G(k)$ under ι. Then according to Proposition 1.4 (iv), for $v \notin S$, the closure P_v of Λ in $G(k_v)$ is a parahoric subgroup of $G(k_v)$, and $\Lambda = G(k) \cap \prod_{v \notin S} P_v$. Let $\Theta_v (\subset \Delta_v)$ be the type of P_v and $\Theta = \prod_{v \notin S} \Theta_v$.

For all but finitely many v, P_v is a hyperspecial parahoric subgroup of $G(k_v)$ and so is of maximum volume ([41: 3.8.2]). Let T be the smallest set of places containing S such that for $v \notin T$, the parahoric subgroup P_v is of maximum volume. Then for all $v \notin T$, as P_v is special (3.1), $\Xi_{\Theta_v} = \{1\}$.

For every $v \in T - S$, we fix a parahoric subgroup P_v^m of $G(k_v)$ of maximum volume such that $P_v^m \cap P_v$ contains an Iwahori subgroup I_v. Let

$$\Lambda^m = G(k) \cap \Big(\prod_{v \in T-S} P_v^m \cdot \prod_{v \notin T} P_v\Big).$$

Then Λ^m is an arithmetic subgroup.

3.3. Using strong approximation, we see at once that

$$\frac{[\Lambda^m : \Lambda^m \cap \Lambda]}{[\Lambda : \Lambda^m \cap \Lambda]} = \prod_{v \in T-S} \frac{[P_v^m : I_v]}{[P_v : I_v]}.$$

Also, for $v \in T - S$,

$$\frac{[P_v^m : I_v]}{[P_v : I_v]} \geqslant \#\Xi_{\Theta_v} \quad \text{(see 3.1)}.$$

Hence,
$$\frac{[\Lambda^m : \Lambda^m \cap \Lambda]}{[\Lambda : \Lambda^m \cap \Lambda]} \geqslant \prod_{v \in V-S} \#\Xi_{\Theta_v}.$$

3.4. For $v \in V_f$, let μ_v (resp. μ'_v) be the Haar measure on $G(k_v)$ (resp. $G'(k_v)$) with respect to which the volume of any Iwahori subgroup of $G(k_v)$ (resp. the volume of the stabilizer in $G'(k_v)$ of any chamber in X_v) is 1.

It is known that $\iota(G(k_v))$ is a closed normal subgroup of $G'(k_v)$ and that $G'(k_v)/\iota(G(k_v))$ is a compact abelian group ([7: 3.19 (i)]). Let I_v be an Iwahori subgroup of $G(k_v)$ and I'_v be the stabilizer in $G'(k_v)$ of the chamber pointwise fixed by I_v. Then $I'_v . \iota(G(k_v)) = G'(k_v)$ and $\iota(I_v) = I'_v \cap \iota(G(k_v))$. Using these facts it is easy to see that μ'_v is the measure determined by the Haar measure on the closed normal subgroup $\iota(G(k_v))$ with respect to which $\iota(I_v)$ has volume 1, and the normalized Haar measure on the compact group $G'(k_v)/\iota(G(k_v))$.

(We note that I'_v is not always an Iwahori subgroup, as defined in Tits [41: 3.7], but it contains a unique such subgroup, necessarily of finite index.)

3.5. For v archimedean, let μ_v (resp. μ'_v) be the Haar measure on $G(k_v)$ (resp. $G'(k_v)$) such that in the induced measure, any maximal compact subgroup of $R_{k_v/\mathbf{R}}(G)$ (**C**) (resp. $R_{k_v/\mathbf{R}}(G')$ $(\mathbf{C})$) has volume 1. In particular, if $k_v = \mathbf{R}$ and G is anisotropic over k_v, then $\mu_v(G(k_v)) = 1 = \mu'_v(G'(k_v))$.

3.6. Let $\mu_{\mathscr{S}}$ (resp. $\mu'_{\mathscr{S}}$) denote the product measure $\prod_{v\in\mathscr{S}} \mu_v$ (resp. $\prod_{v\in\mathscr{S}} \mu'_v$) on $G_{\mathscr{S}}$ (resp. $G'_{\mathscr{S}}$) as well as the induced measure on their quotients by discrete subgroups. Then

$$\mu'_{\mathscr{S}}(G'_{\mathscr{S}}/\Gamma') = [\Gamma' : \Lambda']^{-1}.\mu'_{\mathscr{S}}(G'_{\mathscr{S}}/\Lambda'),$$

and it follows, using the alternate description of the Haar measure μ'_v given in 3.4, that

$$\mu'_{\mathscr{S}}(G'_{\mathscr{S}}/\Lambda') \geqslant \mu_{\mathscr{S}}(G_{\mathscr{S}}/\Lambda) = \frac{[\Lambda^m : \Lambda^m \cap \Lambda]}{[\Lambda : \Lambda^m \cap \Lambda]}\, \mu_{\mathscr{S}}(G_{\mathscr{S}}/\Lambda^m).$$

Hence (see 3.3)

$$(1) \qquad \mu'_{\mathscr{S}}(G'_{\mathscr{S}}/\Gamma') \geqslant \frac{\mu_{\mathscr{S}}(G_{\mathscr{S}}/\Lambda)}{[\Gamma' : \Lambda']} \geqslant \frac{\prod_{v\in V-S} \#\Xi_{\Theta_v}}{[\Gamma' : \Lambda']}\, \mu_{\mathscr{S}}(G_{\mathscr{S}}/\Lambda^m).$$

Now as

$$(2) \qquad [\Gamma' : \Lambda'] \leqslant n^{\epsilon\#\mathscr{S}}.\#H^1(k, C)_{\xi}.\prod_{v\in V-S} \#\Xi_{\Theta_v}$$

(cf. 2.10), we conclude that

$$(*) \qquad \mu'_{\mathscr{S}}(G'_{\mathscr{S}}/\Gamma') \geqslant n^{-\epsilon\#\mathscr{S}}(\#H^1(k, C)_{\xi})^{-1}\, \mu_{\mathscr{S}}(G_{\mathscr{S}}/\Lambda^m).$$

3.7. In [31] the volumes of arithmetic quotients of semi-simple groups have been computed. We shall now describe the result. We begin by observing that since for $v \in S - \mathscr{S}$, G is anisotropic over k_v, $G(k_v)$ is compact and $\mu_v(G(k_v)) = 1$, and hence for any S-arithmetic subgroup Λ of $G(k)$, $\mu_S(G_S/\Lambda) = \mu_{\mathscr{S}}(G_{\mathscr{S}}/\Lambda)$; where μ_S is the measure on G_S/Λ induced by the product measure $\prod_{v\in S} \mu_v$ on G_S.

We recall that r is the absolute rank of G, and, for $v \in V_f$, r_v the rank of G over the maximal unramified extension $\hat{k}_v$ of k_v. Let $\mathscr{G}$ be the unique quasi-split, simply connected inner k-form of G. If G is not a k-form of type $^6D_4$, let ℓ be the smallest Galois extension of k over which $\mathscr{G}$ splits. If G is a k-form of type $^6D_4$, let ℓ be a fixed cubic extension of k contained in the Galois extension, of degree 6, over which $\mathscr{G}$ splits.

Let $\mathfrak{s} = \mathfrak{s}(\mathscr{G}) = 0$ if $\mathscr{G}$ splits over k; if $\mathscr{G}$ does not split over k (i.e. if G is an *outer* form of a split group), then let $\mathfrak{s} = \frac{1}{2}(r-1)(r+2)$ if G is an outer form of type A_r with r odd, $\mathfrak{s} = \frac{1}{2}r(r+3)$ if G is an outer form of type A_r with r even, $\mathfrak{s} = 2r - 1$ if G is an outer form of type D_r (r arbitrary), and $\mathfrak{s} = 26$ if G is an outer form of type E_6; see [31: 0.4]. In particular, we have

$$\mathfrak{s}(\mathscr{G}) \geqslant \begin{cases} 5 \text{ if } \mathscr{G} \text{ does not split over } k \\ 7 \text{ if } \mathscr{G} \text{ is an outer form of type } D_r\ (r \geqslant 4). \end{cases}$$

Let m_i $(1 \leqslant i \leqslant r)$ be the exponents of the compact simply connected real-analytic Lie group of the same type as G; see [31: 1.5]. Note that $r + 2\sum_1^r m_i = \dim G$.

Let $\tau_k(G)$ be the *Tamagawa number* of G/k (see, for example, [31: 3.3]).

With these notations we have ([31: Theorem 3.7]): Let $P = (P_v)_{v \in V_f - S}$ be a coherent collection of parahoric subgroups and Λ the principal S-arithmetic subgroup determined by P (1.5). Then

$$\begin{aligned}\mu_{\mathscr{S}}(G_{\mathscr{S}}/\Lambda) &= \mu_S(G_S/\Lambda)\\ &= D_k^{\frac{1}{2}\dim G}(D_\ell/D_k^{[\ell:k]})^{\frac{1}{2}s}\left(\prod_{v \in V_\infty}\left|\prod_{i=1}^{r}\frac{m_i!}{(2\pi)^{m_i+1}}\right|_v\right)\tau_k(G)\ \mathscr{E}(P),\end{aligned}$$

where $\mathscr{E}(P) = \prod_{v \in S_f} e(I_v) . \prod_{v \in V - S} e(P_v)$; the $e(I_v)$ and $e(P_v)$ are positive real numbers computable in terms of P, the structure of G/k and the Bruhat-Tits theory. For $v \in S_f$ (resp. $v \in V - S$), $e(I_v)$ (resp. $e(P_v)$) is the inverse of the volume of any Iwahori subgroup of $G(k_v)$ (resp. of P_v) with respect to the Haar measure $\gamma_v\,\omega_v^*$; where γ_v is defined in §1.3 and ω_v^* in §2.1 of [31]. In this paper we need the following information, see [31: 3.10, 2.10, 2.11] (the unexplained notation is as in [31]):

(1) $\quad$ for all $v \in S_f$, $e(I_v) > 1$ and for all $v \in V - S$, $e(P_v) > 1$;

(2) $$e(I_v) = (\#\overline{T}_v(\mathfrak{f}_v))^{-1} . q_v^{(r_v + \dim \overline{\mathscr{M}}_v)/2} \geqslant (q_v + 1)^{-r_v} . q_v^{r_v(r_v+3)/2};$$

(3) $$e(I_v) = (q_v - 1)\,(q_v^{d_v} - 1)^{-(r+1)/d_v}\, q_v^{r(r+3)/2} > (q_v - 1)\, q_v^{r(r+1)/2 - 1}$$

if $G(k_v) = SL_{(r+1)/d_v}(\mathfrak{D}_v)$, where $\mathfrak{D}_v$ is a central division algebra of degree $d_v < (r+1)$ over k_v, and $v \in S_f$.

(4) $$e(P_v) = q_v^{(\dim \overline{M}_v + \dim \overline{\mathscr{M}}_v)/2} . (\#\overline{M}_v(\mathfrak{f}_v))^{-1} \quad (v \in V_f - S).$$

Moreover:

(5) $$e(P_v) \geqslant (q_v + 1)^{-1} . q_v^{r_v + 1}$$

if $v \in V_f - S$ and either G is not quasi-split over k_v, or P_v is not special, or G splits over $\hat{k}_v$ but P_v is not hyperspecial. Also,

(6) $$e(P_v) \geqslant (q_v - 1)\, q_v^{(r^2 + 2r - (r+1)^2 d_v^{-1} - 1)/2}$$

if $G(k_v) = SL_{(r+1)/d_v}(\mathfrak{D}_v)$, where $\mathfrak{D}_v$ is a central division algebra of degree $d_v \leqslant r + 1$ over k_v, and

(7) $$e(P_v) \geqslant q_v^{(r+1)/2}$$

if G is an outer form of type A_r, r odd, of k_v-rank $(r-1)/2$, which does not split over $\hat{k}_v$.

In the sequel, we write e_v *for* $e(I_v)$ *and* e_v^m *for* $e(P_v)$, *where* P_v *is a parahoric subgroup of* $G(k_v)$ *of maximal volume.*

3.8. As every arithmetic subgroup of $G'_{\mathscr{S}}$ is contained in a maximal one (1.4 (iii)), combining the bound ($*$) of 3.6 and the formula for the volume of $G_{\mathscr{S}}/\Lambda$ given above, we obtain the following:

$$\mu'_{\mathscr{S}}(G'_{\mathscr{S}}/\Gamma') \geqslant n^{-s\#\mathscr{S}}(\# H^1(k, C)_{\xi})^{-1} D_k^{\frac{1}{2}\dim G}(D_\ell/D_k^{[\ell:k]})^{\frac{1}{2}s} \left(\prod_{v \in V_\infty} \left| \prod_{i=1}^{r} \frac{m_i!}{(2\pi)^{m_i+1}} \right|_v \right) \tau_k(G)\, \mathscr{E},$$

where, in the notation of 3.7,

$$\mathscr{E} = \prod_{v \in S_f} e_v \cdot \prod_{v \in V_f - S} e_v^{m}.$$

This shows that *the volumes* $\mu'_{\mathscr{S}}(G'_{\mathscr{S}}/\Gamma')$ *have a strictly positive lower bound, as* Γ' *runs through the arithmetic subgroups of* $G'_{\mathscr{S}}$. This is then, of course, true with respect to any Haar measure on $G'_{\mathscr{S}}$.

In §5 we shall give an upper bound for the order of $H^1(k, C)_{\xi}$.

3.9. *Proposition. — Let* K' *be a compact open subgroup of the restricted product* G'_{V-S} *of the groups* $G'(k_v)$ $(v \in V - S)$. *Then the number of double cosets* $G'(k)\backslash G'(A)/(G'_S K')$ *is finite.*

This is the finiteness of the class number of G' (at any rate for G'_S non-compact, which is a standing assumption in this paper). It is well-known in the number field case [2], but we do not know of a reference in the function field case (except when G' is anisotropic over k, where S may be taken empty [14: 2.2.7 (iii)]).

We fix a Haar measure ν on $G'(A) = G'_S \times G'_{V-S}$. It is a product of Haar measures ν_S and ν_{V-S} on G'_S and G'_{V-S} respectively. The double cosets mod $G'_S K'$ and $G'(k)$ correspond bijectively to the orbits of $G'_S K'$ on $G'(k)\backslash G'(A)$, which are all open. Since $G'(k)\backslash G'(A)$ has finite Haar measure, it is enough to show that the volumes of these orbits have a strictly positive lower bound. The double cosets are represented by elements of G'_{V-S}; it suffices therefore to consider the orbit of the image of an element $x \in G'_{V-S}$. It is isomorphic to $\Gamma_x\backslash G'_S\, xK'\, x^{-1}$, where $\Gamma_x = G'(k) \cap G'_S\, xK'\, x^{-1}$. Let Γ'_x be the projection of Γ_x into G'_S, with respect to the decomposition $G'(A) = G'_S \times G'_{V-S}$. Then $\nu(G'(k)\backslash G'(k)\, xG'_S\, K') = \nu_S(\Gamma'_x\backslash G'_S) \cdot \nu_{V-S}(K')$. As $xK'\, x^{-1}$ is a compact open subgroup of G'_{V-S}, Γ'_x is an S-arithmetic subgroup. Then the next to last assertion in 3.8 yields our claim.

3.10. *Proposition. — Let* $\mathscr{R}$ *be a finite subset of* V *containing* S, *such that* G *is quasi-split over* k_v *and splits over* $\hat{k}_v$ *for all* $v \notin \mathscr{R}$. *Then the set of arithmetic subgroups* Γ' *of* G'_S *associated to coherent collections* $(P_v)_{v \notin S}$ *of parahoric subgroups (see* 1.5) *which are hyperspecial for* $v \notin \mathscr{R}$ *form finitely many classes with respect to* $\overline{G}(k)$*-conjugacy.*

In view of the construction of the Γ' (see 1.5), it is equivalent to show that the P's in which P_v is hyperspecial for all $v \notin \mathscr{R}$ form finitely many classes under $\overline{G}(k)$-conjugacy. For any $v \in V_f$, the parahoric subgroups of $G(k_v)$ form finitely many conjugacy classes under $G(k_v)$, hence a fortiori under $\overline{G}(k_v)$. It suffices therefore to consider the P's in which P_v belongs to a given conjugacy class of parahoric subgroups in $G(k_v)$ for $v \in \mathscr{R} - S$. Let P and P′ be two such coherent collections. Of course, $P_v = P'_v$ for almost all v's. For $v \notin \mathscr{R}$, any two hyperspecial subgroups of $G(k_v)$ are conjugate under $\overline{G}(k_v)$ [41: 2.5]. There exists then $g \in \overline{G}(A)$ such that ${}^gP = P'$. Let $\overline{P}_v$ be the stabilizer of P_v in $\overline{G}(k_v)$ $(v \notin S)$. Then $\overline{P} = \Pi_{v \notin S} \overline{P}_v$ is a compact open subgroup of $\overline{G}_{V-S}$ and $\overline{G}_S \overline{P}$ is the stabilizer of P in $\overline{G}(A)$. The $\overline{G}(k)$-conjugacy classes of the P's satisfying our conditions correspond therefore to the double cosets of $\overline{G}(A)$ mod $\overline{G}(k)$ and $\overline{G}_S \overline{P}$. They are finite in number by 3.9 and the proposition follows.

4. Euler-Poincaré characteristic of arithmetic groups

We assume in this section that if k is of positive characteristic, then the k-rank of G is zero. Then any arithmetic subgroup of $G'_{\mathscr{S}}$ has a torsion free subgroup of finite index ([33: Theorem 4]) and there exists a $G_{\mathscr{S}}$-invariant measure $\mu^{EP}_{\mathscr{S}}$ on $G_{\mathscr{S}}$ such that, for any arithmetic subgroup Γ of $G_{\mathscr{S}}$,

$$|\chi(\Gamma)| = \mu^{EP}_{\mathscr{S}}(G_{\mathscr{S}}/\Gamma),$$

where $\chi(\Gamma)$ is the Euler-Poincaré characteristic of Γ in the sense of C. T. C. Wall (see [33: §§1.8, 3]).

4.1. It follows from [33: Proposition 25] that, up to sign, $\mu^{EP}_{\mathscr{S}}$ is the product of the Euler-Poincaré measures on the groups $G(k_v)$ $(v \in \mathscr{S})$ introduced in [33: §3], and to be denoted here by μ^{EP}_v. Also, for any nonarchimedean v, μ^{EP}_v is a non-zero multiple $a_v \mu_v$ of the Tits measure μ_v defined in 3.4; here

$$a_v = \mu^{EP}_v(I_v) = (-1)^{s_v} (W_v(\mathbf{q}^{-1}))^{-1},$$

where I_v is an Iwahori subgroup of $G(k_v)$, s_v is the k_v-rank of G and $W_v(\mathbf{q})$ is the Poincaré series associated with the Tits system on $G(k_v)$ whose " B " is an Iwahori subgroup (of $G(k_v)$) and " N " is the group of k_v-rational elements of the normalizer of a suitable maximal k_v-split torus of G ([33: Theorem 6]).

If $v \in \mathscr{S}_\infty$, μ^{EP}_v is non-zero if and only if $G(k_v)$ contains a compact Cartan subgroup ([33: Proposition 23]). Thus if k is a global function field, then $\mu^{EP}_{\mathscr{S}}$ is non-zero; if k is a number field, and $\mu^{EP}_{\mathscr{S}} \neq 0$, then k is necessarily totally real.

4.2. For $v \in \mathscr{S}_\infty$, the Hirzebruch proportionality principle ([33: §3.2]) at once implies that if $G(k_v)$ contains a compact Cartan subgroup, then, up to sign, μ^{EP}_v equals $c_v \mu_v$, where μ_v is the Haar measure on $G(k_v)$ defined in 3.5 and c_v is the Euler-Poincaré characteristic of the compact dual of the symmetric space associated with $G(k_v)$ (i.e., the quotient of a suitable maximal compact subgroup of $G(\mathbf{C})$ by a maximal compact subgroup of $G(k_v)$). The constant c_v is therefore a non-zero integer.

4.3. Assume that $G_{\mathscr{S}_\infty} = \prod_{v \in \mathscr{S}_\infty} G(k_v)$ has a compact Cartan subgroup. Then, combining the above observations, we conclude that for any arithmetic subgroup Γ of $G_{\mathscr{S}}$,

$$|\chi(\Gamma)| \geqslant \prod_{v \in \mathscr{S}_f} |W_v(\mathbf{q}^{-1})|^{-1} \mu_{\mathscr{S}}(G_{\mathscr{S}}/\Gamma).$$

4.4. *A lower bound for* $|W_v(\mathbf{q}^{-1})|^{-1}$. As before, for $v \in V_f$, let $\hat{k}_v$ be the maximal unramified extension of k_v. Let σ_v denote the Frobenius automorphism of $\hat{k}_v/k_v$. Then there is a natural action of σ_v on the affine Weyl group of $G/\hat{k}_v$ and the subgroup of the fixed points is the affine Weyl group of G/k_v. Now the results contained in 1.10.1, 1.11 and 3.3.1 of Tits [41] together with those in 1.30, 1.32, 1.33 and 3.10 of Steinberg [39] imply that

$$(W_v(\mathbf{q}^{-1}))^{-1} = \prod_{j=1}^{r_v} \frac{(1 - \varepsilon_j^v q_v^{1-d_v(j)})(1 - \varepsilon_{0j}^v q_v^{-1})}{(1 - \varepsilon_j^v q_v^{-d_v(j)})},$$

where the $d_v(j)$'s are certain positive integers $\geqslant 2$, and ε_j^v, ε_{0j}^v are certain roots of unity (see Steinberg [39: Theorem 3.10]).

From the above expression for $(W_v(\mathbf{q}^{-1}))^{-1}$, it is obvious that as the $d_v(j)$'s and the q_v's are $\geqslant 2$,

$$(1) \qquad |W_v(\mathbf{q}^{-1})|^{-1} \geqslant \left(\frac{(1 - q_v^{-1})^2}{1 + q_v^{-2}}\right)^{r_v} = \left(\frac{(q_v - 1)^2}{q_v^2 + 1}\right)^{r_v} \quad (\geqslant 5^{-r_v}).$$

As a consequence, we have in particular

$$(2) \qquad |\chi(\Gamma)| = \mu^{\mathrm{EP}}(G_{\mathscr{S}}/\Gamma) \geqslant 5^{-c(S,G)} \mu_{\mathscr{S}}(G_{\mathscr{S}}/\Gamma), \qquad (c(S, G) = \sum_{v \in \mathscr{S}_f} r_v).$$

4.5. For the proof of Theorem 7.3, we need to know $|W_v(\mathbf{q}^{-1})|^{-1}$ explicitly for certain G and v. Using Proposition 24 and Theorem 6 of [33] and the Bruhat-Tits theory, $|W_v(\mathbf{q}^{-1})|^{-1}$ can be easily computed; the values are given in Appendix C, as they are needed.

4.6. Now let Γ', Λ' and Λ be as in 3.2. Then

$$|\chi(\Gamma')| = [\Gamma' : \Lambda']^{-1} |\chi(\Lambda')|$$

and it is obvious that $|\chi(\Lambda')| \geqslant |\chi(\Lambda)|$. Therefore, under the hypothesis of 4.3 we have

$$\begin{aligned} |\chi(\Gamma')| &\geqslant [\Gamma' : \Lambda']^{-1} |\chi(\Lambda)| \\ &\geqslant [\Gamma' : \Lambda']^{-1} \prod_{v \in \mathscr{S}_f} |W_v(\mathbf{q}^{-1})|^{-1} \mu_{\mathscr{S}}(G_{\mathscr{S}}/\Lambda) \\ &\geqslant n^{-\varepsilon \# \mathscr{S}} (\# H^1(k, C)_\xi)^{-1} \prod_{v \in \mathscr{S}_f} |W_v(\mathbf{q}^{-1})|^{-1} \mu_{\mathscr{S}}(G_{\mathscr{S}}/\Lambda^m) \end{aligned}$$

(cf. 3.6), where Λ^m is as in 3.2. By 3.7,

$$\mu_{\mathscr{S}}(G_{\mathscr{S}}/\Lambda^m) = D_k^{\frac{1}{2}\dim G} (D_\ell / D_k^{[\ell:k]})^{\frac{1}{2}s} \left(\prod_{v \in V_\infty} \left| \prod_{i=1}^{r} \frac{m_i!}{(2\pi)^{m_i+1}} \right|_v \right) \tau_k(G)\, \mathscr{E},$$

where $\mathscr{E}$ is as in 3.8; therefore we get the following bound:

$$|\chi(\Gamma')| \geqslant n^{-\varepsilon \# \mathscr{S}} (\# \mathrm{H}^1(k, \mathrm{C})_\xi)^{-1} \prod_{v \in \mathscr{S}_f} |\mathrm{W}_v(\mathbf{q}^{-1})|^{-1} \mathrm{D}_k^{\frac{1}{2}\dim \mathrm{G}} (\mathrm{D}_\ell / \mathrm{D}_k^{[\ell:k]})^{\frac{1}{2}s}$$
$$\cdot \left(\prod_{v \in \mathrm{V}_\infty} \left| \prod_{i=1}^{r} \frac{m_i!}{(2\pi)^{m_i+1}} \right|_v \right) \tau_k(\mathrm{G})\, \mathscr{E}$$

for every arithmetic subgroup Γ'.

5. An upper bound for the order of $\mathrm{H}^1(k, \mathrm{C})_\xi$

In this section we shall give a "good" upper bound for the order of the group $\mathrm{H}^1(k, \mathrm{C})_\xi$, introduced in 2.10.

As in 3.7, let $\mathscr{G}$ be the unique simply connected, quasi-split inner k-form of G. Then the center C of G is k-isomorphic to that of $\mathscr{G}$. We shall begin by considering the case where $\mathscr{G}$ is k-split, i.e. G/k is an inner k-form (of a k-split group). As recalled in 2.6, C is k-isomorphic, in this case, to μ_n^ε, where ε, n and μ_n^ε are as in 2.6. We identify C with μ_n^ε in terms of a fixed k-isomorphism. This then provides an identification of $\mathrm{H}^1(\mathrm{K}, \mathrm{C})$ with $(\mathrm{K}^\times/\mathrm{K}^{\times n})^\varepsilon$ for any field extension K of k. For $x \in (\mathrm{K}^\times)^\varepsilon$, we denote by $\bar{x}$ the element of $\mathrm{H}^1(\mathrm{K}, \mathrm{C})$ which it determines. For $v \in \mathrm{V}$, $\bar{x}_v$ will denote the cohomology class in $\mathrm{H}^1(k_v, \mathrm{C})$ determined by $x \in (k^\times)^\varepsilon$.

Each $v \in \mathrm{V}_f$ gives a homomorphism $(k^\times)^\varepsilon \to \mathbf{Z}^\varepsilon$, which will be denoted again by v. Let now T be the (finite) set of places $v \notin \mathrm{S}$ such that G does not split over k_v. Then in view of Lemma 2.3, it follows from Proposition 2.7 that for $v \notin \mathrm{S} \cup \mathrm{T}$ and $x \in (k^\times)^\varepsilon$, $\xi_v(\bar{x}_v)$ is trivial if and only if $v(x) \in (n\mathbf{Z})^\varepsilon$. From this we conclude that $\mathrm{H}^1(k, \mathrm{C})_\xi \cap (k_n/k^{\times n})^\varepsilon$ is a subgroup of $\mathrm{H}^1(k, \mathrm{C})_\xi$ of index $\leqslant n^{\varepsilon \#(\mathrm{S}_f \cup \mathrm{T})}$, where k_n is the subgroup of $k^\times$ consisting of the elements x such that $v(x) \in n\mathbf{Z}$ for all nonarchimedean v. As $\#(k_n/k^{\times n}) \leqslant h_k\, n^{a(k)}$ (Proposition 0.12), this implies the following:

5.1. *Proposition. — If* G/k *is an inner form of a split group,*

$$\# \mathrm{H}^1(k, \mathrm{C})_\xi \leqslant h_k^\varepsilon\, n^{\varepsilon a(k) + \varepsilon \#(\mathrm{S}_f \cup \mathrm{T})}.$$

5.2. In the rest of this section we treat the case where G/k is an outer form. Then $\mathscr{G}$ is a non-split, quasi-split group of type A_r, or D_r, or E_6. Let ℓ be as in 3.7. Note that ℓ is a separable quadratic extension of k except when G is a triality form of type D_4 in which case it is a separable (but not necessarily Galois) extension of k of degree 3. For v nonarchimedean, let $\ell_v = \ell \otimes_k k_v$. If ℓ_v is a field, let $\tilde{v}$ denote its normalized additive valuation (i.e. the additive valuation whose set of values is $\mathbf{Z}$). Its restriction to k_v is a multiple of v. If v splits over ℓ, let $\tilde{v}_i$ ($i = 1$, 2 and possibly 3) be the normalized additive valuations of ℓ "lying" over v (i.e. whose restriction to $k^\times$ is a multiple of v); in this case ℓ_v is a direct sum of 2 or 3 local fields.

5.3. Let n be as in 2.6 and μ_n be the kernel of the endomorphism $x \mapsto x^n$ of GL_1. Then, except in the case where G/k is a form of type ${}^2D_r$ with r even, the center C of G is k-isomorphic to the kernel of the norm map

$$N_{\ell/k} : R_{\ell/k}(\mu_n) \rightarrow \mu_n.$$

If G/k is of type ${}^2D_r$ with r even, then C is k-isomorphic to $R_{\ell/k}(\mu_2)$.

Assume first that G/k is not of type ${}^2D_r$ with r even. Using the above description of the center C, we get the following commutative diagram:

$$\begin{array}{ccccccc}
\mu_n(k)/N_{\ell/k}(\mu_n(\ell)) & \longrightarrow & H^1(k, C) & \longrightarrow & \ell^\times/\ell^{\times n} & \xrightarrow{N_{\ell/k}} & k^\times/k^{\times n} \\
\downarrow & & \downarrow & & \downarrow & & \downarrow \\
\mu_n(k_v)/N_{\ell/k}(R_{\ell/k}(\mu_n)\,(k_v)) & \longrightarrow & H^1(k_v, C) & \longrightarrow & (\ell \otimes_k k_v)^\times/(\ell \otimes_k k_v)^{\times n} & \xrightarrow{N_{\ell/k}} & k_v^\times/k_v^{\times n} \\
\downarrow & & \downarrow & & \downarrow & & \downarrow \\
\mu_n(\widehat{k}_v)/N_{\ell/k}(R_{\ell/k}(\mu_n)\,(\widehat{k}_v)) & \longrightarrow & H^1(\widehat{k}_v, C) & \longrightarrow & (\ell \otimes_k \widehat{k}_v)^\times/(\ell \otimes_k \widehat{k}_v)^{\times n} & \xrightarrow{N_{\ell/k}} & \widehat{k}_v^\times/\widehat{k}_v^{\times n}
\end{array}$$

in which the rows are exact. Note that the order of $\mu_n(k)/N_{\ell/k}(\mu_n(\ell))$ is at most 2, and this group is trivial if either n is odd or $[\ell : k] = 3$. This is evident from the fact that $N_{\ell/k}(\mu_n(\ell))$ contains $\mu_n(k)^{[\ell:k]}$, and if $[\ell : k] = 3$, then (G/k is a triality form of type D_4 and) $n = 2$. Next we assert that if v is a nonarchimedean place such that G splits over $\widehat{k}_v$, then the image of $\mu_n(k_v)/N_{\ell/k}(R_{\ell/k}(\mu_n)\,(k_v))$ in $H^1(k_v, C)$ acts trivially on Δ_v. To prove this, in view of Lemma 2.3 (i), it suffices to note that if G splits over $\widehat{k}_v$, then $\ell \otimes_k \widehat{k}_v$ is a direct sum of $[\ell : k]$ ($\geqslant 2$) copies of $\widehat{k}_v$, hence $N_{\ell/k}(R_{\ell/k}(\mu_n)\,(\widehat{k}_v)) = \mu_n(\widehat{k}_v)$ and so the image of $\mu_n(k_v)/N_{\ell/k}(R_{\ell/k}(\mu_n)\,(k_v))$ in $H^1(\widehat{k}_v, C)$ is trivial.

Assume now that G/k is of type ${}^2D_r$ with r even. Then $n = [\ell : k] = 2$. In this case C is k-isomorphic to $R_{\ell/k}(\mu_2)$ and will be identified with it in terms of a fixed k-isomorphism. For any field extension K of k, the group $H^1(K, C)$ is canonically isomorphic to $(\ell \otimes_k K)^\times/(\ell \otimes_k K)^{\times 2}$. In particular, $H^1(k, C) = \ell^\times/\ell^{\times 2}$, and $\ell^\times$ acts on Δ_v and $\widehat{\Delta}_v$ through the quotient $\ell^\times/\ell^{\times 2}$; we shall denote the induced homomorphism $\ell^\times \rightarrow \Xi_v (\subset \mathrm{Aut}\,\Delta_v)$ by ξ_v in the sequel.

5.4. *Lemma.* — *Let* $v \in V_f$.

(i) *Assume* G/k *is not of type* ${}^2D_r$ *with* r *even. Let* $L = \{ x \in \ell^\times \mid N_{\ell/k}(x) \in k^{\times n} \}$ *and* $x \in L$. *If* v *does not split over* ℓ, *then* $\widetilde{v}(x) \in n\mathbf{Z}$ *if* v *is ramified in* ℓ, *or if one of* n, $[\ell : k]$ *is odd. In particular, if* v *does not split over* ℓ, $\widetilde{v}(x) \in 2\mathbf{Z}$ *if* G *is a triality form of type* D_4.

(ii) *Assume* G *to be split over* $\widehat{k}_v$ *and quasi-split over* k_v. *Then* $\widetilde{v}(x) \in n\mathbf{Z}$ *if* v *does not split over* ℓ, *and* $\widetilde{v}_i(x) \in n\mathbf{Z}$ *for all* i *if* v *splits over* ℓ, *where* $x \in \ell^\times$ *if* G *is of type* ${}^2D_r$ *with* r *even, and* $x \in L$ *otherwise and* $\xi_v(x) = 1$.

(If G is not of type ${}^2D_r$ with r even, then the image of $H^1(k, C)$ in $\ell^\times/\ell^{\times n}$ is $L/\ell^{\times n}$, see 5.3. At any nonarchimedean place v such that G splits over $\widehat{k}_v$, ξ_v induces a homomorphism of L into Ξ_v ($\subset \mathrm{Aut}\,\Delta_v$) which we have also denoted by ξ_v.)

Proof. — If v does not split over ℓ, and ℓ_v is a ramified extension of k_v, then for $x \in \ell^\times, \tilde{v}(x) = v(\mathrm{N}_{\ell/k}(x))$; if ℓ_v is an unramified extension of k_v, then $\tilde{v}(x) = v(\mathrm{N}_{\ell/k}(x))/[\ell : k]$, so for $x \in \mathrm{L}$, it is an integral multiple of $n/[\ell : k]$. Now assertion (i) of the lemma is obvious. Note that if $[\ell : k]$ is odd, then G is a triality form of type D_4 and $n = 2$.

The second assertion of the lemma follows from 2.3 and 2.7.

5.5. Let $\mathscr{R}$ (resp. T) be the set of places $v \notin \mathrm{S}$ such that G does not split over $\hat{k}_v$ (resp. splits over $\hat{k}_v$ but is not quasi-split over k_v). Both $\mathscr{R}$ and T are finite. Let S_f^0 be the subset of S_f consisting of all places v such that either v splits over ℓ or ℓ_v is an unramified extension of k_v.

Let ℓ_n be the subgroup of $\ell^\times$ consisting of the elements x such that $\tilde{v}(x) \in n\mathbf{Z}$ for every normalized nonarchimedean valuation $\tilde{v}$ of ℓ. Then 5.4 implies that if G/k is not of type ${}^2\mathsf{D}_r$ with r even, then the subgroup $\mathrm{H}^1(k, \mathrm{C})_\xi$ mapping into $\ell_n/\ell^{\times n}$ has index $\leqslant n^{\#(\mathrm{S}_f^0 \cup \mathrm{T})}$ if G is not a triality form, and index $\leqslant 2^{\#\mathscr{R} + 2\#(\mathrm{S}_f^0 \cup \mathrm{T})}$ if G is a triality form. It also implies that if G is of type ${}^2\mathsf{D}_r$ with r even, then $\mathrm{H}^1(k, \mathrm{C})_\xi \cap (\ell_2/\ell^{\times 2})$ is a subgroup of $\mathrm{H}^1(k, \mathrm{C})_\xi$ of index at most $2^{\#\mathscr{R} + 2\#(\mathrm{S}_f^0 \cup \mathrm{T})}$.

By 0.12, the order of $\ell_n/\ell^{\times n}$ is $\leqslant h_\ell\, n^{a(\ell)}$. Moreover, $2^{\#\mathscr{R}} \leqslant \mathrm{D}_\ell/\mathrm{D}_k^{[\ell:k]}$, see [31: Appendix], and as we saw in 5.3, if G is not of type ${}^2\mathsf{D}_r$ with r even, the kernel $\mu_n(k)/\mathrm{N}_{\ell/k}(\mu_n(\ell))$ of the homomorphism $\mathrm{H}^1(k, \mathrm{C}) \to \ell^\times/\ell^{\times n}$ is trivial if G is a triality form and is of order at most 2 in all other cases. Combining all this information, we get:

5.6. *Proposition.* — *Assume that* G *is an outer form (of a split group). Then*

(i) *If* G *is of type* D_r *with* r *even (including the triality forms of type* D_4*),*

$$\#\mathrm{H}^1(k, \mathrm{C})_\xi \leqslant h_\ell\, 2^{a(\ell) + 2\#(\mathrm{S}_f^0 \cup \mathrm{T})}\, \mathrm{D}_\ell/\mathrm{D}_k^{[\ell:k]}.$$

(ii) *In all the other cases,*

$$\#\mathrm{H}^1(k, \mathrm{C})_\xi \leqslant 2h_\ell\, n^{a(\ell) + \#(\mathrm{S}_f^0 \cup \mathrm{T})}.$$

6. A number theoretic result

In this section we shall assume that k is a number field and prove the following proposition, which is needed for the proof of the finiteness theorems in §7.

Let ε, n be as in 2.6 and $m_1 \leqslant \ldots \leqslant m_r$ be the exponents of G (3.7). Recall that $n^\varepsilon \leqslant r + 1$ and $\varepsilon \leqslant 2$. As before, $a(k)$ will denote the number of archimedean places of k.

6.1. *Proposition.* — *Given a positive real number* c *and a nonnegative integer* a, *there exist effectively computable positive integers* $\mathfrak{m}_c$, $\mathfrak{m}_{c,a}$ *and* $\mathfrak{n}_c$, $\mathfrak{n}_{c,a}$ *such that*

(i) if either $r > \mathfrak{m}_c$ *or* $\mathrm{D}_k > \mathfrak{n}_c$, *then*

$$\text{(i)} \qquad \mathrm{D}_k^{\frac{1}{2}\dim \mathrm{G}} \left(\prod_{i=1}^{r} \frac{m_i!}{(2\pi)^{m_i+1}} \right)^{[k:\mathbf{Q}]} > c;$$

(ii) if G is an inner k-form (of a split group) of type other than A_1 and A_2 and either $r > \mathfrak{m}_{e,a}$ or $\mathrm{D}_k > \mathfrak{n}_{e,a}$, then

(ii) $$n^{-2\varepsilon a(k)} h_k^{-\varepsilon} \mathrm{D}_k^{\frac{1}{2}\dim G} \left(\prod_{i=1}^{r} \frac{m_i!}{(2\pi)^{m_i+1}}\right)^{[k:\mathbf{Q}]} > cn^{\varepsilon a};$$

(iii) if G/k is an outer form of type other than A_2 and either $r > \mathfrak{m}_{e,a}$ or $\mathrm{D}_k > \mathfrak{n}_{e,a}$, then

(iii) $$n^{-([\ell:k]+\varepsilon)a(k)} h_\ell^{-1} \mathrm{D}_k^{\frac{1}{2}\dim G} (\mathrm{D}_\ell/\mathrm{D}_k^{[\ell:k]}) \left(\prod_{i=1}^{r} \frac{m_i!}{(2\pi)^{m_i+1}}\right)^{[k:\mathbf{Q}]} > cn^{\varepsilon a};$$

(iv) if $\mathrm{D}_k > \mathfrak{n}_{e,a}$, then

(iv) $$(2^4 \pi^5)^{-[k:\mathbf{Q}]} 3^{-2a(k)} h_k^{-1} \mathrm{D}_k^4 > 3^a c.$$

(v) if k is totally real and $\mathrm{D}_k > \mathfrak{n}_{e,a}$, then

(v) $$(2^3 \pi^2)^{-[k:\mathbf{Q}]} h_k^{-1} \mathrm{D}_k^{3/2} > 2^a c.$$

(vi) There is a positive integer $\mathfrak{n}'_{e,a}$ such that if $\mathrm{D}_\ell > \mathfrak{n}'_{e,a}$, then

(vi) $$(2^4 \pi^5)^{-[k:\mathbf{Q}]} 3^{-a(k)-a(\ell)} h_\ell^{-1} \mathrm{D}_k^4 (\mathrm{D}_\ell/\mathrm{D}_k^2)^2 > 3^a c.$$

Proof. — In the proof of assertions *(v)* and *(vi)* of this proposition we shall use some ideas of [10].

For a number field K, let D_K, h_K, R_K be respectively the absolute value of its discriminant, its class number and regulator. Let $\zeta_\mathrm{K}(s)$ $(= \prod_\mathfrak{p}(1-(\mathrm{N}\mathfrak{p})^{-s})^{-1})$ be its Dedekind zeta-function. Recall that $\zeta_\mathrm{K}(s)$ has a simple pole at $s=1$ and the residue is $2^{r_1(\mathrm{K})}(2\pi)^{r_2(\mathrm{K})} h_\mathrm{K} \mathrm{R}_\mathrm{K}/w_\mathrm{K} \mathrm{D}_\mathrm{K}^{1/2}$, where $r_1(\mathrm{K})$ (resp. $r_2(\mathrm{K})$) is the number of real (resp. complex) places of K and w_K is the order of the finite group of roots of unity in K. Let

$$\mathrm{Z}_\mathrm{K}(s) = -\zeta'_\mathrm{K}(s)/\zeta_\mathrm{K}(s) = \sum_\mathfrak{p} \log(\mathrm{N}\mathfrak{p})/((\mathrm{N}\mathfrak{p})^s - 1)$$

be the negative of the logarithmic derivative of $\zeta_\mathrm{K}(s)$.

According to the Brauer-Siegel theorem ([35: Hilfssatz 1]), for all real $s > 1$,

(1) $$h_\mathrm{K} \mathrm{R}_\mathrm{K} \leqslant w_\mathrm{K} s(s-1) 2^{-r_1(\mathrm{K})} \Gamma\left(\frac{s}{2}\right)^{r_1(\mathrm{K})} \Gamma(s)^{r_2(\mathrm{K})} (2^{-2r_2(\mathrm{K})} \pi^{-[\mathrm{K}:\mathbf{Q}]} \mathrm{D}_\mathrm{K})^{s/2} \zeta_\mathrm{K}(s).$$

R. Zimmert [47] has given the following lower bound for the regulator:

(2) $$\begin{aligned} \mathrm{R}_\mathrm{K} &\geqslant .02 w_\mathrm{K} \exp(.46 r_1(\mathrm{K}) + .1 r_2(\mathrm{K})) \\ &\geqslant .02 w_\mathrm{K} \exp(.1 a(\mathrm{K})), \end{aligned}$$

where $a(\mathrm{K}) = r_1(\mathrm{K}) + r_2(\mathrm{K})$ is the number of archimedean places of K.

A. Odlyzko ([27: Theorem 1]; see also [29]) has provided the following lower bound for D_K:

(3) $$\text{If } [\mathrm{K}:\mathbf{Q}] > 10^5, \text{ then } \mathrm{D}_\mathrm{K} \geqslant (55)^{r_1(\mathrm{K})} (21)^{2r_2(\mathrm{K})}.$$

Moreover, it follows from his results that there exist absolute positive constants (i.e. constants not depending on K) $c_1, c_2\ (\leqslant 2)$ such that for all $s \in (1, 1 + c_2)$

$$\text{(4)} \qquad D_K \geqslant (55)^{r_1(K)} (21)^{2r_2(K)} \exp(2Z_K(s) - 2(s-1)^{-1} - c_1).$$

Since the absolute value of the logarithmic derivative of the Gamma-function is bounded above in the interval $\left[\frac{1}{2}, 2\right]$, there exists a constant c_3 such that, for $1 \leqslant s \leqslant 2$,

$$\text{(5)} \qquad \begin{cases} \Gamma\left(\frac{s}{2}\right) \leqslant \pi^{1/2} \exp(c_3(s-1)) \\ \Gamma(s) \quad \leqslant \exp(c_3(s-1)). \end{cases}$$

Also, it follows at once from [26: Lemma 2] that there is an absolute constant c_4 such that for all $s > 1$,

$$\text{(6)} \qquad \zeta_K(s) \leqslant \exp(Z_K(s) + c_4(s-1)\, a(K)).$$

Taking $s = 2$ in (1) and using (2) and (3) we obtain

$$\text{(7)} \qquad h_K \leqslant 10^2 \left(\frac{\pi}{12}\right)^{[K:Q]} D_K$$

$$\left(\text{as } \zeta_K(2) \leqslant (\zeta_Q(2))^{[K:Q]} = \left(\frac{\pi^2}{6}\right)^{[K:Q]}\right).$$

We shall now prove the assertions *(i)*, *(ii)* and *(iii)* of the proposition. We begin by recalling that for at most one i, $m_i = m_{i+1}$ (and if $m_i = m_{i+1}$ for some i, then G is of type $\mathbf{D}_r$ with r even) and $m_r \to \infty$ with $r \to \infty$; see [31: 1.5]. From this it is clear that, as $m! \gg (2\pi)^{m+1}$ for all $m \gg 0$, there exist positive integers $\mathfrak{m}_c \leqslant \mathfrak{m}_{c,a}$ such that if $r \geqslant \mathfrak{m}_c$ (resp. $r \geqslant \mathfrak{m}_{c,a}$), then

$$\prod_{i=1}^{r} \frac{m_i!}{(2\pi)^{m_i+1}} > c + 1 \qquad \left(\text{resp. } (r+1)^{-(5+2a)} \prod_{i=1}^{r} \frac{m_i!}{(2\pi)^{m_i+1}} > 10^2(c+1)\right).$$

Now as D_k is a positive integer, inequality (i) holds if $r \geqslant \mathfrak{m}_c$. Also as

$$\dim G > 2 \max(\varepsilon, [\ell : k]),$$

$n^{\varepsilon} \leqslant r + 1$ and both $\varepsilon a + 2\varepsilon a(k)$ and $\varepsilon a + ([\ell : k] + \varepsilon)\, a(k)$ are $\leqslant (5 + 2a)\,[k : \mathbf{Q}]$, the inequalities (ii) and (iii) evidently hold for $r \geqslant \mathfrak{m}_{c,a}$ in view of the bound for the class number given by (7). Let us now assume that $2 \leqslant r < \mathfrak{m}_{c,a}$. We observe that, if $r \geqslant 2$ and G is not an inner or outer form of type $\mathbf{A}_2$, then

$$n^{-2\varepsilon} \prod_{i=1}^{r} m_i! \geqslant \frac{3}{4}, \qquad n^{-[\ell:k]-\varepsilon} \prod_{i=1}^{r} m_i! \geqslant \frac{3}{16},$$

and

$$n^{-\varepsilon} \prod_{i=1}^{r} m_i! > 1, \qquad n^{-([\ell:k]+\varepsilon)/2} \prod_{i=1}^{r} m_i! > 1.$$

Now recall that $\dim G = r + 2\Sigma_{i=1}^{r} m_i$. Using this it is easy to see that if G is not of type A_1 or A_2, then

$$r + \sum_{i=1}^{r} m_i \leqslant \frac{18}{11}\left(\frac{1}{2}\dim G - \max(\varepsilon, [\ell : k])\right).$$

Also $r + \Sigma m_i \geqslant 6$. Let $t_{\varepsilon, a}$ be the smallest positive integer such that

$$((2\pi)^{-18/11}.21)^{t_{\varepsilon,a}} > (10^4\, \mathfrak{m}^a_{\varepsilon, a}\, c)^{3/11},$$

then using (3) and (7), for $K = k$ and ℓ, we conclude that if $[k : \mathbf{Q}] \geqslant \max(t_{\varepsilon, a}, 10^5)$, then (ii) (hence also (i)) and (iii) hold. On the other hand, using (7) it is seen that there is a positive integer $u_{\varepsilon, a}$ such that if $[k : \mathbf{Q}] \leqslant \max(t_{\varepsilon, a}, 10^5)$ and $r \leqslant \mathfrak{m}_{\varepsilon, a}$, then for $D_k > u_{\varepsilon, a}$, *(i)*, *(ii)* and *(iii)* hold. Let $\mathfrak{n}_\varepsilon = u_{\varepsilon, 0}$. We shall later choose an integer $\mathfrak{n}_{\varepsilon, a} \geqslant u_{\varepsilon, a}$.

We shall now prove that, for all sufficiently large D_k, the inequality (iv) holds. For this we note that using (7) (for $K = k$), we have

$$(2^4\, \pi^5)^{-[k:\mathbf{Q}]}\, 3^{-2a(k)}\, h_k^{-1}\, D_k^4 \geqslant 10^{-2}(((2^2.3)^{1/3}\, \pi^2)^{-r_1(k)}(2^{2/3}\, \pi^2)^{-2r_2(k)}\, D_k)^3,$$

so if $[k : \mathbf{Q}] > 10^5$, in view of (3),

$$\begin{aligned}(2^4\, \pi^5)^{-[k:\mathbf{Q}]}\, 3^{-2a(k)}\, h_k^{-1}\, D_k^4 &\geqslant 10^{-2}\left(\left(\frac{55}{(2^2.3)^{1/3}\,\pi^2}\right)^{r_1(k)}\left(\frac{21}{2^{2/3}\,\pi^2}\right)^{2r_2(k)}\right)^3 \\ &> 10^{-2}((2.4)^{r_1(k)}\,(1.3)^{2r_2})^3 \\ &\geqslant 10^{-2}(1.3)^{3[k:\mathbf{Q}]},\end{aligned}$$

which implies that there is a positive integer $n_{\varepsilon, a} > 10^5$ such that for $[k : \mathbf{Q}] > n_{\varepsilon, a}$,

$$(2^4\, \pi^5)^{-[k:\mathbf{Q}]}\, 3^{-2a(k)}\, h_k^{-1}\, D_k^4 > 3^a\, c.$$

It is obvious that we can find a positive integer $u'_{\varepsilon, a}$ such that the inequality

$$(((2^2.3)^{1/3}\, \pi^2)^{-r_1(k)}\, (2^{2/3}\, \pi^2)^{-2r_2(k)}\, D_k)^3 > 10^2.3^a\, c$$

holds for all k with $[k : \mathbf{Q}] \leqslant n_{\varepsilon, a}$ and $D_k > u'_{\varepsilon, a}$. Hence for all k with $D_k > u'_{\varepsilon, a}$, the inequality (iv) holds.

We shall now prove that there is a positive integer $u''_{\varepsilon, a}$ such that if $D_k > u''_{\varepsilon, a}$, (v) holds.

(1) and (2) for $K = k$ give us the following (recall that in *(v)*, k is assumed to be totally real):

$$h_k \leqslant 50s(s-1)\, 2^{-r_1(k)}\, \Gamma\left(\frac{s}{2}\right)^{r_1(k)} (\pi^{-[k:\mathbf{Q}]}\, D_k)^{s/2}\, \zeta_k(s)\exp(-\,.1a(k)).$$

This, along with (4), (5) and (6) imply that if $[k : \mathbf{Q}] > 10^5$,

$$\begin{aligned}(2^3\, \pi^2)^{-r_1(k)}\, h_k^{-1}\, D_k^{3/2} \geqslant\; &.02(s(s-1))^{-1}\left(\frac{(55)^{(3-s)/2}}{2^2\, \pi^{(5-s)/2}}\right)^{r_1(k)} \\ &.\exp\Big((2-s)\, Z_k(s) - \frac{1}{2}c_1(3-s) - (3-s)\,(s-1)^{-1} \\ &\quad + (.1 - (c_3 + c_4)\,(s-1))\, r_1(k)\Big).\end{aligned}$$

Now observe that $55/2^2\pi^2 > 1.3$, and $\exp((2-s)\,Z_k(s)) \geqslant 1$ if $s < 2$. So by choosing $s(>1)$ sufficiently close to 1, the above gives the following bound:

There is an absolute constant c_5, such that

(8) $$(2^3\pi^2)^{-r_1(k)}\,h_k^{-1}\,D_k^{3/2} \geqslant c_5(1.3)^{r_1(k)} \quad \text{for all } k \text{ with } [k:\mathbf{Q}] > 10^5.$$

On the other hand, using (7) we find that

$$(2^3\pi^2)^{-r_1(k)}\,h_k^{-1}\,D_k^{3/2} \geqslant 10^{-2}(2\pi^3/3)^{-r_1(k)}\,D_k^{1/2}.$$

From this and (8) it is obvious that there is a positive integer $u''_{c,a}$ such that for all k with $D_k > u''_{c;a}$, the inequality (v) holds.

Take $\mathfrak{n}_{c,a} = \max(u_{c,a}, u'_{c,a}, u''_{c,a})$.

We now finally prove that there exists a positive integer $\mathfrak{n}'_{c,a}$ such that if $D_\ell > \mathfrak{n}'_{c,a}$, then (vi) holds. Since

$$a(k) \leqslant [k:\mathbf{Q}] = \frac{1}{2}(r_1(\ell) + 2r_2(\ell)),$$

it suffices to prove that there is a positive integer $\mathfrak{n}'_{c,a}$ such that if $D_\ell > \mathfrak{n}'_{c,a}$, then

$$(2^4\pi^5)^{-[\ell:\mathbf{Q}]/2}\,3^{-3r_1(\ell)/2-2r_2(\ell)}\,h_\ell^{-1}\,D_k^4(D_\ell/D_k^2)^2$$
$$= (2^2.3^{3/2}.\pi^{5/2})^{-r_1(\ell)}\,(2^4.3^2.\pi^5)^{-r_2(\ell)}\,h_\ell^{-1}\,D_\ell^2 > 3^a c.$$

Now using (1), (2), (4), (5) and (6) for $K = \ell$, we conclude that if $[\ell:\mathbf{Q}) > 10^5$, then

$$(2^2.3^{3/2}.\pi^{5/2})^{-r_1(\ell)}\,(2^4.3^2.\pi^5)^{-r_2(\ell)}\,h_\ell^{-1}\,D_\ell^2$$
$$\geqslant .02(s(s-1))^{-1}\left(\frac{(55)^{(4-s)/2}}{2.3^{3/2}.\pi^{(6-s)/2}}\right)^{r_1(\ell)}\left(\frac{(21)^{(4-s)}}{2^{(4-s)}.3^2.\pi^{(5-s)}}\right)^{r_2(\ell)}$$
$$.\exp\left((3-s)\,Z_\ell(s) - \frac{1}{2}(4-s)\,c_1 - (4-s)\,(s-1)^{-1} + (.1-(c_3+c_4)\,(s-1))\,a(\ell)\right)$$

Now as

$$(55)^{3/2}/2.3^{3/2}.\pi^{5/2} > 2.2, \qquad (21)^3/2^3.3^2.\pi^4 > 1.3,$$

and $$\exp((3-s)\,Z_\ell(s)) \geqslant 1 \quad \text{if } s < 2,$$

by choosing $s(>1)$ sufficiently close to 1, we infer that there is an absolute constant c_6 such that

(9) $$(2^2.3^{3/2}.\pi^{5/2})^{-r_1(\ell)}\,(2^4.3^2.\pi^5)^{-r_2(\ell)}\,h_\ell^{-1}\,D_\ell^2 \geqslant (2.2)^{r_1(\ell)}\,(1.3)^{r_2(\ell)}\,c_6,$$

for all ℓ with $[\ell:\mathbf{Q}] > 10^5$. Also, using (7) for $K = \ell$, we find that

(10) $$(2^2.3^{3/2}.\pi^{5/2})^{-r_1(\ell)}\,(2^4.3^2.\pi^5)^{-r_2(\ell)}\,h_\ell^{-1}\,D_\ell^2 \geqslant 10^{-2}(3^{1/2}.\pi^{7/2})^{-r_1(\ell)}\,\pi^{-7r_2(\ell)}\,D_\ell.$$

From (9) and (10) it is clear that there exists a positive integer $\mathfrak{n}'_{c,a}$ such that for all k and ℓ with $D_\ell > \mathfrak{n}'_c$, the inequality (vi) holds.

19

7. The finiteness theorems

This section is devoted to the proof of the main results of this paper (Theorems 7.2, 7.3, 7.8 and 7.11).

7.1. Let $\mathscr{C}$ be a set of pairs (k, G) consisting of a global field k and an absolutely almost simple, simply connected algebraic group G defined over k such that (i) there is a non-zero lower bound τ for the Tamagawa numbers $\tau_k(G)$ for $(k, G) \in \mathscr{C}$ and (ii) if k is a global function field of genus zero, then G is anisotropic over k i.e. k-rank $G = 0$. We recall here that over a global function field, any absolutely almost simple anisotropic group is necessarily an inner or outer form of type A ([15: §3, Korollar 1]).

It was conjectured by A. Weil that the Tamagawa number of any simply connected semi-simple group, defined over an arbitrary global field, is 1. This conjecture has recently been proved over number fields ([18]; see also [31: 3.3]). The Tamagawa number of any simply connected group of inner type A over an arbitrary global function field is known to be 1 (see [46]). However, whether this is the case in general over a global function field is not yet known.

In view of the above results, we may assume $\mathscr{C}$ to contain all pairs (k, G) such that either k is a number field and G is arbitrary, or k is a global function field and G is of inner type A.

7.2. *Theorem. — Let c be a positive integer and let $\mathscr{C}_c$ be the subset of $\mathscr{C}$ consisting of the pairs (k, G) such that* (i) *if k is a global function field, its genus is* > 0; (ii) G *is anisotropic over k and* $G_\infty := \prod_{v \in V_\infty} G(k)$ *is compact*; (iii) *the class number*

$$\mathfrak{c}(P) := \#(G_\infty \prod_{v \in V_f} P_v \backslash G(A)/G(k))$$

of G/k *with respect to some coherent collection of parahoric subgroups* $(P_v)_{v \in V_f}$ *is* $\leqslant c$. *Then (up to natural equivalence)* $\mathscr{C}_c$ *is finite.*

We recall that a collection $P = (P_v)_{v \in V_f}$ of parahoric subgroups P_v of $G(k_v)$ is said to be coherent if $\prod_{v \in V_\infty} G(k_v) \cdot \prod_{v \in V_f} P_v$ is an open subgroup of the adèle group $G(A)$.

7.3. *Theorem. — Let c be a positive real number and $\mathscr{V}_c$ be the set of triples* (k, G, S) *such that* (i) $(k, G) \in \mathscr{C}$ *and the absolute rank of* G *is at least* 2 *(i.e.* G *is not a form of* SL_2*)*, (ii) S *is a finite set of places of k containing all the archimedean ones so that for all nonarchimedean* $v \in S$, G *is isotropic at v and the subset* S(G) *of* S *consisting of the places where* G *is isotropic is nonempty*, (iii) *there is a k-group G' which is centrally k-isogenous to* G *and an arithmetic subgroup* Γ' *of* $G'_{S(G)}$ *such that either* $\mu'_{S(G)}(G'_{S(G)}/\Gamma') < c$, *or* Γ' *is virtually free**, $0 \neq |\chi(\Gamma')| < c$ *and* G *is not of type* A_2, *where* $\mu'_{S(G)}$ *is as in* 3.6 *and* $\chi(\Gamma')$ *is the Euler-Poincaré characteristic of* Γ' *in the sense of C. T. C. Wall. Then (up to natural equivalence)* $\mathscr{V}_c$ *is finite.*

We shall prove these theorems together.

---

* Equivalently, G is anisotropic over k if the latter is a number field.

7.4. Let $(k, \mathrm{G}) \in \mathscr{C}$. As before, let r denote the absolute rank of G and r_v its rank over the maximal unramified extension $\hat{k}_v$ of k_v if $v \in \mathrm{V}_f$. Let ℓ, $\mathscr{G}$, and $\mathfrak{s}(\mathscr{G})$ be as in 3.7 and let $\overline{\mathscr{G}}$ be the adjoint group of $\mathscr{G}$.

a) Let $(k, \mathrm{G}) \in \mathscr{C}_c$. Then since G_∞ is assumed to be compact, if k is a number field, it is totally real. Let $\mathrm{P} = (\mathrm{P}_v)_{v \in \mathrm{V}_f}$ be a coherent collection of parahoric subgroups such that the class number $\mathfrak{c}(\mathrm{P})$ of G with respect to P is $\leqslant c$. It follows from [31: 4.3, 2.10 and 2.11] that

(1) $$c \geqslant \mathfrak{c}(\mathrm{P}) \geqslant \mathrm{C}(\mathscr{G}/k)\tau\zeta(\mathrm{P}),$$

where

(2) $$\mathrm{C}(\mathscr{G}/k) = \mathrm{D}_k^{\frac{1}{2}\dim\mathscr{G}}(\mathrm{D}_\ell/\mathrm{D}_k^{[\ell:k]})^{\frac{1}{2}\mathfrak{s}(\mathscr{G})} \prod_{v \in \mathrm{V}_\infty} \left| \prod_{i=1}^{r} \frac{m_i!}{(2\pi)^{m_i+1}} \right|_v$$

and $\zeta(\mathrm{P}) = \Pi_{v \in \mathrm{V}_f} e(\mathrm{P}_v)$, with

(3) $$e(\mathrm{P}_v) = q_v^{(\dim \overline{\mathrm{M}}_v + \dim \overline{\mathscr{M}}_v)/2} . (\# \overline{\mathrm{M}}_v(\mathfrak{f}_v))^{-1} > 1 \quad (v \in \mathrm{V}_f).$$

(The unexplained notation is as in [31].) We have

(4) $$e(\mathrm{P}_v) \geqslant (q_v + 1)^{-1}\, q_v^{r_v+1}$$

if either G is not quasi-split over k_v, or P_v is not special, or G splits over $\hat{k}_v$ but P_v is not hyperspecial (3.7 (5)), and

(5) $$e(\mathrm{P}_v) \geqslant (q_v - 1)\, q_v^{(r^2+2r-(r+1)^2 d_v^{-1}-1)/2}$$

if $\mathrm{G}(k_v) = \mathrm{SL}_{(r+1)/d_v}(\mathfrak{D}_v)$, where $\mathfrak{D}_v$ is a central division algebra of degree d_v over k_v (3.7 (6)).

(1) and (3) yield

(6) $$\mathrm{C}(\mathscr{G}/k) \leqslant c/\tau$$

or, more generally,

(6)′ $$\mathrm{C}(\mathscr{G}/k) \prod_{v \in \mathscr{R}} e(\mathrm{P}_v) \leqslant c/\tau \quad (\mathscr{R} \subset \mathrm{V}_f).$$

b) If $(k, \mathrm{G}, \mathrm{S}) \in \mathscr{V}_c$, then from the result stated in 3.7, 3.8 and the bounds obtained in §§4, 5, we get that

(7) $$\text{either } c \geqslant \mathrm{B}(\mathscr{G}/k)\tau\mathscr{F} \quad \text{or} \quad c \geqslant \mathrm{B}(\mathscr{G}/k)\tau\mathscr{F}^{\mathrm{EP}},$$

where

(8) $$\mathrm{B}(\mathscr{G}/k) = 2^{-1} n^{-\varepsilon a(k) - \varepsilon' a(\ell)} h_\ell^{-\varepsilon'} \mathrm{D}_k^{\frac{1}{2}\dim\mathscr{G}}(\mathrm{D}_\ell/\mathrm{D}_k^{[\ell:k]})^{\mathfrak{s}'(\mathscr{G})} \prod_{v \in \mathrm{V}_\infty} \left| \prod_{i=1}^{r} \frac{m_i!}{(2\pi)^{m_i+1}} \right|_v .$$

The constants n, ε are as in 2.6, $\varepsilon' = \varepsilon$ if $\mathscr{G}$ is k-split, $\varepsilon' = 1$ otherwise, and

$$\mathfrak{s}'(\mathscr{G}) = \begin{cases} s(\mathscr{G})/2 - 1 & \text{if } \mathscr{G}/k \text{ is an outer form of type } \mathsf{D}_r,\ r \text{ even}, \\ s(\mathscr{G})/2 & \text{otherwise}, \end{cases}$$

(9) $$\mathscr{F} = \prod_{v \in \mathrm{V}_f} f_v, \qquad \mathscr{F}^{\mathrm{EP}} = \prod_{v \in \mathrm{V}_f} f_v^{\mathrm{EP}},$$

with

(10) $$f_v = e_v n^{-s}, \qquad f_v^{EP} = e_v n^{-s} \mid W_v(\mathbf{q}^{-1}) \mid^{-1} = f_v \mid W_v(\mathbf{q}^{-1}) \mid^{-1}$$

if $v \in S_f$ and $\ell_v = \ell \otimes_k k_v$ is a ramified field extension of k_v,

(11) $$f_v = e_v n^{-2s}, \qquad f_v^{EP} = e_v n^{-2s} \mid W_v(\mathbf{q}^{-1}) \mid^{-1} = f_v \mid W_v(\mathbf{q}^{-1}) \mid^{-1}$$

otherwise ($v \in S_f$),

(12) $$f_v = f_v^{EP} = e_v^{\mathfrak{m}} n^{-s}$$

if $v \in T(G)$, where $T(G)$ is the set of places $v \notin S$ such that G splits over $\hat{k}_v$ but is not quasi-split over k_v, and finally

(13) $$f_v = f_v^{EP} = e_v^{\mathfrak{m}} \quad \text{if} \quad v \notin S \cup T(G).$$

(The e_v and $e_v^{\mathfrak{m}}$'s are as in 3.7). Also recall from 4.4 that

(14) $$\mid W_v(\mathbf{q}^{-1}) \mid^{-1} > ((q_v - 1)^2 (q_v^2 + 1)^{-1})^{r_v} (\geqslant 5^{-r_v}) \quad (v \in V_f).$$

Now we claim that, for $v \in V_f$,

(15) $$f_v > 1,$$
$$f_v^{EP} > 1, \quad \text{unless G is of type } \mathsf{A}_2 \text{ and } q_v \leqslant 3.$$

If $r_v > 4$, this already follows from the previous inequalities. It will be checked in all cases in Appendix C.

We get then from (7) and (15)

(16) $$B(\mathscr{G}/k) < c/\tau$$

or again, more generally, for any subset $\mathscr{R}$ of V_f

(16)′ $$B(\mathscr{G}/k) \prod_{v \in \mathscr{R}} f_v < c/\tau, \qquad B(\mathscr{G}/k) \prod_{v \in \mathscr{R}} f_v^{EP} < c/\tau.$$

c) Next we remark that $D_\ell/D_k^{[\ell:k]} \geqslant 1$. This follows, e.g., from Theorem A in the Appendix of [31].

d) Let now k be a number field. Then we deduce from 6.1 (for $a = 0$) the existence of integers $\mathfrak{m}_0$, $\mathfrak{n}_0$ and $\mathfrak{n}_0'$ such that

(17) $$C(\mathscr{G}/k) \geqslant c \quad (k \text{ is assumed to be totally real when } r = 1),$$
$$B(\mathscr{G}/k) \geqslant c \quad (r \geqslant 2),$$

if either $r > \mathfrak{m}_0$ or $D_k > \mathfrak{n}_0$ or $D_\ell > \mathfrak{n}_0'$.

e) Assume now k to be a function field. We want to prove a similar assertion. If G is of type A, which is necessarily the case if G is anisotropic over k by [15: §3, Kor. 1], we let $A(G)$ denote the set of places v of k where G is a non-split inner form of type A. We now claim that there exist positive integers $\mathfrak{g}_0$, $\mathfrak{g}_0'$, $\mathfrak{m}_0$ and $\mathfrak{q}_0$ such that:

(i) *If k is a global function field of genus > 1 and either $g_k > \mathfrak{g}_0$ or $g_\ell > \mathfrak{g}_0'$, or the absolute rank of $\mathscr{G}$ is greater than $\mathfrak{m}_0$, or $q_\ell > \mathfrak{q}_0$, then $C(\mathscr{G}/k) \geqslant B(\mathscr{G}/k) > c/\tau$.*

(ii) *If k is a global function field of genus 1 and* G *is anisotropic over k, then* $C(\mathcal{G}/k).\Pi_{v\in A(G)}\, e(P_v) > c/\tau$ *if either* $g_\ell > \mathfrak{g}'_c$, *or the absolute rank of* G *is greater than* $\mathfrak{m}_c$, *or* $q_\ell > \mathfrak{q}_c$.

(iii) *If k is a global function field of genus 1, then both* $B(\mathcal{G}/k).\Pi_{v\in S} f_v$ *and* $B(\mathcal{G}/k).\Pi_{v\in S} f_v^{EP}$ *are greater than* c/τ *if either* $g_\ell > \mathfrak{g}'_c$, *or the absolute rank of* G *is greater than* $\mathfrak{m}_c$, *or* $q_\ell > \mathfrak{q}_c$.

(iv) *If k is a global function field of genus zero, and* G *is anisotropic over k, then both* $B(\mathcal{G}/k).\Pi_{v\in S\cup A(G)} f_v$ *and* $B(\mathcal{G}/k).\Pi_{v\in S\cup A(G)} f_v^{EP}$ *are greater than* c/τ *if either* $g_\ell > \mathfrak{g}'_c$, *or the absolute rank of* G *is greater than* $\mathfrak{m}_c$, *or* $q_\ell > \mathfrak{q}_c$.

(If the genus of k is $\leqslant 1$, then $D_k \leqslant 1$ and (6), (16) do not allow one to limit dim G and therefore r. But the point of (ii), (iii) and (iv) is to show that we can compensate for that by multiplying $B(\mathcal{G}/k)$ or $C(\mathcal{G}/k)$ by some of the factors $e(P_v)$ or f_v, f_v^{EP}, which is allowed in view of (6)′, (16)′.)

(i) and (iii) follow easily, we only need to use the upper bound for the class number given in 0.8 (1) and the estimate for e_v provided by 3.7 (2).

We already pointed out that in (ii), (iv), G is a form of type A_r. If it is an inner one, then there is a central division algebra $\mathfrak{D}$ of degree $r+1$ over k such that $G = SL_1(\mathfrak{D})$. It is well-known from class field theory that if d_v is the order of $\mathfrak{D}\otimes_k k_v$ in the Brauer group, then $d_v = 1$ for all but finitely many v's, $r+1$ is the least common multiple of the d_v's and the local invariants m_v/d_v of $\mathfrak{D}$, where m_v is an integer prime to d_v, add up to zero mod 1. This implies that one of the following three conditions is fulfilled:

(•) The number of places where $d_v = r+1$, i.e. where $\mathfrak{D}_v = \mathfrak{D}\otimes_k k_v$ is a division algebra, or, equivalently, where G is anisotropic, is at least two.

(••) $r \geqslant 5$. There is exactly one place where $d_v = r+1$, at least another one where $d_v \geqslant 2$ and a third one where $d_v \geqslant 3$.

(•••) $r \geqslant 5$. There are as least one place where $d_v \geqslant 2$ and two other places where $d_v \geqslant 3$.

If G is an outer form (of type A_r), then there exists a central division algebra $\mathcal{D}$ over a separable quadratic extension ℓ of k and an involution σ of $\mathcal{D}$, of the second kind, such that $G(k) = \{ d \in \mathcal{D}^\times \mid d\sigma(d) = 1 \text{ and } \mathrm{Nrd}(d) = 1 \}$. The local invariant of $\mathcal{D}$ at any place of ℓ which is fixed under the Galois conjugation of ℓ/k is zero. On the other hand, the sum of the local invariants of $\mathcal{D}$ at any two conjugate places of ℓ is zero. This implies that one of the following conditions is fulfilled:

(•) There is a place v of k where G/k_v is an anisotropic inner form of type A_r, i.e. $G(k_v) = SL_1(\mathfrak{D}_v)$, where $\mathfrak{D}_v$ is a central division algebra of degree $r+1$ over k_v.

(••) There are two places v_1, v_2 of k which split over ℓ, such that $G(k_{v_i}) = SL_{(r+1)/d_i}(\mathfrak{D}_i)$, where $\mathfrak{D}_i$ is a central division algebra of degree d_i over k_{v_i} and $d_1 \geqslant 2$, $d_2 \geqslant 3$.

The assertions (ii) and (iv) can now be proved using the estimates for e_v given by 3.7 (2), (3), (6), and the upper bound for the class number given in 0.8 (1). Note that for $v \in S_f$ if $G(k_v) = SL_{(r+1)/d_v}(\mathfrak{D}_v)$, where $\mathfrak{D}_v$ is a central division algebra of degree d_v over k_v, then

$$f_v^{\mathrm{EP}} = (r+1)^{-2}\, q_v^{r(r+3)/2}(q_v - 1)\,(q_v^{r+1} - 1)^{-1}.$$

f) It was proved by Hermite and Minkowski (see [20: Chapter V, Theorem 5]) that there are only finitely many number fields k and ℓ such that $D_k \leqslant \mathfrak{n}_0$ and $D_\ell \leqslant \mathfrak{n}'_0$. Also it follows from Proposition 0.9 and Lemma 0.11 that there are only finitely many global function fields k, each of them having only finitely many separable extensions ℓ of degree $\leqslant 3$, such that $g_k \leqslant \mathfrak{g}_0$, $g_\ell \leqslant \mathfrak{g}'_0$ and $q_\ell \leqslant \mathfrak{q}_0$. Since $\mathscr{G}$, being quasi-split, is uniquely determined by its absolute type and the fields k, ℓ, we now conclude that there is a finite set $\mathscr{Q}_0$ of pairs $(k, \mathscr{G})$ consisting of a global field k and an absolutely almost-simple, simply connected, quasi-split k-group $\mathscr{G}$ such that if either $(k, G) \in \mathscr{C}_0$ or $(k, G, S) \in \mathscr{V}_0$, then G is an inner k-form of $\mathscr{G}$ for some $(k, \mathscr{G})$ in this finite set. Over the finite set $\mathscr{Q}_0$ both $B(\mathscr{G}/k)$ and $C(\mathscr{G}/k)$ have a strictly positive lower bound.

Fix $(k, \mathscr{G}) \in \mathscr{Q}_0$. Then n and r are fixed and $r_v \leqslant r$. It is then clear from (4) and (14) that $e(P_v)$, f_v and f_v^{EP} tend to infinity with q_v if G is not quasi-split over k_v. Therefore we conclude that if there is an inner k-form G of $\mathscr{G}$ such that $(k, G) \in \mathscr{C}_0$ or $(k, G, S) \in \mathscr{V}_0$ for some S, then the cardinality of the residue fields at all non-archimedean places where G fails to be quasi-split is bounded by a constant depending only on k, $\mathscr{G}$, c; moreover, in the latter case, the cardinality of the residue fields at places contained in S_f is also bounded in view of 3.7 (2). Since the set of places of a global field where the cardinality of the residue field is less than a given integer is finite, we see now that there are only a finite number of possibilities for S and that there exists a finite subset $\mathscr{R}$ of V such that G is quasi-split outside $\mathscr{R}$, hence such that the element of $H^1(k, \overline{\mathscr{G}})$ which defines the inner k-form G of $\mathscr{G}$ belongs to the kernel of the natural map

$$\lambda_{\mathscr{R}} : H^1(k, \overline{\mathscr{G}}) \to \prod_{v \in V - \mathscr{R}} H^1(k_v, \overline{\mathscr{G}}).$$

But this kernel is known to be finite, see Appendix B. This shows that there are only finitely many possibilities for G and concludes the proof of 7.2 and 7.3.

7.5. In order to complete the proofs of Theorem A and B of the introduction, there still remains to prove a finiteness assertion for the P's in 7.2 and the Γ''s in 7.3. In view of these theorems, it suffices to show this for one group. Note that, as long as we deal with one group, some of the restrictions made in 7.2 and 7.3 are not necessary.

The group $(\mathrm{Aut}\, G')(A)$ operates canonically on $G'(A)$ and similarly $(\mathrm{Aut}\, G')_S$ operates on G'_S. In particular $\overline{G}(A)$, $\overline{G}_S$ and $\overline{G}(k_v)$ act on $G'(A)$, G'_S, $G'(k_v)$ respectively. This will be referred to as $\overline{G}(A)$ or $\overline{G}_S$ or $\overline{G}(k_v)$-conjugacy.

7.6. *Theorem. — Assume* G *is anisotropic over* k *and* G_∞ *is compact. Let* $c > 0$. *Then, up to* $\overline{G}(A)$*-conjugacy, there are only finitely many coherent collections* $P = (P_v)_{v \in V_f}$ *of parahoric subgroups such that* $\mathfrak{c}(P) \leqslant c$.

Let $c_0 = C(\mathscr{G}/k)\,\tau$. Then $\mathfrak{c}(P) > c_0$ for any P (see 7.4 (1), (3)), therefore we may assume $c > c_0$.

There is a finite subset $\mathscr{R}$ of V with the following properties: (i) $\mathscr{R} \supset V_\infty$; (ii) for $v \notin \mathscr{R}$, G is quasi-split over k_v and splits over $\hat{k}_v$; (iii) $q_v > 3c/2c_0$.

Let P be a coherent collection of parahoric subgroups. Assume that, for some $v \notin \mathscr{R}$, the group P_v is not hyperspecial. Then, since $e(P_v) \geqslant (q_v + 1)^{-1} q_v^{r_v+1}$ (see 7.4 (4)) and $q_v \geqslant 2$, we have $e(P_v) \geqslant c/c_0$, whence

$$\mathfrak{c}(P) > c.$$

As a consequence, if $\mathfrak{c}(P) \leqslant c$, then P_v is hyperspecial for $v \notin \mathscr{R}$. Since any two hyperspecial subgroups of $G(k_v)$ are conjugate under $\overline{G}(k_v)$, $(v \notin \mathscr{R})$, [41: 2.5], it follows that $(P_v)_{v \in V - \mathscr{R}}$ is determined uniquely up to $\overline{G}(A)$-conjugacy. But for a given $v \in \mathscr{R}$ there are only finitely many possibilities for P_v up to conjugacy in $G(k_v)$, whence the theorem.

7.7. *Theorem.* — *Fix* (k, G, S), *a central isogeny* $\iota : G \to G'$ *and* $c > 0$. *We assume* $S \supset V_\infty$ *and* G_S *is not compact. Let* $\mathscr{S}$ *be the subset of* S *consisting of all places where* G *is isotropic. Then, up to* $\overline{G}(k)$*-conjugacy,* $G'_{\mathscr{S}}$ *contains only finitely many finitely generated arithmetic subgroups (for the* k*-structure defined by* G/k*) such that either* $\mu'_{\mathscr{S}}(G'_{\mathscr{S}}/\Gamma') \leqslant c$, *or* Γ' *is virtually torsion-free and* $0 \neq |\chi(\Gamma')| \leqslant c$.

As there is a constant $\mathfrak{e}$ such that we have, for every Γ', $|\chi(\Gamma')| = \mathfrak{e}\mu'_{\mathscr{S}}(G'_{\mathscr{S}}/\Gamma')$, it suffices, in order to prove the theorem, to show that there are only finitely many finitely generated arithmetic subgroups Γ' with $\mu'_{\mathscr{S}}(G'_{\mathscr{S}}/\Gamma') < c$.

Since a finitely generated group contains only finitely many subgroups of a given finite index, it suffices, in view of 1.4 (iii), to prove that $G'_{\mathscr{S}}$ has only finitely many maximal arithmetic subgroups Γ' such that $\mu'_{\mathscr{S}}(G'_{\mathscr{S}}/\Gamma') \leqslant c$. Let then Γ' be maximal. According to 1.4 (iv), there exists a coherent collection $P = (P_v)_{v \in V - S}$ of parahoric subgroups such that Γ' is the normalizer of $\iota(\Lambda)$ where $\Lambda = G(k) \cap \Pi_v P_v$. For $v \in V - S$, let Θ_v be the type of P_v and Ξ_{Θ_v} be as in 2.8.

It follows from the first inequality of 3.6 (1), 3.6 (2) and the formula for the volume $\mu_{\mathscr{S}}(G_{\mathscr{S}}/\Lambda)$ given in 3.7 that there is a constant C depending only on G, k and S such that

$$(*) \qquad \mu'_{\mathscr{S}}(G'_{\mathscr{S}}/\Gamma') \geqslant C \prod_{v \in V - S} (\#\Xi_{\Theta_v})^{-1} . e(P_v),$$

where, for $v \in V - S$, $e(P_v)$ is as in 3.7. Now let e_v^m be as in 3.7. Then the inequalities 3.1 $(*)$ and 3.7 (1) imply at once:

$$(1) \qquad (\#\Xi_{\Theta_v})^{-1} . e(P_v) \geqslant e_v^m > 1.$$

Let T be the smallest subset of V containing S such that, for all $v \notin T$, the group G is quasi-split over k_v and splits over $\hat{k}_v$. If for a $v \notin T$, P_v is not hyperspecial, then

$$(2) \qquad e(P_v) \geqslant (q_v + 1)^{-1} q_v^{r_v+1}.$$

Let $\mathscr{R}$ be a finite subset of V, containing T, such that

(3) $$q_v > 3c(r+1)/2C \quad \text{for } v \notin \mathscr{R}.$$

If for some $v \notin \mathscr{R}$, P_v is not hyperspecial, then $e(P_v) > c(r+1)/C$, as easily follows from (2) and (3); now since $\#\Xi_{\Theta_v} \leqslant r+1$ for every v, we conclude from (*) that $\mu'_{\mathscr{S}}(G'_{\mathscr{S}}/\Gamma') > c$. Thus if $\mu'_{\mathscr{S}}(G'_{\mathscr{S}}/\Gamma') \leqslant c$, then P_v is hyperspecial for $v \notin \mathscr{R}$. The finiteness of the number of $\overline{G}(k)$-conjugacy classes of the Γ''s now follows from 3.10.

7.8. *Theorem.* — a) *Under the assumptions of* 7.2, *the set of* (k, G, P) *such that* $\mathfrak{c}(P) \leqslant c$ *is finite under natural equivalence.*

b) *Under the assumptions of* 7.3, *the set of* (k, G, S, G', Γ') *such that* $\mu_{S(G)}(G_{S(G)}/\Gamma') \leqslant c$ *(resp.* Γ' *is virtually torsion-free,* $0 < |\chi(\Gamma')| \leqslant c$ *and* G *is not of type* A_2*) is finite under natural equivalence.*

Theorem 7.2 reduces the proof of a) to the consideration of the possible P's for a given (k, G), in which case it follows from 7.6. Similarly, 7.3 reduces the proof of b) to the case of one system (k, G, S) and, since G has only finitely many centrally isogeneous groups, of the arithmetic subgroups of one G', which is settled by 7.7.

7.9. *Remark.* — In characteristic zero, the arithmeticity results of Margulis [23] allow us to express the previous finiteness results in a different way:

We consider the 4-tuples (S, k_S, H_S, Γ), where $S = S_\infty \cup S_f$ is a finite set, k_S stands for a collection k_s of local fields of characteristic zero which are archimedean for $s \in S_\infty$ and non-archimedean for $s \in S_f$, H_S is a product of groups $H_s(k_s)$, where H_s is an absolutely almost simple k_s-group and Γ an irreducible discrete subgroup of finite covolume of H_S. Assume moreover that the groups H_s are isotropic for $s \in S_f$ and that the sum of the k_s-ranks of the H_s $(s \in S)$ is at least two. If Γ is not cocompact, then [23] shows that Γ is S-arithmetic for a suitable choice of k having the completions k_s and of a k-group G' isomorphic to H_s over k_s for $s \in S$. If Γ is cocompact, then we may have possibly to enlarge S_∞ and use a k-group G' which is anisotropic at the new archimedean places. It follows that 7.8 implies the finiteness of the 4-tuples (S, k_S, H_S, Γ) under natural equivalence.

In positive characteristic, we deduce from [43] a similar result if we assume moreover Γ to be finitely generated.

7.10. *Corollary.* — *We keep the assumptions of* 7.3 *and assume moreover that* G *is anisotropic over* k, *isotropic over* k_v *for* $v \in S_f$ *and, in case* k *is a number field, that* $G(k_v)$ *is compact for* v *archimedean. Fix an integer* $c > 0$. *Let* X_S *be the product of the Bruhat-Tits buildings* X_v *of* G *over* k_v $(v \in S_f)$. *Then, up to natural equivalence, there exist only finitely many* 5*-tuples* (k, G, G', S_f, Γ') *such that* Γ' *has at most* c *orbits on the set of chambers of* X_S.

Let I'_S be the stabilizer of a chamber in X_S. The number of orbits of Γ' in the set of chambers is also the number of orbits of Γ' on G'_S/I'_S, which, in turn, is equal to the number of orbits of I'_S on G'_S/Γ'. By definition $\mu'_S(I'_S) = 1$, therefore each orbit of I'_S in G'_S/Γ' has volume $\leqslant 1$. Moreover, if k is a number field and v is archimedean, then $G'(k_v)$ is compact by assumption, hence $\mu'_v(G'(k_v)) = 1$ by definition (cf. 3.5). Consequently $\mu'_S(G'_S/\Gamma') \leqslant c$, and the corollary now follows from 7.8.

Remarks. — (1) If the discrete subgroup Γ' of G'_S has finitely many orbits on the set of chambers of X_S, then, as pointed out above, the stability group of a chamber has finitely many orbits on G'_S/Γ' and the latter quotient is necessarily compact. Therefore the supplementary assumptions made here on G are necessary. In the function field case, they imply that G is of type **A**.

(2) Assume now that Γ' consists of special automorphisms of X_S and has finitely many orbits on the set of facets of some given type. Then, as before, we see that the stability group of one such facet has finitely many orbits on G'_S/Γ', hence the latter quotient is compact. However its volume is not bounded by a universal constant, and tends to infinity with the relative ranks at S_f (if the facet is not a chamber). But if the number of elements of S_f is bounded, then the growth of the volume is sufficiently slow so that a minor modification of the previous arguments will again yield a finiteness theorem:

7.11. *Theorem.* — *Let a, c be two positive integers. Then up to natural equivalence, there exist only finitely many pairs* (k, G) *consisting of a number field k and an absolutely almost simple, simply connected k-group* G *such that* (i) G *is anisotropic at all the archimedean places of k,* (ii) *there is a k-group* G′ *k-isogenous to* G, *a finite set* $\mathcal{S}$ *of nonarchimedean places of k of cardinality a and an arithmetic subgroup* Γ' *of* $G'_{\mathcal{S}}$ *which acts by special automorphisms on the product* $X_{\mathcal{S}} = \prod_{v\in\mathcal{S}} X_v$ *of the Bruhat-Tits buildings* X_v *of* G/k_v, $v \in \mathcal{S}$, *with at most c orbits in the set of facets conjugate to some facet* $F = \prod_{\mathcal{S}} F_v$. *Moreover, up to natural equivalence, there are only finitely many* 5*-tuples* (k, G, S, G', Γ') *such that* $\# \mathcal{S} = a$ *and* Γ' *has at most c orbits in the set of facets conjugate to some facet* $F = \prod_{\mathcal{S}} F_v$, *where none of the* F_v*'s is a vertex.*

Proof. — Let k, G, G′, $\mathcal{S}$ be such that $\#\mathcal{S} = a$ and $G'_{\mathcal{S}}$ contains an arithmetic subgroup Γ' which acts by special automorphisms on the product $X_{\mathcal{S}}$ of the Bruhat-Tits buildings X_v of G/k_v, $v \in \mathcal{S}$, with at most c orbits in the set of facets conjugate to some facet $F = \prod_{\mathcal{S}} F_v$. Let C be a chamber of $X_{\mathcal{S}}$ containing F. Then C is a product $\prod_v C_v$, where C_v is a chamber of X_v and F_v a facet of C_v, $v \in \mathcal{S}$. Let I_v (resp. P_v) be the stabilizer of C_v (resp. F_v) in $G(k_v)$. Let $\mu'_{\mathcal{S}}$ be the product of the Tits measures on $G'(k_v)$, $v \in \mathcal{S}$. We want to show first

$$(1) \qquad \mu'_{\mathcal{S}}(G'_{\mathcal{S}}/\Gamma') \leqslant c \cdot \prod_{v\in\mathcal{S}} [P_v : I_v].$$

Let I'_v (resp. P'_v) be the stabilizer of C_v (resp. F_v) in $G'(k_v)$ and $G'_{v,0}$ be the subgroup of $G'(k_v)$ operating on X_v by special automorphisms. The latter is open of finite index in $G'(k_v)$ and contains $\iota(G(k_v))$. Let $P'_{v,0} = G'_{v,0} \cap P'_v$ and $I'_{v,0} = G'_{v,0} \cap I'_v$; we have $G'(k_v) = \iota(G(k_v)) . I'_v$, $P'_{v,0} = \iota(P_v) . I'_{v,0}$, hence

$$(2) \qquad [P'_{v,0} : I'_{v,0}] = [P_v : I_v].$$

20

The group $G'_{\mathscr{S},0} := \prod_{v\in\mathscr{S}} G'_{v,0}$ is transitive on the facets of any given type and $\Gamma' \subset G'_{\mathscr{S},0}$. We have therefore, by assumption,

$$(3) \qquad \#(\Gamma'\backslash G'_{\mathscr{S},0}/P'_{\mathscr{S},0}) \leqslant c, \quad \text{where } P'_{\mathscr{S},0} = \prod_{v\in\mathscr{S}} P'_{v,0},$$

which implies

$$(4) \qquad \mu'_{\mathscr{S}}(G'_{\mathscr{S},0}/\Gamma') \leqslant c.\mu'_{\mathscr{S}}(P'_{\mathscr{S},0}) = c.\prod_{v\in\mathscr{S}} [P'_{v,0} : I'_{v,0}]\,[I'_v : I'_{v,0}]^{-1},$$

and as

$$\mu'_{\mathscr{S}}(G'_{\mathscr{S}}/\Gamma') = [G'_{\mathscr{S}} : G'_{\mathscr{S},0}]\,\mu'_{\mathscr{S}}(G'_{\mathscr{S},0}/\Gamma') = \prod_{v\in\mathscr{S}} [I'_v : I'_{v,0}].\mu'_{\mathscr{S}}(G'_{\mathscr{S},0}/\Gamma')$$

we conclude that

$$(5) \qquad \mu'_{\mathscr{S}}(G'_{\mathscr{S}}/\Gamma') \leqslant c.\prod_{v\in\mathscr{S}} [P'_{v,0} : I'_{v,0}],$$

so that (1) follows from (2) and (5). Proceeding as in 7.4 and taking (1) into account we get first (recall that $\tau_k(G) = 1$ as k is a number field)

$$(6) \qquad c \geqslant B'(\mathscr{G}/k).\Big(\prod_{v\in\mathscr{S}} e_v[P_v : I_v]^{-1}\Big).\prod_{v\in V_f-\mathscr{S}} f_v,$$

where $\quad B'(\mathscr{G}/k) = B(\mathscr{G}/k)\,n^{-2sa}$

and e_v, f_v, $B(\mathscr{G}/k)$ are as in 7.4. In the notation of [31: 2.2],

$$(7) \qquad e_v[P_v : I_v]^{-1} = q_v^{(\dim \overline{M}_v + \dim \overline{\mathscr{M}}_v)/2}(\#\overline{M}_v(\mathfrak{f}_v))^{-1}.$$

By definition $e_v[P_v : I_v]^{-1}$ is the $e(P_v)$ of 7.4, therefore it satisfies

$$(8) \qquad e_v[P_v : I_v]^{-1} > 1 \quad (v \in \mathscr{S})$$

and

$$(9) \qquad e_v[P_v : I_v]^{-1} \geqslant (q_v + 1)^{-1}\,q_v^{r_v+1}$$

if either G is not quasi-split over k_v, or P_v is not maximal (i.e. if F_v is not a vertex). Since $f_v > 1$ (Appendix C), we deduce first from 6.1, (6) and (8), as in 7.4, that there are only finitely many possibilities for k, $\mathscr{G}$. Then from (9) and 3.7 (5), we see that, for given $(k, \mathscr{G})$, there are only finitely many $v \in V_f$ where G may not be quasi-split, whence the finiteness of the G's (hence also of G''s). Moreover, as P_v is maximal if and only if F_v is a vertex, if for no $v \in \mathscr{S}$, F_v is a vertex, then we conclude from (9) that the cardinality of the residue fields at all v in $\mathscr{S}$ is bounded by a constant depending only on k, G and c, which implies the finiteness of the possible $\mathscr{S}$'s; the finiteness of the possible Γ' now follows from 7.8 *b)*.

7.12. The following example shows the necessity of the restriction imposed by (ii) in 7.1, namely that if k is a global function field of genus zero, then G is anisotropic.

Let $n \geqslant 2$ be an integer, q be a power of a prime and F_q be the finite field with q elements. Let k_q be the global function field $F_q(t)$. It is of genus zero and its zeta-function is

$$\zeta_q(s) = (1 - q^{-s})^{-1}.(1 - q^{1-s})^{-1}.$$

Let $\Gamma_{n,q} = \mathrm{SL}_n(\mathrm{F}_q[t^{-1}])$. Then $\Gamma_{n,q}$ is an arithmetic subgroup of $\mathrm{SL}_n(\mathrm{F}_q((t)))$, and with respect to the Tits measure on the latter, its covolume is

$$q^{-(n^2-1)}(\#\mathrm{SL}_n(\mathrm{F}_q))\,(q-1)^{-(n-1)}\,q^{-\frac{1}{2}n(n-1)}\prod_{m=1}^{n-1}\zeta_q(m+1).$$

(See, for example, [31: Theorem 3.7].) Since $q^{(n^2-1)} > \#\mathrm{SL}_n(\mathrm{F}_q)$, we find that the covolume of $\Gamma_{n,q}$ in $\mathrm{SL}_n(\mathrm{F}_q((t)))$ tends to zero if either n or $q \to \infty$.

Similarly, 7.11 is not valid in general over a function field. In fact, [42] provides infinitely many examples of arithmetic subgroups of anisotropic forms of A_2 which are transitive on the edges of a given type on the building of SL_3 over $k((y))$, where k runs through finite fields.

8. Upper bound for the order of finite subgroups and a lower bound for the covolumes of discrete subgroups

In this section we shall sketch an alternative approach to get a lower bound for the covolume of discrete subgroups of the group of rational points of a connected semi-simple isotropic group defined over a nonarchimedean local field of characteristic zero and finite products of groups of this form. This approach was announced in [4]. For arithmetic subgroups, it does not give bounds as sharp as those obtained earlier, and it does not allow one to vary the ground field arbitrarily. However it applies to arbitrary (i.e., not necessarily arithmetic) discrete subgroups and it does not require any information on Tamagawa numbers. It depends on the following two results (Propositions 8.1, 8.2) on upper bounds for the order of finite subgroups, which may be of some independent interest.

Let K be a finite extension of the field $\mathbf{Q}_p$ of p-adic numbers. Let e be its ramification index over $\mathbf{Q}_p$ and q be the cardinality of its residue field.

8.1. *Proposition.* — (i) *The order of any finite abelian subgroup of* $\mathrm{SL}_n(\mathrm{K})$ *is less than* $(2e+1)^n\,(q+1)^{n-1}$.

(ii) *There exists an absolute constant c, not depending on* K *or n, such that the order of any finite subgroup of* $\mathrm{SL}_n(\mathrm{K})$ *is less than* $2^{cn^2/\log n}(2e+1)^n\,(q+1)^{n-1}$.

8.2. *Proposition.* — *Let* G *be a simply connected semi-simple* K*-subgroup of* SL_n. *Let r be the rank of* G *over the maximal unramified extension of* K *and w be the order of its absolute Weyl group. Then*

(i) *The order of any finite abelian subgroup of* G(K) *is less than* $w(2e+1)^n\,(q+1)^r$.

(ii) *There exists a constant d, depending on n but not on* K, *such that the order of any finite subgroup of* G(K) *is less than* $dw(2e+1)^n\,(q+1)^r$.

We will prove these two propositions together.

Let A be a finite abelian subgroup of either G(K) or $\mathrm{SL}_n(\mathrm{K})$. Let A_p be the

p-primary component of A and A$'$ be the sum of prime-to-p-primary components of A. Then $A = A_p \oplus A'$.

Let $\mathfrak{o}$ be the ring of integers of K, $\mathfrak{p}$ the unique maximal ideal of $\mathfrak{o}$ and $F = \mathfrak{o}/\mathfrak{p}$ be the residue field. After replacing A by a conjugate under an element of $GL_n(K)$, we may (and do) assume that A is contained in the maximal compact subgroup $SL_n(\mathfrak{o})$ of $SL_n(K)$. Now since the kernel of the "reduction mod $\mathfrak{p}$" $SL_n(\mathfrak{o}) \to SL_n(F)$ is a pro-p group, this map is injective on A$'$. Let $\bar{A}'$ denote the image of A$'$, let V be the natural n-dimensional representation of $SL_n(F)$ and let $V = \bigoplus_{1 \leqslant i \leqslant a} V_i$ be the decomposition of V as a direct sum of irreducible $F[\bar{A}']$-submodules. Set $\dim V_i = m_i$. Then $\Sigma_{i=1}^a m_i = n$. It is clear that

(1) $$\#A' = \#\bar{A}' \leqslant (q-1)^{-1}.(\prod_{i=1}^{a}(q^{m_i}-1)) \leqslant (q+1)^{n-1}.$$

Now assume that A is a finite abelian subgroup of G(K) and let P be a maximal parahoric subgroup of G(K) containing A$'$. Since G is simply connected, the "reduction mod $\mathfrak{p}$" of P is a connected linear algebraic group defined over the residue field F, see [41: §§3.4, 3.5]. Let M be the quotient of this linear algebraic group by its unipotent radical. Then M is a reductive F-group of absolute rank $\leqslant r$, and the order of its absolute Weyl group is as most w. As A$'$ is a finite abelian group of order prime to p, the natural homomorphism of P into M(F), maps it isomorphically onto an abelian subgroup $\bar{A}'$ of M(F). Now according to a result of Springer and Steinberg [37: Chapter II, Theorem 5.16], $\bar{A}'$ normalizes a maximal F-torus T of M(F) and hence (see [31: Lemma 2.8])

(2) $$\#A' = \#\bar{A}' \leqslant w\#T(F) \leqslant w(q+1)^r.$$

We shall now estimate the order of A_p. For this purpose we consider a maximal commutative semi-simple K-subalgebra $\mathscr{A}$ of the matrix algebra $M_n(K)$ containing A_p. Being semi-simple, $\mathscr{A}$ is a direct sum of certain field extensions K_i of K; $1 \leqslant i \leqslant b$. Let $[K_i : K] = n_i$. Then $\Sigma_{i=1}^b n_i = n$, and so, in particular, $1 \leqslant b \leqslant n$. Now recall that any finite subgroup of the multiplicative group of a field is cyclic, and let c_i be the largest positive integer such that K_i contains a primitive p^{c_i}-th root of unity. Then it is obvious that $\#A_p \leqslant \prod_{i=1}^b p^{c_i}$. On the other hand, the field extension obtained by adjoining a primitive p^{c_i}-th root of unity to $\mathbf{Q}_p$ has ramification index $p^{c_i-1}(p-1)$ over $\mathbf{Q}_p$, and the ramification index of K_i over $\mathbf{Q}_p$ is at most en_i; hence, $p^{c_i-1}(p-1) \leqslant en_i$, which implies that $p^{c_i} \leqslant ep(p-1)^{-1} n_i$. Therefore,

$$\#A_p \leqslant \prod_{i=1}^{b} p^{c_i} \leqslant (ep(p-1)^{-1})^b \prod_{i=1}^{b} n_i \leqslant (ep(p-1)^{-1})^b (n/b)^b$$

(since $\Sigma_{i=1}^b n_i = n$). Now it is easily seen, by computing the maxima of the function $f(x) = (ep(p-1)^{-1})^x (n/x)^x$ in the range $[1, n]$, that

(3) $$\#A_p < (2e+1)^n.$$

The assertions 8.1 (i) and 8.2 (i) now follow from (1), (2) and (3).

Let now $\mathscr{F}$ be a (not necessarily abelian) finite subgroup of $SL_n(K)$. Then, as K is embeddable as a subfield of the field of complex numbers, the quantitative version of a theorem of Jordan proved by Frobenius (see [36: §70, Satz 200]), implies that $\mathscr{F}$ contains an abelian normal subgroup A whose index is $\leqslant n!\,12^{n(\pi(n+1)+1)}$, where $\pi(n+1)$ is the number of positive primes $\leqslant (n+1)$. Using now the bound for the order of finite abelian subgroups obtained above, we conclude that

(4) $$\#\mathscr{F} \leqslant n!\,12^{n(\pi(n+1)+1)}(2e+1)^n\,(q+1)^{n-1},$$

and if $\mathscr{F}$ is a finite subgroup of G(K), that

(5) $$\#\mathscr{F} \leqslant n!\,12^{n(\pi(n+1)+1)}\,w(2e+1)^n\,(q+1)^r.$$

Thus 8.2 (ii) is satisfied with $d = n!\,12^{n(\pi(n+1)+1)}$.

To prove the second assertion of Proposition 8.1, we note that according to the prime number theorem, $\pi(n+1)\log(n+1)/(n+1) \to 1$ as $n \to \infty$. Moreover, $n! < n^n$ and, for every i, $(\log n)^i/n \to 0$ as $n \to \infty$. There exists therefore an absolute constant c such that $n!\,12^{n(\pi(n+1)+1)} < 2^{cn^2/\log n}$. Together with (4), this proves 8.1 (ii).

8.3. Let us now assume that G is a simply connected semi-simple K-subgroup of SL_n. Let μ be the Tits measure on G(K) i.e. the Haar measure with respect to which every Iwahori subgroup of G(K) has volume 1. Let Γ be a discrete subgroup of G(K) and P be a parahoric subgroup of maximum volume. The G(K)-invariant measure on $G(K)/\Gamma$ induced by μ will also be denoted by μ. The group $P \cap \Gamma$, being compact and discrete, is finite. Also, $\mu(P) = [P:I]$. As the natural inclusion of P in G induces an injective map $P/P\cap\Gamma \to G/\Gamma$, we conclude that

$$\mu(G/\Gamma) \geqslant \mu(P)\,.\,(\#(P\cap\Gamma))^{-1} = [P:I]\,(\#(P\cap\Gamma))^{-1}.$$

Using the "reduction mod $\mathfrak{p}$" and the Bruhat-Tits theory (see [41: §§3.5, 3.7]) it is easy to give a good lower bound for $[P:I]$ and Propositions 8.1, 8.2 provide an upper bound for the order of finite subgroups of G(K). Combining these we get a lower bound for the volume of G/Γ. For example, if G is an absolutely simple group of type E_8, then G is K-split, P is hyperspecial and $[P:I] > q^{120}$ (recall that the root system of type E_8 has 240 roots), and considering the embedding of G in SL_{248} given by the adjoint representation, we find from Proposition 8.2 that there is a constant c, which does not depend on K, such that the order of any finite subgroup of G(K), and so in particular of $P\cap\Gamma$, is less than $c(2e+1)^{248}\,(q+1)^8$. Hence

$$\mu(G/\Gamma) > c^{-1}(2e+1)^{-248}\,(q+1)^{-8}\,q^{120}.$$

Note that for a fixed e, the above lower bound goes to infinity with q.

Appendix A: Volumes of parahoric subgroups

This section provides in particular the proofs of two assertions made in 3.1. The arguments are minor modifications of those communicated to us by J. Tits.

A.1. We let K be a non-archimedean local field, q the order of its residue field, H an isotropic absolutely almost simple simply connected K-group, and X the Bruhat-Tits building of H(K). We fix an apartment **A** of X, a chamber C in **A** and let Δ be the set of vertices of C. As usual, the elements of Δ represent either a basis of the affine root system Φ_{af} of H/K or the vertices of the local Dynkin diagram $\mathscr{D}$. Let I be the stability group of C in H(K). It is an Iwahori subgroup. Let **W** be the affine Weyl group of H/K. The isotropy group of an element $c \in \overline{C}$ in H(K) (resp. **W**) is denoted P_c, (resp. $\mathbf{W}_c$). We let μ_T be the Tits measure on H(K). Therefore if the parahoric subgroup P contains I, then $\mu_T(P) = [P : I]$.

A.2. In the classification tables of [41], each vertex β of $\mathscr{D}$ is equipped with a positive integer $d(\beta)$ (written explicitely only if it differs from 1). If r_β is the fundamental reflection associated to $\beta \in \Delta$, i.e., to the wall of $\overline{C}$ opposite β, then $q^{d(\beta)} = \#(Ir_\beta I/I)$ [41: 3.3.1]. Moreover, if w is in the affine Weyl group **W** and $w = r_1 \ldots r_t$ is a reduced decomposition of w, where the r_i are fundamental reflections, then

$$(1) \qquad \#(IwI/I) = q_w = \prod_i q^{d(\beta_i)},$$

where $\beta_i \in \Delta$ is the vertex representing r_i (loc. cit.). This also shows that $q_{w.w'} = q_w \cdot q_{w'}$ if $\ell(w.w') = \ell(w) + \ell(w')$.

A.3. We have to refine and reformulate this. Let T be the maximal K-split torus in H such that T(K) stabilizes **A**. We let Φ^{nd} be the system of non-divisible roots in the relative root system $\Phi = \Phi(H, T)$. We view it as a subset of $X^*(T) \otimes \mathbf{R}$, which, in turn, is identified with the dual ${}^v\mathbf{A}^*$ of the space of translations ${}^v\mathbf{A}$ of **A**. Given an affine root α, there is a unique element $\bar{\alpha} \in \Phi^{nd}$ such that $\bar{\alpha}$ is a positive rational multiple of the vector part of α, and any $a \in \Phi^{nd}$ occurs in this way. For $a \in \Phi^{nd}$, let $\Gamma_a = \cup\, \alpha^{-1}(0)$, where the union is over all the affine roots with vector part proportional to a. It is a union of parallel hyperplanes in **A**. If $c \in \overline{C} \cap \Gamma_a$, we let $r_{a,c}$ be the reflection in the hyperplane of Γ_a containing c. It belongs to **W**. Our previous r_β is then $r_{\bar{\beta},b}$ for any b in the interior of the closed facet of codimension one of $\overline{C}$ not containing β. If now $\alpha \in \Delta$ is such that $c \in \Gamma_{\bar{\alpha}}$, then

$$(1) \qquad \#(Ir_{\bar{\alpha},c}\, I/I) = q^{d(\alpha, c)}, \quad \text{where } 1 \leqslant d(\alpha, c) \leqslant d(\alpha).$$

In fact, $d(\alpha, c)$ can take at most two values as c varies (besides zero when $c \notin \Gamma_{\bar{\alpha}}$). The group $\mathbf{W}_c$ is generated by the $r_{\bar{\beta},c}$, where β runs through the set $\Delta_{(c)}$ of vertices of Δ defining the type of the facet of $\overline{C}$ containing c. Then $\bar{\Delta}_c = \{\bar{\beta} \mid \beta \in \Delta_{(c)}\}$ is a basis of the sub-root system Φ_c of Φ^{nd} given by

$$(2) \qquad \Phi_c = \{\bar{\alpha} \mid c \in \Gamma_{\bar{\alpha}}, (\alpha \in \Phi_{af})\}.$$

By the Bruhat decomposition we have $P_c = \coprod_{w \in \mathbf{W}_c} IwI$, whence

$$(3) \qquad [P_c : I] = \sum_{w \in \mathbf{W}_c} q(w, c), \quad \text{where } q(w, c) := \#(IwI/I).$$

As above, if $w = r_1 r_2 \dots r_t$ is a reduced decomposition of w in $\mathbf{W}_c$, where r_i is one of the $r_{\bar\beta, c}(\beta \in \Delta_{(c)})$, then

$$(4) \qquad q(w, c) = \prod_i q(r_i, c).$$

A.4. For the purpose of this discussion, we shall say that a special vertex c of $\bar{C}$ is *very special* if $d(c)$ has the smallest possible value among the $d(b)$'s for b special. (There are in fact at most two possible values.) We have $d(b) = 1$ if b is hyperspecial, as is shown by inspection of the tables in [41], or could be deduced from 3.8.1 there, hence any hyperspecial point is very special. The parahoric subgroup P_c is said to be very special if c is so.

In the sequel we fix a very special vertex c_0. We have

$$(1) \qquad d(\beta, c_0) = d(\beta) \quad (\beta \in \Delta - \{c_0\}),$$

hence

$$(2) \qquad d(\beta, c_0) \geqslant d(\beta, c) \quad (\beta \in \Delta - \{c_0\},\ c \in \bar{C}).$$

The $\bar\beta$'s for $\beta \in \Delta - \{c_0\}$ form a basis Δ_0 of Φ and $-\bar{c}_0$ is the dominant root (with respect to Δ_0). We identify $\mathbf{W}_{c_0}$ in this way with the Weyl group W of Φ. For $c \in \bar{C}$, we now identify $\mathbf{W}_c$ with the subgroup W_c of W generated by the reflections $r_{\bar\beta}$ $(\beta \in \Delta_{(c)})$. Then Φ_c is the subroot system of Φ^{nd} generated by the corresponding roots and W_c is the Weyl group of Φ_c. If $\alpha, \beta \in \Delta_{(c)}$ are transformed into one another by an element of W_c, then $d(\alpha, c)$ and $d(\beta, c)$ are equal. We may therefore extend the definition of $d(\alpha, c)$ to all α such that $\bar\alpha \in \Phi_c$ and $c \in \Gamma_{\bar\alpha}$ by requiring that it be W_c-invariant. We fix the ordering on Φ^{nd} defined by the basis Δ_0 and, for $c \in \bar{C}$, let Φ_c^+ (resp. Φ_c^-) be the set of roots in Φ_c which are positive (resp. negative) under this ordering. In view of the relation between reduced decompositions and positive roots transformed into negative ones, we can also write A.3 (4) as

$$(3) \qquad q(w, c) = \prod_{\bar\alpha \in \Phi_c^+,\ w\bar\alpha < 0} q^{d(\alpha, c)}.$$

A.5. *Proposition.* — *Let* μ *be a Haar measure on* H(K).

(i) $\mu(P_c)$ $(c \in \bar{C})$ *is maximal among the volumes of parahoric subgroups of* H(K) *if and only if* c *is very special.*

(ii) *Assume* $c \in \bar{C}$ *is not special. Then*

$$(1) \qquad \mu(P_{c_0}) \geqslant \mu(P_c).(1 + q([W : W_c] - 1)).$$

In the proof we may assume that $\mu = \mu_T$. Let first c be special. In view of A.3 (3) and A.4 (3), we have $q(w, c_0) \geqslant q(w, c)$ for all $w \in W$. But, if c is not very special, we have a strict inequality for at least one w, therefore, by A.3 (3), $\mu(P_{c_0}) > \mu(P_c)$. This shows (i) for c special. On the other hand, the second factor on the right hand side of (1) is $\geqslant 2$. Therefore (ii) implies (i) for nonspecial c's. There remains to prove (ii), which we now proceed to do.

Assume c to be nonspecial. Let

$$W^c = \{ w \in W \mid w(\Phi_c^+) \subset \Phi^+ \}.$$

This is a set of representatives for the left cosets W/W_c. Let $u \in W^c$, $w \in W_c$ and $a \in \Phi_c^+$. If $w.a < 0$, then $w.a \in \Phi_c^-$ and therefore $uw.a < 0$, hence

$$w^{-1}\Phi^- \cap \Phi_c^+ = (uw)^{-1}\Phi^- \cap \Phi_c^+.$$

In view of A.4 (3), this shows that

(2) $$q(uw, c_0) \geqslant q(w, c)$$

and

(3) $$q(uw, c_0) \geqslant q.q(w, c) \quad \text{if } (uw)^{-1}\Phi^- \cap \Phi^+ - w^{-1}\Phi^- \cap \Phi^+ \neq \emptyset.$$

If $(uw)^{-1}\Phi^- \cap \Phi^+ = w^{-1}\Phi^- \cap \Phi_c^-$, then

$$\{ a \in \Phi_c^+, w.a < 0 \} = \{ a \in \Phi^+, uw.a < 0 \}.$$

Given w, this determines uw, hence can happen for at most one $u \in W^c$; therefore

(4) $$\sum_{u \in W^c} q(uw, c_0) \geqslant q(w, c)\,(1 + q([W : W_c] - 1)).$$

Then, in view of A.3 (3), the assertion (ii) follows from (4) by summing over $w \in W_c$.

A.6. As in 2.4, we let Ξ be the group of automorphisms of Δ defined by (Ad H) (K). For $c \in \bar{C}$, let Θ_c be the type of the face of C containing c, i.e., the subdiagram of Δ whose vertices correspond to the faces of codimension one of $\bar{C}$ containing c, and Ξ_c be the subgroup of Ξ leaving Θ_c stable. Then we have the following corollary, which is 3.1 (*) in a different notation.

A.7. *Corollary.* — $\mu(P_{c_0}) \geqslant \mu(P_c)\,(\#\Xi_c)$.

If c is special, then $\Xi_c = 1$ and the assertion follows from A.5 (i). Let now c be non-special. In view of A.5 (ii), it suffices to show that

(1) $$1 + q([W : W_c] - 1) \geqslant \#\Xi_c.$$

The left-hand side being $\geqslant 3$, we have only to consider the cases where $\#\Xi_c \geqslant 4$. Then H is either an inner form of type A_r of K-rank r ($r \geqslant 2$), or a K-split form of type D_r ($r \geqslant 4$). In the former case, Ξ is a cyclic group of order $r + 1$. Since $\Xi_c \neq 1$, it is a cyclic group of some order m dividing $r + 1$ and Φ_c is isomorphic to the direct product of m copies of the Weyl group of A_s for some $s \leqslant d$, where $d + 1 = (r + 1)/m$. Then W_c has order $\leqslant ((d + 1)!)^m$, therefore

$$[W : W_c] \geqslant r + 1,$$

and hence the left-hand side of (1) is at least $2r + 1$.

In the second case, Ξ_c is of order 4. It is easily verified that no subgroup of index 2 of W is the Weyl group of a subroot system. Hence the left-hand side of (1) is > 4.

Appendix B: A theorem in Galois cohomology

At the end of the proof of 7.2 and 7.3, we have used a finiteness theorem in Galois cohomology which is well-known in the number field case, but for which we do not know of a reference in the function field case. The purpose of this appendix is to supply a proof. The groups G and G′ are as before.

B.1. *Theorem. — The fibres of the canonical map*

$$\lambda^1_{G'} : H^1(k, G') \to \prod_{v \in V} H^1(k_v, G') \tag{1}$$

are finite.

[In other words, $\lambda^1_{G'}$ is proper with respect to the discrete topology.]

If k is a number field, this follows from Theorem 7.1 in [5]. From now on k is a function field. Let N be the (scheme theoretic) kernel of the central isogeny $\iota : G \to G'$. It is a finite group scheme of multiplicative type, contained in any maximal torus of G. By definition, we have an exact sequence

$$1 \to N \to G \to G' \to 1 \tag{2}$$

and, similarly, if T is a maximal k-torus of G and $T' = \iota(T)$, an exact sequence

$$1 \to N \to T \to T' \to 1. \tag{3}$$

By [15] and [8: III],

$$H^1(k, G) = 0 = H^1(k_v, G) \quad (v \in V). \tag{4}$$

From this and the exact sequence associated to (2),

$$\ldots \to H^1(k, G) \to H^1(k, G') \xrightarrow{\delta} H^2(k, N), \tag{5}$$

it follows that δ is injective. At first, it shows only that $\delta^{-1}(0)$ is the zero element. But the case of an arbitrary fibre of δ is reduced to the previous one by the familiar trick of twisting by a cocycle c representing a given element of $H^1(k, G')$ and replacing the original exact sequence (2) by

$$1 \to N \to G_c \to G'_c \to 1,$$

noting that G_c is also semisimple and simply connected. See e.g. [5: 1.10], in the Galois cohomology case, i.e., if N is reduced. But all this formalism is also available in the flat cohomology case, as is shown in much greater generality in [13: IV, 4.3.4].

Similarly, $\delta_v : H^1(k_v, G') \to H^2(k_v, N)$ is injective. Since $H^2(k_v, N)$ is finite (see Proposition 78 in [34]), this shows that $H^1(k_v, G')$ is finite. [This had already been pointed out by J.-C. Douai, *C. R. Acad. Sci. Paris*, **280** (1975), 321-323, who has showed moreover that δ_v is bijective, but we shall not need this result.]

As a consequence, we are reduced to showing that the fibres of the analogous map

(6) $$\lambda^2_N : H^2(k, N) \to \prod_{v \in V} H^2(k_v, N)$$

are finite. But we now deal with commutative groups, so this amounts to proving that ker λ^2_N is finite. We consider the following commutative diagram with exact rows associated to the exact sequence (3):

$$\begin{array}{ccccccc} H^1(k, T) & \xrightarrow{\alpha} & H^1(k, T') & \xrightarrow{\beta} & H^2(k, N) & \xrightarrow{\gamma} & H^2(k, T) \\ \downarrow{\scriptstyle \lambda^1_T} & & \downarrow{\scriptstyle \lambda^1_{T'}} & & \downarrow{\scriptstyle \lambda^2_N} & & \downarrow{\scriptstyle \lambda^2_T} \\ \prod_v H^1(k_v, T) & \xrightarrow{\tilde{\alpha}} & \prod_v H^1(k_v, T') & \xrightarrow{\tilde{\beta}} & \prod_v H^2(k_v, N) & \xrightarrow{\tilde{\gamma}} & \prod_v H^2(k_v, T). \end{array}$$

By [28: IV, 2.7], the kernel of λ^2_T is finite. This reduces our task to proving that $M = \ker \gamma \cap \ker \lambda^2_N$ is finite. An element $x \in M$ is the image of some element $y \in H^1(k, T')$ such that $\lambda^1_{T'}(y)$ belongs to the kernel of $\tilde{\beta}$, hence to the image of $\tilde{\alpha}$. Recall that for a connected smooth group scheme, the image of the localization map λ^1 belongs to the subset of elements all but finitely many components of which are zero; following [28] we denote it by $\coprod_v$. By §2.6 in [28: IV], the kernels and cokernels of

$$\lambda^1_T : H^1(k, T) \to \coprod_v H^1(k_v, T) \quad \text{and} \quad \lambda^1_{T'} : H^1(k, T') \to \coprod_v H^1(k_v, T')$$

are finite. By diagram chasing, we see that the set of possible y's is finite modulo the image of α and the (finite) kernel of $\lambda^1_{T'}$. Its image under β is therefore finite, as was to be proved.

B.2. *Corollary. — Let $\mathscr{R}$ be a finite subset of* V. *Then the kernel of the map*

$$\lambda^1_{G', \mathscr{R}} : H^1(k, G') \to \prod_{v \notin \mathscr{R}} H^1(k_v, G')$$

is finite.

This follows from B.1 and the fact that $H^1(k_v, G')$ is finite (see [5] in characteristic zero, and the previous proof otherwise).

Appendix C: Verification of the inequalities $f_v > 1$ and $f_v^{EP} > 1$

C.1. In this appendix, we use the notation of §7 freely. Our goal is to check the assertion 7.4 (15), namely

(1) $$f_v > 1 \quad (v \in V_f),$$

(2) $$f_v^{EP} > 1 \text{ unless G is of type } A_2 \text{ and } q_v \leqslant 3 \quad (v \in V_f).$$

If $v \notin S_f \cup T(G)$, then (see 7.4 (13)) f_v and f_v^{EP} are both equal to e_v^m, which is > 1 by 3.7 (1). If $v \in T(G)$, then $f_v = f_v^{EP}$ by 7.4 (12). We have therefore to consider f_v for $v \in S_f \cup T(G)$ and f_v^{EP} for $v \in S_f$.

C.2. *Proof of* (1) *for* $v \in T(G)$. In that case

$$f_v = f_v^{EP} = e_v^{m} n^{-\varepsilon}.$$

Now recall from 3.7 (5) that

$$e_v^{m} \geqslant (q_v + 1)^{-1} q_v^{r_v+1}.$$

(i) G *is of type* B, C, or E_7: Then $r_v = r$, $n^\varepsilon = 2$ and

$$f_v \geqslant (q_v + 1)^{-1} q_v^{r_v+1} 2^{-1} > 1.$$

(ii) G *is of type* E_6: Then $n^\varepsilon = 3$ and $r_v = 6$, hence

$$f_v \geqslant (q_v + 1)^{-1} q_v^{7} 3^{-1} > 1.$$

(iii) G *is of type* D_r: Then $r_v = r \geqslant 4$ and $n^\varepsilon = 4$, so

$$f_v \geqslant (q_v + 1)^{-1} q_v^{r_v+1} 4^{-1} \geqslant (q_v + 1)^{-1} q_v^{r_v-1} > 1.$$

(iv) G *is a form of type* A_r *(which splits over* $\hat{k}_v$ *since* $v \in T(G)$*)*: Then $r_v = r$, $n^\varepsilon = r + 1$ and

$$f_v \geqslant (q_v + 1)^{-1} q_v^{r+1}(r + 1)^{-1} > q_v^{r-1}(r + 1)^{-1} \geqslant 1 \quad \text{if } r \geqslant 3.$$

If $r = 2$, then, as $v \in T(G)$, G/k_v is anisotropic and $G(k_v) \cong SL_1(\mathfrak{D}_v)$, where $\mathfrak{D}_v$ is a division algebra of degree 3. By the inequality in 3.7 (6),

$$e_v^{m} \geqslant (q_v - 1)\, q_v^2,$$

so

$$f_v \geqslant (q_v - 1)\, q_v^2\, 3^{-1} > 1.$$

C.3. Now let us assume that $v \in S_f$. Then $f_v = e_v n^{-\varepsilon}$ if $\ell_v = \ell \otimes_k k_v$ is a ramified field extension of k_v, and $f_v = e_v n^{-2\varepsilon}$ otherwise.

Recall from 3.7 (2) that

$$e_v \geqslant (q_v + 1)^{-r_v} q_v^{r_v(r_v+3)/2}.$$

(i) G *is of type* B, C *or* E_7: Then $n^\varepsilon = 2$, and $r_v = r \geqslant 2$. So

$$e_v n^{-2\varepsilon} \geqslant (q_v + 1)^{-r} q_v^{r(r+3)/2} 2^{-2}$$
$$> q_v^{r(r-1)/2} 2^{-2} > 1 \quad \text{if } r \geqslant 3.$$

If $r = 2$, then G is of type B_2 and we need to use the exact value of e_v:

$$e_v = (q_v - 1)^{-2} q_v^6 \quad \text{if G splits over } k_v,$$
$$e_v = (q_v^2 - 1)^{-1} q_v^6 \quad \text{if G is of rank 1 over } k_v.$$

In both cases $e_v > q_v^4$ and

$$f_v = e_v 2^{-2} > q_v^4 2^{-2} > 1.$$

(ii) G *is of type* E_6: Then $n^\varepsilon = 3$, r_v equals 4 or 6 and

$$f_v \geqslant (q_v + 1)^{r_v} q_v^{r_v(r_v+3)/2} 3^{-2} > 1.$$

(iii) G *is of type* D_r: Then $r_v \geq 2$.

a) If $r_v = 2$, $G/\hat{k}_v$ is a triality form of type D_4. In this case,

$$e_v = (q_v - 1)^{-2} q_v^8$$

and
$$f_v \geq e_v n^{-2s} = (q_v - 1)^{-2} q_v^8 4^{-2} > 1.$$

b) If $G/\hat{k}_v$ is not a triality form, then $r_v \geq 3$ and, using 3.7 (2), we get

$$f_v \geq e_v n^{-2s} > (q_v + 1)^{-r_v} q_v^{r_v(r_v+3)/2} 4^{-2},$$

which is > 1 if $r_v > 3$. On the other hand, if $r_v = 3$, then G is of type D_4 and

$$e_v = (\#\overline{T}_v(\mathfrak{f}_v))^{-1} q_v^{(r_v + \dim \overline{\mathscr{M}}_v)/2} \geq (q_v + 1)^{-3} q_v^{12}$$

(note that $\overline{\mathscr{M}}_v$ is a group of type B_3; therefore, its dimension is 21) and

$$f_v \geq e_v n^{-2s} > (q_v + 1)^{-3} q_v^{12} 4^{-2} > 1.$$

(iv) *a)* G *is of type* A_r *and splits over* $\hat{k}_v$: Then $r_v = r$, $n^s = r + 1$ and, by 3.7 (2),

$$e_v \geq (q_v + 1)^{-r} q_v^{r(r+3)/2},$$

so
$$f_v \geq e_v n^{-2s} \geq (q_v + 1)^{-r} q_v^{r(r+3)/2} (r + 1)^{-2} > 1 \quad \text{if } r > 2.$$

Let now $r = 2$, G/k_v must be isotropic since $v \in S_f$. Then

$$e_v = (q_v - 1)^{-2} q_v^5 \quad \text{if } G/k_v \text{ is of inner type } A_2,$$
$$e_v = (q_v^2 - 1)^{-1} q_v^5 \quad \text{if } G/k_v \text{ is of outer type } A_2,$$

and in both cases,

$$f_v = e_v n^{-2s} = e_v 3^{-2} > 1 \quad \text{for all } q_v.$$

(iv) *b)* G *is of type* A_r, *not splitting over* $\hat{k}_v$: $f_v = e_v (r + 1)^{-1}$ in this case. $r_v = r/2$ if r is even, and equals $(r + 1)/2$ if r is odd. By the inequality in 3.7 (2),

$$f_v \geq (q_v + 1)^{-r_v} q_v^{r_v(r_v+3)/2} (r + 1)^{-1},$$

and it can be easily checked that the number on the right-hand side is greater than 1 if $r_v \geq 3$. Let now $r_v = 2$. Then $G/\hat{k}_v$ is an outer form of type A_2, A_3 or A_4. Making use of the fact that $\#\overline{T}_v(\mathfrak{f}_v) \leq (q_v + 1)^{r_v}$ ([31: 2.8]), we see from the equality in 3.7 (2) that, if $G/\hat{k}_v$ is of type ${}^2A_3$ or ${}^2A_4$, then

$$e_v \geq (q_v + 1)^{-2} q_v^6$$

so
$$f_v \geq (q_v + 1)^{-2} q_v^6 5^{-1} > 1 \quad \text{for all } q_v.$$

If $G/\hat{k}_v$ is of type ${}^2A_2$, then

$$e_v = (q_v - 1)^{-1} q_v^2$$

and
$$f_v = (q_v - 1)^{-1} q_v^2 3^{-1} > 1 \quad \text{for all } q_v.$$

C.4. We now take up the verification of $f_v^{\mathrm{EP}} > 1$ for $v \in S_f$ and G not of type A_2.

(i) G *is of type* B, C, or E_7: Then $r_v \geqslant 2$, $n^{\natural} = 2$, and

$$f_v^{\mathrm{EP}} \geqslant e_v\, 2^{-2}\, |\, \mathrm{W}_v(\mathbf{q}^{-1})\,|^{-1},$$

where $$e_v = (\#\overline{\mathrm{T}}_v(\mathfrak{f}_v))^{-1}\, q_v^{(r_v + \dim \overline{\mathscr{M}}_v)/2}.$$

Now recall that $\#\overline{\mathrm{T}}_v(\mathfrak{f}_v) \leqslant (q_v + 1)^{r_v}$ ([31: 2.8]), $|\, \mathrm{W}_v(\mathbf{q}^{-1})\,|^{-1} \geqslant 5^{-r_v}$ (4.4) and $\overline{\mathscr{M}}_v$ is an absolutely almost simple group of the same type as G. As G is of type B, C, or E, we conclude that $\dim \overline{\mathscr{M}}_v \geqslant r_v(2r_v + 1)$, and so

$$f_v^{\mathrm{EP}} \geqslant (q_v + 1)^{-r_v}\, q_v^{r_v(r_v+1)}\, 2^{-2}\, 5^{-r_v} > 1 \quad \text{if } r_v \geqslant 4.$$

Let $r_v = 3$. If G/k_v is split, then

$$e_v = (q_v - 1)^{-3}\, q_v^{12}$$

and, as $$|\, \mathrm{W}_v(\mathbf{q}^{-1})\,|^{-1} \geqslant \left(\frac{(q_v - 1)^2}{q_v^2 + 1}\right)^3 \quad (4.4),$$

we get $$f_v^{\mathrm{EP}} \geqslant (q_v - 1)^{-3}\, q_v^{12}\, 2^{-2} \left(\frac{(q_v - 1)^2}{q_v^2 + 1}\right)^3 = \frac{2^{-2}\, q_v^{12}(q_v - 1)^3}{(q_v^2 + 1)^3} > 1.$$

If G/k_v is a form of B_3 of relative rank 2, then

$$\#\overline{\mathrm{T}}_v(\mathfrak{f}_v) = (q_v - 1)^2\, (q_v + 1),$$

and so $$f_v^{\mathrm{EP}} \geqslant (q_v - 1)^{-2}\, (q_v + 1)^{-1}\, q_v^{12}\, 2^{-2} \left(\frac{(q_v - 1)^2}{q_v^2 + 1}\right)^3 = \frac{2^{-2}\, q_v^{12}(q_v - 1)^4}{(q_v^2 + 1)^3\, (q_v + 1)} > 1$$

for all q_v.

If G/k_v is a form of type C_3 of relative rank 1, then

$$\#\overline{\mathrm{T}}_v(\mathfrak{f}_v) = (q_v + 1)^2\, (q_v - 1),$$

and $$|\, \mathrm{W}_v(\mathbf{q}^{-1})\,|^{-1} = (q_v^5 - 1)\, (q_v^3 + 1)^{-1}\, (q_v^2 + 1)^{-1}$$

so $$f_v^{\mathrm{EP}} = \frac{q_v^{12}(q_v^5 - 1)}{4(q_v^3 + 1)\, (q_v^2 + 1)\, (q_v + 1)^2\, (q_v - 1)}$$

$$= \frac{q_v^{12}(q_v^4 + q_v^3 + q_v^2 + q_v + 1)}{4(q_v^3 + 1)\, (q_v^2 + 1)\, (q_v + 1)^2} > 1.$$

Let G now be of type B_2. If it is split over k_v, then

$$\#\overline{\mathrm{T}}_v(\mathfrak{f}_v) = (q_v - 1)^2,$$

$$|\, \mathrm{W}_v(\mathbf{q}^{-1})\,|^{-1} = (q_v^3 - 1)\, (q_v - 1)\, (q_v^2 + 1)^{-1}\, (q_v + 1)^{-2},$$

and if it is a form of type B_2 of k_v-rank 1,

$$\#\overline{\mathrm{T}}_v(\mathfrak{f}_v) = (q_v^2 - 1)$$

$$|\, \mathrm{W}_v(\mathbf{q}^{-1})\,|^{-1} = (q_v^3 - 1)\, (q_v^2 + 1)^{-1}\, (q_v + 1)^{-1}$$

and so $$f_v^{\mathrm{EP}} = 2^{-2}\, q_v^6(q_v^2 + q_v + 1)\, (q_v^2 + 1)^{-1}\, (q_v + 1)^{-2} > 1,$$

in both cases.

(ii) G *is of type* D_r:

a) $G/\hat{k}_v$ is a triality form. Then

$$e_v = (q_v - 1)^{-2} q_v^8,$$
$$|W_v(\mathbf{q}^{-1})|^{-1} = (q_v^5 - 1)(q_v - 1)^2 (q_v^6 - 1)^{-1} (q_v + 1)^{-1},$$

so
$$f_v^{EP} \geqslant 4^{-2} q_v^8 (q_v^5 - 1)(q_v^6 - 1)^{-1} (q_v + 1)^{-1} > 1.$$

b) If $G/\hat{k}_v$ is not a triality form, then $r_v \geqslant 3$, $\overline{T}_v(\mathfrak{f}_v) \leqslant (q_v + 1)^{r_v}$ ([31: 2.8]); $\overline{\mathscr{M}}_v$ is of type D_r if G splits over $\hat{k}_v$ and in this case the dimension of $\overline{\mathscr{M}}_v$ is $r(2r-1)$. If G is not split over $\hat{k}_v$, $r_v = r - 1$ and $\overline{\mathscr{M}}_v$ is of type B_{r-1}, its dimension is $(2r-1)(r-1)$.

We take up first the case where G (is of type D_r and) splits over $\hat{k}_v$. Then

$$e_v \geqslant (q_v + 1)^{-r} q_v^{r^2}$$

and so
$$f_v^{EP} \geqslant q_v^{r^2} (q_v + 1)^{-r} 4^{-2} 5^{-r} > 1 \quad \text{if } r \geqslant 5.$$

Let us assume now that $r = 4$. Then there are the following possibilities for G/k_v.

(1) G splits over k_v. In this case $\# \overline{T}_v(\mathfrak{f}_v) = (q_v - 1)^4$ and

$$e_v = (q_v - 1)^{-4} q_v^{16},$$

therefore,

$$f_v^{EP} \geqslant q_v^{16} (q_v - 1)^{-4} 4^{-2} 5^{-4} > 1$$

for all q_v.

(2) G/k_v is of type ${}^2\mathsf{D}_{4,3}^{(1)}$ (and it splits over $\hat{k}_v$). In this case

$$\# \overline{T}_v(\mathfrak{f}_v) = (q_v^2 - 1)(q_v - 1)^2,$$

so
$$e_v = (q_v^2 - 1)^{-1} (q_v - 1)^{-2} q_v^{16},$$

and hence, for all q_v,

$$f_v^{EP} \geqslant q_v^{16} (q_v^2 - 1)^{-1} (q_v - 1)^{-2} 4^{-2} 5^{-4} > 1.$$

(3) G/k_v is of type ${}^1\mathsf{D}_{4,2}^{(1)}$ (and it splits over $\hat{k}_v$). Then $\# \overline{T}_v(\mathfrak{f}_v) = (q_v^2 - 1)^2$, so $e_v = (q_v^2 - 1)^{-2} q_v^{16}$. In this case we need to know the precise value of $|W_v(\mathbf{q}^{-1})|^{-1}$, which is

$$(q_v^5 - 1)(q_v^3 - 1)(q_v^3 + 1)^{-1} (q_v^2 + 1)^{-2} (q_v + 1)^{-1}.$$

Hence

$$f_v^{EP} = 4^{-2} q_v^{16} (q_v^5 - 1)(q_v^3 - 1)(q_v^4 - 1)^{-2} (q_v^3 + 1)^{-1} (q_v + 1)^{-1} > 1$$

for all q_v.

(4) G/k_v is of type ${}^3\mathsf{D}_{4,2}$ (and it splits over $\hat{k}_v$). Then $\# \overline{T}_v(\mathfrak{f}_v) = (q_v^3 - 1)(q_v - 1)$, hence

$$e_v = (q_v^3 - 1)^{-1} (q_v - 1)^{-1} q_v^{16}.$$

In this case

$$|W_v(\mathbf{q}^{-1})|^{-1} = (q_v^9 - 1)(q_v^5 - 1)(q_v^2 + 1)(q_v - 1)^2 (q_v^6 + 1)^{-1} (q_v^6 - 1)^{-2}.$$

Hence, for all q_v,

$$f_v^{\mathrm{EP}} = \frac{q_v^{16}(q_v^9 - 1)(q_v^5 - 1)(q_v^2 + 1)(q_v - 1)}{4^2(q_v^6 + 1)(q_v^6 - 1)^2(q_v^3 - 1)} > 1.$$

(5) G/k_v is of type ${}^2\mathrm{D}^{(2)}_{4,1}$ (and it splits over $\hat{k}_v$). In this case $\#\overline{\mathrm{T}}_v(\mathfrak{f}_v) = (q_v^4 - 1)$

$$e_v = (q_v^4 - 1)^{-1} q_v^{16}$$

$$|\mathrm{W}_v(\mathbf{q}^{-1})|^{-1} = (q_v^5 - 1)(q_v^4 + 1)^{-1}(q_v + 1)^{-1}$$

and so, for all q_v,

$$f_v^{\mathrm{EP}} = 4^{-2} q_v^{16}(q_v^5 - 1)(q_v^8 - 1)^{-1}(q_v + 1)^{-1} > 1.$$

c) Let us assume that G is a form of type D_r which does not split over $\hat{k}_v$. Then $r_v = r - 1$ and $\overline{\mathscr{M}}_v$ is of type B_{r-1}. Therefore

$$\dim \overline{\mathscr{M}}_v = (2r - 1)(r - 1) \quad \text{and} \quad e_v \geqslant (q_v + 1)^{-(r-1)} q_v^{r(r-1)},$$

so
$$f_v^{\mathrm{EP}} \geqslant 4^{-1} q_v^{r(r-1)}(q_v + 1)^{-(r-1)} 5^{-(r-1)}.$$

From this it is easily seen that if $r \geqslant 5$, then $f_v^{\mathrm{EP}} > 1$. Let $r = 4$. Then

$$\#\overline{\mathrm{T}}_v(\mathfrak{f}_v) = (q_v - 1)^3 \quad \text{if G is of rank 3 over } k_v$$

and
$$\#\overline{\mathrm{T}}_v(\mathfrak{f}_v) = (q_v^2 - 1)(q_v + 1) \quad \text{if G is of rank 1 over } k_v.$$

In the first case

$$f_v^{\mathrm{EP}} \geqslant 4^{-1}(q_v - 1)^{-3} q_v^{12} 5^{-3} > 1 \quad \text{for all } q_v.$$

In the second case, we need to know the value of $|\mathrm{W}_v(\mathbf{q}^{-1})|^{-1}$ which is

$$(q_v^5 - 1)(q_v^3 + 1)^{-1}(q_v^2 + 1)^{-1}.$$

We get

$$f_v^{\mathrm{EP}} \geqslant 4^{-1} q_v^{12}(q_v^5 - 1)(q_v^4 - 1)^{-1}(q_v^3 + 1)^{-1}(q_v + 1)^{-1} > 1 \quad \text{for all } q_v.$$

(iii) *Let G/k_v be an inner form of type A_r*: Let $\mathfrak{D}_v$ be the central division algebra such that $G(k_v) \cong \mathrm{SL}_{n_v}(\mathfrak{D}_v)$ and d_v be its degree. Then $d_v n_v = r + 1$,

$$\#\overline{\mathrm{T}}_v(\mathfrak{f}_v) = (q_v^{d_v} - 1)^{n_v}(q_v - 1)^{-1}$$

so
$$e_v = (q_v^{d_v} - 1)^{-n_v}(q_v - 1) q_v^{r(r+3)/2}.$$

As
$$|\mathrm{W}_v(\mathbf{q}^{-1})|^{-1} = (q_v^{d_v} - 1)^{n_v}(q_v^{r+1} - 1)^{-1},$$

we have

$$f_v^{\mathrm{EP}} = (r + 1)^{-2} q_v^{r(r+3)/2}(q_v - 1)(q_v^{r+1} - 1)^{-1},$$

which is easily seen to be greater than 1 if either $r > 2$ or $q_v \geqslant 3$, and less than 1 if $r = 2$ and $q_v = 2$.

(iv) *Let G/k_v be an outer form of type A_r which splits over $\hat{k}_v$*: In this case $r_v = r$ and

$$e_v \geqslant (q_v + 1)^{-r} q_v^{r(r+3)/2}.$$

So
$$f_v^{\mathrm{EP}} \geqslant (r + 1)^{-2} 5^{-r}(q_v + 1)^{-r} q_v^{r(r+3)/2}.$$

This implies that $f_v^{EP} > 1$ if $r > 6$. If $r = 6$, then $\#\overline{T}_v(\mathfrak{f}_v) = (q_v^2 - 1)^3$,

$$e_v = (q_v^2 - 1)^{-3} q_v^{27}$$

and

$$f_v^{EP} \geqslant 7^{-2}\, 5^{-6} (q_v^2 - 1)^{-3} q_v^{27} > 1 \quad \text{for all } q_v.$$

If $r = 5$, $\#\overline{T}_v(\mathfrak{f}_v) = (q_v^2 - 1)^2 (q_v - 1)$ or $(q_v^2 - 1)^2 (q_v + 1)$ depending on whether the k_v-rank of G is 3 or 2;

$$e_v = (q_v^2 - 1)^{-2} (q_v - 1)^{-1} q_v^{20}$$

in the first case, and

$$e_v = (q_v^2 - 1)^{-2} (q_v + 1)^{-1} q_v^{20}$$

in the second case. As

$$f_v^{EP} \geqslant 6^{-2}\, 5^{-5}\, e_v,$$

we conclude that in the first case $f_v^{EP} > 1$. In the second case, the value of $|\, W_v(\mathbf{q}^{-1}) |^{-1}$ is

$$(q_v^8 - 1)(q_v^5 - 1)(q_v^5 + 1)^{-1}(q_v^3 + 1)^{-2}(q_v^2 + 1)^{-1}.$$

It is now simple to see that, for all q_v,

$$f_v^{EP} = 6^{-2} e_v \,|\, W_v(\mathbf{q}^{-1}) |^{-1} > 1.$$

Let now $r = 4$. Then

$$\#\overline{T}_v(\mathfrak{f}_v) = (q_v^2 - 1)^2, \qquad |\, W_v(\mathbf{q}^{-1}) |^{-1} = \frac{(q_v^4 + 1)(q_v^3 - 1)(q_v^2 - 1)}{(q_v^5 + 1)(q_v^3 + 1)(q_v + 1)}$$

and

$$e_v = (q_v^2 - 1)^{-2} q_v^{14}.$$

So

$$f_v^{EP} = \frac{5^{-2} q_v^{14} (q_v^4 + 1)(q_v^3 - 1)}{(q_v^5 + 1)(q_v^3 + 1)(q_v^2 - 1)(q_v + 1)} > 1.$$

We assume now that $r = 3$. Then $\#\overline{T}_v(\mathfrak{f}_v) = (q_v^2 - 1)(q_v - 1)$ if k_v-rank $G = 2$ and $\#\overline{T}_v(\mathfrak{f}_v) = (q_v^2 - 1)(q_v + 1)$ if k_v-rank $G = 1$. In the first case

$$|\, W_v(\mathbf{q}^{-1}) |^{-1} = (q_v^3 - 1)(q_v - 1)(q_v^3 + 1)^{-1}(q_v + 1)^{-1}$$

and in the second case it is equal to $(q_v^3 - 1)(q_v^3 + 1)^{-1}$. So, in both cases,

$$f_v^{EP} = 4^{-2} q_v^9 (q_v^3 - 1)(q_v^2 - 1)^{-1}(q_v^3 + 1)^{-1}(q_v + 1)^{-1} > 1.$$

Let $r = 2$. Then $\#\overline{T}_v(\mathfrak{f}_v) = (q_v^2 - 1)$, $|\, W_v(\mathbf{q}^{-1}) |^{-1} = (q_v^2 + 1)(q_v - 1)(q_v^3 + 1)^{-1}$ and

$$f_v^{EP} = 3^{-2} q_v^5 (q_v^2 + 1)(q_v^3 + 1)^{-1}(q_v + 1)^{-1}$$

which is > 1 if $q_v \geqslant 3$ and < 1 if $q_v = 2$.

(v) *Let* G/k_v *be an outer form of type* A_r *which does not split over* $\hat{k}_v$ *and* $r > 2$:

If $r = 2n$, $\bar{\mathscr{M}}_v$ is an absolutely almost simple group of type B_n, its dimension is $n(2n+1)$,

$$e_v = (q_v - 1)^{-n}\, q_v^{n(n+1)}$$

and $$f_v^{\mathrm{EP}} \geqslant (2n+1)^{-1}\,(q_v - 1)^{-n}\, q_v^{n(n+1)}\, 5^{-n},$$

so $f_v^{\mathrm{EP}} > 1$ if $n \geqslant 3$. If $n = 2$,

$$|\,W_v(\mathbf{q}^{-1})\,|^{-1} = (q_v^3 - 1)\,(q_v - 1)\,(q_v^2 + 1)^{-1}\,(q_v + 1)^{-2},$$

so, in this case, for all q_v,

$$f_v^{\mathrm{EP}} = 5^{-1}\, q_v^6(q_v^2 + q_v + 1)\,(q_v^2 + 1)^{-1}\,(q_v + 1)^{-2} > 1.$$

Now let $r = 2n + 1$ $(n \geqslant 1)$. In this case the group $\bar{\mathscr{M}}_v$ is of type C_{n+1}, its dimension is $(n+1)\,(2n+3)$. Moreover $\#\bar{T}_v(\mathfrak{f}_v) = (q_v - 1)^{n+1}$ if k_v-rank $(G) = n + 1$ and $\#\bar{T}_v(\mathfrak{f}_v) = (q_v - 1)^n\,(q_v + 1)$ if k_v-rank $(G) = n$. So

$$f_v^{\mathrm{EP}} \geqslant (2n+2)^{-1}\,(q_v - 1)^{-n-1}\, q_v^{(n+1)(n+2)}\, 5^{-n-1}$$

in the first case; in the second case

$$f_v^{\mathrm{EP}} \geqslant (2n+2)^{-1}\,(q_v - 1)^{-n}\,(q_v + 1)^{-1}\, q_v^{(n+1)(n+2)}\, 5^{-n-1}.$$

In both cases, $f_v^{\mathrm{EP}} > 1$ if $n \geqslant 2$. So let us assume that $n = 1$. Then, G/k_v is an outer form of type A_3 which does not split over $\hat{k}_v$, and we have

$$|\,W_v(\mathbf{q}^{-1})\,|^{-1} = (q_v^3 - 1)\,(q_v - 1)\,(q_v^2 + 1)^{-1}\,(q_v + 1)^{-2}$$

if k_v-rank $(G) = 2$, and

$$|\,W_v(\mathbf{q}^{-1})\,|^{-1} = (q_v^3 - 1)\,(q_v^2 + 1)^{-1}\,(q_v + 1)^{-1}$$

if k_v-rank $G = 1$. So

$$f_v^{\mathrm{EP}} = 4^{-1}\, q_v^6(q_v^2 + q_v + 1)\,(q_v^2 + 1)^{-1}\,(q_v + 1)^{-2} > 1$$

in both cases.

(vi) If G is of type E_6, E_8, F_4 or G_2, the verification of $f_v^{\mathrm{EP}} > 1$ is easy.

REFERENCES

[1] E. Bombieri, Counting points on curves over finite fields (d'après S. A. Stepanov), Séminaire Bourbaki, *Springer L.N.M.*, 383 (1974).

[2] A. Borel, Some finiteness properties of adele groups over number fields, *Publ. Math. I.H.E.S.*, 16 (1963), 5-30,

[3] A. Borel, On the set of discrete subgroups of bounded covolume in a semi-simple group, *Proc. Indian Acad. Sci. (Math. Sci.)*, 97 (1987), 45-52.

[4] A. Borel et G. Prasad, Sous-groupes discrets de groupes *p*-adiques à covolume borné, *C. R. Acad. Sc. Paris*, 305 (1987), 357-362.

22

[5] A. Borel et J.-P. Serre, Théorèmes de finitude en cohomologie galoisienne, *Comment. Math. Helv.*, **39** (1964), 111-164.

[6] A. Borel et J.-P. Serre, Cohomologie d'immeubles et de groupes S-arithmétiques, *Topology*, **15** (1976), 211-232.

[7] A. Borel et J. Tits, Homomorphismes « abstraits » de groupes algébriques simples, *Ann. Math.*, **97** (1973), 499-571.

[8] F. Bruhat et J. Tits, Groupes réductifs sur un corps local, I, *Pub. Math. I.H.E.S.*, **41** (1971), 5-251; II, *ibid.*, **60** (1984), 5-184; III, *J. Fac. Sci. Univ. Tokyo (Sec. IA)*, **34** (1987), 671-698.

[9] C. Chevalley, Introduction to the theory of algebraic functions of one variable, *Math. Surveys*, Number 6, A.M.S. (1951).

[10] T. Chinburg, Volumes of hyperbolic manifolds, *J. Diff. Geom.*, **18** (1983), 783-789.

[11] M. Deuring, Lectures on the theory of algebraic functions of one variable, *Springer L.N.M.*, **314** (1973).

[12] M. Fried and M. Jarden, *Field arithmetic*, Berlin, Springer-Verlag (1986).

[13] F. Giraud, Cohomologie non abélienne, *Grund. Math. Wiss.*, Springer-Verlag, 1971.

[14] G. Harder, Minkowskische Reduktionstheorie über Funktionenkörpern, *Invent. Math.*, **7** (1969), 33-54.

[15] G. Harder, Über die Galoiskohomologie halbeinfacher algebraischer Gruppen III, *J. Reine Angew. Math.* **274/275** (1975), 125-138.

[16] N. Iwahori and H. Matsumoto, On some Bruhat decompositions and the structure of the Hecke rings of the p-adic Chevalley groups, *Publ. Math. I.H.E.S.*, **25** (1965), 5-48.

[17] W. M. Kantor, R. A. Liebler and J. Tits, On discrete chamber-transitive automorphism groups of affine buildings, *Bull. A.M.S.*, **16** (1987), 129-133.

[18] R. E. Kottwitz, Tamagawa numbers, *Ann. Math.*, **127** (1988), 629-646.

[19] M. Kneser, Galoiskohomologie halbeinfacher algebraischer Gruppen über p-adischen Körpern, I, II, *Math. Z.*, **88** (1965), 250-272.

[20] S. Lang, *Algebraic number theory*, Addison-Wesley, Reading, Mass. (1968).

[21] S. Lang, *Algebra*, Addison-Wesley, Reading, Mass. (1965).

[22] G. A. Margulis, Cobounded subgroups of algebraic groups over local fields, *Funct. Anal. Appl.*, **11** (1977), 45-57.

[23] G. A. Margulis, Arithmeticity of the irreducible lattices in the semi-simple groups of rank greater than 1, *Invent. Math.*, **76** (1984), 93-120.

[24] G. A. Margulis and J. Rohlfs, On the proportionality of covolumes of discrete subgroups, *Math. Ann.*, **275** (1986), 197-205.

[25] J. S. Milne, *Étale Cohomology*, Princeton U. Press (1980).

[26] A. M. Odlyzko, Some analytic estimates of class numbers and discriminants, *Invent. Math.*, **29** (1975), 275-286.

[27] A. M. Odlyzko, Lower bounds for discriminants of number fields, *Acta Arith.*, **29** (1976), 275-297.

[28] J. Oesterlé, Nombres de Tamagawa et groupes unipotents en caractéristique p, *Invent. Math.*, **78** (1984), 13-88.

[29] G. Poitou, Minorations de discriminants (d'après A. M. Odlyzko), Sém. Bourbaki, *Springer L.N.M.*, **567** (1977).

[30] G. Prasad, Strong approximation, *Ann. Math.*, **105** (1977), 553-572.

[31] G. Prasad, Volumes of S-arithmetic quotients of semi-simple groups, *Publ. Math. I.H.E.S.*, **69** (1989), 91-117.

[32] J. Rohlfs, Die maximalen arithmetisch definierten Untergruppen zerfallender einfacher Gruppen, *Math. Ann.*, **244** (1979), 219-231.

[33] J.-P. Serre, Cohomologie des groupes discrets, in *Ann. of Math. Studies*, **70**, Princeton University Press (1971).

[34] S. S. Shatz, Profinite groups, arithmetic and geometry, *Ann. of Math. Studies*, **67**, Princeton University Press, 1972.

[35] C. L. Siegel, Über die Classenzahl quadratischer Zahlkörper, *Acta. Arith.*, **1** (1936), 83-86.

[36] A. Speiser, Die Theorie der Gruppen von endlicher Ordnung, *Grund. Math. Wiss.*, **5**, Springer-Verlag.

[37] T. A. Springer and R. Steinberg, Conjugacy classes, in *Springer L.N.M.*, **131** (1970), 167-294.

[38] R. Steinberg, Regular elements of semi-simple algebraic groups, *Publ. Math. I.H.E.S.*, **25** (1965), 49-80.

[39] R. Steinberg, Endomorphisms of linear algebraic groups, *Mem. A.M.S.*, **80** (1968).

[40] J. Tits, Classification of algebraic semi-simple groups, *Proc. A.M.S. Symp. Pure Math.*, **9** (1966), 33-62.

[41] J. Tits, Reductive groups over local fields, *Proc. A.M.S. Symp. Pure Math.*, **33**, Part 1 (1979).

[42] J. Tits, Buildings and group amalgamations, in *Groups St. Andrews 1985*, Cambridge University Press, 1987, 110-127.
[43] T. N. Venkataramana, On superrigidity and arithmeticity of lattices in semi-simple groups, *Invent. Math.*, **92** (1988), 255-306.
[44] H. C. Wang, Topics on totally discontinuous groups, in *Symmetric spaces* (ed. by W. M. Boothby and G. Weiss), New York, Marcel Dekker (1972), 459-487.
[45] A. Weil, *Basic number theory*, Berlin, Springer-Verlag.
[46] A. Weil, Adeles and algebraic groups, *Progress in Math.*, **23** (1982), Birkhäuser, Boston.
[47] R. Zimmert, Ideale kleiner Norm in Idealklassen und eine Regulatorabschätzung, *Invent. Math.*, **62** (1981), 367-380.

The Institute for Advanced Study
Princeton N.J. 08540
U.S.A.

Tata Institute of Fundamental Research
Homi Bhabha Road
Bombay 400005
India

Manuscrit reçu le 20 décembre 1988.

Addendum to

Finiteness Theorems for Discrete Subgroups of Bounded Covolume in Semi-simple Groups

by Armand Borel and Gopal Prasad

Publ. Math. I.H.E.S. **69** (1989), 119–171

1. If G is an anisotropic outer form of type **A**, defined over a global function field k, then there is a separable quadratic extension ℓ of k and a central division algebra $\mathcal{D}$, over ℓ, with an involution σ of the second kind such that either

$$G(k) = \{ d \in \mathcal{D}^\times \mid d\sigma(d) = 1 \text{ and } \mathrm{Nrd}(d) = 1 \}$$

or there exists an anisotropic hermitian form h on $\mathcal{D}^2$, defined in terms of the involution σ, such that $G = SU(h)$. Due to an oversight, in the proof of Theorem 7.3 of [1] the groups k-isomorphic to $SU(h)$, where h is as above, were not considered. The following construction shows that in fact these groups need to be excluded from the purview of the theorem.

Let q be a positive integral power of an odd prime and $\mathbf{F}_q$ be the field with q elements. Let $k = \mathbf{F}_q(t)$ and $\ell = \mathbf{F}_q(t^{1/2})$. We fix a place v of k which splits over ℓ and which is such that the residue field of k_v is the field with q elements. We also fix two places v_1, v_2 of k, which ramify over ℓ, and where the residue fields are of cardinality q. Now let $\mathcal{D}$ be the central division algebra over ℓ, of degree r, whose local invariant at one of the two places lying over v is m/r, at the other place lying over v it is $-m/r$, where m is an integer prime to r, and zero at all other places of ℓ. Let σ be an involution of $\mathcal{D}$, of second kind, which fixes k pointwise. Let h be an anisotropic hermitian form on $\mathcal{D}^2$, defined in terms of the involution σ, and $G = SU(h)$. We assume that h is so chosen that G is quasi-split at every place, different from v_1, v_2, which splits over ℓ, and of rank $r-1$ over k_{v_i} for $i = 1, 2$. Such a hermitian form exists: in fact, given any $\delta \in k^\times$ such that $(-1)^r \delta$ is a norm from $\ell \otimes_k k_w$ at all places w of k except v_1 and v_2, any hermitian form on $\mathcal{D}^2$ whose determinant is δ will have the desired property. (See [5: Chapter 10, Theorem 6.9].)

Both k and ℓ are fields of genus zero and

$$\zeta_k(s) = (1 - q^{-s})^{-1} . (1 - q^{1-s})^{-1} = \zeta_\ell(s).$$

Hence, for all s,

$$L_{\ell/k}(s) = 1.$$

G/k_v is of rank 1: in fact, there is a central division algebra D_v over k_v, of degree r and invariant m/r, such that $G(k_v)$ is isomorphic to $SL_2(D_v)$. Let (P_w) be a coherent collection of parahoric subgroups such that for every w, P_w is a parahoric subgroup of $G(k_w)$ of maximum volume, and let $\Gamma = G(k) \bigcap_{w \neq v} P_w$. Then, in its natural embedding in $G(k_v)$, Γ is a discrete cocompact subgroup of $G(k_v)$. Using Theorem 3.7 of [4], we find that the volume of $G(k_v)/\Gamma$, with respect to the Tits measure on $G(k_v)$, is precisely $\tau_k(G)$. Thus, if the Weil conjecture on Tamagawa numbers (i.e., the Tamagawa number of any simply connected semi-simple group, defined over a global field, is 1) holds, then for any central division algebra D over a local field K of *positive* characteristic, the group $SL_2(D)$ contains an arithmetically defined cocompact subgroup of covolume 1.

2. In the finiteness assertions of Theorems 7.3 and 7.8 of [1] about virtually torsion-free arithmetic subgroups Γ' with $0 \neq |\chi(\Gamma')| < c$, we have excluded the groups of type $\mathbf{A}_2$. We will now show that using some of the results contained in §§ 4.6, 3.6 (1) and 2.10 of [1], the finiteness assertions can be proved also for groups of type $\mathbf{A}_2$. We shall use the notation of [1].

Using certain observations in 4.6, 3.6 (1) and 2.10 of [1], we conclude that

$$\begin{aligned} |\chi(\Gamma')| &\geqslant [\Gamma' : \Lambda']^{-1} \prod_{v \in \mathcal{S}_f} |W_v(\mathbf{q}^{-1})|^{-1} \mu_{\mathcal{S}}(G_{\mathcal{S}}/\Lambda) \\ &\geqslant \frac{\prod_{v \in V-S} \# \Xi_{\Theta_v}}{[\Gamma' : \Lambda']} \prod_{v \in \mathcal{S}_f} |W_v(\mathbf{q}^{-1})|^{-1} \mu_{\mathcal{S}}(G_{\mathcal{S}}/\Lambda^m) \\ &\geqslant (\# \prod_{v \in \mathcal{S}} C'(k_v) . \# H^1(k, C)'_{\xi})^{-1} \prod_{v \in \mathcal{S}_f} |W_v(\mathbf{q}^{-1})|^{-1} \mu_{\mathcal{S}}(G_{\mathcal{S}}/\Lambda^m). \end{aligned}$$

Now we note that if G' is simply connected,

$$\# H^1(k, C)'_{\xi} \leqslant 2h_{\ell}\, 3^{a(\ell) + \# T},$$

and if G' is adjoint, then

$$\# H^1(k, C)'_{\xi} \leqslant 2h_{\ell}\, 3^{a(\ell) + \#(S_f^0 \cup T)},$$

where S_f^0 is the subset of places $v \in S_f$ such that G splits over $\hat{k}_v$ and T is the finite set of places $v (\notin S)$ of k such that G is anisotropic over k_v but splits over $\hat{k}_v$. The above bounds follows from the arguments in [1 : § 5] once we observe that if G' is simply connected, $\delta_v\, \varphi'(G'(k_v))$ is trivial for all v, where δ_v is as in [1 : 2.5].

Now since

$$\mu_{\mathcal{S}}(G_{\mathcal{S}}/\Lambda^m) \geqslant D_k^4 (D_\ell / D_k^{[\ell : k]})^{5/2} \left(\prod_{v \in V_\infty} |2^4 \pi^5|_v^{-1} \right) \tau_k(G)\, \mathscr{E},$$

where $\mathscr{E}$ is an in [1 : 3.8], we conclude that

$$c \geqslant B(\mathscr{G}/k)\, \tau \Phi^{EP},$$

where $B(\mathcal{G}/k)$ is as in [1 : 7.4 (12), (13)] and

$$\Phi^{EP} = \prod_{v \in S_f} \varphi_v \prod_{v \in V-S} f_v,$$

for $v \in V - S$; here, f_v is as in [1 : 7.4]; for $v \in S_f^0$,

$$\varphi_v = 3^{-1} e_v \,|\, W_v(\mathbf{q}^{-1})|^{-1},$$

and for $v \in S_f - S_f^0$,

$$\varphi_v = (\# C'(k_v))^{-1} e_v \,|\, W_v(\mathbf{q}^{-1})|^{-1}.$$

We claim that

$$(D_\ell/D_k^{[\ell:k]})^{1/2} \prod_{v \in S_f} \varphi_v > 1.$$

Once the claim is established, Proposition 6.1 (iv), (vi) and the arguments used in the proof of Theorem 7.3 can be employed to establish the finiteness assertions of Theorems 7.3 and 7.8 of [1] for groups of type $\mathbf{A}_2$ also.

To prove the claim, the following case analysis suffices.

(1) If G/k_v is of inner type

$$\varphi_v = 3^{-1} q_v^5(q_v - 1)\,(q_v^3 - 1)^{-1} > 1$$

for all q_v.

(2) If G/k_v is an outer form which splits over $\hat{k}_v$, then

$$\varphi_v = 3^{-1} q_v^5(q_v^2 + 1)\,(q_v^3 + 1)^{-1}\,(q_v + 1)^{-1} > 1$$

for all q_v.

(3) If G/k_v is an outer form which does not split over any unramified extension of k_v, then $|\,W_v(\mathbf{q}^{-1})|^{-1} = (q_v - 1)\,(q_v + 1)^{-1}$, and it is easy to see that $\varphi_v > 1$ whenever $q_v > 3$. If $q_v = 2$, $\ell_v := \ell \otimes_k k_v$, being a ramified extension, does not contain a 3rd root of unity, and so $C'(k_v)$ is trivial, $\varphi_v = 4/3$. Finally, if $q_v = 3$,

$$\varphi_v \geqslant 3^{-1} q_v^2(q_v - 1)^{-1}\,(q_v - 1)\,(q_v + 1)^{-1} = 3/4 > 3^{-1/2}.$$

Now Theorem A of the Appendix of [4] implies the claim.

3. *Errata to* [1]. We take this opportunity to make the following two corrections in [1]. In the statement of Theorem 7.3, " Γ *is virtually free* " should be replaced by " Γ *is virtually torsion-free* " and in the footnote on p. 146 " G is anisotropic over k if the latter is a number field " should be replaced by " G is anisotropic over k if the latter is a global function field ".

4. It has been pointed out to us by E. A. Nisnevič that Theorem B.1 of Appendix B of [1] was proved by him in [2] for all connected reductive groups. His method is different from the one used in [1]. For the sake of completeness, we present a proof of the more general result in our framework.

Theorem (Borel-Serre, Nisnevič). — Let G *be a connected reductive group defined over a global field k. Then the fibers of the canonical map*

$$\lambda^1_G : H^1(k, G) \to \coprod_{v\in V} H^1(k_v, G)$$

are finite.

Let T be the connected component of the identity of the center of G and $G' = G/T$. We know that the fibers of the canonical map

$$\lambda^1_{G'} : H^1(k, G') \to \coprod_{v\in V} H^1(k_v, G')$$

are finite [1 : Theorem B.1]. Now using the standard twisting trick, it suffices to prove that the subset of $H^1(k, G)$ consisting of the elements which are mapped onto the trivial element in both $H^1(k, G')$ as well as in $\coprod_{v\notin S} H^1(k_v, G)$, where S is any fixed finite set of places of k, is finite.

Let S be a finite set of places of k containing all the archimedean places and containing at least one place where G′ is isotropic. Let A be the ring of adèles of k and A_S be the ring of S-adèles, i.e. the restricted direct product of the k_v's for $v \notin S$. For every place v of k, there is a natural homomorphism $\partial_v : G'(k_v) \to H^1(k_v, T)$, and so there is a homomorphism

$$\partial : G'(A_S) \to \prod_{v\notin S} H^1(k_v, T).$$

The kernel of ∂ contains a compact open subgroup K′ of $G'(A_S)$; in fact the image of $G(A_S)$ in $G'(A_S)$, under the natural homomorphism $G(A_S) \to G'(A_S)$ is an open subgroup of $G'(A_S)$, see, for example, [3 : I, 3.6]. In particular, the image of ∂ is contained in $\coprod_{v\notin S} H^1(k_v, T)$. Thus we have a commutative diagram with exact rows

$$\begin{array}{ccccccc}
G'(k) & \longrightarrow & H^1(k, T) & \longrightarrow & H^1(k, G) & \longrightarrow & H^1(k, G') \\
\downarrow & & \downarrow & & \downarrow & & \downarrow \\
G'(A_S) & \longrightarrow & \coprod_{v\notin S} H^1(k_v, T) & \longrightarrow & \coprod_{v\notin S} H^1(k_v, G) & \longrightarrow & \coprod_{v\notin S} H^1(k_v, G').
\end{array}$$

Now the above theorem follows from (1) the finiteness of the kernel of the homomorphism $H^1(k, T) \to \coprod_{v\notin S} H^1(k_v, T)$ [3 : IV, 2.6], (2) the finiteness of the number of double cosets $G'(k)\backslash G'(A)/(G'_S K')$ [1 : Proposition 3.9], and a diagram chase.

REFERENCES

[1] A. Borel and G. Prasad, Finiteness theorems for discrete subgroups of bounded covolume in semi-simple groups, *Publ. Math. I.H.E.S.*, **69** (1989), 119-171.
[2] E. A. Nisnevič, Theorem of properness of Galois cohomology maps over function fields (in Russian), *Dokl. Acad. Nauk BSSR*, **23** (1979), 1065-1068.
[3] J. Oesterlé, Nombres de Tamagawa et groupes unipotents en caractéristique *p*, *Invent. Math.*, **78** (1984), 13-88.
[4] G. Prasad, Volumes of S-arithmetic quotients of semi-simple groups, *Publ. Math. I.H.E.S.*, **69** (1989), 91-117.
[5] W. Scharlau, *Quadratic and hermitian forms*, Springer-Verlag, Heidelberg (1985).

Institute for Advanced Study
Princeton N.J. 08540
U.S.A.

Tata Institute of Fundamental Research
Homi Bhabha Road
Bombay 400005
India

Manuscrit reçu le 5 juin 1990.

140.

Correction and Complement to the Paper: Regularization Theorems in Lie Algebra Cohomology. Applications

Duke Math. Jour. **60** (1990), 299–301

1. In that paper, to be referred to by [2], it is proved that the cohomology of an arithmetic subgroup Γ of the group G of real points of a connected reductive $\mathbb{Q}$-group without nontrivial rational character defined over $\mathbb{Q}$, with coefficients in a finite dimensional complex G-module E, can be computed as the relative Lie algebra cohomology with coefficients in the space $C^\infty_{umg}(\Gamma\backslash G) \otimes E$, where $C^\infty_{umg}(\Gamma\backslash G)$ denotes the space of smooth functions on $\Gamma\backslash G$ of uniform moderate growth [2: 3.1]. This is deduced, by an application of the regularization theorem 2.1 of [2], from a similar statement, where $C^\infty_{umg}(\Gamma\backslash G)$ is replaced by the space $C^\infty_{mg}(\Gamma\backslash G)$ of smooth functions which, together with their $U(\mathfrak{g})$-derivatives have moderate growth, but where the bound on the growth may depend on the derivative (whereas there is a uniform one for elements of $C^\infty_{umg}(\Gamma\backslash G)$). For this latter fact, I referred to 3.4 of [1]. However, as Michael Harris pointed out to me, the notion of differential form of moderate growth used there is not the same as the one just mentioned, whence a gap. The purpose of this Note is to fill it. We assume full familiarity with [1], [2], and use the notation there without further comment. In addition, we let $C^\infty_{(mg)}(\Gamma\backslash G)$ be the space of smooth functions of moderate growth on $\Gamma\backslash G$.

2. In [2], we consider the complex

(1) $$A^\infty_{mg}(\Gamma\backslash G; \tilde{E}) = C^*(\mathfrak{g}, K; C^\infty_{mg}(\Gamma\backslash G) \otimes E)$$

of smooth differential forms with moderate growth and the natural inclusion

(2) $$A^\infty_{mg}(\Gamma\backslash X; \tilde{E}) \to A^\infty((\Gamma\backslash X); \tilde{E})$$

or, equivalently,

(3) $$i: C^*(\mathfrak{g}, K; C^\infty_{mg}(\Gamma\backslash G) \otimes E) \to C^*(\mathfrak{g}, K; C^\infty(\Gamma\backslash G) \otimes E)$$

and need to know

Received September 2, 1989.

THEOREM 1. *The cohomology morphism $H^*(i)$ induced by the natural inclusion i of* (3) *is an isomorphism.*

It is for this statement that I referred to 3.4 of [1]. But the complex under consideration there is the complex $\Omega^*_{mg}(\Gamma\backslash X; \tilde{E})$ of smooth differential forms which, together with their exterior derivative, have moderate growth. In the present notation, it can be defined as

$$\{\omega \in C^*(\mathfrak{g}, K; C^\infty(\Gamma\backslash G) \otimes E) | \eta, d\eta \in C^*(\mathfrak{g}, K; C^\infty_{(mg)}(\Gamma\backslash G) \otimes E)\}$$

(Note that $C^*(\mathfrak{g}, K; C^\infty_{(mg)}(\Gamma\backslash G) \otimes E)$ is just a subspace; it is not stable under d.) However, the proof that $H^*(i)$ is an isomorphism follows the same pattern:

First, there is the usual reduction to normal subgroups of finite index, which allows us to assume that Γ is neat. In the compactification $\Gamma\backslash\bar{X}$ of $\Gamma\backslash X$ by corners, let $\mathscr{A}_{mg}$ be the presheaf assigning to an open subset U the space $A^\infty_{mg}(U \cap (\Gamma\backslash X); \tilde{E})$ of smooth differential forms on $U \cap (\Gamma\backslash X)$ which, together with their $U(\mathfrak{g})$ derivatives, have moderate growth. This is a sheaf, also denoted $\mathscr{A}_{mg}$ whose sections on $\Gamma\backslash\bar{X}$ form our complex $A^\infty_{mg}(\Gamma\backslash X; \tilde{E})$. It suffices therefore to show that $\mathscr{A}_{mg}$ is a fine resolution of $\tilde{E}$.

In the corner which contains the closure of $\mathfrak{S}_{t,\omega}$ we take the same local coordinates β^i $(1 \leqslant i \leqslant l)$ and x^j $(l < i < m)$ as in [1: 3.4]. As on p. 33 of [1], we shall also write, for uniformity of notation, x^i for β^i $(1 \leqslant i \leqslant l)$. Let $\mathscr{F}$ be the space of functions on $\mathfrak{S}_{t,\omega}$ which are finite sums of expressions

$$(x^1)^{\lambda_1} \cdot \dots \cdot (x^l)^{\lambda_l} \cdot Q(x^{l+1}, \dots, x^m),$$

where Q is a smooth function of the x^i's $(l < i < m)$ and the λ_i are real numbers. The elements of $\mathscr{F}$ have all moderate growth. This space is obviously stable under the derivatives $\partial\backslash\partial x^j$. On the other hand, it is readily seen, for instance by using a matrix representation of G in which A is diagonal, that the left-invariant vector fields, restricted to $\mathfrak{S}_{t,\omega}$ are linear combinations of the $\partial\backslash\partial x^j$ with coefficients in $\mathscr{F}$. Therefore $\mathscr{F}$ is also stable under $U(\mathfrak{g})$-derivatives, hence belongs to $C^\infty_{mg}(\mathfrak{S}_{t,\omega})$ and we also have

$$\mathscr{F} \cdot C^\infty_{mg}(\mathfrak{S}_{t,\omega}) \subset C^\infty_{mg}(\mathfrak{S}_{t,\omega}). \tag{4}$$

This makes it clear that the arguments given in [1: 3.4] extend to the present case:

First, if φ is a smooth function with support in the corner and $\eta \in A_{mg}(\mathfrak{S}_{t,\omega}: \tilde{E})$, then $\varphi\eta$ and $d(\varphi\eta)$ also belong to $A_{mg}(\mathfrak{S}_{t,\omega}; \tilde{E})$, hence $\mathscr{A}_{mg}$ is fine.

The sections of $\tilde{E}$ are constant on simply connected open subsets of $\mathfrak{S}_{t,\omega}$. It follows then also that they belong to $A^0_{mg}(\mathfrak{S}_{t,\omega}, E)$ whence $H^0(\mathscr{A}_{mg}) = \tilde{E}$.

The ω^i's are left-invariant one-forms, hence are linear combinations of the dx^i's with coefficients in $\mathscr{F}$. Thus, (3) and the relation (6) of [1: 3.4] imply that the homotopy operator A of the Poincaré lemma preserves the condition of moderate growth for all $U(\mathfrak{g})$-derivatives. Hence the Poincaré lemma is valid for $\mathscr{A}_{mg}$.

3. In [2], we also consider the complex, denoted there $\Omega^*_{fd}(X; E)^\Gamma$, of smooth differential forms which, together with their exterior derivatives, are fast decreasing. It contains the complex $\Omega^*_c(X, E)^\Gamma$ of compactly supported smooth forms and we prove that this inclusion is a cohomology isomorphism [1: 5.2]. Now $H^*_c(\Gamma\backslash X; \tilde{E})$ may be identified to the cohomology of the relative Lie algebra complex $C^*(\mathfrak{g}, K; C^\infty_c(\Gamma\backslash G) \otimes E)$ and there is again a companion to Theorem 1, in which $C^\infty_{mg}(\Gamma\backslash G)$ is replaced by the space $C^\infty_{fd}(\Gamma\backslash G)$ of smooth functions on $\Gamma\backslash G$ which, together with all their $U(\mathfrak{g})$-derivatives, are fast decreasing.

THEOREM 2. *The inclusion of relative Lie algebra complexes*

$$C^*(\mathfrak{g}, K; C^\infty_c(\Gamma\backslash G) \otimes E) \to C^*(\mathfrak{g}, K; C^\infty_{fd}(\Gamma\backslash G) \otimes E) \tag{5}$$

is a cohomology isomorphism.

The proof is as before, but in the slightly different context of [1: 5.2], and we describe it briefly.

Let $\mathscr{A}_c$ (resp. $\mathscr{A}_{fg}$) by the presheaf on $\Gamma\backslash\bar{X}$ which assigns to an open set U the smooth forms on $U \cap (\Gamma\backslash X)$ with relatively compact support in $\Gamma\backslash X$ (resp. with coefficients in $C^\infty_{fd}(U \cap (\Gamma\backslash X)) \otimes E$). It is a sheaf, which is fine. This is standard for $\mathscr{A}_c$. It is seen as above for $\mathscr{A}_{fg}$, taking into account the relation

$$C^\infty_{mg}(\Gamma\backslash G) \cdot C^\infty_{fd}(\Gamma\backslash G) \subset C^\infty_{fd}(\Gamma\backslash G) \tag{6}$$

and the fact that the left-invariant vector fields are linear combinations of the derivatives with respect to the local coordinates with coefficients of moderate growth, which is seen as above.

The space of sections over $\Gamma\backslash\bar{X}$ of $\mathscr{A}_c$ (resp. $\mathscr{A}_{fd}$) is the left-hand (resp. right-hand) side of (5). To establish Theorem 2 it suffices therefore, as in [1: 5.2], to show that the inclusion $\mathscr{A}_c \to \mathscr{A}_{fd}$ induces an isomorphism of the derived sheaves. As in *loc. cit.*, p. 47, this amounts to proving that $H^i(\mathscr{A}_{fd})_x = 0$ for all i's and all $x \in \Gamma\backslash\bar{X} - \Gamma\backslash X$. The argument is the one used on p. 47 of [1], once it is noticed that the $U(\mathfrak{g})$-derivatives of $c_I(y)$ in (13) are fast decreasing if those of f are. This in turn follows from (6) and the property of the left-invariant vector fields already mentioned after (6).

REFERENCES

1. A. BOREL, *Stable real cohomology of arithmetic groups II*, Progress in Math. **14**, 21–55, Birkhäuser-Boston, 1981.
2. A. BOREL, *Regularization theorems in Lie algebra cohomology. Applications*, Duke Math. Jour. **50**(1983), 605–623.

INSTITUTE FOR ADVANCED STUDY, SCHOOL OF MATHEMATICS, PRINCETON, NEW JERSEY 08540.

141.

(with A. Ash)

Generalized Modular Symbols

in *Cohomology of arithmetic groups and automorphic forms*
(J-P. Labesse and J. Schwermer ed.), Lect. Notes Math. **1447** (1990), 57–75, Springer-Verlag

Introduction

Let G be a connected semi-simple Q-group, X or $X_{G(\mathbb{R})}$ the space of maximal compact subgroups of the group $G(\mathbb{R})$ of real points of G and Γ an arithmetic subgroup of $G(Q)$. Let H be a connected Q–subgroup and $e \in X$ be fixed under a maximal compact subgroup L of $H(\mathbb{R})$. Then $h \mapsto h \cdot e$ defines a closed embedding of $X_H = H(\mathbb{R})/L$ into X whence also a natural map

$$\lambda_{e,\Gamma,H} : (\Gamma \cap H) \setminus X_H \longrightarrow \Gamma \setminus X,$$

whose image is to be denoted by $[H(\mathbb{R}).e]_\Gamma$. If H is reductive or has no non-trivial rational character defined over Q, $\lambda_{e,\Gamma,H}$ is known to be proper (See e.g. 2.7 in [A]; it is stated there only for Levi subgroups, but the proof is valid without change for any connected reductive Q–subgroup.) We are interested in cases where it is injective (for suitable e, Γ), $[He]_\Gamma$ is orientable and its fundamental class is not homologous to zero in $\Gamma \setminus X$, in singular homology or homology with closed supports, as the case may be. We are mainly concerned with the case where H is a Levi Q-subgroup M of a parabolic Q-subgroup P of G (2.6, 2.8), then $[H(\mathbb{R}).e]_\Gamma$ is a "generalized modular symbol", or the unipotent radical N of P. Our chief tool will be the fact that, after suitable extension k of the groundfield, M is the fixed point set of a finite Q-group of automorphisms all defined over k (2.1, 2.2).

By 4.1 in [A], the dual cohomology classes of the generalized modular symbols restrict nontrivially to the corresponding faces $e'(P)$ of the compactification by corners of [BS] and are not square integrable. A variant of the proof will also lead in §4 to some classes whose restriction to $e'(P)$ are not zero (more precisely are dual to images of interior cohomology classes of the base of $e'(P)$) and which are non-square integrable (§5).

The results on cohomology classes associated to modular symbols are in substance the same as those of [A]. However, the proofs there made use of results in §10 of [BS], which are not true in the generality stated there, as already pointed out on p. 704 of [B], whence the necessity of another proof.

We keep the notation and conventions of the introduction. Unexplained notation pertaining to reduction theory and corners is as in [BS].

§1. INJECTIVITY OF $\lambda_{e,\Gamma,H}$.

1.0. We recall some known facts about X, to be used later without further reference. Let slightly more generally H be a real semi-simple Lie group with finitely many connected components, whose identity component H^o has a finite center and X be the space of maximal compact subgroups of H. It is acted upon transitively by H via conjugation and more generally by Aut$H$. It is homeomorphic to Euclidean space and admits a Riemannian metric invariant under any automorphism of H with respect to which it is complete, with negative curvature. Any two points $a, b \in X$ are joined by a unique geodesic, which is therefore pointwise fixed under any isometry, leaving a and b fixed. Consequently, the fixed point set of a subgroup L of H is a totally geodesic submanifold homeomorphic to Euclidean space. Finally, by E. Cartan's fixed point theorem, any compact group of isometries of X has a fixed point. To a maximal compact subgroup L of H is associated uniquely an involution θ_L of H (the Cartan involution) whose fixed point set is L. If a closed subgroup Q of H is stable under θ_L, then Q is reductive, the restriction of θ_L to Q is a Cartan involution of Q and $L \cap Q$ is a maximal compact subgroup of Q.

1.1. As before, P denotes a parabolic $\mathbb{Q}$-subgroup of G, M a Levi $\mathbb{Q}$-subgroup of P and N its unipotent radical. Let S be a maximal $\mathbb{Q}$-split subgroup of the center of M and $A = S(\mathbb{R})^o$ the identity component of the group of real points of S. Then $M = \mathcal{Z}(S)$ or also $M = \mathcal{Z}(A)$. As usual, let ${}^oM = \cap_{\chi \in X(P)_{\mathbb{Q}}} \ker \chi^2$. Then $P(\mathbb{R}) = {}^oM(\mathbb{R}).A.N(\mathbb{R})$ (semi-direct). We let $\Phi(P, A)$ be the set of roots of P with respect to A and $\Delta(P, A)$ or Δ the set of simple elements of $\Phi(P, A)$. For $t > 0$, let $A_t = \{a \in A | a^\alpha \geq t (\alpha \in \Delta)\}$.

For brevity, we shall sometimes say that $a \in A$ is ***sufficiently far out*** if $a \in A_t$ for some sufficiently big t.

Recall that if Γ is neat, then $\Gamma \cap P \subset {}^oM(\mathbb{R}).N(\mathbb{R})$. We also write Γ_P for $\Gamma \cap P$, Γ_N for $\Gamma \cap N$.

1.2. PROPOSITION: *Assume Γ to be neat. Then there exists t_0 such that $\lambda_{a.e,\Gamma,N}$ is injective for $a \in A_t, t \geq t_0$.*

Proof. We have to show that if t is big enough, a relation

$$\gamma.n.a.e = n'.a.e \qquad (\gamma \in \Gamma; n, n' \in N(\mathbb{R}), a \in A_t) \tag{1}$$

implies that $\gamma \in N$. Let C be a compact subset of $N(\mathbb{R})$ containing a fundamental domain for $\Gamma \cap N$ in $N(\mathbb{R})$. The equality (1) implies the existence of $\gamma_1 \in (\Gamma \cap N).\gamma.(\Gamma \cap N)$ and $c, c' \in C$ such that

$$\gamma_1.c.a.e = c'.a.e. \tag{2}$$

By (2), there exists $k \in K_e$, the isotropy group of e, such that $\gamma_1.c.a = c'.a.k$. By standard reduction theory, there exists t_0 such that if $t \geq t_0$, then (2) implies $\gamma_1 \in \Gamma \cap P$. We have then $k \in P(R)$. Let π be the natural projection of P onto $P/N \cong M$. We have $\pi(\gamma_1)\pi(a) = \pi(a)\pi(k)$. But $\pi(\Gamma \cap P)$ is torsion free, since Γ is neat, $\pi(k)$ belongs to a compact subgroup of $M(\mathbb{R})$ and $\pi(a)$ is central on M. Consequently, $\pi(\gamma_1) = 1$, $\gamma_1 \in N$ and therefore $\gamma \in N$.

1.3. The following lemma is a straightforward generalization of a well-known lemma of H. Jaffee (see [M]), which is the special case where Σ is of order two.

Let Σ be a finite group of Q-automorphisms of G. It operates canonically on X, and has at least one fixed point (i.e., $G(\mathbb{R})$ has at least one maximal compact subgroup stable under Σ). We choose one as our base point, to be denoted e.

LEMMA. *We keep the previous assumptions and let $H = G^{\Sigma}$. Assume that Γ is Σ-stable and torsion free. Then $\lambda_{e,\Gamma,H}$ is injective.*

Proof. Let $u, v \in H(\mathbb{R})$ and $\gamma \in \Gamma$ be such that $\gamma.u.e. = v.e$. We have to show that $\gamma \in H$, hence that ${}^{\sigma}\gamma = \gamma$ for all $\sigma \in \Sigma$. Applying σ to the relation $\gamma.u.e. = v.e$ we get ${}^{\sigma}\gamma.u.e = v.e$; therefore $\gamma^{-1}.{}^{\sigma}\gamma$ belongs to the isotropy group of $u.e$, which is compact, and on the other hand to Γ, which is torsion free, whence $\gamma^{-1}.{}^{\sigma}\gamma = 1$.

1.4. PROPOSITION. *Let m be equal either to four or to an odd prime p; let ζ_m be a primitive m-th root of 1 and $k = Q(\zeta_m)$. Let Σ be a finite group of k-automorphisms of $G \times k$ which is defined over Q. Assume that $H \times k = (G \times k)^{\Sigma}$. Then, for suitable $e \in X$ and Γ, the map $\lambda_{e,\Gamma,H}$ is injective.*

The proof will be given in 1.6, after some preliminaries. The field k is a purely imaginary quadratic extension of a totally real subfield ℓ. Let V_∞ be the set of archimedean places of k. The restriction to ℓ defines a bijection of V_∞ onto the set of archimedean places of ℓ. We let l_v and k_v be the completions of ℓ and k at $v \in V_\infty$. Thus $\ell_v \cong \mathbb{R}$ and $k_v \cong C$ and we have a commutative diagram

$$\begin{array}{ccc} G(\ell_v) & \longrightarrow & G(k_v) \\ \uparrow{\scriptstyle i_v} & & \uparrow{\scriptstyle j_v} \\ G(\mathbb{R}) & \longrightarrow & G(C) \end{array} \tag{1}$$

where i_v, j_v are isomorphisms and the horizontal arrows natural inclusions. Let τ be the generator of $\mathrm{Gal}(k/\ell)$. It extends to the complex conjugation τ_v of k_v with respect to ℓ_v and goes over by j_v to the complex conjugation τ_o of C. The group Σ defines a group $\sum_v$ of automorphisms of $G(k_v)$, which is stable under τ_v, since it is defined over Q. Let $\sum_v'$ be the semi-direct product of $\sum_v$ and $\langle\tau_v\rangle$. Its fixed point set is $i_v(H(\mathbb{R}))$.

1.5. Lemma. *We keep the previous notation. Fix a maximal compact subgroup L of $H(\mathbb{R})$. Then there exist a maximal compact subgroup K of $G(\mathbb{R})$ and for every $v \in V_\infty$ a maximal compact subgroup K_v of $G(k_v)$ stable under $\sum_v'$ such that $\iota_v(K) = G(\ell_v) \cap K_v$.*

Proof. Consider 1.4(1) for two places $v, w \in V_\infty$. Let $j_{v,w} = j_w \circ j_v^{-1}$. Then $j_{v,w}$ is an isomorphism of $G(k_v)$ onto $G(k_w)$ leaving $G(k)$ stable, whose restriction to $G(k)$ is an element γ of $\mathrm{Gal}(k/Q)$ which, necessarily, leaves $G(\ell)$ stable and commutes with τ. The element γ also operates on $\mathrm{Aut}(G(k))$ and leaves Σ invariant, hence γ transforms $\sum_v'$ onto $\sum_w'$. As a result, $\sum_o = j_v^{-1}(\sum_v)$ is independent of v, and so is $\sum_o' = j_v^{-1}(\sum_v')$, which is the semi-direct product of $\langle\tau_o\rangle$ and $\sum_o$.

Let X_c be the symmetric space of maximal compact subgroups of $G(C)$. The fixed point set of τ_o on X_c may be identified with X. Let F be the fixed point set of L in X_c. It is a totally geodesic submanifold (1.0), whose points represent the maximal compact subgroups of $G(C)$ containing L. Since L is contained in the fixed point set of $\sum_o'$, the latter leaves F stable. It has at least one fixed point on F (1.0), say e. It is fixed under τ_o, hence belongs to X and defines a maximal compact subgroup K_o of $G(C)$ whose intersection K with $G(\mathbb{R})$ is maximal compact in $G(\mathbb{R})$. Since $\sum_v' = j_v(\sum_o')$ for all v's, it follows that the groups $K_v = j_v(K_o)$ satisfy our condition.

1.6. *Proof of 1.4:* We take for $e \in X$ the fixed point of K. Let $\widetilde{\Gamma} \subset G(k)$ be an arithmetic subgroup of $G \times k$, which is torsion free and stable under Σ. Its intersection Γ with $G(Q)$ is an arithmetic subgroup of $G(Q)$. We claim that, with these choices, $\lambda_{e,\Gamma,H}$ is injective.

Let $G' = R_{k/Q}G$ and $H' = R_{k/Q}H$. We have

$$G'(R) = \prod_{v \in V_\infty} G(k_v), \qquad H'(\mathbb{R}) = \prod_{v \in V_\infty} H(k_v).$$

We may view G as a Q-subgroup of G'. The product ι of the ι_v's defines a diagonal embedding of $G(\mathbb{R})$ into $G'(\mathbb{R})$ and an embedding of X into $\prod_v X_{G(k_v)}$. By construction $\iota_v(e) = e_v$ is the fixed point of a Σ_v-stable maximal compact subgroup K_v of $G(k_v)$.

We have the canonical isomorphism $G(k) \xrightarrow{\sim} G'(Q)$ and Σ may be viewed as a group of Q-automorphisms of G', whose fixed point set is H'. Let $\gamma \in \Gamma$, $h, h' \in H(\mathbb{R})$ be such that $\gamma.h.e = h'.e$. We have to show that $\gamma \in H$. We have

$$\iota_v(\gamma)\iota_v(h).e_v = \iota_v(h')e_v \qquad (v \in V_\infty).$$

By 1.3, applied to H' and G', this yields that

$$\iota(\gamma) \in H'.$$

But $\iota(\gamma) \in \iota(G)$. Since $H' \cap \iota(G) = \iota(H)$ it follows that $\gamma \in H$, as was to be proved.

§2. Generalized modular symbols.

Let P and $P(\mathbb{R}) = {}^\circ M(\mathbb{R}).A.N(\mathbb{R})$ be as in 1.1. The following proposition shows that we can apply 1.4. to the modular symbol defined by $M(\mathbb{R}) = {}^\circ M(\mathbb{R}).A$.

2.1. Proposition: *Assume G to be of adjoint type. Let m be either an odd prime or equal to four, and strictly greater than the coefficients of the highest Q-root (as expressed as a linear combination of simple Q-roots). Let $k = Q(\zeta_m)$, where ζ_m is a primitive m-th root of 1. Then every Levi Q-subgroup in G is the centralizer of a finite commutative subgroup of $G(k)$, product of cyclic groups of order m.*

For any extension k' of Q, the group $P \times k'$ is parabolic in $G \times k'$ and $M \times k'$ is a Levi k'-subgroup of $P \times k'$. Moreover, each coefficient of the highest k'-root, expressed as a linear combination of simple k' roots, is majorized by one coefficient of the highest Q-root. Also, if G is adjoint, so is $G \times k'$. Therefore, 2.1. follows from the following proposition.

2.2. Proposition: *Let k be a number field and H a semi-simple k-group of adjoint type. Assume that $k \supset Q(\zeta_m)$, where m is either an odd prime or four, strictly greater than the coefficients of the highest k-root, and ζ_m a primitive m-th root of one. Then any Levi k-subgroup in H is the centralizer of a finite commutative subgroup of $H(k)$, product of cyclic subgroups of order m.*

Proof. Let S be a maximal k-split torus in H and Φ the set of k-roots of H with respect to S. Choose an ordering in Φ and let Δ be the set of simple roots. Then the group of k-characters $X(S)$ is generated by Δ, since H is adjoint. Let $Y(S)$ be the group of 1-parameter subgroups of S defined over k, which will have the dual basis $\alpha\check{}$ $(\alpha \in \Delta)$. For any $\alpha \in \Delta$, let $\varphi_\alpha : \mathbf{GL}_1 \to S$ define the 1-parameter subgroup denoted

$\alpha\check{}$. Set $S_\alpha = Im(\varphi_\alpha)$. Then $S = \prod S_\alpha$ (adjoint situation). Also, for any $\beta \in X(S)$, $\varphi_\alpha(t)^\beta = t^{<\alpha\check{},\beta>}$.

Let μ_m denote the group of the m-th roots of 1, and set $\mu_\alpha = \varphi_\alpha(\mu_m)$. For any subset $I \subset \Delta$, set $S_I = (\cap_{\beta\in I}\ker(\beta))^o = \prod_{\alpha\in I'} S_\alpha$, where I' denotes the complement of I in Δ. Then $\mathcal{Z}_H(S_I)$ is a Levi k-subgroup of a parabolic k-subgroup P_I of H, and every Levi k-subgroup in H arises in this way, up to a conjugacy under $H(k)$.

Set $\mu_{I'} = \prod_{\alpha\in I'}\mu_\alpha$. We claim that $\mathcal{Z}_H(S_I) = \mathcal{Z}_H(\mu_{I'})$.

First we show their identity components are equal by using the Lie algebras. We know that $\mathcal{Z}_H(S_I)$ is connected. Let $\mathfrak{h}$ denote Lie(H) and $\mathfrak{h}_\alpha$ the various root spaces. Let $\langle I\rangle$ denote the span of the elements of I in Φ. Then $\mathfrak{h} = \text{Lie}\,\mathcal{Z}_H(S_I) \oplus \oplus_{\beta\notin\langle I\rangle}\mathfrak{h}_\beta$, while Lie $\mathcal{Z}_H(S_I) = \text{Lie}\,\mathcal{Z}_H(S) \oplus \oplus_{\beta\in\langle I\rangle}\mathfrak{h}_\beta$.

If $\beta \in \langle I\rangle$ and α is not in I, $\langle\beta,\alpha\check{}\rangle = 0$, which implies that $\mathcal{Z}_H(S_I) \subset \mathcal{Z}_H(\mu_{I'})$. To see the opposite inclusion, we must show if β is not in $\langle I\rangle$, that Lie $\mathcal{Z}_H(\mu_{I'}) \cap \mathfrak{h}_\beta = (0)$. Here we use the assumption on m. Write $\beta = c\gamma + \ldots$, with $c \neq 0$, $c \in \mathbb{Z}$, and $\gamma \in I'$. Let $t = \varphi_\gamma(\zeta_m) \in \mu_\gamma$. Then $t^\beta = \zeta_m^c \neq 1$ by hypothesis. Thus any nontrivial element of $\mathfrak{h}_\beta$ does not commute with μ_γ, and a fortiori is not in Lie $\mathcal{Z}_H(\mu)$.

So the identity components of $\mathcal{Z}_H(S_I)$ and $\mathcal{Z}_H(\mu)$ are equal. The problem now is to show that $\mathcal{Z}_H(\mu)$ is connected. Take any $x \in \mathcal{Z}_H(\mu)$. Then x normalizes S_I because S_I is the maximal k-split torus in the center of $\mathcal{Z}_H(\mu)^o$. So x permutes the roots of H with respect to S. For $\gamma \in I'$, denote by $\overline{\gamma}$ its restriction to S_I. The $\overline{\gamma}$'s form a basis of $X(S_I)$ and the elements of $\Phi(S_I, H)$ are (positive or negative) integral linear combinations of the $\overline{\gamma}$ ($\gamma \in I'$). It suffices to show that g induces the trivial permutation of $\Phi(S_I, H)$ for, since $\Phi(S_I, H)$ generates $X(S_I)$, then Int x centralizes S_I.

For any $t \in S_I$ and any simple root $\gamma \in I'$, $(x^{-1}.t.x)^{\overline{\gamma}} = t^{x\overline{\gamma}}$. If $t \in \mu_I$, $x^{-1}tx = t$ and we have $t^{\overline{\gamma}} = t^{x\overline{\gamma}}$. As already noted, we can write $x\overline{\gamma} = \sum_{\alpha\in I'} c(\alpha)\overline{\alpha}$ with $c(\alpha) \in \mathbb{Z}$. We want to show $x\overline{\gamma} = \overline{\gamma}$.

Suppose there is some $\alpha \neq \gamma$ with $c(\alpha) \neq 0$. Set $t = \varphi_\alpha(\zeta_m)$. Then $t^{\overline{\gamma}} = 1$, since $\alpha \neq \gamma$. On the other hand, $t^{x\overline{\gamma}} = \zeta^{c(\alpha)} \neq 1$. This is a contradiction.

The only case left is when $x\overline{\gamma} = c(\gamma)\overline{\gamma}$ for all $\gamma \in I'$. We claim that $c(\gamma) = \pm 1$. To see this apply x to an arbitrary root: $x\sum_{\alpha\in I'} d(\alpha)\overline{\alpha} = \sum_{\alpha\in I'} d(\alpha)c(\alpha)\overline{\alpha}$. Since x permutes the roots, we must recover each simple root this way, so that $d(\alpha)c(\alpha) = 1$ for an appropriate choice of $d(\alpha)$. Hence $c(\alpha) = \pm 1$ for all $\alpha \in I'$.

Now suppose $c(\gamma) = -1$ for some $\gamma \in I'$. Then if $t = \varphi_\gamma(\zeta_m)$, we have $\zeta_m = t^{\overline{\gamma}} = t^{x\overline{\gamma}} = t^{-\overline{\gamma}} = \zeta_m^{-1}$, implying $\zeta_m^2 = 1$, a contradiction since $m > 2$ by assumption.

2.3. *Remarks.* 1) If $G = \mathbf{SL}_n$, then every Levi Q-subgroup in G is the fixed point set of an elementary abelian 2-group of Q-automorphisms (not necessarily interior). In fact, up to Q-conjugacy, M is the group $S(\mathbf{GL}_{n_1} \times \cdots \times \mathbf{GL}_{n_s})$ of elements of determinant one in a direct product of diagonal blocks $\mathbf{GL}_{n_i}$ $(i = 1, ..., s; n = n_1 + \ldots + n_s)$. Then M is the centralizer in $\mathbf{GL}_n$ of the group of elements $\mathrm{diag}(\pm I_{n_1}, \ldots, \pm I_{n_s})$, (some of which will have determinant -1 if some n_i is odd).

2) If the system of Q-roots of G is of classical type, or non-reduced, then the biggest coefficient of the highest root is two. In this case we may therefore take $m = 3, 4$.

2.4. *Orientations.* Choose a base point $e \in X$ whose isotropy group K intersects $M(\mathbb{R})$ in a maximal compact subgroup. Then, automatically, $K \cap M = K \cap {}^\circ M(\mathbb{R})$. The map $p \mapsto p.e$ defines a homeomorphism of $X_{M(\mathbb{R})} \times N(\mathbb{R})$ onto X. We view X as a fibre bundle over $X_{\circ M(\mathbb{R})}$ with typical fibre $N(\mathbb{R})$. Fix orientations of X, $X_{M(\mathbb{R})}$ and $N(\mathbb{R})$ compatible with this homeomorphism. For each fibre $mN(\mathbb{R})$, choose the orientation of $mN(\mathbb{R})$ so that $[X_M]$ and $[mN(\mathbb{R})]$ define $[X]$. Of course, translations by elements of ${}^\circ M(\mathbb{R})$ permute the fibres. Note that the identity component of ${}^\circ M(\mathbb{R}).A$ preserves the orientations just chosen, and is transitive on $X_{M(\mathbb{R})}$.

In the sequel, it is necessary that $\Gamma_P = \Gamma \cap P$ preserves the chosen orientations of X, $X_{M(\mathbb{R})}$ and the fibres. We claim it suffices for this that Γ be neat and contained in $G(\mathbb{R})^\circ$. Indeed, any element of $G(\mathbb{R})^\circ$ preserves the orientation of X. Moreover, for $p \in P$, $p \mapsto \det Ad\, p|n$ is a rational Q-character, hence contains any neat arithmetic subgroup in its kernel. Under our conditions therefore $\gamma \in \Gamma_P$ preserves the orientations of X and of the fibres, hence of $X_{M(\mathbb{R})}$, too.

Remark. Similar orientation questions are discussed in [A], 3.2, 3.3. However the argument in 3.3 is not correct since the character χ defined there does not come from a Q-morphism of algebraic groups and it is therefore not clear that it is trivial on any neat arithmetic subgroup.

2.5. THEOREM: *Let P be a parabolic Q-subgroup of G and $P(\mathbb{R}) = {}^\circ M(\mathbb{R}).A.N(\mathbb{R})$ a Langlands decomposition of $P(\mathbb{R})$ (see 1.1). Then there exist $e \in X$, a neat subgroup Γ' of finite index of Γ contained in $G(\mathbb{R})^\circ$ and $t > 0$ such that*

(i) The maps $\lambda_{a.e,\Gamma',N}(a \in A_t)$ and $\lambda_{e,\Gamma',M}$ are injective. Their images $[N(\mathbb{R})ae]_{\Gamma'}$ and $[M(\mathbb{R})e]_{\Gamma'}$, are closed oriented submanifolds of $\Gamma'\backslash X$.

(ii) The submanifolds $[N(\mathbb{R})ae]_{\Gamma'}$ and $[M(\mathbb{R})e]_{\Gamma'}$ intersect transversally at finitely many points, each with intersection number one.

Proof: (a) We assume first that there is a finite group Σ of Q-automorphisms of G such that $M = G^{\Sigma}$. We then choose Γ' of finite index in Γ, and stable under Σ. This is always possible: Select first Γ_1 satisfying our first two conditions and then take the intersection of its conjugates under the elements of Σ.

Next we choose a fixed point e of Σ on X and denote by K its isotropy group in $G(\mathbb{R})$. The Cartan involution of $G(\mathbb{R})$ with respect to K commutes with every element of Σ, therefore leaves $M(\mathbb{R})$ stable and induces a Cartan involution on it; hence $K \cap M(\mathbb{R})$ is a maximal compact subgroup of $M(\mathbb{R})$, therefore also of $P(\mathbb{R})$. We may use e to define the modular symbol associated to M. By 1.2 and 1.3, there exists $t > 0$ such that *(i)* is satisfied. There remains to prove *(ii)*.

Note first that, since $\sigma \in \Sigma$ fixes A pointwise, its differential leaves the root subspaces of A in $\mathfrak{g}$ stable; in particular, σ leaves $N(\mathbb{R})$ stable. Also Σ operates naturally on $\Gamma'\backslash X$ and leaves $[M(\mathbb{R}).e]_{\Gamma'}$ pointwise fixed. Assume that $m.e = n.a.e \bmod \Gamma'$ for some $m \in M(\mathbb{R})$, $n \in N(\mathbb{R})$ and $a \in A_t$. Then we also have ${}^{\sigma}n.a.e = n.a.e \bmod \Gamma'$, and therefore, by 1.2, $\bmod \Gamma'_N$. There exists therefore $\tau_\sigma \in \Gamma'_N$ such that

$$ {}^{\sigma}n = \tau_\sigma.n. \qquad (\sigma \in \Sigma) \qquad (1)$$

Starting again with

$$ m.e = \gamma.n.a.e \qquad (m \in M(\mathbb{R}), n \in N(\mathbb{R}), a \in A_t \text{ and } \gamma \in \Gamma'), \qquad (2)$$

we get

$$ {}^{\sigma}\gamma.{}^{\sigma}n.a.e = \gamma nae. $$

Together with (1), this yields

$$ \gamma^{-1} \cdot {}^{\sigma}\gamma.\tau_\sigma.nae = nae, $$

and shows that $\gamma^{-1}.{}^{\sigma}\gamma.\tau_\sigma$ belongs to the isotropy group of $na.e$, which is compact. Since on the other hand it is contained in the torsion free group Γ', it is equal to 1, hence

$$ {}^{\sigma}\gamma = \gamma.\tau_\sigma^{-1} \qquad (\sigma \in \Sigma) \qquad (3)$$

$$ {}^{\sigma}(\gamma.n) = {}^{\sigma}\gamma.{}^{\sigma}n = \gamma.\tau_\sigma^{-1}.\tau_\sigma.n = \gamma.n \qquad (\sigma \in \Sigma); $$

therefore $\gamma n \in M(\mathbb{R})$ and

$$ \gamma \in \Gamma'_P. \qquad (4)$$

(b) That was the main point. The rest of the argument does not use the special assumption made in (a). To make this explicit for later reference, we derive *(ii)* only assuming *(i)* and that (2) implies (4). In other words, if we denote by $[M(\mathbb{R})e]_{\Gamma'\cap P}$ and $[N(\mathbb{R})ae]_{\Gamma'\cap P}$ the projections of $M(\mathbb{R})e$ and $N(\mathbb{R})ae$ in $(\Gamma'\cap P)\backslash X$, we assume that the natural projection $(\Gamma'\cap P)\backslash X \to \Gamma'\backslash X$ is a bijection of $[M(\mathbb{R})e]_{\Gamma'\cap P} \cap [N(\mathbb{R})ae]_{\Gamma'\cap P}$ onto $[M(\mathbb{R})e]_{\Gamma'} \cap [N(\mathbb{R})ae]_{\Gamma'}$, which reduces us to study the former.

Γ'_P contains $(\Gamma'\cap M)\cdot(\Gamma'\cap N)$ as a subgroup of finite index. The projection $\pi : P \to P/N$ is a Q-isomorphism of M onto P/N. It maps Γ'_P onto an arithmetic subgroup Γ'_M of P/N, which contains $\pi(\Gamma\cap M)$ as a subgroup of finite index. Let $\widetilde{\Gamma}'_M$ be the image of Γ'_M in $M(Q)$ under the inverse of the restriction of π to M. Then $\widetilde{\Gamma}'_M$ contains $\Gamma'\cap M$ as a subgroup of finite index, and $\widetilde{\Gamma}'_M/(\Gamma'\cap M) \cong \Gamma'_P/(\Gamma'\cap M)(\Gamma'\cap N)$. We can write uniquely $\gamma = \gamma_m.\gamma_n$ with $\gamma_m \in \widetilde{\Gamma}'_M, \gamma_n \in N(Q)$ and get

$$m.e = \gamma_m\gamma_n n.ae.$$

Therefore there exists $k \in K$ such that

$$mk = \gamma_m\gamma_n n.a. \tag{5}$$

This implies $k \in P(\mathbb{R})$. But $K \cap {}^\circ M(\mathbb{R})$ is a maximal compact subgroup of $M(\mathbb{R})$, hence of $P(\mathbb{R})$ and therefore $k \in {}^\circ M(\mathbb{R})$. It follows that $a^{-1}(\gamma_n n)a = 1$, hence also $\gamma_n n = 1$ and (5) yields

$$m.e = \gamma_m.a.e.$$

Conversely, let $\tau \in \widetilde{\Gamma}'_M$. Then there exists $n \in N(Q)$ such that $n.\tau \in \Gamma'_P$. Therefore $\tau a.e$ has the same image in $\Gamma'\backslash X$ as $n^{-1}.a.e$, hence it maps to an intersection point; By (i), the elements $\tau.a.e$ and $\tau'.a.e$ have the same image if and only if they belong to the same coset mod $(\Gamma'\cap M)$. As a consequence, the intersection points are the distinct images of the points $\tau.a.e$ $(\tau \in \widetilde{\Gamma}'_M/(\Gamma'\cap M))$. Since $\Gamma'_M \subset \Gamma'_P.N(\mathbb{R})$, its elements preserve the orientation conventions of 2.5, hence the index of each intersection point is one.

(c) We now consider the general case. Let L, K, K_v, ι_v, ι, e, e_v be as in 1.5 (for $H = M$). The map ι defines embeddings of $M(\mathbb{R})$ into $M'(\mathbb{R})$ and of $N(\mathbb{R})$ into $N'(\mathbb{R}) = (R_{k/Q}N)(\mathbb{R})$. As we shall see below (2.6):

(*) There exists a maximal Q–split torus T in G' such that $\iota(A) \subset A'$, where $A' = T(\mathbb{R})^\circ$ and an ordering on roots such that $\Phi(P,A)$ is the set of restrictions of elements in $\Phi(P',A')$; in particular, the simple roots in $\Phi(P',A')$ are the restrictions of the simple roots in $\Phi(P',A')$, and therefore $A_t \subset A'_t$.

We assume this here. By (a) and (b), there exists $t > 0$ and a neat arithmetic subgroup $\widetilde{\Gamma}$ of G' such that *(i)* and *(ii)* are true for $\widetilde{\Gamma}$, M' and N' in G'. Let $\Gamma' \subset G(\mathbb{R})$

be such that $\iota(\Gamma') = \iota(G(\mathbb{R})^\circ) \cap \widetilde{\Gamma}$. It is a neat arithmetic subgroup of G. We claim that *(i)* and *(ii)* are satisfied with the choices made for e, Γ' and t. Using ι, we see first that *(i)* is also true for M and N in G. Assume now we have a relation

$$\gamma.m.x = n.a.x \qquad (\gamma \in \Gamma', m \in M(\mathbb{R}), n \in N(\mathbb{R}), a \in A). \tag{6}$$

As pointed out in b), in order to prove *(ii)*, it suffices to show that (6) implies $\gamma \in \Gamma' \cap P(\mathbb{R})$. Applying ι to (6) we get

$$\iota(\gamma).\iota(m)\iota(x) = \iota(n)\iota(a)\iota(x).$$

By (*) we have, $\iota(a) \subset A'_t$. Therefore (a) shows that $\iota(\gamma) \in P'$. But $(G(\mathbb{R})) \cap P' = \iota(P(\mathbb{R}))$, whence $\gamma \in P(\mathbb{R})$, and $\gamma \in \Gamma' \cap P(\mathbb{R})$.

2.6. There remains to justify (*). This is standard but we give some details for the sake of completeness. The notation and framework is that of [BT], see in particular 5.20, 5.21 there.

Let ${}_{\mathbb{Q}}S$ be a maximal $\mathbb{Q}$–split torus of G contained in M. Choose an ordering on ${}_{\mathbb{Q}}\Phi = \Phi({}_{\mathbb{Q}}S, G)$ and let I be the set of simple $\mathbb{Q}$–roots such that $M = \mathcal{Z}(S_I)$ and that $\Phi(S_I, P)$ consists of positive roots. Then $A = S_I(\mathbb{R})^\circ$. Let now ${}_kS$ be a maximal k–split torus of $G \times k$ contained in M, and containing ${}_{\mathbb{Q}}S \times k$. Choose an ordering on ${}_k\Phi = \Phi({}_kS, G \times k)$ compatible with the ordering on $\Phi({}_{\mathbb{Q}}S, G)$. There exists a set J of simple k–roots such that $M \times k = \mathcal{Z}({}_kS_J)$ and that ${}_kS_J$ is the maximal k–split torus in the center of $M \times k$. Then ${}_{\mathbb{Q}}S \times k \subset {}_kS$. We also have $M \times k = \mathcal{Z}({}_{\mathbb{Q}}S_I \times k)$ therefore ${}_{\mathbb{Q}}S$ contains regular elements of ${}_kS$. In particular every element of $\Phi({}_kS, G \times k)$ restricts non-trivially to $S \times k$, hence to an element of ${}_{\mathbb{Q}}\Phi$. Since the orderings are compatible, it follows that every simple k–root in $\Phi({}_kS, P \times k)$ restricts to a simple element of $\Phi({}_{\mathbb{Q}}S, P)$ and every such root is obtained in this way (maybe more than once, but this is not relevant here). Let T be the maximal $\mathbb{Q}$–split subtorus of $R_{k/\mathbb{Q}}\ {}_kS$. Then $\dim T = \dim {}_{\mathbb{Q}}S$ and there is a canonical isomorphism between ${}_k\Phi$ and ${}_{\mathbb{Q}}\Phi$ (see 5.20 and 5.21 in [BT]). The embedding of $T(\mathbb{R})$ into $(R_{k/\mathbb{Q}}{}_kS)(\mathbb{R})$ is described there, and it is clear that ι defines an embedding of A into $T(\mathbb{R})$ which is compatible with the restriction of relative roots, whence our assertion.

2.7. *Remark.* In the course of the proof, we have seen that a relation

$$m.a.e = \gamma.n.a.e \qquad (\gamma \in \Gamma', m \in M(\mathbb{R}), n \in N(\mathbb{R}), a \in A_t, t >> 0)$$

implies that $\gamma \in \Gamma \cap P$. Moreover, the intersection points of $[M(\mathbb{R}).e]_{\Gamma'}$ and $[N(\mathbb{R}).a.e]_{\Gamma'}$ are the (distinct) images of the elements $\tau.a.e$ where τ runs through a set of representatives of $\widetilde{\Gamma}'_M/(\Gamma' \cap M)$, in the notation of 2.5(b).

§3. Many classes.

In this section we show that by shrinking Γ' we can generate arbitrarily many linearly independent modular symbols in its real homology. The proof is the same as in [A], but is included here for ease of reference.

3.1. Theorem: *We keep the assumptions of 2.5 and choose Γ' and t such that (i) and (ii) are satisfied. Let Γ'' be a normal subgroup of finite index of Γ' and let R be a set of representatives of the double cosets $\Gamma''\backslash\Gamma'/(\Gamma' \cap P)$. Then for $b \in A_t$ the fundamental classes of the submanifolds $[\delta N(\mathbb{R})b.e]$, $\delta \in R$, are linearly independent in $H_d(\Gamma''; \mathbb{R})$, where $d = \dim N(\mathbb{R})$.*

Proof. The intersection pairing between ordinary homology and homology with closed supports shows that it suffices to prove that

$$[\delta N(\mathbb{R}).be]_{\Gamma''} \quad \text{and} \quad [\tau M(\mathbb{R})e]_{\Gamma''} \qquad (\delta, \tau \in R)$$

have a non-void intersection if and only if $\tau = \delta$, in which case their intersection number is > 0.

Assume then there exist $n \in N(\mathbb{R})$, $m \in M(\mathbb{R})$ and $\gamma \in \Gamma''$ such that

$$\gamma\delta nb.e = \tau m.e\ .$$

By 2.8, this implies that $\tau^{-1}\gamma\delta \in \Gamma' \cap P$. In view of the identity $\delta = \gamma^{-1}\tau\tau^{-1}\gamma\delta$, we see that $\delta = \tau$ in R.

Assume now that $\delta = \tau$. The left translation by δ on X induces a homeomorphism of $\Gamma''\backslash X$, (since Γ'' is normal in Γ'). We have then

$$[\delta N(\mathbb{R}).b.e]_{\Gamma''}.[\delta.M(\mathbb{R}).e]_{\Gamma''} = \delta([N(\mathbb{R}).b.e]_{\Gamma''}.[M(\mathbb{R}).e]_{\Gamma''})$$

and this is > 0 by 2.5.

§4. Lifting of some interior cohomology classes of a face.

4.1. In the first four sections, we recall some known facts and notions, mainly to fix the notation. Let Z be a connected oriented manifold, n its dimension and E a commutative field. The image of the natural homomorphism $j : H^\bullet_c(Z;E) \to H^\bullet(Z;E)$ of the cohomology with compact supports into ordinary cohomology is called interior cohomology (with coefficients in E), and is denoted $H^\bullet_!(Z;E)$. The perfect pairing between $H^i_c(Z;E)$ and $H^{n-i}(Z;E)$ defined by the cup product induces a perfect pairing between $H^i_!(Z;E)$ and $H^{n-i}_!(Z;E)$ $(i \in \mathbb{Z})$.

Similarly, we may define the interior homology group $H^!_\bullet(Z;E)$ of Z as the image of the natural homomorphism $j_\bullet : H_\bullet(Z;E) \to H^{c\ell}_\bullet(Z;E)$, where $H^{c\ell}_\bullet(Z;E)$ refers to homology with closed supports. We let Δ_Z be the Poincaré duality isomorphism. It induces isomorphisms

$$H^i(Z;E) \simeq H^{c\ell}_{n-i}(Z;E), \qquad H^i_c(Z;E) \simeq H_{n-i}(Z;E).$$

Again, the perfect pairing between $H_i(Z;E)$ and $H^{c\ell}_{n-i}(Z;E)$ defined by intersection induces one between $H^!_i(Z;E)$ and $H^!_{n-i}(Z;E)$ $(i \in \mathbb{Z})$.

The map Δ_Z induces an isomorphism of $H^i_!(Z;E)$ onto $H^!_{n-i}(Z;E)$ which is compatible with the above pairings.

A continuous proper map preserves interior homology or cohomology.

If Z is the interior of a compact manifold $\overline{Z}$ with boundary $\partial\overline{Z}$, then $H^\bullet(Z;E) = H^\bullet(\overline{Z};E)$ and $H^\bullet_!(Z;E)$ is the kernel of the restriction map $H^\bullet(\overline{Z};E) \to H^\bullet(\partial\overline{Z};E)$. Similarly, $H_\bullet(Z;E) = H_\bullet(\overline{Z};E)$ and $H^!_\bullet(Z;E)$ is the kernel of the boundary map $H_\bullet(\overline{Z};E) \to H_{\bullet-1}(\partial\overline{Z};E)$.

4.2. Returning to our previous situation, we let $\overline{X}$ denote the completion by corners of X, defined in [BS], set $Y = \Gamma\backslash X$, $\overline{Y} = \Gamma\backslash\overline{X}$. We assume Γ to be neat. Then $\overline{Y}$ is a compact manifold with corners and ∂Y is a disjoint union of faces $e'(P)$, where P runs through the proper parabolic Q–subgroups of G, modulo conjugacy under Γ [BS]. Choose P and let $M, {}^\circ M, A, N$ be as in 1.1. Let $\mathcal{P}_M$ be the set of parabolic Q–subgroups having one Levi subgroup equal to M and $\mathcal{B}(P)$ the set of proper parabolic Q–subgroups containing an element of $\mathcal{P}_M$. It is finite. We fix a maximal compact subgroup K of $G(\mathbb{R})$ whose associated Cartan involution leaves $M(\mathbb{R})$ stable. Then $K \cap M(\mathbb{R})$ is a maximal compact subgroup of $M(\mathbb{R})$, or $P(\mathbb{R})$, contained in ${}^\circ M(\mathbb{R})$. Each $Q \subset \mathcal{B}(P)$ has a Levi Q–subgroup of the form $\mathcal{Z}(B)$, where $B \subset A$. Since θ_K is the inversion on A, it leaves invariant B, and hence also $\mathcal{Z}(B)(\mathbb{R})$. Thus every $Q \in \mathcal{B}(P)$ has a Levi Q–subgroup M_Q containing M, whose group of real points is stable under θ_K, hence intersects K in a maximal compact subgroup. We denote by $Q(\mathbb{R}) = {}^\circ M_Q(\mathbb{R}).A_{\mathbb{R}}.N_Q(\mathbb{R})$ the corresponding Langlands decomposition.

We let e be the fixed point of K in X and take it as our base point.

Recall that $e'(P)$ is fibered over $Z_M = \Gamma_M \backslash X_{^\circ M(\mathbb{R})}$ with typical fibre $F = (\Gamma \cap N)\backslash N(\mathbb{R})$. As in 2.5(b), Γ_M is the image of $\Gamma \cap P$ under the natural projection $P \to P/N$. We let m and f be the dimensions of Z_M and F respectively. Therefore $\dim e'(P) = m + f$; of course $\dim X = \dim A + \dim e'(P)$. We assume again $\Gamma \subset G(\mathbb{R})^\circ$, so that it preserves the orientation of X. Then we can fix orientations of $Z_M, A, N(\mathbb{R})$ which add up to the orientation of X and will also be preserved by $\Gamma \cap P$.

We say that Γ separates the P–cusps if $\gamma Q \gamma^{-1} = R$ ($\gamma \in \Gamma; Q, R \in \mathcal{B}(P)$) implies that $Q = R$, hence $\gamma \in Q$. Any Γ contains a subgroup of finite index which separates the P–cusps.

In the sequel we assume that Γ separates the P–cusps, and moreover that $\lambda_{e,\Gamma,M}$ is injective, which can also be achieved by going over to a subgroup of finite index (1.4).

4.3. From now on, $E = \mathbb{R}$ and we do not mention the coefficients anymore. Let $\pi : e'(P) \to Z_M$ be the canonical projection. It is known that this fibration is homologically trivial (see §2.6, 2.7 in [S]). Thus $H^\bullet(e'(P)) = H^\bullet(Z_M) \otimes H^\bullet(F)$, as vector spaces, not necessarily multiplicatively. Still, the natural $H^\bullet(Z_M)$–module structure of $H^\bullet(e'(P))$ goes over to the multiplication on the first factor of the tensor product. Let $a, b \in \mathbb{N}$ be such that $a + b = m$. Then $H_!^a(Z_M)$ and $H_!^b(Z_M)$ are in perfect duality with respect to the cup product. It follows then that $H_!^b(Z_M) \otimes H^f(F)$ may be identified to a subspace E of $H_!^{b+f}(e'(P))$ in perfect duality with $\pi^\bullet(H_!^a(Z_M))$. More intrinsically, we may also describe E as the subspace of $H_!^{b+f}(e'(P))$ such that $\pi_\bullet$ maps $\Delta_{e'(P)}(E)$ isomorphically onto $\Delta_{Z_M}(H_!^b(Z_M))$.

Similarly, $H_\bullet(e'(P)) = H_\bullet(Z_M) \otimes H_\bullet(F)$. The subspace E' of $H_{b+f}^!(e'(P))$ defined by $H_b^!(Z_M) \otimes H_f(F)$ may be characterized as the image of $H_b^!(Z_M)$ under a "Umkehrungshomomorphismus", namely under $\Delta_{e'(P)} \circ \pi^\bullet \circ \Delta_{Z_M}$. Geometrically, given a compact cycle ξ whose homology class $[\xi]$ belongs to $H_b^!(Z_M)$, the corresponding element in E' is represented by the compact cycle $\pi^{-1}(\xi)$. We denote it $[\xi \times F]$.

The face $e'(P)$ is the quotient by $\Gamma \cap P$ of the space $e(P) = A \backslash X$ (cf. [BS]). Let $\sigma : e(P) \to e'(P)$ be the natural projection. The $^\circ M(\mathbb{R})$–orbit of e in $e(P)$ goes over under σ onto a closed submanifold homeomorphic to $Y_M = (\Gamma \cap M)\backslash {}^\circ M(\mathbb{R})$. We have a commutative diagram

$$\begin{array}{ccc} Y_M = (\Gamma \cap M)\backslash {}^\circ M(\mathbb{R}) & \xrightarrow{\nu} & e'(P) \\ & \searrow^{\mu} \quad \swarrow^{\pi} & \\ & Z_M & \end{array}$$

where ν is a closed embedding and μ is an unramified finite covering, (with fibres $\simeq \Gamma_M/\pi(\Gamma \cap M)$). Therefore $\mu_\bullet$ is surjective. Being moreover proper, it maps interior homology onto interior homology.

4.4. The compactification $\overline{e'(P)}$ of $e'(P)$ in $\overline{Y}$ is also a compact manifold with corners [BS]. Using the identification of homology or cohomology of $e'(P)$ (resp. Y) and of $\overline{e'(P)}$ (resp. $\overline{Y}$) we get homomorphisms

$$r_P^\bullet : H^\bullet(Y) \to H^\bullet(e'(P)) \text{ and } r_\bullet^P : H_\bullet(e'(P)) \to H_\bullet(Y),$$

associated to the inclusion $\overline{e'(P)} \hookrightarrow \overline{Y}$. If $\langle\, ,\rangle$ denotes the usual pairing between homology and cohomology of the same dimension, we have

$$\langle r_P^\bullet(\alpha), \beta\rangle = \langle \alpha, r_\bullet^P(\beta)\rangle \qquad (\alpha \in H^q(Y);\ \beta \in H_q(e'(P))). \tag{1}$$

Let ξ be a compact cycle in $e'(P)$ and $\tilde{\xi}$ be a compact subset of $e(P)$ mapped onto ξ by σ. Then, if $a \in A$ is sufficiently far out, the image of $\tilde{\xi}.a.e$ in Y is homeomorphic to ξ. We denote it $[\xi.a.e]_\Gamma$. It is a cycle whose homology class is $r_\bullet^P([\xi])$.

Consider now the situation at the end of 4.3. Assume ξ represents a class of $H_b^!(Z_M)$. Then for a far out the image of $N.\tilde{\xi}.a.e$ in Y is homeomorphic to $\pi^{-1}(\xi)$. It will be denoted $[N\xi a.e]_\Gamma$. Its homology class in $H_\bullet^{c\ell}(Y)$ is $r_\bullet^P([\xi \times F])$.

4.5. THEOREM: *The image of r_P^{f+q} contains a subspace in duality with $\pi^\bullet(H_!^q(Z_M))$ $(q \in \mathbb{N})$.*

Proof. We recall first that, by Poincaré duality, $H_!^q(Z_M)$ is isomorphic to $H_{m-q}^!(Z_M)$. Let $\overline{\xi} \in H_{m-q}^!(Z_M)$. By 4.3, there exists a compact cycle in $\tilde{\xi} \in Y_M$ such that $(\pi \circ \nu)(\tilde{\xi})$ represents $\overline{\xi}$. Let $\xi = \nu(\tilde{\xi})$. The modular symbol $[M(\mathbb{R})e]_\Gamma$ is homeomorphic to $A \times Y_M$, in view of our assumptions. In $A \times Y_M$, we have the (non-compact) cycle $A \times \tilde{\xi}$, which represents a non-trivial element of $H_{\ell+m-q}^{c\ell}(A \times Y_M)$ $(\ell = \dim A)$. We denote the image of $A \times \tilde{\xi}$ in $[M(\mathbb{R})e]_\Gamma$ by $[A\xi e]_\Gamma$. Let now η be a compact cycle in $e'(P)$, of dimension $f+q$, and $[\eta]$ the interior homology class it defines.

As in 4.4, we associate to η a cycle $[\eta a.e]_\Gamma$ homeomorphic to η (for a sufficiently far out). Note that $[A\xi e]_\Gamma$ and $[\eta ae]_\Gamma$ are cycles in complementary dimensions, the latter compact, the former not. We claim:

(*) *For $a \in A_t$, t big enough, we have*

$$[A\xi e]_\Gamma.[\eta ae]_\Gamma = [\xi].[\eta]. \tag{1}$$

Assume this for the moment. Let $\widetilde{\xi}_i$ $(i \in I)$ be compact cycles in Y_M such that the classes of the cycles $\pi\nu(\widetilde{\xi}_i)$ form a basis for the $(m-q)$–th interior homology group of Z_M. and let $\xi_i = \nu(\widetilde{\xi}_i)$. We can find compact cycles η_i in $e'(P)$, of dimension $f+q$, whose interior homology classes satisfy

$$[\xi_i] \cdot [\eta_j] = \delta_{ij} \qquad (i,j \in I).$$

Let $\mu_i \in H^{f+q}(Y)$ be the image of the class of $[A\xi_i e]_\Gamma$ under the Poincaré duality isomorphism $(i \in I)$. We have then, by 4.4 and $(*)$, for $i,j \in I$:

$$\langle r_P^\bullet(\mu_i), [\eta_j]\rangle = \langle \mu_i, r_\bullet^P([\eta_j])\rangle = [A\xi_i e]_\Gamma \cdot r_\bullet^P([\eta_j])$$
$$\langle r_P^\bullet(\mu_i), [\eta_j]\rangle = [A\xi_i e]_\Gamma \cdot [\eta_j ae]_\Gamma = [\xi_i] \cdot [\eta_j] = \delta_{ij},$$

therefore, the span of the μ_i's satisfy our conclusion. There remains to prove $(*)$.

4.6. *Proof of $(*)$.* We shall use a chopping up of A analogous to the one on p. 444 of [A]. However, we shall in this proof only use sets of the form $A_{Q,t} \times C$, with C compact in ${}^\circ Q(\mathbb{R})$.

Fix a metric on $\overline{Y}$. Since η is compact in $e'(P)$, there exists $\delta > 0$ such that η is at distance at least 3δ from the complement of $e'(P)$ in $\partial\overline{Y}$. Note that, since δ separates the P-cusps, η is then at distance $\geq 3\delta$ from any $e'(Q)$, $(Q \in \mathcal{B}(P), G \neq P)$.

There exists $s(P)$ such that $[\eta.ae]_\Gamma$ is in a δ–neighborhood of $e'(P)$ for $a \in A_{s(P)}$.

For $Q \in \mathcal{P}_M$, $Q \neq P$, there exists $t(Q) > 0$ such that the image $[A_{Q,t(Q)}.\xi.e]_\Gamma$ of $A_{Q,t(Q)}.\xi.e$ in Y is in a δ–neighborhood of $e'(Q)$. This follows from reduction theory, since ξ is compact and belongs to ${}^\circ Q(\mathbb{R})$, in fact to ${}^\circ M_Q(\mathbb{R})$. Recall that for $R \supsetneq Q$, hence $R \in \mathcal{B}(P)$, we have a canonical decomposition $A = A_{Q,R} \times A_R$, where $A_{Q,R} = M_R \cap A$ [BS]. It is elementary that, using descending induction on the parabolic rank prk(R) of R (i.e., dim A_R), given strictly positive real numbers $t(R)(R \in \mathcal{B}(P))$, where $t(R) \geq t(R')$ if prk$(R) <$ prk(R'), $(R, R' \in \mathcal{B}(P))$, we can find compact subsets $C_{Q,R} \subset A_{Q,R}$ such that the union U of the $C_{Q,R} \times A_{R,t(R)}$ has a relatively compact complement in $A = A_P$. Moreover, the choice of $C_{Q,R}$ depends only on those made for parabolic subgroups of parabolic rank strictly bigger than that of R. In view of this last point and reduction theory, we may further choose the $t(Q)$ and $C_{Q,R}$, again by descending induction on the parabolic rank, so that $[\xi.C_{Q,R}.A_{R,t(R)}.e]_\Gamma$ is in a δ–neighborhood of $e'(R)$. Therefore, for $R \neq P$, it is at distance at least δ from $[\eta.A_{s(P)}.e]_\Gamma$.

Now let C be a compact subset of A containing the complement of U. Then $[C.\xi.e]_\Gamma$ is compact in Y. There exists then $t > s(P), t(P)$ such that $[C.\xi.e]_\Gamma$ does not intersect $[\eta.a.e]_\Gamma$ if $a \in A_t$.

Let now $u \in \xi$, $v \in \eta$, $b \in A$, $a \in A_t$ and $\gamma \in \Gamma$ be such that

$$u.b.e = \gamma.v.a.e. \tag{1}$$

Thus, the image of ube in Y belongs to $[\eta.a.e]_\Gamma$. In view of our constructions, this forces $b \in A_{t(P)}$. Moreover, by the Siegel property, only finitely many γ's are possible.

Since the cycles $[\eta ae]_\Gamma$ are homologous to one another, we are only interested in γ's for which (1) is true for a arbitrarily far out. Therefore $\gamma P \gamma^{-1} = P$, hence $\gamma \in \Gamma \cap P$. Let $k \in K$ be such that $ubk = \gamma va$. This implies $k \in P(\mathbb{R})$, hence $k \subset {}^\circ M(\mathbb{R})$, and therefore $b = a$. As a consequence

$$[A\xi e]_\Gamma \cap [\eta ae]_\Gamma$$

is the homeomorphic image of $\xi \cap \eta$, and $(*)$ follows.

§5. Non–square integrability.

The argument in section 4 of [A] proves the non-square integrability of the Poincaré duals of modular symbols. A modification of it will establish here the same result for the classes discussed in §4 of this paper.

5.1. Theorem. *Let ξ be a compact cycle in Z_M representing an element of $H^!_q(Z_M)$ and $\xi' = \pi^{-1}(\xi)$ its inverse image in $e'(P)$. Let $\eta(a) = [N.\xi.a.e]_\Gamma$ where a is far out (see 4.4) and ω a closed $(f+q)$-form on Y whose integral on $\eta(a)$ is one. Then ω is not square integrable with respect to a metric ds^2_Y on Y induced by a $G(\mathbb{R})$-invariant metric ds^2_X on X.*

[The cycles $\eta(a)$ $(a \in A_t, t$ large) are homologous to one another, therefore the integral of a closed form on $\eta(a)$ is independent of a.]

Proof. We have the diffeomorphisms

$$\mu : A \times X_{{}^\circ M(\mathbb{R})} \times N(\mathbb{R}) \to X \quad \text{and} \quad \gamma : X_{{}^\circ M(\mathbb{R})} \times N(\mathbb{R}) \to e(P)$$

induced by the decomposition 1.1 of $P(\mathbb{R})$. The former induces a differentiable mapping $\mu_\Gamma : A \times e'(P) \to Y$. If C is open relatively compact in $e'(P)$ and t big enough, then μ_Γ induces a homeomorphism of $A \times C$ onto an open subset of Y. It suffices therefore to show that, for suitable t and C, ω is not square integrable on $(A_t \times C)$ with respect to ds^2_Y, hence that $\mu^*_\Gamma(\omega)$ is not square integrable with respect to the metric $ds^2 = \mu^*_\Gamma(ds^2_Y)$.

On $e(P)$, we consider the metric de^2 obtained via ν from the restriction of ds^2 to $\{1\} \times X_{\circ M(\mathbb{R})} \times N(\mathbb{R})$. It is left-invariant and we denote in the same way the induced metric on $e'(P)$.

Let α be a closed differential $(m-q)$-form on $e'(P)$ representing the Poincaré dual of the class $[\xi \times F]$ of ξ'. If τ is a closed $(q+f)$-form on $e'(P)$, we have

$$\int_{\xi'} \tau = \int_{e'(P)} \tau \wedge \alpha. \tag{1}$$

But $(\tau \wedge \alpha)_x = (\tau, *\alpha)_x dv_x$, where $*$ is the star operator for de^2 and dv_x (resp. $(\, , \,)_x$) the volume element (resp. scalar product) defined at $x \in e'(P)$ by de^2. By the Cauchy–Schwarz inequality, (1) implies

$$\left| \int_{\xi'} \tau \right| \leqq \|\tau\|_{e'(P)} \cdot \|\alpha\|_{e'(P)}, \tag{2}$$

where $\| \cdot \|_{e'(P)}$ refers to the square norm with respect to de^2 for differential forms on $e'(P)$. We have

$$\int_{\eta(a)} \omega = \int_{\{a\} \times \xi'} \omega = 1. \tag{3}$$

Write $\omega = \omega_1 + \omega_2$, where ω_1 has no da's and the maximum number of dn's. We can view ω_1 as a function on A with values in the differential $(f+q)$-forms on $e'(P)$. Then, in view of (2), (3):

$$\int_{\eta(a)} \omega = \int_{\xi'} \omega_1(a) = 1 \leqq \|\omega_1(a)\|_{e'(P)} \|\alpha\|_{e'(P)}, \quad (a \in A_t, t \gg 0). \tag{4}$$

Therefore there exists a constant $c > 0$ such that

$$\|\omega_1(a)\|_{e'(P)} \geq c > 0, \text{ for } a \text{ sufficiently far out.} \tag{5}$$

On the other hand, ω_1 and ω_2 are orthogonal with respect to the metric ds^2_Y, so that

$$\|\omega\|^2 \geqq \|\omega_1\|^2.$$

It suffices therefore to show that $\widetilde{\mu}^*_\Gamma(\omega_1)$ is not square integrable on some set $A_t \times C$, with C open relatively compact in $e'(P)$. We take for C an open relatively compact neighborhood of ξ'. In fact, it will be more convenient to carry out the computation in X. We can write the complement C' in C of a suitable closed set of measure zero as a disjoint union of finitely many open sets C_j $(j \in J)$ such that $X_{\circ M(\mathbb{R})} \times N(\mathbb{R})$ contains an open set D_j mapped homeomorphically onto C_j under the composition

$$X_{\circ M(\mathbb{R})} \times N(\mathbb{R}) \xrightarrow{\nu} e(P) \to e'(P),$$

and we may assume each D_j to be contained in a coordinate chart. Let D be the union of the D_j's. We are reduced to showing that the form $\widetilde{\omega}_1$ on X lifted from ω_1 has a non-square integrable restriction to $A_t \times D$ for some t. Note that (5) implies

$$\sum_j \int_{D_j} |\widetilde{\omega_1}(a)|_x^2 dv_x \geqq c > 0, \quad (a \in A_t, t \gg 0), \tag{6}$$

where the norm and the volume element are those associated to de^2. As recalled in [A: 4.3], we have:

$$\mu^*(ds_Y^2) = da^2 \oplus dz^2 \oplus \oplus_\beta a^{-2\beta} \cdot h_\beta(z), \tag{7}$$

$$\gamma^*(de^2) = dz^2 \oplus \oplus_\beta h_\beta(z), \tag{8}$$

where da^2 and dz^2 are left-invariant metrics on A and $X_{{}^\circ M(\mathbb{R})}$ respectively, β runs through the roots of P with respect to A and $h_\beta(z) = (Ad\ z)^*(h_\beta)$ for some metric h_β on the root space $\mathfrak{u}_\beta$ of β in the Lie algebra of $\mathcal{R}_u P$. [Note that β in [A: 4.3] is replaced by $-\beta$, since Γ acts on the left rather than on the right.] As usual, we let

$$2\rho = \sum_\beta (\dim \mathfrak{u}_\beta) \cdot \beta. \tag{9}$$

On ${}^\circ M(\mathbb{R})$, the square of every rational Q-character of M is trivial, by definition, therefore ${}^\circ M(\mathbb{R})$ acts trivially, up to sign, on the volume form dv_N of $N(\mathbb{R})$ and so dv_N is the product of the volume forms of the h_β, up to sign. On D_j, we can find an orthonormal basis dz_I of q-forms and write

$$\widetilde{\omega}_1 = \sum_I f_I(a,z,n) dz_I \wedge dv_N \tag{10}$$

whence

$$\int_{D_j} |\widetilde{\omega}_1(a)|_x^2 = \sum_I \int_{D_j} f_I^2(a,z,n) dv_M \wedge dv_N, \tag{11}$$

$$\int_{A_t \times D_j} |\widetilde{\omega}_1|_x^2 = \int_{A_t \times D_j} \sum_I f_I^2(a,z,n) a^{2\rho} dv_A \wedge dv_M \wedge dv_N, \tag{12}$$

where dv_A and dv_M are the invariant volume elements on A and $X_{{}^\circ M(\mathbb{R})}$ respectively. By Fubini's theorem, we have

$$\int_{A_t \times D_j} |\widetilde{\omega}_1|_x^2 = \int_{A_t} a^{2\rho} \cdot dv_A \left(\int_{D_j} \sum_I f_I^2(a,z,n) dv_M \wedge dv_N \right). \tag{13}$$

Together with (6) and (11), this yields, for $t \gg 0$,

$$\|\widetilde{\omega}_1\|_{A_t \times D}^2 \geqq c \cdot \int_{A_t} a^{2\rho} \cdot dv_A = \infty, \tag{14}$$

and the theorem is proved.

5.2. COROLLARY. *The classes μ_i constructed in 4.5 are not square integrable.*

It was shown in 4.5 that μ_i is not zero on a class $r^P_{\cdot}([\eta_i])$. But $[\eta_i]$ is of the form $[\xi \times F]$ considered in 5.1. Therefore 5.1 holds for μ_i.

A.A.: Ohio State University, Columbus, OH 43210, USA and University of Cambridge, England

A.B.: Institute for Advanced Study, Princeton, NJ 08540, USA

References

[A] A. Ash, *Non-square integrable cohomology of arithmetic groups*, Duke Math. Jour. **47**(1980), 435–449.

[B] A. Borel, Collected Papers, Vol. III, Springer–Verlag.

[BS] A. Borel and J-P. Serre, *Corners and arithmetic groups*, Comm. Math. Helv. **48**(1974), 244–297.

[BT] A. Borel et J. Tits, *Groupes réductifs*, Publ. Math. I.H.E.S., **27**(1965), 55–150.

[M] J. Millson, *On the first Betti number of a constant negatively curved manifold*, Annals of Math. **104**(1976), 235–247.

[S] J. Schwermer, Kohomologie arithmetisch definierter Gruppen und Eisensteinreihen, Springer LNM **988**, (1983).

143.

The Work of Chevalley in Lie Groups and Algebraic Groups

Proc. Hyderabad Conference on algebraic groups 1989, Manoj Prakashan, Madras, (1991) 1–22;
and Claude Chevalley, Collected Works, Vol. 3, Springer-Verlag (to appear in 2001)

1 The first major research interests of Chevalley pertain to algebraic number theory, in particular class field theory. He had set for himself two main goals, to develop first a local class field theory, independently from the global theory and second an algebraic treatment of the latter, free from analysis. His Thesis [C1] realizes the first one and makes a first step towards the second one, which was achieved about two years later [C2] [C3]. It is on this occasion that he introduced the concept of idele, which soon became fundamental in algebraic number theory.

In the thirties, the Julia Seminar in Paris, started in 1934 as a successor to the Hadamard Seminar, was a meeting ground for a number of younger mathematicians, including some of those who had founded Bourbaki in 1934. At the suggestion of A. Weil, it was agreed to devote it in 1936-37 to the work of E. Cartan on Lie groups and Lie algebras. Chevalley participated, giving one lecture on the representation theory of Lie algebras. This seminar was the starting point of his interest in Lie groups, which soon grew to a major one. Already in 1938, some Bourbaki projects included a report by Chevalley on Lie groups, which was to follow one by C. Ehresmann on manifolds, Chevalley promising to "algebraize it to death". Soon the war came, Bourbaki dispersed, Chevalley went to Princeton, and these plans were left in abeyance as far as Bourbaki was concerned, but Chevalley apparently decided to pursue them in part on his own. In 1941 he published his first two papers on Lie groups. In one [C4], he determines the topological structure of a connected solvable group, showing, in analogy with a well-known result of E. Cartan on semi-simple Lie groups, that a connected solvable group is topologically (or even differentiably) isomorphic to the product of a compact subgroup (a torus) by a euclidean space. Later A. Malcev and K. Iwasawa proved independently a similar result (and the conjugacy of maximal compact subgroups) for a general connected Lie group. (The paper of Iwasawa [I] also introduces what is now known as the *Iwasawa decomposition*, but Iwasawa himself points out that his original proof was valid only for complex semisimple groups and that the one he gives is due to Chevalley).

The paper [C5] is much more a harbinger of the future. Let $\mathfrak{g}$ be a complex Lie algebra and X a regular element (i.e. the multiplicity of the eigenvalue 0 of ad X is the smallest possible for $X \in \mathfrak{g}, X \neq 0$) and $\mathfrak{n}_0$ be the nilspace of ad X (the space of elements of $\mathfrak{g}$ annihilated by some power of ad X). It is a nilpotent subalgebra which plays a basic role in E. Cartan's study of $\mathfrak{g}$. Chevalley proposed to call it a *Cartan subalgebra* of $\mathfrak{g}$ and his main result in that paper is the conjugacy of Cartan subalgebras. It had already been proved for $\mathfrak{g}$ semi-simple by H. Weyl in his Math. Zeitschrift papers [Wy1]. There the conjugacy was under the adjoint group Ad $\mathfrak{g}$ of inner automorphism of $\mathfrak{g}$. However, Chevalley assumes more generally the groundfield to be algebraically closed, of characteristic zero, so that he does not have a ready-made analogue of the adjoint group to perform the conjugacy. Let n and ℓ be the dimension of $\mathfrak{g}$ and $\mathfrak{n}_0$ respectively. In the Grassmannian of ℓ-planes in $\mathfrak{g}$, Chevalley constructs an irreducible variety N, stable under any automorphism of $\mathfrak{g}$, and containing a Zariski-open subset whose points represent the Cartan subalgebras of $\mathfrak{g}$. The generalized eigenspaces of ad X corresponding to the non-zero eigenvalues consist of nilpotent elements. If Y belongs to one of them (or is any nilpotent element of $\mathfrak{g}$ for that matter), then the exponential series

$$\exp \mathrm{ad}Y = \sum_{k \geq 0} (k!)^{-1}.(adY)^k \tag{1}$$

is a finite sum and represents an automorphism of $\mathfrak{g}$. Using a functional determinant argument, Chevalley shows that N has dimension $n - \ell$ and that the conjugates of $\mathfrak{n}_0$ in N under products of such elements fill a Zariski-open subset. The conjugacy then follows from the fact that any two non-empty Zariski open subsets of N have a non-empty intersection [(1)]. Therefore conjugacy is achieved under the group generated by the elements (1) for Y nilpotent. Chevalley himself saw in this argument the beginning of his work in algebraic geometry. It steers away from the traditional analytic approach to Lie groups, towards an algebraico-geometric setting, so it is hardly surprising that he soon turned his attention to linear algebraic groups.

The published work of Chevalley's in Lie groups and algebraic groups extends from 1941 to 1961. It divides rather naturally it into two parts, with 1954 as a cut off date. The second one is devoted entirely to algebraic groups, and consists of a few landmark papers. The first one is more tentative. In it, Lie groups and algebraic groups are closely intertwined, the former playing a dominant role at first, but gradually receeding into the background, in favor of algebraic groups, which became more and more the primary objects. In this account, I shall try to follow the chronological order, but since Chevalley often pursued concurrently various goals, in different directions, I will have at times to deviate from it and to backtrack.

Chevalley's first major achievement in the direction of Lie groups is his Princeton book [C10], to be referred to as Lie I, which became quickly and for many years the basic reference, not only for Lie groups, but also for some foundational material on real analytic manifolds. The latter included in particular a new presentation of tangent spaces, exterior differential forms, E. Cartan's exterior differential calculus, and a global discussion of integral manifolds of a completely integrable system of Pfaffian equations. Although there are hardly any forerunners in the literature, it should not be surmised that all of it was due to him. It is rather likely that some of this material had emerged in discussions and writings within Bourbaki. But the strong algebraic flavor bears his imprint; in particular, the definition of a tangent vector at a point as a derivation of the local ring at that point is most probably his. The exposition of Lie groups is systematically global. The theory of Lie groups had taken a global turn from H. Weyl's papers [Wy1] on, and there had been some expositions of various global aspects of the theory, in particular in [Ca3] [Ca4] and in the last chapters of Pontryagin's book [P], but this was the first textbook exposition. The last chapter contains a proof of the Peter-Weyl theorem and also a version of the Tannaka duality which offers a link between Lie groups and algebraic groups: Given a compact Lie group G, Chevalley introduces the "representative ring" $R(G)$ of G, i.e. the ring of complex valued functions on G generated by the coefficients of the finite dimensional representations of G. He then shows that the set of homomorphisms of $R(G)$ into $\mathbb{C}$ (the maximal spectrum of $R(G)$) admits a natural structure of complex Lie group G_c containing G as a maximal compact subgroup. G appears as the fixed point set of an involution of G_c (for the underlying real group structure) defined by complex conjugation. He also recovers the decomposition of G_c as the product of G by a Euclidean space.

During these years, Chevalley had also been involved with linear algebraic groups. To put his work into context, let me first say a few words about earlier contributions to this topic, all dating from the nineteenth century and due to L. Maurer and E. Picard, (except for two papers by E. Cartan and G. Fano respectively [Ca1] and [F], which were not known to Chevalley). E. Picard had chiefly in mind a Galois theory of linear differential equations, in which the Galois group would be a linear algebraic group, whence his interest in those, to which he devoted some short papers and a chapter in Vol.3 of his Traité d'Analyse. This led to the Picard-Vessiot theory and later to the work of Ritt and E. Kolchin. Of more direct interest to us is the work of L. Maurer (essentially four papers published between 1888 and 1894, the last one [M] being the main one). Initially, Maurer had considered a subgroup G of $GL_n(\mathbb{C})$ defined by one invariant, i.e. consisting of all the invertible linear transformations of $\mathbb{C}^n$ leaving invariant a given rational homogeneous function on $\mathbb{C}^n$. This is a special case of a linear algebraic group, and Maurer soon realized he might as well study any such group, which he called a "regular group", whether it was characterized by one invariant or not. Given a linear algebraic group G, Maurer first investigated the Lie algebra $\mathbf{g}$ of G. If $X \epsilon \mathbf{g}$, and

(2) $X = X_s + X_n$ (X_s semisimple, X_n nilpotent, $X_s.X_n = X_n.X_s$)

is its Jordan decomposition, he noted first that X_s, X_n also belong to $\mathbf{g}$. Let now X be semi-simple, put in diagonal form and let $\lambda_i (1 \leq i \leq n)$ be its eigenvalues. Set

$$C(X) = \{z = (z_i) \epsilon \mathbb{Z}^n \Big| \sum_i z_i.\lambda_i = 0\}$$

and let $\mathbf{h}_X$ be the space of diagonal matrices diag(μ_i), where (μ_i) runs through the n-tuples of complex numbers satisfying the relations

(3) $$\sum_i z_i.\mu_i = 0 \qquad (z = (z_i) \epsilon\, C(X)).$$

This is a diagonal subalgebra having a basis consisting of matrices with integral eigenvalues. Maurer shows that $\mathbf{h}_X$ also belongs to $\mathbf{g}$. (In present day terminology, he had characterized the Lie algebra of the smallest (algebraic) torus in $GL_n(\mathbb{C})$ whose Lie algebra contains X.) In a later part of the paper, he shows that G is a rational variety, a point to which I shall come back later.

However, this was almost fifty years old, some proofs were not complete, some natural questions had not been addressed, so another presentation was in order. Chevalley, first by himself, then in collaboration with H.-F. Tuan, proceeded to provide one and to generalize the theory to an arbitrary groundfield of characteristic zero. To that effect he introduced in [C6] a new algebraic notion, that of a *replica* of a linear transformation X of a finite dimensional vector space V over a field K. Let $V_{p,q}$ be the tensor product of p copies of V and of q copies of its dual $V'(p, q \epsilon \mathbb{N})$. Extended as a derivation of the tensor algebra of V, the transformation X defines an endomorphism $X_{p,q}$ of $V_{p,q}$. Then $Y \epsilon$ End V is a *replica* of X if for any $p, q \epsilon \mathbb{N}$, any element of $V_{p,q}$ annihilated by $X_{p,q}$ is also annihilated by

$Y_{p,q}$. Let now K be perfect. Then the Jordan decomposition (2) holds over K. It is proved that X_s, X_n belong to the space Rep$X$ of replicas of X and that Rep$X$ is the sum of RepX_s and RepX_n. If X is semi-simple and diagonalized over some extension of K, Chevalley shows that Rep X is the space denoted $\mathfrak{h}_X$ above. If K is of characteristic zero and X is nilpotent, then Rep $X = K.X$. Thus Maurer's result is equivalent to the statement that if $\mathfrak{g}$ is the Lie algebra a complex linear algebraic group, then it contains the replicas of all of its elements. In [C9] Chevalley and Tuan announce a converse, the proof of which is given in [C16]. The converse had been formulated over $\mathbb{C}$ only because there was at that time no theory of algebraic groups over other fields, but Chevalley soon began to develop one, to which he devoted [C19]. The title of [C19] certainly conveys the idea that it is meant as a sequel to Lie I, but, in fact, it is hardly so, as can be gathered from the beginning of the introduction:

"The present work constitutes to some extent a sequel to my work 'Theory of Lie groups I', published by Princeton University Press. However, the topics treated here are very different from those considered in 'Theory of Lie Groups', and the proofs of the main theorems contained in this volume do not depend on the general theory of Lie groups."

The field K is always assumed to be infinite. Chevalley defines the notion of algebraic subgroups of $GL(V)$, without any reference to an algebraically closed extension of K. As a result, this notion is not quite the same as that of the group of K-rational points of an algebraic group defined over K, in current terminology, and this has been a source of some confusion. To be more precise, in [C19] a subgroup G of $GL(V)$ is *algebraic* if there exists a set P of polynomials on End V such that G is the set of all elements of $GL(V)$ whose coefficient annihilate the elements of P. The ideal $I(G)$ of G is the set of all polynomials on End V which are zero on G. The group G is irreducible as a variety if $I(G)$ is a prime ideal, in which case the quotient field $K(G)$ of $K[V]/I(G)$ is by definition the field of rational functions of G. Let now L be an algebraically closed extension of K. Let $V^L = V \otimes_K L$ and G^L the smallest algebraic subgroup of $GL(V^L)$ containing G. Its ideal is $I(G) \otimes_K L$, so G^L is in the usual terminology an algebraic group defined over K; its group of rational K-points is indeed G, and, by definition, G is Zariski dense in G^L. It also follows that $L(G^L)$ is a regular extension of K. Conversely, start from an algebraic group $H \subset GL(V^L)$ which is "defined over K" (i.e. $I(H)$ is generated by elements with coefficients in K). Then $H(K) = H \cap GL(V)$ is not necessarily Zariski-dense in H. It can be viewed as an algebraic group in its own right, but its ideal $I(H(K))$ may be strictly bigger than $I(H) \cap K[V]$, hence $H(K)^L$ may be of strictly smaller dimension than H. While reading Lie II, or any paper written in the framework of that book, one has to remember that G is by definition Zariski-dense in G^L, which runs against the usual conventions in algebraic geometry. One drawback of this point of view is that it does not allow for a convenient definition of quotients: If G is as before and N is an algebraic subgroup of $GL(V)$ which is normal in G, then the quotient group G/N cannot be identified in general with an algebraic subgroup of $GL(W)$, where W is some vector space over K, in Chevalley's sense. (If $H \subset GL(V^L)$ is as above, defined over K and Q is normal in H, also defined over K, then H/Q is in a natural way an algebraic group defined over K, but in general, $(H/Q)(K)$ is not equal to $H(K)/Q(K)$.) The difficulty would be even worse if one should try to define in his context an algebraic structure on a coset space, and no attempt is made in Lie II to do so, even in the case of a quotient group.(2)

A number of generalities are developed over any (infinite) K. They include the characterization of an algebraic group by a finite set of semi-invariant polynomials with the same weight or of rational invariants (proved first in [C18]), and the definition of the Lie algebra $L(G)$ of G as the space of $X \epsilon$ **gl**(V) for which the derivation of the field of rational functions on End V associated to the right translation by X leaves the ideal $I(G)$ of G stable. This is so if and only if for any extension L of K, any $s \epsilon G^L$ and any $P \epsilon I(G)$, the differential of P at s, in the direction $s.X$, is zero. The notion of rational representation $\rho : G \rightarrow GL(U)$, where U is a finite dimensional vector space over K, is defined, as well as the differential $d\rho : L(G) \rightarrow L(H)$ for any algebraic subgroup H of $GL(U)$ containing $\rho(G)$.

The main results however are established for K of characteristic zero. This restriction is forced upon Chevalley because his main tool is a formal exponential, defined only in characteristic zero, which allows him to go from $L(G)$ to G in more or less the familiar way. The smallest algebraic subgroup of $GL(V)$ whose Lie algebra contains a given diagonal matrix $X \epsilon$ **gl**(V) is described. Its Lie algebra is the space of replicas of X. By definition a Lie algebra is *algebraic* if it is the Lie algebra of an algebraic group. The Chevalley-Tuan characterization in terms of replicas is proved in this more general context. It is also shown that the derived algebra of any Lie subalgebra of **gl**(n, K) is algebraic (a result which, unknown to Chevalley, had already been announced by E. Cartan over $\mathbb{C}$ in [Ca1]). The very last theorem of Lie II is the existence of a multiplicative Jordan decomposition in G: If $g \epsilon G$ and

$$(4) \qquad g = g_s \cdot g_u \qquad (g_s \text{ semisimple}, g_u \text{ unipotent}, g_s \cdot g_u = g_u \cdot g_s)$$

is its multiplicative Jordan decomposition, then $g_s, g_u \in G$.

I have now to backtrack again to cover the contributions of Chevalley to Lie groups in the late forties. The joint paper with S. Eilenberg [C12] introduces the notion of cohomology of Lie algebras and discusses some applications and open problems. Its origin is a paper of E. Cartan [Ca2], in which the author shows how to reduce the determination of the Betti numbers of the coset space G/H of a compact connected Lie group G modulo a closed subgroup H to an algebraic problem, assuming some theorems he conjectured on that occasion and which were soon proved by G. de Rham. Granted those, Cartan had in substance proved (without the terminology) that the cohomology ring $H^{\cdot}(G/H; \mathbb{R})$ is isomorphic to the cohomology algebra of the algebra of G-invariant differential forms on G/H, with respect to exterior differentiation. Since the latter algebra may be identified with the algebra $(\Lambda^{\cdot} T^*(G/H)_0)^H$ of elements in the exterior algebra of the cotangent space $T^*(G/H)_0$ at the origin of G/H which are invariant under H, acting via the isotropy representation, this was indeed a reduction to an algebraic problem. It was further simplified for symmetric spaces, the main case of interest to Cartan, because then the exterior differential is identically zero on G-invariant forms, so $H^{\cdot}(G/H; \mathbb{R})$ may be identified to $(\Lambda^{\cdot} T^*(G/H)_0)^H$ itself. Therefore Cartan had had no need to make explicit the differentiation in the general case. This is carried out in [C12], and leads to the definition of the cohomology space $H^{\cdot}(\mathfrak{g}; V)$ of a Lie algebra $\mathfrak{g}$ over a field K of characteristic zero, with coefficients in a $\mathfrak{g}$-module V, where V is a vector space over K, and more generally of the relative cohomology space $H^{\cdot}(\mathfrak{g}, \mathfrak{h}; V)$ of $\mathfrak{g}$ modulo a subalgebra $\mathfrak{h}$, with coefficients in V. Both spaces are algebras if $V = K$ is the trivial one-dimensional $\mathfrak{g}$-module. Another incentive to develop this theory was the realization that if $\mathfrak{g}$ is semi-simple, then the vanishing of $H^1(\mathfrak{g}; V)$ implies the complete reducibility of the finite dimensional representations of $\mathfrak{g}$ and the vanishing of $H^2(\mathfrak{g}; V)$ is equivalent to a lemma proved by J.H.C. Whitehead to give a new proof of the Levi theorem. The theorems of H. Hopf and H. Samelson on the cohomology of a compact connected Lie group G could be translated into properties of $H^{\cdot}(\mathfrak{g}; \mathbb{R})$, where $\mathfrak{g}$ is the Lie algebra of G, but there remained the problems of finding direct algebraic proofs. This program, as well a more systematic approach to the cohomology of G/H via cohomology of Lie algebras, was soon carried out by J-L. Koszul [Kz], and then in joint work of H. Cartan, C. Chevalley, J-L. Koszul and A. Weil. I shall not pursue this [(3)] , except however to discuss one problem, namely the determination of the Betti numbers of a compact connected Lie group G. In that case, Cartan's method shows that $H^{\cdot}(G; \mathbb{R})$ may be identified to the invariants of G, acting by the adjoint representation on $\Lambda^{\cdot}\mathbf{g}$. The determination of those had been carried out for the classical groups by R. Brauer. On the other hand, E. Cartan [Ca2] had already found the Poincaré polynomial (4) of the exceptional group $\mathbf{G}_2$. There still remained the four exceptional groups $\mathbf{E}_6$, $\mathbf{E}_7$, $\mathbf{E}_8$ and $\mathbf{F}_4$. Meanwhile Hopf had shown that $H^{\cdot}(G;\mathbb{R})$ is an exterior algebra on $\ell = \text{rank } G$ generators, say of degrees $2m_i - 1 \quad (i = 1,\ldots,\ell)$, so that the problem was reduced to finding the m_i's, often called *the exponents* of G. This was first done by Yen Chih Ta [Y], in a rather roundabout way, by computing the cohomology of certain symmetric spaces of G and then getting back to G by means of a formula of Hirsch (still somewhat conjectural at that time) expressing the Poincaré polynomial of G/H in terms of those of G and H when G and H have the same rank. In [C14], Chevalley proposes another method: Let $S^*(\mathbf{g})$ be the algebra of polynomials over $\mathbf{g}$. It is operated upon by G via the adjoint representation. The joint work alluded to above implies that the algebra I_G of invariants of G in $S^*(\mathbf{g})$ has ℓ algebraically independent homogeneous generators, of degrees $2m_i (i = 1,\ldots,\ell)$, with m_i as above, so that the Poincaré polynomial of G is in principle determined by the structure of I_G. Let now T be a maximal torus of G, $\mathbf{t}$ its Lie algebra, N the normalizer of T in G and $W = N/T$ the Weyl group of G. (By the way, Chevalley asserts there that the terminology " Weyl Group" had been proposed by him. I presume this had been orally, because, as far as I know, it is indeed its first occurence in print.) Then Chevalley states that the restriction to $\mathbf{t}$ induces an isomorphism of I_G onto the algebra I_W of invariants of W acting on $S^*(\mathbf{h})$. In this particular case, this implies that I_W has ℓ algebraically independent generators, but Chevalley also claims that this can be established more generally for any finite Euclidean reflection group. The proof was published only later [C27] (5). This then reduces the determination of the exponents to that of the invariants of a finite group. In that paper, Chevalley announces it can be carried out and confirms the results of Yen Chih Ta. Neither [Y] nor [C14] give all the details. In order to fill this gap, Chevalley and I wrote a paper [C26] in which the exponents are determined with a minimum of computations, using the Hirsch formula and also, for E_8, invariants of the Weyl group of low degree.

In a completely different direction, [C13] gives algebraic *a priori* proofs of two known theorems pertaining to a complex simple Lie algebra $\mathbf{g}$. First the existence of a simple Lie algebra corresponding to a given Cartan matrix (the existence part of the classification, which had been until then carried out case by case), second is the existence of an irreducible representation with a given highest weight. (H. Weyl had already given a general proof [Wy1], [Wy2], but it used transcendental methods. Earlier E. Cartan had relied on an explicit construction of $\ell = \text{rank}\ \mathbf{g}$ fundamental representations, which had to be done case by case.) This Note gives only a sketch of the proofs, but enough indications to make one realize that Chevalley's main constructions, or related ones, occur in many later treatments. (For instance, the infinite dimensional Lie algebra $\tilde{L}$ and the e-extreme $\tilde{L}$-modules in Jacobson's book on Lie algebras are already there.) It also brings somehow to mind infinite dimensional representations and even Kac-Moody algebras. It is therefore rather intriguing that, probably around that time, Chevalley had asserted that the exceptional Lie algebras also belonged to infinite series of simple Lie algebras, like the classical ones, except that they were the only finite dimensional members of those series. I heard this from H. Cartan, around 1950 I think, and later from A. Weil. In retrospect, it is a bit puzzling that nobody took him up on that and asked for more details. Apparently, this seems to be the case, and I do not know what he had in mind.

At about that time, more precisely during the academic year 1947-48, Chevalley saw in the audience of his course on Lie groups and Lie algebras a new student coming from England, who did not seem to know much at first but who caught his undivided attention after some months, when he produced a new proof of Ado's theorem. This was his first contact with Harish-Chandra, who had come to Princeton to study Lie groups under him. In 1948, Harish-Chandra had also found an algebraic proof of the second theorem just stated. Under the influence of [C13], he modified his argument to incorporate a proof of the first one as well. It was published in [HC].

At the end of the introduction to Lie II, Chevalley had announced that Lie I and II were to be part of an exposition of Lie groups and algebraic groups, which would include four more books, but he published only the next one [C24]. Besides the general theory and the classification of semisimple Lie algebras, this treatise was to include the cohomology of Lie algebras and the topology of Lie groups. As far as I know, those chapters were never written, but it may be as a preparation that he had made two further contributions of an expository nature to the global theory of Lie groups. One is a proof of the surjectivity of the exponential map in a compact connected Lie group, which is self-contained and simpler than any other proof I know. The other is a proof of the conjugacy of maximal compact subgroups in a semisimple Lie group not using any Riemannian geometry (in contrast with E. Cartan's famous argument). Both were included in various internal manuscripts of Bourbaki. The former was eventually published in [Bk], but the latter has not yet found its way out of the archives of Bourbaki.

Lie III [C24], notwithstanding its title "Algebraic Groups", is a mixture of algebraic groups (in characteristic zero, essentially a standing assumption), Lie groups and Lie algebras. Lie groups and algebraic groups are studied concurrently, the former being however often subsumed to the latter, for instance by reducing questions on Lie algebras or linear Lie groups to the algebraic case by going over to the algebraic hulls or Zariski closures. Apart from some generalities on linear representations, there are two main parts. The first and main one (Chapters III, IV, V) is concerned with general properties of Lie algebras. It presupposes Lie I and II. Furthermore, Chevalley was at the time of the opinion that the theory of semisimple algebras should precede that of solvable ones. Strict adherence to such principles is not always conducive to an economical presentation. For instance, his proof of Lie's theorem on solvable Lie algebras presupposes the theory of reductive Lie algebras and Cartan's characterization of semisimple algebras. Nevertheless, the main structure theorems are covered, including Ado, Levi-Malcev, a discussion of the universal enveloping algebra and of invariant differential operators.

The second part (Chapter VI) is devoted to Cartan subalgebras and Cartan subgroups and begins with the introduction of the Zariski topology, which is systematically used. A *Cartan subalgebra* of the Lie algebra $\mathfrak{g}$ is, by definition here, a nilpotent subalgebra which is equal to its normalizer. The equivalence with the original definition [C5] is proved; this implies in particular that Cartan subalgebras so defined do exist. As in [C5], a main goal is the conjugacy theorem over an algebraically closed field. The basic idea is the same as there, but Chevalley takes advantage of the fact that he has now a natural framework to express it. Instead of a group generated by the exponentials of nilpotent inner derivations, he uses the irreducible algebraic subgroup of Aut $\mathfrak{g}$ whose Lie algebra is the derived algebra of ad $\mathfrak{g}$ (which he knows to be algebraic by Lie II). He also shows that the conjugacy holds for $\mathfrak{g}$ solvable, even if the groundfield is not algebraically closed. A completely novel element is a definition of Cartan subgroups which is meaningful in any group G. A subgroup C of G is a *Cartan subgroup* if it is maximal nilpotent and every subgroup of finite index of C is of finite index in its normalizer in G. If G is algebraic, these groups are shown to exist and to be the irreducible algebraic subgroups whose Lie algebras are the Cartan subalgebras. If G is a connected Lie group, these are closed subgroups, but they are not necessarily connected. In [C22] Chevalley expects that such a characterization of Cartan subgroups will be useful to study the properties of an algebraic group as an abstract group, an expectation fully confirmed by subsequent developments. The case of compact connected Lie groups, where the Cartan subgroups are the maximal tori, is also considered, and the conjugacy of Cartan subgroups and the surjectivity of the exponential mapping are proved. It is also shown that a compact Lie group is real algebraic, a fact which had been proved in substance in Lie I, in the discussion of Tannaka duality, but had not been stated explicitly there.

From now on we shall be concerned only with algebraic groups. The paper [C23] investigates the structure of the field of rational functions $K(G)$ of an irreducible linear algebraic group G over K, where K is of characteristic zero. It is shown that $K(G)$ is contained in a purely transcendental extension of K (i.e. G is unirational) and is itself purely transcendental (i.e. G is a rational variety) if K is algebraically closed. In his Traité d'Analyse, E. Picard had claimed the last result (over $\mathbb{C}$, of course) but Chevalley points out that Picard's argument only yields the unirationality. In this paper, Chevalley proves in fact more: He constructs the variety M of Cartan subgroups, shows that it is rational, that G is birationally isomorphic to the product of M by a Cartan subgroup C and that C is unirational. Moreover, C is rational if all the elements $\mathrm{ad}\, c\,(c \epsilon C)$ have their eigenvalues in K.[(6)]

From the point of view of algebraic geometry, linear algebraic groups over an algebraically closed groundfield are special case of "algebraic group varieties" i.e. algebraic varieties, endowed with a product structure such that product and inverse are morphisms of algebraic varieties. One class, besides the linear groups was well-known, namely the abelian varieties, which can be characterized as irreducible algebraic groups whose underlying variety is complete. In 1954, Chevalley showed that any irreducible algebraic group G contains a biggest normal linear algebraic subgroup, which is the kernel of a morphism of G onto an abelian variety. Thus abelian varieties and linear algebraic groups are the two building blocks of general algebraic groups. He did not publish it. The same result had been proved shortly afterwards and independently by M. Rosenlicht, using some foundational material supplied by A. Weil. His proof was published in [R], a paper which contains a number of other important results on algebraic groups. [C29] is also concerned with general group varieties. It gives a new proof of a theorem of Barsotti's stating that a group variety is quasi-projective.

In the Notice on his own work [C22], written in Japan probably during the winter 1953-54, Chevalley gives the following motivation for studying algebraic groups over fields other than the complex numbers:

The principal interest of algebraic groups seems to me is that they es-

tablish a synthesis, at least partial, between the two main parts of group theory, namely the theory of Lie groups and the theory of finite groups".

Whether this view was already his at the beginning I do not know, but it became foremost in his mind in the late forties. The model here was L. Dickson who, taking advantage of the fact that the classical groups have an algebraic definition valid over general fields, had constructed new series of finite simple groups over finite fields. He had also done that for the exceptional group $\mathbf{G}_2$ (as well as for $\mathbf{E}_6$, but this was pretty much forgotten at the time and unknown to Chevalley). The task Chevalley set for himself was then to find models of the four other exceptional groups which would make sense over arbitrary fields and again lead to new simple groups. His joint paper with R.D. Schafer [C15] on $\mathbf{F}_4$ and $\mathbf{E}_6$ and his Comptes Rendus Notes on $\mathbf{E}_6$ [C20, 21] are first steps in that direction. In [C22] Chevalley asserts that in the Summer of 1953, he had found new algebraic definitions of $\mathbf{F}_4$, $\mathbf{E}_6$ and $\mathbf{E}_7$, by making use of the triality principle, and that these groups generate infinite series of simple groups, the first new ones in fifty years. No doubt he intended at that time to publish the proofs. In fact, he states in [C22] that in his small book on spinors [C25] he carries out a synthesis of the methods developed by H. Weyl and E. Cartan and generalizes their results over arbitrary groundfields, a "generalization which was necessary in view of the study of the new finite groups I have discovered". But he never did. There was apparently a breakthrough shortly afterwards and Chevalley saw how to carry this out in a uniform, classification-free manner. This leads us to the first major achievement in the second part of Chevalley's work (in the division proposed above), the very influential and justly famous "Tôhoku" paper [C28].

Let $\mathbf{g}$ be a complex semisimple Lie algebra, $\mathbf{h}$ a Cartan subalgebra of $\mathbf{g}$, R the set of roots of $\mathbf{g}$ with respect to $\mathbf{h}$, and S the set of simple roots with respect to some ordering of R. Let $P \subset \mathbf{h}'$ be the lattice of weights, Q the sublattice generated by the roots, $P^\vee$ and $Q^\vee$ their duals in $\mathbf{h}$ (the lattices of coweights and coroots respectively). First, by a careful, largely new, analysis of the constants of structure of $\mathbf{g}$, Chevalley shows one can choose roots vectors X_r $(r \epsilon R)$ and a basis H_s $(s \epsilon S)$ of P which span a $\mathbb{Z}$-form $\mathbf{g}_{\mathbb{Z}}$ of $\mathbf{g}$ so that, for any field K, the elements exp ad tX_r $(t \epsilon K)$ form a unipotent one-parameter group $x_r(K)$ of automorphisms of $\mathbf{g}_K = \mathbf{g}_{\mathbb{Z}} \otimes_{\mathbb{Z}} K$. Chevalley also defines a group of automorphisms H_K (resp. H'_K) of $\mathbf{g}_K$ associated to $\mathbf{h}$ and to the homomorphisms of Q (resp. P) into K^*. Let G_K (resp. G'_K) be the subgroup of Aut $\mathbf{g}_K$ generated by the $x_r(K)$ and H_K (resp. the $x_r(K)$). In G_K, the quotient by H, of the normalizer of H, is the Weyl group W of $\mathbf{g}$ with respect to $\mathbf{h}$. Similarly, G'_K contains H'_K and Norm $H'_K/H'_K = W$.

The group G'_K is normal in G_K and $G_K/G'_K = H_K/H'_K$. Both G_K and G'_K are shown to admit Bruhat decompositions. Let now $\mathbf{g}$ be simple. The main result of the paper is that, except in a few cases where K has two or three elements, G'_K is the derived group of G_K and is a simple group.

Formally, there are no algebraic groups in that paper, though, at the end, Chevalley conjectures that, for $\mathbf{g}$ classical, G_K is isomorphic to the corresponding (split) classical group modulo its center and asks whether G_K is algebraic in general. A first answer was supplied by T. Ono [O]. It became clear later, in the light of [C33], that [C28] gives a construction of the scheme over $\mathbb{Z}$ for the adjoint group of type $\mathbf{g}$ and of the image in the adjoint group of the group of rational points of the simply connected group of the given split type. Altogether, this paper provides a striking illustration for the programmatic statement of the Notice [C22] quoted above.

The next publication of Chevalley is the no less famous Paris Seminar [C30]. There, the framework and point of view are completely different from those of Lie II, III. The Lie algebra appears only briefly, in the last two lectures, and there is no exponential mapping. They are replaced by global arguments in algebraic geometry valid over an algebraically closed groundfield of arbitrary characteristic. Since I am responsible for this change of scenery, I'll digress a little and discuss briefly my own work at that time and how it relates to Chevalley's.

I had been aware of [C9] and of Lie II early on, but from a distance. I got closer to algebraic groups during the first AMS Summer Institute in 1953, devoted to Lie groups and Lie algebras, through my joint work with G.D. Mostow [BM] and a series of lectures by Chevalley on Cartan subalgebras and Cartan subgroups of algebraic groups (the future Chapter VI of Lie III). He was not pleased with it, though, and toward the end said he felt that it was too complicated and there should be a more natural approach valid in any characteristic. Another topic which came up in discussions was a claim by V.V. Morozov, to the effect that maximal solvable subalgebras of a complex semisimple Lie algebra are conjugate. Nobody understood his argument, but I found a simple global proof, using Lie's theorem and the flag variety. In 1954-55, in Chicago, I made a deliberate effort to get away from characteristic zero. Two papers by E. Kolchin [**K1,2**], the first ones to prove substantial results on linear algebraic groups by methods insensitive to the characteristic of the groundfield, led quickly to a structure theory of connected solvable groups. Then I saw how to extend my conjugacy proof of maximal connected solvable subgroups to arbitrary characteristics. That was the decisive step. From then on, it was comparatively smooth sailing and the other results of [B1] followed rather quickly. I lectured on this work and talked about it with Chevalley shortly afterwards, in February 1955 I think, at a Conference in Urbana, Illinois. In the summer of 1955, before leaving the States, I gave him a copy of the manuscript of my forthcoming Annals paper. We did not discuss the subject until the Summer 1956 Bourbaki Congress, where I was told (not by him) that he had classified the simple algebraic groups over any algebraically closed groundfield. When asked, he confirmed it and agreed to give us an informal lecture about this work, at which time he announced he would propose to call "Borel subgroup" a maximal closed connected solvable subgroup of a linear algebraic group. He also told me that, after having read my paper, his first goal had been to prove the normalizer theorem: "A Borel subgroup of a connected linear algebraic group is its own normalizer", after which, "the rest followed by analytic continuation".

The first part of [C30] covers some foundational material and [B1]. The next goal is the normalizer theorem. It is then used, after a rather long analysis, to show that in a connected linear algebraic group G, the intersection of the Borel subgroups containing a given maximal torus T is the radical RG of G. This entails a first discussion of the singular subtori T of codimension one, i.e. the maximal subtori of T whose centralizers are strictly bigger than that of T (which is a Cartan subgroup by [B1]). Assume now G to be semisimple, i.e. $RG = \{1\}$. The next goal is to associate to G and T a root system in $X(T)_{\mathbb{Q}}$, where $X(T)$ is the group of rational characters of T and $X(T)_{\mathbb{Q}} = X(T) \otimes_{\mathbb{Z}} \mathbb{Q}$, whose Weyl group may be identified to Norm T/T acting on $X(T)_{\mathbb{Q}}$ via inner automorphisms. The previous results already imply that the centralizer $\mathcal{Z}(S)$ of a singular torus S of codimension one, modulo its radical, is simple of dimension three. The existence of a reflection amounts to showing that Norm $T \neq T$ in $\mathcal{Z}(S)$, and follows from a fixed point theorem: A torus acting on a projective variety of dimension ≥ 1 has at least two fixed points [(7)]. On the other hand, a one-dimensional connected unipotent group U may be identified with the additive group of the given (universal) algebraically closed groundfield K and then any automorphism of U is given by multiplication by an element of K^*. In particular, if such a subgroup of G is invariant under T, then the action of T on U by inner automorphisms is described by a rational character. The roots are introduced in this way, and the identity components of their kernels are the maximal singular subtori contained in T. Chevalley proves that the roots so defined form a reduced root system in $X(T)_{\mathbb{Q}}$. He associates to (G,T) the pair (Φ, Γ), which I shall call a *root diagram*, where $\Phi \subset X(T)_{\mathbb{Q}}$ is a root system and Γ is a lattice in $X(T)_{\mathbb{Q}}$ intermediary between the lattice $P(\Phi)$ of weights of Φ and the lattice $Q(\Phi)$ generated by the roots, namely $X(T)$ itself. Moreover, Φ is shown to be irreducible if and only if G is simple as an algebraic group. Next the irreducible rational representations of G are constructed. Let B be a Borel subgroup of G containing T. It corresponds to the negative roots for a suitable ordering on Φ. Let λ be a dominant weight contained in $X(T)$. In characteristic zero, the irreducible representation of G with highest weight λ is afforded by the space of regular sections of the line bundle of G/B defined by λ, (viewed as a character of B). This space can also be defined in arbitrary characteristic and is again a finite dimensional vector space on which G acts canonically. It is shown to be non-zero (because λ is dominant), but it is not necessarily irreducible as a G-module. However Chevalley shows that it contains a unique invariant irreducible subspace and establishes in this way again a bijection between dominant weights and irreducible representations. The last seven chapters of the Seminar are chiefly devoted to a uniqueness theorem: An isomorphism of the root diagrams (Φ', Γ') and (Φ, Γ) associated to two groups G', G and maximal tori T', T is induced by an isomorphism of G onto G'. The extremely difficult proof consists in first checking this for Φ of type $\mathbf{A}_n$ or of rank two. The proof in general is then reduced to the rank one and two cases by consideration of the centralizers of singular tori of codimension one and two. In fact, Chevalley proves a more functorial statement. Assume that μ is an isomorphism of $X(T')_{\mathbb{Q}}$ onto $X(T)_{\mathbb{Q}}$ which maps Γ' into Γ and induces a bijection μ_0 of Φ' onto Φ. It is said to be *special* if moreover there exists a bijection $\mu : \Phi \to \Phi'$ such that $\mu_0(\nu(\alpha)) = q(\alpha)\alpha$, $(\alpha \in \Phi)$ where $q(\alpha)$ is a power of the characteristic exponent of K. This condition is easily seen necessary for μ to be associated to an isogeny and Chevalley proves that conversely, any special μ is associated to an isogeny $\iota(\mu) : G \to G'$. If the $q(\alpha)$ are equal to one, that is if μ is an isomorphism of root systems, then $\iota(\mu)$ is a central isogeny. Using his results on representations, Chevalley shows that if (Φ, Γ) is the root diagram associated to (G, T), then there is always $\tilde{G}$ associated to $(\Phi, P(\Phi))$, which admits therefore a central isogeny onto G. (On the other hand, the adjoint group of G corresponds to $\Gamma = Q(\Phi)$.) There are however a few cases, in characteristic two or three, in which G and G' are simple and the function q takes two distinct values. The existence theorem then produces "exceptional isogenies" which, except in the case of $\mathbf{B}_n$ and $\mathbf{C}_n$ in characteristic two, were unknown before.[(8)]

The undated paper [C31] is no doubt contemporary with the seminar. Although it is unpublished and copies are hard to come by, its subject matter and main results are rather well-known. It is concerned with the closures of the orbits of B in G/B, the "Schubert cells" of G/B. The orbits themselves are affine subspaces $C(w)$ $(w \epsilon W)$, the "Bruhat cells", parametrized by the Weyl group W of G. Chevalley first gives a combinatorial description of the cells contained in the closure $\overline{C(w)}$ of $C(w)$ in terms of a reduced decomposition of w. He also shows that $\overline{C(w)}$ is non-singular in codimension one and then conjectures that it is always non-singular, a somewhat surprising conjecture in view of the fact that examples of singular classical Schubert cells in Grassmannians were known, though Chevalley might not have been aware of it. At any rate, these singularities have since been studied and remarkable connections with representation theory, via the Kazhdan-Lusztig polynomials, have been uncovered. Next Chevalley makes an important step towards the determination of the Chow ring $A(G/B)$ of G/B by computing the intersection of $\overline{C(w)}$ with any Schubert cell of codimension one. The formulas are given in terms of roots and weights and depend in fact only on the root diagram (Φ, Γ) associated to G in [C30]. This suffices to show that $A(G/B)$ is independent of the characteristic of the groundfield, so that it may be identified with the similar object attached to the complex analogue of G.

Over $\mathbb{C}$, it is well-known that the map $\mu : (G,T) \mapsto (\Phi(G,T), X(T))$ yields a bijection between isomorphism classes of complex semisimple groups and isomorphism classes of root diagrams. The root system characterizes the Lie algebra of G, and $X(T)$ the various locally isomorphic groups with the given Lie algebra. The group is simply connected if $X(T) = P(\Phi)$ and of adjoint type if $X(T) = Q(\Phi)$. In general $X(T)/Q(\Phi)$ is the center of G and $P(\Phi)/X(T)$ its fundamental group. As pointed out earlier, Chevalley had set up in [C30] a similar mapping in any characteristic and shown that it is also injective. There remained to prove the surjectivity, in order to complete the classification. In the last lecture of [C30], Chevalley remarks that [C28] gives the existence of G of adjoint type, i.e. when $\Gamma = Q(\Phi)$ and adds that with the methods developed to prove the isogeny theorem, it would be easy to associate to any Γ a covering group of the adjoint group, thus establishing the surjectivety of μ. But in [C33] he comes back to this question and handles it in a deeper manner, which has opened new vistas, by a direct generalization of the construction of [C28], now expressed in the language of schemes. This procedure has also built in a natural explanation of why the classification is independent of the characteristic of the groundfield.

Given (Φ, Γ), let G be the associated complex group, $\mathfrak{g}$ its Lie algebra and T a maximal torus of G. The construction of $\mathfrak{g}_{\mathbb{Z}}$ in [C28] allows one to introduce a $\mathbb{Q}$-structure of split group on G. Fix a faithful linear representation (σ, E) of G. It may also be defined over $\mathbb{Q}$. Chevalley asserts the existence of an "admissible" lattice in $E(\mathbb{Q})$, i.e. a lattice spanned by eigenvectors of T and stable under the automorphisms $\exp t\sigma(X_r)$ $(t\in\mathbb{Z})$, where the X_r are part of the basis of $\mathfrak{g}$ constructed in [C28]. The subring of the coordinate ring $\mathbb{C}[G]$ generated by the coefficients of σ with respect to a basis of an admissible lattice is shown to generate a $\mathbb{Z}$-form $\mathbb{Z}[G]$ of $\mathbb{C}[G]$. It represents a scheme over $\mathbb{Z}$ associated to (Φ, Γ). It is then claimed that $\mathbb{Z}[G] \otimes_{\mathbb{Z}} K$ is the coordinate ring of an irreducible K-group $G_{(K)}$ defined and split over the prime field of K, with root diagram (Φ, Γ), whence the existence.(9)

This fundamental paper turned out to be the last (published) research paper of Chevalley's. In the following years, Chevalley devoted his seminar to finite groups and developed an active and successful school in that area, but did not publish anything himself.

Notes

(1) For brevity, I use the language of the Zariski topology, though it was not available at the time.

(2) In §7 of [B2], I recover most of the results of Lie II in a much shorter way, the main reason being, it seems to me, that I can avail myself of the existence of a canonical structure of algebraic variety on a coset space G/H. To see concretely how this works, compare the proofs of the fact that if M and N are algebraic subgroups of G, then the Lie algebra of $M \cap N$ is the intersection of those of M and N.

(3) A detailed exposition may be found in [GHV]. To the references to earlier publications given there, one should now add the letters of A. Weil, written in 1949 and published for the first time in his Collected Papers [Wi].

(4) Recall that the Poincaré polynomial of a space X with finite dimensional real homology is the polynomial

$$P(X,t) = \sum_{i\geq 0} b_i(X).t^i,$$

where $b_i(X)$ is the i-th Betti number of X.

(5) Without Chevalley's knowledge, under rather unusual, if not unique, circumstances. Chevalley had written down his proof in detail but his manuscript did not have all the trimmings of a full-fledged paper and could not be published *ne varietur*. The theorem is proved in characteristic zero, but I needed a complement over a perfect field in some work on the cohomology mod p of compact Lie groups, eventually published in [B0].

In order to prove it, I had to quote not only the statement, but also the proof of his theorem and so needed a publication to refer to. When apprised of this, Chevalley still did not agree to publish a paper and told me I could do whatever I wanted with his manuscript ("Fais-en ce que tu veux"). A. Weil was at the time one of the managing editors of the *American Journal of Mathematics.* We agreed to interpret Chevalley's authorization broadly, as allowing me to complete his manuscript under his name and submit it for publication in the *American Journal of Mathematics.* There was a theoretical possibility that Chevalley might become aware of it since, as an associate editor, he regularly received reports on the activities of the Journal, but we were rather confident that he hardly ever looked at them. Indeed, he learned about the paper only when he saw it in the issue of the Journal containing it. Baffled at first, he quickly realized what had happened and did not mind. I had included the refinement I needed directly in the paper, as a lemma to be used by A. Borel in a forthcoming paper, and I warmly thanked Chevalley in [B0] for having shown his manuscript to me.

(6) As already pointed out, Maurer also claims in [M] that a complex irreducible linear algebraic group is a rational variety. Since this paper is about the only reference to earlier work given in Lie II, I once asked Chevalley why he did not quote it in [C23]. His answer was that since there were mistakes in the first part of [M] he had never bothered to look at the rest of it. In fact, although Maurer does not carry out all details of the proof, his approach is essentially valid, and in a way more direct than Chevalley's, in the sense that it does not use Levi's theorem (which was anyhow not available at the time). Briefly, his idea is the following: First, if G is unipotent, the exponential mapping provides a biregular isomorphism of the Lie algebra $\mathbf{g}$ of G onto G. If not, then $\mathbf{g}$ contains semi-simple elements hence also, by the analysis leading to the definition of $\mathbf{h}_X$ above, a regular semi-simple element X such that ad X has integral eigenvalues. Then $\mathbf{g}$ is the direct sum of a nilpotent algebra $\mathbf{n}^-$ consisting of nilpotent elements corresponding to the strictly negative eigenvalues of ad X and of a solvable algebra $\mathbf{b}$ corresponding to the positive eigenvalues, and G is birationally isomorphic to the product of the corresponding groups N^- and B. Moreover, $\mathbf{b}$ is the direct sum of the nilspace $\mathbf{h}$ of ad X and of the nilpotent ideal $\mathbf{n}$ corresponding to the strictly positive eigenvalues of ad X. Then B is the semi-direct product of the corresponding groups H and N, so that one is reduced to investigate the (nilpotent) algebraic group H.

(7) More generally, a torus operating on a projective variety of dimension n has at least $n+1$ fixed points. After I had noticed this (quite obvious) generalization, I realized it had already been proved over $\mathbb{C}$ by G. Fano in 1896 [F].

(8) A proof not involving case by case consideration has been given later by M. Takeuchi [T]. For central isogenies, it is also presented in [J].

(9) It was pointed out to me several years ago by J. Tits and D. Verma, independently, that this last claim, which amounts to saying that the scheme over $\mathbb{Z}$ defined by $\mathbb{Z}[G]$ has good reduction everywhere, is not quite true. But it becomes so if π is replaced by its direct sum with the adjoint representation. See also 3.9.3 in F. Bruhat et J. Tits, Publ. Math. I. H. E. S. **60** (1983), 197 - 376. For a systematic exposition of split groups with (resp. without) schemes, see [DG] (resp. [S]).

References

[B0] A. Borel, *Sur la torsion des groupes de Lie*, J. Math. pures appl. (9) **35** (1955) 127–139.

[B1] A. Borel, *Groupes linéaires algébriques*, Ann. Math. (2) **64** (1956) 20–82.

[B2] A. Borel, Linear algebraic groups, (Notes by H. Bass), Benjamin (1969), second enlarged edition, GTM 126, Springer-Verlag(1991)

[BM] A. Borel and G. D. Mostow, *On semi-simple automorphisms of Lie algebras*, Ann. Math. (2) **61** (1955) 389–405.

[Bk] N. Bourbaki, Groupes et algèbres de Lie, Chap. 9: Groupes de Lie réels compacts, Masson, Paris 1982.

[Ca1] E. Cartan, *Sur certains groupes algébriques*, C.R. Acad. Sci. Paris **120** (1894) 544–548.

[Ca2] E. Cartan, *Sur les invariants intégraux de certains espaces homogènes clos et les propriétés topologiques de ces espaces*, Ann. Soc. Polonaise Math. **8** (1929) 181–225.

[Ca3] E. Cartan, La théorie des groupes finis et continus et l'Analysis Situs, Mem. Sc. Math. XLII, Gauthier-Villars, Paris 1930.

[Ca4] E. Cartan, *La topologie des espaces représentatifs des groupes de Lie*, Ens. Math. **35** (1936) 177–200.

Books and papers by C. Chevalley:

[C1] *Sur la théorie du corps de classes dans les corps finis et les corps locaux*, J. Fac. Sci. Tokyo Univ. **2** (1933) 364–476.

[C2] (with H. Nehrkorn) *Sur les démonstrations arithmétiques dans la théorie du corps de classes*, Math. Annalen **111** (1935) 364–371.

[C3] *La théorie du corps de classes*, Ann. Math. (2) **41** (1940) 394–417.

[C4] *On the topological structure of solvable groups*, Ann. Math (2) **42** (1941) 668–675.

[C5] *An algebraic proof of a property of Lie groups*, Amer. J. Math **63** (1941) 785–793.

[C6] *A new kind of relationship between matrices*, Amer. J. Math **65** (1943) 521–531.

[C7] *On groups of automorphisms of Lie groups*, Proc. Nat. Acad. Sci **30** (1944) 274–275.

[C8] *La théorie des groupes de Lie*, Proc. First Canadian Math. Congress 1945, Univ. of Toronto Press (1946) 338–354.

[C9] (with H.F. Tuan) *On algebraic Lie algebras*, Proc. Nat. Acad. Sci. USA **51** (1946) 195–196.

[C10] Theory of Lie groups I, Princeton Univ. Press (1946).

[C11] *Algebraic Lie algebras*, Ann. Math. (2) **48** (1947) 91–100.

[C12] (with S. Eilenberg) *Cohomology theory of Lie groups and Lie algebras*, Trans. Amer. Math. Soc. **63** (1948) 85–124.

[C13] *Sur la classification des algèbres de Lie simples et de leurs représentations*, C.R. Acad. Sci. Paris **227** (1948) 1136–1138.

[C14] *The Betti numbers of the exceptional simple Lie groups*, Proc. ICM 1950 Harvard **2**, 21–24.

[C15] (with R.D. Schafer) *The exceptional simple Lie algebras F_4 and E_6*, Proc. Nat. Acad. Sci. USA **36** (1950) 137–141.

[C16] (with H-F. Tuan) *Algebraic Lie algebras and their invariants*, J. Chinese Math. Soc. **1** (1951) 215–242.

[C17] *On a theorem of Gleason*, Proc. Amer. Math. Soc **2** (1951) 122–125.

[C18] (with E.R. Kolchin) *Two proofs of a theorem on algebraic groups*, *ibid.*, 126–137.

[C19] Théorie des groupes de Lie II. Groupes algébriques, Hermann, Paris 1951.

[C20] *Sur le groupe exceptionnel (E_6)*, C.R. Acad. Sci. Paris **232** (1951) 1991–1993.

[C21] *Sur une variété algébrique liée à l'étude du groupe (E_6)*, *ibid.*, 2168–2170.

[C22] *Notice sur les travaux scientifiques de Claude Chevalley*, (unpublished).

[C23] *On algebraic group varieties*, J. Math. Soc. Japan **6** (1954) 303–324.

[C24] Théorie des groupes de Lie III. Groupes algébriques, Hermann, Paris (1954).

[C25] The algebraic theory of spinors, Columbia University Press, New York (1954).

[C26] (with A. Borel) *The Betti numbers of the exceptional Lie groups*, Memoirs Amer. Math. Soc **14** (1955) 1–9.

[C27] *Invariants of finite groups generated by reflections*, Amer. J. Math. **77** (1955) 778–782.

[C28] *Sur certains groupes simples*, Tôhoku Math. J. **7** (1955) 14–66.

[C29] *Plongement projectif d'une variété de groupe*, Proc. Int. Symposium on algebraic number theory, Tokyo-Nikko (1956) 131–138.

[C30] Classification des groupes de Lie algébriques, 2 vol., Notes polycopiées, Inst. H. Poincaré (1956-58).

[C31] *Sur les décompositions cellulaires des espaces G/B*, (unpublished).

[C32] *La théorie des groupes algébriques*, Proc. ICM 1958 Edinburgh, Cambridge Univ. Press (1960) 53–68.

[C33] *Certains schémas de groupes semi-simples*, Sém Bourbaki 1960/61, Exp. **219**.

[DG] M. Demazure et A. Grothendieck, Schémas en groupes I, II, III, Springer LNM **151**, **152**, **153** (1970).

[F] G. Fano, *Sulle varietà algebriche con un gruppo continuo non integrabile di transformazioni proiettive in sè*, Mem. Reale Accad. d. Sci. di Torino (2) **46** (1896) 187–218.

[GHV] W. H. Greub, S. Halperin and J.R. Vanstone, Connections, curvature and cohomology, Vol 3, Academic Press, New York (1976).

[HC] Harish-Chandra, *On some applications of the universal algebra of a semi-simple Lie algebra*, Trans. Amer. Math. Soc. **70** (1951) 28–96.

[I] K. Iwasawa, *On some types of topological groups*, Ann. Math. (2) **50** (1949) 507–558.

[J] J.C. Jantzen, Representations of algebraic groups, Academic Press, New York (1987).

[K1] E.R. Kolchin, *Algebraic matric groups and the Picard-Vessiot theory of homogeneous linear differential equations*, Ann. Math. (2) **49** (1958) 1–42.

[K2] E.R. Kolchin, *On certain concepts in the theory of algebraic matric groups*, *ibid.*, 774–789.

[Kz] J-L. Koszul, *Homologie et cohomologie des algèbres de Lie*, Bull. Soc. Math. France **78** (1950) 65–127.

[M] L. Maurer, *Zur Theorie der continuirlichen, homogenen und linearen Gruppen*, Sitzungsber. d. math. phys. Klasse der Kgl. Bayerischen Akad. d. Wiss. zu München (8) **24** (1894) 297–341.

[O] T. Ono, *Sur les groupes de Chevalley*, J. Math. Soc. Japan **10** (1958) 307–313.

[P] L. Pontrjagin, Topological groups, (translated by E. Lehmer) Princeton Univ. Press (1939)

[R] M. Rosenlicht, *Some basic theorems on algebraic groups*, Amer. J. Math. **78** (1956) 401–443.

[S] R. Steinberg, Lectures on Chevalley groups, (Notes by J. Faulkner and R. Wilson) Yale University (1967).

[T] M. Takeuchi, *A hyperalgebraic proof of the isomorphism and isogeny theorems for reductive groups*, J. Algebra **85** (1983) 179–196.

[Wi] A. Weil, *Géométrie différentielle des espaces fibrés*, Collected Papers, Vol I, 422–436.

[Wy1] H. Weyl, *Theorie des Darstellung kontinuerlicher halbeinfacher Gruppen durch lineare Transformationen*, I. II. III and Nachtrag: I: Math. Zeitschr. **23** (1925) 271–309; II: Math. Zeitschr. **24** (1926) 377–395; III: Math. Zeitschr. **24** (1926) 377–395. Nachtrag: Math. Zeitschr. **24** (1926) 789–791

[Wy2] H. Weyl (mit F. Peter), *Die Vollständigkeit der primitiven Darstellungen einer geschlossenen kontinuerlichen Gruppe*, Math. Annalen **97** (1927) 737–755.

[Y] Yen Chih Ta, *Sur les polynomes de Poincaré des groupes simples exceptionnels*, C.R. Acad. Sci. Paris **228** (1949) 628–630.

Institute for Advanced Study,
Princeton, NJ 08540,
USA

144.

(with G. Prasad)

Values of Isotropic Quadratic Forms at S-Integral Points

Compositio Mathematica **83** (1992), 347–372

1 Introduction

Let F be a non-degenerate indefinite quadratic form on $\mathbf{R}^n$ ($n \geqq 3$). A. Oppenheim conjectured and G. A. Margulis proved [M] that if F is not a multiple of a rational form, then $F(\mathbf{Z}^n)$ is not discrete around the origin. In this paper, we are concerned with a generalization of this result in a S-arithmetic setting.

In the sequel, k is a number field and $\mathfrak{o}$ the ring of integers of k. For every normalized absolute value $|\,.\,|_v$ on k, let k_v be the completion of k at v. Let S be a finite set of places of k containing the set S_∞ of archimedean ones, k_S the direct sum of the fields k_s ($s \in S$) and $\mathfrak{o}_S$ the ring of S-integers of k (i.e. of elements $x \in k$ such that $|x|_v \leqq 1$ for all $v \notin S$).

Let F be a quadratic form on k_S^n. Equivalently, F can be viewed as a collection F_s ($s \in S$), where F_s is a quadratic form on k_s^n. The form is non-degenerate if and only if each F_s is non-degenerate. We shall say that F is isotropic if each F_s is so, i.e. if there exists for each $s \in S$ an element $x_s \in k_s^n - \{0\}$ such that $F_s(x_s) = 0$. If s is a real place, this condition is equivalent to F_s being indefinite (since it is non-degenerate). The form F will be said to be *rational* (over k) if it is a multiple of a form on k^n, i.e. if there exists a form F_o on k^n and λ invertible in k_S such that $F = \lambda . F_o$, and *irrational* otherwise.

Endowed with the product topology, k_S is, with respect to the addition, a locally compact group and $\mathfrak{o}_S$ is a discrete cocompact subgroup. Similarly, $\mathfrak{o}_S^n$ is a cocompact lattice in k_S^n. If F is rational, then $F(\mathfrak{o}_S^n)$ is discrete in k_S, since we can write $F = \lambda . F_o$ ($\lambda \in k_S^*$) and may even assume that F_o has coefficients in $\mathfrak{o}$, whence $F(\mathfrak{o}_S^n) \subset \lambda \mathfrak{o}_S$. As a generalization of the Oppenheim conjecture we shall prove that if F is irrational, isotropic non-degenerate and $n \geqq 3$, then $F(\mathfrak{o}_S^n)$ is non-discrete around the origin of k_S. In fact, we shall establish a somewhat stronger statement:

THEOREM A. *Let F be as above. Assume F to be non-degenerate, isotropic and*

*Supported by the NSF at the Institute for Advanced Study, Princeton, during 1987—88.

$n \geqq 3$. *Then the following conditions are equivalent:*

(i) *F is irrational.*

(ii) *Given* $\varepsilon > 0$, *there exists* $x \in \mathfrak{o}_S^n$ *such that*

$$0 < \max_{s \in S} |F_s(x)|_s < \varepsilon. \tag{1}$$

(iii) *Given* $\varepsilon > 0$ *there exists* $x \in \mathfrak{o}_S^n$ *such that*

$$0 < |F_s(x)|_s < \varepsilon \quad \textit{for all } s \in S. \tag{2}$$

We just remarked that (ii) ⇒ (i) and it is obvious that (iii) ⇒ (ii). So the interest lies in the implications (i) ⇒ (ii) ⇒ (iii). If $k = \mathbf{Q}$ and $S = S_\infty$, then (ii) and (iii) are identical and (i) ⇒ (ii) is the Oppenheim conjecture. Thus our (i) ⇒ (ii) is a direct generalization of it, while (i) ⇒ (iii) is a natural strengthening, which should of course be true if, as is expected, $F(\mathfrak{o}_S^n)$ is dense in k (see Section 6).

To prove Theorem A we shall first handle two main special cases:

(I) The implication (i) ⇒ (ii) when $S = S_\infty$. The argument there is patterned after that of Margulis.[1]

(II) The implication (i) ⇒ (iii) when at least one of the F_s is multiple of a k-rational form. The proof uses strong approximation in algebraic groups, some elementary geometry of numbers and is quite different from that of (I).

The implication (i) ⇒ (iii) in the general case then follows easily from (I) and (II).

Margulis deduced the Oppenheim conjecture from a theorem about closures of orbits of $\mathbf{SO}(2, 1)$ in $\mathbf{SL}_3(\mathbf{R})/\mathbf{SL}_3(\mathbf{Z})$. It is easily seen that conversely the Oppenheim conjecture implies such an orbit theorem. There is a similar equivalence in the general case (Section 1). In (I), we follow Margulis by proving first an orbit theorem but, in (II), we proceed directly to the Oppenheim conjecture, so that Theorem A yields an assertion about closures of orbits in the S-arithmetic case.

The paper is organized as follows: Section 1 contains some preliminary results, in particular a reduction of the proof of Theorem A to the case $n = 3$ and a discussion of the relation, for $n = 3$, between Theorem A and assertions about closures of orbits in spaces of lattices. As a preparation to (I), we give in Section 2 more algebraic geometric proofs of some lemmas of [M] on actions of unipotent groups, so that the consequences drawn in [M] for $\mathbf{SL}_3(\mathbf{R})$ are also valid for $\mathbf{SL}_3(\mathbf{C})$. We then treat (I) in Section 3, (II) in Section 4, and the general case in Section 5. In section 6, we add some remarks and questions about a still open problem, namely whether $F(\mathfrak{o}_S^n)$ is dense when F is irrational.

The generalization from the original case to that of a number field (with

[1] This implication also follows from the validity of the Raghunathan conjecture on orbit closures, recently proved by M. Ratner [Rt], (cf. §7).

$S = S_\infty$), was proposed first in [RR], where some partial results are obtained. We have added finite places following a suggestion of G. Faltings. We thank G. A. Margulis for a simplification in the proof of I which arose in a discussion with one of us (G. P.).

The main results of this paper have been announced, with sketches of some proofs, in [BP].

1. Preliminaries

In this section, we show that it suffices to establish Theorem A for $n = 3$ and then prove that it is equivalent to some statements about closures of orbits in $\mathbf{SL}_3(k_S)/\mathbf{SL}_3(\mathfrak{o}_S)$. We first fix some notation and conventions:

1.0. In the sequel, the notation preceding the statement of Theorem A is used without further reference. We also let $S_f = S - S_\infty$.

For any subring A of k, we view A^n as diagonally embedded in k_S^n. If V is a vector subspace of k^n, then $V_S := V \otimes_k k_S$ is viewed as a k_S-submodule of k_S^n.

If G is a locally compact group, G^o denotes the connected component of the identity in G. An element $d \in G$ is said to *contract* a subgroup U if it normalizes U and $\lim_{n \to \infty} d^n.u.d^{-n} = 1$ for every $u \in U$.

1.1. PROPOSITION. *Let G be a semi-simple algebraic group defined over* $\mathbf{Q}$ *and* $\Gamma \subset G(\mathbf{Q})$ *an arithmetic subgroup. Let E be a subgroup of* $G(\mathbf{R})$ *generated by unipotent elements and assume that* $\overline{E.\Gamma} = R\Gamma$, *where R is a closed connected subgroup of* $G(\mathbf{R})$ *such that $R \cap \Gamma$ has finite covolume in R. Then* $R = \tilde{E}(\mathbf{R})^o$, *where $\tilde{E}$ is the smallest* $\mathbf{Q}$*-subgroup of G whose group of real points contains E.*

Proof. We note first that $\tilde{E}$ is connected in the Zariski topology and, more precisely, that $E \subset \tilde{E}(\mathbf{R})^o$. Indeed, if $u \in E$ then some power u^m of u belongs to $\tilde{E}(\mathbf{R})^o$. If u is unipotent and $\neq 1$, then the whole one-parameter unipotent subgroup $\exp(t \log u^m)$ ($t \in \mathbf{R}$), in particular u itself, belongs to $\tilde{E}(\mathbf{R})^o$.

Next we claim that the Levi subgroups of $\tilde{E}$ are semi-simple. Let C be the quotient $\tilde{E}/\mathscr{R}_u\tilde{E}$ of $\tilde{E}$ by its unipotent radical. The image of E in C is generated by unipotent elements, hence belongs to the derived group $\mathscr{D}C$ of C. The inverse image of $\mathscr{D}C$ in $\tilde{E}$ is defined over $\mathbf{Q}$ and its group of real points contains E, so it coincides with $\tilde{E}$. It follows that $\tilde{E}$ has no non-trivial rational character defined over $\mathbf{Q}$, which, as is known, implies that $\tilde{E}(\mathbf{R})^o.\Gamma$ is closed [*A*], hence also that $R \subset \tilde{E}(\mathbf{R})^o$. On the other hand, by [D: §4], the Zariski-closure $\mathscr{A}(R \cap \Gamma)$ of $R \cap \Gamma$ contains all unipotent elements of R, hence E and therefore $\tilde{E}$. But the inclusion $R \subset \tilde{E}(\mathbf{R})^o$ shows that $\tilde{E}$ contains the smallest $\mathbf{Q}$-subgroup $\tilde{R}$ containing R, hence also $\mathscr{A}(R \cap \Gamma)$. As a result

$$\tilde{E} = \mathscr{A}(\Gamma \cap R) = \tilde{R},$$

and R is contained and Zariski-dense in $\tilde{E}$. Being connected, in the ordinary topology, it must be normal in $\tilde{E}(\mathbf{R})^o$. Its image in $C(\mathbf{R})^o$ is normal, connected, hence closed (C is semi-simple) and Zariski-dense. Therefore the image of R is the whole of $C(\mathbf{R})^o$. It also follows that the image in C of a maximal connected semi-simple subgroup M of R is equal to $C(\mathbf{R})^o$, whence $R = M.(\mathscr{R}_u\tilde{E} \cap R)$. The group M being semi-simple, linear, is of finite index in the group of real points of an algebraic $\mathbf{R}$-group. The group $\mathscr{R}_u\tilde{E} \cap R$ is connected, unipotent, hence also algebraic. As a consequence, R itself is of finite index in the group of real points of an algebraic group [B: §7]. Since R is Zariski-dense in $\tilde{E}$, the proposition is proved.

1.2. PROPOSITION. *Let $m, n \in \mathbf{N}$. For each $s \in S$, let $\mathscr{H}_s$ be a connected almost simple and isotropic k_s-subgroup of $\mathbf{SL}_n$, of dimension m. Let $H_s = \mathscr{H}_s(k_s)$ and H_S the product of the H_s $(s \in S)$, viewed as a subgroup of $\mathbf{SL}_n(k_S)$. Assume that $\Gamma = H_S \cap \mathbf{SL}_n(\mathfrak{o}_S)$ is of finite covolume in H_S. Then there exists a connected k-subgroup M of $\mathbf{SL}_n/k$ such that $M(k_S) = H_S$.*

Proof. Let

$$H_\infty = \prod_{s \in S_\infty} H_s, \qquad H_f = \prod_{s \in S_f} H_s, \tag{1}$$

U be a compact open subgroup of H_f and $\Gamma_U = (U \times H_\infty) \cap \Gamma$. The group Γ_U is of finite covolume in $U \times H_\infty$ hence the projection Γ'_U of Γ_U in H_∞, which is discrete since U is compact, is of finite covolume. We view the product of the $\mathbf{SL}_n(k_s)$ $(s \in S_\infty)$ as the group of real points of the $\mathbf{Q}$-group $L = R_{k/\mathbf{Q}}\mathbf{SL}_n/k$ obtained from $\mathbf{SL}_n$, viewed as a k-group, by restriction of scalars from k to $\mathbf{Q}$. Then Γ'_U is contained in an arithmetic subgroup of L_∞ and, being of finite covolume, is Zariski-dense, since H_∞ is by assumption a product of simple non-compact Lie groups. Therefore H_∞ is the group of real points of a $\mathbf{Q}$-group. By assumption, it is the product of its projections on the factors $\mathbf{SL}_n(k_s)$, therefore it is itself of the form $R_{k/\mathbf{Q}}M(\mathbf{R})$, where M is a connected k-subgroup of $\mathbf{SL}_n$, of dimension m. This already shows that $H_s = M(k_s)$ for all $s \in S_\infty$. Now Γ'_U is to be viewed as an arithmetic subgroup of $M(k)$. Let M_f be the product of the $M(k_s)$ for $s \in S_f$. It is known that Γ'_U, diagonally embedded in M_f, is dense in an open subgroup. Therefore $M_f \subset H_f$ and then $M_f = H_f$ for dimensional reasons.

1.3. PROPOSITION. *Assume F to be irrational and $n \geqq 3$. Then there exists a three-dimensional subspace V of k^n such that the restriction of F to V_S is non-degenerate, isotropic and irrational.*

Proof. Fix $s \in S$. Let H_s be the orthogonal group of F_s. Let M be a subspace of k_s^n on which the restriction of F_s is non-degenerate. It is well-known that the $H_s(k_s)$-orbit of M in the Grassmannian $G_{m,n}$ of m-planes ($m = \dim M$) in k_s^n is open (for the analytic topology) in $G_{m,n}(k_s)$. This follows from the fact that if l_s is an algebraically closed extension of k_s, then $H_s(l_s)(M)$ is open in $G_{m,n}(l_s)$ and from the existence of local cross-sections (in the analytic topology) for the fibration of

$H_s(k_s)$ by the isotropy group of M. In particular, if F_s is isotropic on M, then F_s is non-degenerate and isotropic on any m-dimensional subspace of k_s^n sufficiently close to M.

Assume now $n > 3$. We want to prove the existence of a subspace V of k^n of codimension one, such that the restriction of F to V_S is non-degenerate, isotropic and irrational. For each $s \in S$, fix $M_s \subset k_s^n$ of codimension one, on which F_s is non-degenerate and isotropic. By weak approximation in k_S, we may find $M' \subset k^n$, of codimension one, such that $M' \otimes_k k_s$ is arbitrarily close to M_s for each s, therefore such that the restriction of F to M'_S is non-degenerate and isotropic. There remains to show that there exists such an M' on which F is in addition irrational. Fix $e \in M'(k)$ on which no F_s is zero. Again, this exists by weak approximation. After having multiplied F by a unit in k_S we may assume that $F_s(e) = 1$ for all $s \in S$. Let $\mathscr{M}$ be the set of $(n-1)$-dimensional subspaces of k^n containing e and such that the restriction of F to M_S is non-degenerate and isotropic. Assume that for no $M \in \mathscr{M}$, $F \mid M_S$ is irrational. Then $F(x) \in k$ for all $x \in M(k)$, $M \in \mathscr{M}$. For a given s, the map $x \mapsto F_s(x)$ is a regular function on l_s^n. Since it takes rational values on the union of the $M(k)$, which is obviously Zariski-dense in l_s^n, it is defined over k, hence $F_s(x) \in k$ for all $x \in k^n$ and this implies that F_s is rational over k. Since $F_s(x)$ is independent of s for $x \in M(k)$, $M \in \mathscr{M}$, it follows that F is rational, contradiction. Therefore the restriction of F to some element in $\mathscr{M}$ is irrational, and the proposition follows by descending induction on n.

1.4. LEMMA. *Fix $s \in S$. Let F_s be a non-degenerate isotropic quadratic form on k_s^n. Given a neighborhood U of the origin in k_s^n, there exists $\varepsilon > 0$ such that $F_s(U - 0)$ contains all elements of k_s with absolute value $\leqslant \varepsilon$.*

Proof. The space k_s^n is the direct sum of a hyperbolic plane for F_s and of its orthogonal complement. This reduces us to the case $n = 2$ and $F_s = x.y$, for which our assertion is obvious.

1.5. We now come back to the situation of Theorem A for $n = 3$ and introduce some further notation. We let $G_s = \mathbf{SL}_3(k_s)$ and $G_S = \mathbf{SL}_3(k_S) = \Pi_{s \in S} G_s$. The standard S-arithmetic subgroup of $\mathbf{SL}_3$ is $\Gamma_S = \mathbf{SL}_3(\mathfrak{o}_S)$. We let $\Omega = G_S/\Gamma_S$. It is the space of free $\mathfrak{o}_S$-submodules of k_S^3 of maximal rank and determinant one. We write Λ_o for $\mathfrak{o}_S^3$. For a non-degenerate quadratic form $F = (F_s)$ on k_S^3, we let H_F be the product $\Pi_{s \in S} \mathrm{SO}(F_s)(k_s)$.

1.6. PROPOSITION. *Let $\mathscr{F}$ be the set of non-degenerate isotropic quadratic forms on k_S^3. Then the following two assertions are equivalent*:

(a) *Any $F \in \mathscr{F}$ which does not represent zero rationally either is rational or satisfies* (ii) *of Theorem A.*

(b) *If $F \in \mathscr{F}$ and $z \in \Omega$ are such that $H_F . z$ is relatively compact, then $H_F . z$ is compact.*

Proof. We first assume (b) for $z = \Lambda_o$. Note first that, in view of Mahler's criterion and the fact that H_F consists of elements of determinant one, the assumption $H_F.\Lambda_o$ relatively compact is equivalent to

(1) *There exists a neighborhood U of the origin in k_S^3 such that $H_F.\Lambda_o \cap U = \{0\}$.*

Assume that (ii) of Theorem A does not hold for some $\varepsilon > 0$. Fix U such that $\max |F_s(u_s)|_s < \varepsilon$ for $u = (u_s) \in U$. Then (1) is satisfied, hence $H_F.\Lambda_o$ is compact. By Arens theorem, $H_F.\Lambda_o$ is homeomorphic to $H_F/(H_F \cap \Gamma_S)$, hence $H_F \cap \Gamma_S$ is cocompact in H_F. By 1.2, we can find a k-subgroup M of $\mathbf{SL}_3$ such that $H_F = M(k_S)$ and this implies that F is rational.

Assume now (a). Let $F \in \mathscr{F}$ and $z \in \Omega$ be such that $H_F.z$ is relatively compact. Let $g \in G_S$ be such that $g^{-1}.z = \Lambda_o$ and let $F' = {}^t g.F.g$. Then $H_{F'}.\Lambda_o$ is relatively compact so that (1) holds for F'. In view of 1.4 and Witt's theorem, this implies that F' does not take arbitrarily small values, in absolute value, on $\Lambda_o - \{0\}$, hence also that it does not represent zero rationally; therefore, it does not fulfill (ii). By (a), F' is rational and then $H_{F'} = \mathrm{SO}(F')(k_S)$. It follows that $H_F.\Lambda_o$ is closed and therefore compact. As a consequence, $H_F.z$ is compact, too.

1.7. LEMMA. *Let $n = 3$. Then F is rational and isotropic over k if and only if it is zero on infinitely many lines in k^3.*

Proof. If F is rational over k, then $\mathrm{SO}(F)$ is defined over k. If now F is zero on some rational line, then it is zero on all its transforms under $\mathrm{SO}(F)(k)$, and those are infinite in number since $\mathrm{SO}(F)(k)$ is Zariski-dense in SO(F).

Assume now F to be zero on infinitely many rational lines in k^3 (viewed as embedded diagonally in k_S^3). Take as a universal field an algebraically closed extension K of k containing the fields k_s $(s \in S)$. Then the cones $F_s = 0$ $(s \in S)$ in K^3 have in common a Zariski-dense set of k-rational points. Hence they are identical, all defined over k, and then F is rational and isotropic over k.

1.8. Since $k \subset k_S$, the k_S-module k_S^n can be viewed as a vector space over k. A subset $D \subset k_S^n$ will be called a k-line if it is a one-dimensional subspace for this structure, i.e. if $D = k.x$, for some $x \in k_S^n - \{0\}$. It is rational if $D \subset k^n$. If $n = 1$, and all components x_s of $x \in k_S$ are not zero, then $\mathfrak{o}_S.x$ is discrete in k_S. However, if at least one component x_r is zero, then the set $\mathfrak{o}_S.x$ is not discrete, more precisely contains a sequence of non-zero elements accumulating to 0, since there is a sequence $u_n \in \mathfrak{o}_S - \{0\}$, $n = 1, 2 \ldots$ such that $u_{n,s} \to 0$ for all $s \neq r$.

Let now $D = k.x$ be a rational line in k_S^n. Then $D \cap \mathfrak{o}_S^n$ is a finitely generated $\mathfrak{o}_S$-module, therefore if $F_s(x) \neq 0$ for all $s \in S$, then $F(D \cap \mathfrak{o}_S^n)$ is discrete in k_S. It follows that if F satisfies (iii) of Theorem A, then it takes arbitrarily small values, no component of which is zero, on S-integral points belonging to infinitely many rational lines.

If on the other hand there exist $r, s \in S$ such that $F_r(x) = 0$ and $F_s(x) \neq 0$ for some $x \in \mathfrak{o}_S^n$, then F is obviously not rational, and moreover by the remark in the previous paragraph, it satisfies (ii). In this case therefore, the Oppenheim condition is already fulfilled for $x \in D \cap \mathfrak{o}_S^n$.

1.9. PROPOSITION. *Let $\mathscr{F}$ be the set of all non-degenerate isotropic quadratic forms on k_S^3, and* (a), (b), (c) *the three statements*:

(a) *Any $F \in \mathscr{F}$ is either rational or satisfies* (ii) *of Theorem A.*

(b) *Any $F \in \mathscr{F}$ is either rational or satisfies* (iii) *of Theorem A.*

(c) *If $F \in \mathscr{F}$ and $z \in \Omega$ are such that we can find an open neighborhood U of the origin in k_S^3 and a finite union C of k-lines (see* 1.8) *satisfying the condition*

$$H_F(z - z \cap C) \cap U = \{0\}, \tag{*}$$

then $H_F . z$ is compact.

Then (b) $\Rightarrow$ (c) $\Rightarrow$ (a).

Proof. We show first that (b) $\Rightarrow$ (c). Since $g \in G_S$ is k-linear, the statement (c) is invariant under conjugation by G_S and we may assume without loss of generality that $z = \mathfrak{o}_S^3$ in (*). Under (iii), F takes arbitrary small values on elements of $\mathfrak{o}_S^3$ belonging to infinitely many rational lines (1.8); we see therefore that (*) prevents (iii) to hold, and F is rational. If it were isotropic over k, then it would be zero on infinitely many rational lines (1.7), again contradicting (*). Therefore F is rational, anisotropic over k, hence $H_F . \mathfrak{o}_S^3$ is compact.

We now prove that (c) $\Rightarrow$ (a). Let $F \in \mathscr{F}$ and assume it is not rational. Then it can be zero on only finitely many rational lines (1.7). Let C be their union. Assume (ii) is not true for some $\varepsilon > 0$. There exists a neighborhood U of the origin in k_S^3 such that $|F_s(u_s)| \leqq \varepsilon/2$ for all $s \in S$ and $u = (u_s) \in U$. Then (*) of (c) is fulfilled for this choice of U, C and $z = \mathfrak{o}_S^3$. By (c), $H_F . \mathfrak{o}_S^3$ is compact, and then $H_F = M(k_S)$ for some k-subgroup of $\mathbf{SL}_3$ (1.2), which is equivalent to saying that F is rational.

1.10. REMARK. By definition here, F is isotropic if each F_s is so. This is stronger than the usual requirement: $F(x) = 0$ for some non-zero vector in k_S^n. However, it is necessary here because if we allow one F_s to be anisotropic, Theorem A cannot hold. To see this, observe first that if F_s and F'_s are anisotropic over k_s^n, then $|F_s(x)|_s / |F'_s(x)|_s$ is bounded from above and away from zero on any compact subset of k_s^n not containing the origin, whence the existence of a constant $c > 1$ such that

$$c^{-1}|F_s(x)|_s \leqq |F'_s(x)|_s \leqq c . |F_s(x)|_s \quad \text{for all } x \in k_s^n. \tag{1}$$

Let now F be a rational form; assume F_r to be anisotropic over k_r^n for some $r \in S$, and let $F' = (F'_s)$ be such that $F'_s = F_s$ for $s \neq r$ but F'_r is an anisotropic form on k_r^n which is not a multiple of F_r. Then F' is irrational, but in view of (1), $F'(\mathfrak{o}_S^n)$ is still discrete, since $F(\mathfrak{o}_S^n)$ is so.

2. Geometric lemmas

In this section, unless otherwise stated, the field K is either **R** *or* **C**.

2.1. Let U be a connected unipotent K-group, d its dimension. As a variety, U is K-isomorphic to the d-dimensional affine space under "log". Let u_i be coordinates in the latter and use multiexponential notation for monomials, i.e. if $\alpha=(\alpha_1,\ldots,\alpha_d)\in\mathbf{N}^d$, then u^α stands for $u_1^{\alpha_1}\cdots u_d^{\alpha_d}$ and $|\alpha|=\Sigma\,\alpha_i$. If $u=(u_i)$, then λu is the element with coordinates λu_i $(\lambda\in K)$.

LEMMA. *Let U be as before, E a finite dimensional vector space defined over K and $\sigma\colon U\to \mathrm{GL}(E)$ a rational representation of U in E, defined over K. Let F be the fixed point set of $\sigma(U)$ in E. Choose $c_o\in F(K)$ and let $c_i\in E(K)-F(K)$ $(i=1,2,\ldots)$ be a sequence of elements tending to c_o as $i\to\infty$. Then there exists a non-constant K-morphism of varieties $\varphi\colon U\to F$, mapping the identity onto c_o, such that $\varphi(U(K))$ belongs to the closure of the union of the orbits $\sigma(U(K)).c_i$ $(i=1,2,\ldots)$. If U is commutative, then there exist a sequence $\lambda_i\in K^*$ and a subsequence $\{c_{r_i}\}$ of $\{c_i\}$, such that $|\lambda_i|\to\infty$ and $\varphi(u)=\lim_i\sigma(\lambda_i.u).c_{r_i}$ for all $u\in U$.*

Proof. Fix a basis (e_i) of $E(K)$. Then $\sigma(u)=(\sigma(u)_{p,q})$ is represented by a matrix whose coefficients are polynomials on U, with coefficients in K. Let $(c_{i,q})$ be the coordinates of c_i. The pth coordinate of $\sigma(u).c_i$ is then

$$(\sigma(u).c_i)_p=\sum_q\sigma(u)_{p,q}.c_{i,q}=\sum_\alpha a_{p,i,\alpha}u^\alpha\quad(a_{p,i,\alpha}\in K). \tag{1}$$

For a given i, the $a_{p,i,\alpha}(|\alpha|\neq 0)$ are not all zero since, by assumption, c_i is not fixed under $\sigma(U)$. We can find $\lambda_i\in K^*$ such that

$$|a_{p,i,\alpha}\lambda_i^{|\alpha|}|\leqq 1\quad\text{for all } p,\ q,\ \alpha\quad\text{with } |\alpha|\neq 0 \tag{2}$$

$$|a_{p_o,i,\alpha_o}\lambda_i^{|\alpha_o|}|=1\ \text{for some choice } p_o,\ \alpha_o \text{ of } p\quad\text{and}\quad\alpha\ (|\alpha_o|\neq 0). \tag{3}$$

Passing to a subsequence, if necessary, we may arrange that p_o, α_o are independent of i. Again going over to a subsequence, we may assume that the $a_{p,i,\alpha}\lambda_i^{|\alpha|}$ converge for $|\alpha|\neq 0$. But this is automatic if $|\alpha|=0$, since the sequence c_i converges. Let then

$$a_{p,\alpha}=\lim_{i\to\infty}a_{p,i,\alpha}\lambda_i^{|\alpha|}. \tag{4}$$

We define $\psi\colon U\to E$ by

$$\psi(u)_p=\sum_\alpha a_{p,\alpha}u^\alpha. \tag{5}$$

i.e. by

$$\psi(u)=\lim_{i\to\infty}\sigma(\lambda_i.u).c_i. \tag{6}$$

If $u = 1$, it is represented by zero in our coordinates, hence $\lambda_i . u = u$ and

$$\psi(1) = \lim_i c_i = c_o. \tag{7}$$

The map ψ is not constant since by our construction $|a_{p,\alpha}| = 1$ for some choice of p, $\alpha(|\alpha| \neq 0)$. This implies that $|\lambda_i| \to \infty$, because otherwise a subsequence of $\{\lambda_i . u\}$ would converge to some element $v(u) \in U$ and $\lim_i \sigma(\lambda_i . u) . c_{r_i}$ would be equal to $v(u) . c_o$, hence to c_o since $c_o \in F$, and ψ would be a constant map.

Assume now U to be commutative. We claim that in this case $\psi(U) \subset F$, hence that $\varphi = \psi$ fulfills our conditions. We have to show

$$v . \psi(u) = \psi(u) \quad \text{for all } u, v \in U. \tag{8}$$

We have $v . \psi(u) = \lim_i v . \sigma(\lambda_i . u) . c_{r_i}$ hence also, since U is assumed to be commutative

$$v . \psi(u) = \lim_i \sigma(\lambda_i . u + v) . c_{r_i}. \tag{9}$$

But $\lambda_i . u + v = \lambda_i . (u + v/\lambda_i)$ and, by (1),

$$(\sigma(\lambda_i . (u + v/\lambda_i) c_{r_i}))_p = \sum_\alpha a_{p,i,\alpha} \lambda_i^{|\alpha|} (u + v/\lambda_i)^\alpha. \tag{10}$$

Since $|\lambda_i| \to \infty$ we see that (8) follows from (4) and (5).

We now drop the assumption that U is commutative and write c_i for c_{r_i}. To prove the first assertion, we proceed by induction on the codimension of F in E. There is nothing to prove if $F = E$, so we assume the first assertion established if the c_i belong to a proper U-stable subspace defined over K of E and containing F.

Let now L be a K-subspace of E of codimension one, stable under U and containing F. Since U is unipotent, this always exists and moreover U acts trivially on E/L. Let $\pi: E \to E/L$ be the canonical projection. The relation $c_i \to c_o$ implies $\pi(c_i) \to 0$, hence also $\pi(\sigma(\lambda_i . u) . c_i) \to 0$, which shows that $\psi(U) \subset L$. If now some neighborhood of the identity in $U(K)$ has its image in $F(K)$, then $\psi(U) \subset F$. Otherwise, there is a sequence $c'_j \in L(K) - F(K)$, belonging to $\psi(U(K))$, which converges to c_o. By the induction assumption, there is a non-constant K-morphism of varieties $\varphi: U \to F$, sending 1 onto c_o, such that $\varphi(U(K))$ is in the closure of the union of the orbits $\sigma(U(K)) . c'_j$. By construction (see (6)), c'_j belongs to the closure of the union of the orbits $\sigma(U(K)) . c_i$. Then so does $\sigma(U(K)) . c'_j$, hence also $\varphi(U(K))$.

2.2. In the sequel, there is some interplay between the Zariski topology and the analytic topology. We first settle a minor technical point.

Let G be a K-group, U_i $(i \in I)$ a finite set of smooth irreducible K-varieties, $u_i \in U_i(K)$ and $f_i: U_i \to G$ a K-morphism sending u_i onto the identity $(i \in I)$. It is standard that the $f_i(U_i)$ and their inverses generate a K-subgroup L [B: 2.2].

Assume now that $U_i(K)$ is Zariski-dense in $U_i (i \in I)$. We claim that if V_i is any neighborhood of u_i in $U_i(K)$ (in the K-topology), then the topological identity component $L(K)^o$ of $L(K)$, viewed as a Lie group over K, is generated by the $f_i(V_i)$ and their inverses.

By [B: 2.2], there exist $n \in \mathbf{N}$, a sequence $i(j)$ of elements in I $(j = 1, \ldots, n)$ and $\varepsilon_j = \pm 1$ such that the image of $Y = \Pi_j U_{i(j)}$ under the map

$$\mu: (x_1, \ldots, x_n) \mapsto f_{i(1)}(x_1)^{\varepsilon_1} \cdots f_{i(n)}(x_n)^{\varepsilon_n}$$

is the K-group L. We may assume the V_i's to be open. Then $Z = \Pi_j V_{i(j)}$ contains a point at which d_μ is surjective, since the set of such points is Zariski-open in Y, so the image of Z contains a non-empty open set in $L(K)$ and the group it generates contains $L(K)^o$.

2.3. PROPOSITION. *Let G be a connected K-group and U a unipotent K-subgroup. Let M be a subset of $G(K) - \mathcal{N}_G(U)(K)$ whose closure contains the identity. Then there is a K-morphism of varieties $\psi: U \to \mathcal{N}_G(U)$, mapping the identity onto the identity, and a Zariski K-open neighborhood V of 1 in U such that $\psi(V(K)) \subset \overline{U(K).M.U(K)}$ and $\psi(V(K))$ meets any left-coset of $U(K)$ in at most one point. There is in $\mathcal{N}_G(U)(K)$, viewed as a Lie group over K, a connected Lie subgroup L, containing $U(K)$ strictly and belonging to the subgroup generated by $\mathcal{N}_G(U)(K) \cap \overline{U(K).M.U(K)}$.*

Proof. There exists a finite dimensional vector space E defined over K with a line C and a rational representation $\sigma: G \to \mathrm{GL}(E)$ defined over K such that U is the subgroup of G leaving fixed any point of C. Fix $c_o \in C$. We have then $G.c_o \cong G/U$ and since U is unipotent, also $G(K).c_o \cong G(K)/U(K)$. Let F be the fixed point set of U in E. It contains C and it is elementary that

$$F \cap G(K).c_o = \mathcal{N}_G(U)(K).c_o.$$

Let now (m_i) $(i = 1, 2, \ldots)$ be a sequence of elements in M tending to the identity and let $c_i = m_i.c_o$. Then $c_i \in E(K) - F(K)$ and $c_i \to c_o$. We may apply 2.1 to get a non-constant K-morphism of varieties $\varphi: U \to F$ such that $\varphi(1) = c_o$ and $\varphi(U(K))$ is contained in the intersection of $F(K)$ with the closure of $\bigcup_i \sigma(U(K)).c_i$, hence, a fortiori, in

$$C = F(K) \cap \overline{\sigma(U(K).M).c_o}. \tag{1}$$

The orbit $G.c_o$ is Zariski-open in its closure, which contains C. Therefore U contains a Zariski K-open subset V such that $\varphi(V(K))$ belongs to $C \cap G.c_o$.

Since U is unipotent, there exists a section $s: G/U \to G$, defined over K, of the fibration of G by U. Then $\psi = s_o \varphi$ is a non-trivial K-morphism $U \to G$ mapping 1 onto 1, whose image meets any left coset of U in at most one point. The set $\psi(V(K))$ is contained in the inverse image of C, which is obviously equal to $\mathcal{N}_G(U)(K) \cap \overline{U(K).M.U(K)}$. This proves our first assertion.

The second assertion now follows from 2.2, applied to the case where $\{V_i\}$ consists of U and V.

2.4. NOTATION. We recall and adapt to our framework some notation of [M]. The field K being understood, we let $G = \mathbf{SL}_3(K)$,

$$v_1(s) = \begin{pmatrix} 1 & s & s^2/2 \\ 0 & 1 & s \\ 0 & 0 & 1 \end{pmatrix}, \quad v_2(s) = \begin{pmatrix} 1 & 0 & s \\ 0 & 1 & 0 \\ 0 & 0 & 1 \end{pmatrix}, \quad d(t) = \begin{pmatrix} t & 0 & 0 \\ 0 & 1 & 0 \\ 0 & 0 & t^{-1} \end{pmatrix} \tag{1}$$

($s \in K$, $t \in K^*$), V_1 (resp. V_2, resp. D) the group generated by the $v_1(s)$ (resp. $v_2(s)$, resp. $d(t)$), $V = V_1 . V_2$ and if $K = \mathbf{R}$,

$$V_2^+ = \{v_2(s) \mid s > 0\}, \quad V_2^- = \{v_2(s) \mid s < 0\}, \quad D^o = \{d(t) | t > 0\}. \tag{2}$$

Thus our notation deviates from [M] only in that our D^o is D there. If $K = \mathbf{C}$, then $D^o = D$.

As pointed out in [M], it is easily checked that

$$\mathcal{N}_G(V_1) = D . V. \tag{3}$$

$H \subset \mathbf{SL}_3(K)$ denotes the special orthogonal group of the form $2x_1x_3 - x_2^2$. The group $D . V_1$ is then maximal solvable in H.

2.5. PROPOSITION. *Let $M \subset G - \mathcal{N}_G(V_1)$ be a subset whose closure contains the identity. Then the subgroup C generated by $\mathcal{N}_G(V_1) \cap V_1 . M . V_1$ contains either V or a subgroup of the form $v . D^o . V_1 . v^{-1}$ for some $v \in V$.*

Proof. By 2.3, the subgroup C has a connected Lie subgroup L containing V_1 strictly. Since $\mathcal{N}_G(V_1) = D . V$, the quotient $\mathcal{N}_G(V_1)/V_1$ is isomorphic to $D . V_2$ (i.e. to the affine group of the line). It is elementary that its only connected Lie subgroups over K are V_2 and the V_2-conjugates of D^o, whence the proposition.

2.6. PROPOSITION. *Let M be a subset of $G - H$ whose closure contains the identity. Then $\overline{H . M . D^o . V_1}$ contains V_2 if $K = \mathbf{C}$ and either V_2^+ or V_2^- if $K = \mathbf{R}$.*

Proof. For $K = \mathbf{R}$, this is Lemma 7 in [M]. Our proof is basically the same and we only point out the modification allowing us to include the case $K = \mathbf{C}$.

There is no change in the argument of p. 394 in [M] until the last five lines. There, instead of Lemma 13(i), we invoke 2.1 above and conclude that there is a non-constant K-morphism of varieties of V_1 into $\mathfrak{v}_2$ whose image contains the origin. If $K = \mathbf{C}$, it is then surjective. If $K = \mathbf{R}$, its image contains at least one of the half-lines $\mathfrak{v}_2^+$, $\mathfrak{v}_2^-$. Then the relation on the last line of p. 394 in [M] shows that $\overline{H . M . D^o . V_1}$ contains V_2 if $K = \mathbf{C}$ and either V_2^+ or V_2^- if $K = \mathbf{R}$, under the assumption that $M \cap V_2$ is empty. If it is not, we see from the argument on p. 395 that $H . M . D^o . V_1$ contains all elements $v_2(s) \in V_2$ with $s = t^2 . s_o$, for some $s_o \neq 0$

and all $t \in K^*$. If $K = \mathbf{C}$, this contains $V_2 - 0$, hence the closure contains V_2. In the real case, it contains either V_2^+ or V_2^-.

2.7. LEMMA. *Assume $K = \mathbf{R}$. Let X be a second countable locally compact space on which V_2 operates continuously and $x \in X$. If $V_2^+ . x$ (resp. $V_2^- . x$) is relatively compact, then its closure C contains a non-empty subset invariant under V_2.*
Proof. Let $\varphi(x)$ be the set of limit points of sequences $v_2(t_n).x$, where $t_n \to \infty$ (resp. $t_n \to -\infty$). It is contained in C, and is not empty since C is compact. We claim that it is invariant under V_2. Let $y \in \varphi(x)$ and write it as $\lim_{n\to\infty} v_2(t_n).x$, where $\{t_n\}$ is as before. For $t \in \mathbf{R}$, we have

$$v_2(t).y = \lim_{n\to\infty} v_2(t).v_2(t_n).x = \lim_{n\to\infty} v_2(t+t_n).x.$$

But $t + t_n$ tends to ∞ or to $-\infty$ with t_n, hence $v_2(t).y \in \varphi(x)$.

2.8. REMARKS. (1) 2.1 and 2.3 are valid over a non-archimedean local field K and the proof are identical, with one minor modification to take into account the fact that the set of absolute values is discrete, more precisely the set of powers of some real number $c \in (0, 1)$. Now in 2.1 (2), (3) we let the absolute value in K be the normalized absolute value, keep (2) and replace (3) by

$$|a_{p_0,i,\alpha_0} \lambda_i^{|\alpha_0|}| \geq c^{|\alpha_0|}. \tag{3}$$

(2) In [M], Lemmas 1 and 4 are for general locally compact second countable groups and we shall be able to use them without any modification. Our 2.3 is a replacement for Lemma 5 of [M] and our 2.1 one for Lemma 13(i) of [M]. The formulation of the latter was suggested to one of us by P. Deligne. We have already pointed out that 2.6 generalizes Lemma 7 of [M]. Over $\mathbf{R}$, Proposition 2.5 is a weaker version of Lemma 8(ii) of [M], weaker in the sense that we assume $M \subset G - \mathcal{N}_G(V_1)$ rather than in $G - V_1$. But thanks to 1.1 and 3.1, we shall not need this stronger statement in the proof of 3.4. Finally, 2.7 is just a more general formulation of the argument in Lemma 12 of [M].

3. The implications (i) ⇒ (ii) in the archimedean case

From 3.4 on, it is assumed in this section that $S = S_\infty$.

3.1. LEMMA. *Let T be a finite set, Q_t ($t \in T$) a locally compact second countable group, Q the product of the Q_t, U a closed subgroup of Q and pr_t the projection of Q onto Q_t. Fix $s \in T$ and $q \in Q$. Let Γ be a discrete subgroup of Q on which pr_s is injective. Assume D_s is a subgroup of Q_s containing an element d which contracts $pr_s U$ and such that the image of $D_s . q$ in Q/Γ is relatively compact. Then $U \cap q.\Gamma.q^{-1} = \{1\}$.*

Proof. The projection pr_s is also injective on $q.\Gamma.q^{-1}$, so we may (and do) replace Γ by $q.\Gamma.q^{-1}$. As D_s is now relatively compact modulo Γ, there exists a sequence of elements $\gamma_n \in \Gamma$ such that the set $\{d^n.\gamma_n\}$ $(n \in \mathbf{N})$ is bounded in Q. Let $\gamma \in U \cap \Gamma$ and $\gamma_t = pr_t(\gamma)$. By assumption,

$$\lim_{n\to\infty} d^n.\gamma_s.d^{-n} = 1,$$

therefore $d^n.\gamma.d^{-n}$ converges to some element $\delta \in Q$ such that $pr_s\delta = 1$ and $pr_t\delta = \gamma_t$ for $t \neq s$. But we can write

$$d^n.\gamma.d^{-n} = (d^n.\gamma_n).(\gamma_n^{-1}.\gamma.\gamma_n).(d^n.\gamma_n)^{-1},$$

and, as $d^n\gamma_n$ has a convergent subsequence, we see that a subsequence of $\{\gamma_n^{-1}.\gamma.\gamma_n\}$ converges to an element of Q whose image under pr_s is the identity. Since Γ is discrete and $\gamma_n^{-1}.\gamma.\gamma_n \in \Gamma$, the subsequence is eventually constant, therefore $pr_s(\gamma_n^{-1}.\gamma.\gamma_n) = 1$ for some n, whence $pr_s\gamma = 1$ and $\gamma = 1$ in view of our assumption on Γ.

3.2. We now revert to the setup of Theorem A, assume $n = 3$ and introduce some further notation, extending that of 2.4 to the S-arithmetic case.

We let $G_s = \mathbf{SL}_3(k_s)$ and G or G_S be the product of the G_s $(s \in S)$. Let pr_s be the projection of G onto G_s $(s \in S)$. For $g \in G$, the element $pr_s(g)$ is also called the s-component of g. For $s \in S$, we let $V_{1,s}$, $V_{2,s}$, V_s and D_s be the groups defined as V_1, V_2, V, D in 2.4, except that now the entries are in k_s for the first three groups, in k_s^* for the last one.

We let $\Gamma = \mathbf{SL}_3(\mathfrak{o}_S)$ and view it as a subgroup of G via the diagonal embedding. It is discrete, of finite covolume, but not cocompact. Let $\Omega = G/\Gamma$.

Given a quadratic form $F = \{F_s\}$ on k_S^3 we let $H_s = \mathrm{SO}(F_s)(k_s)$ and H_F be the product of the H_s's. Recall that F is always assumed to be non-degenerate and isotropic.

If $S = S_\infty$, then, as in 1.2, we view G as the group of real points of the $\mathbf{Q}$-group $L = R_{k/\mathbf{Q}}\mathbf{SL}_3$ obtained from $\mathbf{SL}_3$ by restriction of scalars from k to Q.

3.3. PROPOSITION. *Let $r \in S$ and $y \in \Omega$. Then $D_r^o.V_r.y$ and, for r real, $D_r^o.V_{1,r}.V_{2,r}^+.y$ and $D_r^o.V_{1,r}.V_{2,r}^-.y$, are not relatively compact.*

We first consider $D_r^o.V_r.y$ (over $\mathbf{R}$ or $\mathbf{C}$) and choose $g \in G$ such that $g.\mathfrak{o}^3 = y$.

Given a plane E in k_r^3, there exists a sequence $b_n \in \mathfrak{o}^3$ such that the s-component $b_{n,s} \to 0$ for $s \neq r$ while the r-component $b_{n,r}$ is not contained in E. Therefore we can find a sequence $a_n \in g.\mathfrak{o}^3$ such that $a_{n,s} \to 0$ for $s \neq r$ and that the third coordinate $a_{n,r}^3$ of $a_{n,r}$ is not zero. It is then elementary to choose first $v_n \in V_r$ such that the first two coordinates of $v_n.a_{n,r}$ are zero and then $d_n \in D_r^o$ such that $d_n.v_n.a_{n,r} \to 0$, whence also $d_n.v_n.a_n \to 0$. The lemma for $D_r^o.V_r$ now follows from Mahler's criterion.

Let r be real and assume that the closure C of $D_r^o.V_{1,r}.V_{2,r}^+.y$ (resp. $D_r^o.V_{1,r}.V_{2,r}^-.y$) is compact. By 2.7, C contains a non-empty subset Q invariant under V_{2r}. Then $D_r^o V_r . Q$ is contained in C. Since C is assumed to be compact, this contradicts what has already been proved.

REMARK. 3.3 extends to our case Lemmas 11 and 12 of [M]. We have used a simplification of the proof of Lemma 11 proposed by B. J. Birch.

We recall that, up to the end of this section, we assume that $S = S_\infty$.

3.4. THEOREM. *Let $z \in \Omega$ and assume that $H_F . z$ is relatively compact. Then $H_F . z$ is compact.*

Proof. This statement is invariant under conjugation in G. Since a non-degenerate isotropic quadratic form on k_s^3 is equivalent to a multiple of the standard form $2x_1x_3 - x_2^2$, there is no loss in generality in assuming that F_s is equal to the latter form, hence that $H_s = SO(2x_1x_3 - x_2^2)$ for all s. We shall do so. We write H for H_F.

Let $Z = \overline{Hz}$. Let $X \subset Z$ be a minimal H-invariant closed subset and $Y \subset X$ a minimal V_1-invariant closed subset of X (these exist since Z is compact). We shall first show that Y is D^o-invariant.

Fix $r \in S$. Let Y_o be a minimal closed $V_{1,r}$-invariant subset of Y and let R be the identity component of the group of elements $g \in \mathcal{N}_G(V_{1,r})$ such that $g . Y_o = Y_o$. Choose $y \in Y_o$. We claim that $R \cap G_y$ is not cocompact in R. Assume to the contrary that it is. Then $R . y$ is closed, homeomorphic to $R/(R \cap G_y)$, contained in Y_o and containing $V_{1,r} . y$. The latter being dense in Y_o, we see that $Y_o = R . y$ and we are in the situation of 1.1, up to conjugacy, but with the present group L (see 3.2) playing the role of G there. Let $g \in G$ be such that $g . y = o^3$. Then, by 1.1, $g . R . g^{-1} = \tilde{E}(\mathbf{R})^o$, where $\tilde{E}$ is the smallest $\mathbf{Q}$-subgroup of L containing $g . V_{1,r} . g^{-1}$. The group $pr_r(R)$ is contained in the G_r-normalizer of $V_{1,r}$, which is equal to $D_r . V_r$, hence solvable, so $\tilde{E}$ is solvable. Since it is the smallest $\mathbf{Q}$-subgroup containing the unipotent group $g . V_{1,r} \cdot g^{-1}$, it is in fact unipotent. Since any unipotent subgroup of $D_r . V_r$ is contained in V_r, we see that $pr_r(R) \subset V_r$. Consequently, D_r^o contains an element d which contracts $pr_r(R)$. The discrete group G_y is conjugate to Γ, hence pr_r is injective on G_y. Moreover, $D_r^o . y$, being contained in Z, is relatively compact. Therefore 3.1 holds and shows that $R \cap G_y = \{1\}$, a contradiction which implies that $R \cap G_y$ is not cocompact in F. Lemma 4 of [M] now shows that

$$M_o = \{g \in G - R \mid gY_o \cap Y_o \neq \emptyset\}$$

contains 1 in its closure. On the other hand, since $M_o \cap R$ is empty, Lemma 2 in [M] implies that $M_o \subset G - \mathcal{N}_G(V_{1,r})$. By Lemma 3 of [M], Y_o is stable under the subgroup Q generated by $\mathcal{N}_G(V_{1,r}) \cap \overline{V_{1,r} . M_o . V_{1,r}}$. As

$$M_o \subset G - \mathcal{N}_G(V_{1,r}) = G - \mathcal{N}_{G_r}(V_{1,r}) \prod_{s \neq r} G_s,$$

it follows from 2.5 that Q contains either V_r or $v . D_r^o . V_{1,r} . v^{-1}$ for some $v \in V_r$. An elementary computation shows that if $v \in V_r - V_{1,r}$ then

$D_r^o . V_{1,r} v . D_r^o v^{-1}$

contains $D_r V_r$ if r is complex and either $D_r^o . V_{1,r} . V_{2,r}^+$ or $D_r^o . V_{1,r} . V_{2,r}^-$ if r is real. The group Q leaves Y_o stable. Therefore, if either $Q \supset V_r$ or $v \in V_r - V_{1,r}$ then X contains an orbit of $D_r . V_r$ if r is complex and the set of transforms of a point under $D_r^o . V_{1,r}, V_{2,r}^+$ or $D_r^o . V_{1,r} . V_{2,r}^-$ if r is real. Since X is compact, this contradicts 3.3. Consequently $v \in V_{1,r}$ and Q contains $D_r^o V_{1,r}$, i.e. Y_o is stable under $D_r^o . V_{1,r}$. As Y is minimal closed invariant under V_1 and D_r^o normalizes V_1, Lemma 2 of [M] shows that Y is stable under D_r^o. This being true for every $r \in S$, we see that Y is D^o-invariant, as claimed.

Choose $y \in Y$ and let

$$M = \{g \in G - H \mid g . y \in Z\}. \tag{1}$$

Assume first that $1 \in \bar{M}$. For some $s \in S$ we can find a sequence $\{m_i\} \subset M$ tending to 1, such that $pr_s m_i \in G_s - H_s$ for all i's. Since H, D, V_1 are products of their projections on the G_s's, it follows from 2.6 that $\overline{HMD^o V_1}$ contains $V_{2,s}$ if s is complex, and either $V_{2,s}^+$ or $V_{2,s}^-$ if s is real. But Y is a minimal closed $D^o . V_1$ invariant set. By Lemma 1 of [M], $g . Y \subset Z$ for any $g \in \mathcal{N}_G(V_1) \cap \overline{HMD^o V_1}$. Therefore we have $V_{2,s} Y \subset Z$ if s is complex and either $V_{2,s}^+ . Y \subset Z$ or $V_{2,s}^- . Y \subset Z$ if s is real. As a consequence, for $y \in Y$,

$$D_s^o . V_s . y = D_s^o . V_{1,s} V_{2,s} . y \subset Z \quad \text{if } s \text{ is complex}$$

and

$$\text{either } D_s^o . V_{1,s} . V_{2,s}^+ . y \subset Z \quad \text{or} \quad D_s^o . V_{1,s} . V_{2,s}^- . y \subset Z \quad \text{if } s \text{ is real.}$$

Since Z is compact this contradicts 3.3 and shows that $1 \notin \bar{M}$. As $y \in X$ and X is a minimal compact H-invariant set, Lemma 4 of [M] implies that $H/(H \cap G_y)$ is compact. We claim that $z \in Hy$. By assumption $y \in Z$, therefore y is a limit of elements $h_n . z$ $(h_n \in H)$. We can write $h_n . z = c_n . y$ with $c_n \in G$, $c_n \to 1$. But then $c_n . y \in Z$ i.e. $c_n \in M$. Since $1 \notin \bar{M}$, we have $c_n \in H$ for n big enough, whence our contention. But then $H . z = H . y$ is compact.

3.5. We shall say that a sequence $t_n = (t_{n,s})_{s \in S}$ in k_S $(t_{n,s} \in k_s)$, tends to infinity if $|t_{n,s}|_s \to \infty$ for s complex and $t_{n,s} \to \infty$ for s real.

Given $y \in \Omega$, we let $\phi(y)$ be the set of limit points of sequences $v_1(t_n) . y$, where $t_n \to \infty$ in k_S.

LEMMA. *Let y, $z \in \Omega$. Assume that $H_F . z$ is compact, $y \notin H_F . z$ and $V_1 . y$ is relatively compact. Then $\phi(y) \not\subset H_F . z$.*

For $k = \mathbf{Q}$, this is Lemma A of [M]. The proof in the present case is so similar that we shall not repeat it in detail. Assume this assertion to be false, we find as in *loc. cit.* $p \in G/H$, different from the origin $q = H$ in G/H such that $q \in \overline{V_1 . p}$. Now G/H is the product of the G_s/H_s $(s \in S)$ and $V_1 . p$ the product of the $V_{1,s} . p_s$, so that for each s, the point $q_s = H_s$ should be in the closure of $V_{1,s} . p_s$. Since $p_s \neq q_s$ for at least one s this is impossible because G_s/H_s is an affine variety (or at any rate open and closed in one if s is real), and any orbit of a unipotent group in an affine variety is closed, hence $V_{1,s} . p_s$ is also closed in G_s/H_s in the ordinary topology.

3.6. THEOREM. *Let $z \in \Omega$. Assume there exist a finite union C of k-lines* (cf. 1.8) *and an open neighborhood U of the origin in k_S^3 such that*

$$H_F(z - (z \cap C)) \cap U = \{0\}. \tag{1}$$

Then $H_F . z$ is compact.

Proof. Let $g \in G$ be such that $g . \mathfrak{o}^3 = z$. After having replaced F by ${}^t g . F . g$ and U, C by $g^{-1} . U$, $g^{-1} . C$, we are reduced to the case where $z = \mathfrak{o}^3$. Of course, $\mathfrak{o}^3 \cap k . x$ $(x \in k_S^3)$ is $\neq \{0\}$ if and only if $x \in k^n$. We may assume that C consists of finitely many rational lines.

In H_{F_s} we choose a unipotent one-dimensional subgroup Q_s which does not fix any line in C and let Q be the product of the Q_s.

We claim there exists a neighborhood U' of the origin in k_S^3 such that

$$Q . x \cap U' = \varnothing \quad \text{for every non-zero} \quad x \in C \cap \mathfrak{o}^3. \tag{2}$$

Let $x \in C \cap \mathfrak{o}^3 - \{0\}$. For every s, $x_s \neq 0$ and, by our choice of Q, the element x_s is not fixed under Q_s. Therefore $q \mapsto q . x_s$ is injective. Since Q_s is unipotent, $Q_s . x_s$ is closed in k_s^3 and does not contain the origin, hence $Q . x$ does not meet some polydisc $U_d = \{u = (u_s) \mid |u_s|_s < d\}$.

But then the same is true for all elements $\lambda . x$ $(\lambda \in \mathfrak{o} - \{0\})$. Indeed, we have

$$|q_s . \lambda . x|_s = |\lambda|_s |q_s . x_s|_s.$$

and at least one of $|\lambda_s|$ is $\geqq 1$, since $\lambda \in \mathfrak{o}$, $\lambda \neq 0$.

As we remarked in 1.8, if L is a rational line, then $L \cap \mathfrak{o}^3$ is a finitely generated module. There exists therefore a finite subset E of $C \cap \mathfrak{o}^3 - \{0\}$ such that $C \cap \mathfrak{o}^3 = \mathfrak{o} . E$. The claim (2) now follows by choosing a constant d suitable for every $e \in E$. In the sequel, we assume, as we may, that U in (1) is chosen small enough to be contained in U'. In view of the assumption (1), we have now

$$Q . \mathfrak{o}^3 \cap U = \{0\}, \tag{4}$$

therefore $Q . \mathfrak{o}_3$ is relatively compact in Ω by Mahler's criterion. From then on, the argument proceeds as in Section 4 of [M]. Since our framework and formulation are somewhat different, we describe it for the sake of completeness.

We identify k_s with Q_s, hence k_S with Q and let $q(t)$ be the element of Q corresponding to $t = \{t_s\} \in k_S$. Let Y be the set of accumulation points of sequences $q(t_n) . \mathfrak{o}^3$, where $t_n \to \infty$ in the sense of 3.5. Since $Q . \mathfrak{o}^3$ is relatively compact, Y is compact and not empty. The argument of 2.7 shows that it is invariant under Q. We claim that it is connected. Assume it is not. Then it is the union of two disjoint non-empty compact subsets Y_0, Y_1. There exists on Ω a continuous compactly supported real valued function, with values in $[0,1]$, equal to 0 on Y_0 and to 1 on Y_1. Let $a \in Y_0$ and $b \in Y_1$. Then $a = \lim_n q(a_n).\mathfrak{o}^3$ and $b = \lim_n q(b_n).\mathfrak{o}^3$, where $\{a_n\}$ and $\{b_n\}$ are sequences in k_S tending to infinity (3.5). We may assume.

$$f(q(a_n)(\mathfrak{o}^3)) < 1/2, \qquad f(q(b_n).(\mathfrak{o}^3)) > 1/2.$$

We can find a curve $C_n = \{c_n(t) \mid t \in [0,1]\}$ in k_S such that $c_n(0) = a_n$, $c_n(1) = b_n$ and

$$c_{n,s}(t) \geqq \min(a_{n,s}, b_{n,s}) \quad (t \in [0,1]) \text{ if } s \text{ is real,}$$

$$|c_{n,s}(t)| \geqq \min(|a_{n,s}|, |b_{n,s}|) \quad (t \in [0,1]) \text{ if } s \text{ is complex.}$$

There exists then $c_n \in C_n$ such that $f(q(c_n).\mathfrak{o}^3) = 1/2$. The elements $q(c_n).\mathfrak{o}^3$ have an accumulation point $x \in \Omega$ by compactness. But, clearly, $c_n \to \infty$ in the sense of 3.5, hence $x \in Y$. Since f takes the value 1/2 on x this is a contradiction, proving that Y is connected.

Let now $y \in Y$. We want to show that $H.y$ is relatively compact. By construction, there exists a sequence $\{t_n\} \subset k_S$ tending to infinity in the sense of 3.5 such that $q(t_n).\mathfrak{o}^3 \to y$. Let x be an element of the lattice y. There is then a sequence $x_n \in \mathfrak{o}^3$ such that $q(t_n).x_n \to x$ in k_S^3. We claim that for n big enough, $x_n \notin C$. Assume the contrary. Passing to a subsequence, we may further assume, in the above notation, that $x_n = \lambda_n.e$ for some fixed element $e \in E$ and $\lambda_n \in \mathfrak{o}$. For each $s \in S$, the orbit $Q_s.e_s$ is closed and homeomorphic to Q_s. Therefore $q_s(t_{n,s}).e_s$ diverges. Again passing to a subsequence, we may assume that $|\lambda_n|_s \geqq 1$ for some fixed s and all n's, and then it is clear that

$$q_s(t_{n,s}).x_{n,s} = \lambda_n.q_s(t_{n,s}).e_s$$

diverges, too. Hence $q(t_n).x_n$ diverges. Therefore $x_n \notin C$ for n big enough. Then, for $h \in H_F$, we have

$$h.x = \lim_n h.q(t_n).x_n \quad (x_n \notin C).$$

By (1), $h.q(t_n).x_n \notin U$, hence $h.x \notin U$. This shows that $H_F.y \cap U = \{0\}$ hence, by Mahler's criterion, that $H_F.y$ is relatively compact for every $y \in Y$. By 3.4, $H_F.y$ is then compact. Let us show that H_F has at most countably many compact orbits in Ω. Assume that $H_F.a.\mathfrak{o}^3$ is compact for some $a \in G$. Then $a^{-1}.H_F.a.\mathfrak{o}^3$ is compact and, by 1.2, $a^{-1}.H_F.a$ is the group of k_S-points of some k-subgroup of $\mathbf{SL}_3$. But $\mathbf{SL}_3$ has only countably many k-subgroups. Therefore the groups $a^{-1}.H_F.a$, where $a \in G$ is such that $H_F.a.\mathfrak{o}^3$ is compact, form a countable set. But H_F is its own normalizer in G, therefore if $a^{-1}.H_F.a = b^{-1}.H_F.b$ $(a, b \in G)$, then $a \in H_F.b$ and $H_F.a.\mathfrak{o}^3 = H_F.b.\mathfrak{o}^3$, whence our statement. It now follows that Y is the union of at most countably many disjoint closed sets of the form $Y \cap H_F . y$. Since it is compact and connected, as we saw earlier, it must be contained in one of them, say $Y \subset H . z$, for some $z \in Y$. Then the lemma in 3.5 shows that $\mathfrak{o}^3 \in Y$; therefore $H_F . \mathfrak{o}^3$ is compact, and the theorem is proved.

3.7. COROLLARY. *The conditions* (i) *and* (ii) *of Theorem A for* $S = S_\infty$ *are equivalent.*

By 1.3, it suffices to prove this for $n = 3$. Our assertion now follows from 3.6 and the implication (c) $\Rightarrow$ (a) of 1.9.

4. Proof of Theorem A when F is rational for some $s \in S$

4.1. In this section we assume $|S| \geqq 2$, F irrational, the existence of a form F_o on k^n and of $q \in S$ such that $F_q = c_q . F_o$ for some $c_q \in k_q^*$. Let S' be the set of $s \in S$ such that $F_s = c_s . F_o$ with $c_s \in k_s^*$. The set S' contains q and, since F is irrational, $S' \neq S$. Let $T = S - S'$.

Let Q_o be the k-variety defined by $F_o = 0$ and $Q_{o,s} = Q_o \times k_s$. Similarly let Q_s be the k_s-variety defined by $F_s = 0$. It characterizes F_s up to a multiple. We have $Q_{o,s} = Q_s$ for $s \in S'$ but $Q_{o,t} \neq Q_t$ for $t \in T$. In particular, for $t \in T$, $\dim(Q_{o,t} \cap Q_t) < n - 1$ and $Q_{o,t}(k_t) \neq Q_t(k_t)$. [Recall that F_t is isotropic, hence $Q_t(k_t)$ is Zariski dense in Q_t.]

For $t \in T$, let L_t be a fixed open subgroup of $\mathrm{SO}(F_o)(k_t)$ and M_t be the set of vectors in $k_t^n - Q_o(k_t)$, which can be mapped into $Q_t(k_t)$ by an element of L_t. Then M_t is a non-empty open subset of k_t^n and clearly $k_t^* . M_t = M_t$. We fix an element $e_{t,1}$ of M_t. Then the line $k_t . e_{t,1}$ is in M_t, except for the origin. Complete $e_{t,1}$ to a basis $(e_{t,i})$ of k_t^n. Let us denote by $D_{t,r}$ the disc $|x|_t \leqslant r$ in k_t ($r > 0$, real).

For $r > 0$, let $B_{t,r}$ and $C_{t,r}$ be the products of the discs $D_{t,r} . e_{t,j}$ ($j = 2, \ldots, n$), and $D_{t,r} . e_{t,j}$ ($j = 1, 2, \ldots, n$) respectively. We choose $a \geqslant b > 0$ so small that for all $t \in T$,

(i) $e_{t,1} + C_{t,a}$ is contained in M_t. (We note here that then $(D_{t,m} - D_{t,1}) e_{t,1} + C_{t,a}$ is contained in M_t for all $m > 1$).

(ii) The sum of any $|T|$ elements of $C_{t,b}$ is contained in $C_{t,a}$.

We shall now use elementary geometry of numbers to prove:

(*) *Given a polydisc* $U = \Pi_{s \in S'} U_s$ *centered on the origin in* $k_{S'}^n$, *there exists* $x \in \mathfrak{o}_S^n$ *such that* $x_s \in U_s$ *for* $s \in S'$ *and* $x_t \in M_t$ *for* $t \in T$.

Proof. Let $V = \Pi_{s \in S'} V_s$ be a bounded polydisc centered on the origin such that the sum of any $2\,|T|$ elements of V is contained in U. Let $t \in T$. We shall show

$(*)_t$ *There exists* $y(t) \in \mathfrak{o}_S^n$ *and* $m \geqslant 2$ *such that* $y(t)_s \in V_s + V_s$, *if* $s \in S'$, $y(t)_t \in (D_{t,m} - D_{t,1})e_{t,1} + B_{t,b}$ *and* $y(t)_{t'} \in C_{t',b}$ *if* $t' \in T - \{t\}$.

Assume this to be established. Then in view of the conditions (i) and (ii), the sum of the $y(t)$, $t \in T$, satisfies $(*)$.

We now fix $t \in T$ and prove $(*)_t$. For a positive real number r, let

$$\Omega_r = V \times (D_{t,r} \cdot e_{t,1} + B_{t,b/2}) \prod_{t' \in T - \{t\}} C_{t',b/2}.$$

On k_S^n take the product of the usual Haar measures on the k_s^n and let c be the volume of $k_S^n/\mathfrak{o}_S^n$. The set $\mathfrak{o}_S^n$ is discrete in k_S^n, *hence* $(\Omega_1 + \Omega_1) \cap \mathfrak{o}_S^n$ is a finite set, say with q elements. There exists $m \geqslant 2$ such that the volume of $\Omega_{m/2}$ is $> (q+1)c$. Therefore at least one fibre of the projection onto $k_S^n/\mathfrak{o}_S^n$, restricted to $\Omega_{m/2}$, has more than $(q+1)$ elements, so we can find $y_o, \ldots, y_{q+1} \in \Omega_{m/2}$ such that the differences $x_i = y_o - y_i$ $(i = 1, \ldots, q+1)$ are distinct elements of $\mathfrak{o}_S^n$ and at least one of these, say x_1 will then be outside $\Omega_1 + \Omega_1$. Let $y(t) = x_1$, then $y(t)$ satisfies $(*)_t$.

4.2. We now prove Theorem A assuming $|S| \geqq 2$ and at least one of the F_s to be rational. From the Introduction and Section 1, we know that it suffices to show that (i) $\Rightarrow$ (iii) when $n = 3$. We use the notation and assumptions of 4.1 and write H_o for $SO(F_o)$; let $\tilde{H}_o$ be its universal covering and $\sigma\colon \tilde{H}_o \to H_o$ the canonical isogeny. Let $\Gamma \subset H_o(k)$ be the stabilizer of $\mathfrak{o}_S^3$. It is an S-arithmetic subgroup, which we view, as usual, as a discrete subgroup of $H_o(k_S)$. Let Γ_T be its projection on $H_o(k_T)$. We claim first there are open subgroups of finite index $L_t \subset H_o(k_t)$ $(t \in T)$ such that the product L_T of the L_t's is contained in the closure of Γ_T. Let $\tilde{\Gamma}$ be an S-arithmetic subgroup of $\tilde{H}_o(k)$. By assumption, $\tilde{H}_o$ and H_o are isotropic over k_s for $s \in S'$. Therefore, by strong approximation [P], $\tilde{\Gamma}_T$ is dense in an open subgroup of $\tilde{H}_o(k_T)$, hence $\sigma(\tilde{\Gamma}_T)$ is dense in an open subgroup of $H_o(k_T)$. As $\sigma(\tilde{\Gamma}_T)$ is commensurable with Γ_T, our assertion is clear.

Let $\varepsilon > 0$. We choose first the U_s $(s \in S')$ small enough so that $|F_s(y)|_s \leqq \varepsilon$ for $y \in U_s$. Then, using the group L_t just defined, we construct M_t as in 4.1 $(t \in T)$. With respect to those choices, we have seen in 4.1 we may find $x \in \mathfrak{o}_S^n$ satisfying the conditions of $(*)$ there. There exists then $g_t \in L_t$ such that $y_t = g_t \cdot x_t \in Q_t(k_t)$ for every $t \in T$; we can find $z_t \in L_t \cdot x_t$ close to y_t such that $0 < |F_t(z_t)|_t \leqslant \varepsilon/2$. The closure of Γ_T contains $L_T = \Pi_{t \in T} L_t$, we can therefore find $\gamma \in \Gamma$ such that

$$0 < |F_t((\gamma \cdot x)_t)|_t \leqslant \varepsilon \quad (t \in T).$$

By assumption, $F_{0,t}(x_t) \neq 0$. But $F_{0,s}(x_s)$ is conjugate to $F_{0,t}(x_t)$, $(s \in S')$ hence is also $\neq 0$. By construction, for $s \in S'$, $|F_{0,s}(x_s)|_s = |F_s(x_s)|_s \leqslant \varepsilon$ and we have just seen that it is not zero. Since $\gamma \in H_0$, we have $F_s((\gamma \cdot x)_s) = F_s(x_s)$ hence $|F_s(\gamma \cdot x)_s)|_s \leqslant \varepsilon$ for $s \in S'$. Thus the element $\gamma \cdot x$ satisfies our conditions.

5. Proof of Theorem A

We now revert to the assumptions of Theorem A. As pointed out in the Introduction and Section 1, to prove Theorem A, it suffices to show that (i) $\Rightarrow$ (iii) when $n=3$.

By 4.2, this is true if at least one F_s is rational. There remains to consider the case where no F_s is so. Then, by 1.7, the set of $x \in k^n$ for which $F_s(x)=0$ for some $s \in S$ is contained in finitely many lines.

Our present assumption also implies that $F \mid k^3 \otimes_{\mathbf{Q}} \mathbf{R}$ is irrational hence it satisfies (ii) for $x \in \mathfrak{o}^n$ by 3.7. More precisely 3.6 and 3.7 show that (ii) is satisfied by elements $x \in \mathfrak{o}^3$ which belong to infinitely many lines. In view of the above remark, it follows that (iii) is fulfilled, also by elements belonging to infinitely many lines. In particular (ii) $\Rightarrow$ (iii) is proved if $S = S_\infty$.

Let now $S \neq S_\infty$. We have just seen that (iii) is true if $s \in S_\infty$ for $x \in \mathfrak{o}^n$ on infinitely many lines.

Let $s \in S_f$. The set $\mathfrak{o}^n$ is bounded in k_s^n. Therefore given any neighborhood U of the origin in k_s^n, we can find $a_s \in \mathfrak{o}_s$ (integers of k_s) such that $a_s . \mathfrak{o}^n \subset U$. As a consequence, given $\varepsilon > 0$, there exists $a_s \in \mathfrak{o}_s$ such that $|F_s(a_s \cdot x)|_s \leqslant \varepsilon$ for all $x \in \mathfrak{o}^n$.

By strong approximation, there exists $a \in \mathfrak{o}$ such that $|a|_s = |a_s|_s$ for $s \in S_f$. Let $b = \max_{s \in S_\infty} |a|_s$. We can find $x \in \mathfrak{o}^n$ such that

$$0 < |F_s(x)|_s \leqslant \varepsilon/b^2 \quad (s \in S_\infty).$$

We have then

$$0 < |F_s(a \cdot x)|_s \leqslant |a|_s^2 \varepsilon/b^2 \leqslant \varepsilon \quad \text{if } s \in S_\infty$$

and

$$|F_s(a \cdot x)|_s = |a|_s^2 |F_s(x)|_s = |a_s|_s^2 |F_s(x)|_s = |F_s(a_s \cdot x)|_s \leqslant \varepsilon \quad \text{if } s \in S_f.$$

This is true for x belonging to infinitely many lines. On the other hand, given $s \in S$, the form $F_s(x)$ can be zero on only finitely many rational lines, under our present assumptions. Therefore (iii) is already fulfilled by elements of $\mathfrak{o}^n$ in this case. This completes the proof of (i) $\Rightarrow$ (iii).

REMARK. Combined with 1.6 and 1.9, Theorem A yields some assertions about orbits in $\mathbf{SL}_3(k_S)/\mathbf{SL}_3(\mathfrak{o}_S)$. For instance, we see from 1.6 that if F is a non-degenerate isotropic quadratic form on k_S^3 which does not represent zero rationally, then any relatively compact orbit of H_F in $\mathbf{SL}_3(k_S)/\mathbf{SL}_3(\mathfrak{o}_S)$ is compact.

6. Some remarks on the density problem

6.1. The quotient $q_s = k_s^*/k_s^{*^2}$ is of order one if $k_s = \mathbf{C}$, of order two if $k_s = \mathbf{R}$ and of order 4 or 8 if k_s is non-archimedean. Let us consider the following strengthening of (ii):

(iv) Let $\varepsilon > 0$ and $\bar{c}_s \in q_s$. Then there exists $x \in \mathfrak{o}_S^n$, which satisfies (iii) and such that $F_s(x) \in \bar{c}_s$ for all $s \in S$.

We note first that (iv) is equivalent to

(v) The set $F(\mathfrak{o}_S^n)$ is dense in k_S.

Proof. It is obvious that (v) $\Rightarrow$ (iv). Assume now (iv) to hold. Let $r = (r_s) \in k_S$; let $\bar{c}_s$ be the class of r_s in q_s if $r_s \neq 0$, and be any element of q_s otherwise. We let c_s stand for a representative of $\bar{c}_s$. Note that if $x \in \bar{c}_s$, then $c_s^{-1}.x$ is a square. We denote by $(c_s^{-1}.x)^{1/2}$ any square root. Let $x \in \mathfrak{o}_S^n$ be such that $F_s(x) \in \bar{c}_s$ for all $s \in S$. Since $\mathfrak{o}_s$ is a lattice in k_S, there exists a universal constant d and $u(x) \in \mathfrak{o}_S$ such that

$$|(c_s^{-1}r_s/c_s^{-1}F_s(x))^{1/2} - u(x)|_s \leqslant d \quad (s \in S), \tag{1}$$

which can be written

$$|(c_s^{-1}r_s)^{1/2} - u(x)_s(c_s^{-1}F_s(x))^{1/2}|_s \leqslant d\,|(c_s^{-1}F_s(x))^{1/2}|_s. \tag{2}$$

By (iv) we can find a sequence $x_j \in \mathfrak{o}_S^n$ such that

$$F(x_j)_s \in \bar{c}_s \quad (s \in S), \qquad F(x_j) \to 0.$$

Therefore

$$u(x_j)_s(c_s^{-1}F_s(x_j))^{1/2} \to (c_s^{-1}r_s)^{1/2}.$$

Taking the squares and dividing by c_s^{-1}, we get

$$\lim_{j\to\infty} u(x_j)_s^2 F_s(x_j) = r_s,$$

hence also

$$\lim_{j\to\infty} F(u(x_j).x_j) = r.$$

6.2. It is of course conjectured that (i) $\Rightarrow$ (v), hence that the five conditions (i)–(v) are equivalent. In view of Theorem A and the above, this would follow from (iii) $\Rightarrow$ (iv).

It was already known in the 1950s that in the original case ($k = \mathbf{Q}$, $S = S_\infty$), the truth of the Oppenheim conjecture implies the density of $F(\mathbf{Z}^n)$. A stronger result, namely that the values on the primitive vectors are dense, was proved in [DM]. The truth of the Raghunathan conjecture [Rt] also implies the density of the values on primitive vectors, in particular that (i) $\Rightarrow$ (v) for any k if $S = S_\infty$ (§7). The argument of Section 4 above proves it if the set S' there consists of complex places. Whether (i) $\Rightarrow$ (v) is true in general seems to be open at present. Of course, it would again be true in a stronger form if the Raghunathan conjecture would hold when non-archimedian places are allowed.

7. Density in the Archimedean case

(Added January 1991)

In this section, we want to show how a recent result of M. Ratner [Rt] yields the implication (i) $\Rightarrow$ (v), in fact a much stronger statement (see 7.9), in the archimedean case. We shall use a special case of the following theorem:

7.1. THEOREM (M. Ratner [Rt]). *Let $\mathscr{G}$ be a connected semisimple* **Q**-*group, $\Gamma \subset \mathscr{G}(\mathbf{Q})$ an arithmetic subgroup and $\Omega = \mathscr{G}(\mathbf{R})/\Gamma$. Let H be a closed connected subgroup of $\mathscr{G}(\mathbf{R})$ generated by unipotent elements and $z \in \Omega$. Then there exists a closed connected subgroup L of $\mathscr{G}(\mathbf{R})$ such that $\overline{H.z} = L.z$ and $L/(L \cap \mathscr{G}_z)$, where $\mathscr{G}_z$ is the isotropy group of z in $\mathscr{G}(\mathbf{R})$, has finite volume.*

This is in turn a special case of Corollary B in [Rt], but it will suffice for our needs. From this result and 1.1 we derive first:

7.2. COROLLARY. *Let $g \in \mathscr{G}(\mathbf{R})$ be such that $z = g.o$, where o is the coset Γ in Ω. Then $g^{-1}.L.g = \mathscr{H}_g(\mathbf{R})^o$, where $\mathscr{H}_g$ is the smallest* **Q**-*subgroup in $\mathscr{G}$ whose group of real points $\mathscr{H}_g(\mathbf{R})$ contains $g^{-1}.H.g$.*

Proof. After having replaced H and L by $g^{-1}.H.g$ and $g^{-1}.L.g$, respectively, we may assume that $z = o$, $g = 1$. Since Γ is the isotropy group of o, the assumption is equivalent to $L.\Gamma = \overline{H.\Gamma}$. Moreover, $\Gamma \cap L$ is of finite covolume in L by 7.1. Therefore we may apply 1.1 and 7.2 follows.

7.3. In the sequel, $S = S_\infty$, therefore $\mathfrak{o}_S = \mathfrak{o}$. Also $k_S = k \otimes_{\mathbf{Q}} \mathbf{R}$ and, to make it clear that we deal with the case $S = S_\infty$ only, we shall from now on write $k_{\mathbf{R}}$ for k_S. As in 1.2, we view the group

$$G = \mathbf{SL}_n(k_{\mathbf{R}}) = \prod_{s \in S_\infty} \mathbf{SL}_n(k_s)$$

as the group of real points of the **Q**-group $\mathscr{G} = R_{k/\mathbf{Q}}\mathscr{G}'$, where $\mathscr{G}' = \mathbf{SL}_n$, viewed as a k-group.

PROPOSITION. *Let H_s be a closed subgroup of $G_s = \mathbf{SL}_n(k_s)$ $(s \in S_\infty)$ and H the product of the H_s. Then the smallest* **Q**-*subgroup $\mathscr{L}$ of $\mathscr{G}$ whose group of real points contains H is of the form $\mathscr{L} = R_{k/\mathbf{Q}}\mathscr{L}'$, where $\mathscr{L}'$ is a connected k-subgroup of $\mathscr{G}'$.*

This is a simple property of the restriction of scalars, for which we have unfortunately no ready reference. A proof will be given at the end of this section (7.12).

Let now $n \geqq 3$ and $F = (F_s)$ be as in Theorem A. Let H_s be the special orthogonal group of F_s and $H_F = \prod_s H_s$. Then we have:

7.4. COROLLARY. *If $H = H_F$ then either F is rational or the smallest* **Q**-*subgroup containing H is $\mathscr{G}$.*

Proof. We note first that H_s is maximal among proper closed connected subgroups of G_s. This amounts to say that the representation of the Lie algebra $L(H_s)$ of H_s in $L(\mathbf{SL}_n(k_s))/L(H_s)$ is irreducible. This representation is the differential of the natural representation of an orthogonal group in the space of symmetric matrices of trace zero. Its irreducibility (for $n \geqq 3$) is classical. It is another way to state that $\mathbf{SL}_n(k_s)/H_s$ is an irreducible symmetric space, also a standard fact.

Let now $\mathscr{L}$ be the smallest **Q**-subgroup of G such that $\mathscr{L}(\mathbf{R}) \supset H$. By 7.2, it is of the form $R_{k/\mathbf{Q}}\mathscr{L}'$ where $\mathscr{L}'$ is a k-subgroup of $\mathscr{G}'$. In particular $\mathscr{L}(\mathbf{R}) = \prod \mathscr{L}'(k_s)$ is the product of its intersections with the G_s's. Of course $\dim \mathscr{L}'(k_s)$ is independent of s and $\mathscr{L}'(k_s) \supset H_s$. Therefore we have either $\mathscr{L}'(k_s)^o = H_s$ or $\mathscr{L}'(k_s) = G_s$ for every s. In the former case, $\mathscr{L}' = \mathrm{SO}(F_o)$ for a suitable quadratic form F_o on k^n and then $F = c.F_o$ for some $c \in k_{\mathbf{R}}^*$, since a quadratic form is determined up to a multiple by the group leaving it invariant, or its identity component.

From this elementary remark and 7.2 we now deduce immediately:

7.5. THEOREM. *Let F be irrational. Then the orbit $H_F.o$ of the origin in Ω is dense in Ω.*

In fact, if F is irrational, then in 7.3 we have $\mathscr{L}(\mathbf{R}) = G$ hence $\overline{H_F.o} = \Omega$.

7.6. An element $x \in \mathfrak{o}^n$ is primitive if a relation $a.y = x$ ($a \in \mathfrak{o}$, $y \in \mathfrak{o}^n$) implies that a is a unit in $\mathfrak{o}$. A m-tuple $(x_1, \ldots, x_m)(m \leqq n)$ of elements in $\mathfrak{o}^n$ will be said to be primitive if it is part of a basis of $\mathfrak{o}^n$ over $\mathfrak{o}$ or, equivalently, if it spans a direct summand of $\mathfrak{o}^n$. This is the case if and only if there exists $g \in \mathbf{GL}_n(\mathfrak{o})$ such that $g.e_i = x_i (1 \leqq i \leqq m)$, where (e_i) is the canonical basis of $\mathfrak{o}^n$. If $m < n$, we can then also assume $g \in \mathbf{SL}_n(\mathfrak{o})$. Any primitive m-tuple consists of primitive vectors.

It is clear that a subset $(y_1, \ldots, y_m)$ $(m \leqq n)$ of $k_{\mathbf{R}}^n$ is free (over $k_{\mathbf{R}}$) if and only for each s, the s-components $y_{1,s}, \ldots, y_{m,s}$ are linearly independent over k_s. In particular, any free subset is part of a basis. The n-tuple $(y_1, \ldots, y_n)$ is a basis if and only if there exists $g \in \mathbf{GL}_n(k_{\mathbf{R}})$ such that $g.e_i = y_i$ $(i = 1, \ldots, n)$. We shall say that the basis (y_i) is *unimodular* if such a g can be chosen in $\mathbf{SL}_n(k_{\mathbf{R}})$. If $m < n$, any free m-tuple is part of a unimodular basis.

In the next corollaries, we let B_F be the bilinear symmetric form associated to F. We have then

$$2.B_F(x, y) = F(x + y) - F(x) - F(y), \tag{1}$$

$$F(x) = B_F(x, x) \qquad (x, y \in k_{\mathbf{R}}^n).$$

7.7. COROLLARY. *Assume F to be irrational. Let $(y_1, \ldots, y_n)$ be a unimodular basis of $k_{\mathbf{R}}^n$ over $k_{\mathbf{R}}$. Then there exists a sequence $(x_{j,1}, \ldots, x_{j,n})$ $(j = 1, 2, \ldots)$ of bases of $\mathfrak{o}^n$ over $\mathfrak{o}$ such that*

$$B_F(y_a, y_b) = \lim_{j \to \infty} B_F(x_{j,a}, x_{j,b}) \quad (1 \leqq a, b \leqq n). \tag{1}$$

Proof. By assumption, there exists $g \in G$ such that $g . e_i = y_i$ $(1 \leqq i \leqq n)$. Since F is irrational, $H_F . o$ is dense in Ω by 7.5. This is equivalent to saying that $H_F . \mathbf{SL}_n(\mathfrak{o})$ is dense in G, (in the product of the Lie group topologies). There exist therefore sequences $h_j \in H_F$ and $\gamma_j \in \mathbf{SL}_n(\mathfrak{o})$ such that $h_j . \gamma_j \to g$. Since h_j preserves F, we have

$$\begin{aligned} B_F(y_a, y_b) = B_F(g . e_a, g . e_b) &= \lim_j B_F(h_j . \gamma_j . e_a, h_j . \gamma_j . e_b) \\ &= \lim_j B_F(\gamma_j . e_a, \gamma_j . e_b). \end{aligned}$$

Since $(\gamma_j e_1, \ldots, \gamma_j e_n)$ is a basis of $\mathfrak{o}^n$ for all j's, this proves our assertion, with

$$x_{j,i} = \gamma_j . e_i \quad (i = 1, \ldots, n; j = 1, 2, \ldots).$$

7.8. COROLLARY. *Assume F to be irrational. Let $m < n$ and $y_1, \ldots, y_m$ be a free subset of $k_{\mathbf{R}}^n$. Then there exists a sequence $(x_{j,1}, \ldots, x_{j,m})$ $(j = 1, \ldots)$ of primitive m-tuples of $\mathfrak{o}^n$ such that*

$$B_F(y_a, y_b) = \lim_{j \to \infty} B_F(x_{j,a}, x_{j,b}) \quad (1 \leqq a, b \leqq m) \tag{1}$$

This follows from 7.7, once it is noted that $(y_1, \ldots, y_m)$ is part of a unimodular basis.

REMARK. 7.5 for $n = 3$ is already proved in [DM], where the consequence 7.8 for $m = 2$ is also drawn.

7.9. COROLLARY. *Assume F to be irrational. Let $\lambda_1, \ldots, \lambda_{n-1} \in k_{\mathbf{R}}$. Then there exists a sequence of primitive $(n-1)$-tuples $(x_{j,1}, \ldots, x_{j,n-1})$ $(j = 1, 2, \ldots)$ in $\mathfrak{o}^n$ such that*

$$\lambda_i = \lim_{j \to \infty} F(x_{j,i}) \quad (i = 1, \ldots, n-1). \tag{1}$$

In particular the set of values of F on the primitive elements of $\mathfrak{o}^n$ is dense in $k_{\mathbf{R}}$.

Proof. We note first that, since F_s is non-degenerate, its special orthogonal group is an irreducible linear subgroup of $\mathbf{SL}_n(k_s)$. In particular, it does not leave any hyperplane invariant. The form F_s, being moreover isotropic, assumes all values in k_s. Let $\lambda_{i,s}$ be the s-component of λ_i. The previous remarks show that the level surface $F_s(x) = \lambda_{i,s}$ $(x \in k_s^n)$ is not contained in any hyperplane. It is then elementary that we can find a free subset $(c_{s,1}, \ldots, c_{s,n-1})$ of k_s^n such that

$$F_s(c_{s,i}) = \lambda_{i,s} \quad (i = 1, \ldots, n-1,\ s \in S_\infty). \tag{2}$$

The vectors $c_i = (c_{s,i})$ $(i = 1, \ldots, n-1)$ then form a free subset of $k_{\mathbf{R}}^n$ and we have

$$F(c_i) = \lambda_i \quad (i = 1, \ldots, n-1). \tag{3}$$

7.9 now follows from 7.8 for $m = n - 1$, $a = b$, and 7.6(1).

7.10. There remains to justify 7.3. We assume familiarity with the functor of restriction of scalars [W: §1]. Let E be a perfect field and F a finite separable extension of E. Let $\sum$ be the set of embeddings of F in $\bar{E}$. Let $\mathscr{G}'$ be a connected F-group and $\mathscr{G} = R_{F/E}\mathscr{G}'$ be the E-group obtained by restriction of scalars from $\mathscr{G}'$. The Galois group Γ_E of $\bar{E}$ over E operates on $\sum$ by $\sigma \mapsto \gamma \circ \sigma$. We have

$$\mathscr{G}(\bar{E}) = \prod_{\sigma \in \Sigma} {}^{\sigma}\mathscr{G}'(\bar{E}),$$

where ${}^{\sigma}\mathscr{G}'$ is the group obtained from $\mathscr{G}'$ by the base change $F \to {}^{\sigma}F$.

7.11. PROPOSITION. *Let $\mathscr{L}$ be a connected E-subgroup of $\mathscr{G}$. Then there exists a connected F-subgroup $\mathscr{M}'$ of $\mathscr{G}'$ such that $\mathscr{M} = R_{F/E}\mathscr{M}' \subset \mathscr{L}$ and ${}^{\sigma}\mathscr{M}' = \mathscr{L} \cap {}^{\sigma}\mathscr{G}'$ for every $\sigma \in \Sigma$.*

Proof. Let $\mathscr{M}_\sigma = \mathscr{L} \cap {}^{\sigma}\mathscr{G}'$ and $\mathscr{M}$ be the product of the $\mathscr{M}_\sigma$. We claim that it is defined over E. The group $\mathscr{M}_\sigma$ is defined over ${}^{\sigma}F$, clearly. If $\gamma \in \Gamma$ and $\sigma \in \Sigma$, then $\sigma' = \gamma \circ \sigma \in \Sigma$. Since ${}^{\gamma}\mathscr{L} = \mathscr{L}$, we have therefore ${}^{\gamma}(\mathscr{L} \cap {}^{\sigma}\mathscr{G}') = \mathscr{L} \cap {}^{\sigma'}\mathscr{G}'$, hence $\mathscr{M}$ is a $\bar{E}$-subgroup whose group of $\bar{E}$-points is stable under Γ. Therefore it is defined over E [B: AG 14.4]. By construction, it is the product of its intersections with the ${}^{\sigma}\mathscr{G}'$, therefore $\mathscr{M} = R_{F/E}\mathscr{M}'$, with $\mathscr{M}' \subset \mathscr{G}'$ defined over F [BT: 6.18].

7.12. *Proof of* 7.3. Let $\mathscr{L}$ be the smallest $\mathbf{Q}$-subgroup of $\mathscr{G}$ such that $\mathscr{L}(\mathbf{R}) \supset H$. By 7.11, there exists a k-subgroup $\mathscr{M}'$ of $\mathscr{G}' = \mathbf{SL}_n$ such that $\mathscr{M} = R_{k/\mathbf{Q}}\mathscr{M}' \subset \mathscr{L}$ and ${}^{\sigma}\mathscr{M}' = \mathscr{L} \cap {}^{\sigma}\mathscr{G}'$ for all $\sigma \in \Sigma$. Let Σ be the set of embeddings of k into $\bar{\mathbf{Q}}$, fix an embedding of $\bar{\mathbf{Q}}$ into $\mathbf{C}$ and view the elements of Σ as embeddings of k into $\mathbf{C}$. We may identify S_∞ to a subset of Σ consisting of the real embeddings and one representative of each pair of complex conjugate embeddings. Then ${}^{s}\mathscr{G}'(k_s)$ is our group G_s, and $\mathscr{M}(\mathbf{R})$ is the product of the groups $M_s = {}^{s}\mathscr{M}'(k_s)$, $(s \in S_\infty)$.

By assumption $H_s \subset \mathscr{L}(\mathbf{R}) \cap G_s = M_s$, $(s \in S_\infty)$, therefore $H \subset \mathscr{M}(\mathbf{R})$ and then $\mathscr{L} = \mathscr{M}$ since by assumption, $\mathscr{L}$ is the smallest $\mathbf{Q}$-subgroup whose group of real points contains H. This proves 7.3.

References

[A] A. Ash: Non-square-integrable cohomology of arithmetic groups, *Duke Math. J.* 47 (1980), 435–449.

[B] A. Borel: *Linear Algebraic Groups*, Benjamin, New York 1969; 2nd edn., GTM 126, Springer-Verlag 1991.

[BP] A. Borel et G. Prasad: Valeurs de formes quadratiques aux points entiers, *C. R. Acad. Sci. Paris* 307 (1988), 217–220.

[D] S. G. Dani, A simple proof of Borel's density theorem, *Math. Zeitschr.* 174 (1980), 81–94.

[DM] S. G. Dani and G. A. Margulis: Values of quadratic forms at primitive integral points, *Inv. Math.* 98 (1989), 405–425.

[M] G. A. Margulis: Discrete groups and ergodic theory, in *Number Theory, Trace Formulas and Discrete Groups*, Symposium in honor of A. Selberg, Oslo 1987, Academic Press (1989), 377–398.

[P] V. P. Platonov: The problem of strong approximation and the Kneser-Tits conjecture, *Math. USSR Izv.* 3 (1969), 1139–1147; *Addendum*, ibid. 4(1970), 784–786.

[RR] S. Raghavan and K. G. Ramanathan: On a diophantine inequality concerning quadratic forms, *Göttingen Nachr. Mat. Phys. Klasse* (1968), 251–262.

[Rt] M. Ratner: Raghunathan's topological conjecture and distribution of unipotent flows *Duke Math. J.* 63 (1991), 235–280.

[W] A. Weil, Adeles and algebraic groups, *Progress in Mathematics* 23, Birkhäuser, Boston, 1982.

145.

Deane Montgomery (1909 – 1992)

Notices of the American Math. Soc. **39** (1992) 684–686;
Proc. Amer. Philosophical Society **137** (1993) 453–456

Deane Montgomery was born on September 2, 1909, in Weaver, Minnesota. He received a B.A. from Ramline University in 1929, a M.S. in 1930 and a Ph.D. in 1933 from the University of Iowa.

After having held various fellowships at Harvard University, Princeton University, and the Institute for Advanced Study, he went to Smith College, where he was successively assistant professor (1935–1938), associate professor (1938–1941), and professor (1941–1946). During that period, he was also a Guggenheim fellow at the Institute and a visiting associate professor at Princeton University. After two years at Yale University as an associate professor (1946–1948), he came to the Institute, where he was a permanent member from 1948 to 1951 and a professor from 1951 to 1980, at which time he became emeritus.

His thesis adviser had been E. W. Chittenden, and he had a solid background in real analysis and point set topology. His initial research interests focused on the latter, to which he devoted his first four papers. In the tradition of L. E. J. Brouwer and "Polish topology", they already show considerable technical strength and expertise. As soon as he came to Harvard and Princeton, he broadened his interests, first to algebraic or geometric topology (initially on his own and in a private study group including N. Steenrod and Garrett Birkhoff), and then gradually to transformation groups, which became his major interest for the rest of his career.

His first papers in that area, many written in collaboration with Leo Zippin, were in part in the spirit of earlier work of Brouwer and Kerejarkto, aiming at characterizing groups of familiar euclidean motions such as translations or rotations by topological conditions. They were motivated by questions on the foundations of geometry and, foremost, by Hilbert's fifth problem. In the broad sense, the latter asks, given a locally euclidean topological group acting effectively (i.e., no element $\neq 1$ acts trivially) on an analytic manifold, whether coordinates can be introduced to make the group and the operation analytic (the answer is no). In its narrow sense, it asks whether a locally euclidean topological group is, after a suitable change of coordinates, a(n analytic) Lie group. Variants of the first problem and the second one became points of major interest in the next fifteen years or so, but not of sole interest, though.

Among Deane's contributions to the first question, let me mention the following results pertaining to a (separable metric) locally compact group G acting effectively on a manifold M: (i) If G is compact, M analytic, and each transformation is analytic, then G is a Lie group (1945); (ii) (with S. Bochner, 1946). If M is C^2 and every transformation is C^2, and no element $\neq 1$ leaves pointwise fixed a nonempty open subset, then G is a Lie group; (iii) (with S. Bochner, 1947). If M is a compact complex analytic manifold and G the group of automorphisms of M, then G is a complex Lie group acting holomorphically. On the fifth problem proper, after a series of papers with L. Zippin, Deane gave a positive solution in dimension three (1948). Then came shortly afterwards the decisive results proved jointly with L. Zippin: The existence of a closed subgroup isomorphic to $\mathbb{R}$ in a locally compact, noncompact, connected, separable metric group of strictly positive finite dimension (1951) (also established by A. Gleason) and then the reduction to groups without small subgroups (1952). Since A. Gleason had just proved that such a group is a Lie group, that gave a positive answer to Hilbert's fifth problem. In fact, the whole investigation had been carried out for separable metric finite-dimensional locally compact groups and it was shown more generally that such a group is a "generalized Lie group", i.e., possesses an open subgroup that is a projective limit of Lie groups, hence is a Lie group if it is locally connected. The assumption of finite dimensionality was soon removed by H. Yamabe, who was Deane's assistant at the time.

This was the climax of a major effort and, as I remember it, some people were mildly curious to see where Deane would turn, now that this big problem had been solved. But he did not have to look around at all. Apart from writing with L. Zippin a systematic exposition of the work on the fifth problem (1955), he just went back full time to what was really his main interest (and is already the subject matter of the last chapter of that book): Lie groups (especially compact Lie groups) of transformations on manifolds, so that, in the context of his whole work, the contributions of the fifth problem appear almost as a digression, albeit a most important one.

Even during that hot pursuit, Lie transformation groups were very much on his mind, and he brought a number of interesting contributions, in particular in joint works with L. Zippin and with H. Samelson. In fact, two papers with H. Samelson on compact Lie groups transitive on spheres or tori (1943) have a special place in my memory: When I was an assistant in Zurich, H. Hopf once gave me copies of them, and I could generalize and sharpen some of their results. This led to my first single author paper, which I submitted for publication in the *Proceedings of the AMS* to Deane, then an editor; my first contact with him.

The general problem in transformation groups is, roughly, to relate the structures of the group G, the manifold M operated upon, the orbits, fixed points, and the quotient space. At the time, there was one body of special, but deep, work, that of P. A. Smith on homeomorphisms of prime power order of homology spheres or acyclic spaces. Very little was known otherwise, and Deane was a prime mover in the development of a general theory, which he pushed in many directions. He and various collaborators proved a number of foundational results, as well as more special ones, which often opened up fruitful directions for others. A survey of these contributions and of work they led to is given by F Raymond and R. Schultz in the Proceedings of a Conference honoring Deane on his 75th birthday (*Contemporary Mathematics*, vol. 36 (1983)), and I shall not try to duplicate it. It ranges from basic results such as the existence of a slice (with C. T. Yang, 1957), a powerful tool to study a group action near an orbit, the existence of a principal type of orbits (with C. T. Yang, 1958), to more special ones, such as actions on euclidean space or spheres with orbits of small codimension or the existence of smooth actions of SO_3 on euclidean space without fixed points (with P. E. Conner, 1962). In a first phase, the emphasis was on continuity, i.e., on topological properties, but Deane kept up with the great advances of differential topology and soon veered more and more to differentiable actions, adapting techniques and points of view of differential topology. This led to his last major effort, a long series ofjoint papers with C. T. Yang on free or semi-free (i.e., free outside the fixed point set) actions of the circle group on homotopy 7-spheres, which produced notably many interesting examples of homotopy complex projective 3-spaces (1966–1973).

During his tenure as a professor at the Institute, Deane was at the center of activity in topology (algebraic, geometric, differential), one of the highlights in the life of the School, first by his seminar, a perennial feature and a meeting ground for topologists in the Princeton community, but also in more informal ways. He frequently organized seminars in his office, usually with some younger members with whom he would go through some recent developments. He was always seeking out and encouraging young mathematicians. He and his wife Kay would regularly and very warmly receive the visiting members at their home. Maybe remembering his own beginnings in an out of the way place, he had a special interest, and talent, in finding out people with considerable potential among some applicants from rather isolated places about whom not much information was available.

His concern for the Institute went far beyond his immediate scientific interests and was all encompassing. He had a very high view of the role the Institute should play and served this ideal with unwavering and thoroughly unselfish loyalty. In day to day contacts, he was very kind, informal, full of understanding, always ready to help, and struck one as a very mild person, but deceptively so for anyone who, in his eyes, would threaten the Institute's standards, and who would then soon see rising an iron-willed and formidable opponent. His care for the highest standards at the Institute, later gratefully acknowledged in citations by the Trustees, was not always universally understood or shared at the time, so that he and like-minded colleagues had to weather some rather stormy moments, during which he was totally unshakable.

His abiding interest in the welfare of mathematics also led him to accept a number of official positions. In particular, he was Vice President (1952–1953), elected Trustee (1955–1961) and President (1960–1963, includes terms as President-Elect and Ex-President) of the AMS, where he also served on a number of committees, and President of the International Mathematical Union (1974–1978).

Honors, too, came his way: Honorary Doctor of Science from Hamline University (1954), Yeshiva University (1961), the University of Illinois (1977), and the University of Michigan (1986), as well as a Doctor of Laws degree from Tulane University (1967); election to the National Academy of Sciences in 1955, to the American Academy of Arts and Sciences, and the American Philosophical Society in 1958; and receipt of the Steele Prize of the AMS in 1988.

Deane was an early riser and it was a rare event for anyone to be at the Institute before him. Being very gregarious, he talked to practically everybody working in any capacity at the Institute, which won him the respect and affection of members and staff alike and gave him an exhaustive knowledge of the Institute. Through O. Veblen, to whom he had been very close during the latter's late years, it reached to the very beginnings of the Institute so that he was a walking encyclopedia on all aspects of the Institute's history and operations.

In 1988, he and Kay moved to Chapel Hill, NC to be close to their daughter and granddaughters. That prospect did not fully compensate for the severance of the ties with an institution which had meant and still meant so much to him, and it was altogether a rather sad occasion, the sadness of which was hardly mitigated by promises to keep in touch. Being myself a fairly early riser, I often started my day by knocking at his door, sure to find him, to have a chat, mostly about mathematics, mathematicians, and Institute affairs. That I have not been able to do so after his departure has left for me a void which could not be filled.

Deane died in his sleep in Chapel Hill, on March 15, 1992. He is survived by his wife, his daughter Mary Heck, and two granddaughters.

146.

Values of Quadratic Forms at S-integral points

in *Algebraic groups and number theory*, V. Platonov and A. S. Rapinchuk ed.
Uspehi Mat. Nauk. 47, No. 2, (1992) 117–141, (p. 118–120),
Russian Math. Surveys 47, No. 2, (1992) 133–161 (p. 134–136)

This lecture gave a survey of the work done in recent years on the Oppenheim conjecture and on the results about unipotent flows used in dealing with it.

1. Let F be a non-degenerate indefinite quadratic form on $\mathbf{R}^n$ $(n \geqq 3)$. It will be said to be *rational* if a non-zero real multiple has rational coefficients, *irrational* otherwise. The Oppenheim conjecture (made in 1929, though not at first in a fully precise form) states that F is irrational if and only if $F(\mathbf{Z}^n)$ is not discrete around the origin. The interesting part is the "only if", and it suffices to prove it for $n = 3$. It was known early on that its truth would imply that $F(\mathbf{Z}^n)$ is dense in $\mathbf{R}^n$ if F is irrational. The first general proof was given by G.A.Margulis [5] who, following a suggestion of M.S.Raghunathan, deduced it for $n = 3$ from a property of flows on $\Omega_3 = \mathbf{SL}_3(\mathbf{R})/\mathbf{SL}_3(\mathbf{Z})$. In particular if F does not represent zero it follows from (in fact is equivalent to) the fact that a relatively compact orbit of the special orthogonal group $SO(F)$ of F on Ω_3 is closed, one of the main theorems proved in [5]. This last result is itself a very special case of a conjecture of Raghunathan (see below) which also implies that any orbit of $SO(F)$ on Ω_3 is either closed or dense. That was shown by S.G.Dani and Margulis [3] and implies already that the values of an irrational F on the primitive elements of $\mathbf{Z}^n$ are dense.

2. One version of the Raghunathan conjecture asserts that if G is a semisimple $\mathbf{Q}$-group, Γ an arithmetic subgroup and H a closed connected subgroup of $G(\mathbf{R})$ generated by unipotent elements, then the closure of any orbit of H in $G(\mathbf{R})/\Gamma$ is the orbit of a closed subgroup $L \supseteqq H$. A much more general statement was proved by M.Ratner [6], who also showed that the possible L's, as H varies among closed subgroups generated by unipotent elements, form a countable set. After having stated it, I sketched how M.Ratner reduces the proof to the case of a one-dimensional unipotent subgroup H (using in an essential way the countability assertion) and referred to her lecture for more details on this crucial case.

This theorem yields a remarkable strengthening of the Oppenheim conjecture:

Theorem. *Let F be irrational. Given $c_1 \dots c_{n-1} \in \mathbf{R}$ and $\epsilon > 0$ there exists a $(n-1)$-tuple $(x_1, \dots, x_{n-1})$ of primitive elements in $\mathbf{Z}^n$, which are part of a basis of $\mathbf{Z}^n$, such that*

$$|F(x_i) - c_i| < \epsilon \qquad (i = 1, \dots, n-1).$$

(See [2:7.9], which is proved more generally in the setting of the next section, for $S = S_\infty$.)

3. The next part of the talk was devoted to a generalization of the Oppenheim conjecture studied jointly with G.Prasad [1], [2]. Let k be a number field, S a finite set of places of k containing the set S_∞ of archimedean ones, $\mathfrak{o}_S$ the ring of S-integers in k, k_s the completion of k at $s \in S$, and k_S the direct sum of the k_s $(s \in S)$. A quadratic form F on k_S^n is a collection $(F_s)_{s \in S}$, where F_s is a quadratic form on k_s^n. It is said to be *rational* if there exists a unit $\lambda = (\lambda_s) \in k_S^*$ and a form F_o on k^n such that $F_s = \lambda_s F_o$ for all $s \in S$, *irrational* otherwise. We assume each F_s to be non-degenerate, isotropic and $n \geqq 3$. As a slight strengthening of the direct analogue of the Oppenheim conjecture, we show ([2], Theorem A) that if F is irrational, given $\epsilon > 0$ there exists $x \in \mathfrak{o}_S^n$ such that $0 < |F_s(x)| < \epsilon$ for all $s \in S$. It is deduced from two main special cases: (I). $S = S_\infty$, where it follows either from Ratner's result or from a generalization of the original Margulis argument (but proving slightly less, allowing some $F_s(x)$, but not all, to be zero). (II). One of the F_s at least is rational. This is dealt with in a completely different manner by some geometry of numbers and strong approximation in algebraic groups.

4. During the Conference, Prasad and I noticed a very simple consequence of Theorem A which provides a characterization of irrational forms for $n \geqq 2$, without any condition of isotropy.

Proposition. *Let F be a non-degenerate quadratic form on k_S^n $(n \geqq 2)$. Then F is irrational if and only if given $\epsilon > 0$ there exist $x, y \in \mathfrak{o}_S^n$ such that*

$$0 < |F_s(x) - F_s(y)| < \epsilon \quad (s \in S).$$

To see this, one just applies Theorem A of [2] to the quadratic form $F \oplus -F$ on k_S^{2n}, which is possible, since that last form is obviously non-degenerate, isotropic at all places $s \in S$, and irrational if F is so. If $S = S_\infty$, the theorem in §2, in its S_∞-version, implies more strongly that these differences (for x, y primitive even) are dense in k_S if F is irrational.

5. The lecture ended with some open problems: Is $F(\mathfrak{o}_S^n)$ dense if finite places are allowed? If so, is one allowed to use only primitive vectors? More ambitiously, does the analogue of Ratner's theorem hold in this general setting?

Assume now again $k = \mathbf{Q}$, $S = S_\infty$, but F *positive* non-degenerate. Then $F(\mathbf{Z}^n)$ is obviously a discrete set in $\mathbf{R}$. If F is rational, then the difference between any two distinct values is bounded away from zero in absolute value. Let now F be irrational. The previous proposition shows there are arbitrary small such differences. In [4], D.J.Lewis has asked whether the successive differences tend to zero uniformly. This is still an open question.

Institute for Advanced Study, Princeton, NJ 08540, USA

REFERENCES

[1]. A.Borel et G.Prasad, *Valeurs de formes quadratiques aux points entiers*, C.R.Acad. Sci. Paris **307** (1988), 217-220.

[2]. A.Borel and G.Prasad, *Values of isotropic quadratic forms at S-integral points*, to appear in Compos.Math.

[3]. S.G.Dani and G.A.Margulis, *Values of quadratic forms at primitive integral points*, Inv.Math. **98** (1989), 405-425.

[4]. D.J.Lewis, *The distribution of the values of real quadratic forms at integer points*, Proc.Symp.pure math. XXIV, 159-174, AMS 1973.

[5]. G.A.Margulis, *Discrete subgroups and ergodic theory*, Symposium in honor of A.Selberg, Oslo 1987, Academic Press (1989), 377-398.

[6]. M.Ratner, *Raghunathan's topological conjecture and distribution of unipotent flows*, preprint 1990 (to appear).

147.

(avec F. Bien)

Sous-groupes épimorphiques de groupes linéaires algébriques I

C. R. Acad. Sci. Paris **315**, Sér. I, (1992) 649–653

Résumé – Un sous-groupe fermé H d'un groupe algébrique linéaire connexe G est dit *épimorphique* si tout morphisme de G dans un groupe algébrique est déterminé par sa restriction à H. Cette Note est consacrée à des propriétés générales et des exemples de sous-groupes épimorphiques. Nous donnons aussi quelques résultats partiels concernant une propriété de finitude des représentations induites et nous formulons une conjecture générale. Une Note ultérieure liera les sous-groupes épimorphiques à des problèmes de génération finie d'invariants, de compactifications et de multiplicités de représentations.

Epimorphic subgroups of linear algebraic groups I

Abstract – *A closed subgroup* H *of a connected linear algebraic group* G *is said to be* epimorphic *if any morphism of* G *into an algebraic group is determined by its restriction to* H. *This Note is devoted to general properties and examples of epimorphic subgroups. We also give partial results pertaining to a finiteness property of induced representations, and we formulate a general conjecture. A subsequent Note will relate epimorphic subgroups to problems of finite generation of invariants, compactifications and multiplicities of representations.*

1. Dans cette Note, k est un corps commutatif qui, sauf au n° 10, est supposé algébriquement clos, p sa caractéristique, G un k-groupe linéaire connexe et H un k-sous-groupe fermé propre de G. Si X est une variété algébrique, $k[X]$ est la k-algèbre des fonctions régulières sur X, à valeurs dans k. Un k-espace vectoriel E sur lequel G opère par automorphismes est un G-module rationnel s'il est réunion de sous-espaces de dimension finie invariants par G sur lesquels la représentation donnée de G est rationnelle. On note E^H l'espace des invariants de H. En particulier, $k[G]^H$ désigne l'espace des invariants de H dans $k[G]$ par rapport aux translations à droite. C'est un G-module rationnel, G opérant par translations à gauche, qui est égal à $k[G/H]$.

Définition. – H est *épimorphique* dans G si l'inclusion $i: H \to G$ est un épimorphisme dans la catégorie des groupes algébriques linéaires, *i.e.* pour tous morphismes r et s de G dans un k-groupe algébrique linéaire, l'égalité $r \circ i = s \circ i$ entraîne $r = s$.

Théorème 1. – *Les conditions suivantes sont équivalentes :* (i) H *est épimorphique dans* G; (ii) $k[G/H] = k$; (iii) $k[G/H]$ *est de dimension finie;* (iv) *pour tout* G-*module rationnel* E, *on a* $E^H = E^G$; (v) *si un* G-*module rationnel* V *est somme directe de deux sous-espaces* X, Y *invariants par* H, *alors* X *et* Y *sont invariants par* G.

Esquisse de démonstration. – Comme $k[G]^G = k$, il suffit, pour établir que (i) ⇒ (ii), de faire voir que si D est une droite dans $k[G]^H$, elle est aussi fixe, point par point, par G. Elle est contenue dans un sous-espace E de dimension finie de $k[G]$ invariant à droite par G ([3], (5.1). On considère le produit semi-direct $G' = E \rtimes G$ par rapport au produit

$$(x, g).(x', g') = (x + g.x', g.g') \qquad (x, x' \in E,\ g, g' \in G).$$

Soit $d \in D - 0$. Alors $r, s: G \to G'$ définis par $r(g) = (0, g)$ et $s(g) = (g.d - d, g)$ sont deux morphismes qui coïncident sur H, donc sur G, et par suite d est fixe par G. Que (ii) ⇒ (iii) est clair. Si E est un G-module rationnel et $d \in E$ est fixe par H, mais pas par G, alors

---

Note présentée par Armand Borel.

0764-4442/92/03150649 $ 2.00 © Académie des Sciences

l'image de l'application orbitale $g \mapsto g.d$ est au moins de dimension un, ce qui entraîne que $k[G/H]$ est de dimension infinie, d'où (iii) ⇒ (iv). Si l'on est dans la situation de (v), avec V supposé de dimension finie, ce qui loisible, alors la projection de E sur X commute à H, donc à G d'après (iv) appliqué à End E, d'où (iv) ⇒ (v). Pour établir que (v) ⇒ (i) il suffit de considérer deux morphismes r, s de G dans GL(V), où V est de dimension finie. Leur somme directe est un morphisme de G dans $GL(V \oplus V)$. L'espace $V \oplus V$ est somme directe du premier facteur et de la diagonale Z. Si $r = s$ sur H ces deux sous-espaces sont stables par H, donc par G, d'où $r = s$.

Remarque. — Dans un manuscrit non publié [1], communiqué à l'un de nous en 1975, G. Bergman étudie la notion similaire de sous-algèbre épimorphique dans diverses catégories d'algèbres de Lie. La terminologie nous a été suggérée par ce travail, de même que le théorème 1, où l'équivalence de (i), (ii), (iv) et (v) est l'analogue de 3.1, 3.2 dans [1]. L'équivalence de (ii) et (iii) figure aussi dans [14]. Comme nous allons le voir, la catégorie des groupes algébriques linéaires contient beaucoup d'inclusions strictes épimorphiques. Un tel phénomène ne se produit pas dans la catégorie de tous les groupes, par exemple.

2. Nous énumérons ici quelques conséquences simples du théorème 1 ou des définitions. Pour tout k- groupe L, on note X(L) le groupe des caractères de dimension un de L.

(*a*) Tout sous-groupe parabolique est épimorphique.

(*b*) Si L est un sous-groupe fermé connexe de G contenant H et si H est épimorphique dans L et L dans G, alors H est épimorphique dans G.

(*c*) H est épimorphique dans G si et seulement si sa composante connexe de l'identité H^0 ou un sous-groupe de Borel de H^0 l'est.

(*d*) Si G est presque simple, alors H est épimorphique si et seulement si le radical de H^0 est épimorphique.

(*e*) Soit $(G_i)_{i \in I}$ une famille de sous-groupes fermés de G qui engendrent G. Si $H \cap G_i$ est épimorphique dans G_i pour tout $i \in I$, alors H est épimorphique dans G.

3. Rappelons que H est dit *observable* dans G s'il satisfait aux conditions équivalentes suivantes : (i) toute représentation rationnelle de dimension finie de H se prolonge en une telle représentation de G; (ii) G/H est quasi-affine; (iii) $k[G/H]$ sépare les points de G/H; (iv) il existe un G-module rationnel de dimension finie V et $v \in V$ tel que $H = \{g \in G \mid g.v = v\}$ (*cf.* [2]).

L'intersection des sous-groupes observables de G contenant un sous-groupe fermé donné M est observable, c'est l'*enveloppe observable* de M. Elle est connexe si M l'est. La proposition suivante montre que cette notion est en un certain sens l'opposée de celle de sous-groupe épimorphique.

Proposition 1. — *Supposons* H *connexe. Soit* L *le plus grand sous-groupe de* G *tel que* $k[G]^H = k[G]^L$. *Alors* L *est à la fois l'enveloppe observable de* H *dans* G *et le plus grand sous-groupe de* G *dans lequel* H *est épimorphique.*

En particulier, H est épimorphique dans G si et seulement si son enveloppe observable est G.

4. Les remarques du n° 2 montrent que l'étude des sous-groupes épimorphiques se ramène au cas des sous-groupes résolubles connexes. Dans ce numéro, nous supposons que H l'est. Notons S un tore maximal de H et $U = \mathcal{R}_u H$ son radical unipotent. Le groupe H est donc le produit semi-direct de U par S. Pour tout k-groupe L contenant S,

on note $\Phi(S, L)$ l'ensemble des racines de S dans L et, suivant l'usage, $\langle \Psi \rangle$ le sous-groupe engendré par une partie Ψ de $X(S)$.

PROPOSITION 2. — *Supposons* G *réductif et* H *épimorphique dans* G. *Alors* $\langle \Phi(S, H) \rangle = \langle \Phi(S, G) \rangle$. *La réciproque est vraie si* S *est un tore maximal de* G *et si soit* $p \neq 2, 3$, *soit* H *n'est contenu dans aucun sous-groupe réductif propre de* G.

Lorsque S est un tore maximal, on retrouve un résultat de K. Pommerening ([9], 3.6).

5. Supposons G presque simple. Il suit de 2.(*d*) que le radical de tout sous-groupe parabolique propre est épimorphique. Nous allons maintenant construire des exemples de sous-groupes épimorphiques, minimaux en un certain sens. Soient T un tore maximal de G et Ψ un ensemble de racines de T dans G, fermé par addition. On désigne par U_Ψ le sous-groupe de G engendré par les groupes radiciels à un paramètre U_α, $\alpha \in \Psi$.

(*a*) Toute base de $\Phi(T, G)$ peut s'écrire comme union disjointe $I \cup J$, où les éléments de I (resp. J) sont orthogonaux. Soit $\Psi = I \cup (-J)$, et U_Ψ le sous-groupe unipotent (et commutatif) de G sous-tendu par les groupes radiciels à un paramètre U_α ($\alpha \in \Psi$). Alors $H = T . U_\Psi$ est épimorphique. En utilisant de plus le lemme ci-dessous, on voit que l'on peut remplacer T par un sous-tore convenable de dimension un. Par suite, G contient un sous-groupe épimorphique de dimension $l+1$, normalisé par un tore maximal, où l est le rang de G.

(*b*) Supposons $p=0$. Alors G contient un sous-groupe épimorphique de dimension trois. Pour le voir, on part d'un sous-groupe de dimension trois principal L de G et d'un sous-groupe de Borel B_L de L. Si $L \neq G$, il existe un sous-groupe résoluble connexe H de G de dimension trois contenant B_L. Si L est propre maximal, 2.(*e*) montre que H est épimorphique. Sinon, on sait que L est contenu dans un seul sous-groupe propre maximal M et on peut toujours construite un tel H qui ne soit pas contenu dans M. Il est alors aussi épimorphique. En vue de résultats de G. Seitz ([10], [11]) et D. Testerman [12], une construction similaire vaut encore en caractéristique p suffisamment grande. Lorsque G n'est pas de type A_1, on peut montrer qu'un sous-groupe de dimension $\leqq 2$ n'est pas épimorphique si $p=0$ et, en utilisant [13], si $p \geqq 7$.

6. Étant donnée une représentation rationnelle (σ, E) de dimension finie de H, on note $I_H^G(\sigma)$ ou $I_H^G(E)$ la représentation induite de H à G à partir de σ. Par définition $I_H^G(\sigma) = (k[G] \otimes E)^H$, vu comme module sur G par l'action à gauche sur $k[G]$, et $k[G]$ étant vu comme module sur H via les translations à droite. C'est donc aussi l'espace $\mathrm{Hom}_H(E^*, k[G])$ d'entrelacement de la représentation contragrédiente E^* de E et de $k[G]$, sur lequel H opère par translations à droite. On rappelle que le foncteur induction est transitif par rapport aux inclusions de groupes et est exact à gauche [7].

$I_H^G(E)$ est un module sur l'algèbre $k[G]^H$, laquelle s'identifie, en tant que G-module, à la représentation induite de la représentation triviale de H. Comme $k[G]$ n'a pas de diviseur de zéro non nul, le théorème 1 entraîne que H est épimorphique dans G si et seulement s'il existe un H-module rationnel E tel que $I_H^G(E)$ soit de dimension finie non nulle.

Soient $N = \mathscr{R}_u G$ le radical unipotent de G et B un sous-groupe de Borel de G. Nous proposons d'appeler G-*modules principaux* les modules induits $E_\lambda = I_B^G(w_0(\lambda))$, où $\lambda \in X(B)$ et w_0 est le plus long élément du groupe de Weyl de G/N. Le module E_λ est non nul si et seulement si λ est un poids dominant, par rapport à l'ordre défini par B. Dans ce cas, E_λ est irréductible de plus haut poids λ si $p=0$; sinon il contient un seul sous-module irréductible, et le poids dominant de ce dernier est λ. Les G-modules principaux sont les duaux des modules de Weyl utilisés dans [7].

7. Supposons dorénavant H connexe. On note Λ_H, ou si nécessaire Λ_H^G, l'ensemble des $\lambda \in X(H) = X(S)$ tels que $I_H^G(\lambda) \neq 0$. Λ_H contient zéro et est stable par addition, vu que la multiplication $I_H^G(\lambda) . I_H^G(\mu) \to I_H^G(\lambda+\mu)$ est non-triviale pour $\lambda, \mu \in \Lambda_H$. On voit facilement que $\Lambda_H^G = \Lambda_H^L$, où L est l'enveloppe observable de H dans G. Si S est de dimension un, ou si H est normalisé par un tore maximal de G, alors Λ_H est un monoïde de type fini. Nous ne savons pas si c'est toujours le cas.

Pour tout sous-monoïde C du groupe additif de $\mathbb{R}$, notons $C[\Lambda_H]$ l'ensemble des combinaisons linéaires d'éléments de Λ_H à coefficients dans C. C'est un sous-monoïde de $X(S)_{\mathbb{R}} = X(S) \otimes_{\mathbb{Z}} \mathbb{R}$. Si $C = \mathbb{Q}, \mathbb{R}$, on pose $C_+ = \{c \in C \mid c \geqq 0\}$ et $C_- = -C_+$. Le cône $\mathbb{R}_+[\Lambda_H]$ est toujours de dimension maximale dans $X(H)_{\mathbb{R}}$. Notons r l'application de restriction $X(G) \to X(H)$.

LEMME. – *Supposons* H *épimorphique.*

(i) $\mathbb{R}_+[\Lambda_H] \cap \mathbb{R}_-[\Lambda_H] = r(X(G))_{\mathbb{R}}$.

(ii) *Soit* S′ *un sous-tore de* S. *Alors* S′.U *est épimorphique dans* G *si et seulement si tout élément non nul de* Λ_H *a une restriction non triviale à* S′.

(iii) *Si* $\mathbb{Q}_+(\Lambda_H)$ *est de type fini sur* $\mathbb{Q}_+$, *le tore* S *contient un sous-tore* S′ *de dimension un tel que* U.S′ *soit épimorphique.*

(iv) *Si* G *est réductif et* S *est un tore maximal de* G, *alors*

$$\Lambda_H = \{\lambda \in X(H) \mid \lambda(\beta^\vee) \leqq 0, \quad \forall \beta \in \Phi(S, H)\}.$$

8. Nous dirons que H *a la propriété* F *dans* G *si* $I_H^G(E)$ est de dimension finie quel que soit le H-module rationnel de dimension finie E. Vu le théorème 1, cela implique que H est épimorphique. Nous conjecturons que la réciproque est vraie. Elle est bien connue si H = P est parabolique puisque G/P est alors une variété projective. Il suffit de la vérifier pour E irréductible. On montre que H a la propriété F dans G si et seulement si un sous-groupe de Borel de H^0 la possède, ce qui nous ramène à nouveau aux sous-groupes résolubles connexes.

PROPOSITION 3. – *Supposons* G *réductif et* H *résoluble, épimorphique.*

(i) H *a la propriété* F *si et seulement si, quel que soit* $\lambda \in X(H)$, *il n'existe qu'un nombre fini de* G-*modules principaux contenant une droite invariante par* H *de poids* λ.

(ii) *Soit* L *un sous-groupe connexe de* G *dans lequel* H *est normal. Alors* H *a la propriété* F *dans* G *si et seulement si* L *l'a.*

En caractéristique zéro, (i) est facile à prouver en utilisant le théorème de Peter-Weyl algébrique. En caractéristique non-nulle, (i) est une conséquence d'un théorème de F. Grosshans ([6]; th. 8) affirmant que $k[G]$ possède une G-filtration telle que l'algèbre graduée associée s'injecte dans la somme directe de tous les G-modules principaux pris avec certaines multiplicités. Pour (ii), on se ramène aux cas où L est résoluble connexe et L/H est soit un tore, soit unipotent. Dans le premier, on prouve que la restriction de Λ_L à Λ_H est propre; dans le second, on montre l'égalité $\Lambda_L = \Lambda_H$ et on utilise (i).

9. Nous continuons à supposer que H est résoluble, connexe, épimorphique. Admettons aussi que le normalisateur connexe $M = \mathcal{N}_G(U)^\circ$ du radical unipotent U de H contienne un sous-groupe L de la forme $H \ltimes V$, où V est un sous-groupe unipotent de dimension un normalisé, mais pas centralisé par S, et non contenu dans H. On note γ le poids de S dans l'algèbre de Lie de V.

PROPOSITION 4. — *Supposons que* $\mathbb{Q}_+[\Lambda_H]$ *est de type fini et que* $\gamma \notin \mathbb{Q}_-[\Lambda_H]$.

(i) L *a la propriété* F *si et seulement s'il en est de même pour* H.

(ii) *Un tel sous-groupe* V *existe lorsque* M *n'est pas résoluble.*

Ce résultat, joint à proposition 2 (ii), permet quelquefois de démontrer la propriété F par récurrence descendante sur la dimension de H. Par exemple, elle implique que si U contient le radical unipotent $N = \mathscr{R}_u P$ d'un sous-groupe parabolique propre P de G, alors H a la propriété F dans G. (Cela peut aussi se déduire du fait que $k[G]^N$ est une algèbre de type fini [5], point sur lequel nous reviendrons dans une Note ultérieure; *voir* [14], 2.2, dans le cas $H = T.\mathscr{R}_u P$, où T est un tore maximal de G.) Notons que (ii) est évidente en caractéristique zéro. Sinon, cette assertion résulte de [4], 9.16 et [8].

10. Dans cette section, nous supposons k parfait, mais pas nécessairement algébriquement clos, par exemple $k = \mathbb{R}$. On rappelle qu'un k-groupe résoluble connexe est déployé sur k s'il possède une suite de composition formée de k-sous-groupes connexes dont les quotients successifs sont k-isomorphes à $\mathbf{G}_a$ ou $\mathbf{GL}_1$. Les k-sous-groupes résolubles déployés sur k maximaux d'un k-groupe connexe L sont conjugués sous $L(k)$ ([3], 15.14). Le plus grand k-sous-groupe déployé $\mathscr{R}_d L$ du radical $\mathscr{R} L$ de L est normal dans L et est appelé son radical déployé.

THÉORÈME 2. — *Supposons* G *engendré par ses k-sous-groupes unipotents connexes. Alors* H *est épimorphique si et seulement s'il contient un k-sous-groupe résoluble déployé sur k et épimorphique dans* G. *Si* G *de plus est semi-simple et l'image de* H *dans tout quotient propre de* G *par un k-sous-groupe normal est un sous-groupe propre, alors* H *est épimorphique si et seulement si son radical déployé l'est.*

11. (*a*) Un sous-groupe épimorphique de G ne contient pas toujours le radical unipotent de G, ni le centre de G.

(*b*) Si l'on admet des groupes algébriques non-linéaires, on peut construire à l'aide de courbes elliptiques des groupes G tels que $k[G] = k$, mais qui n'ont pas la propriété F.

Note remise le 26 juin 1992, acceptée le 6 juillet 1992.

RÉFÉRENCES BIBLIOGRAPHIQUES

[1] G. BERGMAN, *Epimorphisms of Lie algebras* (preprint, ca. 1970).

[2] A. BYALYNICKI-BIRULA, G. HOCHSCHILD et G. MOSTOW, Extensions of representations of algebraic linear groups, *Amer. J. Math.*, 85, 1963, p. 131-144.

[3] A. BOREL, Linear algebraic groups, 2nd enlarged Edition, *Graduate Texts in Math.*, 126, Springer-Verlag, 1991.

[4] A. BOREL et T. SPRINGER, Rationality properties of linear algebraic groups II, *Tôhoku Math. J.*, (2), 20, 1968, p. 443-497.

[5] F. D. GROSSHANS, The invariants of unipotent radicals of parabolic subgroups, *Invent. Math.*, 73, 1983, p. 1-9.

[6] F. D. GROSSHANS, Contractions of the actions of reductive algebraic groups in arbitrary characteristic, *Invent. Math.*, 107, 1992, p. 127-133.

[7] J. C. JANTZEN, Representations of algebraic groups, *Pure Appl. Math.*, 131, Academic Press, 1987.

[8] T. NAKAMURA, A remark on unipotent subgroups of characteristic $p>0$, *Kodai Math. Sem. Rep.*, 23, 1971, p. 127-130.

[9] K. POMMERENING, Observable radizielle Untergruppen von halbeinfachen algebraischen Gruppen, *Math. Z.*, 165, 1979, p. 243-250.

[10] G. SEITZ, The maximal subgroups of classical algebraic groups, *Mem. Amer. Math. Soc.*, 365, 1987.

[11] G. SEITZ, Maximal subgroups of exceptional algebraic groups, *Mem. Amer. Math. Soc.*, 441, 1991.

[12] D. M. TESTERMAN, Irreducible subgroups of exceptional algebraic groups, *Mem. Amer. Math. Soc.*, 390, 1988.

[13] D. M. TESTERMAN, A_1*-type overgroups of elements of order p in semisimple algebraic groups and the associate finite groups* (preprint 1982).

[14] D. C. VELLA, A cohomological characterization of parabolic subgroups of reductive algebraic groups, *J. Algebra*, 121, 1989, p. 281-300.

F. B. : *Fine Hall, Princeton University, Princeton, NJ* 08544, *U.S.A.*;
A. B. : *Institute for Advanced Study, Princeton, NJ* 08540, *U.S.A.*

148.

(avec F. Bien)

Sous-groupes épimorphiques de groupes linéaires algébriques II

C. R. Acad. Sci. Paris **315**, Sér. I, (1992) 1341–1346

Résumé – Cette Note fait suite à [1]. Étant donné un groupe linéaire algébrique connexe G sur un corps algébriquement clos k et un sous-groupe connexe fermé H, de radical unipotent U, on introduit et discute deux conditions d'engendrement fini pour des anneaux de représentations induites, notés FG et SFG. Si H est épimorphique, la première implique la propriété F de [1] et l'existence d'un plongement projectif normal équivariant de G/H à bord de codimension $\geqq 2$. La deuxième entraîne la première et équivaut au fait que $k[G/U]$ est de type fini.

Epimorphic subgroups of linear algebraic groups II

Abstract – *This Note is a sequel to* [1]. *Given a connected linear algebraic group* G *over an algebraically closed field k and a connected closed subgroup* H, *with unipotent radical* U, *we introduce and discuss two conditions of finite generation for rings of induced representations. If* H *is epimorphic, the first one implies the property* F *of* [1] *and the existence of an equivariant projective normal embedding of* G/H *with boundary of codimension* $\geqq 2$. *The second one implies the first one and is equivalent to the fact that* $k[G/U]$ *is finitely generated.*

1. Nous utilisons librement les notations et définitions de [1]. Nous rappelons que k est un corps algébriquement clos, p sa caractéristique, G un k-groupe linéaire connexe, B un sous-groupe de Borel de G et H un sous-groupe fermé de G. Dans cette Note nous supposons H propre, connexe, notons U le radical unipotent et B_H un sous-groupe de Borel de H.

Une k-algèbre A est appelée une G-*algèbre rationnelle* si G opère sur A par automorphismes de k-algèbre de manière à ce que la représentation linéaire associée sur l'espace vectoriel sous-jacent à A soit rationnelle. A est dite de type fini si elle est engendrée en tant qu'algèbre par un nombre fini d'éléments. Rappelons (*cf.* appendix A au chap. I de [10] pour des références) :

(*) *Si* G *est réductif et* A *une* G-*algèbre rationnelle de type fini, alors* A^G *est de type fini.*

2. Étant donné un G-module rationnel C, on considère $k[G]\otimes C$ comme un $G\times H$ module, G opérant sur $k[G]$ par translations à gauche, sur C par la représentation donnée, et H agissant sur $k[G]$ par translations à droite. En particulier $(k[G]\otimes C)^G$ est un H-module rationnel.

Proposition 1 (transfert). – *L'application k-linéaire* $\varphi: k[G]\otimes C \to C: f\otimes c \mapsto f(1)c$ *pour* $f\in k[G]$ *et* $c\in C$ *induit un isomorphisme* H-*équivariant de* $(k[G]\otimes C)^G$ *sur* C, *qui est un isomorphisme d'algèbres si* C *est une* G-*algèbre rationnelle.*

Dans cet énoncé, l'action de H sur l'image C de φ est la restriction de celle de G. Ce principe, qui revient à dire que $I_G^G C = C$ et est donc presque évident, entraîne que plusieurs propriétés de $k[G]$ comme H-module par rapport aux translations à droite se transportent

Note présentée par Armand Borel.

0764-4442/92/03151341 $ 2.00

à C. Nous dirons qu'elles sont obtenues par transfert. En particulier, on a

(1) $$(k[G]^H \otimes C)^G \cong C^H,$$

résultat dû à Popov [14], Th. 9, et :

COROLLAIRE 1. — (i) *Si* E *est un* H-*module rationnel, alors* $(I_H^G E \otimes C)^G = (C \otimes E)^H$.
(ii) [4], 1.2. *Si* G *est réductif,* C *une* G-*algèbre rationnelle de type fini et* $k[G]^H$ *de type fini, alors* C^H *est de type fini.*

3. Rappelons [1], n° 6, que si $\lambda \in w_0(\Lambda_B^G) = -\Lambda_B^G$, le module induit $I_B^G w_0(\lambda)$ est le G-module principal E_λ. Remarquons que $-\Lambda_B^G = \Lambda_{B^-}^G$, où B^- est un sous-groupe de Borel opposé à B. Le dual E_λ^* de E_λ est *un module de Weyl* [7]. Il est cyclique, engendré par une droite B-invariante de poids $i(\lambda)$, où i est l'involution d'opposition. Il a un unique quotient irréductible, et le poids dominant de ce dernier est $i(\lambda)$. Tout G-module rationnel cyclique engendré par une droite B-invariante de poids $i(\lambda)$ est un quotient de E_λ^* et toute image de E_λ^* dans un G-module par un G-morphisme est de ce type. Si M est un H-module, on a les relations

(1) $$(M \otimes E_\lambda)^H = \mathrm{Hom}_H(E_\lambda^*, M) = \mathrm{Hom}_G(E_\lambda^*, I_H^G M)$$

d'où en particulier,

(2) $$E_\lambda^H = \mathrm{Hom}_G(E_\lambda^*, k[G]^H).$$

4. SOUS-GROUPES EXEMPTS DE MULTIPLICITÉS. — On dira que H est exempt de multiplicités dans G si pour tout module principal E_λ, l'espace $\mathrm{Hom}_G(E_\lambda^*, k[G]^H)$, ou ce qui revient au même E_λ^H, est de dimension $\leqq 1$. Si $p=0$, les G-modules principaux sont toutes les représentations irréductibles de G, à équivalence près, et notre condition revient à demander que chacune intervienne au plus une fois dans $k[G]^H$. Les sous-groupes épimorphiques sont évidemment exempts de multiplicités.

Rappelons que H est dit sphérique dans G si B possède une orbite ouverte dans G/H ou, ce qui revient au même, si H a une orbite ouverte dans G/B. Le critère suivant est dû à M. Brion lorsque G est réductif et $p=0$.

PROPOSITION 2. — H *est exempt de multiplicités si et seulement si son enveloppe observable est sphérique dans* G.

5. On dira qu'un plongement affine ou projectif d'une variété irréductible X a un petit bord si le complément de l'image de X dans son adhérence de Zariski est partout de codimension $\geqq 2$. Il est clair que si G/H admet un prolongement projectif à petit bord, H est épimorphique dans G.

Supposons H observable dans G. Alors, d'après [3], G/H admet un plongement affine à petit bord si et seulement si $k[G]^H$ est de type fini. En utilisant ce fait et la proposition 3.2, p. 97 de [15], on montre :

PROPOSITION 3. — *Supposons que* $k[G]^H$ *soit de type fini et que* H *ait la propriété* F *dans son enveloppe observable* L [1], *n*° 3. *Alors, pour tout* H-*module rationnel* E *de dimension finie,* $I_H^G E$ *est un module de type fini sur* $k[G]^H$.

Par suite, si les sous-groupes épimorphiques ont la propriété F, alors il est vrai plus généralement que si $k[G]^H$ est de type fini, $I_H^G E$ est de type fini sur $k[G]^H$ pour tout H-module rationnel E de dimension finie.

6. PROPRIÉTÉ FG ET COMPACTIFICATIONS. — On dira que H *a la propriété* FG *dans* G si $k[G/H]$ est de type fini et si pour tout caractère $\lambda \in X(H)$ non nul l'anneau

$I_H^G \mathbb{N}\lambda := \bigoplus_{n \geqq 0} I_H^G n\lambda$ est de type fini. En développant par rapport à n, on a

(1) $$I_H^G \mathbb{N}\lambda = k[G]^H \oplus I_H^G \lambda \oplus I_H^G 2\lambda \oplus \ldots \oplus I_H^G n\lambda \oplus \ldots$$

ce qui entraîne immédiatement le :

Lemme 1. — *Si* H *a la propriété* FG, *alors le module* $I_H^G \lambda$ *est de type fini sur* $k[G]^H$, $(\lambda \in X(H))$.

D'après un théorème de Chevalley ([2], 5.1), il existe un G-module rationnel V de dimension finie et une droite D de V tels que H (resp. l'algèbre de Lie de H) soit le stabilisateur de D dans G (resp. l'algèbre de Lie de G). Soit λ le caractère de H dans D. En passant à l'espace projectif associé $\mathbb{P}(V)$, on obtient un plongement G-équivariant de G/H, que nous appellerons un *plongement de Chevalley*. Les sections hyperplanes se remontent sur G/H en sections du fibré associé à λ, donc $\lambda \in \Lambda_H^G$.

Théorème 1. — *Supposons* H *épimorphique dans* G. (i) *Si* H *a la propriété* FG, *alors* G/H *possède un plongement projectif normal à petit bord.* (ii) *Si* G/H *admet un plongement projectif à petit bord, alors* H *a la propriété* F *dans* G.

Equisse de démonstration. — (i) Soit $f: G/H \to \mathbb{P}(V)$ un plongement de Chevalley. Le faisceau inversible $\mathscr{L}$ des germes de sections du fibré défini par le caractère de H associé au plongement est l'image réciproque de $\mathcal{O}(1)$. Appelons $A = \bigoplus_{n \in \mathbb{Z}} \Gamma(G/H, \mathscr{L}^n)$ l'anneau des sections de $\mathscr{L}$. Puisque H est épimorphique et que $\dim \Gamma(G/H, \mathscr{L}^n) \geqq 2$ pour $n \geqq 1$, il résulte du lemme de [1] que $\Gamma(G/H, \mathscr{L}^n) = 0$ si $n < 0$. L'anneau A est gradué par $\mathbb{N}$, normal et finiment engendré puisque H a la propriété FG. Le noyau H_λ de λ est observable et $\tilde{X} = \operatorname{Spec} A$ fournit un plongement affine normal G-équivariant de G/H_λ. Comme $A = k[G/H_\lambda]$, il a un petit bord et il s'ensuit que $X := \operatorname{Proj} A$ donne un plongement projectif normal G-équivariant de G/H à petit bord.

(ii) Soit $f: G/H \to \mathbb{P}(V)$ un plongement projectif G-équivariant de G/H. Quitte à le remplacer par sa normalisation, on peut supposer l'adhérence X de $f(G/H)$ normale. Soit M un H-module rationnel de dimension finie. Il donne lieu à un fibré homogène de fibre type M sur G/H et le faisceau $\mathscr{M}$ de ses germes de sections régulières est localement libre. Comme le plongement f est à petit bord, la proposition 3.2, p. 97 de [15] déjà citée montre que l'image directe $f_*(\mathscr{M})$ de $\mathscr{M}$ est encore cohérente, donc

$$I_H^G M = \Gamma(G/H, \mathscr{M}) = \Gamma(X, f_* \mathscr{M})$$

est de dimension finie.

7. (*a*) Supposons G réductif et $p=0$. On peut alors montrer qu'un sous-groupe sphérique a la propriété FG. Le théorème 1 (i) reste donc valable sous ces hypothèses si l'on y remplace « a la propriété FG » par « est sphérique ». Ce résultat a d'abord été prouvé par M. Brion, et sa démonstration a suggéré la nôtre.

Rappelons que la complexité de H dans G est le minimum des codimensions des orbites de B dans G/H. Les sous-groupes sphériques sont donc de complexité zéro. En utilisant un résultat de F. Knop [8], on peut montrer plus généralement qu'un sous-groupe de complexité $\leqq 1$ a la propriété FG.

(*b*) M. Nagata [11] a prouvé qu'un groupe N commutatif unipotent de dimension 13, plongé génériquement dans le produit $G = (SL_2)^{16}$, est tel que l'algèbre $k[G]^N$ n'est pas finiment engendrée. Soit S le tore de dimension un diagonal dans G; il normalise N. Posons $H = N.S$. On voit facilement que H est épimorphique dans G et a complexité 2.

Il n'a pas la propriété FG. Néanmoins, à l'aide de la description explicite des invariants donnée dans [11] et de la proposition 3 (i) de [1], on montre que H a la propriété F.

8. Nous notons F_λ le H-module principal de poids dominant $\lambda \in -\Lambda_{B_H}^H$ et posons $F_\lambda = \{0\}$ si $\lambda \notin -\Lambda_{B_H}^H$. Pour toute partie Λ de $X(B_H)$, on désigne par F_Λ la somme directe des F_λ $(\lambda \in \Lambda)$. C'est une sous-algèbre de $k[H]$ si Λ est un sous-semigroupe.

Lemme 2. — *Soit* Λ *un sous-semigroupe de* $-\Lambda_{B_H}^H$. *Alors* F_Λ *est une algèbre de type fini si et seulement si* Λ *est de type fini. L'algèbre* F_Λ *est de type fini si* Λ *est un sous-groupe de* $X(B_H)$.

La première assertion se déduit facilement de la relation

$$F_\lambda . F_\mu = F_{\lambda+\mu}, \qquad (\lambda, \mu \in -\Lambda_{B_H}^H) \tag{3}$$

qui résulte de 4.21 dans [7], II, appliqué à H/U. Si Λ est un sous-groupe de $X(B_H)$, on a évidemment $F_\Lambda = F_\Psi$, où $\Psi = \Lambda \cap -\Lambda_{B_H}^H$. Comme $\Lambda_{B_H}^H$ est un cône de type fini, Ψ est de type fini et on est ramené au premier cas.

Définition. — Le sous-groupe H *a la propriété* SFG *dans* G si $I_H^G F_\Lambda$ est de type fini pour tout sous-groupe Λ de $X(B_H)$.

Théorème 2. — *Soit* $V = \mathcal{R}_u B_H$. *Les quatre conditions suivantes sont équivalentes :*
(i) $k[G]^V$ *est de type fini;* (ii) $k[G]^U$ *est de type fini*; (iii) H *a la propriété* SFG *dans* G; (iv) B_H *a la propriété* SFG *dans* G. *Si elles sont satisfaites,* $\Lambda_{B_H}^G$ *est de type fini.*

Esquisse de démonstration. — Le groupe réductif H/U opère par translations à droite sur $k[G]^U$. Comme $k[G]^V = (k[G]^U)^{V/U}$ et que V/U est unipotent maximal dans H/U, l'implication (i) ⇒ (ii) résulte du théorème 9 de [6]. Que (ii) ⇒ (iii) suit du lemme 2, de (*) et de la relation

$$I_H^G F_\Lambda = (k[G] \otimes F_\Lambda)^H = (k[G]^U \otimes F_\Lambda)^{H/U}. \tag{4}$$

L'équivalence de (iii) et (iv) résulte des définitions et de la transitivité de l'induction. L'implication (iv) ⇒ (i) est conséquence de l'égalité

$$k[G]^V = \bigoplus_{\lambda \in X(B_H)} I_V^G \lambda.$$

L'algèbre $k[G]^V$ est la somme directe des $I_H^G F_\lambda$, $(\lambda \in -\Lambda_{B_H}^G)$. Si elle est de type fini, il existe une partie finie Ψ de $-\Lambda_{B_H}^H$ telle que $I_H^G F_\Psi$ engendre $k[G]^V$. Alors Ψ engendre $-\Lambda_{B_H}^G$.

Corollaire 3. — *On suppose que* H *a la propriété* SFG. *Soit* Λ *un sous-semigroupe de type fini de* $-\Lambda_{B_H}^G$. *Alors* $k[G]^H$ *et* $I_H^G F_\Lambda$ *sont de type fini.*

Comme F_Λ est de type fini (lemme 2), cela résulte de (4).

Remarque. — Le corollaire montre que SFG implique FG (donc aussi F si H est supposé épimorphique). Nous ne connaissons pas d'exemple de sous-groupe satisfaisant à FG mais pas à SFG. Ces deux propriétés sont visiblement équivalentes si les tores maximaux de H sont de dimension un.

9. Remarques. — (i) Si H est réductif, il a la propriété SFG. Si H est unipotent, la propriété SFG équivaut à : $k[G]^H$ est de type fini.

(ii) En utilisant (*) et le transfert, on voit que si H a la propriété SFG dans un sous-groupe fermé réductif connexe de G, alors H a la propriété SFG dans G.

(iii) D'après [4], $k[G]^N$ est de type fini si N est le radical unipotent d'un sous-groupe parabolique. Tenant compte de (ii), on voit que si U est le radical unipotent d'un sous-groupe parabolique d'un sous-groupe réductif de G, alors H a la propriété SFG dans G.

(iv) Supposons G réductif. Rappelons que K. Pommerening et V. L. Popov ont conjecturé que $k[G]^N$ est de type fini si N est unipotent connexe et normalisé par un tore maximal de G ([5]; [13]). Vu [4], cela est en tout cas vrai si N est le radical unipotent d'un sous-groupe parabolique. Cette conjecture a aussi été vérifiée dans le groupe exceptionnel $\mathbb{G}_2$ par F. Grosshans et Lin Tan ([5], [16]). Grosshans nous a fait remarquer qu'il suffit de la démontrer en supposant H = T.N épimorphique, comme on le voit en passant à l'enveloppe observable.

10. Soient (σ, M) un H-module rationnel de dimension finie et Λ un sous-semigroupe de type fini de $-\Lambda_H^G$. Alors $I_H^G(F_\Lambda \otimes M)$ est un module sur l'algèbre $I_H^G F_\Lambda$ de façon naturelle. Plus généralement, si A est une G-algèbre rationnelle, $(A \otimes F_\Lambda \otimes M)^H$ est un module sur $(A \otimes F_\Lambda)^H$.

Proposition 4. – *Supposons que* G *soit réductif*, A *de type fini et que* H *ait la propriété* SFG. *Alors* $(A \otimes F_\Lambda \otimes M)^H$ *est un module de type fini sur* $(A \otimes F_\Lambda)^H$. *En particulier* $I_H^G(F_\Lambda \otimes M)$ *est un module de type fini sur* $I_H^G F_\Lambda$.

Il suffit de l'établir pour M irréductible, donc en particulier quand U agit trivialement sur M. Dans ce cas, on part de la relation

$$(A \otimes F_\Lambda \otimes k[M^*])^H = (A^U \otimes F_\Lambda \otimes k[M^*])^{H/U}$$

qui, vu (∗) et le corollaire 1, montre que le membre de gauche est une algèbre de type fini et on utilise un argument de [9] : Soit $k[M^*]_i$ l'espace des polynômes homogènes de degré i sur M^* ($i \in \mathbb{N}$). Il est égal à k (resp. M) si $i=0$ (resp. $i=1$), d'où la relation

$$(A \otimes F_\Lambda \otimes k[M^*])^H = (A \otimes F_\Lambda)^H \oplus (A \otimes F_\Lambda \otimes M)^H \oplus \bigoplus_{i \geqq 2} (A \otimes F_\Lambda \otimes k[M^*]_i)^H,$$

qui implique facilement la proposition.

Note remise le 12 octobre 1992, acceptée le 20 octobre 1992.

Références bibliographiques

[1] F. Bien et A. Borel, Sous-groupes épimorphiques de groupes algébriques linéaires I, *C. R. Acad. Sci. Paris*, 315, série I, 1992, p. 649-653.

[2] A. Borel, *Linear algebraic groups*, 2nd enlarged edition, G.T.M., 126, Springer-Verlag, 1991.

[3] F. D. Grosshans, Observable groups and Hilbert's fourteenth problem, *Amer. J. Math.*, 95, 1973, p. 229-253.

[4] F. D. Grosshans, The invariant of unipotent radicals of parabolic subgroups, *Invent. Math.*, 73, 1983, p. 1-9.

[5] F. D. Grosshans, Hilbert's fourteenth problem for non-reductive groups. *Math. Zeitschrift*, 193, 1986, p. 95-103.

[6] F. D. Grosshans, Contractions of the actions of reductive algebraic groups in arbitrary characteristic, *Invent. Math.*, 107, 1992, p. 127-133.

[7] J. C. Jantzen, *Representations of algebraic groups*, Pure and applied math., 131, Academic Press, 1987.

[8] F. Knop, *Ueber Hilberts vierzehntes Problem für Varietäten mit Kompliziertheit eins*, preprint 1991.

[9] H. Kraft, *Geometrische Methoden in der Invariantentheorie*, Aspekte der Mathematic Viehweg, 1984.

[10] D. Mumford et J. Fogarty, *Geometric Invariant Theory*, Ergebn. d. Math. u.i. Grenzgeb., 34, Springer-Verlag, 1992.

[11] M. Nagata, On the fourteenth problem of Hilbert, *Proc. I.C.M.*, 1958, Cambridge University Press, 1960, p. 459-462.

[12] K. Pommerening, Observable radizielle Untergruppen von halbeinfachen algebraischen Gruppen, *Math. Zeitschrift*, 165, 1979, p. 243-250.

[13] K. Pommerening, Invarianten unipotenter Gruppen, *Math. Zeitschrift*, 176, 1981, p. 359-374.

[14] V. L. Popov, Contraction of the actions of reductive algebraic groups, *Math. U.S.S.R. Sbornik Sb.*, (2), 58, 1987, p. 311-335.

[15] *Séminaire de Géométrie algébrique du Bois-Marie*, 2, 1962, Advanced Studies in Pure Math., 2, North-Holland, Amsterdam, Masson, Paris.

[16] L. Tan, On the Popov-Pommerening conjecture for groups of type G_2, *Algebras, Groups and Geometries*, 5, 1988, p. 421-432.

F. B. : *Department of Mathematics, University of Utah, Salt Lake City, UT* 84112, *E.U.*;

A. B. : *Institute for Advanced Study, Princeton, NJ* 08540, *E.U.*

149.

Réponse à la remise du prix Balzan 1992

Monsieur le Président,
Messieurs les membres de la Fondation Balzan,
Mesdames et Messieurs,

Il y a bientôt trente-cinq ans, le Président de l'Union Internationale des Mathématiciens, chargé de remettre deux médailles Fields, rappela tout d'abord qu'elles sont attribuées traditionnellement à des mathématiciens jeunes, en reconnaissance de leurs travaux, bien sûr, mais aussi pour les encourager à de futurs accomplissements. Il ajouta ensuite qu'un ami avait remarqué que dans la situation actuelle des mathématiques, ce sont les vieux plutôt que les jeunes qui ont besoin d'encouragement! A mesure que les ans passent, cette remarque me paraît de plus en plus pertinente, aussi est-ce à un double titre que j'apprécie très profondément le grand honneur qui m'est fait. Certes, il constitue un précieux témoignage d'estime scientifique pour mes travaux mais je veux y voir aussi un encouragement à les poursuivre plutôt qu'à me borner à les contempler avec plus ou moins de complaisance et satisfaction. Je voudrais aussi partager cette distinction avec la discipline quelque peu mystérieuse à laquelle ils appartiennent, la ou les mathématiques pures.

La mathématique est une construction intellectuelle gigantesque, difficile, sinon impossible, à cerner dans son ensemble. Quelquefois j'aime à la comparer à un iceberg, trouvant que, comme ce dernier, elle comprend une petite partie visible et une grande partie invisible. Par visible j'entends ce qui sert dans le monde extérieur, en technologie, physique, sciences naturelles, astronomie, ordinateurs, etc., et dont l'utilité et la justification sociale sont hors de doute. Les problèmes pratiques ont sans doute été dans l'antiquité à l'origine même des mathématiques. Cependant, à mesure que cette discipline se développait, elle acquerrait une vie autonome et les mathématiciens se sont de plus en plus intéressés à des problèmes purement mathématiques, sans se préoccuper nécessairement d'applications en dehors des mathématiques elles-mêmes. C'est la partie invisible, c'est-à-dire invisible ou en tout cas difficilement saisissable pour le non mathématicien, la mathématique pure.

Ce n'est pas à dire que ces recherches n'auront pas d'applications externes, que l'invisible ne deviendra pas visible. L'expérience montre au contraire, et de plus en plus fréquemment, que même les parties les plus abstraites des mathématiques en trouvent une fois ou l'autre, souvent totalement imprévues. Mais elles ne préoccupent pas en général le mathématicien pur, qui opère dans un monde de formes intellectuelles ayant ses lois propres, ses directives internes, souvent guidé par des critères de nature esthétique. Dans le climat économique actuel, il est facile pour les organismes chargés de financer la recherche d'ignorer ou de faire peu de cas de cette spéculation intellectuelle apparemment gratuite qui, pour autant même qu'elle soit perçue, semble être un luxe, pour donner la priorité à la partie visible, appliquée, dont on peut attendre un gain concret à brève échéance. Aussi, en tant que mathématicien pur, je suis reconnaissant à la Fondation Balzan pour l'intérêt constant qu'elle porte à une discipline si ésotérique, si peu visible, si fermée au non professionnel, intérêt marqué non seulement par le prix que je reçois aujourd'hui, mais aussi par ceux attribués à mes deux prédécesseurs en mathématique, et amis de longue date, Enrico Bombieri et Jean-Pierre Serre.

Mon activité a été et est pour moi aussi bien ma profession que mon hobby de prédilection. Ses directions successives, le choix des questions à étudier, ont été influencés par ces deux points de vue, du reste très souvent confondus. A maintes reprises, j'ai été guidé par un sens de l'architecture de cet édifice auquel nous ajoutons incessamment des étages ou des ailes, tout en rénovant parfois les parties déjà construites, par le sentiment que certains problèmes devaient être étudiés en priorité, pour ouvrir de nouvelles perspectives ou établir une base pour des construction futures. C'est là le point de vue professionnel, mais là souvent, on peut dire que l'utile se joignait à l'agréable, car ces problèmes étaient justement ceux qui m'attiraient le plus. D'autres fois, cependant, je n'aurais pu invoquer de tels motifs, étant mû tout simplement par la curiosité, le besoin de connaître la réponse à une énigme, qu'elle parût importante ou non dans un contexte plus général.

Il y avait tout de même un espoir sous-jacent que l'ensemble de ces travaux constituerait une contribution utile aux mathématiques. En me décernant ce prix, le Comité des Prix affirme qu'il en est bien ainsi et aussi que leur domaine est important. Venant d'un Comité qui examine et reconnaît des contributions si diverses à tant d'aspects de la culture et des sciences, une telle assurance est pour moi, je ne le cacherai pas, une source de profonde satisfaction.

150.

Quelques réflexions sur les mathématiques en général et la théorie des groupes en particulier

Orientamenti et Attività dei premi Balzan 1992, 3–11,
Fondazione Internazionale Balzan, Milan

En feuilletant les *Orientamenti e attività* des lauréats Balzan des deux années précédentes, je ne peux m'empêcher d'envier leurs auteurs. Chacun, quel que soit son domaine, peut entrer directement *in medias res* et se référer à des notions, faits, oeuvres ou personnalités qu'il estime être en droit de supposer connus, afin de situer ou décrire ses propres contributions. Il n'en est hélas pas de même pour le mathématicien "pur" que je suis. Pratiquement rien de ce qui a été fait en mathématiques durant ces deux ou trois derniers siècles n'appartient à la culture de l'"honnête homme", et peut me servir de référence ou de point de comparaison, même vague.

Par exemple, G. Ligeti (prix Balzan 1991) peut sans autre forme de procès faire allusion à Beethoven, Bach, Bartók, Schönberg et à leurs oeuvres et admettre que ses lecteurs, et pas seulement des musiciens professionnels, savent de qui et de quoi il parle. Mais si j'affirme, disons, que certains travaux de H. Weyl et E. Cartan sur les groupes et algèbres de Lie semi-simples ont joué initialement un rôle important dans la genèse des miens, cela n'aura de signification que pour des mathématiciens professionnels engagés dans la recherche, et même pas pour tous, en fait.

Ma tâche aurait sans doute été plus simple à la fin du 18ème siècle. Les grands progrès des mathématiques se centraient alors principalement sur l'analyse et ses applications à des problèmes en mécanique, astronomie, etc., qui les avaient en partie motivés et qu'ils avaient permis de résoudre. En insistant sur ces applications, j'aurais pû en donner une idée, sans devoir m'engager dans une description du calcul infinitésimal ou du calcul des variations. Peut-être aurais-je été récompensé pour des travaux non pas en analyse, mais en théorie des nombres ou en géométrie. Ces deux domaines, surtout le premier, étant déjà plus internes aux mathématiques, ma tâche aurait été plus difficile, mais tout de même pas inabordable: après tout, les nombres entiers et les figures géométriques nous sont familiers, et leur étude remonte aux Grecs, ce qui leur donne déjà une certaine respectabilité.

Mais les mathématiques ont depuis lors pris un essor prodigieux, procédant quelquefois par des sauts brusques, introduisant des concepts de plus en plus abstraits, de plus en plus éloignés de l'intuition quotidienne, qui n'ont trouvé d'écho guère que chez les spécialistes. Mon propre point de départ était déjà le résultat d'un formidable empilement d'abstractions successives, et ce processus n'a fait que s'accélérer depuis. Ma tâche serait en somme de sauter par dessus tous ces développements pour donner une idée de travaux récents en termes relativement familiers il y a deux siècles déjà; une entreprise intéressante, peut-être, mais quelques pages ne sauraient y suffire. Aussi cet essai se bornera à quelques remarques sur les mathématiques et n'abordera qu'incidemment mes contributions.

La prise de conscience de la mathématique comme une activité autonome, non nécessairement liée à des applications au monde extérieur, a été progressive. Peu à peu, les mathématiciens se sont rendu compte qu'ils édifiaient une réalité intellectuelle, sans doute issue initialement de problèmes pratiques, mais de plus en plus indépendante, qu'ils pouvaient légitimement explorer en elle-même. Cette évolution n'a pas toujours été aisée. On peut déjà en voir un exemple dans les difficultés rencontrées en cherchant à donner un sens aux racines carrées de nombres réels négatifs. Durant la Renaissance, des mathématiciens italiens étaient parvenus à exprimer les solutions d'une équation algébrique de degré trois (ou même quatre) par de formules générales portant sur les coefficients de l'équation et faisant intervenir des racines carrées ou cubiques. Dans un cas, appelé *casus irreducibilis,* ces formules contenaient des racines carrées de nombres réels négatifs, ce qui, à première vue était complètement absurde, puisque la règle des signes implique que le carré d'un nombre réel positif ou négatif est toujours positif. Descartes, par exemple, ne voulait pas en entendre parler. Mais on réalisa cependant que si l'on maniait ces symboles en suivant certaines règles formelles, on trouvait effectivement trois solutions réelles. On passait donc de nombres réels (les coefficients) à d'autres nombres réels (les solutions) par l'intermédiaire de nombres "imaginaires", "non existants", "impossibles". Il a fallu plus de deux siècles pour résoudre ce paradoxe apparent. La solution a été de définir un système de nombres plus généraux, les nombres complexes, contenant à la fois les nombres réels et les racines carrées des nombres négatifs. Ces nombres complexes sont représentés par des paires de nombres réels ou, pour parler géométriquement, par les points d'un plan, entre lesquels on définit des opérations analogues, et généralisant, l'addition et la multiplication. On a donc affaire ici à un objet intellectuel, qui généralise et inclut une notion qui semblait enracinée dans le monde qui nous entoure, à savoir les nombres réels.

Une fois acceptés, les nombres complexes, qui s'étaient introduits à propos d'un problème d'algèbre, pénétrèrent peu à peu dans d'autres domaines, en particulier en analyse, où se développa une théorie des fonctions d'une variable complexe. Là aussi, on trouve des exemples de propriétés de fonctions de variables réelles qui ne peuvent se comprendre qu'en passant par l'intermédiaire du complexe, même s'il disparaît de l'énoncé final. Tout cela semblait bien théorique, cependant, et en fait, dans la première partie du 19ème siècle, le mathématicien C.F. Gauss estimait cette théorie justifiée, voire indispensable, en premier lieu au nom de l'harmonie interne des mathématiques. Même lui ne prévoyait pas qu'elle deviendrait fondamentale pour l'ingénieur, en électricité par exemple et ensuite dans bien d'autres domaines.

Une difficulté apparentée se rencontra, même à un plus haut degré, en géométrie. En principe, le but de cette dernière était de rendre compte des propriétés de figures géométriques de l'espace, et il était accepté que sa base était la géométrie euclidienne, décrite par une axiomatique issue de la réalité et la reflétant. Quand certains mathématiciens construisirent des géométries non-euclidiennes, i.e. dans lesquelles un des axiomes d'Euclide (celui des parallèles) n'était pas satisfait, ils se sont heurtés à des réserves mathématiques et philosophiques. Gauss a lui-même affirmé avoir été en possession d'une telle théorie pendant trente ans sans oser la publier "par peur des Béotiens". Une telle géométrie, ne correspondant pas à la réalité, courait le risque d'être qualifiée de jeu gratuit sans portée, indigne d'être appelé géométrie. Il a fallu bien des années pour que l'on en arrive à considérer ces géométries comme des constructions mentales aussi valables l'une que l'autre, qu'elles s'appliquent au monde où nous vivons ou non; un point de vue qui s'imposa définitivement lorsqu'il fut établi qu'elles sont logiquement cohérentes si et seulement si la géométrie euclidienne l'est.

Ce sens de l'autonomie des mathématiques pures n'a fait que s'accentuer au cours des années, aussi la légitimité de celle-ci n'est plus mise en doute, même si elle reste si mystérieuse au non-mathématicien, cela d'autant plus que l'on a constaté à maintes et maintes reprises que des théories développées en mathématique pure pour des raisons internes à la discipline ont trouvé, quelquefois bien plus tard, des applications aussi imprévues que fondamentales aux sciences naturelles ou à la technologie. Cela ne veut pas du tout dire que la mathématique appliquée, i.e. visant dès le départ des applications pratiques, soit pour autant négligée. Bien au contraire elle aussi connaît un développement spectaculaire, aidé par celui des ordinateurs. Mais mon activité se situe dans la première.

La notion de groupe, qui joue un rôle central dans mes travaux, est née vers 1830, dans un mémoire d'un mathématicien de vingt ans, Evariste Galois. Il s'agissait, entre autres, de montrer qu'il n'existait pas de formule universelle exprimant les solutions d'une équation algébrique générale de degré au moins cinq à l'aide de radicaux portant sur les coefficients, contrairement à ce qui était connu pour les degrés plus petits, comme je l'ai rappelé plus haut. Galois a fait la découverte extraordinaire que la différence entre ces deux cas tenait à la structure d'un ensemble de permutations des solutions de l'équation, appelé depuis son groupe de Galois. On dit qu'un ensemble de transformations d'une configuration mathématique est un groupe si cet ensemble contient l'inverse de ses éléments et le composé de deux éléments. On s'est aperçu peu à peu qu'il y avait des groupes de transformations un peu partout en mathématiques et que la structure de ces groupes, considérés en eux-mêmes, était souvent la clef de propriétés des objets transformés, indépendamment de leur nature. L'étude des symétries d'une configuration mathématique se ramène à celle du groupe des transformations la laissant invariante. Par exemple, la géométrie euclidienne traduit essentiellement les propriétés du groupe des déplacements (rotations, translations) de l'espace euclidien. En fait, en 1872, le mathématicien F. Klein formula un vaste programme (Erlanger Programm) qui ramenait en principe la géométrie, dans un sens très général, à la théorie des groupes.

Les groupes de Galois étaient formés d'un nombre fini d'éléments, mais on rencontra dans la suite des groupes infinis. Dans certains, comme le groupe des déplacements de l'espace, les éléments dépendent de certains paramètres variant de façon continue (par exemple axe et angle pour une rotation). Le mathématicien norvégien Sophus Lie jeta les bases d'une théorie de tels groupes, appelés plus tard groupes de Lie, à partir de 1873. Son but initial était de développer une théorie analogue à celle de Galois, mais pour les équations différentielles. Il faut bien dire qu'elle n'eut à cet égard qu'un succès limité, mais plus que compensé par ceux qu'elle a connus ailleurs. En particulier, il était clair dès le début que ce serait un outil précieux dans le programme d'Erlangen de Klein, et S. Lie lui-même l'utilisa abondamment non seulement en analyse, mais aussi en géométrie.

La théorie de Lie elle-même établissait une équivalence entre des problèmes d'analyse et d'algèbre. Par souci d'exactitude, je devrais ajouter que le point de vue de Lie était en partie local; il y était contraint par les outils d'analyse dont il disposait. Ses groupes étaient des "groupes locaux" dans la terminologie actuelle, mais il est bien inutile que j'entre dans ces distinctions: dès que les concepts adéquats furent introduits, la théorie devint globale, principalement sous l'impulsion de H. Weyl et E. Cartan, et c'est ainsi qu'elle se présentait quand j'en abordai l'étude.

Elle avait aussi connu un changement conceptuel: le passage aux groupes "abstraits". A l'origine, pour Galois comme pour Klein ou Lie, les "groupes" étaient des groupes de transformations d'objets mathématiques préexistants. Mais on voulait déduire des propriétés de ces derniers par l'examen de la structure d'un groupe en soi, en faisant plus ou moins abstraction de la nature des objets transformés. Dans un tel cas, la tendance du mathématicien est de définir directement un objet mathématique ne mettant en jeu que la structure à étudier. Cela mena à la notion de groupe abstrait: un ensemble d'éléments entre lesquels est seulement définie une loi de composition satisfaisant à certains axiomes. Cette transition représentait justement un de ces sauts vers une plus grande abstraction auxquels je faisais allusion plus haut, et elle rencontra une certaine résistance. Même F. Klein exprima des réserves. Pour lui, la réalité mathématique intéressante était formée des objets transformés. En faire abstraction pouvait mener à un exercice mental peut-être subtil mais sans valeur, à lâcher la proie pour l'ombre. En fait, les groupes abstraits sont devenus peu à peu eux aussi des objets à part entière de la réalité mathématique, et forment depuis longtemps le point de départ de tout exposé de la théorie des groupes. Du reste, un des succès les plus impressionnants de ces trente dernières années en mathématiques est la solution d'un problème en théorie des groupes abstraits qui a paru longtemps quasiment inaccessible: la classification des groupes finis simples.

En 1912, l'année de sa mort, Henri Poincaré écrivait (dans un rapport sur les travaux de E. Cartan): "la théorie des groupes est, pour ainsi dire, la Mathématique entière, dépouillée de sa matière et réduite à une forme pure. Cet extrême degré d'abstraction a rendu mon exposé un peu aride; pour faire apprécier chacun des résultats, il m'aurait fallu pour ainsi dire lui restituer la matière dont il avait été dépouillé; mais cette restitution peut se faire de mille façons différentes; et c'est cette forme unique que l'on retrouve ainsi sous une foule de vêtements divers, qui constitue le lien commun entre des théories mathématiques qu'on s'étonne souvent de trouver si voisines."

Aujourd'hui, on ne songerait plus guère à parler d'un degré extrême d'abstraction, tant la notion de groupe est devenue fondamentale et usuelle en mathématique. Mais cette citation de Poincaré met par ailleurs en évidence deux aspects complémentaires de l'activité mathématique. La matière première de nos travaux consiste en une myriade de problèmes, souvent très particuliers, et notre tâche est de les résoudre. Mais cela ne suffit pas. On n'est pas satisfait tant que l'on n'a pas trouvé le domaine naturel de validité d'un raisonnement, d'une preuve ou d'une conclusion. La résolution des problèmes est accompagnée d'un effort constant pour développer des théories et méthodes générales permettant de déduire beaucoup de phénomènes particuliers à partir de principes généraux. Les deux sont indispensables. La mathématique ne se réduit ni à une collection de problèmes isolés, ni à un ensemble de théories axiomatiques vides d'applications.

Les groupes de Lie, ou algébriques, ou arithmétiques, présentent une combinaison de ces deux aspects des mathématiques qui m'a irrésistiblement attiré. Les groupes de Lie se construisent à partir de groupes dits simples. Ces derniers se divisent en quatre classes infinies de groupes dits classiques et cinq groupes exceptionnels. Les premiers se rencontrent en beaucoup d'endroits et étaient connus avant même que les groupes de Lie aient été définis. Souvent, des propriétés que l'on espère communes se vérifient d'abord pour les groupes classiques, en utilisant le contexte de leur définition. Même si on parvient à les établir pour les cinq groupes exceptionnels, ce qui s'avère quelquefois plus difficile, la tâche n'est pas vraiment terminée tant que l'on n'a pas trouvé une raison de portée générale, une démonstration valable directement pour un groupe de Lie simple général. Presque invariablement, un tel argument jette de la lumière même sur les cas classiques qui paraissaient si familiers et si bien compris, et sera la base de nouveaux progrès.

Cela s'est avéré être le cas non seulement pour les groupes de Lie, mais aussi pour un grand nombre de leurs espaces homogènes (i.e. espaces sur lesquels le groupe opère de manière à ce que tous leurs points sont équivalents, en ce sens qu'étant donnés deux d'entre eux, il existe une transformation du groupe amenant l'un sur l'autre), dont beaucoup étaient familiers en géométrie ou géométrie algébrique, antérieurement à la naissance de la théorie des groupes de Lie. Les groupes de Lie ou algébriques et leurs espaces homogènes interviennent dans tellement de domaines des mathématiques que Jean Dieudonné a été amené à écrire: “Les groupes de Lie sont devenus le centre des mathématiques; on ne peut rien faire de sérieux sans eux”. Malgré mon intérêt pour eux, je n'irai tout de même pas aussi loin. Les mathématiques sont bien trop complexes, me semble-t-il, pour n'avoir qu'un seul centre. Les liens entre différents sujets des mathématiques forment un réseau extrêmement compliqué. Plus nombreux sont les liens aboutissant à un sujet donné, ou en partant, plus central apparaît le rôle joué par ce dernier. Je ne peux bien entendu que m'associer à Dieudonné en mettant les groupes de Lie parmi ces domaines privilégiés. Je me permettrai même d'y ajouter deux autres jalons de mon activité, les groupes algébriques et les groupes arithmétiques.

Les groupes de Lie et certains de leurs espaces homogènes, en particulier ceux que E. Cartan a appelés symétriques, sont un point de rencontre d'analyse, algèbre, analyse harmonique, géométrie différentielle et topologie. La topologie est par définition l'étude des propriétés d'objets géométriques qui sont invariantes par déformation continue. La topologie algébrique est le domaine par excellence qui fournit des invariants permettant de distinguer entre des objets géométriques qui, localement, sont identiques (exemple le plus simple: un ruban ordinaire et un ruban de Möbius). Mes premiers travaux avaient largement pour but d'appliquer des méthodes très puissantes qui venaient d'être développées en topologie algébrique à l'étude de la topologie des groupes de Lie et espaces homogènes et d'en tirer des conséquences générales en topologie algébrique. Ces méthodes envahirent rapidement d'autres domaines, en particulier la géométrie algébrique, et il s'imposa de développer une théorie de groupes analogues aux groupes de Lie, mais dans le cadre de la géométrie algébrique: ce furent les groupes algébriques. Il s'avéra ensuite que les groupes de Lie ou algébriques, et certains groupes qu'ils contiennent, les groupes arithmétiques, fournissaient le domaine naturel pour une généralisation d'un des plus beaux ornements des mathématiques du 19ème siècle: la théorie des formes modulaires ou automorphes, édifiée par Klein et Poincaré. C'est devenu en effet le cadre d'une vaste synthèse, en cours, entre pratiquement tout de ce que je viens de mentionner et la théorie des nombres.

Dans le paragraphe précédent, je n'ai pu m'empêcher d'utiliser des termes familiers seulement aux mathématiciens et ai ainsi, j'en ai peur, lassé tout autre lecteur qui m'aurait patiemment suivi jusque-là. Mais tout ce que j'espère de ce déluge de termes mathématiques est seulement qu'il donne une idée du rôle, central en effet, joué par l'entité groupes de Lie-groupes algébriques-groupes arithmétiques dans une partie de plus en plus vaste de la mathématique.

C'est par là aussi un agent unificateur, comme Poincaré le dit déjà si élégamment dans le texte précité, ce qui m'amène à toucher un mot de la croyance, qui nous est chère, en l'unité profonde de la mathématique. La croissance vertigineuse de la production, la vision d'une discipline qui se développe dans tant de directions à la fois, dont les spécialistes souvent ne se comprennent pas, tant différentes sont leurs techniques, peut faire craindre que "la" mathématique n'éclate définitivement en plusieurs parties sans relations. Mais on voit périodiquement cette expansion effrénée contrebalancée par des contractions et simplifications: des murs tombent, des fossés se comblent, des fusions inattendues montrent que deux domaines jusque-là considérés comme très différents sont en fait liés pour des raisons profondes qui avaient échappé auparavant. Tout cela nous fait croire envers et contre tout en l'unité de la mathématique, et le rôle joué par mon domaine de prédilection dans ce processus d'unification en est pour moi un des grands attraits.

J'ai déjà mentionné le groupe des déplacements de l'espace euclidien comme un exemple de groupe de Lie. Le groupe de Lorentz en est un autre. Aussi n'est-il pas surprenant que certains groupes de Lie aient été des instruments d'échanges fructueux (de part et d'autre) entre les mathématiques et la physique théorique. Les relations entre ces deux disciplines ont connu de fortes fluctuations au cours de ce siècle, oscillant entre le mariage et l'indifférence, presque le divorce. Très fortes au début du siècle, dans le cadre de la relativité, puis vers les années trente dans celui de la mécanique quantique, elles déclinèrent graduellement dans la suite. D'une part, la mathématique pure prenait un envol extraordinaire, motivé presqu'uniquement de l'intérieur, qui mena à une série de succès spectaculaires. D'autre part, certaines théories physiques paraissaient avoir un caractère si provisoire qu'il ne semblait guère valoir la peine de chercher à les mettre sous forme mathématiquement rigoureuse. Un physicien me l'a dit une fois en s'aidant d'une comparaison avec la théorie de l'atome de Bohr: on savait bien que cette dernière était temporaire, contradictoire dans ses principes, et que l'on ne résoudrait pas ses difficultés simplement en améliorant la présentation mathématique; il fallait de nouvelles idées physiques. (Nous parlions de la physique des particules élémentaires. Il ajoutait aussi que d'autres sujets moins à la mode, dans la physique du solide par exemple, paraissaient par contre tout à fait mûrs pour une attaque par des mathématiciens.) Le contact s'est à nouveau resserré dans les années soixante, par l'intermédiaire de groupes de Lie, notamment par la découverte d'une nouvelle particule élémentaire (Ω^-), dont l'existence avait été conjecturée après examen d'une configuration mathématique qui illustrait, dans un cas très particulier, une théorie de E. Cartan datant de près d'un demi-siècle, et ensuite par la théorie des "quarks". Ces échanges ont plus récemment acquis une nouvelle intensité dans le cadre, en physique, de la théorie conforme des champs et de celle des cordes. Il s'agit là de contacts au plus haut niveau, mettant en jeu des sujets extrêmement difficiles, à la frontière de la recherche, et qui ouvrent des perspectives nouvelles de part et d'autre. Ce fut une grande surprise pour moi de voir quelques physiciens utiliser même des groupes de Lie exceptionnels en théorie des particules élémentaires, groupes dont je ne pouvais m'imaginer qu'ils attireraient l'attention en dehors de la mathématique pure. En ce qui concerne les groupes, cependant, ce sont surtout des variantes de dimension infinie des groupes (ou algèbres) de Lie: groupes ou algèbres de Kac-Moody, groupes de lacets, algèbre de Virasoro, qui jouent un rôle dans ces travaux.

Les mathématiques, que ce soit en elles-mêmes ou dans leurs contacts avec d'autres disciplines, me paraissent être dans un état florissant, et, j'en suis persuadé, ne feront que prospérer, pour autant toutefois qu'elles soient pratiquées par un nombre suffisant de mathématiciens doués, une condition bien évidemment nécessaire, mais dont on commence à craindre qu'elle ne reste pas toujours remplie, à moins que les perspectives d'avenir pour ceux qui voudraient y consacrer le meilleur de leur temps ne s'améliorent. Les tendances de certains organismes chargés d'organiser et de financer l'éducation supérieure et la recherche fondamentale ne paraissent guère rassurantes. De plus en plus, on insiste sur l'utilité, on entend dire que nous avons suffisamment cultivé la recherche fondamentale et qu'il est temps de se consacrer à ses applications. Si une telle politique est adoptée, on s'apercevra sans doute une fois ou l'autre qu'elle tarira la source même des applications pratiques que l'on cherche à favoriser; pas dans l'immédiat, certainement, car la recherche en mathématique possède un tel élan que rien ne pourra l'empêcher de continuer sur sa lancée pour un certain temps. Je pourrais donc me consoler en me disant que, si déclin il y a, je n'en serai pas témoin, mais ce serait une bien maigre consolation pour quelqu'un qui ne se lasse pas d'admirer la richesse et la beauté des mathématiques dans leur état actuel et est convaincu que celles à venir ne leur céderaient en rien.

151.

Some Recollections of Harish-Chandra

Current Science **65**, (1993) 919–920, Bangalore, India,
reproduced in *Current trends in mathematics and physics. A tribute to Harish-Chandra*,
Narosa Publishing House 1995, 210–215

1 I met Harish-Chandra for the first time in Summer 1949, in Zurich. He had come there to study German and maybe also to see W. Pauli, whom he had known while still a physicist. His conversion from theoretical physics to pure mathematics was fairly recent, so that he would still be asked for its reason. One answer, as I heard at the time, was that now he could believe in what he was doing. This should not be viewed as a disparaging comment on theoretical physics, quite the contrary. He had a very high regard for theoretical physics, but felt that a very special intuition, not mathematical, some sort of 'sixth sense', was needed to make progress in it. Later I heard him on various occasions stating that 'no mathematician has ever made a dent in physics' (this is a quote), which, I think, is too sweeping a statement, in particular unfair to H. Weyl. He could still dabble in physics, though. While we were walking along the lake around sunset, the reflection of light struck him as somewhat peculiar. The following morning, when I picked him up in his hotel room, I saw a sheet or two of paper full of computations, with some drawings of reflected light rays.

At that time he had already published some papers on Lie groups and Lie algebras. Since my main interests were Lie groups and topology there was a considerable overlap and it was natural for us to talk shop.

I still remember these contacts rather vividly since he had such a striking personality. I had hardly ever met someone thinking and speaking so fast, although later I realized that this feature is not so uncommon among my Indian friends, or anyone so intense, with such a technical proficiency and concentration power. His working habits, as he described them to me, were also somewhat out of the ordinary: he would usually get up rather late, then have breakfast, read the newspaper, so that he would really get going around ten or eleven, but then it would be 'straight to midnight'. In fact, it seems the habit of getting up rather late had been pretty much a constant in his life. Lily once told me that when he was living with his family, people would often sleep out of doors. In the morning everyone would get up and prepare for the daily work, but not Harish, who instead would go inside the house to sleep some more. His intense work, straight to midnight, may already have been taxing on him since he felt the need to stay away from it for about three months a year. However, contrary to getting up late, this habit which, I presume, would have been later welcomed by his family, was not maintained. The self-imposed pressure from his work became greater and greater, and the time allotted for vacation shrank to practically nothing, until overwork took its toll and the

---

Text of a talk given during the unveiling ceremony of Harish-Chandra's bust in Allahabad, 10 October 1993.

doctor prescribed one month of vacation per year. His visit to Zurich was indeed part of a trip, he was coming from Paris and spoke effusively of P. Gauguin, of whom there was a big exhibit at the time there, in particular of his marvelous sense of color.

As I already said, Lie groups were a common interest. However, Harish told me he felt he had worked enough in Lie groups and their representations and he wanted to do something else. He was planning to go to Harvard for at least a year, to study algebraic geometry with O. Zariski: 'this will be my next venture', as he put it. But it did not lead to a serious involvement, as it turned out. In fact, the next I heard of him was in Paris, in the following year, through letters of Chevalley to some friends, speaking of new extraordinary results on universal enveloping algebras and infinite dimensional representations. This heralded the famous Transactions AMS paper, for which Harish later received the Cole prize of the AMS, and the beginning of his 'long march', of about 25 years, to the explicit determination of the Plancherel measure for semisimple groups, in a long and amazing series of papers which displayed extraordinary technical power and resourcefulness. Later, I heard several times André Weil say that he knew only two mathematicians for whom technical difficulties simply did not exist, namely Chevalley and Harish-Chandra. However, this was technical power at the service of the highest goals. I would be hard put to give another example of a work by a single author extending over 2000 pages at such a consistently high level, or should I say relentlessly high level, devoted to one main goal, very broadly conceived. At the beginning, he was much influenced by the work of others, notably of some mathematicians in the Soviet Union (Gelfand, Naimark, Graev), but then he forged ahead, devised his own strategy and weaponry and became the unquestioned leader in that area. In 1966, at the Moscow Congress, Gelfand told me that at the beginning, they had had a somewhat dim view of his work. They were looking basically at special cases rather than at a general semi-simple group, but they felt that they had more insight and that Harish-Chandra was essentially a technician generalizing their ideas by using the theory of semisimple groups. However, they saw now, especially after having heard his lecture at the Congress and talked with him, that he was far ahead.

Harish-Chandra was a highly principled man, for whom one's life had to have a purpose. In his view, the main one of his own was no doubt to prove the hardest and most fundamental theorems accessible to him. He could look with much satisfaction at his monumental work on the Plancherel formula, which R. P. Langlands, at a memorial service, later likened to 'a Gothic cathedral, heavily buttressed below, but, in spite of its great weight, light and soaring in its upper reaches, coming as close to heaven as mathematics can'. I hope he did, but even assuming it, I have some doubt that he felt he had fully discharged his self-imposed duty. He did not view representation theory as an end in itself, his vision of mathematics was much broader and, all over the years, I detected in him a wish to branch out. But each attempt appears to have been cut short by the occurrence of new ideas pertaining to his main topic, which brought him back, if not forced him back, to it. I already mentioned his plan in 1949 to get into algebraic geometry. His next attempt started during his stay in Paris in 1957–58. It resulted in lasting contributions to the theory of automorphic forms, arithmetic groups, reduction theory, and is better documented, but still not fully, I think. There are also some unfulfilled wishes. Allow me for a minute to turn to the mathematicians here to summarize what he told me at a party, in 1960 I think, maybe 1961, but not later. We happened to be by ourselves for a moment and I asked him what his philosophy on automorphic forms was. He immediately outlined a program (at any rate for split groups): start from an Eisenstein series, attached to some maximal solvable group. It converges in some half-space of a complexified Cartan subalgebra. Extend it meromorphically to the associated positive Weyl chamber. To go from one chamber to another there should be a functional equation attached to a Weyl group element. These functional equations should involve meromorphic functions admitting Euler product expansions, which should contain a tremendous amount of number theoretic information; to get hold of them would be 'hitting the jackpot' (his expression). The specialists will have recognized in very rough outline a preliminary sketch of some features of the program realized later by R. P. Langlands (totally independently), at any rate of its analytic part. The connection with number theory, as envisioned later by Langlands, turned out to be much more sophisticated. However, somewhat to my surprise, he did not pursue this, did not go for the jackpot. I presume he got at the time some of the ideas which led eventually to the determination of the discrete series, arguably the highest peak in his work, and that, of course, had to be given priority. The wish to branch out surfaced again in 1970. Before going on leave to the I.H.E.S. near Paris, during the fall term 1970, he told me he wanted to 'take off' and get involved with number theory. But instead he came back full of projects on representations of p-adic groups, fired up by some new ideas of H. Jacquet.

Once the eminent analyst Solomon Bochner told me, half-jokingly, that he was really a frustrated arithmetician. I have somehow the feeling that there was a little of that in Harish. Number theory, in particular class field theory, had always attracted, if not fascinated, him. In 1961, while in a rather depressed state, he wrote to me that algebraic number theory was the most beautiful topic he had ever come across and that the sole consolation in his misery was his lecturing on class field theory. I submit there are only very few people who would find relief from suffering by coming to grips with a theory as forbidding as class field theory was at the time.

This was indeed the kind of mathematics he admired most: the main results are of great scope, of great aesthetic beauty, but the proofs are technically extremely hard, a description which applies also very well to his own work. In fact, as is often the case with a mathematician having a strong personality, he was naturally attracted to mathematicians with an approach akin to his. Speaking of exceptional technical power at the service of high goals brings to mind C. L. Siegel, whose work Harish-Chandra held indeed in great esteem. Once, in a conversation, A. Weil and I realized that, for a number of years, we had both viewed Harish-Chandra as a kind of spiritual heir to Siegel by his mathematical style, power and concentration on very difficult, basic, questions, and we both felt it was Siegel's loss never to have realized it.

Underlying this tremendous productivity were very strict, almost ascetic, discipline and routine. In his exposition he was extremely meticulous, giving complete proofs (even in his personal seminar notes, which could almost have gone to a printer). It was of course in his nature to do so, but there was also a resolve stemming from a bad experience he had had early, when he discovered a gap in the proofs of results announced in a Note in the Proceedings of the U.S. National Academy of Sciences. He then vowed never to announce a result in print without having written down full proofs in all details. He had 'burnt his fingers' once, as he told me, and did not want this to happen again. However it did, though not with published results, because there was another side to his personality, a juvenile, almost childish enthusiasm when a new idea seemed to work, which belied his stern style of exposition. Sometimes he was so eager to share these new results that he would lecture or announce a lecture without having submitted their proofs to his exacting scrutiny, and a gap might appear unexpectedly. In the last occurrence, it led to a feverish month in which he did fill the gap, but at a high price for his health.

He wrote everything by hand, in an extremely regular writing. In 1966, during the Moscow Congress, a mathematician from the Soviet Union (maybe Kirillov or Kostrikin) remarked to me that a surprising number of papers by Harish-Chandra had either 33 or 66 pages and was mildly wondering why. Somewhat baffled, I could not offer any explanation at the moment but later, while I was looking at some of Harish's manuscripts, a simple one occurred to me. At practically any time, Harish had a big backlog of material to write up. Somehow, he had decided that a convenient way to parcel it out for publication was to write papers of either fifty or one hundred handwritten pages. He managed to do so fairly often and the writing was so regular that the contraction factor from the handwritten to the printed pages was one-third.

His lecturing style reflected faithfully his personality: very precise, complete, clearly written on the blackboard, but technically very demanding and fast. Later, he attempted to slow down, and could do so at least in the first part of a lecture. It was also delivered with much elegance. Once, when he was starting to write at the blackboard, standing erect, half turned to the blackboard, holding the chalk at some distance, as usual, my neighbour turned to me and whispered: 'he really looks like a prince.' But then, in the course of a lecture, he would often get so involved with his material, so excited, that the pace became faster and the old speed came back. That was really his natural tempo to think and speak.

He felt that, because his exposition was so systematic and complete, his papers were easy to read. He was a bit miffed when, at a party in his home, this statement was greeted with a big laugh. We outlined some of the hurdles facing a prospective reader, notably the cascades of references: in a given paper, the reader would be told that the notation of such and such papers was used. Then, if he would look at those works, he might again read the same, referring to still earlier papers, an almost infinite regress. We finally agreed that a given paper, though by no means easy to read, could be understood with a reasonable amount of effort by a fastidious reader, provided he would know thoroughly all of Harish's previous work. So, in a way, there was little difference of opinion because, for a logical, systematic and powerful mind such as his, this last *proviso* was a rather obvious prerequisite for anyone seriously interested in his work.

The sense of purpose Harish gave to his life had some spiritual, even religious underpinning. His religion was not a traditional one with the usual paraphernalia of stories, rituals, prayers and direct intervention of a personal god. Rather it was on an abstract, philosophical level, a yearning for some universal principle, transcending our lives, which would give a sense to the universe. Mathematics was maybe for him a way to approach it this life. He often said that semi-simple Lie groups are so perfect that they must have a divine origin. How seriously this was meant I really did not know for a long time. However, once in a seminar, he stated that a problem had occupied him for years, namely: 'Why has God created the exceptional series?' Together with the previous statement, this seems to me to express a very logical and genuine concern: without the exceptional series, harmonic analysis on a semi-simple group would be essentially well understood and the whole theory, to a large extent based on his work, would be definitive, of great elegance and harmony. The exceptional series complicates matters considerably. Now, if you start from the assumption that semi-simple groups are perfect, you are led to wonder how the exceptional series, which, at this time, seems to us to obscure the theory, will ultimately enhance it. The rather serious tone in which he stated his question makes it appear to me as expressing a frustration that some essential and undoubtedly beautiful feature of the theory remained hidden from him, after so many years of efforts.

In mathematics, Harish's life was indeed a search for fundamental general theorems, with the belief that they should be beautiful, and combine to harmonious theories. He pursued this quest with awesome single-mindedness, persistency, power and success.

ARMAND BOREL

Tata Institute of Fundamental Research
Homi Bhabha Road
Bombay 400 005, India

152.

Introduction to Middle Intersection Cohomology and Perverse Sheaves

Proc. Symposia pure math. **56** Part I, Amer. Math. Soc. (1994) 25–52

0. Introduction

In this paper we review some basic notions and facts on middle intersection cohomology and perverse sheaves. Our main goal is to supply some background material and references for some of the lectures at this Summer Institute dealing with various aspects of the representation theory of reductive algebraic groups.

0.1. In 1977 Goresky and MacPherson [GMP1] introduced new topological invariants on certain singular spaces, called pseudomanifolds, by modifying the usual definition of (PL or simplicial) homology or cohomology groups.

A topological space X (always assumed to be locally compact, Hausdorff, countable at infinity) is an *n-dimensional pseudomanifold* if first it admits a *stratification*, i.e., a decreasing filtration

(1) $$X = X_n \supset X_{n-1} \supset \cdots \supset X_1 \supset X_0 \supset X_{-1} = \emptyset$$

by closed subspaces such that

(2) $$S_{n-k} = X_{n-k} - X_{n-k-1} \qquad (0 \leq k \leq n)$$

is either empty or a $(n-k)$-manifold, called a stratum. Second, it is assumed that

(3) $$X_{n-1} = X_{n-2}, \ S_n \text{ is dense in } X$$

and that a regularity or equisingularity condition is satisfied in the neighborhood of each stratum (see 0.3).

The pseudomanifold is PL (piecewise linear) if it is endowed with a class of triangulations which have common refinements and for which the X_j are subcomplexes.

1991 *Mathematics Subject Classification.* Primary 55N35, 14F32, 32-02, 32S60.
This paper is in final form, and no version of it will be submitted for publication elsewhere.

These homology groups are more precisely assigned to X, a local system E on S_n and a *perversity* **p**. By definition, **p** is a function from the integers in $[0, n]$ to **N** which satisfies the conditions

(4) $\mathbf{p}(0) = \mathbf{p}(1) = \mathbf{p}(2) = 0, \qquad \mathbf{p}(k) \leq \mathbf{p}(k+1) \leq \mathbf{p}(k)+1 \quad (2 \leq k \leq n-1).$

The theory was developed successively on three levels of increasing sophistication:

I. PL intersection homology,

II. sheaf-theoretic intersection cohomology,

III. perverse sheaves.

For our present purposes, it would suffice to review II and III. However, I feel it necessary to touch upon I for motivation and to give an idea of the historical development of the theory. On the other hand, except for some side remarks, I shall concentrate on the so-called *middle perversity* **m** (the most important one anyhow), defined when all strata have *even* codimension by the conditions:

(5) $\mathbf{m}(0) = \mathbf{m}(1) = 0 \quad \text{and} \quad \mathbf{m}(2k) = \mathbf{m}(2k+1) = k-1 \qquad (2 \leq k \leq n).$

[Note that the general perversity (4) always satisfies $0 \leq \mathbf{p}(k) \leq k-2$—in particular, $0 \leq \mathbf{p}(2k) \leq 2k-2$—so that the values of the middle perversity are indeed in the middle of the possible values.]

0.2. *Examples.* (i) Let X be a projective variety (or a complex analytic space) of (necessarily even) real dimension n. Let $S(X)$ be the singular locus of X. Then we have a stratification defined by

(1) $$X_{n-2} = S(X), \quad X_{n-2k} = S(X_{n-2k+2}) \qquad (1 \leq 2k \leq n).$$

In general, it need not satisfy the regularity condition (0.3) but a refinement will [V].

(ii) Let X again be a complex analytic space endowed with a finite partition $X = \bigcup_{w \in W} C(w)$, where $C(w)$ is a locally closed complex analytic submanifold such that $\overline{C(w)}$ and $\overline{C(w)} - C(w)$ are analytic. Let $m(w)$ be the complex dimension of $C(w)$, and assume, moreover, that

(2) $$\overline{C(w)} \subset C(w) \cup \bigcup_{\substack{v \in W \\ m(v) < m(w)}} C(v).$$

Then X has a stratification defined by $X_{2k} = \bigcup_{w \in W,\, m(w) \leq k} C(w)$, whose strata are the unions of the cells $C(w)$ of a given dimension. Again, the regularity condition need not be satisfied, but it will be on some refinement [V].

0.3. Regularity assumptions. Let X be stratified as in 0.1. Then the regularity condition alluded to there is the following:

Let $x \in S_{n-k}$. Then x has a fundamental set of neighborhoods U of the form

(1) $$U = B \times \mathring{c}(L)$$

where B is a neighborhood of x in S_{n-k} and $\overset{\circ}{c}(L)$ is an open cone on a $(k-1)$-dimensional subspace L, called a link at x. Moreover, if X_{n-l} $(l \leq k)$ contains B in its closure, then $X_{n-l} \cap \overset{\circ}{c}(L)$ is a cone over $X_{n-1} \cap L$ and those intersections define again a stratification of L. If the strata in X have even codimension, then the same is true for the induced stratification of L (but dim X-dim L is odd). It is also assumed that L, endowed with the induced stratification, is a (dim L)-PL-pseudosubmanifold so that, in fact, the notion of n-dimensional PL-pseudomanifold is defined by induction on n. A neighborhood of the form (1) will be called distinguished.

In addition, we shall assume that S_n is orientable and oriented. It is sometimes convenient to assume that X is normal, a condition introduced in **[GMP2**, 4.1] and shown there to be equivalent to have all links connected. X admits a canonical normalization $\tilde{X}$, which is a finite ramified covering, and the projection $\tilde{X} \to X$ induces an isomorphism in intersection homology **[GMP2**, 4.1 and 4.2]. If X is a complex algebraic variety, then $\tilde{X}$ is homeomorphic to the normalization of X as an algebraic variety (loc. cit.).

0.4. The basic papers for intersection homology are **[GMP2]**, **[GMP3]**. For convenience, I shall refer mostly to **[B2]**, where the reader will find the original references and further literature. For a history of intersection homology, perverse sheaves, and applications, see **[K]**.

The theory has been extended in various ways, either by weakening the assumptions on X or on the perversity. See **[HS]** for the most recent generalization and references to earlier literature.

I thank L. Scott for some comments, corrections, and a suggestion (see 8.10) which led to some improvements of this paper.

In §§1 to 7, X is an n-dimensional PL-pseudomanifold and $\mathscr{X}$ the underlying stratification. Unless otherwise stated, the strata have even codimension and **m** *is the middle perversity. R denotes a commutative noetherian ring of coefficients and E a local system on S_n (i.e., a locally constant sheaf of finitely generated R-modules). From §5 on, R is a field.*

I. PIECEWISE LINEAR INTERSECTION HOMOLOGY

1. Middle intersection cohomology

1.1. If σ is a simplex in the given PL-structure, then $\sigma \cap S_n$, if not empty, is contractible; hence, the restriction of E to the latter is constant. It makes sense therefore to consider PL-chains of simplices not contained in X_{n-2}, with coefficients in E. Let C_i or $C_i(E)$ denote the R-module of locally finite such i-dimensional PL-chains. We say that $\zeta \in C_i$ satisfies the condition $(\mathbf{m}, i)$ if

$$\text{(1)} \qquad \dim |\zeta| \cap X_{n-2k} \leq i - 1 - k \qquad (0 \leq 2k \leq n+1),$$

and we set

(2) $I^{\mathbf{m}}C_i(E) = \{\zeta \in C_i(E),\ \zeta \text{ satisfies } (\mathbf{m}, i) \text{ and } \partial\zeta \text{ satisfies } (\mathbf{m}, i-1)\}$.

This is obviously a subcomplex of the PL-complex. Its homology is by definition the *middle intersection homology group* $I^{\mathbf{m}}H.(X; E)$ of X with coefficients in E:

(3) $$I^{\mathbf{m}}H_i(X; E) = H_i(I^{\mathbf{m}}C.(E), \partial) \qquad (i \in \mathbf{Z}).$$

It is equal to zero for $i \notin [0, n]$.

The ith middle intersection *cohomology* R-module $I_{\mathbf{m}}H^i(X; E)$ is defined by

(4) $$I_{\mathbf{m}}H^i(X; E) = I^{\mathbf{m}}H_{n-i}(X; E) \qquad (i \in \mathbf{Z}).$$

1.2. THEOREM. *The R-modules $I_{\mathbf{m}}H^i(X; E)$ are topological invariants and are finitely generated if X is compact. If R is a field and X is compact, there exists a canonical perfect Poincaré duality pairing*

$$I_{\mathbf{m}}H^i(X; E) \times I_{\mathbf{m}}H^{n-i}(X; E^*) \to R \qquad (i \in \mathbf{Z}),$$

where E^* is the local system dual to E.

(If X is not compact, one has to replace one of the two groups by the middle intersection cohomology with compact supports, i.e., defined by finite chains.)

Sections 3–5 provide some indications on the proof in the sheaf-theoretic setting, given first in **[GMP3]** and for which we also refer to **[B2**, V]. For a given PL-structure, it had been proved in **[GMP2]**, at any rate for constant coefficients.

1.3. REMARKS. (a) In order to understand the meaning of $(\mathbf{m}, i)$, note that we have

(1) $$i - k - 1 = (i - 2k) + \mathbf{m}(2k).$$

Now if two affine subspaces A, B of respective dimensions a, b in an n-dimensional vector space are in general position, their intersection has dimension $a + b - n$, i.e., $\dim A - \operatorname{codim} B$. The first term on the right-hand side gives, therefore, the dimension of $|\zeta| \cap X_{n-2k}$, if ζ is in general position with respect to X_{n-2k}. The "perversity" is to allow more. In the general case of an arbitrary perversity, the condition $(\mathbf{p}, i)$ is indeed

(2) $$\dim |\zeta| \cap X_{n-k} \le i - k + \mathbf{p}(k).$$

Assume E is the constant sheaf R_X with stalk R and X to be normal. If $\mathbf{p}$ is the zero perversity, then $I_{\mathbf{p}}H^i$ is just the ith ordinary cohomology group. If $\mathbf{p}$ is the biggest possible perversity (i.e., $\mathbf{p}(k) = k - 2$, as already remarked), then $(\mathbf{p}, i)$ is essentially vacuous, and $I^{\mathbf{p}}H_i$ is the ith homology group (with closed supports). If $\mathbf{p} \le \mathbf{q}$ (i.e., $\mathbf{p}(k) \le \mathbf{q}(k)$ for all k's) then $I^{\mathbf{p}}C_i \subset I^{\mathbf{q}}C_i$ obviously, whence a natural map $I^{\mathbf{p}}H_i \to I^{\mathbf{q}}H_i$. In particular, we have natural homomorphisms

(3) $$H^i(X; R) \to I_{\mathbf{m}}H^i(X; R) \to H_{n-i}(X; R)$$

(see **[B2**, V, §2]).

(b) In the case of a general perversity the group $I_{\mathbf{p}}H^i$ is also topological invariant and if $\mathbf{p}$ and $\mathbf{q}$ are "complementary", i.e., $\mathbf{p}(k) + \mathbf{q}(k) = k - 2$ for all k's and R is a field, then there is a Poincaré duality pairing between $I^{\mathbf{p}}H^i(X; E)$ and $I^{\mathbf{q}}H^{n-i}(X; E^*)$ (again for X compact, if not replace one of two cohomology spaces by intersection cohomology with compact supports).

(c) If $E = R_X$ and X is smooth, then $I_{\mathbf{m}}H^{\cdot}$ (or $I_{\mathbf{p}}H^{\cdot}$) is just ordinary cohomology. If X has singularities, then, in general, $I_{\mathbf{m}}H^{\cdot}$ is a new topological invariant, since ordinary cohomology usually fails to obey Poincaré duality. These groups may also be new if X is smooth as a manifold but singular with respect to E, i.e., when E has nontrivial monodromy when going around points in the proper strata. A particularly interesting case is the one where X is a smooth projective variety, X_{n-2} a smooth divisor with normal crossings, and E is defined by a variation of Hodge structure.

(d) Intersection homology does not change under normalization [GMP2, 4.2].

1.4. For illustration, we discuss briefly the simplest non-trivial case—that of isolated singularities, for trivial coefficients and X compact. We assume that $n = 2r$ is even and that X_{n-2} consists of finitely many points. Therefore

(1) $$X_{n-2} = X_0 = S_0 \quad \text{and} \quad X = S_n \cup S_0.$$

The conditions $(\mathbf{m}, i)$ are of interest only for $x \in S_0$. For ζ to be in $I^{\mathbf{m}}C_i$ we must therefore have

(2) $$\dim(|\zeta| \cap X_0) \leq i - r - 1,$$

(3) $$\dim(|\partial\zeta| \cap X_0) \leq i - r - 2.$$

If $i \leq r$, the right-hand side of (2) is < 0, hence ζ does not meet X_0. On the other hand, if $i > r$, the condition (2) is vacuous. In dimension r, a chain not meeting S_0 which is the boundary of any $(r+1)$-chain is allowed. Altogether, we see easily that

(4) $$I^{\mathbf{m}}H_i(X) = \begin{cases} H_i(X - X_0), & i < r, \\ \operatorname{Im}(H_i(X - X_0) \to H_i(X)), & i = r, \\ H_i(X), & i > r. \end{cases}$$

In cohomology this reads

(5) $$I_{\mathbf{m}}H^i(X) = \begin{cases} H^j(X), & j < r, \\ \operatorname{Im}\left(H^j_c(X - X_0) \to H^j(X - X_0)\right), & j = r, \\ H^j_c(X - X_0), & j > r. \end{cases}$$

In this case, Poincaré duality follows therefore from the Poincaré duality on the manifold $X - X_0$ and the groups $I_{\mathbf{m}}H^i(X)$ may be expressed in terms of ordinary cohomology groups. However, already in the case where there is only one proper stratum, but of strictly positive dimension, the reader will easily extend the previous discussion and see that the conditions $(\mathbf{m}, i)$ lead to new groups.

1.5. As a preparation for II, let us also compute the local intersection (co)homology groups $\mathscr{I}_{\mathbf{m}}\mathscr{H}^i(X)_x$ at $x \in S_0$, with coefficients in R. By this we mean, of course, the homology of the complex $\mathscr{C}_x^{\cdot}$ of germs at x of PL-chains satisfying the conditions $(\mathbf{m}, i)$. For $i \leq r$, such a chain does not meet x, as already remarked and, hence, defines the zero germ so that

$$(1) \qquad H_i(\mathscr{C}_x^{\cdot}) = 0 \qquad (i \leq r).$$

For $i > r$, all chains are allowed; hence ,

$$(2) \qquad H_i(\mathscr{C}_x^{\cdot}) = H_i^{\mathrm{cl}}(\overset{\circ}{c}(L)) \qquad (i > r),$$

where the superscript cl refers to homology with closed supports. By the long exact sequence in that homology, this group is equal to $H_i^{\mathrm{cl}}(\overset{\circ}{c}(L) - \{x\})$ and, hence, to $H_i^{\mathrm{cl}}(\mathbf{R} \times L)$. By the Künneth rule, this last group is equal to $H_{i-1}(L)$ (recall that $H_1^{\mathrm{cl}}(\mathbf{R}) = R$ and $H_i^{\mathrm{cl}}(\mathbf{R}) = 0$ for $i \neq 1$). Here L is assumed to be a compact oriented manifold of dimension $2r - 1$; hence, $H_{i-1}(L)$ is isomorphic to $H^{2r-i}(L)$. If we now shift to cohomology, we get

$$(3) \qquad \mathscr{I}_{\mathbf{m}}\mathscr{H}^i(X)_x = \begin{cases} H^i(L) & (i \leq r - 1 = \mathbf{m}(2r)), \\ 0 & (i \geq r). \end{cases}$$

II. SHEAF THEORETIC MIDDLE INTERSECTION COHOMOLOGY

2. Sheaf theory

Familiarity with sheaf theory is assumed. For convenience, we collect here some notions and results in sheaf theory to be used later. For references, see **[B2**, V, §1]. Sections 2.1 and 2.2 are basic for all that follows, but the rest is needed only at some specific places, so that the remainder of §2 should be viewed as a reference to come back to as the need arises.

2.1. Let Y be a topological space (as in 0.1), $\mathscr{S}$ a sheaf of R-modules on Y, and $\mathscr{S} \to \mathscr{J}^{\cdot}$ an injective resolution of S. Then, by definition, the cohomology $H^{\cdot}(Y; \mathscr{S})$ of Y with coefficients in $\mathscr{S}$ is

$$(1) \qquad H^{\cdot}(Y; \mathscr{S}) = H^{\cdot}(\Gamma(\mathscr{J}^{\cdot})),$$

where $\Gamma(\ )$ denotes the functor of continuous sections. Let, more generally, $\mathscr{S}^{\cdot}$ be a differential graded sheaf (DGS), which we always assume to be bounded (i.e., $\mathscr{S}^i = 0$ for $|i|$ large enough). It also has an injective resolution $\mathscr{S}^{\cdot} \to \mathscr{J}^{\cdot}$, by definition, $H^{\cdot}(\Gamma(\mathscr{J}^{\cdot}))$ is the hypercohomology $\mathbb{H}^{\cdot}(Y; \mathscr{S}^{\cdot})$ of Y with respect to $\mathscr{S}^{\cdot}$.

The derived sheaf $\mathscr{H}^{\cdot}\mathscr{S}^{\cdot}$ of $\mathscr{S}^{\cdot}$ is the sheaf associated to the presheaf $U \mapsto H^{\cdot}(S^{\cdot}(U))$ with zero differential. Its stalk at $y \in Y$ is the cohomology $H\mathscr{S}^{\cdot}_y$ of the stalk $\mathscr{S}^{\cdot}_y$ of $\mathscr{S}^{\cdot}$ at y. We have

(2) $$(\mathscr{H}^{\cdot}\mathscr{S}^{\cdot})_y = \lim_{U} H^{\cdot}(\Gamma_U(\mathscr{S}^{\cdot})) = H^{\cdot}\mathscr{S}^{\cdot}_y,$$

where U runs through a fundamental set of open neighborhoods of y and Γ_U refers to continuous sections with support in U.

2.2. The main theorem of sheaf theory asserts the existence of a spectral sequence (E_r) which abuts to $\mathbb{H}^{\cdot}(Y; \mathscr{S}^{\cdot})$ and in which

(1) $$E_2^{p,q} = H^p(Y; \mathscr{H}^q\mathscr{S}^{\cdot}) \qquad (p, q \in \mathbf{N}).$$

Two DGS's $A^{\cdot}$ and $B^{\cdot}$ are *quasi-isomorphic* (q.i.) if there exists a DGS $\mathscr{C}^{\cdot}$ and morphisms $\mathscr{A}^{\cdot} \leftarrow \mathscr{C}^{\cdot} \rightarrow \mathscr{B}^{\cdot}$ which induce isomorphisms of the derived sheaves. If so, the main theorem shows that the hypercohomology groups of Y with respect to $\mathscr{A}^{\cdot}$ and $\mathscr{B}^{\cdot}$ are isomorphic. Therefore, in the derived category of bounded DGS, hypercohomology is well defined. In the sequel, we shall assume some familiarity with derived categories (see [**B2**, V, §5] for an elementary introduction and references). However, although it is essential in III, it is largely window dressing in II, and we shall not use that language before §5.

2.3. The attachment map. Let Z be a closed subspace of Y, $U = Y - Z$, and

(1) $$i : U \rightarrow Y, \qquad j : Z \rightarrow Y$$

be the natural inclusions. For a DGS $\mathscr{S}^{\cdot}$ on Y, the composition of the natural homomorphisms

(2) $$\mathscr{S}^{\cdot} \rightarrow i_* i^* \mathscr{S}^{\cdot} \rightarrow Ri_* i^* \mathscr{S}^{\cdot},$$

where i^* (resp. i_*) is the inverse (resp. direct) image and Ri_* the right derived functor of i_*, is called the attachment map. It is obviously a q.i. at every point of U. To say that it is a q.i. at $z \in Z$, in some dimension i, amounts to the condition

(3) $$H^i(\mathscr{S}^{\cdot}_z) = \varinjlim_V H^i(V - (V \cap Z); \mathscr{S}^{\cdot}),$$

where V runs through a fundamental set of neighborhoods of z in Y. The derived map of the attachment map is part of a long exact sequence involving hypercohomology with support in Z, as will be recalled presently.

2.4. Cohomology with support in a closed subset. We go on with the previous setup. Given the DGS $S^{\cdot}$ on Y, we denote by $\gamma_Z S^{\cdot}$ the DGS of germs of continuous sections of $S^{\cdot}$ with support in Z. It is defined by the presheaf

(1) $$V \mapsto \Gamma_{Z \cap V}(S^{\cdot}|V)$$

(V open in Y). The right derived functor of γ_Z is denoted $j^!$. We have $j_z^! \mathcal{S}^\cdot = \gamma_Z \mathcal{J}^\cdot$, where $\mathcal{J}^\cdot$ is an injective resolution of $\mathcal{S}^\cdot$. The hypercohomology of Y with support in Z, with respect to $\mathcal{S}^\cdot$, is then by definition

$$\mathbb{H}_Z^\cdot(X;\mathcal{S}^\cdot) = H^\cdot(\Gamma(\gamma_Z \mathcal{J}^\cdot)). \tag{2}$$

For any sheaf $\mathcal{S}$ on Y, the kernel of the natural map $\mathcal{S} \to i_* i^* \mathcal{S}$ consists of the sections with support in Z. Moreover, this map is surjective if $\mathcal{S}$ is injective (or flabby). Therefore, we get an exact sequence

$$0 \to j_! \gamma_Z \mathcal{J}^\cdot \to \mathcal{J}^\cdot \to i_* i^* \mathcal{J}^\cdot \to 0, \tag{3}$$

where $j_!$ denotes extension by zero. The long exact sequence of the derived sheaves at $z \in Z$ is then

$$\cdots \to H^i((j_Z^! S^\cdot)_z) \to H^i(\mathcal{S}_z^\cdot) \xrightarrow{\alpha_i} H^i((Ri_* i^* \mathcal{S}^\cdot)_z) \to \cdots, \tag{4}$$

whose global counterpart is

$$\cdots \to \mathbb{H}_Z^i(X;\mathcal{S}^\cdot) \to \mathbb{H}^i(X;\mathcal{S}^\cdot) \to \mathbb{H}^i(U;\mathcal{S}^\cdot) \to \cdots. \tag{5}$$

We shall also sometimes write

$$\mathcal{H}_Z^\cdot \mathcal{S}^\cdot \text{ for } \mathcal{H}^\cdot(j_Z^! \mathcal{S}^\cdot), \text{ in particular, } H_{\{z\}}^\cdot S^\cdot \text{ for } \mathcal{H}^\cdot(j_z^! \mathcal{S}^\cdot). \tag{6}$$

The support of the function $z \mapsto H_{\{z\}}^i(\mathcal{S}^\cdot)$ is the set of points where it is $\neq 0$.

2.5. We shall need this notion in particular when Z is a point. If Y satisfies some local conditions, as is the case for our space X, then we have

$$H^i(j_z^! \mathcal{S}^\cdot)_z = \varprojlim \mathbb{H}_c^i(U;\mathcal{S}^\cdot) \qquad (i \in \mathbf{N}) \tag{1}$$

where U runs through a fundamental system of neighborhoods of z. For pseudomanifolds, see [**B2**, V, 3.10], where it is also shown that the inverse system on the right-hand side is constant on distinguished neighborhoods.

This provides another definition of the left-hand side. Viewed in this way, that construction, for $\mathcal{S}^\cdot = R_X$, when R is a field, goes back to P. Alexandroff and R. L. Wilder, the dimension of the left-hand side being their ith local Betti number *around* x.

2.6. Truncation. Let $\mathcal{S}^\cdot$ be a DGS on Y and $m \in \mathbf{N}$. The *truncation* $\tau_{\leq m}\mathcal{S}^\cdot$ of $\mathcal{S}^\cdot$ is the DGS defined by

$$\tau_{\leq m}\mathcal{S}^i = \begin{cases} \mathcal{S}^i, & i < m, \\ \ker d_i : \mathcal{S}^i \to \mathcal{S}^{i+1}, & i = m, \\ 0, & i > m. \end{cases}$$

2.7. There are several "constructibility" conditions which have to be considered in a thorough discussion of intersection cohomology, for which we refer to [**B2**, V, §§2, 3]. We will keep this here to a minimum and will use mainly the following notion:

Let Y be a topological space endowed with a stratification $\mathscr{Y}$. Then a DGS $\mathscr{S}^{\cdot}$ on Y is said to be $\mathscr{Y}$-clc ($\mathscr{Y}$-cohomologically locally constant) if the derived sheaf $\mathscr{H}^{\cdot}\mathscr{S}^{\cdot}$ is locally constant along the strata of Y. It is $\mathscr{Y}$-cc ($\mathscr{Y}$-cohomologically constructible) if in addition the stalks of the derived sheaf are finitely generated R-modules [**B2**, V, 3.3(ii)].

2.8. Finally, we recall a convention which will be used throughout from §5, mainly for DGS.

If $A^{\cdot} = \{A^i\}_i$ is a **Z**-graded object in some additive category $\mathscr{C}$ and $m \in \mathbf{Z}$, then $A^{\cdot}[m]$ is $A^{\cdot}$, translated by $-m$, i.e.,

(1) $$(A^{\cdot}[m])^i = A^{m+i} \quad (i \in \mathbf{Z}).$$

For instance, if $A^{\cdot}$ is concentrated in degree 0, then $A^{\cdot}[m]$ is concentrated in degree $-m$. Recall that if $B^{\cdot}$ is also graded, then the graded $\mathrm{Hom}^{\cdot}(A^{\cdot}, B^{\cdot})$ is defined by

(2) $$\mathrm{Hom}^i(A^{\cdot}, B^{\cdot}) = \bigoplus_j \mathrm{Hom}(A^j, B^{i+j}).$$

Therefore, for $m \in \mathbf{Z}$,

(3) $$\mathrm{Hom}^{\cdot}(A^{\cdot}, B^{\cdot})[m] = \mathrm{Hom}^{\cdot}(A^{\cdot}, B^{\cdot}[m]) = \mathrm{Hom}^{\cdot}(A^{\cdot}[-m], B^{\cdot}).$$

3. The intersection cohomology sheaf. First axiomatic characterization

3.1. We first fix some notation. Let

(1) $$U_k = X - X_{n-k} \quad (1 \le k \le n+1).$$

The U_k's form an increasing sequence of open subsets. We have

(2) $$U_1 = U_2 = S_n, \quad U_{k+1} = U_k \cup S_{n-k}, \quad U_{n+1} = X.$$

We let i_k and j_k be the inclusions

(3) $$U_k \xrightarrow{i_k} U_{k+1} \xleftarrow{j_k} S_{n-k}.$$

If $S^{\cdot}$ is a DGS on X, then $S_k^{\cdot}$ denotes its restriction on U_k. The attachment map (2.3) is now the composition

(4) $$S_{k+1}^{\cdot} \to i_{k*}S_k^* \to Ri_{k*}S_k^*$$

and will be denoted α_k.

By our standing assumption, the strata have even codimension, therefore

(5) $$U_{2k} = U_{2k+1} \quad (1 \le 2k \le n+1).$$

3.2. Proposition. *Let $\mathcal{I}_{\mathbf{m}}\mathcal{C}^{\cdot}(E)$ be the sheaf of germs of locally finite PL-chains with coefficients in E. Then the DGS $\mathcal{I}_{\mathbf{m}}\mathcal{C}^{\cdot}(E)$ is $\mathcal{X}$-cc and its derived sheaf at $x \in S_{n-2k}$ is given by*

$$(1) \qquad H^i(\mathcal{I}_{\mathbf{m}}\mathcal{C}^{\cdot}(E))_x = \begin{cases} I_{\mathbf{m}}H^i(L;E), & i < k, \\ 0, & i \geq k, \end{cases}$$

where the isomorphism for $i < k$ is given by the derived attachment map $(0 \leq k \leq n)$.

The proof (see 6.1 in [**B2**, II], where it is stated in homology) presents some technical difficulties, but the result is intuitively rather plausible. Let us try to say why, writing $\mathcal{S}^{\cdot}$ for $\mathcal{I}_{\mathbf{m}}\mathcal{C}^{\cdot}(E)$. Recall that x has a fundamental system of neighborhoods U of the form $B \times \overset{o}{c}(L)$ and that the stratification is compatible with that decomposition (0.3).

A Künneth rule in intersection homology allows one to write

$$(2) \qquad I^{\mathbf{m}}H_i(U;E) = I^{\mathbf{m}}H_{i-2k}(\overset{o}{c}(L);E) \qquad (i \in \mathbf{Z}),$$

(using the fact that, in homology with closed supports, $H_i(\mathbf{R}^q;R)$ is equal to R for $i = q$, to zero for $i \neq q$). Passing to cohomology (cf. 1.1(4)), this yields

$$(3) \qquad I_{\mathbf{m}}H^j(U;E) = I_{\mathbf{m}}H^j(\overset{o}{c}(L);E) \qquad (j \geq 0).$$

A discussion similar to that of 1.5 then shows that the right-hand side of (3) is equal to the right-hand side of (1). There remains to see why the isomorphism for $i < k$ is induced by the attachment map. Without trying to prove it, let us at least say why one should expect that

$$(4) \qquad H^i(i_{k*}\mathcal{S}^{\cdot}_k)_x = I_{\mathbf{m}}H^i(L) \qquad (i \in \mathbf{Z};\, x \in S_{n-2k}).$$

To determine the left-hand side we have to compute the hypercohomology of $U - (U \cap S_{n-2k})$ with coefficients in the restriction of $\mathcal{S}^{\cdot}$. This space is equal to the product of B by $\overset{o}{c}(L)$ with the vertex removed and, hence, to $\mathbf{R}^{n-2k+1} \times L$. If we again take the Künneth rule for granted, then we get

$$(5) \qquad \mathbb{H}^i(U - (U \cap S_{n-2k});\, \mathcal{S}^{\cdot}) = I_{\mathbf{m}}H^i(L) \qquad (i \in \mathbf{Z}),$$

an equality that will remain valid if we pass to the limit over U.

3.3. We denote by $AX\ 1$ or more precisely $\mathrm{AX}\ 1_{\mathcal{X},E}$ the following set of conditions on a DGS $\mathcal{S}^{\cdot}$ on X:

(i) $\mathcal{H}^i\mathcal{S}^{\cdot} = 0$ for $i < 0$ and $\mathcal{S}^{\cdot}|U_2$ is q.i. to E.

(ii) For $x \in S_{n-2k'}$ we have $H^i(\mathcal{S}^{\cdot}_x) = 0$ for $i \geq k$.

(iii) The attachment map α_k is a q.i. up to $k-1$.

Remark. For our present purposes, it would do no harm to assume in (i) the stronger condition $\mathcal{S}^i = 0$ for $i < 0$, as is done in [**B2**, 2.2.3]. As remarked there in 2.7(b) any $\mathscr{S}^{\cdot}$ satisfying our present AX $1_{\mathscr{X},E}$ is q.i. to one fulfilling the stronger condition. We could also do with the latter in §4, but it would be too restrictive from §5 on. Of the two, it is anyhow the only one which makes sense in the derived category.

3.4. THEOREM. *The intersection cohomology sheaf* $\mathscr{I}_{\mathbf{m}}\mathscr{C}^{\cdot}(X;E)$ *satisfies AX* $1_{\mathscr{X},E}$. *Any DGS* $\mathscr{S}^{\cdot}$ *satisfying it is q.i. to* $\mathscr{I}_{\mathbf{m}}\mathscr{C}^{\cdot}(X;E)$ *and is* $\mathscr{X}$*-cc. If* X *is compact,* $I_{\mathbf{m}}H^{\cdot}(X;E)$ *is finitely generated.*

The first assertion follows from 3.2, the uniqueness, up to q.i., from 2.5 in [**B2**, V]. That $\mathscr{S}^{\cdot}$ is $\mathscr{X}$-cc (in fact more) is proved in 3.12 *loc. cit.* For the last assertion, see 3.4 there.

3.5. One can define directly a DGS satisfying AX 1 without reference to a PL-structure, namely, Deligne's sheaf, to be denoted $\mathscr{P}^{\cdot}(E)$. Its restriction $\mathscr{P}^{\cdot}(E)_k$ to U_k is defined by induction on k by the following rules:

(1) $\mathscr{P}^{\cdot}(E)_2$ is q.i. to E on U_2.
(2) $\mathscr{P}^{\cdot}(E)_{k+1} = \tau_{\leq \mathbf{m}(k)} Ri_{k*}\mathscr{P}^{\cdot}(E)_k$, $(k \geq 1)$.

Then $\mathscr{P}^{\cdot}(E) = \mathscr{P}^{\cdot}(E)_{n+1}$. The proof of 3.4 consists in fact in showing first that $\mathscr{P}^{\cdot}(E)$ satisfies AX 1, and then that any $\mathscr{S}^{\cdot}$ which does so is q.i. to $\mathscr{P}^{\cdot}(E)$.

3.6. An example. The favorite one at this Summer Institute is no doubt the flag variety G/B, where G is a connected semisimple group (over $\mathbb{C}$ here, but also over a finite field) and B a closed subgroup "whose name I have the honor to bear", to use a formula going back to P. Montel when he was referring to his normal families. The orbits of the unipotent radical U of B form a partition

$$G/B = \bigcup_{w \in W} C(w) \tag{1}$$

as in 0.2(ii), where W is the Weyl group and the "Bruhat cells" or "Schubert cells" $C(w)$ are complex affine spaces. Their closures, the "Schubert varieties", satisfy condition (2) there and $C(v) \subset \overline{C(w)}$ if and only if $v \leq w$ in the Bruhat-Chevalley ordering. Consider $\mathscr{I}_{\mathbf{m}}\mathscr{C}^{\cdot}(\overline{C(w)})$ and its derived sheaf. By homogeneity, $\mathscr{I}_{\mathbf{m}}\mathscr{H}^{\cdot}(\overline{C(w)})_y$ is constant on the U-orbit of y in $\overline{C(w)}$ and, hence, only depends on an element $v \leq w$ in W. It is known to be zero in odd degrees (cf. [**KL**, 4.2]). The Poincaré polynomial (in the variable $t^{1/2}$)

$$P_{v,w}(t) = \sum_i \dim \mathscr{I}_{\mathbf{m}}\mathscr{H}^{2i}(\overline{C(w)})_y . t^i \qquad (y \in C(v),\, v \leq w) \tag{2}$$

is the Kazhdan-Lusztig polynomial.

4. Topological invariance. Second axiomatic characterization

4.1. Recall that the support of a DGS is the set of points where its stalk is $\neq 0$.

We claim first that for a DGS $\mathscr{S}^\cdot$ which is $\mathscr{X}$-clc, (3.3)(ii) is equivalent to

(ii)′ $\dim \operatorname{supp} \mathscr{H}^i\mathscr{S}^\cdot < n - 2i$ $(i \in \mathbf{N},\ i \neq 0)$.

That (ii) $\Rightarrow$ (ii)′ is obvious. On the other hand, $\mathscr{H}^\cdot\mathscr{S}^\cdot$ is $\mathscr{X}$-clc; therefore, if it is not zero at some point of S_{n-k}, then its support has dimension $\geq n-k$. From this the reverse implication is clear.

The interest of (ii)′ is that it is closer to a topological condition, the stratification occurring only by the constructibility condition $\mathscr{X}$-clc. We want now to find a similar equivalent condition to (iii).

We first note that 2.4(3), applied to $Z = S_{n-k}$, $Y = U_{k+1}$, yields the long exact sequence

(1) $$\cdots \to H^i(j_k^!\mathscr{S}^\cdot)_x) \to H^i(\mathscr{S}_x^\cdot) \xrightarrow{\alpha_k^i} H^i((Ri_{k*}\mathscr{S}_k^\cdot)_x) \to \cdots.$$

From this, using the fact that $H^{\mathbf{m}(k)+1}(\mathscr{S}^\cdot)_x = 0$, we see that (iii) is equivalent to

(2) $$H^i(j_k^!\mathscr{S}^\cdot)_x = 0 \quad \text{for } i \leq \mathbf{m}(k)+1.$$

Let f_x (resp. l_x) be the inclusion of x in X (resp. S_{n-k}). Then $f_x = j_k \circ l_x$. This implies $f_x^! = l_x^! \circ f_k^!$. Since S_{n-k} is a manifold, $l_x^!$ is just a shift by $n-k$. Altogether (see [**B**, V, p. 87]) it follows that we can replace (2) by

(3) $$H^i(f_x^!\mathscr{S}^\cdot) = 0 \quad \text{for } i < n - \mathbf{m}(k)\ (x \in S_{n-k}).$$

It can also be shown that if $\mathscr{S}^\cdot$ is $\mathscr{X}$-clc, then the assignment $x \mapsto H^\cdot(f_x^!\mathscr{S}^\cdot)$ is locally constant along the strata of $\mathscr{X}$ [**B2**, V, 3.10]. Then the above argument shows that if $\mathscr{S}^\cdot$ is $\mathscr{X}$-clc, we may in AX 1 replace (ii) by (ii)′ and (iii) by

(iii)′ $\dim\{x \in X \mid H^i(f_x^!\mathscr{S}^\cdot) \neq 0\} < 2i - n$ $(i < n)$.

4.2. We now consider the following set of conditions AX 2 or AX $2_E^\cdot$.

(i) $\mathscr{H}^i\mathscr{S}^\cdot = 0$ for $i < 0$; $\mathscr{S}^\cdot$ is $\mathscr{X}$-clc for some PL-pseudomanifold stratification $\mathscr{Y}$ of X; there exists an open dense submanifold U, whose complement has codimension ≥ 2, and a local system E on U such that $\mathscr{S}^\cdot$ is q.i. to E on U.

(ii) $\dim \operatorname{supp} \mathscr{H}^i\mathscr{S}^\cdot < n - 2i$, $(i > 0)$.

(iii) $\dim\{x \in X \mid H^j(f_x^!\mathscr{S}^\cdot) \neq 0\} < 2j - n$, $(j < n)$.

4.3. Theorem. *The set of conditions $AX2_E$ determines $\mathscr{S}^\cdot$ up to q.i. and is satisfied by the middle intersection cohomology sheaf defined with respect to any PL-pseudomanifold structure.*

This proves in particular the topological invariance of (middle) intersection cohomology. (see [**B2**, V, §4]).

4.4. Since U is now variable and not a priori related to a stratification as in AX 1, some comments on the meaning of the uniqueness up to q.i. are in order.

First, given a local system E on an open subset U with complement of codimension ≥ 2, it is easily seen that there is a biggest open subset $V \supset U$ on which E extends to a local system [**B2**, V, 4.11, 4.12], call it the biggest domain of definition $U(E)$ of E. Second, if $\mathcal{S}^{\cdot}$ satisfies AX 2_E, then we may assume that $U(E)$ contains the biggest stratum U_2 of $\mathcal{Y}$ (*loc. cit.*, 4.14b). Then 4.3 can be stated more precisely as:

4.5. THEOREM. *Let $\mathcal{S}^{\cdot}$ (resp. $\mathcal{T}^{\cdot}$) satisfy AX2 with respect to data $(U, E, \mathcal{X})$ (resp. $V, F, \mathcal{Y}$) and assume that E is isomorphic to F on $U \cap V$. Then $\mathcal{S}^{\cdot}$ and $\mathcal{T}^{\cdot}$ are q.i. to one another and to the middle intersection cohomology sheaves defined with respect to E and $\mathcal{X}$ on the one hand, to F and $\mathcal{Y}$ on the other.*

4.6. The proof of 4.3 is more complicated than that of 3.4 but similar in spirit. Let $\tilde{U} = U(E)$. The idea is to construct by induction on k a stratification $\tilde{\mathcal{X}}$, for which $\tilde{U} = \tilde{U}_2$ and a sheaf $\tilde{\mathcal{P}}^{\cdot}(E)$ associated to $\tilde{\mathcal{X}}$ by Deligne's procedure so that $\tilde{\mathcal{P}}^{\cdot}(E)$ is $\tilde{\mathcal{X}}$-clc and satisfies AX 2 [**B2**, V, 4.15], and then to show any $\mathcal{S}^{\cdot}$ satisfying AX 2 is q.i. to $\mathcal{P}^{\cdot}(E)$ (*loc. cit.*, 4.17).

4.7. PROPOSITION. *Let Y be a pseudomanifold and $p : Y \to X$ the projection of a locally trivial fibration, the fibres of which are manifolds. Then $p^*\mathcal{I}_{\mathbf{m}}\mathcal{C}^{\cdot}(X, E) = \mathcal{I}_{\mathbf{m}}\mathcal{C}^{\cdot}(Y; p^*E)$.*

Here E is a local system defined on an open smooth subset U of X with complement of codimension ≥ 2. Then p^*E is a local system on $p^{-1}U$, which is also smooth, open in Y, with complement of codimension ≥ 2.

This assertion is local on X. We may therefore assume $Y = X \times F$, where F is a manifold and p is the first projection. Let $\mathcal{P}^{\cdot}_X(E)$ (resp. $\mathcal{P}^{\cdot}_Y(p^*E)$) be the Deligne sheaf on X (resp. Y) with coefficients in E (resp. p^*E) (see 3.5). Then we have

$$p^*\mathcal{P}^{\cdot}_X(E) = \mathcal{P}^{\cdot}_Y(p^*E) \tag{1}$$

by 3.14 in [**B2**, V], whence the proposition.

5. Verdier duality and Poincaré duality. Third axiomatic characterization

We recall that, from now on, R is a field. We view the DGS as elements of the bounded derived category $D^{\mathrm{b}}(X)$. Therefore, equality stands for q.i. The constant sheaf on a space Y with stalk R is denoted R_Y. The dual of a vector space V over R is denoted V^*.

So far, no explanation has been given as to why middle intersection cohomology satisfies Poincaré duality, as claimed in 1.2. We take up this point here in the sheaf-theoretic framework. For the PL-theory, see [**GMP2**]. The dualizing sheaf, Verdier duality, and biduality can be discussed without assuming R to be a field, at the cost of some technical complications (see **[B2, V, §§7–9]**), but the case of a field suffices for the purposes of this paper.

5.1. The dualizing sheaf [B2, V, §7]. Let $\mathscr{K}^{\cdot}$ be a c-soft flat resolution of the constant sheaf R_X. The presheaf

$$(1) \qquad U \mapsto (\Gamma_c \mathscr{K}^i | U)^*$$

is a sheaf which is injective **[B2, V, 7.6]** and already flabby if $\mathscr{K}^{\cdot}$ is only assumed to be c-soft **[BM]**, called the dualizing sheaf and denoted $\mathscr{D}_X^{\cdot}$. By definition then

$$(2) \qquad \mathscr{D}_X(U) = (\Gamma_c \mathscr{K}^{-i} | U)^*,$$

therefore, D_X^i is zero for $i \notin [-n, 0]$. We have

$$(3) \qquad \mathbb{H}^i(U; \mathscr{D}_X^{\cdot}) = H^i(\Gamma_U(\mathscr{D}_X^{\cdot})) = (H_c^{-i}(U; R))^* = H_{-i}^{\mathrm{cl}}(U; R),$$

where H_j^{cl} refers, as before, to homology with closed supports. In the present case it is the homology of U with respect to locally finite PL-chains or the singular homology with closed supports. In general, it is the homology introduced in **[BM]**.

Let $x \in X$ and $f_x : x \to X$ the inclusion map. We have already remarked (2.5) that $H_c^j(U; R)$ is constant when U runs through a fundamental system of distinguished neighborhoods. By going over to the limit, we get, using 2.5(1)

$$(4) \qquad (\mathscr{H}^i \mathscr{D}_X^{\cdot})_x = H^{-i}(f_x^! R_X) \qquad (i \in \mathbf{Z};\ x \in X).$$

In **[BM]** the DGS $\mathscr{E}^{\cdot}$ defined by $\mathscr{E}^i = \mathscr{D}_X^{-i}$ is the fundamental homology sheaf, and we can also write

$$(5) \qquad H^j(f_x^! R_X) = \lim_{\substack{\to \\ U}} H_j^{\mathrm{cl}}(U; R).$$

(Note that, for homology with closed supports, there is a natural restriction map from an open set to a smaller one.)

If X is a manifold (oriented, as always), we have therefore

$$(6) \qquad \mathscr{H}^{-n}(\mathscr{D}_X^{\cdot}) = R_X \qquad \mathscr{H}^i(\mathscr{D}_X^{\cdot}) = 0 \quad (i \neq -n).$$

Consequently, if X is a manifold, then $\mathscr{D}_X^{\cdot}[-n]$ is an injective resolution of R_X, i.e., we have, in $D^{\mathrm{b}}(X)$

$$(7) \qquad \mathscr{D}_X^{\cdot} = R_X[n].$$

5.2. The dual of a DGS. The Verdier dual $D_X \mathscr{S}^{\cdot}$ of a DGS $\mathscr{S}^{\cdot}$ on X is

$$(1) \qquad D_X \mathscr{S}^{\cdot} = R\,Hom^{\cdot}(\mathscr{S}^{\cdot}, \mathscr{D}_X^{\cdot}) = Hom^{\cdot}(\mathscr{S}^{\cdot}, \mathscr{D}_X^{\cdot}).$$

(Recall that $\mathscr{D}_X^{\cdot}$ is injective.) In particular, if $\mathscr{S}^{\cdot}$ is concentrated in the interval $[a, b]$, then $D_X\mathscr{S}^{\cdot}$ is concentrated in $[-n-b, -a]$. It is equal to the dual complex of $\mathscr{S}^{\cdot}$ as defined in **[BM]** (see **[B2**, V, 7.8]) and is also sometimes called the Verdier-Borel-Moore dual of $\mathscr{S}^{\cdot}$. The functor $\mathscr{S}^{\cdot} \mapsto D_X\mathscr{S}^{\cdot}$ is exact and preserves q.i. as well as the conditions $\mathscr{X}$-clc and $\mathscr{X}$-cc (*loc. cit.*, 8.7). Moreover, if $\mathscr{S}^{\cdot}$ is $\mathscr{X}$-cc, then $D_X D_X \mathscr{S}^{\cdot} = \mathscr{S}^{\cdot}$ in the derived category $D^{\mathbf{b}}_{\mathscr{X}}(X)$ of DGS which are $\mathscr{X}$-cc **[B2**, V, 8.10]. This is the *biduality*; it can be proved under more general assumptions (see **[B2**, V, 8.11] for references).

If U is an open subset of X, then the definition of $\mathscr{D}_X^{\cdot}$ implies immediately

$$(2) \qquad \mathscr{D}_X^{\cdot}|U = \mathscr{D}_U^{\cdot}, \qquad (D_X\mathscr{S}^{\cdot})|U = D_U(\mathscr{S}^{\cdot}|U).$$

Using 2.8(3), we see that

$$(3) \qquad (D_X\mathscr{S}^{\cdot})[m] = D_X(\mathscr{S}^{\cdot}[-m]) \qquad (m \in \mathbf{Z}).$$

In the sequel, $D_X\mathscr{S}^{\cdot}[m]$ will stand for the left-hand side of (3), but the brackets will be reinstated if there is a danger of ambiguity.

5.3. Assume that X is a manifold. The relation 5.1(8) shows that

$$(1) \qquad D_X\mathscr{S}^{\cdot}[-n] = R\,Hom^{\cdot}(\mathscr{S}^{\cdot}, R_X).$$

Assume now that $\mathscr{S}^{\cdot}$ is a local system E concentrated in degree 0. Then the right-hand side of (1) is just the local system E^* dual to E, hence

$$(2) \qquad D_X E = E^*[n],$$

[B2, V, 7.10] and we get indeed

$$\begin{aligned} D_X D_X E &= Hom^{\cdot}(E^*[n], \mathscr{D}_X^{\cdot}) \\ &= Hom^{\cdot}(E^*, \mathscr{D}_X^{\cdot}[-n] = R\,Hom^{\cdot}(E^*, R_X) = E^{**} = E. \end{aligned}$$

5.4. As a generalization of 5.1(3), we have

$$(1) \qquad \mathbb{H}^i(U; D_X\mathscr{S}^{\cdot}) = \mathbb{H}_c^{-i}(U; \mathscr{S}^{\cdot})^* \qquad (U \text{ open in } X;\ i \in \mathbf{Z})$$

(see 7.7(3) in **[B2**, V], noting that $E_2^{p,q} = 0$ for $p \neq 0$, since R is assumed to be a field, and 7.8). For later reference, we write it as

$$(2) \quad \mathbb{H}^i(U; D_X\mathscr{S}^{\cdot}[-n]) = \mathbb{H}_c^{n-i}(U; \mathscr{S}^{\cdot})^* \qquad (U \text{ open in } X;\ i \in \mathbf{Z}).$$

In view of 2.5(1), it yields

$$(3) \qquad H^i((D_X\mathscr{S}^{\cdot})[-n])_x = H^{n-i}(f_x^!\mathscr{S}^{\cdot}) \qquad (x \in X,\ i \in \mathbf{Z}).$$

This shows that

$$(4) \qquad \mathscr{H}^i(D_X\mathscr{S}^{\cdot}[-n]) = 0 \quad \text{for } i < 0,$$

and that (iii) in 4.2 is equivalent to

(iii)″ dim supp $\mathcal{H}^j(D_X\mathcal{S}^{\cdot}[-n]) < n-2j\ (j>0)$.

Hence AX 2_E is equivalent to AX 3_E:

(i) $\mathcal{H}^i\mathcal{S}^{\cdot} = 0$ *for* $i<0$. $\mathcal{S}^{\cdot}$ *is* $\mathcal{Y}$*-cc for some PL-pseudomanifold stratification* $\mathcal{Y}$ *of* X*; there exists a dense open submanifold* U*, whose complement has codimension* ≥ 2 *and a local system* E *on* U *such that* $\mathcal{S}^{\cdot} = E$ *on* U;

(ii) dim supp $\mathcal{H}^i(\mathcal{S}^{\cdot}) < n-2i\ (i>0)$.

(iii) dim supp $\mathcal{H}^i(D_X\mathcal{S}^{\cdot}[-n]) < n-2i\ (i>0)$.

Theorem 4.3 then implies the first assertion of:

5.5. THEOREM. *The set of conditions AX* 3_E *determines* $\mathcal{S}^{\cdot}$ *in* $D^b(X)$ *uniquely and is satisfied by the middle intersection cohomology sheaf with respect to* E *and any PL-pseudomanifold structure on* X. *If* $\mathcal{S}^{\cdot}$ *satisfies AX* 3_E*, then* $D_X\mathcal{S}^{\cdot}[-n]$ *satisfies AX* 3_{E^*}.

Let $\mathcal{T}^{\cdot} = D_X\mathcal{S}^{\cdot}[-n]$. We claim first

$$D_X\mathcal{T}^{\cdot}[-n] = \mathcal{S}^{\cdot}. \tag{1}$$

In fact, $D_X\mathcal{T}^{\cdot}[-n] = RHom^{\cdot}(D_XS^{\cdot}[-n], \mathcal{D}^{\cdot}_X[-n])$, whence (see 2.8(3))

$$D_XD_X\mathcal{S}^{\cdot}[-n] = RHom^{\cdot}(D_X\mathcal{S}^{\cdot}, \mathcal{D}^{\cdot}_X) = D_XD_X\mathcal{S}^{\cdot} = \mathcal{S}^{\cdot},$$

the last equality following from biduality. But then it is clear that (ii) (resp. (iii)) for $\mathcal{S}^{\cdot}$ is the same as (iii) (resp. (ii)) for $\mathcal{T}^{\cdot}$.

There remains to see that $\mathcal{T}^{\cdot}$ satisfies (i) of AX 3_{E^*}. The relation $\mathcal{H}^i\mathcal{T}^{\cdot} = 0$ for $i<0$ is 5.4(4). By 5.2(2), $\mathcal{T}^{\cdot}|U$ is equal to $D_U\mathcal{S}^{\cdot}[-n]|U$. Since $\mathcal{S}^{\cdot}|U = E$ by assumption, the relation $\mathcal{T}^{\cdot}|U = E^*$ follows from 5.3(2).

5.6. COROLLARY (Poincaré duality). *There is a natural isomorphism*

$$I_{\mathbf{m}}H^i(X;\, E^*) = I_{\mathbf{m}}H_c^{n-i}(X;\, E)^* \qquad (i \in \mathbf{Z}). \tag{1}$$

Let $\mathcal{S}^{\cdot}$ satisfy AX 3_E. By 5.5

$$\mathcal{S}^{\cdot} = \mathcal{I}_{\mathbf{m}}\mathcal{C}^{\cdot}(X;\, E), \qquad D_X\mathcal{S}^{\cdot}[-n] = \mathcal{I}_{\mathbf{m}}\mathcal{C}^{\cdot}(X;\, E^*). \tag{2}$$

The corollary follows therefore from 5.4(2).

REMARK. Assume that the local system E is isomorphic to its contragredient, which is the case for instance if E is constant. Then (2) shows that a DGS satisfying AX 2_E or, equivalently, AX 3_E is self-dual up to a shift in grading. For n even, once we go over to the grading introduced in the next section, it will be a self-duality without further qualification (see end of 6.4).

III. PERVERSE SHEAVES

From now on, the dimension n *of* X *is even and* $r = n/2$.

6. Change of grading

6.1. The definition of a so-called perverse sheaf (actually, not a sheaf, but an element of $D^{b}(X)$) can be given now, namely, any DGS satisfying AX 3 in which the strict inequalities in (ii) and (iii) are replaced by weak ones (cf 8.3, Remark). However, this concept is introduced in **[BBD]** with respect to a different grading convention, which has a number of formal advantages. We first discuss this new grading, to be used henceforth, and reformulate the main previous results using it.

6.2. If $\mathcal{S}^{\cdot}$ is a DGS on X, then $\mathcal{H}^{i}\mathcal{S}^{\cdot}$ can be nonzero only for $i \in [0, n]$. We want to translate this interval to $[-r, r]$, i.e., to replace $\mathcal{S}^{\cdot}$ by $\mathcal{S}^{\cdot}[r]$. Let us write $\tilde{S}^{\cdot}$ for $\mathcal{S}^{\cdot}[r]$. If $\mathcal{S}^{\cdot}$ satisfies AX 2_{E}, then

(1) $$\tilde{S}^{\cdot}|U = E[r],$$

(2) $$\dim \operatorname{supp} \mathcal{H}^{i}\tilde{\mathcal{S}}^{\cdot} < -2i \qquad (i > -r),$$

(3) $$\dim \operatorname{supp}\{x \mapsto H^{i}_{\{x\}}\tilde{\mathcal{S}}^{\cdot}\} < 2i \qquad (i < r).$$

So far, dim always refers to the real dimension. From now on, a set of even real dimension $2i$ will also be said to have complex dimension $\dim_{\mathbf{C}}$ equal to i. Then 4.3, 4.5 yield:

6.3. THEOREM. *Let $\mathcal{S}^{\cdot}$ be a DGS on X. Assume it satisfies the conditions:*
(i) *$\mathcal{S}^{\cdot}$ is $\mathcal{Y}$-cc for some pseudomanifold stratification $\mathcal{Y}$ of X. We have $\mathcal{H}^{i}\mathcal{S}^{\cdot} = 0$ for $i < -r$. There exists a dense open submanifold U of X, whose complement has codimension ≥ 2, and a local system E on U such that $\mathcal{S}^{\cdot}|U = E[r]$.*
(ii) $\dim_{\mathbf{C}} \operatorname{supp} \mathcal{H}^{i}\mathcal{S}^{\cdot} < -i \ (i > -r)$.
(iii) $\dim_{\mathbf{C}} \operatorname{supp}\{x \mapsto \mathcal{H}^{i}_{\{x\}}\mathcal{S}^{\cdot}\} < i \ (i < r)$.
Then $\mathcal{S}^{\cdot}[-r]$ is q.i. to the middle intersection cohomology sheaf of X with respect to E and any pseudomanifold stratification of X.

Note that by 5.2(3), we have

(1) $$D_{X}\mathcal{S}^{\cdot}[-n] = (D_{X}(\mathcal{S}^{\cdot}[r]))[-r].$$

Therefore, (iii) in AX 3_{E} in 5.4 is equivalent to

(2) $$\dim_{\mathbf{C}} \operatorname{supp} \mathcal{H}^{i}(D_{X}(\mathcal{S}^{\cdot}[r])) < -i \qquad (i > -r),$$

whence

6.4. PROPOSITION. *Theorem* 6.3 *remains valid if* (iii) *is replaced by*
(iii)′ $\dim_{\mathbf{C}} \operatorname{supp} \mathcal{H}^{i} D_{X}\mathcal{S}^{\cdot} < -i \ (i > -r)$.

In view of biduality, $\mathcal{S}^{\cdot}$ satisfies (ii) and (iii) if and only $D_{X}\mathcal{S}^{\cdot}$ does so. If $\mathcal{S}^{\cdot}$ satisfies (i), (ii), and (iii′) for E, then $D^{\cdot}_{X}\mathcal{S}^{\cdot}$ satisfies those axioms for E^{*} and we have therefore

(1) $$\mathcal{S}^{\cdot} = \mathcal{I}_{\mathbf{m}}\mathcal{C}^{\cdot}(X; E)[r], \qquad D_{X}\mathcal{S}^{\cdot} = \mathcal{I}_{\mathbf{m}}\mathcal{C}^{\cdot}(X; E^{*})[r].$$

In particular, if E is isomorphic to E^*, then $\mathcal{S}^\cdot = D_X\mathcal{S}^\cdot$.

6.5. The trend in §§3–5 has been to get away from the stratification. But it will also be useful for the discussion of perverse sheaves to go back to conditions emphasizing a stratification, using however the present grading. The condition 3.3(ii) can also be expressed as

(1) $$\mathcal{H}^i\mathcal{S}^\cdot|S_{n-2k} = 0 \quad \text{for } i \geq k,$$

which amounts to

(2) $$H^i\tilde{S}^\cdot|S_{n-2k} = 0 \quad \text{for } i \geq k - r = -\dim_{\mathbf{C}}(S_{n-2k}),$$

i.e., to

(3) $$\mathcal{H}^i\tilde{\mathcal{S}}^\cdot|S = 0 \quad \text{for } i \geq -\dim_{\mathbf{C}} S \ (S \text{ any stratum of } \mathcal{X}).$$

In the notation of 2.4(6), we can also write 4.1(2) as

(4) $$\mathcal{H}^i_S\mathcal{S}^\cdot = 0 \quad \text{for } i \leq k \text{ and } S = S_{n-2k}.$$

(To be strict we have recalled in 2.4 the definition of the cohomology with supports only for a closed subset. It can be also given for a locally closed one. If Z is open in its closure $\overline{Z}$, then, as can be expected

$$\mathcal{H}^\cdot_Z\mathcal{S}^\cdot = \mathcal{H}^\cdot_{\overline{Z}}\mathcal{S}^\cdot|Z.$$

We take here the right-hand side as definition of $\mathcal{H}^\cdot_Z\mathcal{S}^\cdot$.) From (4), we get

(5) $$\mathcal{H}^i_S\tilde{\mathcal{S}}^\cdot = 0 \quad \text{for } i \leq \dim_{\mathbf{C}} S \ (S \text{ stratum of } X),$$

so that the translated middle intersection cohomology sheaf $\mathcal{I}_m\mathcal{C}^\cdot(X; E)[r]$ can also be characterized in the derived category of bounded $\mathcal{X}$-cc complexes by the conditions:

(i) $\mathcal{H}^i\mathcal{S}^\cdot = 0$ *for* $i < -r$ *and* $S^\cdot|U$ *is q.i. to* $E[r]$.
(ii) $\mathcal{H}^i\mathcal{S}^\cdot|S = 0$ *for* $i \geq -\dim_{\mathbf{C}} S$ (S *stratum of* $\mathcal{X}$),
(iii) $\mathcal{H}^i_S\mathcal{S}^\cdot = 0$ *for* $i \leq -\dim_{\mathbf{C}} S$ (S *stratum of* $\mathcal{X}$).

As in 6.4, using Poincaré duality, in particular 5.4(3), we obtain an equivalent set of conditions if we replace (iii) by

(iii)′ $\mathcal{H}^i D_X\mathcal{S}^\cdot|S = 0$ *for* $i \geq -\dim_{\mathbf{C}} S$ (S *stratum of* $\mathcal{X}$).

7. Perverse sheaves on pseudomanifolds

7.1. The perversity **m** (or the general one **p** introduced in 0.1) was defined on the set of integers in $[0, n]$. Actually this integer was always the codimension of a stratum, so that we could also have viewed the perversity as a function from the strata to the integers. This is the point of view of **[BBD]**, which we shall now adopt. There, many perversities are considered, more general in fact than those of **[GMP2]**, hence of **[B2]**, but again, we limit ourselves to one, the selfdual perversity $\mathbf{p} = \mathbf{p}_{1/2}$, defined by

(1) $\mathbf{p}(S) = -\dim_{\mathbf{C}} S \quad (S \text{ stratum of } \mathscr{X}).$

7.2. Definition. An element $\mathscr{S}^{\cdot} \in D^{\mathrm{b}}(X)$ is $\mathscr{X}$-perverse if it is $\mathscr{X}$-cc and satisfies the conditions:

(i) $\mathscr{H}^i\mathscr{S}^{\cdot}|S = 0$ for $i > \mathbf{p}(S)$
(ii) $\mathscr{H}^i D_X\mathscr{S}^{\cdot}|S = 0$ for $i > \mathbf{p}(S)$

for any stratum S of $\mathscr{X}$.

Following tradition, we shall say that $\mathscr{S}^{\cdot}$ is an $\mathscr{X}$-perverse sheaf, although it is rarely concentrated in one degree!

7.3. The discussion of the equivalence of certain conditions in §§3, 4, and 6, in which strict inequalities are replaced by weak ones and vice-versa, shows that if $\mathscr{S}^{\cdot} \in D^{\mathrm{b}}(X)$ is $\mathscr{X}$-cc, then it is $\mathscr{X}$-perverse if and only if it satisfies one of the two following conditions

(a) $\mathscr{H}^i\mathscr{S}|S = 0$ for $i > \mathbf{p}(S)$ and $\mathscr{H}^i_S\mathscr{S}^{\cdot} = 0$ for $i < \mathbf{p}(S)$, (S stratum of $\mathscr{X}$),
(b) $\dim_{\mathbf{C}} \operatorname{supp} \mathscr{H}^i\mathscr{S}^{\cdot} \leq -i$ and $\dim_{\mathbf{C}} \operatorname{supp}\left\{x \mapsto H^i_{\{x\}}(\mathscr{S}^{\cdot}) = 0\right\} \leq i$, for any $i \in \mathbf{Z}$.

The category of bounded $\mathscr{X}$-perverse sheaves will be denoted $D^{\mathrm{b}}_{\mathscr{X},c}$. It is clearly stable under D_X.

7.4. Section 2 of **[BBD]**, from which all this is taken, introduces perverse sheaves associated to general perversities so that there is always a superscript p floating around. With respect to **[BBD]** we have interchanged 7.2 and 7.3(a) (cf 2.1.5 and 2.1.16 there). Moreover, we have from the start imposed a constructibility condition, which is not the case in *loc. cit.* A $\mathscr{X}$-perverse sheaf there is an element of $D^{\mathrm{b}}(X)$ which satisfies 7.3(a). Let $D^{\mathrm{b}}_{\mathscr{X}}(X)$ be the category of $\mathscr{X}$-perverse sheaves in this sense. It is already an abelian category. This follows in **[BBD]** from general considerations on triangulated categories with a t-structure. Since they underline the whole treatment of derived (or triangulated) categories there, we discuss them briefly here, to give some idea of this context. We use a bit more of the language of derived or triangulated categories than before. Most of what we need is recalled in [**B2**, V, §5]. For details and references, see [**BBD**, §1].

7.5. *t*-Structures. We shall always work inside a derived category. If $\mathscr{C}$ is a derived category, $\mathscr{E}$ a full subcategory, and $m \in \mathbf{Z}$, then $\mathscr{E}[m]$ denotes the category of complexes $A^{\cdot}[m]$ $(A \in \mathscr{E})$. In a triangulated category the functor $A^{\cdot} \mapsto A^{\cdot}[m]$ is the $|m|$th iterate of the functor of translation $A^{\cdot} \mapsto A^{\cdot}[1]$ or of the inverse functor $A^{\cdot} \mapsto A^{\cdot}[-1]$ depending on whether m is > 0 or < 0.

Let $\mathscr{D}$ be a triangulated category. A t-structure on $\mathscr{D}$ is the datum of two strictly full subcategories $\mathscr{D}^{\leq 0}$ and $\mathscr{D}^{\geq 0}$ satisfying the conditions:

(i) $Hom(X, Y) = 0$ if $X \in \mathscr{D}^{\leq 0}$ and $Y \in \mathscr{D}^{\geq 0}[-1]$,

(ii) $\mathscr{D}^{\leq 0} \subset \mathscr{D}^{\leq 0}[-1]$ and $\mathscr{D}^{\geq 0} \supset \mathscr{D}^{\geq 0}[-1]$,

(iii) given $X \in \mathscr{D}$, there exists a distinguished triangle (A, X, B) with $A \in \mathscr{D}^{\leq 0}$ and $B \in \mathscr{D}^{\geq 0}[-1]$.

The *heart* of the t-structure is $\mathscr{D}^{\leq 0} \cap \mathscr{D}^{\geq 0}$.

(Cf. 1.3.1 in **[BBD]**.) If $\mathscr{D} = D(A)$ is the derived category of an abelian category A, the *natural t-structure* on $\mathscr{D}$ is given by letting $\mathscr{D}^{\leq 0}$ (resp. $\mathscr{D}^{\geq 0}$) be the category of complexes $C^{\cdot}$ such that $H^i C^{\cdot} = 0$ for $i > 0$ (resp. $i < 0$). Then the heart is A, identified to complexes concentrated in degree 0. In general the heart is an abelian subcategory, even an admissible one, meaning that it has some nice properties spelled out there, which is moreover stable under extensions **[BBD**, 1.3.6].

7.6. Let now $\mathscr{D} = D^{\mathrm{b}}_{\mathscr{X}, c}(X)$. Then there is a t-structure on $\mathscr{D}$ associated to the selfdual perversity, in which $\mathscr{D}^{\leq 0}$ (resp. $\mathscr{D}^{\geq 0}$) is the set of $\mathscr{S}^{\cdot}$ satisfying 7.2(i) (resp. 7.2(ii)) **[BBD**, 2.1.3]. Its heart is the category of $\mathscr{X}$-perverse sheaves, to be denoted $\mathfrak{P}_{\mathscr{X}}(X)$. By 2.1.14 and 2.1.15 in **[BBD]** we have

PROPOSITION. *Let $\mathscr{Y}$ be a refinement of $\mathscr{X}$. Then*

(1) $$D^{\mathrm{b}}_{\mathscr{X}, c}(X)^{\leq 0} \subset D^{\mathrm{b}}_{\mathscr{Y}, c}(X)^{\leq 0} \quad \textit{and} \quad D^{\mathrm{b}}_{\mathscr{X}, c}(X)^{\geq 0} \subset D^{\mathrm{b}}_{\mathscr{Y}, c}(X)^{\geq 0}.$$

In particular,

(2) $$\mathfrak{P}_{\mathscr{X}}(X) \subset \mathfrak{P}_{\mathscr{Y}}(X).$$

8. Perverse sheaves on algebraic varieties

From now on, X is an algebraic variety, over **C** unless otherwise stated, of pure dimension (as an algebraic variety) denoted r.

8.1. We have already pointed out in 0.2 that X admits at least one structure of $2r$-dimensional pseudomanifold, defined by a stratification which is a refinement of the one given by the iterated singular loci. Here, we shall not single one out but consider only stratifications associated to decreasing sequences of Zariski-closed subspaces. The theorems of **[V]** already quoted imply that if Z is a closed irreducible subvariety of X and $\mathscr{X}$ a stratification, then $\mathscr{X}$ has a refinement $\mathscr{Y}$ such that Z is the Zariski-closure of a connected component S of a stratum of $\mathscr{Y}$, which is Zariski-open in Z. As a consequence, any two stratifications admit a common refinement.

A DGS $\mathscr{S}^{\cdot}$ will be said to be cc (cohomologically constructible) if it is $\mathscr{X}$-cc for some stratification $\mathscr{X}$. If so, the results just stated imply that any stratification $\mathscr{Y}$ has a refinement $\mathscr{Y}'$ such that $\mathscr{S}^{\cdot}$ is $\mathscr{Y}'$-cc.

The subcategory of $D^{\mathrm{b}}(X)$ consisting of cohomologically constructible complexes will be denoted $D^{\mathrm{b}}_c(X)$. By definition, it is the union of the

$D^b_{\mathscr{X},c}(X)$ as $\mathscr{X}$ runs through the (algebraic) stratifications of X. Clearly, $D^b_{\mathscr{X},c}(X) \subset D^b_{\mathscr{Y},c}(X)$ if $\mathscr{Y}$ refines $\mathscr{X}$.

The set of stratifications, partially ordered by refinement, is left directed (8.1) hence the $D^b_{\mathscr{X},c}$, form a right directed set with respect to inclusion.

The definition of the autodual perversity $\mathbf{p} = \mathbf{p}_{1/2}$ is extended to all irreducible subvarieties Z by

(1) $$\mathbf{p}(Z) = -\dim Z.$$

Here and in the sequel, dim refers to the dimension in algebraic geometry.

8.2. Definition. An element $\mathscr{S}^{\cdot} \in D^b_c(X)$ is *perverse* if every irreducible closed subvariety Z' has a Zariski-open dense subset Z such that

(1) $$\mathscr{H}^i\mathscr{S}^{\cdot}|Z = \mathscr{H}^i D_X\mathscr{S}^{\cdot}|Z = 0 \quad \text{for} \quad i > \mathbf{p}(Z).$$

Here too, we shall say that $\mathscr{S}^{\cdot}$ is a perverse sheaf.

8.3. Proposition. *Let $\mathscr{S}^{\cdot} \in D^b_c(X)$. Then the following three conditions are equivalent:*

(a) *$\mathscr{S}^{\cdot}$ is perverse.*

(b) *Any irreducible closed subvariety Z' has a dense Zariski-open subset Z such that*

(1) $$\mathscr{H}^i\mathscr{S}^{\cdot}|Z = 0 \ \text{for}\ i > \mathbf{p}(Z) \ \text{and}\ \mathscr{H}^i_Z\mathscr{S}^{\cdot} = 0 \ \text{for}\ i < \mathbf{p}(Z).$$

(c) *$\mathscr{S}^{\cdot}$ is $\mathscr{X}$-perverse* (7.2) *for some stratification $\mathscr{X}$ of X.*

Proof. Assume (a). There exists $\mathscr{X}$ such that $\mathscr{S}^{\cdot}$ is $\mathscr{X}$-cc. Then 7.2(1) (i), (ii) are special cases of 8.2(1); hence, $\mathscr{S}^{\cdot}$ satisfies (c). Assume now the latter, and let Z' be a closed irreducible subvariety of X. There exists a stratum S of $\mathscr{X}$ such that $Z' \cap S$ contains a dense Zariski-open subset Z of Z'. Then 7.2 (i) (ii) imply

$$\mathscr{H}^i\mathscr{S}^{\cdot}|Z = \mathscr{H}^i D_X\mathscr{S}^{\cdot}|Z = 0 \quad \text{for} \quad i > \mathbf{p}(S).$$

Since, obviously, $\mathbf{p}(S) \leq \mathbf{p}(Z)$, this yields (a) hence (a) $\Leftrightarrow$ (c).

Assume (b), and let $\mathscr{X}$ be such that $\mathscr{S}^{\cdot}$ is $\mathscr{X}$-cc. Then, in view of 7.3, $\mathscr{S}^{\cdot}$ is $\mathscr{X}$-perverse; hence, (b) implies the equivalent conditions (a) and (c). Assume now (c), and let Z be an irreducible closed subvariety of X. As recalled in 8.1, there exists a refinement $\mathscr{Y}$ of $\mathscr{X}$ such that Z is the closure of a component S of a stratum of $\mathscr{Y}$. By 7.6, $\mathscr{S}^{\cdot}$ is also $\mathscr{Y}$-perverse. But then 7.3(a) for S implies (1) here for Z, whence (b).

Remark. If $\mathscr{S}^{\cdot}$ is perverse, then there exists a Zariski-open subset U of X and a local system E on U such that $\mathscr{S}^{\cdot}|U = E[r]$. To see this, note that for Z open in X, $\mathscr{H}^{\cdot}_Z\mathscr{S}^{\cdot}$ is just the restriction of $\mathscr{H}^{\cdot}\mathscr{S}^{\cdot}$ to Z. Therefore (1) implies the existence of such a Z on which $\mathscr{H}^i\mathscr{S}^{\cdot} = 0$ for $i \neq -r$. Thus, $\mathscr{S}^{\cdot}$ reduces on Z to a sheaf concentrated in degree $-r$; but it is also $\mathscr{X}$-cc for some stratification $\mathscr{X}$, so $\mathscr{S}^{\cdot} = \mathscr{H}^{-r}\mathscr{S}^{\cdot}$ is locally constant on some subset U which is dense and Zariski-open in Z, hence also in X. Therefore, a perverse sheaf always satisfies (i) of 6.3, and the conditions imposed on a perverse sheaf differ from those characterizing $\mathcal{I}_{\mathbf{m}}\mathcal{C}^{\cdot}(X; E)[r]$ in 6.3 only by a weakening of (ii) and (iii). In particular, the latter is perverse.

8.4. We let $\mathfrak{P}(X)$ denote the category of perverse sheaves on X. It is also stable under D_X. By 8.3, it is the inductive limit under inclusions of the categories $\mathfrak{P}_{\mathcal{X}}(X)$. By 7.6, the t-structures on the various $D^{\mathrm{b}}_{\mathcal{X},c}(X)$ are compatible with inclusion; therefore, there is on $D^{\mathrm{b}}_c(X)$ a t-structure in which

$$(1) \qquad D^{\mathrm{b}}_c(X)^{\leq 0} = \bigcup_{\mathcal{X}} D^{\mathrm{b}}_{\mathcal{X},c}(X)^{\leq 0}, \quad D^{\mathrm{b}}_X(X)^{\geq 0} = \bigcup_{\mathcal{X}} D^{\mathrm{b}}_{\mathcal{X},c}(X)^{\geq 0},$$

the heart of which is $\mathfrak{P}(X)$. In particular, (7.5) $\mathfrak{P}(X)$ is an abelian subcategory of $D^{\mathrm{b}}_c(X)$, stable under extensions. In fact, it is much more (4.3.1, 4.3.2 in **[BBD]**):

8.5. Theorem. (i) *The category* $\mathfrak{P}(X)$ *is abelian, stable under* D_X *and extensions, noetherian and artinian. In particular, every object is of finite length.*

(ii) *Let* Z *be an irreducible closed subvariety,* U *a Zariski open smooth subset of* Z, E *a simple local system on* E, *and* $i : Z \to X$ *the inclusion. Then* $i_*(\mathcal{I}_{\mathbf{m}}\mathcal{C}^{\cdot}(Z; E)[\dim Z])$ *is a simple object of* $\mathfrak{P}(X)$ *and every simple object of* $\mathfrak{P}(X)$ *is of this form.*

(A local system is simple if it is defined by an irreducible representation of the fundamental group.) To be strict, the simple objects of $\mathfrak{P}(X)$ are defined differently in **[BBD**, 4.3], see the end of the next section.

8.6. Intermediate extension. Let $A \subset B$ be locally closed subvarieties of X and $i : A \to B$ the inclusion. **[BBD]** defines a functor $i_{!*}$, *the intermediate extension*, which to a perverse sheaf $\mathcal{S}^{\cdot}$ on A associates a perverse extension of $\mathcal{S}^{\cdot}$ to B. If A is closed in B, this is just the extension by zero. Assume A is open in B. Then $\mathcal{T}^{\cdot} = i_{!*}\mathcal{S}^{\cdot}$ is characterized by the condition:

(1) Every closed subvariety Z' of $B - A$ has a dense Zariski-open subvariety Z such that

$$(1) \qquad \mathcal{H}^i\mathcal{T}^{\cdot}|Z = 0 \text{ for } i \geq \mathbf{p}(Z) \quad \text{and} \quad \mathcal{H}^i_Z\mathcal{T}^{\cdot} = 0 \text{ for } i \leq \mathbf{p}(Z).$$

Again, as in 8.5, if B has a stratification such that A is union of strata, it suffices to require that condition for the strata not in A. (Note that in (1) we have weak inequalities rather than strict ones as in 8.3, so that these conditions are stronger.) This functor has also a transitivity property with respect to the inclusions $A \to B \to X$ **[BBD**, 2.1.7.1].

The construction of 3.5 is a special case. To describe it, let us now take $B = X$ and fix a stratification $\mathcal{X}$ of X.

For $k \in \mathbf{Z}$ let U_k be the union of the strata of dimension $\geq -k$. It is open, and we have

(1) $$U_k = \varnothing \quad \text{if } k < -r, \quad U_k = X \quad \text{if } k \geq 0,$$

(2) $$U_k \subset U_{k+1} \quad \text{and} \quad U_{k+1} = U_k \cup S_{-k-1} \quad (k \in \mathbf{Z}),$$

where S_j denotes the (total) stratum of dimension j. Note also that U_{-r} is the open set on which E is defined, denoted U_2 in 3.5. Let $i_k : U_k \to U_{k+1}$ be the inclusion. In 3.5 we introduced Deligne's operation to extend a DGS on U_k to one on U_{k+1}. In our present notation and grading convention, it is

(3) $$\tau_{\leq k} Ri_{k*} \qquad (k \in [-r, -1]).$$

This is $i_{k!*}$. Deligne's sheaf $\mathscr{P}^{\cdot}(E)$ is obtained from E on U_{-r} by applying

(4) $$\tau_{\leq -1} \circ Ri_{-1*} \circ \tau_{\leq -2} \circ Ri_{-2*} \circ \cdots \circ \tau_{\leq -r} Ri_{-r*}$$

which, by (3) and the transitivity of the operation of intermediate extension, is equal to $i_{!*}$, where $i = U_{-r} \to X$ is the inclusion. Therefore,

(5) $$\mathscr{P}^{\cdot}(E)[r] = i_{!*}(E[r]).$$

Let now U be a smooth locally closed irreducible subvariety of X, d its dimension, Z its Zariski closure, and E a local system on U. Let $k : U \to Z$ and $j : Z = X$ be the canonical inclusions and $i = j \circ k$. Then $k_{!*}(E[d]) = \mathscr{I}_{\mathbf{m}}\mathscr{C}^{\cdot}(Z; E)[d]$, as just explained. On the other hand, $j_{!*} = j_*$ is the extension by zero and $i_{!*} = j_{!*} \circ k_{!*}$. Therefore,

(6) $$j_*(\mathscr{I}_{\mathbf{m}}\mathscr{C}^{\cdot}(Z; E)[d]) = i_{!*}(E[d]).$$

For E simple, we can replace in 8.5(ii) the left-hand side of (6) by the right-hand side, which yields the description of the simple objects of $\mathfrak{P}(X)$ given in [BBD].

8.7. Morphisms. If $f : Y \to X$ is a proper morphism of irreducible algebraic varieties, then the all-important *decomposition theorem* is valid [BBD, 6.2.5]. It asserts, in particular, that $Rf_*(\mathscr{I}_{\mathbf{m}}\mathscr{C}^{\cdot}(Y; \mathbf{Q})[s])$ ($s = \dim_{\mathbf{C}} Y$) is a direct sum of objects of the form $j_{!*}(E[d])$, where E is a local system (over $\mathbf{Q}$) on a locally closed smooth subvariety U and $j : U \to X$ the inclusion, but where d is not necessarily the dimension of U.

This last *proviso* reflects the fact that $Rf_*(\mathscr{I}_{\mathbf{m}}\mathscr{C}^{\cdot}(Y; \mathbf{Q})[s])$ is not necessarily perverse. It is, however, if Y is smooth (or rationally smooth), dim $Y = \dim X$, and f is *semismall.* The latter condition means that for $i \in \mathbf{N}$, the set of $x \in X$ for which $\dim f^{-1}(x) \geq i$ has dimension $\leq r - 2i$ [BMP].

If Y is smooth, then $\mathscr{I}_{\mathbf{m}}\mathscr{C}^{\cdot}(Y; \mathbf{Q}) = \mathbf{Q}_Y$. If f is moreover a resolution of singularities of X, then $\mathscr{I}_{\mathbf{m}}\mathscr{C}^{\cdot}(X; \mathbf{Q})$ occurs among the summands of $Rf_*\mathbf{Q}_Y$; therefore, $I_{\mathbf{m}}H^{\cdot}(X; \mathbf{Q})$ is a direct summand of $H^{\cdot}(Y; \mathbf{Q})$.

In [S], M. Saito proves a similar decomposition theorem for a proper morphism of irreducible complex analytic varieties, assuming that the domain space is the image of a smooth Kähler manifold by a proper surjective morphism.

8.8. The Riemann-Hilbert correspondence. It underlies the use of $\mathscr{D}$-modules in representation theory (cf. Milicic's lectures). For the terminology on $\mathscr{D}$-modules we refer to those lectures or **[B3]**.

Let X be quasi-projective. Then the de Rham functor DR establishes an equivalence of categories between the bounded derived category $D^{\mathrm{b}}_{\mathrm{rh}}(X)$ of (algebraic) regular holonomic systems and the category $D^{\mathrm{b}}_{\mathrm{c}}(X)$, which induces an equivalence between the categories of regular holonomic modules and that of perverse sheaves **[B3, VIII]**.

The proof of this last assertion **[B3**, §22] shows more precisely that the de Rham functor maps the natural t-structure on $D^{\mathrm{b}}_{\mathrm{rh}}(X)$ onto the t-structure defined on $D^{\mathrm{b}}_{c}(X)$ by the self-dual perversity (cf. 7.5) and, hence, yields an equivalence between the hearts of these two t-structures, which are respectively the regular holonomic modules and the perverse sheaves.

8.9. Sections 8.1 to 8.7 carry over, mutatis mutandi, in positive characteristic when X is a variety over the algebraic closure of a finite field, in the étale topology, and using l-adic sheaves **[BBD]**. In fact, the decomposition theorem was derived there from results in characteristic p (the purity theorem).

8.10. The last sections arose from a suggestion by L. Scott, namely, to include 8.14 in the case where P is solvable and Q is subminimal, to be referred to in the contribution of E. Cline, B. Parshall and L. Scott to this volume.

From now on Y is an algebraic variety, $f: Y \to X$ is a morphism which is the projection map of a locally trivial (in the Zariski topology) algebraic fibration with irreducible *smooth* fibres, and d is the dimension of a typical fibre F. We choose a stratification $\mathscr{X}$ of X. Then the inverse images of the strata of $\mathscr{X}$ are smooth and are the strata of a stratification $\mathscr{Y}$ of Y.

We let $f^*[d]: D^{\mathrm{b}}(X) \to D^{\mathrm{b}}(Y)$ be the functor which assigns $f^*\mathscr{S}^{\cdot}[d]$ to $\mathscr{S}^{\cdot}$. Thus

$$(2) \qquad f^*[d](\mathscr{S}^{\cdot}) = f^*\mathscr{S}^{\cdot}[d] = f^*(\mathscr{S}^{\cdot}[d]), \qquad (\mathscr{S}^{\cdot} \in D^{\mathrm{b}}(X)).$$

8.11. PROPOSITION. *We keep the notation and assumptions of* 8.10. *If* $\mathscr{S}^{\cdot}$ *is a simple perverse sheaf on* X, *then* $f^*[d]\mathscr{S}^{\cdot}$ *is one on* Y. *In particular,* $f^*[d]$ *maps* $\mathfrak{P}(X)$ *into* $\mathfrak{P}(Y)$ *and if* f *is proper,* $f^*[d] \circ Rf_*$ *satisfies the decomposition theorem.*

PROOF. We recall that f^*, hence $f^*[d]$, is exact and in particular commutes with the formation of a derived sheaf.

Assume the first assertion has been proved. Since f^* is exact, 8.5 then implies that $f^*[d]$ maps $\mathfrak{P}(X)$ into $\mathfrak{P}(Y)$. Also, it shows that if $\mathcal{S}^\cdot$ is a direct sum of simple elements of $\mathfrak{P}(X)$, up to shifts in degrees, then $f^*[d]\mathcal{S}^\cdot$ is so in $D^b(Y)$. Now if f is proper, Rf_* satisfies the decomposition theorem (8.7) and the last assertion follows. There remains to show the first one.

Let Z be a closed irreducible subvariety of X, E a simple locally constant system on a dense Zariski-open subset U of Z, s the dimension of Z, and $i: Z \to X$ the inclusion map. Then $i_*(\mathcal{I}_{\mathbf{m}}\mathcal{C}^\cdot(Z; E)[s])$ is a simple object in $\mathfrak{P}(X)$ and every simple perverse sheaf is of this type (8.5). It suffices therefore to show that $f^*(i_*(\mathcal{I}_{\mathbf{m}}\mathcal{C}^\cdot(Z; E)))[s+d]$ is a simple object in $\mathfrak{P}(Y)$. Note that by 4.5 there is a certain leeway in the choice of U. We may therefore assume, after shrinking U if necessary, that the fibration is trivial over U. Then $f^{-1}(U) \simeq U \times F$. Since F is irreducible, it is clear that f^*E is a simple local system on $f^{-1}U = V$. Let W be the Zariski closure of V and $j: W \to Y$. Since W has dimension $s+d$, it suffices to show, by 8.5 again

(1) $$f^*(i_*(\mathcal{I}_{\mathbf{m}}\mathcal{C}^\cdot(Z; E)))[s+d] = j_*(\mathcal{I}_{\mathbf{m}}\mathcal{C}^\cdot(W; f^*E))[s+d].$$

The variety W is the inverse image of Z, and both i_* and j_* are the extension by zero. We are therefore reduced to proving

(2) $$f^*(\mathcal{I}_{\mathbf{m}}\mathcal{C}^\cdot(Z; E)) = \mathcal{I}_{\mathbf{m}}\mathcal{C}^\cdot(W; f^*E),$$

but this follows from 4.7.

8.12. The counterpart of 8.11 in the l-adic setting is proved in **[BBD]**; see 4.2.5 and 4.2.6.2 there.

In view of 8.6(6), a more elegant statement including 8.11(1) is that f^* commutes with intermediate extensions. A sketch of a proof is given in *loc. cit.*, bottom of p. 110. This can also be carried out in our framework. The intermediate extension is a succession of elementary steps of the type 8.6(3), i.e., direct image followed by truncation. The problem is local on the base. In our situation, we are therefore reduced to the following problem:

Let U be an open subset of X over which the fibration is trivial. It is viewed as stratified by its intersections with the strata of $\mathscr{X}$. Let $V \subset V'$ be open in U, where V is a union of strata and $V' = V \cup S$, with S a stratum, say of dimension m, not contained in V. The inverse images of V, V', S are isomorphic to the products $V \times F$, $V' \times F$, $S \times F$ respectively, and $S \times F$ is a stratum of dimension $m+d$ for the induced stratification on $f^{-1}U$. Moreover, f restricts on each of these products to the projection on the first factor. Consider the following commutative diagram of maps

$$\begin{array}{ccc} V \times F & \xrightarrow{i'} & V' \times F \\ \downarrow f & & \downarrow f' \\ V & \xrightarrow{i} & V' \end{array}$$

where i, i' are inclusion and f, f' the projections on the first factor. Then we have to show

(1) $$f'^*[d](\tau_{\leq -m-1} Ri_*) = \tau_{\leq -m-d-1} Ri'_* f^*[d] \quad \text{on } D^b(V).$$

By 3.13 in [B2, V], we have

(2) $$Ri'_* f^* = f'^* Ri_* \quad \text{on } D^b(V),$$

hence

(3) $$Ri'_* \circ f^*[d] = f'^*[d] \circ Ri_* \quad \text{on } D^b(V).$$

So it suffices to show

(4) $$f'^*[d] \circ \tau_{\leq s} = \tau_{\leq s-d} \circ (f'^*[d]) \quad \text{on } D^b(V') \ (s \in \mathbf{Z}).$$

This follows from the definition of truncation (2.6) and the exactness of f'^*.

8.13. In order to reduce the proof of 8.11 to that of the first assertion, we have used the full force of 8.5. However, the fact that $f^*[d]$ maps $\mathfrak{P}(X)$ into $\mathfrak{P}(Y)$ can also be checked directly. We sketch the argument. We shall have to use the relation

(1)
$$f^* R\,Hom^{\cdot}(\mathcal{S}^{\cdot}, \mathcal{T}^{\cdot}) = R\,Hom^{\cdot}(f^*\mathcal{S}^{\cdot}, f^*\mathcal{T}^{\cdot}) \qquad (\mathcal{S}^{\cdot}, \mathcal{T}^{\cdot} \in D^b(X)).$$

This is a local statement on X, in the analytic topology. Since the fibration is locally trivial, the proof of (1) is reduced to the case where $Y = X \times F$ and f is the first projection. It then follows from 10.21 in [B, V].

Next, we claim

(2) $$f^*[d]\mathcal{D}^{\cdot}_X = \mathcal{D}^{\cdot}_Y[-d].$$

This assertion is local on X, so we may again assume $Y = X \times F$ and f to be the projection on the first factor. By 10.26 in [B, V], we have

(3) $$f^*\mathcal{D}^{\cdot}_X \overset{L}{\otimes} g^*\mathcal{D}^{\cdot}_F = \mathcal{D}^{\cdot}_Y,$$

where $g : Y \to F$ is the second projection and $\otimes^L$ refers to the left derived functors of $\otimes$ (i.e., the tensor product of one factor by a flat resolution of the other [B2, V, 6.2]). Since F is a manifold of real dimension 2d, 5.1(7) gives $\mathcal{D}^{\cdot}_F = \mathbb{C}_F[2d]$, hence $g^*\mathcal{D}^{\cdot}_F = \mathbb{C}_Y[2d]$. It is free, so that $\otimes^L$ is just the ordinary tensor product and (3) now reads

(4) $$f^*\mathcal{D}_X[2d] = \mathcal{D}^{\cdot}_Y,$$

which is obviously equivalent to (2).

In view of the relation between $\mathscr{X}$ and $\mathscr{Y}$, it is clear that if $\mathscr{S}^{\cdot}$ is $\mathscr{X}$-cc, then $f^*[d]\mathscr{S}^{\cdot}$ is $\mathscr{Y}$-cc. We have, using (1) and (4):

$$f^*D_X\mathscr{S}^{\cdot}[d] = f^*(R\,Hom^{\cdot}(\mathscr{S}^{\cdot}, \mathscr{D}^{\cdot}_X[d]))$$
$$= R\,Hom^{\cdot}(f^*\mathscr{S}^{\cdot}, f^*\mathscr{D}^{\cdot}_X[d]) = R\,Hom^{\cdot}(f^*\mathscr{S}^{\cdot}, \mathscr{D}^{\cdot}_Y[-d]).$$

Together with 2.8(3), this yields

$$(5)\qquad f^*D_X\mathscr{S}^{\cdot}[d] = D_Y(f^*\mathscr{S}^{\cdot}[d]).$$

Assume now $\mathscr{S}^{\cdot}$ to be perverse. It is then $\mathscr{X}$-perverse, and it suffices to show that $f^*\mathscr{S}^{\cdot}[d]$ is $\mathscr{Y}$-perverse (8.3). Since it is $\mathscr{Y}$-cc, as already remarked, there remains to check that it satisfies (i) and (ii) of 7.2. Let S be a stratum of $\mathscr{Y}$ and $T = f(S)$. By construction, it is a stratum of $\mathscr{X}$ and $S = f^{-1}(T)$. By assumption, there is a dense Zariski-open subset T' of T such that

$$(6)\qquad \mathscr{H}^i\mathscr{S}^{\cdot}|T' = \mathscr{H}^iD_X\mathscr{S}^{\cdot}|T' = 0 \text{ for } i > -\dim T.$$

(Recall that dim now refers to the complex dimension.) We have

$$\mathscr{H}^i(f^*\mathscr{S}^{\cdot}[d]) = \mathscr{H}^{i+d}(f^*\mathscr{S}^{\cdot}) = f^*\mathscr{H}^{i+d}\mathscr{S}^{\cdot}.$$

Since $\dim S = \dim T + d$, (6) gives

$$(7)\qquad \mathscr{H}^i(f^*\mathscr{S}^{\cdot}[d])|S' = 0 \quad \text{for } i > -\dim S$$

where $S' = f^{-1}T'$. This proves (i) of 7.2. Using (5), we get

$$(8)\qquad \mathscr{H}^iD_Y(f^*\mathscr{S}^{\cdot}[d]) = \mathscr{H}^{i+d}f^*D_X\mathscr{S}^{\cdot} = f^*\mathscr{H}^{i+d}D_X\mathscr{S}^{\cdot};$$

so that (6) implies

$$(9)\qquad (\mathscr{H}^iD_Y(f^*\mathscr{S}^{\cdot}[d]))|S' = 0 \text{ for } i > -\dim S,$$

which is (ii) of 7.2.

REMARK. (4) and (5) can be written

$$f^*\mathscr{D}^{\cdot}_X[\dim_{\mathbf{R}} F] = \mathscr{D}^{\cdot}_Y, \qquad f^*D_X\mathscr{S}^{\cdot}[\dim_{\mathbf{R}} F] = D_Yf^*\mathscr{S}^{\cdot} \quad (\mathscr{S}^{\cdot} \in D^b_c(X)).$$

In this form, they are valid more generally if X and Y are pseudomanifolds and f the projection of a locally trivial fibration, the fibres of which are connected manifolds (possibly of odd dimension). Similarly, (1) also holds under those assumptions. The proofs are the same.

In the last section, we shall assume some familiarity with the theory of linear algebraic groups (see, e.g., [**B1**]).

8.14. PROPOSITION. *Let G be a connected affine algebraic group (over $\mathbb{C}$), $P \subset Q$ parabolic subgroups of G, $f: G/P \to G/Q$ the canonical projection,* *and* d *the dimension of* Q/P. *Then* $f^*[d]$ *and* $f^*[d] \circ Rf_*$ *satisfy the conclusion of* 8.11.

PROOF. It suffices to prove that $Y = G/P$, $X = G/Q$, and f satisfy the assumptions of 8.11. First, dividing out by the radical of G, we are reduced to the case where G is semisimple. Let U be the unipotent radical of a parabolic subgroup opposite to Q. Then the projection $p : G \to G/Q$ maps U isomorphically onto its image C, which is Zariski open in G/Q [**B2**, 21.20]. The fibration of G by Q, which is principal, is then trivial over C, hence over any translate of C, and is therefore locally trivial in the Zariski topology. *A fortiori*, the fibration of G/P over G/Q, with typical fibre Q/P and projection f, is locally trivial. Since Q/P is an irreducible smooth projective variety, the assumptions of 8.11 are indeed fulfilled.

In this review, I have of course followed the existing literature, progressing from IC to perverse sheaves. But the reader should be warned that this point of view may soon become passé. I understand that in his Colloquium Lectures at the January, 1991 AMS meeting R. MacPherson proposed to go the other way: The perverse sheaves, being the most beautiful objects of the theory, should be the starting point; intersection homology should come later and be derived from the theory of perverse sheaves.

REFERENCES

[**BBD**] A. A. Beilinson, J. Bernstein, et P. Deligne, *Faisceaux pervers*, Astérisque **100** (1982).

[**B1**] A. Borel, *Linear algebraic groups*, Second enlarged edition, Graduate Texts in Math., vol. 126, Springer-Verlag, New York, 1991.

[**B2**] A. Borel et al., *Intersection cohomology*, Progr. Math., vol. 50, Birkhäuser, Boston, 1984.

[**B3**] ——, *Algebraic D-modules*, Perspect. Math., vol. 2, Academic Press, Boston, 1987.

[**BM**] A. Borel and J. C. Moore, *Homology theory for locally compact spaces*, Michigan Math. J. 7 (1960), 137-159.

[**BMP**] W. Borho and R. MacPherson, *Partial resolutions of nilpotent varieties*, Astérisque **101–102** (1983), 23–74.

[**GMP1**] M. Goresky et R. MacPherson, *La dualité de Poincaré pour les espaces singuliers*, C. R. Acad. Sci. Paris Sér. A I Math **284** (1977), 1549–1551.

[**GMP2**] ——, *Intersection homology theory*, Topology **19** (1980), 135–162.

[**GMP3**] ——, *Intersection homology*. II, Invent. Math. **71** (1983), 77–129.

[**HS**] N. Habegger and L. Saper, *Intersection cohomology of cs-spaces and Zeeman's filtration*, Invent. Math. **105** (1991), 247–272.

[**KL**] D. Kazhdan and G. Lusztig, *Schubert varieties and Poincaré duality*, Proc. Sympos. Pure Math., vol. 36, Amer. Math. Soc., Providence, RI, 1980, pp. 185–203

[**K**] S. L. Kleiman, *The development of intersection homology theory*, Century of Mathematics in America, vol. II, Amer. Math. Soc., Providence, RI, 1989, pp. 543–585.

[**S**] M. Saito, *Decomposition theorem for proper Kähler morphisms*, Tôhoku Math. J. **42** (1990), 127–148.

[**V**] J.-L. Verdier, *Stratifications de Whitney et théorème de Bertini-Sard*, Invent. Math. **36** (1976), 295-312.

INSTITUTE FOR ADVANCED STUDY, PRINCETON, NJ 08540, USA
E-mail address: borel@math.ias.edu

153.

On the Place of Mathematics in Culture[0]

in *Duration and Change, fifty years at Oberwolfach*, (M. Artin, H. Kraft, R. Remmert eds.), Springer-Verlag (1994) 139–158

The place of mathematics in our society strikes me as a rather peculiar one. It is certainly somewhere in the catalogue of our activities, even generally, if sometimes ruefully, perceived as an item of growing importance, it being obvious that the evermore encroaching technology we have to deal with has a mathematical underpinning. But where should it be classified? Is it viewed as part of culture? Hardly so: a "cultured person" is expected to have a knowledge and appreciation of literature, history, art, music, past civilizations, etc, but not necessarily of science and not at all of mathematics. To qualify for such a label, it is mandatory to know something about such towering figures as Plato, Michel-Angelo, Goethe, V.Hugo, Peter the Great, Napoleon, etc, but it is no liability to be blissfully ignorant about the greatest mathematicians, such as L.Euler, C.F.Gauss, D.Hilbert or H.Poincaré, if not a source of some pride. In his book: "*The Two Cultures*", C.P.Snow argued indeed that the concept of culture should comprise a scientific culture besides the traditional humanistic one, on equal footing with it. But even there, mathematics is cursorily treated and I, for one, could not see whether it was in his mind part of either.

A great variety of opinions has been (and is being) held about mathematics, from high praise to downright contempt: "Queen of Science", reflecting the often held view that mathematics is the language of science and that in fact no discipline is properly speaking a science until it can be expressed mathematically, or "Servant of Science", implying that its main, if not sole, function and justification is the solution of problems raised in natural sciences, technology, etc, to help us deal with the real world. It has, however,

[0] Faculty lecture given at the Institute for Advanced Study on October 16, 1991. It overlaps with an earlier lecture published first in German: *Mathematik: Kunst und Wissenschaft*, Themenreihe XXXIII, C.F.v.Siemens Stiftung, München 1982 and then, in a slightly modified form, in an English translation: *Mathematics: Art and Science*, Mathematical Intelligencer 5, n° 4 (1983), 9–17, Springer-Verlag. I shall allow myself to borrow from it without further references, a course of action which would be frowned upon in the case of a mathematical paper, and for which I apologize to anyone who has read it and still remembers it.

also been maintained that it has an artistic component. On the occasion of a Thesis defense (in 1845), the candidate asserted that mathematics was "art and science", to which an opponent retorted it was "only art, not science". A long line of thinkers has also viewed the intensive practice of mathematics with some suspicion, finding it had a numbing effect on the mind, making it ill prepared for nobler pursuits such as theology or philosophy (a criticism I can live with). A few have not hesitated to view mathematics as a mere collection of tautologies: after all, the argument goes, you start from self-evident truths, called axioms, and manipulate them with elementary logical operations. So what can you expect to get?[(1)]

Needless to say, I shall try to present a more positive view of mathematics than that last one. As a starter, I would like to compare it with an iceberg. For people with no lost love for mathematics, this may seem to evoke remoteness, coldness, a forbidding and threatening object to stay away from. I have in mind a different point of comparison, however, namely that it consists of a small visible part and a much bigger invisible one. By visible part I mean the one which interacts with other activities of obvious importance, such as natural sciences, technology, computer science, etc, and is clearly useful, worth financial support, even by tax payers' money. There is no doubt that this visible part is at the origin of mathematics, which was born in antiquity from the need to solve very practical problems, such as measuring of land or quantities, book accounting, keeping track of the movements of celestial bodies, then engineering, building canals, bridges, etc. But as mathematics grew, it started to acquire a life of its own. Mathematicians began to think about problems regardless of whether they had any applications. In short, they engaged in "pure mathematics" or "mathematics for mathematics' sake". This is what I refer to as the invisible part of mathematics, i.e. invisible to the layman, sorry, to the layperson. When apprised of that distinction, the outsider is often surprised and wonders as to how worthwhile such an activity can be: are there really still problems to be solved? How do you find them? Are they just mind teasers, such as you see in some Sunday papers, or is there a hierarchy? Are some problems more important than others? If so, what are your criteria? Our first answer is that we indeed have criteria, and they are mostly aesthetical in nature. We weave patterns of certain ideas, as a painter weaves patterns of forms and colors, a composer of sounds, a poet of words, and we are acutely sensitive to elegance, harmony in proofs, in statements, and the handsome development of a theory. Mathematicians have often waxed eloquent on this, so, for instance, Poincaré in 1897:

> In addition to this it provides its disciples with pleasures similar to painting and music. They admire the delicate harmony of the numbers and the forms; they marvel when a new discovery opens up to them an unexpected vista; and does the joy that they feel not have an aesthetic character even if the senses are not involved at all?...

> For this reason I do not hesitate to say that mathematics deserves to be cultivated for its own sake, and I mean the theories which cannot be applied to physics just as much as the others.(2)

and also earlier:

> If we work, it is less to obtain those positive results the common people think are our only interest, than to feel that aesthetic emotion and communicate it to those able to experience it.(3)

This point of view does not present the whole picture, however, as we shall see; but let us accept it for the moment and view pure mathematicians as artists. Then we are artists with a singular privilege: we write only for our peers, without attempting to reach a broader public. Hardly what is expected from a painter, sculptor or composer. I know this difference did not escape at least one well-known Princeton composer, Milton Babbitt, who, over thirty years ago, in a since then famous article: "*Who cares if you listen*?"(4), requested precisely that freedom for himself and fellow composers. He felt contemporary music had reached such a level of complexity that the layperson cannot have access to it without special preparation, as is the case with modern science. So why not follow the example of science and free the composer of contemporary music from the obligation to try to reach a wider public, who is anyhow reluctant and does in any case have "its own music... to read by, eat by, dance by and to be impressed by". Still, this point of view does not seem to have prevailed, even among composers. But our writing is indeed mostly addressed to fellow workers, which has of course contributed mightily to our invisibility.

However, we do not live solely on theorems and thin air. We also need some money and other amenities for more mundane needs. Now the people or institutions "with the power of the purse" might wish to separate the visible from the invisible. They might tell us, we do not really mind your weaving patterns of ideas accessible only to people of our ilk, but why should we subsidize you for doing so? Why can't we just pay for services rendered to the visible part, whose usefulness we can gauge, at least up to a point, and leave it up to you to devote some or all or none of your free time to "pure mathematics" as a hobby?

The answer is a fundamental fact, whose importance can hardly be overemphasized: *it is not possible to separate the two.* At a given moment, it is of course easy to point out that a particular part of mathematics has no outside use whatsoever, but to predict it never will is very hazardous, and many who did had later to eat their own words, even if only posthumously. John v.Neumann, who certainly cannot be accused of having ignored the applications of mathematics, once said in an address to Princeton alumni(5):

> But still a large part of mathematics which became useful developed with absolutely no desire to be useful, and in a situation where

> nobody could possibly know in what area it would become useful: and there were no general indications that it even would be so.... This is true of all science. Successes were largely due to forgetting completely about what one ultimately wanted, or whether one wanted anything ultimately; in refusing to investigate things which profit, and in relying solely on guidance by criteria of intellectual elegance....
>
> And I think it extremely instructive to watch the role of science in everyday life, and to note how in this area the principle of laissez faire has led to strange and wonderful results.

This surprising relevance of the apparently irrelevant has often been commented upon, for instance by our first Director, A.Flexner in an article: "*Usefulness of useless knowledge*"[6], a theme recently revived by our present Director, or the physicist E.Wigner in an often quoted paper: "*The unreasonable effectiveness of mathematics in the natural sciences*"[7]. This theme could easily be the topic of a full, even entertaining, lecture but I shall have to be satisfied with a few examples. In sketching them, I shall use some mathematical concepts only vaguely defined, if at all, but a full understanding is not necessary. All I want to convey is a sense of the unexpectedness, or unreasonableness if you will, of the application to physics of a mathematical theory created for purely internal reasons. For the first two, I shall follow the steps of E.Witten in his own Faculty Lecture, since they are always the first ones to come to mind.

The most venerable one is the use of the conics by J.Kepler. The conics are the plane curves obtained by intersecting a cone over a circle with a plane. Apart from some degenerate cases, they are the familiar ellipses, parabolae and hyperbolae, the theory of which was developed by the Greeks in about the 4th century B.C., as a chapter of geometry.

At the beginning of the 17th century, Kepler had at his disposal a lot of observational data on the movements around the sun of the planets known at the time (five of them), and he was trying to find a general law describing their main features. According to a view generally accepted for over a thousand years, the universe, having been created by God, had to obey perfect laws, which should then be expressible in terms of the geometric figures hailed as the most perfect ones, namely circles, spheres and the five Platonic solids. Kepler did try to construct very complicated models using those, but they did not fit the data. Finally, he cast down his eyes on the lowly ellipses, which had no claim to such divine perfection, but it worked. He postulated that a planet moved in a plane on an ellipse, having the sun as one of its focal points, a hypothesis which turned out to agree with the observations and became the first Kepler law. Two other properties of ellipses allowed him to describe other features of the orbits of the planets, leading to the second and third Kepler laws[8]. Later, I.Newton derived these laws from his law of gravitation, providing a striking confirmation of the latter.

In that example, the time lag between the mathematical theory and the application is about two thousand years. In our next one, Riemannian geometry and general relativity, it amounts to about sixty years. In fact the story starts a bit earlier, with a fundamental contribution by C.F.Gauss, published in 1827, to "differential geometry", i.e. the study of curves and surfaces in three-space by the tools of analysis. Imagine two-dimensional mathematicians living on a surface, who have no feeling for a third dimension, though they may conceive it intellectually, in other words, who are in the same relationship with a third dimension as we are with a fourth spatial dimension. Assume this surface to be made of flexible material. It may therefore take various positions in space and be more or less curved. For instance, if it is a piece of my handkerchief, it can be flat, or very curved, say if I wrap it around my fist. Our mathematicians cannot see the difference just by measuring lengths and angles on the surface, since we precisely assume they are not altered by these deformations. However, Gauss showed the existence of a quantity they can compute at every point and which bears some relation to curvature, call it the Gaussian curvature. It would still not help them to distinguish between the various shapes I can give to my handkerchief. However, if I would ask: "Do you live on my handkerchief or on a portion of a sphere?" they could answer because the Gauss curvature is zero in the former case, but not in the latter. This discovery, which Gauss found striking enough to call "Theorema egregium", led to the division of the differential geometry of surfaces in space into two parts: "Inner geometry", dealing with the properties of surfaces depending only on measures of lengths and angles on the surface, and "Outer geometry", describing how surfaces sit in space. Shortly thereafter came the discovery of non-euclidean geometry, in which Euclid's parallel axiom is not satisfied, and in 1854, starting from Gauss' discovery, B.Riemann presented an extraordinarily bold generalization of the known geometries. It deals not only with surfaces, or three-space, but with continua of any dimension. In such a continuum a "metric" is given, which allows one to measure lengths and angles. Inspired by the Theorema egregium he defines a notion of curvature: at a given point, he considers the Gaussian curvature there of all surfaces going through it. This "Riemannian curvature" is now a function of the point, as before, and moreover of the direction of the surface. This opened up the possibility of vast generalizations, though it remained to be seen how substantial they would be. Once the paper of Riemann was published (1868) mathematicians pursued these ideas and saw quickly that they led indeed to a far reaching broadening of geometry.

By the way, these two episodes give a glimpse into two of the ways a mathematical mind can work. Gauss' theorem is not that hard to prove, once noticed, but it took remarkable insight to look for it and to realize its importance. The second one displays the power of abstraction and generalization: Starting from Gauss' observation, Riemann takes flight to unsuspected heights and opens up completely new vistas.

I now come to the application. From about 1908 on, A.Einstein was trying to develop a general theory of gravitation, not restricted to uniform motions, as special relativity was; but he was bogged down by the lack of mathematical tools. In 1912, he became professor at the E.T.H. in Zurich, where he found among his new colleagues a good friend and former classmate, the mathematician Marcel Grossmann. He appealed to him, telling him: "*Grossmann, you have to help me, otherwise I'll go crazy*"[(9)]. Einstein explained his difficulties, and Grossmann led him to Riemannian geometry as well as to subsequent developments such as tensor calculus and covariant derivative. This help was indeed what Einstein needed, and the first paper on general relativity was a joint one, Grossmann writing the mathematical part and Einstein the physical. It still took Einstein three more years of extraordinary effort to complete his theory, but it remained (and still is) firmly embedded in four-dimensional Riemannian geometry, without which it cannot even be formulated, forcing the physicists to learn a theory they could safely ignore until then.

These two examples illustrate the more or less usual relationship between physics and mathematics: The physicist has a problem, needs a theorem or a framework and the mathematician either makes it to order or, as in these cases, points out, maybe with some cockiness, that what is needed is available and has been in stock for a long time. But for me, as a pure mathematician, it is even more exciting when mathematics leads to completely new and fundamental physical insights. As an example, let me mention the discovery of the positron. In 1928 P.A.M.Dirac had set up equations for the motion of the free electron which were compatible with the requirements of relativity and quantum mechanics. However, these equations also admitted solutions with the same mass as the electron but with the opposite electric charge. As no such particle had been observed, and as it was assumed that all elementary particles were known, these "physically meaningless" solutions were viewed as a flaw of the theory, and to modify it so as to eliminate them became a matter of great concern. All attempts failed; so, after about two years, Dirac made up his mind that maybe the mathematics was right and then postulated the existence of such a particle. It was discovered later, christened the positron, and provided the first example of an antiparticle or of antimatter. Since then, as far as I know, it is still a dogma in nuclear physics that with each particle there should be an anti-particle.

I have confined myself to theoretical physics, but these unexpected uses of mathematics have been ubiquitous in science or technology, some of the most recent ones in areas such as cryptography or error-correcting codes, which until then had felt rather safe from the latest developments in pure mathematics. As tempting as it is to pursue this theme further, I shall leave the visible part of my iceberg for most of the rest of this lecture and try to make the invisible one a bit more concrete, more visible.

As the quotations from Poincaré make clear, the feeling that mathematics is an art is strong. The mathematician G.H.Hardy once wrote that mathematics, if socially justifiable, could only be so as an art[10]. This analogy is strengthened when we think about the way we work. You should not believe that mathematicians are just thinking machines who always proceed in steps clearly planned with implacable logic. This impression is often given by papers. Those are organized for maximal efficiency of the exposition; omitting all the false leads, they often proceed in an order inverse to that which led to the discovery. In the case of a very clear-headed and far-seeing mathematician, such efficiency may indeed reflect the usual way of his thinking. But frequently the going is rougher: at the start, we often do not know whether a given proposition is likely to be true or false, whether we should try to prove or disprove it, do not see a clear way to go about it, make various attempts without knowing whether the goal is reachable with the means at our disposal, as if trying to find our way in deep fog in unfamiliar surroundings. In such situations, considerations of elegance are secondary and pragmatism reigns. We want to find the solution or the proof by whatever available means. Then, if we are lucky, all of a sudden a breakthrough occurs, sometimes so unexpected that the word which comes to mind is "inspiration", not unlike that of a composer or poet. We have well-known statements by Gauss and Poincaré on these unexpected flashes of insight. After having found the solution of a problem he had been working off and on for ten years, Gauss wrote[11]:

> Finally, just a few days ago, success - but not as a result of my laborious search, but only by the grace of God I would say. Just as it is when lightning strikes, the puzzle was solved; I myself would not be able to show the threads which connect that which I knew before, that with which I had made my last attempt, and that by which it succeeded.

More recently, the Japanese mathematician K.Oka, reflecting on the genesis of his great discoveries (in the function theory of several complex variables), said that for a long time he was at a loss as to how to proceed, did not know where to go, and added that one should be patient: if one waits long enough, the solution which lies in our subconscious is likely eventually to emerge[12]. This unforeseeable character is also present elsewhere in science. In his Faculty Lecture here last year, J.Bahcall, speaking about one of the great puzzles facing astrophysicists, said we do not really know what we are looking for, but we will know it when we have found it. As he was saying this, I was reminded of an almost identical statement made I think by the French sculptor A.Rodin "What I am looking for I know after I have found it".

This dependence on ideas, the flow or occurrence of which we cannot control, makes us share with creative artists, composers, poets, painters, a gnawing worry of drying up. For instance, in 1913, when Einstein was about to leave the E.T.H. for Berlin, to head a Max Planck Institute, he confided to a friend:

The gentlemen from Berlin speculate on me as on a prized egg-laying hen. I do not know whether I can still lay eggs.[(13)]

For us mathematicians, this worry is worsened by the overwhelming evidence of an unfortunate correlation between aging and drying up. When I think of my colleagues in Historical Studies, I am green with envy, and greener by the year. There you seem to get started around forty, hit your stride at 50 or 60, and from then on you enjoy a majestic *crescendo*. One of my colleagues in that School, upon retiring at seventy, showed me his files and told me he had there material for twenty volumes. Well, for us, it is not quite the same. True, life does not stop after 35, as I used to think, many, many years ago. Enough mathematicians have done extremely deep work later, say around 45-50, to give some comfort. First rate achievements may even come later but less and less frequently, so much so that an eminent emeritus colleague once wrote, also, many, many years ago, that such achievements [of elder mathematicians] "fill us each time with astonishment and admiration". Altogether the correct overall musical marking to describe life for mathematicians after sixty, say, would be *diminuendo* rather than *crescendo* though it may be *sostenuto* for some privileged ones.

So far pure mathematics has been depicted solely as an art, but this analogy goes only so far and does not account for some of its other features. Though it is the work of individuals, it is also a collective effort in which the contributions of these individuals strongly depend on one another. Art and music have developed in various cultures in many widely different ways, but there is essentially only one mathematics. Often, people discover a theorem or establish a new theory quite independently at about the same time. A given problem has usually just one solution. Even when artists or composers face a common problem, say what should take the place of the object in painting or of traditional harmony in music, they come up with very different solutions. Also, in mathematics, many papers or even books are written jointly. One rarely hears about joint works in painting or music. Haydn and Mozart knew one another socially, performed music together, admired one another, composed in related styles, but never collaborated on a piece of music. There are of course the Haydn quartets by Mozart, but no Haydn-Mozart quartet or symphony. Similarly, it would be hard for me to imagine two mathematicians as close to one another as Braque and Picasso in the six years preceding World War I, seeing one another almost daily, looking at and discussing their work, borrowing from one another, painting at that time in styles so similar, that it is sometimes difficult at first sight to decide who had painted what, and not eventually producing a joint work... unless they became embroiled in a priority fight.

There is also a notion of progress. Mathematics is strictly cumulative. Once a theorem has been proved and the proof is accepted, it is there forever. Future developments may change views about its interest, supersede it, absorb it in a more general one, it still remains. The work of a generation will add to and improve upon that of the previous ones, even if it is not a comparable intellectual achievement. To come back to my second example, Gauss and Riemann are both giants and I would not want to attempt a comparison: still I would maintain that Riemannian geometry is of a greater scope and impact on mathematics than Gauss' theory of surfaces; and I could go on describing further developments, such as H.Weyl's first example of a gauge theory or E.Cartan's theory of connections, which have further increased the power of the theory and the range of its applications. The old saying, quoted by I.Newton, which goes back to Bernard de Chartres early in the twelfth century, namely, that by standing on the shoulders of giants, one may see more and further than they did, is definitely true in mathematics. It is not even always necessary to stand on the shoulders of giants: by standing on those of two worthy predecessors, it is often possible to see further than they did simply by relating their works in a previously unsuspected way.

I know that an idea of progress has also existed in art and music. The notion that a given generation, having the benefit of the experiences of earlier ones, would be in a position to do better, was rather commonly accepted, notably around the end of last century. In fact, the Bernard de Chartres aphorism has been quoted under various guises and attributions in many contexts. For instance, Renaissance painters, equipped with the laws of perspective, were (mistakenly) rated higher than their "primitive" predecessors. Or a history of music would show a steady progress from modal music, to polyphony and to the crowning achievement, harmony, relegating modal music as complex as Indian or Balinese music to the rank of the expression of a primitive or semi-civilized culture. But the stormy developments in the twentieth century, a better knowledge and appreciation of the past and of many world cultures have pretty much done away with such views. In fact, the opposite opinion has been held, too. How often new developments, departing from accepted norms, have been viewed with alarm, as a sign of decadence. We need only remember how various schools of paintings were greeted at first or look at Slonimsky's "*Lexicon of Musical Invective*" to see that many of the now widely respected composers did not fare well at all at first. Saying this makes me realize however that my depiction of the progress in mathematics has been a bit too rosy and one-sided. New is not always better or not always perceived as such. Also, we are not exempt from controversies (some rather bitter, I shall mention one later) about the value of a new theory or point of view. Some were also denounced as dangerous trends, leading to the degeneration of our beloved mathematics, but these controversies are usually rather short lived, and make way to a consensus, mainly for a good reason I shall come to in a moment.

It would be difficult to account for these features of mathematics by viewing it only as an art, with some additional guidelines stemming from the needs of other disciplines. Rather, they are among the attributes of a science and also point out to the existence of a world of mathematical concepts, problems, theorems, to which mathematicians constantly add, collectively or individually. Saying this however leads one right into a question which has been debated for ages and will presumably go on as long as there are mathematicians: namely, what is the locus of that mathematical reality? Do we create mathematics step by step, or does it preexist us and we merely discover it little by little, as if exploring an unknown country? Both views have had and still have their advocates. Those of the latter may appeal to a religious belief or to a philosophical tradition going back to Plato. The Platonic view has been held by many, for instance, by G.H.Hardy(14):

> I believe that mathematical reality lies outside us, that our function is to discover or observe it, and that the theorems which we prove, and which we describe grandiloquently as our "creations", are simply our notes of our observations. This view has been held, in one form or another by many philosophers of high reputation, from Plato onwards....

K.Gödel was also of that opinion and more recently R.Penrose expressed similar feelings(15):

> Is mathematics invention or discovery? When mathematicians come upon their results are they just producing elaborate mental constructions which have no actual reality, but whose power and elegance is sufficient simply to fool even their inventors into believing that these mere mental constructions are 'real'? Or are mathematicians really uncovering truths which are, in fact, already 'there' - truths whose existence is quite independent of the mathematicians' activities? I think that, by now, it must be quite clear to the reader that I am an adherent of the second, rather than the first, view, at least with regard to such structures as complex numbers and Mandelbrot set.

For a religious mathematician, the locus of that mathematical reality will usually be God's mind. Such were the points of view, e.g., of C.Hermite(16):

> There exists, if I am not mistaken, an entire world which is the totality of mathematical truths, to which we have access only with our mind, just as a world of physical reality exists, the one like the other independent of ourselves, both of divine creation.

Or of G.Cantor, who even pushed this belief much further, viewing himself as a messenger of God: the theory he had published had been revealed to him by God and it was his mission to be a good messenger and spread it out. Maybe he was moved to such an extreme position not only by his deep faith but also in some measure by the predicament he was in: he had developed a very daring, even revolutionary, theory, that of transfinite numbers, which dealt with actual infinities (as opposed to the usual "potential infinities"), even introducing different orders of actual infinities. His theory met with considerable opposition from many mathematicians, one of them attacking him savagely, and was also viewed with suspicion by some philosophers and theologians. The objections of the official Church mattered a great deal to Cantor; so he took great pains to discuss them, and he was greatly relieved when it was pronounced that his theory of infinity did not contradict accepted religious doctrine(17).

In this connection, it is difficult not to mention C.S.Ramanujan, who attributed to the family goddess his mathematical gifts and some of his formulae, communicated to him in dreams(18). More recently, a colleague of mine, in an introduction to a series of lectures on his own work, pointed out he had been preoccupied by a certain problem for many years and then startled his audience by stating: "*Why has God created the exceptional series?*"

But even a mathematician who believes mathematics is purely a human creation, as I do, has the obscure feeling that it exists somewhere out there. And I catch myself time and again talking as if it really does. To come to terms with that impression, I shall simply take the view, which is quite common, that this is a cultural phenomenon: if a concept is such that we are convinced it exists in the minds of others in the same way as it does in ours, so that we can discuss it, argue about it, then this very fact translates to a feeling of an objective existence, outside, and independent of, a particular individual. This attitude is not peculiar to mathematics, but a common experience in many aspects of our lives. So there are many religions in the world and a person will as a rule believe in at most one; in fact, the stronger the belief, usually the more exclusive it will be. For that person, the preferred religion will of course be an example of a collection of concepts, thoughts, stories reflecting an objective reality of non-human origin, while all the other religions just illustrate what I referred to, namely, a mental construct by a group of people, which they believe (erroneously) to have an objective existence. We also encounter this often in literature. Aren't we inclined to think that Sherlock Holmes has actually lived? After all, there is a Sherlock Holmes Museum, a Sherlock Holmes Society in England, among whose activities is the organization of trips, or should I say pilgrimages, to the site of the fight between S.Holmes and his arch-enemy, Dr. Moriarty, at Reichenbach Falls, in Switzerland. A few years ago, at a meeting of the learned American Philosophical Society in Philadelphia, held, as are all its meetings "to promote useful knowledge", there was a lecture on the tastes and achievements of S.Holmes in music. This phenomenon of a mental creation acquiring an objective reality is apparently also often experienced by writers, who create a character and then view themselves more or less as observers, having little or no control on the thoughts and acts of that character. For instance, Erskine Caldwell once said in an interview(19):

> I have no influence over them. I'm only an observer, recording. The story is always being told by the characters themselves. In fact,

> I'm often critical, or maybe ashamed, of what some of them say and do – their profanity or their immorality. But I have no control over it.
>
> "But you do at least understand their motivations?"
>
> I'm not an oracle by any means. I'm often at a loss to explain the desires and the motivations of my people. You'll have to find your explanation in them. They're their own creations.

Now it is very easy in mathematics, with its very precise language, to create a new object or conçept, which will make sense in the same way to its creator as to whomever may be interested in it. Some of the properties of this mathematical object derive easily from the definitions, but others not at all. It may require tremendous efforts, over a long time, to pry them out, and then how can we escape the feeling that this object was there before and we just stumbled upon it? If moreover, the interest in those problems and the efforts to solve them are shared by others, the feeling of an objective existence becomes practically irresistible. It can still be argued as to whether there is underlying it a higher degree of reality, whether we create, or observe or reminisce. It seems to me that such a discussion can go on ad *infinitum*, without any prospect of a final convincing answer, and I leave it at that.[20]

Whatever its origin or locus, we have then at our disposal this enormous amount of concepts, theorems, open and solved problems, theories, a "mathematical reality", which has been amassed over more than two thousand years. It has for us as much objective reality as the natural world for a physicist or natural scientist. This analogy helps me to complete the first answer given to a question raised earlier: if one leaves out the applications, what are the internal guidelines of mathematics, in the invisible part of the iceberg? As already said, those of aesthetic nature are of paramount importance, but are not quite the only ones I think. The structure of this mathematical world brings a hierarchy between the open questions and makes a problem or a theory more important at a given time, and helps us to single out some of greater interest. After all, we accept easily the view that the investigation of nature imposes such criteria upon the natural scientist. That quest for knowledge is an unending one. Many years ago I read of a comparison of the amount of knowledge in science at a given time with a ball immersed in the sea of the unknown: When its radius increases, then so does its surface of contact with the unknown. It was, I believe, attributed to Poincaré, but I have not been able to trace it back. At any rate, this comparison applies perfectly well to our mathematical world, even though it is an intellectual one, in which we operate essentially with intellectual tools.

The feeling of a structure in this mathematical world is strengthened by what I would like to call our "belief in a myth", namely, the fundamental unity of mathematics. It is a myth because mathematics is much too big to be comprehended by one person and, at a given time, consists of many seemingly unrelated parts. But time and again totally unsuspected connections appear, two different topics, the respective experts of which had so far little in common, all of a sudden become part of a bigger one, under the impact of a new, usually more abstract, point of view. This counterbalancing of unbridled expansion by contraction through unification makes many of us strong believers in the ultimate coherence of the whole of mathematics.

The analogy with natural sciences can even be pushed further. Pure mathematics also has an experimental side besides its theoretical one. The latter is the one usually associated with it: we strive to set up general theories, in which certain key theorems have many consequences. But we need first to build up the material to organize, or to explain. As already pointed out, when facing a new object or proposition, we often do not know in which direction to go, what should be proved or disproved and, when we do, what method might work. The only way, for most of us, to gain such intuition is to look at special cases and this is the experimental side. Although the exposition of mathematics usually goes from the general to the particular, the way to the discovery is often the opposite one. These experiments were traditionally performed with pencil and paper, but computers are used more and more, increasing considerably their scope and their impact on mathematics, also making the analogy with laboratory work closer.

There are even findings about our objects which surprise us enormously and make an impression not unlike that made on physicists by the discovery of an elementary particle. A few years ago, it was shown that the euclidean space in four dimensions carries several differentiable structures. I shall not try to define those terms. Let me just say that euclidean space is one of the most basic structures in mathematics. There is one in each dimension and it was known to have a unique differentiable structure except possibly in dimension 4. It was quite a shock when it was shown that this last case was indeed exceptional. The feeling of many mathematicians may have been akin to that of the physicist I.Rabi who, when apprised of the discovery of a new particle, the muon, somewhat unwelcome since it was unexpected and did not fit into any existing theory, exclaimed: "*Who ordered that*?". So our world takes a very concrete form for us. It does not need much of a provocation for me even to maintain that a definite mathematical object, say the n-dimensional sphere, "exists" at least as much as an elementary particle which lives all of 10^{-22} seconds, can be detected only by extremely complicated experiments and by a sophisticated interpretation of the experimental data.

By the way, I cannot resist pointing out that the discovery of these new differentiable structures provides a striking boost for the belief in the ultimate coherence of mathematics. It came about by comparing, and drawing consequences from, two theorems proved in completely different frameworks, mathematical physics and geometric topology, by totally unrelated techniques. In fact, nobody at the time was familiar with both.

This double or triple aspect of mathematics makes it also easier to approach the question of "aesthetics" in mathematics. It is often remarked there is quicker consensus in mathematics about the relative importance of theorems or of some of its parts than in other disciplines, in particular in art, so that it would be somewhat surprising if it were based purely on aesthetic feelings, since there usually are so many conflicting opinions in art about recent work. When we speak of beauty, elegance, we indeed express first of all an aesthetic judgment on the ideas, how they are put together and how original they are. But, combined with them, often implicitly, are judgments closer to a "bottom line" mentality. In considering a theorem, we are not only interested in its proof, but also in its consequences and applications, in its power. This also contributes to a feeling of beauty. It may even have a retroactive effect. For instance, to come back to Gauss' "Theorema Egregium", it is indeed beautiful in its own right, but the feeling of beauty is heightened by the knowledge that it was the starting point of Riemannian geometry. Strictly thinking, one should distinguish between beauty and importance, but we often lump them together as "aesthetic" judgments. So "success" is often implicitly part of beauty and, since it is difficult to argue with success, a consensus on "beauty" in this broad sense is more easily reached: if a new theorem solves an old outstanding problem, gives more power to a theory, brings new fruitful viewpoints, it will have to be accepted and will even eventually command admiration.

In stressing usefulness within mathematics I should not go too far and run the danger of being accused of adopting for mathematics a "profit oriented" attitude I had decried earlier. Such relative judgments are unavoidable, if only to guide one's own activity, but should not be absolute. It would also be easy to describe cases in which a very special topic, of interest at first only to a handful of specialists, sometimes even scorned by other mathematicians, turned out to have unexpected applications in a broader context. (Cantor's theory of transfinite numbers, alluded to earlier, is a case in point.) The eventual relevance of the apparently irrelevant also takes place within mathematics. The freedom for mathematicians with regard to science advocated by J.v.Neumann in my earlier quote has to be granted to them within pure mathematics, too.

In talking about pure mathematics, I have taken for granted that it is a legitimate object of study, regardless of applications. This view is now widely accepted, though not universally or unconditionally, but it took some time to emerge. The focus of mathematics in the late seventeenth and in the eighteenth centuries was analysis and its applications to mechanics. Number theory, originally the study of properties of the natural integers, the paragon of pure mathematics, was practiced by only a few (though outstanding) mathematicians and was rather commonly viewed as a minor topic, even though it had a distinguished pedigree, going back to the Greeks. The mathematician L.Euler, who worked on all aspects of pure mathematics and its applications, even writing a book on ship building, did publish several papers on number theory, but he felt it necessary to state in his introductory remarks that this was as justified as research on more applied topics, because it added to our knowledge, and in fact might eventually be useful even from a more practical point of view. In the nineteenth century, C.F.Gauss did not express himself publicly on such matters but only in a few letters, from which one sees that he valued mathematics way over some of the applied work his functions required him to do:

> ... all the measurements in the world do not compare with one Theorem, through which the science of the eternal truths is brought further.[21]

And he valued number theory above all:

> The higher I put and have always put this part of mathematics above all the others - the more painful it is to me that - directly or indirectly, through external circumstances - I am so far from my favorite occupation.[22]

Curiously, the first statement in print claiming autonomy for pure mathematics may be due to someone who knew some mathematics, but was not at all a professional mathematician, namely, Johann Wolfgang von Goethe, who wrote:

> Mathematics must declare itself, however, independently of everything external, must go its own great intellectual way and develop more purely than can happen when, as hitherto, it is linked up with the empirical and is aiming at gaining something from it or adjusting itself to it.

in an aphorism published in 1829 I shall come back to in a moment. Later, some mathematicians also came out of the closet to make similar claims. For instance, W.Hankel, in a book on complex numbers and quaternions published in 1867, wrote:

> Needed to establish a general arithmetic is therefore a purely intellectual mathematics, free from intuition, a pure science of forms, in which not quantities or their representatives, the numbers, are related, but intellectual objects, objects of thoughts, which may correspond to actual objects or relations, but do not have to.[23]

That claim was however still confined to one part of mathematics, arithmetic or algebra, but a bit later, G.Cantor made more sweeping ones, to the effect that any mathematical concept is a legitimate object of study provided its definition is logically consistent. He viewed that freedom as the essence of mathematics and, to stress it, remarked he would prefer to speak of "free mathematics" rather than "pure mathematics", if he had the choice[24].

This view became more and more widespread, but not always without some reservation. Even H.Poincaré, in the 1897 lecture quoted earlier, issued a warning

> If I may be allowed to continue my comparison with the fine arts, then the pure mathematician who would forget the existence of the outside world could be likened to the painter who knew how to combine colors and forms harmoniously, but who lacked models. His creative power would soon be exhausted.(25)

In order to make his point, Poincaré takes for granted, as an unquestionable truth, that a main goal of painting is representation of the outside world or, more broadly, that the latter is the main source of a painter's inspiration. But, in fact, this very tenet had been more and more questioned by some painters. In order to broaden the discussion, I would like to extend the distinction between invisible and visible to painting and music by letting the former refer to a practice based on the use of purely internal criteria and motivation and the latter to one in which the practitioners also obey imperatives from the outside, whose fulfillment may be more easily detected and appreciated by the outsider. In painting, those would be the duty to represent nature, people, idealized beauty, religious subjects, to exalt the mind, etc. It is that duty, viewed traditionally as the aim of painting, which had been more and more under attack. About twenty five years earlier, for instance, James Whistler had called the famous portrait of his mother an "arrangement in grey and black", and proposed that it should stand or fall on its merits as an "arrangement", since the identity of the portrait, though of interest to him, could hardly matter to the public. A bit later, he stated

> As music is the poetry of sound, so is painting the poetry of sight, and the subject matter has nothing to do with harmony of sound or color.(26)

In 1890, Maurice Denis began his first theoretical writing on painting with the since then famous sentence:

> It is well to remember that a painting, before being a warhorse, a naked woman or some anecdote is essentially a plane surface covered with colors assembled in a certain order.(27)

In 1897, the very year of Poincaré's lecture, a painter in Munich, August Endell, was prophesying an art with forms that mean nothing, represent nothing, will be able to excite our souls as only music has been able to do with sounds(28).

By hindsight, we see that these artists were groping for one form or another of nonfigurative painting, developed a bit later, from about 1910 on, by Kandinsky, Kupka, Malevich and others(29). For some, it became the only way to paint; others had a more balanced view, or kept with the traditional ideals, and at present we have the whole spectrum, representing all these outlooks.

In two of the above quotations, the artists refer to music as the model to emulate, the paragon of a "pure art" in which the primary guidelines are internal ones, i.e., in the language I have adopted here, consisting mainly of the invisible part. But here too this had not always been a prevalent view. In the Europe of the seventeenth and eighteenth centuries, composers had patrons and had to produce, on order, lots of functional music; if some of the great ones let themselves be carried away by their genius, there were often grumblings, not excluding even Bach or Mozart, about their music being too much of a display of technical virtuosity, ignoring feelings or melody, etc. Mozart himself lamented during his last illness that he had to leave when he saw prospects of improvement in his financial situation which would have allowed him to be no longer a slave to fashion, no longer chained to speculators and to follow his spirit freely and independently and write as his heart dictated[30]. His successors, Beethoven, Schubert and others did indeed take that right into their hands of course, but it seems that this view of music was presented systematically and explicitly first by the music critic Eduard Hanslick in 1854 in his treatise: "Vom Musikalisch-Schönen"[31]; there he stresses that music is an arrangement of sounds which may evoke feelings but is not obliged to express one, and he quotes many earlier statements of others insisting on the contrary that the primary duty of music is to express human feelings.

Such analogies have their limitations and I do not want to push them too far. Nevertheless it seems interesting that the growing feeling of the specificity of mathematics, not only in its techniques, but also in its goals, was paralleled by similar developments elsewhere. I have singled out painting and music, but it reflects an even broader cultural phenomenon of the time.

At this point, "pure mathematics", the invisible part of my iceberg, emerges as a discipline which is at the same time an art and a science of the mind, an intellectual science, both experimental and theoretical. It is now time to take again the visible one into consideration. I will do this by confessing a sin and trying to atone for it. It may have been a surprise that I could enlist Goethe as an ally in promoting pure mathematics. I did this by using a device not unknown to people with an axe to grind, namely, selective or out-of-context quotation. I read to you only the second half of an aphorism, the first half of which states:

> Physics must show itself as separate from mathematics. The former must exist in a decided independence and must endeavour with all loving, respecting and pious strength to penetrate into nature and nature's holy life, wholly without concern for what mathematics achieves and performs for its part.[32]

and gives it a quite different meaning. On the one hand, it would seem I should welcome it as one more instance of the claims for specificity I was just quoting, but unfortunately it is emphatically not one I can go along with. I have indeed pleaded for the right to do pure mathematics, without worrying about applications, but have never intended to assert one should do so. Of course, if challenged, I might give a quite impressive list of achievements in pure mathematics in these last fifty years or so motivated by purely internal considerations without any visible output from outside. But it would be a very futile enterprise. A moment ago, I took some pride in relating an occasion in which pure mathematics led to new insights in physics, but physicists could point to many examples of the converse. In fact the interaction between physics and mathematics is at present at an all time high, as a two-way street, and I would be particularly ill-advised to deny it here, since this place is a focal point for it. Using for a last time the comparison with the iceberg, I should point out that the separation exists only when looking at it from the outside but not within the iceberg. Furthermore, this analogy, like all analogies, has its limitations. A severe one is its failure to give any account of the constant shifting and exchange between its two main parts. To complete my description of mathematics in the context of science I would rather say that it is both a queen and a servant of science or, more democratically speaking, an equal partner, freely exchanging ideas, theories, and problems with it for mutual benefit.

Mathematics appears now as a very complex, many faceted, structure. In fact, most of the opinions I mentioned at the beginning appear to carry some part of the truth: Queen of Science, servant of science, art, as well as an experimental and theoretical mental science in its own right, a gigantic, awesome product of the collective human mind. As a professional mathematician, I must confess I feel rather uncomfortable in just philosophizing about it, beating around the bush, rather than getting to some of its actual contents and saying more specifically why we are in awe of it. But, as two members of the first Institute Faculty, Marston Morse and John von Neumann, said over fifty years ago in a report on the School of Mathematics to the Board of Institute Trustees, mathematics is written in a unique language which cannot be translated into any other, so that one has to learn it to be in a position to understand mathematics. The situation seems to be improving though. Efforts to reach out, to make some cracks in the language barrier, are increasing. Some recent essays do manage, I think, to give to an interested reader a good idea of the substance and goals of mathematics, or at least of parts of it(33). However, my purpose here was somewhat different. Rather, I tried to show that, in spite of its esoteric character, a number of features of its evolution and development, and of questions about its essence, trends, aesthetics, goals have their counterparts, sometimes with similar answers, in science or other human endeavors more commonly viewed as part of our culture. I hope to have in this way made more plausible my belief that mathematics is an integral and important part of culture, whether scientific, humanistic or artistic.

NOTES

(1) For a discussion of, and references to, such dim views of mathematics, see A.Pringsheim, *Ueber den Wert und angeblichen Unwert der Mathematik*, Jahresbericht d.Deutschen Mathematiker-Vereinigung **13** (1904), 357–382.

(2) "Et surtout leurs adeptes y trouvent des jouissances analogues à celles que donnent la peinture et la musique. Ils admirent la délicate harmonie des nombres et des formes; ils s'émerveillent quand une découverte nouvelle leur ouvre une perspective inattendue; et la joie qu'ils éprouvent ainsi n'a-t-elle pas le caractère esthétique, bien que les sens n'y prennent aucune part?..."

"C'est pourquoi je n'hésite pas à dire que les mathématiques méritent d'être cultivées pour elles-mêmes et que les théories qui ne peuvent être appliquées à la physique doivent l'être comme les autres." [Address to the First International Congress of Mathematicians in Zürich: *Sur les rapports de l'analyse pure et de la physique*, (which he could not deliver in person), see the Proceedings of that Congress 81–90, Verlag von B.G.Teubner, Leipzig 1898; also reproduced under the title: *L'analyse et la physique*, in *La Valeur de la Science*, pp. 137–151, E.Flammarion, Paris 1905.]

(3) "Si nous travaillons, c'est moins pour obtenir ces résultats auxquels le vulgaire nous croit uniquement attachés, que pour ressentir cette émotion esthétique et la communiquer à ceux qui sont capables de l'éprouver." [*Notice sur Halphen*, Jour. Ecole Polytechnique, 60ème cahier, 1890; Oeuvres de G.H.Halphen, Vol. **1**, XVII-XLIII, Gauthier-Villars, Paris 1916.]

(4) High Fidelity **8**, N° 2, 1958, pp. 30–40, 126–127; also reproduced in: *Contemporary composers on contemporary music*, pp. 243–250, E.Schwartz and B.Childs, ed., Da Capo Press, New York 1967.

(5) J.v.Neumann, "*The role of mathematics in the science and in society*", address to Princeton Graduate Alumni, June 1954. Cf. *Collected Works*, 6 Vol., Pergamon, New York, 1961, Vol. VI, pp. 477–490.

(6) Harper's **179**, October 1939, pp. 544–552.

(7) Communications on Pure and Applied Mathematics **13** (1960), pp. 1–14.

(8) See e.g. M.Kline, *Mathematical Thought from Ancient to Modern Times*, Oxford University Press, N.Y. 1972, Chapter 12, §5, pp. 243-245.

(9) "*Grossmann, Du musst mir helfen, sonst werd' ich verrückt*" [cf. L.Kollros, *Albert Einstein en Suisse. Souvenirs*, in *Fünfzig Jahre Relativitätstheorie*, Helvetica Physica Acta, Supplementum IV, Birkhäuser Verlag, Basel 1956, pp. 271–281.]

(10) G.H.Hardy, *A Mathematician's Apology*, Cambridge University Press, 1940; new printing with a foreword by C.P.Snow, pp. 139–140.

(11) "Endlich vor ein Paar Tagen ist's gelungen – aber nicht meinem mühsamen Suchen, sondern bloß durch die Gnade Gottes möchte ich sagen. Wie der Blitz

einschlägt, hat sich das Räthsel gelöst; ich selbst wäre nicht im Stande, den leitenden Faden zwischen dem, was ich vorher wußte, dem, vomit ich die letzten Versuche gemacht hatte, - und dem, wodurch es gelang, nachzuweisen..." [In a letter to H.W.M.Olbers, September 3, 1805, *Gesammelte Werke* 10_I, p. 23.]

(12)K.Oka, *Ten Essays in Spring Evenings* (in Japanese). See K.Oka, *Collected Essays*, Gakushu Kenkya-sha, 1969. I thank T.Oda, of Sendai University, who drew my attention to them and translated some excerpts for me.

(13)"Die Herren Berliner spekulieren mit mir wie mit einem prämierten Leghuhn. I weiss nicht, ob ich noch Eier legen kann" [*loc. cit.*(9), pp. 123–124.]

(14)*loc. cit.*(10).

(15)R.Penrose, *The Emperor's New Mind*, Oxford University Press 1989, p. 96, reproduced here by permission of Oxford University Press.

(16)"Il existe, si je ne me trompe, tout un monde qui est l'ensemble des vérités mathématiques, dans lequel nous n'avons accès que par l'intelligence, comme il existe un monde des réalités physiques, l'un et l'autre indépendants de nous, tous deux de création divine." [G.Darboux, *La vie et l'Oeuvre de Charles Hermite*, Revue du Mois **10**, January 1906, p. 46.]

(17)For all this, see W.J.W.Dauben, *Georg Cantor. His Mathematics and Philosophy of the Infinite*, Harvard University Press, in particular pp. 132-146, 232–239.

(18)R.Kanigel, *The Man Who Knew Infinity*, C.Scribner's Sons, New-York 1991. *See* notably pp. 36, 64–67.

(19) The Paris Review N° 86, 1982, 127–157, see p. 132. The first paragraph is also reproduced in *The Writer's Chapbook*, G.Plimpton ed. Viking 1989, p. 194.

(20) As was pointed out, it is rather natural for a believer in God to conclude that mathematics preexists us, in the mind of God, but such a religious belief may also lead to the opposite view. At any rate, it did for R.Dedekind, who maintained in a letter to H.Weber (January 24, 1888) that we are of divine race and possess without any doubt a creative power, not only in material things (trains, telegraphs), but especially in spiritual things. Let me add, for the mathematician, that Dedekind made this statement in a discussion of the notion of number and of his definition of irrational numbers by what became known as Dedekind cuts. Weber held the view that the irrational numbers were the cuts themselves, whereas for Dedekind they were something new, created by the mind, corresponding to a cut, but not to be identified with it:

"so möchte ich doch rathen, unter der Zahl (Anzahl, Cardinalzahl) lieber nicht die Classe (das System aller einander ähnlichen endlichen Systeme) selbst zu verstehen, sondern etwas Neues (dieser Classe Entsprechendes), was der Geist erschafft. Wir sind göttlichen Geschlechtes und besitzen ohne jeden Zweifel schöpferische Kraft nicht blos in materiellen Dingen (Eisenbahnen, Telegraphen), sondern ganz besonders in geistigen Dingen. Es ist dies ganz dieselbe Frage, von der Du am Schlusse Deines Briefes bezüglich meiner Irrational-Theorie sprichst, wo Du sagst, die Irrationalzahl sei überhaupt Nichts anderes als der Schnitt selbst, während ich es vorziehe, etwas N e u e s (vom Schnitte Verschiedenes) zu erschaffen, was dem Schnitte entspricht, und wovon ich sage, daß es den Schnitt hervorbringe, erzeuge. Wir haben das Recht, uns eine solche Schöpfungskraft zuzusprechen, und außerdem ist es der Gleichartigkeit aller Zahlen wegen viel zweckmäßiger, so zu verfahren"" [R.Dedekind, *Gesammelte Werke*, Vol. III, p.489, Viehweg and Sohn, Braunschweig 1932.]

(21)"...Alle Messungen der Welt wiegen nicht ein Theorem auf, wodurch die Wissenschaft der ewigen Wahrheiten wahrhaft weiter gebracht wird." [*Briefwechsel zwischen Gauss und Bessel*, Verlag W.Engelmann, Leipzig 1880, letter 143, March 14, 1824, p. 428.]

(22)"Denn je höher ich diesen Theil der Mathematik über alle andern setze und von jeher gesetzt habe, um so schmerzhafter ist es mir, dass – unmittelbar oder mittelbar durch die äussern Verhältnisse - ich so sehr von meiner Lieblingsbeschäftigung entfernt bin." [Letter to G.Lejeune Dirichlet, February 2, 1838, *Gauss' Werke* **12**, pp.309–312.]

(23)"Die Bedingung zur Aufstellung einer allgemeinen Arithmetik ist daher eine von aller Anschauung losgelöste, rein intellectuelle Mathematik, eine reine Formenlehre, in welcher nicht Quanta oder ihre Bilder, die Zahlen verknüpft werden, sondern intellectuelle Objecte, Gedankendinge, denen actuelle Objecte oder Relationen solcher entsprechen können, aber nicht müssen. [H.Hankel, *Vorlesungen ueber die complexen Zahlen und ihre Funktionen*, I.Teil, *Theorie der complexen Zahlensysteme*, L.Voss, Leipzig, 1867, p.10.]

(24)G.Cantor, *Ueber unendlich lineare Punktmannigfaltigkeiten*, 5. Fortsetzung, Math. Annalen **21** (1883), 545–586; G.W. 165-208, §8.

(25)"Si l'on veut me permettre de poursuivre ma comparaison avec les beaux-arts, le mathématicien pur qui oublierait l'existence du monde extérieur serait semblable à un peintre qui saurait harmonieusement combiner les couleurs et les formes, mais à qui les modèles feraient défaut. Sa puissance créatrice serait bientôt tarie."

(26)J.A.McNeill Whistler, *The Gentle Art of Making Enemies*, G.P.Putnam and Sons, New-York 1890, see p.126 (May 22, 1878).

(27)"Se rappeler qu'un tableau – avant d'être un cheval de bataille, une femme nue, ou une quelconque anecdote est essentiellement une surface plane recouverte de couleurs en un certain ordre assemblées. [Art et Critique, Paris 1890;

also in Théories (1890-1910), Hermann, Paris 1964. Translated in H.B.Chipp, *Theories of Modern Art*, University of California Press 1968.]

(28)A.Endell, *Formenschönheit und dekorative Kunst*, Dekorative Kunst I, N° 2, Nov. 1897, p.75, see also P.Weiss, *Kandinsky in Munich*, Princeton University Press 1979, p. 34.

(29)I have limited myself to some signposts, without any attempt at comprehensiveness. They point to a trend towards "Art for Art's Sake", an idea which had also been entertained, in various forms, by some writers in the first half of the nineteenth century. In fact, that expression itself seems to appear first in the diary of Benjamin Constant, who wrote on February 10, 1804, during a visit in Weimar:

"L'art pour l'art et sans but, tout but dénature l'art. Mais l'art atteint un but qu'il n'a pas."

(Art for art's sake and without a goal, every goal falsifies art. But art achieves a goal it does not have.) a statement which strikes me as not devoid of similarity with the one of J.v.Neumann quoted earlier.

(30)F.X.Niemetschek, *Lebensbeschreibung des k.k.Kappellmeisters Wolfgang Amadeus Mozart*, 1798. Reprint VEB Deutscher Verlag für Musik, p. 54. Translation by H.Mautner: *Life of Mozart*, L.Hyman, London 1956, p. 45.

(31) *Vom Musikalisch-Schönen*, 1854, translated by G.Cohen, *The Beautiful in Music*, Ewer and Co. N.Y. 1891, reprinted by Da Capo Press, N.Y.

(32) The full aphorism reads in the original: "Als getrennt muß sich darstellen: Physik von Mathematik. Jene muß in einer entschiedenen Unabhängigkeit bestehen und mit allen liebenden, verehrenden, frommen Kräften in die Natur und das heilige Leben derselben einzudringen suchen, ganz unbekümmert, was die Mathematik von ihrer Seite leistet und tut. Diese muß sich dagegen unabhängig von allem Äußern erklären, ihren eigenen großen Geistesgang gehen und sich selber reiner ausbilden, als es geschehen kann, wenn sie wie bisher sich mit dem Vorhandenen abgibt und diesem etwas abzugewinnen oder anzupassen trachtet." [See: *Betrachtungen in Sinne der Wanderer*, Goethes Werke, Propyläen Ausgabe, Berlin, Vol. **41** p. 380.]

This collection of aphorisms (the one quoted here being number 134) was published in 1829, as a supplement to "*Wilhem Meisters Wanderjahre*", Bd 2, but is likely to have been written earlier.

Goethe's advocacy of a strict separation between mathematics and physics may not be unrelated to the conflict between his theory of colors and that of Newton. In his opinion, the latter was mistaken, and it had gained support chiefly because of Newton's prestige as a mathematician. *See Ferneres ueber Mathematik und Mathematiker*, ibid. Vol. **39**, pp. 89 and 439.

(33) notably *The Mathematical Experience*, by P.J.Davis and R.Hersch, Birkhäuser, Boston, 1980.

154.

(with J. Yang)

The Rank Conjecture for Number Fields

Math. Res. Letters 1 (1994) 689–699

0. Introduction

The rank conjecture states that two filtrations (the γ-filtration and the rank filtration) of the K_n-group $K_n(k)$ of an infinite field k are complementary (see [11, 4.1] or [8]). Let E denote a field of characteristic zero. In the case of a number field, and in view of known results on the γ-filtration, the rank conjecture is implied by two assertions pertaining to the homomorphism

$$j_{i,m,n} : H_i(\mathbf{GL}_m(k); E) \longrightarrow H_i(\mathbf{GL}_n(k); E) \quad (m \le n)$$

induced by the standard inclusion

$$j_{m,n} : \mathbf{GL}_m(k) \to \mathbf{GL}_n(k) \quad (m \le n),$$

namely,

$$(R1) \qquad \operatorname{Im} j_{2m-1,m,n} \supset PH_{2m-1}(\mathbf{SL}_n(k); E),$$
$$(R2) \qquad \operatorname{Im} j_{2m-1,m-1,n} \cap PH_{2m-1}(\mathbf{GL}_n(k); E) = \{0\}.$$

for $n \gg m \ge 2$, where PH_i is the space of primitive elements (1.4). The goal of [11] is to prove (R1) and (R2), but the argument is incomplete (see Section 6). More precisely, the proof of (R1) for quadratic fields in [11] is incorrect and the method there does not apply to $k = \mathbb{Q}$. Our first objective is to provide a complete proof (Section 5). To that effect, we first show (Theorem 2.1) that if G is a simply connected, almost simple and isotropic k-group G, then $H^\bullet(G(k); \mathbb{R})$ is canonically isomorphic to the continuous (Eilenberg-MacLane) cohomology space $H^\bullet_{\mathrm{ct}}(G(k \otimes_{\mathbb{Q}} \mathbb{R}); \mathbb{R})$. This, in turn, is an immediate consequence of the main theorem of [1] on the cohomology of S-arithmetic congruence subgroups of $G(k)$. The rank conjecture for number fields follows easily from this theorem, applied to $\mathbf{SL}_n$ viewed as a k-group and from some known facts about the continuous cohomology of $\mathbf{SL}_n(\mathbb{R})$ or $\mathbf{SL}_n(\mathbb{C})$ (Section 4).

Received September 19, 1994.

Most of the work on this note was carried out while the first named author enjoyed the hospitality of the Max-Planck-Institut für Mathematik in Bonn, Germany, as an A. v. Humboldt awardee. The second named author was supported in part by the NSF grant DMS-9401411.

Finally, in Section 7, we describe $H_\bullet(\mathbf{GL}_n(k); E)$ and show that $j_{i,m,n}$ is injective for all i's and surjective for $i \leq 2m-1$.

1. Notations and Review

1.1. In this note, k always stands for a number field, V_∞ (resp. V_f) for the set of archimedean (resp. non-archimedean) places of k and $V = V_\infty \cup V_f$. For $v \in V$, the completion of k at v is denoted by k_v. If $S \subset V$, then define $S_f = S \cap V_f$.

1.2. If G is an affine k-group, and S a finite subset of V, then $r(S, G)$ is the sum of the k_v-ranks of G for $v \in S$. Moreover, as usual, we let

$$G_\infty = G(k \otimes_{\mathbb{Q}} \mathbb{R}) = \prod_{v \in V_\infty} G(k_v),$$

which is viewed as a real Lie group.

1.3. If E is a commutative ring and Γ a group, then $H^i(\Gamma; E)$ and $H_i(\Gamma; E)$ are the ith cohomology and homology group of Γ with coefficients in the trivial Γ-module E, in the sense of Eilenberg-MacLane, and $H^\bullet(\Gamma; E)$ (resp. $H_\bullet(\Gamma; E)$) is the direct sum of the $H^i(\Gamma; E)$ (resp. $H_i(\Gamma; E)$).

If M is a real Lie group and E is either $\mathbb{R}$ or $\mathbb{C}$, then $H^i_{\mathrm{ct}}(M; E)$ denotes the ith continuous cohomology group of M with respect to E, i.e., cohomology based on continuous (or differentiable) Eilenberg-MacLane cochains (see e.g. IX in [7]). Let M^δ be M viewed as a discrete group (forget the topology). We let

$$f : H^\bullet_{\mathrm{ct}}(M; E) \longrightarrow H^\bullet(M^\delta; E) \tag{1}$$

be the natural map.

1.4. All fields in this paper are assumed to be commutative. Let E be a field and

$$A^\bullet = \bigoplus_{i=0}^{\infty} A^i$$

be a graded algebra over E which is connected, i.e., $A^0 = E$. We let $IA = \sum_{i \geq 1} A^i$. Then $DA^i = A^i \cap (IA \cdot IA)$ is the space of decomposable elements in A^i and $PA^i = A^i / DA^i$ the space of primitive elements in degree i. Let A_i be the dual space to A^i over E and let

$$A_\bullet = \bigoplus_{i=0}^{\infty} A_i.$$

Then, by definition, the space PA_i of primitive elements in A_i is the subspace on which DA^i is zero. The duality between A^i and A_i defines one between PA^i and PA_i.

2. Theorem 2.1

We first draw a consequence of the main result of [1].

Theorem 2.1. *Let G be a connected, simply connected, almost absolutely simple k-group of strictly positive k-rank $r_k(G)$. Then the natural homomorphism*

$$\mu : H^{\bullet}_{ct}(G_\infty;\mathbb{R}) \longrightarrow H^{\bullet}(G(k);\mathbb{R}) \tag{1}$$

which is a composition of f with the restriction map

$$H^{\bullet}(G^{\delta}_{\infty};\mathbb{R}) \longrightarrow H^{\bullet}(G(k);\mathbb{R})$$

is an isomorphism.

Proof. In this proof, S is a finite subset of V containing V_∞ and at least one finite place. Let Γ_S be a congruence S-arithmetic subgroup of $G(k)$. For example, if we fix an embedding $G \hookrightarrow \mathbf{GL}_N$ over k, we may take

$$\Gamma_S = G(\mathcal{O}_S) := G(k) \cap \mathbf{GL}_N(\mathcal{O}_S),$$

where $\mathcal{O}_S$ is the ring of S-integers of k. Then Theorem 1 in [1] asserts:

$$H^{\bullet}(\Gamma_S;\mathbb{R}) = H^{\bullet}_{ct}(G_\infty;\mathbb{R}) \oplus H^{\bullet}_{cusp}(\Gamma_S;\mathbb{R}). \tag{2}$$

The second summand on the right-hand side is the so-called cuspidal cohomology ([1], [6]). The only property of $H^{\bullet}_{cusp}(\Gamma_S;\mathbb{R})$ relevant here is that it vanishes in degrees less than $r(S_f;G)$. If S runs through a strictly increasing sequence of subsets of V, then $r(S_f;G) \to \infty$. Therefore, given $i \in \mathbb{N}$, we have

$$H^{i}(\Gamma_S;\mathbb{R}) \cong H^{i}_{ct}(G_\infty;\mathbb{R}) \tag{3}$$

if S is sufficiently large (e.g. if $|S_f| > i/r_k(G)$). The field k is the inductive limit of the rings $\mathcal{O}_S$ ($S \subset V$), hence $G(k)$ is the inductive limit of the subgroups $G(\mathcal{O}_S)$. Therefore

$$H_{\bullet}(G(k);\mathbb{R}) = \varinjlim_{S} H_{\bullet}(G(\mathcal{O}_S);\mathbb{R}).$$

By duality, equation (3) implies that for a fixed i,

$$H_i(G(\mathcal{O}_S);\mathbb{R}) \cong H_i(G(k);\mathbb{R})$$

for sufficiently large S (depending on i). Dualizing the statement again, we see that

$$H^i(G(k);\mathbb{R}) \xrightarrow{\sim} H^i(G(\mathcal{O}_S);\mathbb{R})$$

for sufficiently large S, whence an isomorphism

$$H^i(G(k);\mathbb{R}) \xrightarrow{\sim} H^i_{\mathrm{ct}}(G_\infty;\mathbb{R}).$$

That it can be described as in the theorem can be seen by a discussion analogous to that in Section 4.8 of [11]. □

3. Some Remarks

3.1. For G anisotropic over k, the isomorphism (2) and Theorem 2.1 are proved in [7, XIII, 3.5, 3.9], except that the formulation of (2) is slightly different, since all the cohomology is cuspidal in that case. The argument used to deduce the theorem from (1) is the same as above.

3.2. The first named author proved in [3] the existence of a constant $c(G)$ such that

$$H^i_{\mathrm{ct}}(G_\infty;\mathbb{R}) \longrightarrow H^i(\Gamma;\mathbb{R})$$

is an isomorphism for $i \leq c(G)$ and any arithmetic subgroup Γ of $G(k)$. In [4], the result is extended to S-arithmetic subgroups and then to $G(k)$ with the same range. The main theorem of [1] shows that in fact, for S-arithmetic groups, the range increases with S, whence the improvement over the results of [4].

3.3. The proof of equation 2.1(2) consists of two main steps. First it is shown that $H^\bullet(\Gamma_S;\mathbb{R})$ is equal to the L^2-cohomology of Γ_S. This relies on [9], which is not yet published. The second and easier step consists in computing the L^2-cohomology. It is also carried out in [6].

3.4. The assumption that G is simply connected is probably not essential. In fact, the second step in [6] does not need that assumption. However, the theorem does not extend to connected reductive groups. For instance, $H^\bullet(\mathbf{GL}_1(k);\mathbb{R})$ is dual to an exterior algebra over an infinite dimensional vector space, while $H^\bullet_{\mathrm{ct}}(G_\infty;\mathbb{R})$ is finite dimensional.

4. Application to $\mathbf{SL}_m(k)$

In this section, we study the effect on homology and cohomology of the standard inclusion

$$\jmath_{m,n} : \mathbf{SL}_m(k) \longrightarrow \mathbf{SL}_n(k).$$

4.1. Let $G_n = \mathbf{SL}_n$ be viewed as an algebraic group over k. Then

$$(1) \qquad H^\bullet_{\mathrm{ct}}(G_{n,\infty};\mathbb{R}) \cong \bigotimes_{v\in V_\infty} H^\bullet_{\mathrm{ct}}(\mathbf{SL}_n(k_v);\mathbb{R}).$$

It is known that

$$(2) \qquad H^\bullet_{\mathrm{ct}}(\mathbf{SL}_n(k_v);\mathbb{R}) = \begin{cases} H^\bullet(\mathbf{SU}_n/\mathbf{SO}_n;\mathbb{R}) & v \text{ real},\\ H^\bullet(\mathbf{SU}_n;\mathbb{R}) & v \text{ complex}.\end{cases}$$

The values of the right-hand side, computed in [2], are

$$(3) \quad H^\bullet(\mathbf{SU}_n/\mathbf{SO}_n;\mathbb{R}) = \begin{cases} \Lambda\langle x_{n,5}, x_{n,9}, \ldots, x_{n,2n-1}\rangle & n \text{ odd},\\ \Lambda\langle x_{n,5}, x_{n,9}, \ldots, x_{n,2n-3}, e_n\rangle & n \text{ even}.\end{cases}$$

$$(4) \qquad H^\bullet(\mathbf{SU}_n;\mathbb{R}) = \Lambda\langle x_{n,3}, x_{n,5}, \ldots, x_{n,2n-1}\rangle$$

where $\Lambda\langle\ \rangle$ denotes a graded exterior algebra over the generators listed in $\langle\ \rangle$, the second subscript of $x_{n;j}$ denotes the degree and e_n is of degree n.

4.2. One can easily study the effect of $j^\bullet_{m,n}$, $(m < n)$ using (3), (4). For consistent choices of the generators, we have

$$(5) \qquad j^i_{m,n}(x_{n,i}) = x_{m,i} \quad (i \le 2m-1),$$
$$(6) \qquad j^i_{m,n}(x_{n,i}) = 0 \quad (i > 2m-1),$$
$$(7) \qquad j^n_{m,n}(e_n) = 0 \quad (n \text{ even}).$$

Therefore

$$(8) \qquad j^i_{m,n} : PH^i_{\mathrm{ct}}(G_{n,\infty};\mathbb{R}) \longrightarrow PH^i_{\mathrm{ct}}(G_{m,\infty};\mathbb{R})$$

is an isomorphism if i is odd and $\le 2m-1$.

As a corollary, we obtain the following

Theorem 4.3. *For any field E of characteristic 0,*

(a) $j^i_{m,n} : PH^i(SL_n(k);E) \to PH^i(SL_m(k);E)$ and
(b) $j_{i,m,n} : PH_i(SL_m(k);E) \to PH_i(SL_n(k);E)$ are isomorphisms for i odd, $i \le 2m-1$.
(c) $j^{2m-1}_{m-1,n}(PH^{2m-1}(SL_n(k);E)) = \{0\}$.

The assertion (a) is a consequence of (8) and 2.1, the assertion (b) follows by duality and (c) from (2.1) and (6), with m replaced by $m-1$.

5. The Rank Conjecture

Theorem 5.1. *The rank conjecture is true for all number fields.*

As recalled in the introduction, we have to prove the two assertions (R1) and (R2) stated there.

By duality and 3.9 in [5], the inclusion $\mathbf{SL}_n(k) \to \mathbf{GL}_n(k)$ induces an isomorphism

$$(1) \quad PH_{2m-1}(\mathbf{SL}_n(k); E) \xrightarrow{\sim} PH_{2m-1}(\mathbf{GL}_n(k); E), \quad (n \gg m \geq 2).$$

Therefore, it suffices to prove (R1) with **GL** replaced by **SL**. Then it follows from 4.3(a).

By duality and (1), (R2) is equivalent to

$$j^{2m-1}_{n,m}(PH^{2m-1}(\mathbf{SL}_n(k); E)) = 0 \quad (n \geq m \geq 2),$$

which follows from 4.3(c).

REMARK: We have given a recent reference for (1) or the dual assertion, but it was known before. For instance, it can be derived from the homology stability of $\mathbf{GL}_n$ proved by Suslin [10] and the fact that (1) is true with n replaced by ∞, since both sides are then naturally isomorphic to $K_{2m-1}(k) \otimes E$ by the Milnor-Moore theorem.

6. Erratum to [11]

The main goal of [11] is to prove (R1), (R2) above for $k \neq \mathbb{Q}$. The proof of (R2) given there is valid for all k's, but it implicitly uses the fact that

$$PH^{2n-1}(\mathbf{GL}_n(k); \mathbb{R}) \cong PH^{2n-1}(\mathbf{SL}_n(k); \mathbb{R})$$

discussed above. Also the proof there made use of Proposition 4.9, which is unnecessary. Once it is known that the map

$$PH^{2n-1}(\mathbf{GL}(k); \mathbb{R}) \to H^{2n-1}(\mathbf{GL}(k); \mathbb{R}) \to H^{2n-1}(\mathbf{GL}_m(k); \mathbb{R})$$

is trivial for $m < n$, the lower rank conjecture follows by duality.

The proof of (R1) for quadratic fields in [11] is not valid, however. First of all, the estimate for $\alpha(J)$ on p. 305 in the imaginary quadratic field case is incorrect. Hence the injectivity assertion for j^{2n-1} on top of p. 305 is not proved for imaginary quadratic fields, and in fact is clearly false for $n = 2$. Even when the injectivity on top of p. 305 holds, the argument in 4.12 in [11], based on 4.9, does not quite give (R1). There is some confusion on the notion of primitive elements. More precisely, a subspace DH_i of $H_i(\mathbf{SL}_n(k); \mathbb{R})$ is defined, which is a supplement to $PH_i(\mathbf{SL}_n(k); \mathbb{R})$. The argument in [11] shows that

$$j_{2m-1,m,n}(H_{2m-1}(\mathbf{SL}_m(k); \mathbb{R}))$$

contains a supplement to DH_i, but not necessarily $PH_{2m-1}(\mathbf{SL}_n(k);\mathbb{R})$ itself.

7. Homology of $\mathbf{GL}_n(k)$ In Characteristic Zero

7.1. Given a field F and $n \in \mathbb{N}$, we identify $x \in F^*$ to the diagonal matrix $d(x) \in GL_n(F)$ with all diagonal entries equal to 1, except for the first one, which is equal to x. The group $\mathbf{GL}_n(k)$ is the semi-direct product of the subgroup

$$Z = \{ d(x) \mid x \in F^* \}$$

and the normal subgroup $\mathbf{SL}_n(F)$. The group Z operates by inner automorphisms on the cohomology (or homology) of $\mathbf{SL}_n(F)$ and we let $\delta(x)$ be the automorphism so defined. In case F is a topological field, we have similarly an operation $\delta(x)$ on the continuous cohomology. We first determine the Z-invariants in $H^\bullet(\mathbf{SL}_n(F);\mathbb{R})$, when $F = k, k_v$ $(v \in V_\infty)$.

We refer to Section 4 (2), (3), (4) for the description of $H^\bullet_{\rm ct}(\mathbf{SL}_n(k_v);\mathbb{R})$.

Lemma 7.2. *We keep the previous notation. Then*

(1)

$$H^\bullet_{\rm ct}(\boldsymbol{SL}_n(k_v);\mathbb{R})^{k_v^*} = \begin{cases} H^\bullet_{\rm ct}(\boldsymbol{SL}_n(k_v);\mathbb{R}) & (v \textit{ complex or } n \textit{ odd}) \\ \Lambda_{\mathbb{R}}\langle x_{n,5}, x_{n,9}, \dots, x_{n,2n-3}\rangle & (v \textit{ real, } n \textit{ even}). \end{cases}$$

Proof. The spaces $\mathbf{SU}_n/\mathbf{SO}_n$ and $\mathbf{SU}_n$ are deformation retracts of $\mathbf{SL}_n(\mathbb{C})/\mathbf{SL}_n(\mathbb{R})$ and $\mathbf{SL}_n(\mathbb{C})$ respectively. We can therefore replace $\mathbf{SU}_n$ and $\mathbf{SO}_n$ by $\mathbf{SL}_n(\mathbb{C})$ and $\mathbf{SL}_n(\mathbb{R})$ in (2), (3), (4) of Section 4. Note that on the left-hand side of these relations we have the singular cohomology groups of some manifolds, so any connected group of homeomorphisms acts trivially on them. If v is complex, then k_v^* is connected, whence the lemma in that case. The same is true for the group $\mathbb{R}^*_+$ of strictly positive real numbers if v is real so that, in that case, we are reduced to studying the effect of $\delta(-1)$. Since it belongs to $\mathbf{O}_m$, we may again use the compact picture.

The map

$$p^\bullet : H^\bullet(\mathbf{SU}_n/\mathbf{SO}_n;\mathbb{R}) \to H^\bullet(\mathbf{SU}_n;\mathbb{R})$$

induced by the projection

$$p : \mathbf{SU}_n \longrightarrow \mathbf{SU}_n/\mathbf{SO}_n$$

is injective on odd dimensional generators (cf. [2]). Since $p^\bullet$ commutes with $\delta(-1)$, this implies that $\delta(-1)$ acts trivially on the subalgebra of $H^\bullet(\mathbf{SU}_n/\mathbf{SO}_n;\mathbb{R})$ generated by the $x_{n,i}$. For n odd, this concludes the proof.

If n is even, there remains to be seen that

$$\delta(-1)(e_n) = -e_n.$$

The class e_n is the image by transgression in the bundle

$$\mathbf{SO}_n \to \mathbf{SU}_n \to \mathbf{SU}_n/\mathbf{SO}_n,$$

of a class $c_{n-1} \in H^{n-1}(\mathbf{SO}_n;\mathbb{R})$ and $\delta(-1)$ commutes with transgression. So we are reduced to showing that

$$\delta(-1)(c_{n-1}) = -c_{n-1}.$$

The class c_{n-1} is the image of a generator d_{n-1} of $H^{n-1}(\mathbf{S}^{n-1};\mathbb{R})$ under the map

$$q^\bullet : H^\bullet(\mathbf{S}^{n-1};\mathbb{R}) \longrightarrow H^\bullet(\mathbf{SO}_n;\mathbb{R})$$

induced by the projection

$$q : \mathbf{SO}_n \longrightarrow \mathbf{S}^{n-1}.$$

The projection q commutes with $\delta(-1)$ acting by inner automorphism on $\mathbf{SO}_n$ and via $d(-1)$ on $\mathbb{R}^n$. Since $d(-1)$ changes the orientation, our assertion follows. □

7.3. For each $v \in V_\infty$, let σ_v be an associated embedding. When $v \in V_{\mathbb{R}}$, σ_v is unique. If v is complex, then take σ_v to be one of the two conjugate embeddings associated to v. The homomorphism μ of Section 2 (1) may be written as a tensor product

$$\mu = \bigotimes_{v\in V_\infty} \mu_v,$$

where μ_v is the composition

$$H^\bullet_{\mathrm{ct}}(\mathbf{SL}_n(k_v);\mathbb{R}) \to H^\bullet(\mathbf{SL}_n(k_v)^\delta;\mathbb{R}) \xrightarrow{\sigma_v^\bullet} H^\bullet(\mathbf{SL}_n(k);\mathbb{R}).$$

Let

$$\begin{aligned} S^\bullet_{n,v} &= \mu_v(H^\bullet_{\mathrm{ct}}(\mathbf{SL}_n(k_v);\mathbb{R})^{k_v^*}) \\ S^\bullet_n &= \bigotimes_{v\in V_\infty} S^\bullet_{n,v}. \end{aligned}$$

Then 2.1(1) may be written

$$H^\bullet(\mathbf{SL}_n(k);\mathbb{R}) \cong S^\bullet_n \otimes \Lambda_{\mathbb{R}}\langle \mu(e_{n,1}),\ldots,\mu(e_{n,r})\rangle,$$

where $e_{n,i}$ is the Euler class in the factor associated to v_i, and $V_{\mathbb{R}} = \{v_1,\ldots,v_r\}$ is the set of real places, or equivalently real embeddings, of k.

Lemma 7.4. *We have the equalities*

(a) $H^\bullet(SL_n(k);\mathbb{R})^{k^*} = S_n^\bullet$,
(b) $S_m^\bullet = j_{m,n}^\bullet(H^\bullet(SL_n(k);\mathbb{R}))\quad (n > m \geq 2)$.

Moreover,

(c) $j_{m,n}^i\colon S_n^i \to S_m^i$ *is an isomorphism for* $i \leq 2m-1$.

Proof. We note first that

$$\mu_v \circ \delta(\sigma_v(x)) = \delta(x) \circ \mu_v \quad (v \in V_\infty; x \in k^*) \tag{2}$$

on $H^\bullet_{\mathrm{ct}}(\mathbf{SL}_n(k_v);\mathbb{R})$, as follows from the definition. Let

$$N = \{\, x \in k^* \mid \sigma_v(x) > 0,\ v \in V_{\mathbb{R}} \,\}, \qquad C = k^*/N. \tag{3}$$

By (2) and the initial remark in the proof of 7.2, N acts trivially on $H^\bullet(\mathbf{SL}_n(k);\mathbb{R})$, hence

$$H^\bullet(\mathbf{SL}_n(k);\mathbb{R})^{k^*} = S_n^\bullet \otimes (\Lambda_{\mathbb{R}}\langle \mu(e_{n,1}),\ldots,\mu(e_{n,r})\rangle)^C. \tag{4}$$

By weak approximation, $C = (\mathbb{Z}/2\mathbb{Z})^r$, i.e., given any pattern of signs $\{\, \epsilon_{\sigma_v} \mid v \in V_{\mathbb{R}} \,\}$, there exists $x \in k^*$ such that sign $\sigma_v(x) = \epsilon_{\sigma_v}$, for $v \in V_{\mathbb{R}}$. Then (2) and 7.2 show that the second factor on the right-hand side of (4) reduces to the constants, and (a) is proved. Now (b) and (c) follow from Section 4 (5), (6), (7). □

REMARK: In view of 7.4(b), S_m has a natural $\mathbb{Q}$-structure $S_{m,\mathbb{Q}}$, namely the image of the rational cohomology of $\mathbf{SL}_n(k)$. We let

$$S_{m,E} = S_{m,\mathbb{Q}} \otimes_{\mathbb{Q}} E. \tag{5}$$

Theorem 7.5. *We have the isomorphisms*

$$H^\bullet(k^*; H^\bullet(SL_n(k);E)) = H^\bullet(k^*;E) \otimes S_{n,E}^\bullet = H^\bullet(GL_n(k);E).$$

Proof. The algebra $H^\bullet(k^*; H^\bullet(\mathbf{SL}_n(k);E))$ is the abutment of a spectral sequence associated to the group extension (see (3))

$$1 \to N \to k^* \to C \to 1 \tag{6}$$

in which, as usual,

$$E_2^{p,q} = H^p(C; H^q(N; H^\bullet(\mathbf{SL}_n(k);E))). \tag{7}$$

Since N acts trivially on $H^\bullet(\mathbf{SL}_n(k);E)$, we have

$$H^q(N; H^\bullet(\mathbf{SL}_n(k);E)) \cong H^q(N;E) \otimes H^\bullet(\mathbf{SL}_n(k);E). \tag{8}$$

The group C is finite and E is of characteristic zero, hence

$$E_2^{p,q} = \begin{cases} (H^q(N;E) \otimes H^\bullet(\mathbf{SL}_n(k);E))^C, & p=0, \\ 0, & p \neq 0. \end{cases} \tag{9}$$

But k^* is commutative, hence acts trivially on the cohomology of N and we get

$$E_2^{0,q} = H^q(N;E) \otimes (H^\bullet(\mathbf{SL}_n(k);E))^C. \tag{10}$$

Therefore by 7.4 and the remark to 7.4,

$$H^\bullet(k^*; H^\bullet(\mathbf{SL}_n(k);E)) = H^\bullet(N;E) \otimes S_{n,E}. \tag{11}$$

The same computation shows that the spectral sequence of (6) for the trivial k^*-module E collapses and yields

$$H^\bullet(N;E) = H^\bullet(k^*;E), \tag{12}$$

and the first equality follows from (11) and (12).

To prove the second equality, we consider the commutative diagram of extensions

$$\begin{array}{ccccccccc} 1 & \longrightarrow & \mathbf{SL}_n(k) & \longrightarrow & \mathbf{GL}_n(k) & \longrightarrow & k^* & \longrightarrow & 1 \\ & & \downarrow & & \downarrow & & \| & & \\ 1 & \longrightarrow & \mathbf{SL}(k) & \longrightarrow & \mathbf{GL}(k) & \longrightarrow & k^* & \longrightarrow & 1 \end{array} \tag{13}$$

where as usual, $\mathbf{SL}(k)$ and $\mathbf{GL}(k)$ are defined to be the inductive limits of $\mathbf{SL}_n(k)$ and $\mathbf{GL}_n(k)$ respectively, and the vertical arrows are natural inclusions. It is known that the spectral sequence of the second row of (13) is trivial, hence

$$H^\bullet(\mathbf{GL}(k);E) = H^\bullet(k^*;E) \otimes H^\bullet(\mathbf{SL}(k);E)$$

(see 3.9 in [5]). Lemma 7.4(b) and stability imply that the image of

$$j^\bullet_{n,\infty} : H^\bullet(\mathbf{SL}(k);E) \to H^\bullet(\mathbf{SL}_n(k);E)$$

is $S_{n,E}$.

Let $\{E'_r\}$ be the spectral sequence of the first row of (13). Then the previous remark and the first part of 7.5 show that

$$j_{n,\infty} : E_2^{p,q} \to {E'}_2^{p,q}$$

is surjective. Since the first spectral sequence collapses at E_2, the same is true for the second one, whence the second part of 7.5.

A priori, there might be an extension problem since we have proved this only for $E_\infty = E_2$. However, since the "fibre factor" $S_{n,E}$ is an exterior algebra on odd dimensional generators, hence free among graded anti-commutative algebras over E, this is not an issue. □

Corollary 7.6. *Let $m, n \in \mathbb{N}$, $m \leq n$. Then*

(a) $j^i_{m,n} : H^i(GL_n(k); E) \to H^i(GL_m(k); E)$
is surjective for all i's and injective for $i \leq 2m-1$;

(b) $j_{i,m,n} : H_i(GL_m(k); E) \to H_i(GL_n(k); E)$
is injective for all i's and surjective for $i \leq 2m-1$.

Proof. The first assertion follows from 7.4, 7.5, the structure of $S^\bullet_{n,E}$ given by 7.2 and 4.2. The second one follows from the first by duality. □

Remark 7.7. We see in particular that $j_{i,m,n}$ is an isomorphism for $i \leq 2m-1$, which gives a sharper version of the "upper" rank conjecture. We recall that by 3.4 in [10], this is true for any infinite field and any trivial coefficients if $i \leq m$. We also understand that Suslin had conjectured the injectivity of $j_{i,m,n}$ for all i's for infinite fields.

References

1. D. Blasius, J. Franke, F. Grunewald, *Cohomology of S-arithmetic groups in the number field case*, Invent. Math. **116** (1994), 75–93.
2. A. Borel, *Sur la cohomologie des espaces fibrés principaux et des espaces homogènes de groupes de Lie compacts*, Ann. Math. **57** (1953), 115–207.
3. ———, *Stable real cohomology of arithmetic groups*, Ann. Sci. Ecole Norm. Sup. (4) **7** (1974), 235–272.
4. ———, *Stable real cohomology of arithmetic groups II*, Prog. Math. **14** (1981), 21–55, Birkhäuser-Boston.
5. ———, *Values of zeta-functions at integers, cohomology and polylogarithms*, preprint.
6. A. Borel, J.-P. Labesse, J. Schwermer, *On the construction of cuspidal cohomology of S-arithmetic subgroups of reductive groups*, Preprint.
7. A. Borel, N. Wallach, *Continuous cohomology, Discrete Subgroups, and Representations of Reductive Groups*, Princeton University Press, Princeton, 1980.
8. J-L. Cathelineau, *Homologie du groupe linéaire et polylogarithmes*, Sém. Bourbaki, 1992–93, Exp. **772**, Astérisque **216**, 1993, 311–341, Soc. Math. France.
9. J. Franke, *Harmonic analysis in weighted L_2-spaces*, in preparation.
10. A. A. Suslin, *Homology of GL_n, characteristic classes and Milnor K-theory*, LNM, **1046** (1984), 357–375, Springer-Verlag.
11. J. Yang, *On the real cohomology of arithmetic groups and the rank conjecture for number fields*. Ann. Scient. Éc. Norm. Sup. **25** (1992), 287–306.

INSTITUTE FOR ADVANCED STUDY, PRINCETON, NJ 08540
E-mail address: borel@math.ias.edu

DEPARTMENT OF MATHEMATICS, DUKE UNIVERSITY, DURHAM, NC 27708-0320
E-mail address: yang@math.duke.edu

155.

Response to "Theoretical Mathematics: Toward a Cultural Synthesis of Mathematics and Theoretical Physics, by A. Jaffe and F. Quinn"

Bull A.M.S. N.S. **30** (1994), 179–181

Armand Borel
Institute for Advanced Study
Princeton, NJ 08540
borel@math.ias.edu

Some comments on the article by A. Jaffe and F. Quinn:

There are a number of points with which I agree, but they are so obvious that they do not seem worth such an elaborate discussion. On the other hand, I disagree with the initial stand and with much of the general thrust of the paper, so that I shall not comment on it item by item, but limit myself to some general remarks.

First the starting point. I have often maintained, and even committed to paper on some occasions, the view that mathematics is a science, which, in analogy with physics, has an experimental and a theoretical side, but operates in an intellectual world of objects, concepts and tools. Roughly, the experimental side is the investigation of special cases, either because they are of interest in themselves or because one hopes to get a clue to general phenomena, and the theoretical side is the search of general theorems. In both, I expect proofs of course, and I reject categorically a division into two parts, one with proofs, the other without.

I also feel that what mathematics needs least are pundits who issue prescriptions or guidelines for presumably less enlightened mortals. Warnings about the dangers of certain directions are of course nothing new. In the late forties, H. Weyl was very worried by the trend towards abstraction, exemplified by the books of Bourbaki or that of Chevalley on Lie groups, as I knew from M. Plancherel. Later, another mathematician told me he had heard such views from H. Weyl in the late forties but, then, around 1952 I believe, i.e. after the so-called French explosion, H. Weyl told him: "I take it all back."

In fact, during the next quarter century, we experienced a tremendous development of pure mathematics, bringing solutions of one fundamental problem after the other, unifications, etc., but during all that time, there was in some quarters some whining about the dangers of the separation between pure math and applications to sciences, and how the great nineteenth century mathematicians cultivated both (conveniently ignoring some statements by none other than Gauss which hardly support that philosophy). [To avoid any misunderstanding, let me hasten to add that I am not advocating the separation between the two, being quite aware of the great benefits on both sides of interaction, but only the freedom to devote oneself to pure mathematics, if so inclined.]

Of course, I agree that no part of mathematics can flourish in a lasting way without solid foundations and proofs, and that not doing so was harmful to Italian algebraic geometry for instance. I also feel that it is probably so for the Thurston program, too. It can also happen that standards of rigor deemed acceptable by the practitioners in a certain area turn out to be found wanting by a greater mathematical community. A case in point, in my experience, was E. Cartan's work on exterior differential forms and connections, some of which was the source of a rather sharp exchange between Cartan and Weyl. Personally, I felt rather comfortable with it but later, after having been exposed to the present points of view, could hardly understand what I had thought to understand. We all know about Dirac diagonalizing any self-adjoint operator and using the Dirac function. And there are of course many more examples. But I do believe in the self-correcting power of mathematics, already expressed by D. Hilbert in his 1900 address, and all I have mentioned (except for Thurston's program) has been straightened out in due course. Let me give another example of this self-correcting power of mathematics. In the early fifties, the French explosion in topology was really algebraic topology with a vengeance. Around 1956, I felt that topology as a whole was going too far in that direction and I was wishing that some people would again get their hands dirty by using more intuitive or geometric points of view (I did so for instance in a conversation with J. C. Whitehead at the time, which he reminded me of shortly before his death, in 1960; I had forgotten it.) But shortly after came the developments of PL-topology by Zeeman and Stallings, of differential topology by Milnor and Smale, and there was subsequently in topology a beautiful equilibrium between algebraic, differential and PL points of view.

But this was achieved just because some gifted people followed their own inclinations, not because they were taking heed of some solemn warning.

In advocating freedom for mathematicians, I am not innovating at all. I can for instance refer to a lecture by A. Weil (*Collected Papers* II, 465–469) praising disorganization in mathematics and pointing out that was very much the way Bourbaki operated. As a former member of Bourbaki, I was of course saddened to read that all that collective work, organized or not, ended up with the erection of a bastion of arch-conservatism. Not entertaining pyramids of conjectures? Let me add that Weil was not ostracized for his conjectures, nor was Grothendieck for his standard conjectures and the theory motives, nor Serre for his "questions".

F. Quinn is not making history in raising questions about the Research Announcements in the *Bulletin*, as you know. At some point, they were functioning poorly and their suppression was suggested by some. To which I. Singer answered that people making such proposals did not know what the AMS was about (or something to that effect) and offered to manage that department for a few years. He did so and it functioned very well during his tenure. Also, the *Comptes Rendus* have a very long history of R.A.s. There were ups and downs of course, but for the last twenty years or so, it seems to me to have been working well on the whole. All this to say that the problems seen by F. Quinn are not new, have been essentially taken care of in the past, and I do not see the need for new prescriptions.

156.

Values of Indefinite Quadratic Forms at Integral Points and Flows on Spaces of Lattices

Bull. A.M.S. **32** (1995), 184–204

This mostly expository paper centers on recently proved conjectures in two areas:

A) A conjecture of A. Oppenheim on the values of real indefinite quadratic forms at integral points.

B) Conjectures of Dani, Raghunathan, and Margulis on closures of orbits in spaces of lattices such as $\mathbf{SL}_n(\mathbf{R})/\mathbf{SL}_n(\mathbf{Z})$.

At first sight, A) belongs to analytic number theory and B) belongs to ergodic and Lie theory, and they seem to be quite unrelated. They are discussed together here because of a very interesting connection between the two pointed out by M. S. Raghunathan, namely, a special case of B) yields a proof of A).

The first main goal of this talk is to describe the Oppenheim conjecture and various refinements and to derive them from one statement about closures of orbits in the space of unimodular lattices in $\mathbf{R}^3$ (see Proposition 2 in §2.3). In §§3 and 4 we put this statement in context and describe more general conjectures and results on orbit closures and invariant probability measures on quotients of Lie groups by discrete subgroups. §5 gives some brief comments on the proofs and further developments. §§6, 7, and 8 are devoted to the so-called S-arithmetic setting, where we consider products of real and p-adic groups. §6 is concerned with a generalized Oppenheim conjecture; §7 with a generalization of the orbit closure theorem proved by M. Ratner [R8]; and §8 with applications to quadratic forms. Since the subject matter of that last section has not been so far discussed elsewhere, we take this opportunity to present proofs, obtained jointly with G. Prasad. Finally, §9 gives the proof of a lemma on symmetric simple Lie algebras, a special case of which is used in §8.

I am glad to thank M. Ratner and G. Prasad for a number of remarks on, and corrections to, a preliminary version of this paper, thanks to which many typos and inaccuracies have been eliminated.

I. Values of indefinite quadratic forms

1. The Oppenheim conjecture.

1.1. In the sequel F denotes a *non-degenerate* quadratic form on $\mathbf{R}^n$ which is *indefinite*, i.e. $F(x) = 0$ for some $x \in \mathbf{R}^n - \{0\}$ or equivalently $F(\mathbf{R}^n) = \mathbf{R}$. It may be written

$$F(x) = \sum_{1 \leq i,\, j \leq n} f_{ij} x_i x_j \qquad (f_{ij} = f_{ji} \in \mathbf{R},\ \det(f_{ij}) \neq 0).$$

1991 *Mathematics Subject Classification.* Primary 22-02, 73K12.

This paper is an outgrowth of a Progress in Mathematics lecture given at the joint AMS-CMS meeting, Vancouver, B.C., Canada, August 15–19, 1993.

Unless otherwise stated we assume $n \geqq 3$. We are concerned with $F(\mathbf{Z}^n)$.

Definition. F is said to be *rational* if $F(x)/F(y) \in \mathbf{Q}$ whenever $x, y \in \mathbf{Q}^n$ and $F(y) \neq 0$, and *irrational* otherwise.

F is rational if and only if there exists $c \in \mathbf{R}^*$ such that $F = c.F_o$, where F_o has rational coefficients (with respect to a basis of $\mathbf{Q}^n$). We may then also arrange F_o to have integral coefficients. Therefore

$$F(\mathbf{Z}^n) = c.F_o(\mathbf{Z}^n) \subset c.\mathbf{Z}$$

is discrete. The Oppenheim conjecture states that, conversely, if F is irrational, then $F(\mathbf{Z}^n)$ is not discrete around the origin. More precisely, consider the two conditions:

(i) *F is irrational.*

(ii) *Given $\epsilon > 0$, there exists $x \in \mathbf{Z}^n$ such that $0 < |F(x)| < \epsilon$.*

We just saw that (ii) $\Rightarrow$ (i). The *Oppenheim conjecture is that* (i) $\Rightarrow$ (ii).

1.2. Historically, this is a bit of an oversimplification. In 1929, A. Oppenheim stated that the following is very likely to be true: if F is irrational and $n \geqq 5$, then $|F(x)|$ takes arbitrary small values on $\mathbf{Z}^n$ [O1, O2]. Formally, this may be written

(ii)′ Given $\epsilon > 0$, there exists $x \in \mathbf{Z}^n$ such that $|F(x)| < \epsilon$,

and is automatically satisfied if F "represents zero rationally"; i.e. if there exists $x \in \mathbf{Q}^n - \{0\}$, hence also $x \in \mathbf{Z}^n - \{0\}$, such that $F(x) = 0$. But Oppenheim had clearly (ii) in mind, and he made it explicit in [O3], still for $n \geqq 5$, though; but [O4, O5] show that he was wondering whether it might be true for $n \geqq 3$ already (it was well known to be false for $n = 2$; an example is given in (1.4)). Then, later, the conjecture (i) $\Rightarrow$ (ii)′ became erroneously known as "Davenport's conjecture", though Davenport referred to Oppenheim. The implication (i) $\Rightarrow$ (ii)′ is obviously equivalent to the Oppenheim conjecture (i) $\Rightarrow$ (ii) for forms not representing zero rationally. The bound $n \geqq 5$ had been suggested to A. Oppenheim by a theorem of A. Meyer according to which a rational indefinite quadratic form in $n \geqq 5$ variables always represents zero rationally (see e.g. [S, IV, §3]). He felt that, in the irrational case, it should take values close to zero on $\mathbf{Z}^n$.

1.3. The condition (ii) leaves open the possibility that $F(\mathbf{Z}^n)$ accumulates to zero only on one side, but Oppenheim showed in [O3] that this cannot happen for $n \geqq 3$, our standing assumption (but that it can for $n = 2$). It then follows by a very elementary argument that (ii) implies:

(iii) $F(\mathbf{Z}^n)$ is dense in $\mathbf{R}$,

so that the conjectural dichotomy was in fact

(A)
F rational $\Leftrightarrow$ $F(\mathbf{Z}^n)$ discrete;
F irrational $\Leftrightarrow$ $F(\mathbf{Z}^n)$ dense.

1.4. To conclude this section, we give a simple counterexample for $n = 2$, borrowed from [G].

Let $F(x, y) = y^2 - \theta^2.x^2$, where θ is quadratic, irrational > 0, and θ^2 is irrational. The form F is irrational. As is well known, there exists $c > 0$ such that

$$(1) \qquad |\theta - y/x| \geqq c.x^{-2} \qquad (x, y \in \mathbf{Z}, x \neq 0).$$

For $x \neq 0$, we can write

$$(2) \qquad F(x, y) = x^2(y/x + \theta)(y/x - \theta).$$

We have to prove that $|F(x, y)|$ has a strictly positive lower bound for $x, y \in \mathbf{Z}$ not both zero. This is clear if one of them is equal to zero. So let $x, y \neq 0$. We may assume them to be > 0. Then $|\theta + y/x| \geqq \theta$. Together with (1) and (2), this yields $|F(x, y)| \geqq c.\theta$.

2. Results.

2.1. The first partial results on the Oppenheim conjecture were obtained in the framework of analytic number theory. It was shown to be true for diagonal forms in $n \geqq 9$ variables [C], in $n \geqq 5$ variables [DH] and for general forms in $n \geqq 21$ variables [DR]. Oppenheim himself proved it when F represents zero rationally, for $n \geqq 5$ in [O4] and for $n = 4$ in [O5]. In both papers he stated his belief it should be true for $n = 3$, though it was obviously false for $n = 2$.

It is easy to see that if F is irrational, then there exists a three-dimensional subspace $V \subset \mathbf{Q}^n$ such that the restriction of F to $V \otimes_{\mathbf{Q}} \mathbf{R}$ is non-degenerate, indefinite and irrational. Therefore it suffices to prove the Oppenheim conjecture for $n = 3$.

2.2. Around 1980, M. S. Raghunathan made a conjecture on closures of orbits in spaces of lattices (see 3.2). He noticed further that a very special case, namely Proposition 1 below, would readily imply the "Davenport conjecture". G. A. Margulis, following this strategy, then proved Proposition 1 and deduced from it that (i) $\Rightarrow$ (ii)$'$ [M1, M2]. When informed (by the author, October 1987) of the fact that the Oppenheim conjecture was a slightly stronger one, he quickly completed his argument and established:

Theorem 1 (Margulis [M3]). *The Oppenheim conjecture is true.*

2.3. The statement on closures of orbits proved and used by Margulis to show that (i) $\Rightarrow$ (ii)$'$ is:

Proposition 1. *Let $n = 3$. Then any relatively compact orbit of $SO(F)$ on $\Omega_3 = \mathbf{SL}_3(\mathbf{R})/\mathbf{SL}_3(\mathbf{Z})$ is compact.*

It is of course equivalent to the same assertion for the topological identity component $SO(F)^o$ of $SO(F)$ (which has index two).

A slight extension of it (see Theorem 1$'$ in [M3]) allowed him to establish the full Oppenheim conjecture. Rather than explaining how, I shall sketch a derivation of this conjecture from the following assertion, stronger than Proposition 1 (but still a very special case of the topological conjecture; see §3), proved shortly afterwards for this purpose by S. G. Dani and G. A. Margulis [DM1]:

Proposition 2. *We keep the previous notation and assumptions. Then any orbit of $SO(F)^o$ on Ω_3 either is closed and carries an $SO(F)^o$-invariant probability measure or is dense.*

(In this paper, all measures are Borel measures.)

They deduced directly from it that (i) implies not only (iii), but also that the set of values of F on the *primitive vectors* in $\mathbf{Z}^n$ is dense [DM1]. Recall that $x \in \mathbf{Z}^n$, $x \neq 0$, is *primitive* if it is not properly divisible in $\mathbf{Z}^n$. The set of primitive vectors is $\mathbf{SL}_n(\mathbf{Z}).e_1$, where e_1 is the first canonical basis vector. This is a sharpening of (iii) which, as far as I know, had never been considered before.

Note that, since Ω_3 is not compact, Proposition 2 obviously implies Proposition 1.

2.4. We now sketch the proof of the Oppenheim conjecture, or rather of its strengthening just mentioned, using Proposition 2.

We let $G = \mathbf{SL}_3(\mathbf{R})$, $H = SO(F)^o$ and $\Gamma = \mathbf{SL}_3(\mathbf{Z})$.

Let o be the origin in Ω_3, i.e. the coset Γ. Then by Proposition 2, $H.o$ is either closed with finite invariant measure or dense. Assume first it is dense. Then its inverse image $H.\Gamma$ in G is also dense. Fix $c \in \mathbf{R}$. There exists $x \in \mathbf{R}^n$, $x \neq 0$ such that $F(x) = c$. Let $e_1, \ldots, e_n$ be the canonical basis of $\mathbf{R}^n$. There exists $g \in \mathbf{SL}_3(\mathbf{R})$ such that $g.e_1 = x$. Since $H.\Gamma$ is dense in G, we can find sequences $\gamma_j \in \Gamma$ and $h_j \in H$ $(j = 1, \ldots)$ such that $h_j.\gamma_j \to g$. Then we have

$$c = F(g.e_1) = \lim_{j\to\infty} F(h_j.\gamma_j.e_1) = \lim_{j\to\infty} F(\gamma_j.e_1);$$

therefore c is in the closure of $F(\mathbf{SL}_3(\mathbf{Z}).e_1)$.

Assume now that $H.o$ is closed and supports an H-invariant probability measure. Let $\Gamma_o = H \cap \Gamma$. Then H/Γ_o is homeomorphic to $H.o$ and has therefore finite invariant volume. By a general result, this implies that Γ_o is "Zariski-dense" in H, i.e. is not contained in any algebraic subgroup. However, in this case, it can be checked more directly. In fact, the only important property of real algebraic groups relevant here is that they have only finitely many connected components, in the usual topology. It suffices therefore to show that Γ_o is not contained in any closed subgroup of H having finitely many connected components. Let M be one. Then H/M also carries an invariant probability measure (define the measure of an open set U in H/M as equal to that of its inverse image in H/Γ_o under the natural projection $H/\Gamma_o \to H/M$). By a standard fact (see e.g. [Bu], VII, §2, no. 6), this implies that M, hence also M^o, is unimodular. Also, since H/M^o is a finite covering of H/M, it carries an invariant probability measure, too, and the same is true if we replace H by its twofold covering $\mathbf{SL}_2(\mathbf{R})$ and M^o by the identity component of its inverse image there. Being unimodular, M^o is not a maximal connected solvable subgroup of H. It is then either conjugate to the subgroup A of diagonal matrices with positive entries or to the subgroup N of upper triangular unipotent matrices. Recall the Iwasawa decomposition $H = K.A.N$, where $K = \mathbf{SO}(2)$. If $M^o = A$, then there is an N-equivariant diffeomorphism of H/M^o onto $N \times K$. If $x \in N$, $x \neq 1$, then we can find a neighborhood of 1 in $N \times K$, the translates of which by the powers of x are disjoint, hence H/M^o has infinite invariant measure. If $M^o = N$, then there is similarly an A-equivariant diffeomorphism of H/M^o onto $A \times K$, and we see in the same way that an invariant measure has infinite volume. Thus, the existence of M leads to a contradiction, which shows that $\Gamma \cap H$ is Zariski-dense in H. (In both cases, it would suffice in fact to note that H/M^o is not compact, in view of a theorem of G. D. Mostow [M, Theorem

7.1] which states that if the quotient of a connected Lie group by a closed subgroup with finitely many connected components carries an invariant measure, it is compact. However this result is closely related to the Zariski-density theorem I was trying to prove directly.)

Consider the natural action of H on the space of 3×3 real symmetric matrices. The only invariants of H are the multiples cF of F $(c \in \mathbf{R})$. Since Γ_o is Zariski-dense in H, these are also the only invariants of Γ_o. But invariance under Γ_o translates into a system of linear equations for the coefficients of F, with integral coefficients. It has therefore a rational solution; i.e. cF has rational coefficients for some $c \neq 0$, hence F is rational by definition.

2.5. The results of M. Ratner recalled in §3 imply that Proposition 2 is valid for all $n \geqq 3$, (see 3.6). As we saw, the case $n = 3$ suffices for the results on quadratic forms discussed so far, but Proposition 2 for arbitrary n also yields a stronger approximation theorem for quadratic forms. To state it, let us say that a subset $(x_1, \ldots, x_m)$ of $\mathbf{Z}^n$ $(m \leqq n)$ is primitive if it is part of a basis of $\mathbf{Z}^n$. If $m < n$, this condition is equivalent to the existence of $g \in \mathbf{SL}_n(\mathbf{Z})$ such that $g.e_i = x_i$ $(i = 1, \ldots, m)$. Then we have [BP2, 7.9].

Theorem 2. *Let $c_i \in \mathbf{R}$ $(i = 1, \ldots, n-1)$. Assume F to be irrational. Then there exists a sequence $(x_{j1}, \ldots, x_{jn-1})$ $(j = 1, 2, \ldots)$ of primitive subsets of $\mathbf{Z}^n$ such that*

$$\text{(1)} \qquad \lim_{j \to \infty} F(x_{ji}) = c_i \qquad (i = 1, \ldots, n-1).$$

Proposition 2 already implies it for two values c_1, c_2, as was shown in [DM1]. The proof is an easy extension of the first argument in 2.4. In fact [BP, 7.9] is concerned more generally with the S-arithmetic case (see §6) but without finite places. The proof in the general S-arithmetic case will be given in 8.4.

2.6. Propositions 1 and 2 are very special cases of the general results of M. Ratner outlined in the next section. However, Proposition 1 had earlier been given a comparatively elementary proof in [M1, M3], and [DM3] provides also an elementary proof of a theorem weaker than Proposition 2 but stronger than those of [M1, M3] and already sufficient to show that if F is irrational, its set of values at primitive vectors is dense in $\mathbb{R}$. The main theorems on flows, in particular the property of "uniform distribution" (see 3.9), have led to some quantitative refinements of the Oppenheim conjecture. We state here the simplest one

1 **Proposition 3.** *Given $a, b > 0$ with $a < b$, there exists constants r_o, $c > 0$ such that*

$$Card\{x \in \mathbf{Z}^n \cap B_r | a \leqq |F(x)| \leqq b\} \geqq c.r^{n-2} \qquad (r \geqq r_o).$$

Here, B_r denotes the euclidean ball in $\mathbf{R}^n$ of radius r with center the origin. This was proved first, independently, by S. G. Dani and S. Mozes on the one hand and M. Ratner on the other (unpublished). A more general statement, valid for certain compact sets of quadratic forms, is proved in [DM4]; see Corollary on page 95.

2.7. The truth of the Oppenheim conjecture also yields a characterization of arbitrary non-degenerate rational quadratic forms in $n \geqq 2$ variables, as was noted by G. Prasad and me [B]:

Proposition 4. *Let $n \geqq 2$ and E be a non-degenerate quadratic form on $\mathbf{R}^n$. Then E is irrational if and only, given $\epsilon > 0$, there exists $x, y \in \mathbf{Z}^n$ such that*

$$0 < |E(x) - E(y)| < \epsilon. \tag{1}$$

This follows by applying Theorem 1 to $E \oplus -E$ on $\mathbf{R}^{2n}$. In fact, in view of 2.4, we may also find primitive vectors x, y satisfying (1). Let now E be *positive* non-degenerate. Then $E(\mathbf{Z}^n)$ is discrete in $\mathbf{R}$, but one can still ask questions about the difference between two consecutive values. (1) shows that the lower bound of the non-zero differences $|F(x) - F(y)|$ is zero. This is a small step towards a conjecture made by D. J. Lewis [L], namely: given $\varepsilon > 0$, there exists $R(\varepsilon) > 0$ such that if the norm of x is $\geqq R(\varepsilon)$, then there exists $y \in \mathbf{Z}^n$ so that (1) is satisfied. In other words, if we go far enough, these successive differences are uniformly bounded and tend to zero.

II. Flows on spaces of lattices

In this part, G is a connected Lie group and Γ is a discrete subgroup which, unless otherwise stated, has finite invariant covolume; i.e. $\Omega = G/\Gamma$ carries a G-invariant probability measure, and H is a closed subgroup of G.

If L is a Lie group, L^o denotes the connected component of the identity in L. If L is a group operating on a space X, and $x \in X$, then L_x is the isotropy group of x, i.e. the subgroup of all elements of L leaving x fixed.

We recall that the assumption on Ω forces G to be unimodular. The orbits of H on Ω define a foliation. We are concerned with the closures of the leaves and the supports of H-invariant ergodic probability measures for certain classes of subgroups H.

The most important case here is the one where

$$G = \mathbf{SL}_n(\mathbf{R}),\ \Gamma = \mathbf{SL}_n(\mathbf{Z}) \text{ and } H = SO(F)^o, \tag{MC}$$

to be referred to as our main special case (MC). Then $\Omega = G/\Gamma$ may be identified with the space of unimodular lattices in $\mathbf{R}^n$.

3. The topological conjecture.

3.1. The prototype here is the *horocycle flow* on a Riemann surface of finite area. Let then $G = \mathbf{PSL}_2(\mathbf{R})$ and X be the upper half-plane. Assume, to avoid ramification, that Γ is torsion free. Then X/Γ is a compact Riemann surface or the complement of finitely many points in one, and G/Γ may be identified to the unit tangent bundle of X/Γ. Take for H the group of matrices

$$\begin{pmatrix} 1 & c \\ 0 & 1 \end{pmatrix} \qquad (c \in \mathbf{R}).$$

Then the orbits of H in G/Γ are the orbits of the horocycle flow. By results of Hedlund [H] any orbit of H in G/Γ is either compact or dense, and the former does not occur if G/Γ is compact.

In [D2], S. G. Dani states a far-reaching conjectural generalization of the previous theorem, proposed by M. S. Raghunathan (Conjecture II, p. 358).

3.2. **Conjecture** (M.S. Raghunathan). *Assume G to be reductive (as a Lie group, i.e. $\mathfrak{g}$ reductive). Let H be an Ad-unipotent (see below) one-parameter subgroup of G and $x \in \Omega$. Then there exists a connected closed subgroup L of G containing H such that $\overline{H.x} = L.x$.*

In [M1, M3], Margulis extended the conjecture to the case where G is a connected Lie group and H is a subgroup generated by elements which are Ad-unipotent in G.

An element $g \in G$ is Ad-unipotent if its image Ad g in the adjoint representation of G in its Lie algebra $\mathfrak{g}$ is unipotent (all eigenvalues equal to one). Similarly, $y \in \mathfrak{g}$ is ad-nilpotent if ad y is a nilpotent endomorphism of $\mathfrak{g}$. If G is semi-simple, linear, then $g \in G$ is Ad-unipotent if it is either unipotent as a matrix or central in G, and $y \in \mathfrak{g}$ is ad-nilpotent if and only if it is a nilpotent matrix. If $y \in \mathfrak{g}$, then the one-parameter subgroup $\exp \mathbf{R}.y$ is Ad-unipotent (i.e. consists of Ad-unipotent elements) if and only if y is ad-nilpotent.

Various special cases of this conjecture were established, notably for horospherical subgroups of reductive groups [D3], until M. Ratner proved it in full generality or, rather, obtained a stronger conclusion under more general assumptions:

3.3. **Theorem 3** (M. Ratner [R4]). *Let H^o be the identity component of H. Assume that H/H^o is finitely generated, H^o is generated by Ad-unipotent elements and each coset $h.H^o$ ($h \in H$) contains an Ad-unipotent element. Let $x \in \Omega$. Then there exists a closed subgroup L of G containing H such that $\overline{H.x} = L.x$ and $L.x$ supports a L-invariant probability measure ergodic for the action of H.*

Remark. The subgroup L is not necessarily unique. For instance, a bigger closed subgroup L' of the same dimension, such that $L'/(L' \cap G_x) = L/(L \cap G_x)$, would also do. However, the Lie algebra $\mathfrak{l}$ of L is uniquely determined: the differential μ_1 at 1 of the map $g \mapsto g.x$ is an isomorphism of $\mathfrak{g}$ onto the tangent space to Ω at x, and $\mathfrak{l}$ is the subspace of $\mathfrak{g}$ mapped by μ_1 onto the tangent space to $\overline{H.x}$ at x (which is well defined, since $\overline{H.x}$ is a submanifold by the theorem). Therefore $L^o.H$ is the smallest possible choice for L and is unique. It will be denoted $L(x, H)$. If H is connected, then $L(x, H)$ is the only connected, subgroup satisfying the conclusion of the theorem.

This normalization is introduced in [R3, p. 546].

As an obvious consequence of the above, we have the

3.4. **Corollary.** *Assume that H is connected and maximal among proper closed connected subgroups of G. Then any orbit of H in Ω is either closed and supports an H-invariant probability measure or dense. In particular, if Ω is not compact, a relatively compact orbit is closed and supports an invariant probability measure.*

3.5. The main point in [R4] is to show 3.3 for H connected, one-dimensional, in which case Ratner proves a stronger result (see 3.9), and the following complement:

(*) *For a given $x \in \Omega$, the set of subgroups $L(x, H)$ occurring in 3.3, when H runs through all the connected closed Ad-unipotent one-dimensional subgroups, is countable.*

3.6. In this section we consider our main special case (MC). Since F is indefinite and $n \geqq 3$, the group H is generated by connected one-dimensional unipotent subgroups. Moreover $SO(F)$ is the fixed point set of an involutive automorphism of G; therefore, as is well known and easy to prove, H is maximal among proper closed connected subgroups, so that 3.4 holds and yields in particular Proposition 2 in any dimension $n \geq 3$.

We now give some indications of how to prove $(*)$ and reduce the proof of 3.3 for connected H's to one-dimensional ones, in the case under consideration.

We note first (this is completely general) that if 3.3 and 3.5 are true for H and x, then they also hold for $g.H.g^{-1}$ and $g.x (g \in G)$. We may therefore assume x to be the origin o of Ω.

(a) If $L.o$ is closed and carries an invariant probability measure, then $L \cap \Gamma$ is of finite covolume in L. This implies, rather easily, that L is defined over $\mathbf{Q}$ (see Proposition 1.1 in [BP2]). However $\mathbf{SL}_n(\mathbf{R})$ contains only countably many real algebraic subgroups defined over $\mathbf{Q}$, whence $(*)$.

(b) We now assume 3.3 to be true for all one-dimensional connected Ad-unipotent subgroups of H. We want to prove 3.3 for H.

Let $\mathcal{N}$ be the variety of nilpotent elements in the Lie algebra $\mathfrak{h}$ of H. For $y \in \mathcal{N}$, we let U_y denote the unipotent (or, equivalently, Ad-unipotent) one-dimensional subgroup $\exp \mathbf{R}.y$.

$\mathcal{N}$ is an algebraic variety, invariant under Ad H. It is a finite union of irreducible subvarieties V_j $(j \in J)$. Since $\mathcal{N}$ is invariant under Ad H and H is connected, H leaves each V_j invariant, and the subgroup M_j generated by the subgroups U_y $(y \in V_j)$ is normal in H. We claim that $M_i = H$ for some $i \in I$. This is obviously the case for any i if H is simple. The group H is simple except when $n = 4$ and F has signature $(2, 2)$. Then $\mathfrak{h}$ is the direct sum of two copies of the Lie algebra of $\mathbf{SL}_2(\mathbf{R})$. In that case, we may take for V_i the Zariski closure of the orbit of any element $y \in \mathcal{N}$ whose projections on the two factors are not zero.

For $y \in \mathcal{N}$, let L_y be the closed subgroup such that $L_y.o$ is the closure of $U_y.o$. Let $\mathcal{L}$ be the set of the subgroups L_y. For $L \in \mathcal{L}$, let M_L be the set of $y \in V_i$ such that $L_y.o \subset L.o$. It is obviously closed. Since $\mathcal{L}$ is countable, there exists at least one L such that M_L contains a non-empty open subset of V_i. Choose one such L of smallest possible dimension. Then let Y be a non-empty open subset of U_i such that $\overline{U_y.o} \subset L.o$ for $y \in Y$. Let R be the subgroup generated by the groups U_y $(y \in Y)$. Then any $r \in R$ leaves $L.o$ stable, hence $R \subset L$ and $\overline{R.o} \subset L.o$. Let $\mathfrak{r}$ be the Lie algebra of R. Then $\mathfrak{r} \cap V_i$ is an algebraic subset which contains a non-empty open subset of V_i, hence is equal to V_i. Therefore $R = H$ and $L.o$ is the closure of $H.o$.

3.7. *Remarks.* (i) The argument in (a) is valid if G is a linear algebraic group defined over $\mathbf{Q}$ and Γ an arithmetic subgroup.

(ii) The proof in (b) is valid without change in the general case, once the existence of V_i is proved. It can be easily deduced from the fact that H is the semi-direct product of a normal nilpotent subgroup, all of whose elements are Ad-unipotent, and of a semisimple group without compact factors. For another argument, see [R4].

3.8. The crucial difference between the case $n = 2$, where the Oppenheim conjecture is false, and $n \geqq 3$ lies in the fact that $SO(F)^o$ is generated by unipotent elements for $n \geqq 3$ but does not contain any (except 1) for $n = 2$. In fact, in that last case, the flow defined by H is the geodesic flow, and it is well known that its orbits may be neither dense nor closed and may have closures which are not manifolds.

3.9. Let $H = \exp \mathbf{R}.y$ with $y \in \mathfrak{g}$ ad-nilpotent be an Ad-unipotent one-parameter group. In this case, M. Ratner establishes a further property of $H.x$, namely, that $H.x$ is *uniformly distributed* in its closure [R4, Theorem B]:

Let $L = L(x, H)$ be as in the remark to Theorem 3 and $d\nu_L$ the L-invariant probability measure with support Lx. Then

$$(1) \qquad \lim_{t\to\infty} t^{-1} \int_0^t f((\exp s.y).x)ds = \int f.d\nu_L$$

for every bounded continuous function f on Ω.

This had been proved for $G = \mathbf{SL}_2(\mathbf{R})$ in [DS] and for G nilpotent in [P]. An extension of Ratner's result to connected unipotent subgroups has been given by N. Shah [Sh, Corollary 1.3].

4. The measure theoretic conjecture.

We have so far emphasized the topological conjecture because of its relevance to the proof of the Oppenheim conjecture. However, it appeared comparatively recently in ergodic theory (motivated by the "Davenport conjecture" in fact), in the context of activity centering on a basic problem in ergodic theory: given a group L acting on a measure space X, classify the ergodic L-invariant probability measures. To complete the picture, I will now discuss one aspect of this problem. From the point of view of the applications to the Oppenheim conjecture, it is not strictly needed, as pointed out in 2.6. However, it is an essential (and the hardest) step in the work of M. Ratner leading to 3.3 (and 3.9).

4.1. The starting point here is again the horocyclic flow on a Riemann surface of finite area (see 3.1; we use the same notation). When G/Γ is compact, H. Furstenberg [F] proved, as a strengthening of the fact that all orbits of H are dense, that the horocycle flow is "uniquely ergodic" (only one H-invariant ergodic probability measure). The existence of closed orbits when G/Γ is not compact shows this is not so in general. But the results of [D1] imply that an H-invariant ergodic probability measure on G/Γ is either G-invariant or supported by a closed orbit of H. This led to the following conjecture, to be called here the "measure theoretic conjecture".

4.2. **Conjecture** (Dani [D2], Margulis [M1, M3]). *Let H be Ad-unipotent and μ an ergodic H-invariant probability measure on Ω. Then there exists a closed subgroup L of G containing H, a point $x \in \Omega$ such that $L.x$ is closed and μ a L-invariant measure with support $L.x$.*

To be more precise, this is conjectured in [D2] when G is reductive (as in 3.2) and H one-dimensional. Moreover, it is proved in [D2] when H is a maximal horospherical subgroup of G.

In her papers M. Ratner refers to this, or rather to a variant of it (see below) as the "measure theoretic Raghunathan conjecture" because it is the measure theoretic analogue of the topological conjecture. It seems to me this is somewhat of a misnomer, since, as far as I know, Raghunathan did not consider the measure theoretic case at all.

4.3. To state Ratner's theorem, we use a definition introduced in [R1, R2, R3]: let μ be an H-invariant probability measure on Ω. Denote by $\Lambda = \Lambda(\mu)$ the set of $g \in G$ which leaves it invariant. It is a closed subgroup [Rl, Proposition 1.1]. Then μ is said to be *algebraic* if there exists $x \in \Omega$ such that $G_x \cap \Lambda$ is of finite covolume in Λ and Λx is the support of μ. In particular Λx is closed [R, Theorem 1.13].

Theorem 4 (M. Ratner). *We drop the assumption that Γ has finite covolume. Let H be as in Theorem 3. Then any ergodic H-invariant probability measure μ on Ω is algebraic. If H is connected, it contains a one-parameter subgroup, Ad-unipotent in G, which acts ergodically on (Ω, μ).*

The first assertion is proved in [R3, Theorem 3], after having been established for G solvable in [R1] and for G semisimple, Γ cocompact, in [R2]. The second one is Proposition 5.2 in [R3].

To reduce the proof to the one-dimensional case, [R3] also provides a countability statement (Theorem 2 there), namely:

Theorem 5. *Fix $x \in \Omega$. Let $\Phi_x(G, \Gamma)$ be the set of closed connected subgroups L of G with the following property: $G_x \cap L$ has finite covolume in L, and L contains a connected subgroup of G generated by Ad-unipotent elements of G which acts ergodically on $(L.x, \nu_L)$, where ν_L is the L-invariant probability measure on $L.x$. Then $\Phi_x(G, \Gamma)$ is countable.*

5. Some remarks on the proofs and further developments.

5.1. We shall not try to describe the proofs, which take over 200 pages, and limit ourselves to some comments, all the more since we can refer to Ratner's survey [R9] for further information.

There is so far only one proof of Theorem 4, given in [R1, R2, R3]. It is also described for $\mathbf{SL}_2(\mathbf{R})$ in [R5] and sketched in the general case in [R6]. The assumption Ad-unipotent for one-parameter subgroup H is used in two crucial ways. First, the adjoint representation $\mathrm{Ad}_{\mathfrak{g}} : h \mapsto \mathrm{Ad}_{\mathfrak{g}} h$ of H on the Lie algebra $\mathfrak{g}$ of G is given by a polynomial mapping of H into $\mathrm{End}(\mathfrak{g})$, and all orbits of H there are closed. Second, if G is semisimple, the Lie algebra $\mathfrak{h}$ of H belongs to a "$\mathfrak{sl}_2$-triple"; i.e. there exists a homomorphism $\varphi : M \to G$ of a covering M of $\mathbf{SL}_2(\mathbf{R})$ such that H is the (isomorphic) image of the identity component of the inverse image of the group of upper triangular unipotent matrices of $\mathbf{SL}_2(\mathbf{R})$. The image A under φ of the identity component of the inverse image of the group of diagonal matrices in $\mathbf{SL}_2(\mathbf{R})$ then normalizes H, and $\mathrm{Ad}_{\mathfrak{g}} A$ is diagonalisable (over the reals). M. Ratner then says that A is diagonal, or is a diagonal, for H. This allows one in particular to use the representation theory of $\mathbf{SL}_2(\mathbf{R})$ to describe the actions of A and H on $\mathfrak{g}$ by the adjoint representation.

The first fact yields some control of some orbits, in particular of the time passed in certain subsets. The starting point of such estimates is the following property of polynomials on the line:

Let $\mathcal{P}(n)$ be the set of polynomials on $\mathbf{R}$ *of degrees* $\leqq n$. *Then there exists* $\eta \in (0, 1)$ *with the following property: if* $P \in \mathcal{P}(n)$ *is such that for given* $t, \theta > 0$, *it satisfies the condition*

$$\max_{s\in[0,t]}|P(s)| = |P(t)| = \theta,$$

then $\theta/2 < |P(s)| \leq \theta$ *for* $s \in [(1-\eta)t, t]$.

5.2. The passage from Theorem 4 to the uniform distribution theorem 3.9 is carried out in [R4], described for $\mathbf{SL}_2(\mathbf{R})$ in [R5] and sketched for the general case in [R6]. The proof is by induction on $\dim G$ so that it may be assumed that there is no proper closed connected subgroup M of G containing H such that $M \cap G_x$ is of finite covolume in M. In this case, it has to be shown that $H.x$ is uniformly distributed with respect to the G-invariant probability measure $d\nu_G$ on Ω. This will also imply that $H.x$ is dense in G. For $t > 0$, let $T_{x,t}$ be the measure on the space $C_o(G)$ of bounded continuous functions on G defined by

$$T_{x,t} = t^{-1}\int_0^t f(\exp sy).x)ds.$$

Then it has to be shown that

$$d\nu_G = \lim_{t\to\infty} T_{x,t}$$

in the weak sense, i.e. $\int f d\nu_G = \lim T_{x,t}(f)$ for $f \in C_o(G)$. The measures $T_{x,t}$ are obviously $\leqq 1$ in norm. Since a bounded set of measures in the weak* topology is relatively compact, the set $M(x, H)$ of measures which are limit points of sequences T_{x,t_j} $(t_j \to \infty)$ is not empty. All these measures are H-invariant. One has to prove eventually that $M(x, H)$ consists solely of $d\nu_G$. Let $\mu \in M(x, H)$ and Y be its support. It is shown first that $\mu(\Omega) = 1$. By a general fact, μ admits a decomposition into ergodic H-invariant probability measures, i.e. there exists a family of $\xi = \{\mu_y\}$ of ergodic H-invariant probability measure so that the supports $C(y)$ of the μ_y form a partition of Y and a measure ν_ξ on the quotient Y/ξ such that $\mu(f) = \int \mu_y(f|C(y))\nu_\xi$. (All this is not really true as stated, but only up to sets of measure zero for the measures under consideration.) Now by Theorem 4, each μ_y is algebraic, i.e. there exists for $y \in Y$ a closed subgroup Λ_y containing H such that $C(y) = \Lambda_y^o.y$, the intersection $\Lambda_y^o \cap G_y$ has finite covolume in Λ_y^o and μ_y is the Λ_y^o-invariant probability measure on $\Lambda_y^o.y$. Then the main part of the proof consists in showing, under our initial assumption, that the μ-measure of the union of the $C(y)$ which are $\neq \Omega$ is zero. This is established for G semisimple in Theorem 2.1 and in the general case in Corollary 3.1 of [R4].

5.3. Another way to go from Theorem 4 to 3.9 is described in [DM4]. The authors also prove their own variant of a countability theorem (Theorem 5.1):

Fix a right invariant Riemannian metric on G, *whence also a Riemannian metric on* Ω. *Given* $c > 0$, *let* $\mathcal{V}_c$ *be the set of closed connected subgroups* H *such that* $H\Gamma/\Gamma$ *is closed and has volume* $\leqq c$. *Then the set of intersections* $H \cap \Gamma$ *for* $H \in \mathcal{V}_c$ *is finite. Let further* $\rho : G \to GL(V)$ *be a finite dimensional representation of* G *with kernel central in* G. *Then the set of* $H \in \mathcal{V}_c$ *for which* $\rho(H \cap \Gamma)$ *is Zariski dense in* $\rho(H)$ *is finite.*

In this theorem, Γ need not have finite covolume.

5.4. In fact, [DM4] proves a generalization of 3.9, also independently obtained in [R6, Theorem 7], involving a sequence of Ad-unipotent subgroups. Let U_n ($n \in \mathbf{N}$) and U be Ad-unipotent one-parameter subgroups of G. The relation $U_n \to U$ means, by definition, that $U_n(t) \to U(t)$ for every $t \in \mathbf{R}$. We refer to 3.3 for the definition of $L(x, U)$. In view of Theorem 3, $L(x, U) = G$ if and only if $U.x$ is dense in Ω.

Theorem 6. *Let U be a one-dimensional Ad-unipotent subgroup of G such that $L(x, U) = G$ and $x \in \Omega$. Let $x_n \in \Omega$ tend to x and U_n be a sequence of one-dimensional Ad-unipotent subgroups of G tending to U. Then for any sequence $t_n \to \infty$ and any bounded continuous function f on Ω we have*

$$\lim_{t_n \to \infty} t_n^{-1} \int_0^{t_n} f(U_n(s).x_n).ds = \int_\Omega f d\nu_G.$$

5.5. In [MS] the authors consider a sequence of Ad-unipotent one-dimensional subgroups $\{U_n\}$ (not necessarily convergent) and a convergent sequence of measures $\mu_n \to \mu$, where μ_n is ergodic and U_n-invariant. They show, among other things, that μ is algebraic, invariant and ergodic for some one-dimensional Ad-unipotent subgroup.

5.6. The assumption that H^o is generated by Ad-unipotent subgroups is of course essentially used. Some assumption on H is certainly needed since, for instance, some orbits of the group of diagonal matrices A of $\mathbf{SL}_2(\mathbf{R})$ in $\mathbf{SL}_2(\mathbf{R})/\mathbf{SL}_2(\mathbf{Z})$ have closures which are not even manifolds. Nevertheless, Theorems 3 and 4 (resp. Theorem 4) have been extended to some more general classes of groups by M. Ratner (resp. S. Mozes).

a) [R6, Theorem 9]. H is connected, generated by a closed connected subgroup M, itself generated by Ad-unipotent one-dimensional subgroups, and by subgroups A_i $(1, \dots, m)$ where A_i is diagonal with respect to some one-dimensional Ad-unipotent subgroup U_i of M (see 5.1 for that notion).

b) [Mo]. G has a connected semi-simple subgroup L without compact factors containing H, and H is connected, epimorphic in L.
Here epimorphic is in the sense of [BB]. This is equivalent to requiring that the regular functions on L/H be only the constants.

There is an overlap between these two classes. For instance, both contain the parabolic subgroups of a connected semisimple subgroup L of G without compact factors.

III. The S-arithmetic case

The Oppenheim conjecture gives a criterion for an indefinite quadratic form to be "rational", meaning rational with respect to $\mathbf{Q}$. In [RR], the authors initiated the consideration of an analogous question over a number field k. It involved looking at quadratic forms over the archimedean completions of k. This was taken up and generalized in [BP1] and [BP2], where finite places are also included (following a suggestion of G. Faltings). Subsequently, extensions of some of (resp. all) the results on flows have been obtained in [MT2] (resp. [R8]). To complete the picture, I describe some of these generalizations in this section, assuming familiarity with some basic concepts in algebraic number theory and also, in §§7 and 8, with the theory of linear algebraic groups.

6. The generalized Oppenheim conjecture.

6.1. In the sequel, k is a number field and $\mathfrak{o}$ the ring of integers of k. For every normalized absolute value $|\cdot|_v$ on k, let k_v be the completion of k at v. In the sequel S is a finite set of places of k containing the set S_∞ of the archimedean ones, k_S the direct sum of the fields k_s $(s \in S)$ and $\mathfrak{o}_S$ the ring of S-integers of k (i.e. of elements $x \in k$ such that $|x|_v \leqq 1$ for $v \notin S$). For s non-archimedean, the valuation ring of k_s is denoted $\mathfrak{o}_s$.

Let F be a quadratic form on k_S^n. Equivalently, F can be viewed as a family (F_s) $(s \in S)$, where F_s is a quadratic form on k_s^n. The form F is non-degenerate if and only each F_s is non-degenerate. We say that F is isotropic if *each* F_s is so, i.e. if there exists for every $s \in S$ an element $x_s \in k_s^n - \{0\}$ such that $F_s(x_s) = 0$. The form F is said to be *rational* (over k) if there exists a quadratic form F_o on k^n and a unit c of k_S such that $F = c.F_o$, *irrational* otherwise.

6.2. The following theorem reduces to the truth of the Oppenheim conjecture if $k = \mathbf{Q}$ and $S = S_\infty$ consists of the infinite place. That (i) and (ii) are equivalent is the generalized Oppenheim conjecture.

Theorem 7 ([BP2], Theorem A). *Let $n \geqq 3$ and F be an isotropic non-degenerate quadratic form on k_S^n. Then the following two conditions are equivalent:*

(i) *F is irrational.*

(ii) *Given $\epsilon > 0$, there exists $x \in \mathfrak{o}_S^n$ such that $0 < |F_s(x)|_s < \epsilon$ for all $s \in S$.*

Here again (ii) $\Rightarrow$ (i) is obvious, and the main point is (i) $\Rightarrow$ (ii). For $S = S_\infty$, the proof is patterned after that of Margulis [M3]; it is based on a generalization of Proposition 1 (and of Theorem 1′ in [M3]) to the case where

$$G = \prod_{s \in S} \mathbf{SL}_3(k_s), \ \Gamma = \mathbf{SL}_3(\mathfrak{o}_S), \ H = \prod_{s \in S} SO(F_s). \tag{1}$$

To treat the general case, this result is combined with an argument using strong approximation in algebraic groups and some geometry of numbers.

Remarks. 1) We have assumed that each F_s is isotropic over k_s. If this is not so, it is easily seen that Theorem 7 cannot hold (see 1.10 in [BP2]).

2) In the original case $k = \mathbf{Q}$ and $S = S_\infty$, we already pointed out in 1.3 that the truth of the Oppenheim conjecture and one theorem of Oppenheim imply that $F(\mathbf{Z}^n)$ is dense in $\mathbf{R}$ when F is irrational. In the general case, going from Theorem 7 to the density appears to be more difficult. At this time, it has to make use of the orbit closure theorem, which then yields again a much stronger statement (see §8).

7. Closures of orbits.

7.1. The results of §6 and their proofs led one naturally to ask whether the results on flows reviewed in II would extend to a framework including the one of 6.2, where G would be a product of real and p-adic groups. This generalization was carried out by M. Ratner [R7, R8, R9] for all of her results and, independently, by G.A. Margulis and G. Tomanov [MT1, MT2] for the measure theoretic conjecture in the setting of algebraic groups. We focus here on the orbit closure theorem and then, in the next section, deduce from it an extension of Proposition 2, hence also of Theorem 2, to the S-arithmetic case.

7.2. Again, the framework of Ratner's work is Lie group theory, rather than algebraic groups. We refer to [Bu1] or [S1] for the notion of Lie group over a local field E of characteristic zero (i.e. $\mathbf{R}$, $\mathbf{C}$ or a finite extension of $\mathbf{Q}_p$ or, equivalently, the fields k_s of 6.2, for variable number fields k). If $\mathcal{G}$ is an algebraic group defined over E, then the group $\mathcal{G}(E)$ of rational points of $\mathcal{G}$ is in a natural way a Lie group over E, and the Lie algebra of $\mathcal{G}(E)$, as a Lie group, is the space of rational points over E of the Lie algebra of $\mathcal{G}$, as an algebraic group.

7.3. Let k, S, k_s, o_s, k_S be as in 6.1. For each $s \in S$, there is given a Lie group G_s over k_s and a closed subgroup H_s generated by Ad-unipotent one-dimensional subgroups over k_s. The product G of the G_s is then in a natural way a locally compact topological group, and the product H of the H_s is a closed subgroup.

As usual, we identify $G_t\,(t \in S)$ to the subgroup of G consisting of the elements $(g_s)_{s \in S}$ such that $g_s = 1$ for $s \neq t$.

Two slight restrictions are imposed on G_s if s is non-archimedean. First, the kernel of the adjoint representation is the center $Z(G_s)$ of G. (If G_s is the group of rational points of an algebraic group which is connected in the Zariski topology, this is automatic, otherwise the kernel could be bigger.) Second, it is required that the orders of the finite subgroups of G_s are bounded. This is always true if G_s is linear. An argument is given in [S1, IV, Appendix 3], proof of Theorem 1. We sketch it: the maximal compact subgroup of $\mathbf{GL}_n(k_s)$ are all conjugate to $\mathbf{GL}_n(\mathfrak{o}_s)$, [S1, Theorem 1, p. 122], so it suffices to consider the finite subgroup of $\mathbf{GL}_n(\mathfrak{o}_s)$. Since $\mathbf{GL}_n(\mathfrak{o}_s)$ is compact, it follows from [S1, Theorem 5, p. 119] that it contains a torsion-free normal open subgroup N. Then $\mathbf{GL}_n(\mathfrak{o}_s)/N$ is a finite group, and any finite subgroup of $\mathbf{GL}_n(\mathfrak{o}_s)$ is isomorphic to a subgroup of that quotient.

In [R8], G_s is said to be Ad-regular if it satisfies the first condition and regular if it satisfies both. In particular, we see from the above that if G_s is the group of rational points of a connected linear algebraic group defined over k_s, it is regular.

In Theorem 8 below, G_s is assumed to be regular for $s \in S$ non-archimedean.

Theorem 8 [R8, Theorem 2]. *Let G, H be as above. Let M be a closed subgroup of G containing H and Γ a discrete subgroup of finite covolume of M. Let $x \in M/\Gamma$. Then M contains a closed subgroup L such that $L.x$ is the closure of $H.x$ and $L \cap M_x$ has finite covolume in L.*

This is stated and proved by M. Ratner for $k = \mathbf{Q}$, but this is no loss in generality. Let $\mathbf{Q}_s$ be the completion of $\mathbf{Q}$ in k_s. It is therefore equal to $\mathbf{R}$ if $k_s = \mathbf{R}, \mathbf{C}$ and to a field $\mathbf{Q}_p$ for some prime p if k_s is non-archimedean. Then a Lie group over k_s, of dimension m, may be viewed in a natural way as a Lie group over $\mathbf{Q}_s$, of dimension $m[k_s : \mathbf{Q}_s]$, in the same way as a complex Lie group can be viewed as a real Lie group of twice the dimension. Let us denote by G'_s the group G_s thus endowed with a structure of Lie group over $\mathbf{Q}_s$. Then the identity map of the product G' of the G'_s onto G is an isomorphism of topological groups. Moreover, if U is a one-dimensional Ad-unipotent group over k_s, then U' is a direct sum of $[k_s : \mathbf{Q}_s]$ one-dimensional Ad-unipotent subgroup over $\mathbf{Q}_s$ of G'_s. Therefore it is clear that Theorem 8 for G' implies it for G.

One advantage of the shift to Lie groups over $\mathbf{Q}_s$ is the fact (due to E. Cartan over the real numbers) that any closed subgroup of a Lie group over $\mathbf{Q}_s$ is a Lie group over $\mathbf{Q}_s$ [Bu1, Chapter 3, §8 no. 2; S1, Chapter V, §9].

7.4. The steps in the proof of Theorem 8 are similar to those for real Lie groups. There is first a theorem proving that H-invariant ergodic probability measures are algebraic, also established in [MT2] when the G_s are groups of rational points of linear algebraic groups. Then a countability statement [R8, Theorem 1.3] allows one to reduce the proof of Theorem 8 to the case where H is one-dimensional Ad-unipotent, contained in one factor, in which case a uniform distribution theorem is also proved [R8, Theorem 3].

8. **Applications to quadratic forms.** We now generalize the density theorems of 2.3 and 2.5 to the present case, using Theorem 8, in the same way as was done in the case $S = S_\infty$ in §7 of [BP2]. As stated in the introduction, we include the proofs, obtained jointly with G. Prasad.

8.0. We shall use the following lemma. It should be known in the theory of affine symmetric spaces. For lack of a reference, we have included a proof in the appendix.

Lemma. *Let E be a field of characteristic zero, $\mathfrak{g}$ a simple Lie algebra over E, $\sigma \neq 1$ an involutive automorphism of $\mathfrak{g}$ and $\mathfrak{k}$ the fixed point set of σ. Assume that $\mathfrak{k}$ is semi-simple. Then any $\mathfrak{k}$-invariant subspace of $\mathfrak{g}$ containing $\mathfrak{k}$ is equal to $\mathfrak{k}$ or $\mathfrak{g}$. In particular, $\mathfrak{k}$ is a maximal proper subalgebra of $\mathfrak{g}$.*

8.1. We revert to the notation of 6.2. Moreover, let

$$G_s = \mathbf{SL}_n(k_s),\ H_s = SO(F_s) \qquad G = \prod_{s\in S} G_s, \quad H = \prod_{s\in S} H_s,$$

$\mathscr{G}_s$ be $\mathbf{SL}_n$ viewed as algebraic group over k_s and $\mathscr{H}_s$ the algebraic group over k_s such that $\mathscr{H}_s(k_s) = H_s$.

Following a notation of [BT], we let H_s^+ denote the subgroup of H_s generated by one-dimensional unipotent (hence Ad-unipotent) subgroups. We claim that it is a closed and open normal subgroup of finite index of H_s. If $k_s = \mathbf{C}$, this is immediate, since H_s is semisimple and connected in the usual topology, and in fact $H_s = H_s^+$. If $k_s = \mathbf{R}$, then H_s^+ is the topological identity component of H_s and has index two. Now let k_s be non-archimedean. Let $\tilde{\mathscr{H}}_s$ be the universal covering of $\mathscr{H}_s$, i.e. the spinor group of F_s, and $\mu\colon \tilde{\mathscr{H}}_s \to \mathscr{H}_s$ the central isogeny. Let $\tilde{H}_s = \tilde{\mathscr{H}}_s(k_s)$. It is known that $\tilde{H}_s = \tilde{H}_s^+$ is generated by one-dimensional unipotent subgroups [BT, 6.15], that $\mu(\tilde{H}_s^+) = H_s^+$ [BT, 6.3] and that $\mu(\tilde{H}_s)$ is a normal open and closed subgroup of finite index of H_s [BT, 3.20], whence our assertion in that case.

We note that H_s^+ is not compact, since it is of finite index in H_s and the latter, being the orthogonal group of an *isotropic* form, is not compact.

We let $\mathfrak{h}_s$ be the Lie algebra of H_s and N_s the normalizer of $\mathfrak{h}_s$ in G_s, i.e.

$$N_s = \{g \in G_s | \operatorname{Ad} g(\mathfrak{h}_s) = \mathfrak{h}_s\}.$$

We claim that N_s is also the normalizer of H_s or of H_s^+. In fact, both groups, viewed as Lie subgroups of G_s, have $\mathfrak{h}_s$ as their Lie algebra, therefore any element $g \in G_s$ normalizing H_s or H_s^+ belongs to N_s. Conversely, let $g \in N_s$. Since $\mathfrak{h}_s$ is the space of rational points of the Lie algebra of $\mathscr{H}_s$ and is of course Zariski-dense in it, the automorphism $\operatorname{Int} g : x \mapsto g.x.g^{-1}$ of $\mathscr{G}_s$ leaves $\mathscr{H}_s$ stable, hence g normalizes H_s and therefore also H_s^+.

Lemma. (i) *H_s^+ has finite index in N_s.* (ii) *Let M be a subgroup of G_s containing H_s^+. Then either $M = G_s$ or $M \subset N_s$.*

(i) Since H_s^+ has finite index in H_s, it suffices to show that H_s has finite index in N_s. The only quadratic forms on k_s^n invariant under H_s are the multiples of F_s. If $x \in N_s$, then ${}^t x.F_s.x$ is invariant under H_s, hence of the form $c.F_s$ $(c \in k_s^*)$. It has the same determinant as F_s; hence $c^n = 1$, and therefore N_s/H_s is isomorphic to a subgroup of the group of n-th roots of unity.
(ii) Identify F_s to a symmetric, invertible, matrix. Then the map

$$\sigma : x \mapsto F_s.{}^t x^{-1}.F_s^{-1} \qquad (x \in G_s)$$

is an automorphism of G_s, obviously of order two, and H_s is the fixed point set of σ. The differential $d\sigma$ of σ at the origin is an involutive automorphism of $\mathfrak{g}_s$ with fixed point set $\mathfrak{h}_s$. The group $\mathscr{G}_s$ (resp. $\mathscr{H}_s$) is simple (resp. semisimple) as an algebraic group; therefore $\mathfrak{g}_s$ (resp. $\mathfrak{h}_s$) is a simple (resp. semisimple) Lie algebra. By 8.0, any $\mathfrak{h}_s$-invariant subspace of $\mathfrak{g}_s$ containing $\mathfrak{h}_s$ is equal to $\mathfrak{h}_s$ or to $\mathfrak{g}_s$.

Now let M be a subgroup of G_s containing H_s^+ but not contained in N_s. We have to show that $M = G_s$. Let $\mathfrak{g}$ be the subspace generated by the subalgebras $\operatorname{Ad} m(\mathfrak{h}_s)$, $(m \in M)$. It is normalized by M, obviously, and in particular by H_s^+. Therefore it is $\mathfrak{h}_s$-invariant. It is $\neq \mathfrak{h}_s$, since M is not in N_s. By the remark just made, $\mathfrak{g} = \mathfrak{g}_s$. There exists therefore a finite set of elements $m_i \in M (1 \leq i \leq a)$ such that $\mathfrak{g}_s$ is the sum of the subalgebras $\mathfrak{h}_i = \operatorname{Ad} m_i(\mathfrak{h}_s)$. The Lie algebra $\mathfrak{h}_i$ is the Lie algebra of $H_i = m_i.H_s^+.m_i^{-1}$. Let $Q = H_1 \times \ldots \times H_d$ be the product of the H_i and $\mu : Q \to G_s$ be the map which assigns to $(h_1, \ldots, h_a)$ $(h_i \in H_i)$ the product of the h_i's. It is a morphism of k_s-manifolds, whose image is contained in M. The tangent space at the identity of Q is the direct sum of the $\mathfrak{h}_i$'s. Therefore the differential $d\mu$ of μ at the identity maps the tangent space to Q onto $\mathfrak{g}_s$. This implies that $\mu(Q)$ contains an open neighborhood of the identity in G_s (see [S1, III, 10.2]). Since it belongs to M, the latter is an open subgroup of G_s. It contains H_s^+, which is not compact, as noted in 8.1, hence is noncompact. Moreover, it is elementary that $G_s = \mathbf{SL}_n k_s$ is generated by the group of unipotent upper triangular matrices and its conjugates. It then follows from Theorem (T) in [Pr] that $M = G_s$.

8.3. Let $\Gamma = \mathbf{SL}_n(\mathfrak{o}_S)$. It is viewed as a discrete subgroup of G via the embeddings $\mathbf{SL}_n(k) \to \mathbf{SL}_n(k_s)$. The quotient $\Omega = G/\Gamma$ has finite volume. We let o be the coset Γ in Ω.

Theorem 9. *If F is irrational, the orbit $H.o$ is dense in Ω.*

Let H^+ be the product of the groups H_s^+ (see 8.1). Since H_s^+ has finite index and is normal, open and closed in H_s, the same is true for H^+ in H, and it is equivalent to prove that $H^+.o$ is dense in Ω.

By Theorem 8 (with $M = G$), there exists a closed subgroup L of G such that $L.o$ is the closure of $H^+.o$ and $L \cap \Gamma$ has finite covolume in L.

Let $M_s = L \cap G_s$. It is a closed normal subgroup of L which contains H_s^+. By 8.2, we have either $M_s \subset N_s$ or $M_s = G_s$. Now let P_s be the projection of L into G_s. It normalizes M_s and contains it. Assume $M_s \subset N_s$. Then $\mathrm{Ad}\, g$ ($g \in P_s$) leaves invariant the Lie algebra of M_s, which is the same as that of H_s, hence g belongs to N_s. In particular, P_s is closed and open in N_s. If $M_s = G_s$, then $P_s = G_s$. Therefore the product M of the M_s is normal, closed and open, of finite index in the product P of the P_s, and P is closed. We have of course $M \subset L \subset P$. As a consequence, M is normal, open and closed, of finite index, in L.

Now define Q_s by the rule: $Q_s = H_s$ if $M_s \subset N_s$, and $Q_s = G_s$ if $M_s = G_s$, and let Q be the product of the Q_s. Then $Q \cap L$ is open and closed, of finite index, in both L and Q. Therefore $Q \cap \Gamma$ has finite covolume in Q. By Proposition 1.2 in [BP2], there exists a k-subgroup $\mathcal{Q}$ of $\mathbf{SL}_n$ such that $\mathcal{Q}(k_s) = Q_s$ for every $s \in S$. This shows first of all that either $Q_s = G_s$ for all $s \in S$ or $Q_s = H_s$ for all $s \in S$. In the first case, $L = G$ and $H^+.o$ is dense. We have to rule out the second one. In that case $H.o$ is closed, $L = H^+$ and $H \cap \Gamma$ has finite covolume in H. Moreover $\mathcal{Q}$ is the orthogonal group of a form F_o on k^n, and there exists a unit c of k_S such that $F = c.F_o$, i.e. F is rational over k, contradicting our assumption.

8.4. We can now generalize 2.5.

A subset $(x_1, \ldots, x_m)$ $(m \leqq n)$ of $\mathfrak{o}_S^n$ is said to be primitive if it is part of a basis of $\mathfrak{o}_S^n$ over $\mathfrak{o}_S$. If $m < n$, it is so if and only if there exists $g \in \mathbf{SL}_n(\mathfrak{o}_S)$ such that $x_i = g(e_i)$ $(i = 1, \ldots, m)$, where e_i is the i-th canonical basis element of k_S^n.

Corollary. *Assume F to be irrational. Let $\lambda_i \in k_s$ $(i = 1, \ldots, n-1)$. Then there exists a sequence of primitive $(n-1)$-tuples $(x_{j,1}, \ldots, x_{j,n-1})$ $(j = 1, 2 \ldots)$ in $\mathfrak{o}_S^n$ such that*

$$\lambda_i = \lim_{j \to \infty} F(x_{j,i}) \qquad (i = 1, \ldots, n-1).$$

In particular, the set of values of F on primitive elements of $\mathfrak{o}_S^n$ is dense in k_S.

The argument is the same as in 7.9 in [BP2]. We repeat it for the sake of completeness.

Let $\lambda_{i,s}$ be the component of λ_i in k_s $(s \in S)$. The form F_s, being isotropic, takes all values in k_s. The representation of H_s in k_s^n is irreducible, and no level surface $F_s = c$ is contained in a hyperplane; hence, given $s \in S$, we can find linearly independent vectors $y_{s,i} \in k_s^n$ such that $F_s(y_{s,i}) = \lambda_{i,s}$. There exists then $g_s \in G_s$ such that $g_s(e_i) = y_{s,i}$ $(i = 1, \ldots, n-1)$. Let $g = (g_s)$ and $y_i = (y_{s,i}) \in k_S^n$. Then

$$F(y_i) = F(g.e_i) = \lambda_i \qquad (i = 1, \ldots, n-1).$$

By Theorem 8, $H.o$ is dense in Ω, hence $H.\Gamma$ is dense in G. There exist therefore elements $h_j \in H$, $\gamma_j \in \Gamma$ $(j = 1, 2, \dots)$ such that $h_j.\gamma_j \to g$. Then

$$\lambda_i = F(g.e_i) = \lim_{j\to\infty} F(h_j.\gamma_j.e_i) = \lim_{j\to\infty} F(\gamma_j.e_i) = \lim_{j\to\infty} F(x_{j,i}),$$

where $x_{j,i} = \gamma_j.e_i$ $(j = 1, 2, \dots; i = 1, \dots, n-1)$. For each j, $(x_{j,1}, \dots, x_{j,n-1})$ is a primitive subset of $\mathfrak{o}_s^n$, whence the corollary.

8.5. Errata to [BP2]. In the proof of 7.4, the difference between H^+ and H has been overlooked. In the archimedean case the subgroup of H_F generated by unipotent elements is the topological identity component H_F^o of H_F, and its index is twice the number of real places. The corrections on page 369 are:

Line 2: After H_s add: Let H_s^o be the connected component of the identity in H_s and H_F^o the product of the H_s^o.
Line 3: Replace H_F by H_F^o.
Line 5: Replace H_s by H_s^o.
Line 12: Replace G by $\mathscr{G}$ and 7.2 by 7.3.
Line 23: Replace 7.3 by 7.4.

9. **Appendix: A Lemma on symmetric Lie algebras.** In this appendix, we prove the lemma in 8.0. (We recall that, if $E = \mathbf{R}$ and $\mathfrak{k}$ is maximal compact or $\mathfrak{g}$ is compact, this is true without restriction on $\mathfrak{k}$ and is due to E. Cartan.)

The case where $\mathfrak{g}$ is one-dimensional is left to the reader, so we assume $\mathfrak{g}$ to be also semisimple. Let $\mathfrak{p}$ be the (-1)-eigenspace of σ in $\mathfrak{g}$. Then

$$(1) \qquad \mathfrak{g} = \mathfrak{k} \oplus \mathfrak{p} \qquad [\mathfrak{k}, \mathfrak{k}] \subset \mathfrak{k}, [\mathfrak{k}, \mathfrak{p}] \subset \mathfrak{p} \qquad [\mathfrak{p}, \mathfrak{p}] \subset \mathfrak{k}$$

as usual. If $\mathfrak{m}$ is a $\mathfrak{k}$-invariant subspace of $\mathfrak{p}$, then $\mathfrak{k} \oplus \mathfrak{m}$ is a subalgebra, so that in fact the last clause of the lemma is equivalent to the lemma itself.

Let B be the Killing form of $\mathfrak{g}$. It is non-degenerate and $B(\mathfrak{k}, \mathfrak{p}) = 0$; hence the restrictions of B to $\mathfrak{k}$ and to $\mathfrak{p}$ are non-degenerate. We recall that, by invariance

$$(2) \qquad B([a, b], c) = B(a, [b, c]) \qquad (a, b, c \in \mathfrak{g}).$$

The proof is divided into a number of steps:

a) Let $\mathfrak{m} \subset \mathfrak{p}$ be a $\mathfrak{k}$-invariant subspace. Then $[\mathfrak{m}, \mathfrak{m}]$ is an ideal of $\mathfrak{k}$ and $[\mathfrak{m}, \mathfrak{m}] \oplus \mathfrak{m}$ a subalgebra normalized by $\mathfrak{k}$.

This follows by straightforward application of the Jacobi identity and (1).

b) We have $[\mathfrak{p}, \mathfrak{p}] = \mathfrak{k}$.

In fact, $[\mathfrak{p}, \mathfrak{p}] \oplus \mathfrak{p}$ is a subalgebra normalized by $\mathfrak{k}$ in view of a), hence a non-zero ideal, hence equal to $\mathfrak{g}$.

c) Let $\mathfrak{m}$, $\mathfrak{n}$ be $\mathfrak{k}$-invariant subspaces of $\mathfrak{p}$ and assume $B(\mathfrak{m}, \mathfrak{n}) = 0$. Then $[\mathfrak{m}, \mathfrak{n}] = 0$.

Let $m \in \mathfrak{m}$, $n \in \mathfrak{n}$ and $k \in \mathfrak{k}$. Then

$$B(k, [m, n]) = B([k, m], n) \subset B(\mathfrak{m}, \mathfrak{n}) = 0;$$

therefore $[m, n]$ belongs to the radical of the restriction of B to $\mathfrak{k}$. Since the latter is non-degenerate, this proves $[m, n] = 0$.

d) There is no proper $\mathfrak{k}$-invariant subspace $\mathfrak{m}$ of $\mathfrak{p}$, on which the restriction of B is non-degenerate.

Let $\mathfrak{m}$ be one and $\mathfrak{n}$ its orthogonal complement. Then $\mathfrak{p} = \mathfrak{m} \oplus \mathfrak{n}$. By c)

$$(3) \qquad [\mathfrak{m}, \mathfrak{n}] = 0.$$

By b), $\mathfrak{a} = [\mathfrak{m}, \mathfrak{m}] \oplus [\mathfrak{m}]$ and $\mathfrak{b} = [\mathfrak{n}, \mathfrak{n}] + \mathfrak{n}$ are subalgebras, and (3) implies that $[\mathfrak{a}, \mathfrak{b}] = 0$. By b) and (3), $\mathfrak{k} = [\mathfrak{m}, \mathfrak{m}] + [\mathfrak{n}, \mathfrak{n}]$; hence $\mathfrak{g} = \mathfrak{a} + \mathfrak{b}$ and $\mathfrak{a}$, $\mathfrak{b}$ are distinct non-zero ideals of $\mathfrak{g}$, a contradiction.

e) There is no proper $\mathfrak{k}$-invariant subspace $\mathfrak{m}$ of $\mathfrak{p}$ on which the restriction of B is degenerate but non-zero.

Assume $\mathfrak{m}$ is such a subspace. Let $\mathfrak{r}$ be the radical of $B|\mathfrak{m}$. It is non-zero, $\neq \mathfrak{m}$, invariant under $\mathfrak{k}$. There exists a $\mathfrak{k}$-invariant supplement $\mathfrak{n}$ to $\mathfrak{r}$ in $\mathfrak{m}$, and $\mathfrak{n} \neq 0$. Then $B|\mathfrak{n}$ is non-degenerate, and we are back to d).

f) There is no proper $\mathfrak{k}$-invariant subspace $\mathfrak{m}$ of $\mathfrak{p}$ which is isotropic for B.

Let $\mathfrak{m}$ be one. If $\dim \mathfrak{m} < \dim \mathfrak{p}/2$, then the orthogonal subspace $\mathfrak{n}$ to $\mathfrak{m}$ has dimension $> \dim \mathfrak{p}/2$ and is invariant under $\mathfrak{k}$, and the restriction of B to $\mathfrak{n}$ is non-zero. We are back to e). There remains to consider the case where $\dim \mathfrak{p}$ is even, $\dim \mathfrak{m} = \dim \mathfrak{p}/2$, $B|\mathfrak{p}$ is hyperbolic, and $\mathfrak{m}$ is maximal isotropic. Moreover, the representation of $\mathfrak{k}$ in $\mathfrak{m}$ is irreducible; otherwise we would be back to the case just treated. There exists a supplement $\mathfrak{n}$ to $\mathfrak{m}$ in $\mathfrak{p}$ which is $\mathfrak{k}$-invariant. Then, again, it has to be maximal isotropic. By c),

$$[\mathfrak{m}, \mathfrak{m}] = [\mathfrak{n}, \mathfrak{n}] = 0; \tag{4}$$

hence by b)

$$\mathfrak{k} = [\mathfrak{m}, \mathfrak{n}]. \tag{5}$$

It follows from 1) and 4) that $\mathfrak{m}$ and $\mathfrak{n}$ consist of nilpotent matrices. The normalizer of $\mathfrak{n}$ in $\mathfrak{m}$ is $\mathfrak{k}$-invariant, hence reduced to zero in view of (5) and the irreducibility of the representation of $\mathfrak{k}$ in $\mathfrak{m}$. Therefore $\mathfrak{n}$ is the nilpotent radical of $\mathfrak{k} \oplus \mathfrak{n}$, and the latter is the normalizer of $\mathfrak{n}$. By Theorem 2 in [Bu1, 8, §10], $\mathfrak{k} \oplus \mathfrak{n}$ is parabolic. But $\mathfrak{k}$ is assumed to be semisimple, whence a contradiction. This concludes the proof.

Remarks. 1) In this paper, 8.0 is only needed in case $E = k_s$, $\mathfrak{g} = \mathfrak{g}_s$ and $\mathfrak{k} = \mathfrak{h}_s$. If it holds after extension of the groundfield, it is clearly already true in the original situation. Since k_s may be embedded into $\mathbf{C}$, this reduces us to the case where $E = \mathbf{C}$, $\mathfrak{g} = \mathbf{sl}_n\mathbf{C}$ and $\mathfrak{k}$ is the Lie algebra of $\mathbf{SO}_n(\mathbf{C})$. A reader who would have a direct argument in that last case could then avoid any recourse to 8.0.

2) As the fixed point set of an involution, $\mathfrak{k}$ is always reductive. The assumption $\mathfrak{k}$ semisimple has been used only in the last step of the proof. Some restriction is necessary, since otherwise $\mathbf{sl}_2$ already provides a counterexample, with σ having the diagonal matrices as fixed point set. Over $\mathbf{C}$, other counterexamples are given by the complexifications of hermitian symmetric pairs. In fact, the proof leads to a complete description of the cases where $\mathfrak{k}$ is not proper maximal, namely,

$$\mathfrak{g} = \mathfrak{k} \oplus \mathfrak{n} \oplus \mathfrak{m}$$

where $\mathfrak{n}$, $\mathfrak{m}$ are commutative, are the nilpotent radicals of two parabolic subalgebras with maximal reductive subalgebra $\mathfrak{k}$, the representation of $\mathfrak{k}$ in $\mathfrak{n}$ and $\mathfrak{m}$ are contragredient of one another, irreducible, the split center of $\mathfrak{k}$ is one-dimensional and acts by dilations on $\mathfrak{m}$ and $\mathfrak{n}$. Conversely, given such a decomposition of $\mathfrak{g}$, the map of $\mathfrak{g}$ onto itself which is the identity on $\mathfrak{k}$ and $-$Id. on $\mathfrak{m} \oplus \mathfrak{n}$ is obviously an automorphism.

References

[BB] F. Bien and A. Borel, *Sous-groupes épimorphiques de groupes algébriques linéaires*. I, C. R. Acad. Sci. Paris Sér. I Math. **315** (1992), 649–653.

[B] A. Borel, *Values of quadratic forms at S-integral points*, Algebraic Groups and Number Theory (V. Platonov and A. S. Rapinchuk, eds.), Uspekhi Mat. Nauk. **47** (1992), 118–120; Russian Math. Surveys **47** (1992), 134–136.

[BP1] A. Borel and G. Prasad, *Valeurs de formes quadratiques aux points entiers*, C.R. Acad. Sci. Paris Sér. I Math. **307** (1988), 217–220.

[BP2] ———, *Values of isotropic quadratic forms at S-integral points*, Compositio Math. **83** (1992), 347–372.

[BT] A. Borel and J. Tits, *Homomorphismes "abstraits" de groupes algèbriques simples*, Ann. of Math. (2) **97** (1973), 499–571.

[Bu] N. Bourbaki, *Intégration*, Chap. 7, 8, Hermann, Paris, 1963.

[Bu1] ———, *Groupes et algèbres de Lie*, Chap. 2, 3, Hermann, Paris, 1972.

[C] S. Chowla, *A theorem on irrational indefinite quadratic forms*, J. London Math. Soc. **9** (1934), 162–163.

[D1] S. G. Dani, *Invariant measures of horospherical flows on non-compact homogeneous spaces*, Invent. Math. **47** (1978), 101–138.

[D2] ———, *Invariant measures and minimal sets of horospherical flows*, Invent. Math. **64** (1981), 357–385.

[D3] ———, *Orbits of horospherical flows*, Duke Math. J. **53** (1986), 177–188.

[DM1] S. G. Dani and G. A. Margulis, *Values of quadratic forms at primitive integral points*, Invent. Math. **98** (1989), 405–424.

[DM2] ———, *Orbit closures of generic unipotent flows on homogeneous spaces of* $SL(3, \mathbf{R})$, Math. Ann. **286** (1990), 101–128.

[DM3] ———, *Values of quadratic forms at integral points: An elementary approach*, Enseign. Math. (2) **36** (1990), 143–174.

[DM4] ———, *Limit distributions of orbits of unipotent flows and values of quadratic forms*, Adv. Soviet Math. **6** (1993), 91–137.

[DS] S. G. Dani and J. Smillie, *Uniform distribution of horocyclic orbits for Fuchsian groups*, Duke Math. J. **51** (1984), 185–194.

[DH] H. Davenport and H. Heilbronn, *On indefinite quadratic in five variables*, J. London Math. Soc. **21** (1946), 185–193.

[DR] H. Davenport and H. Ridout, *Indefinite quadratic forms*, Proc. London Math. Soc. **9** (1959), 544–555.

[F] H. Furstenberg, *The unique ergodicity of the horocycle flow*, Recent Advances in Topological Dynamics, Lecture Notes in Math., vol. 318, Springer, New York, 1972, 95–115.

[G] E. Ghys, *Dynamique des flots unipotents sur les espaces homogènes*, Astérisque **206** (Sém. Bourbaki 1992–93, no. 747) (1992), 93–136.

[H] G. Hedlund, *Fuchsian groups and transitive horocycles*, Duke Math. J. **2** (1936), 530–542.

[L] D. J. Lewis, *The distribution of values of real quadratic forms at integer points*, Proc. Sympos. Pure Math., vol. XXIV, Amer. Math. Soc., Providence, RI, 1973, pp. 159–174.

[M1] G.A. Margulis, *Lie groups and ergodic theory*, Algebra Some Current Trends (L. L. Avramov, ed.) Proc. Varna 1986, Lecture Notes in Math., vol. 1352, Springer, New York, 130–146.

[M2] ———, *Indefinite quadratic forms and unipotent flows on homogeneous spaces*, Banach Center Publ., vol. 23, Polish Scientific Publishers, Warsaw, 1989.

[M3] ———, *Discrete subgroups and ergodic theory*, Number Theory, Trace Formulas and Discrete Groups (symposium in honour of A. Selberg), Academic Press, San Diego, CA, 1989, pp. 377–398.

[MT1] G. A. Margulis and G. Tomanov, *Measure rigidity for algebraic groups over local fields*, C. R. Acad. Sci. Paris Sér. I. Math. **315** (1992), 1221–1226.

[MT2] ———, *Invariant measures for actions of unipotent groups over local fields on homogeneous spaces*, Invent. Math. **116** (1994), 347–392.

[M] G. D. Mostow, *Homogeneous spaces with finite invariant measure*, Ann. of Math. **75** (1992), 17–37.

[Mo] S. Mozes, *Epimorphic subgroups and invariant measures*, preprint.

[MS] S. Mozes and N. Shah, *On the space of ergodic invariant measures of unipotent flows*, Ergodic Theory and Dynamical Systems (to appear).

[O1] A. Oppenheim, *The minima of indefinite quaternary quadratic forms of signature* 0 , Proc. Nat. Acad. Sci. U.S.A **15** (1929), 724–727.

[O2] ———, *The minima of indefinite quaternary quadratic forms*, Ann. of Math. **32** (1931), 271–298.

[O3] ———, *Values of quadratic forms.* I, Quart. J. Math. Oxford Ser. (2) **4** (1953), 54–59.

[O4] ———, *Values of quadratic forms.* II, Quart. J. Math. Oxford Ser. (2) **4** (1953), 60–66.

[O5] ———, *Values of quadratic forms.* III, Monatsh. Math. **57** (1953), 97–101.

[P] W. Parry, *Ergodic properties of affine transformations and flows on nilmanifolds*, Amer. J. Math. **91** (1969), 757–771.

[Pr] G. Prasad, *Elementary proof of a theorem of Bruhat-Tits-Rousseau and of a theorem of Tits*, Bull. Soc. Math. France **110** (1982), 197–202.

[RR] S. Raghavan and K.G. Ramanathan, *On a diophantine inequality concerning quadratic forms*, Nachr. Akad. Wiss. Göttingen Math.-Phys. Kl. II (1968), 251–262.

[R] M. S. Raghunathan, *Discrete subgroups of Lie groups*, Ergeb. Math. Grenzgeb. (3), vol. 68, Springer-Verlag, Berlin, 1992.

[R1] M. Ratner, *Strict measure rigidity for unipotent subgroups of solvable groups*, Invent. Math. **101** (1990), 449–482.

[R2] ———, *On measure rigidity of unipotent subgroups of semisimple groups*, Acta Math. **165** (1990), 229–309.

[R3] ———, *On Raghunathan's measure conjecture*, Ann. of Math. **134** (1991), 545–607.

[R4] ———, *Raghunathan's topological conjecture and distributions of unipotent flows*, Duke Math. J. **63** (1991), 235–280.

[R5] ———, *Raghunathan's conjectures for* $\mathbf{SL}_2(\mathbf{R})$, Israel J. Math. **80** (1992), 1–31.

[R6] ———, *Invariant measures and orbit closures for unipotent actions on homogeneous spaces*, Geom. Funct. Anal. **4** (1994), 236–257.

[R7] ———, *Raghunathan's conjectures for p-adic Lie groups*, Internat. Math. Res. Notices **5** (1993), 141–146.

[R8] ———, *Raghunathan's conjectures for cartesian products of real and p-adic groups*, Duke Math. J. (to appear).

[R9] ———, *Interactions between ergodic theory, Lie groups and number theory*, Proc. ICM 94 (to appear).

[S] J-P. Serre, *A course in arithmetic*, Graduate Texts in Math., vol. 7, Springer, New York, 1973.

[S1] ———, *Lie algebras and Lie groups* (second ed.), Lecture Notes in Math., vol. 1500, Springer, New York, 1992.

[Sh] N.A. Shah, *Limit distributions of polynomial trajectories on homogeneous spaces*, preprint.

School of Mathematics, Institute for Advanced Study, Princeton, New Jersey 08540
E-mail address: `borel@math.ias.edu`

157.

Values of Zeta-Functions at Integers, Cohomology and Polylogarithms

in *Current trends in mathematics and physics. A tribute to Harish-Chandra*, S. D. Adhikari ed., 1–44, Narosa Publishing House, Bombay 1995

In this paper, mostly expository, I shall survey the main relations, proven or conjectured, between the values of the Dedekind zeta-function ζ_F of a number field F at integers, various cohomological or homological invariants of certain classical arithmetic groups and polylogarithms.

Let, as usual, r_1 (resp. r_2) be the number of real (resp. pairs of complex conjugate) embeddings of F into $\mathbf{C}$. We recall that the multiplicity d_m of the zero of ζ_F at $1-m$ ($m \in \mathbf{N}$, $m \neq 0$) is equal to $r_1 + r_2 - 1$ if $m = 1$, to r_2 if m is even and to $r_1 + r_2$ if m is odd ≥ 3.

After having recalled some facts about ζ_F in §1 we consider in §2 the case where $d_m = 0$. This happens if and only if F is totally real and m is even. Then $\zeta_F(1-m) \in \mathbf{Q}$ by a theorem of Siegel and Klingen [K] . We indicate briefly the derivation of this result by use of the Euler-Poincaré characteristic of the modular symplectic groups $\mathbf{Sp}_{2n}(\mathfrak{o}_F)$, where $\mathfrak{o}_F$ is the ring of integers of F, and how it can be applied to relate the denominator of $\zeta_F(1-m)$ and the orders of the finite subgroups of $\mathbf{Sp}_{2n}(\mathfrak{o}_F)$.

The main concern of this survey is the case $d_m \neq 0$. The starting point of this discussion is the fact that the integer d_m ($m \geqq 1$) is the dimension of the space $PH_{2m-1}(\mathbf{GL}_N\mathfrak{o}_F; \mathbf{C})$ of primitive elements in the (Eilenberg-MacLane) complex homology of $\mathbf{GL}_N\mathfrak{o}_F$ for N large (see §3).

This theorem is proved by setting up, via invariant differential forms, an isomorphism α^m between the space of primitive elements $PH^{2m-1}(X_{N,u}; \mathbf{C})$ in the complex cohomology of a certain compact symmetric space $X_{N,u}$ and the dual space of the homology space mentioned above. Since the dimension of $PH^{2m-1}(X_{N,u}; \mathbf{C})$ is easily seen to be equal to d_m, this proves the result.

These two cohomology spaces have natural $\mathbf{Q}$-structures, defined by the cohomology with coefficients in $\mathbf{Q}$. However, α^m does not seem to map one to the other, and this leads one to consider its *regulator*: if $e_1, \dots, e_{d_m}$ and $f_1, \dots, f_{d_m}$ are bases of the rational cohomology in these two spaces, the regulator is by definition the determinant of $(\alpha^m(e_i))$ with respect to (f_i). So defined, it is an element of $\mathbf{C}^*/\mathbf{Q}^*$ (see 5.1). It is known (see 5.5(5)) that

$$R'_m \sim D_F^{1/2}\zeta_F(m).(\pi i)^{-m.d} \tag{1}$$

where $d = [F : Q]$, D_F is the discriminant of F over $\mathbf{Q}$ and $\sim$ means up to a factor in $\mathbf{Q}^*$. We have assumed $d_m \neq 0$, but the proof also covers the case $d_m = 0$, and gives back the theorem of Siegel-Klingen, or, equivalently, shows that the right hand side of (1) is rational (see 5.6(17)). We recall that in all other cases, nothing is known about the arithmetic nature of that expression.

For any field L of characteristic zero, the space $PH_{2m-1}(\mathbf{GL}_N\mathfrak{o}_F; L)$ may be identified canonically with $K_{2m-1}\mathfrak{o}_F \otimes_{\mathbf{Z}} L$. Here K_iA is the i-th K-group of the ring A in the sense of Quillen (see §6). Moreover $K_i\mathfrak{o}_F \otimes L$ is naturally isomorphic to $K_iF \otimes L$ for $i > 1$. The previous results can then be formulated in the context of the algebraic K-theory of number fields. The dual map to α^m yields a map of $K_{2m-1}(F) \otimes \mathbf{R}$ onto a homology space with real coefficients of the same dimension d_m. For various reasons, it is better to switch to twisted coefficients. Recall that if A is a subring of $\mathbf{C}$, and $j \in \mathbf{N}$, then $A(j)$ denotes $(2\pi i)^j.A$, viewed as an A-submodule of $\mathbf{C}$. One is then led to consider, instead of the dual map to α^m, an isomorphism

$$\Lambda_{F,m} : K_{2m-1}F \otimes \mathbf{R} \tilde{\rightarrow} \mathbf{R}(m-1)^{d_m}. \tag{2}$$

These two spaces have natural $\mathbf{Q}$-structures defined by $K_{2m-1}F \otimes \mathbf{Q}$ and $\mathbf{Q}(m-1)$ respectively, so that we can again consider the regulator $R(\Lambda_{F,m}) = R_m$ of $\Lambda_{F,m}$. The previous theorem implies

$$R_m \sim D_F^{1/2}.\zeta_F(m).(\pi i)^{-m.d_{m-1}} \tag{3}$$

(see 7.3 and §10).

For $m = 1$, the regulator is basically the Dedekind-Dirichlet regulator of algebraic number theory (cf. 5.2), which is a determinant with entries the logarithms of absolute values of the various conjugates of a fundamental system of units. This and a number of, at first surprising, connections between the dilogarithm and K-theory have led D.Zagier to conjecture that R_m can be expressed as a determinant involving higher polylogarithms. In §9 we give various forms of these conjectures, and review the main results supporting them.

Much of the work on this paper was carried out during visits at the Tata Institute of Fundamental Research in Bombay, India and later at the Max-Planck-Institut für Mathematik in Bonn, Germany, as an A. v. Humboldt awardee. I thank both institutions for their hospitality.

§1. Notation. The zeta-function of F

Throughout the paper, F is a number field, $\mathfrak{o}_F$ the ring of integers of F, d and D_F the degree and discriminant of F over $\mathbf{Q}$, Δ_F the absolute value of D_F, r_1 (resp. r_2) the number of real (resp. pairs of complex conjugate embeddings) of F into $\mathbf{C}$. We recall that the zeta function ζ_F of F is defined by

$$\zeta_F(s) = \sum_{\mathfrak{a}} N\mathfrak{a}^{-s} = \prod_{\mathfrak{p}} (1 - N_{\mathfrak{p}}^{-s})^{-1},$$

where $\mathfrak{a}$ runs through the ideals, $\mathfrak{p}$ through the prime ideals, of $\mathfrak{o}_F$, and $N\mathfrak{a}$ is the norm of $\mathfrak{a}$. It converges for $Re\, s > 1$, has an analytic continuation to $\mathbf{C}$

which is holomorphic, except for a simple pole at $s = 1$. The values at s and $1-s$ are related by a functional equation which can be written

$$(1) \qquad \zeta_F(-s) = \frac{\Delta_F^{s+1/2}\Gamma(s+1)^d}{\pi^{d(s-1)}2^{ds+r_2}}(\sin\frac{\pi s}{2})^{r_1}(\sin \pi s)^{r_2}\zeta_F(s+1).$$

We are only interested in its values for $s \in \mathbf{N}$. We let d_m be the multiplicity of the zero at $1-m$ ($m \in \mathbf{N}$, $m \neq 0$). From (1) and the fact that $\zeta_F(s) \neq 0$ for $Re\, s > 1$ and has a simple pole at $s = 1$, one sees easily

$$(2) \qquad d_m = \begin{cases} r_1 + r_2 - 1 & m = 1 \\ r_2 & m \text{ even}, m \neq 0 \\ r_1 + r_2 & m \text{ odd}, m \neq 1 \end{cases}$$

We set

$$(3) \qquad \xi_F(m) = \frac{\Delta_F^{1/2}\zeta_F(m)}{\pi^{d.m}}$$

$$(4) \qquad \zeta_F(1-m)^* = \lim_{s\to 1-m} \xi_F(s)(s+m-1)^{-d_m}.$$

We shall write for a, $b \in \mathbf{C}^*$

$$(5) \qquad a \sim b \text{ if } a/b \in \mathbf{Q}^*.$$

From the functional equation (1), one sees

$$(6) \qquad \zeta_F(1-m)^* \sim \xi_F(m)\pi^{d_m}.$$

§2. Non-zero values and Euler characteristics

2.1 From (2) in §1 we see that

$$(1) \qquad d_m = 0 \Leftrightarrow m \text{ even}, r_2 = 0 \qquad (m \in \mathbf{N}, m > 1).$$

Then, by (6) above

$$(2) \qquad \zeta_F(1-m) \sim \xi_F(m).$$

As recalled in the introduction, the theorem of Siegel-Klingen (cf. [K]), asserts that in this case $\zeta_F(1-m) \in \mathbf{Q}$. Actually, for $k = \mathbf{Q}$, this was proved by L.Euler who showed more precisely that

$$\zeta(1-m) = B_{m/2}/m$$

where B_j is the j-th Bernoulli number.

2.2 For a commutative ring A with identity we let $\mathbf{Sp}_{2n}A$ be the group of symplectic $2n \times 2n$ matrices, with coefficients in A. The group $\mathbf{Sp}_{2n}\mathfrak{o}_F$ belongs to a class of groups Γ which have a finite Euler-characteristic $\chi(\Gamma)$ in the sense of C.T.C. Wall, which is a rational number. By induction on n, the theorem of Siegel-Klingen follows therefore from the following theorem of Harder [Ha].

2.3 Theorem. *(Harder [Ha]) For $n = 1, 2, \dots$, we have*

1 (1)
$$\prod_{m=1}^{n} \zeta_F(1-2m) = 2^{n(d-1)}\chi(Sp_{2n}\mathfrak{o}_F).$$

It is also known that $\mathbf{Sp}_{2n}\mathfrak{o}_F$ contains only finitely many conjugacy classes of finite subgroups (cf. [BS] e.g.). If q is a common multiple of their orders, then

2
$$q.\zeta_F(1-m) \in \mathbf{Z}.$$

This fact has been used in [S] [Br 1] to relate the orders of finite subgroups of $\mathbf{Sp}_{2n}\mathfrak{o}_F$ and the denominators of $\zeta_F(1-m)$. More generally [Br 2], if S is a finite set of finite places of F, there are similar relations between the orders of the finite subgroups of $\mathbf{Sp}_{2n}\mathfrak{o}_{F,S}$, where $\mathfrak{o}_{F,S}$ is the ring of S-integers of F, and the partial zeta function $\zeta_{F,S}$

$$\zeta_{F,S}(s) = \zeta_F(s).\prod_{\mathfrak{p}\in S}(1 - N\mathfrak{p}^{-s}).$$

§3. Homological interpretation of d_m

3.0 Let X be a space, $H^i(X;L)$, (resp. $H_i(X;L)$) its ith-cohomology (resp. homology) group with respect to a field of coefficients L. We let DH^i denote the space of decomposable elements in H^i, i.e. the subspace spanned by the products of elements of strictly smaller degree, in the cohomology ring $H^{\cdot}(X;L)$. The quotient H^i/DH^i is the space $PH^i(X;L)$ of primitive elements in cohomology and the subspace $PH_i(X;L)$ of elements in $H_i(X;L)$ annihilated by $DH^i(X;L)$ is the space of primitive elements in homology. In all cases of interest here these spaces are finite dimensional. Then $PH^i(X,L)$ and $PH_i(X;L)$ are the duals of one another.

3.1 For any $\mathbf{N} \geq 1$, the derived group of $\mathcal{D}\mathbf{GL}_N\mathfrak{o}_F$ of $\mathbf{GL}_N\mathfrak{o}_F$ is $\mathbf{SL}_N\mathfrak{o}_F$ and the quotient is $\mathfrak{o}_F^*$. On the other hand, the first integral homology group of a discrete group Γ is equal to $\Gamma/\mathcal{D}\Gamma$. Therefore if L is a field of characteristic zero

(1)
$$H_1(\mathbf{GL}_N\mathfrak{o}_F;L) = \mathfrak{o}_F^* \otimes_{\mathbf{Z}} L$$

so that

(2)
$$\text{rank } \mathfrak{o}_F^*/\text{torsion} = \dim H_1(\mathbf{GL}_N\mathfrak{o}_F;L) \qquad (N \geq 1).$$

In dimension one, every element is primitive, so that (1) may be viewed as the case $j = 1$ of the following theorem

3.2 Theorem. *Let $j \in \mathbf{N}$ and $N > 2j$. Then*

$$\dim PH_j(\mathbf{GL}_N\mathfrak{o}_F;\, L) = \begin{cases} 0 & \text{if } j \text{ is even} \\ d_m & \text{if } j\text{=}2m\text{-}1. \end{cases}$$

The theorem is proved in [Bo 1], for $j \geq 2$ in a slightly different formulation (see 3.3).

In the next sections, we introduce the framework of the proof.

3.3 The paper [Bo 1] uses SL_N rather then GL_N. We first recall how to go from one to the other.

Let Z be the subgroup of diagonal matrices in $\mathbf{GL}_{N+1}\,\mathfrak{o}_F$ with all diagonal entries equal to one, except for the first one. It is isomorphic to $\mathfrak{o}_F^*$ and $\mathbf{GL}_{N+1}\mathfrak{o}_F$ is the semi-direct product of Z and $\mathbf{SL}_{N+1}\mathfrak{o}_F$. Identify $\mathbf{SL}_N\mathfrak{o}_F$ with the lower right $N \times N$ corner, completed by zeroes on the first row and column, except for one at the top left corner. The subgroup M of $\mathbf{GL}_{N+1}\mathfrak{o}_F$ generated by Z and $\mathbf{SL}_N\mathfrak{o}_F$ is their direct product. The inclusion $M \to \mathbf{GL}_{N+1}\mathfrak{o}_F$ induces a morphism of extensions

$$\begin{array}{ccccccccc} 1 & \longrightarrow & \mathbf{SL}_N\mathfrak{o}_F & \longrightarrow & M & \longrightarrow & Z & \longrightarrow & 1 \\ & & \downarrow & & \downarrow & & \| & & \\ 1 & \longrightarrow & \mathbf{SL}_{N+1}\mathfrak{o}_F & \longrightarrow & \mathbf{GL}_{N+1}\mathfrak{o}_F & \longrightarrow & Z & \longrightarrow & 1 \end{array}$$

whence a homomorphism $(E_r) \to ('E_r)$ of the spectral sequence (E_r) of the bottom extension into that, $('E_r)$, of the top one. We have

$$E_2^{p,q} = H^q(Z;\, H^q(\mathbf{SL}_{N+1}\mathfrak{o}_F,\, L))$$
$$'E_2^{p,q} = H^p(Z;\, L) \otimes H^q(\mathbf{SL}_N\mathfrak{o}_F;\, L).$$

Since $H^q(\mathbf{SL}_{N+1}\mathfrak{o}_F;\, L) \to H^q(\mathbf{SL}_N\mathfrak{o}_F;\, L)$ is an isomorphism in the stable range, Z acts trivially on $H^q(\mathbf{SL}_{N+1}\mathfrak{o}_F;\, L)$ at least up to that range. Hence

$$E_2^{p,q} = H^p(Z;\, L) \otimes H^q(\mathbf{SL}_{N+1}\mathfrak{o}_F;\, L)$$

at least for $q < N/2$, hence $E_2^{p,q} \to {'E_2^{p,q}}$ is an isomorphism for $q < N/2$ and any p. This implies readily that it induces isomorphisms $E_r^{p,q} \to {'E_r^{p,q}}$ for all $r \geqq 2$ and p, q such that $p+q < N/2$. Since the top extension in (1) is trivial, we see that

(2) $$H^{\cdot}(\mathbf{GL}_{N+1}\mathfrak{o}_F;\, L) = H^{\cdot}(\mathfrak{o}_F^*;\, L) \otimes H^{\cdot}(\mathbf{SL}_{N+1}\mathfrak{o}_F;\, L)$$

at least up to the total degree $N/2$.

The first factor on the right-hand side is the cohomology of the free abelian group $\mathfrak{o}_F^*$/torsion, hence an exterior algebra over $H^1(\mathfrak{o}_F^*;\, L)$. Therefore

(3) $$PH^1(\mathbf{GL}_N\mathfrak{o}_F;\, L) = H^1(\mathfrak{o}_F^*;\, L),$$

(4) $$PH^j(\mathbf{GL}_N\mathcal{O}_F; L) = PH^j(\mathbf{SL}_N\mathcal{O}_F; L) \qquad (j \geq 2)$$

and we can in 3.2, substitute $\mathbf{SL}_N$ to $\mathbf{GL}_N$ for $j \geq 2$. This is how 3.2 is formulated in [Bo 1]. For a K-theoretic interpretation of 3.2 and (4), see §6.

3.4 In the sequel, we let

(1) $$G_N = \mathcal{R}_{F/\mathbf{Q}}(\mathbf{SL}_N \times F)$$

be the (Weil) restriction of scalars from F to $\mathbf{Q}$ of $\mathbf{SL}_N$, viewed as a group defined over F. It is defined over $\mathbf{Q}$. If A is an algebra over $\mathbf{Q}$

(2) $$G_N(A) = \mathbf{SL}_N(F \otimes_{\mathbf{Q}} A).$$

Let V_∞ be the set of archimedean places of F, F_v the completion of F at $v \in V_\infty$ and $\sum$ the set of embeddings of F into $\mathbf{C}$. They have respectively $r_1 + r_2$ and d elements. We have from (2)

(3) $$G_N(\mathbf{R}) = \prod_{v \in V_\infty} \mathbf{SL}_N F_v \qquad G_N(\mathbf{C}) = (\mathbf{SL}_N\mathbf{C})^{\Sigma}.$$

We shall also write

(4) $$G_N(\mathbf{R}) = \prod_{v \in V_\infty} G_{N,v} \qquad G_N(\mathbf{C}) = \prod_{\sigma \in \Sigma} G_{N,\sigma}$$

where therefore

(5) $$G_{N,v} = \mathbf{SL}_N F_v \qquad G_{N,\sigma} = \mathbf{SL}_N\mathbf{C} \quad (v \in V_\infty, \sigma \in \Sigma).$$

Let $K_{N,v}$ be the standard maximal compact subgroup of $\mathbf{SL}_N F_v$ and $K_N = \prod_{v \in V_\infty} K_{N,v}$. Thus $K_{N,v} = \mathbf{SO}_N$ if v is real, $K_{N,v} = \mathbf{SU}_N$ if v is complex. The space X_N of maximal compact subgroups of $G_N(\mathbf{R})$ may be identified with $G_N(\mathbf{R})/K_N$. We have

(6) $$X_N = \prod_{v \in V_\infty} X_{N,v}, \text{ where } X_{N,v} = \mathbf{SL}_N F_v / K_{N,v}.$$

3.5 If M is a smooth manifold, we denote by $\Omega^{\cdot}_M$ the graded algebra of smooth real valued differential forms on M. If C is a $\mathbf{R}$-submodule of $\mathbf{C}$, then $\Omega^{\cdot}_M \otimes_{\mathbf{R}} C$ is the vector space of C-valued differential forms (an algebra if C is a subring). As usual, if H is a group of diffeomorphisms of M, we let $\Omega^{\cdot H}_M$ be the algebra of H-invariant differential forms. We let

(1) $$I^{\cdot}_N = \left(\Omega^{\cdot}_{X_N}\right)^{G_N(\mathbf{R})}.$$

We have

(2) $$I^{\cdot}_N = \otimes_{v \in V_\infty} I^{\cdot}_{N,v} \qquad \left(I^{\cdot}_{N,v} = \left(\Omega^{\cdot}_{X_{N,v}}\right)^{G_{N,v}}\right).$$

We shall denote by $E(\{x_i,\, j_i\}_{i\in I})$ an exterior algebra over $\mathbf{R}$, with generators x_i $(i \in I)$, j_i being the degree of x_i. It is known that

$$(3) \qquad I^{\cdot}_{N,v} = \begin{cases} E(\{x_i,\, 2i-1\}_{1<i\leq N}) & \text{if } v \text{ is complex} \\ E(\{x_i,\, 4i+1\}_{1\leq i\leq N/4}) & \text{if } v \text{ is real and } N \text{ is odd.} \end{cases}$$

The assumption N odd is for convenience. If N is even and v is real, the previous equality is still valid in the stable range $j < N/2$, which is all that matters for us, but the assumption N odd brings with it some mild simplifications in various statements. (3) implies

$$(4) \qquad \dim PI^j_N = \begin{cases} 0 & j \text{ even },\, j \neq 0 \\ r_2 & j = 2m-1,\, m \text{ even}, \leq N \\ r_1 + r_2 & j = 2m-1,\, m \text{ odd},\, 3 \leq m \leq N. \end{cases}$$

3.6 The space X_N is contractible and $\mathbf{SL}_N\mathfrak{o}_F$ acts properly on it. For cohomology with respect to a field of characteristic zero, the quotient $Y_N = SL_N\mathfrak{o}_F\backslash X_N$ may be viewed as an Eilenberg-MacLane space for $\mathbf{SL}_N\mathfrak{o}_F$, i.e.

$$H^{\cdot}(Y_N;\, L) = H^{\cdot}(\mathbf{SL}_N\mathfrak{o}_F;\, L).$$

The elements of $I^{\cdot}_N$ may be viewed as differential forms on X_N or Y_N. They are closed, whence a map

$$j^{\cdot} : I^{\cdot}_N \to H^{\cdot}(Y_N;\, \mathbf{R}).$$

In view of 3.5(3), Theorem 3.2 follows from the following result, proved in [Bo 1].

3.7 Theorem. *The map* $j^{\cdot} : I^{\cdot}_N \to H^{\cdot}(Y_N;\, \mathbf{R})$ *is an isomorphism in degrees* $j < 2N$.

3.8 It is more elegant for the statements, though not helpful for the proofs so far, to go over to the limit with respect to N.

For a commutative ring A with unit, we let $\mathbf{GL}\,A = \varinjlim \mathbf{GL}_N A$, the limit being taken with respect to the standard inclusions $GL_N A \to GL_{N+1}A$. Similarly $\mathbf{SL}\,A = \varinjlim \mathbf{SL}_N A$. If $A = \mathfrak{o}_F,\, F$, then $\mathcal{D}\mathbf{GL}\,A = \mathbf{SL}\,A$ and $\mathbf{GL}\,A/\mathbf{SL}\,A = A^*$. The previous statements implies a stability property, namely, the restriction map $H^j(\mathbf{GL}_{N+1}\mathfrak{o}_F;\, L) \to H^j(\mathbf{GL}_N\mathfrak{o}_F;\, L)$ is an isomorphism if $N > 2j$. It follows that $H^j(\mathbf{GL}\mathfrak{o}_F;\, L) = H^j(\mathbf{GL}_N\mathfrak{o}_F;\, L)$ for $N > 2j$, and similarly for $\mathbf{SL}_N$.

On the other hand X_N is embedded as a totally geodesic closed submanifold in X_{N+1} therefore the restriction of forms to X_N defines a map: $I^{\cdot}_{N+1} \to I^{\cdot}_N$. By 3.5 it is an isomorphism in low dimensions, we can go to the limit and consider

$$(1) \qquad I^{\cdot}_\infty = \varprojlim I^{\cdot}_N$$

we see that

$$I^{\cdot}_{\infty} = \otimes_{v \in V_\infty} I^{\cdot}_{\infty, v} \tag{2}$$

where

$$I^{\cdot}_{\infty, v} = \begin{cases} E(\{x_i,\ 4i+1\}_{i=1,\dots}) & \text{for } v \text{ real} \\ E(\{x_i,\ 2i-1\}_{i=2,3\dots}) & \text{for } v \text{ complex} \end{cases}$$

and in view of 3.3, Theorem 3.2 for all N's is equivalent to the assertion that

$$j^{\cdot}_{\infty} : I^{\cdot}_{\infty} \tilde{\to} H^{\cdot}(\mathbf{SL}\mathfrak{o}_F;\ \mathbf{R}) \tag{3}$$

is an isomorphism.

3.9 It can be proved (see [Bo 3]) that the restriction map

$$H^{\cdot}(\mathbf{SL}_N F;\ \mathbf{Q}) \to H^{\cdot}(\mathbf{SL}_N \mathfrak{o}_F;\ \mathbf{Q}) \tag{1}$$

is an isomorphism. Again this goes over to the limit and is valid for $\mathbf{SL}\,F$ and $\mathbf{SL}\mathfrak{o}_F$. In view of 3.2, we also have, for any field L of characteristic zero:

$$PH^j(\mathbf{GL}_N F;\ L) = PH^j(\mathbf{GL}_N \mathfrak{o}_F;\ L) \qquad (j \geq 2) \tag{2}$$

hence

$$\begin{aligned} PI^j_{\infty} &= PH^j(\mathbf{GL}\,F;\ L) = PH^j(\mathbf{GL}_N\,F;\ L) = \\ &= PH^j(\mathbf{SL}\,F;\ L) = PH^j(\mathbf{SL}_N\,F;\ L) \quad (j \geq 2,\ N > 2j). \end{aligned} \tag{3}$$

The relation (2) also follows from K-theory, as will be recalled in §6.

Note that by going over to the limit in 3.3(2), we get

$$H^{\cdot}(\mathbf{GL}\mathfrak{o}_F;\ L) = H^{\cdot}(\mathfrak{o}^*_F) \otimes H^{\cdot}(\mathbf{SL}\mathfrak{o}_F;\ L) \tag{4}$$

Similarly, we have

$$H^{\cdot}(\mathbf{GL}F;\ L) = H^{\cdot}(F^*;\ L) \otimes H^{\cdot}(\mathbf{SL}F;\ L) \tag{5}$$

Both can be proved by the argument used in 3.3: for $A = \mathfrak{o}_F$, F, erase $N+1$ and replace $\mathbf{SL}_N A$ by the copy of $\mathbf{SL}A$ consisting of the matrices with zeroes on the first row and column, except for one at the top left entry.

§4. Some interpretations of $I^{\cdot}_N$ and some symmetric spaces

The algebra $I^{\cdot}_N$ has several interpretations which play an important role in the sequel, also involving two other spaces $X_{N,\mathbf{C}}$ and $X_{N,u}$. In this section, we discuss various isomorphisms and canonical identifications.

We let $\mathfrak{g}_N(\mathbf{R})$ and $\mathfrak{k}_N$ be the Lie algebras of $G_N(\mathbf{R})$ and K_N respectively.

4.1 We have the Cartan decomposition $\mathfrak{g}_N(\mathbf{R}) = \mathfrak{k}_N \oplus \mathfrak{p}_N$, where $\mathfrak{p}_N$ is the orthogonal complement of $\mathfrak{k}_N$ in $\mathfrak{g}_N(\mathbf{R})$ with respect to the Killing form. Therefore

(1) $$\mathfrak{p}_N = \oplus_{v \in V_\infty} \mathfrak{p}_{N,v}$$

where $\mathfrak{p}_{N,v}$ is the space of real symmetric (resp. complex hermitian) $N \times N$ matrices of trace zero. The space $\mathfrak{p}_N$ can be (and will be) identified to the tangent space $T(X_N)_0$ to X_N at the coset K_N. The restriction of differential forms on X_N to $T(X_N)_0$ then yields an isomorphism

(2) $$I_N^{\cdot} \tilde{\rightarrow} (\Lambda^{\cdot} \mathfrak{p}_N^*)^{K_N} = H^{\cdot}(\mathfrak{g}_N(\mathbf{R}), \mathfrak{k}_N; \mathbf{R}),$$

where the last term is the relative Lie algebra cohomology of $\mathfrak{g}_N(\mathbf{R})$ modulo $\mathfrak{k}_N$.

4.2 By a theorem of van Est, the last term on 4.1(2) is naturally isomorphic to the algebra $H_{ct}^{\cdot}(G_N(\mathbf{R}); \mathbf{R})$ of *continuous Eilenberg-MacLane cohomology* of $G_N(\mathbf{R})$ (see e.g. [BW:IX]). An explicit realization of that isomorphism has been given by J. Dupont [D]. We recall it briefly. Fix a point $x \in X_N$. Given $g_0, \dots, g_n \in G_N(\mathbf{R})$, we consider a "geodesic simplex" (which may be degenerate) $\Delta(g_o, \dots, g_n)$ with vertices $g_o x, \dots, g_n x$ defined inductively: if $\Delta(g_{i+1}, \dots, g_n)$ is defined, then $\Delta(g_i, \dots, g_n)$ is the union of all geodesic segments joining $g_i x$ to the points of $\Delta(g_{i+1}, \dots, g_n)$. If now $\omega \in I_N^n$, one associates to it the real valued cochain

$$(g_o, \dots, g_n) \mapsto \int_{\Delta(g_o, \dots, g_n)} \omega . dv,$$

where dv is an invariant measure on X_N. This cochain is a cocycle and the resulting map $I_N^{\cdot} \to H_{ct}^{\cdot}(G_N(\mathbf{R}); \mathbf{R})$, which is independent of $x \in X_N$, is an isomorphism.

4.3 An embedding $\sigma : F \to \mathbf{C}$ defines a place of F, to be denoted $v(\sigma)$. If $\sigma \in \Sigma$, then $\overline{\sigma}$ denotes the complex conjugate embedding $x \mapsto \overline{\sigma(x)}$, $(x \in F)$. We have $v(\sigma) = v(\overline{\sigma})$.

We let τ be the complex conjugation of $G_N(\mathbf{C})$ with respect to $G_N(\mathbf{R})$. If v is real, the inclusion

(1) $$G_{N,v(\sigma)} = \mathbf{SL}_N(\mathbf{R}) \subset G_{N,\sigma} = \mathbf{SL}_N(\mathbf{C})$$

is the natural one and the restriction of τ to $G_{N,\sigma}$ is the complex conjugation. If σ is complex, the inclusion

(2) $$G_{N,v(\sigma)} = \mathbf{SL}_N(\mathbf{C}) \subset G_{N,\sigma} \times G_{N,\overline{\sigma}} = \mathbf{SL}_N(\mathbf{C}) \times \mathbf{SL}_N(\mathbf{C})$$

is the map

(3) $$x \mapsto (x, \overline{x}) \qquad (x \in \mathbf{SL}_N(\mathbf{C}))$$

and the restriction of τ to $G_{N,\sigma} \times G_{N,\bar{\sigma}}$ is

(4) $$(x, y) \mapsto (\bar{y}, \bar{x}) \qquad (x, y \in \mathbf{SL}_N(\mathbf{C})).$$

Let $K_{N,u} = \mathbf{SU}_N$ be the standard maximal compact subgroup of $G_N(\mathbf{C})$ and $X_{N,u} = K_{N,u}/K_N$. The space $X_{N,u}$ is the compact dual symmetric space to X_N.

4.4 At the Lie algebra level, we have

(1) $$\mathfrak{k}_{N,u} = \oplus_{\sigma\in\sum} \mathfrak{k}_{N,u,\sigma} \qquad \mathfrak{k}_{N,u,\sigma} = \mathfrak{sl}_N(\mathbf{C}).$$

The Cartan decomposition of $\mathfrak{k}_{N,u}$ with respect to $\mathfrak{k}_N$ is

(2) $$\mathfrak{k}_{N,u} = \mathfrak{k}_N \oplus \mathfrak{p}_{N,u}, \text{with } \mathfrak{p}_{N,u} = i.\mathfrak{p}_N,$$

so that we have an identification

(3) $$T(X_{N,u})_0 = \mathfrak{p}_{N,u} = i\mathfrak{p}_N$$

where $i\mathfrak{p}_N$ is viewed as a real subspace of $\mathfrak{p}_{N,\mathbf{C}} = \mathfrak{p}_N \otimes_{\mathbf{R}} \mathbf{C}$.

Let $I^{\cdot}_{N,u}$ be the algebra of real-valued $K_{N,u}$-invariant differential forms on $X_{N,u}$. As is well known

(4) $$I^{\cdot}_{N,u} \tilde{\rightarrow} H^{\cdot}(X_{N,u}; \mathbf{R}).$$

Clearly, in analogy with 4.1(2) we have

(5) $$I^{\cdot}_{N,u} \cong (\Lambda^{\cdot}\mathfrak{p}^*_{N,u})^{K_N}.$$

There are natural injective $\mathbf{R}$-linear maps

(6) $$(\Lambda^{\cdot}\mathfrak{p}^*_{N,u})^{K_N} \xrightarrow{\lambda} (\Lambda^{\cdot}\mathfrak{p}^*_{N,\mathbf{C}})^{K_N(\mathbf{C})} \xleftarrow{\mu} (\Lambda^{\cdot}\mathfrak{p}^*_N)^{K_N}$$

as well as restriction maps

(7) $$(\Lambda^{\cdot}\mathfrak{p}^*_{N,u})^{K_N} \xleftarrow{\lambda'} (\Lambda^{\cdot}\mathfrak{p}_{N,\mathbf{C}})^{K_N(\mathbf{C})} \xrightarrow{\mu'} (\Lambda^{\cdot}\mathfrak{p}^*_N)^{K_N}.$$

In view of (3), 4.1(2), (5), the composition $\nu = \lambda' \circ \mu$ provides isomorphisms

(8) $$\nu : I^j_N \tilde{\rightarrow} i^j I^j_{N,u} \qquad (j \in \mathbf{N}).$$

In particular

(9) $$\nu : PI^{2m-1}_N \tilde{\rightarrow} iPI^{2m-1}_{N,u}.$$

If we now combine (9), with 3.7 but ignore for the moment the factor i, we get the form of 3.7 stated in the introduction, at any rate in absolute value.

4.5 These isomorphisms can be interpreted more geometrically in terms of invariant differential forms.

Let $K_N(\mathbf{C})$ be the complexification of K_N in $G_N(\mathbf{C})$. Then

$$K_N(\mathbf{C}) \cap K_{N,u} = K_N(\mathbf{C}) \cap G_N(\mathbf{R}) = K_N \tag{1}$$

consequently $X_{N,u}$ and X_N are embedded in $X_N(\mathbf{C}) = G_N(\mathbf{C})/K_N(\mathbf{C})$, as orbits of the coset $K_N(\mathbf{C})$. The spaces $X_{N,u}$ and X_N are real forms of $X_N(\mathbf{C})$, with real dimension equal to the complex dimension of $X_N(\mathbf{C})$. A $G_N(\mathbf{C})$-invariant form on $X_N(\mathbf{C})$ restricts on X_N (resp. $X_{N,u}$) to a form invariant under $G_N(\mathbf{R})$ (resp. $K_{N,u}$). Conversely a form on X_N (resp. $X_{N,u}$) invariant under $G_N(\mathbf{R})$ (resp. $K_{N,u}$) is the restriction of a unique $G_N(\mathbf{C})$-invariant $\mathbf{C}$-valued form on $X_N(\mathbf{C})$. The map ν may therefore also be defined in the following way: given a $G_N(\mathbf{R})$-invariant $\mathbf{R}$-valued form on X_N, let $\tilde{\omega}$ be its unique extension to $X_N(\mathbf{C})$, as a $G_N(\mathbf{C})$-invariant $\mathbf{C}$-valued form on $X_N(\mathbf{C})$. Then $\nu(\omega)$ is the restriction of $\tilde{\omega}$ to $X_{N,u}$.

§5. Regulators

3 **5.1** Let A and B be two finite dimensional vector spaces over $\mathbf{Q}$. Assume that there is given an isomorphism $\lambda : A \otimes_{\mathbf{Q}} \mathbf{C} \tilde{\to} B \otimes_{\mathbf{Q}} \mathbf{C}$. If $e_1, \ldots, e_q$ (resp. $f_1, \ldots, f_q$) is a basis of A (resp. B) over $\mathbf{Q}$, then the regulator $R(\lambda)$ is the determinant of the coordinates of $\lambda(e_1), \ldots, \lambda(e_q)$ with respect to the basis $f_1, \ldots, f_q$. A change of either basis multiplies $R(\lambda)$ by a factor in $\mathbf{Q}^*$ so that $R(\lambda)$ is to be viewed as an element of $\mathbf{C}^*/\mathbf{Q}^*$. If A and B have natural $\mathbf{Z}$-structures $A_{\mathbf{Z}}$ and $B_{\mathbf{Z}}$ and we use only bases of $A_{\mathbf{Z}}$ and $B_{\mathbf{Z}}$, then $R(\lambda)$ is determined up to sign.

If C is another vector space over $\mathbf{Q}$ and $\mu : B \otimes_{\mathbf{Q}} \mathbf{C} \to C \otimes_{\mathbf{Q}} \mathbf{C}$ an isomorphism, then obviously

$$R(\mu \circ \lambda) \sim R(\mu).R(\lambda), \tag{1}$$

where $a \sim b$ ($a, b \in \mathbf{C}^*$) means that $a/b \in \mathbf{Q}^*$. In particular

$$R(\lambda)R(\lambda^{-1}) \sim 1. \tag{2}$$

If we identify $A \otimes_{\mathbf{Q}} \mathbf{C}$ with $B \otimes_{\mathbf{Q}} \mathbf{C}$, we may also formulate the definition of a regulator in a slightly different way: given a finite dimensional complex vector space E and two $\mathbf{Q}$-structures A, B on E, then the determinant of a basis of A with respect to basis of B may be viewed as the regulator of the identity map of E, endowed with the $\mathbf{Q}$-structure A onto E, endowed with the $\mathbf{Q}$-structure B.

5.2 An example, which is at the origin of the terminology, is the Dirichlet-Dedekind regulator R_F of number theory. Take for $A_{\mathbf{Z}}$ the group $A_{\mathbf{Z}} = o_F^*/\text{tors} \oplus \mathbf{Z}$. Let $c = r_1 + r_2$, $\{e_1, \ldots, e_{c-1}\}$ be a basis of the first summand and e_c a generator of the second one. Let $B_{\mathbf{Z}} = \mathbf{Z}^{V_\infty}$ be the free abelian group with basis elements f_v ($v \in V_\infty$). Consider the map λ defined by:

$$\lambda e_i = \sum c_{i,v} f_v, \quad (c_{i,v} = \log |e_i|_v; (i < c)$$

$$\lambda(e_c) = \sum c_{c,v} f_v, \quad (c_{c,v} = 1, v \in V_\infty).$$

This defines an isomorphism of $A_{\mathbf{Z}} \otimes \mathbf{R}$ onto $B_{\mathbf{Z}} \otimes \mathbf{R}$. Its regulator is

$$R(\lambda) = \det(c_{i,v})$$

by the definition above. Since $\sum_v c_{i,v} = 0$ $(1 \leq i < c)$ have,

$$|R(x)| = c.R_F,$$

where R_F is the Dirichlet-Dedekind regulator.

5.3 Our next goal is to review the results of [Bo 2]. We shall formulate them in a slightly different way in order to keep track of the powers of i which occur naturally; moreover we want to use twisted coefficients.

If A is a subring of $\mathbf{C}$, (in the sequel $A = \mathbf{Z}$, $\mathbf{Q}$, $\mathbf{R}$) and $j \in \mathbf{N}$, we let $A(j) = A(2\pi i)^j$, viewed as an A-submodule of $\mathbf{C}$. If X is a space, then $a \mapsto a(2\pi i)^j$ induces an isomorphism $\iota(j)$ of A-modules of $H^m(X;\, A)$ onto $H^m(X;\, A(j))$ for every $m \in \mathbf{Z}$.

Let $A = \mathbf{Q}$ and assume $H^m(X;\, \mathbf{Q})$ to be finite dimensional. Then $H^m(X;\, \mathbf{Q}(j))$, may be identified to $H^m(X;\, \mathbf{Q}) \otimes_{\mathbf{Q}} \mathbf{Q}(j)$ and both $H^m(X;\, \mathbf{Q})$, $H^m(X;\, \mathbf{Q}(j))$ define $\mathbf{Q}$-structures on $H^m(X;\, \mathbf{C})$. If $(e_1, \dots, e_q)$ is a basis of $H^m(X;\, \mathbf{Q})$, then the $f_a = (2\pi i)^j e_a$ form a basis of $H^m(X; \mathbf{Q}(j))$. The regulator of the identity map $\sim$ (relating $H^m(X; \mathbf{Q})$ to $H^m(X;\, \mathbf{Q}(j))$ is therefore

$$R(\iota(j)) \sim (2\pi i)^{-j.q}. \tag{1}$$

Assuming that $H^m(X;\, \mathbf{Z})$ is finitely generated, for simplicity, we may view similarly $H^m(X;\, \mathbf{Z})$/torsion and $H^m(X;\, \mathbf{Z}(j))$/torsion as defining $\mathbf{Z}$-structures on $H^m(X;\, \mathbf{C})$. The corresponding regulator is then $(2\pi i)^{-jq}$, up to sign.

Note that if $A = \mathbf{R}$, then $\mathbf{R}(j) = i^j.\mathbf{R}$ so that we have a natural isomorphism

$$\mathbf{R}(j+1) = \mathbf{C}/\mathbf{R}(j) \qquad (j \in \mathbf{N}). \tag{2}$$

In the sequel, I shall follow [G1], [G2] and other literature and use the notation $\mathbf{R}(m)$, $\mathbf{R}(m-1)$ rather than $i^m\mathbf{R}$ and $(\mathbf{C}/i^m\mathbf{R})$, though the latter is formally preferable.

5.4 If we associate to $\omega \in I_N^{2m-1}$ the differential form on $X_{N,u}$ which is equal to $i\omega$ on the tangent space at the origin, we get an isomorphism

$$I_N^{2m-1} \xrightarrow{\sim} H^{2m-1}(X_{N,u}; \mathbf{R})$$

which maps PI_N^{2m-1} onto $PH^{2m-1}(X_{N,u}; \mathbf{R})$. Using 5.3 and 3.6 we have a chain of isomorphisms

$$\begin{aligned} PH^{2m-1}(X_{N,u}; \mathbf{R}(m)) &\overset{\iota(m)}{\xrightarrow{\sim}} PH^{2m-1}(X_{N,u}; \mathbf{R}) \overset{\beta m}{\xrightarrow{\sim}} PI_N^{2m-1} \\ &\overset{\gamma_m}{\xrightarrow{\sim}} PH^{2m-1}(\mathbf{SL}_N \mathcal{O}_F; \mathbf{R}) \end{aligned} \tag{1}$$

($m \geq 2$, $N > 4m$). All the terms except for PI_N^{2m-1}, have natural **Q**-structures, defined by cohomology with coefficients in $\mathbf{Q}(m)$ and $\mathbf{Q}$, with respect to which we can consider regulators.

5.5 Theorem. *The regulator $R(j_F^m)$ of the isomorphism*

$$(1) \quad j_F^m : PH^{2m-1}(X_{N,u}; \mathbf{R}(m)) \tilde{\to} PH^{2m-1}(\mathbf{SL}_N o_F; \mathbf{R}),\ (m \geq 2,\ N > 4m),$$

composition of the isomorphisms in 5.4(1), is

$$(2) \qquad R(j_F^m) \sim D_F^{1/2} \zeta_F(m)\, (\pi i)^{-m.d_{m-1}}.$$

Since

$$(3) \qquad D_F^{1/2} \in i^{r_2}.\mathbf{R}$$

as is well known, we see that

$$(4) \qquad R(j_F^m) \in \begin{cases} \mathbf{R} & \text{if } m \text{ is odd and} > 1 \\ i^{r_2}.\mathbf{R} & \text{if } m \text{ is even and} \geq 2. \end{cases}$$

The regulator of

$$PH^{2m-1}(X_{N,u}; \mathbf{R}(m)) \tilde{\to} PH^{2m-1}(X_{N,u}; \mathbf{R})$$

is $(2\pi i)^{m.d_m}$ by 5.3(1), so (2) is equivalent to

$$(5) \qquad R(\alpha^m) \sim D_F^{1/2}.\zeta_F(m).(\pi i)^{-d.m}$$

where α^m is the composition of β^m and γ^m in 5.3(1).

This is Theorem 6.1 of [Bo 2], except that the regulator there is the absolute value $\Delta_F^{1/2}.\zeta_F(m).\pi^{-d.m}$ of the present regulator. However, as we shall indicate below, the proof actually yields the value given in (5). [To relate the present statements with those of [Bo 2], note also that we consider here $2m-1$ rather than $2m+1$ and that d_m there is d_{m-1} here.]

5.6 The space I_N has no natural **Q**-structure. It could be given one if K_N were defined over **Q**, but this is not the case in general. To obviate that, we go from symmetric spaces and relative Lie algebra cohomology over to groups and Lie algebra cohomology.

We let $p_1 : K_{N,u} \to X_{N,u}$ be the projection of the principal fibration of $K_{N,u}$ with structural group K_N. The group $\mathbf{SL}_N o_F$ acts properly and freely on $G_N(\mathbf{R})$ by left translations and there is a fibration

$$(1) \qquad p_3 : Z_N := \mathbf{SL}_N o_F \backslash G_N(\mathbf{R}) \to Y_N.$$

with fibre K_N (in fact, there may be some singular fibres due to the torsion in $\mathbf{SL}_N o_F$ but this does not matter for rational cohomology and we could systematically go over to a torsion-free subgroup of finite index to avoid this complication, as is done in [Bo1], [Bo2]. We shall ignore this here). We let $\{{}_iE_r\}$ be the spectral sequence of p_i in complex cohomology ($i = 1, 3$). We have

$$ {}_1E_2^{p,q} = H^p(X_{N,u};\mathbf{C}) \otimes H^q(K_N;\mathbf{C}) \tag{2}$$

$$ {}_3E_2^{p,q} = H^p(Y_N;\mathbf{C}) \otimes H^q(K_N;\mathbf{C}). \tag{3}$$

There is similarly a spectral sequence $\{{}_2E_r\}$ in Lie algebra cohomology, which abuts to $H^{\cdot}(\mathfrak{g}_N(\mathbf{R});\, \mathbf{R})$ and in which

$$ {}_2E_2^{p,q} = H^p(\mathfrak{g}_N(\mathbf{R}),\, \mathfrak{k}_N;\mathbf{C}) \otimes H^q(\mathfrak{k}_N;\, \mathbf{C}). \tag{4}$$

We recall that $H^{\cdot}(\mathfrak{k}_N;\, \mathbf{C})$ (resp. $H^{\cdot}(\mathfrak{g}_n(\mathbf{R});\, \mathbf{C})$ may be identified with the algebra of biinvariant differential forms on K_N (resp. $G_N(\mathbf{R})$ or $K_{N,u}$) and this leads to isomorphisms

$$ H^{\cdot}(\mathfrak{k}_N;\, \mathbf{C}) \cong H^{\cdot}(K_N;\, \mathbf{C}),\ H^{\cdot}(\mathfrak{g}_N(\mathbf{R});\, \mathbf{C}) \cong H^{\cdot}(K_{N,u};\mathbf{C}) \tag{5}$$

and to a natural homomorphism

$$ H^{\cdot}(\mathfrak{g}_N;\, \mathbf{C}) \rightarrow H^{\cdot}(Z_N;\, \mathbf{C}), \tag{6}$$

so we have now a commutative diagram

$$\begin{array}{ccccc} H^{\cdot}(K_{N,u};\, \mathbf{C}) & \xrightarrow{\tilde\beta^{\cdot}} & H^{\cdot}(\mathfrak{g}_N(\mathbf{R});\, \mathbf{C}) & \xrightarrow{\tilde\gamma^{\cdot}} & H^{\cdot}(Z_N;\, \mathbf{C}) \\ \uparrow p_1^{\cdot} & & \uparrow p_2^{\cdot} & & \uparrow p_3^{\cdot} \\ H^{\cdot}(X_{N,u};\, \mathbf{C}) & \xrightarrow{\beta^{\cdot}} & H^{\cdot}(\mathfrak{g}_N(\mathbf{R}),\, \mathfrak{k}_N;\, \mathbf{C}) & \xrightarrow{\gamma^{\cdot}} & H^{\cdot}(Y_N;\, \mathbf{C}). \end{array} \tag{7}$$

This is of course valid for any N. For N odd however, which we assume, all these spectral sequences degenerate at the E_2-level. By theorems of H. Hopf and H. Samelson, and their analogues in Lie algebra cohomology by J-L. Koszul [K], $H^{\cdot}(K_{N,u};\, \mathbf{C})$ and $H^{\cdot}(\mathfrak{g}_N(\mathbf{R});\, \mathbf{C})$ are exterior algebras over canonically defined subspaces of "primitive elements", so that in those cases we may identify in a natural way the spaces PH^j, which are by definition quotients, to subspaces. Moreover, the images of the injective maps $p_1^{\cdot}$ and $p_2^{\cdot}$ are subalgebras generated by primitive elements and we can write uniquely

$$ H^{\cdot}(K_{N,u};\, \mathbf{C}) = A_1^{\cdot} \otimes B_1,\ H^{\cdot}(\mathfrak{g}_N(\mathbf{R});\, \mathbf{C}) = A_2^{\cdot} \otimes B_2^{\cdot} \tag{8}$$

where $A_i^{\cdot}$ and $B_i^{\cdot}$ are generated by primitive elements,

$$ A_i^{\cdot} = Im\ p_i^{\cdot} \qquad (i = 1,\, 2) \tag{9}$$

and the restriction maps

$$(10) \qquad r_1^{\cdot} : B_1^{\cdot} \to H^{\cdot}(K_N;\, \mathbf{C}) \qquad r_2^{\cdot} : B_2^{\cdot} \to H^{\cdot}(\mathfrak{k}_N;\, \mathbf{C})$$

are isomorphisms.

In view of (10) and the definition of $\tilde{\gamma}^{\cdot}$, it follows that the fibres of p_3 are also totally non-homologous to zero (over $\mathbf{C}$) so that we also have

$$(11) \qquad H^{\cdot}(Z_N;\, \mathbf{C}) = H^{\cdot}(Y_N;\, \mathbf{C}) \otimes H^{\cdot}(K_N;\, \mathbf{C})$$

or more precisely

$$(12) \qquad H^{\cdot}(Z_N;\, \mathbf{C}) = A_3^{\cdot} \otimes B_3^{\cdot}$$

where

$$(13) \qquad A_3^{\cdot} = Im\ p_3^{\cdot}$$

and the restriction map r_3: $H^{\cdot}(Z_N;\, \mathbf{C}) \to H^{\cdot}(K_N;\, \mathbf{C})$ induces an isomorphism of B_3 onto $H^{\cdot}(K_N;\, \mathbf{C})$. In particular, we see that, in the stable range, $\tilde{\gamma}^{\cdot}$ is also an isomorphism and that the composition

$$\tilde{\alpha}^{\cdot} : H^{\cdot}(K_{N,u};\mathbf{C}) = H^{\cdot}(X_{N,u};\mathbf{C}) \otimes H^{\cdot}(K_N;\mathbf{C}) \to H^{\cdot}(Y_N;\mathbf{C}) \otimes H^{\cdot}(K_N;\mathbf{C}) = H^{\cdot}(Z_N;\mathbf{C})$$

may be identified to $\alpha^{\cdot} \otimes Id$.

We now go over to the spaces of primitive elements. Then the isomorphism

$$P\tilde{\alpha}^{\cdot} : PH^{\cdot}(K_{N,u};\, \mathbf{C}) = PH^{\cdot}(X_{N,u};\, \mathbf{C}) \otimes PH^{\cdot}(K_N;\, \mathbf{C}) \tilde{\to} PH^{\cdot}(Y_N;\, \mathbf{C}) \otimes PH^{\cdot}(K_N;\, \mathbf{C}) = PH^{\cdot}(Z_N;\, \mathbf{C}),$$

is given in degrees $< N/4$ by

$$(14) \qquad P\tilde{\alpha}^{\cdot} = P\alpha^{\cdot} \otimes Id.$$

If m is even and F totally real, then

$$r_1^{\cdot} : PH^{2m-1}(K_{N,u};\, \mathbf{C}) \to PH^{2m-1}(K_N;\, \mathbf{C})$$
$$r_2^{\cdot} : PH^{2m-1}(Z_N;\, \mathbf{C}) \to PH^{2m-1}(K_N;\, \mathbf{C})$$

are isomorphisms, so that

$$(15) \qquad R(\alpha^m) \sim 1 \qquad (m \text{ even } \geq 2,\ F \text{ totally real}).$$

The group G_N is defined over $\mathbf{Q}$, hence $\mathfrak{g}_N(\mathbf{R})$ has a $\mathbf{Q}$-form, namely $\mathfrak{g}_N(\mathbf{Q})$, so that $H^{\cdot}(\mathfrak{g}_N(\mathbf{R});\, \mathbf{C})$ has a $\mathbf{Q}$-structure, defined by $H^{\cdot}(\mathfrak{g}_N(\mathbf{Q});\, \mathbf{Q})$, therefore the regulator of α^m is the composition of the regulators of β^m and γ^m. By 5.4, 5.5 in [Bo 2] and the Erratum to the latter

$$(16) \qquad R(\beta^m) \sim D_F^{1/2}(\pi i)^{dm},\ R(\gamma^m) \sim \zeta_k(m).$$

Since $D_F \in \mathbf{Q}^*$, we have $D_F^{1/2} \sim D_F^{-1/2}$ and we get 5.5(5).

Comparing 5.5(5) and (15), and taking 5.4(3) into account, we obtain

$$(17) \qquad D_F^{1/2}\zeta_k(m)\pi^{dm} \in \mathbf{Q}^*,\ (m \text{ even } \geq 2; F \text{ totally real}),$$

which proves anew the Siegel-Klingen theorem.

§6. Algebraic K-groups

6.1 Let G be a group. Assume its derived group $N = \mathcal{D}G$ is perfect ($N = \mathcal{D}N$). D. Quillen has introduced a space BG^+, obtained from the classifying space BG of G by adding two- and three-cells by the so-called plus construction, with the following properties

(i) $\pi_1(BG^+) = G/N$.

(ii) The inclusion $BG \to BG^+$ induces isomorphisms

$$H_i(BG; \mathbf{Z}) \tilde{\to} H_i(BG^+; \mathbf{Z}) \qquad (i \in \mathbf{Z})$$

(iii) The natural isomorphisms $\pi_1(BG) = G$, $\pi_1(BG^+) = G/N$ carry the canonical map $\pi_1(BG) \to \pi_1(BG^+)$ over to the projection $G \to G/N$.

If G is commutative, $BG = BG^+$. If G is perfect, then BG^+ is simply connected. In particular, BN^+ is simply connected.

The natural fibration

$$(1) \qquad BN \to BG \to B(G/N)$$

yields, by the plus construction, a fibration

$$(2) \qquad BN^+ \to BG^+ \to B(G/N)^+ = B(G/N),$$

where the first map is a universal covering map [Ber]. In particular

$$(3) \qquad \pi_i(BN^+) = \pi_i(BG^+) \qquad (i \geq 2).$$

6.2 Let A be a commutative ring with unit. Assume that

$$(1) \qquad \mathcal{D}\mathbf{GL}\, A = \mathbf{SL}\, A$$

and is perfect. Then, by Quillen's first definition

$$(2) \qquad K_iA = \pi_i(B\mathbf{GL}\, A^+) \qquad (i \in \mathbf{N}).$$

In view of 3.1, we also have

$$(3) \qquad K_iA = \pi_i(B\mathbf{SL}\, A^+) \qquad (i \geq 2).$$

The Hurewicz homomorphism yields homomorphisms

(4) $$K_iA \tilde{\to} H_i(\mathbf{GL}\,A;\,\mathbf{Z}) = H_i(B\mathbf{GL}\,A;\,\mathbf{Z}) \qquad (i \in \mathbf{N})$$

(5) $$K_iA \tilde{\to} H_i(\mathbf{SL}\,A;\,\mathbf{Z}) = H_i(B\mathbf{SL}\,A;\,\mathbf{Z}) \qquad (i \geq 2).$$

By a theorem of Milnor-Moore [MM], these homomorphisms give rise to isomorphisms

(6) $$K_iA \otimes L \tilde{\to} PH_i(\mathbf{GL}\,A;\,L) \qquad (i \in \mathbf{N})$$

(7) $$K_iA \otimes L \tilde{\to} PH_i(\mathbf{SL}\,A;\,L) \qquad (i \in \mathbf{N},\, i \neq 1)$$

if L is a field of characteristic zero.

The rings o_F, $o_{F,S}$, F all satisfy the initial condition of 6.1. In particular, (4), (5), (6), (7), for $A = o_F$, provide an alternate proof of 3.3(3). The dual of the relation 3.9(2) can be written, in view of (6), (7)

(8) $$K_iF \otimes_{\mathbf{Z}} L = K_io_F \otimes_{\mathbf{Z}} L \qquad (i \in \mathbf{N};\, i \neq 1)$$

which also follows from the localization exact sequence in K-theory.

In fact, that long exact sequence, combined with D. Quillen's determination of the K-groups of finite fields, show that K_jF is an infinitely generated torsion group if j is even $\geqq 2$ and is a finitely generated group if j is odd and $\geqq 3$.

§7. Universal classes. Regulators in K-theory

7.1 The $\mathbf{Z}$-module $H^{2m-1}(\mathbf{SU_N};\,\mathbf{Z}(m))$ is free and $PH^{2m-1}(\mathbf{SU_N};\,\mathbf{Z}(m))$ is of rank one. We let e_m be the generator of the latter which is mapped onto the m-th Chern class $c_m \in H^{2m}(B\mathbf{SU}_N;\,\mathbf{Z}(m))$ by transgression in the universal bundle for $\mathbf{SU}_N$.

As we shall recall later, in a more general context, there is a canonical isomorphism

(1) $$H^{2m-1}(\mathbf{SU_N};\,\mathbf{R}(m)) \tilde{\to} H^{2m-1}_{ct}(\mathbf{SL}_N\mathbf{C};\,\mathbf{R}(m-1)), \quad (m \geq 2)$$

where the subscript ct refers to continuous Eilenberg-MacLane cohomology (4.2). If M is a Lie group, we let M^δ denote M, viewed as a discrete subgroup. Note that, if L is a local field, then $(\mathbf{SL}_NL)^\delta = \mathbf{SL}_NL^\delta$. The identity map $L^\delta \to L$ is continuous, whence homomorphisms

(2) $$H^{\cdot}_{ct}(M;\,L) \to H^{\cdot}(M^\delta;\,L).$$

In particular, we have a homomorphism

(3) $$H^{2m-1}_{ct}(\mathbf{SL}_N\mathbf{C};\,\mathbf{R}(m-1)) \to H^{2m-1}(\mathbf{SL}_N\mathbf{C}^\delta;\,\mathbf{R}(m-1)).$$

We let b_m be the image of e_m under the composition of (1) and (3). It can also be viewed as a homomorphism

$$H_{2m-1}(\mathbf{SL}_N\mathbf{C}^\delta;\,\mathbf{R}) \to \mathbf{R}(m-1)$$

which, by 6.2(4), defines a homomorphism

$$\lambda_m : K_{2m-1}\mathbf{C} \otimes_{\mathbf{Z}} \mathbf{R} \to \mathbf{R}(m-1),$$

to be called *the m-th regulator map of* **C**.

Remark. It is more usual to state these definitions and facts for $\mathbf{U}_N$ and $\mathbf{GL}_N$. For $m \geqq 2$, the two are equivalent (3.3). If we consider $\mathbf{GL}_N$ then the definition of the regulator map also extends to $m = 1$. In that case $K_1\mathbf{C} = \mathbf{C}^*$ and b_1 is the map $x \mapsto \log|x|$.

7.2 We consider the **R**-vector space $\mathbf{R}(m-1)^\Sigma$, with basis elements

(1) $$c_\sigma(2\pi i)^{m-1} := c_\sigma(m-1) \qquad (\sigma \in \Sigma).$$

The complex conjugation τ operates on $\mathbf{R}(m-1)^\Sigma$ by permuting σ and $\overline{\sigma}$ and by sending i to $-i$. The fixed point set $\mathbf{R}(m-1)^\Sigma_\tau$ is therefore spanned by the elements

(2) $$d_\sigma(m-1) = (c_\sigma(m-1) + (-1)^{m-1}c_{\overline{\sigma}}(m-1))/2$$

and its dimension is d_m. A basis consists of the $d_\sigma(m-1)$, where σ runs through a set of representatives of pairs of complex conjugate complex embeddings, to which one should add, if m is odd, the elements $d_\sigma(m-1)$, σ real. Note that

(3) $$d_{\overline{\sigma}}(m-1) = (-1)^{m-1}d_\sigma(m-1) \qquad (\sigma \in \Sigma).$$

For each $\sigma \in \Sigma$ we have natural homomorphisms

(4) $$K_{2m-1}F \otimes \mathbf{R} \xrightarrow{\iota_\sigma} K_{2m-1}\mathbf{C} \otimes \mathbf{R} \xrightarrow{\lambda_m} \mathbf{R}(m-1).$$

The product of these maps defines a homomorphism

(5) $$\Lambda_{F,m} = \prod \lambda_m \circ \iota_\sigma : K_{2m-1}F \otimes \mathbf{R} \to \mathbf{R}(m-1)^\Sigma.$$

7.3 Theorem. *The image of* $\Lambda_{F,m}$ *belongs to* $\mathbf{R}(m-1)^\Sigma_+$. *The restriction map*

(1) $$K_{2m-1}F \otimes \mathbf{R} \to \mathbf{R}(m-1)^\Sigma_+ \xrightarrow{\sim} \mathbf{R}(m-1)^{d_m}$$

(the m-th regulator map of F) is an isomorphism. Its regulator, with respect to the **Q***-structures defined by* $K_{2m-1}F \otimes \mathbf{Q}$ *and* $\mathbf{Q}(m-1)$ *is*

(2) $$R(\Lambda_{F,m}) \sim D_F^{1/2}\zeta_F(m)(\pi i)^{-d.m}.$$

This regulator is therefore the same as that of j_F^m in 5.5, and [Bo 2] is the reference usually given for 7.3. As far as I can see, it is not completely obvious that 5.5 implies 7.3. We shall sketch a proof of a somewhat stronger result in §10.

§8. Polylogarithms

We give here a brief summary of what is needed later on polylogarithms. For more details about the material in this section, see [O] or [Z] and the references given there.

8.0 If M is a set, we let $\mathbf{Z}[M]$ denote the free abelian group with basis the elements of M. A general element is written in the form of a finite sum

(1) $$c = \sum_x n_x[x] \qquad (x \in M,\, n_x \in \mathbf{Z})$$

Moreover, if E is a commutative field, we let

$$\mathcal{F}(E) = \mathbf{Z}[\mathbf{P}^1(E)]$$

8.1 The m-th logarithm $Li_m z$ is a function of one complex variable defined by the power series

(1) $$Li_m\, z = \sum_{n \geqq 1} z^n . n^{-m} \qquad (m \in \mathbf{N},\, m \geq 1).$$

It converges absolutely for $|z| \leq 1$ if $m \geqq 2$, for $|z| < 1$ if $m = 1$ and admits an analytic continuation to $(\mathbf{C} - [1, \infty[)$. We have

(2) $$Li_1 z = -\log(1 - z)$$

(3) $$Li_m(1) = \zeta_{\mathbf{Q}}(m) \qquad (m \geq 2)$$

(4) $$Li_1' z = (1 - z)^{-1}$$

(5) $$Li_m' z = z^{-1}.Li_{m-1} z \qquad (m \geq 2).$$

It extends to a "multivalued function" on $\mathbf{C} - \{0, 1\}$. This means the following. Let X be the universal covering of $\mathbf{C} - \{0, 1\}$, with base point $1/2$ and $p : X \to \mathbf{C} - \{0, 1\}$ the natural projection. There exists a holomorphic function $\mathbf{Li}_m$ on X such that the germ of $\mathbf{Li}_m$ at $1/2$ is Li_m. If $s : U \to Y$ is a holomorphic section of p on some open disc U, then $\mathbf{Li}_m \circ s : U \to \mathbf{C}$ is a determination of Li_m on U. Given $c \in X$, we denote by $Li_m^{[c]}$ the determination such that $Li_m^{[c]}(p(c)) = \mathbf{Li}_m(c)$.

8.2 We want related functions which are independent of the determination. For $m = 1$, the function $\log|z|$ is one. For $m = 2$, an example is the Bloch-Wigner dilogarithm. After some experimentation by him and others, D. Zagier has proposed the following functions, which were then singled out by motivic considerations by A. Beilinson and P. Deligne.

Start from the function

$$\tilde{P}_m(z) = \sum_{0 \leqq l < m} \frac{b_l}{l!} \log^l |z|^2 Li^{[c]}_{m-1}(z) \qquad (m \geqq 2). \tag{1}$$

The b_l are given by the power series

$$\sum b_l x^l = \frac{2x}{e^{2x} - 1}. \tag{2}$$

Therefore $b_l = 2^l B_l (l!)^{-1}$, where the B_l's are Bernoulli numbers. Let

$$P_m = \begin{cases} iIm\tilde{P}_m, & (m \geq 2, \text{ even}) \\ Re\tilde{P}_m & (m \text{ odd}). \end{cases} \tag{3}$$

Then

$$P_1(z) = ReLi_1(z) = -\log|(1-z)| \tag{4}$$

$$P_2(z) = iD_2(z) \tag{5}$$

where

$$D_2(z) = Im(Li_2^{[c]} z + \log|z| \log^{[c]}(1-z)) \tag{6}$$

is the Bloch-Wigner dilogarithm.

8.3 For $m \geq 2$, the function P_m is real analytic on $\mathbf{C} - \{0, 1\}$, with values in $\mathbf{R}(m-1)$. It extends continuously to $\mathbf{P}^1(\mathbf{C})$ and

$$P_m(0) = P_m(\infty) = 0 \tag{1}$$

$$P_m(1) = \begin{cases} \zeta_{\mathbf{Q}}(m) & (m \text{ odd}, m \geqq 3) \\ 0 & (m \text{ even}, m \geqq 2). \end{cases} \tag{2}$$

Moreover

$$P_m(\overline{z}) = \overline{P_m(z)} = (-1)^{m-1} P_m(z). \tag{3}$$

If $c = \sum_x n_x[x] \in \mathcal{F}(\mathbf{C})$, we let

$$P_m(c) = \sum_x n_x P_m(x). \tag{4}$$

This defines a map

(5) $$q_m : \mathcal{F}(\mathbf{C}) \to \mathbf{R}(m-1).$$

Let

(6) $$Q_{F,m} : \mathcal{F}(F) \to \mathbf{R}(m-1)^{\Sigma}$$

be the product of the maps $q_m \circ \sigma$ ($\sigma \in \Sigma$). It follows from (3) that

(7) $$Im\ Q_{F,m} \subset \mathbf{R}(m-1)^{\Sigma}_{+} \simeq \mathbf{R}(m-1)^{d_m}.$$

§9. Zagier's conjectures

We shall give three forms of these conjectures and summarize what has been proved so far.

9.1 *Conjecture* C_1. We first state it in a rather vague way, to indicate what one is looking for. Let Σ_c be the set of a set of representatives of the pairs of complex conjugates complex embeddings of F and define

(1) $$\Sigma_m = \begin{cases} \Sigma_c \ (m \text{ even, } m \geqq 2) \\ \Sigma_c \cup \{\text{real places}\} \ (m \text{ odd, } m \geqq 3). \end{cases}$$

It has therefore d_m elements. Then there should exist d_m elements

$$c_i = \Sigma_k\ n_{ik}[x_{ik}] \in \mathbf{Z}[F^*] \quad (i = 1, \dots, d_m)$$

such that $\zeta_F(m)$ should be equal, up to the product by a rational number, $D_F^{1/2}$ and some power of π, to

(2) $$\det\left(P_m(\sigma(c_i))\right) \qquad (i = 1, \dots, d_m; \sigma \in \Sigma_m)$$

where, by definition

(3) $$\sigma(c_i) = \Sigma_k\ n_{ik}\ [\sigma(x_{ik})].$$

9.2 *A second form.* The model for these conjectures is for $m = 1$ the formula expressing the residue of $\zeta_F(s)$ at $s = 1$, namely

(1) $$\mathrm{Res}_{s=1}\zeta_F(m) \sim \Delta_F^{1/2}\pi^{r_2}R_F.$$

In this case the elements occurring in the determinant are described rather explicitly. In the same spirit, the second form of the conjecture attempts to indicate more precisely where the c_i's should come from. To this effect, several subgroups and subquotients of $\mathcal{F}(F)$ are introduced.

More generally, if E is any commutative field we define, by induction on $m \geqq 1$, subgroups

(2) $$R_m(E) \quad (m \geqq 1) \text{ and } A_m(E) \quad (m \geqq 2)$$

of $\mathcal{F}(E)$. If $R_m(E)$ is defined, we let

$$(3) \qquad w_m : \mathcal{F}(E) \to \mathcal{F}(E)/R_m(E) := B_m(E)$$

be the natural projection. Let

$$(4) \qquad R_1(E) = \langle [xy] - [x] - [y],\ (x,\ y \in E^*),\ [0],\ [\infty] \rangle\,.$$

Then, it is easily seen that the map $[x] \mapsto x$ $(x \in E^*)$, $[0]$, $[\infty] \to 1$ induces an isomorphism

$$(5) \qquad B_1(E) \xrightarrow{\sim} E^*.$$

We define a map

$$(6) \qquad \delta_2 : \mathcal{F}(E) \to E^* \wedge E^*$$

by the rule

$$(7) \qquad x \in E^* \mapsto w_1(x) \wedge w_1(1-x)$$

let

$$(8) \qquad A_2(E) = \ker \delta_2.$$

Fix $m \geqq 2$. Assume that $A_m(E)$ has been defined for all commutative fields. As usual, $E(t)$ is the field of rational functions in t, with coefficients in E. We let

$$(9) \qquad R_m(E) = \left\langle \sum n_i(f_i(0) - f_i(1)) \middle| \sum n_i f_i \in A_m(E(t)) \right\rangle$$

$$(10) \qquad \delta_{m+1} : \mathcal{F}(E) \to B_m(E) \otimes B_1(E) = B_m(E) \otimes E^*$$

(cf (3), (5)) be the map defined by

$$(11) \qquad \delta_{m+1}(x) = w_m(x) \otimes w_1(x) \qquad (x \in E)$$

and

$$(12) \qquad A_{m+1}(E) = \ker \delta_{m+1}.$$

We have (cf 1.16 in [G 2]):

$$(13) \qquad R_m(E) \subset A_m(E)$$

so that δ_m induces a map

$$\overline{\delta}_m : B_m(E) \to \begin{cases} E^* \wedge E^*, & \text{for } m = 2, \\ B_{m-1}(E) \otimes E^*, & \text{for } m \geqq 3 \end{cases} \tag{14}$$

and

$$\ker \overline{\delta}_m = A_m(E)/R_m(E). \tag{15}$$

It is known that, as a consequence of the functional equations satisfied by the polylogarithms, we have:

$$P_m(R_m(\mathbf{C})) = 0 \tag{16}$$

(*see* 1.15 in [G2]) so that P_m may be viewed as a map from $B_m(\mathbf{C})$ to $\mathbf{R}(m-1)$. We let then

$$P_{F,m} : B_m(F) \to \mathbf{R}(m-1)_+^{\Sigma} \tag{17}$$

be the map defined by the product of the $P_m \circ \sigma$ $(\sigma \in \Sigma)$.

On the other hand, [BD] proves the existence of a homomorphism

$$\mu_m : A_m(F) \to K_{2m-1}F \tag{18}$$

such that

$$P_{F,m} = \Lambda_{F,m} \circ \mu_m. \tag{19}$$

Then C_1 would follow from 7.3 and

Conjecture C_2. *There exist elements* $c_\tau \in A_m(F)$ $(\tau \in \Sigma_m)$ *such that the* $P_{F,m}(c_\tau)$ *are linearly independent.*

If C_2 is true and $N_m(F)$ denotes the kernel of $P_{F,m}$ on $A_m(F)$, then $P_{F,m}$ induces an isomorphism of $A_m(F)/N_m(F) \otimes \mathbf{Q}$ onto $K_{2m-1}F \otimes \mathbf{Q}$. If one had a direct definition of $N_m(F)$, by means of constructions involving only $\mathcal{F}(F)$, this would yield a definition of $K_{2m-1}F \otimes \mathbf{Q}$. The most precise form of the conjecture, here called the *third form*, asserts indeed that $N_m(F) \otimes \mathbf{Q}$ is equal either to $R_m(F) \otimes \mathbf{Q}$ or to a group $C_m(F) \otimes \mathbf{Q}$, where $C_m(F)$ is also defined inductively ([Z], §8).

9.3 Remark. In the general case of a field E, the $B_m(E)$ and $\overline{\delta}_m$ define the Goncharov complex $\Gamma^{\cdot}(E, m)$:

$$\begin{aligned} &B_m(E)_{\mathbf{Q}} \xrightarrow{\overline{\delta}_m} \left(B_{m-1}(E) \otimes E^*\right)_{\mathbf{Q}} \xrightarrow{\overline{\delta}_{m-1} \otimes Id} \left(B_{m-2}(E) \otimes \wedge^2 E^*\right)_{\mathbf{Q}} \to \cdots \\ &\to B_2(E) \otimes \wedge^{m-2} E^*)_{\mathbf{Q}} \xrightarrow{\overline{\delta}_2 \otimes Id} (\Lambda^m E^*)_{\mathbf{Q}}, \end{aligned} \tag{1}$$

where the subscript $\mathbf{Q}$ stands for $\otimes_{\mathbf{Z}}\mathbf{Q}$. By definition, this is a cohomological complex starting in degree one. In particular, if $E = F$, then

$$H^1(\Gamma(F, m); \mathbf{Q}) \cong A_m(F)/R_m(F) \otimes \mathbf{Q}. \tag{2}$$

We should add that Goncharov also considers (1) without tensoring with $\mathbf{Q}$. In that case, $\delta_i \circ \delta_{i-1}$ is zero modulo torsion only. This allows for some variants modulo torsion or involving some specific torsion primes (see [G1], [G2]).

9.4 The conjecture C_2 has been proved in the cases $m = 2, 3$. ([G1], [G2], [Y2]). Actually, in the case $m = 2$, it had been known to follow from [Bo2] and results of Bloch and Suslin. The strategy is different from that underlying 9.3 and has been formalized by Goncharov to a further conjecture, which implies the existence of a natural map $\nu_m : K_{2m-1}F \to A_m(F)/R_m(F)$, i.e. going into the direction opposite to that of μ_m, such that

$$P_{F,m} \circ \nu_m = \Lambda_{F,m}. \tag{1}$$

To my knowledge, the third form has not yet been stated as proved for $m = 2, 3$ so far. According to Goncharov, however, it follows from C_2 and known facts for $m = 2$ and it seems within reach for $m = 3$, in both cases with $N_m(F) \otimes \mathbf{Q} = R_m(F) \otimes \mathbf{Q}$. We now give some indications on Goncharov's approach to C_2 for $m = 2, 3$.

9.5 In this sketch, we make use of the truth of the so-called "rank conjecture" for number fields, which simplifies the argument considerably, and we first discuss it briefly. It pertains to two filtrations of $K_jF \otimes \mathbf{R}$, and is implied by the two statements about the natural maps

$$\varphi_{a,b} : H_{2a-1}(\mathbf{GL}_a F; \mathbf{R}) \to H_{2a-1}(\mathbf{GL}_b F; \mathbf{R}), \qquad (b \gg a) \tag{1}$$

defined by inclusion, namely:

$$\operatorname{Im} \varphi_{a,b} \supset PH_{2a-1}(\mathbf{GL}_b F; \mathbf{R}) \qquad (b \gg a) \tag{2}$$

$$\operatorname{Im} \varphi_{a-1,b} \cap PH_{2a-1}(\mathbf{GL}_b F; \mathbf{R}) = \{0\} \qquad (b \gg a) \tag{3}$$

cf. [Y1]. The proof given there (assuming $F \neq \mathbf{Q}$) is incomplete. More precisely, it yields (3) for all F's and (2) for fields of degree $\geqq 3$ over $\mathbf{Q}$. However a proof of a stronger result will be given in a joint Note of J. Yang and the author [BY], so we take it for granted. For $m = 2$, it can already be derived from known results (see [G1], §2, n° 6) and for $m = 3$, it also follows from Goncharov's argument.

9.6 Consider the commutative diagrams

$$\begin{array}{ccc} H^{2m-1}_{ct}(\mathbf{GL}_N\mathbf{C}; \mathbf{R}(m-1)) & \longrightarrow & H^{2m-1}(\mathbf{GL}_N\mathbf{C}^\delta; \mathbf{R}(m-1)) \\ \downarrow r. & & \downarrow r^{\cdot} \\ H^{2m-1}_{ct}(\mathbf{GL}_m\mathbf{C}; \mathbf{R}(m-1)) & & H^{2m-1}(\mathbf{GL}_m\mathbf{C}^\delta; \mathbf{R}(m-1)) \end{array} \tag{1}$$

and, for $\sigma \in \Sigma$

$$\begin{array}{ccc} H_{2m-1}(\mathbf{GL}_m F; \mathbf{R}) & \xrightarrow{\sigma} & H_{2m-1}(\mathbf{GL}_m\mathbf{C}^\delta; \mathbf{R}) \\ j.\downarrow & & \downarrow j. \\ H_{2m-1}(\mathbf{GL}_N F; \mathbf{R}) & \xrightarrow{\sigma} & H_{2m-1}(\mathbf{GL}_N\mathbf{C}^\delta; \mathbf{R}) \end{array} \tag{2}$$

where $r^{\cdot}$ is induced by the restriction, $j.$ by the inclusion and σ by the obvious embedding.

Remember the definition of λ_m and $\Lambda_{F,m}$ (7.1, 7.2). On an element

$$z \in H_{2m-1}(\mathbf{GL}_N\, \mathbf{C}^{\delta}; \mathbf{R}),$$

the value $\lambda_m(z)$ is the evaluation $\langle b_m, z\rangle$ by a universal class b_m, which is the image of a class in continuous cohomology, denoted by b_m, too. The map $\Lambda_{F,m}$ is a product of maps $\lambda_m \circ \sigma$, i.e. of evaluations $\langle b_m, \sigma(y)\rangle$, with $y \in PH_{2m-1}(\mathbf{GL}_N F; \mathbf{R})$. Let $b'_m = r^{\cdot}(b_m)$. By 9.5, y is the image of a class $y' \in H_{2m-1}(\mathbf{GL}_m\, F; \mathbf{R})$, therefore

$$\langle b_m, \sigma(y)\rangle = \langle b'_m, \sigma(y')\rangle, \tag{3}$$

so that the computation is reduced to $\mathbf{GL}_m$. The proof of C_1 then consists in constructing a cocycle η_m representing a rational multiple of b'_m such that $\langle \eta_m, \sigma(y')\rangle$ is of the required type. So far, this could be done only for $m = 2, 3$.

9.7 *The case* $m = 2$. Given four distinct points $x_i \in \mathbf{P}^1(\mathbf{C})$ $(1 \leqq i \leqq 4)$, we let

$$r_2(x_1, x_2, x_3, x_4) = (x_1 - x_3).(x_2 - x_4)/(x_1 - x_4).(x_2 - x_3) \tag{1}$$

denote their cross-ratio. Fix $a \in \mathbf{Q}^*$. Given four elements $g_i \in \mathbf{GL}_2\, \mathbf{C}$ $(1 \leqq i \leqq 4)$ such that the $g_i.a$ are distinct, we consider the following assignment:

$$\eta_3 : (g_1, \dots, g_4) \mapsto P_2(r_2(g_1 a, \dots, g_4 a)) \in \mathbf{R}(1) = i\mathbf{R}. \tag{2}$$

We claim that this assignment defines a cocycle in its domain of definition, i.e. that if $g_i \in \mathbf{GL}_2\, \mathbf{C}$ are five elements such that the points $g_i a$ $(i = 1, \dots, 5)$ are distinct, then

$$\Sigma_i (-1)^i\, \eta_3(g_1, \dots, \hat{g}_i, \dots, g_5) = 0 \tag{3}$$

where, as usual, $\hat{g}_i$ means that g_i is omitted.

Since r_2 is invariant under $\mathbf{GL}_2\, \mathbf{C}$, it is enough to check this when $(g_1 a, g_2 a, g_3 a) = (0, 1, \infty)$. Then $x = g_4 a$ and $y = g_5 a$ belong to $\mathbf{C}^*$ and (3) is equivalent to the 5-term functional equation for P_2:

$$P_2\left(\frac{1-y^{-1}}{1-x^{-1}}\right) - P_2\left(\frac{1-y}{1-x}\right) + P_2\left(\frac{y}{x}\right) - P_2(y) + P_2(x) = 0. \tag{4}$$

This cocycle is not everywhere defined, but obviously measurable. It is also bounded, since P_2 is, hence locally L^1, and it defines therefore a continuous cohomology class by [Ba]. It is independent of the choice of a.

It is clear that the values of η_3 on homology classes are sums of terms of the type required in the conjecture. There remains then only to show that it represents a rational multiple of b'_2.

Since $H^3_{ct}(\mathbf{GL}_2, \mathbf{R})$ is one-dimensional, this can be done by a direct check, using an imaginary quadratic field. In fact, an argument based on a theorem of J. Dupont, sketched in [G1], §2, n° 5, shows that it represents b'_3 itself. Let λ'_2 be the map $H_3(\mathbf{GL}_2\mathbf{C}^\delta; \mathbf{R}) \to i\mathbf{R}$ defined by

$$\lambda'_2(z) = \langle \eta_3, z\rangle \qquad (z \in H_3(\mathbf{GL}_2\mathbf{C}^\delta; \mathbf{R})).$$

In order to construct ν_2 in 9.4(1), it suffices to show the existence of

$$\nu'_2 : H_3(\mathbf{GL}_2\mathbf{C}^\delta; \mathbf{R}) \to A_2(\mathbf{C})/R_2(\mathbf{C}),$$

such that $\lambda'_2 = P_2 \circ \nu'_2$. Its construction is based on the use of the so-called Grassmannian complex.

9.8 Let E be an infinite field and $n \in \mathbf{N}$. A set of m-points in E^n is said to be general if any subset of $k \leqq m$, n elements is free. Let $\tilde{C}_m(n)$ be a free abelian group over the set of general $(m+1)$-tuples in E^n. It is operated upon by $\mathbf{GL}_nE$ in the obvious way. The sum $\tilde{C}.(n)$ of the $\tilde{C}_m(n)$ is endowed with the differential $d : \tilde{C}_m(n) \to \tilde{C}_{m-1}(n)$ defined by

$$d(x_1, \ldots, x_m) = \Sigma_i(-1)^i(x_1, \ldots, \hat{x}_i, \ldots, x_m). \tag{1}$$

It is easily seen to provide a resolution of the trivial $\mathbf{GL}_nE$-module $\mathbf{Z}$. Let $C.(n)$ be the complex of coinvariants of $\tilde{C}.(n)$. Since any free resolution maps into a given resolution, we get a map

$$\alpha : H.(\mathbf{GL}_n E; \mathbf{Z}) \to H.(C.(n)). \tag{2}$$

Let now $n = 2$ and $E = \mathbf{C}$. The main point is to relate a truncation of $C.(2)$ to the Goncharov complex $\Gamma(F, 2)$ (see 9.3), more precisely to construct maps

$$f_1 : C_3(2) \to B_2(\mathbf{C}), \qquad f_0 : C_2(2) \to \Lambda^2\mathbf{C}^* \tag{3}$$

so that the following diagram is commutative

$$\begin{array}{ccccc} C_4(2) & \xrightarrow{d} & C_3(2) & \xrightarrow{d} & C_2(2) \\ \downarrow & & \downarrow f_1 & & \downarrow f_0 \\ 0 & \longrightarrow & B_2(\mathbf{C}) & \xrightarrow{\delta_2} & \Lambda^2\mathbf{C}^*, \end{array} \tag{4}$$

a construction for which we refer to §2, n^o3 of [G1]. This provides in particular a homomorphism

$$\beta : H_3(C_1(2)) \to A_2(\mathbf{C})/R_2(\mathbf{C}) \otimes \mathbf{Q}.$$

If the homogeneous complex is used to define $H.(\mathbf{GL}_3\mathbf{C}; \mathbf{Z})$ and α, it is easily seen that $\beta \circ \alpha$ is the sought for map ν'_2.

9.9 Remark. If we view $\mathbf{P}^1(\mathbf{C})$ as the two-sphere at infinity of the natural compactification of the 3-dimensional hyperbolic space H_3, then it is known that $\eta_3(g_1, \ldots, g_4)$ is the hyperbolic volume of the ideal tetrahedron with vertices the $g_i.a$ (up to a factor 2/3). This is the starting point of D. Zagier's approach to the conjecture for $m = 2$. However, it does not prove the full conjecture C_1, but only that one can use elements $c_\tau \in \mathcal{F}(E)$, where E runs through extensions of F of bounded degree (see [Z] and the references given there).

9.10 The proof sketched in 9.6–9.8 suggests by itself a line of attack for other m's. We describe it for $m = 3$, the only other case where it could so far be brought to completion.

We first state Goncharov's definition of a generalized cross-ratio on $\mathbf{P}^2(\mathbf{C})$, as given in [G3: §7].

Fix a volume form $\omega \in \Lambda^3\mathbf{C}^3$, $\omega \neq 0$. Given $y_1, y_2, y_3, \in \mathbf{C}^3$ let

$$\Delta(y_1, y_2, y_3) = \langle \omega, y_1 \wedge y_2 \wedge y_3 \rangle. \tag{1}$$

For a point $x \in \mathbf{P}^2(\mathbf{C})$, we denote by x' a non-zero element on the line in $\mathbf{C}^3$ defined by x. Let now $\{x_i\}$ $(1 \leqq i \leqq 6)$ be six points of $\mathbf{P}^2(\mathbf{C})$ in general position. The expression

$$\frac{\Delta(x_1', x_2', x_4').\Delta(x_2', x_3', x_5').\Delta(x_3', x_1', x_6')}{\Delta(x_1', x_2', x_5').\Delta(x_2', x_3', x_6').\Delta(x_3', x_1', x_4')} := r_3'(x_1, \ldots, x_6) \tag{2}$$

is independent of the choice of ω or of the representatives x_i' of the x_i (which justifies the notation on the right-hand side). Then the generalized cross-ratio $r_3(\{x_i\})$ is by definition an element of $\mathcal{F}(\mathbf{C})$ obtained by antisymmetrization of r_3':

$$r_3(\{x_i\}) = \sum_{\tau \in \mathfrak{S}_6} sgn\sigma.[r_3'(x_{\tau(1)}, \ldots, x_{\tau(6)})], \tag{3}$$

where $\mathfrak{S}_6$ is the symmetric group in six letters and, as before, $[u]$ $(u \in \mathbf{P}^1(\mathbf{C}))$ denotes the basis element of $\mathcal{F}(\mathbf{C})$ defined by u. The functional equation for P_3 found by Goncharov implies that if $x_1, \ldots, x_7$ are seven points of $\mathbf{P}^2(\mathbf{C})$ in general position, then

$$\sum_{1 \leqq j \leqq 7} (-1)^j P_3(r_3(\{x_i\})_j) = 0. \tag{4}$$

Fix $a \in \mathbf{P}^2(\mathbf{C})$. This means that on the set of 6-tuples of elements $g_i \in \mathbf{GL}_3(\mathbf{C})$ such that the $g_i.a$ are in general position, the Eilenberg-MacLane cochain

$$(g_1, \cdots, g_6) \mapsto P_3(r_3(\{g_i.a\}))$$

is a cocycle (see Thm. 3.15 in [G2]). This yields first a locally integrable 5-cocycle and then, as in 9.7, an element

$$\eta_5 \in H^5_{ct}(\mathbf{GL}_3(\mathbf{C}); \mathbf{R}(2))$$

(which is independent of the choice of a). The remaining steps are similar to those in the case $m = 2$, but technically more complicated.

9.11 We refer to [G1], §1 for a discussion of various problems and conjectures pertaining to the general case.

§10. Appendix

In this appendix we sketch how to go from 5.5 to 7.3 and also recall the construction of an isomorphism generalizing 7.1(1).

10.1 We first fix some notation. Recall that we have chosen in 7.1 a primitive generator e_m in $H^{2m-1}(\mathbf{SU}_N; \mathbf{Z}(m)$. We write it as $e'_m(2\pi i)^m$, where e'_m is a primitive element in $H^{2m-1}(\mathbf{SU}_N; \mathbf{Z})$. The complex conjugation on $\mathbf{SU}_N$ induces an involution $x \mapsto \overline{x}$ on each cohomology group. We have

$$\overline{e}'_m = (-1)^m e'_m. \tag{1}$$

It suffices to see this for $N = m$. There it follows from the fact that e'_m is also, up to sign, the image of a generator of $H^{2m-1}(\mathbf{S}^{2m-1}; \mathbf{Z})$ under the map $H^{2m-1}(\mathbf{S}^{2m-1}; \mathbf{Z}) \to H^{2m-1}(\mathbf{SU}_m; \mathbf{Z})$ associated to the canonical projection $\mathbf{SU}_m \to \mathbf{S}^{2m-1} \subset \mathbf{C}^m$. Indeed, the projection commutes with complex conjugation in $\mathbf{SU}_m$ and $\mathbf{C}^m$ and the latter changes the orientation of $\mathbf{C}^m$ if and only if m is odd. We have

$$K_{N,u} = (\mathbf{SU}_N)^{\Sigma}, \ G_N(\mathbf{C}) = (\mathbf{SL}_N\mathbf{C})^{\Sigma}. \tag{2}$$

In analogy with 3.4(4) we write

$$K_{N,u} = \prod_{\sigma\in\Sigma} K_{N,\sigma}, \tag{3}$$

where $K_{N,\sigma} \cong SU_N$ is the standard maximal compact subgroup of $G_{N,\sigma}$. The complex conjugation τ of $G_N(\mathbf{C})$ with respect to $G_N(\mathbf{R})$ (see 4.3) leaves $K_{N,u}$ stable and its fixed point set there is K_N.

10.2 *Some generators.*

We denote by $e'_{m,\sigma}$ the primitive generator in $H^{2m-1}(K_{N,\sigma}; \mathbf{Z})$ equal to e'_m. The $e'_{m,\sigma}$ $(\sigma \in \Sigma)$ for a basis of the space $PH^{2m-1}(K_{N,u}; \mathbf{Z})$. The complex conjugation operates on the latter. In view of (1), (3) we see that the elements

$$(e'_{m,\sigma} + (-1)^m e'_{m,\overline{\sigma}}) \qquad (\text{resp.} e'_{m,\sigma} + (-1)^{m-1} e'_{m,\overline{\sigma}}) \tag{1}$$

generate the 1-eigenspace (resp. (-1)-eigenspace), of τ in $PH^{2m-1}(\mathbf{SU}_N; \mathbf{Z})$. Its dimension is d_{m-1} (resp. d_m). The space $A^{\cdot}_1$ (resp. $B^{\cdot}_1$) in 5.6(8) is generated by the primitive elements on which τ acts by $-Id.$ (resp. $Id.$). Recall that $p^{\cdot}_1$ identifies $H^{\cdot}(X_{N,u}; \mathbf{C})$ with $A^{\cdot}_1$. We let $f'_{m,\sigma}$ be the unique element of $PH^{2m-1}(X_{N,u}; \mathbf{Q})$ such that

$$p^{\cdot}_1(f'_{m,\sigma}) = (e'_{m,\sigma} + (-1)^{m-1} e'_{m,\overline{\sigma}})/2. \tag{2}$$

Therefore,

$$(3) \qquad p_1^{\cdot}(f'_{m,\sigma}) = \begin{cases} 0 \text{ if } \sigma \text{ is real, } m \text{ even, } m \geqq 2. \\ e'_{m,\sigma} \text{ if } \sigma \text{ is real, } m \text{ odd, } m \geqq 3. \end{cases}$$

and

$$(4) \qquad f'_{m,\sigma} = (-1)^{m-1} f'_{m,\bar{\sigma}} \quad (\sigma \in \Sigma).$$

Then $\{f'_{m,\sigma}\}$, with running through the set Σ_m defined by 9.1(1), is a basis of $PH^{2m-1}(X_{N,u}; \mathbf{Q})$ and the elements

$$(5) \qquad f_{m,\sigma} = f'_{m,\sigma} \otimes (2\pi i)^m$$

from a basis of $PH^{2m-1}(X_{N,u}; \mathbf{Q}(m))$.

In the sequel, we let for any $j \in \mathbf{N}$

$$(6) \qquad r^{\cdot} : H^{\cdot}_{ct}(G_N(\mathbf{C}), R(j)) \to H^{\cdot}_{ct}(G_N(\mathbf{R}); \mathbf{R}(j))$$

be the natural map induced by restrictions of Eilenberg-MacLane cochains.

10.3 Proposition. *There exist natural isomorphisms*

$$(1) \qquad \tilde{\mu}^m : H^{2m-1}(K_{N,u}; \mathbf{R}(m)) \tilde{\to} H^{2m-1}_{ct}(G_N(\mathbf{C}); \mathbf{R}(m-1))$$

$$(2) \qquad \mu^m : H^{2m-1}(X_{N,u}; \mathbf{R}(m)) \to H^{2m-1}_{ct}(G_N(\mathbf{R}); \mathbf{R}(m-1))$$

such that

$$(3) \qquad r^{\cdot} \tilde{\mu}^m(e_{m,\sigma}) = 2.\mu^m(f_{m,\sigma}).(\sigma \in \Sigma).$$

The relation (3) implies

$$(4) \qquad r^{\cdot}.\tilde{\mu}^m.p_1^{\cdot}(f_{m,\sigma}) = 2.\mu^m(f_{m,\sigma}) \qquad (\sigma \in \Sigma).$$

This means that the diagram

$$(5) \qquad \begin{array}{ccc} PH^{2m-1}(K_{N,u}; \mathbf{R}(m)) & \xrightarrow{\tilde{\mu}^m} & PH^{2m-1}_{ct}(G_N(\mathbf{C}); \mathbf{R}(m-1)) \\ \uparrow p_1^{\cdot} & & r^{\cdot} \downarrow \\ PH^{2m-1}(X_{N,u}; \mathbf{R}(m)) & \xrightarrow{\mu^m} & PH^{2m-1}_{ct}(G_N(\mathbf{R}); \mathbf{R}(m-1)) \end{array}$$

is commutative up to a factor 2. The proof of 10.3 is contained in the next two sections.

10.4 The proofs of 10.3(1), (2) use the usual trick involving $i.\mathfrak{p}$ and $\mathfrak{p}$. We establish (2) first. Recall the Cartan decompositions

(1) $$\mathfrak{g}_N(\mathbf{R}) = \mathfrak{k}_N \oplus \mathfrak{p}_N,\ \mathfrak{k}_{N,u} = \mathfrak{k}_N \oplus i\mathfrak{p}_N$$

(cf. 4.1 and 4.4(2)). This yields, in view of 4.4(5), 4.1 and the van Est theorem (4.2)

(2) $$H^{2m+1}(X_{N,u}; \mathbf{R}(m)) = (\Lambda^{2m-1}(i\mathfrak{p}_N^*))^{K_N} \otimes i^m\mathbf{R},$$

(3) $$H^{2m+1}_{ct}(G_N(\mathbf{R}); \mathbf{R}(m-1)) = (\Lambda^{2m-1}\mathfrak{p}_N^*)^{K_N} \otimes i^{m-1}\mathbf{R}.$$

Since $i^{2m-1} = \pm i$, this gives (2). The proof of (1) combines this argument with an identity pertaining to the complexification of a real group which is itself a complex group viewed as real group. Let

(4) $$H_N = R_{\mathbf{C}/\mathbf{R}} G_N/\mathbf{C}.$$

Then

(5) $$H_N(\mathbf{R}) = G_N(\mathbf{C}) \text{ and } H_N(\mathbf{C}) = G_N(\mathbf{C}) \times G_N(\mathbf{C})$$

with $H_N(\mathbf{R})$ embedded in $H_N(\mathbf{C})$ as

(6) $$H_N(\mathbf{R}) = \{(x, \overline{x}) | x \in G_N(\mathbf{C})\}.$$

As a maximal compact subgroup of $H_N(\mathbf{C})$ we choose

(7) $$H_{N,u} = K_{N,u} \times K_{N,u}$$

and as maximal compact subgroup L_N of $H(\mathbf{R})$ the subgroup

(8) $$L_N = \{(x, \overline{x}) | x \in K_{N,u}\} \simeq K_{N,u}.$$

The subgroups $H_N(\mathbf{R})$ and $H_{N,u}$ of $H(\mathbf{C})$ are two real forms of H, with intersection L_N. We have then the Cartan decompositions, corresponding to (1) above:

(9) $$\mathfrak{h}_N(\mathbf{R}) = \mathfrak{l}_N \otimes \mathfrak{q}_N \qquad \mathfrak{h}_{N,u} = \mathfrak{l}_N \otimes i\mathfrak{q}_N$$

where $\mathfrak{h}_N$ and $\mathfrak{l}_N$ are the Lie algebras of H_N and L_N respectively and $\mathfrak{q}_N$ is the orthogonal complement of $\mathfrak{l}_N$ in $\mathbf{h}_N(\mathbf{R})$ with respect to the Killing form. In analogy with (2), (3), this gives

(10) $$H^{2m-1}(H_{N,u}/L_N; \mathbf{R}(m)) \tilde{\to} (\Lambda^{2m-1} i\mathfrak{q}_n^*)^{L_N} \otimes i^m\mathbf{R}.$$

(11) $$H^{2m-1}_{ct}(G_N(\mathbf{R}); \mathbf{R}(m-1)) \tilde{\to} (\Lambda^{2m-1}\mathfrak{q}^*_N)^{L_N} \otimes i^{m-1}\mathbf{R}.$$

But, and this is the special feature of this situation alluded to above, there is a natural isomorphism

(12) $$H_{N,u}/L_N \tilde{\to} K_{N,u}.$$

The projection of $H_{N,u}$ on its first factor already yields such an isomorphism of varieties. It is better to use the map

(13) $$(x, y) \mapsto x.\overline{y}^{-1}.$$

which is an isomorphism of symmetric spaces for $H_{N,u}$, the group $K_{N,u}$ being viewed as a symmetric space on which $H_{N,u}$ acts by left and right translations. More precisely, the element $(a, b) \in H_{N,u}$ induces the map $x \mapsto a.x.{}^t b$ (x, a, $b \in K_{N,u}$). The involution τ goes over to the inversion $x \mapsto x^{-1}$. We can therefore replace $H_{N,u}/L_N$ in (10) by $K_{N,u}$ and then (1) follows from (10) and (11).

10.5 There remains to prove 10.3 (3). In the diagram (5), all maps are natural, nevertheless the two vertical arrows go in opposite directions. By Dupont's interpretation of the van Est isomorphism (see 4.2) the map $r^{\cdot}$ is induced by restriction of $G_N(\mathbf{C})$-invariant forms on $G_N(\mathbf{C})/K_{N,u}$ to $G_N(\mathbf{R})$-invariant forms on $G_N(\mathbf{R})/K_N$. We would like similarly to replace $p_1^{\cdot}$ by a map

$$H^{\cdot}(K_{N,u}; \mathbf{R}(m)) \to H^{\cdot}(X_{N,u}; \mathbf{R}(m))$$

corresponding to a restriction of differential forms to a submanifold. This can be easily done by means of a construction of E. Cartan, in which a symmetric space is viewed as a submanifold of the group, namely, the space of transvections, rather than as a quotient.

We let $K_{N,u}$ operate on itself by the operations

(1) $$\lambda_g : x \mapsto g.x.\tau(g^{-1}) \qquad (x, g \in K_{N,u}).$$

The orbit

(2) $$S_N = \{g.\tau(g^{-1})\} \qquad (g \in K_{N,u})$$

of the identity is a closed submanifold and the isotropy group of 1 is clearly K_N. The map

(3) $$q : g \mapsto g.\tau(g^{-1})$$

of $K_{N,u}$ into itself induces therefore an isomorphism $\tilde{q}$ of $X_{N,u}$ onto S_N. The map q decomposes into

(4) $$K_{N,u} \xrightarrow{p_1} X_{N,u} \xrightarrow{\tilde{q}} S_N \xrightarrow{i} G_u$$

where i is the inclusion. We are interested in the effect of $i^{\cdot}$ on the primitive elements and how it relates to $p_1^{\cdot}$. In view of (4), we are reduced to $q^{\cdot}$. For the latter, we have the following lemma of B. Harris [Hr]:

$$q^{\cdot}(x) = x - \tau(x) \qquad (x \in PH^j(K_{N,u};\, \mathbf{C});\ j \in \mathbf{N}). \tag{5}$$

This implies therefore

$$i^{\cdot}(x) = 0 \text{ if } \tau(x) = x \quad i^{\cdot}(x) = 2x \text{ if } \tau(x) = -x, \tag{6}$$

hence that $i^{\cdot}$ is an isomorphism of $A_1^{\cdot}$ onto $H^{\cdot}(S_N;\, \mathbf{C})$, the kernel of which is generated, as an ideal, by the primitive elements in $B_1^{\cdot}$ (see 5.4 for the notation).

We now consider the diagram

(7)

$$\begin{array}{ccccc}
H^{2m-1}(K_{N,u};\mathbf{R}(m)) & = & H^{2m-1}(K_{N,u};\mathbf{R}(m)) & \xrightarrow{\tilde{\mu}^m} & H^{2m-1}_{ct}(G_N(\mathbf{C});R(m-1)) \\
\uparrow{\scriptstyle p_1} & & \downarrow{\scriptstyle i^{\cdot}} & & \downarrow{\scriptstyle r^{\cdot}} \\
H^{2m-1}(X_{N,u};\mathbf{R}(m)) & \xrightarrow{\tilde{q}} & H^{2m-1}(S_N;\mathbf{R}(m)) & \xrightarrow{\mu^m \circ \tilde{q}} & H^{2m-1}_{ct}(G_N(\mathbf{R});\mathbf{R}(m-1)).
\end{array}$$

The tangent space to S_N at the origin is $\mathfrak{p}_{N.u}$ and again $(\Lambda\mathfrak{p}^*_{N,u})^{K_N}$ may be identified to the space of $K_{N,u}$-invariant forms of S_N, hence to the cohomology of S_N. From this it follows that the right-hand square is commutative. Then 10.3(3) follows from (6).

10.6 In order to derive 7.3 from 5.5, it suffices to show that $R(j_F^m) = R(\Lambda_{F,m})$. In fact, we prove more strongly that the maps

$$j_F^m,\ \Lambda_{F,m} : PH_{2m-1}(\mathbf{SL}_N o_F;\, \mathbf{Q}) \to \mathbf{R}(m-1)^{\Sigma}_{+}$$

underlying them the same image. We have commutative diagrams:

$$\begin{array}{ccc}
H^{2m-1}_{ct}(G_N(\mathbf{C});\, \mathbf{R}(m-1) & \longrightarrow & H^{2m-1}(G_N(\mathbf{C}^\delta);\, \mathbf{R}(m-1)) \\
\downarrow{\scriptstyle r^{\cdot}} & & \downarrow{\scriptstyle r^{\cdot}} \\
H^{2m-1}_{ct}(G_N(\mathbf{R});\, \mathbf{R}(m-1) & \longrightarrow & H^{2m-1}(G_N(\mathbf{R}^\delta);\, \mathbf{R}(m-1))
\end{array} \tag{1}$$

$$\begin{array}{ccc}
PH_{2m-1}(\mathbf{SL}_N o_F;\, \mathbf{R}) & \longrightarrow & H_{2m-1}(G_N(\mathbf{R}^\delta);\, \mathbf{R}) \\
& \searrow & \downarrow \\
& & H_{2m-1}(G_N(\mathbf{C}^\delta);\, \mathbf{R}).
\end{array} \tag{2}$$

The map $\Lambda_{F,m}$ in 7.2 can equivalently be defined as obtained by evaluation on $K_{2m-1}F \otimes \mathbf{Q} = PH_{2m-1}(\mathbf{SL}_N o_F;\, \mathbf{Q})$ of the elements $\tilde{\mu}^m(e_{m,\sigma})$. Similarly, evaluation by the $f_{m,\sigma}$ defines a map $J_{F,m} : K_{2m-1}F \otimes \mathbf{Q} \to \mathbf{R}(m-1)^{\Sigma}_{\tau}$ which, modulo the isomorphism μ^m, is the map j^m_F of 5.5. But now, 10.3(3) shows that

$$\text{(3)} \qquad \operatorname{Im} J_{F,m} = \operatorname{Im} \Lambda_{F,m},$$

whence, in particular, 7.3.

References

[Ba] P. Blanc, *Sur la cohomologie continue des groupes localement compacts*, Annales Sci. Ec. Norm. Sup. (4) **12**, 1979, 137–168.

[Bei] A. Beilinson, *Higher regulators and values of L-functions,*, Sovr. Probl. Math. **24**, Moscow, VINITI, 1984, 181–238. English translation: Jour Soviet Math. **30**, 1985, 2036–2070.

[Ber] A. J. Berrick, *The plus construction and fibrations*, Quarterly J. Math. Oxford (2) **33**, 1982, 149–157.

[BD] A. Beilinson and P. Deligne, *Polylogarithms and regulators*, Proc. AMS Research Summer Conference on "Motives", to appear.

[Bo 1] A. Borel, *Stable real cohomology of arithmetic groups*, Ann. Sci. Ec. Norm. Super., (4) **7**(1974), 235–272.

[Bo 2] A. Borel, *Cohomologie de SL_n et valeurs de fonctions zeta aux points entiers*, Ann. Sc. Norm. Super. Pisa, Cl. Sci., (4) **4**(1977), 613–636; Correction, ibid. **7**(1980), 373.

[Bo 3] A. Borel, *Stable real cohomology of arithmetic groups II*, Prog. Math., Boston **14**(1981), 21–55.

[BS] A. Borel et J-P. Serre, *Cohomologie d'immeubles et de groupes S-arithmétiques*, Topology **15**(1976), 211–232.

[BW] A. Borel and N. Wallach, *Continuous cohomology, discrete subgroups and representations of reductive groups*, Ann. Math. Stud,"Princeton University Press". **94**(1980).

[BY] A. Borel and J. Yang, *The rank conjecture for number fields*, in preparation.

[Br 1] K. Brown, *Euler characteristics of discrete groups and G-spaces*, Inv. Math. **27**, 1974, 229–264.

[Br 2] K. Brown, *Euler characteristics of groups; the p-fractional part*, Inv. Math. **29**, 1975, 1–5.

[C] J-L. Cathelineau, *Homologie du groupe linéaire et polylogarithmes*, Sém. Bourbaki, 1992–93, Exp. **772**.

[D] J. Dupont, *Simplicial de Rham cohomology and characteristic classes of flat bundles*, Topology **15**, 1976, 233–245.

[G 1] A. B. Goncharov, *Geometry of configurations, polylogarithms and motivic cohomology*, to appear in Adv. in Math.

[G 2] A. B. Goncharov, *Polylogarithms and motivic Galois groups*, Proc. AMS Conference on Motives.

[G 3] A.B. Goncharov, *The classical polylogarithms, algebraic K-theory and $\zeta_f(n)$*, in The Gelfand mathematical seminars, 1990–1992 (L. Corwin, I. Gelfand, J. Lepowsky eds) Birkhäuser, Boston 1993.

[Ha] G. Harder, *A Gauss-Bonnet formula for discrete arithmetically defined groups*, Ann. Sci. E. N. S. **4** (1971), 409–455.

[Hr] B. Harris, *On the homotopy groups of the classical group*, Annals of Math. **74**, 1961, 407–413.

[K] H. Klingen, *Ueber die Werte der Dedekindschen Zetafunktion*, Math. Annalen **114**, 1962, 265–272.

[MM] J. Milnor and J. Moore, *On the structure of Hopf algebras*, Annals of Math. **81**, 1965, 211–264.

[O] J. Oesterlé, *Polylogarithmes*, Sém. Bourbaki 1992/93. Exp. **762**.

[R] M. Rapoport, *Comparison of the regulators of Beilinson and Borel*, in Beilinson's conjectures on special values of L-functions, (M. Rapoport, N. Schappacher, P. Schneider eds.), Perspectives in Mathematics 4, 1988, 169–190, Academic Press.

[S] J-P. Serre, *Cohomologie des groupes discrets*, Prospects in Mathematics, Annals of Math, Studies **70**, Princeton University Press, 1970, 77–169.

[Y 1] J. Yang, *On the real cohomology of arithmetic groups and the rank conjecture*, Annals Sci. E. N. S.: (4) **25**, 1992, 287–306.

[Y 2] J. Yang, *The Hain-MacPherson's trilogarithm, the Borel regulator and the value of Dedekind zeta- functions at 3*, Preprint.

[Z] D. Zagier, *Polylogarithms, Dedekind zeta-functions and the algebraic K-theory of fields*, in Arithmetic algebraic geometry (G.v. Geer, F. Oort, J. Steenbrink, eds), Progress in Math. 89, 1991, 391–430.

Institute for Advanced Study
Princeton, NJ 08540, USA
and
Max-Planck-Institut für Mathematik
Gottfried-Claren-Strasse 26, 53225 Bonn, Germany.

158.

(with F. Bien and J. Kollár)

Rationally Connected Homogeneous Spaces

Inv. Math. 124 (1996), 103–127

to Reinhold Remmert

Introduction

An irreducible algebraic variety X over an algebraically closed field k is said to be *quasi-complete* if the k-algebra $k[X]$ of regular functions on X reduces to the constants. The case of interest in this paper is when $X = G/H$ is the quotient of a connected affine k-group by a closed subgroup, an assumption which is kept throughout this introduction. Then X is quasi-complete if and only if H is epimorphic in G, in the sense of [BB a, b]. A sufficient condition for quasi-completeness is the existence of sufficiently many images of $\mathbb{P}^1(k)$ in X or in quotients by closed subgroups containing H. They are fulfilled in many cases, including most of the examples given in [BB a, b]. Furthermore they imply property F of [BB a]: the space of regular sections of a homogeneous vector bundle is finite dimensional, thus confirming in those cases the conjecture of [BB a] that all epimorphic subgroups have property F. This paper is accordingly devoted first to a discussion of various notions of connectedness by images of rational curves, their relations and properties and then to the applications to several series of examples.

To describe more precisely the contents of this paper, we first introduce some conventions and definitions. A *rational curve* is by definition in this paper a smooth complete curve of genus 0, i.e. a copy of $\mathbb{P}^1(k)$; a *chain of rational curves* is a variety $C = \cup C_i$ of rational curves such that $C_i \cap C_{i+1}$ consists of one point ($i = 1, \dots, n-1$) and there are no other intersection points between the C_i's. A *rational chain in X* is a morphism $f : C \to X$ of a chain of rational curves into X. Its image can be singular. It *connects x and y* if $x \in f(C_1)$ and $y \in f(C_n)$. Section 1 is devoted to smoothings of rational chains by chains of length one. The results are analogous to those proved in [KMM] for smooth projective varieties in characteristic zero. The arguments are in part similar, based on results of [Gb]. In particular, it is shown that if $f : C \to X$ connects x and y and $f^*T_X|_{C_i}$ is ample for each i, where T_X is the tangent bundle to X, then x and y are connected by the image of one rational curve (1.9). This also holds in characteristic zero without the assumption of ampleness (1.10).

We shall say that X is *rationally connected* if any two points are contained in the image of a rational curve, *rationally chain connected* if any two points are connected by a rational chain and *generically rationally connected* if there is an open dense subset $U \subset X \times X$ such that for any $(x, y) \in U$, x and y are connected by the image in X of a rational curve. The main result of Sect. 2 asserts that there is a unique closed subgroup $H' \supset H$ such that any rational chain in X projects to a point in $X' = G/H'$ and that H'/H is generically rationally connected (2.5). In particular, if X is rationally chain connected, then it is generically rationally connected (2.6). In all this, "generically" may be erased in characteristic zero (2.8, 2.9). In Sect. 3, it is shown that the projection $X \rightarrow X'$ preserves the coherence of G-sheaves (3.2). It is this fact which yields Property F in the applications.

Section 4 is devoted to those. First it is proved that if H is epimorphic, if G contains a subgroup L isogeneous to $\mathbf{SL}_2$ such that $L \cap H$ is a Borel subgroup of L and G is generated by H and L, then X is rationally chain connected (4.1), hence generically rationally connected, or rationally connected in characteristic zero (4.2). In the remaining applications, G is semisimple. We show, mainly by a repeated application of 4.1, 4.2, that if H is epimorphic and normalized by a maximal torus, then H has property F. Next we consider, in characteristic zero, the normalizer $\mathcal{N}(Z)$ of a one-dimensional unipotent subgroup Z and prove that X is rationally connected if $H = S \cdot U$, where U is the unipotent radical of $\mathcal{N}(Z)$, S a one-dimensional torus such that $S \cdot Z$ is the Borel subgroup of a subgroup isogeneous to $\mathbf{SL}_2$, and Z is not contained in any proper normal subgroup of G (4.5). In 4.6 we discuss similarly the three-dimensional examples of epimorphic subgroups of simple groups described in [BB a, 5(b)]. In 4.8, G is a product of $\mathbf{SL}_2$. We consider a class of solvable subgroups which contains as a special case the group attached in [BB b, 7(b)] to Nagata's counterexample to the Hilbert 14th problem [N] and show that H not only has property F, but that X is rationally connected. In 4.10, 4.11, an analogous result is proved for semisimple groups.

In Sect. 5, X is assumed to admit an embedding into a smooth projective variety, with boundary of codimension $\geqq 2$. Theorem 1 in [BB b] shows that in that case H has property F. Here it is established more strongly that X is generically rationally connected (5.1). Finally, Sect. 6 first lists some implications between various properties considered here and in [BB a, b], and then formulates some open problems and a conjecture pertaining to converse statements.

In this paper, k is an algebraically closed field. Algebraic groups over k are assumed to be reduced, unless otherwise specified. If L is an algebraic group and M a closed subgroup, then L/M denotes the quotient homogeneous space. In particular, the canonical projection $L \rightarrow L/M$ is separable. L^o is the connected component of e in L. Throughout, G is a connected algebraic k-group and H a closed k-subgroup of G.

A property is said to hold for general points in a variety Z if it holds for all points in an open Zariski-dense subset of Z.

From Sect. 4 on, G is affine.

Notation and definitions

For the ease of references and to fix some notation, we recapitulate here the main definitions given in the introduction.

0.1. An irreducible (reduced) variety X over k is *quasi-complete* if the k-algebra $k[X]$ of regular functions on X reduces to the constants.

0.2. A *rational curve* is a smooth complete curve of genus zero, i.e. a copy of $\mathbb{P}^1(k)$. A *chain of rational curves of length n* is a connected and reduced curve C with irreducible components $C_1, \ldots, C_n$ such that every component C_i is a rational curve, has two marked points x_i^- and x_i^+, and $x_i^+ = x_{i+1}^-$ for $i = 1, \ldots, n-1$. The curve C_i intersects C_{i+1} at x_i^+ if $i < n$, and C_{i-1} at x_i^- if $i = 2, \ldots, n$. There are no other intersection points.

We also write $x_1^- = 0$ and $x_n^+ = \infty$. These are smooth points of C. The canonical bundle ω_C of C is a line bundle which has degree -1 on C_1 and C_n and degree zero on C_j for $1 < j < n$.

Let X be a variety.

0.3. A *rational chain* in X is a morphism $f : C \to Y$ where C is a chain of rational curves. It connects $x, y \in X$ if $f(0) = x$ and $f(\infty) = y$.

No condition on the singularities of $f(C)$ is imposed.

0.4. (i) X is *rationally chain connected* if for any $(x, y) \in X \times X$ there is a rational chain connecting x and y.

(ii) X is *generically rationally connected* if there is a Zariski open set $U \subset X \times X$ such that for any $(x, y) \in U$, the points x, y are contained in the image of a rational curve under a morphism.

(iii) X is *rationally connected* if for any $x, y \in X$ there exists a morphism $f : \mathbb{P}^1 \to X$ whose image contains x and y.

Clearly, (iii) $\Rightarrow$ (ii) $\Rightarrow$ (i). For a study of these properties in the case of smooth projective varieties in characteristic zero, see [KMM a, b, c]. The main complication here is that the varieties of interest are not necessarily complete. However, they are homogeneous and this will allow us to extend some of the results of [KMM a, b, c].

1. Chains of rational curves

We review a special case of the theory of Hilbert schemes studied in [M] and [KMM a, b, c]. Details are worked out there, but the formulations are not given in the generality we need here. In particular, we have to extend the context from projective varieties to quasi-projective varieties.

1.1 Definition. *A* ***family of rational chains*** *on a variety X is a diagram*

$$\begin{array}{ccc} U & \xrightarrow{u} & X \\ {\scriptstyle q}\downarrow & & \\ S & & \end{array}$$

where U and S are algebraic varieties, $q : U \to S$ *is proper and flat, every fiber of q is a chain of rational curves and u is a morphism of varieties. The length of the chains is allowed to vary.*

If we are also given a section $s : S \to U$ of q such that $u \circ s(S) = x \in X$, then we say that the diagram is a *family of rational chains through x.*

1.2 Definition. *Let* $f : C \to X$ *be a rational chain in X. A* ***smoothing*** *of f is a family of rational chains*

$$\begin{array}{ccc} U & \xrightarrow{u} & X \\ {\scriptstyle q}\downarrow & & \\ S & & \end{array}$$

where (S, o) *is a smooth pointed curve such that*

(i) $q^{-1}(o) \simeq C$ *and* $u|_C \simeq f$,

(ii) $q^{-1}(s)$ *is a rational curve for* $s \in S$, $s \neq o$.

Let $f : C \to X$ be a rational chain connecting x_1 and x_2. We say that the above *smoothing fixes* x_1 (resp. x_1 and x_2) if there are sections $t_1 : S \to U$ (resp. $t_1 : S \to U$ and $t_2 : S \to U$) such that

$t_1(o) = 0 \in C \subset U$ (resp. also $t_2(o) = \infty \in C \subset U$),
$u \circ t_1(S) = x_1$ (resp. also $u \circ t_2(S) = x_2$).

1.3 *Example of a smoothing family.* Let (S, o) be a smooth pointed curve and $X_1 = \mathbb{P}^1 \times S$ with projection $\pi_1 : X_1 \to S$ and sections $t_{11} : S \simeq \{0\} \times S \hookrightarrow X_1$ and $t_{12} : S \simeq \{\infty\} \times S \hookrightarrow X_1$.

Assume that X_i, π_i and t_{i1}, t_{i2} are already defined. Let $b : X_{i+1} \to X_i$ be the blow up of X_i at $t_{i2}(o)$. Let $\pi_{i+1} = \pi_i \circ b$ be the projection. Set $t_{(i+1)1} = b^{-1} \circ t_{i1}$ and $t_{(i+1)2} = b^{-1} \circ t_{i2}$. The latter morphism is first defined only on $S - o$, then extended to S by taking the closure of $t_{(i+1)2}(S - o)$ in X_{i+1}. This extension exists. Indeed, a morphism from an open set of a smooth curve to a projective variety always extends, see e.g. [H] I, 6.8. Although X_{i+1} is not complete, it is projective over S, hence the extension property also holds. For instance, one can use the valuative criterion of properness ([H], II.4.7) with $X = X_{i+1}$, $Y = T = S$ and $U = S - o$.

$\pi_n : X_n \to S$ is flat, the central fiber is a chain C of rational curves of length n, while the general fiber is a single rational curve. We will use the two sections t_{n1}, t_{n2} in the sequel.

Let V be a vector bundle on $\mathbb{P}^1$. It is well-known that V splits as a direct sum of line bundles $V = \sum \mathcal{O}(a_i)$, see [Ga].

1.4 Definition. *A vector bundle V on $\mathbb{P}^1$ is called **semipositive** (resp. **ample** or **positive**) if $a_i \geqq 0$ (resp. $a_i \geqq 1$) for every i.*

If V is generated by global sections, then it is semipositive. If $V \otimes \mathcal{O}(-1)$ is generated by global sections, then V is ample.

It is elementary that $H^0(\mathbb{P}^1, \mathcal{O}(a)) = 0$ if and only $a < 0$. Therefore, by III, 5.1 of [H], itself a special case of Serre duality (cf. below),

$$(1) \qquad H^1(\mathbb{P}^1, V) = 0 \textit{ if } V \textit{ is semi-positive}.$$

(In fact, already if $a_i \geqq -1$ for all i's, which is what is used in 1.6).

1.5 Lemma. *Let $X = G/H$. If $f : \mathbb{P}^1 \to X$ is any morphism, then f^*T_X is semipositive.*

Proof. Indeed, the tangent bundle T_X is generated by global sections, namely the vector fields coming from the Lie algebra of G. Thus, f^*T_X is also generated by global sections, and it must be semipositive. □

If $\mathscr{E}$ is a sheaf of sections of a vector bundle on a variety Z, and D is a divisor on X, then $\mathscr{E}(D)$ is the coherent sheaf obtained by tensoring $\mathscr{E}$ with the invertible sheaf $\mathcal{O}(D)$ corresponding to the divisor D.

1.6 Proposition. *Let $C = \cup C_i$ be a chain of rational curves, and E a vector bundle on C. Suppose that $E|_{C_i}$ is semi-positive (resp. ample) for every i. Then $H^1(C, E(-0)) = 0$ (resp. $H^1(C, E(-0 - \infty)) = 0$).*

Proof. The dual bundle E^* restricted to C_i is semi-negative (resp. negative). Since C is a complete intersection, the dualizing sheaf of C coincides with its canonical bundle ω_C, see proof of [H] III.7.12. By Serre duality ([H], III.7.7), we have:

$$H^1(C, E(-0)) = (H^0(C, E^*(0) \otimes \omega_C))^*,$$

$$H^1(C, E(-0 - \infty)) = (H^0(C, E^*(0 + \infty) \otimes \omega_C))^*.$$

Now, ω_C has degree -1 on C_1 and C_n, and has degree 0 on all other components of C. Hence, $E^*(0) \otimes \omega_C$ is semi-negative on $C_1, \ldots, C_{n-1}$, and has negative degree on C_n, (resp. $E^*(0 + \infty) \otimes \omega_C$ is negative on all components C_i). On each component C_i, these bundles split as direct sums of line bundles $\sum \mathcal{O}(a_i)$ where the a_i are at most 0. Therefore, if a section of these bundles vanishes at one point, it must be identically zero. Since all a_i's are negative on C_n, there is only the zero section on C_n. Now, any two consecutive components intersect at one point. Hence, we can propagate the zeroes of the section all the way across to C_1. We obtain that $E^*(0) \otimes \omega_C$ (resp. $E^*(0 + \infty) \otimes \omega_C$) has no non-zero global section on C. □

1.7. We need a general result on the space of morphisms between two schemes. It goes back to A. Grothendieck's foundational work in algebraic geometry

[G b], in which the representability of functors similar to the one defined below is proved.

Let X/S and Y/S be locally Noetherian schemes over a Noetherian scheme S. Let $B \subset X$ be a subscheme, proper over S, and $b : B \to Y$ an S-morphism. Let

$$\mathrm{Hom}_S(X, Y; b) : (\mathrm{LNSch}/S)^o \to (\mathrm{Sets})$$

be the contravariant functor from the category of locally Noetherian S-schemes to the category of sets, defined by:

$$\mathrm{Hom}_S(X, Y; b)(T) = \{T\text{-morphisms } f : X_T \to Y_T \mid f|_{B_T} = b_T\}$$

where $T \in \mathrm{LNSch}/S$. Here, we use the notation $X_T := X \times_S T$ for any scheme X, and $b_T := b \times_S \mathrm{id}_T$.

The following theorem summarizes some properties of the functor $\mathrm{Hom}_S(X, Y; b)$ in the case of interest to us, namely under the following assumptions:

The scheme X/S is reduced, projective, flat and of relative dimension one over S and B/S is a closed subscheme of X, smooth, finite and flat over S. The scheme Y/S is smooth, quasi-projective over S and $b : B \to Y$ a S-morphism. We denote by $I_B \subset \mathcal{O}_X$ the ideal sheaf of B. We tacitly identify $\mathrm{Hom}_S(X, Y; b)$ with a subfunctor of $\mathrm{Hilb}(X \times Y)$ by considering the graph of morphisms.

1.8 Theorem. *We keep the previous assumption and notation. Let $s \in S$ a point and $[f]$ an element of* $\mathrm{Hom}_S(X, Y; b)(s)$, *corresponding to a morphism* $X_s \to Y_s$.

(i) $\mathrm{Hom}_S(X, Y; b)$ *is represented by a scheme* $\mathrm{Hom}_S(X, Y; b)$.

(ii) *The kernel of the map of Zariski tangent spaces*

$$T_{[f]} : \mathrm{Hom}_S(X, Y; b) \to T_s S$$

is isomorphic to

$$H^0(X_s, f^* T_{Y_s} \otimes I_{B_s}) .$$

(iii) *If* $H^1(X_s, f^* T_{Y_s} \otimes I_{B_s}) = 0$, *then* $\mathrm{Hom}_S(X, Y; b)(s) \to S$ *is a smooth morphism at* $[f]$.

For a proof of (i), see [G a, n° 221] or [M, Prop. 1]. (ii) and (iii) are relative versions of Prop. 3 in [M] and are proved by the same method. Some special cases are asserted in [KMM b], Cor. 1.5, but no complete proof is available in the literature at this time. However the theorem follows from II.1.7 in the forthcoming book [K1]. More precisely, 1.7.1 and 1.7.2 there, together with our assumptions, imply that the fiber at $[f_s]$ has the same dimension as its Zariski-tangent space, hence is smooth, and that the assumption of 1.7.3 holds. Smoothness then follows from 1.7.3, which asserts flatness at $[f_s]$, and III.10.2 in [H], which states that a flat morphism with smooth fibers is smooth. The proof of II.1.7 in [K1] is based on I.2.17 and I.5.14 there.

1.9 Corollary. *Let Z be a smooth variety, z_1, z_2 two closed points of Z, and $f : C \to Z$ a rational chain connecting z_1 and z_2. Suppose that $f^*T_Z|_{C_i}$ is semipositive (resp. ample) for every i, where $C = \cup C_i$. Then the chain $f : C \to Z$ admits a smoothing fixing z_1 (resp. z_1 and z_2).*

Proof. Let n be the length of the chain C and let us use the notation $\pi_n : X_n \to S$ defined in Example 1.3. Recall that the fiber of π_n over any $s \in S$, $s \neq o$ is a $\mathbb{P}^1$, and we can identify the fiber of π_n over $o \in S$ with C. Let $B = \operatorname{im} t_{n1}$ (resp. $B = \operatorname{im} t_{n1} \cup \operatorname{im} t_{n2}$) be the image of the sections of π_n. Consider the constant morphisms $b : B \to \{z_1\} \subset X$ (resp. $b : B \to \{z_1\} \cup \{z_2\} \subset X$)). In the above theorem, let X be X_n, Y be $Z \times S$, and choose $s = o$. Then $X_o \simeq C$, $Y_o \simeq Z$, and $B_o = \{0\}$ (resp. $B_o = \{0\} \cup \{\infty\}$; notation of 0.2). Apply 1.6 with $E = f^*T_Z$ and use the hypothesis to get:

$$H^1(C, f^*T_Z \otimes I_{B_o}) = 0\,.$$

Then 1.8 (iii) implies that $\operatorname{Hom}_S(X_n, Z \times S; b) \to S$ is a smooth morphism at $f : C \to Z$. Thus, we can extend f to an S-morphism $X_n \to Z \times S$ over a neighborhood of $o \in S$. For any $s \in S - \{o\}$, we obtain a morphism $\mathbb{P}^1 \to Z$. This provides the desired smoothing of $f : C \to Z$. □

1.10 Proposition. *Let X be any smooth variety, S a variety and $u : \mathbb{P}^1 \times S \to X$ a dominant separable morphism such that $u(\{0\} \times S) = x \in X$. Then $u^*T_X|_{\mathbb{P}^1 \times \{s\}}$ is ample for general $s \in S$.*

Proof. We have a natural map

$$T_{\mathbb{P}^1 \times S} \to u^*T_X$$

which is surjective at a general point, because the morphism is dominant, separable. Pick $s \in S$ general and smooth; let $u_s = u|_{\mathbb{P}^1 \times \{s\}}$, and $m = \dim S$. Its differential yields a morphism

$$p_s : \mathcal{O}_{\mathbb{P}^1}(2) \oplus \mathcal{O}_{\mathbb{P}^1}^m \simeq T_{\mathbb{P}^1 \times S}|_{(\mathbb{P}^1 \times \{s\})} \to u_s^*T_X\,.$$

Write $u_s^*T_X = \oplus_i \mathcal{O}_{\mathbb{P}^1}(a_i)$. Since p_s is surjective at a general point of $\mathbb{P}^1 \times s$, we see that $a_i \geqq 0$ for all i's and that if $a_i = 0$ for some i, then one of the summands $\mathcal{O}_{\mathbb{P}^1}$ of the left hand side maps isomorphically onto $\mathcal{O}_{\mathbb{P}^1}$ on the right hand side. By hypothesis, u contracts the zero section to a point $x \in X$, hence the restriction of p_s to $\mathcal{O}_{\mathbb{P}^1}^m$ is identically zero at $(0, s)$. Thus $a_i \geqq 1$ for every i, hence $u_s^*T_X$ is ample. □

If k has characteristic zero, then a dominant morphism is automatically separable, and the previous result applies. The above lemma fails in the inseparable case.

2. Rational curves on homogeneous spaces

2.0. Given a variety X and a closed point $x \in X$, let

$$\begin{array}{ccc} U & \xrightarrow{u} & X \\ {\scriptstyle q}\downarrow & & \\ S & & \end{array}$$

be a connected family of rational chains through x such that $\dim(\operatorname{im} u)$ is maximal. We will denote by $C(x)$ an irreducible component containing x of the closure of $\operatorname{im} u$ in X, of maximal dimension. In general, $C(x)$ need not be unique, and the dimension of $C(x)$ may vary with x. It is clearly independent of x if X is homogeneous.

Lemma 2.1. *Let $X = G/H$ and $x \in X$. Let $f : \mathbb{P}^1 \to X$ be a morphism. Then $f(\mathbb{P}^1)$ is either contained in $C(x)$ or disjoint from it.*

Proof. Suppose that $f(\mathbb{P}^1) \nsubseteq C(x)$, but $f(\mathbb{P}^1) \cap C(x) \neq \emptyset$. Pick a point $z \in f(\mathbb{P}^1) \cap C(x)$. Composing f with an automorphism of $\mathbb{P}^1$ we may assume that $f(0) = z$. Let $p_z : G \to X$ be the morphism $g \mapsto g(z)$. Then $p_z : G \to X$ is a principal fiber bundle with structure group H_z, the isotropy group of z.

Set $M = \{g \in G | g(z) \in C(x)\}$. Then $M = p_z^{-1}(C(x))$, thus $M \to C(x)$ is a fiber bundle. Therefore if $Z \subset M$ is any irreducible component then $p_z : Z \to C(x)$ is surjective. Let Z be the unique irreducible component of M containing $e \in G$.

Let $C(x)_0 \subset C(x)$ be a dense and open subset such that $C(x)_0 \subset u(U)$. Then $Z_0 = Z \cap p_z^{-1}(C(x)_0)$ is dense in Z.

Consider the action morphism $\varphi : G \times \mathbb{P}^1 \to X$ given by $(g, b) \mapsto g \cdot f(b)$. The closure $\operatorname{cl}(Z_0')$ of the image

$$Z_0' = \{z \cdot f(b) | z \in Z_0, b \in \mathbb{P}^1\}$$

of $Z_0 \times \mathbb{P}^1$ is irreducible and has dimension $> \dim C(x)$, since it contains $C(x)$ properly. We now use the notation q and s of 1.1. Given $w \in U$, let C_w be the smallest chain of rational curves in $q^{-1}(q(w))$ which connects $s(q(w))$ and w. Let D_w be the chain of rational curves obtained by adding to C_w a copy of $\mathbb{P}^1$ by identifying w and 0. Given $(z, b) \in Z \times \mathbb{P}^1$, there is a rational chain in X connecting x and $z \cdot f(b)$, image of D_w, namely $u(C_w) \cup z \cdot f(\mathbb{P}^1)$ where w is such that $u(w) = z(x)$. The closure of the union of these images contains $\operatorname{cl}(Z_0')$, hence has an irreducible component through x of dimension $> \dim C(x)$. This is a contradiction proving the lemma, once it is shown that these rational curves are part of a family of rational curves through x in the sense of 1.1. This amounts to formalize what has just been described.

By base change from the original family $S \xleftarrow{q} U \xrightarrow{u} X$ of rational chains we obtain

$$\begin{array}{ccccc} Z_0 \times U \times_S U & \xrightarrow{\pi_3} & U & \xrightarrow{u} & X \\ {\scriptstyle q_1}\downarrow & & & & \\ Z_0 \times U & & & & \end{array}$$

where π_3 is the third coordinate projection. This family has a section

$$t : Z_0 \times U \to Z_0 \times U \times_S U \quad \text{given by} \quad (z, w) \mapsto (z, w, w) .$$

Let U_2 be the scheme obtained by gluing together $Z_0 \times U \times_S U$ and $Z_0 \times U \times \mathbb{P}^1$ along the two sections $t(Z_0 \times U)$ and $Z_0 \times U \times \{0\}$. Let $q_2 : U_2 \to Z_0 \times U$ be the projection. This is a family of rational chains.

We would like to construct a morphism $u_2 : U_2 \to X$ giving us the required family of rational chains. We can attempt to do this by setting

$$u_2|Z_0 \times U \times_S U = u \circ \pi_3, \quad \text{and} \quad u_2|Z_0 \times U \times \mathbb{P}^1 = (g \circ f)(s),$$

where $g \in Z_0$ and $s \in \mathbb{P}^1$.

These definitions unfortunately do not always agree along the two sections. The above family corresponds to discontinuous chains where first we start along a chain from $q : U \to S$ and continue along a translate of $f(\mathbb{P}^1)$. To get the right family we have to restrict to those chains where these two intersect. Thus set

$$U_3 = \{(g, w) \in Z_0 \times U | (g \circ f)(0) = u(w)\} .$$

U_3 is a closed subscheme of $Z_0 \times U$. Let $q_3 : U_3 \to Z_0 \times U$ be the restriction to U_3 of $q_2 : U_2 \to Z_0 \times U$. The morphism u_2 restricts to a morphism $u_3 : U_3 \to X$ defining a family of rational chains through x which contains those constructed above, thus concluding the proof of the lemma.

2.2. We fix an ample divisor E on X. The degree d_E or simply d of a chain $f : C \to X$ (with respect to E) is the sum of the degrees of f^*E on the irreducible components of C. It is locally constant on families. We let $D_1(d)$ (resp. $D_2(d)$) be the set of pairs $(x, x') \in X \times X$ such that there is a morphism $f : \mathbb{P}^1 \to X$ (resp. a rational chain $f : C \to X$) of degree $d_E \leqq d$ connecting x and y. Clearly $D_1(d) \subset D_2(d)$, and X is generically rationally connected if and only if $D_1(d)$ contains a dense open subset of $X \times X$, for some d.

2.3 Lemma. *Let X be a variety, E an ample divisor on X and $d \in \mathbb{N}$. Let $D_1(d)$ and $D_2(d)$ be as in 2.2.*

(i) *There is a scheme S with finitely many irreducible components and a morphism*

$$\begin{array}{ccc} S \times \mathbb{P}^1 & \xrightarrow{v} & X \\ \downarrow & & \\ S & & \end{array}$$

such that $D_1(d)$ is the image of the morphism

$$\Pi : S \ni s \mapsto (v(s, 0), v(s, \infty)) \in X \times X .$$

(ii) *There is a scheme S with finitely many irreducible components and a family of rational chains*

$$\begin{array}{ccc} U & \xrightarrow{v} & Y \\ {\scriptstyle q}\downarrow & & \\ S & & \end{array}$$

with two sections $t_i : S \to U$*, such that* $D_2(d)$ *is the image of the morphism*

$$\Pi : S \ni s \mapsto (v(t_1(s)), v(t_2(s))) \in X \times X .$$

(iii) *The sets* $D_1(d)$ *and* $D_2(d)$ *are constructible.*

(iv) *If* $D_1(d)$ *is dense in* $X \times X$ *and* $x \in X$ *is a general point then there is an irreducible subvariety* $S_0 \subset S$ *such that* $v(s, 0) = x$ *for every* $s \in S_0$*, and*

$$\pi : S_0 \ni s \mapsto v(s, \infty) \in X$$

is dominant.

Proof. Let us prove first (ii). The proof of (i) is essentially a special case.

The bound $\deg f^*E \leqq d$ implies that the rational chain C has length at most d. Therefore, it is enough to show that the subsets $D_2(d;k)$ of $D_2(d)$, where we use rational chains of fixed length $k \leqq d$, are constructible.

Fix a chain C of length k. Any morphism $f : C \to X$ is determined by its graph $\Gamma(f) \subset C \times X$, which is a reduced algebraic curve. Let E_C be an ample divisor on C. Then $E' = p_C^* E_C + p_X^* E$ is ample on $C \times X$, and the intersection number of $\Gamma(f)$ and E' is at most $\deg E_C + d$, hence bounded.

Now, embed $C \times X$ into some projective space $\mathbb{P}^n$. Let $Y \subset \mathbb{P}^n$ be the closure of the image and $Z = Y \backslash (C \times X)$ the boundary. The family of reduced algebraic curves with bounded degree in Y is bounded in the Hilbert scheme of all reduced algebraic curves in Y. Being disjoint from Z is an open condition. Being the graph of a morphism is again an open condition. Therefore, we obtain an open subset S' of this Hilbert scheme which parametrizes the graphs of morphisms $f : C \to X$ whose image has degree at most d. Let $q' : U' \to S'$ be the universal family over S' and $v' : U' \to X'$ the second projection. This is nearly the family we want, except for the existence of the two sections t_i.

In order to get the sections we set

$$S = U' \times_{S'} U' \quad \text{and} \quad U = U' \times_{S'} U' \times_{S'} U' .$$

The two sections are given by

$$t_1(u_1', u_2') = (u_1', u_2', u_1') \quad \text{and} \quad t_2(u_1', u_2') = (u_1', u_2', u_2') .$$

This proves (ii). Since q is flat, there is an open set $S^0 \subset S$ such that q is smooth over S^0. Let $U^0 = q^{-1}(S^0)$. Then $q : U^0 \to S^0$ is smooth and every geometric fiber is a $\mathbb{P}^1$. Furthermore q has two sections. We claim there is an open cover $S^0 = \cup_i S_i^0 \varepsilon$ of S^0 such that $q^{-1}(S_i^0) = \mathbb{P}^1 \times S_i^0$ for every i. If so, (i) is proved once S (resp. U) is replaced by the disjoint union of the S_i^0 (resp. U_i^0). The previous claim follows from:

(∗) Let $q : U^0 \to S^0$ be a smooth proper morphism with a section $t : S^0 \to U^0$, such that every geometric fiber is a $\mathbb{P}^1$. Then there is an open cover $S^0 = \cup_i S_i^0$ of S^0 such that $q^{-1}(S_i^0) = \mathbb{P}^1 \times S_i^0$ for every i.

Proof of (∗). Consider the invertible sheaf $L = \mathcal{O}_{U^0}(t(S^0))$. It has degree one on each fiber, therefore q_*L is a vector bundle of rank two on S^0 such that $U^0 = \mathrm{Proj}_{S^0} L$. Since a vector bundle is locally trivial in the Zariski topology, (∗) follows.

Let $q : U \to S$ be the third projection composed with q' and $v : U \to X$ be the third projection composed with v'. By a result of Chevalley (see e.g. [H] II.3.19 Exercise), the image of a locally closed set is constructible. Thus (i) and (ii) imply (iii).

In order to see (iv), set $S_0' = \{s \in S | v(s, 0) = x\}$. Then $S_0' \subset S$ is a closed subscheme and $\pi' : S_0' \ni s \mapsto v(s, \infty) \in X$ is dominant. S_0' has only finitely many irreducible components, thus there is at least one irreducible component $S_0 \subset S_0'$ such that $\pi' | S_0$ is dominant. □

2.4 Lemma. *Let $X = G/H$. Then $D_2(d)$ is contained in the closure* $\mathrm{cl}(D_1(d))$ *of $D_1(d)$.*

Proof. Let $(x, y) \in D_2(d)$ and $f : C \to X$ a rational chain of degree $\leq d$ connecting x and y. The space X is homogeneous, therefore (1.5, 1.9) the chain f can be smoothed fixing x, by means of the family $S \leftarrow U \xrightarrow{u} X$ of 1.3. The proof of 1.9 shows that the degree of f^*E is preserved under smoothing. Therefore $(x, u(U)) \subset D_2(d)$ and $(x, u(U - C)) \subset D_1(d)$. Consequently (x, y) belongs to the closure of $D_1(d)$.

Thus we have $D_1(d) \subset D_2(d) \subset \mathrm{cl}(D_1(d))$. In particular, $D_1(d)$ and $D_2(d)$ have the same closure. Both are constructible; therefore, if they are dense in $X \times X$, they contain a dense *open* subset.

2.5 Theorem. *Let $X = G/H$. Then there is a closed subgroup $H' \supset H$ of G with the following properties, where $p : G/H \to G/H'$ is the canonical projection*:

(i) *If $f : \mathbb{P}^1 \to X$ is a morphism, then $f(\mathbb{P}^1)$ is contained in a fiber of p.*
(ii) *Every fiber of p is generically rationally connected* (*see* 0.4).

Note that by the theorem H'/H is irreducible.

Proof. Let $x \in X$ be a point with stabilizer subgroup H. Let $C(x)$ be as in 2.0 and $C(x)_0 \subset C(x)$ be as in the proof of 2.1. Let $g \in G$ be such that $g \cdot x \in C(x)$. We claim that $g \cdot C(x) = C(x)$. Since $C(x)_0$ is dense in $C(x)$, it suffices to show that $g \cdot C(x)_0 \subset C(x)$. Let $y \in C(x)_0$. There exists a rational chain $f : C \to X$ which connects x and y, hence $g \circ f : C \to X$ connects $g \cdot x$ and $g \cdot y$. Write the chain C as in 0.2. Using 2.1, we see by induction on i that

$g \circ f(C_i) \subset C(x)$, whence our assertion. Let H' be the stability group of $C(x)$. The previous argument shows that it is transitive on $C(x)$. Hence $C(x) = H' \cdot x$ and $C(x) = H'/H$. In particular, H'/H is irreducible, H'^o is transitive on $C(x)$ and we can also write $C(x) = H'^o/(H \cap H'^o)$.

Consider the natural projection $p : G/H \to G/H'$. Let $f : \mathbb{P}^1 \to X$ be a morphism and assume that its image is not contained in any fiber of p. Composing f with a suitable translation by $g \in G$ we obtain a morphism $g \circ f : \mathbb{P}^1 \to X$ whose image intersects $C(x)$ but is not contained in it. This is impossible by Lemma 2.1, which proves (i).

All fibers of p are isomorphic to each other, thus in order to prove (ii), we must find a length-one chain connecting two general points of $C(x)$.

There exists an integer c such that any rational chain connecting x to an element of $C(x)_0$ has degree $\leqq c$ (with respect to an ample divisor on $C(x)$, see 2.2). The space $C(x)$ is covered by the translates $g \cdot C(x)_0$, $(g \in H')$ of $C(x)_0$ and any two have a non-empty intersection. Therefore any two points of $C(x)$ are connected by a rational chain of degree $\leqq d = 3 \cdot c$. In the notation of 2.2, this means that $D_2(d) = C(x) \times C(x)$. Then by 2.4 and the end remark to 2.4, $D_1(d)$ contains a non-empty open subset of $C(x) \times C(x)$. This implies our assertion (see 2.2). □

2.6 Corollary. (i) *Every rational chain in X is contained in a fibre of p.*

(ii) *If X is rationally chain connected, it is generically rationally connected.*

(i) follows obviously from 2.5 (i). It implies that if X is chain connected, then a fibre of p is open, hence equal to X, and (ii) follows from 2.5 (ii).

Applying the above theorem to X' and repeating the procedure we obtain:

2.7 Corollary. *Let $X = G/H$. Then there is a unique sequence of quotient spaces $X_i = G/H_i$, where $H_0 = H \subset H_1 \subset \cdots \subset H_n$, such that the sequence of canonical morphisms*

$$p : X = X_0 \xrightarrow{p_0} X_1 \to \cdots \to X_{n-1} \xrightarrow{p_{n-1}} X_n = Z$$

has the following properties:

(i) *If $f : C \to X_i$ is a rational chain in X_i, then $f(C)$ is contained in a fiber of p_{i-1} $(i = 1, \ldots, n)$.*

(ii) *H_i/H_{i-1} is generically rationally connected* $(i = 1, \ldots, n)$.

(iii) *Every morphism $f : \mathbb{P}^1 \to Z$ is constant.*

Note that the H_i/H_{i-1} $(i = 1, \ldots, n)$ are irreducible, although, if H is not connected, the H_i's are not necessarily connected, and that H_n/H is irreducible, quasi-complete.

No examples are known to us where n in the corollary above needs to be bigger than one, i.e. where the target X' in Theorem 2.5 admits non-constant morphisms $f : \mathbb{P}^1 \to X'$. In the projective case, see conjecture 2.11 in [KMM b]. If X is proper and G is affine, i.e. X is a flag space for G, then X' is a point.

2.8 Proposition. *Assume k to be of characteristic zero. Let $X = G/H$ and $x \in X$. Suppose X is rationally chain connected. Then X is rationally connected.*

Proof. Looking at the last step in the proof of Theorem 2.5, we see that there is a family of length-one rational chains $\{f_s : \mathbb{P}^1 \to X | s \in S\}$ which connect x with any point of a dense open subset $X^0 \subset X$. By Lemma 1.10, $f_s^* T_X$ is ample for general $s \in S$, thus by shrinking S we may assume that ampleness holds for every $s \in S$.

Let $x_1, x_2 \in X$. We have to show they are contained in the image of a morphism $f : \mathbb{P}^1 \to X$. Let $g_i \in G$ be such that $g_i(x) = x_i$ $(i = 1, 2)$. Being Zariski-open in X, the sets $g_1(X^0)$ and $g_2(X^0)$ have a non-empty intersection. Let y be a point in it. By construction, there are morphisms $f_i : \mathbb{P}^1 \to X$ such that

$$f_1(0) = x_1, \qquad f_1(\infty) = y = f_2(0), \qquad f_2(\infty) = x_2 .$$

We can glue f_1 and f_2 to get a morphism $h : C \to X$ of a length-two chain such that $h(0) = x_1$ and $h(\infty) = x_2$. By construction $h^* T_X$ is ample on both components of the chain. By 1.9, this chain is smoothable fixing x_1 and x_2; this proves the existence of f.

From the construction, it is clear that the degree of the morphism $f : \mathbb{P}^1 \to X$ is twice the degree of the family of curves f_s, thus bounded independently of $x_1, x_2 \in X$.

2.9 Corollary. *Let* $\operatorname{char} k = 0$. *Then* 2.5(ii) *and* 2.6(ii), *remain true if "generically" is erased.*

3. Preservation of coherence

We continue with the notation of Sect. 2.

3.1 Lemma. *Let Y be a variety over k. Suppose that $D_1(d)$ (see 2.2) is dense in $Y \times Y$ for some $d > 0$. Let $\mathcal{E}$ be a locally free sheaf on Y. Then* $\dim H^0(Y, \mathcal{E}) < \infty$.

Proof. Choose a point $y \in Y$ such that $D_1(d) \cap (\{y\} \times Y)$ is Zariski-dense in $\{y\} \times Y$.

By 2.3, there exists a family of morphisms $\mathbb{P}^1 \to Y$

$$\begin{array}{ccc} U \times \mathbb{P}^1 & \xrightarrow{v} & Y \\ \downarrow & & \\ U & & \end{array}$$

such that v is dominant and $v(U \times \{0\}) = y$.

$v^* \mathcal{E}$ can be viewed as a family of vector bundles on $\mathbb{P}^1$. This family has bounded degrees as explained after (2.8) above, so there is a universal constant

$d = d(\mathscr{E}) > 0$ such that

$$H^0(\mathbb{P}^1, \mathcal{O}_{\mathbb{P}^1}(-d) \otimes v^*\mathscr{E}|_{\{u\}\times\mathbb{P}^1}) = 0 \quad \text{for every } u \in U\,.$$

This implies that any (regular) section of $\mathfrak{m}_y^d \otimes \mathscr{E}$, where $\mathfrak{m}_y \subset \mathcal{O}_Y$ is the ideal sheaf of y, is zero on $u(U)$. The sheaf $\mathfrak{m}_y^d \otimes E$ is locally free, isomorphic to E, outside y. Since $u(U)$ contains a non-empty Zariski-open subset, it follows that $H^0(Y, \mathfrak{m}_y^d \otimes E) = 0$. Then

$$\dim H^0(Y,\mathscr{E}) \leqq \dim H^0(Y,(\mathcal{O}_Y/\mathfrak{m}_y^d)\otimes\mathscr{E}) = \operatorname{rank}\mathscr{E} \cdot \dim H^0(Y,\mathcal{O}_Y/\mathfrak{m}_y^d)$$

is finite. □

Remark. It is easy to see that $d(\mathscr{E}_1 \otimes \mathscr{E}_2) \leqq d(\mathscr{E}_1) + d(\mathscr{E}_2)$. Thus we can also estimate the growth of H^0 in tensor products of vector bundles. The growth we get is the same as one would expect for projective varieties. For line bundles of degree d along the curves in the family v, the above proof shows that the dimension of the space of global sections is at most $\binom{n+d}{n}$, where $n = \dim Y$. This upper bound is achieved by $Y = \mathbb{P}^n$ with the family of lines.

3.2 Theorem. *Let H' be a closed subgroup of G containing H and $p : G/H \to G/H'$ the canonical projection. Assume that the fibers of p are generically rationally connected. Let $\mathscr{F}$ be a coherent G-sheaf on G/H. Then $p_*\mathscr{F}$ is a coherent G-sheaf on G/H'.*

Proof. Fix a point $0 \in G/H'$ and identify the fiber of p over 0 to H'/H. Let $\mathscr{F}_0 = \mathscr{F}|_{p^{-1}(0)}$, viewed as a sheaf on H'/H. We have a commutative diagram

$$\begin{array}{ccccc} H'/H & \xleftarrow{r_0} & H'/H \times G & \xrightarrow{r} & G/H \\ \downarrow{\scriptstyle p_0} & & \downarrow{\scriptstyle p_G} & & \downarrow{\scriptstyle p} \\ 0 & \xleftarrow{q_0} & G & \xrightarrow{q} & G/H', \end{array}$$

where the morphisms are defined by

$$\begin{array}{ccccc} h'H & \xleftarrow{r_0} & (h'H, g) & \xrightarrow{r} & gh'H \\ \downarrow{\scriptstyle p_0} & & \downarrow{\scriptstyle p_G} & & \downarrow{\scriptstyle p} \\ 0 & \xleftarrow{q_0} & g & \xrightarrow{q} & gH'\,. \end{array}$$

Both squares are fiber squares and $r_0^*\mathscr{F}_0 \cong r^*\mathscr{F}$. The morphisms q_0 and q are both flat. Since cohomology commutes with flat base change ([H], III.9.3) we see that

$$q_0^*((p_0)_*\mathscr{F}_0) \cong (p_G)_*(r^*\mathscr{F}) \cong q^*(p_*\mathscr{F})$$

$(p_0)_*\mathscr{F}_0$ is coherent (i.e. finite dimensional) by 3.1, hence so is $q_0^*((p_0)_*\mathscr{F}_0)$ ([H], II, 5.8). Thus $q^*(p_*\mathscr{F})$ is coherent. q is surjective hence also faithfully flat. Therefore a sheaf $\mathscr{E}$ on G/H' is coherent if and only if $q^*\mathscr{E}$ is coherent [Gc:2.5.2]. This shows that $p_*\mathscr{F}$ is coherent. □

4. Applications to epimorphic subgroups

In this section H is connected. We recall that G is affine from now on.

4.1 Proposition. *Assume that G is generated by H and a closed subgroup L isogeneous to* $\mathbf{SL}_2$, *such that* $H \cap L$ *is a Borel subgroup of L. Then there exists* $d \in \mathbb{N}$ *such that any two points of G/H are connected by a chain of rational curves of length* $\leqq d$. *In particular, H is epimorphic and G/H is quasi-complete.*

Proof. It follows from 2.2 in [B] that there exists an integer d such that every element $g \in G$ can be written as a product

$$(1) \quad g = h_1 \cdot l_1 \cdot h_2 \cdot l_2 \cdots \cdot h_m \cdot l_m \quad (h_i \in H,\ l_i \in L,\ i = 1, \ldots, m,\ m \leqq d).$$

Let x_o be the coset H in G/H and put $C_o = L \cdot x_o$. It is a rational curve in G/H. Let $\mathscr{C} = \{gC_o | g \in G\}$. We want to show that gx_o is connected to x_o by a chain of m rational curves in $\mathscr{C}$, the last of which is $gC_o\,(g \in G)$.

This is proved by induction on m. If $m = 1$, then $gx_o = h_1 \cdot l_1 \cdot x_o \in h_1 C_o = gC_o$. In (1), we can write $g = g' \cdot h_m \cdot l_m$, where, by the induction assumption, $g'x_o$ is connected to x_o by a chain of length $m-1$, the last element C'_{m-1} of which is $g' \cdot C_o$. The isotropy group of $g'x_o$ is $g' \cdot H \cdot g'^{-1}$. Let $C_m = {}^{g'}h_m \cdot C'_{m-1}$. This rational curve contains $g' \cdot x_o$, since ${}^{g'}h_m$ fixes it. Moreover

$${}^{g'}h_m C'_{m-1} = ({}^{g'}h_m) \cdot g' \cdot C_o = g' \cdot h_m \cdot C_o = g' \cdot h_m \cdot L x_o = g' \cdot h_m \cdot l_m \cdot C_o = gC_o\,.$$

This provides a chain of rational curves in $\mathscr{C}$ of length m connecting x_o with $g \cdot x_o$, and ending with gC_o, as claimed.

4.2 Corollary. *Under hypothesis* 4.1:

(i) *The space G/H is generically rationally connected.*

(ii) *If* $\mathscr{F}$ *is a coherent G-sheaf on G/H, then* $\dim H^0(G/H; \mathscr{F})$ *is finite.*

(iii) *If k has characteristic zero, then G/H is rationally connected.*

This follows from 4.1 and 2.6, 2.8, 3.2. We do not know whether 4.1 implies rational connectedness in positive characteristic.

4.3 Theorem. *Let G be reductive and H epimorphic, normalized by a maximal torus T of G. Then H has property F in G.*

The group $H' = H \cdot T$ normalizes H and has maximal rank. Therefore H has property F if and only H' does (see the remark below) and we may assume that $T \subset H$. We want to prove that if $H \neq G$ there exists a closed subgroup $H_1 \supsetneqq H$ such that H_1/H is generically rationally connected.

Let $\Phi = \Phi(T, G)$ (resp. $\Psi = \Psi(T, H)$) be the system of roots of G (resp. H) with respect to T. We let U_α be the one-parameter unipotent subgroup normalized by T on which T acts via α (see e.g. [B:13.18]). The group H is generated by T and the U_α ($\alpha \in \Psi$) and in fact its Lie algebra is the direct sum of $\mathfrak{t}$ and of the Lie algebras $\mathfrak{u}_\alpha$ of the U_α [B:13.20]. Assume $H \neq G$. Since it is epimorphic, it is not reductive and its unipotent radical is $\neq \{1\}$. There exists therefore $\alpha \in \Psi$ such that $-\alpha \notin \Psi$. Consider the subgroup L_α generated by U_α, $U_{-\alpha}$. It is the derived group of the group G_α of [B:13.18]. It is three-dimensional and its intersection with H is a Borel subgroup of L_α. Therefore the subgroup H_1 generated by L_α and H is $\neq H$ and H_1/H is generically rationally connected by 4.2. Repeating this construction we see that there exists a sequence of closed connected subgroups

$$H = H_o \subset H_1 \subset \cdots \subset H_n = G$$

such that H_i/H_{i-1} is generically rationally connected ($i = 1, \ldots, n$). In particular the space Z of 2.7 is reduced to a point and property F follows from 3.2.

Remark. Formally, that follows from Prop. 3(ii) in [BB a], but that proposition has to be amended in two ways. First the proof (not given) uses the assumption that Λ_L^G is finitely generated, which should be added. However, it holds whenever L contains a maximal torus of G (cf. [BB a]), hence is applicable to H'. Second contrary to what is done there, H is not assumed to be solvable here, so it should be pointed out that H has property F if and only a Borel subgroup of H does (see [BB a]).

4.4. We assume G reductive, k of characteristic 0 and recall some known facts about the structure of the normalizer $\mathcal{N}_G(Z)$ of a 1-dimensional unipotent subgroup Z, first in Lie algebra terms.

Let z be a generator of $\mathfrak{z} = \mathrm{Lie}(Z)$. By the Jacobson–Morosow theorem ([Bk], VIII, Sect. II, n° 2, Prop. 2), there exists a 3-dimensional simple Lie subalgebra $\mathfrak{l}$ of $\mathfrak{g}$, with basis (s, z, z_-) such that

$$(1) \qquad [s,z] = 2z, \qquad [s,z_-] = -2z_-, \qquad [z,z_-] = s\,.$$

Write $\mathfrak{g}$ as the direct sum of the centralizer $\mathfrak{z}_o$ of $\mathfrak{l}$ and of non-trivial irreducible submodules $\mathfrak{m}_1 = \mathfrak{l}, \mathfrak{m}_2, \ldots, \mathfrak{m}_c$ of $\mathfrak{g}$, viewed as a $\mathfrak{l}$-module via the adjoint representation. Fix a highest weight vector z_i of $\mathfrak{m}_i$ (with respect to $k \cdot s \oplus k \cdot z$), equal to z for $i = 1$. We claim

$$(2) \qquad \mathfrak{n}_\mathfrak{g}(z) = \mathfrak{z}_o \oplus k \cdot s \oplus \mathfrak{u}, \quad \text{and} \quad \mathfrak{z}_\mathfrak{g}(z) = \mathfrak{z}_o \oplus \mathfrak{u}, \quad \mathfrak{u} = \oplus_i k \cdot z_i$$

where $\mathfrak{z}_o$ is reductive, $\mathfrak{u}$ is the nilradical of $\mathfrak{n}_\mathfrak{g}(z)$ and $\mathfrak{h} = k \cdot s \oplus \mathfrak{u}$ is a normal solvable ideal of $\mathfrak{n}_\mathfrak{g}(z)$.

Proof. Since $\mathfrak{z}_o$ is the centralizer of a reductive subalgebra, it is reductive. It normalizes the sum of the isotypic subspaces for the non-trivial irreducible submodules for $\mathfrak{l}$. By the representation theory for $\mathfrak{sl}_2$, the Lie algebra $\mathfrak{u}$ is the centralizer of z in the sum $\mathfrak{m}$ of the $\mathfrak{m}_i$'s. The rest of the claim is clear.

Globally, denoting by the corresponding Roman capital the group with a given Lie algebra, we have

$$(3) \qquad \mathcal{N}_G(Z)^o = Z_o \cdot S \cdot U, \quad \mathcal{Z}_G(Z) = Z_o \cdot U, \qquad U = \mathcal{R}_u(\mathcal{N}_G(Z)^o),$$
$$H = S \cdot U \subset \mathcal{R}(\mathcal{N}_G(Z)^o)$$

where $\mathcal{R}$ (resp. $\mathcal{R}_u$) refers to the radical (resp. unipotent radical). It is known that any two three-dimensional simple subgroups (TDS) containing Z are conjugate by an element centralizing Z, (cf. [Kt] or [SS]). Therefore this decomposition is unique up to conjugation by elements of $\mathcal{Z}_G(Z)$.

4.5 Proposition. *Let* $\operatorname{char} k = 0$ *and* G *be semisimple. Let* Z *be a one-dimensional unipotent subgroup,* $H = S \cdot U$ *a subgroup normalizing* Z*, where* $U = \mathcal{R}_u(\mathcal{N}_G(Z)^o)$ *and* $S \cdot Z$ *is a Borel subgroup of a subgroup* L *isogeneous to* $\mathbf{SL}_2$ *(see 4.4). Assume that the projection of the Lie algebra* $\mathfrak{z}$ *of* Z *on each simple factor of* $\mathfrak{g}$ *is not zero. Then* G/H *is rationally connected.*

Proof. By 4.2, it suffices to show that $G' = \langle L, H\rangle$ is equal to G. We use the notation of 4.4. Since L acts irreducibly on $\mathfrak{m}_i$ and $\mathfrak{g}' \cap \mathfrak{m}_i \neq 0$, the Lie algebra $\mathfrak{g}'$ contains the sum of the $\mathfrak{m}_i$'s. The latter is also the sum of the isotypic subspaces in $\mathfrak{g}$ for the non-trivial irreducible $\mathfrak{l}$-submodules of $\mathfrak{g}$, hence is invariant under $\mathfrak{z}_o$. Therefore so is $\mathfrak{g}'$. But $\mathfrak{g} = \mathfrak{z}_o + \mathfrak{g}'$, therefore $\mathfrak{g}'$ is an ideal of $\mathfrak{g}$, hence a sum of simple factors. Since $\mathfrak{z} \in \mathfrak{g}'$ and $\mathfrak{z}$ projects non trivially on each simple factor, $\mathfrak{g}' = \mathfrak{g}$.

4.6. It may happen that the conclusion already holds for a suitable proper subgroup of H. An example is the three-dimensional epimorphic subgroup of an almost simple group considered in [BB a: 5(b)]:

Assume G to be almost simple and Z to contain *regular* unipotent elements. Then L is a principal TDS (see [Kt]). It is known that L is either proper maximal among connected subgroups of G or properly contained in exactly one proper connected subgroup, say R, of G. There exists $\mathfrak{m}_j$ which is outside the Lie algebra $\mathfrak{r}$ of R. Then $S \cdot Z$ normalizes the one-dimensional unipotent subgroup Z_j with Lie algebra $k \cdot z_j$ and we let $H = S \cdot Z \cdot Z_j$. Then again $G = \langle L, Z\rangle$, since $\langle L, Z\rangle$ is a subgroup which contains properly L and is different from R, in view of the fact that $Z_j \not\subset R$.

In this case, by [Kt], c is the rank of $\mathfrak{g}$, the representations $\mathfrak{m}_i$ have odd degrees, which are equal to the dimensions of the primitive elements in the cohomology of $\mathfrak{g}$. Whenever the highest degree of a primitive element for $\mathfrak{g}$ is strictly greater than the corresponding number for R, we may take for z_j a root vector corresponding to the highest root. A case where this condition is not fulfilled is that of the principal TDS of $\mathbf{SL}_{2n}$. There $R = \mathbf{Sp}_{2n}$ and the common highest degree of a primitive element is $4n - 1$.

4.7. In this section and the next one, k is of characteristic zero. Our goal is to generalize the construction of 7(b) in [BB b], which associates an epimorphic subgroup to Nagata's counterexample to the Hilbert's 14th problem, and to prove that the associated quotients are rationally connected.

We fix a positive integer D and let G be the direct product of D copies G_i of $\mathbf{SL}_2(k)$ $(i = 1,\dots,D)$. Let

$$t_i = \begin{pmatrix} 1 & 0 \\ 0 & -1 \end{pmatrix}, \qquad e_i = \begin{pmatrix} 0 & 1 \\ 0 & 0 \end{pmatrix}, \qquad f_i = \begin{pmatrix} 0 & 0 \\ 1 & 0 \end{pmatrix}$$

be the standard basis of $\mathfrak{g}_i$. Hence

$$(2) \qquad [t_i, e_i] = 2 \cdot e_i, \qquad [t_i, f_i] = -2f_i, \qquad [e_i, f_i] = t_i \qquad (i = 1,\dots,D)$$

and all other commutators between elements of $\{e_i, f_i, t_i\}$ $(i \in [1,D])$ are zero.

We let T_i, N_i, N_i^- be the corresponding one-dimensional subgroups of G_i, consisting respectively of the diagonal elements of determinant 1, the upper triangular and lower triangular unipotent matrices, and T (resp. N, resp. N^-) be the product of the T_i (resp. N_i, resp. N_i^-). The exponential mapping defines an isomorphism of the Lie algebra $\mathfrak{n}$ (resp. $\mathfrak{n}^-$) of N (resp. N^-) onto N (resp. N^-). The group G is viewed as a group defined over $\mathbb{Q}$ in the obvious way. In particular, N_i, N_i^-, N, N^- and T are defined over $\mathbb{Q}$, and the e_i, f_i, t_i form a basis of $\mathfrak{g}$ for the associated $\mathbb{Q}$-structure.

Let T_o be the one-dimensional subtorus of T which is the diagonal of T. Its Lie algebra $\mathfrak{t}_o$ is $k \cdot t_o$, where $t_o = \sum_{i \geq 1} t_i$. If $t \in T_o$, then $\operatorname{Ad} t$ operates by dilations on $\mathfrak{n}$ or $\mathfrak{n}^-$. In the sequel $\mathfrak{u}$ is a u-dimensional subspace of $\mathfrak{n}, d$ its codimension in $\mathfrak{n}$ and $U = \exp \mathfrak{n}$. The group U is normalized by T_o and we let $H = T_o \cdot U$.

4.8 Proposition. *We keep the notation and conventions of* 4.7.

(i) *If* $\mathfrak{u}$ *is not contained in any coordinate hyperplane and* H *is epimorphic, then* G/H *is rationally connected.*

(ii) *If the projection of* $\mathfrak{u}$ *on any coordinate* 2*-plane, with kernel the complementary coordinate subspace, is surjective, then* G/H *is rationally connected.*

Proof. The assumption of (ii) also implies that $\mathfrak{u}$ is not contained in any coordinate hyperplane. We first assume the latter. There exists then

$$z = \sum c_i \cdot e_i \in \mathfrak{u} \text{ with } c_i \neq 0 \text{ for all } i\text{'s}.$$

Let

$$z^- = \sum_i c_i^{-1} \cdot f_i .$$

It follows from 4.7 (2) that

$$(1) \qquad [t_o, z] = 2z, \qquad [t_o, z^-] = -2z^-, \qquad [z, z^-] = t_o$$

therefore t_o, z, z_- span the Lie algebra of a connected subgroup L isomorphic to $\mathbf{SL}_2$ and $L \cap H = S \cdot Z$ is a Borel subgroup of L, where $Z = \exp k \cdot z$.

Let $M = \langle L, H \rangle$ be the subgroup generated by L and H. We show first that M is semisimple. [The argument, simpler than our original one, was suggested by the referee.]

Let $\mathfrak{r}$ be the nilradical of $\mathfrak{m}$. We claim that $\mathfrak{r} = (0)$. It is normalized by $\mathfrak{u}$ and the Lie algebra $\mathfrak{l}$ of L. If $\mathfrak{r} \neq (0)$, there exists $x \in \mathfrak{r}$, $x \neq 0$ such that $[\mathfrak{u}, x] = 0$. The element

$$t = [x, z^-] = \sum x_i c_i^{-1} t_i$$

is semisimple and, on the other hand, belongs to $\mathfrak{r}$, hence is nilpotent. Therefore $t = 0$, and, consequently, $x = 0$, a contradiction. Thus M is reductive. Since it is generated by the unipotent subgroups U and $\exp k \cdot z^-$, it is semisimple.

We now note that the three following conditions are equivalent:

a) H is epimorphic; b) $M = G$; c) G/H is rationally connected.

Indeed, a) $\Rightarrow$ b) by [BB a, n° 3], b) $\Rightarrow$ c) by 4.2 and c) implies that G/H is quasi-complete, hence a).

This concludes the proof of (i) and shows that (ii) is implied by

(ii)′ *If the projection of* $\mathfrak{u}$ *on every coordinate 2-plane, with kernel the complementary coordinate subspace, is surjective, then* $M = G$.

Given a subset A of $[1, D]$, we let $|A| = \operatorname{Card} A$ and

$$G_A = \prod_{i \in A} G_i, \qquad N_A = \prod_{i \in A} N_i, \qquad \mathfrak{n}_A = \bigoplus_{i \in A} \mathfrak{n}_i .$$

The group M is semisimple, with a commutative maximal unipotent subgroup $N_{\mathfrak{m}}$, hence is isogeneous to a product of group of type $\mathbf{SL}_2$ and the number of simple factors is equal to the dimension $n_{\mathfrak{m}}$ of $\mathfrak{n}_{\mathfrak{m}}$. If Q is one such factor, then its image under the projection of G onto G_i is either trivial or the whole group. Therefore if Q, Q' are two such factors, they cannot both project non-trivially on the same factor G_i. There is consequently a partition of $[1, D]$ into $n_{\mathfrak{m}}$ disjoint non-empty subsets I_j $(j = 1, \ldots, n_{\mathfrak{m}})$ such that M is the product of its intersections M_j with the G_{I_j}, where M_j is isogeneous to $\mathbf{SL}_2$.

Let $\mathfrak{m}_j = \mathfrak{n}_m \cap \mathfrak{n}_{I_j}$ $(j = 1, \ldots, n_{\mathfrak{m}})$. It is one-dimensional and $\mathfrak{n}_{\mathfrak{m}}$ is the direct sum of the $\mathfrak{m}_j$'s. Therefore the projection of $\mathfrak{n}_m$ in $\mathfrak{n}_{I_j}$, with kernel the sum of the $\mathfrak{n}_{I_k} (k \neq j)$, is one-dimensional and then so is that of $\mathfrak{u}$. (Note that the projection of z on any coordinate subspace is $\neq 0$). Now $M \neq G$ if and only $\mathfrak{n} \neq \mathfrak{n}_{\mathfrak{m}}$, i.e. if and only there exists j such that $|I_j| \geqq 2$. But if $|I_j| \geqq 2$, the projection of $\mathfrak{u}$ on any coordinate 2-plane contained in $\mathfrak{n}_{I_j}$ is one-dimensional, and (ii)′ follows.

4.9 *Remark.* Let $G_u(\mathfrak{n})$ be the Grassmannian of u-dimensional subspaces of $\mathfrak{n}$. Let $\mathfrak{q}$ be a coordinate 2-plane in $\mathfrak{n}$ and $\mathfrak{q}'$ the complementary coordinate subspace. Then the projection of a u-dimensional subspace $\mathfrak{v}$ on $\mathfrak{q}$ has dimension $\leqq 1$ if and only $\mathfrak{v} \cap \mathfrak{q}'$ has dimension $\geqq u - 1$. The set of such subspaces is a proper Zariski closed subset of $G_u(\mathfrak{n})$, defined over $\mathbb{Q}$. Therefore the $\mathfrak{u}$'s satisfying the condition in 4.8(ii) form a dense Zariski-open subset of $G_u(\mathfrak{n})$. In particular, if $\mathfrak{u}$ defines a generic point of $G_u(\mathfrak{n})$ over $\mathbb{Q}$, then G/H is rationally connected.

If $D = 16$, and $d = 3$, and $\mathfrak{u}$ is "general", then U, acting on the natural representation of G in k^{32}, tensor product of the identity representations of the G_i, is Nagata's counterexample to Hilbert's 14th problem [N]. Here "general" means defined by three linear homogeneous equations, all of whose coefficients are algebraically independent over $\mathbb{Q}$. It fulfills the condition of (ii). In [BB a], n° 7(b), it was pointed out that the associated group H has property F. The proof was based on the explicit determination of the invariants of U by Nagata. 4.8 provides a stronger statement. It would also apply to the group associated similarly to the counterexample where $D = 9, d = 3$, given by R. Steinberg (1991, to be published in [S]).

4.10. We assume G to be semisimple and use some standard facts about the structure of $\mathfrak{g}$ (see e.g. [Bk]). Let $\mathfrak{t}$ be a Cartan subalgebra, $\mathfrak{b} = \mathfrak{t} \oplus \mathfrak{n}$ a Borel subalgebra containing $\mathfrak{t}, R$ the set of roots of $\mathfrak{g}$ with respect to $\mathfrak{t}$ and S the set of simple roots for the ordering defined by $\mathfrak{b}$. For $x \in R$, let e_α be a basis vector for the one-dimensional root subspace

$$\text{(1)} \qquad \mathfrak{g}_\alpha = \{x \in \mathfrak{g}, [t, x] = \alpha(t)x, \quad (t \in \mathfrak{t})\}\ .$$

We let $\mathfrak{n}_1$ be the subspace of $\mathfrak{n}$ spanned by the e_α ($\alpha \in S$). Then $\mathfrak{n} = \mathfrak{n}_1 \oplus \mathscr{D}\mathfrak{n}$. As a consequence $\mathfrak{n}_1$ generates $\mathfrak{n}$ as a subalgebra. The main diagonal of $\mathfrak{t}$ is the line $\mathfrak{t}_o$ on which all simple roots take the same value. We let T_o be the corresponding one-dimensional torus in the maximal torus T with Lie algebra $\mathfrak{t}$. Given

$$\text{(2)} \qquad z = \sum_{\beta \in S} c_\beta \cdot e_\beta \qquad (c_\beta \neq 0,\ \beta \in S)\ ,$$

there exists z^- in the subspace $\mathfrak{n}_1^-$ spanned by the $e_{-\beta}$ ($\beta \in S$) such that t_o, z, z^- span a copy of $\mathbf{sl}_2$ (see [Bk], VIII, Sect. 11, n° 4, Prop. 8). This is a principal $\mathbf{sl}_2$, in the sense of [Kt].

We may assume that the e_α and the $t_\alpha = [e_\alpha, e_{-\alpha}]$ span a $\mathbb{Q}$-form of $\mathfrak{g}$, cf. [Bk]. This is the rational structure used in the sequel.

4.11 Proposition. *We keep the notation and assumptions of* 4.10. *Let U be any closed subgroup of N normalized by T_o and $H = T_o \cdot U$. Assume that the projection of the Lie algebra $\mathfrak{u}$ of U in $\mathfrak{n}_1$ is not contained in any rational hyperplane and contains an element z as in* 4.10(2) *with all $c_i \in \mathbb{Q}^*$. Then G/H is rationally connected.*

Proof. We let the *degree* $d°\alpha$ of a root α be the sum of the coefficients of α, expressed as a linear combination of simple roots. Let $\mathfrak{n}_j$ be the subspace of $\mathfrak{n}$ spanned by the roots of degree j. Then the above $\mathfrak{n}_1$ is indeed the subspace spanned by the degree-one roots and $\mathscr{D}\mathfrak{n} = \bigoplus_{j \geqq 2} \mathfrak{n}_j$. Fix $t_o \in \mathfrak{t}_o$ on which all simple roots are equal to one. Then $[t_o, e_\alpha] = d°\alpha \cdot e_\alpha$. Since $\operatorname{ad} t_o$ leaves $\mathfrak{u}$ invariant, the latter is a direct sum of its intersections with the $\mathfrak{n}_j$. In particular, if π is the projection of $\mathfrak{u}$ into $\mathfrak{n}_1$, then $\pi(\mathfrak{u}) = \mathfrak{u} \cap \mathfrak{n}_1$.

The unique $z^- = \sum_\beta d_\beta e_{-\beta}$ which forms a $\mathbf{sl}_2$-triple with t_o and z has also rational coefficients and the d_β's are all $\neq 0$ (cf. [Bk], loc. cit.) We let L be the subgroup isogeneous to $\mathbf{SL}_2$ with Lie algebra spanned by t_o, z, z^-. Then $L \cap H$ is a Borel subgroup of L. By 4.2, it suffices to prove that $G = \langle L, H \rangle$. Let $M = \langle L, H \rangle$ and $\mathfrak{m}$ its Lie algebra.

We claim now that $\mathfrak{t} \cap \mathfrak{m}$ contains an element s on which the simple roots take *distinct* non-zero values. Let $x = \sum x_\beta \cdot e_\beta$ in $\mathfrak{u} \cap \mathfrak{n}_1$. Since the difference of two distinct simple roots is not a root, we have

$$[x, z^-] = \sum_{\beta \in S} x_\beta \cdot d_\beta \cdot t_\beta .$$

For $\alpha \in S$, $\alpha(t_\beta) = n_{\alpha,\beta}$ is an integer. We have

$$\alpha([x, z^-]) = \sum_{\beta \in S} x_\beta \cdot d_\beta \cdot n_{\alpha,\beta} .$$

The condition $\sum_\beta x_\beta \cdot d_\beta \cdot n_{\alpha,\beta} = 0$ defines a rational hyperplane in $\mathfrak{n}_1$. Moreover, since the Cartan matrix is invertible, the relation

$$(1) \qquad \sum_\beta x_\beta (d_\beta n_{\alpha\beta} - d_\beta n_{\gamma\beta}) = 0 \quad (\alpha, \gamma \in S, \alpha \neq \gamma)$$

also defines a rational hyperplane. By assumption $\mathfrak{u} \cap \mathfrak{n}_1$, i.e. $\pi(\mathfrak{u})$, is not contained in any. We can therefore find x such that the $\alpha([x, z^-])$ are non-zero and distinct for $\alpha \in S$. We let then $s = [x, z^{-1}]$. We have

$$(\operatorname{ad} s)^i(z) = \sum_{\alpha \in S} \alpha(s)^i \cdot c_\alpha \cdot e_\alpha \cdot \quad (i = 1, 2 \ldots)$$

The determinant of the elements $(\operatorname{ad} s)^i(z)$ $(1 \leqq i \leqq \dim \mathfrak{n}_1)$ with respect to the basis $\{c_\beta \cdot e_\beta\}$ of $\mathfrak{n}_1$ is a Vandermonde determinant, hence $\neq 0$ in view of our assumption on s, therefore these elements span $\mathfrak{n}_1$. Similarly the $(\operatorname{ad} s)^i(z^-)$ span $\mathfrak{n}_1^-$. Consequently, $\mathfrak{m}$ contains e_α and $e_{-\alpha}$ $(\alpha \in S)$. But those generate $\mathfrak{g}$, therefore $M = G$. □

Remark. We have assumed that $\pi(\mathfrak{u})$ is not contained in any rational hyperplane of $\mathfrak{n}_1$. However, the proof shows that it suffices to require that $\pi(\mathfrak{u})$ be not contained in any of a finite set of rational hyperplanes, namely, the coordinate hyperplanes and the hyperplanes defined by the equations (1) above.

5. Homogeneous spaces with small compactifications

Let X be an irreducible variety and $X \hookrightarrow Y$ an embedding of X into a projective variety. We say that X has small boundary in Y if X is open and $Z := Y - X$ has codimension $\geqq 2$.

5.1 Theorem. *Suppose that G/H has a small boundary in a smooth projective variety Y. Then G/H is generically rationally connected.*

Proof. Any linear algebraic group is rational; fix a birational map $\rho : \mathbb{P}^n \dashrightarrow G$. The quotient map $G \to G/H \subset Y$ is smooth. Therefore the composition $\phi : \mathbb{P}^n \dashrightarrow G \to Y$ is a dominant rational map which is generically smooth, i.e.

separable. Lines in $\mathbb{P}^n$ connect any two points, thus any two general points of Y are connected by the image of a line in $\mathbb{P}^n$.

Let $p \in \mathbb{P}^n$ be a general point and let

$$\mathbb{P}^{n-1} \overset{q}{\leftarrow} V \overset{v}{\rightarrow} \mathbb{P}^n$$

be the family of lines through p. The map v is birational, thus $\phi \circ v$ is separable. There is an open and dense subset $D \subset \mathbb{P}^{n-1}$ such that $q^{-1}(D) \cong D \times \mathbb{P}^1$ and $\phi \circ v$ is a morphism on $q^{-1}(D)$. By 1.10 we conclude that if $f : \mathbb{P}^1 \to Y$ corresponds to a general line in $\mathbb{P}^n$ through p then $f^* T_Y$ is ample.

We have to show that we can move these images of rational curves away from the boundary $Z := Y - G/H$. We shall use 1.8, in the case where S there is $\operatorname{Spec} k$, hence where S and $T_s S$ are reduced to a point.

Let $U_1 \subset \operatorname{Hom}(\mathbb{P}^1, Y)$ be an irreducible component containing the images of the general lines constructed above. In general, those only give a small subset of U_1. By replacing U_1 with a suitable Zariski open subset $U \subset U_1$, we may assume that $f^* T_Y$ is ample for every $[f] \in U$. In particular, $H^1(\mathbb{P}^1, f^* T_Y) = 0$ for every $[f] \in U$ (see 1.4(1)). Thus, by 1.8

$$\dim U = \dim H^0(\mathbb{P}^1, f^* T_Y) = \deg f^* T_Y + \dim Y \,. \tag{1}$$

Let $U_Z \subset U$ be the set of those morphisms whose image intersects Z. It is Zariski closed. Assume that $U_Z \neq U$. Then the set of $f \in U$ with image in G/H is Zariski open and dense in U, therefore the set of $q \in G/H$ connected to p by the image of a rational curve in G/H contains a dense open set, and the theorem is proved. There remains then to show that $U_Z \neq U$. For $z \in Z$ and $a \in \mathbb{P}^1$, let $b_{a,z} : \{a\} \to \{z\}$ be the constant morphism. Since $f^* T_Y$ is ample, $f^* T_Y(-a)$ is semi-positive, thus 1.6 (applied to $C = \mathbb{P}^1$) implies $H^1(\mathbb{P}^1, f^* T_Y(-a)) = 0$, and 1.8 yields

$$\dim(\operatorname{Hom}(\mathbb{P}^1, Y, b_{a,z}) \cap U) = \dim H^0(\mathbb{P}^1, f^* T_Y(-a)) = \deg f^* T_Y \,. \tag{2}$$

By construction

$$U_Z = \bigcap_{a \in \mathbb{P}^1, z \in Z} (\operatorname{Hom}(\mathbb{P}^1, Y, b_{a,z}) \cap U)$$

hence, taking (1), (2) and the assumption $\dim Z \leqq \dim Y - 2$ into account, we get

$$\dim U_Z \leqq \dim \mathbb{P}^1 + \dim Z + \deg f^* T_Y \leqq \dim Y - 1 + \deg f^* T_Y = \dim U - 1 \,.$$

6. Open problems

6.1. Let $X = G/H$. We have encountered the following conditions

(i) H is epimorphic.

(ii) H has property F.

(iii) there is an increasing sequence of closed subgroups $H = H_o \subset H_1 \subset H \subset H_n = G$ such that H_{i+1}/H_i is generically rationally connected ($i = 0,\ldots,n-1$).

(iv) X is generically rationally connected.

(v) X admits an open embedding into a smooth projective variety Z such that codim $(Z - X) \geqq 2$.

We have

$$(1) \qquad (v) \Rightarrow (iv) \Rightarrow (iii) \Rightarrow (ii) \Rightarrow (i).$$

Indeed, (v) $\Rightarrow$ (iv) by 5.1, (iv) $\Rightarrow$ (iii) is obvious, (iii) $\Rightarrow$ (ii) by 3.2 and (ii) $\Rightarrow$ (i) is elementary (see [BB a, n° 8]).

Little is known about implications in the other direction. It is conjectured in [BB a] that (i) $\Rightarrow$ (ii). This has been established in the present paper in a number of cases, but always by proving that a stronger condition holds, namely (iv) or (iii) for a subgroup normalizing H.

Taking into account the remark at the end of 2.7, we would also like to propose the

Conjecture 6.2. *Assume that H has finite index in its normalizer and is epimorphic. Then X is generically rationally connected.*

6.3. We mention another condition considered in [BB b]:

(vi) H is epimorphic and $k[G]^U$ is finitely generated, where U is the unipotent radical of H.

By Theorems 1, 2 of [BB b], it implies the weakening (v)′ of (v) where "smooth" is replaced by "normal".

Condition (vi) is of interest in connection with the Pommerening–Popov conjecture (loc. cit. 9(iv)), which says that if U is unipotent, normalized by a maximal torus of G (and G is reductive), then $k[G]^U$ is finitely generated.

Assume here that Theorem 5.1 holds with smooth replaced by normal. If the Pommerening–Popov conjecture holds, then for H epimorphic, normalized by a maximal torus, G/H would be generically rationally connected. In view of the results of Grosshans and Lin Tan quoted in [BB b] and of Grosshans in [G], this would show that G/H is generically rationally connected if U is the unipotent radical of a parabolic subgroup, or if G is simple of rank 2. (Only G_2 is mentioned explicitly in these papers, but Theorem 7 in [G] takes care of A_2 and B_2 in the epimorphic case.)

M. Brion has communicated to us examples where G is simple of rank $\geqq 2$, H is epimorphic, normalized by a maximal torus, and G/H admits a normal projective embedding with small boundary, but no smooth one. In particular (i) $\not\Rightarrow$ (v).

6.4. Let G/S be a group scheme over a scheme S and H/S a subgroup scheme such that G/H is flat (and smooth) over S. Assume that there is a closed point $s \in S$ such that the fiber of G/H over s is rationally connected. Then the same holds for fibers over an open set of S. This is even true in the mixed characteristic case, [KMM]. The above conjecture would imply that the same holds for epimorphic subgroups of finite index in their normalizers.

Finally, we raise the following question, which has in fact motivated this paper. A positive answer would yield the implication (i) ⇒ (iii).

6.5. Are the following assertions true?

(i) If every morphism $f : \mathbb{P}^1 \to G/H$ is constant, then G/H is quasi-affine.

(ii) If G/H is quasi-complete (and $G \neq H$), then there exists a non-constant morphism $f : \mathbb{P}^1 \to G/H$.

In fact, (i) and (ii) are equivalent. We sketch the proof of this equivalence. First we reduce it to the case where H is connected. The subgroup H is observable in G if and only if H^o is so (see [BHM], Theorem 7 and n° 5), therefore G/H is quasi-affine if and only G/H^o is. On the other hand H is epimorphic in G if and only H^o is [BB a, n° 2(c)], hence G/H is quasi-complete if and only G/H^o is so. Finally, since $\mathbb{P}^1$ is simply connected, there exists a non-constant morphism of $\mathbb{P}^1$ into G/H if and only if there is one in G/H^o.

Assume now H to be connected. Let L be its observable hull ([BHM], [BB a]). It is the biggest closed subgroup of G in which H is epimorphic [BB a, Prop. 1]. Consider now the construction underlying 2.7 and let us write K for H_n. Since K/H is quasi-complete, H is epimorphic in K, hence $K \subset L$. Now we claim that $K = L$ is equivalent to each of (i) and (ii). In fact, K is the smallest subgroup of G containing H such that every morphism of $\mathbb{P}^1$ in G/K is constant, and G/K is quasi-affine if and only if $K = L$. Therefore (i) is equivalent to $K = L$. On the other hand, K is the biggest subgroup containing H such that the image of any non-constant morphism of $\mathbb{P}^1$ into G/H and intersecting K/H is contained in K/H and L the biggest subgroup containing H such that L/H is quasi-complete. Hence (ii) is also equivalent to $K = L$.

References

[B] A. Borel: Linear algebraic groups, 2nd enlarged edition. GTM **126**, Springer, Berlin Heidelberg New York 1991

[Bk] N. Bourbaki: Groupes et algèbres de Lie, Chap. 7.8. Hermann, Paris, 1975

[BB a] F. Bien, A. Borel: Sous-groupes épimorphiques de groupes linéaires algèbriques I. C. R. Acad. Sci. Paris Ser. I, **315** (1992) 649–653

[BB b] F. Bien, A. Borel: Sous-groupes épimorphiques de groupes linéaires algèbriques II. C. R. Acad. Sci. Paris Ser. I, **315** (1992) 1341–1346

[BHM] A. Byalinicki-Birula, G. Hochschild, G.D. Mostow: Extensions of representations of algebraic linear groups. Am. J. Math. **85** (1963) 131–144

[G] F.D. Grosshans: Hilbert's fourteenth problem for non-reductive groups. Math. Z. **193** (1986) 95–103

[G a] A. Grothendieck: Sur la classification des fibrés holomophes sur la sphère de Riemann. Am. J. Math. **79** (1956) 121–138

[G b] A. Grothendieck: Techniques de construction et théorèmes d'existence en géométrie algébrique IV: Les schémas de Hilbert. Sém. Bourbaki 1961–62 Exp. **221**, Benjamin, NY 1966

[G c] A. Grothendieck: Eléments de Géométrie algèbrique IV, (seconde partie). Publ. Math. I.H.E.S. **24** (1965)

[H] R. Hartshorne: Algebraic Geometry, GTM **52**, Springer, Berlin Heidelberg New York, 1977

[K] J. Kollár: Flips, Flops, Minimal Models, etc. Surv. Differ. Geom. **1** (1991) 113–199

[K1] J. Kollár: Rational curves on algebraic varieties. Erg. Math. Grenzgeb. Springer, Berlin Heidelberg New York (to appear)

[Kt] B. Kostant: The principal three-dimensional subgroup and the Betti numbers of a complex simple Lie group. Am. J. Math. **81** (1959) 973–1032

[KMM a] J. Kollár, Y. Miyaoka, S. Mori: Rational Curves on Fano Varieties. Proc. Alg. Geom. Conf. Trento, LNM **1515** (1992) pp. 100–105, Springer, Berlin Heidelberg New York

[KMM b] J. Kollár, Y. Miyaoka, S. Mori: Rationally Connected Varieties. J. Alg. Geom. **1** (1992) 429–448

[KMM c] J. Kollár, Y. Miyaoka, S. Mori: Rational Connectedness and Boundedness of Fano Manifolds. J. Differ. Geom. **36** (1992) 765–769

[MM] Y. Miyaoka, S. Mori: A Numerical Criterion for Uniruledness. Ann. Math. **124** (1986) 65–69

[M] S. Mori: Projective Manifolds with Ample Tangent Bundles. Ann. Math. **110** (1979) 593–606

[N] M. Nagata: On the fourteenth Hilbert problem. Proc. I.C.M. 1958, Cambridge University Press 1960, pp. 459–462

[SS] T.A. Springer, R. Steinberg: Conjugacy classes, Seminar on algebraic groups and related finite groups. LNM **131**, pp. 167–266, Springer, Berlin Heidelberg New York

[S] R. Steinberg: Nagata's example. LNM Australian Math. Soc., Cambridge University Press (1995)

160.

(with J-P. Labesse and J. Schwermer)

On the Cuspidal Cohomology of S-Arithmetic Subgroups of Reductive Groups over Number Fields

Comp. Math. **102**, 1996, 1–40

The main goal of this paper is to prove the existence of cuspidal automorphic representations for some series of examples of S-arithmetic subgroups of reductive groups over number fields which give rise to non-vanishing cuspidal cohomology classes. In order to detect these cuspidal automorphic representations we combine two techniques, both of which can be seen as special cases of Langlands functoriality. Prior to that, we have to extend to the S-arithmetic case the definition of cuspidal cohomology and to show it is a direct summand in the cohomology. We emphasize that in our framework the class of S-arithmetic groups contains the class of arithmetic groups.

Before describing the contents of the paper we recall some facts about the cohomology of S-arithmetic groups. Let G be a semisimple group over a number field k. As usual S is a finite set of places of k containing all the archimedean ones. Let (ϕ, E) be a finite dimensional complex representation of G_S trivial on G_{S_f} and let Γ be an S-arithmetic subgroup of $G(k)$. The cohomology groups $H^{\cdot}(\Gamma; E)$ of Γ with values in E can be computed using the differential cohomology of G_S in the space of smooth functions on $\Gamma\backslash G_S$:

$$H^{\cdot}(\Gamma; E) = H^{\cdot}_d(G_S; C^\infty(\Gamma\backslash G_S) \otimes E).$$

The space of cusp forms plays a central role; accordingly it is natural to investigate the corresponding space of cohomology called the cuspidal cohomology:

$$H^{\cdot}_{\text{cusp}}(\Gamma; E) := H^{\cdot}_d(G_S; \mathbf{L}^2_{\text{cusp}}(\Gamma\backslash G_S)^\infty \otimes E)$$

where

$$\mathbf{L}^2_{\text{cusp}}(\Gamma\backslash G_S)^\infty$$

is the space of smooth vectors, in the space of square integrable functions, generated by cusp forms. It is a subspace of the cohomology with respect to the discrete spectrum

$$H^{\cdot}_{\text{disc}}(\Gamma; E) := H^{\cdot}_d(G_S; \mathbf{L}^2_{\text{disc}}(\Gamma\backslash G_S)^\infty \otimes E)$$

which is in turn a subspace of the $\mathbf{L}^2$-cohomology

$$H^{\cdot}_{(2)}(\Gamma; E) := H^{\cdot}_{d}(G_S; \mathbf{L}^2(\Gamma\backslash G_S)^{\infty} \otimes E).$$

The natural inclusions

$$\begin{aligned}\mathbf{L}^2_{\text{cusp}}(\Gamma\backslash G_S)^{\infty} &\to \mathbf{L}^2_{\text{disc}}(\Gamma\backslash G_S)^{\infty} \otimes E \\ &\to \mathbf{L}^2(\Gamma\backslash G_S)^{\infty} \otimes E \to C^{\infty}(\Gamma\backslash G_S) \otimes E \end{aligned} \tag{1}$$

yield therefore natural homomorphisms

$$H^{\cdot}_{\text{cusp}}(\Gamma; E) \xrightarrow{\mu} H^{\cdot}_{\text{disc}}(\Gamma; E) \xrightarrow{\nu} H^{\cdot}_{(2)}(\Gamma; E) \xrightarrow{\sigma} H^{\cdot}(\Gamma; E). \tag{2}$$

When the discrete group Γ is cocompact, all the maps in (1) are isomorphisms, hence so are those in (2). A first goal of this paper is to provide some information on (2) in the isotropic case. In particular we show that $\sigma \circ \nu \circ \mu$ is injective. A second goal is to construct, in some cases, non-trivial cohomology classes in $H^{\cdot}_{\text{cusp}}(\Gamma; E)$.

From now on, we assume for convenience in this introduction that G is almost absolutely simple over k. We are only concerned with the isotropic case and assume that $rk_k(G) > 0$, so that, in particular, $G(k_v)$ is not compact for any $v \in S$.

In Part I we prove a decomposition theorem for functions of uniform moderate growth on $\Gamma\backslash G_S$ (see Section 2 for the statement), originally established by R. Langlands in the arithmetic case, i.e. when $S = S_{\infty}$. His argument had so far only been sketched in a letter [Lan2] and elaborated upon in an unpublished preprint of the first named author. We take this opportunity to present a complete proof for any S. In the case $S = S_{\infty}$, another one is contained in [Ca2] (see 4.6).

In Part II we define and study the cuspidal cohomology of Γ. We generalize to the S-arithmetic situation the regularization theorem of [B5]. From this and the decomposition theorem it follows that the cohomology space $H^{\cdot}(\Gamma; E)$ is canonically a direct sum

$$H^{\cdot}(\Gamma; E) = \bigoplus_{\mathcal{P} \in \mathcal{A}} H^{\cdot}_{\mathcal{P}}(\Gamma; E) \tag{3}$$

where $\mathcal{P}$ runs through the set $\mathcal{A}$ of classes of associate parabolic k-subgroups of G (5.4). Of main interest to us is the summand indexed by G, which we shall also denote by $H^{\cdot}_{\text{cusp}}(\Gamma; E)$ and call the cuspidal cohomology, since it can be identified to the space so denoted above. This shows in particular that the homomorphism

$$\sigma \circ \nu \circ \mu : H^{\cdot}_{\text{cusp}}(\Gamma; E) \to H^{\cdot}(\Gamma; E)$$

(see (2)) is injective, a fact already established in the arithmetic case in [B4], by a different argument.

Let r_f be the sum of the k_v-ranks of G/k_v for $v \in S_f$. For $v \in S_f$ the Steinberg representation is the only irreducible unitary representation of $G(k_v)$ with non trivial cohomology, besides the trivial representation. Then arguments similar to those of [BW:XIII] show that for all $i \in \mathbb{Z}$

$$H^{i+r_f}_{\text{cusp}}(\Gamma; E) = \bigoplus_{\pi} H^i_d(G_\infty; I^\infty_\pi \otimes E), \tag{4}$$

where $I_\pi \otimes H_{\pi_{S_f}}$ is the isotypic subspace of $(\pi, H_\pi) = (\pi_\infty \otimes \pi_{S_f}, H_{\pi_\infty} \otimes H_{\pi_{S_f}})$ in $\mathbf{L}^2_{\text{cusp}}(\Gamma\backslash G_S)$ and the sum runs over the set of equivalence classes of irreducible unitary representations $\pi = \pi_\infty \otimes \pi_{S_f}$ for which π_{S_f} is the Steinberg representation of G_{S_f} (6.5).

Assume that S_f is non-empty. A straightforward adaptation of an argument of N. Wallach [W] shows that

$$H^{\cdot}_{\text{disc}}(\Gamma; E) = H^{\cdot}_{\text{cusp}}(\Gamma; E) \oplus H^{\cdot}_d(G_\infty; E). \tag{5}$$

In Section 7 it is shown that (5) gives the whole $\mathbf{L}^2$-cohomology i.e. that ν in (2) is an isomorphism. For congruence subgroups of simply connected groups similar results are contained in [BFG]; moreover it is shown there, by use of the main result of [F], that (5) is equal in that case to the full cohomology of Γ with coefficients in E, i.e. that σ in (2) is an isomorphism under those assumptions. We shall not need this fact.

In Part III, we return to the general case where S_f may be empty, and we produce in Sections 10 and 11 examples of S-arithmetic groups Γ containing a subgroup Γ_1 of finite index for which

$$H^{\cdot}_{\text{cusp}}(\Gamma_1; E) \neq 0. \tag{6}$$

Our basic tool in Section 10 will be the twisted trace formula, the twist being given by a rational automorphism α of G. Some technical preliminaries are carried out in Sections 8 and 9: in Section 8 we construct and study, in the twisted case, analogues of Euler-Poincaré functions due to Clozel–Delorme for real Lie groups and to Kottwitz for $\mathfrak{p}$-adic groups; the twisted analogues will be called Lefschetz functions. In Section 9 we establish in the twisted case a simple form of the trace formula; this is a variant of a theorem due to J. Arthur.

It suffices to find a Γ' commensurable with Γ with property (6). Such a Γ' may be taken to be an arbitrary small congruence subgroup; then we may draw on our understanding of automorphic representations in the adelic setting. The relation (4) above shows that proving (6) amounts to finding cuspidal automorphic adelic representations (π, H_π) such that

$$H^{\cdot}_d(G_S; H^\infty_{\pi_S} \otimes E) \neq 0. \tag{7}$$

Observe that in this new setting, as far as existence (or non vanishing) assertions are concerned, we are free to enlarge S whenever it is convenient, even if one is primarily interested in arithmetic groups. Beyond the classical case of groups G whose archimedean component G_∞ has discrete series, very little is known about such automorphic representations (see [Sch] for a discussion of the state of art in 1989). In order to detect some in some cases, we combine two techniques, both of which can be seen as special cases of Langlands functoriality.

The first one, used in Section 10, is based on a very crude and preliminary form of what should be the stabilization of the twisted trace formula. The non-vanishing of the cuspidal cohomology for a small enough Γ' (or equivalently the existence of cuspidal automorphic representations such that (7) holds) is clear whenever one can prove the non-vanishing of the cuspidal Lefschetz number:

$$\mathrm{Lef}_{\mathrm{cusp}}(\alpha, h, \Gamma'; E) := \sum(-1)^i \operatorname{trace}(\alpha \times h \mid H^i_{\mathrm{cusp}}(\Gamma'; E)), \tag{9}$$

for some suitably chosen Hecke correspondence h for some small enough Γ'. This turns out to be the case if (and only if) the automorphism α of G is such that the S-local Lefschetz number:

$$\mathrm{Lef}(\alpha, G_S; H_{\pi_S} \otimes E) = \sum(-1)^i \operatorname{trace}(\alpha \mid H^i_d(G_S; H^\infty_{\pi_S} \otimes E)) \tag{8}$$

does not vanish identically for representations of G_S (10.4). This is the case if, at archimedean places, α is a 'Cartan-type automorphism' (10.5). Our technique is a variant of the one used by Rohlfs and Speh to exhibit cohomology in some cases [RS1] [RS2]; but we have to refine their argument to get cuspidal cohomology. The twisted trace formula allows us to compute the Lefschetz number in the discrete spectrum. As one may guess from (5) the non-vanishing of the cuspidal Lefschetz number is easier to prove if S_f is non empty; as already observed this particular case is enough for our needs. Particular cases of our result appear in [RS3]. The existence of Cartan-type rational automorphism of G is easily seen when G is split over a totally real field k; also, if G' is defined over a CM-field k' i.e. a quadratic totally imaginary extension k' of a totally real field k, the complex conjugation induces a Cartan-type automorphism of the group $G = \mathrm{Res}_{k'/k} G'$ (this was observed by Clozel) (10.6).

Conjecturally, cuspidal representations with non-vanishing S-local Lefschetz numbers, which contribute non trivially to the twisted trace formula, should be liftings from cuspidal representations, of some twisted endoscopic group, with non-vanishing S-local Euler-Poincaré numbers. Cartan-type automorphisms do not exist in general and one may try to use other Langlands functorialities.

In Section 11 we shall use cyclic base change for $\mathrm{GL}(n)$. Representations that are 'Steinberg' at some finite place are well behaved with respect to this lifting: they remain cuspidal after base change. Base change preserves the non-vanishing of cohomology; combined with (10.6) this allows us to show that $\Gamma = \mathrm{SL}(n, \mathcal{O}_k)$, where $\mathcal{O}_k$ is the ring of integers of k, and k is obtained by a tower of cyclic extensions from a totally real number field k_0, has subgroups of finite index with non-vanishing cuspidal cohomology. This generalizes to all n's a result proved in [LS] for $n = 2$ or 3.

I. A DECOMPOSITION THEOREM FOR FUNCTIONS OF UNIFORM MODERATE GROWTH ON $\Gamma\backslash G_S$

0. Assumptions and notation

0.1. The following notation and assumptions are to be used throughout the paper:

- k is a number field, $\mathcal{V}$ the set of places of k and $\mathcal{V}_\infty$ (resp. $\mathcal{V}_f$) the set of infinite (resp. finite) places of k. We denote by $|\cdot|_v$ the normalized absolute value on the completion k_v of k at v. For $v \in \mathcal{V}_f$, $\mathfrak{O}_v$ is the ring of integers of k_v. For any finite set of places Σ, we let k_Σ be the direct sum of the k_v for $v \in \Sigma$; for $x \in k_\Sigma$ let
$$|x|_\Sigma = \prod_{v\in\Sigma} |x_v|_v.$$
For any connected component L of a k-group, we let
$$L_\Sigma = L(k_\Sigma) = \prod_{v\in\Sigma} L(k_v), \quad L_\infty = \prod_{v\in\mathcal{V}_\infty} L(k_v).$$
- S will denote a finite set of places of k, containing $\mathcal{V}_\infty$. Thus $S = S_\infty \cup S_f$, where $S_\infty = \mathcal{V}_\infty$ and $S_f = S \cap \mathcal{V}_f$.
- G is a connected reductive k-group.
- $\mathcal{U}(\mathfrak{g})$ is the universal enveloping algebra of the Lie algebra of left-invariant vector fields on G_∞.

0.2. We fix a maximal compact subgroup K_v of G_v, assumed to be 'good' for $v \notin \mathcal{V}_\infty$. Let $K_\infty = \prod_{v\in\mathcal{V}_\infty} K_v$, $K_f = \prod_{v\in S_f} K_v$ and $K = K_\infty K_f$.

0.3. We shall use a height $\|\ \|$ on G_S defined by means of a faithful finite dimensional representation of G over k. The height is a product of local heights. For each archimedean place it is a Hilbert-Schmidt norm on endomorphisms; for finite places it is a sup norm on endomorphisms.

0.4. In the sequel Γ denotes an S-arithmetic subgroup of $G(k)$. It is viewed as a discrete subgroup of G_S via the diagonal embedding. The group G is the almost direct product of a central torus Z^0 and of its derived group $\mathcal{D}G$, which is semisimple. Γ is commensurable with the product of the intersections $\Gamma \cap Z^0(k)$ and $\Gamma \cap \mathcal{D}G(k)$, which are S-arithmetic in $Z^0(k)$ and $\mathcal{D}G(k)$ respectively.

0.5. If f, g are strictly positive functions on a set X, we write $f \prec g$ if there exists a constant $c > 0$ such that $f(x) \leqslant cg(x)$ for all $x \in X$, and then say that f is essentially bounded by g.

1. The vector space $\mathfrak{a}_P$ and the function $\hat{r}_P$

1.1. Let P be a connected linear k-group, N its unipotent radical, M a Levi k-subgroup. Denote by $X(P)_k$ the group of k-rational characters of P, and let

$$\mathfrak{a}_P = \mathrm{Hom}_{\mathbb{Z}}(X(P)_k, \mathbb{R});$$

we shall also use its dual

$$\mathfrak{a}_P^* = X(P)_k \otimes_{\mathbb{Z}} \mathbb{R}.$$

We shall denote by

$$H_P : P_S \to \mathfrak{a}_P$$

the map defined as follows: for any $\lambda = \chi \otimes r \in X(P)_k \otimes_{\mathbb{Z}} \mathbb{R}$

$$\mathrm{e}^{\langle \lambda, H_P(x) \rangle} = |\chi(x)|_S^r.$$

The kernel of H_P will be denoted P_S^1; it contains the unipotent radical and all compact or S-arithmetic subgroups of P. To M is associated canonically a section of the homomorphism H_P, the image of which is the connected component $A_M(\mathbb{R})^0$ of the group of real points of the maximal $\mathbb{Q}$-split torus A_M of the center of $\mathrm{Res}_{k/\mathbb{Q}} M$. Not to overburden notation we shall sometimes denote by A_M^0 or even simply A^0 the vector group $A_M(\mathbb{R})^0$. We denote a_P the compositum of the map H_P and of this section; hence any $x \in P_S$ can be written uniquely as

$$x = y.a_P(x)$$

with $y \in P_S^1$. Given $\lambda \in \mathfrak{a}_P^*$ we have

$$a_P(x)^\lambda = \mathrm{e}^{\langle \lambda, H_P(x) \rangle}.$$

1.2. Now assume that P is a parabolic k-subgroup in G. The subgroup M is the centralizer of A and is therefore determined by A, and A runs through the set of maximal $\mathbb{Q}$-split tori in $\mathrm{Res}_{k/\mathbb{Q}} P$. The pair (P, A) is called a p-pair. The set of k_S-points of a parabolic k-subgroup P of G with Levi subgroup M, has a Langlands decomposition

$$P_S = N_S A^0 M_S^1.$$

Moreover one has $G_S = P_S K$. This allows us to extend the function a_P to a function on G_S by the formula $a_P(pk) = a_P(p)$.

There is a natural map $\mathfrak{a}_P \to \mathfrak{a}_G$, the kernel of which will be denoted by $\mathfrak{a}_P^G$. The set $\Delta(P, A)$ of simple roots of P with respect to A may be identified with a subset of $\mathfrak{a}_P^*$. Given a real number $t > 0$ let

$$A_t = \{a \in A^0 | a^\alpha > t, \quad \forall \alpha \in \Delta(P, A)\}.$$

We denote by $\hat{\Delta}(P, A)$ the set of fundamental weights which is the dual basis to the basis in $\mathfrak{a}_P^G$ given by the simple coroots. The cone generated by the dominant regular weights is the positive Weyl chamber; its dual cone in $\mathfrak{a}_P$ is sometimes denoted $^+\mathfrak{a}_P$. As in [A1] we let $\hat{\tau}_P$ be the characteristic function of the latter:

$$\hat{\tau}_P(H) = 1 \iff \varpi(H) > 0, \quad \forall \varpi \in \hat{\Delta}(P, A), \quad H \in \mathfrak{a}_P.$$

1.3. Let ω be a relatively compact open subset in P_S^1. The Siegel set $\mathfrak{S}(P, \omega, t)$ in G_S, relatively to the p-pair (P, A), is by definition

$$\mathfrak{S}(P, \omega, t) = \omega A_t K. \tag{1}$$

If the set ω is a product $\omega_\infty \omega_f$ with $\omega_\infty \subset P_\infty^1$ and $\omega_f \subset P_{S_f}^1$, the Siegel set $\mathfrak{S}(P, \omega, t)$ is the product

$$\mathfrak{S}(P, \omega, t) = (\omega_\infty A_t K_\infty)(\omega_f K_f) \tag{2}$$

of a Siegel set $\omega_\infty A_t K_\infty$ of G_∞ by a relatively compact open subset in G_{S_f}.

1.4. Given a minimal parabolic k-subgroup P_0 there exist a Siegel set $\mathfrak{S}_0$ relative to P_0 and a finite subset $C \subset G(k)$ such that

$$G_S = \Gamma C \mathfrak{S}_0. \tag{1}$$

For G_∞ this is classical ([B2], Section 15). The general case follows from 8.4 and the proof of 8.5 in [B1].

We recall that $C\mathfrak{S}_0$ has the 'Siegel property', namely, it meets only finitely many of its translates under Γ (cf. [B2] Section 15 and [B1] 8.4).

1.5. A function $f \in C^\infty(G_S)$ is of *moderate growth* if there exists $m \in \mathbb{N}$, such that for every $x \in G_S$

$$|f(x)| \prec ||x||^m. \tag{1}$$

Its restriction to any Siegel set $\mathfrak{S}(P, \omega, t)$ satisfies an inequality

$$|f(x)| \prec a_P(x)^\lambda ||a_G(x)||^{m'}$$

for some $\lambda \in \mathfrak{a}_P^*$, which is zero on $\mathfrak{a}_G$, and some $m' \in \mathbb{N}$. Conversely if now f is Γ-invariant and if this condition is satisfied for the functions $x \mapsto f(cx)$ with $c \in C$, in the notation of 1.4 (1), then f is of moderate growth. (If $S = S_\infty$ see [HC] lemma 6 p. 9. The general case again follows from 8.4 and 8.5 in [B1].)

Let f be Γ-invariant; if for any Siegel set $\mathfrak{S}(P, \omega, t)$, any $\lambda \in \mathfrak{a}_P^*$, which is zero on $\mathfrak{a}_G$, and some $m' \in \mathbb{N}$, the function f satisfies an inequality

$$|f(x)| \prec a_P(x)^\lambda ||a_G(x)||^{m'}$$

we say that the function is rapidly decreasing.

2. The decomposition theorem

2.1. A function $f \in C^\infty(G_S)$ is of *uniform moderate growth* (u.m.g.) if it is uniformly locally constant under right translations on G_{S_f} and if there exists $m \in \mathbb{N}$, such that for every $D \in \mathcal{U}(\mathfrak{g})$ and $x \in G_S$

$$|Df(x)| \prec ||x||^m. \tag{1}$$

Let $V_\Gamma = C^\infty_{\text{umg}}(\Gamma \backslash G_S)$ be the space of smooth functions on $\Gamma \backslash G_S$ of uniform moderate growth. By 1.1, applied to the case $P = G$, we have $G_S = G_S^1 \times A_G^0$ and $\Gamma \subset G_S^1$. Similarly we define V_Γ^1 to be the set of functions on $\Gamma \backslash G_S^1$ of uniform moderate growth.

2.2. As usual $\Gamma_P = \Gamma \cap P(k)$, $\Gamma_N = \Gamma \cap N(k)$ but $\Gamma_M = \Gamma_N \backslash \Gamma_P$. If f is a Γ_P-left-invariant function, then its constant term f^P along P is:

$$f^P(x) = \int_{\Gamma_N \backslash N_S} f(nx)\, dn,$$

the Haar measure being normalized so that the quotient has volume one. For $x \in G_S$, the function $f_x^P : m \mapsto f^P(mx)$ $(m \in M_S^1)$ is Γ_M-left-invariant. A function $f \in V_\Gamma$ is said to be *negligible along* P if for all $x \in G_S$ the function f_x^P is orthogonal to the cusp forms on $\Gamma_M \backslash M_S^1$.

Let $\mathcal{A}$ be the set of classes of associate parabolic k-subgroups. For $\mathcal{P} \in \mathcal{A}$, denote by $V_\Gamma(\mathcal{P})$ the space of elements of $f \in V_\Gamma$ which are negligible along Q for every parabolic k-subgroup $Q \notin \mathcal{P}$. We shall also need the space $V_\Gamma^1(\mathcal{P}) = V_\Gamma(\mathcal{P}) \cap V_\Gamma^1$. We recall the

2.3 PROPOSITION. *If a function f is negligible along all parabolic k-subgroups it is zero, and if it is negligible along all proper parabolic k-subgroups, it is cuspidal.*

Proof. We refer to [Lan1] Lemma 3.7 and its corollary p. 58. The statement and the proof extend to the adelic case (see for example [MW], Proposition 1.3.4)

and also to the S-arithmetic case. Note that in the case of congruence subgroups it follows directly from the adelic case by taking functions invariant under a suitable open compact subgroup outside of S.

As a consequence, the sum

$$V_\Gamma(\mathcal{A}) := \sum_{\mathcal{P}\in\mathcal{A}} V_\Gamma(\mathcal{P})$$

is direct. We shall prove in Section 4 that $V_\Gamma = V_\Gamma(\mathcal{A})$. This will complete the proof of the

2.4 THEOREM. *(Langlands)*

$$V_\Gamma = \bigoplus_{\mathcal{P}\in\mathcal{A}} V_\Gamma(\mathcal{P}).$$

In the case $S = S_\infty$, this theorem and a sketch of proof were communicated by R. P. Langlands in a letter to the first named author [Lan 2]. The following proof is a variant of the original one in our slightly more general setting; there are two new ingredients:

(a) The truncation operator in the form given by J. Arthur. This replaces an inductive construction of a (smooth) truncation, and a delicate analysis of the geometry of Siegel sets to prove that the truncated function is rapidly decreasing.

(b) The Dixmier–Malliavin theorem [DM], which allows one to work up to a convolution; this makes the passage from moderate growth to uniform moderate growth easy.

3. Preliminaries on E-series and constant terms

3.1. Let P be a parabolic k-subgroup of G and $P = MN$ a Levi decomposition over k. Let α be a smooth compactly supported function on G_S and $T_P \in \mathfrak{a}_P$.

LEMMA. *Let $f \in C^\infty(\Gamma_P N\backslash G_S)$ and E_{P,f,T_P} the series defined by*

$$E_{P,f,T_P}(x) = \sum_{\gamma\in\Gamma_P\backslash\Gamma} \hat{\tau}_P(H_P(\gamma x) - T_P) f(\gamma x) \quad (x \in G_S). \tag{1}$$

(i) *The series E_{P,f,T_P} converges absolutely and locally uniformly.*

(ii) *Assume that f is of u.m.g and let α be a smooth compactly supported function on G_S. Then $E_{P,f,T_P} * \alpha \in V_\Gamma$.*

Proof. The convergence of the series and the moderate growth of its sum (which already follows if f has moderate growth) is Corollary 5.2 in [A1]. Its convolution with a smooth compactly supported function is then of u.m.g.

3.2. Let $\mathcal{Q}$ be an associate class of parabolic k-subgroups in M. We denote by $\mathcal{Q}^G$ the associate class in $\mathcal{A}$ which consists of the parabolic k-subgroups of G having a Levi k-subgroup conjugate to M_Q, where M_Q is a Levi k-subgroup of an element $Q \in \mathcal{Q}$.

LEMMA. *We keep the assumptions of 3.1. Assume that f is of u.m.g. and that the functions f_x^P belong to $V_{\Gamma_M}(\mathcal{Q})$ for all $x \in G_S$. Then*

$$E_{P,f^P,T_P} * \alpha \in V_\Gamma(\mathcal{Q}^G).$$

Proof. We have to prove that if R is a parabolic k-subgroup of G not belonging to $\mathcal{Q}^G$, the constant term of E_{P,f^P,T_P} along R is negligible:

$$\int_{\Gamma_{M_R}\backslash M^1_{R,S}} \psi(m).(E_{P,f^P,T_P})^R(my)\,\mathrm{d}m = 0, \tag{1}$$

for all cusp forms ψ on $\Gamma_{M_R}\backslash M^1_{R,S}$ and $y \in G$. This vanishing result is a variant for our E-series of the classical vanishing result for constant terms of Eisenstein series constructed from cusp forms. In the case of arithmetic subgroups see [Lan1] Lemma 4.4 or also [HC], pp. 34–39, in particular the proof of Corollary 3 p. 39. In the adelic case see [MW] II.1.8 and II.2.1. The proof in our case uses the same formal manipulations, which is allowed since our E-series are absolutely convergent.

4. Proof of the theorem

4.1. We note first that V_Γ may be viewed as a union of differentiable G_S-modules $V_{G,m} (m \in \mathbb{N})$, where $V_{G,m}$ is the set of $f \in V_\Gamma$ whose derivatives Df are essentially bounded by $||x||^m$, endowed with the semi-norms $\sup_x |Df(x)|.||x||^{-m}$. Therefore, since at finite places our functions are uniformly locally constant, it follows from the Dixmier–Malliavin theorem [DM] that any $f \in V_\Gamma$ is a finite linear combination of terms $h * \alpha$, with $h \in V_\Gamma$ and $\alpha \in C_c^\infty(G_S)$ (with support in a prescribed neighborhood of 1). In other words,

$$V_\Gamma = \bigcup_{\alpha \in C_c^\infty(G_S)} V_\Gamma * \alpha. \tag{1}$$

4.2. We have the decomposition

$$\mathbf{L}^2(\Gamma\backslash G_S^1) = \bigoplus_{\mathcal{P}\in\mathcal{A}} \mathbf{L}^2_{\mathcal{P}}(\Gamma\backslash G_S^1), \tag{1}$$

where $\mathbf{L}^2_{\mathcal{P}}(\Gamma\backslash G_S^1)$ is the space of $\mathbf{L}^2$-functions on $\Gamma\backslash G_S^1$ which are negligible along the parabolic k-subgroups $Q \notin \mathcal{P}$. This follows from 2.3 (see also [MW] II.2.4). We have then

$$\mathbf{L}^2(\Gamma\backslash G_S^1)^\infty = \bigoplus_{\mathcal{P}\in\mathcal{A}} \mathbf{L}^2_{\mathcal{P}}(\Gamma\backslash G_S^1)^\infty. \tag{2}$$

But, again by [DM], the elements of $\mathbf{L}^2_{\mathcal{P}}(\Gamma\backslash G^1_S)^\infty$ are finite sums of terms $f * \alpha$ and hence have u.m.g. (cf. [BJ] 1.6, which refers to [HC] I.3, Lemma 9 and its corollary p. 10), therefore belong to $V^1_\Gamma(\mathcal{P})$:

$$\mathbf{L}^2_{\mathcal{P}}(\Gamma\backslash G^1_S)^\infty \subset V^1_\Gamma(\mathcal{P}). \tag{3}$$

As a consequence,

$$\mathbf{L}^2(\Gamma\backslash G^1_S)^\infty \subset \bigoplus_{\mathcal{P}\in\mathcal{A}} V^1_\Gamma(\mathcal{P}) = V^1_\Gamma(\mathcal{A}). \tag{4}$$

4.3. By 2.1, $G_S = G^1_S \times A^0_G$ and $\Gamma \subset G^1_S$. Given a function f on $\Gamma\backslash G_S$ and $a \in A^0_G$ we let f_a be the function on $\Gamma\backslash G^1_S$ defined by $x \mapsto f(xa)$.

LEMMA. *Let $f \in V_\Gamma$. Assume that for $a \in A^0_G$, $f_a \in \mathbf{L}^2(\Gamma\backslash G^1_S)$ and that its $\mathbf{L}^2$-norm is slowly increasing, i.e.*

$$\|f_a\|_2 \prec \|a\|^m$$

for some $m \in \mathbb{N}$. Let $\alpha \in C^\infty_c(G_S)$. Then $f * \alpha \in V_\Gamma(\mathcal{A})$.

Proof. We apply 4.2(2) to f_a, with $a \in A^0_G$. There exists then a finite set of functions $(f_{\mathcal{P}})_{\mathcal{P}\in\mathcal{A}}$ on $\Gamma\backslash G_S$, such that

$$f = \sum_{\mathcal{P}\in\mathcal{A}} f_{\mathcal{P}} \tag{1}$$

with $f_{\mathcal{P},a} \in V^1_\Gamma(\mathcal{P})$. We show first that each $f_{\mathcal{P}}$ is locally integrable on G_S. (In an earlier version, we had overlooked this point. N. Wallach drew our attention to it and provided the argument.) By hypothesis f is square integrable on $\Gamma\backslash G_S = \Gamma\backslash G^1_S \times A^0_G$ for the measure

$$d\mu(xa) = dx \otimes \|a\|^{-n}\, da$$

if n is large enough, where dx and da are Haar measures on G^1_S and A^0_G respectively. But $f_{\mathcal{P}}$ is the image of f by the orthogonal projection on

$$\mathbf{L}^2(\Gamma\backslash G^1_S)_{\mathcal{P}} \hat{\otimes} \mathbf{L}^2(A^0_G, \|a\|^{-n}\, da);$$

it is in particular square integrable for the above measure and hence locally integrable for any Haar measure on G_S. It remains to show that

$$f_{\mathcal{P}} * \alpha \in V_\Gamma(\mathcal{P}). \tag{2}$$

The function $f_{\mathcal{P}} * \alpha$ is smooth. Since the formation of constant terms commutes with convolution, $f_{\mathcal{P}} * \alpha$ is negligible outside $\mathcal{P}$. There remains to see that it is of u.m.g.

For each a, $||f_a||_2^2$ is a sum of the $||f_{\mathcal{P},a}||_2^2$. Therefore there exists $m \in \mathbb{N}$ such that

$$||f_{\mathcal{P},a}||_2 \prec ||a||^m. \tag{3}$$

Let $D \in \mathcal{U}(\mathfrak{g})$. We have

$$D(f_{\mathcal{P}} * \alpha)(y) = (f_{\mathcal{P}} * \beta)(y) = \int_{\Gamma\backslash G_S} f_{\mathcal{P}}(x) K_\beta(x,y)\, dx \tag{4}$$

where $D\alpha = \beta$ and

$$K_\beta(x,y) = \sum_{\gamma\in\Gamma} \beta(x^{-1}\gamma y) \quad (x, y \in G_S). \tag{5}$$

We claim that this series converges absolutely, uniformly in x, locally uniformly in y and that there exist $r \in \mathbb{N}$ and a compact set C in A_G^0, which depend only on the support of α, such that

$$|K_\beta(xa, yb)| \prec ||y||^r \tag{6}$$

for $D \in \mathcal{U}(\mathfrak{g})$, $x, y \in G_S^1$ and $a, b \in A_G^0$. Moreover $K_\beta(xa, yb) = 0$ if $ab^{-1} \notin C$. This follows from standard arguments (see for example [HC] Lemma 9, [A1] Lemma 4.3 or [MW] Lemma I.2.4). We can write

$$(f_{\mathcal{P}} * \beta)(yb) = \int_{bC} da \int_{\Gamma\backslash G_S^1} f_{\mathcal{P},a}(x) K_\beta(xa, yb)\, dx$$

for $y \in G_S^1$ and $b \in A_G^0$. In view of (3) and (6)

$$|D(f_{\mathcal{P}} * \alpha)(yb)| = |(f_{\mathcal{P}} * \beta)(yb)| \prec ||y||^r \mathrm{vol}(\Gamma\backslash G_S^1)||b||^m$$

with r and m independent of D. This implies that $f_{\mathcal{P}} * \alpha$ is of u.m.g.

4.4. We now use Arthur's truncation operator Λ^T. The parameter T belongs to $\mathfrak{a}_{P_0}$ where P_0 is a fixed minimal parabolic k-subgroup of G; it will be assumed to be far enough in the positive Weyl chamber. Given a function f on $\Gamma\backslash G_S$, $\Lambda^T f$ is a sum, over a set $\mathcal{X}$ of representatives of Γ-conjugacy classes of parabolic k-subgroups of G, of E-series of the type studied in 3.1:

$$\Lambda^T f = \sum_{P\in\mathcal{X}} (-1)^{\mathrm{prk}(P)-\mathrm{prk}(G)} E_{P, f^P, I_P(T)}$$

where $I_P : \mathfrak{a}_{P_0} \to \mathfrak{a}_P$ is the linear map defined in [Mü] p. 488 and prk(P) the parabolic k-rank of P. Given a function f on $\Gamma\backslash G_S$ of u.m.g., the fundamental property of the truncation operator is that $\Lambda^T f$ is rapidly decreasing on $\Gamma\backslash G_S^1$.

More precisely, if $\mathfrak{S}_0$ is a Siegel set in G_S for a minimal parabolic k-subgroup P_0, let $\mathfrak{S}_0^1 = \mathfrak{S}_0 \cap G_S^1$. Then

$$\sup_{(x,a)\in\mathfrak{S}_0^1\times A_G^0} |\Lambda^T f(xa)|\, ||x||^n ||a||^{-m} \prec \sup_{D\in\mathcal{U}(\mathfrak{g}),y\in G_S} |Df(y)|\,.\,||y||^{-r}$$

for all n. In the adelic setting, this a variant of Lemma 1.4 page 95 of [A2], where we have replaced the $\mathbf{L}^1$-norm over some parameter space S for a measure $d\sigma$ by the sup-norm over A_G^0. In the arithmetic case, this is theorem 5.2 of [OW]; a proof may be found in Section 7 of [OW]. Again in view of 8.4 and 8.5 of [B1], it extends to the S-arithmetic case. This shows that $(\Lambda^T f)_a$ is square integrable on $\Gamma\backslash G_S^1$ and that the $\mathbf{L}^2$-norm of $(\Lambda^T f)_a$ is a slowly increasing function on A_G^0.

4.5. Combined with 4.2 and 4.3, the last assertion of 4.4 shows that $(\Lambda^T f) * \alpha$ belongs to $V_\Gamma(\mathcal{A})$. To finish the proof we proceed by induction on the semisimple k-rank. By construction, $f - \Lambda^T f$ is an alternated sum of E-series E_{P,f^P,T_P}, where $T_P = I_P(T)$, for P in a set of representatives of Γ-conjugacy classes of proper parabolic k-subgroups of G. By Lemma 3.2 we know that $E_{P,f^P,T_P} * \alpha$ belongs to $V_\Gamma(\mathcal{A})$ since the constant term f^P is of u.m.g. and the induction assumption implies that the functions $m \mapsto f^P(mh)$ belong to $V_{\Gamma_M}(\mathcal{A}_M)$ for all $h \in K$.

4.6. REMARK. W. Casselman has shown the existence of a similar decomposition theorem in the arithmetic case ($S = S_\infty$) for the Schwartz space of $\Gamma\backslash G_S$, i.e. the space of functions which are uniformly rapidly decreasing (u.r.d.) (i.e., all $\mathcal{U}(\mathfrak{g})$-derivatives are rapidly decreasing in the sense of 1.5) [Ca2]. The above proof also yields this, in the S-arithmetic case, once it is noted that the argument of 3.1 also shows that if f is u.r.d., then $E_{P,f^P,T_P} * \alpha$ is u.r.d. Conversely 2.4 for $S = S_\infty$ follows from 1.16 and 4.7 in [Ca2].

II. COHOMOLOGY OF S-ARITHMETIC GROUPS

5. Decomposition of cohomology

Let (ϕ, E) be a finite dimensional irreducible complex rational representation of G_∞. We view E as a representation of G_S which is trivial on G_{S_f}. For convenience, we assume it to be irreducible. It is therefore a tensor product of representations E_v, where E_v is an irreducible representation of $G(k_v)$ ($v \in \mathcal{V}_\infty$).

5.1. Assume that $S = \mathcal{V}_\infty$. It is shown in [B5] that there is a canonical isomorphism

$$H^{\cdot}(\Gamma; E) = H^{\cdot}(\mathfrak{g}, K; V_\Gamma \otimes E). \qquad (1)$$

In 2.4, the direct summands are obviously $(\mathfrak{g}, K)$ submodules, therefore we get

5.2 THEOREM. *Assume that* $S = \mathcal{V}_\infty$. *Then there is a canonical direct sum decomposition*

$$H^\cdot(\Gamma; E) = \bigoplus_{\mathcal{P}\in A} H^\cdot_{\mathcal{P}}(\Gamma; E), \tag{1}$$

where

$$H^\cdot_{\mathcal{P}}(\Gamma; E) = H^\cdot(\mathfrak{g}, K; V_\Gamma(\mathcal{P}) \otimes E). \tag{2}$$

REMARK. For $\mathcal{P} = \{G\}$

$$V_\Gamma(\{G\}) = \mathbf{L}^2_{\{G\}}(\Gamma\backslash G)^\infty := \mathbf{L}^2_{\mathrm{cusp}}(\Gamma\backslash G)^\infty \tag{3}$$

is the space of cuspidal functions. The corresponding summand is therefore the cuspidal cohomology $H^\cdot_{\mathrm{cusp}}(\Gamma; E)$. Consequently (1) proves again that the inclusion

$$\mathbf{L}^2_{\mathrm{cusp}}(\Gamma\backslash G)^\infty \subset C^\infty(\Gamma\backslash G)$$

yields an injective map

$$H^\cdot_{\mathrm{cusp}}(\Gamma; E) \to H^\cdot(\Gamma; E)$$

[B4]. Moreover, it exhibits a natural complement to the cuspidal cohomology.

5.3. We now return to the case of a general S-arithmetic group and want to extend the foregoing in the framework of continuous (or differentiable) cohomology, for which we refer to [BW:IX]. By [BW:XIII, 1.1]

$$H^\cdot(\Gamma; E) = H^\cdot_d(G_S; C^\infty(\Gamma\backslash G_S) \otimes E). \tag{1}$$

5.4 PROPOSITION. *The inclusion*

$$\iota : V_\Gamma \to C^\infty(\Gamma\backslash G_S)$$

induces an isomorphism

$$\iota^\cdot : H^\cdot_d(G_S; V_\Gamma \otimes E) \tilde{\to} H^\cdot_d(G_S; C^\infty(\Gamma\backslash G_S) \otimes E). \tag{1}$$

Proof. We can operate either in differentiable cohomology, denoted $H^\cdot_d$ in [BW] or, by passing to K-finite vectors, in the variant $H^\cdot_e$ of [BW:X, 5]. We use the former. The spaces $H^{\cdot}_d(G_S; V_\Gamma \otimes E)$ and $H^{\cdot}_d(G_S; C^\infty(\Gamma\backslash G_S) \otimes E)$ are the abutments of spectral sequences (E_r) and (E'_r) [BW:IX, 4.3] in which

$$E_2^{p,q} = H^p_d(G_{S_f}; H^q_d(G_\infty; V_\Gamma \otimes E)) \tag{2}$$

and

$$E'^{p,q}_2 = H^p_d(G_{S_f}; H^q_d(G_\infty; C^\infty(\Gamma\backslash G_S) \otimes E)). \tag{3}$$

The inclusion ι induces a homomorphism $(E_r) \to (E'_r)$ of spectral sequences. It suffices therefore to show that $E_2 \tilde{\to} E'_2$ is an isomorphism and for this, it is enough to prove that the homomorphism

$$\iota^\cdot : H^{\cdot}_d(G_\infty; V_\Gamma \otimes E) \to H^{\cdot}_d(G_\infty; C^\infty(\Gamma\backslash G_S) \otimes E) \tag{4}$$

is an isomorphism. The group G_{S_f} operates by right translations on these cohomology groups and $\iota^\cdot$ is G_{S_f}-equivariant. By going over to the K_∞-finite elements in V_Γ and $C^\infty(\Gamma\backslash G_S)$, we may replace the differentiable cohomology by the relative Lie algebra cohomology $H^\cdot(\mathfrak{g}_\infty, K_\infty; \cdot)$. The latter commutes with inductive limits. The two spaces of functions under consideration are inductive limits of the subspaces of elements fixed under a compact open subgroup K'_f of G_{S_f}. We are then reduced to showing that the natural homomorphism

$$H^\cdot\left(\mathfrak{g}_\infty, K_\infty; V_\Gamma^{K'_f}\right) \to H^\cdot(\mathfrak{g}_\infty, K_\infty; C^\infty(\Gamma\backslash G_S/K'_f)), \tag{5}$$

is an isomorphism for every compact open subgroup K'_f of G_{S_f}. But the space $\Gamma\backslash G_S/K'_f$ is isomorphic to a finite disjoint union of arithmetic quotients:

$$\Gamma\backslash G_S/K'_f = \coprod_c \Gamma_c\backslash G_\infty$$

[BJ:4.3]. We are thus reduced to the case dealt with in [B5].

5.5. Using 5.3, 5.4 and the decomposition Theorem 2.4, we can therefore write $H^\cdot(\Gamma; E)$ as a direct sum

$$H^\cdot(\Gamma; E) = \bigoplus_{\mathcal{P}\in\mathcal{A}} H^\cdot_{\mathcal{P}}(\Gamma; E), \tag{1}$$

where, by definition,

$$H^\cdot_{\mathcal{P}}(\Gamma; E) = H^\cdot_d(G_S; V_\Gamma(\mathcal{P}) \otimes E). \tag{2}$$

In particular, $V_\Gamma(\{G\})$ is the space of cuspidal functions and the corresponding summand is the cuspidal cohomology of Γ, also to be denoted $H^\cdot_{\text{cusp}}(\Gamma; E)$.

6. Cohomology with coefficients in the discrete spectrum

6.1. Let F be a non-archimedean local field, L a connected reductive F-group and P_0 a minimal parabolic F-subgroup of L. We shall denote by $\mathrm{St}(L)$ the *Steinberg representation* of $L(F)$. We recall that it can be defined as the natural representation of $L(F)$ in the quotient of $C^\infty(P_0(F)\backslash L(F))$ by the invariant subspace generated by the functions which are left-invariant under a parabolic subgroup $Q \supsetneqq P_0$. It is the space of C^∞-vectors of an irreducible representation of $L(F)$ which is square integrable modulo the split component of the center of $L(F)$ (see for example [B3]). It is the trivial representation if $L(F)$ is compact or if L is a torus. The center of $L(F)$ belongs to its kernel. If L is an almost direct product of two F-subgroups L_1, L_2 then $\mathrm{St}(L) = \mathrm{St}(L_1) \otimes \mathrm{St}(L_2)$. From 4.7 in [BW:X] we see then that

$$H^i_d(L(F); \mathrm{St}(L)) = \begin{cases} 0, & i \neq \mathrm{rk}_F(\mathcal{D}L) \\ \mathbb{C}, & i = \mathrm{rk}_F(\mathcal{D}L) \end{cases} \qquad (i \in \mathbb{Z}), \tag{1}$$

where $\mathcal{D}L$ is the derived group of L. Moreover, if L is almost absolutely simple over F, we deduce from results of W. Casselman (see [Ca1] or [BW:XI, 3.9]) that the only irreducible admissible unitarizable representations (π, H) of $L(F)$ for which $H^{\cdot}_d(L(F); H) \neq 0$ are the Steinberg representation and the trivial representation. In the latter case, we have

$$H^i_d(L(F); \mathbb{C}) = \begin{cases} \mathbb{C}, & i = 0 \\ 0 & i \neq 0 \end{cases} \qquad (i \in \mathbb{Z}). \tag{2}$$

6.2. We now come back to the S-arithmetic groups. In the remainder of Section 6 we shall assume that G is semisimple, almost absolutely simple over k of strictly positive k-rank. Let $\widetilde{G}$ be the universal covering of G and τ the canonical isogeny $\widetilde{G} \to G$. It induces a morphism $\tau_S : \widetilde{G}_S \to G_S$ with finite kernel and cokernel.

6.3 LEMMA. *We keep the previous assumptions. Let (π, H) be an irreducible unitary representation of G_S which occurs discretely in $\mathbf{L}^2(\Gamma\backslash G_S)$. Assume that π has a non-compact kernel. Then π is the trivial representation if either $G = \widetilde{G}$ or*

$$H^{\cdot}_d(G_S; H^\infty \otimes E) \neq 0.$$

Proof. Assume first that G is simply connected. By our assumptions, G is absolutely almost simple over k and G_v is not compact for every $v \in S$. Therefore G_v is simple modulo its center as an abstract group [T]. There exists then $v \in S$ such that G_v is in the kernel of π. By hypothesis, H^∞ is realized as a space of functions on G_S right-invariant under G_v and left-invariant under Γ. Since G_v is normal, they are also left-invariant under G_v. The group G being assumed simply connected, strong approximation is valid and implies that $G_v.\Gamma$ is dense in G_S. Therefore the elements of H^∞ are left-invariant under G_S, hence are constant functions.

We now drop the assumption $G = \tilde{G}$. The group $G' = \tau_S(\tilde{G}_S)$ is normal of finite index in G_S. As a $\tilde{G}_S$-module, H is the direct sum of finitely many irreducible G'-modules $H_i (i \in I)$ which are permuted transitively by G_S (cf. [Sil] or Lemma A.2 (ii) in the Appendix). Because of this last fact, their kernels are isomorphic, and in particular either all compact or all non-compact. Since G' is open of finite index in G_S, its intersection with $\ker \pi$ is not compact, hence $\ker \pi_i$ is not compact and the previous argument shows that H_i is a trivial G'-module $(i \in I)$, i.e. that $G' \subset \ker \pi$. Consequently, H is an irreducible representation of the finite group G_S/G', in particular is finite dimensional. Write it as a tensor product $(\pi, H) = \otimes_{s \in S}(\pi_s, H_s)$, where (π_s, H_s) is an irreducible finite dimensional representation of $G(k_s)$. Since $H_d^{\cdot}(G_S; H^\infty \otimes E) \neq 0$ and the Künneth rule holds [BW:XIII, 2.2], we have

$$H_d^{\cdot}(G(k_s); H_s \otimes E) \neq 0, \ (s \in \mathcal{V}_\infty), \ H_d^{\cdot}(G(k_s); H_s) \neq 0 \quad (s \in S_f).$$

If $s \in S_f$ this forces H_s to be trivial (6.1). Now let s be archimedean. We are dealing with relative Lie algebra cohomology with coefficients in a finite dimensional representation, hence $H_s \otimes E$ must contain the trivial representation, i.e. E must be the contragredient representation to H_s. In particular it must have a kernel of finite index. Since E is a rational representation, it must be trivial, and then so is $H_s \, (s \in \mathcal{V}_\infty)$.

REMARK. The first part of the proof is just a variant of 3.4 in [BW:XIII] and 2.2 in [BW:VII]. The second part of the previous argument allows one to suppress the assumption $G = \tilde{G}$ in [BW:XIII, 3.4], hence also in 3.5 there.

6.4. An irreducible unitary representation (π, H) of G_S can be written uniquely as $(\pi_\infty, H_\infty) \otimes (\pi_{S_f}, H_f)$, where (π_∞, H_∞) (resp. (π_{S_f}, H_f)) is an irreducible unitary representation of G_∞ (resp. G_{S_f}). If T is a finite subset of $\mathcal{V}_f$ we let

$$\mathrm{St}(G_T) := \bigotimes_{v \in T} \mathrm{St}(G_v)$$

be the Steinberg representation of G_T. We let $\hat{G}_{\mathrm{cusp}}(S, \Gamma, \mathrm{St})$ (resp. $\hat{G}_{\mathrm{disc}}(S, \Gamma, \mathrm{St})$) be the set of equivalence classes of irreducible unitary representations of G_S which occur in the cuspidal (resp. $\mathbf{L}^2$-discrete) spectrum of $\Gamma \backslash G_S$ and in which π_{S_f} is the Steinberg representation. Assume that S_f is non-empty, then

$$\hat{G}_{\mathrm{cusp}}(S, \Gamma, \mathrm{St}) = \hat{G}_{\mathrm{disc}}(S, \Gamma, \mathrm{St}). \tag{1}$$

Since the Steinberg representation is tempered, this follows from an argument of Wallach's [W], which asserts that an irreducible unitary representation $\pi = \otimes_{s \in S} \pi_s$

belonging to the discrete spectrum and tempered at one place belongs to the cuspidal spectrum. Wallach's argument is carried out only at infinity (i.e. for $S = \mathcal{V}_\infty$), but it extends very easily to the S-arithmetic case, in the same way as it was done by Clozel [Cl3] in the adelic case.

Recall that if $A^\cdot$ is a module graded by $\mathbb{Z}$, given $m \in \mathbb{Z}$, then $A^\cdot[m]$ denotes the graded module defined by $(A^\cdot[m])^i = A^{m+i}$ $(i \in \mathbb{Z})$.

6.5 THEOREM. *Let r_f be the sum over $s \in S_f$ of the k_s-ranks of the groups G/k_s. Let us denote by $I_\pi \otimes H_{\pi_{S_f}}$ the isotypic subspace of $(\pi, H_\pi) = (\pi_\infty \otimes \pi_{S_f}, H_{\pi_\infty} \otimes H_{\pi_{S_f}})$ in $\mathbf{L}^2_{\text{cusp}}(\Gamma\backslash G_S)$. Under the assumptions of 6.2,*

$$H^\cdot_{\text{cusp}}(\Gamma; E) = \bigoplus_\pi H^\cdot_d(G_\infty; I^\infty_\pi \otimes E)[-r_f], \tag{1}$$

where π runs through $\hat{G}_{\text{cusp}}(S; \Gamma, \text{St})$. If moreover S_f is non-empty, the discrete part

$$H^\cdot_{\text{disc}}(\Gamma; E) := H^\cdot_d(G_S; \mathbf{L}^2_{\text{disc}}(\Gamma\backslash G_S)^\infty \otimes E)$$

of $H^\cdot_{(2)}(\Gamma; E)$ is given by

$$H^\cdot_{\text{disc}}(\Gamma; E) = H^\cdot_{\text{cusp}}(\Gamma; E) \oplus H^\cdot_d(G_\infty; E). \tag{2}$$

Proof. First we show that $H^\cdot_d(G_S; \mathbf{L}^2_{\text{disc}}(\Gamma\backslash G_S)^\infty \otimes E)$ is finite dimensional.
(a) Assume that G is simply connected. We shall use the variant $H^\cdot_e$ of the cohomology introduced in [BW:X, 5.1]. Let $V = \mathbf{L}^2_{\text{disc}}(\Gamma\backslash G_S)^\infty$; denote by V_f the space of K-finite vectors in V. By [BW:XII, 2.5]

$$H^\cdot_e(G_S; V_f \otimes E) = H^\cdot_d(G_S; V \otimes E).$$

By [BW:X, 5.3], there is a spectral sequence abutting to $H^\cdot_e(G_S; V_f \otimes E)$, in which

$$E_2^{p,q} = H^p_e(G_{S_f}; H^q_e(G_\infty; V_f \otimes E)). \tag{3}$$

The space V_f is the inductive limit of the spaces $V^{K'_f}$, where K'_f runs through the compact open subgroups of G_{S_f}. Let Y be the Tits building of G_{S_f}. The cohomology of G_{S_f} can be computed as that of a simplicial sheaf on a chamber $C \simeq G_{S_f}\backslash Y$ which associates to a face σ of C the vector space.

$$H^q_d(G_\infty; \mathbf{L}^2_{\text{disc}}(\Gamma\backslash G_S)^\infty \otimes E)^{K_\sigma}, \tag{4}$$

where K_σ is the isotropy subgroup of σ in G_{S_f} [BW:X, 2.5]. Since the representations of K_σ are fully reducible, taking fixed points commutes with the formation of cohomology, hence (4) can also be written as

$$H_d^q(G_\infty; \mathbf{L}^2_{\mathrm{disc}}(\Gamma\backslash G_S/K_\sigma)^\infty \otimes E). \tag{5}$$

As already pointed out in 5.4, $\Gamma\backslash G_S/K_\sigma$ is a disjoint union of finitely many quotients $\Gamma_c\backslash G_\infty$, where Γ_c is arithmetic; the $\mathbf{L}^2$-discrete spectrum is the direct sum of the $\mathbf{L}^2$-discrete spectra of the quotients $\Gamma_c\backslash G_\infty$. But it is proved in [BG] that the relative Lie algebra cohomology with coefficients in the $\mathbf{L}^2$-discrete spectrum of an arithmetic group is finite dimensional. Therefore $E_2^{\cdot,q}$ is the cohomology of G_{S_f} with coefficients in a finite dimensional space, which is in particular an admissible G_{S_f}-module. It is then finite dimensional [BW:X, 6.3]. As a consequence, E_2 is finite dimensional, therefore so is the abutment of the spectral sequence and our assertion is proved in case (a).

(b) Let $\tilde{G}, \tau, G'$ be as in 6.2, 6.3 and $N = \ker \tau_S$. The group N is finite and acts trivially on $\mathbf{L}^2_{\mathrm{disc}}(\Gamma\backslash G_S)^\infty \otimes E$, therefore the spectral sequence of $\tilde{G}_S$ modulo N yields an isomorphism

$$H_d^{\cdot}(G'; \mathbf{L}^2_{\mathrm{disc}}(\Gamma\backslash G_S)^\infty \otimes E) \simeq H_d^{\cdot}(\tilde{G}_S; \mathbf{L}^2_{\mathrm{disc}}(\Gamma\backslash G_S)^\infty \otimes E). \tag{6}$$

On the other hand, the spectral sequence of G_S modulo G' degenerates to an isomorphism

$$H_d^{\cdot}(G_S; \mathbf{L}^2_{\mathrm{disc}}(\Gamma\backslash G_S)^\infty \otimes E) \simeq H_d^{\cdot}(G'; \mathbf{L}^2_{\mathrm{disc}}(\Gamma\backslash G_S)^\infty \otimes E)^{G_S/G'}, \tag{7}$$

which, together with (6), provides the reduction to case (a).

This implies of course also that $H_e^{\cdot}(G_S; \mathbf{L}^2_{\mathrm{cusp}}(\Gamma\backslash G_S)^\infty \otimes E)$ is finite dimensional, but this already follows from the fact that it injects into $H^{\cdot}(\Gamma; E)$ (5.5), since the latter is known to be finite dimensional [BS].

Let (π, H_π) be an irreducible representation of G_S contained in the $\mathbf{L}^2$-discrete spectrum. By the Künneth rule

$$H_d^{\cdot}(G_S; H_\pi^\infty \otimes E) = H_d^{\cdot}(G_\infty; H_{\pi_\infty}^\infty \otimes E) \otimes H_d^{\cdot}(G_{S_f}; H_{\pi_{S_f}}^\infty). \tag{8}$$

If π is trivial, then by 6.1(2)

$$H_d^{\cdot}(G_S; H_\pi^\infty \otimes E) = H_d^{\cdot}(G_\infty; E). \tag{9}$$

Let now π be non-trivial. By 6.1, 6.3, the left-hand side of (8) can be non-zero only if π_{S_f} is the Steinberg representation, hence if $\pi \in \hat{G}_{\mathrm{disc}}(S, \Gamma, \mathrm{St})$. Moreover, its contribution to cohomology is equal to

$$H_d^{\cdot}(G_\infty; H_{\pi_\infty}^\infty \otimes E)[-r_f], \tag{10}$$

in view of the Künneth rule (8) and 6.1 (1). If we write $\mathbf{L}^2_{\mathrm{disc}}(\Gamma\backslash G_S)$ as a Hilbert direct sum of G_S-invariant irreducible subspaces, then only finitely many can contribute to the cohomology. Let then V be the Hilbert direct sum of those with respect to which the cohomology of G_S is zero. The space $H^{\cdot}(G_S; V^\infty)$ is finite dimensional in view of our initial argument. Then it is equal to zero by [BW:XIII, 1.6]. In view of (10) and 6.4(1), this concludes the proof of (1) and also shows that

$$H^{\cdot}_{\mathrm{disc}}(\Gamma; E) = H^{\cdot}_d(G_\infty; E) \oplus \bigoplus_\pi H^{\cdot}_d(G_\infty; I^\infty_\pi \otimes E)[-r_f], \tag{11}$$

where π runs through $\hat{G}_{\mathrm{disc}}(S, \Gamma, \mathrm{St})$, so that (2) now follows from (1) and 6.4(1).

6.6. REMARK. In the anisotropic case, $H^{\cdot}_{\mathrm{disc}}(\Gamma; E) = H^{\cdot}(\Gamma; E)$. The equality (11) above so modified is proved in [BW], XIII, 3.5, for G simply connected, but, as already remarked in 6.3, the argument there allows one to suppress that restriction.

7. $\mathbf{L}^2$-cohomology

We prove the vanishing of the cohomology with coefficients in direct integrals (1) below when S_f is not empty and $\mathfrak{a}_P \neq 0$. This implies that the $\mathbf{L}^2$-cohomology of Γ is reduced to $H^{\cdot}_{\mathrm{disc}}(\Gamma; E)$, provided that the complement of the discrete spectrum in the $\mathbf{L}^2$-spectrum is the sum, over Γ-conjugacy classes of classes of Levi k-subgroups of proper parabolic k-subgroups of G, of direct integrals of discrete spectra for Levi subgroups. It is likely that Langlands' proof [Lan1] extends to the S-arithmetic case but we do not know of a reference in this more general case. For congruence subgroups this follows easily from the corresponding adelic result (cf. [MW] VI.2.1).

7.1. Let P be a proper parabolic k-subgroup of G, N its unipotent radical and M a Levi k-subgroup. Recall (2.1) that $M_S = M^1_S \times A^0$. Let V be the $\mathbf{L}^2$-discrete spectrum of $\Gamma_M\backslash M^1_S$. It is viewed as a representation of P^1_S trivial on N_S. Let $I(P, V, \mu)$ be the induced representation from $V \otimes \mathbb{C}_{\rho+i\mu}$, where $\rho = \rho_P$ is as usual and $\mathbb{C}_\nu$ denotes $\mathbb{C}$ on which A^0 acts by ν ($\nu \in \mathfrak{a}^*_P \otimes \mathbb{C}$). Let $I_{P,V}$ be the direct integral

$$I_{P,V} = \int^{\oplus} I(P, V, \mu)\,\mathrm{d}\mu \qquad \text{(over the positive Weyl chamber in } \mathfrak{a}^*_P\text{)}. \tag{1}$$

We want to prove that

$$H^{\cdot}_d(G_S; I_{P,V} \otimes E) = 0 \quad \text{if } S_f \neq \emptyset. \tag{2}$$

7.2. As in [BC:3.4] we use Shapiro's lemma and are reduced to consider

$$H_d^{\cdot}\left(P_S; E \otimes V \otimes \int^{\oplus} \mathbb{C}_{\rho+i\mu}\, \mathrm{d}\mu\right). \tag{3}$$

It is the abutment of a spectral sequence in which

$$E_2^{p,q} = H_d^p\left(M_S; H_d^q(N_S; E) \otimes V \otimes \int^{\oplus} \mathbb{C}_{\rho+i\mu}\, \mathrm{d}\mu\right). \tag{4}$$

We have

$$H_d^{\cdot}(N_S; E) = H_d^{\cdot}(N_\infty; E) \otimes H_d^{\cdot}(N_{S_f}; \mathbb{C}). \tag{5}$$

The archimedean factors are given by Kostant's theorem, as in *loc. cit.* For $v \in S_f$, $H_d^{\cdot}(N_v; \mathbb{C})$ is reduced to $\mathbb{C}$ in dimension 0 by [BW:X, 4.1].

The center of M contains a non-trivial k-split torus T. Consider a finite place $v \in S_f$. To compute the cohomology of M_S we use the spectral sequence of M_S modulo T_v. Its term E_2 is

$$H_d^{\cdot}\left(M_S/T_v; H_d^{\cdot}\left(T_v; \mathbb{C}_\rho \otimes V \otimes \int^{\oplus} \mathbb{C}_{i\mu}\, \mathrm{d}\mu\right) \otimes H^{\cdot}(N_S; E)\right),$$

since T_v acts trivially on $H^{\cdot}(N_S; E)$; the fact that we can factor out $H^{\cdot}(N_S; E)$ is the essential difference with the archimedean case and the main point here. Let C_v be the maximal compact subgroup of T_v. Using the spectral sequence of T_v modulo C_v and remembering that the cohomology with respect to a compact subgroup reduces to the invariants in dimension 0, we get

$$H_d^{\cdot}\left(T_v; V \otimes \mathbb{C}_\rho \otimes \int^{\oplus} \mathbb{C}_{i\mu}\, \mathrm{d}\mu\right) = H_d^{\cdot}\left(T_v/C_v; V^{C_v} \otimes \mathbb{C}_\rho \otimes \int^{\oplus} \mathbb{C}_{i\mu}\, \mathrm{d}\mu\right)$$

which is equal to

$$V^{C_v} \otimes H_d^{\cdot}\left(T_v/C_v; \int^{\oplus} \mathbb{C}_{\rho+i\mu}\, \mathrm{d}\mu\right)$$

Now T_v/C_v is a finitely generated free abelian group. We may identify its cohomology with the Lie algebra cohomology of a commutative Lie algebra with coefficients in a direct integral. It is zero by [BC:3.2] since $\rho \neq 0$.

III. CONSTRUCTION OF CUSPIDAL COHOMOLOGY CLASSES

In this part α is an automorphism of G defined over k, of finite order ℓ. Denote by $\widetilde{L}$ the semi-direct product $G \rtimes \langle\alpha\rangle$; this is a non-connected algebraic group whose identity component equals G. Here, $\langle\alpha\rangle$ is viewed as a k-group all elements of which are rational over k and $\widetilde{L}$ as a semi-direct product over k. In particular $\widetilde{L}(k)$ is Zariski-dense in $\widetilde{L}$. Let L be the coset defined by α. If A is a k-algebra, we denote by $L(A)^+$ the group generated by $L(A)$. Note that in general $L(A)^+ \subsetneqq \widetilde{L}(A)$.

8. Lefschetz functions for automorphisms of finite order

8.1. Let F be the completion of k at some place. Let E be a finite dimensional representation of $L(F)^+$, assumed to be trivial if F is non-archimedean. Let (π, H_π) be an admissible irreducible representation of $L(F)^+$. A description of admissible irreducible representations of $L(F)^+$ in terms of admissible irreducible representations of $G(F)$ is given in the Appendix. Consider an element β in $L(F)$. By abuse of notation we denote again by β the automorphism of the differentiable cohomology groups $H_d^{\cdot}(G(F); H_\pi \otimes E)$ induced by β. The Lefschetz number of β with respect to $H_\pi \otimes E$ is by definition

$$\mathrm{Lef}(\beta, G(F); H_\pi \otimes E) = \sum (-1)^i \,\mathrm{trace}(\beta \mid H_d^i(G(F); H_\pi^\infty \otimes E)).$$

Since $G(F)$ acts trivially on the differentiable cohomology groups, this number is independent of the choice of β in the coset $L(F)$, and we shall sometimes denote it by $\mathrm{Lef}(\alpha, G(F); H_\pi \otimes E)$ instead of $\mathrm{Lef}(\beta, G(F); H_\pi \otimes E)$.

Fix a minimal parabolic F-subgroup P_0 of G over F with Levi decomposition $P_0 = M_0 N_0$ over F. Since all minimal parabolic subgroups and Levi decompositions over F are conjugate under $G(F)$ we may choose β_0 in the coset defined by α such that P_0 and M_0 are β_0-stable: $\beta_0 P_0 \beta_0{}^{-1} = P_0$ and $\beta_0 M_0 \beta_0{}^{-1} = M_0$. Let P be a β_0-stable parabolic F-subgroup with a β_0-stable Levi decomposition $P = MN$. We denote by $P(F)^+$ (resp. $M(F)^+$) the subgroup generated by $P(F)$ and β_0 (resp $M(F)$ and β_0).

When dealing with representations induced from representations of a parabolic subgroup we shall use normalized induction (as in 7.1): it differs from ordinary induction by a shift by the square root of the modulus function of the parabolic subgroup, so that it preserves unitarity. We shall first exhibit some properties of Lefschetz numbers for F non-archimedean.

8.2 PROPOSITION. *Let F be non-archimedean.*

(1) $\mathrm{Lef}(\alpha, G(F); H_\pi) = 0$ *whenever the restriction of π to $G(F)$ is a constituent of a representation induced from a unitary representation of a proper parabolic subgroup, trivial on the unipotent radical.*

(2) *The map* $\pi \mapsto \mathrm{Lef}(\alpha, G(F); H_\pi)$ *does not vanish identically.*

Proof. We recall that, according to a theorem due to W. Casselman, admissible irreducible representations of $G(F)$ with non-trivial cohomology are the irreducible subquotients of the right regular representation in the space of smooth functions on $P_0(F)\backslash G(F)$ (cf. [BW:X, 4.12]). Hence, if π has non-trivial cohomology, π is a subrepresentation of $i_{P_0}^G(\delta_{P_0}^{1/2})$, the semi-simplification of the representation of $G(F)$ induced from $\delta_{P_0}^{1/2}$. On the other hand let $P = MN$ be a proper parabolic F-subgroup of G and assume that π is a constituent of the representation of $G(F)$ induced from an irreducible unitary representation σ of $M(F)$ extended trivially to the unipotent radical $N(F)$. By Frobenius reciprocity σ is a subrepresentation of the semi-simplification of the Jacquet module $r_P(i_{P_0}^G(\delta_{P_0}^{1/2}))$. The semi-simplification of this Jacquet module is a sum of representations of the form $i_{M\cap P_0}^M(\delta_{P_0}^{1/2} \circ w)$ of $M(F)$, where w runs over a subset of the Weyl group and where $\delta_{P_0}^{1/2} \circ w$ is considered as a representation of $M_0(F)$, extended trivially to $N_0(F) \cap M(F)$ (cf. [BDK:5.4]). Hence the unitary representation σ itself should be a subrepresentation of $i_{M\cap P_0}^M(\delta_{P_0}^{1/2} \circ w)$ for some w. But if $M \neq G$ the representation $i_{M\cap P_0}^M(\delta_{P_0}^{1/2} \circ w)$ has a non-unitary central character; this is a contradiction. This proves assertion (1).

To prove assertion (2) we first observe that since P_0 is β_0-stable

$$P_0(F)\backslash G(F) = P_0(F)^+\backslash L(F)^+$$

and hence I_{P_0} has a canonical extension to $L(F)^+$. Now the Steinberg representation $\mathrm{St}(G(F))$ of $G(F)$ is the unique irreducible quotient of I_{P_0}; hence the Steinberg representation is β_0-stable and has a canonical extension to a representation $\mathrm{St}(L(F)^+)$ of $L(F)^+$. By [BW:X, 4.7], we know that $H_d^q(G(F); \mathrm{St}(G(F))) = 0$ unless $q = \mathrm{rk}\, G(F)$, in which case the cohomology space is canonically isomorphic to $H^q(G(F); I_{P_0})$ and is one-dimensional. The automorphism induced by β_0 acts by 1 on the non-trivial cohomology space, and one has

$$\mathrm{Lef}(\beta_0, G(F); \mathrm{St}(L(F)^+)) = \mathrm{Lef}(\alpha, G(F); \mathrm{St}(L(F)^+)) = (-1)^{\mathrm{rk}\, G(F)}.$$

8.3. Let $K \subset G(F)$ be a maximal compact subgroup. If F is an archimedean field, K may and will be chosen α-stable; it is well known that it is possible and can be seen as follows: the symmetric space attached to $G(F)$ is the set of maximal compact subgroups of $G(F)$; it is a complete simply connected Riemannian manifold of negative curvature; α acts on this space and generates a finite group of isometries, which has a fixed point, say K, by a well known theorem of É. Cartan ([Hel] Chap. I Th.13.5 p.75). If F is non-archimedean, K is assumed to be special.

Let f be a smooth K-finite function with compact support on $G(F)$; we shall denote by f^* the function on $L(F)$ defined by $f^*(x \rtimes \alpha) = f(x)$. We have to recall two definitions. According to J. Arthur (cf. [A5], Section 7, p. 538) a function f^* on $L(F)$ is said to be *cuspidal* if the trace of $\pi(f^*)$ vanishes whenever π is a representation of $L(F)^+$ induced from an irreducible representation (τ, H_τ) of $P(F)^+$ trivial on $N(F)$, where $P = MN$ is a β_0-stable proper parabolic F-subgroup. A smooth compactly supported function f^* is said to be *very cuspidal* if for any β_0-stable proper parabolic F-subgroup P with Levi decomposition $P = MN$, the constant term with respect to P (or along P)

$$f_P^*(m) = \delta_P(m)^{1/2} \int_K \int_{N(F)} f^*(k^{-1}mnk)\, dn\, dk$$

vanishes for all $m \in M(F)\beta_0$. As suggested by the terminology very cuspidal functions are cuspidal, since trace $\pi(f^*)$ = trace $\tau(f_P^*)$ if π is induced from a representation τ of $P(F)^+$ trivial on $N(F)$.

8.4 PROPOSITION. *(1) There exist K-finite, cuspidal, compactly supported smooth functions $f_{\alpha,E}$ on $G(F)$ such that for any admissible representation π of $L(F)^+$ of finite length*

$$\text{trace}\, \pi(f^*_{\alpha,E}) = \text{Lef}(\alpha, G(F); H_\pi \otimes E).$$

(2) If F is archimedean, the functions $f^*_{\alpha,E}$ may be chosen to be very cuspidal.

Such functions are called *Lefschetz functions.*

Proof. Assume first that F is non-archimedean; then E is assumed to be trivial. We have to generalize a construction due to Kottwitz [Ko]. We use the notation of [BW:X, 2]. The automorphism α acts on the Tits building Y associated to $G(F)$. Let s be a face of Y; denote by $L(F)_s^+$ (resp. $G(F)_s$) the stabilizer of s in $L(F)^+$ (resp. $G(F)$). Denote by sign_s the function on $L(F)^+$ equal to the signature of the permutation of the vertices of s induced by x if $x \in L(F)_s^+$ and equal to zero otherwise. Choose a chamber C in Y; since $G(F)$ acts transitively on the chambers there is a $\beta_1 \in L(F)$ which fixes C. If H_π is the space of an admissible representation π of $L(F)^+$, the differentiable cohomology $H_d^{\cdot}(G(F); H_\pi)$ is isomorphic to the cohomology of a complex whose terms are:

$$C^q(Y; H_\pi)^{G(F)} = \bigoplus_{\dim(s)=q} H_\pi^s$$

where H_π^s is the largest subspace of H_π on which $G(F)_s$ acts by the character $sign_s$, and the sum is over the faces s of C ([BW:X, 2.5]); β_1 acts on this complex. Consider the function on $G(F)$

$$x \mapsto f_\alpha(x) = \sum (-1)^{\dim(s)} \text{meas}(G(F)_s)^{-1} \text{sign}_s(x \rtimes \alpha),$$

where the sum is over the faces s of C such that $\beta_1(s) = s$. We have

$$\mathrm{trace}(\pi(f_\alpha^*)) = \sum (-1)^{\dim(s)} \mathrm{sign}_s(\beta_1)\mathrm{trace}(\beta_1 | H_\pi^s)$$

and hence

$$\mathrm{trace}(\pi(f_\alpha^*)) = \mathrm{Lef}(\beta_1, G(F); H_\pi) = \mathrm{Lef}(\alpha, G(F); H_\pi).$$

To prove that such functions are cuspidal we have to show that

$$\pi \mapsto \mathrm{Lef}(\alpha, G(F); H_\pi) = \mathrm{Lef}(\beta_0, G(F); H_\pi)$$

vanishes on representations (π, H_π) of $L(F)^+$ induced from a representation (τ, H_τ) of $P(F)^+$, where $P = MN$ is a proper β_0-stable parabolic F-subgroup, and τ is irreducible and trivial on the unipotent radical $N(F)$. By [BW:X, 4.2] we have

$$\mathrm{Lef}(\beta_0, G(F); H_\pi) = \mathrm{Lef}(\beta_0, M(F); H_{\tau_1})$$

where $\tau_1 = \tau \otimes \delta_P^{1/2}$. Since all cohomology groups $H_d^q(M(F); H_{\tau_1})$ vanish unless the center of $M(F)$ acts trivially, we are reduced to prove the vanishing of the Lefschetz numbers if this center acts trivially. If $M \neq G$, the center of M^+ contains a non-trivial split torus A on which β_0 acts trivially. Let $A(F)^1$ be the maximal compact subgroup of $A(F)$. Using the Hochschild-Serre spectral sequence associated to the exact sequence

$$1 \to A(F)/A(F)^1 \to M(F)^+/A(F)^1 \to M(F)^+/A(F) \to 1$$

and since β_0 acts trivially on A, we see that

$$\mathrm{Lef}(\beta_0, M(F); H_{\tau_1}) = \mathrm{Lef}(\beta_0, M(F)/A(F); H_{\tau_1})\mathrm{Lef}(1, A(F)/A(F)^1; \mathbb{C}),$$

where $A(F)$ acts trivially on $\mathbb{C}$. Now $A(F)$ is the group of F-points of a non-trivial F-split torus, thus the group $A(F)/A(F)^1$ is isomorphic to $\mathbb{Z}^n$ for some $n > 0$ and its Euler-Poincaré characteristic vanishes: $\mathrm{Lef}(1, A(F)/A(F)^1; \mathbb{C}) = 0$. This implies the vanishing of all Lefschetz numbers for $M(F)^+$ and concludes the proof of assertion (1) for non-archimedean fields.

Assume now that F is archimedean. For $\alpha = 1$ the result is known: Euler-Poincaré functions were first constructed by Clozel and Delorme using their trace Paley-Wiener theorem [CD]; then Laumon, in a letter to J. Arthur, has shown that a more direct construction, due to N. Wallach and which uses 'multipliers', yields very cuspidal Euler-Poincaré functions (see [Lab]). For arbitrary α, some functions $f_{\alpha,E}$ have been constructed by this latter method in [Lab] Proposition 12. This takes care of assertion (1). To prove assertion (2) we have to check that one can extend to the twisted case the arguments in [Lab] Section 4. Let us denote by $\mathfrak{g}$ and $\mathfrak{k}$ the Lie algebras of $G(F)$ and K respectively. Since K is α-stable, the same is true for the Cartan decomposition $\mathfrak{g} = \mathfrak{k}+\mathfrak{p}$. Let $\mathfrak{a}_0$ be a maximal abelian subspace in $\mathfrak{p}$. Choose a set of simple coroots $\check{\Delta}_0$ for $(\mathfrak{g}, \mathfrak{a}_0)$. The pairs $(\mathfrak{a}_0, \check{\Delta}_0)$ are all conjugate under K and hence we may choose $\beta_1 = k_1 \rtimes \alpha$ in $K \rtimes \alpha$ which preserves $\mathfrak{a}_0$ and the set of simple coroots. This defines a β_1-stable minimal parabolic subgroup Q_0. Since P_0 and Q_0 are conjugate under K there exists $k_0 \in K$ such that $\beta_1 = k_0^{-1}\beta_0 k_0$. A β_1-stable subset $\check{\Delta} \subset \check{\Delta}_0$ defines a β_1-stable subalgebra $\mathfrak{a} \subset \mathfrak{a}_0$. The dimension of the subalgebra of β_1-fixed vectors $\mathfrak{a}^{\beta_1} \subset \mathfrak{a}$ is the number of β_1-orbits in $\check{\Delta}$. Any β_0-stable proper parabolic subgroup P is K-conjugate to a parabolic subgroup Q with Levi decomposition $Q = MN$, where M is the centralizer of an abelian β_1-stable subalgebra $\mathfrak{a} \subset \mathfrak{a}_0$ as above. In fact, M is also the centralizer of the subalgebra $\mathfrak{a}^{\beta_1}$. Now, if t belongs to $(K \cap M).\beta_1$ then $\mathrm{Ad}(t)$ acts trivially on the non-trivial subalgebra $\mathfrak{a}^{\beta_1} \subset \mathfrak{p}$, and hence

$$\sum(-1)^i \operatorname{trace}(\mathrm{Ad}(t) \mid \Lambda^i\mathfrak{p}) = \det(1 - \mathrm{Ad}(t) \mid \mathfrak{p}) = 0.$$

Let us denote by $\chi_{\alpha,E}$ the function on K defined by

$$x \mapsto \operatorname{trace}(x \rtimes \alpha|\check{E}) \sum(-1)^i \operatorname{trace}(\mathrm{Ad}(x \rtimes \alpha) \mid \Lambda^i\mathfrak{p})$$

where $\check{E}$ is the contragredient of E. Let $\mu_{\alpha,E}$ be the measure supported on K which is the product of $\chi_{\alpha,E}$ by the normalized Haar measure of K (cf. [Lab] p. 616). According to [Lab] Section 4 the constant term along Q of the measure $\mu_{\alpha,E}$ is the product of the function on $(K \cap M)$: $m \mapsto \chi_{\alpha,E}(m.k_1)$ by the Haar measure on $(K \cap M)$ and hence it vanishes. In other words $\mu_{\alpha,E}$ is very cuspidal. The function $f_{\alpha,E}$ is defined by applying a multiplier to $\mu_{\alpha,E}$. Since taking constant terms commutes with multipliers [A3] we see that $f^*_{\alpha,E}$ is very cuspidal. This proves assertion (2).

Remark: The second statement is likely to hold also for non-archimedean fields. For $G = \mathrm{GL}(n)$ and $\alpha = 1$ such functions have been constructed by Laumon using results of Waldspurger ([Lau] Chapter 5).

8.5. The previous construction immediately extends to groups over S. Let E be the tensor product over S of finite dimensional representations E_v of L_v^+, trivial for finite places. Consider $f_{\alpha,E} = \otimes_{v\in S} f_{\alpha,E_v}$ where f_{α,E_v} is a Lefschetz function for each $v \in S$. Given an admissible representation π_S of $L_S^+ = G_S \rtimes \langle\alpha\rangle$ which is the restriction of a tensor product $\otimes_{v\in S}\pi_v$ of representations of G_v^+ one has

$$\begin{aligned} \operatorname{trace} \pi_S(f^*_{\alpha,E}) &= \operatorname{Lef}(\alpha, G_S; H_{\pi_S} \otimes E) = \prod_{v\in S} \operatorname{Lef}(\alpha, G_v; H_{\pi_v} \otimes E_v) \\ &= \sum(-1)^i \operatorname{trace}(\alpha \mid H^i_d(G_S, H_{\pi_S} \otimes E)) \end{aligned}$$

Such functions are called *Lefschetz functions for α and E (over S).*

9. A simple form of the trace formula

9.1. Let F be a global field; we denote by $\mathbb{A}_F$ the ring of adèles of F. Let $\mathbb{A}_k^S$ be the subring of adèles of k with null component in S. There is a natural action ρ of the semi-direct product

$$L(\mathbb{A}_k)^+ = G(\mathbb{A}_k) \rtimes \langle\alpha\rangle$$

on $\mathbf{L}^2(G(k)\backslash G(\mathbb{A}_k))$:

$$(\rho(y \rtimes \alpha^r)f)(x) = f(\alpha^{-r}(xy)).$$

The discrete spectrum $\mathbf{L}^2_{\text{disc}}(G(k)\backslash G(\mathbb{A}_k))$ is invariant under this action. Let ρ_{disc} denote the representation of $L(\mathbb{A}_k)^+$ on the discrete spectrum. Denote by $m_{\text{disc}}(\pi)$ the multiplicity with which the irreducible representation π of $L(\mathbb{A}_k)^+$ occurs in the discrete spectrum. We have

$$\operatorname{trace} \rho_{\text{disc}}(f) = \sum m_{\text{disc}}(\pi) \operatorname{trace} \pi(f).$$

Here we use the fact that $\rho_{\text{disc}}(f)$ is of trace class [Mü]; in particular the above expansion is absolutely convergent, and a partial summation indexed by absolute values of norms of the imaginary part of infinitesimal characters, as in [A5] Theorem 4.4 and in 9.2 below, is not necessary here.

We want to compute the trace of $\rho_{\text{disc}}(f)$ for functions supported on $L(\mathbb{A}_k)$ of the form $f = f^*_{\alpha,E} \otimes h^*$ where $f_{\alpha,E}$ is a Lefschetz function for α over S and $h \in C_c^\infty(G(\mathbb{A}_k^S))$. We assume here Card$(S) \geqslant 2$ and we would like to use the 'simple form of the trace formula' ([A5] Theorem 7.1 p. 538) which should be valid for functions $f = \otimes f_v$ such that f_v is cuspidal for at least two places. This simple form is established by J. Arthur using the invariant form of the trace formula. The proof of this invariant trace formula ([A5], Theorems 3.3 and 4.4) relies on the twisted trace Paley-Wiener theorem. Unfortunately, while this theorem has been proved for non-archimedean fields [Ro], it is still not known to be true for archimedean fields in full generality (as far as we know it is proved in the non-twisted case [CD], and in the case of base change [D]).

However one can bypass this difficulty: mimicking the proof of Theorem 7.1 in [A5], we shall get in 9.5 an unconditional proof of the simple form of the trace formula we need, using the ordinary – non-invariant – trace formula and functions $f = \otimes f_v$ satisfying slightly more stringent assumptions: the functions f_v will be assumed to be very cuspidal at one place and cuspidal at another one.

The proof is quite technical and the reader may skip Section 9 if he is willing to take 9.5 for granted. We shall assume the reader to be familiar with the contents of Section 7 in [A5]. In particular, we shall use the notation adopted there without further explanation. Fix a minimal parabolic k-subgroup P_0^0 with Levi decomposition $P_0^0 = M_0^0 N_0$ over k. A parabolic k-subgroup P^0 containing M_0^0 has a unique Levi k-subgroup M^0 containing M_0^0. Denote by $\widetilde{P}$ the normalizer of P^0 in $\widetilde{L}$ and by $\widetilde{M}$ the normalizer of M^0 in $\widetilde{P}$. Let $P = \widetilde{P} \cap L$ and $M = \widetilde{M} \cap L$; they are called a parabolic subset and a Levi subset respectively. Since all minimal parabolic subgroups and Levi decompositions over k are conjugate under $G(k)$, the cosets P_0 and M_0 have a common rational point over k; but for an arbitrary parabolic subgroup P^0 the sets $P(k)$ and $M(k)$ may be empty. Let $\mathcal{L}(L)$ denote the set of Levi subsets containing M_0. As usual, upper indices are often omitted if $L' = L$ in the various distributions $J_M^{L'}$ that will show up.

9.2 PROPOSITION. *Let $f = \otimes f_v$ be a function in $C_c^\infty(L(\mathbb{A}_k))$. Assume that*

(a) *At a place v_0 the function f_{v_0} is very cuspidal.*
(b) *At a place $v_1 \neq v_0$, the function f_{v_1} is cuspidal.*

Then there exists a finite set $S(f)$ of places of k such that, if Σ is a finite set of places of k containing $S(f)$, the trace formula can be written:

$$\begin{aligned} J(f) &= \sum_{\gamma \in (L(k))_{G,\Sigma}} a^L(\Sigma, \gamma) J_L(\gamma, f) \\ &= \sum_{t \geqslant 0} \sum_{\pi \in \Pi_{\mathrm{disc}}(L,t)} a^L_{\mathrm{disc}}(\pi) J_L(\pi, f). \end{aligned}$$

Proof. Given f, for a large enough finite set of places Σ, the ordinary trace formula can be written (cf. [A5], p. 508 and 521):

$$\begin{aligned} J(f) &= \sum_{M \in \mathcal{L}(L)} \frac{|W^M|}{|W^L|} \sum_{\gamma \in (M(k))_{M,\Sigma}} a^M(\Sigma, \gamma) J_M^L(\gamma, f) \\ &= \sum_{t \geqslant 0} \sum_{M \in \mathcal{L}(\mathcal{L})} \frac{|W^M|}{|W^L|} \int_{\Pi(M,t)} a^M(\pi) J_M^L(\pi, f)\, d\pi. \end{aligned}$$

The proposition is easily proved combining this formula and the following lemma.

9.3 LEMMA. *Let $f = \otimes f_v$ be a function in $C_c^\infty(L(\mathbb{A}_k))$ which satisfies assumptions* (a) *and* (b) *in 9.2 above. Then $J_M^L(*, f) = 0$ if $M \neq L$; moreover $J_L(\pi, f) = 0$ if π is induced from a representation of a proper parabolic subgroup, trivial on the unipotent radical.*

Proof. Since the distributions $J_M^L(*, f)$ arise from a (L, M)-family one may use the descent and splitting formulas ([A4], Proposition 7.1 p. 357 and Corollary 7.4 p. 358). The splitting formula ([A4] Corollary 7.4) shows that $J_M^L(*, f)$ can be expressed as a sum of terms indexed by pairs of Levi subsets (L_1, L_2); to each such pair of Levi subsets is attached a pair of parabolic subsets (Q_1, Q_2) via a 'section' (cf. [A4] p. 356–357) so that:

$$J_M^L(\,*\,, f) = \sum \mathrm{d}_M^L(L_1, L_2) J_M^{L_1}(\,*\,, f_{v_0, Q_1}) J_M^{L_2}(\,*\,, f_{Q_2}^{v_0}).$$

One should note that the constant terms f_{Q_i} along parabolic subsets Q_i with Levi subsets L_i that show up in the above formula cannot in general be replaced by their invariant avatars f_{L_i} unless the distributions are invariant; recall that unless $M = L$ the distributions $J_M^L(*, f)$ are non-invariant.

Since f_{v_0} is very cuspidal, $f_{v_0,Q} = 0$ unless $Q = L$ and since $d_M^L(L, L') = 0$ unless $M = L'$, all terms vanish except if $L_1 = L$ and $L_2 = M$:

$$J_M^L(*, f) = J_M^L(*, f_{v_0}) J_M^M(*, f_M^{v_0}).$$

We have used the equality $d_M^L(L, M) = 1$ and the fact that J_M^M is an invariant distribution.

By assumption there is a place $v_1 \neq v_0$ where f_{v_1} is cuspidal, so that $f_M^{v_0} = 0$ if $M \neq G$, and hence $J_M^L(*, f) = 0$ if $M \neq L$, as expected. It remains to see that $J_L(\pi, f) = 0$ if π is an induced representation, but this follows from the cuspidality of f_v for at least one place v.

9.4. If at one place, say w, the support of f_w is inside the set of regular elements (in particular they are semisimple), the summation over the set $(L(k))_{G,\Sigma}$ in the geometric side of the trace formula can be replaced by a sum over $(L(k))_{G,\mathrm{reg}}$, the set of regular $G(k)$-conjugacy classes in $L(k)$. We recall that

$$J_L(\gamma, f) = \int_{G_\gamma(\mathbb{A}_k)\backslash G(\mathbb{A}_k)} f(x^{-1}\gamma x)\, \mathrm{d}x,$$

where G_γ is the identity component of the centralizer of γ in G. If γ is regular then $a^L(\Sigma, \gamma)$ equals 0 unless γ is elliptic in which case it equals the volume of $G_\gamma(k)\backslash G_\gamma(\mathbb{A}_k)$ divided by the order of $G_\gamma(k)$ in the centralizer of γ in $G(k)$; this expression is independent of Σ and will be denoted $a^L(\gamma)$.

9.5 PROPOSITION. *Let $f = \otimes f_v$ be a function in $C_c^\infty(L(\mathbb{A}_k))$. Assume that f satisfies assumptions* (a) *and* (b) *in 9.2 and*

(c) *There is a place v such that* trace $\pi_v(f_v) = 0$, *whenever π_v is a constituent of a representation induced from a unitary representation of a proper parabolic subgroup, trivial on the unipotent radical.*
(d) *At one place w the support of f_w is inside the set of regular elements.*

Then

$$\operatorname{trace} \rho_{\text{disc}}(f) = \sum_{\gamma \in (L(k))_{G,\text{reg}}} a^L(\gamma) J_L(\gamma, f).$$

Proof. As in the proof of Corollary 7.3 in [A5] we see that if there is a place v such that $\operatorname{trace} \pi_v(f_v) = 0$, whenever π_v is a constituent of a representation induced from a unitary representation of a proper parabolic subgroup, trivial on the unipotent radical (by the way this implies that f_v is cuspidal), then

$$a^L_{\text{disc}}(\pi) J_L(\pi, f) = m_{\text{disc}}(\pi) \operatorname{trace} \pi(f).$$

We conclude by using 9.2.

10. Non-vanishing results

In this section we assume, as in 6.2, that G is an almost absolutely simple connected algebraic group over k of strictly positive k-rank.

10.1. We shall say that the cuspidal cohomology of G over S, with coefficients in E, does not vanish if every S-arithmetic subgroup has a subgroup of finite index with non-zero cuspidal cohomology with respect to E. If $E = \mathbb{C}$, the coefficients will not be mentioned.

To get information about the non-vanishing of the cuspidal cohomology of deep enough S-arithmetic subgroups of $G(k)$, we shall first study the group $H^{\cdot}_{\text{cusp}}(G, S; E)$, which we shall call the *cuspidal cohomology group for G over S with values in E*, which is by definition the inductive limit over *congruence* S-arithmetic subgroups Γ of $G(k)$ of the cuspidal cohomology groups:

$$H^{\cdot}_{\text{cusp}}(G, S; E) = \varinjlim H^{\cdot}_{\text{cusp}}(\Gamma; E).$$

This inductive limit is isomorphic to the cohomology of the adelic cuspidal spectrum:

$$H^{\cdot}_{\text{cusp}}(G, S; E) = H^{\cdot}_d(G_S; \mathbf{L}^2_{\text{cusp}}(G(k)\backslash G(\mathbb{A}_k))^\infty \otimes E).$$

The non-vanishing of the cuspidal cohomology for G over S, which will be proved below for some groups G, implies the non-vanishing of the cuspidal cohomology for deep enough S-arithmetic subgroups Γ of $G(k)$, not necessarily of congruence type:

10.2 PROPOSITION. *Assume that $H^p_{\text{cusp}}(G, S; E)$ is non-trivial for some p, then any S-arithmetic subgroup Γ of $G(k)$, contains a subgroup Γ_1 of finite index such that the cuspidal cohomology $H^p_{\text{cusp}}(\Gamma_1; E)$ is non-trivial.*

Proof. By hypothesis there exists a congruence S-arithmetic subgroup Γ_0 for which $H^p_{\text{cusp}}(\Gamma_0; E)$ is non-trivial. The subgroup $\Gamma_0 \cap \Gamma$ is of finite index in both Γ and Γ_0 and contains a subgroup Γ_1 of finite index, invariant in Γ_0. The usual argument to show that the cohomology of Γ_0 injects in the cohomology of Γ_1 also applies to the cuspidal cohomology: we have

$$\mathbf{L}^2_{\text{cusp}}(\Gamma_0 \backslash G_S)^\infty = \left(\mathbf{L}^2_{\text{cusp}}(\Gamma_1 \backslash G_S)^\infty\right)^{\Gamma_0/\Gamma_1},$$

and since the formation of cohomology commutes with taking fixed points with respect to finite groups this implies

$$H^p_{\text{cusp}}(\Gamma_0; E) = H^p_{\text{cusp}}(\Gamma_1; E)^{\Gamma_0/\Gamma_1}$$

so that the restriction map

$$H^p_{\text{cusp}}(\Gamma_0; E) \to H^p_{\text{cusp}}(\Gamma_1; E)$$

is injective (this also follows from the representation theoretic description of the cuspidal cohomology groups in theorem 6.5) and our claim follows.

We shall use the trace formula to prove an $\mathbf{L}^2$-Lefschetz formula and then apply the latter to show the non-vanishing of the cuspidal cohomology in some cases.

For α as in Section 9 and $h \in C_c^\infty(G(\mathbb{A}_k^S))$, the $\mathbf{L}^2$-Lefschetz number on the discrete part of the $\mathbf{L}^2$-spectrum is, by definition,

$$\begin{aligned}&\mathrm{Lef}_{\text{disc}}(\alpha, h, G; E)\\ &:= \sum(-1)^i \operatorname{trace}(\alpha \times h \mid H^i_d(G_S; \mathbf{L}^2_{\text{disc}}(G(k) \setminus G(\mathbb{A}_k))^\infty \otimes E).\end{aligned}$$

Remark that only the irreducible representations π of $L(\mathbb{A}_k)^+$ whose restriction σ to $G(\mathbb{A}_k)$ remain irreducible contribute non-trivially to the α-twisted trace (see Corollary A.4 in the appendix). If π is one, we may decompose σ into a tensor product $\sigma = \sigma_S \otimes \sigma^S$ of representations of G_S and $G(\mathbb{A}_k^S)$ respectively; according to A.3 the representations σ_S and σ^S may be extended to representations π_S and π^S of L_S^+ and $L(\mathbb{A}_k^S)^+$ respectively, in such a way that π is the restriction of $\pi_S \otimes \pi^S$ to $L(\mathbb{A}_k)^+$; (observe that $L(\mathbb{A}_k)^+$ is of index ℓ in $L_S^+ \times L(\mathbb{A}_k^S)^+$). The spectral decomposition (Theorem 6.5) and 6.5(8) show that

$$\mathrm{Lef}_{\text{disc}}(\alpha, h, G; E) = \sum m_{\text{disc}}(\pi) \, \mathrm{Lef}(\alpha, G_S; \pi_S \otimes E) \operatorname{trace} \pi^S(h^*).$$

We have shown in Section 8 that there exist Lefschetz functions $f_{\alpha,E}$ for α and E over S. Since

$$\mathrm{Lef}(\alpha, G_S; \pi_S \otimes E) = \operatorname{trace} \pi_S(f^*_{\alpha,E})$$

we have

$$\mathrm{Lef}_{\mathrm{disc}}(\alpha, h, G; E) = \mathrm{trace}\Big(\rho(\alpha)\rho(f_{\alpha,E} \otimes h) \mid \mathbf{L}^2_{\mathrm{disc}}(G(k)\backslash G(\mathbb{A}_k))\Big)$$

10.3 PROPOSITION. *Assume that S contains at least one finite place v and that at some place $w_1 \notin S$ the support of $h^*_{w_1}$ is inside the set of elliptic regular elements. Then the Lefschetz number is given by*

$$\mathrm{Lef}_{\mathrm{disc}}(\alpha, h, G; E) := \sum_{\gamma \in (L(k))_{G,\mathrm{reg}}} a^L(\gamma)\, J_L(\gamma, f^*_{\alpha,E} \otimes h^*)$$

where

$$J_L(\gamma, f^*_{\alpha,E} \otimes h^*) = \int_{G_\gamma(k_S)\backslash G(k_S)} f^*_{\alpha,E}(x^{-1}\gamma x)\, \mathrm{d}x \int_{G_\gamma(\mathbb{A}_k^S)\backslash G(\mathbb{A}_k^S)} h^*(x^{-1}\gamma x)\, \mathrm{d}x.$$

Proof. We shall apply Proposition 9.5; we have to check that the various assumptions are satisfied. By Proposition 8.4 the functions f^*_{α,E_v} may be chosen to be cuspidal for $v \in S_f$ and very cuspidal for $v \in V_\infty$. Since $S \neq S_\infty$ assumptions 9.2 (a) and (b) are satisfied. By 8.2(1) the assumption 9.5(c) is fulfilled for $v \in S_f$; finally 9.5(d) is part of our hypothesis.

Let us define the cuspidal Lefschetz number by

$$\mathrm{Lef}_{\mathrm{cusp}}(\alpha, h, G; E) := \mathrm{trace}\Big(\rho(\alpha)\rho(f_{\alpha,E} \otimes h) \mid \mathbf{L}^2_{\mathrm{cusp}}(G(k)\backslash G(\mathbb{A}_k))\Big).$$

10.4 THEOREM. *Assume that the Lefschetz number at infinity*

$$\pi_\infty \mapsto \mathrm{Lef}(\alpha, G_\infty; H_{\pi_\infty} \otimes E)$$

where $(\pi_\infty, H_{\pi_\infty})$ varies through the set of equivalence classes of irreducible unitary representations of the Lie group G_∞, is not identically zero. Then the cuspidal cohomology $H^{\cdot}_{\mathrm{cusp}}(G, S; E)$ does not vanish.

Proof. Recall that we have assumed G to be almost absolutely simple over k and that the k-rank of G is at least 1. To prove the theorem we are free to enlarge S arbitrarily since, according to Theorem 6.5, if $S \subset \Sigma$

$$H^{\cdot}_d(G_\Sigma, \mathbf{L}^2_{\mathrm{cusp}}(G(k)\backslash G(\mathbb{A}_k))^\infty \otimes E)[-r_f(G_\Sigma)]$$

injects in

$$H^{\cdot}_d(G_S, \mathbf{L}^2_{\mathrm{cusp}}(G(k)\backslash G(\mathbb{A}_k))^\infty \otimes E)[-r_f(G_S)].$$

It suffices to show that for S big enough and some $h \in C_c^\infty(G(\mathbb{A}_k^S))$ the cuspidal Lefschetz number $\mathrm{Lef}_{\mathrm{cusp}}(\alpha, h, G;\, E)$ does not vanish. We shall assume that S contains at least one finite place v. Theorem 6.5 also shows that only the trivial representation may contribute to the non-cuspidal discrete spectrum. From this we get that

$$\mathrm{Lef}_{\mathrm{disc}}(\alpha, h, G; E) = \mathrm{Lef}_{\mathrm{cusp}}(\alpha, h, G; E) + \mathbf{1}(f_{\alpha,E} \otimes h).$$

By assumption for archimedean places and by 8.2.(2) for non-archimedean ones the Lefschetz number is not identically zero; hence there is a representation π_S of G_S such that the Lefschetz number $\mathrm{Lef}(\alpha, G_S; \pi_S \otimes E)$ is not zero. But

$$\mathrm{Lef}(\alpha, G_S; \pi_S \otimes E) = \mathrm{trace}\, \pi_S(f^*_{\alpha,E})$$

can be computed, using Weyl's integration formula, as an integral of orbital integrals $J_L(x, f^*_{\alpha,E})$ against the character of π_S for some measure on the set of G_S-conjugacy classes in L_S, recalling that the character of π_S is given by a locally integrable function on L_S. This fact – due to Harish-Chandra for connected reductive groups over local fields of characteristic zero – has been proved for non-connected reductive real Lie groups by Bouaziz [Bou], and for non-connected reductive groups over non-archimedean local fields of characteristic zero by Clozel [Cl1].

The non-vanishing of the Lefschetz number over S implies that for some semisimple element $x_0 \in L_S$, the orbital integral of the Lefschetz function $J_L(x_0, f^*_{\alpha,E})$ does not vanish. The same will be true for some, close enough, rational regular elliptic element $\gamma_0 \in L(k)$, which can always be found since these elements are dense in L_S. If h^* has a small enough support in a neighborhood of γ_0, the Lefschetz formula in Proposition 10.3 reduces to:

$$\mathrm{Lef}_{\mathrm{cusp}}(\alpha, h, G; E) + \mathbf{1}(f_{\alpha,E} \otimes h) = a^L(\gamma_0) J_L(\gamma_0, f^*_{\alpha,E}) J_L(\gamma_0, h^*).$$

Fix the function h^*_v at all places $v \notin S$ but one, say w, with a non-vanishing orbital integral for γ_0; we take h^*_w to be, up to scalars, the characteristic function of a decreasing sequence of (sufficiently small) neighborhoods of γ_0 so that

$$J_L(\gamma_0, h^*) = c_1 \int_{G_{\gamma_0}(k_w)\backslash G(k_w)} h^*_w(x^{-1}\gamma_0 x)\, \mathrm{d}x \equiv 1,$$

while

$$\mathbf{1}(f_{\alpha,S} \otimes h) = c_2 \int_{G(k_w)} h_w(x)\, \mathrm{d}x$$

goes to zero when the neighborhood decreases. Such sequences of neighborhoods and scalars exist since the variety $G_{\gamma_0}(k_w)\backslash G(k_w)$ has a strictly smaller dimension than $G(k_w)$. To finish the proof we simply have to recall that for γ_0 regular elliptic the number $a^L(\gamma_0)$ is the volume of $G_{\gamma_0}(k)\backslash G_{\gamma_0}(\mathbb{A}_k)$ divided by some integer and, in particular, it is non-zero.

10.5. To apply 10.4 to a given pair G, E we have to exhibit an automorphism α of G defined over k such that E extends to a representation of L^+_∞ and such that the Lefschetz number for α at infinity $\pi_\infty \mapsto \mathrm{Lef}(\alpha, G_\infty; \pi_\infty \otimes E)$ is not identically zero. We say that G over k admits a Cartan-type automorphism if there is a k-rational automorphism α of G such that on the Lie group G_∞, for some $x \in G_\infty$, the automorphism $\theta = \alpha \circ \mathrm{Ad}(x)$ is a Cartan involution.

10.6 THEOREM. *Let G be an absolutely almost simple group G over k that admits a Cartan-type automorphism. When $E = \mathbb{C}$, the cuspidal cohomology over S does not vanish.*

Proof. As observed by Rohlfs and Speh ([RS1], Proposition 4.3 p. 493), the Lefschetz number for a Cartan involution θ is never identically zero: in fact since θ acts by -1 on $\mathfrak{g}/\mathfrak{k}$, the Lefschetz number for θ in the $\mathfrak{g}$-K_∞-cohomology of the trivial representation of L^+_∞ is strictly positive. But on the differentiable cohomology at infinity θ and α have the same Lefschetz number. Hence we may apply Theorem 10.4.

This theorem bears some resemblance to some of the results of Rohlfs and Speh in [RS1], where they prove non-vanishing results for Lefschetz numbers on the total cohomology of arithmetic groups. Here we work with cuspidal cohomology and we allow S-arithmetic discrete subgroups for arbitrary large S. More general finite dimensional representations E could be used (see [RS1] and [LS:5.4]) but from now on we shall restrict ourselves to $E = \mathbb{C}$.

10.7 COROLLARY. *Assume that G is k-split and k totally real or $G = \mathrm{Res}_{k'/k}G'$ where k' is a CM-field. Then the cuspidal cohomology of G over S does not vanish.*

Proof. If G is split semisimple over a totally real field k, a Cartan involution of G_∞ induces, up to an inner automorphism, an automorphism of the Dynkin diagram which is itself induced by a rational automorphism α of finite order. Let G' be defined over a CM-field k' ; it is a quadratic totally imaginary extension of a totally real field k; the complex conjugation induces a Cartan-type automorphism of the group $G = \mathrm{Res}_{k'/k}G'$.

11. A base change construction

11.1. Let k/k_0 be a cyclic extension of number fields, of prime degree. Consider an irreducible cuspidal automorphic representation $\pi = \otimes\pi_v$ of $\mathrm{GL}_n(\mathbb{A}_{k_0})$. J. Arthur and L. Clozel have shown ([AC] Chap. 3 Theorem 4.2 (c) and 5.2) that there exists a unique representation $\Pi = b_{k/k_0}(\pi)$ of $\mathrm{GL}_n(\mathbb{A}_k)$, which is locally everywhere the base change lift of π. The representation Π is not necessarily cuspidal, but it is always induced from a cuspidal representation of a Levi subgroup of a parabolic subgroup, extended trivially to the unipotent radical. Now, suppose that a local component π_v of π at a finite place v of k_0 is the Steinberg representation. Then, for any place w above v the component Π_w is again the Steinberg representation (cf. [AC], Chap. 1, Lemma 6.12 or p. 56) and cannot be properly induced. Thus, Π is cuspidal in that case.

More generally, consider a finite extension k/k_0 such that there is a tower

$$k = k_m \supset k_{m-1} \supset \cdots \supset k_1 \supset k_0 \tag{1}$$

of intermediate fields so that k_{i+1}/k_i is a cyclic extension of prime degree. The above base change maps b_{k_{i+1}/k_i} can be composed to define a map b_{k/k_0}.

From this, a base change which associates to a given irreducible cuspidal automorphic representation σ of $\mathrm{SL}_n(\mathbb{A}_{k_0})$ an L-packet $b_{k/k_0}(\sigma)$ of automorphic representations of $\mathrm{SL}_n(\mathbb{A}_k)$ has been defined in [LS] 3.6 and 4.5. We recall that it is constructed by means of a bijection between representations of GL_n up to twists by characters and L-packets of representations of SL_n [LS:3.6], (see also [Cl2] p. 136–138).

This can be summed up in the following proposition.

11.2 PROPOSITION. *Let k/k_0 be as above. If σ is an irreducible cuspidal automorphic representation of $\mathrm{SL}_n(\mathbb{A}_{k_0})$ such that for some finite place v of k_0 the local component σ_v of σ is the Steinberg representation, then the L-packet $b_{k/k_0}(\sigma)$, contains a representation which is cuspidal and the local component of which at any place above v is the Steinberg representation.*

We may now state and prove the following generalization to SL_n of [LS:6.3] for extensions of the type 11.1(1) in which k_0 is totally real.

11.3 THEOREM. *Let k_0 be a totally real field and let k/k_0 be as in 11.1(1). Then the cuspidal cohomology for SL_n/k over S with trivial coefficients does not vanish.*

Proof. By 6.5 and 10.7 there is a cuspidal automorphic representation σ occurring in the cuspidal spectrum $\mathrm{L}^2_{\mathrm{cusp}}(\mathrm{SL}_n(k_0)\backslash\mathrm{SL}_n(\mathbb{A}_{k_0}))$ which is a Steinberg representation at $v \in S_f$ and the infinite component σ_∞ of which has non-trivial differentiable cohomology. On the other hand the base change lift for SL_n preserves the non-vanishing of differentiable cohomology: at finite places this is clear since Steinberg representations lift to Steinberg representations; at infinity it is checked in [LS] Section 5 (a more sophisticated reader might look in [J]). Therefore the base change lift b_{k/k_0} provides us with an L-packet which contains cuspidal automorphic representations that contribute non-trivially to the cuspidal cohomology.

11.4. For more general groups a proof of the existence of cyclic base change, at least for cuspidal representations that are Steinberg representations at some finite places, should be soon available. Using [J] one would be able to extend Theorem 11.3 to all simple split groups over a tower of cyclic extensions of a totally real number field. The use of some other Langlands functoriality (non-cyclic base change; comparison with inner forms...) should produce other non-vanishing results (as in [LS]). In particular it seems likely that if G is a split simple group over any global field k, and if $E = \mathbb{C}$, the cuspidal cohomology over S does not vanish.

Appendix

A.1. Let H^+ be a locally compact topological group with a closed invariant subgroup H, such that H^+/H is compact abelian. We consider a category of continuous representations in complex vector spaces of H and similarly one for H^+, called *good* representations, for which Schur's lemma is valid (e.g. unitary representations or admissible representations of $\mathfrak{p}$-adic groups) and compatible with the restriction from H^+ to H. Let π be a good irreducible representation of H^+ and let $X(\pi)$ be the group of characters χ of the abelian group H^+/H such that $\pi \otimes \chi \simeq \pi$. For any character χ of H^+/H we choose an intertwining operator I_χ between π and $\pi \otimes \chi$; if $\chi \in X(\pi)$ we may and will choose I_χ invertible, otherwise define I_χ to be zero. Of course I_χ is a self-intertwining operator for the restriction σ of π to H, and the various I_χ for $\chi \in X(\pi)$ are linearly independent operators. Now let U be a self-intertwining operator for σ; for any character χ of the abelian group H^+/H we define an intertwining operator between π and $\pi \otimes \chi$:

$$U_\chi = \int_{H^+/H} \chi(x)\pi(x)U\pi(x)^{-1}\,d\dot{x}.$$

Here $d\dot{x}$ is the normalized Haar measure on the compact abelian group H^+/H. Since U_χ intertwines π and $\pi \otimes \chi$ and π is irreducible, there are scalars c_χ such that $U_\chi = c_\chi I_\chi$.

A.2 LEMMA. *Let σ be the restriction to H of an irreducible representation π of H^+.*

(i) *The representation σ is reducible if and only if there exists a non-trivial character χ of H^+/H such that $\pi \otimes \chi$ is equivalent to π.*
(ii) *Assume that $X(\pi)$ is finite. The algebra $I(\sigma)$ of self-intertwining operators of σ is finite dimensional: its dimension equals the order of the group $X(\pi)$. It is semisimple and the representation σ is a finite direct sum of irreducible representations.*
(iii) *If $X(\pi)$ is cyclic of order ℓ, the algebra $I(\sigma)$ is commutative; the representation σ is a direct sum of ℓ inequivalent irreducible representations.*

Proof. If $X(\pi)$ is of finite order, say ℓ, for each $\chi \in X(\pi)$ the operator I_χ can be chosen such that $I_\chi^\ell = 1$; moreover we have by Fourier inversion,

$$U = \sum_{\chi \in X(\pi)} c_\chi I_\chi.$$

In other words the operators I_χ $(\chi \in X(\pi))$ form a basis of the algebra $I(\sigma)$ of self-intertwining operators of σ. Hence, its dimension equals the order of the group $X(\pi)$; this is the first part of assertion (ii). In particular if $X(\pi)$ is trivial, $I(\sigma)$ is reduced to scalars. This last remark and Schur's lemma imply (i). To prove (iii) observe that the algebra $I(\sigma)$ is generated over $\mathbb{C}$ by I_χ if χ generates the group $X(\pi)$. We still have to conclude the proof of (ii). Denote by $H(\pi)$ the intersection of the kernels of characters $\chi \in X(\pi)$, and note that $X(\pi)$ is a product of finite cyclic groups; by using induction on the number of cyclic factors in the quotient group $H^+/H(\pi)$ and the previous remarks, we see readily that the representation σ is a finite direct sum of irreducible representations and that the algebra $I(\sigma)$ is semisimple.

A.3. Assume now that H^+/H is cyclic of order ℓ. We shall denote by α an element of H^+ whose class modulo H generates the quotient group; in particular α^ℓ belongs to H. Let σ be a good representation of H and σ^α the representation $x \mapsto \sigma(\alpha x \alpha^{-1})$. The representation σ is called α-invariant if σ^α is equivalent to σ, i.e. if there exists an operator A in the space of σ such that

$$\sigma^\alpha(x) = A\,\sigma(x)\,A^{-1}.$$

For example, the restriction σ to H of a representation π of H^+ is α-invariant.

If σ is α-invariant and irreducible, by Schur's lemma, $\sigma(\alpha^{-\ell})A^\ell$ is a scalar. We may choose A such that $A^\ell = \sigma(\alpha^\ell)$ and define an extension π of σ to H^+, by letting for $x \in H$

$$\pi(x\alpha^i) = \sigma(x)A^i.$$

The representation π is uniquely defined by σ up to tensor products with characters χ of the cyclic group H^+/H. Since π restricted to H is irreducible, the above lemma implies that $\pi \otimes \chi$ can be equivalent to π only if $\chi = 1$.

More generally, let r be the smallest integer such that $\sigma^{\alpha^r} \simeq \sigma$; then σ can be extended to a representation π_r of the subgroup H_r generated by H and α^r; then π_r, induced from H_r to H^+, yields an irreducible representation π of H^+. We have $\pi \otimes \chi \simeq \pi$ if and only if $\chi(\alpha)$ is a root of unity of order r.

The following corollary is elementary but very useful. We say that the character of π exists as a distribution if $\phi \mapsto \operatorname{trace} \pi(\phi)$ is a continuous linear form on $C_c^\infty(H^+)$.

A.4 COROLLARY. *Assume that the character of π exists as a distribution. If π is irreducible but its restriction to H is reducible, the character vanishes on the subspace of smooth functions compactly supported on αH.*

Proof. By A.2(i), $X(\pi)$ is non-trivial. For $\chi \in X(\pi)$ we have $\pi \otimes \chi \simeq \pi$ and hence

$$\text{trace}\, \pi(\phi) = \text{trace}(\chi \otimes \pi)(\phi) = \chi(\alpha)\text{trace}\, \pi(\phi),$$

hence trace $\pi(\phi)$ vanishes if $\chi(\alpha) \neq 1$.

This applies in particular to the pair $G(\mathbb{A}_k)$, $G(\mathbb{A}_k)^+$ considered in 10.2.

Acknowledgements

The first named author thanks the Humboldt foundation for its support during two visits at Eichstätt. The 'Prix Alexandre Humboldt pour la coopération scientifique franco–allemande', in particular the École Normale Supérieure, Paris, as host institution, supported the third named author during a certain period of this work. He is grateful for this support. We would like to thank as well the Deutsche Forschungsgemeinschaft for support on various occasions.

References

[A1] Arthur, J.: A trace formula for reductive groups I: Terms associated to classes in $G(\mathbf{Q})$, *Duke Math. J.* 45 (1978) 911–952.

[A2] Arthur, J.: A trace formula for reductive groups II: Applications of a truncation operator, *Compos. Math.* 40 (1980) 87–121.

[A3] Arthur, J.: A Paley-Wiener theorem for real reductive groups, *Acta Mathematica* 150 (1983) 1–89.

[A4] Arthur, J.: The invariant trace formula. I. Local theory, *J. Amer. Math. Soc.* 1 (1988) 323–383.

[A5] Arthur, J.: The invariant trace formula. II. Global theory, *J. Amer. Math. Soc.* 1 (1988) 501–554.

[AC] Arthur, J. and Clozel, L.: Simple Algebras, Base Change, and the Advanced Theory of the Trace Formula, *Annals of Math. Studies* 120, Princeton Univ. Press, 1989.

[BDK] Bernstein, J., Deligne, P. and Kazhdan, D.: Trace Paley-Wiener theorem for reductive p-adic groups, *Journal d'Analyse Mathématique* 47 (1986) 180–192.

[BFG] Blasius, D., Franke, J. and Grunewald, F.: Cohomology of S-arithmetic groups in the number field case, *Invent. Math.* 116 (1994) 75–93.

[B1] Borel, A.: Some finiteness properties of adele groups over number fields, *Publ. Math. I.H.E.S.* 16 (1963) 5–30.

[B2] Borel, A.: Introduction aux Groupes Arithmétiques, Hermann, Paris, 1969.

[B3] Borel, A.: Admissible representations of a semisimple group over a local field with vectors fixed under an Iwahori subgroup, *Invent. Math.* 35 (1976) 233–259.

[B4] Borel, A.: Stable real cohomology of arithmetic groups II. Manifolds and Lie groups, J. Hano et al. (eds.), *Progress in Math.* 14, 21–55, Birkhäuser, Boston, Basel, Stuttgart, 1981.

[B5] Borel, A.: Regularization theorems in Lie algebra cohomology. Applications, *Duke Math. J.* 50 (1983) 605–623.

[BC] Borel, A. and Casselman, W.: $\mathbf{L}^2$-Cohomology of locally symmetric manifolds of finite volume, *Duke Math. J.* 50 (1983) 625-647.

[BG] Borel, A. and Garland, H.: Laplacian and discrete spectrum of an arithmetic group, *Amer. J. Math.* 105 (1983) 309–335.

[BJ] Borel, A. and Jacquet, H.: Automorphic Forms and Automorphic Representations in *Automorphic Forms, Representations and L-functions Proc. of Symp. in Pure Math.* 33(1), 189–202, A.M.S., Providence, 1979.

[BS] Borel, A. and Serre, J.-P.: Cohomologie d'immeubles et de groupes S-arithmétiques, *Topology* 15 (1976) 211–232.

[BT] Borel, A. and Tits J.: Homomorphismes 'abstraits' de groupes algébriques simples, *Annals of Math.* 97 (1973) 499–571.

[BW] Borel, A. and Wallach, N.: Continuous Cohomology, Discrete Subgroups, and Representations of Reductive Groups, *Annals of Math. Studies* 94, Princeton Univ. Press, Princeton, 1980.

[Bou] Bouaziz, A.: Sur les caractères des groupes de Lie réductifs non connexes, *Journal of Funct. Anal.* 70 (1987) 1–79.

[Ca1] Casselman, W.: A new non-unitary argument for $\mathfrak{p}$-adic representations, *J. Fac. Sci. Univ. Tokyo* 28 (1982) 907–928.

[Ca2] Casselman, W.: Introduction to the Schwartz space of $\Gamma\backslash G$, *Can. J. Math* XLI (1989) 285–320.

[Cl1] Clozel, L.: Characters of non-connected p-adic groups, *Canad. J. Math.* 39 (1987) 149–167.

[Cl2] Clozel, L.: Représentations galoisiennes associées aux représentations automorphes autoduales de $\mathrm{GL}(n)$, *Publ. Math. IHES* 73 (1991) 97–145.

[Cl3] Clozel, L.: On the cohomology of Kottwitz's arithmetic varieties, *Duke Math. J.* 72 (1993) 757–795.

[CD] Clozel, L. and Delorme, P.: Le théorème de Paley-Wiener invariant pour les groupes de Lie réductifs II, *Ann. Sci. Ec. Norm. Sup. 4e serie* 23 (1990) 193–228.

[D] Delorme, P.: Théorème de Paley-Wiener invariant tordu pour le changement de base $\mathbf{C}/\mathbf{R}$, *Compositio Math.* 80 (1991) 197–228.

[DM] Dixmier, J. and Malliavin, P.: Factorisations de fonctions et de vecteurs indéfiniment différentiables, *Bull. Sci. Math.* 102 (1978) 307–330.

[F] Franke, J.: Harmonic analysis in weighted $\mathbf{L}^2$-spaces. In preparation.

[HC] Harish–Chandra Automorphic Forms on Semisimple Lie Groups, *Lect. Notes in Math.* 62, Springer-Verlag, Berlin, Heidelberg, New York, 1968.

[Hel] Helgason, S.: Differential Geometry and Symmetric Spaces, *Pure and Applied Math.* 12, Academic Press, 1962.

[J] Johnson, J. F.: Stable base change $\mathbf{C}/\mathbf{R}$ of certain derived functor modules, *Math. Annalen* 287 (1990) 467–493.

[Ko] Kottwitz, R. E.: Tamagawa numbers, *Annals of Math.* 127 (1988) 629–646.

[Lab] Labesse, J.-P.: Pseudo-coefficients très cuspidaux et K-theorie, *Math. Ann.* 291 (1991) 607–616.

[LS] Labesse, J.-P. and Schwermer, J.: On liftings and cusp cohomology of arithmetic groups, *Invent. Math.* 83 (1986) 383–401.

[Lan1] Langlands, R. P.: On the Functional Equations Satisfied by Eisenstein Series, *Lect. Notes in Math.* 544, Springer, Berlin, Heidelberg, New York, 1976.

[Lan2] Langlands, R. P.: Letter to A. Borel dated October 25, 1972.

[Lau] Laumon, G.: Cohomology with compact support of Drinfeld modular varieties. Part I. Publication de l'Université de Paris-Sud, 1991.

[MW] Mœglin, C. and Waldspurger, J-L.: *Décomposition spectrale et séries d'Eisenstein*, Progress in Math. 113, Birkhäuser, Boston, Basel, Berlin, 1994.

[Mü] Müller, W.: The trace class conjecture in the theory of automorphic forms, *Annals of Math.* 130 (1989) 473–529.

[OW] Osborne, M. S. and Warner, G.: The Selberg trace formula II: Partition, reduction, truncation, *Pacific J. Math.* 106 (1983) 307–496.

[Ro] Rogawski, J. D.: Trace Paley Wiener theorem in the twisted case, *Transactions AMS* 309 (1988) 215–229.

[RS1] Rohlfs, J. and Speh, B.: Automorphic representations and Lefschetz numbers, *Ann. Sci. Ec. Norm. Sup. 4e série* 22 (1989) 473–499.

[RS2] Rohlfs, J. and Speh, B.: Lefschetz numbers and twisted stabilized orbital integrals, *Math. Ann.* 296 (1993) 191–214.

[RS3] Rohlfs, J. and Speh, B.: On the cuspidal cohomology of arithmetic groups and cyclic base change, *Math. Nach.* 158 (1992) 99–108.

[Sch] Schwermer, J.: Cohomology of arithmetic groups, automorphic forms and *L*-functions, Cohomology of Arithmetic Groups and Automorphic forms, *Lect. Notes in Math.* 1447, Springer, Berlin, Heidelberg, New York, 1990.

[Sil] Silberger, A.: Isogeny restriction of irreducible admissible representations are finite direct sums of irreducible admissible representations, *Proc. AMS* 73 (1979) 263–264.

[T] Tits, J.: Algebraic and abstract simple groups, *Annals of Math.* 80 (1964) 313–329.

[VZ] Vogan, D. and Zuckerman, G. J.: Unitary representations with non-zero cohomology, *Compositio Math.* 53 (1984) 51–90.

[W] Wallach, N.: On the constant term of a square-integrable automorphic form, Operator algebras and group representations II, *Monographs and Studies in Math.* 17 (1984) 227–237, Pitman, London.

161.

Class Functions, Conjugacy Classes and Commutators in Semisimple Lie Groups

in *Algebraic groups and Lie groups*, G. Lehrer ed., Australian Math. Soc. Lecture Series **9**, Cambridge University Press 1997, 1–19

To the memory of Roger Richardson

Introduction

Let G be a connected Lie group, $\pi : H \to G$ a covering of G. Let us say that the conjugacy class $C(g) = \{x.g.x^{-1} | x \in G\}$ of $g \in G$ splits in H if $\pi^{-1}(C(g))$ is the disjoint union of the conjugacy classes in H of the elements of $\pi^{-1}(g)$. J. Milnor pointed out that if $G = \mathbf{SL}_2(\mathbb{R})$ every conjugacy class splits completely in any covering. As a generalization and sharpening, we prove (see 1.2 for a slightly stronger statement and 1.3).

Theorem I. *Let G be a connected semisimple Lie group with finite center and $\pi : H \to G$ a covering.*

(i) *If $\pi_1(G)$ is torsion free, any conjugacy class in G splits in H.*
(ii) *If $\pi_1(G)$ has torsion and H is the universal covering of G, there is a conjugacy class in G which does not split.*

The first condition is fulfilled notably in the following case

(*) *G is linear, has no compact factor, its complexification is simply connected, and its quotient G/K by a maximal compact subgroup K is a bounded symmetric domain.*

(see 1.4). This holds in particular for the symplectic group $\mathbf{Sp}_{2n}(\mathbb{R})$ $(n \geqq 1)$. In that case, Milnor had also established the splitting of conjugacy classes in the universal covering by constructing a continuous class function on $\mathbf{Sp}_{2n}(\mathbb{R})$ which extends the determinant function of $K = \mathbf{U}(n)$. This, and a similar construction by G. Lusztig on an indefinite special unitary group, led to ask whether any continuous class function on K extends to one on G. We shall prove it more generally for linear reductive groups of inner type (3.5).

Theorem II. *Let G be a linear real reductive group of inner type (see 2.1), K a maximal compact subgroup and M a topological space. Then any M-valued continuous class function f on K extends to one on G.*

This extension is not necessarily unique, but the proof will exhibit one constructed canonically, with the same set of values. To describe it, we use a slight sharpening of the (multiplicative) Jordan decomposition in a real linear algebraic group, discussed in §2, which states that any $g \in G$ can be written uniquely as a product of three commuting elements g_e, g_d and g_a where g_e is "elliptic" (i.e. contained in a compact subgroup), g_d is contained in the identity component of a split torus, and g_u is unipotent. In fact, this is carried out slightly more generally for the class of "almost algebraic groups" (see 2.1), which includes the reductive groups of inner type.

The canonical extension F of f is then defined by $F(g) = f(C(g_e) \cap K)$. It is well defined since $C(g_e) \cap K$ is known to consist of exactly one conjugacy class in K (see 3.4) and the main point is to show that it is continuous. This follows directly from 3.3, which is the main technical result of this paper.

It is known that a connected semisimple linear Lie group G has finite *commutator length*, i.e. there exists an integer N such that any element in G is a product of at most N commutators (4.2). Milnor pointed out (see [20]) that this is not true anymore in the universal covering of $\mathbf{SL}_2(\mathbb{R})$. As a generalization, we shall show in §4 (see 4.2, 4.4):

Theorem III. *Let G be a connected semisimple Lie group. Then G has finite commutator length if and only if its center is finite.*

As can be seen from the above, the theorems proved here were mostly suggested by remarks of Milnor on $\mathbf{SL}_2(\mathbb{R})$ or $\mathbf{Sp}_{2n}(\mathbb{R})$. Those of §§1, 2, 3 were presented at an Oberwolfach meeting in 1976, in a somewhat less general form, but were not until now written up for publication.

0. Notation and conventions

0.1. Let G be a group. Then $\mathscr{D}G$ denotes its derived group and $\mathscr{C}G$ its center. If A is a subset of G, then $\mathscr{Z}_G(A)$ or $\mathscr{Z}(A)$ is its centralizer and $\mathscr{N}_G(A)$ or $\mathscr{N}(A)$ its normalizer:

$$\mathscr{Z}_G(A) = \{g \in G | g.a = a.g \ (a \in A)\}, \ \mathscr{N}_G(A) = \{g \in G | g.A = A \cdot g\}.$$

The inner automorphism $x \mapsto g.x.g^{-1}$ of G defined by g is denoted Int g. We also write ${}^xg$ for $x.g.x^{-1}$ and, if $A \in G$, ${}^GA$ for $\{gag^{-1} | g \in G, \ a \in A\}$.

The conjugacy class of g in G is denoted $C(g)$ or $C_G(g)$.

0.2. If G is a Lie group, then G^o denotes its connected component of the identity. If G is an algebraic group over $\mathbb{C}$, it is also the identity component of G in the Zariski topology. In this paper, all algebraic groups are linear, over $\mathbb{C}$. If $G \subset \mathbf{GL}_n(\mathbb{R})$, we let G_c be the complexification of G, or, equivalently, the smallest complex Lie subgroup of $\mathbf{GL}_n(\mathbb{C})$ containing G.

The Lie algebra of a Lie group is usually denoted by the corresponding German letter, and similarly for an algebraic group.

As usual, Ad_G is the adjoint representation of G in its Lie algebra. It associates to g the differential at 1 of Int g.

0.3. Let G be a Lie group with finitely many connected components. We recall that any compact subgroup of G is contained in a maximal one, the maximal ones are conjugate under G^o and G is topologically the product of any one of them by a euclidean space [13].

1. Inverse images of conjugacy classes

In this section, G and H are groups and $\pi : H \to G$ an epimorphism with kernel $N \subset \mathscr{C}H$.

Lemma 1.1. *Let $g \in G$ and $Q_g = \pi^{-1}(\mathscr{Z}_G(g))$.*

(a) *The following conditions are equivalent*

(i) *$Q_g = \mathscr{Z}_H(h)$ for some $h \in \pi^{-1}(g)$.*

(ii) *$Q_g = \mathscr{Z}_H(h)$ for all $h \in \pi^{-1}(g)$.*

(iii) *$C_G(g)$ splits in H.*

(iv) *π induces a bijection of $C_H(h)$ onto $C_G(g)$ for every $h \in \pi^{-1}(g)$.*

(b) *Let $h \in \pi^{-1}(g)$. The restriction to Q_g of the commutator map*

$$c_h : x \mapsto x.h.x^{-1}.h^{-1}$$

is a homomorphism of Q_g into N, with kernel $\mathscr{Z}_H(h)$.

Proof. Obviously, $\mathscr{Z}_H(h) = \mathscr{Z}_H(h.c)$ for all $h \in H$ and $c \in \mathscr{C}H$, hence (i) $\Leftrightarrow$ (ii) and

$$\mathscr{Z}_H(h) \subset Q_g \quad (h \in \pi^{-1}(g)). \tag{1}$$

On the other hand

$$x.h.x^{-1} = h.n \quad (x \in H,\ h \in \pi^{-1}(g),\ n \in N) \Leftrightarrow x \in Q_g \tag{2}$$

and

$$n \neq 1 \text{ in } (2) \Leftrightarrow x \notin \mathscr{Z}_H(h). \tag{3}$$

This shows that *(i)* $\Leftrightarrow$ *(iii)*. Let $x, y \in H$ and $h \in \pi^{-1}(g)$. Then

$$x.h.x^{-1} = y.h.y^{-1} \Leftrightarrow x \in y.\mathscr{Z}_H(h) \tag{4}$$

$$\pi(x.h.x^{-1}) = \pi(y.h.y^{-1}) \Leftrightarrow x \in y.Q_g, \tag{5}$$

whence *(i)* $\Leftrightarrow$ *(iv)*. This proves (a).

(b) The map c_h is constant on the coset $x.\mathscr{Z}_H(h)$. If $x \in Q_g$, then $c_h(x) \in N$ by (2) and then it is clear that c_h is a homomorphism of Q_g into N, with kernel $\mathscr{Z}_H(h)$.

Theorem 1.2. *Assume that G and H are connected semisimple Lie groups and that $\mathscr{C}G$ is finite.*

(i) *If N is torsion-free, then $C_G(g)$ splits in H for every $g \in G$.*

(ii) *If N has torsion and the maximal compact subgroups of H are semisimple then there exists $g \in G$ such that $C_G(g)$ does not split in H.*

Proof. (i) Let $g \in G$, let $h \in \pi^{-1}(g)$ and $Q_g = \pi^{-1}\big(\mathscr{Z}_G(g)\big)$. By 1.1, it suffices to show that $Q_g/\mathscr{Z}_H(h)$ is a finite group. Both groups contain N, hence $Q_g/\mathscr{Z}_H(h) \cong \mathscr{Z}_G(h)/\pi(Z_H(h)\big)$. By 1.1(b), $Q_g^\circ = \mathscr{Z}_H(h)^\circ$, hence $\big(\pi(\mathscr{Z}_H(h))\big)^\circ = \mathscr{Z}_G(g)^\circ$. Therefore $Q_g/\mathscr{Z}_H(h)$ is isomorphic to a quotient of $\mathscr{Z}_G(g)/\mathscr{Z}_G(g)^\circ$ and it suffices to prove that $\mathscr{Z}_G(g)$ has finitely many connected components.

Let G' be the adjoint group of G and $\sigma : G \to G'$ the canonical projection. Its kernel is $\mathscr{C}G$ and is finite by assumption. It suffices therefore to show that $\mathscr{Z}_{G'}(\sigma(g))$ has finitely many connected components. This reduces us to the case where G is linear. Let G_c be the complexification of G. It is a connected semisimple algebraic group and $G = G_c(\mathbb{R})^o$. The group $\mathscr{Z}_G(g)$ is equal to the group of real points of the algebraic group $\mathscr{Z}_{G_c}(g)$, hence has finitely many connected components ([19] or [3], 6.4). The same is then true of $\mathscr{Z}_{G_c}(g) \cap G = \mathscr{Z}_G(g)$.

(ii) Assume that N contains an element $z \neq 1$ of finite order. Any maximal compact subgroup K of H contains z. By assumption, K is semisimple. It is connected, since H is. By [8] there exists $a, b \in K$ such that $a.b.a^{-1}.b^{-1} = z$. We have then $z.b = a.b.a^{-1}$, which shows that $C_G(\pi(b))$ does not split in H (see 1.1).

1.3. Proof of Theorem I. Let $\tilde{\pi} : \tilde{G} \to G$ be the universal covering of G. Then $\pi_1(G) = \ker \tilde{\pi}$. If H is a covering of G, then it also has $\tilde{G}$ as a universal covering and $\pi_1(H)$ is a subgroup of $\pi_1(G)$. Assume the latter is torsion-free. Then so is $\pi_1(H)$. By 1.2(i) any conjugacy class in H or G is the bijective image of a conjugacy class in $\tilde{G}$ under the natural projection. The same is then true for $\pi : H \to G$. In view of 1.1, this implies (i). The maximal compact subgroups of $\tilde{G}$ are simply connected (0.3), in particular semisimple. Therefore (ii) is a special case of 1.2(ii).

Proposition 1.4. *Assume that G is a connected semisimple linear Lie group without compact factor, whose quotient by a maximal compact subgroup K is a bounded symmetric domain and that the complexification G_c of G is simply connected. Then $\pi_1(G)$ is torsion-free.*

Proof. Let S be the identity component of the center of K. The assumptions on G and G/K imply that $K = \mathscr{Z}_G(S)$. Furthermore, K is a maximal compact subgroup of the centralizer M of S in G_c. and $\mathscr{D}K$ a maximal compact subgroup of $\mathscr{D}M$. Since G_c is simply connected, it is known that $\mathscr{D}M$ is simply connected, hence so is $\mathscr{D}K$. But K is homeomorphic to the product of $\mathscr{D}K$ and of a torus, [1, 3.2], as already recalled. Hence $\pi_1(K)$ has no torsion.

2. Jordan decomposition in almost algebraic Lie groups

2.1. Let G be a Lie group with finitely many connected components. It is said to be *almost real algebraic* if there is given a linear algebraic $\mathbb{R}$-group H and a morphism of Lie groups $\sigma : G \to H(\mathbb{R})$ with finite central kernel, and image open in $H(\mathbb{R})$.

We view H and σ as part of the data, although the notation may not always indicate it. If (G', σ', H') is another such triple, a morphism $\varphi : G \to G'$ consists in fact of a morphism of Lie groups $\varphi : G \to G'$, and of a $\mathbb{R}$-morphism of algebraic groups $\psi : H \to H'$ such that $\sigma' \circ \varphi = \psi \circ \sigma$.

The Lie group G is *reductive* if it is almost algebraic and H^o is a reductive algebraic group, i.e. has a unipotent radical reduced to $\{1\}$. It is *reductive of inner type* i.e. if $\mathrm{Ad}_H \sigma(G) \subset \mathrm{Ad}_H H^o$. The notion of reductive group of inner type is the same as that of ([18], 2.11, 2.28) and essentially equivalent to that of reductive group in [10, §3], also adopted in [17].

Our object of interest is G so, replacing H by a subgroup of finite index if needed, we may always assume $\sigma(G)$ to be Zariski dense in H. Note that, since H is of inner type, $\mathscr{C}G^o \subset \mathscr{C}G$ and $\mathscr{C}H^o \subset \mathscr{C}H$; moreover, $\mathscr{Z}_H(H^o)$ meets every connected component of H. The subgroup $\mathscr{Z}_H(H^o)$ is normal in H, defined over $\mathbb{R}$ and $H/\mathscr{Z}_H(H^o)$ is defined over $\mathbb{R}$, connected, semisimple of adjoint type. The group $\big(\mathscr{Z}_G(G^o)\big)^o$ is a real form of $\mathscr{Z}_H(H^o)^o$, the group $\sigma\big(\mathscr{Z}_G(G^o)\big)$ is of finite index in $\mathscr{Z}_H(H^o)(\mathbb{R})$, the map σ induces a morphism $\sigma' : G/\mathscr{Z}_G(G^o) \to H/\mathscr{Z}_H(H^o)$ and $\big(G/\mathscr{Z}_G(G^o), H/\mathscr{Z}_H(H^o), \sigma'\big)$ is an almost algebraic semisimple group of inner type.

2.2. Let G be almost algebraic and $g \in G$.

(a) A closed connected subgroup A is a *split toral subgroup* if $\sigma(A) = S(\mathbb{R})^o$, where S is a $\mathbb{R}$-split torus in H. Clearly, A is the identity component of $\sigma^{-1}\big(S(\mathbb{R})\big)^o$ and σ is an isomorphism of A onto its image $S(\mathbb{R})^o$.

(b) g is said to be *split positive* if it belongs to a split toral subgroup. Then g is the only split positive element in $\sigma^{-1}\big(\sigma(g)\big)$. The element g is split positive if and only if $\sigma(g)$ is diagonalisable over $\mathbb{R}$ and has strictly positive eigenvalues.

(c) g is *unipotent* if either $g = 1$ or $\sigma(g)$ is unipotent and $\neq 1$. Let X be the unique element in $\mathfrak{g}$ such that $\sigma(g) = \exp\sigma(X)$. Then $g = \exp X$ and is the unique unipotent element in $\sigma^{-1}\big(\sigma(g)\big)$.

(d) g is *elliptic* if it belongs to a compact subgroup of G, and semisimple if $\sigma(g)$ is so in H.

2.3. Many of the properties of semisimple groups (Cartan decomposition, Iwasawa decomposition etc.) extends to reductive groups of inner type, as is shown in those references given in 2.1. In particular we shall use the following:

The maximal compact subgroups of G are conjugate under G^o. Let K be one. There is a unique involution θ_K of G with fixed point set K, called the Cartan involution of G with respect to K. Let $\mathfrak{p}$ be the (-1)-eigenspace of in $\mathfrak{g}$. Then $(k, X) \mapsto k.\exp X$ is an isomorphism of manifolds of $K \times \mathfrak{p}$ onto G. In particular $\exp\mathfrak{p} = P$ is closed. Any split toral subgroup of G is conjugate under K to a split toral subgroup contained in P. The maximal ones are conjugate under K. If G is commutative, then P is the biggest split toral subgroup of G and $G = K \times P$.

Let M be a subgroup of G which is stable under θ_K. Then M is reductive, and the restriction of θ_K to M is a Cartan involution of M. In particular $M \cap K$ is a maximal compact subgroup of M, and we have the Cartan decomposition $M = (M \cap K).\exp(\mathfrak{m} \cap \mathfrak{p})$. For this, see [11].

Proposition 2.4. *Let G be an almost algebraic group and $g \in G$.*

(i) *There exist unique elements g_s, g_u such that $g = g_s.g_u = g_u.g_s$ and g_s (resp. g_u) is semisimple (resp. unipotent).*

(ii) *Let g be semisimple. There exist unique elements g_e and g_d in G such that $g = g_e.g_d = g_d.g_e$ and g_e (resp. g_d) is elliptic (resp. split positive).*

If G is linear, g is elliptic (resp. split positive) if and only its eigenvalues are of modulus one (resp. real and > 0).

Proof. (i) Let $g' = \sigma(g)$. Then $g' = g'_s.g'_u$ with g'_u unipotent, g'_s semisimple and $g'_s g'_u = g'_u g'_s$ by the usual Jordan decomposition ([2], 4.4). If $g'_u \neq 1$, there exists a unique unipotent g_u in the inverse image of g'_u. It commutes with g. Then $g_s = g.g_u^{-1}$ commutes with g, and is semisimple since $\sigma(g_s) = g'_s$. This proves the existence of one decomposition. If $g = \tilde{g}_s.\tilde{g}_u$ is another one, then $\sigma(\tilde{g}_s) = g'_s$ and $\sigma(\tilde{g}_u) = g'_u$ by the uniqueness of the Jordan decomposition in H, whence $\tilde{g}_u = g_u$ and then $\tilde{g}_s = g_s$.

(ii) Let g be semisimple and let M be the smallest algebraic subgroup of H containing $\sigma(g)$. If $M(\mathbb{R})$ is compact, then so is its inverse image and g is elliptic. Assume it is not. It is defined over $\mathbb{R}$. and is of multiplicative type, i.e. it consists of semisimple elements and is commutative. Let S be the greatest $\mathbb{R}$-split torus contained in M and $A' = S(\mathbb{R})^o$. The group $M(\mathbb{R})$ is the direct product of A' and its greatest compact subgroup K'. Then $\sigma^{-1}\big(M(\mathbb{R})\big) = K \times A$, where K is compact, A is the identity component of $\sigma^{-1}(A')$ and is a split toral subgroup. Then $g = g_e.g_d$ with $g_e \in K$, hence elliptic, $g_d \in A$, hence split positive, and $g_e.g_d = g_d.g_e$. In view of its construction, $\sigma^{-1}\big(M(\mathbb{R})\big)$ is invariant under any automorphism of G leaving g and $\ker\sigma$ stable. In particular, any element of G commuting with g also commutes with g_e and g_d. Let now $g = g'_e.g'_d$ be another such decomposition. Then g_e, g_d, g'_e, g'_d commute pairwise. Assume first that $G \subset H(\mathbb{R})$. Then $g'_e g_e^{-1} = g_d {g'}_d^{-1}$. The right-hand side is diagonalisable over $\mathbb{R}$, with strictly positive eigenvalues, whereas the left-hand side is diagonalisable over $\mathbb{C}$, with eigenvalues of modulus one. Hence both sides are equal to 1, which proves the uniqueness of the decomposition in that case. Coming back to the general case, we see that $\sigma(g'_d) = \sigma(g_d)$, whence $g'_d = g_d$ (see 2.2 (a)) and the uniqueness. This proves the first assertion. If G is linear, then σ is the identity, and the second assertion follows, too.

Remark. A decomposition similar to (ii) is given in [17], part II, p. 41.

Corollary 2.5. *Given $g \in G$, there exist unique commuting elements g_e, g_d, g_u such that $g = g_e.g_d.g_u$, where g_e (resp. g_d, resp. g_u) is elliptic (resp. split positive, resp. unipotent).*

This decomposition will be referred to as the *refined Jordan decomposition.* It is invariant under morphisms (as defined in 2.1).

2.6. Let G be reductive of inner type. Let N be a maximal unipotent subgroup of G, necessarily contained in G^o, A a maximal split toral subgroup normalizing N and K a maximal compact subgroup. Then $k \times a \times n \mapsto k.a.n$ is a diffeomorphism of $K \times A \times N$ onto G (Iwasawa decomposition, see e.g. [18], 2.1.8). Moreover $M.A.N$, where $M = \mathcal{Z}_K(A)$, is a minimal parabolic subgroup of G and any split toral subgroup (resp. unipotent subgroup) of G is conjugate to a subgroup of A (resp. N).

Let $g \in G$. As a consequence of the above, we see that if $g_e \in \mathcal{C}G$, then g is conjugate to an element $h \in M.A.N$ such that $h_e = g_e$, $h_d \in A$ and $h_u \in N$.

3. Extension of certain class functions in linear reductive groups of inner type

The following lemma is undoubtedly well-known. Not knowing of a reference, we give a proof for the sake of completeness.

Lemma 3.1. *Let L be a Lie group, $g \in L$ such that* $\mathrm{Ad}\, g$ *is semisimple,* $Z = \mathscr{Z}_L(g)$ *and* $\mathfrak{m}$ *a supplement to the Lie algebra* $\mathfrak{z}$ *of* $\mathscr{Z}_L(g)$ *in* $\mathfrak{g}$ *which is stable under* $\mathrm{Ad}\, g$. *Let* $\tau : \mathfrak{m} \times Z \to G$ *be the map defined by* $\tau(X, z) = \mathrm{Ad} \exp X(z)$ $(X \in \mathfrak{m},\, z \in Z)$. *Then its differential* $d\tau_0$ *at* $(0, g)$ *is invertible.*

Proof. The tangent space to $\mathfrak{m} \times Z$ (resp L) at $(0, g)$ (resp. g) is $\mathfrak{m} \oplus \mathfrak{z}.g$ (resp. $\mathfrak{g}.g$). Let $X \in \mathfrak{m}$, $Y \in \mathfrak{z}$. Then

$$\tau(sX, e^{tY}) = e^{sX}.g.e^{tY}.e^{-sX}.g^{-1}.g \qquad (s, t \in \mathbb{R}).$$

From

$$d\tau_0(X, Y) = \frac{d^2\tau}{dsdt} e^{sX}.e^{tY}.ge^{-sX}.g^{-1}.g\Big|_{s=t=0}$$

we get

$$d\tau_0(X, Y) = (I - \mathrm{Ad}\, g)X + Y.$$

Since $\mathrm{Ad}\, g - I$ is invertible on $\mathfrak{m}$, this proves the lemma.

Lemma 3.2. *Let G be a reductive linear group, $a \in \mathscr{C}G$ an elliptic element, $b \in G$ a semisimple element, $x_n, g_n \in G$ $(n = 1, 2\ldots)$ such that $x_{n,d} = 1$, $x_{n,e} \to a$, $g_n.x_n g_n^{-1} \to b$. Then $a = b$.*

Let K be a maximal compact subgroup of G containing a. The eigenvalues of b, in the given linear embedding of G, are limits of those of x_n, hence of $x_{n,e}$ and are therefore of modulus one. Consequently, b is elliptic (2.4) and there exists $g \in G$ such that $b' = g.b.g^{-1} \in K$. Let H be the complexification of G. Let L be a maximal compact subgroup of H containing K. It is a real form of H. By the Peter-Weyl theorem, the conjugacy classes in L are separated by the characters of the irreducible finite dimensional continuous (or real analytic) representations of L. By Tannaka duality (see [6], Chapter VI) the characters of the real analytic representations of L are the restrictions of linear combinations of those of the holomorphic or antiholomorphic finite dimensional irreducible representations of H. If σ is one and χ its character, then

$$\chi(b) = \lim_{n\to\infty} \chi(g_n.x_n.g_n^{-1}) = \lim_{n\to\infty} \chi(g_n.x_{n,e}.g_n^{-1}) = \lim_{n\to\infty} \chi(x_{n,e}) = \chi(a)$$

therefore a and b are conjugate in L. Since a is central also in H, this implies $b = a$.

The following theorem is the main technical result of this paper. Lemma 3.2 is a special case except for the fact that G is not assumed there to be of inner type.

Theorem 3.3. *Let G be a real linear reductive Lie group of inner type (see 2.1) and K a maximal compact subgroup of G. Let $a \in K$, $y \in G$ and $b = y_e$. Let g_n, x_n $(n = 1, 2\ldots)$ be sequences of elements in G such that $x_{n,e} \to a$ and $g_n.x_n.g_n^{-1} \to y$. Then a and b are conjugate in G.*

If G is compact, then $y = b$, and $x_n = x_{n,e}$. We may assume that $g_n \to g_o$ and get $b = g_o.a.g_o^{-1}$, which proves the theorem. From now on, G is assumed to be non compact. We let H be the complexification of G. Then G is Zariski dense in H and of finite index in $H(\mathbb{R})$.

The proof consists in a reduction in several steps to a special case where H is semisimple, connected, in which we can avail ourselves of a theorem of [9].

a) *We may assume* $b \in K$. Let $s \in G$ be such that ${}^s b \in K$. We have $({}^s y)_e = {}^s y_e = {}^s b$. Moreover

$${}^s y_n = (sg_n).x_n.(sg_n)^{-1} \to {}^s y.$$

The theorem, assumed to be proved if $b \in K$, shows that a is a conjugate to ${}^s b$, hence also to b.

b) *We may assume in addition that* $y = y_s$ *and* $g_n \in G^o$.

Let $\{V_j\}(j = 1, 2\dots)$ be a decreasing sequence of neighborhoods of 1 converging to $\{1\}$. We claim we can find $z_j \in \mathscr{Z}_G(y_s)$ such that $z_j.y_u.z_j^{-1} \in V_j$ $(j = 1, 2\dots)$.

If $y_u = 1$ there is nothing to prove. So assume $y_u \neq 1$. There exists then a unique $X \in \mathfrak{g}$ such that $\exp X = y_u$. The element X is nilpotent and belongs to the Lie algebra of the derived group of $\mathscr{Z}(g_s)$, which is semisimple. By the Jacobson-Morosow theorem, we can find in $\mathscr{D}\big(\mathscr{Z}(y_s)\big)$ a three-dimensional $\mathbb{R}$-split simple group containing $\exp \mathbb{R}X$ and a split toral subgroup D normalizing $\exp \mathbb{R}X$. The elements of D act by dilations on X under the adjoint representation. For any strictly positive real number r there is $d \in D$ such that $\mathrm{Ad}\, d$ is the dilation by r on X; the existence of the z_j's follows. Then ${}^{z_j}y \to y_s$. Choose a neighborhood W_j of y such that ${}^{z_j}W_j \subset y_s.V_j$. If $n(j)$ is such that

$$g_{n(j)}.x_{n(j)}\, g_{n(j)}^{-1} \in W_j.$$

Then

$$z_j.g_{n(j)}.x_{n(j)}.\, g_{n(j)}^{-1}.z_j^{-1} \to y_s.$$

Changing the notation, we may assume that our original sequence is this subsequence. So we have $g_n.x_n.g_n^{-1} \to y_s$. The group G has finitely many connected components. Passing again to a subsequence, we may assume that there exists $c \in G$ such that $g_n = c.g'_n$ with $g'_n \in G^o$. Then

$$g'_n.x_n.g'^{\,-1}_n \to c^{-1}.y_s.c.$$

The proposition, assumed to be proved under the assumption b), shows that a is conjugate to $c^{-1}.b.c$, hence also to b.

c) Assume G^o to be *commutative*. Then it is central in G, as already pointed out. Therefore we have $x_{n,e} \to a$, and $x_{n,d}.x_{n,e} \to y$. Also (see 2.3), G^o is the direct product of a compact subgroup K and a split toral subgroup A, therefore $x_{n,d} \to y_d$ and $x_{n,e} \to y_e = b$, whence $a = b$. In particular, this proves 3.3 when $\dim G \leqq 2$. From now, we assume the theorem established in dimensions $< \dim G$.

d) *We may assume in addition that* $\mathscr{C}G$ *is compact.*

Assume $\mathscr{C}G$ is not compact. Since G is of inner type, $\mathscr{C}G^o$ and $\mathscr{C}G$ have the same connected component of the identity, and the same is true for $H(\mathbb{R})$ in the ordinary topology and for H in the Zariski topology.

Let S be the greatest split torus of the center of H. Since $\mathscr{C}H(\mathbb{R})$ is not compact, $\dim S \geqq 1$. Then $A = S(\mathbb{R})$ is the greatest central split toral subgroup of G. We claim that there exists a normal subgroup G_1 of G such that $G = G_1 \times A$. Let L be the greatest compact subgroup of $\mathscr{C}G^o$. Then $\mathscr{C}G^o = L \times A$. The algebra $\mathfrak{g}$ being the direct sum of its derived algebra and of its center, we see that the group G_1^o generated by L and $\mathscr{D}G^o$ is the almost direct product of these groups and $G^o = G_1^o \times A$. The group G_1^o is invariant in G. We let $G_1 = K.G_1^o$; then G_1^o is indeed the identity component of G_1. Since $\mathscr{C}G_1^o \subset \mathscr{C}G^o$ we have $\mathscr{C}G_1^o \subset \mathscr{C}G_1$. The group G_1^o is normal in G^o, hence its Zariski closure H_1^o in H is normal in H. For $h \in H^o$ the restriction of $\mathrm{Ad}\, h$ to the Lie algebra of H_1 belongs to $\mathrm{Ad}\, H_1^o$. It follows that G_1 is reductive, of inner type, and of course linear. We can write uniquely $y = y' \cdot z$ with $y' \in G_1$ and $z \in A$. Then $y'_e = y_e = b$. Similarly we have

$$x_n = x'_n.z_n \qquad (x'_n \in G_1,\, z_n \in A,\, x'_{n,e} = x_{n,e}, n = 1, 2, \ldots)$$
$$g_n = g'_n.q_n \qquad (g'_n \in G_1,\, q_n \in A,\, n = 1, 2 \ldots)$$

hence

$$x'_{n,e} \to a,\; g'_n.x'_n.g'^{-1}_n \to y',$$

so that the theorem, applied to G_1, as is allowed by our induction assumption, shows that a is conjugate to b already in G_1.

e) *We may assume in addition that* $x_{n,d} = 1$.

Assume that $x_{n,d} \neq 1$ for infinitely many n's, hence, going over to a subsequence, for all n's. Let S be a maximal $\mathbb{R}$-split torus of H. Since $\mathscr{C}G^o$ is compact, S belongs to $\mathscr{D}H^o$. Then $A = S(\mathbb{R})^o$ is a maximal split toral subgroup of G, belongs to $\mathscr{D}G^o$ and $A \cap \mathscr{C}G = \{1\}$.

Let θ be the Cartan involution of G with respect to K and $\mathfrak{p}$, P be as in 2.3. As recalled there we may, possibly after conjugation by an element of K, assume $A \subset P$. Let $Z_n = \mathscr{Z}_G(x_{n,e})$. It is stable under θ, since $x_{n,e} \in K$, reductive and contains x_n, $x_{n,d}$. The restriction of θ to Z_n is a Cartan involution and

$$Z_n = (K \cap Z_n).(P \cap Z_n)$$

is a Cartan decomposition, i.e. $K_n = K \cap Z_n$ is maximal compact in Z_n and $P \cap Z_n = \exp.\mathfrak{p}_n$, where $\mathfrak{p}_n = \mathfrak{p} \cap \mathfrak{z}_n$. There exists $u_n \in K_n$ such that ${}^{u_n}x_{n,d} \in P$ and $k_n \in K$ such that $k_n.u_n.x_{n,d}.u_n^{-1}.k_n^{-1} \in A$ (see 2.3).

Passing to a subsequence, we may assume that $k_n \to k_o \in K$. Let $g'_n = g_n.k_n^{-1}.u_n^{-1}$ and $x'_n = k_n.u_nx_n.u_n^{-1}.k_n^{-1}$. Then

$$x'_{n.e} = {}^{k_nu_n}x_{n,e} \to {}^{k_o}a, \quad x'_{n,d} = {}^{k_nu_n}x_{n,d} \in A.$$

If b is conjugate to ${}^{k_o}a$, then it is conjugate to a as well. Therefore we are reduced to the case where $x_{n,d} \in A$ for all n's.

Let $\Phi = \Phi(S;\, H)$ be the set of roots of H^o with respect to S. It may also be viewed as the set of roots of G with respect to A, or of $\mathfrak{g}$ with respect to $\mathfrak{a}$. For a subset Ψ of Φ, let A_Ψ be the intersection the kernels of the $\alpha \in \Psi$. There exists Ψ such that for infinitely many n's, $x_{n,d} \in A_\Psi$ and does not belong to $A_{\Psi'}$ for any bigger set Ψ'. Passing to a subsequence, we assume this is the case for all n's. Let $A' = A_\Psi$ and S' the Zariski closure of A' in H. It is a split torus contained in S. The group A belongs to $\mathscr{D}G$, as already remarked. Therefore the differentials of the restrictions of the roots to the Lie algebra $\mathfrak{a}'$ of A' span the dual of $\mathfrak{a}'$. As a consequence, A' is the split center of its centralizer and $\mathscr{Z}_G(A')$ is of finite index in its normalizer, which is equal to the normalizer $\mathscr{N}_G(A')$ of A'. In view of the construction, $\mathscr{Z}_G(x_{n,d}) = \mathscr{Z}_G(A')$, hence $x_n \in \mathscr{Z}_G(A')$ $(n = 1, 2, \ldots)$.

Going over to a subsequence, we may assume that the $x_{n,d}$ belong to the same Weyl chamber in A', i.e. to the same connected component C of the complement of the union of the subgroups $\ker\alpha\,(\alpha \in \Phi - \Psi)$. There exists then a parabolic subgroup $P \subset H^o$ with Levi subgroup $\mathscr{Z}_H(S')$ and unipotent radical $\mathscr{R}_uP$ corresponding to the positive roots for the order defined by C ([2], 20.4). Let $U = \mathscr{R}_uP(\mathbb{R})$ and $Q = \mathscr{Z}_G(A').U$.

Let $G' = Q/U$ and $\pi : G \to G'$ the canonical projection. The group G' is linear, reductive and of inner type ([18], 2.2.8).

We have $G = K.Q^o$ as follows from the Iwasawa decomposition (2.6), and can write $g_n = k_n.q_n$ with $k_n \in K$ and $q_n \in Q^o$. Passing to a subsequence, we may assume that $k_n \to k_o \in K$. Then

$$q_n.x_n.q_n^{-1} \to k_o^{-1}.y.k_o,$$

therefore $k_o^{-1}.y.k_o \in Q$. The induction assumption, applied to G', shows the existence of $v \in \mathscr{Z}_G(A')$ such that

$${}^{\pi(v)}\pi(a) = \pi(k_o^{-1}bk_o).$$

The element $k_o^{-1}.b.k_o$ is semisimple, belongs to Q. Therefore ([12], §7) there exists $u \in U$ such that

$$u.k_o^{-1}.b.k_o.u^{-1} \in \mathscr{Z}_G(A').$$

This element has the same image, under π, as $k_o^{-1}.b.k_o$. Since π is an isomorphism of $\mathscr{Z}_G(A')$ onto G', we see that a is conjugate to $u.k_o^{-1}.b.k_o.u^{-1}$, hence to b. This proves e).

f) We have now reduced the proof of 3.3 to the following case:

$$(*)\ b \in K,\ y = y_s,\ \mathscr{C}G \text{ compact},\ g_n \in G^o \text{ and } x_{n,d} = 1,\ (n = 1, 2 \ldots).$$

We first claim that

$$y = y_e = b. \tag{1}$$

The kernel of Ad_G is $\mathscr{C}G$, hence is compact. It contains therefore no split positive element, except for the identity. Hence a semisimple element $x \in G$ is elliptic if and only if the eigenvalues of $\mathrm{Ad}_G\, x$ are of modulus one. The eigenvalues of $\mathrm{Ad}_G\, y$ are the limits of those of the elements $\mathrm{Ad}_G\, g_n.x_n.g_n^{-1}$, hence of $\mathrm{Ad}_G\, x_{n,e}$ since $x_{n,s} = x_{n,e}$ by $(*)$, which are of modulus one. Therefore y, which is semisimple by $(*)$, is elliptic, which proves (1).

The element y is then conjugate to one in K, so it is enough to consider the case where

$$y = y_e = b \in K. \tag{2}$$

Let L be a maximal compact subgroup of H containing K. The argument in 3.2, based on the Peter-Weyl theorem and Tannaka duality, shows that a and b are conjugate in L. This implies first that either

$$G^o \subset \mathscr{Z}_G(a) \cap \mathscr{Z}_G(b) \tag{3}$$

or

$$G^o \not\subset \mathscr{Z}_G(a) \qquad G^o \not\subset \mathscr{Z}_G(b). \tag{4}$$

Assume (3). The element $x_{n,u}$ belongs to G^o, hence $x_n \in G^o.a$ and then $b \in G^o.a$, too. We may replace G by the subgroup generated by G^o and a, hence assume that $a \in \mathscr{C}(G)$. Then 3.2 implies that $a = b$, which proves the theorem in that case.

g) There remains to establish 3.3 under the assumptions $(*)$, (1) and (4).

We first show that we can assume, in addition

$$x_{n,u} = 1 \qquad \text{for all } n\text{' s.} \tag{5}$$

By 3.1, we can find sequences of elements $q_n \in G$ and $s_n \in \mathscr{Z}_G(b)$ such that

$$q_n \to 1,\ s_n \to b,\ g_n.x_n.g_n^{-1} = q_n.s_n.q_n^{-1} \qquad (n \geqq 0). \tag{6}$$

The element b belongs to K, hence $\mathscr{Z}_G(b)$ is stable under the Cartan involution θ_K associated to K and (see 2.3) $K \cap \mathscr{Z}_G(b)$ is a maximal compact subgroup of $\mathscr{Z}_G(b)$. There exists therefore elements $r_n \in \mathscr{Z}_G(b)$ such that

$$r_n.s_{n,e}.r_n^{-1} \in K \cap \mathscr{Z}_G(b) \qquad (n \geqq n_o).$$

Passing to a subsequence, we may assume that $r_n.s_{n,e}.r_n^{-1} \to a' \in K \cap \mathscr{Z}_G(b)$. We now have

$$r_n^{-1}.(r_n.s_n.r_n^{-1})r_n = s_n \to b$$
$$(r_n s_n r_n^{-1})_e = r_n.s_{n,e}.r_n^{-1} \to a'.$$

Then 3.2 shows that $a' = b$. Now we have, taking (6) into account

$$r_n.q_n^{-1}.g_n.x_{n,e}.g_n^{-1}.q_n.r_n^{-1} = r_n.s_{n,e}.r_n^{-1} \to b$$

and of course still $x_{n,e} \to a$. This reduces us to the case where $x_n = x_{n,e}$ $(n \geqq 1)$, i.e. $x_n \in K$, as claimed.

By 3.1, applied to K, there exist sequences $q_n \in K$, $s_n \in \mathscr{Z}_G(a) \cap K = \mathscr{Z}_K(a)$ such that

$$q_n \to 1,\ s_n \to a,\ x_n = q_n.s_n.q_n^{-1} \in K.$$

We have $s_n = s_{n,e}$ and

$$g_n.q_n^{-1}.s_n.q_n.g_n^{-1} = g_n.x_n g_n^{-1} \to b.$$

Replacing g_n by $g_n.q_n^{-1}$ and x_n by s_n, we are reduced to the case where

$$x_n = x_{n,e} \in \mathscr{Z}_G(a) \cap K = \mathscr{Z}_K(a).$$

Let T be a maximal torus of $\mathscr{Z}_K(a)^\circ \cap K$. Then any element of $\mathscr{Z}_K(a)^\circ.a$ is conjugate under $\mathscr{Z}_K(a)^\circ$ to an element of $T.a$. [If $x \in \mathscr{Z}_K(a)^\circ.a$, there exists $k \in \mathscr{Z}_K(a)^\circ$ such that $k.x.a^{-1}.k^{-1} \in T$, whence $k.x.k^{-1} \in T.a$.] Let then $k_n \in \mathscr{Z}_K(a)^\circ$ be such that $k_n.x_n.k_n^{-1} \in T.a$.

We may assume that $k_n \to k_o \in \mathscr{Z}_K(a)^\circ$. We have

$$k_n.x_n.k_n^{-1} \to k_o.a.k_o^{-1} = a$$

$$g_n.k_n^{-1}.(k_n.x_n.k_n^{-1}).(g_n.k_n^{-1})^{-1} = g_n.x_n.g_n^{-1} \to b.$$

Replacing x_n by $k_n.x_n.k_n^{-1}$ and g_n by $g_n.k_n^{-1}$, we see that we can arrange to have $x_n \in T.a$.

h) We may now assume $(*)$, (1), (4) and

$$x_n \in T.a, \text{ where } T \text{ is a maximal torus of } \mathscr{Z}_G(a)^\circ \cap K. \tag{7}$$

We claim

$(**)$ *there exists a compact set C in G such that we can write*

$$g_n = u_n.v_n \qquad (u_n \in C, v_n \in \mathscr{Z}_G(a), n = 1, \dots).$$

We first conclude the proof assuming $(**)$. Passing to a subsequence, we may arrange that $u_n \to u$. Then

$$v_n.x_n.v_n^{-1} = u_n^{-1}.g_n.x_n.g_n^{-1}.u_n \to u^{-1}.b.u = b'$$

and therefore $b' \in \mathscr{Z}_G(a)$. Then $a = b'$ by 3.2, hence a is conjugate to b.

There remains to establish $(**)$. We use the construction outlined at the end of 2.1, and write R for $\mathscr{Z}_G(G^\circ)$ and S for $\mathscr{Z}_H(H^\circ)$. Then R is of finite index in $S(\mathbb{R})$. We have $R^\circ = (\mathscr{C}G^\circ)^\circ$, which is compact under our present assumptions. Therefore R is compact. Let $\mu : G \to G' = G/R$ be the canonical projection and $a' = \mu(a)$.

Then $\mu(x_n) \to a'$, $\mu(g_n).\mu(x_n).\mu(g_n^{-1}) \to \mu(b)$. Assume that $(**)$ holds in G'. We want to deduce $(**)$ in G. There exists a compact set $C' \subset G'$ and elements $u'_n \in C'$, $v'_n \in \mathscr{Z}_{G'}(a')$ such that $\mu(g_n) = u'_n.v'_n$. Since R is compact, $\mu^{-1}(C') = C$ is compact. We may find $\tilde{u}_n \in C$ mapping onto u'_n, hence such that $\tilde{v}_n = \tilde{u}_n^{-1}.g_n \in \mu^{-1}\big(\mathscr{Z}_G(a')\big)$. We claim that $\mathscr{Z}_G(a)$ has finite index in $\mu^{-1}\big(\mathscr{Z}_G(a')\big)$. The Lie algebra $\mathfrak{z}$ of $\mathscr{Z}_G(a)$ is the 1-eigenspace of Ad a and contains the Lie algebra of R $\big($since $\mathscr{C}G^\circ \subset \mathscr{C}G\big)$. The Lie algebra $\mathfrak{z}'$ of $\mathscr{Z}_{G'}(a')$ is the 1-eigenspace of $\mathrm{Ad}_{G'}\, a'$. Since these elements are semisimple,

$$d\mu(\mathfrak{z}) = \mathfrak{z}' \text{ and } d\mu^{-1}(\mathfrak{z}') = \mathfrak{z}$$

so that

$$\mu\big(\mathscr{Z}_G(a)^\circ\big) = \mathscr{Z}_{G'}(a')^\circ, \ \big(\mu^{-1}\big(\mathscr{Z}_{G'}(a)\big)\big)^\circ = \mathscr{Z}_G(a)^\circ.$$

Since these groups, being almost algebraic, have finitely many connected components, this proves our claim. Passing to a subsequence, we can find $c \in \mu^{-1}\big(\mathscr{Z}_{G'}(a')\big)$ such that $v'_n = c.v_n$ with $v_n \in \mathscr{Z}_G(a)$. We let then $u_n = \tilde{u}_n.c$. This proves (**) in G.

We may now replace G by G', hence H by $H' = H/S$. This reduces us to the case where H is connected semisimple, and even of adjoint type although we do not need that last fact. The normalizations (5), (7) still hold.

The centralizer $\mathscr{Z}_H(T)$ of T is connected ([2], 11.12) and contains a. There is a Cartan subgroup C of $\mathscr{Z}_H(T)$ containing T and a, and defined over $\mathbb{R}$. We now have

$$x_n \in C(\mathbb{R}) \cap \mathscr{Z}_G(a) \cap K, \qquad (n = 1, 2, \dots). \tag{8}$$

We want to deduce (**) from a theorem of [9], applied to H. To this effect, we need one further preliminary remark. Let $X = H/\mathscr{Z}_H(a)$ and $\nu : H \to X$ be the canonical projection. By 6.4 in [3], $X(\mathbb{R})$ is the union of finitely many orbits of $H(\mathbb{R})$, which are open and closed in $X(\mathbb{R})$, hence closed in X (in the ordinary topology). Since G has finite index in $H(\mathbb{R})$, this also holds for the orbits of G. In particular $\nu(G)$ is closed in X. The equality $G \cap \mathscr{Z}_H(a) = \mathscr{Z}_G(a)$ then shows that the projection μ induces an isomorphism of $G/\mathscr{Z}_G(a)$ onto a closed subset of X.

By Theorem 1 of [9], there exists a compact neighborhood B of a in C with the following property: given a compact subset $\omega \in H$, there is a compact subset $\omega^* \in X$ such that if $h \in H$, $x \in B$ are such that $h.x.h^{-1} \in \omega$, then $\nu(h) \in \omega^*$. We now apply this to our sequence. We may assume $x_n \in B$ and $g_n.x_n.g_n^{-1}$ in some compact neighborhood of b. Then the elements $\nu(g_n)$ belong to the intersection of $\nu(G)$ with a compact subset of X. The subset $\nu(G)$ is closed in X, isomorphic to $G/\mathscr{Z}_G(a)$, therefore the $\nu(g_n)$ belong to a compact subset of $G/\mathscr{Z}_G(a)$. Since any compact subset of $G/\mathscr{Z}_G(a)$ is the image of some compact subset of G, this proves (**).

Corollary 3.4. *Let $C(K)$ be the quotient of K be itself, acting by inner automorphisms and $\sigma_K : K \to C(K)$ the canonical projection. Then there exists a continuous map $\sigma_G : G \to C(K)$ which is constant on conjugacy classes and extends σ_K.*

It is known that two elements of K which are conjugate in G are already conjugate in K (see [2], 24.7, Prop. 2 for a more general statement). As a consequence, if $x \in G$ is elliptic, then $C_G(x) \cap K$ consists of one conjugacy class in K. We then define σ_G by $\sigma_G(g) = \sigma_K\big(C(g_e) \cap K\big)$. This is a map which extends σ_K and is constant on conjugacy classes. By definition then, $\sigma_G(g) = \sigma_G(g_e)$. There remains to show that σ is continuous.

Let $y \in G$ and $\{y_n\}$, $(n = 1, \dots,)$, a sequence of elements in G tending to y. We have to show that $\sigma_G(y_n) \to \sigma_G(y)$. There exists $g_n \in G$ such that

$$(g_n^{-1}.y_n.g_n)_e \in K.$$

Let $x_n = g_n^{-1}.y_n.g_n$. Then $x_{n,e} \in K$ and, passing to a subsequence, we may assume that $x_{n,e} \to a \in K$. Then

$$\sigma_G(y_n) = \sigma_G(x_n) = \sigma_K(x_{n,e}) \to \sigma_K(a) = \sigma_G(a). \quad (1)$$

On the other hand $\sigma_G(y) = \sigma_G(y_e)$ by definition. We have

$$x_{n,e} \in K,\ x_{n,e} \to a,\ y_n = g_n.x_n g_n^{-1} \to y.$$

By 3.3, a and y_e are conjugate in G, hence $\sigma_G(a) = \sigma_G(y_e) = \sigma_G(y)$. Together with (1), this shows that $\sigma_G(y_n) \to \sigma_G(y)$.

Corollary 3.5. *Let M be a topological space and $f : K \to M$ a continuous M-valued class function. Then f extends to a continuous M-valued class function on G.*

By definition, f can be written as $f = f' \circ \sigma_K$, with $f' : C(K) \to M$ continuous. Then $F' = f' \circ \sigma_G$ satisfies our conditions.

Corollary 3.6. *Let $u, v \in K$. Assume there exist sequences $x_n, y_n \in G$ such that $x_n \to u$, $y_n \to v$ and, for each n, x_n is conjugate to y_n in G. Then u and v are conjugate in K.*

For each n, $\sigma_G(x_n) = \sigma_G(y_n)$, hence $\sigma_G(u) = \sigma_G(v)$ and

$$\sigma_K(u) = \sigma_G(u) = \sigma_G(v) = \sigma_K(v)$$

which shows that u and v are conjugate in K.

3.7. Given f as in 3.5, we let F_f be the function constructed in proving 3.5, to be called the *canonical extension of f*.

Let S, T be tori, $\nu : S \to T$ and $\mu : K \to S$ continuous homomorphisms. Then ν and $\mu \circ \mu$ are continuous class functions on K. We claim that

$$F_{\nu \circ \mu} = \nu \circ F_\mu.$$

Let $g \in G$, and $k \in C(g_e) \cap K$. Let φ be either μ or $\nu \circ \mu$. By construction $F_\varphi(g) = \varphi(k)$. This implies our assertion.

3.8. We indicate briefly how to prove Theorem I(i) using 3.5 when G is linear. As in 1.3, it suffices to consider the case where H is the universal covering of G. Let K be a maximal compact subgroup of G. It is the semidirect product of its derived group $\mathscr{D}K$ by a torus S ([1], 3.2). Since $\pi_1(K)$ is isomorphic to $\pi_1(G)$, hence is torsion-free, the inclusion $S \to G$ induces an isomorphism $\pi_1(S) \tilde{\to} \pi_1(G)$. Let ν be the inverse isomorphism. We identify these groups to the kernels of the projections π and $\pi' : V \to S$, where V is the universal covering of S. Let f be the projection of K on S, with kernel $\mathscr{D}K$ and $F : G \to S$ its canonical extension. There is a unique lifting $\tilde{F} : \tilde{G} \to V$ of F which maps 1 onto 1 and satisfies the conditions $F(\pi(g)) = \pi'(\tilde{F}(g))$ and $\tilde{F}(g.n) = \tilde{F}(g).\nu(n)$ ($g \in G$, $n \in N$). The first relation implies that $\tilde{F}$ is a class function. Since $\nu(N)$ acts freely on V, the second one shows that if $n \neq 1$ $\tilde{F}(g.n) \neq \tilde{F}(g)$, therefore g and $g.n$ are not conjugate. In view of 1.1, this proves $I(i)$.

4. Commutator length in semisimple Lie groups

In this section, G is a connected semisimple Lie group.

4.1. Let L be a group which is equal to its derived group $\mathscr{D}L$. Then every element $g \in L$ is a product of finitely many commutators $(a, b) = a.b.a^{-1}.b^{-1} (a, b \in L)$. The commutator length $c\ell_L(g)$ or simply $c\ell(g)$ of g is the minimum of the number of factors over all such expressions. The *commutator length* $c\ell(L) \in \mathbb{N} \cup \infty$ of L is the lower upper bound of the $c\ell(g), (g \in L)$.

Let $\nu : L' \to L$ be a covering, and $N = \ker \nu$. It follows immediately from the definitions that

$$c\ell(L) \leqq c\ell(L') \leqq c\ell(L).\big(\sup{}_{n \in N} c\ell_L(n)\big). \tag{1}$$

In particular, if N is finite, $c\ell(L)$ is finite if and only if $c\ell(L')$ is so.

The following proposition is certainly known, even though I do not know of a reference for it.

Proposition 4.2. *Assume that G has a finite center. Then $c\ell(G)$ is finite.*

The Lie algebra $\mathfrak{g}$ is equal to its derived algebra, hence is spanned by elements $[x, y]$ $(x, y \in \mathfrak{g})$. This implies that the image of the map $G^{2n} \to G$ which associates to $(g_1, \dots, g_{2n})$ the product of the commutators (g_i, g_{n+i}) $(i = 1, \dots, n; n = \dim G)$ contains a neighborhood of the origin. Therefore $c\ell_G(g)$ has a finite upper bound if g runs through any compact subgroup. Let K be a maximal one and $G = K.P$ be the corresponding Cartan decomposition (2.3). It is known that any element in P is a commutator [7] (the argument is recalled below), whence the proposition.

4.3. Much more precise information is available in many cases. We review some of it, without claiming completeness. In [8] it is shown that if G is compact, then $c\ell(G) = 1$. To see this it suffices to prove that any element in a maximal torus T is a commutator. But if $w \in \mathscr{N}(T)$ represents an element of the Weyl group with finite fixed point set (e.g. a Coxeter transformation), then it is readily seen that $t \mapsto (w, t)$ is a surjective homomorphism of T onto itself.

Let now G be non-compact and $G = K.P$ as above. Let A be a maximal split toral subgroup contained in P. Then the same argument, using an element $w \in \mathscr{N}_G(A) \cap K$ representing a Coxeter transformation of the Weyl group of G with respect to A, proves that every element $a \in A$ is of the form (w, b), $(b \in A)$. By conjugation, this shows that any $p \in P$ is a commutator. Using that last fact and Goto's result, we see that if K is semisimple, then $c\ell(G) \leq 2$, as pointed out in [7]. On the other hand, if G is a complex semisimple group, then $c\ell(G) = 1$. This was proved first for $\mathbf{SL}_n$ over any algebraically closed field in [16], in general over $\mathbf{C}$ in [14] and in general over any algebraically closed groundfield in [15].

For further results on specific groups, see [7]. The precise value of $c\ell(G)$ for G of adjoint type, say, seems not to have been determined in general. I do not know of an example where it is > 2.

Proposition 4.4. *Assume G has an infinite center. Then $c\ell(G)$ is infinite.*

Proof. Let $G_o = G/N$ and $\tau : G \to G_o$ the canonical projection,where $N = \mathscr{C}G$. Then $\mathscr{C}G_o = \{1\}$. Let K_o be a maximal compact subgroup of G_o. The group N is commutative, finitely generated, hence product of a finite group N_t by a free commutative subgroup of some rank $d > 0$. It suffices to prove that $c\ell(G/N_t)$ is infinite and we may therefore assume N to be free abelian. It is contained in $\tilde{K} = \tau^{-1}(K_o)$. In view if the known structure of coverings of compact Lie groups (see e.g. [5], §1, n° 1, Prop. 5), $\tilde{K} = K' \times V$, where K' is a compact group on which τ is injective and V a vector group of rank d. Then K_o is the almost direct product of $\tau(K')$ and of a d-dimensional torus S_1. Let $S = K_o/\tau(K')$ and $f_o : K_o \to S$ the canonical projection. S is a d-dimensional torus, image of S_1. Let N' be the fundamental group of S and $\pi' : V' \to S$ the universal covering of S. Then f_o lifts to a surjective homomorphism $\rho : K' \times V \to V'$ with kernel K', which maps N onto a subgroup of finite index of N'. The map f_o may be viewed as a continuous S-valued class function on K_o. Let F_o be its canonical extension (3.7). It lifts uniquely to a continuous function $F = F_G : G \to V$ which maps 1 to 0 and satisfies the conditions

$$F(x.n) = F(x).\rho(n) \quad \pi'\big(F(x)\big) = F_o\big(\pi(x)\big) \quad (x \in G,\, n \in N).$$

In particular it is surjective.

For $m \in \mathbb{N}$, let us denote by $G_{(m)}$ the set of elements in G of commutator length $\leqq m$. Then 4.4 is a consequence of

(∗) *$F(G_{(m)})$ is bounded in V ($m \in \mathbb{N}$).*

There remains to prove (∗). In order to avoid some technical complications in the last part of the proof we first show that we can assume G to be the quotient G_1 of its universal covering $\tilde{G}$ by the torsion subgroup of the center of $\tilde{G}$. The group K_o is the almost direct product of its derived group $\mathscr{D}K_o$ by the identity component C_o of its center. The latter has dimension c equal to the rank of $\mathscr{C}G_1$. The group $\mathscr{D}K_o$ belongs to the kernel of f_o and S_1 belongs to C_o. Therefore $f_o = \nu \circ f_1$, where f_1, ν

$$K_o \xrightarrow{f_1} K_o/\mathscr{D}K_o = C \xrightarrow{\nu} S$$

are the canonical projections. Let F_1 be the canonical extension of f_1 (3.7). It has a canonical lifting $F_{G_1} : G_1 \to V_C$ mapping 1 to 0.

Let $\tilde{\nu}$ be the map $V_C \to V$ covering ν. The map $\tilde{\nu} \circ F_{G_1} : G_1 \to V$ is constant on the left cosets of $\ker(G_1 \to G)$, hence defines a continuous map $G \to V$. We claim it is F_G. This follows from the fact that the diagram

$$\begin{array}{ccccc}
G_1 & \xrightarrow{F_{G_1}} & V_C & \xrightarrow{\tilde{\nu}} & V \\
\downarrow & & \vert & & \Vert \\
G & & \xrightarrow{\quad F_G \quad} & & V \\
\downarrow & & \downarrow & & \downarrow \\
G_0 & \xrightarrow{F_1} & C & \xrightarrow{\nu} & S
\end{array}$$

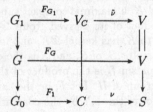

is commutative because $\tilde{\nu} \circ F_{G_1}$ is the lifting of F_o to G_1 by 3.7. It suffices then to prove that $F_{G_1}(G_{1(m)})$ is bounded in V_G. So we assume that $G = G_1$, hence that S is a quotient of $\mathscr{C}K_o$ by a finite group, under ν.

We consider the diagram

$$\begin{array}{ccccccc} G^{2m} & \xrightarrow{\gamma} & G^m & \xrightarrow{\mu} & G & \xrightarrow{F} & V \\ \sigma\downarrow & & \downarrow & & \downarrow & & \downarrow\pi \\ G_o^{2m} & \xrightarrow{\gamma_o} & G_o^m & \xrightarrow{\mu_o} & G_o & \xrightarrow{F_o} & S \end{array} \tag{1}$$

where γ, γ_o are commutator maps, which assign to $(g_1, \dots, g_{2m})$ the element with components (g_i, g_{m+i}) $(i = 1, \dots, m)$, the maps μ, μ_o are product maps and the vertical maps natural projections. Let $\varphi = F_o \circ \mu_o \circ \gamma_o$. We claim

(**) *there is a finite cover* $\mathscr{U} = \{U_i\}(i \in I)$ *of* S *by contractible closed subsets such that* $\varphi^{-1}(U_i)$ *has finitely many connected components* $(i \in I)$.

Assuming (**) provisorily, we prove (*).

The kernel of σ is N^{2m}. Since

$$(x.z,\, y.z') = (x,\, y) \quad (x, y \in G,\ z,\ z' \in N),$$

the map γ is constant of the cosets of N^{2m}, hence factors through σ, i.e. there exists a continuous map $\rho : G_o^{2m} \to G^m$ such that $\gamma = \rho \circ \sigma$. If we add ρ to (1), it clearly remains commutative. We have $\mu \circ \gamma(G^{2m}) = G_{(m)}$ and G^{2m} is the union of the $\sigma^{-1}\varphi^{-1}(U_i)$. Therefore $F(G_{(m)})$ is the union of the sets

$$V_i = (F \circ \mu \circ \rho)\left(\varphi^{-1}(U_i)\right) \quad (i \in I).$$

Let Z be a connected component of $\varphi^{-1}(U_i)$. Then $F.\mu.\rho(Z)$ is a *connected* subset of $\pi^{-1}(U_i)$. But U_i is closed, contractible, hence its inverse image is a disjoint union of compact subsets, all homeomorphic to U_i under π. Therefore $F.\mu.f(Z)$ is bounded, hence so is V_i and (*) is proved.

We still have to establish (**). We shall use the following fact: in ordinary topology, a real algebraic variety X and the complement $U = X - Z$ of a real algebraic subvariety Z in X have finitely many connected components (see [19], Theorems 3, 4). We shall call U a difference of real algebraic sets.

The group G_o is the product of simple groups with center reduced to $\{1\}$. A maximal compact subgroup of such a factor either is semisimple or has a one-dimensional center (in which case the corresponding symmetric space is isomorphic to a bounded symmetric domain). Let $G_{o,1}, \dots, G_{o,c}$ be the simple factors of G_o with non-semisimple maximal compact subgroup, H_o the product of the remaining factors and K_i a maximal compact subgroup of $G_{o,i}$. The group K_i is the almost direct product of its derived group $\mathscr{D}K_i$ and of a circle group C_i. The restriction f_i of f_o to K_i has kernel $\mathscr{D}K_i$ and induces an isomorphism of $C_i/C_i \cap \mathscr{D}K_i$ onto a direct factor S_i of S, and S is the direct product of the S_i's $(1 \leqq i \leqq c)$. The homomorphism f_o is the product of the f_i's by the trivial map of a maximal compact subgroup of H. Let F_i be the canonical extension of f_i to $G_{o,i}$. Then F_o is the product of the F_i's by the constant map of H_o into 1. The group G is the product of groups G_i $(i = 1, \dots, c)$ by H, where G_i is a covering of $G_{o,i}$ with fundamental group $\mathbb{Z}$. The subset $G_{(m)}$ is contained in the product of the $G_{i(m)}$ by H and (1) may be written as a product of similar diagrams for the G_i's with F_o replaced by F_i, and F by the lifting of F_i mapping G_i to the universal covering V_i of S_i, and of a similar diagram for H where F and F_o are replaced by constant maps. From this it is clear that it suffices to prove (**) for each G_i. We are therefore reduced to the case where G_o is simple, $(\mathscr{C}K_0)^o = C$ and S are one-dimensional.

Let $s \in S - \{1\}$ and $c \in C$ mapping onto S. Then $\mathscr{Z}_G(s) = K_o$, since, on one hand $\mathscr{Z}_G(s)$ contains K_o and on the other hand it is $\neq G_o$ and K_o is proper maximal in G_o. Therefore, if $F_o(g) = s$ $(g \in G_o)$, then g is elliptic and contained in ${}^G(\mathscr{D}K_o.c)$.

As a consequence, if Z is a proper subset of $S - \{1\}$ and Z' its inverse image in C, then

$$F^{-1}(Z) = {}^G(\mathscr{D}K_o.Z'). \tag{2}$$

Let $\psi = \mu_o \circ \gamma_o$. It commutes with G_o acting on G_o by inner automorphisms and on G_o^{2m} by

$${}^g(g_1, \dots, g_{2m}) = ({}^g g_1, \dots, {}^g g_{2m}) \qquad (g, g_i \in G_o; i = 1, \dots, 2m) \tag{3}$$

$\psi^{-1}(K_o)$ and $\psi^{-1}(DK_o.c)$ $(c \in C)$ are algebraic subsets in G_o^{2m}. Let E be a finite set of points on S containing 1 and $\{Z_j\}$ $(j \in J)$ the connected components of $S - E$. The set $\psi^{-1}(K_o) - \psi^{-1}(\mathscr{D}K_o.E)$ is a difference of algebraic sets, hence has finitely many connected components. It is the disjoint union of the $\mathscr{D}K_o.Z_j$, which are relatively open and closed subsets, hence consist each of finitely many connected components. Using (2) and (3) we see that

(+)$\varphi^{-1}(Z)$ *has finitely many connected components if* Z *is an interval in* $S - \{1\}$, *either open or closed in* $S - \{1\}$.

There remains to examine $F_o^{-1}(1)$. We want to show the existence of an algebraic subset $Y \subset G_o$ such that $F_o^{-1}(1) = {}^G Y$.

We note first that the set of subgroups $\mathscr{Z}_G(k)$, $k \in \mathscr{D}K_o$, form finitely many conjugacy classes in G. To see this we may assume that k is in a maximal torus T of $\mathscr{D}K_o$. Then the Lie algebra of $\mathscr{Z}_G(k)$ is described by the roots of G with respect to T which are equal to one on k, which gives only finitely many possibilities. In addition there are finitely many possibilities for the component group. Let $\{Q_i\}(i \in I)$ be a set of representatives of these conjugacy classes. For each Q_i, choose an Iwasawa decomposition $Q_i = K_i.A_i.N_i$, where $K_i = K_o \cap Q_i$ (see 2.3, 2.6). Let $Y_i = (\mathscr{D}K_o \cap \mathscr{C}Q_i).A_i.N_i$. The intersection $\mathscr{D}K_o \cap \mathscr{C}Q_i$ is a subgroup of $\mathscr{Z}_{Q_i}(A_i) \cap K_i$, hence normalizes $A_i.N_i$, so that Y_i is an algebraic subgroup.

Let Y be the union of the Y_i's. We want to show that $F_o^{-1}(1) = {}^G Y$. If $g \in Y_i$, then g_e is central in Y_i and belongs to $\mathscr{D}K_o$, hence $F_o(g) = 1$, which shows that ${}^G Y \subset F_o^{-1}(1)$. Conversely, let $g \in G$ be such that $F_o(g) = 1$. After conjugation, we may assume that $g_e \in K_o$ and therefore $g_e \in \mathscr{D}K_o$. The group $\mathscr{Z}_G(g_e)$ is conjugate to some Q_i, which reduces us to the case where $g \in Q_i$

and $Q_i = \mathscr{Z}_G(g_e)$. Then (see 2.6) g is conjugate to an element of Y_i, whence $F_o^{-1}(1) \subset {}^GY$. Then $\psi^{-1}(Y)$ is algebraic and

$$\psi^{-1}\left(F_o^{-1}(1)\right) = {}^G\left(\psi^{-1}(Y)\right)$$

has finitely many connected components.

Now (**) follows readily. Take for $\mathscr{U}$ the union of two closed half circles covering S and intersecting only at their end points, one of which contains 1 in its interior. To the one not containing 1, we can apply (+). The other interval is the union of $\{1\}$, to which we apply the remark made above, and of two intervals for which (+) is valid.

References

1. A. Borel, *Sous-groupes commutatifs et torsion des groupes de Lie compacts connexes*, Tôhoku Math. J. **13** (1961), no. 2, 216–240.
2. ______, *Linear algebraic groups*, Springer-Verlag, 1991, 2nd enlarged edition, GTM **126**.
3. A. Borel and J-P. Serre, *Théorèmes de finitude en cohomologie galoisienne*, Comment. Math. Helv. **39** (1964), 111-164.
4. A. Borel and J. Tits, *Groupes réductifs*, Publ. Math. I.H.E.S. **27** (1965), 55–150.
5. N. Bourbaki, *Groupes et algèbres de Lie*, Masson, Paris, 1982, Chap. 9.
6. C. Chevalley, *Theory of Lie groups I*, Princeton University Press, 1946.
7. D. Z. Djokovic, *On commutators in real semisimple Lie groups*, Osaka J. Math **23** (1986), 223–228.
8. M. Goto, *A theorem on compact semi-simple groups*, J. Math. Soc. Japan **1** (1949), 270–272.
9. Harish-Chandra, *A formula for semisimple Lie groups*, Amer. J. Math. **79** (1957), 733–760.
10. ______, *Harmonic analysis on real reductive groups I. the theory of the constant term*, J. Functional Analysis **19** (1975), 104–204.
11. G. D. Mostow, *Self-adjoint group*, Annals of Math. **62** (1955), 44–55.
12. ______, *Fully reducible subgroups of algebraic groups*, Amer. J. Math. **78** (1956), 200–221.
13. ______, *Covariant fiberings of Klein spaces, II*, Amer. J. Math. **84** (1962), 465–474.
14. S. Paciencer and H. C. Wang, *Commutators in a semi-simple Lie group*, Proc. A.M.S. **13** (1962), 907–913.
15. R. Ree, *Commutators in semi-simple algebraic groups*, Proc. A.M.S. **15** (1964), 457–460.
16. K. Shoda, *Einige Sätze über Matrizen*, Japanese J. Math. **13** (1936), 361–365.
17. V. S. Varadarajan, *Harmonic analysis on real reductive groups*, Springer-Verlag, 1977, LNM **576**.
18. N. Wallach, *Real reductive groups I*, Academic Press, 1988.
19. H. Whitney, *Elementary structure of real algebraic varieties*, Annals of Math. **66** (1957), 545–566.
20. J. Wood, *Bundles with totally disconnected structure group*, Comment. Math. Helv. **46** (1971), 257–273.

INSTITUTE FOR ADVANCED STUDY, OLDEN LANE, PRINCETON, NJ 08540, USA

162.

On the Work of E. Cartan on Real Simple Lie Algebras

letter, Notices of the A.M.S. 44, 1997, 430–431

In [2], A. J. Coleman asserts that E. Cartan had obtained in [1] (which, by the way, is [38] in his *Collected Papers*, not [39]), not only a classification of real forms of complex simple Lie algebras, but also of their Cartan subalgebras, a fact he had never seen referred to in the literature. I was asked to comment on this assertion, and it was later suggested that I write to the *Notices* about it, whence this letter.

First, let me dispose of a very minor point. By classification of Cartan subalgebras, we mean nowadays up to inner automorphisms. The distinction between inner and outer automorphisms does not seem to me to occur in E. Cartan's work before 1927, so that it would have been a classification up to automorphisms. However, had he obtained one, this would be minor quibbling, but I do not believe he had, even implicitly. Let me explain why.

By index of a nondegenerate quadratic form on $\mathbb{R}^n$, Cartan means the number of positive squares minus the number of negative squares, once the form is diagonalized. Let $\mathfrak{g}_c$ be a complex simple Lie algebra, $\mathfrak{h}_c$ a Cartan subalgebra of $\mathfrak{g}_c$, and r its rank. The character δ of a real form $\mathfrak{g}$ of $\mathfrak{g}_c$ is the index of its Killing form. Given a Cartan subalgebra $\mathfrak{h}$ of $\mathfrak{g}$, Cartan denotes by δ_0 the index of the restriction of the Killing form of $\mathfrak{g}$ to $\mathfrak{h}$. It may depend on $\mathfrak{h}$, of course. The integers δ and δ_0 play a fundamental role in [1]. It turns out that "in general" (see below) the index characterizes the real form, up to isomorphisms, and this dictates Cartan's strategy. He starts from a real form $\mathfrak{h}$ of $\mathfrak{h}_c$ and tries to construct, by analyzing the constants of structure and the restrictions of roots to $\mathfrak{h}$, a real form $\mathfrak{g}$ having $\mathfrak{h}$ as a Cartan subalgebra. This gives him a certain number of possibilities for δ and δ_0. Furthermore, early in his discussion, he divides the possibilities into two categories for some types of $\mathfrak{g}_c$'s. Then, within one category, he proves that two real forms $\mathfrak{g}, \mathfrak{g}'$ with the same δ are isomorphic. To this end he first shows that $\mathfrak{g}$ and $\mathfrak{g}'$ contain Cartan subalgebras $\mathfrak{h}, \mathfrak{h}'$ with the same δ_0. [If $\delta = r$ (split form; he says normal form), he may try $\delta_0 = r$ (split Cartan subalgebra). If he finds $\mathfrak{h}, \mathfrak{h}'$ with $\delta_0 = -r$ (compact Cartan subalgebras), he often uses those; this would of course be the only possibility if $\delta = -\dim \mathfrak{g}$ (compact form).] Then the argument consists in establishing an isomorphism of $\mathfrak{g}$ onto $\mathfrak{g}'$ bringing $\mathfrak{h}$ onto $\mathfrak{h}'$. This is indeed a conjugacy assertion in a given $\mathfrak{g}$ for Cartan subalgebras with the chosen δ_0, but it is only a first step towards a classification. As a second one along those lines it would be necessary, given $\mathfrak{g}$, to find all possible values of δ_0. As far as I can see, Cartan does not do it, nor does he seem interested. There would then be the problem of the conjugacy of Cartan subalgebras with a given δ_0. If they were conjugate, one might hope that some generalization of Cartan's procedure might prove it, but this is not always true. Since δ_0 is the only invariant of $\mathfrak{h}$ considered in the paper, this rules out a priori the possibility for this paper to contain such a classification. There is indeed only one conjugacy class if $\delta_0 = r, -r$, as Cartan shows in many cases, but even for those he does not make a general statement.

To conclude, I note that Cartan's tables give a quantitative meaning to the above "in general": $\mathfrak{g}_c$ may have two nonisomorphic real forms with the same character only if it is of type $A_r, r = m^2 - 3$ $(r > 1,$ odd), or D_r, $r = m^2$, where m is a positive integer.

In order not to lengthen this letter, I shall not discuss the other assertions of [2] about Killing and Cartan, though they seem to me somewhat misleading and inaccurate.

References

[1] E. Cartan, *Les groupes réels, simples, finis et continus,* Ann. Sci. École Norm. Sup. **31**, (1914), 263-365.

[2] A. J. Coleman, *Groups and physics, dogmatic opinions of a senior citizen,* Notices Amer. Math. Soc. **44**, number 1 (1997), 8-17.

Armand Borel
Institute for Advanced Study

(Received January 31, 1997)

164.

Jean Leray and Algebraic Topology

in J. Leray, Selected papers, Vol. I, 1–21, Springer 1998

I. 1933–1939

1. Leray's first contributions to mathematics belong to fluid dynamics. ([1931a] to [1933a]). The joint paper with Juliusz Schauder [1933b] marks his first involvement with algebraic topology. It follows the same pattern as earlier work of Schauder: proof of a theorem of algebraic topology in Banach spaces and applications to the existence of solutions of certain P.D.E. We first summarize briefly the parts of it most relevant here.

Schauder's results in algebraic topology pertain to a transformation of a Banach space B into itself of the form

$$\Phi(x) = x - F(x) \tag{1}$$

where F is a completely continuous map (i.e. transforms bounded sets into relatively compact ones) defined on B or sometimes only on a bounded subset. He extends to this situation two fundamental results of L.E.J. Brouwer in the finite dimensional case, namely

a) *A fixed point theorem:* if F is defined on, and maps into itself, the closure of an open non-empty bounded convex set, then it has a fixed point in it.

This is first proved in [13] (in a slightly different form) and used to show the existence of solutions for certain hyperbolic equations.

b) *Invariance of the domain:* if F is defined on B and Φ is bijective, then Φ maps any open set onto an open set, hence is bicontinuous.

It is applied to certain elliptic equations. The type of theorem obtained is roughly the following: if for some initial choice of data there is only one solution and if for nearby data there is at most one solution, then there is indeed one solution for data sufficiently close to the initial ones (*see* [14] for the precise statements).

In both cases, the proof of the topological theorem is a reduction to Brouwer's case by means of suitable finite-dimensional approximations.

2. The first part of the joint paper [1934c] with J. Schauder, announced in [1933b], is devoted to the definition and basic properties of a "topological degree" of Φ, again in analogy with Brouwer's work. Schauder had already used the Brouwer index, but here the goal is to have a definition valid in a Banach space, for a transformation of the type (1) above, when F is defined on the closure $\bar{\omega}$ of an open bounded set ω (but does not necessarily leave it stable). Given a point b not on the image of the boundary ω' of $\bar{\omega}$, they define a topological index $d(\Phi, \omega, b)$ with the following natural properties:

1) If $\omega = \omega_1 \bigcup \omega_2$ where ω_1 and ω_2 are two disjoint bounded open sets, then

$$d(\Phi, \omega, b) = d(\Phi, \omega_1, b) + d(\Phi, \omega_2, b).$$

(ii) If $d(\Phi, \omega, b) \neq 0$, then $b \in \Phi(\omega)$.

(iii) The degree remains constant when ω and F vary continuously in such a way that b never meets $\Phi(\omega')$.

They also introduce an index $i(\Phi, a)$ of Φ at a point a which is isolated in its fiber $\Phi^{-1}(b)$, where $b = \Phi(a)$. It is an integer which, under some further technical assumptions on F, is equal to ± 1. If $\Phi^{-1}(b)$ consists of finitely many points a_j, then $d(\Phi, \omega, b)$ is the sum of the $i(\Phi, a_j)$.

As in [13], [14], these results are proved by reduction to the finite-dimensional case. They are applied to a family of transformations

$$\Phi(x, k) = x - F(x, k) \tag{2}$$

depending on a parameter k varying in a closed interval K of the real line, where for each $k \in K$, the transformation $F(x, k)$ is as above, defined on $\overline{\omega(k)}$, and the union of the $\overline{\omega(k)}$ is assumed to be bounded in $B \times K$. The goal is to investigate the fixed points of $F(x, k)$, i.e. to find those x and k for which $\Phi(x, k) = 0$. To this end, the index $d(\Phi(x, k), \omega(k), 0)$ is examined. It is assumed that for some value k_0 of k in K, $F(x, k_0)$ has finitely many isolated fixed points, all in $\omega(k_0)$ and that $d(\Phi(x, k_0), \omega(k_0), 0)$ is not zero. If it is known on the other hand that $F(x, k)$ has at most finitely many fixed points, contained in some bounded set independently of k, then, under some further technical assumptions which I shall not state, it is shown that $F(x, k)$ has at least one fixed point for every $k \in K$, and that some of these fixed points form a family depending continuously on k. This result is then applied to a variety of functional or partial differential equations.

3. In [1935a] Leray brings a complement to the topological part of [1934c]. He gives a formula for the topological degree of the composition of two maps (of type (1)) and deduces from it first the invariance of domain, under assumptions somewhat more general than those of Schauder, and second a theorem about the number of bounded connected components of the complement of a bounded closed subset: it is the same for two closed bounded subsets F_1, F_2 if there exists a homeomorphism φ of F_1 onto F_2 such that the differences $\varphi(x) - x$ ($x \in F$) belong to some relatively compact subset.

The paper [1934c] reduces the proof of the existence of solutions to *a priori* majorations and also shows that, under suitable circumstances, local uniqueness implies global uniqueness. Leray's publications until 1939 provide many applications of these principles to fluid dynamics and P.D.E., for which I refer to Part II of these Selecta.

II. 1940–1945

4. The Second World War broke out in 1939 and J. Leray was made prisoner by the Germans in 1940. He spent the next five years in captivity in an officers' camp, Oflag XVIIA in Austria. With the help of some colleagues, he founded a university there, of which he became the Director ("recteur"). His major mathematical interests had been so far in analysis, on a variety of problems which, though theoretical, had their origins in, and potential applications to, technical problems in mechanics or fluid dynamics. Algebraic topology had been only a minor interest, geared to applications to analysis(1). Leray feared that if his competence as a "mechanic" ("mécanicien", his word) were known to the German authorities in the camp, he might be compelled to work for the German war machine, so he converted his minor interest to his major one, in fact to his essentially unique one, presented himself as a pure mathematician and devoted himself mainly to algebraic topology(2).

The first major outcome of this work is the series of three papers [1945a,b,c], the three parts of a "course in algebraic topology taught in captivity", announced in part in [1942a, b, c, d].

In describing them and subsequent work, I shall use the current terminology, which has been standard for the last 45 years or so, but also indicate the one proposed by Leray. This should not hide the fact that many of these concepts were completely new at the time and underwent some variations before the present formulations were arrived at.

As was pointed out, the theorems of algebraic topology used by Schauder and Leray-Schauder reviewed above were all proved by reduction to the finite-dimensional case by suitable approximations. The first main goal of Leray was to build up a theory of equations and transformations directly applicable to more general topological spaces. This required a new definition of homology. Leray was also keen not to use any subdivision of complexes, simplicial approximations, orientability assumptions and not to assume the spaces to be quasilinear ([1945a], p.97–98). Before getting to these papers, I shall first describe Leray's starting point, as outlined in *loc.cit.*

5. Until about 1935, the basic objects of algebraic topology were the homology groups, usually defined for simplicial complexes, though more general concepts had been introduced by Vietoris and E. Čech, mainly for compact spaces. Around 1935, it was discovered by several people (J. Alexander, E. Čech, Kolmogorov, H. Whitney) that a product adding degrees could be introduced on complexes dual to those defining homology, without assuming the underlying space to be a manifold (in which case a product could be defined by Poincaré duality from the intersection product in homology). The de Rham theorems, expressing the homology of compact smooth manifolds in terms of differential forms, had already shown that this product could be defined directly by means of the exterior product of differential forms. With this example in mind, Alexander realized that also in the general case these new complexes could be defined directly, not as duals to some pre-existing ones and that homology groups could in turn be viewed as dual objects to the new groups. His definitions of the new complex, boundary operator and product were indeed inspired by exterior differential calculus [1].

Leray adopted Alexander's point of view and minimized, almost suppressed from 1946 on, the use of the traditional homology groups. They occur mainly for comparison purposes or for the treatment of a generalization of manifolds and of Lefschetz numbers. He never lost sight of the analogy with E. Cartan's exterior differential calculus, of which he had acquired first-hand knowledge by writing up for publication the notes of a course given by E. Cartan [5]. In the introduction to [1945b], he points out that his "forms on a space" (*see* **6**) obey most of the rules of the calculus of Pfaffian forms and states that the main interest of that paper seems to him to be its treatment of a problem in topology, alien to any assumption of differentiability, by computations of that nature.

Early on, H. Whitney had proposed to call cohomology groups and cup product the new groups and the product [17], a suggestion which was soon rather widely adopted, but not by Leray until 1953. Prior to that, the word cohomology occurs only in two C.R. Notes. Leray kept to homology otherwise, prefacing several of his later papers, including the main ones, by stating that he would call homology what is usually referred to as cohomology since he will deal exclusively with the latter. I shall use cohomology.

Another requirement for the new cohomology ring to be defined was that it should allow one to carry over to more general spaces the proof of the theorem of H. Hopf, to the effect that a compact connected manifold endowed with a continuous product satisfying certain conditions (for instance, defining a group structure) has the same rational homology as a product of odd-dimensional spheres.

6. In [1945a] Leray first defines a notion of *abstract* complex over a ring L (either $\mathbb{Z}$, or $\mathbb{Z}/m\mathbb{Z}$ or $\mathbb{Q}$): a graded free finitely generated L-module, with a coboundary operator d increasing the degree by one (n° 7). It is called a simplex if it is acyclic. A *concrete* complex K on a space E (I shall simply say complex on E) is an abstract complex, to each basis element X of which is assigned a non-empty subset $|X|$ of E, its support. The support of a linear combination is then, by definition, the union of the supports of the basis elements occurring in it and it is required that $|dX| \subset |X|$. If F is a subspace of E, the intersection $F.K$ of F and K is the quotient of K by the submodule of elements with support not meeting F, the support of the image $F.X$ of X being $F \cap |X|$. Let K' be another complex on E. For $x \in E$ there is a natural homomorphism $r_x : K \otimes K' \to xK \otimes xK'$. The support of $Z \in K \otimes K'$ is then the set of $x \in E$

for which $r_x(Z) \neq 0$. The *intersection* $K \circ K'$ of K and K' is, by definition, the quotient of $K \otimes K'$ by the submodule of elements with empty support, endowed with the obvious supports.

Next, Leray introduces (the first version of) an important notion in all of his work in topology, that of "couverture": a complex K on E with *closed* supports such that xK is acyclic for all $x \in E$ and the sum of the basis elements of degree zero is a cocycle, called the unit cocycle of K. A linear combination of elements of K, with coefficients in L, as above, is a "form on E". I do not know of any translation of couverture in the mathematical literature. In later presentations of the theory, beyond Leray's work, it appears in such a disguised form that I shall neither need nor venture a translation, and simply use the French word [in sheaf theory, the sheaf associated to a couverture would be a resolution of the constant sheaf with stalk L].

The notion of couverture is stable under product, intersection, intersection with a closed subset and inverse image. The stability under product or intersection is a consequence of an algebraic argument we shall come to in **9**, which is fundamental for the whole paper and later developments.

Let E be a normal space. The union of the couvertures on E, with coefficients in L, is shown to be an L-algebra, with respect to sum and a product defined via the above intersection product. Its cohomology is, by definition, the cohomology ring of E, with coefficients in L, to be denoted here $H^*(E; L)$. It is not compared in this paper with the definitions of Alexander and Kolmogorov. [In [1945b], Leray states he cannot do so for lack of documentation.] It is mostly used for compact spaces (with an extension to differences of such, *see* **7**).

A *cover* of E is a collection of subsets, the union of which is E. To a finite closed cover of E is associated an abstract complex, its nerve, which is made into a complex on E by assigning to a simplex as support the intersection of the subsets represented by its vertices. This is a couverture, the couverture generated by the given cover. To compute $H^*(E; L)$ it suffices to consider a cofinal family of finite closed covers and all the couvertures obtained by iterated intersections from those defined by their nerves. More precisely, the constructions in n°s 16 to 18 present the cohomology ring as a direct limit of cohomology rings of couvertures associated to the nerves of suitable finite closed covers. They could be replaced by slightly bigger open finite open covers with the same nerves, so that, for compact spaces, the cohomology is essentially equivalent to Čech cohomology. For normal, not locally compact spaces, it does not seem to me that this type of cohomology has been considered elsewhere. It will also not occur later in Leray's work. Except on one point (*see* **9**) I shall not discuss technical details at all, since modifications of the definitions often led to simpler and more powerful arguments. Leray then establishes many properties of his cohomology ring. A first immediate consequence of the construction is Theorem 12, p. 122, according to which every cohomology class of strictly positive dimension is nilpotent. A compact space is said to be simple if it is acyclic. A fundamental result (Theorem 6, p. 126) asserts that a compact space which is a deformation retract of one of its points is simple. If a couverture has simple supports, then the cohomology groups of E are those of the underlying abstract complex (Theorem 12, p. 138). This shows in particular that the cohomology of a finite polyhedron is dual to the usual homology. If the non-empty intersections of the elements in a cover are simple, in which case the cover is said to be "convexoïd", then there is a finite procedure to determine the cohomology *ring* (n° 37).

§22 is devoted to a Künneth rule and §§23 to 25 to generalizations of the theorems of Hopf mentioned earlier and of Samelson. Theorem 2 quoted above allows one to adapt to compact connected spaces with a product the argument of Hopf's, to the effect that a homogeneous indecomposable element (Leray says "maximal cycle") has odd dimension.

Let E be a compact connected space endowed with a convexoïd finite cover by closed subsets, and ξ a continuous map of E into itself (a representation of E into itself in Leray's terminology). Chapter III associates to ξ a Lefschetz number Λ_ξ. It is defined simplicially, using the couverture defined by the given cover and its dual complex. It is the same for two homotopic continuous maps (Theorem 16, p. 162). The space E is said to be convexoïd if it has a fundamental set of closed neighborhoods which are simple as well as all their non-empty intersections [this is the definition given in [1953a], the one here is slightly different in formulation, but equivalent.] It is shown that if $\Lambda_\xi \neq 0$, then ξ has at least one fixed point (Theorem 17, p. 163). If E is moreover a topological group, then n° 44 generalizes a theorem of Hopf on the degree of the k-th power map.

7. The first part of [1945b], Chapter IV, is concerned with relations between the cohomology of a normal space E, a closed subspace F and the difference $U = E - F$. By definition, the latter is the "cohomology of the interior of U", i.e. the cohomology of the sub-complex of the complex defining $H^*(E; L)$ consisting of elements with support in U. It is not quite a topological invariant, since the supports are subsets of U which are closed in E. However, it is in the (main) case where E is compact, because these subspaces are just the compact subsets of U so they have an internal characterization. He then shows that $H^*(E; L)$, $H^*(F; L)$ and $H^*(U; L)$ are related by a long exact cohomology sequence (not in this language, but by proving three times that a kernel is an image). This leads to a generalization of Alexander duality, modulo an identification of the duals of Leray's cohomology groups with the usual homology groups, discussed in sections 35 of [1945a] and 66 of the present paper. Next, n°s 56–59 consider the case where E is the union of two closed subsets and establish the existence of a Mayer–Vietoris sequence.

For a locally compact space, the cohomology introduced by Leray will turn out to be equivalent to the Alexander–Spanier cohomology with compact supports. As Leray points out, it is non-trivial for the line and so it is not true that a non-compact space and a deformation retract have in general the same cohomology. To remedy this, he introduces in section IV another type of cohomology groups, based on the notion of "pseudo-cycle". As far as I know it plays a role in the 1945 paper but disappears from his later treatment, so, again, I shall just keep his terminology. A "pseudo-cycle" on the normal space E is an operator which assigns to each compact subset of B of E a cohomology class of B, the assignment being compatible with intersection by a closed subset. They can be added, intersected and form a graded L-algebra. [It is in fact the projective limit of the cohomology of the compact subsets of E, with respect to inclusions.] If any two elements of E are contained in a compact connected subset, then $H^0 \cong L$. Leray also defines the "pseudocycles in the interior of U", in case U is open in E and proves the existence of a long exact cohomology sequence. With respect to that cohomology, Euclidean space is acyclic and more generally E and a deformation retract have isomorphic cohomology rings.

Chapter V is devoted to manifolds, Poincaré duality, computes the cohomology of projective spaces and discusses the relations between the cohomology of a closed subset of the n-sphere and of its complement, in particular proves the Jordan–Brouwer theorem.

8. The first part of [1945c], Chapter VI, generalizes the Leray-Schauder theory in the framework of the cohomology theory developed in [1945, a, b].

Let E be a convexoïd space (*see* **6**), O an open subset of E and ξ a continuous map in E of a closed subset F of E. Leray defines an index $i(O)$, related to the fixed points of ξ contained in O. It is an integer, equal to the Lefschetz number Λ_ξ of ξ if $O = E$. It is defined if the closure $\overline{O}$ of O belongs to F and there is no fixed point in $\overline{O} - O$. It is zero if there is no fixed point in O and is invariant under continuous deformation (Theorem 22, p. 212). If E is moreover acyclic, it depends only on the restriction of ξ to $\overline{O} - O$. In fact, n° 88 provides a direct definition using only the restriction of ξ to $\overline{O} - O$.

Next the definition and properties of $i(O)$ are extended to different, or apparently different, situations, in particular to the following one: E is a topological space (not necessarily convexoïd), F a closed subspace, T a convexoïd space and $\xi = \varphi \circ \tau$ is a composition of continuous maps

$$\xi : F \xrightarrow{\tau} T \xrightarrow{\varphi} E.$$

The total index $i(O)$ of ξ is then by definition the index $i(\varphi^{-1}(O))$ of $\tau \circ \varphi$ (n°s 81, 82, pp. 223-225).

As is pointed out on p. 213, these results contain the Leray-Schauder theory, and the applications include the theorems of existence and/or uniqueness of solutions in [1934c] as well as in [1933c], [1936a], [1939].

Some of the main results on $i(O)$ and on its relation with the Lefschetz number were reproved and generalized in [1959c]. Notably, the assumption made several times in the present paper that the cohomology of F is finitely generated is dropped[(3)].

Chapters VII and VIII are devoted to algebraic topology. In particular, the following situation is considered: E and E' are topological spaces, T an acyclic convexoïd space of homeomorphisms of E into E', F a closed subspace of E and $\tau : F \to T$ a continuous map. Let F' be the set of points $\tau(x).x$ ($x \in F$). Then it is shown that the group of pseudocycles of $E - F$ and $E' - F'$ are isomorphic (Theorem 35, p. 245). If $E = E'$ is euclidean space and $\tau(x)$ is the translation bringing x to $\tau(x).x$, this yields the Alexander theorem and the invariance of the domain. Finally, a generalization of the Fredholm alternative is also proved.

9. These three papers first of all fulfill Leray's initial main goals, namely, to set up a cohomology theory (Chapters I, II, IV) and use it as a framework for a theory of equations encompassing the one of Leray-Schauder (Chapters III, V, VI). In addition they prove and generalize a number of theorems in algebraic topology, mostly known in some form, though sometimes derived in quite novel ways and greater generality (chapters VII, VIII). However, Leray realized that he could go much further, as hinted in a footnote to [1945c] p. 201. This led to the work announced in [1946a, b], which broke entirely new ground.

The starting point is an argument which occurs repeatedly in [1945a]. Its first goal was to prove that the "forms on a space" (*see* **6**) obey some of the rules of exterior differential calculus (cf. the introductory remarks in [1945b] quoted above in **5**). According to [1950a] p. 9 or [1959c], p.10, it is the analysis of this argument which led Leray to the cohomological invariants of a continuous map, described initially in [1946b]. Its first occurrence is in lemma 2, n° 4: given two abstract complexes C and C', where C is acyclic and has a unit cocycle U, it asserts that the cohomology of the product is naturally isomorphic to that of C'. Let z be a cochain of degree $m > 0$. It is a sum of expressions $u^a \times v^b$, where u^a is a cochain of degree a of C and v^b a cochain of degree b of C' ($a + b = m$). It is said to be of weight q if q is the maximum of a. Assume $q \geqq 1$. If z is a cocycle, then the sum of the terms of weight $q+1$ in dz is zero, and it follows that z is cohomologous to a cocycle of weight $q-1$ hence, by induction on q, to a cocycle of weight zero. As to those, it is easily seen that the map $c' \mapsto u \times c'$ induces an isomorphism of $H^*(C')$ onto the space of cocycles of weight zero modulo coboundaries, which proves the lemma. It is first used to prove that a product of couvertures is again one and then applied to more complicated situations ([1945a], n°s 17, 27, 32). Theorem 12, p. 138, quoted in **6**, is also an application of that principle, to be referred to as the fundamental argument or fundamental lemma.

Prop. 10.4 in [1950a] provides a translation in terms of spectral sequences and it is only in that form that it is used there(4). For the reader familiar with the notion of filtration and spectral sequences, we sketch the proof of lemma 2 in those terms (*see* also remark 2, p. 8, 9 of Exp. VI in [3]).

Let $S = C \times C'$. Filter S by $S^{-p} = \bigoplus_{q \geq p} C^q \times C'$.

Then $E^p_{-1} = S^{-p}/S^{-p-1}$, the differential d_{-1} is the partial differential with respect to C. The induction on the weight shows that $E^p_0 = 0$ for $p \neq 0$, hence $E^*_0 = E^*_\infty = GH(S) = E^0_0 = H(C')$, and the second part.

III. 1946–1950

10. We now come to the two seminal Notes [1946a, b]. The first one introduces sheaves, cohomology with respect to a sheaf and the cohomology ring of a continuous closed map $\pi : E \to E^*$ of normal spaces. Leray wanted to associate to π a cohomology ring of E^* with respect to the "variable coefficients" $H^*(\pi^{-1}y)$, $(y \in E^*)$(5). This led formally to the notions just listed.

A *sheaf* $\mathcal{B}$ on E is a functor which associates to every *closed* subset F of E a module (or ring, as the case may be) $\mathcal{B}(F)$, which is zero if F is empty, and to each inclusion $F' \subset F$ a homomorphism $r_{F'F} : \mathcal{B}(F) \to \mathcal{B}(F')$ with the usual transitivity properties. It is called normal if $\mathcal{B}(F)$ is the inductive limit of the $\mathcal{B}(F')$ $(F \subset F')$.

Let $b \in \mathcal{B}(F)$. It is said to be reducible if there exists a finite cover $\{F_i\}_{i \in I}$ of F such that $r_{F_i,F} b = 0$ for $i \in I$. The reduced sheaf defined by $\mathcal{B}$ associates to F the quotient of $\mathcal{B}(F)$ by the module of reducible elements.

In a way, sheaves had already implicitly occured in [1945a]: let K be a complex on E. To a closed subspace F, there is associated the section $F.K$ of K by F (*see* **6**), i.e. the quotient of K by the submodule of elements with support not meeting F. Clearly, the map $F \mapsto F.K$ defines a sheaf, which is normal. The stalk $x.K$ at $x \in E$ already played a considerable role in the theory, as we saw. Another important example of a normal sheaf is the q-th cohomology sheaf $\mathcal{B}^q : F \mapsto H^q(F; L)$.

A form on E with coefficients in $\mathcal{B}$ is a finite linear combination $\sum_i b_i X_i$, where the X_i's are basis elements of some couverture and $b_i \in \mathcal{B}(|X_i|)$. If B is the constant sheaf $\mathcal{L} : B(F) = L$, these are the forms on E of [1945a] and it is asserted that the constructions and results there generalize, whence the definition of the cohomology group (or ring) $H^*(E; \mathcal{B})$ of E with respect to the normal sheaf $\mathcal{B}$ of modules (or rings).

These two examples show why it is natural in the present setup to view sheaves as functors from closed rather than open subspaces. Recall that the cohomology here is with compact supports, so that the assignment to open sets of their cohomology would lead to a "cosheaf", where the natural maps go in the same direction as the inclusions.

By definition the sheaf $\pi(\mathcal{B})$ on E^* associates $\mathcal{B}\pi^{-1}(F^*)$ to $F^* \subset E^*$.

The last part of [1946a] introduces the cohomology ring of π. Let again $\mathcal{B}^q$ be the q-th cohomology sheaf of E. The transform $\pi(\mathcal{B}^q)$ associates to F^* the q-th cohomology ring of $\pi^{-1}(F^*)$. Then the (p,q)-cohomology group of π is $H^p(E^*; \pi(\mathcal{B}^q))$ and the cohomology ring of π is the direct sum of these groups, endowed with the product inherited from those in the cohomology of E and of the closed subsets of E(6).

11. The Note [1946b] is devoted to the "structure of the cohomology ring of π". By that is meant a construction allowing one to relate it to the cohomology of E. It is a first version of what became later known as the spectral sequence of π. Not all the features of the latter appear explicitly, but several essential ingredients already do.

We let $P_1^{p,q} = H^p(E^*; \pi(\mathcal{B}^q))$, and P_1 be the direct sum of the $P_1^{p,q}$. We shall call p the base-degree. The structure in question is defined by a sequence of submodules

$$(1) \qquad 0 = Q_{-1}^{p,q} \subset Q_0^{p,q} \subset \ldots \subset Q_{p-1}^{p,q} \subset P_{p+1}^{p,q} \subset \cdots \subset P_2^{p,q} \subset P_1^{p,q}$$

of $P_1^{p,q}$, of submodules

$$(2) \qquad 0 = E^{-1,p+1} \subset E^{0,p} \subset \cdots \subset E^{p-1,1} \subset H^p(E;L) = E^{p,0}$$

of $H^p(E;L)$, and isomorphisms

$$(3) \qquad \Delta_r : P_r^{p,q} | P_{r+1}^{p,q} \tilde{\rightarrow} Q_r^{p-r,q+r+1} / Q_{r-1}^{p-r,q+r+1}$$

$$(4) \qquad \Gamma_{p,q} : P_{p+1}^{p,q} / Q_{q-1}^{p,q} \tilde{\rightarrow} E^{p,q} / E^{p-1,q+1}$$

allowing to get information on the successive quotients of the composition series (2) by successive approximations, starting from P_1.

These modules and Δ_r are defined in terms of couvertures and of the action of the coboundary operator on these. The construction is quite intricate, only sketched there and I can only try to give some idea of it.

Fix $r \in [2, p+1]$. Let $x \in P_1^{p,q}$ represented by a form on E^*, i.e. a finite sum $\sum_\alpha z^{p,\alpha} C^{q,\alpha}$ where $z^{p,\alpha}$ is a cocycle on $\pi^{-1}(|X^{q,\alpha}|)$. Assume there is a form $L^{p,q}$ on E which can be written

$$L^{p,q} = \sum_\alpha L'^{p,\alpha} \pi^{-1}(X^{q,\alpha}) + \sum_{s>0;\lambda} L^{p-s,\lambda} \pi^{-1}(X^{q+s,\alpha})$$

such that $L'^{p,\alpha}.\pi^{-1}(|X^{q,\alpha}|) \sim z^{p,\alpha}$ and that $dL^{p,q}$ can be written similarly, but with p replaced by $p-r$ and q by $q+r+1$, where the terms of base degree $p-r$ in $dL^{p,q}$ represent a class y in $P_1^{p-r,q+r+1}$. By definition, $P_r^{p,q}$ is generated by those x's, $Q_r^{p-r,q+r+1}$ by the y's and Δ_r associates y to x. By definition, Δ_r annihilates $P_{r+1}^{p,q}$. There is an ambiguity in the choice of $L^{p,q}$, which leads to view Δ_r as a map of $P_r^{p,q}/P_{r+1}^{p,q}$ onto $Q_r^{p-r,q+r+1}/Q_{r+1}^{p-r,q+r+1}$.

If $L^{p,q}$ can be chosen to be a cocycle, then $x \in P_{p+1}^{p,q}$ (and conversely), the class $[x]$ belongs to $E^{p,q}$ by definition and $\Gamma_{p,q}x$ is its image in $E^{p,q}/E^{p-1,q+1}$.

In short, $E^{p,q}/E^{p-1,p+1}$ is a subquotient of P_1. It is arrived at by successive approximations, by means of a descending induction on the base-degree. This is reminiscent of the fundamental lemma, of course, but going from the latter to this construction is obviously a "giant step". It is only sketched in this Note, was never described in more detail, so that it is hardly possible, at least for this writer, to see how it could effectively be used. The results announced in this and the following Note, proved before Leray arrived at the next, and final, formulation of his theory show that he did.

12. The last part of [1946b] gives some applications. First an analog in this context of a theorem of Vietoris: if E^* is compact and $\pi^{-1}(x)$ is acyclic for all $x \in E^*$, then π induces an isomorphism of $H^*(E^*; L)$ onto $H^*(E; L)$. If π is the projection of a locally trivial fiber bundle, with typical fiber F and E^* is simply connected, the Betti numbers of E are majorized by those of $E^* \times F$. Finally, if E is a compact simply connected Lie group, F a closed one-parameter subgroup, L a field of characteristic zero, then $H^*(E/F; L)$ is obtained from $H^*(E; L)$, which is an exterior algebra on odd degree generators, by replacing one factor $\wedge x_{2i+1}$ by $L[x]/(x^{i+1})$, where x has degree two. (In fact, $i = 1$, [11b]).

13. In the following year, the theory underwent a number of changes, partly under the influence of contributions by J-L. Koszul and H. Cartan. In [11a], Koszul gives a purely algebraic definition of the construction underlying [1946b], introducing what is now known as the spectral sequence of a filtered differential algebra A (in the case of a decreasing filtration). A (decreasing) filtration on an algebra A is defined by a sequence of two-sided ideals A^p ($p \in \mathbb{Z}$) such that

$$A^p \supset A^q \text{ if } p \leqq q,\ \cup_p A^p = A,\ \cap_p A^p = \{0\}. \tag{1}$$

$$A^p . A^q \subset A^{p+q} \tag{2}$$

and, if A is differential, $dA^p \subset A^p$ ($p \in \mathbb{Z}$). In that case, $H(A)$ is endowed with the filtration defined by the $H(A)^p$, where $H(A)^p$ denotes the subgroup of cohomology classes represented by a cocycle in A^p. The spectral sequence relates the grading ring $Gr A = \oplus_p A^p/A^{p+1}$ to $Gr\ H(A) = \oplus_p H(A)^p/H(A)^{p+1}$ by means of a sequence of graded algebras E_r ($r \geqq 0$), where $E_0 = Gr\ A$ and E_{r+1} is the cohomology of E_r with respect to a differential d_r. If A is graded and d increases the degree by one, then E_r is bigraded and d_r increases the filtration degree by r, decreases a complementary degree by $r - 1$. Moreover E_r tends to $Gr\ H(A)$, and is equal to it if the filtration is bounded.(7)

In the notation of **11**, $E_r^{p,q} = P_{r-1}^{p,q}/Q_{r-2}^{p,q}$ [11b].

Early in 1947, H. Cartan noticed a formal similarity between the fundamental lemma of [1945a] and a proof of the de Rham theorems contained in a letter of A. Weil [15].(8) This was his starting point towards an axiomatic cohomology theory, quite different from his previous approach (Comm. Math. Helv. **18** (1945), 1–15) which was much more in the mainstream of algebraic topology at the time.

14. Cartan and Leray lectured at a Colloquium in Paris, June 26 – July 2, 1947, but the Proceedings, published in 1949 only, do not contain their original communications. The article of Leray [1949a] is "different in title and contents" from the oral lecture and summarizes a lecture given in November 1947 and a course given at the Collège de France in 1947-48 (*see* the footnote on p. 61). Cartan withdrew his communication and replaced it by a short text, written in 1949, stating that his views had changed considerably, partly under the influence of [1946a] and of Leray's lecture at the Colloquium and he was preparing a full-fledged exposition (the subject matter of [7]).

The theory outlined in [1949a] is basically the final form, as can be seen from the systematic exposition [1950a]. It starts with algebraic notions: differential algebra, filtered ring and spectral sequence attached to a differential filtered ring, essentially as introduced in [11a], or [6a] with some technical differences, though: the subgroups defining the filtration need not be two-sided ideals and, in the differential case, are not necessarily stable under the differential. The spectral sequence (E_r) may therefore have non-trivial terms with negative index, which in a sense tend to $Gr\ A$ as $r \to -\infty$. If the filtration and grading are bounded, then $E_r = Gr\ A$ (resp. $E_r = GrH(A)$) for r sufficiently small (resp. big). The fundamental lemma is now embedded in some spectral sequence statements. [Initially, Leray allows a filtration by the real numbers $\mathbb{R}$, but uses only $\mathbb{Z}$ when defining the spectral sequence and in his subsequent papers. Filtrations and spectral sequences indexed by $\mathbb{R}$ were considered later by R. Deheuvels (Annals of Math. **61** (1995), 13–72), upon Leray's suggestion, in connection with the calculus of variations.]

Spaces are always locally compact. The notion of sheaf is as in [1946a], except that the condition "normal" is replaced by "continuous". It is the same if the space X is compact, but stronger otherwise(9). For instance, given a ring L, the constant sheaf which assigns L to every closed subset F, the transition homomorphisms being the identity, is continuous if and only if X is compact. If it is not, the sheaf associating L to compact subsets, and zero to non-compact ones, is continuous; it is called the "sheaf identical to L". The notion of sheaf is further extended to that of differential filtered sheaf.

Next complexes are defined. The original definition (*see* **6**), in which the supports are now assumed to be closed, is modified in two ways, proposed by H. Cartan in his lecture: a complex is not necessarily a free module and is moreover assumed to be endowed with a product adding the degrees, i.e. it is a differential graded ring (with closed supports). The most important innovation however is the introduction of *fine* complexes. This was done at the Colloquium lecture already, while Cartan proposed a similar notion in his own (*see* the already quoted footnote on p. 61): a complex K on X is fine if , given a finite open cover $\{U_\alpha\}$ of X by subsets which are relatively compact or with compact complements, there exists endomorphisms r_α of K, for the additive structure only, such that $\operatorname{supp} r_\alpha k \subset U_\alpha$ for all $k \in K$ and the sum of the r_α is the identity. This then replaces a union of complexes with arbitrarily small supports (*see* **6**). The intersection $K \circ \mathcal{B}$ of a complex K with the sheaf $\mathcal{B}$ is defined. It is a complex, which is fine if K is so. Similarly, the intersection $K \circ K'$ of two complexes K, K' is fine if one of them is so. The sheaf is assumed to be continuous, which forces the elements of $K \circ \mathcal{B}$ to have compact supports. The cohomology ring $H^{\cdot}(K^{\cdot} \circ \mathcal{B})$ of X with respect to $\mathcal{B}$ is is by definition $H^{\cdot}(K^{\cdot} \circ \mathcal{B})$, where $K^{\cdot}$ is a fine couverture. It has of course to be shown to be independent of the choice of $K^{\cdot}$, up to natural isomorphisms. To this effect the fundamental lemma, or some variant, is used to show that if $K^{\cdot}$ and $M^{\cdot}$ are fine couvertures, then $K^{\cdot} \circ M^{\cdot}$ is also one and the natural maps

$$K^{\cdot} \circ \mathcal{B} \to K^{\cdot} \circ M^{\cdot} \circ \mathcal{B}, \quad M^{\cdot} \circ \mathcal{B} \to K^{\cdot} \circ M^{\cdot} \circ \mathcal{B}$$

induce isomorphisms in cohomology. The construction of Alexander, modified by Čech, the initial inspiration for Leray (*see* 5) gives rise to a fine couverture, showing that when $\mathcal{B}$ is the "sheaf identical to a ring L" (*see* above), the cohomology $H^{\cdot}(X \circ \mathcal{B})$ is the Alexander-Čech, also called Alexander-Spanier, cohomology of X with compact supports, coefficients in L.

The case of a differential filtered sheaf is also considered (n° 23) and the homology sheaf $\mathcal{HB} : F \mapsto H(\mathcal{B}(F))$, denoted there $\mathcal{FB}$, is introduced. The group $H^{\cdot}(K^{\cdot} \circ \mathcal{B})$, computed with respect to a total differential (the "hypercohomology" with respect to $\mathcal{B}$ in case the filtration is associated to a grading), is naturally filtered and is the abutment of a spectral sequence in which one term is $H^{\cdot}(K^{\cdot} \circ \mathcal{HB})$. In the present set up, this is the *fundamental theorem of sheaf theory*. Again, the hypercohomology and the spectral sequence are independent of the fine couverture $K^{\cdot}$ and define topological invariants.

A familiar consequence, not drawn here or in [1950a], but apparently in the original text[(10)], pertains to homomorphisms of differential filtered sheaves. Let $\mu : \mathcal{B} \to \mathcal{C}$ be one. If it induces an isomorphism of $H(\mathcal{B}(x))$ onto $H(\mathcal{C}(x))$ for all $x \in X$, then it induces an isomorphism of $H^{\cdot}(K^{\cdot} \circ \mathcal{B})$ onto $H^{\cdot}(K^{\cdot} \circ \mathcal{C})$.

All this is valid in fact only under suitable boundedness assumptions on the degrees and filtrations under consideration, which I have ignored (e.g., it suffices that they be bounded in both directions).

Let π: $X \to Y$ be a continuous map and $K^{\cdot}_X$ (resp. $K^{\cdot}_Y$) a fine couverture of X (resp. Y). Then $\pi^{-1}K^{\cdot}_Y \circ K^{\cdot}_X$ is a fine couverture of X. The spectral sequence of $Z = \pi^{-1}K^{\cdot}_Y \circ K^{\cdot}_X \circ \mathcal{B}$ with respect to the filtration defined by the degree in $K^{\cdot}_Y$ is by definition the spectral sequence of π. It relates the (hyper)cohomology of Y with respect to $\pi_*(K^{\cdot}_X \circ \mathcal{B})$ to the cohomology of $H^{\cdot}(X \circ \mathcal{B})$. In fact, given integers ℓ, m, Leray defines a filtration of Z, using m-times the degree in $K^{\cdot}_Y$ and ℓ times the degree in $K^{\cdot}_X$, whence a spectral sequence for each choice of ℓ and m, but they are not essentially different. In studying fibre bundles Leray uses mostly the ones corresponding to $\ell = 0, m = 1$, or $\ell = -1, m = 0$. In the sequel, I shall always stick to the former. The r-th term of a spectral sequence is denoted $\mathcal{H}_r$ by Leray. The index depends on the filtration: $\mathcal{H}_r$ for the filtration $\ell = 0, m = 1$ is $\mathcal{H}_{r-1}$ for the filtration $\ell = -1, m = 0$. [The construction of [1946b], *see* **11**, is a precursor of the spectral sequence assigned to $\ell = -1, m = 0$.] As before, I shall use E_r.

The spectral sequence replaces the construction of [1946b]. The underlying idea is the same, but more easily described, notably because it starts with a complex on X, rather than the equivalent of the E_2-term. Let

$$Z^{p,q} = \pi^{-1}K^p_Y \circ K^q_X \circ \mathcal{B} \text{ and } F_pZ = \sum_{i \geqq p,\, q \geqq 0} Z^{i,q}.$$

The F_pZ define the filtration $\ell = 0$, $m = 1$. Let $z \in Z$. Its filtration degree is the biggest p such that $z \in F_pZ$. Cocycles are arrived at by successive approximations: one looks at $z \in F_pZ$ such that $dz \in F_{p+r}Z$ ($r \geqq 0$). These elements form C^p_r. The latter contains C^{p+1}_{r-1} and $D^p_r := dC^{p-r}_r$; by definition

$$E_r = C^p_r/(C^{p+1}_{r-1} + D^p_r)$$

and d_r is induced by d. As r gets bigger, z is closer to a cocycle, and its actually one if r is greater than its total degree.

Except in [1950b], no groundring is specified and Leray speaks of filtered rings and spectral rings. He shifts to filtered algebras and spectral algebras in [1950c].

15. Assume now that a discrete group G acts freely and properly on X and that $Y = X/G$ is the quotient space. The map π has discrete fibres and the spectral sequence cannot give much information. However, Leray indicated in his lecture how to associate a spectral sequence to that situation when G is finite. This led to a joint paper with H. Cartan [1949b], which defines the spectral sequence of a finite regular covering of a locally compact space, relating Eilenberg-MacLane cohomology of G with coefficients in $H^{\cdot}(X; L)$ to the cohomology of Y, in cohomology with compact supports, and to two C. R. Notes of Cartan, where the restrictions that G be finite and the supports be compact are lifted [6].

16. The paper [1950a], based on courses given at the Collège de France in 1947-48 and 1949-50, provides a comprehensive exposition of the theory. The overall plan is the same as that of [1949a] with many technical refinements I shall not go into, depending for instance on various assumptions made on the complexes under consideration and on whether filtrations are bounded or not. I content myself to mention some items not occurring in [1949a]. If X has finite cohomological dimension, then it carries a fine couverture with degrees bounded by the dimension n°40. The Mayer-Vietoris sequence attached to a cover by two closed sets is established in n°49. n°67 considers the effect of retractions on cohomology and discusses homotopic maps. n°s.69 to 73 are devoted to locally constant systems in Steenrod's sense and their relations with the fundamental group. Determination of the cohomology when X has a finite convexoïd cover n°74, or is more particularly a finite polyhedron n°75, spectral sequence of a simplicial map between polyhedra n°77. The last two sections show that the spectral sequence of a map is not necessarily a homotopy invariant and give some indications on how to define homotopy invariants by means of these constructions.

This paper is the last one devoted by Leray to his theory of cohomology with compact supports of a locally compact space with respect to a sheaf and to the general properties of the spectral sequence of a continuous map. The former was considerably generalized by H. Cartan [9][(11)]. From the start, Leray applied the latter to fibre bundles, in particular to the study of the relations between the cohomology rings of a compact connected Lie group, a closed subgroup U and the quotient G/U. We now turn to these applications, backtracking a little since they began in 1946 already.

17. The work of Leray on fibre bundles and homogeneous spaces is contained in six C.R. Notes and two papers. Four of the C.R. Notes announce without proofs results established, often in greater generality, in one of the two papers. I shall therefore treat them rather briefly. The cohomology is usually with respect to a field C, of characteristic zero when homogeneous spaces are discussed.

Given a space with finitely generated cohomology, its Poincaré polynomial $P(E,t)$ (with respect to C) is, by definition

$$P(E,t) = \sum_{i\geqq 0} \dim H^i(E;C).t^i$$

[1946c] considers first a map $\pi : E \to E^*$ as in [1946b] and describes some relations between Poincaré polynomials of E, E^* and the invariants of π defined in [1946b].

The remaining part of the Note is concerned with a locally trivial fibre bundle (E,B,F,p) with total space E, base B, typical fibre F and projection p. First some consequences of Poincaré duality are drawn when E,B,F are orientable compact connected manifolds. The last section gives sharper relations between the Poincaré polynomials of E,B,F when the cohomology rings of the fibres form a constant system. [1946d] describes the Poincaré polynomial of G/T, when G is a simple compact connected classical Lie group and T a maximal torus of G. This is pursued further in [1949e], where the rational cohomology of G/U is determined when G is locally isomorphic to a product of classical groups by a torus and U a closed subgroup of maximal rank. If U is connected $P(G/U,t)$ is given by a formula conjectured by G. Hirsch. If it is not and U^o is its identity component, then $H^{\cdot}(G/U;C)$ may be identified to the invariants of U/U^o in $H^{\cdot}(G/U;C)$, the operation of U/U^o being defined by right translations.

From 1949 on, the cohomology ring of a compact space X, with respect to coefficients which are clear from the context, is denoted $\mathcal{H}_X$. The Note [1949f] has three parts: the first one extends some of the results of the previous one to a compact space X which is a principal bundle for G. In particular, the projection $X/T \to X/G$ induces an isomorphism of $H^{\cdot}(X/G)$ onto the invariants of the Weyl group $\mathcal{N}T/T$ in $H^{\cdot}(X/T)$. If X is a group containing G as a subgroup, this reduces the study of $H^{\cdot}(X/G)$ to that of $H^{\cdot}(X/T)$. The second one (Theorem 2) determines the cohomology of G/S, where G is simple, classical, and S a singular subtorus of codimension one in a maximal torus T of G. It is the tensor product of an algebra of even dimensional elements by an exterior algebra with one generator of degree equal to the maximum of the degrees of the primitive generators of $H^{\cdot}(G;\mathbb{Q})$. The third part is devoted to sphere fibrations and describes how the Gysin exact sequence relates to the spectral sequence of the projection.

18. The two Notes [1949c] and [1949d] are somewhat apart and concerned with a topic Leray did not come back to (but is taken up again in [4]). There homology and cohomology (in characteristic zero) do occur and the standard terminology is used. By theorems of Hopf and Samelson, $H^{\cdot}(G)$ and the homology algebra $H_{.}(G)$, where the product is the Pontrjagin product, are exterior algebras $\wedge P^{\cdot}$ and $\wedge P.$ over spaces of primitive elements, in natural duality. Assume G operates on a locally compact space X by means of a map $q : G \times X \to X$ and let $m : G \times G \to G$ be the product map. Then, by definition

$$q \circ m = q \circ q : G \times G \times X \to X.$$

By consideration of the corresponding maps in cohomology it is shown that a primitive homogeneous element $x \in P_a$ induces a differential δ_x of $H^{\cdot}(X)$ decreasing the degree by a and this assignment extends to a homomorphisms of $\wedge P_{.}$ into the algebra of graded endomorphisms of $H^{\cdot}(X)$. If X is a compact orientable manifold and $c \in H^p(X)$ is dual to a submanifold of codimension p, then a submanifold dual to $\delta_x.c$ is described geometrically.

In the case $X = G/U$, where G is a compact connected Lie group, U a closed connected subgroup, $X = G/U$, this construction yields a new proof of a theorem of Samelson asserting that $\pi^* H^{\cdot}(G/U)$ is a subalgebra generated by primitive elements, where $\pi : G \to G/U$ is the canonical projection. If U is of maximal rank, the Euler-Poincaré characteristic $X(G/U)$ of U is $\neq 0$, according to a theorem of Hopf and Samelson, hence π^* annihilates $H^{\cdot}(G/U)$ for all $i > 0$.

The next Note considers more generally a projection $\pi : X \to Y$, where G acts on X and Y and commutes with π. Then the operations of $\wedge P_{.}$ on X and Y extend to differentials of the terms E_r of the spectral sequence of π, commuting with the differentials. Various consequences are drawn.

Let G and U be as in the previous Note, but U not necessarily of maximal rank. Let $\mathcal{N}U$ be the normalisator of U in G and M its identity component. Then $\mathcal{N}U/M$ operates freely by right translations on G/U. The Lefschetz number of $n \in \mathcal{N}U$ is $\chi(G/U)$ if $n \in M$, and is zero if $n \notin M$. Therefore the representation of $\mathcal{N}U/M$ induced in $H^{\cdot}(G/U)$ is a multiple of the regular representation if the Betti numbers of G/U in odd degrees are all zero.

19. The paper [1950b] is devoted first to general properties of the spectral sequence of a fibre bundle (E, B, F, π): structure of E_2, case where the cohomology algebras of the fibres form a constant system, interpretation of π^* and of the restriction $r^{\cdot} : H^{\cdot}(E) \to H^{\cdot}(F)$ in the spectral sequence, triviality of the spectral sequence in case $r^{\cdot}$ is surjective (i.e. F is totally non-homologous to zero), various inequalities relating to Poincaré polynomials of E, B, F, etc. The last chapter discusses several special cases: F is a sphere, where generalizations of results of Gysin and of Chern-Spanier are obtained, B is a sphere, where the H.C. Wang exact sequence is proved, F is a product of even dimensional spheres and 2 is divisible in the coefficient ring, in which case F is totally non-homologous to zero, E, B, F are compact orientable manifolds.

20. The paper [1950c] uses the filtration $\ell = -1, m = 0$, for spectral sequences therefore E_r here stands for $\mathcal{H}_{r-1}$ there. Cohomology is always with respect to a field of characteristic zero. G, T and U are as before, and $W = \mathcal{N}T/T$ is the Weyl group of G.

In the first part, U has the same rank as G. Results stated earlier for classical groups are now proved in general. The new ingredients are the theorem of Chevalley on invariants of finite reflection groups and an argument, supplied by this writer, showing that the Betti numbers of G/T vanish in odd degrees. The Hirsch formula giving $P(G/U, t)$ when U is connected is established. Let $\mathcal{P}_T$ be the symmetric algebra over $H^1(T)$, where all the degrees are doubled. The group W operate on it. Let $\mathcal{R}_G$ be the ideal generated by the invariants of W without constant term. It is shown that $H^{\cdot}(G/T) = \mathcal{P}_T/\mathcal{R}_G$ and that $E_3 = E_\infty$ in the spectral sequence of the projection $G \to G/T$. The next section of the paper is devoted to the situation considered in [1949f] and establishes without restriction on G the theorems stated there. This reduces the study of $H^{\cdot}(G/U)$ to that of $H^{\cdot}(G/S)$, where S is a maximal torus of U. It may be assumed to be contained in T and the last section provides a theorem on the E_2-term of the spectral sequence of the projection $G/S \to G/T$, with fibre T/S. It is of course equal to $H^{\cdot}(G/T) \otimes H^{\cdot}(T/S)$. However, using the results of the first part, Leray shows that it is isomorphic, as a differential algebra, with $\mathcal{P}_S \otimes H^{\cdot}(G)$, endowed with an explicitly given differential d. Unpublished computations to prove Theorem 2 of [1949f] indicate that Leray had that picture in mind already then. Here it is particularly interesting because a theorem announced by Cartan in [8] implies that $H^{\cdot}(G/S) = H^{\cdot}(\mathcal{P}_S \otimes H^{\cdot}(G))$. As a consequence $E_3 = E_\infty$, a fact which is clear if $\dim T/S = 1$, but not otherwise.

This is the last paper devoted by Leray to algebraic topology, a topic which had played in his work a minor role in the thirties, a major one in the forties, occurred only incidentally in it after 1950 and was profoundly influenced by Leray's contributions.

Notes

(1) J. Schauder once wrote to Leray that he did not view himself as a topologist per se and commented in another letter: "I am, as you are, a man of the applications" ("Ich bin, so wie Sie, ein Mann der Anwendungen"), a remark quoted by Leray in [1979].

(2) The only exceptions where a course in analysis, based on the Notes of a Cours d'Analyse at the Ecole Polytechnique, brought by some prisoners who had been students there, and a course on special relativity (where "Einstein" became "Albert" whenever some member of the German staff was passing by during a lecture).

(3) In [10], A. Deleanu extends the theory to neighborhood retracts of convexoïd spaces (which are not always convexoïd), so that it also includes Lefschetz's fixed point theorem for absolute neighborhood retracts.

(4) In [3], this lemma (Exp. I, Théorème 6), the main argument of which is called induction on the weight, is also used to give a first proof, without spectral sequences, of a main uniqueness theorem of [1949a] or [1950a], and to compare Leray's cohomology ring with others (Exp. III, IV). It is also a main tool in Cartan's first two versions of the theory (*see* Note(11)).

(5) In the comments to [16], p. 526-27, Vol. II, of his Collected Papers, A. Weil recalls a short conversation in June 1945 with Leray, just back from captivity, in which Leray spoke of a homology theory with variable coefficients depending on the point, an idea he found quite striking and communicated shortly afterwards to H. Cartan.

(6) $\pi(\mathcal{B})$ and $\pi(\mathcal{B}^q)$ are the analogues of the direct image $\pi_*\mathcal{B}$ of $\mathcal{B}$ and of the q-th right derived functor $R^q\pi_*\mathcal{B}$ of the direct image functor in the now standard sheaf theory.

(7) The term filtration is not used there. It was proposed later by H. Cartan, in print for the first time in [6a]. As to the E_r's, Koszul speaks of a sequence of homologies, Leray of a spectral ring, from [1949a] on, and of a spectral algebra in [1950c]. I shall use spectral sequence.

(8) We try here to compare them. Let M be a smooth connected manifold, N the nerve of the open cover constructed in [15] (or [16]), such that all non-empty intersections U_σ (σ simplex of N) are contractible (the analog of a convexoïd cover in [1945a]). Let $A^{p,q}$ be the space of p-cochains of N which assign to a p-simplex σ the smooth differential q-forms on U_σ. The direct sum $A^{\cdot\cdot}$ of the $A^{p,q}$ is a bigraded algebra, endowed with two commuting differentials

$$d : A^{p,q} \to A^{p,q+1} \qquad \delta : A^{p,q} \to A^{p+1,q}$$

stemming from exterior differentiation and from the coboundary operator in N. Let $E^{p,q}$ (resp. $H^{p,q}$) be the subspace of $A^{p,q}$ spanned by the elements annihilated by $d\delta$ (resp. d or δ). (Weil's notation is different, his two superscripts are the total degree $m = p + q$ and q). Weil establishes isomorphisms

$$F^{0,m}/H^{0,m} = H^m_{DR}(M),\ F^{m,0}/H^{m,0} = H^m(N),$$

$$F^{p,q}/H^{p,q} = F^{p+1,q-1}/H^{p+1,q-1} \qquad (0 \leqq q \leqq m)$$

where $H^m_{DR}(M)$ refers to de Rham cohomology, which, by composition, yield an isomorphism of $H^m_{DR}(M)$ onto $H^m(N)$. Each step is quite similar to the key argument in the fundamental lemma, though there is no reason to believe that Weil was aware of it. On the other hand, it seems rather plausible (also to Weil) that the definition of the $A^{p,q}$ had been suggested in part by the idea of cohomology with variable coefficients. In fact, apart from the fact that Weil deals with an open rather than closed cover, $A^{p,q}$ is, in the framework of [1946a], the space of p-forms of the couverture N with coefficients in the sheaf of differential q-forms. However, the global strategy of the proof is different from that of Leray to establish uniqueness theorems, which amounts to compare the two objects under consideration to a third one, their intersection, rather than directly to one another. The algebra $A^{\cdot\cdot}$ admits a total differential $d - \delta$ (which Weil does not consider explicitly, but the sequences of coelements of total degree m satisfying (I) in [16] are cocycles with respect to it). Then a descending induction on p and q would show that $H^m(A^{\cdot\cdot})$ is isomorphic to $H^m_{DR}(M)$ and to $H^m(N)$. In [16], written later, the argument is further simplified by the use of homotopy operators, which even allow one to define directly maps in both directions between simplicial cochains and differential forms. Weil also shows in the same way that $H^m(N)$ is isomorphic to the m-th cohomology space of M in singular cohomology.

(9) Let X be not compact and $\overline{X}$ its one-point compactification. Given the sheaf $\mathcal{B}$ on X, define the sheaf $\overline{\mathcal{B}}$ on $\overline{X}$ by the rule $\overline{\mathcal{B}}(F) = \mathcal{B}(F \cap X)$, ($F$ closed in $\overline{X}$). Then, by definition $\mathcal{B}$ is continuous on X if $\overline{\mathcal{B}}$ is normal on $\overline{X}$.

(10) In [2], n° 2, I state this is so. I do not remember whether I had seen the original text or had only been informed by Leray.

(11) This is the third version of Cartan's work on this topic. The first one [12], which is likely to be rather close to the oral lecture at the 1947 Colloquium, is also based on the notion of complex with supports. The main change with respect to the definition in [1945a] is the introduction of differential graded complexes, called gratings in analogy with a terminology of J. Alexander, which are graded algebras and fine. There are no sheaves as such but, as in [1945a], given a complex K, the functor assigning to a closed subset F the complex $F.K$ plays an important role. There are no spectral sequences. The uniqueness theorem is established for compact spaces, by means of an analogue of the consequence

of the fundamental theorem of sheaf theory mentioned in **14**: a homomorphism $K^{\cdot} \to L^{\cdot}$ of fine differential graded gratings which induces an isomorphism of $H^{\cdot}(xK^{\cdot})$ onto $H^{\cdot}(xL^{\cdot})$ for all $x \in X$ induces an isomorphism in cohomology (again under suitable boundedness conditions), in the special case where $H^{i}(xK)$ and $H^{i}(xK')$ are zero for $i > 0$.

The main argument to establish it, p. 159–165, is patterned after the fundamental one of [1945a], outlined here in **9**. [No reference is indicated there, but this is acknowledged in the next version [7], Exp. XV, n° 7.] Applications to the de Rham theorems and the singular cohomology of HLC spaces are also given.

The second stage [7] is still devoted to locally compact spaces, but cohomology with closed supports is included (if the space is also paracompact). The basic notion in [7] is that of sheaf, defined as in [1946a], a condition similar to normality being embedded into the definition. Those sheaves correspond in fact to "presheaves" in current terminology. The distinction between presheaves and sheaves becomes important if cohomology with closed supports of non-compact spaces is to be included, and Cartan introduces the *completion* of the given sheaf, which would now be called the sheaf associated to, or defined by, a presheaf.

The notion "fine" is carried over to sheaves. A sheaf in which all transition homomorphisms are surjective (which is in fact the sheaf associated to the complex of sections on the whole space) is called a carapace, and the cohomology is defined by means of fine carapaces, in which $H^{\cdot}(B(x))$ is acyclic for all $x \in X$, the counterpart of a fine couverture. Numerous examples are given and, once the uniqueness theorem is proved, many consequences are drawn, including Poincaré duality on manifolds, for cohomology with closed supports or with compact supports.

In [9] the theory is developed in much greater generality, with a stronger use of homological algebra, which Cartan was developing at the time with S. Eilenberg. X is only assumed to be regular. A sheaf is now defined as a functor on open subsets and injective resolutions are introduced. Cohomology is defined with respect to a family Φ of supports and the spectral sequence of a continuous map is also defined in that context. The fundamental theorem of sheaf theory (XIX, Thm. 3) is proved in full generality.

This exposition and Cartan-Eilenberg's "Homological Algebra" (Princeton University Press, 1956) paved the way for the treatment of sheaf theory and spectral sequences in the framework of homological algebra by A. Grothendieck: *Sur quelques points d'algèbre homologique*, Tôhoku M. J. **9**, 1957, 119–221 and R. Godement: "Topologie algébrique et théorie des faisceaux", Hermann, Paris 1958.

References

The references to J. Leray's papers are to the bibliography at the end of this volume.

Further references

1. J. Alexander, *On the connectivity ring of an abstract space*, Annals of Math. **37** (1936), 698–708.
2. A. Borel, *Remarques sur l'homologie filtrée*, J.Math. Pures Appl. (9) **29**, (1950), 313–322; Collected Papers I, 57–66, Springer.
3. A. Borel, Cohomologie des espaces localement compacts, d'après J. Leray, mimeographed Notes, E.P.F. Zurich, 1951; 3rd edition: LNM **2**, 1964, Springer.
4. A. Borel, *Sur l'homologie et la cohomologie des groupes de Lie compacts connexes*, Amer. J. Math. **76** (1954), 273–342; Collected Papers I, 322–391, Springer.
5. E. Cartan, La méthode du repère mobile, la théorie des groupes continus et les espaces généralisés, Notes written by J. Leray, Hermann, Paris 1935; Oeuvres Complètes III_2, 1259–1320.
6. H. Cartan, *Sur la cohomologie des espaces où opère un groupe.* a) *Notions algébriques préliminaires*, C. R. Acad. Sciences Paris **226** (1948), 148–150; b) *Etude d'un anneau différentiel où opère un groupe*, **ibid.** 303–305; Oeuvres III, 1226–1228, 1229–1231, Springer.
7. H. Cartan, Séminaire de topologie algébrique de l'E.N.S. 1948-49, Exp. XII to XVII.
8. H. Cartan, *La transgression dans un groupe de Lie et dans un espace fibré principal*, Colloque de Topologie C.B.R.M. Bruxelles 1950, 57–71; Oeuvres III, 1268–1282, Springer.
9. H. Cartan, Séminaire de topologie algébrique de l'E.N.S. 1950-51, Exp. XVI to XX.
10. A. Deleanu, *Théorie des points fixes sur les rétractes de voisinages des espaces convexoïdes*, Bull. Soc. Math. France **87** (1959), 235–243.
11. J-L. Koszul, a) *Sur les opérateurs de dérivation dans un anneau*, C.R. Acad. Sciences Paris **225** (1947), 217–219; b) *Sur l'homologie des espaces homogènes*, **ibid.** 477–479.
12. H. Pollack and G. Springer, Algebraic topology, based upon lectures by H. Cartan at Harvard University, (Spring 1948), mimeographed Notes, Harvard University 1949.
13. J. Schauder, *Der Fixpunktsatz in Funktionalräumen*, Studia Mathematica **2** (1930), 170–179; Oeuvres, 168–176, Polish Scientific Publishers, Warsaw 1978.
14. J. Schauder, *Ueber den Zusammenhang zwischen der Eindeutigkeit und der Lösbarkeit partieller Differentialgleichungen zweiter Ordnung von elliptischen Typus*, Math. Annalen **106** (1932), 667–721; Oeuvres, 235–297.
15. A. Weil, *Lettre à Henri Cartan* (Jan. 1947), Collected Papers II, 45–47, Springer.
16. A. Weil, *Sur les théorèmes de de Rham*, Comm. Math. Helv. **26** (1952), 119–145; Collected Papers II, 17–43, Springer.
17. H. Whitney, *On products in a complex*, Annals of Math. **39** (1938), 397–432; Collected Papers II, 294–329, Birkhäuser.

Institute for Advanced Study, School of Math., Princeton, NJ 08540, USA

165.

Twenty-five Years with Nicolas Bourbaki, 1949–1973*

Notices of the AMS 45 (1998), 373–380.
Mitteilungen der Deutschen Mathematiker Vereinigung, Heft 1, 1998, 8–15

1. The choice of dates is dictated by personal circumstances: they roughly bound the period in which I had inside knowledge of the work of Bourbaki, first through informal contacts with several members, then as a member for twenty years, until the mandatory retirement at 50.

Being based largely on personal recollections, my account is frankly subjective. Of course, I checked my memories against the available documentation, but the latter is limited in some ways: not much of the discussions about orientation and general goals has been recorded.[1] Another member might present a different picture.

2. To set the stage, I shall briefly touch upon the first fifteen years of Bourbaki. They are fairly well documented[2] and I can be brief.

In the early thirties, the situation of mathematics in France, at the university and research levels, the only ones of concern here, was highly unsatisfactory. World War I had essentially wiped out one generation. The upcoming young mathematicians had to rely for guidance on the previous one, including the main and illustrious protagonists of the so-called 1900 school, with strong emphasis on analysis. Little information was available about current developments abroad, in particular about the flourishing German school (Göttingen, Hamburg, Berlin), as some young French mathematicians (J. Herbrand, C. Chevalley, A. Weil, J. Leray) were discovering during visits to those centers.[3]

3. In 1934, A. Weil and H. Cartan were Maîtres de Conférences (the equivalent of assistant professors) at the University of Strasbourg. One main duty was, of course, the teaching of differential and integral calculus. The standard text was the Traité d'Analyse of E. Goursat, which they found wanting in many ways. Cartan was frequently bugging Weil with questions on how to present this material, so that, at some point, to get it over with once and for all, Weil suggested they write themselves a new "Traité d'Analyse". This suggestion was spread around and soon a group of about 10 mathematicians

---

* Lecture given at the University of Bochum, Germany, October 1995, in a Colloquium in honor of R. Remmert, and at the International Center for Theoretical Physics, Trieste, Italy, September 1996.

[1] The Archives of Bourbaki at the Ecole Normale Supérieure, Paris, contain reports, surveys, successive drafts or counterdrafts of the chapters, remarks on those resulting from discussions, and proceedings of the Congresses, called "Tribus". Those provide mainly a record of plans, decisions, commitments for future drafts, as well as jokes, sometimes poems.

[2] See [2; 3; 6; 7; 8; 14].

[3] For this, see pp. 134–136 of [8].

began to meet regularly to plan this treatise. It was soon decided that the work would be collective, without any acknowledgment of individual contributions. In summer 1935, the pen name Nicolas Bourbaki was chosen.[4]

The membership varied over the years; some people in the first group dropped out quickly, others were added and later there was a regular process of additions and retirements. I do not intend to give a detailed account. At this point, let me simply mention that the true "founding fathers", those who shaped Bourbaki and gave it much of their time and thoughts until they retired are:

Henri Cartan
Claude Chevalley
Jean Delsarte
Jean Dieudonné
André Weil

born respectively in 1904, 1909, 1903, 1906, 1906, all former students at the Ecole Normale Supérieure in Paris.[5]

A first question to settle was how to handle references to background material. Most existing books were found unsatisfactory. Even B. v. d. Waerden's Moderne Algebra, which had made a deep impression, did not seem well suited to their needs (besides being in German). Moreover, they wanted to adopt a more precise, rigorous style of exposition than had been traditionally used in France, so they decided to start from scratch and, after many discussions, divided this basic material into six "Books", each consisting possibly of several volumes, namely,

I	Set theory
II	Algebra
III	Topology
IV	Functions of one real variable
V	Topological Vector Spaces
VI	Integration

These books were to be linearly ordered: references at a given spot could be only to the previous text in the same book or to an earlier book (in the given ordering). The title "Eléments de Mathématique" was chosen in 1938. It is worth noting that they chose "Mathématique" rather than the much more usual

[4] See [3] for the origin of the name.

[5] They all contributed in an essential way. For Cartan, Chevalley, Dieudonné and Weil I could witness it at first-hand, but not for Delsarte, who was not really active anymore when I came on board. But his importance has been repeatedly stressed to me by Weil in conversations. See also [14] and comments by Cartan, Dieudonné, Schwartz in [3: pp. 81–83]. In particular, he played an essential role in transforming into a coherent group, and maintaining it so, a collection of strong, some quite temperamental, individuals. Besides, obviously, Book IV, "Functions of one real variable" owes much to him. Some other early members, notably Szolem Mandelbrojt and René de Possel, have also contributed substantially to the work of the group in its initial stages.

"Mathématiques". The absence of "s" was of course quite intentional, one way for Bourbaki to signal its belief in the unity of mathematics.

4. The first volumes to appear were the Fascicle of Results on Set Theory (1939) and then, in the forties, Topology and three volumes of Algebra.

At that time, as a student and later assistant at the E.T.H. (Swiss Federal Institute of Technology) in Z"urich, I read them and learned from them, especially from Multilinear Algebra, for which there was no equivalent anywhere, but with some reservations. I was rather put off by the very dry style, without any concession to the reader, the apparent striving for the utmost generality, the inflexible system of internal references and the total absence of outside ones (except in Historical Notes). For many, this style of exposition represented an alarming tendency in mathematics, towards generality for its own sake, away from specific problems. Among those critics was H. Weyl, whose opinion I knew indirectly through his old friend and former colleague M. Plancherel, who concurred, at a time I was the latter's assistant.

5. In fall 1949 I went to Paris, having received a fellowship at the C.N.R.S. (Centre National de la Recherche Scientifique), benefiting from an exchange convention just concluded between the C.N.R.S. and the E.T.H. I quickly got acquainted with some of the senior members (H. Cartan, J. Dieudonné, L. Schwartz) and, more usefully for informal contacts, with some of the younger ones, notably Roger Godement, Pierre Samuel, Jacques Dixmier and, most importantly, Jean-Pierre Serre, the beginning of intense mathematical discussions and of a close friendship. Of course, I also attended the Bourbaki Seminar, which met three times a year and offered each time six lectures on recent developments.

Those first encounters quickly changed my vision of Bourbaki. All these people, the elder ones of course, but also the younger ones, were very broad in their outlook. They knew so much and knew it so well. They shared an efficient way to digest mathematics, to go to the essential points, and reformulate it in a more comprehensive and conceptual way. Even when discussing a topic more familiar to me than to them, their sharp questions often gave me the impression I had not really thought it through. That methodology was also apparent in some of the lectures at the Bourbaki seminar, such as Weil's on theta functions (Exp. 16, 1949) or Schwartz's on Kodaira's big Annals paper on harmonic integrals (Exp. 26, 1950). Of course, special problems were not forgotten, in fact were the bread and butter of most discussions. The writing of the books was obviously a different matter.

6. Later, I was invited to attend (part of) a Bourbaki Congress and was totally bewildered. Those meetings (as a rule three per year, two of one week, one of about two) were private affairs, devoted to the books. A usual session would discuss a draft of some chapter or maybe a preliminary report on a topic under consideration for inclusion, then or later. It was read aloud line by line by a member and anyone could at any time interrupt, comment, ask questions or criticize. More often than not, this "discussion" turned into a chaotic shouting match. I had often noticed that Dieudonné with his stentorian voice, his propensity for definitive statements, extreme opinions, would automatically raise the decibel level of any conversation he would take part in. Still, I was not prepared for what I saw and heard: "Two or three monologues shouted at top voice, seemingly independently of one another" is how I briefly summarized for myself my impressions that first evening, a description not unrelated to Dieudonné's comments in [8]:

> "Certain foreigners, invited as spectators to Bourbaki meetings, always come out with the impression that it is a gathering of madmen. They could not imagine how these people, shouting – some times three or four at the same time – could ever come up with something intelligent ..."

It is only about ten years ago, reading the text of a 1961 lecture by Weil on organization and disorganization in mathematics [13] that I realized this anarchic character, if not the shouting, was really by design. Speaking of Bourbaki, Weil said, in part (freely translated):

> "... keeping in our discussions a carefully disorganized character. In a meeting of the group, there has never been a president. Anyone speaks who wants to and everyone has the right to interrupt him ..."
>
> "The anarchic character of these discussions has been maintained throughout the existence of the group ..."
>
> "A good organization would have no doubt required that everyone be assigned a topic or a chapter, but the idea to do this never occurred to us ..."
>
> "What is to gather concretely from that experience is that any effort at organization would have ended up with a treatise like any other ..."

The underlying thought was apparently that really new, ground breaking ideas were more likely to arise from confrontation than from an orderly discussion. When they did emerge, Bourbaki members would say: "the spirit has blown" ("l'esprit a soufflé") and it is indeed a fact that it blew much more often after a "spirited" (or should I say stormy) discussion than after a quiet one.

Other rules of operation also seemed to minimize the possibility of a publication in a finite time:

Only one draft was read at a given time and everyone was expected to take part in everything. A chapter might go through six or even more drafts. The first one was written by a specialist, but anyone might be asked to write a later one. Often this was hardly rewarding. Bourbaki could always change his mind. A draft might be torn to pieces and a new plan proposed. The next version, following those instructions, might not fare much better and Bourbaki might opt for another approach or even decide that the former one was preferable, after all, and so on, resulting sometimes in something like a periodicity of two in the successive drafts.

To slow down matters further, or so it seemed, there were no majority votes on publications: all decisions had to be unanimous and everyone had a veto right.

7. However, in spite of all those hurdles, the volumes kept coming out. Why such a cumbersome process did converge was somewhat of a mystery even to the founding members (see [6], [8]), so I do not pretend to be able to fully explain it. Still, I'll venture to give two reasons.

The first one was the unflinching commitment of the members, a strong belief in the worthiness of the enterprise, how distant the goals might seem to be, and the willingness to devote much time and energy to it. A typical Congress day would include three meetings, totaling about seven hours of often hard, at times tense discussions, a rather grueling schedule. Added to this was the writing of drafts, sometimes quite long, which might take a substantial part of several weeks or even months, with the prospect of seeing the outcome heavily criticized, if not dismissed, or even summarily rejected after reading of at most a few pages, or left in abeyance ("put into the refrigerator"). Many, even if read with interest, have not led to any publication. As an example, the "pièce de résistance" on the second Congress I attended was a manuscript by Weil of over 260 pages on manifolds, Lie groups, titled "Brouillon de calcul infinitésimal", based on the idea of "nearby points" ("points proches"), a generalization of Ehresmann's jets. This was followed later by about 150 pages of elaboration by Godement, but Bourbaki never published anything on nearby points.

On the other hand, whatever was accepted would be incorporated without any credit to the author. Altogether, a truly unselfish, anonymous, demanding work, by people striving to give the best possible exposition of basic mathematics, moved by their belief in its unity and ultimate simplicity.

My second reason is the superhuman efficiency of Dieudonné. Although I did not try to count pages, I would expect him to have written more than any two or three other members combined. For about twenty-five years, he would routinely start his day (in mathematics i.e., maybe after an hour of piano playing) by writing a few pages for Bourbaki. In particular, but by far not exclusively, he took care of the final drafts, exercises and preparation for the printer of all the volumes (about thirty) which appeared while he was a member, and even slightly beyond.

This no doubt accounts to a large extent for the uniformity of style of the volumes, frustrating any effort to try to individualize one contribution or the other. But this was not really Dieudonné's style, rather the one he had adopted for Bourbaki. Nor was it the personal one of other Bourbaki members except for Chevalley. Even to Bourbaki, he seemed sometimes too austere and a draft of his might be rejected as being "too abstract". The description "severely dehumanized book ...", given by Weil in his review of a book by Chevalley ([12], p. 397) is one many people would have applied to Bourbaki itself. Another factor contributing to this impersonal, not user-friendly, presentation[6],

[6] Called "abstract, merciless abstract" by E. Artin in his review of Algebra [1], adding however "... the reader who can overcome the initial difficulties will be richly rewarded for his efforts by deeper insights and fuller understanding." (p. 479).

was the very process by which the final texts were arrived at. Sometimes a heuristic remark, to help the reader, would find its way into a draft. While reading it, in this or some later version, its wording would be scrutinized, found to be too vague, ambiguous, impossible to make precise in a few words, and then, almost invariably, thrown out.

8. As a by-product, so to say, the activity within Bourbaki was a tremendous education, a unique training ground, obviously a main source of the breadth and sharpness of understanding I had been struck by in my first discussions with Bourbaki members.

The requirement to be interested in all topics clearly led to a broadening of horizon, maybe not so much for Weil who, it was generally agreed, had the whole plan in his mind almost from the start or for Chevalley, but for most other members, as was acknowledged in particular by Cartan [7:xix]:

> "This work in common with men of very different characters, with a strong personality, moved by a common requirement of perfection, has taught me a lot, and I owe to these friends a great part of my mathematical culture."[7]

and by Dieudonné [8:143–144]:

> "In my personal experience, I believe that if I had not been submitted to this obligation to draft questions I did not know a thing about, and to manage to pull through, I should never have done a quarter or even a tenth of the mathematics I have done."

But the education of members was not a goal per se. Rather, it was forced by one of the mottoes of Bourbaki: "The control of the specialists by the non-specialists". Contrary to my early impressions in Z"urich, related earlier, the aim of the treatise was not the utmost generality in itself, but rather the most efficient one, the one most likely to fill the needs of potential users in various areas. Refinements of theorems which seemed mainly to titillate specialists, without appearing to increase substantially the range of applications, were often discarded. Of course, later developments might show that Bourbaki had not made the optimal choice.[8] Nevertheless this was a guiding principle.

Besides, many discussions took place outside the sessions about individual research or current developments. Altogether, Bourbaki represented an awesome amount of knowledge at the cutting edge, which was freely exchanged.

This made it obvious that for Bourbaki current research and the writing of the Eléments were very different, almost disjoint, activities. Of course, the latter was meant to supply foundations for the former and the dogmatic style, going from the general to the special, best suited for that purpose (see [5]).

---

[7] "Ce travail en commun avec des hommes de caractères très divers, à la forte personnalité, mus par une commune exigence de perfection, m'a beaucoup appris, et je dois à ces amis une grande partie de ma culture mathématique"

[8] For instance, the emphasis on locally compact spaces in Integration, on which P. Halmos had expressed strong reservations in his review [11], indeed did not address the needs of probability theory, and this led to the addition of a Chapter (IX) to Integration.

However, the Eléments were not meant to stimulate, suggest, or be a blueprint for, research (as stressed in [8] p. 144). Sometimes I have wondered whether a warning should not have been included in the "Mode d'emploi".

9. All this bore fruit and the fifties were a period of spreading influence of Bourbaki, both by the treatise and the research of members. Remember in particular the so-called French explosion in algebraic topology, the coherent sheaves in analytic geometry, then in algebraic geometry over ℂ, later in the abstract case, and homological algebra. Although very much algebraic, these developments also reached analysis, via Schwartz's theory of distributions and the work of his students B. Malgrange and J.-L. Lions on PDE. Early in 1955, A. Weinstein, a "hard analyst" had told me he felt safe from Bourbaki in his area. But, less than two years later, he was inviting Malgrange and Lions to his institute, at the University of Maryland.

I am not claiming at all that all these developments were solely due to Bourbaki. After all, the tremendous advances in topology had their origin in Leray's work, and R. Thom was a main contributor. Also, K. Kodaira, D. Spencer and F. Hirzebruch had had a decisive role in the applications of sheaf theory to complex algebraic geometry; but undeniably the Bourbaki outlook and methodology were playing a major role. This was recognized early on by H. Weyl, in spite of the critical comments mentioned earlier. Once R. Bott told me he had heard negative remarks on Bourbaki by H. Weyl in 1949 (similar to those I knew about) but, by 1952, the latter said to him: "I take it all back". Others however (like W. Hurewicz, in a conversation in 1952) would assert that all that had nothing to do with Bourbaki, only that they were strong mathematicians. Of course, the latter was true, but the influence of Bourbaki on one's work and vision of mathematics was obvious to many in my generation. For us H. Cartan was the most striking illustration, almost an incarnation, of Bourbaki. He was amazingly productive, in spite of having many administrative and teaching duties at the Ecole Normale Supérieure. All his work (in topology, several complex variables, Eilenberg–MacLane spaces, earlier in potential theory (with J. Deny) or harmonic analysis on locally compact abelian groups (with R. Godement) did not seem to involve brand new, ground breaking ideas. Rather, in a true Bourbaki approach, it consisted in a succession of natural lemmas and, all of a sudden, the big theorems followed. Once, with Serre, I was commenting on Cartan's output, to which he replied "Oh, well, twenty years of messing around with Bourbaki, that's all". Of course, he knew there was much more to it, but this remark expressed well how we felt Cartan exemplified Bourbaki's approach and how fruitful the latter was. At the time Cartan's influence through his seminar, papers and teaching was broadly felt. Speaking of his generation, R. Bott said of him "He has been truly our teacher", at the Colloquium in honor of Cartan's 70th year [4].

The fifties also saw the emergence of someone who was even more of an incarnation of Bourbaki, in his quest for the most powerful, most general, and most basic, namely, Alexander Grothendieck. His first research interests, from 1949 on, were in functional analysis. He quickly made mincemeat of many problems on topological vector spaces put to him by Dieudonné and Schwartz and proceeded to establish a far reaching theory. Then he turned his attention to algebraic topology, analytic and algebraic geometry and soon came up with a version of the Riemann-Roch theorem which took everyone by surprise, already by its formulation, steeped in functorial thinking, way ahead of anyone else. As major as it was, it turned out to be just the beginning of his fundamental work in algebraic geometry.

10. The fifties were thus outwardly a time of great success for Bourbaki. However, in contrast, it was inwardly one of considerable difficulties, verging on a crisis.

Of course there were some grumblings against Bourbaki's influence. We had witnessed progress in, and a unification of, a big chunk of mathematics, chiefly through rather sophisticated (at the time), essentially algebraic, methods. The most successful lecturers in Paris were Cartan and Serre, who had a considerable following. The mathematical climate was not favorable to mathematicians with a different temperament, a different approach. This was indeed unfortunate, but could hardly be held against Bourbaki members, who did not force anybody to carry on research in their way.[9]

The difficulties I want to discuss were of a different, internal nature, partly engineered by the very success of Bourbaki, tied up with the "second part", i.e. the treatise beyond the first six books. In the fifties, these were essentially finished and it was understood the main energies of Bourbaki would henceforth concentrate on the sequel. It had been in the mind of Bourbaki very early on (after all, there was still no "Traité d'Analyse"). Already in September 1940 (Tribu n° 3), Dieudonné had outlined a grandiose plan in 27 books, encompassing most of mathematics. More modest ones, still reaching beyond the Eléments, also usually by Dieudonné, would regularly conclude the Congresses. Also many reports on, and drafts of, future chapters had already been written. However, mathematics had grown enormously, the mathematical landscape had changed considerably, in part through the work of Bourbaki, and it became clear we could not go on simply following the traditional pattern. Although this had not been intended, the founding members had often carried a greater weight on basic decisions but they were now retiring[10] and the primary

[9] In this connection, I would like to point out that the subtitle "Le choix bourbachique" in [9] is extremely misleading. Bourbaki members gave many talks at the Seminar and had much input in the choice of the lectures, so it is fair to say that most topics discussed were of interest to at least some members of Bourbaki, but many equally interesting ones turned out to be left out, if only because no suitable speaker appeared to be available. So the seminar is by no means to be viewed as a concerted effort by Bourbaki to present a comprehensive survey of all recent research in mathematics of interest to him, and a ranking of contributions. Such conclusions by Dieudonné are solely his own. He says that much in his introduction, p. xi, but it seems worth repeating. Of course, like most mathematicians, Bourbaki members had strong likes and dislikes, but it never occured to them to erect them as absolute judgements by Bourbaki, as a body. Even when it came to his strong belief in the underlying unity of mathematics, Bourbaki preferred to display it by action rather than by proclamation.

[10] It had been apparently agreed early on that the retirement age would be 50 (at the latest). However, when the time came to implement that rule, from 1953 on, there was little mention

responsibility was shifting to younger members. Some basic principles had to be reexamined.

One for instance, was the linear ordering and the system of references. We were aiming at more special topics. To keep a strict linear ordering might postpone unduly the writing up of some volumes. Also when that course had been adopted at the beginning, there was indeed a dearth of suitable references. But Bourbaki had caught on, some new books were rather close to Bourbaki in style and some members were publishing others. To ignore them might lead to a considerable duplication, and waste, of effort. If we did not, how could we take them into account without destroying the autonomous character of the work? Another traditional basic tenet was that everyone should be interested in everything. As meritorious as it was to adhere to, it had been comparatively easy while writing the Eléments, which consist of basic mathematics, part of the baggage of most professional mathematicians. It might however be harder to implement it when dealing with more specialized topics, closer to the frontier. The prospect of dividing up, of entrusting the primary responsibility of a book to a subset of Bourbaki, was lurking but was not one we would adopt lightly. These questions and others were debated, though not conclusively for a while. There were more questions than answers. In short, two tendencies, two approaches, emerged: one (let me call it the idealistic one) to go on building up broad foundations in an autonomous way, in the tradition of Bourbaki, the other, more pragmatic, to get to the topics we felt we could handle, even if the foundations had not been thoroughly laid out in the optimal generality.

11. Rather than remain at the level of vague generalities, I would like to illustrate this dilemma by an example.

At some point a draft on elementary sheaf theory was produced. It was meant to supply basic background material in algebraic topology, fibre bundles, differential manifolds, analytic and algebraic geometry. However, Grothendieck objected[11] we had to be more systematic, and provide first foundations for this topic itself. His counterproposal was to have as the next two books:

Book VII: Homological algebra,
Book VIII: Elementary topology,

the latter to be tentatively subdivided into:

Chap. I: Topological categories, local categories, gluing of local categories, sheaves.
Chap. II: H^1 with coefficients in a sheaf
Chap. III: H^n and spectral sequences
Chap. IV: Coverings

to be followed by

Book IX: Manifolds,

which had already been planned.

of it until 1956, when Weil wrote a letter to Bourbaki announcing his retirement. From then on, it has been strictly followed.

[11] At the March 1957 Congress, later called "Congress of the inflexible functor".

He also added a rather detailed plan for the chapter on sheaves I shall not go into.

This was surely in the spirit of Bourbaki. To oppose it would have been a bit like arguing against motherhood, so it had to be given a hearing. Grothendieck lost no time and presented to the next Congress, about three months later, two drafts:

Chap. 0. Preliminaries to the book on manifolds. Categories of manifolds, 98 pages.

Differentiable manifolds, Chap I: The differential formalism, 164 pages,

and warned that much more algebra would be needed, e.g. hyperalgebras. As was often the case with Grothendieck's papers, they were at points discouragingly general, but at others rich in ideas and insights. However, it was rather clear that if we followed that route, we would be bogged down with foundations for many years, with a very uncertain outcome. Conceived so broadly, his plan aimed at supplying foundations not just for existing mathematics, as had been the case for the Eléments, but also for future developments, to the extent they could be foreseen. If the label "Chapter 0" was any indication, one could fear that the numbering might go both ways, Chapters -1, -2 ... being needed to give foundations to foundations, etc.

On the other hand, many members thought we might achieve more tangible goals in a finite time, not so fundamental maybe, but still worthwhile. There were quite a number of areas (algebraic topology, manifolds, Lie groups, differential geometry, distributions, commutative algebra, algebraic number theory, to name a few) in which we felt the Bourbaki approach might produce useful expositions, without needing such an extensive foundational basis as a prerequisite.

The ideal solution would have been to go both ways, but this exceeded by far our possibilities. Choices had to be made, but which ones? The question was not answered for some time, resulting in a sort of paralysis. A way out was finally arrived a year later, namely, to write a Fascicle of results on differential and analytic manifolds, thus bypassing, at least provisorily, the problem of foundations, at any rate for the main topics we had in mind. After all, as far as manifolds were concerned, we knew what kind of basic material was needed. To state what was required, and prove it for ourselves, was quite feasible (and was indeed carried out rather quickly).

This decision lifted a stumbling block and we could now set plans for a series of books, which, we hoped, would essentially include notably commutative algebra, algebraic geometry, Lie groups, global and functional analysis, algebraic number theory, automorphic forms.

Again, this was too ambitious. Still, in the next fifteen years or so, a sizable number of volumes appeared:

Commutative algebra (9 chapters)

Lie groups and Lie algebras (9 chapters)

Spectral theory (2 chapters)

besides preliminary drafts for several others.

12. In 1958 a decision had also been taken to solve in principle a problem which had been plaguing us for quite a while: additions to the Eléments. On occasions, while writing a new chapter, we would realize that some complement to one of the first six books was in order. How to handle this? Sometimes, if a volume became out of print, it was possible to include these complements in a revised edition. If not, one could conceivably add an appendix to the new chapter. But this threatened to create a lot of confusion for the references. In 1958 it was resolved to revise the Eléments and publish a "final" edition, not to be tampered with for at least fifteen years. Unfortunately, it took more time and efforts than anticipated. It is in fact not quite finished by now, and (I feel) it slowed down progress in the more innovative parts of the treatise. But it was certainly in the logic of Bourbaki, and hardly avoidable.

13. Of the three books listed above, "Commutative Algebra" was obviously well within Bourbaki's purview; it could, and did in fact, proceed independently of the resolution of the dilemma we had faced. But the fascicle of results on manifolds was an essential prerequisite for the book on Lie groups and Lie algebras. The latter also shows that the more pragmatic way could lead to useful work. A good example is provided by Chapters 4, 5, 6, on reflection groups and root systems.

It started with a draft of about 70 pages on root systems. The author was almost apologetic in presenting to Bourbaki such a technical and special topic, but asserted this would be justified later by many applications. When the next draft, of some 130 pages, was submitted, one member remarked that it was all right, but really Bourbaki was spending too much time on such a minor topic, and others acquiesced. Well, the final outcome is well-known: 288 pages, one of the most successful books by Bourbaki. It is a truly collective work, involving very actively about seven of us, none of whom could have written it by himself. Bourbaki had developed a strong technique to elicit a collaboration on a given topic between specialists and people with related interests, looking at it from different angles. My feeling (not unanimously shared) is that we might have produced more books of that type, but that the inconclusive discussions and the controversies, the difficulties in mapping out a clear plan of activity, had created a loss of momentum from which Bourbaki never fully recovered. There is indeed a tremendous amount of unused material in Bourbaki's archives.

This approach was less ambitious than the Grothendieck plan. Whether the latter would have been successful, had we gone fully into that direction, seems unlikely to me, but not ruled out. The development of mathematics does not seem to have gone that way, but implementation of that plan might have influenced its course. Who knows?

Of course, Bourbaki has not realized all its dreams, reached all of its goals, by far. Enough was carried out, it seems to me, to have a lasting impact on mathematics, by fostering a global vision of mathematics and of its basic unity, and also by his style of exposition and choice of notation, but, as an interested party, I am not the one to express a judgment.

What remains most vividly in my mind is the unselfish collaboration, over many years, of mathematicians with diverse personalities toward a common goal, a truly unique experience, maybe a unique occurence in the history of mathematics. The underlying commitment and obligations were taken for granted, not even talked about, a fact which seems to me more and more astonishing, almost unreal, as these events recede into the past.

Bibliography

[1] E. Artin, *"Review of Algebra (I–VII) by N. Bourbaki"*. Bull. Amer. Math. Soc. **59**, (1963), pp. 474–479

[2] L. Baulieu, *"A Parisian café and ten proto – Bourbaki meetings (1934–35)."* Math. Intelligencer **15** (1993), pp. 27–35

[3] L. Baulieu, *"Bourbaki: Une histoire du groupe de mathématiciens français et de ses travaux.* Thèse." Université de Montréal, 1989, 2 Vol

[4] R. Bott, *On characteristic classes in the framework of Gelfand-Fuks cohomology, in* Colloque analyse et topologie en l'honneur de H. Cartan, Astérisque **32–33** (1976), 113–139, Soc. Math. France; Collected papers **3**, 492–558, Birkhäuser 1995

[5] N. Bourbaki, *"L'architecture des mathématiques", in* Les grands courants de la pensée mathématique (F. Le Lionnais, ed.), Cahiers du Sud (1948); English translation: Amer. Math. Monthly **57** (1950), pp. 221–232

[6] H. Cartan, *"Nicolas Bourbaki and contemporary mathematics"*, Math. Intelligencer **2** (1979–80), pp. 175–180

[7] H. Cartan, Oeuvres Vol. 1, Springer 1979

[8] J. Dieudonné, *"The work of Nicholas Bourbaki"*. Amer. Math. Monthly **77** (1970), pp. 134–145

[9] J. Dieudonné, Panorama des mathématiques pures. Le choix bourbachique. Bordas, Paris, 1977

[10] D. Guedj, *"Nicholas Bourbaki, collective mathematician: an interview with Claude Chevalley"*. Math. Intelligencer **7** (1985), no. 2, pp. 18–22

[11] P. Halmos, *"Review of Integration (I–IV) by Bourbaki"*. Bull. Amer. Math. Soc. **59**, (1963), pp. 249–255

[12] A. Weil, *Review of "Introduction to the theory of algebraic functions of one variable by C. Chevalley"*. Bull. Amer. Math. Soc. **57** (1951), pp. 384–398; Oeuvres Scientifiques II, pp. 2–16, Springer 1979

[13] A. Weil, *"Organisation et désorganisation en mathématique"*. Bull. Soc. Franco-Japonaise des Sci. **3** (1961), pp. 23–35; Oeuvres Scientifiques II, pp. 465–469, Springer 1979

[14] A. Weil, *"Notice biographique de J. Delsarte"*. Oeuvres de Delsarte I, pp. 17–28, C.N.R.S., Paris (1971); Oeuvres Scientifiques III, pp. 217–228, Springer 1980

167.

Full Reducibility and Invariants for $SL_2(C)$

Ens. Math. (2) 44, 1998, 71–90

1. Let G be a group, V a finite dimensional vector space over a commutative field k (mostly **C** in this lecture), n the dimension of V and π a representation of G in V i.e. a homomorphism $G \to GL(V)$ of G into the group $GL(V)$ of invertible linear transformations of V. The choice of a basis of V provides an isomorphism of V with k^n, of $GL(V)$ with the group $\mathrm{GL}_n(k)$ of $n \times n$ invertible matrices with coefficients in k, and a realization of π as a matrix representation:

$$g \mapsto \pi(g) = \big(\pi(g)_{ij}\big)_{1 \leq i,j \leq n}. \tag{1}$$

Two main problems pertaining to this situation were considered already in the 19th century, in various special cases, for $k = \mathbf{C}$.

I) Invariants. Let $k[V]$ be the space of polynomials on V with coefficients in k and $k[V]_m$ ($m \in \mathbf{N}$) the space of homogeneous polynomials of degree m. The group G acts via π on $k[V]$ by the rule

$$g \circ P(v) = P\big(\pi(g)^{-1} . v\big) \qquad (v \in V,\ P \in k[V],\ g \in G)$$

leaving each $k[V]_m$ stable. [The argument on the right-hand side will usually be written $g^{-1} . v$ if there is no ambiguity about π.]

Let $k[V]^G$ be the space of polynomials which are invariant under G, i.e. which are constant on the orbits of G. It is an algebra over k and the (first) problem of invariant theory is to know whether it is finitely generated, as a k-algebra.

II) FULL REDUCIBILITY. The representation (π, V) is said to be reducible if there exists a G-invariant subspace $W \neq \{0\}, V$, and fully or completely reducible if any G-invariant subspace has a G-invariant complement. If so, V can be written as a direct sum of G-invariant irreducible subspaces. One is interested in groups having classes of fully reducible representations or in finding families of groups all of whose representations over a given k are fully reducible.

In this lecture, I shall discuss the history of these two problems mainly for one group, namely the group $SL_2(\mathbf{C})$ of 2×2 complex invertible matrices of determinant one, for $k = \mathbf{C}$ and holomorphic representations, i.e. in which the $\pi(g)_{ij}$ in (1) are holomorphic functions in the entries of g. Occasionally, some remarks will be made on other groups, to put certain results in a more general context, or for historical reasons, but our main focus of attention will still be $SL_2(\mathbf{C})$. Even so restricted, this history is surprisingly complicated, in part because the principal contributors were sometimes not aware of other work already done. In one case, it seems even that one of them had forgotten some of his own.

2. The irreducible representations of $SL_2(\mathbf{C})$ were determined by S. Lie. As we know, there is for each $m \in \mathbf{N}$, up to equivalence, one representation of degree $m+1$ in the space, to be denoted V_m, of homogeneous polynomials of degree m on $\mathbf{C}^2$, acted upon via the identity representation of $SL_2(\mathbf{C})$.

In fact, S. Lie formulated his result differently, more geometrically [LE]. For him, a representation is not a linear one, but a projective one, i.e. a homomorphism into the group of projective transformations of some complex projective space $\mathbf{P}_m(\mathbf{C})$. As usual, $\mathbf{P}_m(\mathbf{C})$ is viewed as the quotient of $\mathbf{C}^{m+1} - \{0\}$ by dilations. This identifies the group $\operatorname{Aut} \mathbf{P}_m(\mathbf{C})$ of projective transformations of $\mathbf{P}_m(\mathbf{C})$ with the quotient $GL_{m+1}(\mathbf{C})/\mathbf{C}^*$ of $GL_{m+1}(\mathbf{C})$ by the non-zero multiples of the identity matrix, or also with $PSL_{m+1}(\mathbf{C}) = SL_{m+1}(\mathbf{C})/\text{center}$, i.e. modulo the group of multiples $c.\mathrm{Id}$ of the identity matrix, where $c^{m+1} = 1$. Let B be the group of upper triangular matrices in $SL_2(\mathbf{C})$. The quotient $SL_2(\mathbf{C})/B = C$ is a smooth complete rational curve, i.e. a copy of $\mathbf{P}_1(\mathbf{C})$. The group B is solvable, connected, therefore, by Lie's theorem it has a fixed point in any projective representation and so, if this point is not fixed under G, its orbit is a copy of C. Lie looks for the cases where such a C is "as curved as possible" ("möglichst gekrümmt") meaning, not contained in a proper projective subspace. It is also a fact, implicitly assumed by Lie, that the action of $SL_2(\mathbf{C})$ on such a curve is always induced by projective transformations of the ambient projective space. Therefore the search of smooth rational complete curves in projective spaces which are "as curved as possible", up to projective transformations, is tantamount to the classification of the irreducible holomorphic representations of $SL_2(\mathbf{C})$ (linear or projective, there is no essential difference since $SL_2(\mathbf{C})$ is simply connected), up to equivalence. Given $m \geq 1$, the smooth projective rational curves not contained in a proper projective subspace, of smallest degree (number of intersection points with a generic hyperplane), form in $\mathbf{P}_m(\mathbf{C})$ one family, operated upon transitively by $\mathrm{Aut}\,\mathbf{P}_m(\mathbf{C})$, and the degree is m. The irreducible representations are those in which the G-orbit of a fixed point of B has degree m. It is then the only closed orbit of G. In [LE], p. 785–6, S. Lie reports that E. Study has proved the full reducibility of the representations of $SL_2(\mathbf{C})$ (again, in an equivalent projective formulation I do not recall here, but which will appear in § 13), but he does not describe the proof because it is long, maybe not quite correct, and simplifications are hoped for. He adds it is very likely to be true for representations of $SL_n(\mathbf{C})$, any $n \geq 2$. In fact, Study had made this conjecture in a letter to him, even more generally for semisimple groups.

3. In his Thesis E. Cartan provides a proof of full reducibility [Cr1]. It is algebraic, deals with Lie algebras so establishes in fact the full reducibility of the representations of the Lie algebra $\mathfrak{sl}_2(\mathbf{C})$ of $SL_2(\mathbf{C})$, but this is equivalent. He does not state the theorem explicitly, however. The proof is embedded (pp. 100–2) in that of another one, due to F. Engel, to the effect that a non-solvable Lie algebra always contains a copy of $\mathfrak{sl}_2(\mathbf{C})$. But a statement is given at the beginning of Chapter VII (p. 116) with a reference to the passage just quoted for the proof.

In 1896, G. Fano, who knew about Study's theorem through [LE] and was surely not aware of Cartan's proof, maybe not even of Cartan's Thesis, gave an entirely different one in the framework of algebraic geometry, using the properties of "rational normal scrolls" [F].

He first makes two remarks of an algebraic nature which simplify the argument.

a) An induction on the length of a composition series shows it suffices to carry the proof when the space E of the given representation contains one irreducible G-invariant subspace F such that E/F is also irreducible. In other words, since the V_m's are the irreducible representations, up to equivalence, it suffices to consider the case of an exact sequence

$$0 \to V_m \to E \to V_n \to 0 \qquad (1)$$

(again, in projective language, see 13.1).

b) In (1), it may be assumed that $m \geq n$. If $m < n$, this is seen by going over to the contragredient representations

$$0 \to V_n^* \to E^* \to V_m^* \to 0, \tag{2}$$

noting that E is fully reducible if and only if E^* is, and that for each m, the representation V_m is self-contragredient. This also shows that it suffices to consider the case where $m \leq n$. In fact, this last reduction allows for a considerable simplification in Cartan's proof, whereas the reduction to $m \geq n$ is the one Fano uses. (see § 12, § 13 for more details).

4. Another development came from a different source: the idea of averaging over a finite group. In 1896 it was shown that a finite group G of linear transformations always leaves invariant a positive non-degenerate hermitian form. It was stated by A. Loewy without proof [L] and by E.H. Moore, who announced it at some meeting, communicated his proof to F. Klein, and published it later [Mo]. This argument is the now standard one (and Loewy stated later it was his, too): starting from a positive non-degenerate hermitian form $H(\ ,\ )$ on $\mathbf{C}^n$, he considers the sum $H^o(\ ,\ )$ of its transforms under the elements of G:

$$H^o(x, y) = \sum_{g \in G} H(g^{-1}.x\,,\, g^{-1}.y) \tag{1}$$

which he calls a *universal invariant* for G. It is obviously G-invariant and positive non-degenerate. This construction seems quite obvious, but Klein viewed it as interesting enough to make it the subject matter of a communication to the German Math. Soc. [K]. For Moore it was an application of a "well-known group theoretic process". In [Lo] and [Mo], this fact is used to show that a linear transformation of finite order is diagonalizable (which was known, but with more complicated proofs). A bit later, H. Maschke, a colleague of Moore at Chicago, made use of this universal invariant to establish the full reducibility of linear representations of a finite group [Ma]. The standard argument is of course to point out that if V is a G-invariant subspace, then so is its orthogonal complement with respect to H^o. This is the gist of Maschke's proof, but presented in a rather complicated manner.

5. The idea of averaging was pushed much further by A. Hurwitz in a landmark paper [H]. He was interested in the invariant problem. He starts by saying it is well-known one can construct invariants for a finite linear group by averaging, but he is concerned with certain infinite groups, specifically $SL_n(\mathbf{C})$ and the special complex orthogonal group $SO_n(\mathbf{C})$ $(n \geq 2)$.

Hurwitz recalls first that if G is a finite linear group acting on $\mathbf{C}^n$ and P is a polynomial on $\mathbf{C}^n$ then the polynomial $P^\natural$ defined by

$$P^\natural(x) = N^{-1} . \sum_{g \in G} P(g^{-1} . x) \qquad (x \in \mathbf{C}^n), \tag{1}$$

where N is the order of G, is obviously invariant under G (the factor N^{-1} is inserted so that $P^\natural = P$ if P is invariant). If now G is infinite, the initial idea is to replace the summation in (1) by an integration, with respect to a measure invariant by translations. However, if the group is not compact (Hurwitz says bounded), this integral may well diverge. To surmount that difficulty, A. Hurwitz used a procedure which turned out later to be far reaching, namely to integrate over a compact subgroup G_u, which insures convergence, but choosing it big enough so that invariance under G_u implies the invariance under the whole group, an argument later called the "unitarian trick" by H. Weyl [W1]. This is carried out for $SU_n \subset SL_n(\mathbf{C})$ and $SO_n \subset SO_n(\mathbf{C})$. I describe it for $G = SL_2(\mathbf{C})$ and $G_u = SU_2$. The latter is

$$G_u = SU_2 = \left\{ \begin{pmatrix} a & b \\ -\bar{b} & \bar{a} \end{pmatrix}, \quad a, b \in \mathbf{C}, \quad |a|^2 + |b|^2 = 1 \right\}. \tag{2}$$

Write $a = x_1 + ix_2$, $b = x_3 + ix_4$, with the x_i real. Then SU_2 may be identified to the unit 3-sphere

$$\mathbf{S}^3 = \{(x_1, \ldots, x_4) \in \mathbf{R}^4, \quad x_1^2 + \cdots + x_4^2 = 1\}. \tag{3}$$

It can be parametrized by the Euler angles φ, ψ, θ:

$$\begin{aligned} x_1 &= \cos\psi . \cos\varphi . \cos\theta \\ x_2 &= \cos\psi . \cos\varphi . \sin\theta \\ x_3 &= \cos\psi . \sin\varphi \\ x_4 &= \sin\psi . \end{aligned} \qquad (|\varphi|, |\psi| \le \pi/2, \ \theta \in [0, 2\pi]) \tag{4}$$

The measure

$$dv = \cos\psi . \cos\varphi . d\psi . d\varphi . d\theta \tag{5}$$

is then invariant under translations and the volume of $\mathbf{S}^3$ with respect to dv is 8π. Let $\sigma : G \to GL_N(\mathbf{C})$ be a holomorphic linear representation of G and P be a polynomial on $\mathbf{C}^N$. Integration on G_u yields the polynomial $P^\natural$ given by

$$\begin{aligned} P^\natural(x) &= (8\pi)^{-1} \int_{G_u} P(g^{-1} . x)\, dv \\ &= (8\pi)^{-1} \int_{-\pi/2}^{\pi/2} \cos\psi . d\psi \int_{-\pi/2}^{\pi/2} \sin\varphi . d\varphi \int_0^{2\pi} P(g^{-1} . x)\, d\theta \end{aligned} \tag{6}$$

($x \in \mathbf{C}^N$). It is invariant under the action of G_u and the claim is that it is even invariant under G itself. Given $x \in \mathbf{C}^N$ consider the function μ_x on G given by

$$\mu_x(g) = P^\natural(g^{-1}.x) - P^\natural(x) \qquad (g \in G). \tag{7}$$

It is holomorphic in the entries of g, and is identically zero for $g \in G_u$. To establish that it is identically zero on G, it suffices to show that it is zero on a neighborhood U of the identity. The tangent space to G (resp. G_u) at the identity is the complex (resp. real) vector space $\mathfrak{g}$ (resp. $\mathfrak{g}_u$) of 2×2 complex (resp. skew-hermitian) matrices of trace zero. Take U small enough so that it is the isomorphic image of a neighborhood U_o of the origin in $\mathfrak{g}$ under the exponential mapping. Let $\tilde{\mu}_x$ be the pull back of $\mu_x|_U$ by the inverse mapping. Then $\tilde{\mu}_x$ is a holomorphic function on U_o which is zero on $U_o \cap \mathfrak{g}_u$. But $\mathfrak{g}_u$ is a real form of $\mathfrak{g}$, i.e. as a real vector space, $\mathfrak{g}$ is the direct sum of $\mathfrak{g}_u$ and of the space $i\mathfrak{g}_u$ of hermitian 2×2 matrices of trace zero. Hence $\tilde{\mu}_x$ is identically zero on U_o and our assertion follows.

6. From this Hurwitz deduces that the algebra $\mathbf{C}[\mathbf{C}^N]^G$, to be denoted I_G to simplify notation, of invariant polynomials on $\mathbf{C}^N$ is finitely generated: The projector $P \mapsto P^\natural$ obviously satisfies the relation

$$(P.Q)^\natural = P.Q^\natural, \quad \text{if } P \text{ is } G\text{-invariant}. \tag{1}$$

By Hilbert's finiteness theorem, the ideal I generated by the elements of I_G without constant term is finitely generated. Let $Q_i \in I_G$ $(1 \leq i \leq s)$ be a generating system of this ideal, which may be assumed to consist of homogeneous invariant elements of strictly positive degrees. Let now $Q \in I_G$ be homogeneous. It certainly belongs to I. There exist therefore homogeneous polynomials $A_1, \ldots, A_s$ such that

$$Q = \sum_{1 \leq i \leq s} Q_i . A_i .$$

Then, we have, by (1)

$$Q^\natural = Q = \sum_i Q_i . A_i^\natural .$$

Since the $A_i^\natural$ have strictly lower degrees than Q, it follows by induction on the degree, that I_G is generated, as an algebra, by the Q_i.

In analogy with Maschke's theorem, Hurwitz could have easily given a new proof of the full reducibility of the holomorphic representations of $\mathrm{SL}_2(\mathbf{C})$, and, more generally the first proof for $\mathrm{SL}_n(\mathbf{C})$ $(n \geq 3)$ and $\mathrm{SO}_n(\mathbf{C})$ $(n \geq 4)$.

Indeed if, as in §4, $H(\ ,\ )$ is a positive non-degenerate hermitian form on $\mathbf{C}^N$, the form H^o constructed as in (1), but using integration

$$H^o(x,y) = \int_{G_u} H(g^{-1}.x, g^{-1}.y)\, dv \qquad (x, y \in \mathbf{C}^N)$$

is invariant under G_u and still positive non-degenerate, whence the full reducibility of the (continuous) representations of G_u. There remains to show that every G_u-invariant subspace is G-invariant. Let V be one and W its orthogonal complement with respect to H^o. Fix a basis $(f_1, \ldots, f_N)$ of $\mathbf{C}^N$ whose first $p = \dim V$ elements span V and the last $N-p$ span W. Then the matrix coefficients $\sigma(g)_{ij}$ $(i \leq p,\ j > p)$ are holomorphic functions on G which vanish on G_u hence, by the argument outlined previously, are identically zero on G. Therefore V and W are G-invariant, and full reducibility is proved.

7. I spoke of a "landmark paper". This is only by hindsight because the paper was completely forgotten for about 25 years and, apparently, no specialist of Lie groups or Lie algebras was aware of it and had realized that a proof of Study's conjecture for $SL_n(C)$ was at hand.

Meanwhile, progress was made on two fronts:

a) Character theory for complex representations of finite groups, orthogonality relations, etc (Frobenius, Schur, Burnside, 1896-1906).

b) Construction of all irreducible representations of complex simple Lie algebras by E. Cartan ([Cr2], 1914).

In 1922, I. Schur discovers Hurwitz's paper and uses it to extend the character theory a) to representations of SU_n or SO_n [S]. Two years later, H. Weyl combines b) and the point of view of Hurwitz-Schur to generalize it to all complex or compact semisimple groups [W]. Until he came on the scene, Schur was not aware of Cartan's work nor Cartan of Schur's or Hurwitz's. He also points out a gap in [Cr2]: the construction of irreducible representations makes implicit use of full reducibility, a problem Cartan had not considered at all there. At that point, as a proof, there was then only Weyl's generalization of Hurwitz and Schur, which was highly transcendental. Both Cartan and Weyl felt that an algebraic proof of such a purely algebraic statement was desirable, but viewed it as rather unlikely that one would be forthcoming. Cartan could have pointed out that in the case of $SL_2(C)$ or rather its Lie algebra, one was contained in his Thesis. The fact that he did not makes me think that he had forgotten about it (but not forever, though: it is again given in his book on spinors [Cr3]).

8. In comparing physicists and mathematicians it is often said that the physicists, unlike mathematicians, do not care that much for rigorous proofs. Here, we are dealing with the search for a second proof, in a different framework, of a theorem already established, a problem which would normally seem even less attractive to a physicist. In that case, however, it did attract one, H. L Casimir, whose approach had its origin in the use of group representations in quantum mechanics. Since it involves SO_3 or $SO_3(\mathbf{C})$ rather than SU_2 or $SL_2(\mathbf{C})$, let me recall first that SO_3 (resp. $SO_3(\mathbf{C})$) is the quotient of SU_2 (resp. $SL_2(\mathbf{C})$) by its center, which consists of $\pm\mathrm{Id}$. In particular, $\mathfrak{sl}_2(\mathbf{C})$ may be viewed as the complexification of the Lie algebra $\mathfrak{so}_3$ of SO_3, so that we can take as a basis of it the infinitesimal rotations D_x, D_y, D_z around the three coordinate axes in $\mathbf{R}^3$, where x, y, z are the coordinates:

$$(1)\quad D_x = \begin{pmatrix} 0 & 0 & 0 \\ 0 & 0 & 1 \\ 0 & -1 & 0 \end{pmatrix} \quad D_y = \begin{pmatrix} 0 & 0 & -1 \\ 0 & 0 & 0 \\ 1 & 0 & 0 \end{pmatrix} \quad D_z = \begin{pmatrix} 0 & -1 & 0 \\ 1 & 0 & 0 \\ 0 & 0 & 0 \end{pmatrix}.$$

Viewed as differential operators on functions, these transformations are

$$(2)\qquad D_x = y\,.\,\partial_z - z\,.\,\partial_y, \quad D_y = z\,.\,\partial_x - x\,.\,\partial_z, \quad D_z = y\,.\,\partial_x - x\,.\,\partial_y\,.$$

The application to quantum mechanics makes use of

$$L_x = i^{-1}\,.\,D_x, \quad L_y = i^{-1}\,.\,D_y, \quad L_z = i^{-1}\,.\,D_z$$

called the components of the moment of momentum and of

$$(3)\qquad L^2 = L_x^2 + L_y^2 + L_z^2,$$

the square of the moment of momentum.

The decisive idea is to use L^2. It is a differential operator, also represented by minus the sum of the square of the matrices in (1). It belongs to the associative algebra of endomorphisms of $\mathbf{C}^2$ generated by $\mathfrak{sl}_2(\mathbf{C})$, a quotient of the so-called universal enveloping algebra of $\mathfrak{sl}_2(\mathbf{C})$, but not to the Lie algebra itself.

An elementary computation shows that L^2 commutes with the infinitesimal rotations, hence with $\mathfrak{sl}_2(\mathbf{C})$ itself. A linear representation (σ, V) extends to one of the enveloping algebra and in particular $\sigma(L^2)$ is defined. If σ is irreducible, then $\sigma(L^2)$ is a scalar multiple of the identity (Schur's lemma).

In the representation V_n of degree $n+1$, this scalar is equal to $n(n+2)/4$. It characterizes the representation, up to equivalence [1]).

In order to prove full reducibility, Casimir notes that it suffices to consider the case of the exact sequence (1) in §3. Assume first $m \neq n$, the main case. Then $\sigma(L^2)$ has two eigenvalues, $m(m+2)/4$ and $n(n+2)/4$. The eigenspace W for the latter eigenvalue intersects V_m only at the origin. Since $\sigma(L^2)$ commutes with $\sigma(\mathfrak{sl}_2(C))$, the space W is also invariant under $SL_2(C)$. Its projection in V_n is invariant, non-zero, hence equal to V_n, so W is the sought for complement to V_m. If $m = n$, the existence of an invariant complement is proved by a rather elementary computation, sketched in 12.4.

9. An analog of L^2 had been introduced in 1931 by Casimir for any complex semisimple Lie algebra, later called the Casimir operator. Using it van der Waerden generalized Casimir's argument to give the first algebraic general proof of the full reducibility of finite dimensional representations of complex semisimple Lie algebras [CW].

Later it was realized that the Casimir operator is an element in the center of the universal enveloping algebra (which generates it for $\mathfrak{sl}_2(C)$). The full center was investigated in the late forties by G. Racah, also a physicist, on the one hand, by C. Chevalley and Harish-Chandra on the other, and became a powerful tool in the study of the topology of compact Lie groups and of infinite dimensional representations of semisimple Lie groups.

Racah's motivation was representation theory. From a physicist's point of view, the eigenvalue of L^2 gave a parametrization of an irreducible representation of $SL_2(C)$ by a number with a physical meaning, whereas the highest weight had none. For higher dimensional groups, the eigenvalue of L^2 does not characterize the representation, up to equivalence, which makes the general argument in [CW] quite complicated. Racah's idea was to search for more operators commuting with the Lie algebra (r independent ones if r is the rank of the Lie algebra), the eigenvalues of which would again characterize the irreducible representations. This would then allow one to treat the case of two inequivalent irreducible representations in a short exact sequence in the same way as for $\mathfrak{sl}_2(C)$ and considerably simplify the proof. At that time, the mathematicians were not searching for a new algebraic proof, however, and this was not at all a motivation for Chevalley and Harish-Chandra.

[1]) In the physics literature and in [W2], the irreducible representations of $SL_2(C)$ are parametrized by $(1/2)N$. The representation V_j there is our V_{2j}. It has degree $2j+1$ and the eigenvalue of L^2 is $j(j+1)$. It is a spin representation, i.e. non trivial on the center, if and only j is a half-integer.

10. The paper [CW] was followed shortly by two other algebraic proofs, one by R. Brauer [Br] and one based on a lemma of J. H. C. Whitehead, which is now best expressed in the framework of Lie algebra cohomology, and became the standard algebraic argument for a number of years.

In 1956, a new proof was published by P. K. Raševskiĭ [R]. Consider the group $\mathrm{Aff}(\mathbf{C}^N)$ of *affine* transformations of $\mathbf{C}^N$. It is the semidirect product of the group of translations by the group $\mathrm{GL}_N(\mathbf{C})$. Accordingly, its Lie algebra is the semidirect product $\mathfrak{s} \oplus \mathfrak{t}$ of the space of translations $\mathfrak{t}$ by the Lie algebra $\mathfrak{s}$ of $\mathrm{GL}_N(\mathbf{C})$. The new ingredient is the proof that any representation of a semisimple Lie algebra in the Lie algebra $\mathfrak{aff}(\mathbf{C}^N)$ of $\mathrm{Aff}(\mathbf{C}^N)$ leaves a point of $\mathbf{C}^N$ fixed or, globally speaking, any complex semisimple group of affine transformations of $\mathbf{C}^N$ has a fixed point. Let now σ be a representation of the complex semisimple Lie algebra $\mathfrak{g}$ in $\mathbf{C}^M$ and $V \subset \mathbf{C}^M$ an invariant subspace. Then the set of subspaces W of $\mathbf{C}^M$ complementary to V forms in a canonical way an affine space, with space of translations $\mathbf{C}^M/V$. It is operated upon naturally by $\sigma(\mathfrak{g})$. The existence of a fixed point implies that of a $\mathfrak{g}$-invariant complement to V, whence the full reducibility.

When N. Bourbaki was preparing Volume 1 of the book on Lie groups and Lie algebras, entirely devoted to Lie algebras, an algebraic proof was needed. The cohomological one did not seem really suitable, requiring as it did lots of preliminaries on cohomology of Lie algebras, which it did not seem appropriate to introduce at that early stage of the exposition. Then Bourbaki turned to Raševskiĭ's proof and made it somewhat more algebraic and self-contained. After the book was published in 1961, I stumbled once on a copy of [Br], and realized this argument was the one of [Br], another example of a paper overlooked for over 25 years, the knowledge of which would have saved some work to Bourbaki.

11. This pretty much concludes my story, but as Poincaré once wrote, there are no problems which are completely solved, only problems which are more or less solved. Still considering SL_2 one may ask about the problems I and II for $\mathrm{SL}_2(k)$ where k is an algebraically closed groundfield of positive characteristic p. It is well-known that full reducibility does not necessarily hold. Take for example k of characteristic two and the representation V_2 of degree three on the homogeneous quadratic polynomials. It has $x^2, x.y$ and y^2 as a basis. In characteristic 2, we have the rule $(a+b)^2 = a^2 + b^2$ so the linear combinations of x^2 and y^2 are the squares of the linear forms, and form a two-dimensional invariant subspace V. A complementary subspace is of dimension one therefore, if invariant, would be acted up trivially by $\mathrm{SL}_2(k)$.

However, $SL_2(k)$ does not leave any non-zero element of V_2 stable, so V_2 is not fully reducible.

This does not rule out a positive answer to problem I, but if so, another approach had to be devised. D. Mumford proposed a weaker notion than full reducibility, now called geometric reductivity: if C is an invariant one-dimensional subspace, there exists a homogeneous G-invariant hypersurface not containing C (in the case of full reducibility it could be a hyperplane). Then Nagata showed that this condition indeed implies the finite generation of the algebra of invariants. Later geometric reductivity was proved by C.S. Seshadri for $SL_2(k)$ and by W. Haboush in general.

Even over $\mathbf{C}$, the problems of full reducibility and of the determination of irreducible representations resurfaced not for $SL_2(\mathbf{C})$, but for its generalization as a Kac-Moody Lie algebra, or for the deformation of its Lie algebra as a "quantum group". This has led to further problems and to more contacts with mathematical physics.

APPENDIX: MORE ON SOME PROOFS OF FULL REDUCIBILITY

We give here more technical details on the proofs of full reducibility for $\mathfrak{sl}_2(\mathbf{C})$ or $SL_2(\mathbf{C})$ due to Cartan, Fano and Casimir, assuming some familiarity with Lie algebras and algebraic geometry. We let $\mathfrak{g}$ stand for $\mathfrak{sl}_2(\mathbf{C})$.

12. LIE ALGEBRA PROOF:

12.1. Let

$$h = \begin{bmatrix} 1 & 0 \\ 0 & -1 \end{bmatrix}, \quad e = \begin{bmatrix} 0 & 1 \\ 0 & 0 \end{bmatrix}, \quad f = \begin{bmatrix} 0 & 0 \\ -1 & 0 \end{bmatrix}. \tag{1}$$

be the familiar basis of $\mathfrak{g}$. It satisfies the relations

$$[h,e] = 2e \qquad [h,f] = -2f \qquad [e,f] = -h. \tag{2}$$

The elements h, e, f define one-parameter subgroups ($t \in \mathbf{R}$)

$$e^{th} = \begin{pmatrix} e^t & 0 \\ 0 & e^{-t} \end{pmatrix} \quad e^{te} = \begin{pmatrix} 1 & t \\ 0 & 1 \end{pmatrix} \quad e^{tf} = \begin{pmatrix} 1 & 0 \\ -t & 0 \end{pmatrix}.$$

By letting them act on functions of x, y and taking the derivatives for $t = 0$, we get expressions of h, e, f as differential operators, namely

$$h = x \,.\, \partial_x - y \,.\, \partial_y, \quad e = x \,.\, \partial_y, \quad f = -y \,.\, \partial_x. \tag{3}$$

Let E be a representation space for $\mathfrak{g}$ and E_c ($c \in \mathbf{C}$) the eigenspace for h with eigenvalue c. Then (2) implies

$$(4) \qquad e\,.\,E_c \subset E_{c+2} \qquad f\,.\,E_c \subset E_{c-2}\,.$$

More generally, if $(h - c\,.\,I)^q\,.\,v = 0$ for some $q \geq 1$, then

$$(5) \qquad (h-(c+2)\,.\,I)^q\,.\,e\,.\,v = 0 = (h-(c-2)\,.\,I)^q\,.f\,.\,v = 0\,.$$

12.2. We now consider V_m. It has a basis $x^{m-i}\,.\,y^i$ $(i = 0,\dots,m)$ and $x^{m-i}\,.\,y^i$ is an eigenvector for h, with eigenvalue $m-2i$. Let

$$(1) \qquad v_{m-2i} = \binom{m}{i}\,x^{m-i}\!.\,y^i \qquad (i = 0,\dots,m)\,.$$

The v_{m-2i} form a basis of V_m and we have:

$$(2) \qquad h\,.\,v_{m-2i} = (m-2i)v_{m-2i} \qquad (i = 0,\dots,m)\,.$$

A simple computation, using 12.1(2), (3), yields

$$\begin{aligned} &(3) \qquad f\,.\,v_{m-2i} = -(i+1)\,v_{m-2i-2} \\ & \qquad\qquad\qquad\qquad\qquad\qquad\qquad\qquad (i = 0,\dots,m)\,, \\ &(4) \qquad e\,.\,v_{m-2i} = (m-i+1)\,v_{m-2i+2} \end{aligned}$$

with the understanding that

$$(5) \qquad v_{m+2} = v_{-m-2} = 0\,.$$

(3) and (4) imply

$$(6) \qquad f\,.\,e\,.\,v_{m-2i} = -i(m-i+1)v_{m-2i}$$

$$(7) \qquad e\,.f\,.\,v_{m-2i} = (i+1)(m-i)v_{m-2i}\,.$$

REMARKS. (a) The eigenvalues of h in V_m are integers. By consideration of a Jordan-Hölder series, it follows that this is true for any finite dimensional representation.

(b) In $\mathbf{P}(V_m)$ the rational normal curve occurring in Lie's description of the irreducible projective representations of $\mathrm{SL}_2(\mathbf{C})$ (see § 2) is the orbit of the point representing the line spanned by x^m. This is also the unique fixed point in $\mathbf{P}(V_m)$ of the group U generated by e, i.e. the group of upper triangular unipotent (eigenvalues equal to one) matrices. It is therefore also the locus of the fixed points of the conjugates of U in $\mathrm{SL}_2(\mathbf{C})$, and each such conjugate has a unique fixed point in $\mathbf{P}(V_m)$.

12.3. Note that 12.2(5) is a consequence of 12.2(4) and of the commutation relations 12.1(1). A similar argument shows more generally that if E is a representation of $\mathfrak{g}$ and $v \in E$ satisfies the conditions

$$e \, . \, v = 0, \quad h \, . \, v = c \, . \, v \qquad (c \in \mathbf{C}), \tag{1}$$

then the elements $f^i \, . \, v$ $(i \geq 0)$ span a finite dimensional $\mathfrak{g}$-submodule F. In particular $\mathbf{C} \, . \, v$ is the eigenspace with eigenvalue c and all other eigenvalues of h in F are of the form $c - q$ $(q \in \mathbf{N}, q \geq 1)$.

12.4. *First proof of full reducibility.* We use 12.3, which is contained in [Cr1] and the two remarks a) and b) of §3. This reduces the proof of full reducibility of a $\mathfrak{g}$-module E to the case of a short exact sequence

$$0 \to V_m \to E \xrightarrow{\pi} V_n \to 0 \qquad (m \leq n). \tag{1}$$

Let $m < n$. Then h has an eigenvector $v \in E$ with eigenvalue n, which does not belong to V_m. It is annihilated by e, since there are no weights $> n$ in V_m or V_n, hence in E. By 12.3, it generates a $\mathfrak{g}$-submodule distinct from V_m, which must therefore be a $\mathfrak{g}$-invariant complement to V_m.

Let now $m = n$. Let $\{v_{m-2i}\}$ $(i = 0, \dots, m)$ be the basis of V_m, viewed as subspace of E, constructed in 12.2. Let v'_m be a vector which maps under π onto the similar basis element of the quotient and let $v'_{m-2i} = \binom{m}{i} \, . \, v'_m$. Then the v'_{m-2i} project onto the basis of E/V_m defined in 12.2. There exists $a \in \mathbf{C}$ such that

$$h \, . \, v'_m = m \, . \, v'_m + a \, . \, v_m \, . \tag{2}$$

We claim it suffices to show that $a = 0$. Indeed, in that case, 12.3 again implies that v'_m generates a $\mathfrak{g}$-submodule distinct from V_m, hence a supplement to V_m.

There remains to prove that $a = 0$. We claim first

$$h \, . \, v'_{m-2i} = (m - 2i) v'_{m-2i} + a \, . \, v_{m-2i} \qquad (i = 0, \dots, m). \tag{3}$$

For $i = 0$, this is (2). Assuming it is proved for i, we obtain (3) for $i + 1$ by applying f to both sides and using 12.1(2), 12.2(3).

For $i \geq 1$, we have, by 12.1(2) and 12.2(3)

$$i \, . \, e \, . \, v'_{m-2i} = -e \, . f \, . \, v'_{m-2i+2} = -f \, . \, e \, . \, v'_{m-2i+2} + h \, . \, v'_{m-2i+2} \, . \tag{4}$$

By (3) and 12.2(6), this yields

$$i \, . \, e \, . \, v'_{m-2i} = i(m - i + 1) \, . \, v'_{m-2i+2} + a \, . \, v_{m-2i+2} \, . \tag{5}$$

If we apply (5) for $i = m + 1$, we get $a \, . \, v_{-m} = 0$, hence $a = 0$.

REMARK. This last computation is contained in [CW] and also, unknown to the authors, in [Cr1]. As we saw, the proof for $m < n$ reduces immediately to 12.3, and by b) in §3 it suffices to consider that case when $m \neq n$. A direct computation along the lines of the previous proof is longer if $m > n$ (see 12.5). Cartan performs it even for a Jordan-Hölder series of any length, which leads to a rather complicated argument. By using his operator, Casimir did not have to make any distinction between the cases $m < n$ and $m > n$.

12.5. To give a better idea of Cartan's proof, we discuss the case $m > n$ directly, without reducing to $m < n$.

We let v'_n and v'_{n-2i} $(i \geq 0)$ be as before. Note first that if n and m have different parities, then V_n and V_m have no common eigenvalue for h. In particular h has no element of weight $n+2$ in E and the eigenspace for n is one-dimensional, hence spanned by v'_n. Again, by 12.3, v'_n generates a complementary $\mathfrak{g}$-module. So we assume that $m \equiv n \mod 2$. As before, the whole point is to find v'_n satisfying the condition 12.3(1), for $c = n$.

As above, there is a constant a such that

$$h \,.\, v'_n = n \,.\, v'_n + a \,.\, v_n \,. \tag{1}$$

We want to prove v'_n may be chosen so that $a = 0$. As in 12.4, we see that

$$h \,.\, v'_{n-2i} = (n-2i) \,.\, v'_{n-2i} + a \,.\, v_{n-2i} \qquad (i \geq 0)\,. \tag{2}$$

The weights in V_n are contained in $[n, -n]$, so the projection of $f \,.\, v'_{-n}$ in V_n is zero and we have, for some constant c,

$$f \,.\, v'_{-n} = c \,.\, v_{-n-2} \,. \tag{3}$$

Let $v''_n = v'_n - c \,.\, v_n$ and following 12.2(4), define v''_{n-2i} inductively by the relation

$$v''_{n-2i} = -i \,.\, f \,.\, v''_{n-2i+2} \qquad (i = 1, \ldots, n)\,. \tag{4}$$

By induction on i, we see that

$$h \,.\, v''_{n-2i} = (n-2i) \,.\, v''_{n-2i} + a \,.\, v_{n-2i} \qquad (i = 0, \ldots, n) \tag{5}$$

and also, in view of (3), that

$$f \,.\, v''_{-n} = f \,.\, v'_{-n} - c \,.\, f \,.\, v_{-n} = 0 \,. \tag{6}$$

For $i = n$, the equality (5) gives

$$h \,.\, v''_{-n} = -n \,.\, v''_{-n} + a \,.\, v_{-n} \,. \tag{7}$$

Apply now f to both sides and recall that $f.h = h.f + 2f$. In view of (6) and 12.2(3) for $m = n$, we get

$$(8) \qquad a.(n+1).v_{-n-2} = 0.$$

But $n < m$ so $v_{-n-2} \neq 0$, whence $a = 0$.

We may therefore assume that v'_n is an eigenvector of h. There is no eigenspace for h with eigenvalue $n+2$ in V'_n, hence

$$(9) \qquad e.v'_n = b.v_{n+2} \qquad (b \in \mathbf{C}).$$

By 12.2(4), for $i = (m-n)/2$ (recall that $m \equiv n$ (2)), we get

$$(10) \qquad e.v_n = \big((m+n+2)/2\big).v_{n+2},$$

therefore

$$(11) \qquad w_n = v'_n - b\big((m+n+2)/2\big)^{-1}.v_n$$

satisfies the conditions

$$(12) \qquad h.w_n = n.w_n, \quad e.w_n = 0,$$

so that, by 12.3, $\mathfrak{g}.w_n$ is a copy of V_n complementary to V_m.

13. FANO'S PROOF:

It deals with projective transformations and uses algebraic geometry. Given a finite dimensional vector space F over $\mathbf{C}$, we let $\mathbf{P}(F)$ be the projective space of one-dimensional subspace of F. If F is of dimension n, $\mathbf{P}(F)$ is isomorphic to $\mathbf{P}_{n-1}(\mathbf{C})$.

13.1. The proof is contained in §§7, 8, 9 of [F]. §9 shows how to reduce it to the case considered in §3, a), b), that is, to the case of a short exact sequence 12.4(1) with $m \geq n$, but expressed in projective language, namely:

The space $\mathbf{P} = \mathbf{P}(E)$ contains a minimal irreducible invariant projective subspace $W = \mathbf{P}(V_m)$ of dimension m and the induced projective representation in the space W' of projective $(m+1)$-subspaces containing W is irreducible.

The problem is then to find an invariant projective subspace D not meeting W. If so, it has necessarily dimension n and $\mathbf{P}(E)$ is the join of W and D. Moreover, by the remark b) in §3, it may be assumed that $m \geq n$. Let us write N for the dimension of $\mathbf{P}$. Then $N = m+n+1$ and $m \geq (N-1)/2$.

As in 12.2(b), U is the one-parameter subgroup of $G = SL_2(\mathbf{C})$ generated by e. Its fixed point set is also the subspace E^e of E annihilated by e. Since

U is unipotent, any line invariant under U is pointwise fixed, so that the projective subspace $\mathbf{P}(E^U)$ associated to E^U is also the fixed point set $\mathbf{P}(E)^U$ of U on $\mathbf{P}(E)$. Similarly, it may be identified with the set $\mathbf{P}(E)^e$ of zeros of the vector field on $\mathbf{P}(E)$ defined by the action of U.

In §7, Fano proves that $\mathbf{P}(E)^e$ is a projective line. I am not sure I understand his argument, so I shall revert to the linear setup. As just pointed out, we have to show that E^e is two-dimensional.

In V_m and V_n it is one-dimensional, so the exact sequence 12.4(1) shows that $\dim E^e \leq 2$. As in §3, let E^* be the contragredient representation to E. Then E^{*^e} is the dual space to E/eE so it is equivalent to prove that $\dim E^{*^e} = 2$. Therefore we may assume that $m \leq n$ (our assumption earlier, but not the one of Fano). Fix a vector $v' \in E$ projecting onto a highest weight vector in V_n. It is an eigenvector of h if $m < n$, is annihilated by $(h - n.I)^2$ otherwise, and in both cases is annihilated by e (see 12.1 (4), (5)).

13.2. The next and main part of Fano's argument depends on some properties of the "rational normal scrolls", which we now recall (see [GH], p. 522–527). Assume $N \geq 2$ and let Z be a surface in $\mathbf{P}$, not contained in any projective subspace. Then its degree is at least $N-1$ ([GH], p. 173). Those of degree $N-1$ have been classified, up to projective transformations ([GH], *loc.cit.*). Only one is not ruled, the Veronese embedding of $\mathbf{P}_2(\mathbf{C})$ in $\mathbf{P}_5(\mathbf{C})$.

The others are the *rational normal scrolls* $S_{a,b}$ $(a+b=N-1)$, obtained in the following way: Fix two independent projective subspaces A, B of dimension a, b. Then $\mathbf{P} = A * B$ is the join of A and B. Let C_A (resp. C_B) be a rational normal curve in A (resp. B) and $\varphi : C_A \to C_B$ an isomorphism. Then $S_{a,b}$ is the space of the lines $D(x, \varphi(x))$ $(x \in C_A)$. If $a > 0$, but $b = 0$, then C_B is a point, φ maps C_A onto a point and $S_{a,b}$ is the cone over C_A with vertex C_B. It has a unique singular point, namely C_B and this is the only case where $S_{a,b}$ is not smooth ([GH], p. 525).

A rational curve in $S_{a,b}$ which cuts every line $D(x, \varphi(x))$ in exactly one point is called a directrix. By construction C_A (resp. C_B) is a directrix of degree a (resp. b). The main result used by Fano is that if $a > b$, then C_B is the unique directrix of degree b ([GH], p. 525). Fano deduces this essentially from an earlier result of C. Segre [Se].

If $a = b$, then we may identify A to B by a map φ which takes C_A to C_B. It is clear that in that case $S_{a,b} = S_{a,a} = \mathbf{P}^1(\mathbf{C}) \times \mathbf{P}^1(\mathbf{C})$.

13.3. We now come back to the situation in 13.1. In W there is exactly one rational normal curve C stable under G. The zero set $\mathbf{P}(E)^e$ of e is a line (13.1) and $\mathbf{P}(E)^e \cap C$ consists of one point, namely, W^e. Let Z be the set of transforms $g \cdot \mathbf{P}(E)^e$ of $\mathbf{P}(E)^e$, $(g \in G)$. Since $\mathbf{P}(E)^e$ is stable under the upper triangular group B and G/B is complete (in fact a smooth rational curve), Z is a projective subvariety, a G-stable ruled surface. We first dispose of a special case. The line $g \,.\, P(E)^e$ is the fixed point set of the subgroup ${}^gU = g \,.\, U \,.\, g^{-1}$, conjugate to U by g. Assume that two distinct such lines have a common point. It would then be fixed by two distinct conjugates of U. But it is immediate that two such subgroups generate G, so that there would be a fixed point D of G in $\mathbf{P}(E)$, necessarily outside W. Then $\mathbf{P}(E)$ would be the join of W and D, and we would be through. From now on, we assume that the lines $g \,.\, \mathbf{P}(E)^e$ either coincide or are disjoint. We want to prove that Z has degree $N-1$ in $\mathbf{P}(E)$. First we claim that it is not contained in any hyperplane Y of $\mathbf{P}(E)$. Indeed, if it were, it would be contained in a G-stable proper subspace F, the intersection of the transforms of Y. The subspace F would contain W properly, which would contradict the irreducibility of the quotient representation in $\mathbf{P}(V_n)$. The degree of Z is therefore at least $N-1$ ([GH], p. 173–4). There remains to show that it is $\leq (N-1)$.

Let $C' \subset W'$ be the closed orbit of G, which plays the same role as C in W. In particular, it has degree n. Let now Y be a generic hyperplane of $\mathbf{P}(E)$ among those containing W. Viewed as a hyperplane in W', it cuts C' in n distinct points Q_i $(i = 1, \ldots, n)$. Let U_i be the conjugate of U which fixes Q_i (see 12.2, (b)). The intersection $Z \cap Y$ is a (reducible) curve. We want to prove it has degree $N-1$ in Y. We claim first

$$(1) \qquad Y \cap Z = C \cup D_1 \cup \cdots \cup D_n \qquad (D_i = \mathbf{P}(E)^{U_i}),$$

where the D_i are disjoint projective lines, each intersecting C at exactly one point.

First, by construction, $C \subset Z \cap Y$, in fact $C = W \cap Z \cap Y$. Let $x \in Z \cap Y$, $x \notin W$. It belongs to some line $D_g = g \,.\, \mathbf{P}(E)^e$. The line D_g also contains $g \,.\, W^e$, which belongs to $Z \cap Y$, too. Therefore $D_g \subset Y$, and of course $D_g \subset Z$, hence $Z \cap Y$ is the union of C and some of the lines D_g. The line D_g spans with W a projective subspace of dimension equal to $\dim W + 1$, which represents a point of W', fixed under ${}^gU$. It belongs therefore to Y if and only ${}^gU$ is one of the U_i, i.e. if and only if D_g is one of the D_i's and (1) follows.

Since C has degree $m = \dim W$ in W, it follows that $Z \cap Y$ is a curve of degree $m + n$ in Y, hence Z is a surface of degree at most $m + n = N - 1$ in $\mathbf{P}(E)$.

Thus Z is a ruled surface, not contained in a hyperplane, of smallest possible degree. It is therefore a "rational normal scroll" (13.2). It is isomorphic to $S_{a,b}$ where $a = \dim W = m$ and $b = N - 1 - a = n$.

Recall that we have reduced ourselves to the case $a \geq b > 0$. Assume first $a > b$. Then, (see 13.2), Z contains a *unique* directrix of degree b. It is a normal curve in a b-dimensional subspace, which must be invariant under G, since Z is. This provides the complementary subspace to W.

Let now $m = n$. Then (13.2), $Z = C \times C'$ is a product of two copies of $\mathbf{P}^1(\mathbf{C})$, where C is, as before, a G-stable rational normal curve in W and $C' = \mathbf{P}(E)^e$. The transforms $g\,.\,\mathbf{P}(E)^e$ of C' are the lines $\{c\} \times C'\,(c \in C)$.

The lines $C_y = C \times \{y\}$ $(y \in C')$ are "directrices". We claim that they are all invariant under G. Clearly, the intersection number $C_y \cdot C_z$ is zero if $y \neq z$ $(y, z \in C')$. Let $g \in G$. Since it is connected to the identity, we have then also $(g\,.\,C_y) \cdot C_z = 0$, therefore $g\,.\,C_y \cap C_z = \emptyset$ unless $g\,.\,C_y = C_z$. Since $g\,.\,C_y$ must meet at least one C_z, we have $g\,.\,C_y = C_z$ for some z and we see that G permutes the curves $\{C_y\}\,(y \in C')$. Each such curve contains a fixed point of e, hence of U. Therefore C_y is stable under U. Now the subgroup H of G leaving each curve C_y stable is a normal subgroup, which is $\neq \{1\}$ since it contains U. But G is a simple Lie group, therefore $H = G$, which proves our contention. Any curve C_y is a rational normal curve in a subspace W'_y which is necessarily G-stable. This provides infinitely many G-invariant subspaces and concludes the proof.

REMARK. Let us compare the orders of the steps in the proofs of Cartan and of Fano. In 12.4 and 12.5 the first item of business is to show that the action of h on a certain h-stable two-dimensional subspace is diagonalisable. That space is E^e in 12.4, and subsequently shown to be E^e in 12.5. Once a new eigenvector of h annihilated by e is found, 12.3 can be used. In Fano, the first step is to show that E^e is two-dimensional or rather, equivalently, that $\mathbf{P}(E)^e$ is a projective line. There, the analogue of the first step of Cartan would be to prove the existence of *two* fixed points on $\mathbf{P}(E)^e$ of h, or of the group $H = \{e^{th}\}$ generated by h. One is W^e. In the generic case $m > n$, Fano's argument may also be viewed as a search for this second fixed point: it is the intersection of $\mathbf{P}(E)^e$ with the (unique) directrix C_B. However, since the proof provides directly the G-orbit C_B of that second fixed point, the argument is not phrased in that way.

I am grateful to Thierry Vust for a careful reading of the manuscript, which led to a number of corrections and clarifications.

REFERENCES

[Bo] BOURBAKI, N. *Groupes et algèbres de Lie 7, 8.* Hermann, Paris, 1975.

[Br] BRAUER, R. Eine Bedingung für vollständige Reduzibilität von Darstellungen gewöhnlicher und infinitesimaler Gruppen. *Math. Z. 41* (1936), 330–339. *C.P.* II, 462–471.

[Cr1] CARTAN, E. Sur la structure des groupes de transformation finis et continus. Thèse, Paris, Nony 1894. *O.C.* I, 137–287, Gauthier-Villars, 1952.

[Cr2] —— Les groupes projectifs qui ne laissent invariante aucune multiplicité plane. *Bull. Soc. Math. France 41* (1913), 53–96. O.C.: I_1, 355–398.

[Cr3] —— *Leçons sur la théorie des spineurs* I, II. Hermann, Paris, 1938.

[CW] CASIMIR, H. L. und B. L. V. D. WAERDEN. Algebraischer Beweis der vollen Reduzibilität der Darstellungen halbeinfacher Liescher Gruppen. *Math. Annalen 111* (1935), 1–11.

[CE] CHEVALLEY, C. and S. EILENBERG. Cohomology of Lie groups and Lie algebras. *Trans. AMS 63* (1948), 85–124.

[F] FANO, G. Sulle varietà algebriche con un gruppo continuo non integrabile di trasformazioni proiettive in sè. *Mem. Reale Accad.Sci. di Torino (2) 46* (1896), 187–218.

[GH] GRIFFITHS, P. and J. HARRIS. *Principles of Algebraic Geometry.* Wiley and Sons, New York, 1978.

[H] HURWITZ, A. Ueber die Erzeugung der Invarianten durch Integration. *Nachr. k. Gesellschaft der Wiss. zu Göttingen, Math.-Phys. Klasse* (1897), 71–90. *Math. Werke* II, 546–564. Birkhäuser Verlag, Basel, 1933.

[K] KLEIN, F. Ueber einen Satz aus der Theorie der endlichen (discontinuirlichen) Gruppen linearer Substitutionen beliebig vieler Veränderlichen. *Jahresbericht der Deutschen Math.-Ver. 5* (1890), 57.

[LE] LIE, S. und F. ENGEL. *Theorie der Transformationsgruppen III.* Teubner, Leipzig, 1893.

[Lo] LOEWY, A. Sur les formes quadratiques définies à indéterminées conjuguées de M. Hermite. *C.R. Acad. Sci. Paris 123* (1896), 168–171.

[Ma] MASCHKE, H. Beweis des Satzes, dass diejenigen endlichen linearen Substitutionsgruppen, in welchen einige durchgehends verschwindende Coefficienten auftreten, intransitiv sind. *Math. Annalen 52* (1899), 363–368.

[Mo] MOORE, E. H. A universal invariant for finite groups of linear substitutions: with applications in the theory of the canonical form of a linear substitution of finite order. *Math. Annalen 50* (1898), 213–219.

[R] RAŠEVSKIĬ, P. K. On some fundamental theorems of the theory of Lie groups. *Uspehi Mat. Nauk. (N.S.) 8* (1953), 3–20.

[S] SCHUR, I. Neue Anwendungen der Integralrechnung auf Probleme der Invariantentheorie. 1. Mitteilung. *Sitzungsber. d. Preussischen Akad.d.Wiss., Math.-Phys. Klasse* (1924), 189–208; *Ges.Abh.* II, 440–459, Springer.

[Se] SEGRE, C. Sulle rigate razionali in uno spazio lineare qualunque. *Atti d. Reale Acc. d. Sci. di Torino 19* (1884), 355–373.

[W1] WEYL, H. Theorie der Darstellung kontinuierlicher halbeinfacher Gruppen durch lineare Transformationen I, II, III und Nachtrag, I: *Math. Z. 23* (1925), 271–309; II: *Math. Z. 24* (1926), 328–376; III: *Math. Z.*

24 (1926), 377–395; Nachtrag: *Math Z.* *24* (1926), 789–791. *G.A.* II, 543–647, Springer.

[W2] —— *Gruppentheorie und Quantenmechanik*. S. Hirzel, Leipzig, 1928, 1931. English translation, Dutton, New-York, 1932. Reprinted, Dover publications, 1949.

(Reçu le 12 janvier 1998; version révisée reçue le 2 mars 1998)

Armand Borel
School of Mathematics
Institute for Advanced Study
Princeton, NJ 08540
U.S.A.
e-mail: borel@math.ias.edu

168.

André Weil and Algebraic Topology*

Notices of the AMS **46**, 1999, 422-427,
Gazette des Mathématiciens, n° spécial, 63–74, Soc. Math. France 1999

André Weil is associated more with number theory or algebraic geometry than with algebraic topology. But the latter was very much on his mind during a substantial part of his career. This led him first to contributions to algebraic topology proper, in a differential geometric setting, and then also to the use in abstract algebraic geometry and several complex variables of ideas borrowed from it.

According to [W3], I, p. 562 his first contacts with algebraic topology took place in Berlin, 1927, in long conversations with, and lectures from, Heinz Hopf. The first publication of H. Hopf on the Lefschetz fixed point formula appeared the following year, so it is rather likely that Weil heard about it at the time. At any rate, his first paper involving algebraic topology is indeed an application of that formula to the proof of a fundamental theorem on compact connected Lie groups (which Weil attributes to E. Cartan, but is in fact due to H. Weyl): Let G be a compact connected Lie group. Then the maximal tori (i.e. maximal connected abelian subgroups) of G are conjugate by inner automorphisms and contain all elements of G ([1935c] in [W3], I, 109–111).

The proof is a repeated application of the Lefschetz fixed point formula to translations by group elements on the homogeneous space G/T, where T is a maximal torus. Note that the isotropy groups on G/T are the conjugates of T, so that an element belongs to a conjugate of T if and only if it fixes some point in G/T. Weil first points out that T is of finite index in its normalizer $N(T)$. If $t \in T$ generates a dense subgroup of T, then its fixed points are the same as those of T and a local computation shows their indexes to be simultaneously equal to 1 or to -1. The Lefschetz number of t is then $\neq 0$. But since t is connected to the identity, this number is equal to the Euler-Poincaré characteristic $\chi(G/T)$ of G/T, which is therefore $\neq 0$. As a consequence, any element $g \in G$ has a non-zero Lefschetz number, hence a fixed point, and belongs to a conjugate of T. If T' is another torus, and t' generates a dense subgroup of T', then any torus containing t' will also contain T', whence the conjugacy statement.

This was the first new proof of that theorem, completely different from the original one, which relied on a study of singular elements (cf. H. Weyl,

* Lecture given at the Institute for Advanced Study on January 8, 1999, at a Conference on André Weil's Work and its influence, January 8–9, 1999.
NDLR: reproduit avec l'autorisation de l'auteur et des *Notices of the American Mathematical Society.*

Collected Papers II, 629–633). It was rediscovered independently, about five years later, by H. Hopf and H. Samelson (Comm. Math. Helv. **13**, 1940–41, 240–251).

For about ten years, from 1942 on, topology was present in several works of Weil, often pursued simultaneously, which I first list briefly:

a) In algebraic geometry: foundations, introduction of fibre bundles, formulation of the Weil conjectures.
b) New proof of the de Rham theorems. Together with Leray's work, this was the launching pad for H. Cartan's work in sheaf theory.
c) Characteristic classes for differentiable bundles: Allendoerfer-Weil generalization of the Gauss-Bonnet theorem, theory of connections, the Chern-Weil homomorphism, the Weil algebra.
d) Joint work with Cartan, Koszul and Chevalley on cohomology of homogeneous spaces.
e) A letter to H. Cartan (August 1, 1950) on complex manifolds, advocating the use of analytic fibre bundles in the formulation of problems such as those of Cousin.

There is a last item I would like to add, dating from 1961–2:

f) Local rigidity of discrete cocompact subgroups of semisimple Lie groups.

On the face of it, it does not belong to algebraic topology, but can be fitted under my general title when stated as a theorem on group cohomology. This formulation was originally an afterthought, but turned out to be important to suggest further developments.

Algebraic Geometry

The algebraic geometry, as developed mainly by the Italian School, did not offer a secure framework for the proof of the Riemann hypothesis for curves and other researches of Weil in algebraic geometry. He had to develop new Foundations, with as one of its main goals a theory of intersections of subvarieties. It had also to be over any field. This implied a massive recourse to algebra, but Weil still wanted to keep a geometric language and picture. Until then, only projective, affine or quasi-projective varieties had been considered, i.e. subvarieties of some standard spaces. He wanted a notion of "abstract variety" which would be the analogue of a manifold (albeit with singularities). His first version [W1] is a bit awkward, as acknowledged in the foreword to the second edition, because no topology is introduced. From ([1949c], [W3], I, 411–413) on however, he uses the language of the Zariski topology (introduced in 1944 by O. Zariski), and I shall do so right away. Fix a "universal field", K i.e. an algebraically closed field of infinite transcendence degree over its prime field. Let V be an algebraic subset of K^n i.e. an affine variety. In the Zariski topology, the closed subsets of V are the algebraic subsets. The open sets are, of course, their complements and are quite big. If V is irreducible, any two non-empty ones intersect in a dense open one, so that the topology is decidedly not Hausdorff (unless V is a point), which may explain some reluctance to use it initially. To define an (irreducible) abstract variety V, start from a finite collection (V_i, f_{ji}), $(i, j \in I)$, where V_i is an irreducible affine algebraic set, f_{ji} a birational correspondence from V_i to V_j satisfying certain conditions so that, in particular: f_{ii} is the identity, $f_{ij} = f_{ji}^{-1}$, there exist open subsets $D_{ji} \subset V_i$ such that f_{ji} is a biregular mapping of D_{ji} onto D_{ij} and $f_{ji} = f_{jk} \circ f_{ki}$. Two points $P_i \in D_{ji}$ and $P_j \in D_{ij}$ are equivalent if $f_{ji}(P_i) = P_j$. The "abstract variety" V is by definition the quotient of the disjoint union $\tilde{V}$ of the V_i by that equivalence relation.

Note that V is obtained by gluing together disjoint affine sets. For lack of suitable concepts, it was not possible to start from a topological space and require that it be endowed locally with a given structure, as is done for manifolds, (as was done later by Serre using the notion of ringed space [S2]). As a result, the V_i and f_{ji} are part of the structure, which is rather unwieldy, and requires a somewhat discouraging amount of algebra to be worked with. Nevertheless, Weil develops the theory of such varieties and of the intersection of cycles. For the latter, the analogy with the complex case and the intersection product in the homology of manifolds (on which he had lectured earlier at the Hadamard seminar ([W3] I, p. 563)) is always present. In particular, a key property is the analogue of Hopf's inverse homomorphism (See [W1], Introduction, xi–xii). Weil also introduces an analogue of compact manifolds, the complete varieties, which include the projective ones. [W1] supplied the framework for a detailed proof of the Riemann hypothesis for curves, further work on abelian varieties, and was essentially the only one for algebraic geometry over any field until Grothendieck's theory of schemes (from about 1960 on).

Algebraic topology also underlies the formulation of the conjectures in ([1949b], cf. [W3], I, 399–410), soon to be called the Weil conjectures, which suggests to look for a cohomology theory for complete smooth varieties in which a Lefschetz fixed point formula would be valid. This vision, which turned out to be prophetic, was quite unique at the time.

In ([1949c], cf. [W3], I, 411–12), Weil introduces in algebraic geometry fibre bundles with an algebraic group, say G, as structural group. Given a variety B and a finite open cover $\{V_i\}(i \in I)$ of B, assume given regular maps $s_{ij} : V_i \cap V_j \to G$ $(i, j \in I$; s_{ii} is the constant map to the identity), with the usual transitivity conditions. Let F be a variety on which G operates. Then a fibre bundle E on B, with typical fibre F, is obtained by gluing the products $V_i \times F$ by means of the s_{ij}, as usual. Weil also considers the case of principal bundles ($F = G$, acted upon itself by right translations). In particular, if $G = \mathbb{C}^*$ is the multiplicative group of non-zero complex numbers, the isomorphism classes of such bundles correspond to linear equivalence classes of divisors. It also allowed Weil to interpret in a more conceptual way earlier work on algebraic curves (see [W3], I, p. 531, 541, 570 for comments). A detailed exposition is given in [W2], where the classification of such bundles is studied in some simple cases.

In view of the big size of the neighborhoods on which such a bundle is trivial, it was not a priori clear this would lead to an interesting theory. That it did is one reason why Weil began to gain confidence in the Zariski topology.

Of course, his definition of fibre bundle was greatly generalized later. Already in [1949c], Weil points out it would be desirable to have a notion broad enough so that B could be the set of prime spots of a number field. Later (séminaire Chevalley 1958), J-P. Serre introduced an important generalization of local triviality: a bundle is locally isotrivial if every point has an open neighborhood admitting an unramified covering on which the lifted bundle is trivial. This notion, which includes the fibration of an algebraic group by a closed subgroup, led A. Grothendieck to the definition of étale topology.

The de Rham Theorems

In January 1947, Weil wrote to H. Cartan a letter ([W3], II, 45–47) outlining a new proof of the de Rham theorems, published later in ([1952a], [W3], II, 17–43), the first one since de Rham's Thesis. It is limited to compact manifolds but this restriction is lifted, with very little complication, in the final version.

Given a smooth compact connected manifold M, Weil first shows the existence of a finite open cover $\{U_i\}_{i\in I}$ of M such that any non-empty intersection of some of the U_i's is contractible. Let N be the nerve of this cover and, for each simplex $\sigma \in N$, let U_σ be the intersection of the U_i represented by the vertices of σ. Given $p, q \in \mathbb{N}$, let $A^{p,q}$ be the function assigning to each p-simplex σ of N the space of differential q-forms on U_σ.

The direct sum A of $A^{p,q}$ is endowed with two differentials

$$d : A^{p,q} \to A^{p,q+1}, \; \delta : A^{p,q} \to A^{p+1,q},$$

where d stems from exterior differentiation and δ from the coboundary operator on N, followed by restriction of differential forms. Let $F^{p,q}$ (resp. $H^{p,q}$) be the subspace of $A^{p,q}$ of elements annihilated by $d\delta$ (resp,. d or δ). Then Weil establishes the isomorphisms

$$F^{0,m}/H^{0,m} = H^m_{DR}(M), \qquad F^{m,0}/H^{m,0} = H^m(N). \tag{2}$$

$$F^{p,q}/H^{p,q} = F^{p+1,q-1}/H^{p+1,q-1} \qquad (0 \le q \le m), \tag{3}$$

where $H^m_{DR}(M)$ refers to real de Rham cohomology and $H^m(N)$ to the real cohomology of N. This proves, by induction that $H^m_{DR}(M)$ is isomorphic to $H^m(N)$, hence to $H^m(M)$, since the U_σ are contractible. This last fact is taken for granted in the letter but Weil also shows in [1952a] by similar arguments that $H^m(M)$ is equal to the m-th singular cohomology of M over the reals.

H. Cartan had studied the work of Leray in topology, in particular his war time paper (*J. Math. pure appl.* **29**, 1945, 95–248) and he noticed a similarity with Weil's proof. That was tremendously suggestive to him, and quickly gave rise to a flurry of letters to Weil, in which Cartan initiated his theory of *faisceaux et carapaces* (sheaves and gratings), of which he gave later three versions (see [B] for references), first following Leray rather closely, arriving eventually at a much greater generality.

What Cartan had noticed is an analogy between the proofs of the isomorphisms in (3) and an argument which occurs repeatedly in Leray's paper, to which Leray himself traced later the origin of the spectral sequence (see [B]). However, Weil's argument was completely independent from it: as stated in a slightly later letter to Cartan, Weil did not know that paper and in fact suspected, on the strength of a report by S. Eilenberg in *Math. Reviews*, that it did not bring much new, if anything. On the other hand, it is quite plausible that the definition of the $A^{p,q}$ was in part inspired by a short conversation Weil had with Leray in summer 1945, in which the latter spoke of a cohomology 'with variable coefficients". In fact, an analogue of A in the theory Leray was developing at the time would be a couverture, N, with coefficients in the differential graded sheaf associated to differential forms.

Characteristic Classes

In 1941–42 Weil was for some time at Haverford College, PA, where he met C. Allendoerfer. This led to their joint work on the generalized Gauss-Bonnet theorem [AW]. Given a smooth compact oriented Riemannian manifold M of even dimension m, it expresses the Euler-Poincaré characteristic $\chi(M)$ of M as the integral over M of a differential m-form built from the components of the curvature tensor. Such a formula had already been proved by Allendoerfer and by Fenchel for submanifolds of euclidean space. At the time, the Allendoerfer-Weil theorem was in principle more general since it was not known whether a Riemannian manifold was globally isometrically diffeomorphic to a submanifold of euclidean space, though it had been established locally. Because of that, the nature of their proof forced them to prove a more general statement, though I do not know whether the added generality has led to further applications.

Recall the Gauss-Bonnet formula in the most classical case: P is a relatively compact open subspace on a surface in $\mathbb{R}^3$, bounded by a simple closed curve, union of finitely many smooth arcs. Then the integral of the Gaussian curvature K on P, plus the sum of the integrals of the curvature on the boundary arcs and of the outside angles at the meeting points of those, is equal to 2π. The Allendoerfer-Weil formula gives a generalization of such a formula for a Riemannian polyhedron. It is proved first for polyhedra in euclidean space. The general case then follows by using a polyhedral subdivision, small enough so that its building blocks can be isometrically embedded in euclidean space and by proving a suitable addition formula.

In 1944 S.S. Chern produced a proof of the Allendoerfer-Weil formula for closed manifolds (*Ann. Math.* **45**, 1944, 747–52) which was much simpler and a harbinger of further developments on characteristic classes. On M choose a vector field X with only one zero, of order $|\chi(M)|$ at some point x_0, which is always possible. Let E be the unit tangent bundle to M, $p : E \to M$ the canonical projection and Ω the Gauss-Bonnet form. The key point is that $p^*\Omega = d\Pi$ is the exterior derivative of some explicitly given form Π, the restriction of which to a fibre F represents the fundamental class [F] of the latter. The vector field X defines a submanifold V in E, a copy of $M - \{x_0\}$, with boundary the unit sphere $F_o = p^{-1}(x_0)$, with multiplicity $|\chi(M)|$. The Gauss-Bonnet formula then follows from Stokes theorem, applied to $V \cup F_0$.

The relationship between Ω, Π, and $[F_0]$ is a first example of a notion developed later under the name of *transgression* in a fibre bundle: a cohomology class β of a fibre F is transgressive, if there is a cochain (in the cohomology theory used, here a differential form) on the total space E whose restriction to F is closed, represents β, and whose coboundary belongs to the image of a cohomology class η of the base B, under the map induced by the projection $p : E \to B$. The classes β and η will be said to be related by transgression. This notion, and the terminology, were introduced first by J-L. Koszul in a Lie algebra cohomology setting in his Thesis (Bull. S.M. France **78**, 1950, 65–127).

In *Ann. Math.* **47**, 1946, 85–121, Chern gives several definitions of the characteristic classes $c^i(M) \in H^{2i}(M; \mathbb{C})$, since then called the Chern classes $(1 \leq i \leq m)$. In particular, if M is endowed with a hermitian metric, they can be expressed by closed differential forms which are locally defined in terms of the curvature tensor. Again each one is related by transgression in a suitable bundle to the fundamental class of the fibre.

It is at this point that Weil comes in. He was familiar with the work of Chern, with the theory of fibre bundles, in particular the classification theorem in terms of universal bundles, having written jointly with S. Eilenberg, with some help from N. Steenrod, a report on fibre bundles for Bourbaki (which, incidentally, provided much background material for the second Cartan seminar [C2]). He was also aware of Ehresmann's publications on fibre bundles and on the formulation of E. Cartan's theory of connections in that framework, as well as of Koszul's work towards his Thesis quoted above. All this came together in a series of letters to Cartan, Chevalley, Koszul, of which the first four have been published (almost completely) for the first time thirty years later ([1949e], in [W3], I, 422–36). Some were shown around at the time, however. In particular, the first one is the basis of Chapter III in [C4] and this is how its contents became widely known.

Let G be a compact connected Lie group, ξ a principal G-bundle, E (resp. B) the total space (resp. base) of ξ. A connection on ξ is defined by means of a 1-form on E with values in the Lie algebra g of G, satisfying certain conditions. Let I_G be the algebra of polynomials on g invariant under the adjoint representation and $P \in I_G$ a homogeneous element of degree q. Replacing the variables in P by the components of the curvature tensor of the connection, Weil associates to P a differential $2q$-form on M, which is proved to be closed, hence to define an element $c_P \in H^{2q}(M; \mathbb{R})$. A fundamental theorem asserts that c_P is independent of the connection. The proof is short, but stunning. In the Fall of 1949, in Paris, I read this letter and said once to Cartan that this proof seemed to come out of the blue and I could not trace it back to anything. *"That's genius. You don't explain genius"*, was his answer. The image of I_G under this homomorphism, which became known as the Chern-Weil homomorphism, is then the characteristic algebra of ξ.

At the end of the first letter, Weil states a conjecture relating the primitive generators of $H^*(G; \mathbb{R})$ (recall that it is an exterior algebra with a distinguished set of generators, called primitive), to the characteristic algebra by transgression, soon proved by Chevalley. This already provided a generalization of Chern's treatment of characteristic classes of hermitian bundles, modulo some normalization and plausible identifications. In the third letter (addressed to Koszul, as well as the 4th one), Weil makes the analogy closer. Recall that in the classical case the characteristic classes are the images of cohomology classes of a classifying space (a complex Grassmannian for hermitian bundles), under the homomorphism induced by a classifying map (see [C4] for example). Weil proposes an algebraic analogue of that situation. He introduces an algebra, which, following Cartan [C3], I shall denote $W(g)$ and call the Weil algebra of g. By definition, $W(g) = S(g^*) \otimes \wedge g^*$ is the tensor product of the symmetric algebra $S(g^*)$ by the exterior algebra $\wedge g^*$ of the dual g^* of g. It is graded, anticommutative, an element $x \in g^*$ being given the degree 1 (resp. 2) if it is viewed as belonging to $\wedge g^*$ (resp. $S(g^*)$). The Weil algebra is further endowed with a specific differential. The latter leaves $S(g^*) \otimes \mathbb{R}$ stable, the cohomology of which is isomorphic to I_G. The algebra $\wedge g^*$, endowed with the Lie algebra cohomology differential, is a quotient of $W(g)$. The transgression in $W(g)$ provides a bijection of the space of primitive generators of $H^*(g)$, (which is isomorphic to $H^*(G; \mathbb{R})$) onto a space spanned by independent homogeneous generators of I_G (the latter is, by a theorem of Chevalley, a polynomial algebra). A connection on ξ provides a homomorphism of $W(g)$ onto a subalgebra of differential forms on E which, after having passed to cohomology, yields the Chern-Weil homomorphism. Thus $W(g)$ plays the role of an algebra of differential forms on a universal G-bundle, analogy reinforced by the fact, proved by Cartan [C3], that $W(g)$ is acyclic.

So far, I have focused on characteristic classes. But these letters, combined with Koszul's Thesis, led to further correspondence on the cohomology of homogeneous spaces, and to more results announced by H. Cartan (Colloque de topologie, CBRM Bruxelles 1950, 57–71) and J-L. Koszul (*ibid.* 73–81). A full exposition is given in [GHV].

Complex Manifolds and Holomorphic Fibre Bundles

On August 1, 1950, Weil wrote to H. Cartan a letter about global analysis in several complex variables (unpublished). He first claims that it is high time to stop viewing the object of these investigations as a sort of "domain" spread over n-space or complex projective space. One should look at complex manifolds, noting, of course, that not much can be proved without further assumptions such as compact, Kähler, global existence of holomorphic functions with non-zero Jacobians, etc. Then he points out that analytic fibre bundles underly some classical problems. For instance, the Cousin data for the multiplicative Cousin problem (find a function with a given divisor of zeros and poles) leads to a principal $\mathbb{C}^*$-bundle. For a solution to exist, the bundle should first be topologically trivial. This condition is not always sufficient, but it is on a domain of holomorphy. Pursuing that idea, he conjectures that a complex vector bundle on a polycylinder with structural group a complex Lie group which is topologically trivial should be analytically trivial. Unfortunately, I could find only the first page of this letter in Weil's papers, and the original seems to be lost, or at any rate could not be located. The beginning of the last sentence on the page: *"Once one has taken the habit to look for fibre bundles in these questions, one soon sees them everywhere (or "almost everywhere") and there is an enormous gain ..."*, makes one wish to see the rest.

These remarks were taken into account by H. Cartan (Proceedings I.C.M. 1950, Vol. 1, 152–164), who also pointed out that the first Cousin problem (find a meromorphic function with given polar parts) leads to a principal complex bundle, too, but with fibre the additive group of $\mathbb{C}$.

Local Rigidity

It is well-known that compact Riemann surfaces of higher genus have moduli (non-compact ones, too, but I confine myself to the compact case). Such a surface is a quotient $\Gamma\backslash X$ of the upper half-plane $X = \mathbf{SL}_2(\mathbb{R})/\mathbf{SO}(2)$ by a discrete cocompact subgroup Γ of $\mathbf{SL}_2(\mathbb{R})$. Equivalently, this means that there are small deformations of Γ in $\mathbf{SL}_2(\mathbb{R})$ which are not conjugate to Γ. In the fifties, it began to be suspected that these phenomena were pretty much unique to that case among compact locally symmetric spaces $\Gamma\backslash X$, where $X = G/K$ is the quotient of a non-compact semisimple Lie group with finite center by a maximal compact subgroup K. The question is then to show that the locally symmetric space structure on $\Gamma\backslash X$ is locally rigid (no small deformation which is not an isomorphism) or, equivalently, that Γ is locally rigid: any local deformation of Γ in G is a conjugate of Γ). The first results along those lines were obtained by E. Calabi [C1], E. Calabi-E. Vesentini [CV], from the geometric point of view and A. Selberg [S1], for $G = \mathbf{SL}_n(\mathbb{R})$, from the group theoretical point of view.

The paper [S1] and an unpublished sequel to [C1] were the starting point for the three papers of Weil on that topic ([1960c], [1962b], [1964a] in [W3], II, 449–464, 486–510, 517–525). In the first one, Weil proves, for any connected Lie group, a conjecture of Selberg in [S1], to the effect that if Γ is discrete cocompact, any small deformation of Γ is discrete, cocompact, isomorphic to Γ. To formulate the problem, he introduces the variety $R(\Gamma, G)$ of homomorphisms of Γ into G. The group Γ is finitely presented (as fundamental group of the compact smooth manifold G/Γ). Let $(g_1, \ldots, g_N)$ be a generating subset. Then $R(\Gamma, G)$ may be viewed as the real analytic subvariety of the product $G^{(N)}$ of N copies of G defined by the relations between these generators. Let $x_o = (g_1, \ldots, g_N)$. The theorem is then that x_o has a neighborhood in $R(\Gamma, G)$, all elements of which represent discrete, cocompact subgroups of G isomorphic to Γ.

Assume now that G is semisimple, with finite center (an assumption which is implicit in [1962b], but could be lifted) with no factor which is compact or three dimensional. Then it is shown in [1962b] that Γ is locally rigid, as conjectured in [S1], too, by proving that the orbit

$$G.x_o = \{g.g_1.g^{-1}, \ldots, g.g_N.g^{-1},\ (g \in G)\}$$

contains a neighborhood of x_o in $R(\Gamma, G)$. The theorem is further extended to the case where G has some factors locally isomorphic to $\mathbf{SL}_2(\mathbb{R})$, provided that the projection of Γ on any such factor is not discrete. At the time, it was rumored (and in fact stated in [S1]) that Calabi had proved local rigidity when X is the hyperbolic n-space ($n \geqq 3$), but this was not contained in his only publication on that matter [C1] and Weil kept telling me that an essential idea was still missing. But he found it in Notes by Kodaira of some 1958–59 seminar lectures by Calabi, and then proved the above results within a few days.

The paper [CV] considers first of all the case where X is an irreducible bounded symmetric domain and shows that its complex structure is locally rigid, provided X is not isomorphic to the unit ball in $\mathbb{C}^n$ ($n \geq 2$). Both [C1] and [CV] follow the model of the Kodaira-Spencer theory of deformations of complex structures. Local rigidity follows then from the vanishing of a first cohomology group, with coefficients in germs of Killing vector fields in [C1], of holomorphic tangent vector fields in [CV]. In [1964a], Weil provides similarly a cohomological translation of [1962b] by showing that the proof there implies the vanishing of the 1st cohomology group $H^1(\Gamma; g)$ of Γ with coefficients in the Lie algebra g of G, acted upon by the adjoint representation.

The proof of [1962b] was already cohomological in spirit and is described so by Weil in his comments. It is first reduced to the case of a one-parameter group of deformations, defined by a vector field ξ. Without changing its class modulo inner automorphisms, he replaces ξ by a "harmonic" one, i.e. by the minimum of a suitable variation problem. It is then shown to be G-invariant and a simple Lie algebra computation shows that it is zero if G has no factor which is compact or locally isomorphic to $\mathbf{SL}_2(\mathbb{R})$.

Weil was in fact not a newcomer to the group cohomology. In 1951, he had asked a student, Arnold Shapiro, to prove a certain lemma on the cohomology of finite groups. The later complied and the lemma came up later in countless variations, all known as "Shapiro's lemma" (cf. [W3], I, pp. 498 and 577).

Weil never came back to these questions, but several further developments originated in these papers. If instead of g we take $\mathbb{C}$ acted upon trivially, then $H^1(\Gamma; \mathbb{C})$ is trivial if and only if the commutator subgroup of Γ is of finite index in Γ. The vanishing of $H^1(\Gamma; \mathbb{C})$ was proved in many cases, using an approach similar to Weil's, by Matsushima, who extended it further to determine some higher cohomology groups (*Osaka M.J.* [**14**], 1962, 1–20). Later I generalized Matsushima's theorem to non-cocompact arithmetic groups, which yielded the determination of the rational K-groups of rings of algebraic integers (*Annales Sci. ENS* Paris (4) **7**, 1974, 235–272) and led to the study of higher regulators in algebraic K-theory (*Ann. Sci. Sc. Norm. Sup.* Pisa (4) **4**, 1977, 613–656). In another direction, N. Mok, Y-T. Siu and S-K. Yeung used a non-linear version of Matsushima's argument to establish archimedean superrigidity of cocompact discrete subgroups (*Inv. Math.* **113**, 1993, 57–83).

This concludes my survey of algebraic topology in the work of A. Weil. Viewed as part of his overall output, it is quantitatively minor. Still, it reaches out to an impressive amount of mathematics, has been very influential and testifies to the breadth of his outlook, as well as to his concentration on essential questions.

Bibliography

[A.W] A. Allendoerfer and A. Weil, *The Gauss-Bonnet theorem for Riemannian polyhedra* Trans. A.M.S. (VI) **53**, 1943, 101–129; [W3], I, 299–327

[B1] A. Borel, *Jean Leray and algebraic topology* in Leray's Selected papers I, 1–21, Springer and Soc. Math. France 1997

[C1] E. Calabi, *On compact Riemannian manifolds with constant curvature* I, Proc. Symp. pure math. III, 155–180, AMS 1961

[CV] E. Calabi and E. Vesentini, *On compact locally symmetric Kählerian manifolds* Ann. Math. **71** (1960), 472–507

[C2] H. Cartan et al., *Homotopie et espaces fibrés*, Séminaire E.N.S. 1949–50

[C3] H. Cartan, *Notions d'algèbre différentielle; application aux groupes de Lie et aux variétés où opère un groupe de Lie* Colloque de Topologie C.B.R.M. Bruxelles 1950, 15–27; Collected Papers III, 1255–67

[C4] S.S. Chern, *Topics in differential geometry*, Notes, Institute for Advanced Study, 1951

[GHV] W. Greub, S. Halperin and R. Vanstone, *Connections, curvature and cohomology*, Vol. 3, Pure and Applied math **47**, Academic Press 1976

[S1] A. Selberg, *On discontinuous groups in higher-dimensional symmetric spaces* in Contributions to function theory, Int. Colloquium TIFR Bombay, 1960, 147–164

[S2] J-P. Serre, *Faisceaux algébriques cohérents* Ann. Math. **61**, 1955, 197–278, Collected Papers I, 310–91

[W1] A. Weil, *Foundations of algebraic geometry*, A.M.S. Colloquium public. XXIX, 1946; second edition 1962

[W2] A. Weil, *Fibre spaces in algebraic geometry (Notes by A. Wallace)*, Notes, University of Chicago, 1952

[W3] A. Weil, *Collected Papers*, 3 volumes, Springer 1979

170.

André Weil: quelques souvenirs*

Gazette des Mathématiciens, n° spécial, 37–41, Soc. Math. France 1999

André Weil avait un sens aigu de l'humour, était prompt à voir l'aspect comique ou ridicule d'une situation ou d'une déclaration, ce qui se traduisait rapidement par un commentaire mordant, voire blessant, ou simplement spirituel, amusant. Je voudrais donner deux exemples dans cette veine, ayant trait à des incidents qui se sont produits à l'Institute for Advanced Study.

La participation de Weil animait souvent nos séances de faculté, quelquefois peut-être un peu trop. Une fois, alors que nous discutions des plans pour une nouvelle bibliothèque, un débat extrêmement acrimonieux se développa autour d'un point vraiment mineur, hors de proportion avec la véhémence de la discussion. Le lendemain, en en parlant avec Weil, je lui dis que cela me rappelait le «Le Lutrin» de N. Boileau, une comparaison que l'amusa. Pour moi, ce rapprochement était une fin en soi, mais pas pour lui, car notre directeur, J.R. Oppenheimer, reçut deux ou trois jours plus tard la lettre suivante:

A MONSIEUR
Monsieur Robert Oppenheimer
En son Eschole de Princeton dans le Nouveau Jersey ches les Indiens d'Amérique

Je ne sçais, MONSIEUR, si ce peu de réputation que j'eus de mon vivant sera parvenu jusqu'à vous. Mais je puis bien vous dire qu'ici, au royaume des ombres, il n'est bruit que de vous et des débats, si glorieux pour vous, où vous fustes opposé à quelques savantastres dont le charitable dessein était d'expédier promptement chez nous tout ce qui reste d'hommes sur la terre. On dit aussi que vous sçavez parfaitement les langues tant anciennes que modernes, et mesme celle des brachmanes. C'est ce qui fait que pour vous escrire je m'exprime en françois, ma langue maternelle, plutost qu'en la vostre que j'entends fort bien, depuis que Messieurs Pope et Addison m'en communiquèrent l'usage, mais dont je ne me sers qu'avec un peu de peine.

Vous n'ignorez pas, MONSIEUR, qu'une fois par siècle notre auguste souverain Pluton nous accorde un congé durant lequel nous avons permission de nous réincarner sur terre. J'ay dès longtemps songé à consacrer à Calliope mon prochain séjour parmi les vivans, car j'eus autrefois pour l'histoire un goust fort pronouncé.

Mesme j'eus l'honneur d'estre historiographe du roy. J'advoue que je n'ay rien laissé qui m'acquitast envers la postérité de ce qu'elle eust eu droit d'attendre de moi à ce titre. Du moins osay-je me flatter que mon poème du LUTRIN aura asseuré à mon nom quelque durable renommée parmi les sçavans férus d'histoire ecclésiastique, et que votre docte confrère, M. le Conseiller Aulique Cantorovitche, si versé en celle-ci, ne manquera pas d'en témoigner.

Or notre auguste souveraine Proserpine, constamment informée par les voies les plus sûres de ce qui se passe au monde des vivans, a daigné me faire avertir de la

* Version française rédigée par l'auteur de l'article paru dans les *Notices of the American Mathematical Society* (avril 1999)

grande guerre qui pointe en votre Eschole, sur le sujet de votre librairie ou bibliothèque. Mesme on dit que le canon aurait desjà tonné en vos murs, et que vos mathématiciens s'apprestent à la deffense. Je ne sçaurois certes trouver plus digne sujet pour ma Muse.

Mais, lorsque nous autres morts rendons visite aux vivans, nous devenons sujets, tout comme vous, à de fascheuses nécessités, c'est de manger et de boire. Des gens dignes de foi m'ont asseuré qu'il est en votre pouvoir, après avoir pris avis de vos conseillers ordinaires, d'accorder des pensions ou stipendia à ceux que vous en jugez dignes. Aussi osay-je m'adresser à vous, avec l'asseurance que ma requeste, dont l'effet sera d'immortaliser votre Eschole, ne peut manquer d'estre receue de vous avec faveur. C'est dans cette persuasion, MONSIEUR, que j'ay l'honneur de me dire ici

De votre Magnificence
Le très-humble et très-obéissant serviteur
Nicholas Boileau-Despreaux

Peu de temps après, Weil écrivit une courte note à l'historien E. Kantorowicz, à l'époque chargé des affaires courantes de l'École des études historiques («executive officer» est le titre officiel):

«*I hear that Dr. Oppenheimer has received an application for a stipend from my famous countryman, Monsieur N. Boileau. Obviously this concerns your school. May I nevertheless, without impropriety, put in a word of recommendation? His project seems most promising. It is true that he has been dead for a great many years. But surely, in the eyes of historians and scholars, this should be counted as a feather in his cap.*»

[Cette dernière remarque faisait allusion au fait, dûment noté, que la moyenne d'âge des membres visiteurs était nettement plus élevée en histoire qu'en mathématiques (ou physique). Durant sa première année académique à l'Institute, 1958–59, Weil avait même dessiné les graphes de la distribution des âges chez les visiteurs mathématiciens, physiciens et historiens. En mathématiques, la courbe commençait à 21 ans, avait un maximum à 30–34 ans, puis décroissait et ne montrait qu'un membre dans les intervalles 45–49 et 60–64. Chez les historiens, le membre le plus jeune avait 27–28 ans, il y avait deux maximaux locaux à 50–54 et 60–64 et encore un membre dans la fourchette 80–84.]

Quelques jours plus tard, Oppenheimer offrait à Boileau une position de membre, avec bourse, pour 25 ans. Inutile de dire que les discussions sur la bibliothèque ne furent plus les mêmes dans la suite.

En juillet 1969, Deane Montgomery, qui occupait un bureau au rez-de-chaussée de «Fuld Hall», notre bâtiment principal, trouvait qu'il y faisait anormalement chaud. Bien entendu, on est habitué à un temps chaud et humide à Princeton en juillet, mais cela paraissait vraiment excessif, à tel point que Deane descendit au sous-sol pour voir s'il pouvait trouver une explication. Ce fut le cas: une pièce était chauffée en permanence pour y élever des faisans et autres volailles de choix. Deane demanda simplement à l'administrateur général de les faire enlever, mais la nouvelle s'était évidemment propagée partout. Deane, qui était notre «executive officer» à ce moment, reçut de Weil la lettre suivante:

«*Dear Deane*

It is my understanding that Fuld Hall is being converted into a Pheasant Breeding Farm, and that, for what must appear to everyone as narrowly selfish motives, you have raised objections against this excellent plan.

Had you canvassed your colleagues first (as was your obvious duty), you would have discovered that there is widespread and enthusiastic agreement in favor of the aforesaid project – it being understood, of course, that a bonus of a brace or two of those valuable birds would be distributed at Christmas, Thanksgiving and other suitable occasions, to all members of our Faculty....

I should be obliged if you would formally communicate these views of mine to all those who are in any way concerned with or interested in the Pheasant Project (first and foremost, of course, our Director).»

Tout lecteur des *Souvenirs d'apprentissage* réalise que les intérêts culturels et linguistiques de son auteur étaient vastes. Connaissant plusieurs langues anciennes ou modernes, féru de littérature et de poésie, il aimait entre autres lire dans l'original du latin (sa fille Sylvie m'a une fois raconté qu'à l'époque o^{os} elle préparait le «bac», il lui arrivait de demander à son père la traduction de certains mots qu'elle ne trouvait pas dans son lexique, sans doute pour raisons de décence), du grec, voire du sanscrit. Cependant la mathématique avait la priorité absolue, ce qui fermait à Weil un certain nombre de portes, comme il me l'a dit une fois, en pensant à ses autres intérêts. Il estimait que son «dharma» était de faire des mathématiques (*loc. cit.* p. 132) et sa vie était organisée dans ce sens. En 1955, nous étions à Chicago, vivant dans le même immeuble que les Weil. Tôt un matin j'eus une idée mathématique qui, quoique très simple, était si cruciale pour mon travail et que je ne pouvais attendre de vérifier avec lui si elle était valable. Je téléphonai, expliquant pourquoi je désirais le voir. Son épouse, Eveline, qui m'avait répondu, m'informa qu'il se préparait à partir pour un court voyage. Un peu déçu, je dis que j'attendrais son retour. «Oh non,» répliqua-t-elle après lui avoir parlé, «pour André, les mathématiques viennent toujours en premier. Il trouvera le temps de vous voir avant de partir.» Les Weil vivaient au rez-de-chaussée. La chambre qui servait de bureau à Weil faisait saillie, avec des fenêtres sur tout le tour, ce qui fait que toutes les fois que l'on sortait de la maison où y entrait, on ne pouvait s'empêcher de voir si Weil se trouvait dans son bureau. C'était si souvent le cas que l'on pouvait presque se demander s'il y vivait, travaillant et surtout tapant à la machine presque constamment. Un jour la concierge lui dit: «Vous travaillez vraiment énormément. Si vous continuez ainsi, vous deviendrez célèbre.»

171.

Algebraic Groups and Galois Theory in the Work of Ellis R. Kolchin

in *Selected works of Ellis R. Kolchin*, with commentary,
(H. Bass, A. Buium and P. Cassidy eds.), 505–525, American Math. Soc. 1999

The Galois theory of differential fields, a generalization of the Picard-Vessiot theory, was a major concern of E. Kolchin's during the first thirty years or so of his scientific life. From the beginning, it appeared that these Galois groups would be algebraic groups, or rather be naturally isomorphic to such. However, the theory of algebraic groups was not suitably developed in Kolchin's view for his purpose when he first needed it, so that a minor, but persistent and essential, theme in his work is the theory of algebraic groups.

As implied by the title, this lecture will emphasize both themes, but I still do not claim to provide an even-handed discussion of all of Kolchin's work in this area. More specifically, I shall take for granted whatever is needed from the theory of differential fields, or more generally from differential algebra, to which Kolchin had to add in no small measure, in order to arrive as directly as possible to the two main topics and their relationships.

My starting point in describing Kolchin's work will be his Annals paper on algebraic matric groups and Picard-Vessiot theory [K10]. It already contains important results, but also a program which was to occupy him for about 25 years. In order to put this paper in context, I would like first to make some historical remarks, to indicate how the theory presented itself to him then.

*First Kolchin Memorial Lecture, given at Columbia University on April 16, 1993

1. The idea of using Lie groups to develop some sort of Galois theory for differential equations is as old as the Lie theory itself. It was in the mind of S. Lie when he first developed it, as can already be seen from an 1874 letter to A. Mayer.[1] He had noticed that several classical methods to find the solutions of certain differential equations could be given a unified treatment by using a one-parameter subgroup leaving the equation invariant, whence the idea that if a Lie group leaves invariant a given system of ordinary differential equations this should help one to reduce the search of solutions to the study of a system of lower degree. The existence of such a group would imply certain properties of the solutions, reminding one of the role of the Galois group of an algebraic equation. But the analogy is rather weak: every algebraic equation has a Galois group, the structure of which gives essential insight into the properties of the roots, whereas most differential equations do not admit a non-trivial Lie group of transformations.

2. A point of view closer to the model of the Galois theory was proposed by Emile Picard for homogeneous linear differential equations, in a series of C.R. Notes and papers between 1883 and 1898, see [P2], and in the third volume of his "Traité d'Analyse" [P1]. Consider the ordinary homogeneous differential equation

$$(\mathcal{E}) \qquad \frac{d^n y}{dx^n} + p_1(x)\frac{d^{n-1}y}{dx^{n-1}} + \cdots + p_n(x)y = O$$

where the p_i's belong to some field F of functions meromorphic in a given region of the plane, stable under the taking of derivatives and containing $\mathbf{C}$. It has a "fundamental system of solutions" $y_1, \cdots, y_n$, i.e. a set of solutions linearly independent over $\mathbf{C}$. Then every other solution is a linear combination of the y_i's with constant coefficients. As was well-known, the condition for linear independence is that the Wronskian

$$W(y_1, \cdots, y_n) = \begin{vmatrix} y_1 & \cdots & y_x \\ y_1^{(1)} & \cdots & y_n^{(1)} \\ & \vdots & \\ y_1^{(n-1)} & \cdots & y_n^{(n-1)} \end{vmatrix}$$

[1]Cf the letter of Febr.3, 1874 in vol.V of S. Lie's Collected Papers, p.586.

be $\neq 0$. Here and below, $y_i^{(j)}$ stands for $\frac{d^j y_i}{dx^j}$. Let now

$$A = (a_i^j) \epsilon GL_n(\mathbf{C}).$$

Then

$$\overline{y}_j = \sum_i a_j^i y_i$$

is another fundamental set of solutions. We want to impose on A certain conditions analogous to those defining the Galois group of an algebraic equation. In that case it can be required that A is an automorphism of the splitting field of the equation or, equivalently, without using that notion, that it respects all algebraic relations between the roots. Here we use an analogue of that second formulation. Let

$$Y_i^{(j)} \qquad (i = 1, \cdots, n; j = 0, 1, 2, \cdots)$$

be indeterminates. Let $\mathcal{P}(\mathcal{E})$ be the set of polynomials $P(Y_i^{(j)})$ in the $Y_i^{(j)}$, with complex coefficients, such that $P(y_i^{(j)})$, viewed as a function of x, belongs to F. The condition imposed on A is then that $P(\overline{y}_j^{(j)})$ is the same function of x as $P(y_i^{(j)})$ for all $P \epsilon \mathcal{P}(\mathcal{E})$. These A's form the "transformation group of $(\mathcal{E})$", in E. Picard's terminology, to be denoted $G(\mathcal{E})$, or the "rationality group of $(\mathcal{E})$", in that of F. Klein. The latter also pointed out that if $n = 2$ and the equation has regular singular points, then $G(\mathcal{E})$ is the smallest algebraic group containing the monodromy group of the equation.[2]

To justify the use of "group", Picard pointed out only that $G(\mathcal{E})$ contains the product of any two of its elements. That $A \in G(\mathcal{E})$ implies $A^{-1} \in G(\mathcal{E})$ was shown later by A. Loewy [L].[3]

The group $G(\mathcal{E})$ is first of all a Lie group but Picard pointed out that it is moreover "algebraic" in the sense that the matrix coefficients are algebraic functions of suitably chosen parameters (rather that merely analytic), which led him to study some properties of such groups. In particular, he determined

[2]This is also true of a linear homogeneous equation with regular singular points of any degree, cf. F. Beukers, *Differential Galois theory*, in From number theory to physics, Waldschmidt et al eds. Springer, 1992, 413-439, Thm. 2.5.1. Here it is understood that $F = \mathbf{C}(z)$.

[3]There were precedents in Lie theory where this condition had been forgotten. However, Picard includes it when it comes to defining finite groups of substitutions (cf [P1], p. 452).

3

the structure of those of dimensions 1 or 2 (cf [96] in [P2]) and stated that the previous parameters can be chosen so that the matrix coefficients are rational functions of them ([P1], XVII, n^0 12) (i.e. that the group variety is unirational in present-day language).

The theory was further developed by E. Vessiot (1892 - 1904), see his account in [V], where earlier references are also given. His main result is in substance that if $G(\mathcal{E})$ is connected solvable ("integrable" in Lie's sense) if and only if $(\mathcal{E})$ is solvable by "quadratures". That notion is not precisely or consistently defined, as pointed out by Kolchin in [K10], and we shall come back to it later (see §6). Of course, it means adjoining a primitive but, apparently, the exponential of a primitive was also allowed.

3. The whole theory was somewhat obscure and the necessity of various clarifications or reformulations was felt already at the time. In fact, in giving the definition of $G(\mathcal{E})$ above, I have followed [L] and [V] rather than Picard himself. There was little further development until [K10]. There after summarizing the history of the subject and pointing out various ways in which it was lacking in precision or rigor, Kolchin states:

> *the purposes of the present paper are, first to develop a set of theorems on algebraic matric groups at least adequate to meet the demands of the Picard-Vessiot theory and second, to algebraize, rigorize, round out, and augment that theory*

The framework for the algebraization needed to discuss differential equations was already available, namely the differential algebra, developed by his teacher J. F. Ritt. On the group side, Kolchin felt the need to build a theory of algebraic matric groups, rather than rely on the theory of Lie groups, which he viewed as far deeper than the one to be developed. His paper then has three main parts.

A) algebraic matric groups (Chap. I)
B) Galois theory of differential fields (Chap. III)
C) The Picard-Vessiot theory (Chap. IV, V)

(Chap.II being devoted to preliminaries on differential algebra.)

Before coming to (A), let me mention that there was more at the time to the theory of algebraic matric groups than he implied. In particular L. Maurer had published several interesting papers (1890-94) in which he notably characterized the Lie algebras of algebraic groups and essentially proved that a group variety is rational (rather than unirational). His theory had been revived and generalized by Chevalley and Tuan, and then by Chevalley.

Since B) and C) are in characteristic zero, this might have filled Kolchin's needs. Whether he was aware of it or not I do not know, but the fact that it was not taken into account was ultimately a "good thing" because Kolchin decided that:

> *To emphasize the purely algebraic nature of the subject matter the proofs are carried out in manner valid for fields of non-zero as well as zero characteristic.*

so that Part (A) and the companion paper [K11] constitute in fact the birth certificate of the theory of linear algebraic groups over algebraically closed fields of *arbitrary* characteristic. In the following summary, I shall discuss both together.

4. Let C be an algebraically closed commutative field, p its characteristic. The notion of algebraic matrix group over C introduced by Kolchin is the nowadays usual one: a subgroup G of $GL_n(C)$ is algebraic if it is the set of all invertible matrices the coefficients of which annihilate a given set of polynomials in n^2 indeterminates with coefficients in C, i.e. G is the intersection of $GL_n(C)$ with a closed subvariety of $M_n(C)$, called by Kolchin the underlying manifold of G.

4.1. A first main result is that if G is connected and solvable, it can be put in triangular form, since then known as the Lie-Kolchin theorem. He also shows that G is solvable, as an abstract group, if and only it is solvable as an algebraic group: the series of the smallest algebraic groups containing the successive derived groups stops at $\{1\}$. (In fact, it was proved later that the abstract derived groups are automatically algebraic subgroups, cf. e.g. [B]).

4.2. If G is commutative, then it is the direct product of two subgroups G_s and G_u consisting respectively of the semisimple and unipotent elements in G. [An immediate consequence, not drawn by Kolchin, is that any linear algebraic group is stable under the (multiplicative) Jordan decomposition.]

4.3. Kolchin also introduces the notions of "anticompact" and "quasicompact" groups of matrices, i.e consisting of unipotent and semisimple matrices respectively. He shows that any anticompact group can be put in triangular form. If G is quasicompact, algebraic, connected, then it can be put in diagonal form (i.e. is a torus in the usual terminology). Also he shows that if G is connected, the k-th power map $x \mapsto x^k$ is a dominant morphism, if $(k,p) = 1$ in case $p \neq 0$.

4.4. These papers also contain some general comments about connected components, Jordan-Hölder series and algebraic subgroups. Altogether, they present the first body of theorems on linear algebraic groups with proofs insensitive to the characteristic of the groundfield, carried out in the framework of algebraic geometry, without any recourse to Lie algebras and the usual mechanism of Lie theory. They were influential on my own work, as can be seen from my first paper on this topic (Annals Math. **64** (1956), 20-82), in which these results are all proved again. They are also incorporated in [B] and belong to any systematic exposition of the theory.

5.1. For (B) and (C) the basic notion is that of a *differential field.* A (commutative) field F is differential if it is endowed with a derivation,[4] i.e. self-map δ satisfying

$$\delta(a+b) = \delta(a) + \delta(b), \ \ \delta(a.b) = \delta(a).b + a.\delta(b) \ \ (a, b \epsilon F).$$

Then

$$C_F = \{a \epsilon F, \delta(a) = 0\}$$

is a subfield, the *field of constants* of F. The nth derivative of a is $\delta^n.a$ and is written $a^{(n)}$. An *automorphism* of F is an automorphism of the underlying field commuting with the derivation.

In fact, this is an *ordinary* differential field. Kolchin develops the theory more generally for *partial* differential fields, i.e. fields endowed with a finite set of commuting derivations.

[4]Called in [K10] a derivative. From [K17] on, Kolchin switched to *derivation operator* and let *derivative operators* be the operators on F defined by the powers of δ (similarly for partial differential fields). I shall use derivation and derivatives.

In the sequel, unless otherwise stated, F is a differential field of characteristic zero with algebraically closed field of constants C_F or C. The field F is partial from § 7 on, ordinary before, unless otherwise stated.

5.2. Let $E \supset F$ be a field containing F as a subfield. It is a *differential extension* of F it is a differential field with derivation leaving F stable and inducing on F its given derivation. It is a *finitely generated differential extension* if there exist moreover finitely many elements $\eta_1 \cdots, \eta_l$ in E such that E is generated over F by the η_i's and their derivatives of all orders, in which case we write

$$E = F < \eta_1, \cdots, \eta_n > .$$

It is not necessarily a finitely generated field extension of F, i.e. its transcendence degree $\partial^0(E/F)$ over F may be infinite.

In the present context, the equation $(\mathcal{E})$ takes the form

$$(\mathcal{E}) \qquad L(y) = y^{(n)} + p_1.y^{(n-1)} + \cdots + p_n.y = 0, \ (p_i \epsilon F, i = 1, \cdots, n)$$

and we are looking for solutions y in some differential extension of F. The standard facts about the solutions of $(\mathcal{E})$ in the classical case recalled earlier generalize to the following theorem (see [K4] and [K10], §§ 14, 15).

Theorem. *There exists a finitely generated differential extension E of F containing n solutions $\eta_1 \cdots, \eta_n$ of $(\mathcal{E})$ such that*

$$W(\eta_1, \cdots, \eta_n) = \begin{vmatrix} \eta_1, & \cdots & , \eta_n \\ \eta_1^{(1)}, & \cdots & , \eta_n^{(1)} \\ \eta_1^{(n-1)}, & \cdots & , \eta_n^{(n-1)} \end{vmatrix} \neq 0.$$

In any differential extension E' of E, the η_i are linearly independent over $C_{E'}$ and span over $C_{E'}$, the space of solutions of $(\mathcal{E})$ in E'. No differential extension F' of F contains more than n solutions of $(\mathcal{E})$ linearly independent over the constants.

A set of n linearly independent solutions of $(\mathcal{E})$ is called a *fundamental set of solutions* of $(\mathcal{E})$.

5.3. A differential extension E of F is a *Picard-Vessiot extension* (a P.V. extension for short) of F if it is of the form

$$E = F < \eta_1, \cdots, \eta_n >$$

where the η_i form a fundamental set of solutions of an equation $(\mathcal{E})$ *and* if $C_E = C_F$.

It is clear from $(\mathcal{E})$ that the derivatives of all orders of the η_i's are linear combinations with coefficients in F of the $(n-1)$ first ones. Therefore $\partial^0(E/F)$ is finite.

Kolchin proved later that given $(\mathcal{E})$, one can find a fundamental set of solutions $\eta_1, \cdots, \eta_n$ in some extension such that $F < \eta_1, \cdots, \eta_n >$ has the same field of constants as F (cf [K12] or [K29], Prop. 13, p.412). This is under our standing assumption that C_F is algebraically closed (of char. 0). If not, then Seidenberg has produced an equation $(\mathcal{E})$ such that $C_E \neq C_F$ for all differential field extensions E generated over F by a fundamental set of solutions of that equation (cf. [S] or [K29], Exercise 1, p.413).

5.4. Let E be a P.V. extension of F. Then, by definition, its *Galois group over* F is the group $G = G(E/F) = Aut(E/F)$ of automorphisms of E (as a differential field) leaving fixed each element of F. Let $g \epsilon G$. Then

$$g.\eta_i = \sum_j c_i^j(g).\eta_j$$

where the $c_i^j(g)$ belong to C_E at first, hence to C_F since both are assumed to be equal. The map

$$g \mapsto c(g) = (c_i^j(g))$$

is a faithful linear representation of G into $GL_n(C_F)$.

Before stating the main theorem concerning $G(E/F)$, we note that if E_1 is a differential field intermediate between E and F then its field of constants is equal to C_F, since $C_E = C_F$. Therefore, it is clear from the definition that E is a P.V. extension of E_1 and $G(E/E_1)$ is defined as before.

5.5. Theorem. *Let E be a P.V. extension of F. We keep the previous notation.*

(i) The group $c(G)$ is the group of rational points over C_F of an algebraic subgroup of $GL_n(C_F)$ defined over C_F, of dimension equal to $\partial^0(E/F)$.

(ii) The map which assigns to an algebraic subgroup H of $c(G)$ the subfield E^H of elements fixed under H is a bijection between algebraic subgroups and intermediate differential subfields. The inverse bijection assigns to a differential subfield E_1 the group $G(E/E_1)$ of automorphisms of E over E_1.

(iii) H is a normal subgroup if and only if $\sigma(E^H) = E^H$ for all $\sigma \epsilon G(E/F)$. In that case, the restriction of an automorphism to E^H defines an isomorphism of G/H onto $Aut(E^H/F)$.

See §§ 17 to 20 in [K10]. There F is ordinary. The generalization to partial differential fields is carried out in [K15]. We indicate briefly how Kolchin define P.V. extensions in the general case. To this effect he first remarks that $p_i = \pm W_i/W_0$, where

$$W_i = \det\,(\eta_k^{(j)}), (1 \leq i, k \leq n; 0 \leq j \leq n, j \neq n-i)$$

by Cramer's rule. Given now n elements $y_1, \cdots, y_n$ in a partial differential field E and partial derivatives $\theta_1, \cdots, \theta_n$, let

$$W(\theta_1, \cdots, \theta_n, y_1, \cdots, y_n) = \det\,(\theta_j y_i).$$

If the y_i's are linearly independent over constants, there is always a choice of partial derivatives of orders $< n$, say $\sigma_1, \cdots, \sigma_n$ such that

$$W(\sigma_1, \cdots, \sigma_n; y_1, \cdots, y_n) \neq 0. \tag{1}$$

([K15]), lemma 1). Assume now F to be a partial differential field. A differentiable extension of E of F is a P.V. extension if

$$C_E = C_F, E = F < \eta_1, \cdots, \eta_n >,$$

where the η_i are linearly independent over constants and

$$W(\theta_1, \cdots, \theta_n; \eta_1, \cdots, \eta_n)/W(\sigma_1, \cdots, \sigma_n; \eta_1, \cdots, \eta_n) \epsilon F$$

for all choices $\theta_1, \cdots, \theta_n$ of partial derivatives of orders $\leq n$, the σ_i's being a set of partial derivatives of orders $< n$ such that (1) is satisfied by the η_i's.

The initial remark above shows readily that this definition is equivalent to the original one in the ordinary case. For other characterizations and further discussion of P.V. extensions, see [K29], VI, $n^0 6$, and the exercises following it, pp.409-418.

In [K10], Kolchin does not raise the question as to whether E^H in (*iii*) is a P.V. extension of F. Later on, he provided a positive answer in [K18], III, n^0 3, Corollary to Theorem 2 (see also 9.2 below).

6. The last sections of the paper (§§ 23 to 27) are devoted to a precise version and generalization of Vessiot's theorem about equations solvable by "quadratures". Consider first two special cases:

6.1. The primitives of $a \epsilon F$, i.e. the solutions of $y' = a$, differ by a constant. The equation $y' = a$ is not homogeneous but any solution η_2 and the constant $\eta_1 = 1$ span the space of solutions of

(1) $$y'' - (a'/a).y' = 0.$$

The extension $E = F < \eta_2 >$ is P.V. If $g \epsilon G(E/F)$ then

$$g.\eta_1 = \eta_1, \qquad g.\eta_2 = \eta_2 + c = \eta_2 + c.\eta_1,$$

so that G can be identified to the group of upper triangular unipotent matrices in $\mathbf{GL}_2(C_F)$, i.e. to the additive group $\mathbf{G}_a$ of C_F.

6.2. Let now $E = F < \eta >$, where η is a solution of

$$y' - a.y = 0.$$

Then $\eta = \exp b$, where b is a primitive of a. Any other fundamental solution is a non-zero multiple of η, whence an isomorphism of $G(E/F)$ onto a subgroup of $\mathbf{GL}_1$. There are then two cases:

a) η is transcendental, which is equivalent to $G(E/F) = \mathbf{GL}_1$,

b) η is algebraic. Then $G(E/F)$ is finite cyclic, of some order h, and η is an h-th root of an element of F.

The results of §27 imply conversely that if E is a P.V. extension with a Galois group of one of the three previous types, then we are in the corresponding case.

6.3 From this it follows first that a P.V. extension has a solvable Galois group if and only it can be obtained from F by a succession of steps 6.1, 6.2. In fact, Kolchin carries out the discussion a bit more generally so as to allow for more general algebraic extensions.

An extension $E = F < \alpha_1, \cdots, \alpha_q >$ of F is said to be *Liouvillian* if $C_E = C_F$ and α_{i+1} is either algebraic over $F < \alpha_1, \cdots, \alpha_i >$ or a primitive or an exponential of a primitive of an element in that field ($i = 0, \cdots, q - 1$). Let now E be a P.V. extension of F. Then (§25, Theorem), if E is contained in a Liouvillian extension, then $G(E/F)$ has a solvable identity component. Conversely, the latter property implies that E is Liouvillian. More precisely, Kolchin distinguishes ten types of Liouvillian extensions and ten types of algebraic groups and show that they match in case of P.V. extensions contained in a Liouvillian extension.

7. The above sections 5,6 summarize part C in my initial division of [K10]. Part B is Kolchin's first attempt at a Galois theory of differential fields, not necessarily associated to differential equations and is much more tentative. To this effect, he has to introduce a notion of normal extension of F, but can prove only part of the sought for Galois correspondence. See the introduction of [K16] for a discussion of the "blemishes" of that theory. There he defines the concept of "strongly normal extension", which was to be the basis of all his further work on Galois theory, so I shall directly go over to it. The main results are proved first in [K16], [K18] and in the paper [K19], written jointly with S. Lang. The whole theory was given a comprehensive exposition in [K29].

8.1. Kolchin first introduces a *universal field* in analogy with the universal field in Weil's Foundations of Algebraic Geometry [W].

A differential extension U of F is a *universal extension* if given a finitely generated differential extension F_1 of F in U and a finitely generated differential extension E of F_1, there exists an isomorphism of E into U over F_1. It is algebraically closed and its field of constants C_U is a universal field for C_F in Weil's sense, (i.e. algebraically closed, of infinite transcendence degree over C_F). Then nothing is lost by restricting oneself to finitely generated extensions of F contained in U and we shall do so. (Cf [K16 : I, no.5], and [K29: III, No.7] for an extension to arbitrary characteristics.)

8.2. Let now E be a finitely generated differential extension of F. Wanted is a condition of normality. The most obvious one is that E be stable under any automorphism of U over F. However, Kolchin noticed that this forces E to be algebraic over F, hence is too strong, so that isomorphisms of E into U over F have to be allowed. In order to avoid endless repetitions of "into U", I shall follow Kolchin's convention and call this simply an isomorphism of E over F, (whereas an "automorphism" of E over F is of course an isomorphism of E onto itself fixing F pointwise). However, they must be restricted in some fashion. On the other hand, whatever the definition of normal extension is, a P.V. extension has to be one. Assume then that $E = F < \eta_1, \cdots, \eta_n >$ is a P.V. extension, where the η_i are a fundamental set of solutions of the underlying equation $L(y) = 0$. Let σ be an isomorphism of E over F. Then the $\sigma.\eta_i$ form a fundamental set of solutions in $\sigma(E)$, therefore

$$\sigma.\eta_i = \sum c_i^j \eta_j, (i = 1, \cdots, n)$$

where the c_i^j belong to the constant field of $\sigma(E)$, hence also to C_U. Therefore

$$\sigma(E) \subset E.C_U.$$

Kolchin saw that this was the decisive restriction and he introduced the following definitions:

Definitions.

(i) An isomorphism σ of E over F is *strong* it satisfies the two conditions

(1) $\sigma \mid C_E = Id.$

(2) $\sigma(E) \subset E.C_U$ and $E \subset \sigma(E).C_U$ i.e. $E.C_U = \sigma(E).C_U$.

(ii) E is a *strongly normal extension* of F if it is finitely generated over F (as a differentiable field) and every isomorphism of E over F is strong.

See [K29], VI, n° 2,3. These definitions appear first in [K16] except that (1) is not needed since $C_E = C_F$ by a standing assumption there (on p.772).

If E is strongly normal over F, then it is clearly so over every differential subfield containing F.

In the sequel, $C(\sigma)$ denotes the field of constants of the field $E.\sigma(E)$. The condition (2) may also be written

$$E.C(\sigma) = E.\sigma(E) = \sigma(E).C(\sigma). \tag{3}$$

If E is strongly normal, $C_E = C_F$ ([K29], V, Prop. 9, p.393) and E has finite transcendence degree over F ([K29] VI, Prop.11, p.394; the latter statement is also contained in Theorem 2, III, $n^0 2$ of [K16]).

8.3. *The Galois group.* If E is any finitely generated differential extension of F, then it is linearly disjoint from C_U over its own field of constants C_E of E (see [K16], I, $n^0 4$, Prop.3 and the comment following it, or [K29], II, n^0 1, Cor.1, p.87). As a consequence, a strong isomorphism of E over F extends uniquely to an automorphism of $E.C_U$ fixing C_U pointwise, i.e. to an automorphism of $E.C_U$ over $F.C_U$. Conversely, if τ is an automorphism of $E.C_U$ over $F.C_U$, then its restriction to E is a strong isomorphism of E over F which determines τ completely. Assume now E to be strongly normal. Then every isomorphism of E over F is strong, by definition, whence a canonical bijection

$$\{Isom\ E/F\} = \text{Aut}\ (E.C_U/F.C_U),$$

which allows one to view the left-hand side as a *group*. By definition, this is the Galois group $G(E/F)$ of E over F.

8.4. In the sequel we write C for C_F and C_U is viewed as the universal field for C, i.e. algebraic varieties defined over C are assumed to have their points in C_U. If M is an irreducible one, then the residue field $C(x)$ of $x \epsilon M$ at x is the subfield of C_U generated over C by the values at x of the rational functions on M which are defined at x.

We now give a first, provisory as far as (i) is concerned, formulation of the fundamental theorem of Kolchin's Galois theory of strongly normal extensions:

Theorem. *Let E be a strongly normal differential extension of F.*

(i) There exists an algebraic group $\mathbf{G}$ *defined over C such that $G(E/F)$ may be identified with $\mathbf{G}(C_U)$ so that $C(\sigma)$ is the residue field of $\sigma \epsilon G(E/F)$. For a finitely generated extension K of C in C_U, this isomorphism identifies*

13

$\mathbf{G}(K)$ *with* $\{\sigma \epsilon G(E/F) \mid \sigma(E) \subset E.K\}$. *In particular,* $\mathbf{G}(C_F) = Aut(E/F)$. *The dimension of* $\mathbf{G}$ *is equal to* $\partial^0(E/F)$.

(ii) The map which assigns to an algebraic subgroup H *of* $G(E/F)$ *defined over* C *the field* E^H *of invariants of* H *establishes a bijection between algebraic* C*-subgroups of* $G(E/F)$ *and intermediate differential fields between* F *and* E. *The inverse bijection assigns to such a field* F_1 *the subgroup of* $G(E/F)$ *of elements of* $G(E/F)$ *fixing* F_1 *pointwise.*

(iii) The C*-subgroup* H *is normal if and only if* E^H *is strongly normal over* F. *In that case, the restriction of* $\sigma \epsilon G(E/F)$ *to* E^H *defines an isomorphism of algebraic groups of* $G(E/F)/H$ *onto* $G(E^H/F)$.

For (*ii*) and (*iii*), see [K29] VI, $n^0 4$, Theorem 3,4 respectively. It is also shown in the latter that the conditions in (*iii*) are equivalent to either of:

(a) For each element α of E^H not contained in F, there exists a strong isomorphism of E^H over F not leaving α fixed.

(b) $\sigma(E^H) \subset E^H C_U$ for every $\sigma \epsilon G(E/F)$.

We shall soon come to (*i*) and to what is already proved in [K16] and [K18]. Before doing so, we list some complements to the fundamental theorem. We write G for $G(E/F)$ and view it as an algebraic group over C.

9.1. The group $G(E/F)$ is not necessarily connected. In fact the fixed point set of the identity component G^0 of G is the algebraic closure F^0 of F in E and we have

$$[F^0 : F] = [G : G^0].$$

9.2. In this theorem, the algebraic groups are not necessarily linear. In fact, Kolchin proves that E is P.V. if and only if $G(E/F)$ is linear. The new implication is of course that $G(E/F)$ linear implies E to be P.V. (see [K18], III, n^0 3, Theorem 2, p. 891 or Cor.2 to Theorem 8, p.427 in [K29]). Now, given a linear algebraic group G and a closed normal subgroup N there always exists a rational representation of G with kernel N (a by now standard fact, proved first in [K14]); it follows from this characterization that if E is P.V., any strongly normal extension of F in E is also P.V. ([K18], Cor. to Theorem 2, p.891). A slightly more direct way to derive the corollary from the theorem is also hinted at in exercise 2, p.427 of [K29].

9.3. Assume now that G is connected of dimension one. The two cases where it is linear were discussed in 6.1, 6.2. There remains the one where G is an elliptic curve. This corresponds to the case where E is *Weierstrassian* over F, first discussed in [K16] III $n^0 6$ p. then also in [K18] III. $n^0 3$ and in [K29], VI $n^0 5$. Briefly, $E = F < \alpha >$, with α transcendental and there exist $g_2, g_3 \epsilon C_F, a \epsilon F$ such that

$$g_2^3 - 27g_3^2 \neq 0, (\alpha')^2 = a^2(4\alpha^3 - g_2\alpha - g_3).$$

9.4. By the Theorem, every Galois group is a group variety over C. As a step towards a converse, Kolchin shows that given a connected group variety over C, there exists a differential field with field of constants C and a strongly normal extension of that field with Galois group G. This is still far from a solution to the "inverse problem" of Galois theory, namely, given F, to determine the possible Galois groups of strongly normal extensions of F.[5]

9.5. Let G be connected. Then by the Chevalley-Barsotti structure theorem G has a biggest normal linear subgroup N such that G/N is an abelian variety (see e.g. [C], a proof is also given in [K29], V, n^0 24). Let $E' = E^N$. Then, in view of 9.2 E' is the smallest strongly normal extension of F in E such that E is P.V. over E', and $G(E'/F)$ is an abelian variety.

9.6. in [K19] and in the last section of [K29] the realization of strongly normal extensions by function fields of principal homogenous spaces is studied. Let G be a connected group variety over C. A variety V over F is a principal homogenous space (over F) for G if G acts simply transitively on it. Given $v \epsilon V$, the map $g \mapsto v.g$ is an isomorphism of varieties over $F(v)$. There is always such a v which is algebraic over F. Then G operates over $F(V)$ by right translations. It is shown that a strongly normal extension of F with Galois group G may be so realized.

10.1. We now come back to (i) in the fundamental theorem (8.4) and to the reason why its formulation had been called provisory. It is correct as stated and references will be provided. Still, it misrepresents an important aspect

[5]Kolchin has not contributed further to the inverse problem, which puts it outside the purview of this talk. For a historical survey, see the paper by M. Singer in this volume.

of Kolchin's approach. The assertion "is isomorphic to an algebraic group over C" refers implicitly to the standard notion of algebraic group, namely, an algebraic variety over C endowed with a group structure such that inverse and product are defined by C-morphisms of varieties. In the sequel, I shall refer to those as group varieties over C. But Kolchin proceeds differently: he first defines a structure on $G(E/F)$ analogous to that of an algebraic group, in which the fields $C(\sigma)$ play the role of residue fields, so that (i) is in fact the combination of two statements:

(a) The Galois group $G(E/F)$, endowed with the fields $C(\sigma)$, is an algebraic group in Kolchin's sense.

(b) $G(E/F)$ may be identified to a group variety over C_F so that $C(\sigma)$ is the residue field at σ $(\sigma \epsilon G(E/F))$.

10.2. To describe this, let me first recall, however briefly, some concepts familiar in [W], which play a great role in Kolchin's approach.

Let K be an algebraically closed groundfield. All extensions of K are to be contained in a fixed universal field. Let V be a variety over K. Then, as already recalled, every point $x \epsilon V$ has a residue field $K(x)$, a finitely generated extension of K. The points of V are partially ordered by a notion of specialization over K, noted $x \underset{K}{\rightarrow} x'$ or simply $x \rightarrow x'$. [If x, x' are in an open affine subset, every regular function vanishing at x also vanishes at x']. It implies that $\partial^0(K(x)/K) \geq \partial^0(K(x')/K)$. The specialization is *generic* if there is equality, in which case $x' \rightarrow x$, *non-generic* otherwise. The locus Z_x of x over K is the set of specializations (over K) of x. It is the smallest K-irreducible subset of V containing x, and x is a generic point of Z_x over K.

If W is another variety over K, a map $f : V \rightarrow W$ is a morphism if $f^0(K(f(x)) \subset K(x)$ for all $x \epsilon V$ and if it is compatible with specializations.

10.3. We now return to [K16] and the strongly normal extension E of F, let G^* be the group defined by the strong isomorphisms of E over F (see 8.3) and G the subgroup of automorphisms of E over F. By definition, G^* is the Galois group of E over F, but the main results of [K16] pertain to G.

Kolchin introduces the notion of specialization (over C) for isomorphisms of E over F and shows that the specialization of a strong one σ is strong

(II, n^0 5, Prop. 5, p.774). The strong isomorphism σ is isolated if it not a non-generic specialization of another strong isomorphism. G^* has finitely many isolated elements, up to generic specialization. This, and some further properties of strong isomorphisms allow Kolchin to define on G^* a structure of algebraic group and morphisms of such groups with the usual properties. Then G is endowed with the induced structure of algebraic group, in which the algebraic subgroups are the intersections of G with algebraic subgroups of G^* defined over C, i.e. in fact the groups of rational C-points of such subgroups. It is this notion of algebraic subgroup which underlies the statement of the fundamental theorem in [K16], which deals only with G. (See n^0s 2,3 in III). The question as to whether G^* may be naturally identified with a group variety over C, in Weil's sense, is raised in [K16: p.759] and positively answered in [K18], II, n^0 1, Thm. 1, p.878. More precisely, the fields $C(\sigma)$ are the residue fields in the group variety structure. This then realizes (*a*) and (*b*) in 10.1 and yields a proof of (*i*) in 8.4.

In [K19] this identification is used to put the discussion of realization via principal homogeneous spaces (see 9.6) in the framework of the usual algebraic group theory. But Kolchin comes back in [K29] to his original point of view and develops it much more systematically.

10.4. Kolchin's treatment of "algebraic sets and groups" is the subject matter of Chapter V in [K29]. The references in this section and the two following ones are to that Chapter.

Let K be as in 10.2. The starting notion is that of pre-K set ($n^0 2$). The set V is a pre$-K$ set if to each point $x \epsilon V$ is assigned a finitely generated extension $K(x)$ of K, subject to relations of specializations, satisfying certain natural conditions. V is a K-group if moreover it is a group and we have

$$K(x.y) \subset K(x).K(y) \qquad K(x^{-1}.y) \subset K(x).K(y), \ (x, y \epsilon V)$$

and some further conditions relative to specializations. V is a homogeneous K-space if it is a pre-K-set acted upon transitively by a K-group G, again with certain natural conditions (p.220). A K-set is a pre-K-subset of a homogeneous K-space. Thus, in this algebraic geometry, the algebraic sets (the K-sets) are all contained in homogeneous spaces. Kolchin then develops his theory along lines familiar in other contexts : effect of extensions of the groundfield, Zariski topology, regular extensions and, for K-groups: connected components, products, morphisms, quotients (to be discussed below) linear groups, Galois cohomology, ending up with a proof of the Chevalley-Barsotti theorem mentioned in 9.5.

10.5. We now want to relate those concepts to the corresponding ones in algebraic geometry. There, in any treatment familiar to me, one starts with a notion of affine variety (or scheme) and the varieties (or schemes) are by definition locally affine. Kolchin's approach is *a priori* very different as we saw. However, a product of $\mathbf{G}_a$'s, i.e. the additive group of a vector space over K, is homogeneous and its K-subsets are included in Kolchin's set up. They are essentially affine varieties in disguise and provide the link between the two points of view.

A K-affine set is by definition a K-set which is isomorphic to a K-subset of some $\mathbf{G}_a^n$. The main point is then to show that in a K-set A, any finite set of points is contained in an open affine K-subset, from which it follows that A has a finite open cover by affine K-subsets (n^0 16, Thm. 4 and Cor., p.311). This then implies that Kolchin's K-groups or homogeneous K-spaces may be naturally identified with group or homogeneous varieties (*loc. cit.* Remark p.311-312). [In view of this identification, Thm.4 quoted above is not surprising since it is known that any homogeneous variety is quasi-projective (Chow's theorem).]

10.6. Kolchin achieves in this way autonomy, at the cost of a huge (his word) chapter on algebraic groups. Whether it is worth the effort or whether a shorter treatment of the Galois theory using the standard theory of algebraic groups could be given, I will not try to assess, but I would like to mention two points, one minor, one major, in which Kolchin's approach has paid off. The former is the construction of the quotient space G/H of a K-group G by a closed K-subgroup H. Wanted is a structure of K-variety on G/H with a universal property, such that $\pi : G \to G/H$ is a separable K-morphism commuting with the action of G. For linear groups, there is a simple trick to achieve this (see [B]). For non-linear groups the original approach of Weil [W1] is rather awkward. The usual method nowadays, which also proceeds from the local to the global, is to start from the existence of the quotient of some invariant Zariski-open set, proved by M. Rosenlicht, and to use translations (see [Ss]). From Kolchin's point of view, it is straightforward and directly global. The starting point is Theorem 4, p.240 according to which, given a closed subset A of a homogeneous K-space, there is a smallest field $L \supset K$ such that A is a L-subset. It is finitely generated over K and denoted $K(A)$. Then, Kolchin assigns to $x \epsilon G/H$ the field $K(\pi^{-1}(x))$. If $x' \epsilon G/H$, then $x \to x'$ if there exist $y \epsilon \pi^{-1}(x), y' \epsilon \pi^{-1}(x')$ such that $y \to y'$. It is then rather easily seen that his axioms for a homogeneous K-space are satisfied and that π is separable (n^0 11).

10.7. The second point reaches further, to the last major project of Kolchin: the foundation of a theory of differential algebraic groups i.e., roughly speaking, groups locally defined by algebraic differential equations [K35]. In the seventies, one of affine or linear differential algebraic groups had been developed by P.J. Cassidy, but Kolchin decided to supply the foundations for a general theory. In spite of many technical complications, he could use Chapter V of [K29] almost as a blueprint: his axiomatic treatment there could be adapted to this more general situation and in fact, a number of sections could be taken almost verbatim, with minor changes in notation. It initiates more broadly a "differential algebraic geometry", at any rate for subsets of homogeneous spaces, as in [K29].

In the linear case, the theory was pursued by P.J. Cassidy and led to a classification of semisimple differential groups and Lie algebras [Ca]. On the other hand, A. Buium applied similar ideas to the study of diophantine problems over function fields, thus establishing completely new connections with other topics in mathematics. This whole area is in full development and well worth a report, which I have unfortunately neither the time nor the competence to give.[6] Still, I wanted to mention it as a counterpart to the main topic of my talk: the latter is a well-rounded theory, with statements of a considerable aesthetic beauty, seemingly in final form. The former opens up new directions and points out to work for the future.

[6]but for which I can now refer to the paper by A. Buium and P. Cassidy in this volume.

REFERENCES

The symbol [Kx] refers to item x in the bibliography of E. Kolchin's work, at the end of this volume.

[B] A. Borel, Linear algebraic groups, 2nd enlarged edition, GTM **126**, Springer 1991.

[Ca] P.J. Cassidy, *The classification of the semisimple differential algebraic groups and the linear semisimple differential algebraic Lie algebras*, Jour. Algebra **121** (1988), 169-238.

[C] C. Chevalley, *Une démonstration d'un théorème sur les groupes algébriques*, Jour. Math. pur. appl. (9) **39** (1960), 307–317.

[L] A. Loewy, *Die Rationalitätsgruppe einer linearen homogenen Differentialgleichung*, Math. Annalen **65** (1908), 129-160.

[P1] E. Picard, Traité d'Analyse, t. III, Gauthier-Villars, Paris 1898 or 1908.

[P2] E. Picard, Oeuvres, t.2, éditions du CNRS, Paris, 1979.

[Se] A. Seidenberg, *Contribution to the Picard-Vessiot theory of homogeneous linear differential equations*, Amer. J. Math. **78** (1956), 808-817.

[Ss] C. S. Seshadri, *Some results on the quotient space by an algebraic group of automorphisms*, Math. Annalen **149** (1963), 286-301.

[V] E. Vessiot, *Méthodes d' intégration élémentaires*, Encycl. des sci. math. pures et appliquées, tome II, Vol. 3 (1910), pp.58-170.

[W] A. Weil, Foundations of algebraic geometry, A.M.S. Colloquium Publ. **29**, New-York, 1946.

[W1] A. Weil, *On algebraic groups and homogeneous spaces*, Amer. J. Math. **77** (1955), 493-512.

School of Mathematics, Institute for Advanced Study, Olden Lane, Princeton, NJ 08540

20

173.

Henri Poincaré and Special Relativity

Enseignement Math. **45** (1999), 281–300

1. In the later part of the last century and at the beginning of the present one, physicists had great difficulties in reconciling rational mechanics and electromagnetism. A solution was presented by A. Einstein in 1905, in the framework of a theory which was gradually accepted and given about ten years later (by Einstein) the name of *special relativity.* However, other people had worked and published earlier on these questions, most importantly for us in this paper H. A. Lorentz and H. Poincaré. The relationships between the contributions of Lorentz and Poincaré, or of Lorentz and Einstein are fairly well established, but this is not so for those of Einstein and Poincaré. In fact, for many years, there has been (and still is) some controversy as to how much Poincaré had anticipated or independently developed special relativity. Much has been written on this topic and many opinions have been expressed. To start with, I'll mention three (others will follow later):

2. *a*) In 1953, Sir Edmund Whittaker published the second volume of his *"History of Aether and Electricity"* [W]. It contains a chapter on these matters, entitled "The relativity theory of Lorentz and Poincaré". The 1905 paper of Einstein [E1] is mentioned (p. 40): it sets forth the relativity theory of Lorentz and Poincaré, with some amplifications. It also introduces the postulate of the constancy of the speed of light, which was widely accepted at first and heavily criticized later.

The second 1905 paper by Einstein on relativity [E2], containing the relation $E = mc^2$, is quoted (p. 52) as introducing a special case of that formula, already anticipated by Poincaré, stated in full generality first in 1908 by C. N. Lewis.

b) In his book on Einstein: *"Subtle is the Lord..."* [P1], published in 1982, A. Pais states (p. 21) that *Poincaré never understood the basis of special relativity*. (See footnote [7]) below for a less abrupt statement.)

c) The mathematician and mathematical physicist Shlomo Sternberg, who did not agree with that assessment, wrote in an unpublished report (1983) he had asked for the opinion of some of his physicist friends who have looked into the subject: "*most feel H. Poincaré had indeed all or almost all of the basic concepts of special relativity in advance of or independently of Einstein*".

3. In this lecture, I shall try to sketch the development of these ideas, and to present, in particular, the information known to me on the contributions, views and mutual influences of these three scientists. The literature on this topic is so abundant that practically all opinions have been held and I do not pretend to offer a new one, all the more since the background material is known. Most of it is mentioned or discussed in several places, notably in the books of Pais [P1], A. Fölsing [F2], and A. Miller [M1]. A further difficulty in this discussion stems from the relative rarity of statements by Poincaré on Einstein, and vice versa. In particular, I do not know of any officially recorded one by Poincaré on Einstein and special relativity (or, as he would have said, the *new mechanics*).

4. In the second half of the 19th century, there was general agreement on the wave nature of light: it propagates in a medium, the aether, with the same speed c in all directions for an observer at rest with respect to the aether. However, nobody knew what the aether was, nor how to ascertain to be at rest with respect to it. But, owing to the revolution of the earth around the sun, one could hope to detect a relative uniform motion of a laboratory on earth with respect to the aether. The most decisive experiment in this respect was carried out by Michelson and Morley in 1887; it was negative. It turned out to be impossible to observe any optical effect of the earth's motion relative to the aether, at any rate up to second order in v/c, where v is the assumed speed of the earth with respect to the aether.

Around 1890 the Irish physicist G. Fitzgerald and, somewhat later, independently, Lorentz proposed an explanation: "the contraction hypothesis". Recall that in the Michelson-Morley experiment two rays of light go back and forth on two rods of the same length, ℓ, at right angles, one $\overline{OA}$ moving with speed v with respect to the aether, while the other one $\overline{OB}$ is assumed to be perpendicular to the direction of motion relative to the aether, and then are brought to interfere. According to the Newtonian addition of speeds, there should be a difference between the travelling times taken by these two rays, but the experiment did not show any. However, if it is assumed that the rod $\overline{OA}$ is shortened by the factor $(1 - v^2/c^2)^{1/2}$, then the times are indeed equal. Lorentz introduces and discusses this assumption in §§89–92 of his monograph [L3].

5. In 1898, H. Poincaré published a remarkable paper: "The measure of time" [P6]. Each human being has a feeling of a flow of time, a succession of events, which appears to reflect an absolute time. But how can one define that objective time? Poincaré singles out two difficulties. First, we have no direct intuition of the equality of two intervals of time. However, this had already been pointed out and he even gives some references. But a second point had not attracted much attention so far: to say that two events at different places occur at the same time has no objective meaning. Conventions on the measure of time are needed. Simultaneity is not an absolute concept. It is difficult to separate the qualitative problem of simultaneity from the quantitative problem of the measure of time, no matter whether a chronometer is used, or whether account must be taken of a velocity of transmission, such as that of light, because such a velocity could not be measured without *measuring* a time.

I must say that when I studied special relativity, over fifty years ago, this point eventually appeared to me as a most crucial one (and one which could be grasped without any formula). Apparently, this is indeed what had also happened to Einstein in 1905, as I shall recall in a moment, but the conclusion of Poincaré indicates that he did not attach a particularly fundamental importance to it:

> To conclude: We do not have a direct intuition of simultaneity, nor of the equality of two durations. If we think we have this intuition, this is an illusion. We replace it by the aid of certain rules which we apply almost always without taking account of them.
>
> But what is the nature of these rules? No general rule, no rigorous rule; a multitude of little rules applicable to each particular case.
>
> These rules are not imposed upon us and we might amuse ourselves in inventing others; but they could not be cast aside without greatly complicating the enunciation of the laws of physics, mechanics and astronomy.
>
> We therefore choose these rules, not because they are true, but because they are the most convenient, and we may recapitulate them as follows: "The simultaneity of two events, or the order of their succession, the equality of two durations, are to be so defined that the enunciation of the natural laws may be as simple as possible. In other words, all these rules, all these definitions are only the fruit of an unconscious opportunism."

In the same vein, it seems rather telling to me that when Poincaré included this article in [P10], he did not indicate the original reference. In fact when I read it, also some fifty years ago, I wondered why Einstein was not mentioned at all, not realizing that that particular chapter had already been published in 1898.

6. In 1902, Einstein settled in Berne. After some time, he was hired as a 3rd class expert at the Patent Office. Before, in order to make ends meet, he had given some private lessons. One of his (few) students was a Rumanian, Maurice Solovine, who was more interested in general knowledge, philosophy, than in mathematics or physics *per se.* Their conversations went quickly beyond the framework of private lessons, they became friends and Solovine later proposed to meet regularly to read and discuss together some important books. Together with another friend of Einstein's, the mathematician C. Habicht, they created to that effect what they called the "Akademie Olympia". In later reminiscences on the Akademie Olympia, in the introduction of [S3], Solovine listed some of the books they had read. He singled out one which "deeply impressed us and kept us breathless for weeks on end" ("*un livre qui nous a profondément impressionnés et tenus en haleine pendant de longues semaines*"), namely, a collection of essays published by Poincaré in 1902 [P8]. It consists of various articles dealing with mathematics or physics. There is no precise record of what they actually read, but it is very likely that they devoted much attention to the two chapters most relevant for us. One, "Theories of modern physics", is part of the text of a lecture given at an international congress in Paris, 1900, though this is not indicated there. Poincaré discusses the contraction hypothesis but would like a general principle according to which it would not be possible to detect relative motion with respect to the aether. After describing the hypothetical properties of that medium, he asks whether it really exists.

Another paper they surely read, too, is "Classical Mechanics", at the beginning of which Poincaré stresses four points:

1) There is no absolute space, only relative motions.
2) There is no absolute time. The equality of two time intervals has no meaning by itself.
3) We have no direct intuition of the simultaneity of two events occurring at different locations, and he refers to [P6] for a discussion.
4) Euclidean geometry is only some sort of convention of language. Non-euclidean geometry would be less convenient, but equally legitimate.

Points 1), also stressed by E. Mach, and 4) no doubt left a mark.

Einstein later acknowledged that he owed that insight to the sharp-witted ("*scharfsinnig*") Poincaré. However it does not seem that 2) and 3) struck a chord at that time. At any rate, they were not referred to (by Einstein, or anyone else for that matter), at that time.

7. It seems that these were the only works of Lorentz or Poincaré known to Einstein when he wrote [E1]. Specifically, a letter to C. Seelig written in 1955 ([S2], p. 116) implies that he did not know [L4], [L5], [P9] at the time. ("... *as far as I am concerned, I knew only Lorentz's important work of 1895, but not his later one nor the subsequent work by Poincaré. In that sense, my work was independent.*") In the early fifties, he also told A. Pais that he had never read [P12] (see [P1], p. 171). I shall now discuss briefly the contributions of Lorentz and Poincaré or, rather, only that part which is most relevant here, up to 1906, it being understood that Einstein became aware of them only later.

In 1899 Lorentz published another paper on the contraction [L4]. He realized that for the equations of electromagnetism to remain valid in a frame in uniform relative motion v (in the x-direction), one needed not only to make the change of spatial coordinates

$$(1) \qquad x' = (x - vt)(1 - v^2/c^2)^{-1/2}, \qquad y' = y, \qquad z' = z$$

but also to define a new time in the moving frame. As such, he proposed

$$(2) \qquad t' = t - vx/c^2 .$$

He had already introduced it in [L3], pp. 49–50 and had called it "local time", in opposition to the "general time" t because it depends on the spatial coordinate x, whereas t does not. This *local time* was (and remained) for him a mathematical device to perform computations in the moving frame. There was still only one true time, the absolute time t.

In 1904, in the lecture [P9], Poincaré discusses the local time (an "ingenious idea"). He gives it a physical meaning by prescribing a rule to synchronize clocks, and states a relativity principle, but justified, or explained, by the contraction hypothesis. He also speculates on a new mechanics, in which no speed could exceed c.

In [L5], Lorentz now modifies (2) and introduces the change of coordinates

$$(3) \qquad x' = (x - vt)(1 - v^2/c^2)^{-1/2}, \qquad t' = (t - vx/c^2)(1 - v^2/c^2)^{-1/2}$$

$$(4) \qquad y' = y, \qquad z' = z .$$

and envisages a formulation of physics in which all speeds would be at most c. The contraction has to have a dynamical origin, and is later also called the hypothesis of the deformable electron: electrons were viewed as the ultimate constituents of matter, balls when at rest with respect to the aether, but becoming ellipsoids flattened in the direction of a movement with respect to the aether, whence the contraction.

In June 1905, Poincaré published a *Comptes Rendus* Note [P11] which announced a paper submitted on July 5th and published a year later [P12]: "Sur la dynamique de l'électron".

He introduces the linear change of coordinates of [L5] (cf. (3), (4) above), which he calls a *Lorentz transformation.* He also defines the Lorentz group and proves the invariance of the equations of electromagnetism with respect to Lorentz transformations. Invariance with respect to the latter is the principle of relativity. However, this principle of relativity needs the contraction: the speed of light is c in the aether; it appears to be so to an observer in uniform motion with respect to the aether because of the contraction. He interprets a Lorentz transformation as a rotation in 4-space, with respect to the Lorentz metric, thus initiating the 4-dimensional formalism later developed by Minkowski [M2]. He also deduces the relativistic compositions of (uniform) velocities. Since the principle of relativity is justified by the contraction, he wants to give a dynamical explanation for the latter. He develops a theory of the forces which transform the electron, from a ball when at rest with respect to the aether, to an ellipsoid, flattened in the direction of the movement relative to the aether.

This is the last technical paper by Poincaré on the "new mechanics". There remain three general lectures, and an article [P13], also of a general nature, without formulas.

8. I now come to Einstein. Within three months in 1905 he submitted three papers: on the photoelectric effect, with the hypothesis of light quanta, on the determination of the Avogadro number (his Thesis) and on Brownian motion (submitted May 10). After that he was free to turn to the problems discussed above, which had occupied him off and on since his student days. As he told in a lecture in Kyoto, December 1922, published in English in [E6], Einstein went to his friend and colleague at the Patent Office Michele Besso, saying he wanted to have a battle of ideas about a problem he had been struggling with unsuccessfully. During that discussion, he suddenly saw the key to the problem: "*My solution was an analysis of the notion of time: it is not absolutely defined but only relatively and there is an inseparable connection between time and signal velocity.*"

He explained this point to a few friends, one of whom, J. Souter, also a colleague at the Patent Office, reminisced about it 50 years later [S4]. Einstein described how an observer on the famous clock tower in Berne and one on a tower in a nearby location, Muri, could synchronize clocks and define simultaneity. In the next five to six weeks he wrote his paper [E1], received June 30th and published three months later.

The paper states two postulates as a starting point:

1) The equations of physics are the same for two observers in relative uniform motion.

2) The speed of light is constant, independent of the relative motion of the source.

From this he deduces the Lorentz transformation relating time and space coordinates of two observers in relative uniform motion, which was new to him. The Lorentz-Fitzgerald contraction is thus given a *kinematical* explanation. Einstein proves that the Lorentz transformations form a group and then deduces the relativistic addition of speeds. No frame of reference is singled out and the aether is superfluous. He also draws further dynamical consequences which I need not go into here.

9. The paper did not create right away the splash he had expected or hoped for, according to his sister Maja, but, unknown to him for some time, it aroused much interest in some quarters: Max Planck lectured on it in his seminar (1905–6), and this impressed in particular Max von Laue, who visited Einstein in 1907. W. Wien asked a student, Laub, to report on it. Laub was later the first collaborator of Einstein, and had been very surprised to learn that Einstein was just an 8-hour a day employee in some office.

It should be added that in those first years, there was no clear distinction for most physicists between the theories of Lorentz and of Einstein [0]). They were viewed as prediction equivalent [1]). One spoke of Einstein's version, or generalization, of the Lorentz theory.

At that time however a big cloud was hanging over the theory, the experiments of W. Kaufmann [K1]. There was a competing theory, by

[0]) Not quite for all, though. Recently Professor E. L. Schucking kindly drew my attention to pp. 466–8 in the book *"Lehrbuch der Optik"*, second edition, Hirzel, Leipzig, 1906, by P. Drude where the author briefly describes the views of Lorentz [L5] and Einstein [E1], accepts the latter, and does not find the objection to it by Kaufmann (already announced in 1905) convincing.

[1]) As we know, they are not. Already in [E3] Einstein suggested an experiment to check his prediction about time dilation, which has no analog in the Lorentz theory, an experiment which could be carried out conclusively only in 1937 (Ives-Stilwell, see [M1], 7.2).

Abraham, in which electrons remained rigid balls (and even a third one, by A. H. Bucherer). All predicted an increase of the mass of an electron under a relative uniform motion, but the formulas were different. In fact there were transverse and longitudinal masses. For the former, Lorentz-Einstein predicted

$$m = m_0(\sqrt{1 - v^2/c^2})^{-1/2} \sim m_0(1 + \frac{1}{2}\, v^2/c^2) \tag{1}$$

and Abraham

$$m = m_0(1 + \frac{2}{5}\, v^2/c^2) \tag{2}$$

(see e.g. [M1], 1.10.2, 1.12.3). Kaufmann announced that his experiments confirmed (2). He was quite affirmative, claiming flatly that, unless he had made an error of principle, the Lorentz-Einstein theory was incompatible with his results. Max Planck, in a lengthy report [P4], was much more cautious and thought it was too early to draw definite conclusions. In fact, that same year, he published a first paper on relativity [P3], in which he stated that the principle of relativity, "introduced by Lorentz, and in still greater generality by Einstein", brings such enormous simplifications to electromagnetism that it is worth exploring its consequences in more than one way. He does not want to rule out at that stage that it will eventually be found compatible with experiments.

Lorentz and Poincaré were quite shaken by Kaufmann's results. The former wrote to the latter (March 8, 1908): "*Unfortunately, my hypothesis of the flattening of electrons is in contradiction with Kaufmann's results and I must abandon it. I am, therefore, at the end of my Latin.*" (Cf. [M1], pp. 336–7 for a facsimile of the letter). Poincaré wondered in [P13], section V, whether the relativity principle did not have as general a validity as had been expected. Einstein, on the other hand, was unmoved. He asked for further experiments. Though he points out that the theories of Abraham and Bucherer yield curves which are significantly closer to the one observed by Kaufmann than the curve provided by the relativity theory, he concludes that, in his opinion, *the theories of Abraham and Bucherer have a rather small probability of being correct, because their basic assumptions on the mass of the moving electron are not derived from theoretical systems which encompass a greater complex of phenomena* ([E4], III, § 10). In short, let us wait until the experiments confirm my theory. He did not have to wait long since in 1908, Bucherer announced the results of new experiments, which confirmed the Lorentz-Einstein theory versus Abraham's or his own, [B3].

10. This paved the way for H. Minkowski who, about three weeks later, gave his famous Cologne lecture [M2] with its ringing introduction:

> "The views of space and time I wish to lay before you have sprung from the soil of experimental physics; herein lies their strength. They are radical. Henceforth space by itself and time by itself are doomed to fade away into mere shadows and only a kind of union of the two will preserve an independent reality."

This is also the paper where Minkowski develops the four-dimensional formalism in which we all have learned special relativity, which Einstein viewed initially as superfluous erudition ("*überflüssige Gelehrsamkeit*"), but later praised highly, see [P1], p. 152.

Poincaré was also quick to point out, in a footnote added in proofs to the reproduction of [P13] in [P14], that Bucherer's experiments confirmed "*the views of Lorentz*".

11. How relieved Lorentz was can be seen from his book *"The Theory of electrons"* [L6]. Published in 1909, it was based on lectures given at Columbia University in 1906, but he had delayed publication in order to add recent developments. However, it was written while Kaufmann's results cast a doubt on his theory and he is very cautious, presenting his views as tentative, but in a last Note, added in 1908, he refers to Bucherer's results and concludes:

> "In all probability, the only objection against the hypothesis of the deformable electron and the principle of relativity has now been removed."

The last sections of this book contain a first discussion of Einstein's theory. Let S_0 be a system at rest with respect to the aether, A_0 an observer fixed in it, S a system moving with uniform speed v with respect to S_0 and A an observer in it. Lorentz first describes how, because of the contraction, the speed of light will also be c for A. Then he draws attention to a reciprocity pointed out by Einstein: if the observer A views S_0 as moving at speed $-v$, and performs the same computations, he will get back the coordinates in S_0. Then (§ 194) he concludes in a somewhat ambivalent manner which seems to me worth quoting extensively:

> It will be clear by what has been said that the impressions received by the two observers A_0 and A would be alike in all respects. It would be impossible to decide which of them moves or stands still with respect to the ether, and there would be no reason for preferring the times and lengths measured by the one to those determined by the other, nor for saying that either of them is in possession of the „true" times or the „true" lengths. This is a point which Einstein has laid particular stress on, in a theory in which he starts from what

he calls the principle of relativity, i.e. the principle that the equations by means of which physical phenomena may be described are not altered in form when we change the axes of coordinates for others having a uniform motion of translation relatively to the original system.

I cannot speak here of the many highly interesting applications which Einstein has made of this principle. His results concerning electromagnetic and optical phenomena (leading to the same contradiction with Kaufmann's results that was pointed out in § 179) agree in the main with those which we have obtained in the preceding pages, the chief difference being that Einstein simply postulates what we have deduced, with some difficulty and not altogether satisfactorily, from the fundamental equations of the electromagnetic field. By doing so, he may certainly take credit for making us see in the negative result of experiments like those of Michelson, Rayleigh and Brace, not a fortuitous compensation of opposing effects, but the manifestation of a general and fundamental principle.

Yet, I think, something may also be claimed in favor of the form in which I have presented the theory. I cannot but regard the ether, which can be the seat of an electromagnetic field with its energy and its vibrations, as endowed with a certain degree of substantiality, however different it may be from all ordinary matter. In this line of thought, it seems natural not to assume at starting that it can never make any difference whether a body moves through the ether or not, and to measure distances and lengths of time by means of rods and clocks having a fixed position relatively to the ether.

It would be unjust not to add that, besides the fascinating boldness of its starting point, Einstein's theory has another marked advantage over mine. Whereas I have not been able to obtain for the equations referred to moving axes *exactly* the same form as for those which apply to a stationary system, Einstein has accomplished this by means of a system of new variables slightly different from those which I have introduced...

12. In April 1909, Poincaré gave in Göttingen the "Wolfskehl-lectures". This was the first of an annual series of six lectures in mathematics, physics or mathematical physics. Poincaré gave the first five in German. At the beginning of the sixth he pointed out that he had been able to do so, though in poor German, because he had crutches, namely, the formulas. In the present lecture there will be none, so he will have to give it in French. It was on "*La nouvelle mécanique*" ("The new mechanics") [P15]. It is basically the Lorentz-Poincaré theory. There is an aether, an absolute time, but the contraction makes it impossible to detect a uniform relative motion with respect to it, whence a relativity principle.

A similar, in part identical lecture was given in Lille, in August 1909 [P16]. There is no mention of Einstein at all. However, it seems rather likely that Poincaré must have become aware of Einstein's theory from about that time on: a French translation of [M2] was published in 1909 and one of [E4] in 1910 [E5]. Besides, is it not likely that he received a copy of [L6]?

In 1910 M. Planck published eight lectures given at Columbia University in 1909 [P5]. After describing Einstein's ideas on space and time, he adds:

> It need scarcely be emphasized that this new conception of the idea of time makes the most serious demands upon the capacity of abstraction and the power of imagination of the physicist. It surpasses in boldness everything previously suggested in speculative natural phenomena and even in the philosophical theories of knowledge: non-euclidean geometry is child's play in comparison. And, moreover, the principle of relativity, unlike non-euclidean geometry, which only comes seriously into consideration in pure mathematics, undoubtedly possesses a real physical significance. The revolution introduced by this principle into the physical conceptions of the world is only to be compared in extent and depth with that brought about by the introduction of the Copernican system of the universe.

On October 13, 1910, Poincaré gave a lecture (in French) in Berlin. It is similar in substance, as regards our main topic of interest, to the sixth one in Göttingen, and would hardly be worth mentioning, were it not for what the first biographer of Einstein, the Polish journalist Alexander Moszkowski, wrote about it. The latter published in 1922 a book on Einstein, based largely on their conversations [M3]. At the beginning, he relates that he attended Poincaré's Berlin lecture. This occasion is the first time he heard the name of Einstein and he was so impressed by what he learned on him that he resolved he should get acquainted with him. This took place a few years later and led to the book.

However, the text of Poincaré's lecture is available in German translation [P17] and, again, there is no mention in it of Einstein or of Einstein's point of view. It is therefore reasonable to conjecture that someone brought up Einstein's relativity in the discussion following the lecture. After all, it was widely known and accepted in Germany by that time. If so, it would of course be interesting to know what was said. Maybe it is not overly optimistic to believe that Moszkowski provides at least a glimpse of it, since he writes later in his book (p. 231), "H. Poincaré had confessed, as late as 1910, that it caused him the greatest effort to find his way into Einstein's new mechanics", which, incidentally, would imply conclusively that he knew it.

13. A few days later (October 24–29), it was Lorentz's turn to give the Wolfskehl lectures in Göttingen. The second one was entitled "*The relativity principle of Einstein*". After expressing his pleasure in lecturing at a University where Minkowski had worked, he describes his point of view (aether, absolute time, local time, contraction) and Einstein's, in which all local times are on the same footing. He asserts that both are equivalent, and the choice between them

a question of taste. The note taker was Max Born, then a young Privatdozent, who also wrote up Lorentz's lectures [L7] from his own notes. In a comment added much later (1965) to a letter of Einstein of 1920, he wrote that, as a convinced Einstein disciple, he found that attitude "absurd and reactionary" ([B1], p. 72)[2]).

14. In 1911 appeared the first paper on Einstein's relativity by a French physicist, namely, Paul Langevin, professor at the Collège de France [L1][3]).

It gives a clear description of Einstein's theory, but is not really a complete endorsement. For instance, Langevin did not want to give up the aether:

> But it should not be concluded from that, as has sometimes been done prematurely, that the notion of aether should be abandoned, that aether does not exist, is unreachable through experiment. Only a uniform speed with respect to aether cannot be detected, but every change of speed, every acceleration has an absolute meaning.

---

[2]) Lorentz had the same point of view in [L8]. However in the second edition of [L6], which keeps most of the text of the first one, he adds a note to the page quoted in 11, beginning with:

> If I had to write the last chapter now, I should certainly have given a more prominent place to Einstein's theory of relativity (§ 189) by which the theory of electromagnetic phenomena in moving systems gains a simplicity that I had not been able to attain. The chief cause of my failure was my clinging to the idea that the variable t only can be considered as the true time and that my local time t' must be regarded as no more than an auxiliary mathematical quantity. In Einstein's theory, on the contrary, t' plays the same part as t; if we want to describe phenomena in terms of x', y', z', t' we must work with these variables exactly as we could do with x, y, z, t.

Still, while acknowledging here the superiority of Einstein's point of view, he maintained to the end of his life his preference for his own framework. In a letter written in 1954, quoted in [F1], p. 251, Einstein points out that Lorentz could not give up the aether, but adds that anyone who has lived through these times understands this.

This reluctance to abandon a familiar framework has often been compared to Einstein's own attitude, later, towards quantum mechanics.

[3]) The spreading of special relativity in France was apparently very slow, notwithstanding [E5] and the French translation of [M2], which do not seem to have made much impression, and mistrust towards it persisted for a long time in some quarters. The *Journal de Physique* regularly took up an issue of a foreign periodical, listed the titles of the papers, often with a capsule review. Einstein did not come out well. The summary of [E1] ((4), 5, 1906, 490) is hardly informative:

> The author proposes to establish the electrodynamics of moving bodies by taking into account absolute movement, only assuming light propagation in the vacuum with fixed speed, independent of the source.

In the summary of [P3], it is said that the principle of relativity, just introduced by Lorentz, implies extraordinary simplifications. The summary of [B3] speaks of a confirmation of Lorentz theory. A number of papers by Einstein, or Einstein-Laub, are quoted by title only.

In a lecture at the French Math. Society in 1912 (published in his *Collected Papers*, III_1, p. 179), "*Sur les groupes de transformations de contact et la cinématique nouvelle*", E. Cartan attributes the postulate of the constancy of the speed of light and the new kinematics to Lorentz. In his article: "La cinématique nouvelle et le groupe de Lorentz", published in the *Encyclopédie des Sciences Mathématiques*, 1915, he speaks of the "principle of relativity" and gives as references [P12], [M2], [L5] and a paper of F. Klein. Einstein is not mentioned at all. See also footnote [4]).

and in 1913 [L2], N° 13, after describing the contraction, he adds:

> The idea that such a contraction occurs solely because a body moves seemed at first peculiar. Then Einstein showed that it only corresponds to one of the consequences of the new notions of space and time imposed by the principle of relativity.
>
> From a completely different point of view, an important remark of Poincaré has thrown light on the very mechanism of that contraction.

In other words, the true explanation is the dynamical one of Lorentz-Poincaré, rather than those kinematical ideas about space and time. [It should be added, though, that later on in [L2], Langevin adopts Einstein's point of view completely.]

15. The first Solvay Congress took place in Bruxelles (Oct. 30–Nov. 3, 1911) and is probably the first and only time Poincaré and Einstein met, after which Einstein wrote to his friend H. Zangger: "Poincaré was simply quite antagonistic (with respect to Relativity) and showed little understanding for the situation, in spite of his sharp wit" (see [S1], p. 43). Also, the few recorded exchanges between them in discussions do not show much understanding. Those were on other matters, since the Congress had been convened not to discuss the "new mechanics", on which Einstein's theory was now widely accepted, but rather radiation and quanta [S0]. In a general report ([S0], 407–435), Einstein had also included his own light quanta hypothesis but with little success. The general consensus was apparently that Einstein was wrong on that. It took several years before his views were accepted, and eventually recognized by a Nobel prize (1922). Still he created a great impression, "the strongest apart maybe from Lorentz", wrote F. A. Lindemann, one of the three secretaries of the Congress; another secretary, Maurice de Broglie, reminisced later that Einstein and Poincaré were "in a class by themselves". Einstein was now famous, a recognized leading physicist.

At the time, Einstein was professor in Prague, but people at the E.T.H. (Swiss Federal Institute of Technology) in Zurich were trying to secure a position for him. Letters of recommendation were needed and Poincaré and Marie Curie were each asked to write one. The first sentence of Poincaré's letter is well known: "Mr. Einstein is one of the most original minds I know" and gives the impression that Poincaré now has a high regard for Einstein and that, maybe, Einstein's remarks to Zangger were based on some misunderstanding, but let us read the letter in full (see [P1], 170–1 and [F1], 83–4 for the original):

"Monsieur Einstein is one of the most original minds I have known; in spite of his youth he already occupies a very honorable position among the leading scholars of his time. We must especially admire in him the ease with which he adapts himself to new concepts and his ability to infer all the consequences from them. He does not remain attached to the classical principles and, faced with a physics problem, promptly envisages all possibilities. This is translated immediately in his mind into an anticipation of new phenomena, susceptible some day to experimental verification. I would not say that all his expectations will resist experimental check when such checks will become possible. Since he is probing in all directions, one should anticipate, on the contrary, that most of the roads he is following will lead to dead ends; but, at the same time, one must hope that one of the directions he has indicated will be a good one; and this suffices. This is how one should proceed. The role of mathematical physics is to state well the questions, but it is up to the experiments to answer.

The future will show more and more the value of M. Einstein, and the university which will succeed in attracting this young master is sure to draw much honor from it."

This implies first that Einstein, according to Poincaré, does not have yet any first rate achievement to his credit. It can only be hoped, even expected, that one will occur in the future, among many mistakes. Second, as someone who has had to read and evaluate thousands of letters of recommendation, (besides having written a fair number of them), I am rather wary of letters long on general remarks and short on specifics. Well, this one does not mention even one achievement of Einstein. Replace his name by, say, H. A. Lorentz or J. Doe, and you have a letter of recommendation for H. A. Lorentz or J. Doe. Altogether, not a very informed backing of the initial and final sentences.

16. Poincaré died unexpectedly in July 1912, after an operation which had appeared to be successful, so the lecture he gave in London in early May on "Space and Time" may well be his last word on this topic [P18]. The point of view is still that of his previous lectures,

"the principle of relativity, in its old form, had to be abandoned and to be replaced by the principle of relativity of Lorentz. It is the transformations of the "Lorentz group" which do not alter the differential equations of the dynamics. If you assume that the system is referred not to fixed axes, but to axes moving by a translation, you have to admit that all bodies undergo a deformation, that a sphere, for example, transforms into an ellipsoid, the small axis of which is parallel to the translation of the axes."

but then, there seems to be a shift in his discussion of the local time:

"It is necessary that the time be profoundly modified; here are two observers the first bound to the fixed axes, the second one to the moving ones, but both believing to be at rest. Not only such a figure, which the first one views as a sphere, will appear to the second one as an ellipsoid; but two events viewed by the first one as simultaneous, will not be so anymore for the second one."

"... in the new conception, space and time are no longer distinct entities which can be considered separately, but two parts of the same whole, two parts which are so closely knit that they cannot be easily separated."

which of course echoes Minkowski [M2], see 10 above. Does this mean that he is now leaning towards Einstein's relativity? Hardly, because he concludes:

What shall be our position in view of these new conceptions? Shall we be obliged to modify our conclusions? Certainly not; we had adopted a convention because it seemed convenient and we had said that nothing could constrain us to abandon it. Today some physicists want to adopt a new convention. It is not that they are constrained to do so; they consider this new convention more convenient; that is all. And those who are not of this opinion can legitimately retain the old one in order not to disturb their old habits. I believe, just between us, that this is what they shall do for a long time to come.

Now who can "some physicists" be, who claim the necessity to adopt new conventions about space and time in physics, which seem inconvenient and unnecessary to Poincaré? To me, there is only one answer: Einstein and those who had accepted his theory[4]).

Let us now return to the second and third opinions quoted earlier (2*b*), *c*)).

The foregoing surely leads one in the direction of *b*), except however that in order to avoid taking a stand as to whether Poincaré did or did not understand something, I would rather say: "Poincaré knew Einstein's relativity

---

[4]) There is hardly any doubt that Poincaré discussed these matters with E. Picard, and some of the latter's remarks in [P2] (freely translated) strike me as amplifications of Poincaré's opinion:

We know the importance of the principle of relativity, the starting point of which is the impossibility, claimed on the strength of a few negative experiments, to detect a relative uniform motion of a system by means of experiments in optics or electricity made inside the system. Assuming on the other hand that the idea of Lorentz and his electromagnetic equations are unassailable, one has been led to view as necessary a change of our ideas on *space* and on *time*... The mathematicians interested by a group of transformations leaving invariant the quadratic form $x^2 + y^2 + z^2 - c^2t^2$ (c speed of light) have produced elegant dissertations on this topic and have no doubt contributed to the popularity of the principle of relativity. In other times, before rejecting the traditional ideas on space and time, one might have examined very critically those on the aether and the formulation of the equations of electromagnetism; but the appetite for the new knows no bounds today... Science surely has no dogmas, and it may happen that precise experiments constrain us one day to modify some ideas which are now common sense. But is it already the time to do so?

Poincaré saw the danger of those fads and, in a lecture on the new mechanics, he implored professors not to discard the old and proven mechanics... He has lived long enough to see the main protagonists of the new ideas at least ruin partially their own work.

theory but did not accept it", a conclusion which in substance also agrees with those reached in [F2] or [M1].

17. However this does not prevent the friends quoted by S. Sternberg in 2*c*) from being right in many ways. Let us go back to the period 1898–1906 and see what Poincaré had anticipated: in 1898, the relativity of simultaneity, in 1900 he questions the existence of the aether, finds the hypotheses of Lorentz too special and asks for a principle implying the impossibility of detecting a relative uniform motion with respect to the aether. In 1904, he arrives at one, but tied up to the contraction and proposes a synchronization of "local times". In 1905 he defines the Lorentz group, shows it leaves invariant the equations of electromagnetism, computes the relativistic addition of uniform speeds and introduces a four-dimensional formalism. Moreover, his early writings on this topic surely had some influence on Einstein, beyond the non-necessity of Euclidean geometry. This incessant questioning of the framework and of various implicit assumptions was in tune with Einstein's own approach and the latter also acknowledged that Poincaré, Hume and Mach had influenced him. It certainly can be said that Poincaré had anticipated or independently developed "almost all" concepts of special relativity. But the missing part, even if quantitatively small, is the crux of the matter, what Lorentz called a starting point of fascinating boldness and Planck the boldest idea ever in the philosophy of knowledge. Poincaré viewed the relativity of simultaneity as a source of non-indispensable conventions, not as forcing a fundamental reorganization of our way of apprehending space and time, a point of view quite similar to that of Lorentz, in fact, as we saw[5]).

To conclude, I'll quote one more opinion, that of Louis de Broglie, the father of the wave formulation of quantum mechanics. It is in the introduction to Volume IX of Poincaré's *Collected Papers*, which contains Poincaré's papers on mathematical physics [P21]. Note that the intent of such an introduction is to present the work under consideration in as positive a light as possible. Indeed,

[5]) As we saw, a fundamental difference between Einstein and Lorentz-Poincaré is the fact that the former provides a *kinematical* explanation for the contraction, while Lorentz and Poincaré look for a *dynamical* one. In that respect, I would like to quote from a recent letter of Jürg Fröhlich, commenting on an earlier draft of this paper: "It deserves to be emphasized that the special theory of relativity, although it brought about a revolution in our conception of space and time, is purely kinematical. A consistent formulation of *dynamical laws* for the new mechanics turned out to be *impossible*. A relativistic formulation of the dynamical laws governing the motion of gravitating bodies was only accomplished in the *general theory of relativity*. The insight that relativistic *dynamics* forces one to go beyond the special theory of relativity and the formulation of the general theory are undoubtedly due to Einstein."

after discussing [P12], de Broglie writes that Poincaré had accomplished there a work of capital importance, but then adds:

> ... but at the same time, maybe because he was more an analyst than a physicist, he did not perceive the general point of view, based on a very fine analysis of measures of distances and duration, which the young Albert Einstein had just discovered by a stroke of genius, and which led him to a complete transformation of our ideas on space and time. Poincaré has not taken this decisive step but he is, with Lorentz, the one who contributed most to make it possible[6]).

which seems to me to be a fair short summary of the situation[7]).

REFERENCES

[B1] BORN, M. *Einstein-Born Briefwechsel 1916–1955*. Nymphenburger Verlagsbuchhandlung, München, 1969.

[B2] —— *My life, recollections of a Nobel laureate*. C. Scribner's Sons, New York, 1978.

[B3] BUCHERER, A. H. Messungen an Becquerelstrahlen. Die experimentelle Bestätigung der Lorentz-Einsteinschen Theorie. *Phys. Z. 9* (1908), 755–762.

[E1] EINSTEIN, A. Zur Elektrodynamik bewegter Körper *Ann. Phys. 17* (1905), 891–921; [E7], 276–310.

[E2] —— Ist die Trägheit eines Körpers von seinem Energieinhalt abhängig? *Ann. Phys. 18* (1905), 639–641; [E7], 312–315.

---

[6]) In volume 11 of Poincaré's *Collected Papers*, second part, pp. 62–71, in a paper written on the occasion of Poincaré's 100th birthday, L. de Broglie returns to this question, and elaborates on the likely reason why Poincaré did not take that decisive step, saying in part

> "... il avait une attitude un peu sceptique vis-à-vis des théories physiques, considérant qu'il existe en général une infinité de points de vue différents, d'images variées, qui sont logiquement équivalents et entre lesquels le savant ne choisit que pour des raisons de commodité. Ce nominalisme semble lui avoir parfois fait méconnaître le fait que, parmi les théories logiquement possibles, il en est cependant qui sont plus près de la réalité physique, mieux adaptées en tout cas à l'intuition du physicien et par là plus aptes à seconder ses efforts. C'est pourquoi le jeune Albert EINSTEIN, âgé alors seulement de 25 ans et dont l'instruction mathématique était rudimentaire en comparaison de celle du profond et génial savant français, est cependant arrivé avant lui à la vue synthétique qui, utilisant et justifiant toutes les tentatives partielles de ses devanciers, a balayé d'un seul coup toutes les difficultés. Coup de maître d'un esprit vigoureux guidé par une intuition profonde des réalités physiques!"

[7]) This opinion is rather consonant, except for the inclusion of Lorentz, with one expressed at a meeting of the Société Française de Philosophie held in June 1922, of which A. Pais says on p. 168 of [P1] that it coincides with his own assessment:

> "The solution anticipated by Poincaré was given by Einstein in his memoir of 1905 on special relativity. He accomplished the revolution which Poincaré had foreseen at a moment when the development of physics seemed to lead to an impasse."

[E3] —— Möglichkeit einer neuen Prüfung des Relativitätsprinzips. *Ann. Phys. 23* (1907), 197–198; [E7], 402–403.

[E4] —— Relativitätsprinzip und die aus demselben gezogenen Folgerungen. *Jahrb. Radioakt. 4* (1907), 411–462; *5* (1907), 98–99 (Berichtigungen); [E7], 433–484, 495.

[E5] —— Principe de relativité et ses conséquences dans la physique moderne. *Arch. Sci. Phys. et Nat. 29* (1910), 5–28 and 125–144. (French translation of [E4].)

[E6] —— How I created the theory of relativity. *Physics Today* (Aug. 1982), 45–7.

[E7] —— *Collected Papers*. Vol. 2. Princeton University Press, 1989.

[F1] FLÜCKIGER, M. *Einstein in Bern*. (2nd edition). P. Haupt, Bern, 1974.

[F2] FÖLSING, A. *Einstein, eine Biographie*. Suhrkamp, Frankfurt am Main, 1993.

[K1] KAUFMANN, W. Über die Konstitution des Electrons. *Ann. Phys. 19* (1906), 487–553; Nachtrag zu der Abhandlung: Über die Konstitution des Electrons. *Ann. Phys. 20* (1906), 639–640.

[L1] LANGEVIN, P. L'évolution de l'espace et du temps. *Scientia 10* (1911), 31–54. Reproduced in *La Physique depuis 20 ans*, V. O. Doin, Paris, 1923, 265–300.

[L2] —— L'inertie de l'énergie et ses conséquences. Lecture given at the French Physics Society, 26.1.1913; in *La physique depuis vingt ans*, VII. O. Doin, Paris, 1923, 345–405; *Œuvres*. C.N.R.S., Paris, 1950, 397–426.

[L3] LORENTZ, H. A. *Versuch einer Theorie der elektrischen und optischen Erscheinungen in bewegten Körpern*. Brill, Leiden, 1895; Reprinted in *Collected Papers, 5*, 1–137.

[L4] —— Théorie simplifiée des phénomènes électriques et optiques dans des corps en mouvement. *Verhl. Kon. Akad. Wetensch. (7) 507* (1899). Reprinted in *Collected Papers, 5*, 139–155.

[L5] —— Electromagnetic phenomena in a system moving with any velocity less than that of light. *Proc. R. Acad. Amsterdam (6) 809* (1904). Reprinted in *Collected Papers, 5*, 172–197.

[L6] —— *The Theory of Electrons*. Brill, Leiden, 1909; rev. ed. 1916; Dover, New York, 1952.

[L7] —— Alte und neue Fragen der Physik. *Phys. Z. (11)* (1910), 1234–1257. Reprinted in *Collected Papers, 7*, 205–257.

[L8] —— *Das Relativitätsprinzip, Vorlesungen gehalten in Teylers Stiftung in Haarlem*. Teubner, 1914.

[M1] MILLER, A. *Albert Einstein's special theory of relativity*. Addison-Wesley, 1981.

[M2] MINKOWSKI, H. Raum und Zeit. Lecture delivered to the 80th Naturforscherversammlung at Cologne on 21 September 1908; in *Phys. Z. 20* (1909), 104–111. French translation in *Ann. Sci. École Norm. Sup. (3) 26* (1909), 499–517. *Collected Papers II*, 431–44.

[M3] MOSZKOWSKI, A. *Einstein, Einblicke in seine Gedankenwelt*. Fontane, Berlin, 1922. English translation: *Conversations with Einstein*. Horizon Press, New York, 1970.

[P1] PAIS, A. *Subtle is the Lord... The science and the life of Albert Einstein*. Oxford University Press, New York, 1982.

[P2] PICARD, E. L'Œuvre de Henri Poincaré, *Ann. Sci. École Norm. Sup. (3) 30* (1913), 463–487.

[P3] PLANCK, M. Das Prinzip der Relativität und die Grundgleichungen der Mechanik. *Verh. Phys. Ges. 4* (1906), 136–141.

[P4] —— Die Kaufmannschen Messungen der Ablenkbarkeit der β-Strahlen in ihrer Bedeutung für die Dynamik der Elektronen. *Phys. Z. 7* (1906), 753–761. Published without the discussion session in *Verh. D. Phys. Ges. 8* (1906), 418–432.

[P5] —— *Acht Vorlesungen über theoretische Physik, gehalten an der Columbia University in the city of New York im Frühjahr 1909*. Leipzig, 1910. English translation: *Eight lectures on theoretical physics, delivered at Columbia University 1909*. Columbia University Press, 1915.

[P6] POINCARÉ, H. La mesure du temps. *Rev. Mét. Mor. 6* (1898), 371–384; reproduced in [P10], 35–58.

[P7] —— Sur les rapports de la Physique expérimentale et de la Physique mathématique. Rapports présentés au Congrès international de Physique réuni à Paris en 1900 (4 vols.; Gauthier-Villars, Paris, 1900), vol. 1, 1–29. Reproduced as Chap. IX, X of [P8]

[P8] —— *La Science et l'Hypothèse*. Flammarion, Paris, 1902. English translation in [P19].

[P9] —— L'état actuel et l'avenir de la Physique mathématique. Lecture delivered on 24 September, 1904 to the International Congress of Arts and Science, Saint Louis, Missouri, and published in *Bull. Sci. Math. 28* (1904), 302–324. Reproduced with minor editing, as Chap. VII, VIII and IX of [P10], 91–111.

[P10] —— *La valeur de la Science*. Flammarion, Paris, 1905. English translation in [P19].

[P11] —— Sur la dynamique de l'électron. *C.R. Acad. Sci. 140* (1905), 1504–1508. [P20], 489–493.

[P12] —— Sur la dynamique de l'électron. *Rend. Circ. Mat. Palermo 21* (1906), 129–175. [P20], 494–550.

[P13] —— La dynamique de l'électron. *Rev. Générale Sci. Pures Appl. 19* (1908), 386–402. [P20], 551–586. Reproduced, apart from minor modifications, in [P14], 199–250.

[P14] —— *Science et Méthode*. Flammarion, Paris, 1908. English translation in [P19]

[P15] —— *Sechs Vorträge über ausgewählte Gegenstände aus der reinen Mathematik und mathematischen Physik*. Teubner, Leipzig, 1910.

[P16] —— *La mécanique nouvelle. Conférence, Mémoire et Note sur la théorie de la relativité*. Gauthier-Villars, Paris, 1909. Besides the conference, contains reprints of [P11] and [P12].

[P17] —— *Die neue Mechanik*. Himmel und Erde XXIII, n°3, Berlin, 1910; Teubner, Leipzig, 1912.

[P18] —— L'espace et le temps. Lecture delivered at the University of London, May 4, 1912. Reproduced in [P20], 15–24.

[P19] —— *The foundations of Science*. The Science Press, New York, 1913. Authorized translation of [P8], [P10], [P14] by G.D. Halsted.

[P20] —— *Dernières pensées*. Flammarion, Paris, 1913. English translation: *Mathematics and Science: Last Essays*. Dover, 1963.

[P21] —— *Œuvres*. Vol. IX. Gauthier-Villars, Paris, 1954.

[S0] *La théorie du rayonnement et les quanta, rapport et discussions de la réunion tenue à Bruxelles, du 30 octobre au 3 novembre 1911, sous les auspices de M.E. Solvay, publiés par MM. P. Langevin et M. de Broglie*. Gauthier-Villars, Paris, 1912.

[S1] SEELIG, C. (ed.) *Helle Zeit-Dunkle Zeit, in memoriam Albert Einstein*. Europa Verlag, Zürich, 1956. Reedition, Braunschweig, 1986.

[S2] —— *Albert Einstein, Leben und Werk eines Genies unserer Zeit*. Europa Verlag, Zürich, 1960.

[S3] SOLOVINE, M. *Einstein, lettres à Maurice Solovine*. Gauthier-Villars, Paris, 1956.

[S4] SOUTER, J. *Comment j'ai appris à connaître Einstein*. Radio broadcast, August 6, 1955, in [F1], 154–158.

[W] WHITTAKER, E. *A History of the theories of Aether and Electricity*. Vol. 2. Nelson, London, 1953.

(Reçu le 17 décembre 1998; version révisée reçue le 25 mars 1999)

Armand Borel
School of Mathematics
Institute for Advanced Study
Princeton, NJ 08540
U. S. A.
e-mail: borel@IAS.edu

Commentaires et corrections supplémentaires aux volumes I, II, III

$(x.y)$ (resp. $(x.yb)$) signifie page x, ligne y (resp. page x, ligne y du bas).

Volume I

45.10b: remplacer "1197–1199" par "1129".
34.1b: remplacer "1128" par "1129".
51.5: remplacer "existence" par "extension".

Volume II

154.15: remplacer **SU**(2) par **SU**(3), (trois fois).
365.2b: remplacer "c" par "ç".
366.13b: remplacer "E_x" par "E_a".
371: Ajouter à la section 1.16 la remarque suivante:
Si l'on peut choisir un élément $b \in p^{-1}(c)$ qui appartient à $B^{\mathfrak{g}}$, alors a est le cocycle trivial $a_s = 1$ pour tout $s \in \mathfrak{g}$, donc $\rho_c = 1$. Par suite, l'opération de $C^{\mathfrak{g}}$ passe au quotient $C^{\mathfrak{g}}/p(B^{\mathfrak{g}})$.
504.6b: remplacer k par $\tilde{k}$.
545.17: Il y a ici une erreur qui m'a été signalée par R. Sankaran, due au fait que je cite A. Blanchard incorrectement. La formule (1) est valable dans le cas des variétés de Hopf ($u \neq 0, v = 0$), mais si $u.v \neq 0$, le sous-groupe discret Γ doit être remplacé par un sous-groupe complexe central, isomorphic au groupe additif de $\mathbb{C}$. Il s'ensuit dans les deux cas que $A^o(M_{u,v})$ est engendré par des sous-groupes compacts, et que par conséquent l'argument de 9.3 est valable.
649.12: remplacer "opéra" par "opère".
677.15, 16: remplacer $\subset_\alpha$ par c_α.
684: Réf. 5: remplacer P. Cartier par B. Kostant.
691.11: Supprimer la dernière parenthèse.
779.6b: le titre de l'article de T. Nakamura est: *A remark on unipotent groups of characteristic* $p > 0$.

Volume III

122. En caractéristique zéro, une démonstration du Corollaire 3.2 est aussi contenue dans un article de B. Weisfeiler: *On a class of unipotent subgroups of semisimple algebraic groups*, Uspehi Math. Nauk **21** (1966), 222–3, Math. Rev. **34**, 2634), ce que je n'ai remarqué que beaucoup plus tard.
160. La démonstration du Théorème 2.3 n'est pas correcte, comme me l'a signalé tout d'abord E. Leuzinger, qui en a fourni une dans [1] (voir le Théorème

5.7 et la discussion dans la remarque 5.6). Cette erreur m'a aussi été mentionnée par L. Ji et R. MacPherson et par L. Saper.

[1] E. Leuzinger, *Tits geometries and arithmetic groups*, to appear.

230.17: remplacer 36 par 37

375.20: remplacer "importance" par "importante"

391.18b: dorénavant ne doit pas être caractères gras

391.17b: remplacer "preque" par "presque".

507.14: remplacer 14.3 par 24.10.

507.17: remplacer $r(a)/2$ par $r(h_a)/2$.

515.18: remplacer 15 par 16.

597. Les espaces complexes compacts symétriques au sens de [114] dont le groupe d'automorphismes a une composante neutre semisimple (resp. réductive) ont été classés par D.N. Ahiezer [1] (resp. R. Lehmann [2], [3]).

[1] D.N. Ahiezer, *On algebraic varieties that are symmetric in the sense of Borel*, Sov. Math. Doklady **30** (1984), 579–582.

[2] R. Lehmann, *Complex symmetric spaces*, Ann. Institut Fourier **39** (1989), 373–416.

[3] R. Lehmann, *Singular complex symmetric torus embeddings*, Archiv. d. Math. **56** (1991), 68–80.

633.14: remplacer "elment" par "element".

634.9: insérer "the" devant "volume".

635.12: remplacer y par y' dans le deuxième membre de l'égalité (3).

660.4: remplacer l'exposant 9 dans Ω^9 par un point.

666.17: remplacer $\mathfrak{g}$ par G.

691.14b: remplacer "dense" par "danse".

Commentaires et corrections au volume IV

Les Notes à chaque article sont numérotées consécutivement. Le symbole X.z réfère à la Note z, page X de ce volume.

121. Laplacian and the Discrete Spectrum of an Arithmetic Group (with H. Garland)

2.1 Ce théorème était un premier pas dans la direction de la conjecture centrale du sujet, à savoir: la convolution par une fonction à décroissance rapide est un opérateur à trace dans le spectre L^2 discret d'un groupe arithmetique. Elle a été établie par W. Müller pour une fonction K-finie [2], puis ensuite sans cette hypothèse dans [4], (annoncé dans [3]). Indépendamment, une autre démonstration du résultat général a été donnée par L. Ji [1].

[1] L. Ji, *The trace class conjecture for arithmetic groups*, Jour. Diff. Geometry **48** (1998), 165–203.

[2] W. Müller, *The class conjecture in the theory of automorphic forms*, Annals of Math. **130** (1989), 473–529.

[3] W. Müller, *The trace class conjecture without the K-finiteness assumption*, Comptes Rendus Acad. Sci. **324** (1997), 1333–1228.

[4] W. Müller, *The trace class conjecture in the theory of automorphic forms II*, Geometric and Functional Analysis **8** (1998), 315–55.

La démonstration dépend en partie d'une estimation de la fonction $N_d(\lambda)$ qui donne le nombre de valeurs propres du Laplacien dans le spectre L^2 discret qui sont $\leq \lambda$. Un problème de base est de savoir si elle satisfait à la loi de Weyl (établie par Weyl pour les variétés compactes), ou si au moins cette dernière fournit une borne supérieure. Cela a été démontré dans des cas particuliers variés, mais le problème reste ouvert en général, pour autant que je sache. Pour une exposition de quelques résultats récents et une vue d'ensemble d'autres plus anciens, voir:

L. Ji, *The Weyl upper bound on the discrete spectrum of locally symmetric spaces*, preprint.

21.2 Remplacer X par C_g.

124. Cohomology and Spectrum of an Arithmetic Group

69.1 Dans cet article on s'intéresse à la cohomologie d'un sous-groupe arithmétique Γ d'un $\mathbb{Q}$-groupe semisimple G, à coefficients complexes (ou, plus généralement, dans un module de dimension finie pour le groupe $G(\mathbb{R})$ des points réels de G, mais je me borne ici à $\mathbb{C}$). On sait qu'elle peut s'écrire sous la forme $H^*(\bar{X}/\Gamma;\mathbb{C})$, où $\bar{X}$ désigne la compactification définie dans [98] de l'espace symétrique X des sous-groupes compacts maximaux de $G(\mathbb{R})$. Le quotient $\bar{X}/\Gamma$ est une variété compacte à coins (au moins si Γ est net, hypothèse peu restrictive). Le quotient $\partial\bar{X}/\Gamma$ est réunion d'un nombre fini de faces associées à des $\mathbb{Q}$-sous-groupes paraboliques de G. Dans 4.1, je soulève la question de savoir s'il existé des classes de cohomologie de $\bar{X}/\Gamma$ dont la restriction aux faces est nulle, sans que la restriction au bord tout entier $\partial\bar{X}/\Gamma$ le soit, et propose d'appeler classes fantômes ("ghost classes") de tels éléments.

Un premier exemple a été fourni par G. Harder [4]. D'autres, représentés par des formes $G(\mathbb{R})$-invariantes sur X, ont été trouvés par J. Franke [2] pour $\mathbf{SL}_n$ et ensuite par A. Kewenig et T. Rieband [5] pour beaucoup de groupes classiques.

70.2 $H^*(\Gamma;\mathbb{C})$ peut aussi s'écrire en cohomologie relative d'algèbres de Lie sous la forme

$$H^*(\Gamma;\mathbb{C}) = H^*(\mathfrak{g},K;C^\infty(\Gamma\backslash G(\mathbb{R}))$$

(où $\mathfrak{g}$ est l'algèbre de Lie de $G(\mathbb{R})$ et K un sous-groupe compact maximal de $G(\mathbb{R})$). Dans 4.2, j'annonce que l'on peut remplacer $C^\infty(\Gamma\backslash G(R))$ par l'espace $C^\infty_{umg}(\Gamma\backslash G(\mathbb{R}))$ des fonctions indéfiniment différentiables sur $\Gamma\backslash G(\mathbb{R})$ à croissance modérée uniforme (ce qui est démontré dans [125]). Il ne manque qu'une condition pour qu'une telle fonction soit une forme automorphe, à savoir d'être annulée par un idéal de codimension finie du centre de l'algèbre enveloppante de $\mathfrak{g}$, et je demande si en fait l'inclusion des formes automorphes dans C^∞_{umg} induit un isomorphisme en cohomologie, tout en remarquant que, si c'est le cas, c'est sans doute beaucoup plus difficile à établir que le passage de C^∞ à C^∞_{umg}. C'est en effet le cas pour la démonstration fournie par J. Franke [1]. En fait, Franke se place dans un cadre adèlique, ce qui fait que sa démonstration est limitée aux sous-groupes arithmétiques de congruence, Il n'y a guère doute qu'elle devrait s'étendre aux groupes arithmétiques généraux, mais, pour autant que je sache, cela n'a pas été établi.

G. Harder avait aussi émis une conjecture voisine, mais pas identique, selon laquelle toute classe de cohomologie peut être représentée par une forme harmonique fermée, qu'il a démontrée dans le cas du rang $\mathbb{Q}$-rang égal à 1 [3].

[1] J. Franke, *Harmonic analysis in weighted L_2-spaces*, Annales Sci. Ecole Normale Sup. (4) **31** (1998), 181–279.

[2] J. Franke, *Topological models for some summand of the Eisenstein cohomology of congruence subgroups*, preprint, (in [5]).

[3] G. Harder, *On the cohomology of discrete arithmetically defined groups*, Proc. Intern. Colloquium on Discrete subgroups of Lie groups and applications to moduli, Tata Institute of fundamental research, Bombay 1973, Oxford University Press 1975, 129–160.

[4] G. Harder, *Some results on the Eisenstein cohomology of arithmetic subgroups of GL_n* in Cohomology of arithmetic groups and automorphic forms, SLN **1447** (1989), 85–153.

[5] A. Kewenig und T. Rieband, *Geisterklassen im Bild der Borelabbildung für symplektische und orthogonale Gruppen*, Preprint, University of Bonn, 1997.

130. On Affine Algebraic Homogeneous Spaces

126.1 F. Knop m'a signalé, il a plusieurs années, que le Théorème 1(ii) n'est pas vrai en caractéristique 2. L'exemple la plus simple est déjà fourni par $\mathbf{SL}_2$.

Soient donc k de caractérisque deux et $G = \mathbf{SL}_2$. Per définition, G opère sur k^2, dont on note e_1, e_2 la base canonique. La carré symétrique cette représentation (dite identique) est de degré trois, et contient un sous-espace invariant V de codimension un, formé des carrés des éléments de k^2. Si $x = e_1.e_2$, alors on vérifie aisément que l'orbite $G.x$ de x est égale à l'espace affine $x + V$. Le groupe d'isotropie contient le groupe des matrices diagonales comme sous-groupe d'indice deux et n'est pas connexe. F. Knop m'a aussi signalé des examples analogues formés à partir de la représentation identique de $\mathbf{SO}_{2n+1}$ ou de la représentation de degré 7 groupe exceptionnel $\mathbb{G}_2$.

Cependant le Theor. 1(ii) et sa démonstration sont valables si H est connexe ou si k n'est pas de caractéristique 2.

131. Cohomologie d'intersection et L^2-cohomologie de variétés arithmétiques de rang rationnel deux (avec W. Casselman)

131.1 Le but de cette Note est d'esquisser une démonstration de la conjecture de Zucker lorsque le $\mathbb{Q}$-groupe simple ambiant est de rang deux sur $\mathbb{Q}$, (le cas du rang rationnel 1 est traité dans [122]). Elle dépendait d'un théorème de type Paley-Wiener de Casselman, dont la démonstration n'a pas été publiée jusqu'à présent.

Depuis, la conjecture de Zucker a été prouvée par E. Looijenga [1] et par L. Saper et M. Stern [2].

[1] E. Looijenga, *L^2-cohomology of locally symmetric spaces*, Compos. Math. **67** (1988), 3–20.

[2] L. Saper and M. Stern, *L_2-cohomology of arithmetic varieties*, Ann. Math. **132** (1960), 1–69.

138. The School of Mathematics at the Institute for Advanced Study

214.1 La chronologie ici est un peu vague. Quand ces calculs ont été faits, Hirzebruch n'avait pas encore démontré sa version du théorème de Riemann-Roch, mais il avait conjecturé la formule à établir, que nous avions admise. Il s'agissait de calculer la dimension de l'espace des sections régulières d'un fibré homogène, défini par un poids dominant, sur le quotient d'un groupe simple complexe par un sous-groupe connexe résoluble maximal. Un théorème d'annulation de

Kodaira montrait que le membre de gauche de la formule conjecturée, en principe une caractéristique d'Euler-Poincaré, était en fait égal à la dimension cherchée, et le membre de droite permettait de la calculer en termes de poids et racines.

215.2 Plus précisément, la table qui m'avait été fournie par H. Toda, et publiée ensuite dans les Comptes Rendus, concernait certains groupes d'homotopie des sphères. En utilisant des fibrations bien connues, on pouvait en tirer des renseignements sur quelques groupes d'homotopie de certains groupes de Lie compacts, notamment du groupe exceptionnel $\mathbb{G}_2$. Ils contredisait en partie des résultats que nous avions obtenus de manière entièrement différente. En particulier les valeurs trouvées pour $\pi_{10}(G_2)$ modulo 2-torsion étaient en conflit.

139. Finiteness Theorems for Discrete Subgroups of Bounded Covolume in Semisimple Groups (with G. Prasad)

229.1 Une application aux nombres de classes de groupes algébriques semisimples sur des corps de nombres est donnée dans l'Appendice Al du livre:
V. Platonov and A. Rapinchuk, Algebraic groups and number theory, pure and applied math. series **139**, Academic Press 1994.

143. The Work of Chevalley in Lie Groups and Algebraic Groups

309.1 En fait, le texte présenté est ici une version légèrement révisée de l'article paru initialement.

144. Values of Isotropic Quadratic Forms at S-integral Points (with G. Prasad)

331.1 Cet article, annoncé dans [137], résumé dans [146], s'insère dans la suite des travaux suscités par la démonstration de Margulis d'une conjecture d'Oppenheim concernant les formes quadratiques non-dégénérées isotropes sur $\mathbb{R}^n$. Suivant une suggestion de M. S. Raghunathan, il l'avait déduite d'un cas très particulier, démontré par lui, d'une conjecture de Raghunathan concernant les adhérences de certaines orbites dans un quotient G/Γ d'un groupe semisimple G par un sous-groupe discret Γ de covolume fini. Cela provoqua énormément d'intérêt pour cette conjecture, et pour une conjecture analogue, faite par Dani, généralisée par Margulis, concernant les measures sur G/Γ invariantes par certains sous-groupes (appelée Raghunathan's measure theoretic conjecture by M. Ratner). Les deux ont été établies par M. Ratner dans une série d'articles culminant avec [3].

D'autre par S. Raghavan et K.G. Ramanathan proposèrent dans [2] une généralisation de la conjecture d'Oppenheim sur un corps de nombres k, mettant en jeu des formes quadratiques isotropes associées aux places archimédiennes de k. Dans [144] nous reprenons cette idée mais allons plus loin, en admettant aussi des places finies de k, ce qui revenait en fait à se placer dans un cadre S-arithmétique. La démonstration que nous donnons de la conjecture d'Oppenheim ainsi généralisée laissait penser que les conjectures précitées de Raghunathan et Dani pouvaient être aussi vraies dans cette situation plus générale, ce qui fut en effet établi par M. Ratner [4], même sous des hypothèses encore plus faibles concernant les groupes ambiants. L'article [1] démontre ou esquisse la démonstration de ces conjectures dans le cadre des groupes algébriques.

[1] G. A. Margulis and G. M. Tomanov, *Invariant measures for action of unipotent groups over local fields*, Inv. Math. **116** (1994), 347–392.

[2] S. Raghavan and K. G. Ramanathan, *On a diophantine inequality concerning quadratic forms*, Göttingen Nachr. Math. Phys. Klasse (1968), 252–262.

[3] M. Ratner, *Raghunathan's topological conjecture and distributions of unipotent flows*, Duke Math. J. **63** (1991), 234–280.

[4] M. Ratner, *Raghunathan's conjecture for cartesian products of real and p-adic Lie groups*, Duke Math. J. **77** (1995), 275–382.

Section 8.5 de [155] donne quelques corrections à la démonstration de 7.4

151. Some Recollections of Harish-Chandra

386.1 Ce texte a aussi été reproduit dans [1], malheureusement avec quelques fautes, dont une particulièrement gênante: p. 213, ligne 20, remplacer "grief" par "grips".

[1] Current trends in mathematics and physics. A tribute to Harish-Chandra (S. D. Adhikari ed.), Narosa publishing house, Bombay 1995, 210–215.

156. Values of Indefinite Quadratic Forms at Integral Points and Flows on Spaces of Lattices

458.1 Des résultats beaucoup plus complets sur une version quantitative de la conjecture d'Oppenheim ont été obtenus depuis. Voir en particulier.

A. Eskin, G. A. Margulis and S. Mozes, *Upper bounds and asymptotics in a quantitative version of the Oppenheim conjecture*, Ann. Math. **147** (1998), 93–131.

Un exposé historique systématique des travaux sur la conjecture d'Oppenheim est fourni par G. A. Margulis, *Oppenheim conjecture*, in Fields medallists' lectures, World Science Series, 20th century math. **5**, World Scientific publisher, 1997.

157. Values of Zeta Functions at Integers, Cohomology and Polylogarithms

La formule (1) doit être

478.1 (1) $$\prod_{m=1}^{n} \xi_F(1-2m) = \chi(\mathbf{Sp}_{2n}\underline{\sigma}_F)$$

et la ligne 7

478.2 (2) $$q.\chi(\mathbf{Sp}_{2n}\underline{\sigma}_F) \in \mathbb{Z}.$$

485.3 Étant donné un sous-anneau A de $\mathbb{C}$, on note $A(j)$ le sous-A-module de $\mathbb{C}$ égal à $A(2\pi i)^j$. Le m-ème régulateur est un homomorphisme

$$\lambda_m : K_{2m-1}\mathbb{C} \otimes_{\mathbb{Z}} \mathbb{R} \to \mathbb{R}(m-1).$$

Dans un contexte beaucoup plus général. S. Beilinson a défini un m-ième régulateur à valeurs dans $\mathbb{Z}(m-1)$. En considérant son image dans $\mathbb{R}(m-1)$, dans un cas particulier, il obtient aussi un régulateur

$$\lambda'_m : K_{2m-1}\mathbb{C} \otimes_{\mathbb{Z}} \mathbb{R} \to \mathbb{R}(m-1).$$

Précisant un résultat antérieur de Beilinson, M. Rapoport démontre dans [1] que $\lambda_m/\lambda'_m \in \mathbb{Q}$.

Dans [2], les auteurs conjecturent que ce quotient est égal à deux, et le démontrent modulo une conjecture affirmant que pour un fibré universel plat il y a égalité entre la classe de Chern-Simons et une classe de Chern introduite dans un cadre simplicial par Beilinson.

[1] M. Rapoport, *Comparison of the regulators of Beilinson and Borel*, in Beilinson's conjectures on special values of L-functions, Perspective in Math. **4**, 169–175, Academic Press 1988.

[2] J. Dupont, R. Hain and S. Zucker, *Regulators and characteristic classes of flat bundles*, to appear in the Proceedings of the Banff Conference on algebraic cycles (june 1998).

La réference pour [G1] est: Advances in math. **114** (1995), 197–318 et pour [G2]: Proc. symposia pure math. **55** (1994), part 2, 43–96.

Pour une exposé des travaux sur cette question jusqu'à ce moment, on renvoie à A. B. Goncharov, *Polylogarithms in arithmetic and geometry*, Proc. ICM Zürich 1994, 374–397.

Pour quelques résultats ultérieurs dans la direction de la conjecture de Zagier, voir R. de Jeu, *Zagier's conjecture and wedge complexes in algebraic K-theory*, Compos. Math. **96** (1995), 197–247.

173. Henri Poincaré and Special Relativity

Cet article étant le seul que j'aie publié sur un sujet de physique, on m'a quelquefois demandé ce qui m'y avait amené. La part de Poincaré à la relativité restreinte est un question bien naturelle que j'avais occasionnellement rencontrée dès mes études, sans jamais vraiment l'approfondir avant 1980 et mon impression superficielle était que la contribution de Poincaré était substantielle, peut-être même plus que généralement admis. J'en serais sans doute resté là si Abraham Pais n'avait passé quelques mois à l'Institute durant l'année académique 1980-81, pour terminer son livre sur Einstein [P1], et n'avait suggéré que nous discutions de Poincaré, dont le rôle dans cette question n'était pas clair pour lui.

Son impression initiale était en gros la même que la mienne et je me le rappelle très bien me disant au début, à plusieurs reprises, à propos de la relativité restreinte: "Poincaré knew a great deal", impressions qui furent d'abord renforcées lorsque nous remarquâmes que "La mesure du temps" [P6] avait été publiée en 1898 déjà (nous ne connaissions cet article que par sa reproduction dans [P10], qui ne comporte aucune référence). Durant les semaines suivantes, nous avons lu et discuté tout ce que nous trouvions sur Poincaré ayant trait à la "Mécanique Nouvelle", en particulier ses conférences de 1909 à 1912, pour constater, avec étonnement, que, malgré [P6], Poincaré en était essentiellement resté au point de vue de Lorentz, ce qui a amené Pais aux conclusions énoncées dans [P1], et citées dans mon article. Comme c'était en substance aussi les miennes, même si ma formulation eût été un peu différente, je considérais cette question comme réglée et ne comptais pas y revenir.

Cependant, au cours de nombreuses discussions durant les années suivantes, j'ai constaté que la plupart de mes interlocuteurs n'en connaissaient que certains aspects et il m'a semblé qu'il ne serait pas inutile d'en faire un exposé d'ensemble, d'où cet article, qui a été lui-même précédé d'une conférence donnée à plusieurs reprises. À cette occasion, j'ai revu tous les documents, ai mieux apprécié certains d'entre eux, et suis arrivé à des conclusions semblables, à mon avis encore mieux étayées qu'auparavant, si bien exprimées par L. de Broglie.

Selon ce dernier, si Poincaré n'a pas fait le pas décisif, c'est peut-être parce qu'il était trop mathématicien et pas assez physicien. Mais cela ne signifie pas qu'aucun mathématicien n'aurait pu y parvenir. En fait, il n'est pas exclu que H. Minkowski y soit essentiellement arrivé, indépendamment et à peu près en même temps que Einstein.

En effet, M. Born relate dans l'introduction à ses Mémoires [B2] que Minkowski lui a une fois confié (en automne 1908, sans doute): "it came to him as a great shock when Einstein published his paper in which the equivalence of the different local times of observers moving relative to each other was pronounced: for he had reached the same conclusion independently but did not publish it because he wished first to work out the mathematical structure in all its splendor", et M. Born s'empresse d'ajouter que Minkowski "never made a priority claim and always gave Einstein his full share of credit in this important discovery".

En fait on peut remarquer que l'exposition dans [M2], et aussi dans un article antérieur de 1907, ne semble rien devoir à Einstein. Chaque Mémoire contient une seule référence à Einstein, précisément pour l'équivalence des temps locaux.

Un témoignage de l'admiration de Minkowski pour Einstein (qui avait été son élève à l'Ecole Polytechnique Fédérale (EPF)) peut être aussi glané dans quelques souvenirs [K] de L. Kollros (camarade de classe d'Einstein, plus tard professeur de géométrie à l'EPF), qui parle d'une promenade à Rome en avril 1908 en compagnie de H.A. Lorentz et Minkowski durant laquelle "tous deux reconnaissaient la grande valeur des idées neuves introduites par le jeune savant de 26 ans", ajoutant que "Lorentz renonçait à son hypothèse de la contraction des corps en mouvement pour se rallier à la nouvelle explication rationnelle et cohérente du résultat négatif de l'expérience de Michelson" (ce à quoi il ne s'est finalement pas résolu, comme nous l'avons vu).

[K] L. Kollros, Albert Einstein en Suisse. Souvenirs, in Fünfzig Jahre Relativitätstheorie, Helvetica Physica Acta, Supplementum IV, Birkhäuser Verlag, Basel, 1956, 271–181, traduction en allemand dans [S1].

Bibliographie

La numérotation de gauche suit l'ordre chronologique de parution (de rédaction pour [30] et [87]). Un chiffre romain I, II, III ou IV à droite d'un titre indique dans quel volume l'article en question est contenu. Les titres suivis du signe 0 sont ceux des travaux non reproduits ici.

1. (avec J. de Siebenthal) Sur les sous-groupes fermés connexes de rang maximum des groupes de Lie clos, C. R. Acad. Sci., Paris **226** (1948) 1662–1664. I, 1–2
2. Some remarks about Lie groups transitive on spheres and tori, Bull. Amer. Math. Soc. **55** (1949) 580–587. I, 3–10
3. (avec J. de Siebenthal) Les sous-groupes fermés de rang maximum des groupes de Lie clos, Comment Math. Helv. **23** (1949) 200–221. I, 11–32
4. Groupes d'homotopie des groupes de Lie, Espaces Fibrés et Homotopie, Sém. H. Cartan, E. N. S. Paris 1949–1950, Exp. 12, 13; Notes polycopiées, 2éme éd. Secrét. Mathématique, Soc. Math. France (1955). 0
5. Limites projectives de groupes de Lie, C. R. Acad. Sci., Paris **230** (1950) 1197–1199. I, 33–35
6. Sections locales de certains espace fibrés, C. R. Acad. Sci., Paris **230** (1950) 1246–1248. I, 36–38
7. Le plan projectif des octaves et les sphères comme espaces homogènes, C. R. Acad. Sci., Paris **230** (1950) 1378–1380. I, 39–41
8. Groupes localement compacts, Séminaire Bourbaki, Exp. 29 (1949/50). I, 42–54
9. (avec J-P. Serre) Impossibilité de fibrer un espace euclidien par des fibres compactes, C. R. Acad. Sci., Paris **230** (1950) 2258–2259. I, 55–56
10. Remarques sur l'homologie filtrée, J. Math. Pures Appl., (9) **29** (1950) 313–322. I, 57–66
11. Impossibilité de fibrer une sphère par un produit de sphères, C. R. Acad. Sci., Paris **231** (1950) 943–945. I, 67–69
12. Sous-groupes compacts maximaux des groupes de Lie, Séminaire Bourbaki, Exp. 33(1950/51). I, 70–76
13. Sur la cohomologie des variétés de Stiefel et de certains groupes de Lie, C. R. Acad. Sci., Paris **232** (1951) 1628–1630. I, 77–79
14. La transgression dans les espaces fibrés principaux, C. R. Acad. Sci., Paris **232** (1951) 2392–2394. I, 80–82
15. Sur la cohomologie des espaces homogènes de groupes de Lie compacts, C. R. Acad. Sci., Paris **233** (1951) 569–571. I, 83–85

16. (avec J-P. Serre) Détermination des p-puissances réduites de Steenrod dans la cohomologie des groupes classiques, Applications, C. R. Acad. Sci., Paris **233** (1951) 680–682. I, 86–88
17. Cohomologie des espaces homogènes, Séminaire Bourbaki, Exp. 45 (1950/51). 0
18. Cohomologie des espaces localement compacts d'après J. Leray, Notes. E. P. F. Zürich, 1951; 3ème édition: Lect. Notes Math. **2** (1964), Springer-Verlag. 0
19. (avec A. Lichnérowicz) Groupes d'holonomie des variétés riemanniennes, C. R. Acad. Sci., Paris **234** (1952) 1835–1837. I, 89–91
20. (avec A. Lichnérowicz) Espaces riemanniens et hermitiens symétriques, C. R. Acad. Sci., Paris **234** (1952) 2332–2334. I, 92–94
21. Les espaces hermitiens symétriques, Séminaire Bourbaki, Exp. **62** (1951/52). I, 95–103
22. Les fonctions automorphes de plusieurs variables complexes, Bull. Soc. Math. France **80** (1952) 167–182. I, 104–119
23. Sur la cohomologie des espaces fibrés principaux et des espaces homogènes de groupes de Lie compacts, (Thèse, Paris, 1952) Ann. Math. (2) **57** (1953) 115–207. I, 121–216
24. (avec J-P. Serre) Sur certains sous-groupes des groupes de Lie compacts, Comm. Math. Helv. **27** (1953) 128–139. I, 217–228
25. La cohomologie *mod*2 de certains espaces homogènes, Comm. Math. Helv. **27** (1953) 165–197. I, 229–261
26. (avec J-P. Serre) Groupes de Lie et puissances réduites de Steenrod, Amer. J. Math. **75** (1953) 409–448. I, 262–301
27. Les bouts des espaces homogènes de groupes de Lie, Ann. Math. (2) **58** (1953) 443–457. I, 302–316
28. Homology and cohomology of compact connected Lie groups, Proc. Nat. Acad. Sci. USA **39** (1953) 1142–1146. I, 317–321
29. Sur l'homologie et la cohomologie des groupes de Lie compacts connexes, Amer. J. Math. **76** (1954) 273–342. I, 322–391
30. Représentations linéaires et espaces homogènes kähleriens des groupes simples compacts (inédit, Mars 1954), published in [120], I, 392–396. I, 392–396
31. Kählerian coset spaces of semi-simple Lie groups, Proc. Nat. Acad. Sci., USA **40** (1954) 1147–1151. I, 397–401
32. Topics in the homology theory of fibre bundles, Univ. of Chicago 1954 (Notes by E. Halpern), Lect. Notes Math. **36** (1967). 0
33. Topology of Lie groups and characteristic classes, Bull. Amer. Math. Soc. **61** (1955) 397–432. I, 402–437
34. Nouvelle démonstration d'un théorème de P. A. Smith, Comm. Math. Helv. **29** (1955) 27–39. I, 438–450
35. (with C. Chevalley) The Betti numbers of the exceptional groups, Mem. Amer. Math. Soc. **14** (1955) 1–9. I, 451–459

36. (with G. D. Mostow) On semi-simple automorphisms of Lie algebras, Ann. Math. (2) **61** (1955) 389–405. I, 460–476
37. Sur la torsion des groupes de Lie, J. Math. Pures Appl. (9) **35** (1955) 127–139. I, 477–489
38. Groupes algébriques, Séminaire Bourbaki, Exp. **121** (1955/56). 0
39. Groupes linéaires algébriques, Ann. Math. (2) **64** (1956) 20–82. I, 490–552
40. Transformation groups with two classes of orbits, Proc. Nat. Acad. Sci. USA **43** (1957) 983–985. I, 553–555
41. Travaux de Mostow sur les espaces homogènes, Séminaire Bourbaki, Exp. **142** (1956/57). I, 556–564
42. The Poincaré duality in generalized manifolds, Mich. Math. J. **4** (1957) 227–239. I, 565–577
43. (with F. Hirzebruch) Characteristic classes and homogeneous spaces I, Amer. J. Math. **80** (1958) 458–538. I, 578–658
44. (avec J-P. Serre) Le théorème de Riemann-Roch, d'après Grothendieck, Bull. Soc. Math. France **86** (1958) 97–136. I, 659–698
45. (with F. Hirzebruch) Characteristic classes and homogeneous spaces II, Amer. J. Math. **81** (1959) 315–382. II, 1–68
46. Fixed points of elementary commutative groups, Bull. Amer. Math. Soc. **65** (1959) 322–326. II, 69–73
47. (with F. Hirzebruch) Characteristic classes and homogeneous spaces III, Amer. J. Math. **82** (1960) 491–504. II, 74–87
48. On the curvature tensor of the hermitian symmetric manifolds, Ann. Math., (2) **71** (1960) 508–521. II, 88–101
49. (with J. C. Moore) Homology theory for locally compact spaces, Mich. Math. J. **7** (1960) 137–159. II, 102–124
50. Density properties for certain subgroups of semi-simple groups without compact components, Ann. Math., (2) **72** (1960) 179–188. II, 125–134
51. Commutative subgroups and torsion in compact Lie groups, Bull. Amer. Math. Soc. **66** (1960) 285–288. II, 135–138
52. Seminar on transformation groups, Ann. Math. Stud. **46** (1960), (with contributions by G. Bredon, E. Floyd, D. Montgomery, R. Palias). 0
53. Sous groupes commutatifs et torsion des groupes de Lie compacts connexes, Tôhoku Math. J., (2) **13** (1961) 216–240. II, 139–163
54. (with Harish-Chandra) Arithmetic subgroups of algebraic groups, Bull. Amer. Math. Soc. **67** (1961) 579–583. II, 164–168
55. Some properties of adele groups attached to algebraic groups, Bull. Amer. Math. Soc. **67** (1961) 583–585. II, 169–171
56. (avec A. Haefliger) La classe d'homologie fondamentale d'un espace analytique, Bull. Soc. Math. France **89** (1961) 461–513. II, 172–224
57. (mit R. Remmert) Über kompakte homogene kählersche Mannigfaltigkeiten, Math. Ann. **145** (1962) 429–439. II, 225–235

58. (with Harish-Chandra) Arithmetic subgroups of algebraic groups, Ann. Math., (2) **75** (1962) 485–535. II, 236–286
59. Ensembles fondamentaux pour les groupes arithmétiques, Colloque sur la Théorie des Groupes Algébriques, Buxelles 1962, 23–40. II, 287–304
60. Some finiteness properties of adele groups over number fields, Publ. Math., Inst. Hautes Etud. Sci. **16** (1963) 5–30. II, 305–330
61. Arithmetic properties of linear algebraic groups, Proc. Int. Congr. Mathematicians, Stockholm 1962, Uppsala 1963, 10–22. II, 331–343
62. Compact Clifford-Klein forms of symmetric spaces, Topology **2** (1963) 111–122. II, 344–355
63. (with W. L. Baily Jr.) On the compactification of arithmetically defined quotients of bounded symmetric domains, Bull. Amer. Math. Soc. **70** (1964) 588–593. II, 356–361
64. (avec J-P. Serre) Théorèmes de finitude en cohomologie galoisienne, Comment. Math. Helv. **39** (1964) 111–164. II, 362–415
65. Cohomologie et rigidité d'espaces compacts localement symétriques, Séminaire Bourbaki, Exp. 265 (1963/64). II, 416–423
66. (avec J. Tits) Groupes réductifs, Publ. Math., Inst. Hautes Etud. Sci. **27** (1965) 55–150. II, 424–520
67. Statement of the index theorem. Outline of proof, Chap. I in: Seminar on the Atiyah-Singer index theorem by R. Palais et al., Ann. Math. Stud. **57** (1965) 1-11. II, 521–531
68. A spectral sequence for complex analytic bundles, Appendix Two in: F. Hirzebruch, Topological methods in algebraic geometry, 3rd edition, 202–217, Springer 1966. II, 532–547
69. (with W. L. Baily Jr.) Compactification of arithmetic quotients of bounded symmetric domains, Ann. Math. (2) **84** (1966) 442–528. II, 548–634
70. Density and maximality of arithmetic subgroups, J. Reine Angew. Math. **224** (1966) 78–89. II, 635–646
71. Opérateurs de Hecke et fonctions zêta, Séminaire Bourbaki, Exp. **307** (1965/66). II, 647–661
72. Class invariants, Chap. III, IV in: *Seminar on complex multiplication*, with S. Chowla, C. S. Herz, K. Iwasawa, J-P. Serre, Lect. Notes Math. **21** (1966), Springer-Verlag. 0
73. Linear algebraic groups, Proc. Symp. Pure Math. **9**, Amer. Math. Soc. (1966) 3–19. II, 662–678
74. Reduction theory for arithmetic groups, Proc. Symp. Pure Math. **9**, Amer. Math. Soc. (1966) 20–25. II, 679–684
75. Introduction to automorphic forms, Proc. Symp. Pure Math. **9**, Amer. Math. Soc. (1966) 199–210. II, 685–696
76. (with T. A. Springer) Rationality properties of linear algebraic groups, Proc. Symp. Pure Math. **9**, Amer. Math. Soc. (1966) 26–32. II, 697–703

77. (with R. Narasimhan) Uniqueness conditions for certain holomorphic mappings, Invent. Math. **2** (1967) 247–255. II, 704–712

78. Sur une généralisation de la formule de Gauss-Bonnet, An. Acad. Bras. Cienc. **39** (1967) 31–37. II, 713–719

79. Ensembles fondamentaux pour les groupes arithmétiques et formes automorphes, Notes d'un cours à l'Inst. H. Poincaré 1964, rédigées par H. Jacquet, J.-J. Sansuq et B. Schiffmann, Ecole Normale Supérieure, Paris 1967. 0

80. (with T. A. Springer) Rationality properties of linear algebraic groups II, Tôhoku Math. J., (2)**20** (1968) 443–497. II, 720–774

81. On the automorphisms of certain subgroups of semi-simple Lie groups, Proc. Inter. Colloquium on Algebraic Geometry 1968, Tata Institute, Bombay (1969) 43–73. III, 1–31

82. (avec J. Tits) On 'abstract' homomorphisms of simple algebraic groups, Proc. Colloquium on Algebraic Geometry 1968, Tata Institute, Bombay (1969) 75–82. III, 32–39

83. Injective endomorphisms of algebraic varieties, Arch. Math. **20** (1969) 531–537. III, 40–46

84. *Introduction aux groupes arithmétiques*, Actualités Sci. Ind. no. 1341, Hermann, Paris (1969). 0

85. *Linear algebraic groups* (Notes by H. Bass), Math. Lecture Notes Series, Benjamin, Inc. New York (1969); Traduction russe Moscou MIR (1972). 0

86. Sous-groupes discrets de groupes semi-simples (d'après D. A. Kajdan et G. A. Margoulis), Séminaire Bourbaki, Exp. 358, (1968/69), Lect. Notes Math. **179** (1971) 199–216, Springer-Verlag. III, 47–56

87. On periodic maps of certain $K(\pi, 1)$, (1969), published in [120], III, 57–60. III, 57–60

88. Pseudo-concavité et groupes arithmétiques, Essays on Topology and Related Topics, Mémoires dédiés à G. de Rham, Springer (1970) 70–84. III, 61–75

89. Properties and linear representations of Chevalley groups, in *Seminar on algebraic groups and related finite groups*, Lect. Notes Math. **131** (1970) 1–55, Springer-Verlag. III, 76–108

90. (avec J-P. Serre) Adjonction de coins aux espaces symétriques; Applications à la cohomologie des groupes arithmétiques, C. R. Acad. Sci., Paris **271** (1970) 1156–1158. III, 109–111

91. (avec J-P. Serre) Cohomologie à supports compacts des immeubles de Bruhat-Tits; Applications à la cohomologie des groupes S-arithmétiques, C. R. Acad. Sci., Paris **272** (1971) 110–113. III, 112–115

92. (avec J. Tits) Eléments unipotents et sous-groupes paraboliques de groupes réductifs I, Invent. Math. **12** (1971) 95–104. III, 116–125

93. Cohomologie réelle stable de groupes S-arithmétiques, C. R. Acad. Sci., Paris **274** (1972) 1700–1702. III, 126–128
94. (avec J. Tits) Compléments à l'article: 'Groupes réductifs,' Publ. Math., Inst. Hautes Etud. Sci. **41** (1972) 253–276. III, 129–152
95. Some metric properties of arithmetic quotients of symmetric spaces and an extension theorem, J. Differ. Geom. **6** (1972) 543–560. III,153–170
96. Représentations de groupes localement compacts, Lect. Notes Math. **276** (1972). 0
97. (avec J. Tits) Homomorphismes 'abstraits' de groupes algébriques simples, Ann. Math., (2) **97** (1973) 499–571. III, 171–243
98. (with J-P. Serre) Corners and arithmetic groups. With an appendix by A. Douady and L. Hérault: Arrondissement des Variétés à coins. Comment. Math. Helv. **48** (1973) 436–491. III, 244–299
99. Cohomologie de certains groupes discrets et Laplacien p-adique (d'après H. Garland), Séminaire Bourbaki, Exp. **437** (1973/74), Lect. Notes Math. **431** (1975) 12–35, Springer-Verlag. III, 300–314
100. Stable real cohomology of arithmetic groups, Ann. Sci. Ec. Norm. Super., (4) **7** (1974) 235–272. III, 315–352
101. Cohomology of arithmetic groups, Proc. Int. Congr. of Mathematicians, Vancouver, 1974, Vol. **1** (1975) 435–442. III, 353–360
102. Linear representations of semi-simple algebraic groups, Proc. Symp. Pure Math. **29**, Amer. Math. Soc. (1975) 421–439. III, 361–373
103. Formes automorphes et séries de Dirichlet (d'près R. P. Langlands), Séminaire Bourbaki, Exp. **466** (1974/75), Lect. Notes Math. **514** (1976) 183–222, Springer-Verlag. III, 374–398
104. Cohomologie de sous-groupes discrets et représentations de groupes semi-simples, Astérisque **32–33** (1976) 73–112. III, 399–438
105. (avec J-P. Serre) Cohomologie d'immeubles et de groupes S-arithmétiques, Topology **15** (1976) 211–232. III, 439–460
106. Admissible representations of a semi-simple group over a local field with vectors fixed under an Iwahori subgroup, Invent. Math. **35** (1976) 233–259. III, 461–487
107. (with B. M. Schreiber) p-adic linear groups with ergodic automorphisms, Isr. J. Math. **24** (1976) 199–205. III, 488–494
108. Cohomologie de SL_n et valeurs de fonctions zeta aux points entiers, Ann. Sc. Norm. Super. Pisa, Cl. Sci., (4) **4** (1977) 613–636; Correction, ibid. **7** (1980) 373. III, 495–519
109. (with G. Harder) Existence of discrete cocompact subgroups of reductive groups over local fields, J. Reine Angew. Math. **298** (1978) 53–64. III, 520–531
110. (avec J. Tits) Théorèmes de structure et de conjugaison pour les groupes algébriques linéaires, C. R. Acad. Sci., Paris **287** (1978) 55–57. III, 532–534

111. On the development of Lie group theory, Proc. of the Bicentennial Congr. of the Dutch Math. Soc., Math. Centre Tract 100/101 (1979) 25–37 and Nieuw Archief voor Wiskunde (3) **27** (1979) 13–25; Math. Intell. **2.2** (1980) 67–72. III, 535–547
112. (with H. Jacquet) Automorphic forms and automorphic representations, Proc. Symp. Pure Math. **33**, Part 1, Amer. Math. Soc. (1979) 189–202. III, 548–561
113. Automorphic L-functions, Proc. Symp. Pure Math. **33**, Part 2, Amer. Math. Soc. (1979) 27–61. III, 562–596
114. Symmetric compact complex spaces, Arch. Math. **33** (1979) 49–56. III, 597–604
115. (with N. Wallach) *Continuous cohomology, discrete subgroups and representations of reductive groups*, Ann. Math. Stud. **94** (1980), Princeton U. Press. III, 605–613
116. Stable and L^2-cohomology of arithmetic groups, Bull. Amer. Math. Soc., (N.S.) **3** (1980) 1025–1027. III, 614–616
117. Commensurability classes and volumes of hyperbolic 3-manifolds, Ann. Sc. Norm. Super. Pisa, CL Sci., (4) **8** (1981) 1–33. III, 617–649
118. Stable real cohomology of arithmetic groups II, Prog. Math., Boston **14** (1981) 21–55. III, 650–684
119. Mathematik: Kunst und Wissenschaft, Themen-Reihe der Carl Friedrich von Siemens Stiftung XXXIII München 1982. English translation, The Mathematical Intelligencer Vol. 5, No. 4, Springer-Verlag New York 1983, 9–17. Finnish translation by O. Pekonen in Symbolien Metsässä, Mathemaattisia Esseitä O. Pekonen, ed. Art House Osakeyhtiö, Helsingissä 1992, 11–36. III, 685–701
120. *Oeuvres. Collected Papers (1948–1982)*, 3 vol. Springer-Verlag (1983). 0
121. (with H. Garland) Laplacian and the discrete spectrum of an arithmetic group, Amer. J. Math. **105** (1983) 309–335. IV, 1–27
122. L^2-cohomology and intersection cohomology of certain arithmetic varieties, in *E. Noether in Bryn Mawr*, Springer-Verlag (1983) 119–131. IV, 28–40
123. On free groups of semi-simple groups, Enseign. Math. (2) **29** (1983) 151–164. IV, 41–54
124. Cohomology and spectrum of an arithmetic group, Proc. of a Conference on Operator Algebras and Group Representations, Neptun, Rumania (1980), Pitman (1983) 28–45. IV, 55–72
125. Regularization theorems in Lie algebra cohomology. Applications, Duke Math. J. **50** (1983) 605–623. IV, 73–91
126. (with W. Casselman) L^2-cohomology of locally symmetric manifolds of finite volume, Duke Math. J. **50** (1983) 625–647. IV, 92–114
127. Linear algebraic groups, Lectures at the Mathematical Institute of the Chinese Acad. Sci., Beijing, 1981, Notes by Zhe-Xian Wan, Adv. in Math. **13** (1984) 161–206 (in Chinese). 0

128. Sheaf theoretic intersection cohomology, in *Intersection Cohomology*, A. Borel et al, PM **50** (1984) 47–182, Birkhäuser, Boston. 0

129. The L^2-cohomology of negatively curved Riemannian symmetric spaces, Ann. Acad. Sci. Fenn. Ser. A, I. Math **10** (1985) 95–105. IV, 115–125

130. On affine algebraic homogeneous spaces, Archiv. d. Math. **45** (1985) 74–78. IV, 126–130

131. (with W. Casselman) Cohomologie d'intersection et L^2-cohomologie de variétés arithmétiques de rang rationnel 2, C. R. Acad. Sci. Paris **301** (1985) 369–373. IV, 131–135

132. Hermann Weyl and Lie groups in *Hermann Weyl 1885–1985*. K. Chandrasekharan ed., Springer-Verlag 1986, 53–82. IV, 136–165

133. *Algebraic D-Modules*, A. Borel et al., Perspectives in Mathematics **2**, Academic Press Boston, 1987. 0

134. A vanishing theorem in relative Lie algebra cohomology, in *Algebraic Groups*, Utrecht 1986, Lect. Notes Math. **1271** (1987), 1–16, Springer-Verlag. IV, 166–181

135. (with G. Prasad) Sous-groupes discrets de groupes p-adiques à covolume borné, C. R. Acad. Sci. Paris **305** (1987), 357–362. IV, 182–187

136. On the set of discrete subgroups of bounded covolume in a semisimple group, Proc. Indian Acad. Sci. (Math. Sci.) **97** (1987), 45–52. IV, 188–195

137. (avec G. Prasad) Valeurs de formes quadratiques aux points entiers, C. R. Acad. Sci. Paris **307** (1988), 217–220. IV, 196–199

138. The School of Mathematics at the Institute for Advanced Study, in *A Century of Mathematics in America*, Part III, (editor P. Duren, with the assistance of R. Askey, H. Edwards, U. Merzbach), Amer. Math. Soc., Providence, R.I., (1989) 119–147. IV, 200–228

139. (with G. Prasad) Finiteness theorems for discrete subgroups of bounded covolume in semi-simple groups, Publ. Math. I.H.E.S. **69** (1989), 119–171; Addendum, ibid. **71** (1990), 173–177. IV, 229–286

140. Correction and complement to the paper: Regularization theorems in Lie algebra cohomology. Applications, Duke Math. Jour. **60** (1990), 299–301. IV, 287–289

141. (with A. Ash) Generalized modular symbols, in *Cohomology of arithmetic groups and automorphic forms* (J-P. Labesse and J. Schwermer ed.), Lect. Notes Math. **1447** (1990), 57–75, Springer-Verlag. IV, 290–308

142. *Linear Algebraic Groups*, 2nd enlarged ed., Grad. Text. Math. **126**, Springer-Verlag 1991, 288 + xi pages. 0

143. The work of Chevalley in Lie groups and algebraic groups, *Proc. Hyderabad Conference on algebraic groups 1989*, Manoj Prakashan, Madras, (1991) 1–22. IV, 309–330

144. (with G. Prasad) Values of isotropic quadratic forms at S-integral points, *Compositio Mathematica* **83** (1992), 347–372. IV, 331–356

145. Deane Montgomery (1909–1992), *Notices of the American Math. Soc.* **39** (1992) 684–686; Proc. Amer. Philosophical Society **137** (1993) 453–456. IV, 357–360
146. Values of quadratic forms at S-integral points in *Algebraic groups and number theory*, V. Platonov and A. S. Rapinchuk ed. Uspehi Mat. Nauk. **47**, No 2, (1992) 117–141, (p. 118–120), Russian Math. Surveys 47, No. 2, (1992) 133–161 (p. 134–136). IV, 361–363
147. (avec F. Bien) Sous-groupes épimorphiques de groupes linéaires algébriques I, C.R. Acad. Sci. Paris **315**, Sér. I, (1992) 649–653. IV, 364–368
148. (avec F. Bien) Sous-groupes épimorphiques de groupes linéaires algébriques II, C.R. Acad. Sci. Paris **315**, Sér. I, (1992) 1341–1346. IV, 369–374
149. Réponse à la remise du prix Balzan 1992. IV, 375–376
150. Quelques réflexions sur les mathématiques en général et la théorie des groupes en particulier, Orientamenti et Attività dei premi Balzan 1992, 3–11, Fondazione Internazionale Balzan, Milan. IV, 377–385
151. Some recollections of Harish-Chandra, Current Science **65**, (1993) 919–920, Bangalore, India, reproduced in *Current trends in mathematics and physics. A tribute to Harish-Chandra*, Narosa Publishing House 1995, 210–215. IV, 386–390
152. Introduction to middle intersection cohomology and perverse sheaves, Proc. Symposia pure math. **56** Part I, Amer. Math. Soc. (1994) 25–52. IV, 391–418
153. On the place of mathematics in culture, in *Duration and Change, fifty years at Oberwolfach*, (M. Artin, H. Kraft, R. Remmert eds), Springer-Verlag (1994) 139–158. IV, 419–440
154. (with J. Yang), The rank conjecture for number fields, Math. Res. Letters **1** (1994) 689–699. IV, 441–451
155. Response to "Theoretical mathematics: toward a cultural synthesis of mathematics and theoretical physics, by A. Jaffe and F. Quinn", Bull A.M.S. N.S. **30** (1994), 179–181. IV, 452–453
156. Values of indefinite quadratic forms at integral points and flows on spaces of lattices, Bull. A.M.S. **32** (1995), 184–204. IV, 454–474
157. Values of zeta-functions at integers, cohomology and polylogarithms, in *Current trends in mathematics and physics. A tribute to Harish-Chandra*, S.D. Adhikari ed., 1–44, Narosa Publishing House, Bombay 1995. IV, 475–508
158. (with F. Bien and J. Kollar), Rationally connected homogeneous spaces, Inv. Math. **124** (1996), 103–127. IV, 509–533
159. Introduction to automorphic forms in one variable, Advances in Mathematics (China) **25** (1996), 97–158 (in Chinese). 0
160. (with J-P. Labesse and J. Schwermer), On the cuspidal cohomology of S-arithmetic subgroups of reductive groups over number fields, Comp. Math. 102, 1996, 1–40. IV, 534–573

161. Class functions, conjugacy classes and commutators in semisimple Lie groups, in *Algebraic groups and Lie groups*, G. Lehrer ed., Australian Math. Soc. Lecture Series **9**, Cambridge U. Press 1997, 1–19. IV, 574–592
162. On the work of E. Cartan on real simple Lie algebras, letter, Notices of the A.M.S. **44**, 1997, 430–1. IV, 593–594
163. Automorphic forms on $SL_2(R)$. Cambridge Tracts in Mathematics **130**, Cambridge University Press, 1997, x + 192 p. 0
164. Jean Leray and algebraic topology, in J. Leray, Selected papers, Vol. I, 1–21, Springer 1998. IV, 595–615
165. Twenty-five years with Nicolas Bourbaki, 1949–1973, Notices of the AMS **45** (1998), 373–380. Mitteilungen der Deutschen Mathematiker Vereinigung, Heft 1, 1998, 8–15. IV, 616–627
166. Semisimple groups and Riemannian symmetric spaces, Texts and Readings in Mathematics **16**, Hindustan Book Agency, New Delhi, India, 1998, x + 136 p. 0
167. Full reducibility and invariants for $SL_2(C)$, Ens. Math. (2) **44**, 1998, 71–90. IV, 628–647
168. André Weil and algebraic topology, Notices of the AMS **46**, 1999, 422–27, Gazette des Mathématiciens, n^o spécial, 63–74, Soc. Math. France 1999. IV, 648–657
169. André Weil, Notices of the AMS **46**, 1999, 442–44. 0
170. André Weil: quelques souvenirs, Gazette des Mathématiciens, n° spécial, 37–41, Soc. Math. France 1999. IV, 658–660
171. Algebraic groups and Galois theory in the work of Ellis R. Kolchin, in *Selected works of Ellis R. Kolchin*, with commentary, (H. Bass, A. Buium and P. Cassidy eds), 505–525, American Math. Soc. 1999. IV, 661–680
172. (with N. Wallach), *Continuous cohomology, discrete subgroups, and representations of reductive groups*, 2nd enlarged edition, Math. surveys and monographs **67**, American Math. Soc., 1999. 0
173. Henri Poincaré and special relativity, Enseignement Math. **45** (1999), 281–300. IV, 681–700

À paraître après 1999

(with R. Friedman and J. Morgan), Almost commuting elements in compact Lie groups.

(with G. Henkin and P. Lax), Jean Leray (1906–1998).

Essays in the history of Lie groups and algebraic groups.

(with L. Ji), Compactifications of symmetric and locally symmetric spaces.

Acknowledgements

Springer-Verlag would like to thank the original publishers of Armand Borel's papers for granting permission to reprint them here.

The numbers following each source correspond to the numbering of the articles in the bibliography at the end of each volume.

Reprinted from A Century of Mathematics in America, by P. Duren (ed.), © American Mathematical Society: 138

Reprinted from Algebraic Groups, © Springer-Verlag: 134

Reprinted from Algebraic Groups and Lie Groups by Lehrer (ed.), © Cambridge University Press: 161

Reprinted from Amer. J. Math., © by Johns Hopkins University Press: 26, 29, 43, 45, 47, 121

Reprinted from An. Acad. Bras. Cienc., © by Academia Brasiliera de Ciencias: 78

Reprinted from Ann. Acad. Sci. Fenn., Ser. A I, © Academia Scientiarum Fennica: 129

Reprinted from Ann. Math. Stud., © by Princeton University Press: 67, 115

Reprinted from Ann. Math., (2), © by Princeton University Press: 23, 27, 36, 39, 48, 50, 58, 69, 97

Reprinted from Ann. Sc. Norm. Super. Pisa, Cl. Sci., (4), © by Scuola Normale Superiore, Italy: 108, 117

Reprinted from Ann. Sci. Ec. Norm. Super., (4), © by Editions Bordas-Dunod-Gauthier-Villars: 100

Reprinted from Arch. Math., © by Birkhüser Verlag, Basel: 83, 114

Reprinted from Astérisque. © by Société Mathématique de France: 104

Reprinted from Bull. Am. Math. Soc., © by The American Mathematical Society: 2, 33, 46, 51, 54, 55, 63, 116, 155, 156

Reprinted from Bull. Am. Math. Soc., New Ser., © American Mathematical Society: 155, 156

Reprinted from Bull. Soc. Math. France. © by Editions Bordas-Dunod-Gauthier-Villars: 22, 44, 56

Reprinted from C. R. Acad. Sci., Paris, © by Editions Bordas-Dunod-Gauthier-Villars: 1, 5, 6, 7, 9, 11, 13, 14, 15, 16, 19, 20, 90, 91, 93, 110

Reprinted from C. R. Acad. Sci., Paris, © Editions Elsevier: 131, 135, 137, 147, 148

Reprinted from Cohomology of Arithmetic Groups and Automorphic Forms, by J-P. Labesse and J. Schwermer (eds.), © Springer-Verlag: 141

Reprinted from Comment. Math. Helv., © by The University of Zürich: 3, 24, 25, 34, 64, 98

Reprinted from Compos. Math., © by Kluwer Academic Publishers: 144, 160
Reprinted from Current Science, © Current Science: 151
Reprinted from Current Trends in Mathematics: A Tribute to Harishchandra by Adhikari © 1995 Narosa Publishing House, New Delhi: 157
Reprinted from Duke Math. J., © Duke University Press: 125, 126, 140
Reprinted from Duration and Change, Fifty Years at Oberwolfach, by M. Artin, H. Kraft, R. Remmert (eds.), © Springer-Verlag: 153
Reprinted from E. Noether in Bryn Mawr, © Springer-Verlag: 122
Reprinted from Enseign. Math., II Sér., © Enseignement Mathématique: 123, 167, 173
Reprinted from Essays on Topology and Related Topics, © by Springer-Verlag Berlin Heidelberg New York: 88
Reprinted from Gaz. Math., Soc. Math. Fr., © Société Mathématique de France: 170
Reprinted from Hermann Weyl 1885-1985, © Springer-Verlag: 132
Reprinted from Invent. Math., © by Springer-Verlag Berlin Heidelberg New York: 77, 92, 106, 158
Reprinted from Isr. J. Math., © by Weizmann Science Press: 107
Reprinted from J. Differ. Geom., © by The Lehigh University: 95
Reprinted from J. Leray, Selected Papers vol. I, © Springer-Verlag: 164
Reprinted from J. Math. Pures Appl.. (9), © by Editions Bordas-Dunod-Gauthier-Villars: 10, 37
Reprinted from J. Reine Angew. Math., © by Walter de Gruyter & Co.: 70, 109
Reprinted from Math. Ann.. © by Springer-Verlag Berlin Heidelberg New York: 57
Reprinted from Math. Res. Letters 1, © Mathematical Publishing: 154
Reprinted from Mem. Amer. Math. Soc., © by The American Mathematical Society: 35
Reprinted from Mich. Math. 3., © by The University of Michigan: 42, 49
Reprinted from Mitteilungen der DMV, © Deutsche Mathematiker-Vereinigung e.V.: 165
Reprinted from Notices Am. Math. Soc., © American Mathematical Society: 145, 162, 168, 136
Reprinted from Operator Algebras and Group Representations, vol. 1, by Gr. Arsene (ed.), © Pitman: 124
Reprinted from Orientamenti et Attività dei premi Balzan 1992, © Fondazione Internazionale Premio E. Balzan: 150
Reprinted from Proc. Indian Acad. Sci., Math. Sci., © Indian Academy of Science: 136
Reprinted from Proc. Int. Congr. Mathematicians, Stockholm, 1962. © by The American Mathematical Society: 61
Reprinted from Proc. Int. Congr. Mathematicians, Vancouver, 1974, © by The American Mathematical Society: 101
Reprinted from Proc. Nat. Acad. Sci. USA, © by the National Academy of Sciences USA: 28, 31, 40
Reprinted from Proc. of the Bicentennial Congr. of the Dutch Math. Soc., Math. Centre Tract 100/101, © Mathematisch Centrum, Amsterdam: 111

Reprinted from Proc. Symp. Pure Math., © by The American Mathematical Society: 73, 74, 75, 76, 102, 112, 113, 152

Reprinted from Prog. Math., Boston, © by Birkhäuser Boston, Inc.: 118

Reprinted from Publ. Math., Inst. Hautes Etud. Sci., © by Institute des Hautes Etudes Scientifiques, France: 60, 66, 94, 139

Reprinted from Russ. Math. Surv., © London Mathematical Society: 146

Reprinted from Selected Works of Ellis R. Kolchin, by H. Bass, A. Buium and P. Cassidy (eds.), © American Mathematical Society: 171

Reprinted from Séminaire Bourbaki, © by Association des Collaborateurs de Nicolas Bourbaki: 8, 12, 21, 41, 65, 71, 86, 99, 103

Reprinted from Tata Institute of Fundamental Research, Bombay, © by Tata Institute of Fundamental Research: 81, 82

Reprinted from Tôhoku Math. J., (2), © by The Tôhoku University: 53, 80

Reprinted from Topological Methods in Algebraic Geometry, © by Springer-Verlag Berlin Heidelberg New York: 68

Reprinted from Topology. © by Pergamon Press: 62, 105